Study Guide for

Organic Chemistry

Second Edition

Seyhan N. Eğe
The University of Michigan

Roberta W. Kleinman
Lock Haven University of Pennsylvania

Marjorie L. C. Carter
The University of Michigan

D. C. Heath and Company   LEXINGTON, MASSACHUSETTS / TORONTO

Cover: Photomicrograph of cyclohexanone oxime by Manfred Kage/Peter Arnold, Inc.

Contents

To the Student

The *Study Guide* that accompanies your text has been prepared to help you study organic chemistry. The textbook contains many problems designed to assist you in reviewing the chemistry that you need to know. The *Study Guide* contains the answers to these problems worked out in great detail to help you to develop the patterns of thought and work that will enable you to complete a course in organic chemistry successfully. In addition, notes that clarify points that may give you difficulty are provided in many answers.

Suggestions for the best way to study organic chemistry are given on pages 2 - 4 of the textbook. Before you work with the *Study Guide,* please review those pages in the text. There you will find emphasis on four things necessary for success in organic chemistry:

1. steady consistent studying,
2. working with problems as the most effective way to learn organic chemistry,
3. training your eye to look for the structural features in molecules that determine their chemical and physical properties, and
4. training your hand to draw correct structural formulas for reacting species and products.

This *Study Guide* contains two features to help you to study more effectively. The sections on Problem-Solving Skills in the textbook show you how to analyze problems in a systematic way by asking yourself questions about the structural changes in and the reactivity of the reagents shown in the problem. These same questions are used in arriving at the answers shown in the *Study Guide* for some of the problems. If you follow the reasoning shown in these answers, you will review the thinking patterns that are useful in solving problems.

The *Study Guide* also contains Concept Maps, which are summaries of important ideas or patterns of reactivity presented in a two-dimensional outline form. The textbook has notes in the margins telling you when a Concept Map appears in the *Study Guide*. The Concept Maps are located among the answers to the problems. The Table of Contents of the *Study Guide* on pp. iii- vii will tell you where each Concept Map is. The Concept Maps will be the most useful to you if you use them as a guide to making your own. For example, when you review your lecture notes, you will learn the essential points much more easily if you attempt to summarize the contents of the lecture in the form of a concept map. At a later time you may want to combine the contents of several lectures into a different concept map. Your maps need not look like the ones in the *Study Guide*. What is important is that you use the format to try to see relationships among ideas, reactions, and functional groups in a variety of ways.

The *Study Guide* will be most helpful to you if you make every attempt to solve each problem completely before you look at the answer. Recognition of a correct answer is much easier than being able to produce one yourself, so if you simply look up answers in the book to see whether you "know how to do the problem" or "understand" a principle, you will probably decide that you do. In truth, however, you will not have gained the practice in writing structural formulas and making the step-by-step decisions about reactivity that you will need when faced by similar questions on examinations. Work out all answers in detail, writing correct, complete formulas for all reagents and products. Build molecular models to help you draw correct three-dimensional representations of molecules. Consult the models whenever you are puzzled by questions of stereochemistry.

If you do not understand the answer to a problem, study the relevant sections of the text again, and then try to do the problem once more. The problems will tell you what you need to spend most of your time studying. As you solve the many review problems that bring together material from different chapters, your knowledge of the important reactions of organic chemistry will solidify.

We hope that the *Study Guide* will serve you as a model for the kind of disciplined care that you must take with your answers if you wish to train yourself to arrive at correct solutions to problems with consistency. We hope also that it will help you to develop confidence in your ability to master organic chemistry so that you enjoy your study of a subject that we find challenging and exciting.

Seyhan N. Eğe
Roberta W. Kleinman
Marjorie L. C. Carter

1

An Introduction to Structure and Bonding in Organic Compounds

Concept Map 1.1   Ionic compounds and ionic bonding.

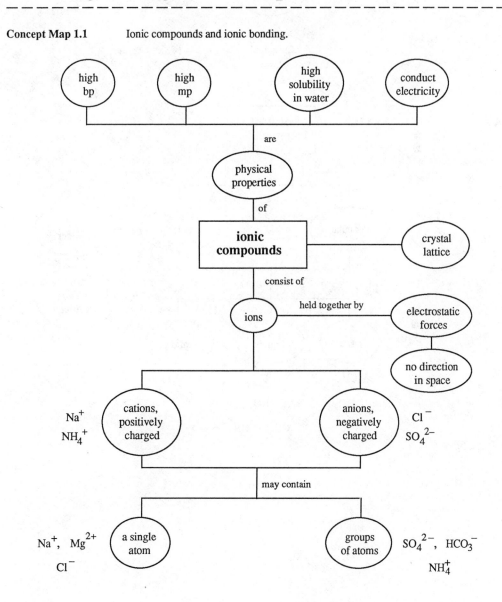

Concept Map 1.2   Covalent compounds.

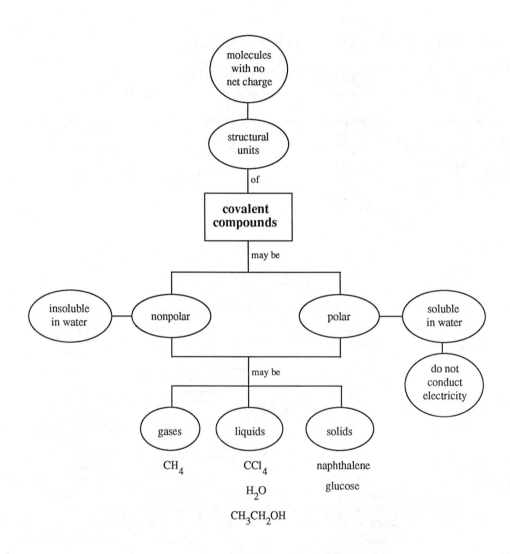

1.1   Formal charge = number of valence electrons − number of nonbonding electrons − $\frac{1}{2}$ number of bonding electrons.

CCl_4

Cl $7 - 6 - \frac{1}{2}(2) = 0$

C $4 - 0 - \frac{1}{2}(8) = 0$

CH_3Br

H $1 - 0 - \frac{1}{2}(2) = 0$

C $4 - 0 - \frac{1}{2}(8) = 0$

Br $7 - 6 - \frac{1}{2}(2) = 0$

$CH_3\overset{+}{O}H_2$

H $1 - 0 - \frac{1}{2}(2) = 0$

C $4 - 0 - \frac{1}{2}(8) = 0$

O $6 - 2 - \frac{1}{2}(6) = +1$

(deficiency of one electron)

$^-NH_2$

H $1 - 0 - \frac{1}{2}(2) = 0$

N $5 - 4 - \frac{1}{2}(4) = -1$

(excess of one electron)

$CH_3\overset{+}{N}H_3$

H $1 - 0 - \frac{1}{2}(2) = 0$

C $4 - 0 - \frac{1}{2}(8) = 0$

N $5 - 0 - \frac{1}{2}(8) = +1$

(deficiency of one electron)

H_2NNH_2

H $1 - 0 - \frac{1}{2}(2) = 0$

N $5 - 2 - \frac{1}{2}(6) = 0$

PH_3

P $5 - 2 - \frac{1}{2}(6) = 0$

H $1 - 0 - \frac{1}{2}(2) = 0$

H_2S

H $1 - 0 - \frac{1}{2}(2) = 0$

S $6 - 4 - \frac{1}{2}(4) = 0$

1.1 (cont)

CH_3CH_2OH

H : C : C : O : H

H—C—C—O—H

H	$1-0-\frac{1}{2}(2)=0$
C	$4-0-\frac{1}{2}(8)=0$
O	$6-4-\frac{1}{2}(4)=0$

$HOCH_2CH_2OH$

H : O : C : C : O : H

H—O—C—C—O—H

H	$1-0-\frac{1}{2}(2)=0$
C	$4-0-\frac{1}{2}(8)=0$
O	$6-4-\frac{1}{2}(4)=0$

$CH_3\overset{+}{O}CH_3$

CH_3

H : C : O : C : H

H—C—O—C—H

H—C—H

H	$1-0-\frac{1}{2}(2)=0$
C	$4-0-\frac{1}{2}(8)=0$
O	$6-2-\frac{1}{2}(6)=+1$

(deficiency of one electron)

1.2 (a)

CH_3

$CH_3—\overset{+}{N}—CH_3$

CH_3

N $5-0-\frac{1}{2}(8)=+1$

(b)

: Br — C — Br :

C $4-2-\frac{1}{2}(4)=0$

(c)

H

$CH_3—\overset{+}{O}—CH_3$

O $6-2-\frac{1}{2}(6)=+1$

(d)

$CH_3—\overset{-}{N}—H$

N $5-4-\frac{1}{2}(4)=-1$

(e)

: Cl :

: Cl — $\overset{-}{C}$ — Cl :

C $4-2-\frac{1}{2}(6)=-1$

(f)

CH_3

$CH_3—\overset{+}{C}—CH_3$

C $4-0-\frac{1}{2}(6)=+1$

1.3

$$
\begin{array}{ccc}
 & :\!\overset{\cdot\cdot}{F}\!: & H \\
 & | & | \\
:\!\overset{\cdot\cdot}{F}\!-\!\!\!&\overset{-}{B}\!-\!\overset{+}{N}\!&-\!H \\
 & | & | \\
 & :\!\overset{\cdot\cdot}{F}\!: & H \\
\end{array}
$$

F $7 - 6 - \frac{1}{2}(2) = 0$

H $1 - 0 - \frac{1}{2}(2) = 0$

B $3 - 0 - \frac{1}{2}(8) = -1$ (excess of one electron)

N $5 - 0 - \frac{1}{2}(8) = +1$ (deficiency of one electron)

1.4 BF_3 has room in its orbitals to accept a pair of electrons. We expect it will react with any neutral or negatively charged species that has nonbonding electrons. Even though the hydronium ion has a pair of nonbonding electrons on it, the positive charge on the ion makes them unavailable for further bonding.

1.5   <u>Condensed Formula</u>   <u>Lewis Structures</u>

(a) CH_3CCH_3 (with O double bonded)

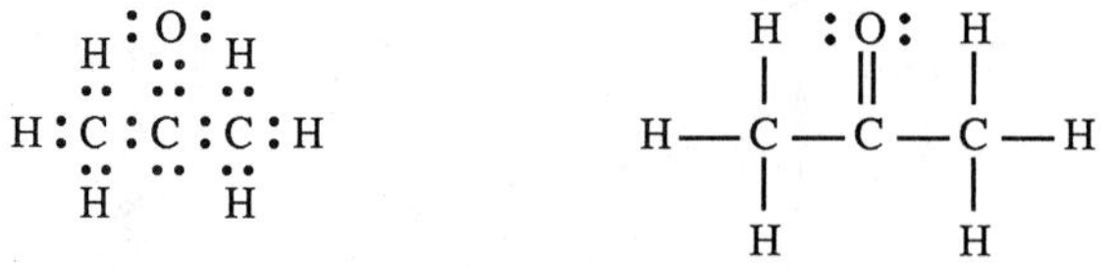

(b) $CH_3C{\equiv}CH$

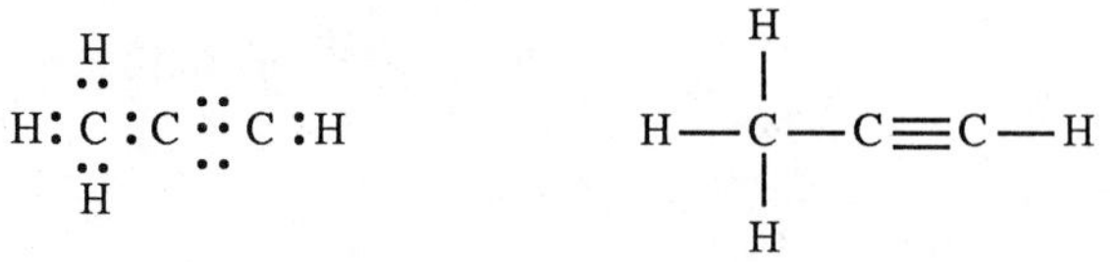

(c) $HCNH_2$ (with O double bonded)

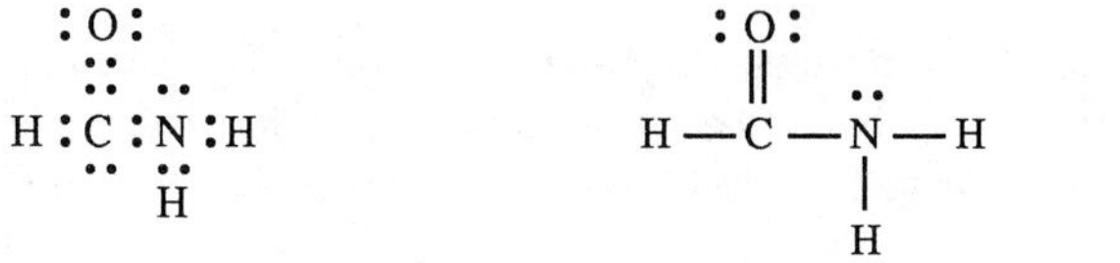

(d) $HCOH$ (with O double bonded)

(e) CH_3COCH_3 (with O double bonded)

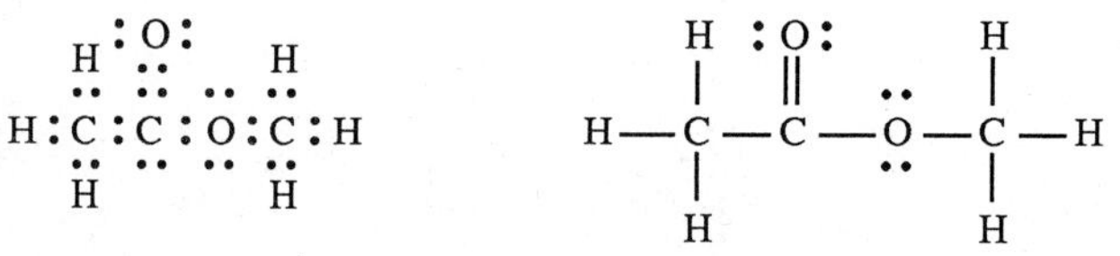

(f) $HON{=}O$

or

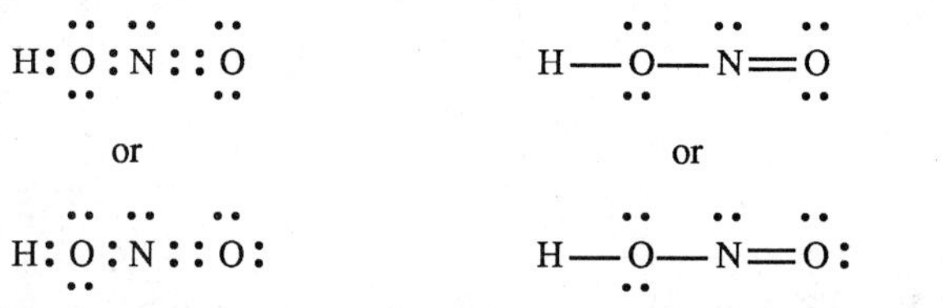

Note that the nonbonding electrons may be placed around a doubly bonded oxygen in two ways, illustrated in the structural formulas given above.

(g) $CH_3N{=}NCH_3$

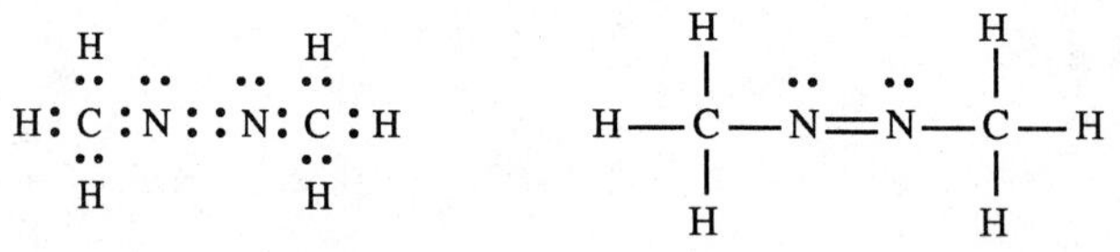

(h) $CH_2{=}CHCl$

Concept Map 1.3   Covalent bonding.

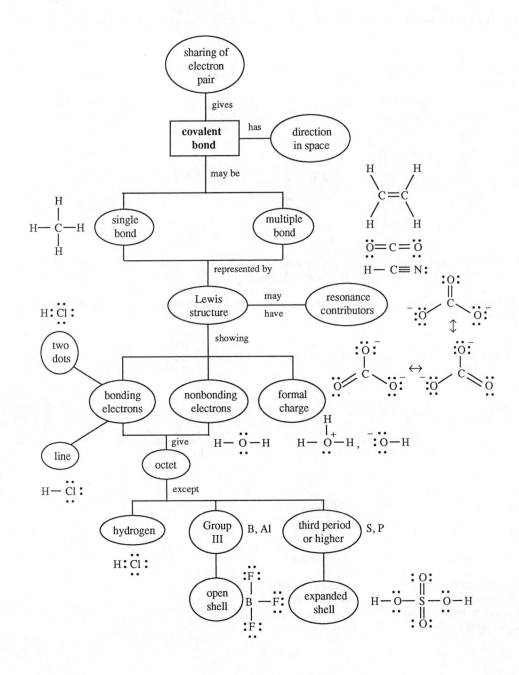

1.6

(a)

$$\text{H—O—C—O}: \leftrightarrow \text{H—O—C=O} \leftrightarrow \text{H—O=C—O}: \leftrightarrow \text{H—O—C—O}:$$

         ①   ②   ③

     major and equivalent   minor   minor

     no separation of charge;   separation of charge   separation of charge;
     complete octets   complete octets   carbon has sextet

(b)

$$\text{H—C—H} \leftrightarrow \text{H—C—H}$$

    ①   ②

   major   minor

no separation of charge;   separation of charge;
complete octets   carbon has sextet

Note that such a minor resonance contributor may nevertheless be significant in the chemical reactivity of a species (see Problems 1.10 and 1.11 for examples).

(c)

$$:\text{O—N=O} \leftrightarrow \text{O=N—O}: \leftrightarrow :\text{O—N—O}:$$

       ①   ②

     major and equivalent   minor

     no separation of charge;   separation of charge
     complete octets   nitrogen has sextet

d)

$$:\text{N≡C—O}:^{-} \leftrightarrow :\text{N=C=O}$$

   ①   ②

   major   minor

no separation of charge;   no separation of charge;
complete octets;   complete octets;
negative charge on more   negative charge on less
electronegative atom   electronegative atom

1.6 (cont)

(e) $\overset{..}{\text{O}}\overset{..}{:}$ ... (see structure)

$$\text{H}-\overset{..}{\underset{..}{\text{O}}}-\overset{\overset{\overset{..}{\text{O}}\overset{..}{:}}{\|}}{\underset{+}{\text{N}}}-\overset{..}{\underset{..}{\text{O}}}{:}^{-} \leftrightarrow \text{H}-\overset{..}{\underset{..}{\text{O}}}-\overset{\overset{\overset{..}{\text{O}}\overset{..}{:}^{-}}{|}}{\underset{+}{\text{N}}}=\overset{..}{\underset{..}{\text{O}}}$$

equivalent resonance contributors

complete octets;
no contributor with complete octets
that does not have separation of
charge can be written

(f) $:\overset{..}{\underset{..}{\text{O}}}-\overset{..}{\underset{+}{\text{O}}}=\overset{..}{\underset{..}{\text{O}}} \leftrightarrow \overset{..}{\text{O}}=\overset{..}{\underset{+}{\text{O}}}-\overset{..}{\underset{..}{\text{O}}}:^{-} \leftrightarrow {}^{-}:\overset{..}{\underset{..}{\text{O}}}-\overset{..}{\underset{..}{\text{O}}}-\overset{..}{\underset{..}{\text{O}}}^{+} \leftrightarrow {}^{+}\overset{..}{\underset{..}{\text{O}}}-\overset{..}{\underset{..}{\text{O}}}-\overset{..}{\underset{..}{\text{O}}}:^{-}$

① ②

major and equivalent	minor and equivalent
complete octets;	one oxygen has sextet;

all contributors have separation of charge

(g) $\overset{..}{\underset{..}{\text{O}}}=\overset{+}{\text{N}}=\overset{..}{\underset{..}{\text{O}}} \leftrightarrow \overset{..}{\underset{..}{\text{O}}}=\overset{..}{\text{N}}-\overset{..}{\underset{..}{\text{O}}}^{+} \leftrightarrow {}^{+}\overset{..}{\underset{..}{\text{O}}}-\overset{..}{\text{N}}=\overset{..}{\underset{..}{\text{O}}}$

① ②

major	minor and equivalent
complete octets; positive charge on less electronegative atom	oxygen has sextet; positive charge on more electronegative atom

1.7   The boron atom in boron trifluoride has only six electrons and can accept an electron pair from the nitrogen atom in ammonia to complete the octet.

$$:\overset{..}{\text{F}}: \quad \text{H}$$
$$\quad | \qquad \quad |$$
$$:\overset{..}{\text{F}}-\text{B} \quad :\text{N}-\text{H}$$
$$\quad | \qquad \quad |$$
$$:\overset{..}{\text{F}}: \quad \text{H}$$

formation of covalent bond

Likewise the reaction between carbon dioxide and hydroxide ion is rationalized by looking at a resonance contributor in which the carbon in carbon dioxide has only six electrons and can accept an electron pair from the oxygen atom of the hydroxide ion to complete the octet.

1.7 (cont)

$$\overset{\cdot\cdot}{\underset{\cdot\cdot}{O}}=\overset{+}{C}-\overset{\cdot\cdot}{\underset{\cdot\cdot}{O}}:\qquad\overset{-}{:}\overset{\cdot\cdot}{\underset{\cdot\cdot}{O}}-H$$

formation of covalent bond

1.8   Ammonia will react with formaldehyde. In the minor resonance contributor of formaldehyde (Problem 1.6b), the carbon atom has a sextet and can accept an electron pair from the nitrogen atom in ammonia to complete the octet.

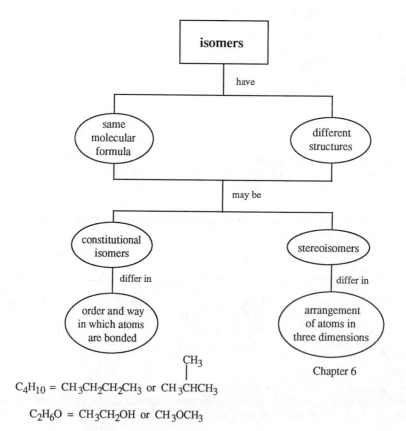

formation of covalent bond

Concept Map 1.4   Isomers.

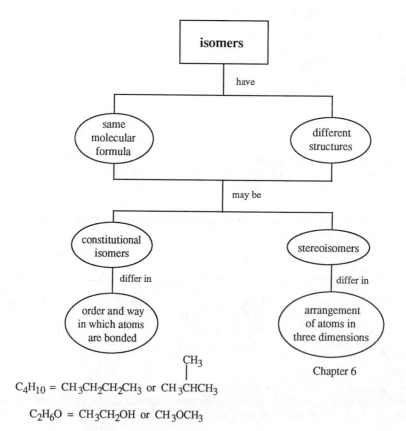

C_4H_{10} = $CH_3CH_2CH_2CH_3$ or $CH_3\overset{\overset{\textstyle CH_3}{|}}{C}HCH_3$

C_2H_6O = CH_3CH_2OH or CH_3OCH_3

1.9 C_3H_7Cl

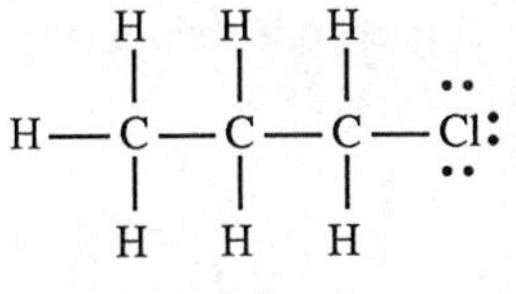

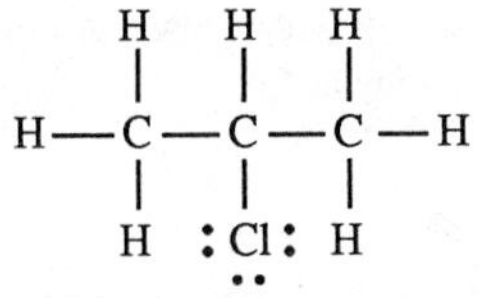

$CH_3CH_2CH_2Cl$

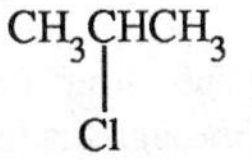

CH_3CHCH_3
|
Cl

C_3H_8O

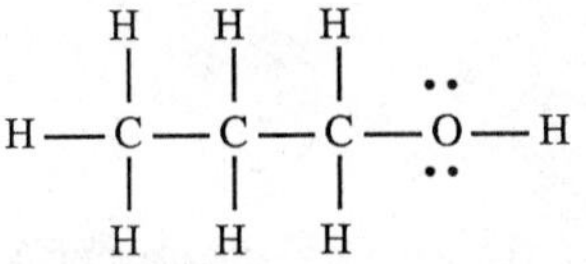

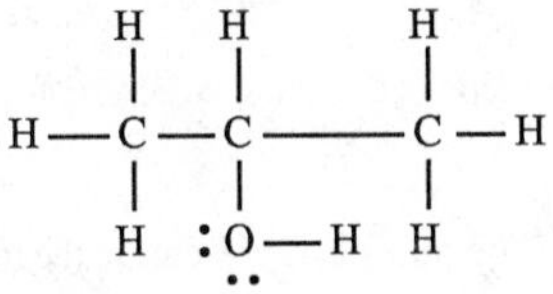

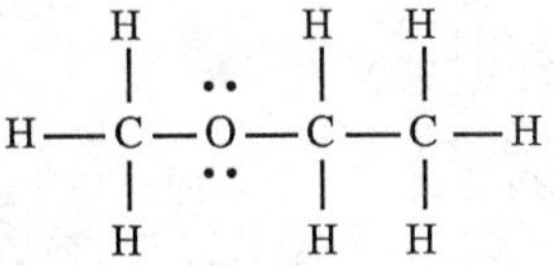

$CH_3CH_2CH_2OH$

CH_3CHCH_3
|
OH

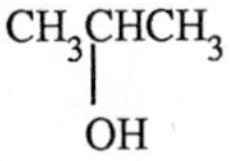

$CH_3OCH_2CH_3$

C_5H_{12}

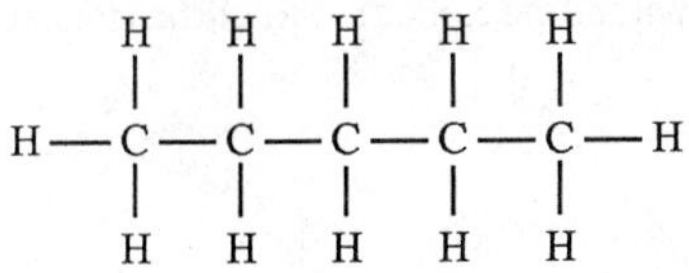

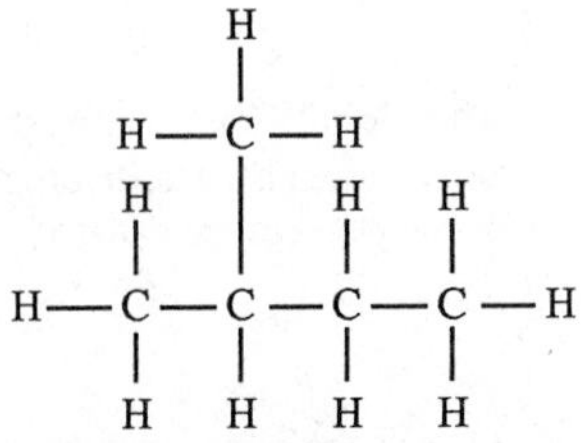

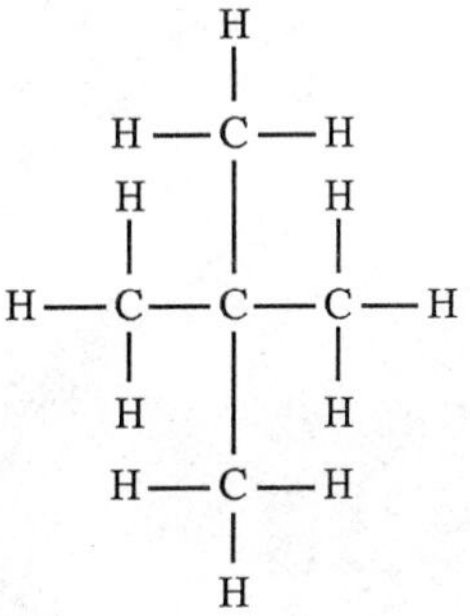

$CH_3CH_2CH_2CH_2CH_3$

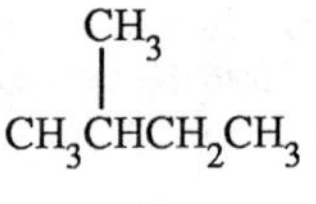

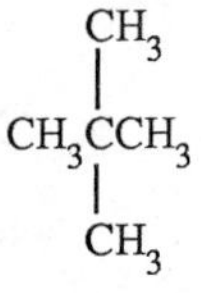

CH_3
|
$CH_3CHCH_2CH_3$

CH_3
|
CH_3CCH_3
|
CH_3

C_2H_4O

O
||
CH_3CH

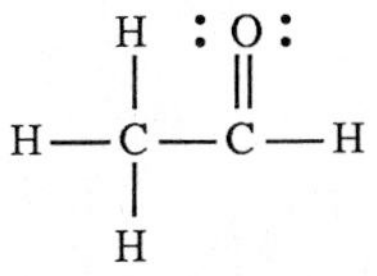

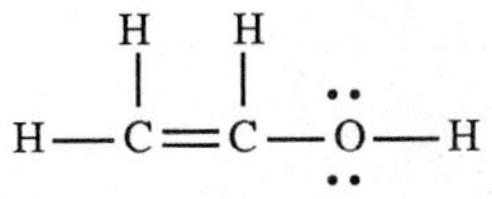

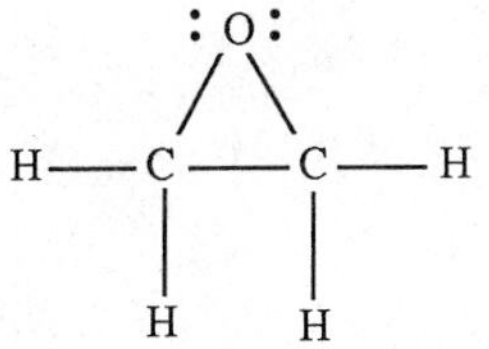

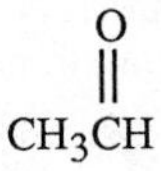

$CH_2=CHOH$

enol

O
/ \
CH_2-CH_2

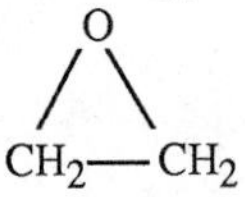

1.10   Different three-dimensional representations of the compounds and the relationships between them are shown below and on the next page. Note that there are many more possibilities for each compound besides those that are shown. It will be helpful to work with molecular models to see the different orientations in space that molecules and parts of molecules may have.

(a)   109.5°

(This is the same structure as

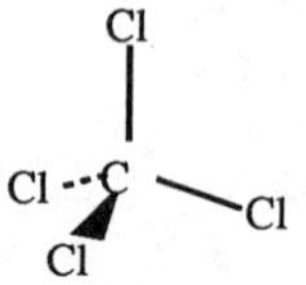

The formula representing the molecule has been rotated around an axis through the top chlorine atom and the carbon atom.)

(b)   ~109.5°

~109.5°

(This is the same structure as   .

The formula representing the molecule has been rotated in the plane of the paper.)

(c)   ~109.5°   1.09 Å

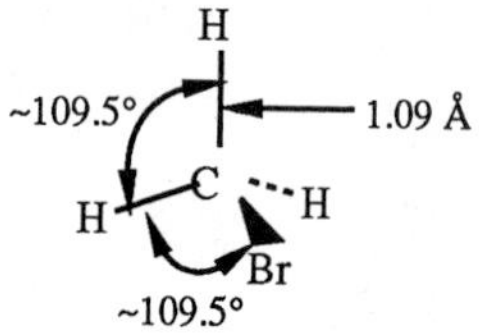

(This is the same structure as   .

The formula representing the molecule was first rotated around an axis going through the top hydrogen atom and the carbon atom and then rotated in the plane of the paper.)

(d)   ~109.5°
109.5°
109.5°

(This is the same structure as   .

The formula representing the molecule was rotated around an axis perpendicular to the carbon-carbon bond.)

(e)   ~105°

~109.5°

(This is the same structure as   .

The formula representing the molecule was changed by rotating the O—H bond around an axis going through the carbon atom and the oxygen atom without rotating the carbon atom. This type of transformation will be discussed in greater detail in Chapter 5.)

1.10 (cont)

(f)   (This is the same structure as .

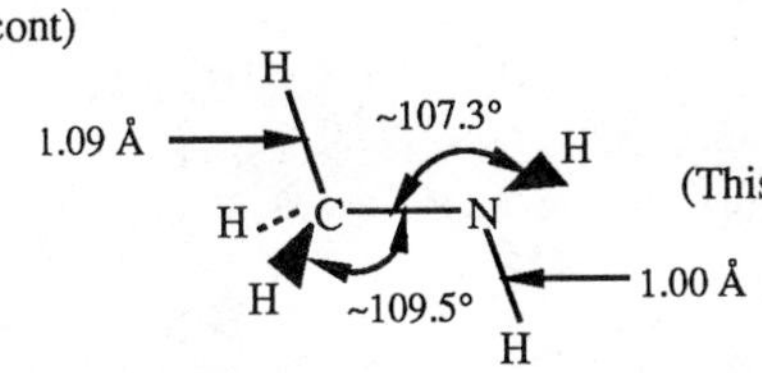

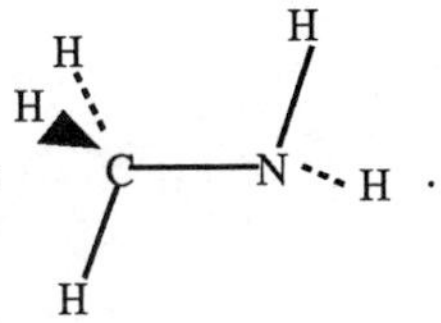

The formula representing the molecule was rotated around an axis going through the carbon atom and the nitrogen atom.)

(g)   (This is the same structure as .

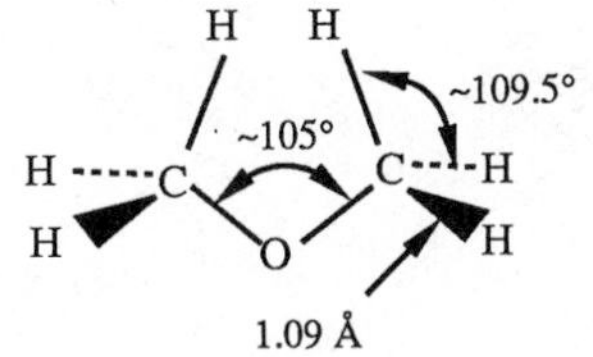

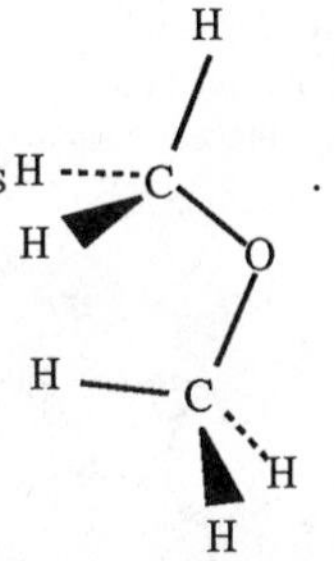

The formula representing the molecule was changed by rotating the CH_3 [methyl] group on the right about an axis through the carbon-oxygen bond on the left.)

(h)   (This is the same structure as .

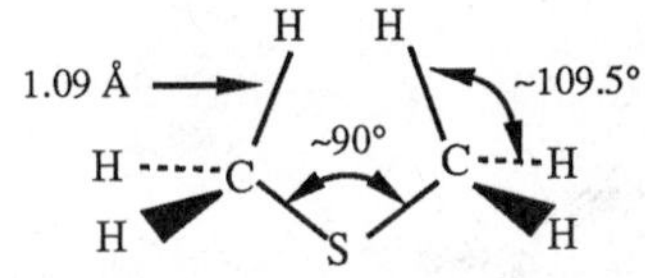

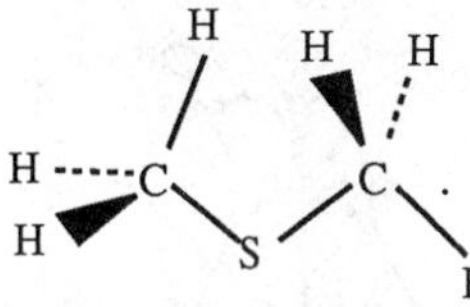

The formula representing the molecule was changed by rotating the CH_3 [methyl] group on the right about an axis through the carbon-sulfur bond on the left.)

1.11   (a)

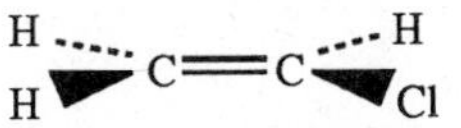

(molecule perpendicular to plane of paper)

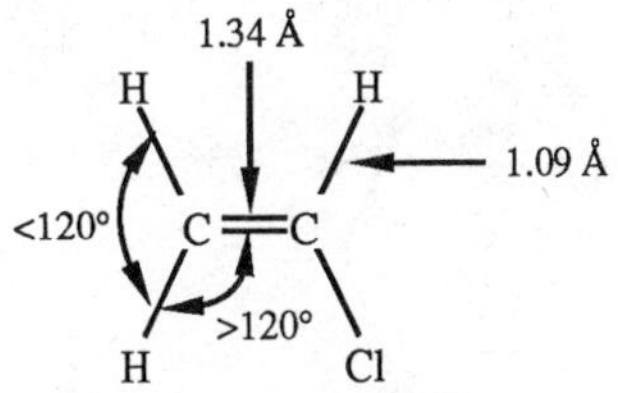

(molecule in plane of paper)

1.11 (cont)

(b)

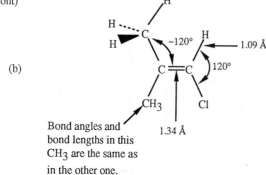

Bond angles and
bond lengths in this
CH₃ are the same as
in the other one.

1.34 Å

(c)

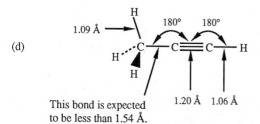

1.34 Å

H ····· C ═ C ····· H (with Cl on left wedges)
Cl   Cl

Chlorine atoms are both on
the same side of the paper.

H ····· C ═ C ····· Cl
Cl   H

Chlorine atoms are on the
opposite sides of the paper.

(d)

1.09 Å →

180°   180°

H

H ···· C — C ≡ C — H
H

1.20 Å 1.06 Å

This bond is expected
to be less than 1.54 Å.

(e)

1.09 Å → ~120°

H
C ═ O
H

(molecule in plane of paper)

H ·····
        C ═ O
H

(molecule perpendicular
to plane of paper)

Concept Map 1.5   Polarity of covalent molecules.

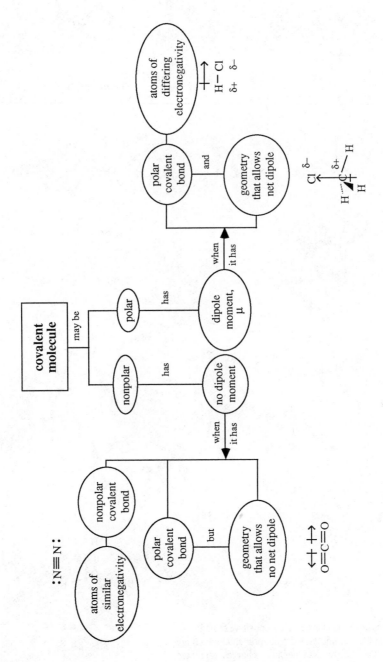

1.12   (a)

H ...... N$^{\delta-}$ — H $^{\delta+}$
       H

(b)

O $^{\delta-}$
H   H $^{\delta+}$

(c)

Br $^{\delta-}$
H ...... C$^{\delta+}$ — H
    H

(d)

$$O=C=O \quad {\scriptstyle \delta+\;\;\delta-}$$

(e)

F
F ...... C$^{\delta+}$ — F $^{\delta-}$
F

(f)

H   Cl $^{\delta-}$
C = C $^{\delta+}$
Cl   H

(g)

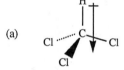

$C^{\delta+}$ — $O^{\delta-}$ H $\delta+$

1.13   (a)

(b)

(c)

I —— Cl

(d)

(e)

(f)

H
Br ...... C — H
Br

ions have full charges, either positive or negative, and thus do not have dipole moments, which are partial positive and negative charges separated within the same molecule

1.13 (cont)

(g)

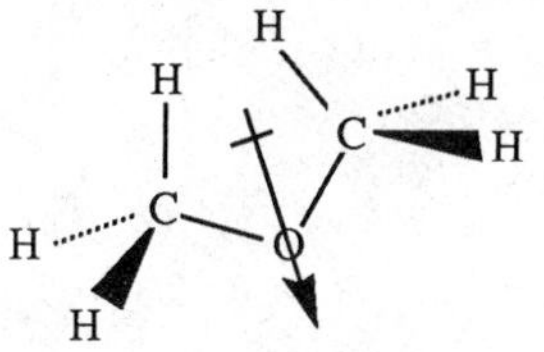

(h)

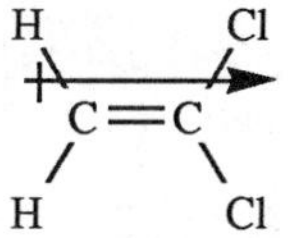

1.14   (a) $CH_3CH_2CH_2CH_2OH$ has the highest boiling point because of hydrogen bonding.

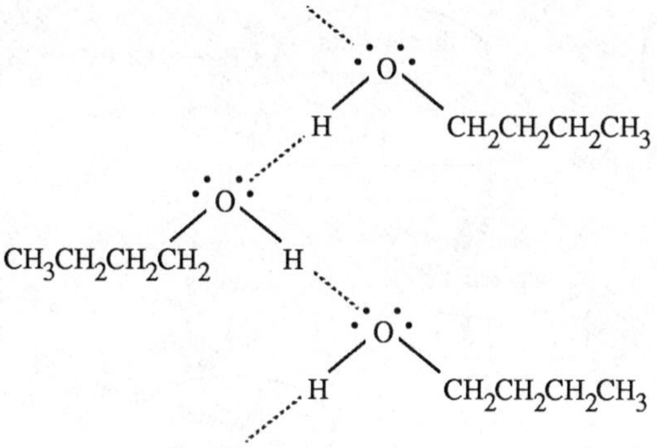

(b) CH_3OH has the highest boiling point because of hydrogen bonding.

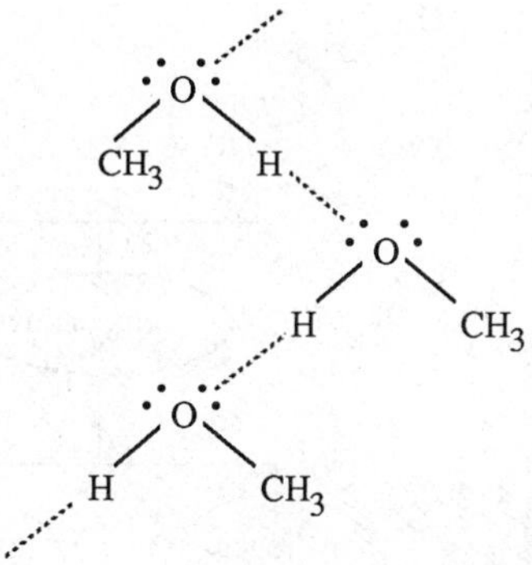

(c) CH_3OH has the highest boiling point [see(b)]. The hydrogen bonding in CH_3SH is not significant (see p. 25 in the text).

Concept Map 1.6   Nonbonding interactions between molecules.

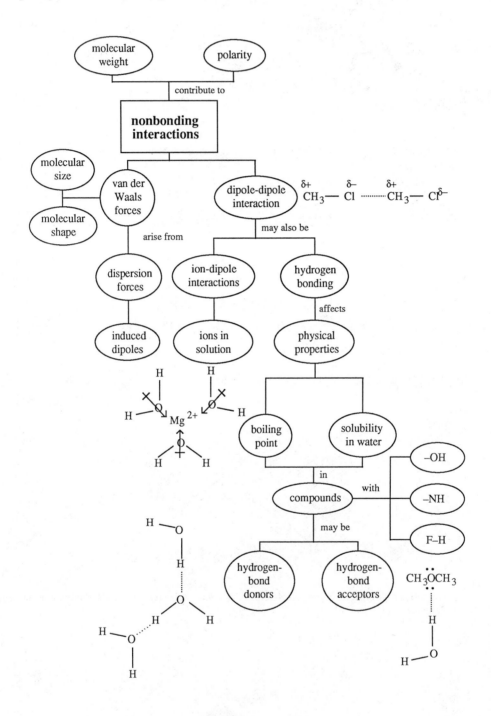

1.15   (a) CH_3CH_2OH

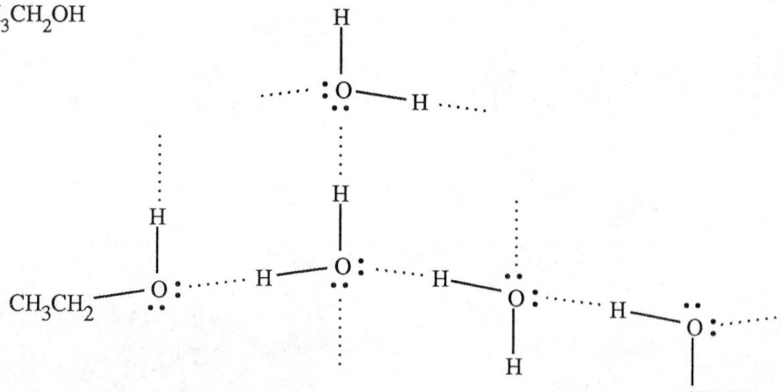

(b) $CH_3CH_2CH_2OH$

see (a) above and substitute $CH_3CH_2CH_2$— for CH_3CH_2—

(c) $HOCH_2CH_2CH_2CH_2CH_2OH$

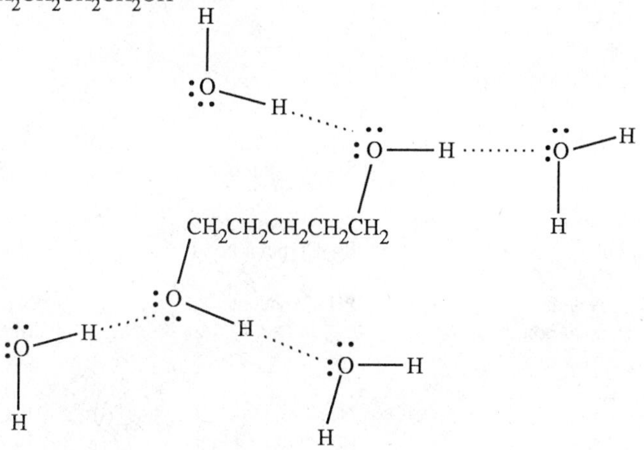

Twice as many hydrogen bonds as $CH_3CH_2CH_2CH_2CH_2OH$

(d) $CH_3CH_2\overset{\displaystyle O}{\overset{\displaystyle \|}{C}}OH$

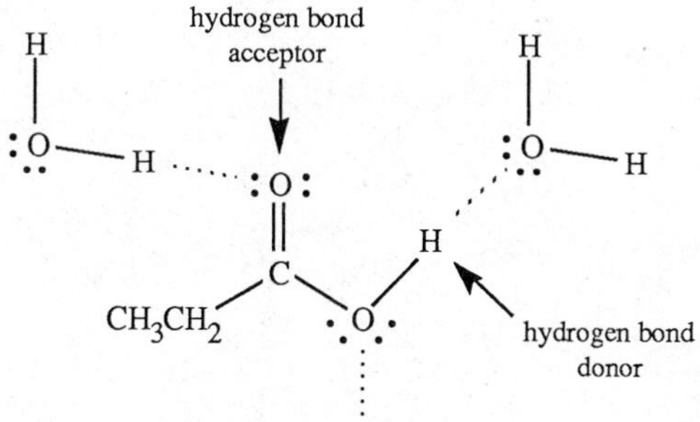

$CH_3\overset{\displaystyle O}{\overset{\displaystyle \|}{C}}OCH_3$ is only a hydrogen bond acceptor.

1.15 (cont)

(e) $CH_3CH_2CH_2NH_2$

$$:\overset{\displaystyle ..}{O}\!\!-\!\!H$$
$$|$$
$$H$$
$$\vdots$$

$$CH_3CH_2CH_2\overset{\displaystyle ..}{N}\!\!-\!\!H\cdots\cdots:\overset{\displaystyle ..}{O}\!\!-\!\!H$$
$$| \qquad\qquad\qquad |$$
$$H \qquad\qquad\qquad H$$
$$\vdots$$

$$:\overset{\displaystyle ..}{O}\!\!-\!\!H$$
$$|$$
$$H$$

$CH_3CH_2CH_2Cl$ does not participate in hydrogen bonding to a significant extent and has little soluble in water.

1.16   (a) MgF_2, ionic

(b) SiF_4, covalent
Si—F, polar bond

(c)   NaH, ionic
(the anion is $H:^-$)

(d)   ClF, covalent
polar bond

(e) SCl_2, covalent
S—Cl, polar bond

(f) OF_2, covalent
O—F, polar bond

(g) SiH_4, covalent
Si—H, polar bond

(h) PH_3, covalent

(i) $NaOCH_3$, ionic
(the anion, $CH_3\!-\!\overset{..}{\underset{..}{O}}:^-$,
contains covalent bonds)

(j) CH_3Na, ionic
(the anion, $^-:CH_3$,
contains covalent bonds)

(k) Na_2CO_2, ionic
(the anion, CO_3^{-2},
contains covalent bonds)

(l)   BrCN, covalent
Br—C and C≡N,
polar bonds

1.17   (a) $:\!\overset{..}{\underset{..}{F}}\!\!-\!\!\overset{..}{N}\!\!-\!\!\overset{..}{\underset{..}{F}}\!:$
$$|$$
$$:\!\overset{..}{\underset{..}{F}}\!:$$

(b) $:\!\overset{..}{\underset{..}{Cl}}\!\!-\!\!Al\!\!-\!\!\overset{..}{\underset{..}{Cl}}\!:$
$$|$$
$$:\!\overset{..}{\underset{..}{Cl}}\!:$$

(c) $H\!\!-\!\!\overset{\displaystyle H}{\underset{\displaystyle H}{C}}\!\!-\!\!\overset{..}{\underset{..}{S}}\!\!-\!\!\overset{\displaystyle H}{\underset{\displaystyle H}{C}}\!\!-\!\!H$

(d) $H\!\!-\!\!\overset{\displaystyle H}{\underset{\displaystyle H}{C}}\!\!-\!\!\overset{..}{N}\!\!-\!\!H$
$$|$$
$$H$$

(e) $H\!\!-\!\!\overset{\displaystyle H}{\underset{\displaystyle H}{C}}\!\!-\!\!\overset{\displaystyle H}{\underset{:\overset{..}{\underset{..}{Cl}}:}{C}}\!\!-\!\!\overset{\displaystyle H}{\underset{\displaystyle H}{C}}\!\!-\!\!H$

(f) $:\!\overset{..}{\underset{..}{O}}\!\!-\!\!H$

1.17 (cont)

(g)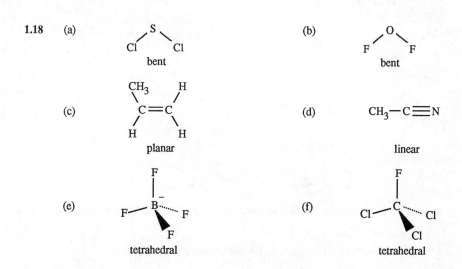

(h)

(i)

(j) $O = S = O$

Note that sulfur, an element in the third row, has 10 electrons around it and still has a zero formal charge.

S: $6 - 2 - {}^8\!/_2 = 0$

(k)

(l)

(m)

(n)

Here sulfur has 12 electrons around it but a zero formal charge.

S: $6 - 0 - {}^{12}\!/_2 = 0$

1.18   (a)

S
Cl   Cl
bent

(b)

O
F   F
bent

(c)

CH_3 H
C = C
H   H
planar

(d) $CH_3 - C \equiv N$

linear

(e)

F
F — B — F
F
tetrahedral

(f)

F
Cl — C — Cl
Cl
tetrahedral

1.18 (cont)

(g)

CH$_3$

CH$_3$—N$^+$·····CH$_3$

CH$_3$

tetrahedral

(h)

H—P·····H

H

pyramidal

(i)

CH$_3$—O$^+$·····H

H

pyramidal

(j)

CH$_3$—N·····CH$_3$

CH$_3$

pyramidal

1.19   (a)

H C=C H / H Cl

(b)

H, H, C—S, C·····H, H, H, H

(c)

H, H, C—C≡C—C, H, H, H, H

does not have a significant dipole moment

(d)

F, Br—C·····Br, Br

C—F bond dipole is much greater than C—Br bond dipoles

(e)

H, H, C—C, O—H, H, H

(f)

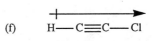

H—C≡C—Cl

(g)

Cl, Cl, C=C, Cl, Cl

individual bond dipoles cancel

1.20   (a) Hydrogen atoms are attached to carbon, not oxygen, nitrogen or fluorine. Hydrogen bonding is not important

      (b) There is a hydrogen atom attached to oxygen, so hydrogen bonding is important.

      (c) No hydrogen atoms are on oxygen, nitrogen or fluorine, so no hydrogen bonding.

1.21 (cont)

    (d) There is a hydrogen atom attached to oxygen, so hydrogen bonding is important.

    (e) Hydrogen atoms are attached to nitrogen, so hydrogen bonding does occur.

    (f) Here, too, hydrogen bonding occurs because of presence of N—H bonds.

    (g) The hydrogen attached to sulfur does not participate significantly in hydrogen bonding.

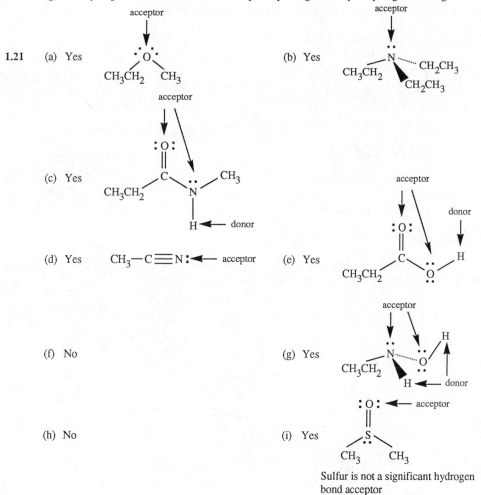

1.21 (a) Yes

(b) Yes

(c) Yes

(d) Yes

(e) Yes

(f) No

(g) Yes

(h) No

(i) Yes

Sulfur is not a significant hydrogen bond acceptor

1.22 (a) NaCl is more soluble (ionic compound, strong ion-dipole interactions)

    (b) HCOH (does not have nonpolar carbon chain)

1.22 (cont)

$$\overset{O}{\overset{\|}{}}$$

(c) HCOH (hydrogen bond acceptor and donor and does not have nonpolar carbon chain)

(d) $CH_3CH_2OCH_2CH_3$ (larger dipole, oxygen more electronegative than sulfur; also hydrogen bond acceptor)

$$\overset{O}{\overset{\|}{}}\qquad\overset{O}{\overset{\|}{}}$$

(e) $HOCCH_2CH_2COH$ (two hydrogen bond donor and four acceptor sites)

1.23

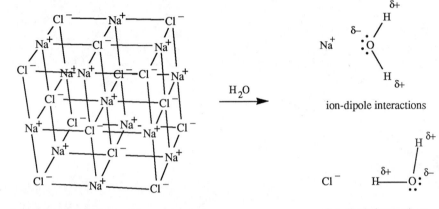

ion-ion interactions in crystal, very strong   ion-dipole interactions

In water solution, the solvation of the ions by the polar water molecules stabilizes the ions and compensates for the loss of the electrostatic interactions present in the crystal.

1.24

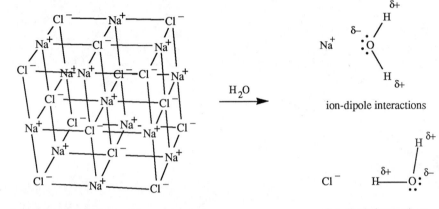

In water, all atoms are involved in hydrogen bonding. A strong, three-dimensional network results.

1.24 (cont)

········ H——$\ddot{\underset{\displaystyle\cdot\cdot}{F}}$: ········ H——$\ddot{\underset{\displaystyle\cdot\cdot}{F}}$: ········ H——$\ddot{\underset{\displaystyle\cdot\cdot}{F}}$: ········

In hydrogen fluoride, only two atoms are involved in hydrogen bonding, giving rise to a weaker linear network.

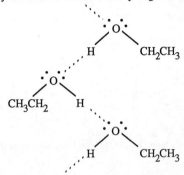

In ethanol, only two atoms are involved in hydrogen bonding. The rest of the molecule is nonpolar and involved in van der Waals interactions, which are much weaker.

1.25   (a)

$$\underset{\displaystyle H}{\overset{\displaystyle H}{H—C—C \equiv N:}}$$ Lewis structure

(b)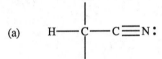

$CH_3—C \equiv N:$ dipole moment

(c) $CH_3—\overset{+}{C} = \ddot{\underset{\displaystyle\cdot\cdot}{N}}:^{-}$ Resonance contributor with separation of charge and negative formal charge on the more electronegative element. A separation of charge increases the dipole moment.

1.26

(a) $:N \equiv \overset{+}{N}—\ddot{\underset{\displaystyle\cdot\cdot}{O}}:^{-}$ ↔ $^{-}:N = \overset{+}{N} = \ddot{\underset{\displaystyle\cdot\cdot}{O}}$ ↔ $\overset{+}{N} = N—\ddot{\underset{\displaystyle\cdot\cdot}{O}}:^{-}$

          ①   ②   ③

complete octets;   complete octets;   nitrogen has sextet;
negative charge on more   negative charge on less   large separation of
electronegative atom   electronegative atom   charge

major   minor

1.26 (cont)

(b)

all equivalent with no separation of charge and at least an octet around each atom; major

↔ (three others);

minor; have separation of charge

(c)

$CH_3—C—O—H$ ↔ $CH_3—C=O—H$ ↔ $CH_3—C—O—H$

① ② ③

complete octets;
no separation of charge

complete octets;
large separation of charge

carbon has sextet;
separation of charge

major

minor

1.27   (a) $:C=O$ $^-:C≡O:^+$ $^+:C—O:^-$

carbon does not have
octet

both atoms have octets
but negative charge is
on less electronegative
atom

carbon does not have an
octet but negative charge
is on more electronegative
atom

(b) Fe^{2+} is an electron pair **acceptor**. The carbon atom of carbon monoxide bonds to iron; it must be an electron pair **donor**. Therefore, the resonance contributor in which carbon has a negative charge and each atom has a complete octet is more important than the other contributors. The resonance contributor with separation of charge and a positive charge on the carbon atom would lead to electrostatic repulsion between Fe^{2+} and the carbon atom and must, therefore, not make a significant contribution to the structure.

1.28

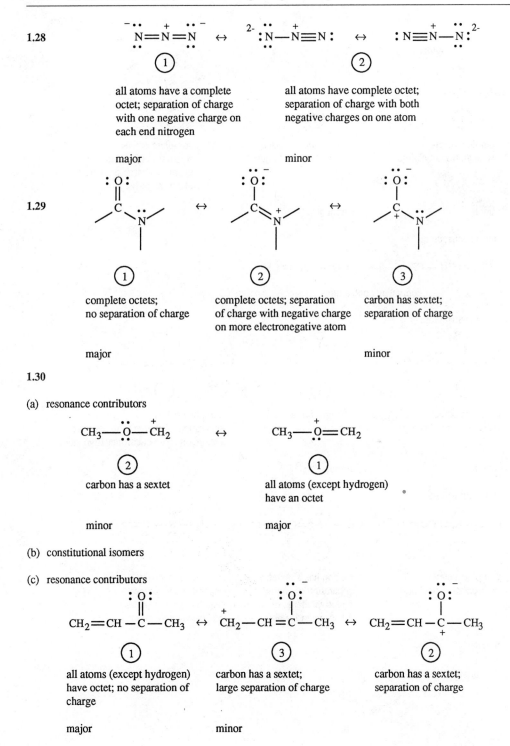

1.29

1.30

(a) resonance contributors

(b) constitutional isomers

(c) resonance contributors

1.30 (cont)

(d) resonance contributors

$$CH_3-\overset{\overset{\displaystyle :O:}{\|}}{C}-\overset{..}{\underset{..}{O}}-CH_3 \;\leftrightarrow\; CH_3-\overset{\overset{\displaystyle :\overset{..}{O}:}{|}}{\underset{+}{C}}-\overset{..}{\underset{..}{O}}-CH_3 \;\leftrightarrow\; CH_3-\overset{\overset{\displaystyle :\overset{..}{O}:^-}{|}}{C}=\overset{+}{\underset{..}{O}}-CH_3$$

                    ①   ③   ②

     complete octets;   carbon has a sextet;   complete octets;
     no separation of charge   separation of charge   separation of charge;
                                                           negative charge on more
                                                            electronegative atom

     major   minor

(e) resonance contributors

$$CH_2=CH-\overset{+}{CH_2} \quad\leftrightarrow\quad \overset{+}{CH_2}-CH=CH_2$$

                     equivalent contributors

(f) resonance contributors

$$CH_3-\overset{\overset{\displaystyle CH_3}{|}}{\underset{..}{N}}-\overset{..}{N}=\overset{..}{\underset{..}{O}} \;\leftrightarrow\; CH_3-\overset{\overset{\displaystyle CH_3}{|}}{\underset{..}{N}}-\overset{..}{\underset{+}{N}}-\overset{..}{\underset{..}{O}}:^- \;\leftrightarrow\; CH_3-\overset{\overset{\displaystyle CH_3}{|}}{\underset{+}{N}}=\overset{..}{N}-\overset{..}{\underset{..}{O}}:^-$$

                    ①   ③   ②

     complete octets;   nitrogen has a sextet;   complete octets;
     no separation of charge   separation of charge   separation of charge with negative
                                                                charge on more electronegative atom

     major   minor

(g) resonance contributors

$$CH_3-\overset{..}{N}=C=\overset{..}{\underset{..}{O}} \;\leftrightarrow\; CH_3-\overset{..}{N}=\overset{+}{C}-\overset{..}{\underset{..}{O}}:^- \;\leftrightarrow\; CH_3-\overset{..}{\underset{..}{N}}{}^{-}-\overset{+}{C}=\overset{..}{\underset{..}{O}}$$

                    ①   ②   ③

     complete octets;   carbon has a sextet;   carbon has a sextet;
     no separation of charge   separation of charge;   separation of charge;
                                           negative charge on more   negative charge on less
                                           electronegative atom   electronegative atom

     major   minor

1.30 (cont)

(h) resonance contributors

complete octets

major

carbon has a sextet

minor

(i)   The first two are resonance contributors, the third is a constitutional isomer of the first two.

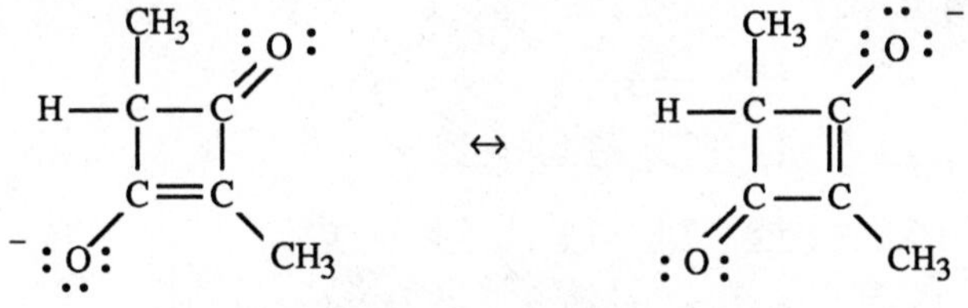

both have complete octets around
each atom (except hydrogen)

essentially equivalent

(j)   resonance contributors

The first structure is a minor contributor (negative charge is on the less electronegative atom); the next four are equivalent and major resonance contributors (negative charge is on the more electronegative atom).

2

Covalent Bonding and Chemical Reactivity

		1s	2s	2p			3s	3p		
2.1	sodium	↑↓	↑↓	↑↓	↑↓	↑↓	↑			
	magnesium	↑↓	↑↓	↑↓	↑↓	↑↓	↑↓			
	aluminum	↑↓	↑↓	↑↓	↑↓	↑↓	↑↓	↑		
	silicon	↑↓	↑↓	↑↓	↑↓	↑↓	↑↓	↑	↑	
	phosphorus	↑↓	↑↓	↑↓	↑↓	↑↓	↑↓	↑	↑	↑
	sulfur	↑↓	↑↓	↑↓	↑↓	↑↓	↑↓	↑↓	↑	↑
	chlorine	↑↓	↑↓	↑↓	↑↓	↑↓	↑↓	↑↓	↑↓	↑
	argon	↑↓	↑↓	↑↓	↑↓	↑↓	↑↓	↑↓	↑↓	↑↓

2.2

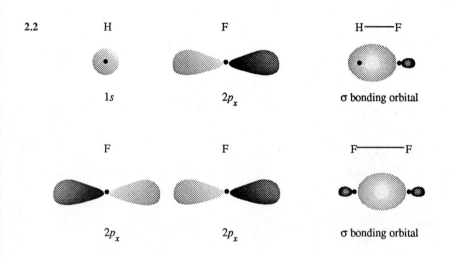

H   F   H——F

1s $2p_x$ σ bonding orbital

F   F   F———F

$2p_x$ $2p_x$ σ bonding orbital

Concept Map 2.1   Molecular orbitals and covalent bonds.

2.3   (a) No overlap possible; wrong orientation of *p* orbital in space. The *s* orbital is approaching the nodal plane of the *p* orbital where there is no electron density.

(b) No overlap possible; wrong orientation of one *p* orbital in space.

(c) No bonding will occur. The orbitals are out of phase, and only an antibonding interaction will occur.

(d) Bonding will occur. The orbitals are in phase and are approaching each other so that overlap is possible.

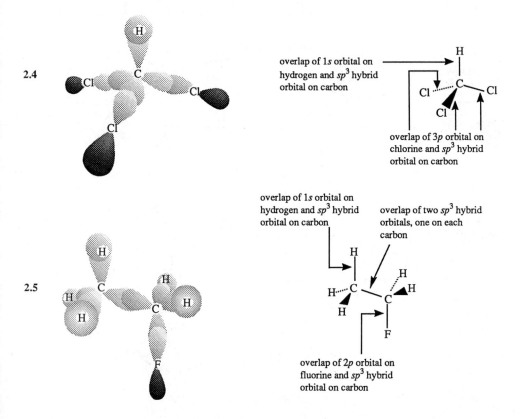

2.4

overlap of 1*s* orbital on hydrogen and *sp*³ hybrid orbital on carbon

overlap of 3*p* orbital on chlorine and *sp*³ hybrid orbital on carbon

2.5

overlap of 1*s* orbital on hydrogen and *sp*³ hybrid orbital on carbon

overlap of two *sp*³ hybrid orbitals, one on each carbon

overlap of 2*p* orbital on fluorine and *sp*³ hybrid orbital on carbon

2.6   The three 2*p* orbitals on nitrogen are 90° apart. Bonds formed by the overlap of these orbitals would have bond angles close to 90°. The actual bond angles in ammonia have been experimentally determined to be 107°.

2.7

nonbonding electrons

2.8

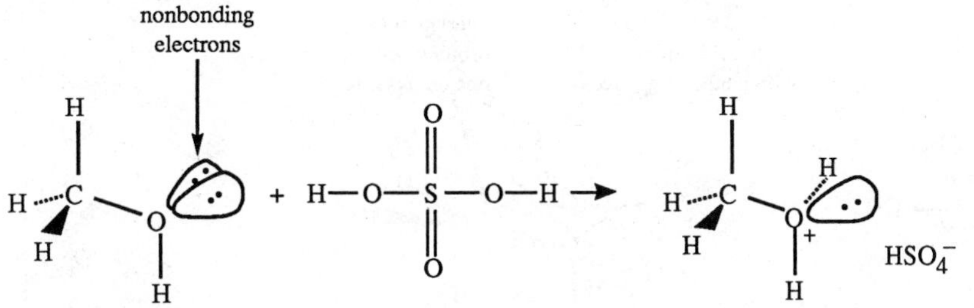

nonbonding electrons

methanol

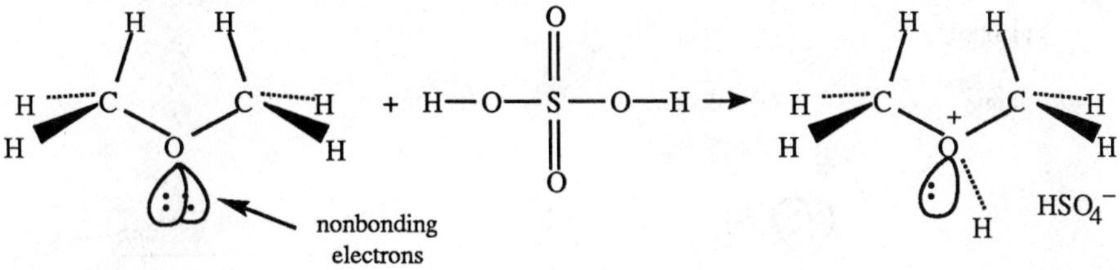

dimethyl ether

nonbonding electrons

2.9

(a)

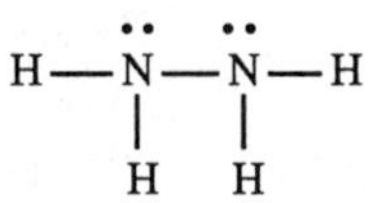

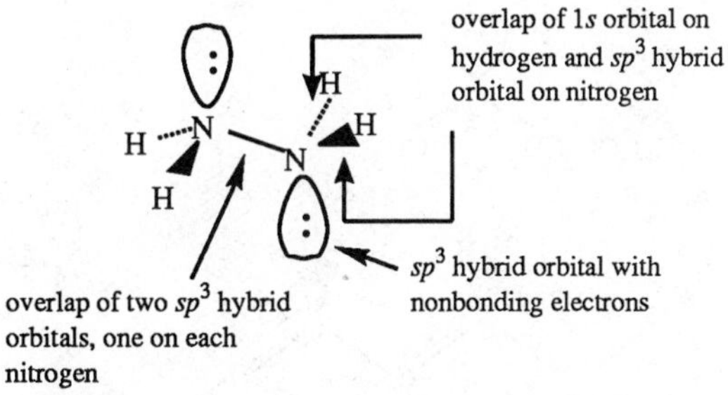

$$H—N—N—H$$

overlap of 1s orbital on hydrogen and sp^3 hybrid orbital on nitrogen

overlap of two sp^3 hybrid orbitals, one on each nitrogen

sp^3 hybrid orbital with nonbonding electrons

(b) BF_4^-

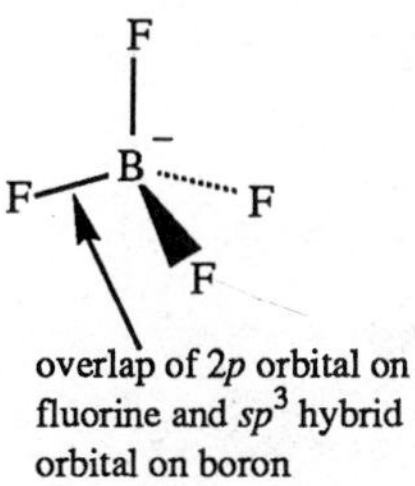

overlap of 2p orbital on fluorine and sp^3 hybrid orbital on boron

overlap of two sp^3 hybrid orbitals, one on carbon and one on nitrogen

(c) $CH_3—\overset{+}{N}—CH_3$

overlap of 1s orbital on hydrogen and sp^3 hybrid orbital on carbon

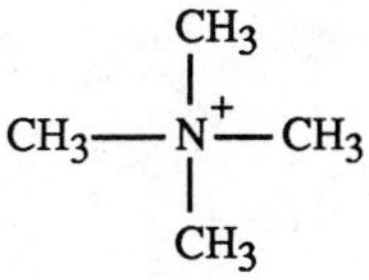

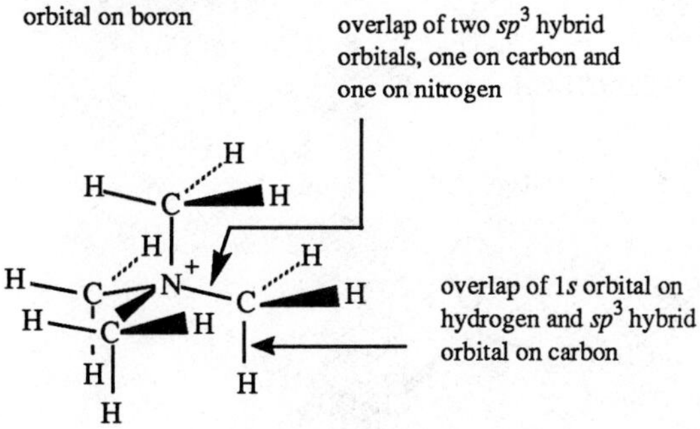

2.9 (cont)

(d) $CH_3 \overset{\underset{\displaystyle CH_3}{|}}{\underset{\displaystyle \cdot\cdot}{O}}\overset{+}{-} CH_3$

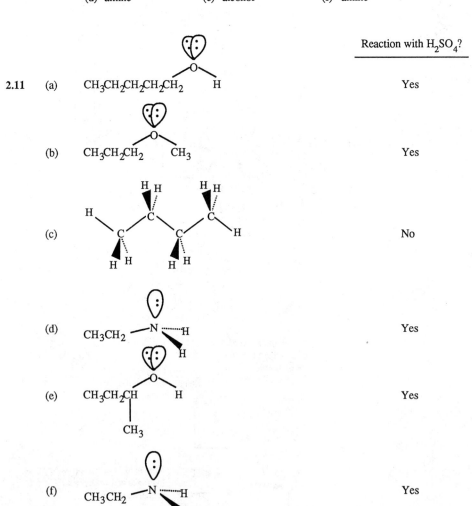

sp^3 hybrid orbital with nonbonding electrons

overlap of two sp^3 hybrid orbitals, one on carbon and one on oxygen

2.10   (a) alcohol   (b) ether   (c) alkane

(d) amine   (e) alcohol   (f) amine

		Reaction with H_2SO_4?
2.11 (a)	$CH_3CH_2CH_2CH_2CH_2$ O H	Yes
(b)	$CH_3CH_2CH_2$ O CH_3	Yes
(c)		No
(d)	CH_3CH_2 — N — H, H	Yes
(e)	CH_3CH_2CH O H, CH_3	Yes
(f)	CH_3CH_2 — N — H, CH_2CH_3	Yes

2.12

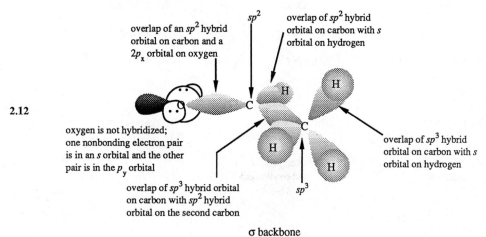

overlap of an sp^2 hybrid
orbital on carbon and a
$2p_x$ orbital on oxygen

sp^2

overlap of sp^2 hybrid
orbital on carbon with s
orbital on hydrogen

overlap of sp^3 hybrid
orbital on carbon with s
orbital on hydrogen

oxygen is not hybridized;
one nonbonding electron pair
is in an s orbital and the other
pair is in the p_y orbital

overlap of sp^3 hybrid orbital
on carbon with sp^2 hybrid
orbital on the second carbon

sp^3

σ backbone

overlap of two $2p_z$ orbitals, one
on carbon and one on oxygen, in
phase with each other

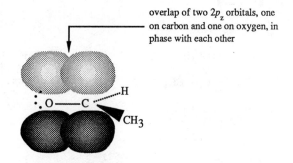

2.13

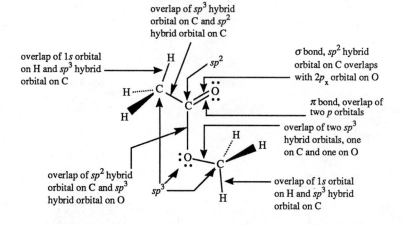

overlap of sp^3 hybrid
orbital on C and sp^2
hybrid orbital on C

overlap of $1s$ orbital
on H and sp^3 hybrid
orbital on C

sp^2

σ bond, sp^2 hybrid
orbital on C overlaps
with $2p_x$ orbital on O

π bond, overlap of
two p orbitals

overlap of two sp^3
hybrid orbitals, one
on C and one on O

overlap of sp^2 hybrid
orbital on C and sp^3
hybrid orbital on O

sp^3

overlap of $1s$ orbital
on H and sp^3 hybrid
orbital on C

2.14   (a) alcohol   (b) alkene   (c) aldehyde

      (d) ketone   (e) amine   (f) alkane

      (g) carboxylic acid   (h) ether   (i) ester

2.15   (a) alcohol and   (b) amine and   (c) ketone and
          carboxylic acid   carboxylic acid   carboxylic acid

      (d) alcohol and   (e) alkene and
          aldehyde   carboxylic acid

2.16   (a)   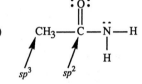   (b)

2.17   (a) alkyne   (b) alkene   (c) ketone

      (d) ester   (e) nitrile   (f) aldehyde

      (g) carboxylic acid   (h) amine

2.18   (a)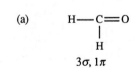
            $3\sigma, 1\pi$ (b) 8σ

Concept Map 2.2   Hybrid orbitals.

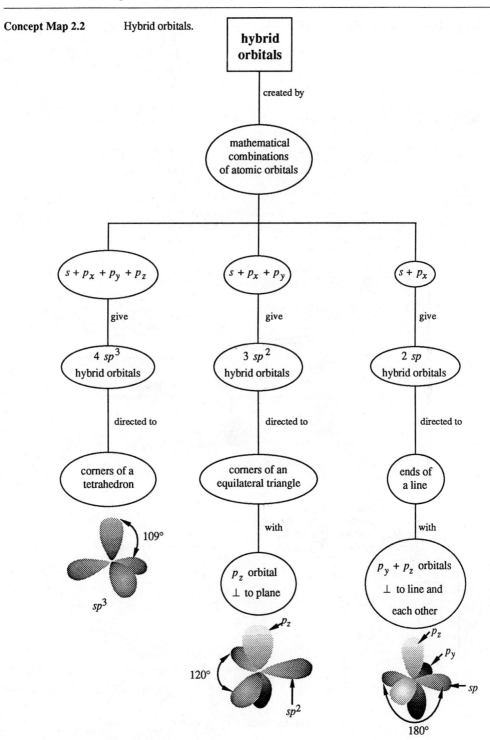

2.18   (c)   H—C≡C—C≡C—H   (d)   H—C=C—C≡N
                                                   | |
                                                   H   H

          5σ, 4π   6σ, 3π

                H   O   H   H   H   H
                | || | | | |
   (e)   H—C—C—O—C—H   (f)   H—C—C—C—H
                | | | | |
                H   H   H   H   H

          10σ, 1π   10σ

overlap of 1s orbital   overlap of two sp hybrid
on H and sp hybrid on C   orbitals, one on N and one on C

2.19

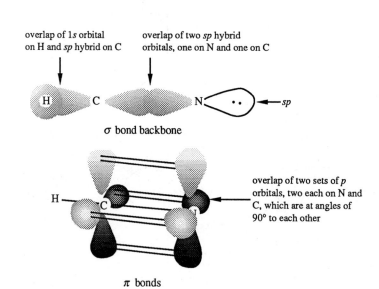

σ bond backbone

overlap of two sets of p
orbitals, two each on N and
C, which are at angles of
90° to each other

π bonds

overlap of sp³ hybrid orbital
on one C and sp hybrid   overlap of 1s orbital
on the other C   on H and sp hybrid orbital on C

2.20

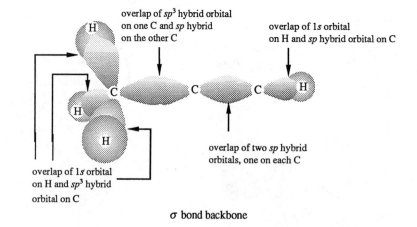

overlap of two sp hybrid
orbitals, one on each C

overlap of 1s orbital
on H and sp³ hybrid
orbital on C

σ bond backbone

2.20 (cont)

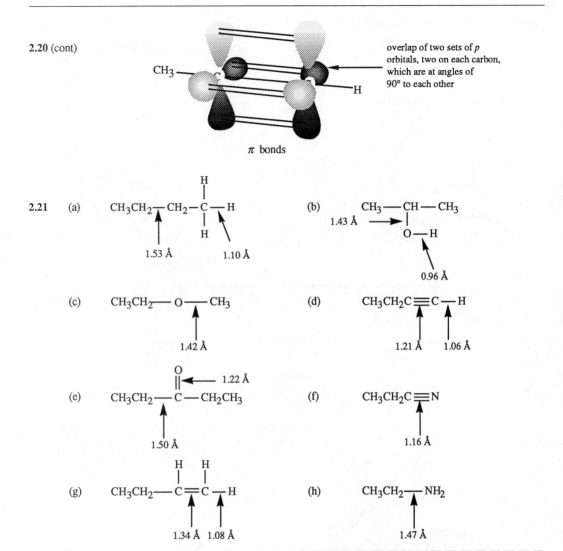

overlap of two sets of *p* orbitals, two on each carbon, which are at angles of 90° to each other

π bonds

2.21 (a) $\text{CH}_3\text{CH}_2\text{—CH}_2\text{—C—H}$ with H above and H below

1.53 Å   1.10 Å

(b) $\text{CH}_3\text{—CH—CH}_3$
      1.43 Å → with O—H below
                      0.96 Å

(c) $\text{CH}_3\text{CH}_2\text{—O—CH}_3$

1.42 Å

(d) $\text{CH}_3\text{CH}_2\text{C}\!\equiv\!\text{C—H}$

1.21 Å   1.06 Å

(e) $\text{CH}_3\text{CH}_2\text{—C—CH}_2\text{CH}_3$ with O double-bonded above ← 1.22 Å

1.50 Å

(f) $\text{CH}_3\text{CH}_2\text{C}\!\equiv\!\text{N}$

1.16 Å

(g) $\text{CH}_3\text{CH}_2\text{—C}\!=\!\text{C—H}$ with H on each C

1.34 Å 1.08 Å

(h) $\text{CH}_3\text{CH}_2\text{—NH}_2$

1.47 Å

Concept Map 2.3   (see p. 40)

2.22

(a) $\text{CH}_3\text{—H} + \text{Cl—Cl} \longrightarrow \text{CH}_3\text{—Cl} + \text{H—Cl}$

Bond energies (kcal/mol)   99   58   81   103

$$\Delta H_r = (99 + 58) - (81 + 103) = -27 \text{ kcal/mol}$$

Concept Map 2.3   Bond lengths and bond strengths.

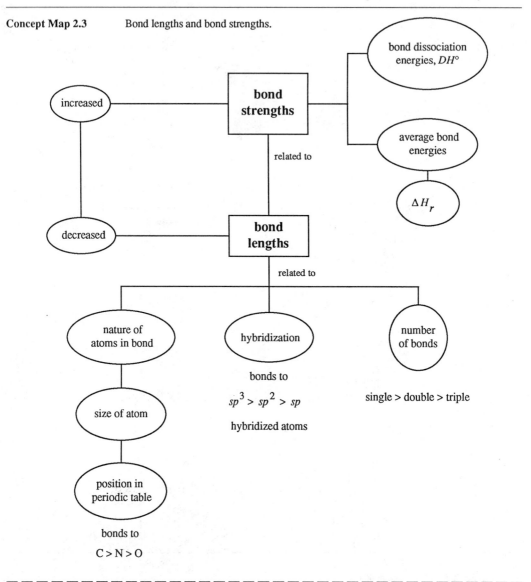

- -

2.22 (cont)

(b)

$$H_2C \Longleftrightarrow O \quad + \quad H \Longleftrightarrow C \equiv N \quad \longrightarrow \quad H \Longleftrightarrow O \Longleftrightarrow CH_2 \Longleftrightarrow C \equiv N$$

Bond energies (kcal/mol)   176   99   111   86   83

$$\Delta H_r = (176 + 99) - (111 + 86 + 83) = -5 \text{ kcal/mol}$$

2.22 (cont)   (Note that to calculate ΔH_r in a molecule in which a π bond is broken, both the π bond and the σ bond are broken and the σ bond is reformed in the product.)

(c)

$$H_2C = CH_2 \quad + \quad H - Br \quad \longrightarrow \quad H - CH_2 - CH_2 - Br$$

146   87   99   83   68

$$\Delta H_r = (146 + 87) - (99 + 83 + 68) = -17 \text{ kcal/mol}$$

(d)

$$CH_3 - CH - CH_2 \quad \longrightarrow \quad CH_3 - CH = CH_2 \quad + \quad H - O - H$$

83 ... 146 ... 111

86 ... OH ... H ... 99

$$\Delta H_r = (86 + 83 + 99) - (146 + 111) = +11 \text{ kcal/mol}$$

2.23

(a)

$$CH_3CH_2CH=CHCH_3 \quad + \quad Br_2 \quad \xrightarrow[\substack{\text{carbon} \\ \text{tetrachloride} \\ \text{dark, 25 °C}}]{} \quad CH_3CH_2CHCHCH_3$$

$$\underset{Br \ Br}{|\ \ |}$$

(b)

$$CH_3C \equiv CCH_3 \quad + \quad 2\,Br_2 \quad \xrightarrow[\substack{\text{carbon} \\ \text{tetrachloride} \\ \text{dark, 25 °C}}]{} \quad CH_3C - CCH_3$$

Br Br above, Br Br below

$$Br_2 \searrow$$

$$\underset{Br}{\overset{CH_3}{C}} = \underset{CH_3}{\overset{Br}{C}} \qquad \nearrow Br_2$$

(c)

$$HC \equiv CCH_2C \equiv CH \quad + \quad 4\,Br_2 \quad \xrightarrow[\substack{\text{carbon} \\ \text{tetrachloride} \\ \text{dark, 25 °C}}]{} \quad HC - CCH_2C - CH$$

Br Br Br Br above; Br Br Br Br below

$$Br_2 \downarrow$$

$$\underset{Br}{\overset{CH_3}{C}} = \underset{CH_2C \equiv CH}{\overset{Br}{C}} \xrightarrow{Br_2} \underset{Br}{\overset{H}{C}} = \underset{\underset{\underset{Br}{C} = \underset{H}{C}}{CH_2}}{\overset{Br}{C}} \xrightarrow{Br_2} \underset{Br}{\overset{HCBr_2CBr_2CH_2}{C}} = \underset{H}{\overset{Br}{C}}$$

$$Br_2 \uparrow$$

2.23 (cont)

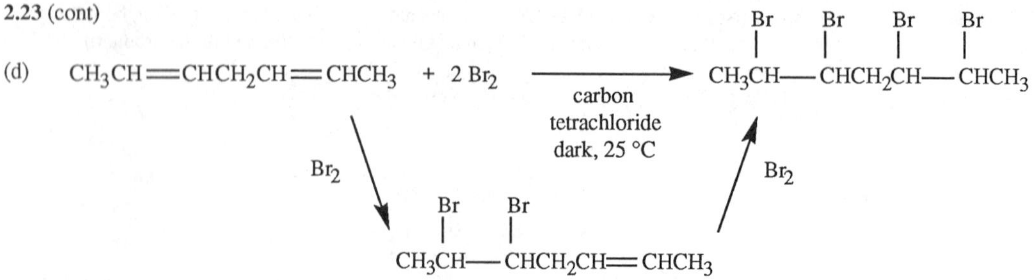

(d)

$$CH_3CH=CHCH_2CH=CHCH_3 + 2\,Br_2 \xrightarrow[\substack{\text{carbon}\\\text{tetrachloride}\\\text{dark, 25 °C}}]{} \underset{Br\ \ \ Br\ \ \ Br\ \ \ Br}{CH_3CH-CHCH_2CH-CHCH_3}$$

Br_2 ↘   ↗ Br_2

$$\underset{Br\ \ \ Br}{CH_3CH-CHCH_2CH=CHCH_3}$$

2.24   Alkanes: Tetrahedral carbon atoms bonded only to other tetrahedral carbon or hydrogen atoms

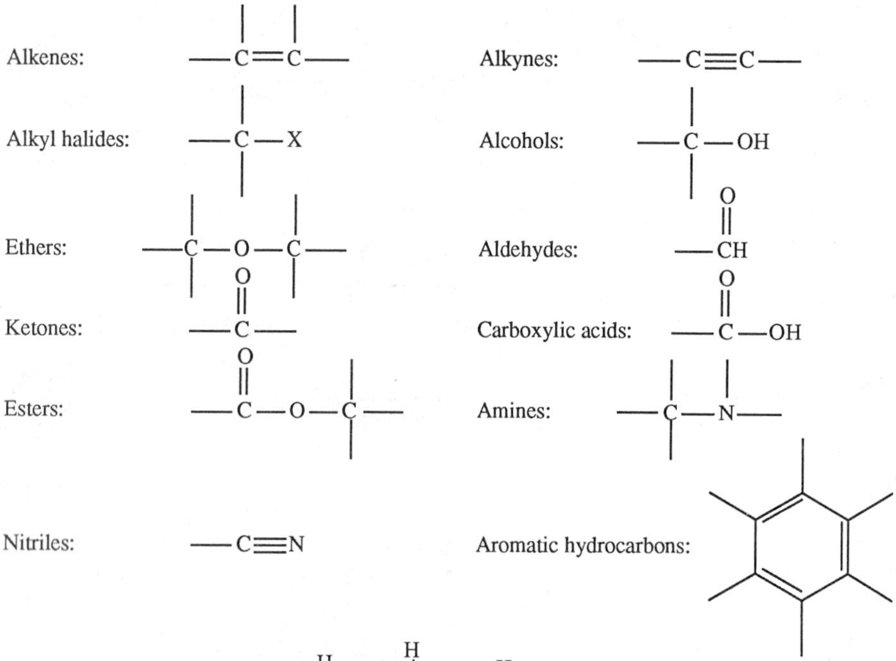

Alkenes: $-C=C-$ Alkynes: $-C\equiv C-$

Alkyl halides: $-C-X$ Alcohols: $-C-OH$

Ethers: $-C-O-C-$ Aldehydes: $-\overset{\overset{\displaystyle O}{\|}}{C}H$

Ketones: $-\overset{\overset{\displaystyle O}{\|}}{C}-$ Carboxylic acids: $-\overset{\overset{\displaystyle O}{\|}}{C}-OH$

Esters: $-\overset{\overset{\displaystyle O}{\|}}{C}-O-C-$ Amines: $-C-N-$

Nitriles: $-C\equiv N$ Aromatic hydrocarbons:

2.25

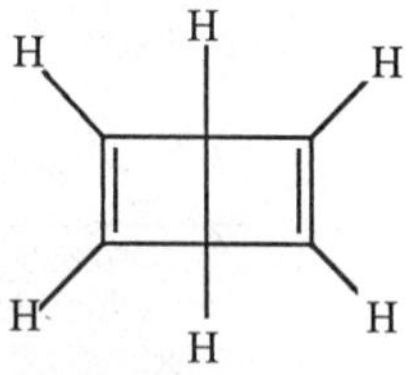

No delocalization of electrons in this structure.
We would expect the compound to behave like an alkene.

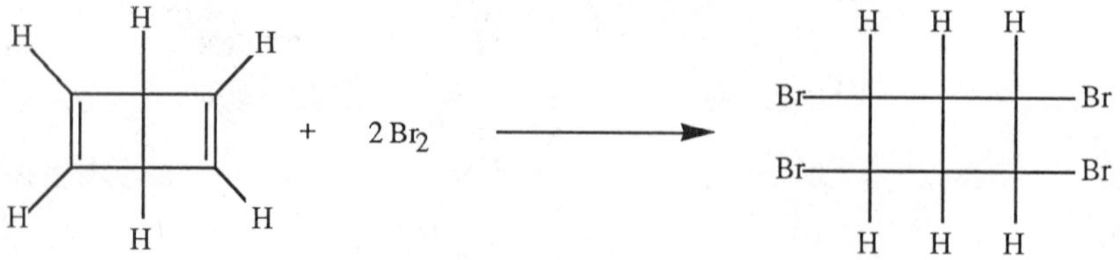

$+$ $2\,Br_2$ $\longrightarrow$

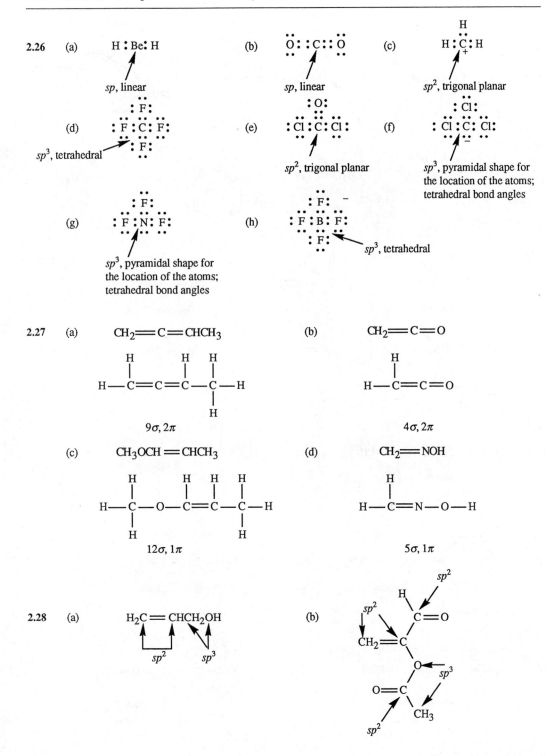

2.26   (a)   H : Be : H

sp, linear

(b)   O : : C : : O

sp, linear

(c)   H : C : H (with H above)

sp², trigonal planar

(d)   : F : C : F : (with F above and below)

sp³, tetrahedral

(e)   : Cl : C : Cl : (with O above)

sp², trigonal planar

(f)   : Cl : C : Cl : (with Cl above)

sp³, pyramidal shape for the location of the atoms; tetrahedral bond angles

(g)   : F : N : F : (with F above)

sp³, pyramidal shape for the location of the atoms; tetrahedral bond angles

(h)   : F : B : F : (with F above and F below)

sp³, tetrahedral

2.27   (a)   CH₂=C=CHCH₃

H—C=C=C—C—H (with H atoms attached)

9σ, 2π

(b)   CH₂=C=O

H—C=C=O (with H attached)

4σ, 2π

(c)   CH₃OCH=CHCH₃

H—C—O—C=C—C—H (with H atoms attached)

12σ, 1π

(d)   CH₂=NOH

H—C=N—O—H (with H attached)

5σ, 1π

2.28   (a)   H₂C=CHCH₂OH

sp²   *sp³*

(b)   (structure with labels *sp²*, *sp²*, *sp³*, *sp²* on CH₂=C, C=O, O, O=C, CH₃)

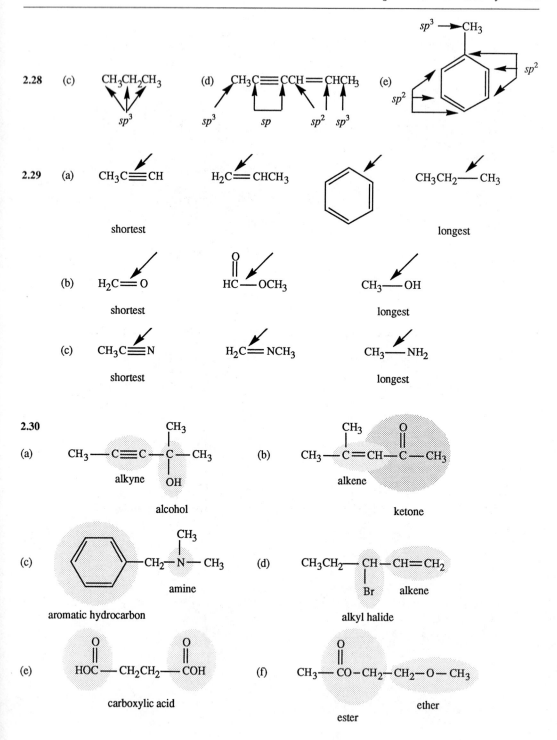

2.28 (c) CH₃CH₂CH₃   (d) CH₃C≡CCH=CHCH₃   (e)

sp^3 sp^3 sp sp^2 sp^3

$sp^3 \longrightarrow CH_3$

sp^2

sp^2

2.29 (a) CH₃C≡CH   H₂C=CHCH₃   CH₃CH₂—CH₃

shortest   longest

(b) H₂C=O   HC—OCH₃ (with =O)   CH₃—OH

shortest   longest

(c) CH₃C≡N   H₂C=NCH₃   CH₃—NH₂

shortest   longest

2.30

(a) CH₃—C≡C—C(CH₃)—CH₃ with OH
          alkyne   alcohol

(b) CH₃—C(CH₃)=CH—C(=O)—CH₃
          alkene   ketone

(c) [benzene ring]—CH₂—N(CH₃)—CH₃
          amine
     aromatic hydrocarbon

(d) CH₃CH₂—CH(Br)—CH=CH₂
          alkyl halide   alkene

(e) HOC(=O)—CH₂CH₂—C(=O)OH
          carboxylic acid

(f) CH₃—CO(=O)—CH₂—CH₂—O—CH₃
          ester   ether

2.30 (cont)

(g)

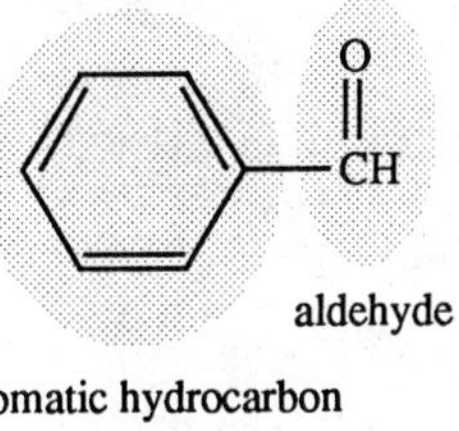

aldehyde

aromatic hydrocarbon

(h) CH_3—CH—CH_2—CH

OH

aldehyde

alcohol

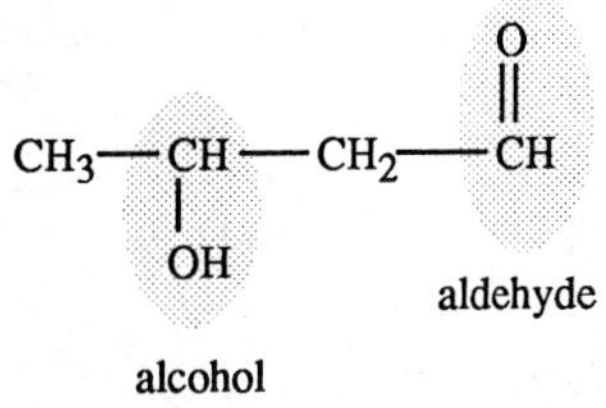

(i) ——CH_2OH

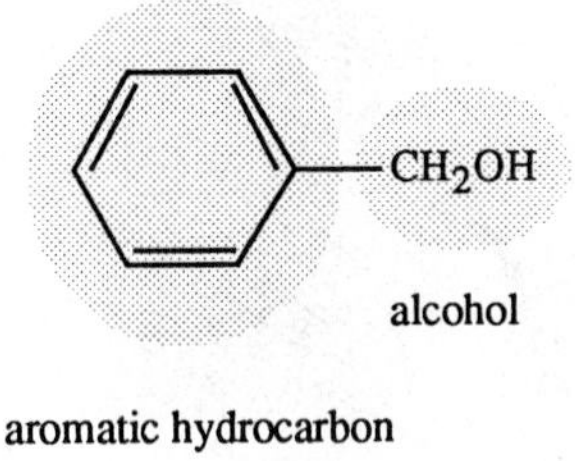

alcohol

aromatic hydrocarbon

(j) $CH_3CH_2CH_2$——C≡N

nitrile

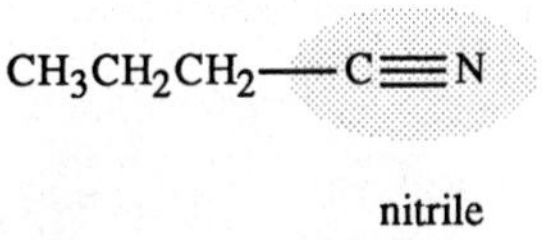

2.31   a)

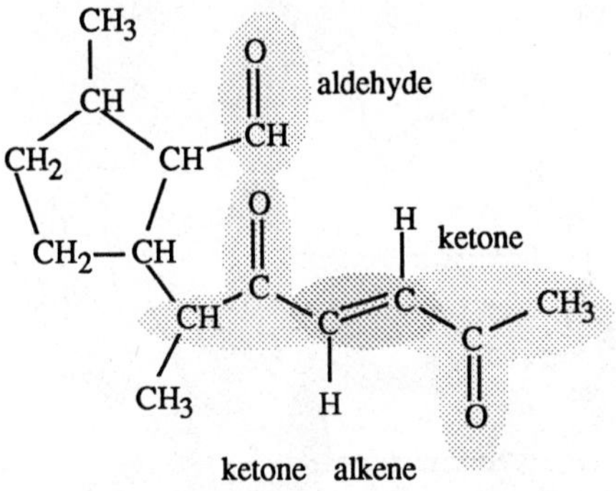

CH₃

CH

CH₂

CH₂—CH

O

CH

aldehyde

CH

O

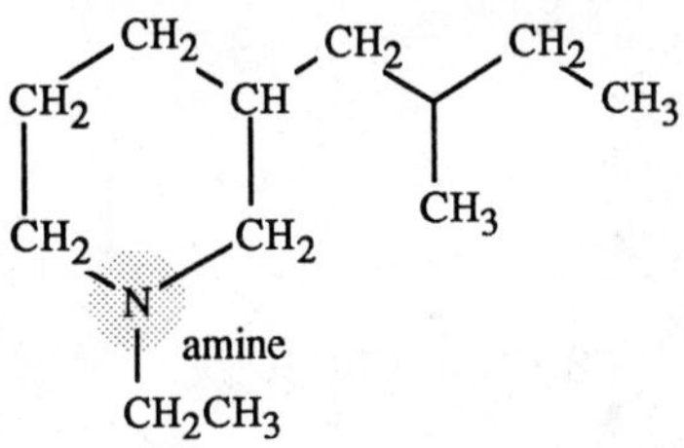

C

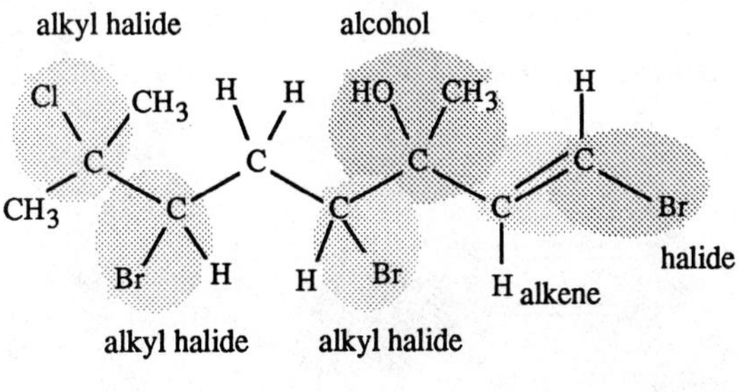

CH

CH₃

H

ketone

C=C

H

CH₃

C

O

ketone alkene

b) CH₂

CH₂

CH₂

CH

CH₂

CH₂

N

CH₂

CH₂

CH₃

amine

CH₃

CH₂CH₃

c)

alkyl halide

Cl   CH₃   H   H

C

CH₃

C

Br   H

alkyl halide

alcohol

HO   CH₃

C

C

H   Br

H

alkyl halide

H

C=C

Br

H alkene

halide

e)

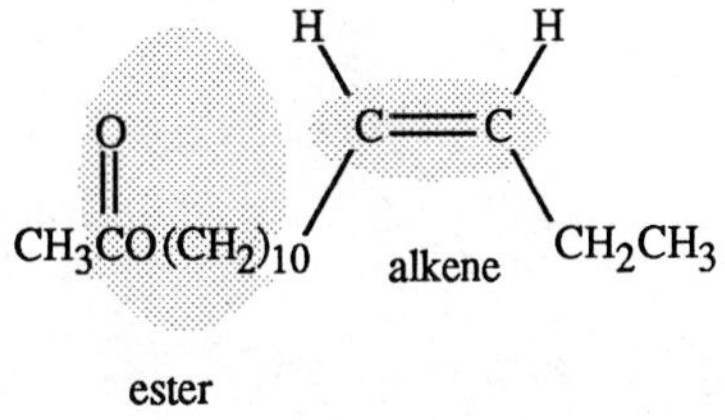

O

CH₃CO(CH₂)₁₀

H

C=C

alkene

H

CH₂CH₃

ester

d)

H

C=C

CH₃(CH₂)₄   alkene

H H

C=C

CH₂   alkene

H

(CH₂)₂C≡C(CH₂)₂COH

alkyne   carboxylic acid

O

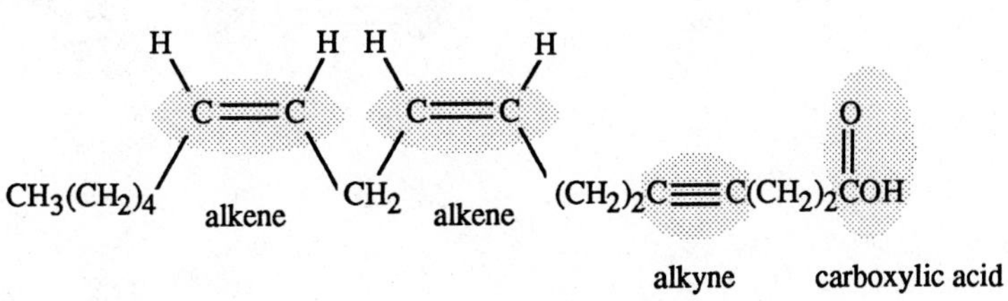

2.32

(a) $H_2C{=\!\!=}O$

sp^2

(b) CH_3NHOH

sp^3

(c) $CH_3CH{=\!\!=}CHCH_3$

sp^3 sp^2 sp^3

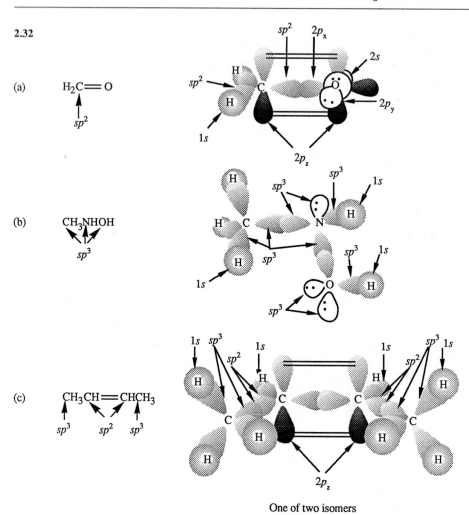

One of two isomers

(d) $CH_3C{\equiv}CCH_3$

sp^3 sp sp^3

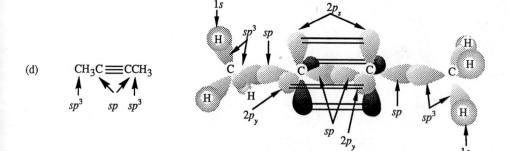

2.32

(a) $CH_3C \equiv N: \ + \ H-OSOH \ \longrightarrow \ CH_3C \equiv \overset{+}{N}-H \quad HSO_4^-$

(the $OSOH$ group bears $=O$ above and below)

(b) $CH_3\overset{:O:}{\underset{}{C}}CH_3 \ + \ H-OSOH \ \longrightarrow \ CH_3\overset{\overset{+}{:O}-H}{\underset{}{C}}CH_3 \quad HSO_4^-$

(c) $CH_3\overset{CH_3}{\underset{}{C}} = \overset{..}{N}CH_3 \ + \ H-OSOH \ \longrightarrow \ CH_3\overset{CH_3}{\underset{}{C}} = \overset{+}{\underset{H}{N}}CH_3 \quad HSO_4^-$

(d) $CH_3\overset{:O:}{\underset{..}{C}}OCH_3 \ + \ H-OSOH \ \longrightarrow \ CH_3\overset{\overset{+}{:O}-H}{\underset{..}{C}}OCH_3 \quad HSO_4^-$

or

$CH_3\overset{:O:}{\underset{..}{C}}\overset{+}{\underset{H}{O}}-CH_3 \quad HSO_4^-$

2.34

(a) $CH_3CH = CHCH_3 \ + \ Br_2 \xrightarrow[\substack{\text{carbon} \\ \text{tetrachloride} \\ \text{dark, 25 °C}}]{} CH_3\overset{}{\underset{Br}{C}}H - \overset{}{\underset{Br}{C}}HCH_3$

(b) $CH_3CH_2CH = CH_2 \ + \ H_2 \xrightarrow[\text{catalyst}]{} CH_3CH_2\overset{}{\underset{H}{C}}H - \overset{}{\underset{H}{C}}H_2$

(c) $CH_3CH_2CH_2CH_3 \ + \ Br_2 \xrightarrow[\substack{\text{carbon} \\ \text{tetrachloride} \\ \text{dark, 25 °C}}]{} \text{no reaction}$

(d) (phenyl)$-CH = CH_2 \ + \ Br_2 \xrightarrow[\substack{\text{carbon} \\ \text{tetrachloride} \\ \text{dark, 25 °C}}]{} \text{(phenyl)}-\overset{}{\underset{Br}{C}}H - \overset{}{\underset{Br}{C}}H_2$

2.34 (cont)

(e) $CH_3C\equiv CH$ + Br_2 (excess) $\xrightarrow[\substack{\text{carbon}\\\text{tetrachloride}\\\text{dark, 25 °C}}]{}$

$$CH_3\underset{\underset{Br}{|}}{\overset{\overset{Br}{|}}{C}}-\underset{\underset{Br}{|}}{\overset{\overset{Br}{|}}{CH}}$$

(f) $CH_3C\equiv CH$ + H_2 (excess) $\xrightarrow[\text{catalyst}]{}$

$$CH_3\underset{\underset{H}{|}}{\overset{\overset{H}{|}}{C}}-\underset{\underset{H}{|}}{\overset{\overset{H}{|}}{CH}}$$

2.35

(a)

$$H_2C{=}CH_2 + Br{-}Br \longrightarrow Br{-}CH_2{-}CH_2{-}Br$$

Bond energies (kcal/mol)   146   46   68   83   68

$$\Delta H_r = (146 + 46) - (68 + 83 + 68) = -27 \text{ kcal/mol}$$

(b)

$$CH_3-\underset{\underset{CH_3}{|}}{\overset{\overset{CH_3}{|}}{C}}{-}OH + H{-}Cl \longrightarrow CH_3-\underset{\underset{CH_3}{|}}{\overset{\overset{CH_3}{|}}{C}}{-}Cl + H{-}OH$$

86   103   81   111

$$\Delta H_r = (86 + 103) - (81 + 111) = -3 \text{ kcal/mol}$$

(c)

$$\overset{\overset{O}{\|}}{179}\; CH_3{-}C{-}CH_3 + \overset{111}{H}{-}O{-}H \longrightarrow \underset{\underset{O-H}{86}}{\overset{\overset{O-H}{86}}{CH_3{-}C{-}CH_3}}$$

$$\Delta H_r = (179 + 111) - (111 + 86 + 86) = +7 \text{ kcal/mol}$$

2.36

$$\overset{\overset{O}{\|}}{CH_3CCH_3}$$

ketone

$$\overset{\overset{O}{\|}}{CH_3CH_2CH}$$

aldehyde

$$CH_2{=}CHCH_2OH$$

alkene and alcohol

2.36 (cont)

$CH_2\!\!=\!\!CHOCH_3$

alkene and ether

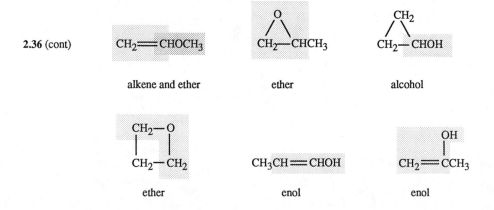

ether

alcohol

$\begin{array}{c} CH_2\!-\!O \\ |\qquad| \\ CH_2\!-\!CH_2 \end{array}$

ether

$CH_3CH\!\!=\!\!CHOH$

enol

$CH_2\!\!=\!\!CCH_3$

enol

2.37

$CH_2\!\!=\!\!C\!\!=\!\!CH_2$

$sp^2 \qquad sp \qquad sp^2$

$1s$ sp^2 sp^2 $1s$

sp

σ bond backbone

π bonds

2.38 CH_3NH_2 + BF_3 $\longrightarrow$ $CH_3\!-\!\overset{+}{N}\!-\!\overset{-}{B}\!-\!F$

sp^3 sp^3

tetrahedral   tetrahedral

2.38 (cont)

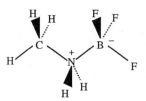

2.39   (a)

         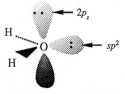

(b) The one in which the oxygen atom is sp^2 hybridized. The electron pair in the p orbital should have a different energy than the electron pair in the sp^2 hybrid orbital. The model of water in which the oxgyen atom is sp^3 hybridized has both pairs of nonbonding electrons in sp^3 hybrid orbitals. These electrons should, therefore, be of the same energy. Experimentally we should not see them as occupying two different energy levels.

(c) Yes, no hybridization at the oxygen atom also results in two different energy levels for the two pairs of nonbonding electrons.

overlap of a $1s$ orbital
on hydrogen with a $2p$
orbital on oxygen

An sp^2-hybridized oxygen atom would be expected to have a H—O—H bond angle of 120°. If oxygen were not hybridized, the H—O—H bond angle would be expected to be 90°. The experimentally determined value for the bond angle is 105°, between 120° and 90°.

(d) The oxygen atom would have sp hybridization.

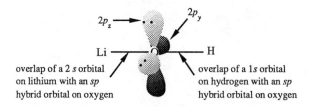

overlap of a 2 s orbital   overlap of a $1s$ orbital
on lithium with an sp on hydrogen with an sp
hybrid orbital on oxygen   hybrid orbital on oxygen

3

Reactions of Organic Compounds as Acids and Bases

Concept Map 3.1   The Brønsted-Lowry theory of acids and bases.

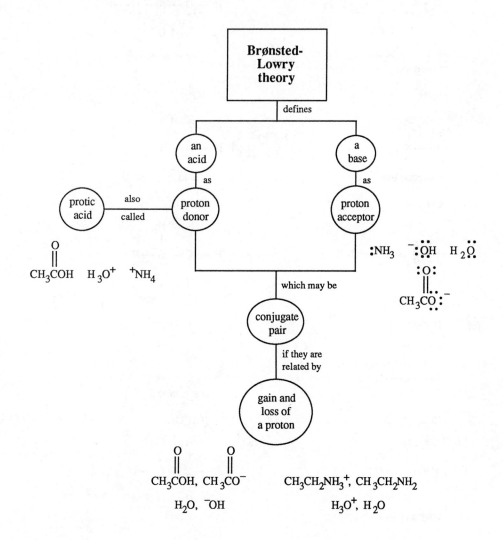

3.1   (a) $(CH_3)_2O$ $[(CH_3)_2OH]^+$

base   conjugate acid

(b) HSO_4^- H_2SO_4

conjugate base   acid
(one resonance contributor)

(c) $^-NH_2$ NH_3

conjugate base   acid

(d) CH_3OH $CH_3OH_2^+$

base   conjugate acid

(e) $H_2C=O$ $H_2C=OH^+$

base   conjugate acid

(f) CH_3O^- CH_3OH

conjugate base   acid

Concept Map 3.2   The Lewis theory of acids and bases.

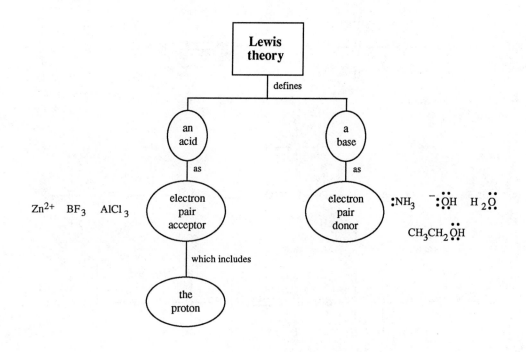

- -

3.2

(a)

3.2 (cont)

(b)

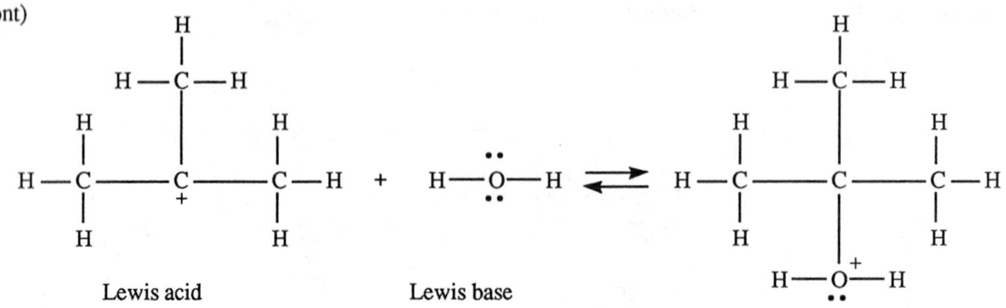

$$
\begin{array}{c}
\text{H} \\
| \\
\text{H—C—H} \\
\end{array}
$$

Lewis acid   Lewis base

(c)

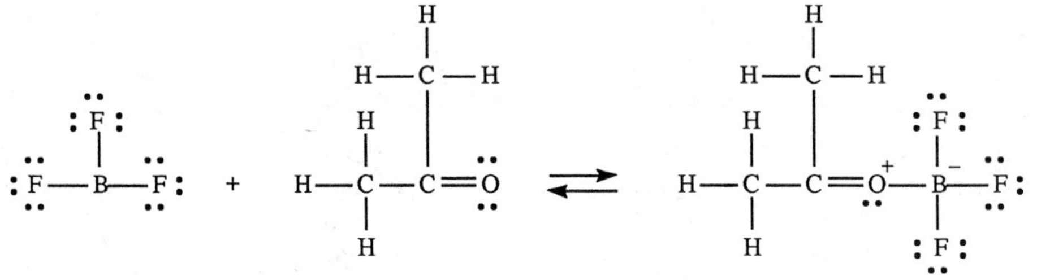

Lewis acid   Lewis base

(d) Hg^{2+} +   H—C—C—S—H   ⇌ $\left[\text{H—C—C—S—H} \right]^{2+}$

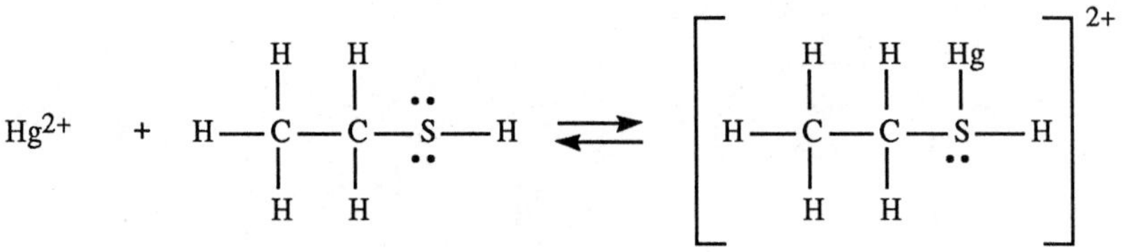

Lewis acid   Lewis base

3.3

(a) CH_3NHCH_3 + $AlCl_3$ ⟶ $CH_3\overset{+}{N}CH_3$ (with $^-AlCl_3$ and H)

   CH_3NHCH_3 + H_2SO_4 ⟶ $CH_3\overset{+}{N}CH_3$ (with H) + HSO_4^-

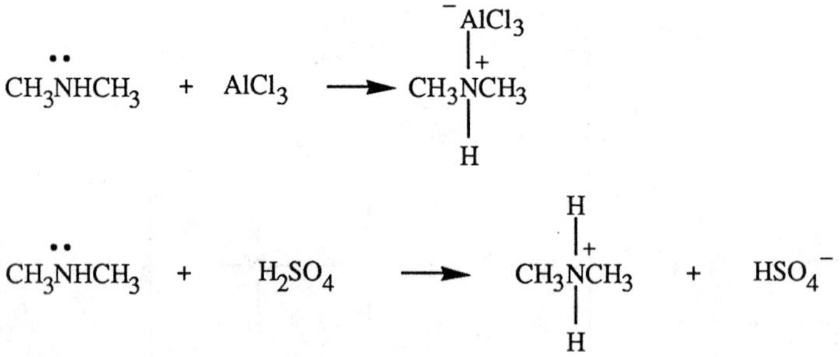

(b)   HCH (with :O:) + $AlCl_3$ ⟶ HCH (with $\overset{+}{O}{-}^-AlCl_3$)

   HCH (with :O:) + H_2SO_4 ⟶ HCH (with $\overset{+}{O}{-}H$) + HSO_4^-

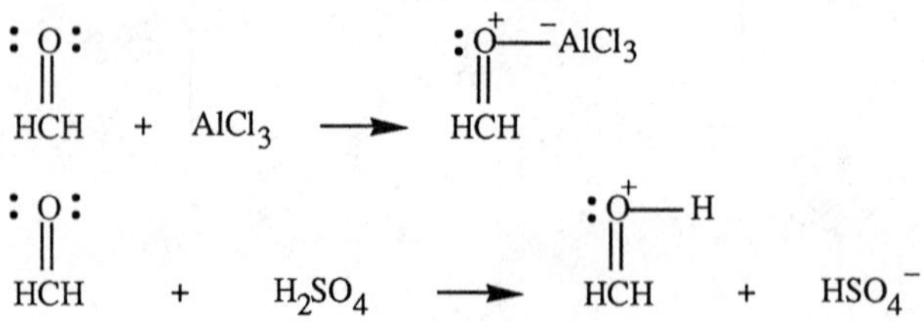

3.3 (cont)

(c) $HC\equiv CH$ + $AlCl_3$ $\longrightarrow$ $H-\overset{+}{C}=\overset{\overset{\displaystyle ^-AlCl_3}{|}}{C}-H$

$HC\equiv CH$ + H_2SO_4 $\longrightarrow$ $H-\overset{+}{C}=\overset{\overset{\displaystyle H}{|}}{C}-H$ + HSO_4^-

(d) $CH_3CH_2-\overset{\overset{\displaystyle CH_2CH_3}{|}}{\underset{\cdot\cdot}{P}}-CH_2CH_3$ + $AlCl_3$ $\longrightarrow$ $CH_3CH_2-\overset{\overset{\displaystyle CH_2CH_3}{|}}{\underset{\underset{\displaystyle ^-AlCl_3}{|}}{\overset{+}{P}}}-CH_2CH_3$

$CH_3CH_2-\overset{\overset{\displaystyle CH_2CH_3}{|}}{\underset{\cdot\cdot}{P}}-CH_2CH_3$ + H_2SO_4 $\longrightarrow$ $CH_3CH_2-\overset{\overset{\displaystyle CH_2CH_3}{|}}{\underset{\underset{\displaystyle H}{|}}{\overset{+}{P}}}-CH_2CH_3$ + HSO_4^-

(e) $CH_3CH_2-\overset{\cdot\cdot}{\underset{\cdot\cdot}{O}}H$ + $AlCl_3$ $\longrightarrow$ $CH_3CH_2-\overset{+}{\underset{\cdot\cdot}{O}}\overset{\overset{\displaystyle ^-AlCl_3}{|}}{H}$

$CH_3CH_2-\overset{\cdot\cdot}{\underset{\cdot\cdot}{O}}H$ + H_2SO_4 $\longrightarrow$ $CH_3CH_2-\overset{+}{\underset{\cdot\cdot}{O}}\overset{\overset{\displaystyle H}{|}}{H}$ + HSO_4^-

(f) $CH_3CH_2\overset{\cdot\cdot}{\underset{\cdot\cdot}{S}}-CH_2CH_3$ + $AlCl_3$ $\longrightarrow$ $CH_3CH_2\overset{+}{\underset{\cdot\cdot}{S}}\overset{\overset{\displaystyle ^-AlCl_3}{|}}{-}CH_2CH_3$

$CH_3CH_2\overset{\cdot\cdot}{\underset{\cdot\cdot}{S}}-CH_2CH_3$ + H_2SO_4 $\longrightarrow$ $CH_3CH_2\overset{+}{\underset{\cdot\cdot}{S}}\overset{\overset{\displaystyle H}{|}}{-}CH_2CH_3$ + HSO_4^-

3.4   (a)   ethyl acetate $H-\overset{\overset{\displaystyle H}{|}}{\underset{\underset{\displaystyle H}{|}}{C}}-\overset{\overset{\displaystyle :O:}{||}}{C}-\overset{\cdot\cdot}{\underset{\cdot\cdot}{O}}-\overset{\overset{\displaystyle H}{|}}{\underset{\underset{\displaystyle H}{|}}{C}}-\overset{\overset{\displaystyle H}{|}}{\underset{\underset{\displaystyle H}{|}}{C}}-H$

$CH_3\overset{\overset{\displaystyle :O:}{||}}{C}-\overset{\cdot\cdot}{\underset{\cdot\cdot}{O}}-CH_2CH_3$ + H_2SO_4 $\longrightarrow$ $CH_3\overset{\overset{\displaystyle :\overset{+}{O}-H}{||}}{C}-\overset{\cdot\cdot}{\underset{\cdot\cdot}{O}}-CH_2CH_3$ + HSO_4^-

$CH_3\overset{\overset{\displaystyle :O:}{||}}{C}-\overset{\cdot\cdot}{\underset{\cdot\cdot}{O}}-CH_2CH_3$ + H_2SO_4 $\longrightarrow$ $CH_3\overset{\overset{\displaystyle :O:}{||}}{C}-\overset{\cdot\cdot}{\underset{\underset{\displaystyle H}{\overset{+}{|}}}{O}}-CH_2CH_3$ + HSO_4^-

3.4 (cont)

(b)   2-(*N*,*N*-dimethylamino)ethanol

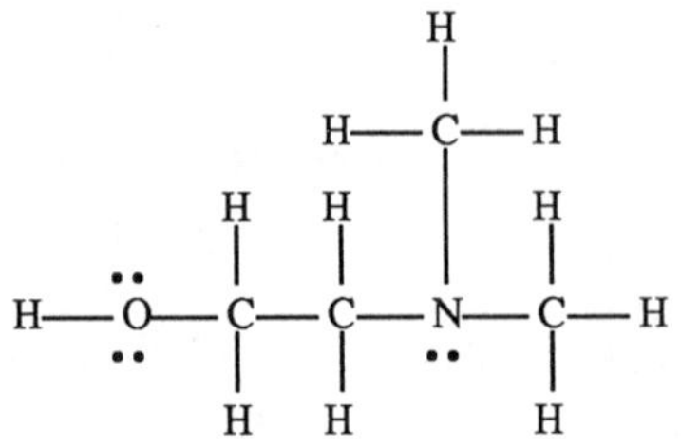

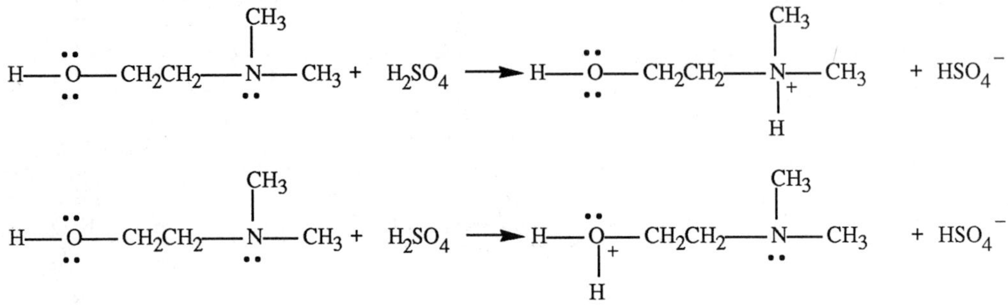

(c)   2-propen-1-ol

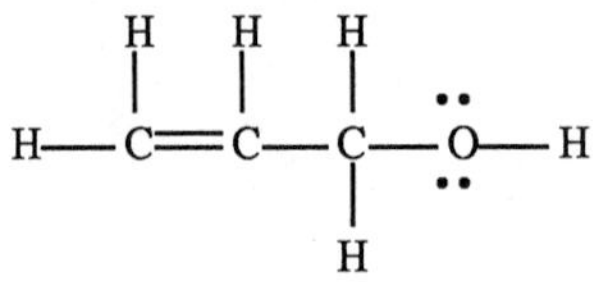

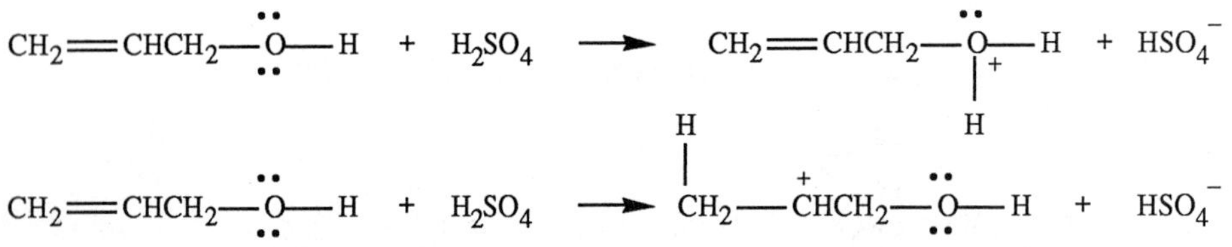

$$CH_2{=}CHCH_2{-}\ddot{O}{-}H \;+\; H_2SO_4 \;\longrightarrow\; CH_2{=}CHCH_2{-}\overset{+}{\underset{H}{\ddot{O}}}{-}H \;+\; HSO_4^{-}$$

$$CH_2{=}CHCH_2{-}\ddot{O}{-}H \;+\; H_2SO_4 \;\longrightarrow\; CH_2{-}\overset{H}{\underset{}{C}}{}^{+}HCH_2{-}\ddot{O}{-}H \;+\; HSO_4^{-}$$

(d)   methyl isocyanate

$$CH_3{-}\ddot{N}{=}C{=}\ddot{O} \;+\; H_2SO_4 \;\longrightarrow\; CH_3{-}\overset{+}{\underset{H}{N}}{=}C{=}\ddot{O} \;+\; HSO_4^{-}$$

$$CH_3{-}\ddot{N}{=}C{=}\ddot{O} \;+\; H_2SO_4 \;\longrightarrow\; CH_3{-}\ddot{N}{=}C{=}\overset{+}{\underset{H}{\ddot{O}}} \;+\; HSO_4^{-}$$

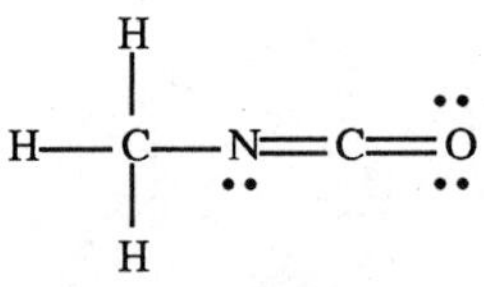

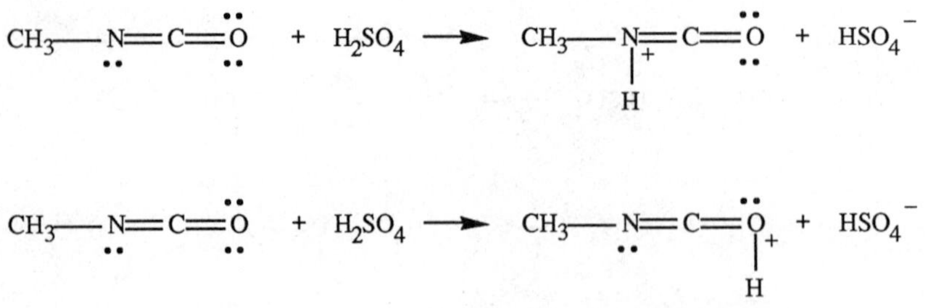

3.5   base   conjugate acid

(a)

$$CH_3CH_2 \quad\quad H$$
$$C=C$$
$$H \quad\quad CH_2CH_3$$

$$CH_3CH_2 \quad H$$
$$\overset{+}{C}-C-H$$
$$H \quad CH_2CH_3$$

(b) $CH_3CH_2-\overset{\cdot\cdot}{\underset{\cdot\cdot}{O}}-CH_2CH_3$

$$CH_3CH_2-\overset{\cdot\cdot}{\underset{+}{O}}-CH_2CH_3$$
$$\overset{|}{H}$$

(c) $CH_3CH_2\overset{\cdot\cdot}{N}CH_2CH_3$
$$\overset{|}{H}$$

$$CH_3CH_2\overset{+}{N}CH_2CH_3$$
$$\overset{|}{H}$$

(d) $CH_3CH_2CH_2-\overset{\cdot\cdot}{\underset{\cdot\cdot}{S}}-CH_3$

$$CH_3CH_2CH_2-\overset{\cdot\cdot}{\underset{+}{S}}-CH_3$$
$$\overset{|}{H}$$

(e)
$$:\overset{\cdot\cdot}{O}:$$
$$\|$$
$$CH_3CH_2CH$$

$$:\overset{+}{O}-H$$
$$\|$$
$$CH_3CH_2CH$$

(f) $CH_3CH_2CH_2-\overset{\cdot\cdot}{\underset{\cdot\cdot}{O}}H$

$$CH_3CH_2CH_2-\overset{\cdot\cdot}{\underset{+}{O}}H$$
$$\overset{|}{H}$$

(g)
$$:\overset{\cdot\cdot}{O}:$$
$$\|$$
$$CH_3CH_2CCH_2CH_3$$

$$:\overset{+}{O}-H$$
$$\|$$
$$CH_3CH_2CCH_2CH_3$$

3.6   Equations from Problem 3.3 rewritten:

(a)

$$CH_3 \quad\quad Cl$$
$$CH_3\overset{|}{N}: \quad\quad \overset{|}{Al}-Cl \quad\longrightarrow\quad CH_3\overset{|}{\overset{+}{N}}-\overset{-}{AlCl_3}$$
$$\overset{|}{H} \quad\quad \overset{|}{Cl} \quad\quad\quad\quad \overset{|}{H}$$

$$CH_3 \quad\quad\quad\quad\quad\quad\quad\quad\quad CH_3$$
$$CH_3\overset{|}{N}: \quad H-\overset{\cdot\cdot}{\underset{\cdot\cdot}{O}}-SO_3H \quad\longrightarrow\quad CH_3\overset{|}{\overset{+}{N}}-H \quad \overset{-}{:}\overset{\cdot\cdot}{\underset{\cdot\cdot}{O}}-SO_3H$$
$$\overset{|}{H} \quad\quad\quad\quad\quad\quad\quad\quad\quad \overset{|}{H}$$

3.6 (cont)

(b)

$$:\overset{..}{O}: \qquad \overset{Cl}{\underset{Cl}{\overset{|}{Al}}}{-}Cl \qquad \overset{+}{:}\overset{..}{O}{-}^{-}AlCl_3$$
$$\underset{HCH}{\parallel} \qquad \qquad \longrightarrow \qquad \underset{HCH}{\parallel}$$

$$:\overset{..}{O}: \qquad H{-}\overset{..}{\underset{..}{O}}{-}SO_3H \qquad \overset{+}{:}\overset{..}{O}{-}H \qquad \overset{..}{:}\overset{..}{\underset{..}{O}}{-}SO_3H$$
$$\underset{HCH}{\parallel} \qquad \qquad \longrightarrow \qquad \underset{HCH}{\parallel}$$

(c)

$$HC{\equiv}CH \qquad \overset{Cl}{\underset{Cl}{\overset{|}{Al}}}{-}Cl \qquad \longrightarrow \qquad H{-}\overset{+}{C}{=}\overset{\overset{-AlCl_3}{|}}{C}{-}H$$

$$HC{\equiv}CH \quad H{-}\overset{..}{\underset{..}{O}}{-}SO_3H \qquad \longrightarrow \qquad H{-}\overset{+}{C}{=}\overset{\overset{H}{|}}{C}{-}H \qquad \overset{..}{:}\overset{..}{\underset{..}{O}}{-}SO_3H$$

(d)

$$CH_3CH_2{-}\overset{\overset{CH_2CH_3}{|}}{\underset{..}{P}}{-}CH_2CH_3 \qquad \overset{Cl}{\underset{Cl}{\overset{|}{Al}}}{-}Cl \longrightarrow \qquad CH_3CH_2{-}\overset{\overset{CH_2CH_3}{|}}{\underset{\underset{AlCl_3}{-}}{\overset{+}{P}}}{-}CH_2CH_3$$

$$CH_3CH_2{-}\overset{\overset{CH_2CH_3}{|}}{\underset{..}{P}}{-}CH_2CH_3 \quad H{-}\overset{..}{\underset{..}{O}}{-}SO_3H \longrightarrow \quad CH_3CH_2{-}\overset{\overset{CH_2CH_3}{|}}{\underset{\underset{H}{|}}{\overset{+}{P}}}{-}CH_2CH_3 \quad \overset{..}{:}\overset{..}{\underset{..}{O}}{-}SO_3H$$

(e)

$$CH_3CH_2{-}\overset{..}{\underset{..}{O}}H \qquad \overset{Cl}{\underset{Cl}{\overset{|}{Al}}}{-}Cl \qquad \longrightarrow \qquad CH_3CH_2{-}\overset{\overset{+}{\underset{..}{O}}}{\underset{\underset{AlCl_3}{-}}{}}H$$

$$CH_3CH_2{-}\overset{..}{\underset{..}{O}}H \quad H{-}\overset{..}{\underset{..}{O}}{-}SO_3H \qquad \longrightarrow \qquad CH_3CH_2{-}\overset{\overset{H}{|}}{\underset{..}{\overset{+}{O}}}H \qquad \overset{..}{:}\overset{..}{\underset{..}{O}}{-}SO_3H$$

(f)

$$CH_3CH_2\overset{..}{\underset{..}{S}}{-}CH_2CH_3 \qquad \overset{Cl}{\underset{Cl}{\overset{|}{Al}}}{-}Cl \longrightarrow \qquad CH_3CH_2\overset{\overset{-AlCl_3}{|}}{\underset{..}{\overset{+}{S}}}{-}CH_2CH_3$$

$$CH_3CH_2\overset{..}{\underset{..}{S}}{-}CH_2CH_3 \quad H{-}\overset{..}{\underset{..}{O}}{-}SO_3H \longrightarrow \quad CH_3CH_2\overset{\overset{H}{|}}{\underset{..}{\overset{+}{S}}}{-}CH_2CH_3 \quad \overset{..}{:}\overset{..}{\underset{..}{O}}{-}SO_3H$$

3.6 (cont)

Equations from Problem 3.4 rewritten:

(a)

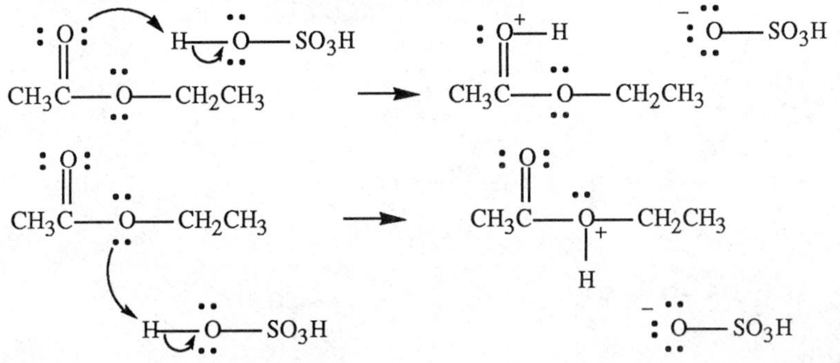

$$CH_3C-O-CH_2CH_3 \quad + \quad H-O-SO_3H \longrightarrow CH_3C-O-CH_2CH_3 \quad + \quad {}^-O-SO_3H$$

$$CH_3C-O-CH_2CH_3 \longrightarrow CH_3C-\overset{+}{O}-CH_2CH_3$$

$$H-O-SO_3H \qquad \qquad {}^-O-SO_3H$$

(b)

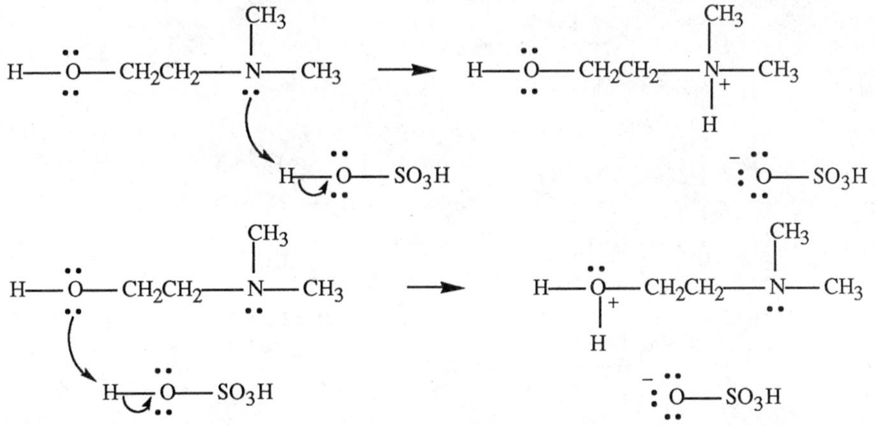

$$H-O-CH_2CH_2-\overset{CH_3}{\underset{}{N}}-CH_3 \longrightarrow H-O-CH_2CH_2-\overset{CH_3}{\underset{H}{\overset{+}{N}}}-CH_3$$

$$H-O-SO_3H \qquad \qquad {}^-O-SO_3H$$

$$H-O-CH_2CH_2-\overset{CH_3}{\underset{}{N}}-CH_3 \longrightarrow H-\overset{+}{O}-CH_2CH_2-\overset{CH_3}{\underset{}{N}}-CH_3$$

$$H-O-SO_3H \qquad \qquad {}^-O-SO_3H$$

(c)

$$CH_2=CHCH_2-O-H \longrightarrow CH_2=CHCH_2-\overset{+}{O}-H$$

$$H-O-SO_3H \qquad \qquad H \qquad {}^-O-SO_3H$$

$$CH_2=CHCH_2-O-H \longrightarrow CH_2-\overset{+}{C}HCH_2-O-H$$

$$H-O-SO_3H \qquad \qquad H \qquad {}^-O-SO_3H$$

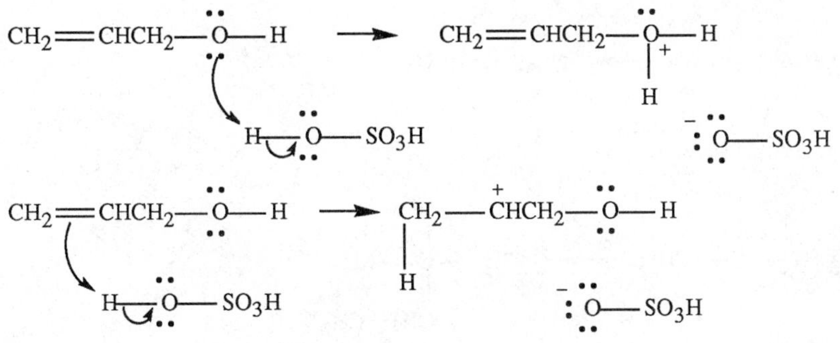

3.6 (cont)

(d) CH_3—N=C=O   → CH_3—N=C=O
                                        |+
                                        H

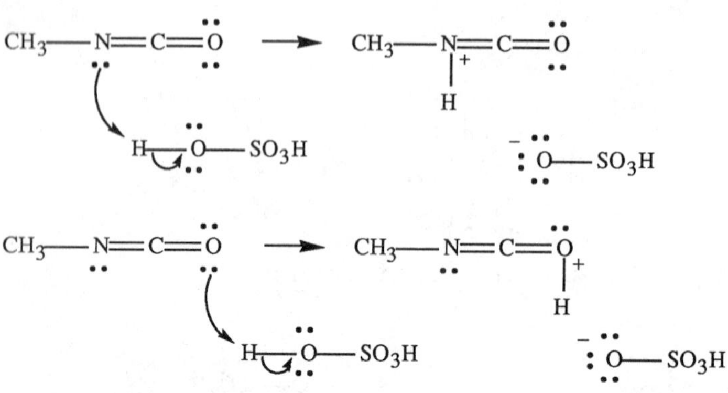

H—O—SO_3H $^-$O—SO_3H

CH_3—N=C=O   → CH_3—N=C=O
                                        |+
                                        H

H—O—SO_3H $^-$O—SO_3H

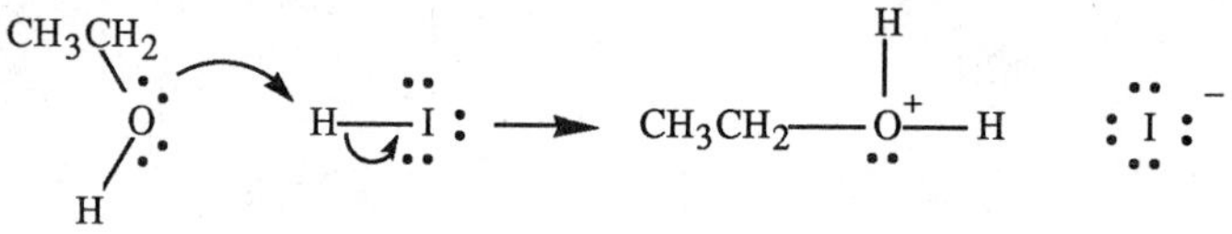

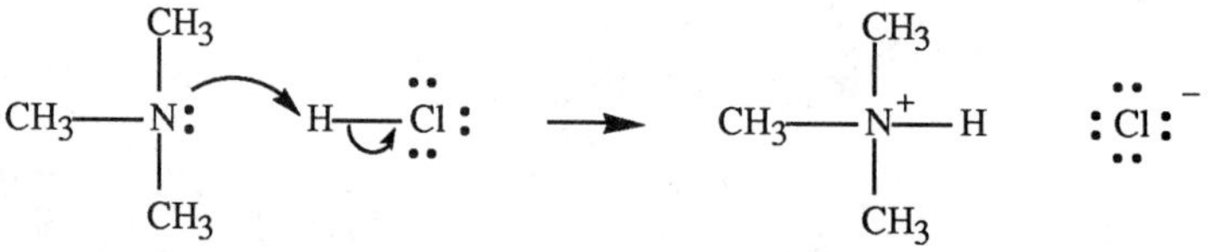

3.7   (a) CH_3CH_2
                    O   H—I: → CH_3CH_2—O$^+$—H   :I:$^-$
                    H |
                                                          H

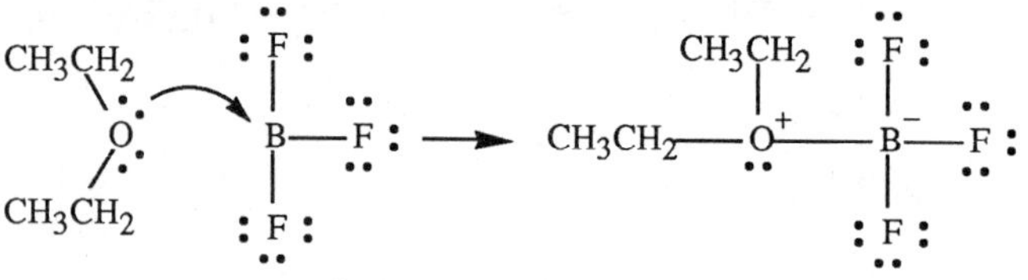

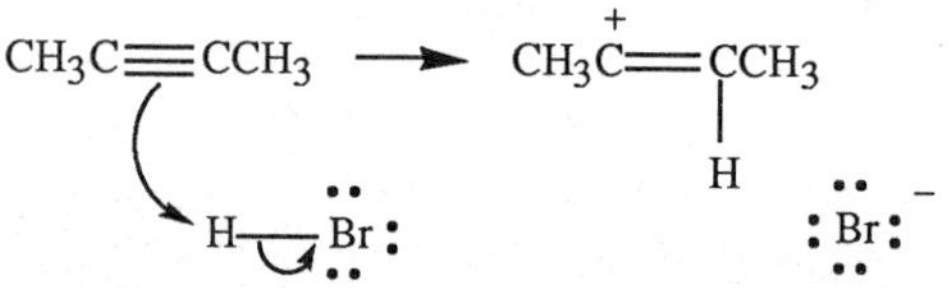

(b) CH_3
             |
   CH_3—N: H—Cl: → CH_3—N$^+$—H   :Cl:$^-$
             | |
            CH_3 CH_3

(c) CH_3CH_2 :F: CH_3CH_2 :F:
             O   B—F: → CH_3CH_2—O$^+$ B—F:
       CH_3CH_2 :F: :F:

(d) $CH_3C\equiv CCH_3$ → $CH_3C=CCH_3$
                                          |
                                          H
       H—Br: :Br:$^-$

(e) CH_3 CH_3
             | |
   CH_3—C$^+$—CH_3   → CH_3—C—CH_3
                                          |
                                    CH_2—$CH_2$$^+$
       CH_2=CH_2

3.8 H_2S is the stronger acid.

$$H_2S + HO^- \rightleftharpoons HS^- + H_2O$$

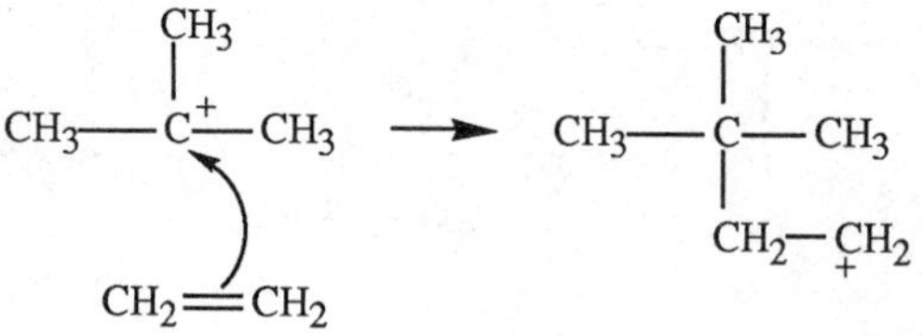

3.8 (cont)

Hydrogen sulfide ionizes more readily than water does because the hydrogen sulfide anion is more stable relative to its conjugate acid than hydroxide ion is. The hydrogen sulfide anion is a weaker base than hydroxide anion for the same reason that methanethiolate anion is a weaker base than the hydroxide anion. The sulfur atom is larger than an oxygen atom. Therefore the negative charge on the hydrogen sulfide anion is more spread out and less likely to attract a proton than the charge on a hydroxide anion.

3.9 $CH_3CH_2SH + CH_3CH_2O^- \xrightleftharpoons{\quad\quad} CH_3CH_2S^- + CH_3CH_2OH$

Ethanethiolate is a weaker base than methoxide ion because the negative charge is more spread out on the sulfur atom than on the oxygen atom. (See the answer to Problem 3.8.) In the reaction of ethanethiol and methoxide ion, the equilibrium lies far to the right.

Concept Map 3.3   Relationship of acidity to position of the central element in the periodic table.

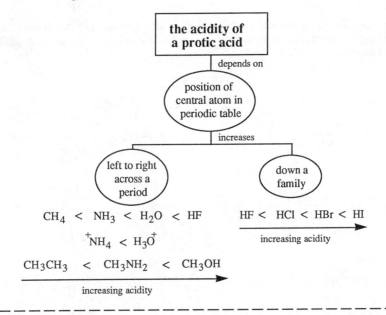

$CH_4 < NH_3 < H_2O < HF$

$^+NH_4 < H_3O^+$

$CH_3CH_3 < CH_3NH_2 < CH_3OH$

increasing acidity

$HF < HCl < HBr < HI$

increasing acidity

3.10   <u>most basic</u>   <u>least basic</u>

(a) CH_3NHCH_3 > CH_3OCH_3 > $CH_3CH_2CH_3$

amine   ether   alkane
(nonbonding electrons   (nonbonding electrons   (no nonbonding electrons)
on nitrogen)   on oxygen)

(b) CH_3^- > NH_2^- > OH^- > F^-

methyl anion   amide ion   hydroxide ion   fluoride ion
a carbanion

3.10 (cont)   <u>most basic</u>   <u>least basic</u>

(c) $CH_3\overset{-}{N}CH_3$ > CH_3O^-

           amide anion   alkoxide anion

3.11   (a) One of the nonbonding electron pairs on the oxygen atom can react with the empty *p*-orbital of the carbo-
           cation.

an oxonium ion

(b) The product formed is a protonated alcohol, an oxonium ion, which has a pK_a ~–2. It will transfer a proton
     to the oxygen atom of water in an equilibrium reaction. With a large excess of water, equilibrium lies
     toward the right.

3.12   <u>least acidic</u>   <u>most acidic</u>

(a) $CH_3CH_2CH_3$ < $CH_3CH_2NH_2$ < CH_3CH_2OH < CH_3CH_2SH

           alkene   amine   alcohol   thiol

(b) CH_3OCH_3 < CH_3NHCH_3 < $CH_3\overset{+}{O}-CH_3$ (with H above)

        (proton on a carbon atom)   (proton on a nitrogen atom)   (proton on the oxygen atom
                                                                              of an oxonium ion)

(c) $CH_3CH_2CH_3$ < $CH_3\overset{+}{N}-CH_3$ (with CH₃ above, H below)   < $CH_3\overset{+}{O}-CH_3$ (with H above)

           alkane   ammonium ion   oxonium ion

3.13   <u>acid</u>   <u>conjugate base</u>

(a)

$$CH_3C(CH_3)(CH_3)\text{---}\overset{..}{\underset{..}{O}}H$$

$$CH_3C(CH_3)(CH_3)\text{---}\overset{..}{\underset{..}{O}}{:}^-$$

(b) $CH_3CH_2CH_2\overset{..}{\underset{..}{S}}H$ $CH_3CH_2CH_2\overset{..}{\underset{..}{S}}{:}^-$

(c)

$$CH_3CH(CH_3)\text{---}\overset{+}{\underset{H}{O}}\text{---}CHCH_3(CH_3)$$

$$CH_3CH(CH_3)\text{---}\overset{..}{\underset{..}{O}}\text{---}CHCH_3(CH_3)$$

(d)

$$CH_3\text{---}\langle\text{ring}\rangle\text{---}\overset{H}{\underset{H}{\overset{+}{N}}}\text{---}H$$

$$CH_3\text{---}\langle\text{ring}\rangle\text{---}\overset{H}{\underset{H}{N}}\text{---}H$$

(e)

$$\langle\text{ring}\rangle\text{---}\overset{..}{\underset{..}{O}}\text{---}H$$

$$\langle\text{ring}\rangle\text{---}\overset{..}{\underset{..}{O}}{:}^-$$

(f)

$$CH_3CH_2\underset{\underset{:F:}{\overset{\overset{:O:}{\|}}{|}}}{C}H\text{---}\overset{..}{\underset{..}{O}}H$$

$$CH_3CH_2\underset{\underset{:F:}{\overset{\overset{:O:}{\|}}{|}}}{C}H\text{---}\overset{..}{\underset{..}{O}}{:}^-$$

3.14   The acidity constant is

$$K_a = \frac{[H^+][{:}B]}{[HB^+]}$$

where :B and HB⁺ are the conjugate base and conjugate acid forms respectively.

pK_a is defined as $-\log K_a$

For formic acid $\quad pK_a = -\log (1.99 \times 10^{-4})$

$$= -(\log 1.99 + \log 10^{-4})$$

$$= -(0.299 - 4)$$

3.14 (cont)

$$pK_a = -(-3.701)$$

$$pK_a = +3.701$$

For acetic acid $pK_a = +4.76$

As the pK_a gets larger (more positive), K_a gets smaller and the amount of dissociation of the acid decreases. Formic acid has the smaller pK_a and thus the larger K_a and is a stronger acid than acetic acid.

3.15

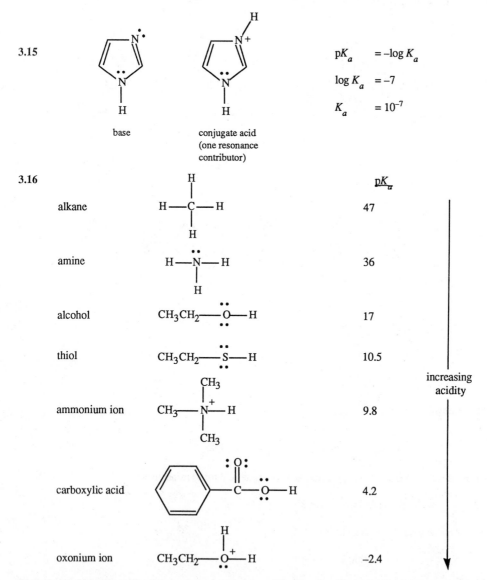

$$pK_a = -\log K_a$$

$$\log K_a = -7$$

$$K_a = 10^{-7}$$

base   conjugate acid
                        (one resonance
                        contributor)

3.16

		pK_a
alkane	$H-CH_3$ (H—C—H with H)	47
amine	$H-NH_2$ (H—N—H with H)	36
alcohol	CH_3CH_2-O-H	17
thiol	CH_3CH_2-S-H	10.5
ammonium ion	$CH_3-N^+(CH_3)_2-H$	9.8
carboxylic acid	$C_6H_5-C(=O)-O-H$	4.2
oxonium ion	$CH_3CH_2-O^+H-H$	-2.4

increasing
acidity

(Note that in this table acidity increases down the table, unlike the pK_a table inside the front cover of the textbook.)

3.17 $\underline{\text{reactants}}$ $\underline{\text{products}}$

$CH_3OH + OH^-$ $CH_3O^- + H_2O$

pK_a 15.5 pK_a 15.7

Approximately equal amounts of both reactants and products are present; methanol and water have about the same acidity.

$CH_3SH + OH^-$ $CH_3S^- + H_2O$

pK_a ~10.5 pK_a 15.7

The reaction goes essentially to completion; methanethiol is a much stronger acid than water..

$CH_3OH + NH_2^-$ $CH_3O^- + NH_3$

pK_a 15.5 pK_a 36

The reaction goes essentially to completion; methanol is a much stronger acid than ammonia

$CH_3NH_3^+ + OH^-$ $CH_3NH_2 + H_2O$

pK_a 10.6 pK_a 15.7

The reaction goes essentially to completion; methylammonium ion is a much stronger acid than water.

$CH_3OH_2^+ + H_2O$ $CH_3OH + H_3O^+$

pK_a ~~1.7 pK_a ~1.7

Approximately equal amounts of starting material and product are present; methyloxonium ion and hydronium ion have about the same acidity.

3.18

(a) CH_3CH_2OH + K^+OH^- $\rightleftharpoons$ $CH_3CH_2O^-K^+$ + H_2O

           pK_a 17 pK_a 15.7

(b) $CH_3CH_2\overset{+}{N}H_2$ Cl^- + Na^+OH^- $\longrightarrow$ CH_3CH_2NH + H_2O + Na^+Cl^-
                   | |
               CH_3CH_2 CH_3CH_2

              pK_a 10 pK_a 15.7

(c) CH_2FCOH + $Na^+HCO_3^-$ ⇌ $CH_2FCO^-Na^+$ + H_2CO_3 ⇌ H_2O + $CO_2\uparrow$

   pK_a 2 pK_a 6.5

(d) CH_3—⟨ ⟩—OH + $Na^+HCO_3^-$ ⟶   no reaction (pK_a for H_2CO_3 is 6.5)

   pK_a 10

(e)   ⟨ ⟩—NH_2 + HCl ⟶ ⟨ ⟩—$\overset{+}{N}H_3$ Cl$^-$

   pK_a –7 pK_a 4.6

(f) $CH_3CH_2CH_2SH$ + $CH_3CH_2O^-$ Na^+ ⟶ $CH_3CH_2CH_2S^-$ Na^+ + CH_3CH_2OH

   pK_a 10.5 pK_a 17

(g) CH_3—⟨ ⟩—COH + Na^+ CN^- ⟶ CH_3—⟨ ⟩—CO^-Na^+ + HCN

   pK_a 4 pK_a 9.1

(h) $CH_3CCH_2CCH_3$ + $CH_3O^-Na^+$ ⟶ $CH_3CCHCCH_3$ + CH_3OH

   pK_a 9.0 Na^+ pK_a 15.5

3.19 $\Delta G° = -2.303\ RT \log K_{eq}$

   -2.59 kcal/mol = $(-2.303)(1.987 \times 10^{-3}$ kcal/mol·K$)(298$ K$)(\log K_{eq})$

   $\log K_{eq}$ = 1.899

   K_{eq} = $10^{1.899}$ = 79.3

   K_{eq} = $\dfrac{K_a\ (ClCH_2CO_2H)}{K_a\ (ClCH_2CO_2H)}$

3.19 (cont)

$$79.3 \;=\; \frac{K_a\,(\text{ClCH}_2\text{CO}_2\text{H})}{(1.75 \times 10^{-5})}$$

$$K_a\,(\text{ClCH}_2\text{CO}_2\text{H}) \;=\; (79.3)(1.75 \times 10^{-5}) \;=\; 1.39 \times 10^{-3}$$

$$pK_a\,(\text{ClCH}_2\text{CO}_2\text{H}) \;=\; -\log K_a \;=\; -\log (1.39 \times 10^{-3})$$

$$pK_a \;=\; -(-3 + 0.143) = 2.86$$

This is very close to the value, 2.8, given in the front cover of the textbook for chloroacetic acid.

Concept Map 3.4   The relationship of pK_a to energy and entropy factors.

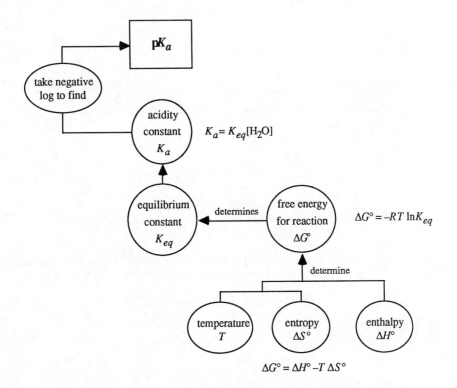

3.20 CF_3CH_2OH ionizes more readily than CH_3CH_2OH because the $CF_3CH_2O^-$ ion is more stable relative to its conjugate acid than the ethoxide ion is. The electron-withdrawing effect (negative inductive effect) of the three fluorine atoms reduces the negative charge on the oxygen atom in 2,2,2-trifluoroethoxide ion relative to that in ethoxide ion.

	pK_a	
$ClCH_2COH$ (with C=O)	2.86	
Cl_2CHCOH (with C=O)	1.30	increasing acidity
CCl_3COH (with C=O)	0.64	

3.21

Inductive effects are additive. The more electron-withdrawing chlorine atoms there are in the molecule, the more stable its conjugate base will be relative to the acid, and, therefore, the more acidic the conjugate acid will be.

3.22

:N≡C—C(H)(H)—C(=O)—O—H

cyanoacetic acid

:N≡C—CH₂—C(=O)—O—H ⟷ ⁻:N=C⁺—CH₂—C(=O)—O—H

⁻:N≡C—CH₂—C(—O⁻)(=O⁺...)—O—H ⟷ :N≡C—CH₂—C=O⁺—H

Cyanoacetic acid is stronger than acetic acid because the atom attached to the α-carbon atom has a positive charge in one important resonance contributor. The cyano group is therefore electron withdrawing.

⁻:O—N⁺(=O)—C(H)(H)—C(=O)—O—H

nitroacetic acid

3.22 (cont)

Nitroacetic acid is a stronger acid than acetic acid because the nitrogen attached to the α–carbon atom is more electronegative than hydrogen and always has a formal positive charge. The nitro group, with one nitrogen atom and two oxygen atoms, is a strongly electron-withdrawing group. Nitroacetic acid is a stronger acid than cyanoacetic acid because the atom attached to the α-carbon atom in nitroacetic acid carries a full positive charge (all resonance contributors have a positively charged nitrogen atom). Only one resonance contributor in cyanoacetic acid has a positive carbon atom attached to the α-carbon atom.

3.23   Triiodomethane (CHI_3) would be expected to be a weaker acid than trichloromethane ($CHCl_3$) because iodine is less electronegative than chlorine and cannot stabilize the conjugate base of triiodomethane as effectively as chlorine stabilizes the trichloromethyl anion.

3.24

(a)

The anion produced by removal of the proton of the carboxyl group is stabilized by resonance. The anion that would be produced by removal of the proton on the hydroxyl group would not be stabilized by resonance.

(b)

The anion produced by the removal of the α-hydrogen atom is stabilized by resonance.

(c)

The removal of the proton from the thiol group is easier than the removal of a proton from the hydroxyl group. The larger size of the sulfur atom means that the negative charge is more spread out and, therefore, better stabilized than if it were on an oxygen atom.

3.24 (cont)

(d)

The hydrogen atom on the α-carbon atom with the bromine atoms is more acidic than the hydrogen on the other α-carbon atom because the electron-withdrawing effect of the bromine atoms, as well as resonance, stabilizes the anion that results from its loss.

3.25   least basic   most basic

$$CH_3\text{—} \bigcirc \text{—}NH_2 \quad < \quad CH_3CH_2NH_2 \quad < \quad (CH_3CH_2)_2NH$$

3.26   The less positive (or more negative) a pK_a is, the stronger the conjugate acid and the weaker its conjugate base. The conjugate acid of diphenylamine, with a pK_a of 0.8, is a strong acid, and diphenylamine is thus a weak base. The ability of a base to remove a proton from an acid depends on how available the electron pair of the base is, i.e., how much electron density is available. In diphenylamine the electron pair on the nitrogen atom is delocalized into both phenyl rings, reducing the amount of electron density on the nitrogen atom.

Concept Map 3.5   The effects of structural changes on acidity and basicity.

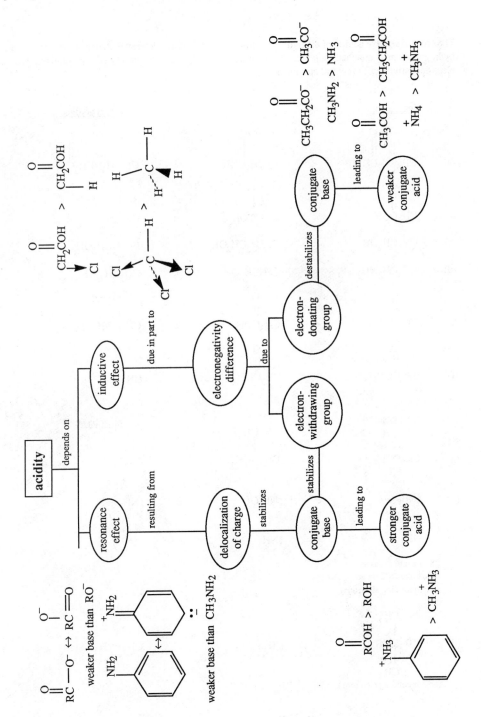

3.27   least acidic   most acidic

$$NaH \quad < \quad AlH_3 \quad < \quad H_2S \quad < \quad HCl$$

The order of relative acidity of the hydrogen atoms increases as the electronegativity of the atom to which the hydrogen atom is attached increases. This is due to the increasing stability of the conjugate base as the electronegativity of the central atom increases.

3.28   least acidic   most acidic

(a) $CH_3CH_2\overset{O}{\overset{\|}{C}}OH$ $<$ $ClCH_2\overset{O}{\overset{\|}{C}}OH$ $<$ $Cl^-\ H_3\overset{+}{N}CH_2\overset{O}{\overset{\|}{C}}OH$

(b) $CH_3\overset{O}{\overset{\|}{C}}OH$ $<$ $CCl_3\overset{O}{\overset{\|}{C}}OH$ $<$ $CF_3\overset{O}{\overset{\|}{C}}OH$

(c) CH_3CH_2OH $<$ $ClCH_2CH_2OH$ $<$ FCH_2CH_2OH

(d) $CH_3CH_2NH_2$ $<$ CH_3CH_2OH $<$ CH_3CH_2SH

(e) $CH_3\overset{O}{\overset{\|}{C}}OCH_3$ $<$ $CH_3\overset{O}{\overset{\|}{C}}OH$ $<$ $CH_3\overset{\overset{+}{O}-H}{\overset{\|}{C}}OH$

(f) $CH_3\overset{O}{\overset{\|}{C}}OCH_3$ $<$ $CH_3\overset{O}{\overset{\|}{C}}NH_2$ $<$ $CH_3\overset{O}{\overset{\|}{C}}OH$

3.29   least basic   most basic

(a) $CH_3\overset{..}{\underset{..}{O}}CH_3$ $<$ $CH_3\overset{CH_3}{\underset{..}{\overset{|}{N}}}CH_3$ $<$ $Cl_3C\!:^-$

(b) $:NF_3$ $<$ $:NH_2OH$ $<$ $:NH_3$

(c) $\overset{+}{N}H_4$ $<$ $:NH_3$ $<$ $\overset{-\ ..}{:NH_2}$

(this cannot act as a
base because there are
no nonbonding electrons)

(d) $CH_3\overset{CH_3}{\underset{CH_3}{\overset{|}{\underset{|}{Si}}}}CH_3$ $<$ $CH_3\overset{..}{\underset{..}{S}}CH_3$ $<$ $CH_3\overset{CH_3}{\underset{..}{\overset{|}{P}}}CH_3$

(no nonbonding electrons)

3.30   Basicity is determined by the availability of the electron pair that will remove the proton from the acid. $CH_3OCH_2CH_2CH_2NH_2$ is a weaker base than $CH_3CH_2CH_2CH_2NH_2$ because the electron-withdrawing oxygen atom of the methoxy group is pulling electron density away from the nitrogen atom. In $(CH_3O)_2CHCH_2NH_2$ there are two electron-withdrawing methoxy groups, which further reduce the availability of the electron pair. In $N{\equiv}CCH_2CH_2NH_2$ the cyano group is even more electron withdrawing than the methoxy group because of the positive charge on the carbon atom of the cyano group in one resonance contributor (see Problem 3.22).

3.31

(a) F^- + BF_3 $\longrightarrow$ BF_4^-

    Lewis base   Lewis acid

(b) Ag^+ + $2\,NH_3$ $\longrightarrow$ $Ag(NH_3)_2^+$

    Lewis acid   Lewis base

(c) $Al(H_2O)_6^{3+}$ + OH^- $\longrightarrow$ $Al(OH)(H_2O)_5^{2+}$ + H_2O

    Brønsted-Lowry   Brønsted-Lowry   conjugate base   conjugate acid
    acid   base

(d) $CH_3CH_2\overset{\overset{\displaystyle O}{\|}}{C}OH$ + OH^- $\longrightarrow$ $CH_3CH_2\overset{\overset{\displaystyle O}{\|}}{C}O^-$ + H_2O

    Brønsted-Lowry   Brønsted-Lowry   conjugate base   conjugate acid
    acid   base

(e) $CH_3CH_2\overset{\overset{\displaystyle CH_2CH_3}{|}}{N}CH_2CH_3$ + $CF_3\overset{\overset{\displaystyle O}{\|}}{C}OH$ $\longrightarrow$ $CH_3CH_2\overset{\overset{\displaystyle CH_2CH_3}{|}}{\underset{\underset{\displaystyle H}{|}}{\overset{+}{N}}}CH_2CH_3$ + $CF_3\overset{\overset{\displaystyle O}{\|}}{C}O^-$

    Brønsted-Lowry   Brønsted-Lowry   conjugate acid   conjugate base
    base   acid

(f) $CH_3CH_2SCH_2CH_3$ + BF_3 $\longrightarrow$ $CH_3CH_2\overset{\overset{\displaystyle {}^-BF_3}{|}}{\underset{+}{S}}CH_2CH_3$

    Lewis base   Lewis acid

(g) $ClCH_2\overset{\overset{\displaystyle O}{\|}}{C}OH$ + HCO_3^- $\longrightarrow$ $ClCH_2\overset{\overset{\displaystyle O}{\|}}{C}O^-$ + H_2CO_3

    Brønsted-Lowry   Brønsted-Lowry   conjugate base   conjugate acid
    acid   base

3.31(cont)

(h) $H_2PO_4^-$ + OH^- $\longrightarrow$ HPO_4^{2-} + H_2O

Brønsted-Lowry   Brønsted-Lowry   conjugate base   conjugate acid
acid   base

(i) $CH_3CH_2CH_2SH$ + $CH_3CH_2O^-$ $\longrightarrow$ $CH_3CH_2CH_2S^-$ + CH_3CH_2OH

Brønsted-Lowry   Brønsted-Lowry   conjugate base   conjugate acid
acid   base

3.32

(a)

+ Na^+OH^- $\longrightarrow$ + H_2O + Na^+ + Cl^-

pK_a ~10.6 pK_a 15.7

(b) H_2O + $CH_3CH_2\overset{+}{O}CH_2CH_3$ $\longrightarrow$ $CH_3CH_2OCH_2CH_3$ + H_3O^+
                              |
                              H

                    pK_a −3.6 pK_a −1.7

(c) $CH_3\overset{O}{\overset{||}{C}}OH$ + $CH_3CH_2S^-Na^+$ $\longrightarrow$ $CH_3\overset{O}{\overset{||}{C}}O^-Na^+$ + CH_3CH_2SH

        pK_a 4.8 pK_a 10.5

(d) $CCl_3\overset{O}{\overset{||}{C}}OH$ + $Na^+ HCO_3^-$ $\longrightarrow$ $CCl_3\overset{O}{\overset{||}{C}}O^-Na^+$ + H_2CO_3 ($\longrightarrow$ H_2O + $CO_2\uparrow$)

        pK_a 0.6 pK_a 6.5

(e) $CH_3CH_2\overset{O}{\overset{||}{C}}OH$ + NH_3 $\longrightarrow$ $CH_3CH_3\overset{O}{\overset{||}{C}}O^-$ + NH_4^+

        pK_a ~4.9 pK_a 9.4

(f) CH_3NO_2 + $CH_3CH_2O^-Na^+$ $\longrightarrow$ $Na^+ {}^-CH_2NO_2$ + CH_3CH_2OH

        pK_a 10.2 pK_a 17

3.32 (cont)

(g) CH_3CH_2SH + $NaNH_2$ ⟶ $CH_3CH_2S^-Na^+$ + NH_3

pK_a 10.5   pK_a 36

(h) $HC≡CH$ + $Na^+HCO_3^-$ ⟶   no reaction (pK$_a$ for H_2CO_3 is 6.5)

pK$_a$ 26

3.33

(a) CH_3CH_2—Br ⟶ CH_3CH_2—$C≡N$ + Br^-

:C≡N:

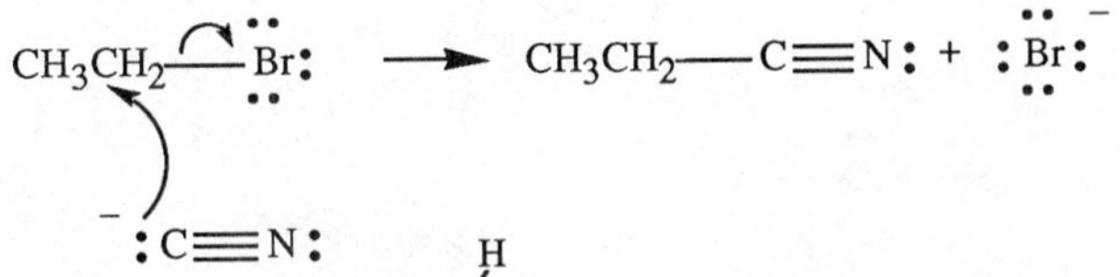

(b)

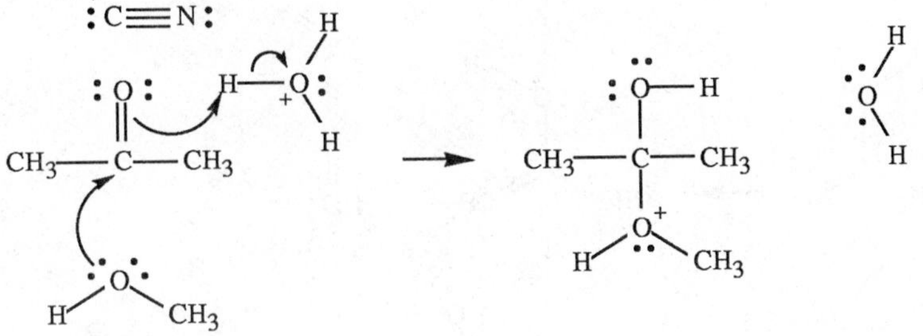

(c) CH_2=CH_2 ⟶ CH_2—CH_3

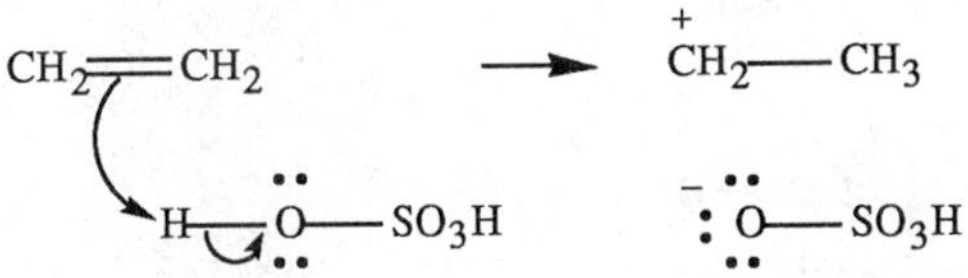

(d)

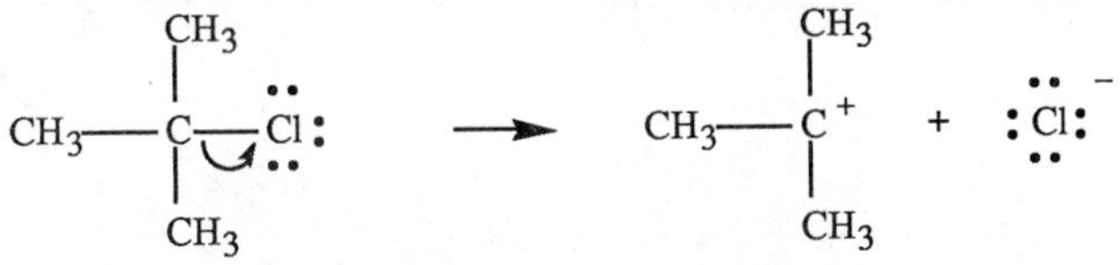

(e)

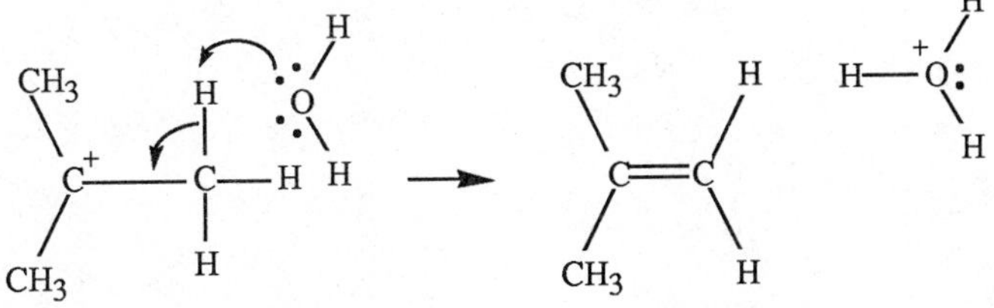

(f)

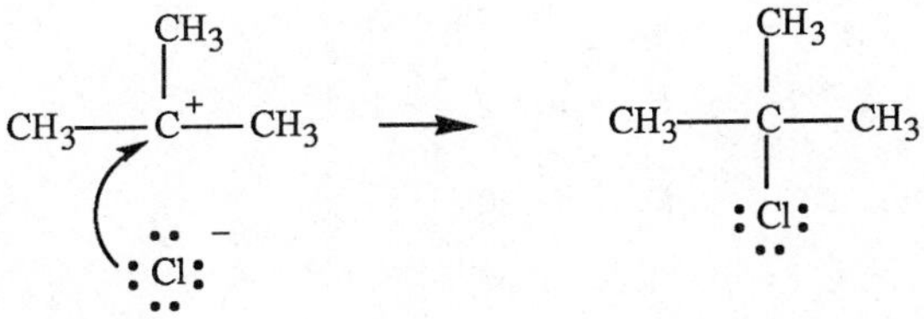

3.33 (cont)

(g)

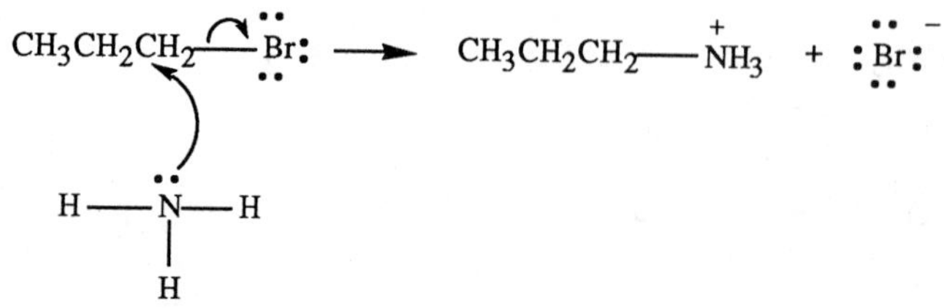

CH₃CH₂CH₂—Br: ⟶ CH₃CH₂CH₂—$\overset{+}{N}$H₃ + :Br:⁻

H—N—H
   |
   H

(h)

H—C—C—CH₃ ⟶ H—C=C—CH₃
   | |
   H   H

H—O—H

:O—H

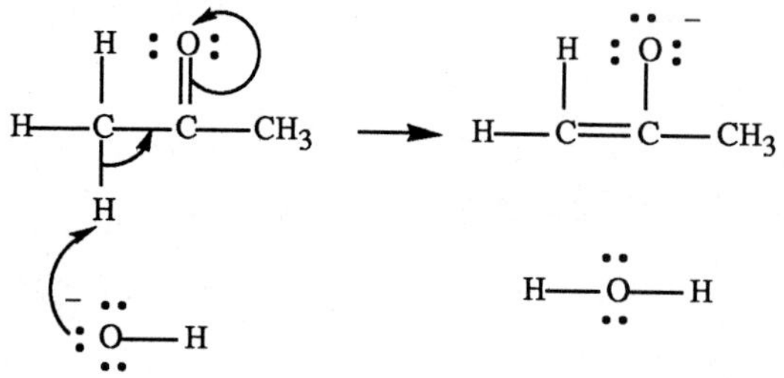

(i)

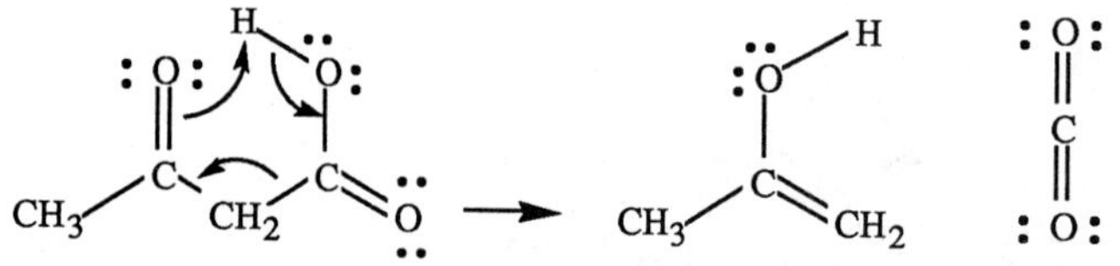

CH₃—C—CH₂—C=O ⟶ CH₃—C=CH₂   :O:
                                   ‖
                                   C
                                   ‖
                                   :O:

(j)

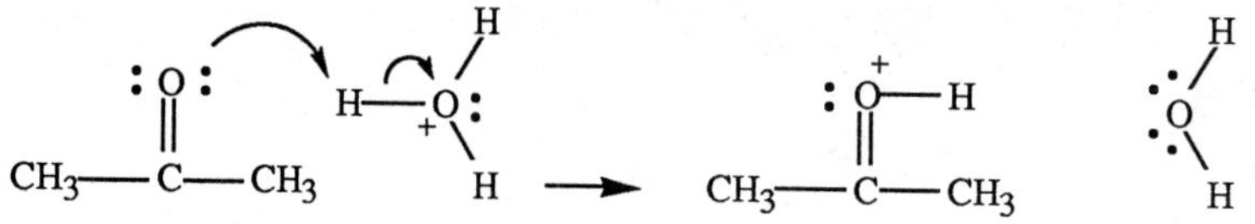

CH₃—C—CH₃   H—O: ⟶   CH₃—C—CH₃   H—O—H
      ‖ | |
      O   H   O—H

(k)

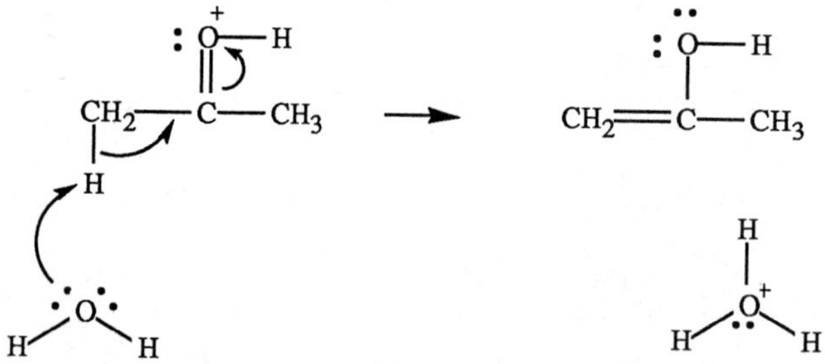

CH₂—C—CH₃   ⟶   CH₂=C—CH₃
   | ‖ |
   H O—H   O—H

H—O—H   H—$\overset{+}{O}$—H
                      |
                      H

3.34

(a)

$$CHCl_2\overset{O}{\underset{\|}{C}}OH$$

$$pK_a = 1.3$$
$$K_a = 10^{-pK_a} = 10^{-1.3}$$
$$= 10^{+0.7} \times 10^{-2}$$
$$K_a = 5 \times 10^{-2}$$

3.34 (cont)

(b) $CH_3\overset{+}{N}H_3$

$pK_a = 10.4$

$K_a = 10^{-pK_a} = 10^{-10.4}$

$\quad\quad = 10^{+0.6} \times 10^{-11}$

$K_a = 4 \times 10^{-11}$

(c) CCl_3CH_2OH

$pK_a = 12.2$

$K_a = 10^{-pK_a} = 10^{-12.2}$

$\quad\quad = 10^{+0.8} \times 10^{-13}$

$K_a = 6 \times 10^{-13}$

3.35

(a) CH_3CH_2OH

$K_a = 10^{-17} = 10^{-pK_a}$

$pK_a = -\log K_a = +17$

(b) CH_3CH_2SH

$K_a = 3.16 \times 10^{-11}$

$pK_a = -\log (3.16 \times 10^{-11})$

$\quad\quad = (\log 3.16 + \log 10^{-11})$

$\quad\quad = (0.5 - 11) = -(-10.50)$

$pK_a = +10.50$

(c) $CH_3CH_2\underset{+}{\overset{H}{O}}CH_2CH_3$

$K_a = 3.98 \times 10^{+3}$

$pK_a = -\log (3.98 \times 10^{+3})$

$\quad\quad = (\log 3.98 + \log 10^{+3})$

$\quad\quad = (0.69 + 3) = -(3.60)$

$pK_a = -3.60$

3.36

$\underset{\overset{|}{\overset{+}{N}H_3}}{\overset{\overset{O}{\|}}{CH_2CO^-}}$

$pK_a \ \overset{+}{RNH_3} \sim 10.6$

$\underset{\overset{|}{NH_2}}{\overset{\overset{O}{\|}}{CH_2COH}}$

$pK_a \ \underset{\overset{\|}{O}}{RCOH} \sim 4.0$

(a)   The carboxylate anion, $-\overset{\overset{O}{\|}}{C}O^-$, is a weaker base than the amino group, $-NH_2$. Glycine is best represented as $\underset{+}{H_3}NCH_2\overset{\overset{O}{\|}}{C}O^-$.

(b)   The high melting point for a compound with such a low molecular weight is indicative of an ionic compound. The high water solubility is also indicative of strong electrostatic forces. Both observations are consistent with the charged structure, $\underset{+}{H_3}NCH_2\overset{\overset{O}{\|}}{C}O^-$, for glycine.

3.37

(a) $CH_3CH_2\overset{..}{\underset{..}{O}}CH_2CH_2\overset{\overset{\displaystyle CH_3}{|}}{\underset{..}{N}}{-}CH_3$ + $H{-}\overset{\overset{\displaystyle H}{|}}{\underset{..}{\overset{+}{O}}}{-}H$ $\longrightarrow$

$CH_3CH_2\overset{..}{\underset{..}{O}}CH_2CH_2\overset{\overset{\displaystyle CH_3}{|}}{\underset{\underset{\displaystyle H}{|}}{\overset{+}{N}}}{-}CH_3$ + $H{-}\overset{..}{\underset{..}{O}}{-}H$ weaker acid than oxonium ion

(b) $CH_3\overset{\overset{\displaystyle :\overset{..}{\underset{..}{O}}:}{\|}}{C}CH_2CH_2CH_2CH_2CH_2\overset{..}{\underset{..}{O}}{-}H$ + $H{-}\overset{\overset{\displaystyle H}{|}}{\underset{..}{\overset{+}{O}}}{-}H$ $\longrightarrow$

$CH_3\overset{\overset{\displaystyle :\overset{..}{\underset{..}{O}}:}{\|}}{C}CH_2CH_2CH_2CH_2CH_2\overset{\overset{\displaystyle H}{|}}{\underset{..}{\overset{+}{O}}}{-}H$ + $H{-}\overset{..}{\underset{..}{O}}{-}H$ weaker acid than protonated carbonyl

(c) $CH_3C{\equiv}CCH_2CH_2\overset{..}{\underset{..}{O}}{-}CH_3$ + $H{-}\overset{\overset{\displaystyle H}{|}}{\underset{..}{\overset{+}{O}}}{-}H$ $\longrightarrow$

$CH_3C{\equiv}CCH_2CH_2\overset{\overset{\displaystyle H}{|}}{\underset{..}{\overset{+}{O}}}{-}CH_3$ + $H{-}\overset{..}{\underset{..}{O}}{-}H$ nonbonding electrons are more easily protonated than π electrons

3.38 (a) $K_a = \dfrac{[A^-]\,[H_3O^+]}{[HA]}$

taking the log of both sides: $\log K_a = \log\left(\dfrac{[A^-]\,[H_3O^+]}{[HA]}\right)$

$$\log K_a = \log [H_3O^+] + \log \dfrac{[A^-]}{[HA]}$$

multiplying both sides by -1

$$-\log K_a = -\log [H_3O^+] - \log \dfrac{[A^-]}{[HA]}$$

rearranging the equation by adding $\log \dfrac{[A^-]}{[HA]}$ to both sides

$$-\log [H_3O^+] = -\log K_a + \log \dfrac{[A^-]}{[HA]}$$

substituting the following relationships

$$pK_a = -\log K_a$$

$$pH = -\log [H^+] = -\log [H_3O^+]$$

3.38 (cont)

we get the Henderson-Hasselbach equation

$$pH = pK_a + \log \frac{[A^-]}{[HA]}$$

(b)   when $pH = pK_a$

$$\log \frac{[A^-]}{[HA]} = 0 \text{ and}$$

$$\frac{[A^-]}{[HA]} = 1$$

or the concentrations of the acid and its conjugate base are equal.

$$[HA] = [A^-]$$

(c) $pH \cong 6.5$

$$pK_a = 7$$

$$pH = pK_a + \log \frac{[A^-]}{[HA]}$$

$$6.5 = 7 + \log \frac{[A^-]}{[HA]}$$

$$\log \frac{[A^-]}{[HA]} = -0.5$$

$$\frac{[A^-]}{[HA]} = 0.316$$

$$[A^-] = 0.316[HA]$$

and $[HA] = \sim 3[A^-]$

There is more protonated than unprotonated imidazole present at pH 6.5.

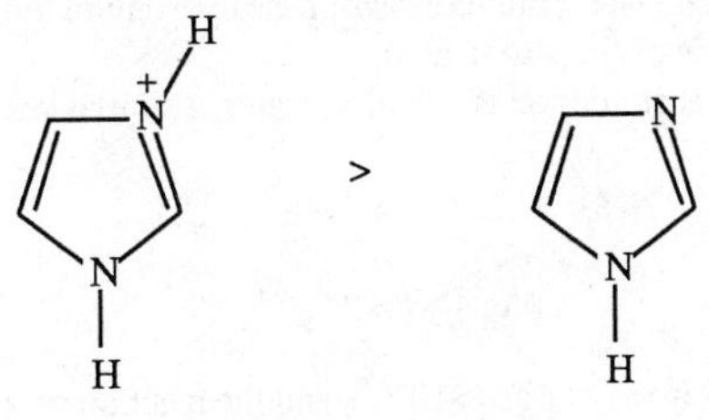

3.39   (a)

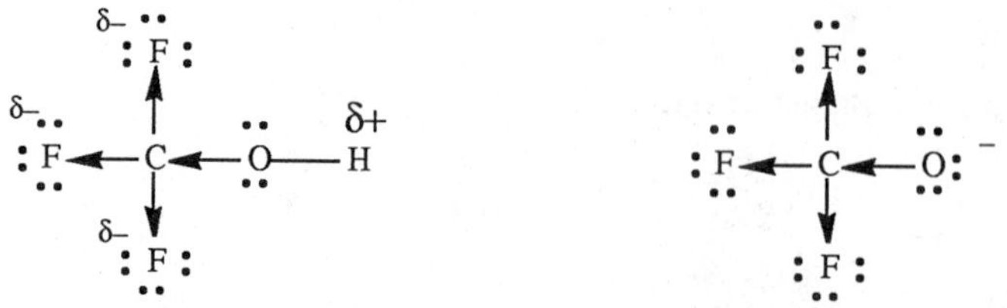

acid   conjugate base

The inductive effect of the fluorines makes the proton on the hydroxyl group more acidic and the anion less basic. The inductive effect of the fluorines also reduces the availability of the electron pairs on the oxygen atom so they are not as as readily involved in hydrogen bonding, decreasing the boiling point.

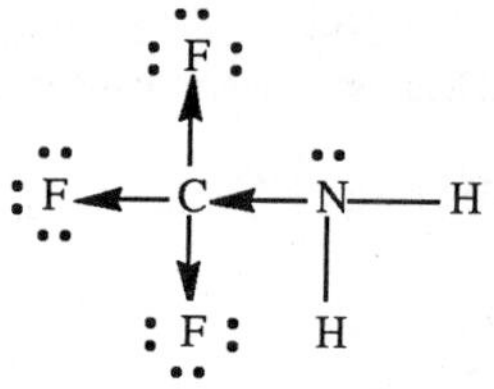

The basicity of the nitrogen atom is also decreased by the electron-withdrawing effect of the fluorines, which makes the electron pair on the nitrogen less available for protonation.

(b)

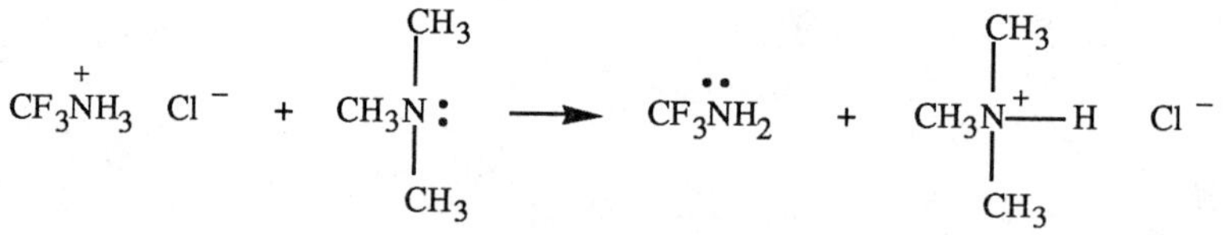

$$CF_3\overset{+}{N}H_3 \ Cl^- \ + \ CH_3\underset{CH_3}{\overset{CH_3}{\underset{|}{\overset{|}{N}}}}\!\!: \ \longrightarrow \ CF_3\overset{..}{N}H_2 \ + \ CH_3\underset{CH_3}{\overset{CH_3}{\underset{|}{\overset{|}{\overset{+}{N}}}}}\!\!-H \quad Cl^-$$

We would expect trimethylamine to be a stronger base than trifluoromethylamine [see part (a)], and to remove a proton from the trifluoromethylammonium ion.

3.40   (a) When a strong acid such as H—Cl is added to water, the following reaction takes place.

$$H\text{---}Cl \ + \ H_2O \ \rightarrow \ Cl^- \ + \ H_3O^+$$

$$pK_a \ -7 \qquad\qquad\qquad pK_a \ -1.7$$

The equilibrium constant for this reaction is greater than 10^{+5}, and the reaction proceeds essentially to completion. Similar reactions take place in water with other strong acids such as sulfuric, nitric, and phosphoric acids. A solution of strong acid in water thus consists of the hydronium ion, the strongest acid which can exist in water) and the conjugate base of the strong acid.

When a strong base such as amide ion is added to water, a similar reaction takes place.

$$NH_2^- \ + \ H_2O \ \rightarrow \ NH_3 \ + \ OH^-$$

$$pK_a \ 15.7 \quad pK_a \ 36$$

Again the equilibrium constant is very large ($>10^{+20}$), and the reaction proceeds essentially to completion. A solution of a strong base in water thus consists of the hydroxide ion (the strongest base that can exist in water) and the conjugate acid of the strong base.

4

Reaction Pathways

Concept Map 4.1   The factors that determine whether a chemical reaction between a given set of reagents is likely.

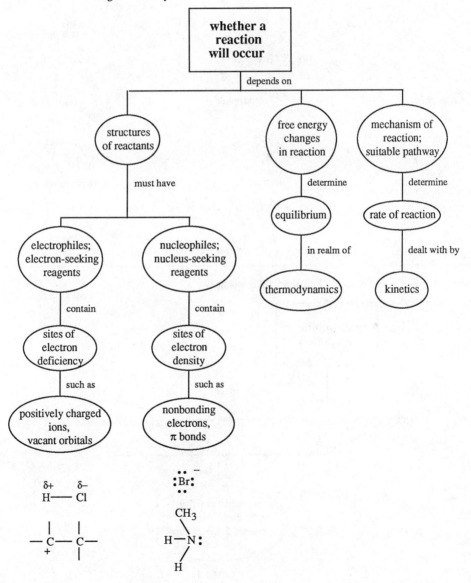

4.1   For those species in which more than one electrophilic or nucleophilic sites is present, the stronger site is indicated.

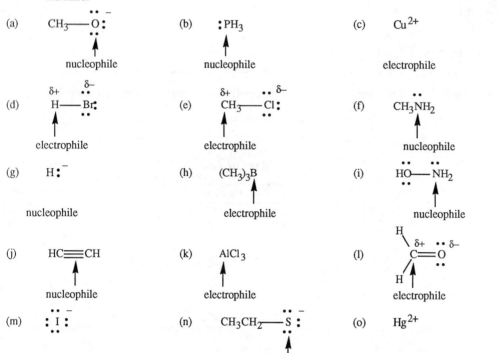

(a) CH_3—$\overset{\cdot\cdot}{\underset{\cdot\cdot}{O}}$: ⁻

nucleophile

(b)   :PH_3

nucleophile

(c) Cu^{2+}

electrophile

(d) $\overset{\delta+}{H}$—$\overset{\delta-}{\underset{\cdot\cdot}{Br}}$:

electrophile

(e) $\overset{\delta+}{CH_3}$—$\overset{\delta-}{\underset{\cdot\cdot}{Cl}}$:

electrophile

(f) CH_3NH_2

nucleophile

(g)   H:⁻

nucleophile

(h) $(CH_3)_3B$

electrophile

(i) $\overset{\cdot\cdot}{HO}$—$\overset{\cdot\cdot}{NH_2}$

nucleophile

(j)   HC≡CH

nucleophile

(k) $AlCl_3$

electrophile

(l) $\overset{H}{\underset{H}{>}}C\overset{\delta+}{=}\overset{\delta-}{\underset{\cdot\cdot}{O}}$

electrophile

(m)   :$\overset{\cdot\cdot}{\underset{\cdot\cdot}{I}}$:⁻

nucleophile

(n) CH_3CH_2—$\overset{\cdot\cdot}{\underset{\cdot\cdot}{S}}$:⁻

nucleophile

(o) Hg^{2+}

electrophile

Concept Map 4.2   A nucleophilic substitution reaction.

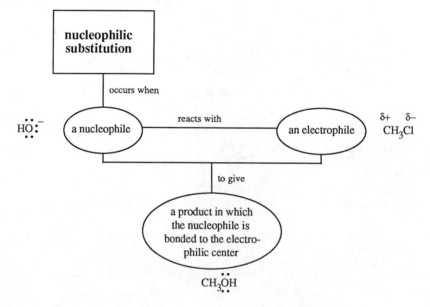

4.2

(a)   electrophile → CH_3 — $\overset{..}{\underset{..}{Br}}$:   → CH_3 — $\overset{..}{\underset{..}{O}}H$ +   : $\overset{..}{\underset{..}{Br}}$: $^-$

nucleophile → $^-$: $\overset{..}{\underset{..}{O}}$ — H

(b)   electrophile → CH_3 — $\overset{..}{\underset{..}{I}}$:   → CH_3 — $\overset{..}{\underset{..}{O}}H$ +   : $\overset{..}{\underset{..}{I}}$: $^-$

nucleophile → $^-$: $\overset{..}{\underset{..}{O}}$ — H

electrophile

(c) CH_3CH_2 — $\overset{..}{\underset{..}{Cl}}$:   → CH_3CH_2 — $\overset{..}{\underset{..}{O}}H$ +   : $\overset{..}{\underset{..}{Cl}}$: $^-$

nucleophile → $^-$: $\overset{..}{\underset{..}{O}}$ — H

(d)   electrophile → CH_3 — $\overset{..}{\underset{..}{I}}$:   → CH_3 — $\overset{+}{N}H_3$: $\overset{..}{\underset{..}{I}}$: $^-$

: NH_3

nucleophile

4.3   Sodium ion, Na^+, has a complete valence shell (see Problem 2.1 for the electronic configuration of elemental sodium), and the energies of the empty orbitals in the $n=3$ shell are too high to be available for accepting an electron pair. Both the mercury(II), Hg^{2+}, and copper(II), Cu^{2+}, ions are transition metals that have lower energy empty orbitals available to accept an electron pair.

Concept Map 4.3   (see p. 84)

Concept Map 4.4   (see p. 84)

4.4   At 311 K, $k_r = 3.55 \times 10^{-5}$ L/mol·sec.

Initial rate $= k_r[CH_3Cl][OH^-]$; when $[OH^-]$ 0.05 M:

initial rate $= (3.55 \times 10^{-5}$ L/mol·sec)(0.003 mol/L)(0.05 mol/L).

$= 5.33 \times 10^{-9}$ mol/L·sec, which is five times the original rate (see p. 114 in the text).

When $[CH_3Cl] = 0.001$ M, the rate will be one-third the rate when $[CH_3Cl] = 0.003$ M, or 3.55×10^{-10} mol/L·sec.

Concept Map 4.3   The factors that determine the equilibrium constant for a reaction.

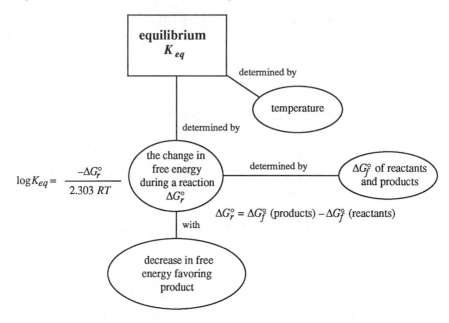

Concept Map 4.4   The factors that influence the rate of a reaction as they appear in the rate equation.

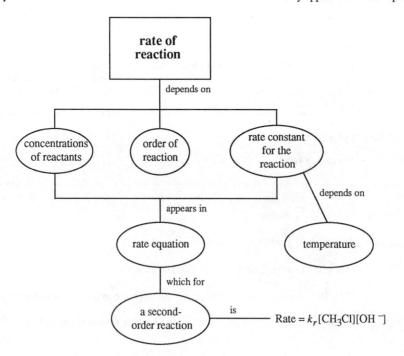

Concept Mapt 4.5   Some of the factors other than concentration that influence the rate of a reaction.

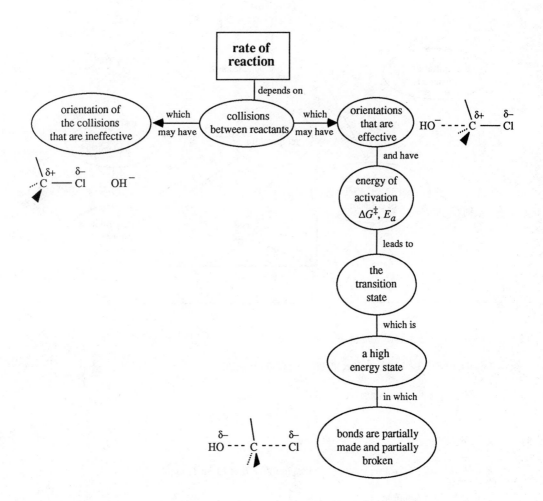

Concept Map 4.6   Factors that influence the rate of a reaction, and the effect of temperature.

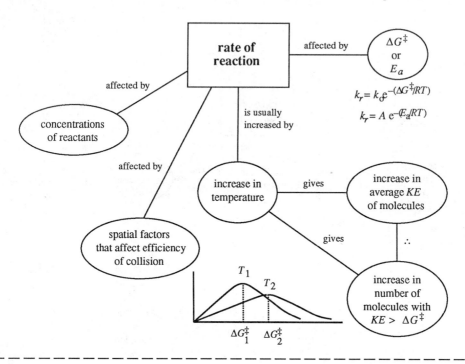

4.5   At 308 K, $k_r = 5.3 \times 10^{-4}$ L/mol·sec.

Initial rate = $(5.3 \times 10^{-4}$ L/mol·sec$)(0.001$ mol/L$)(0.01$ mol/L$)$

           = 5.3×10^{-9} mol/L·sec

4.6 $k_r = Ae^{-E_a/RT}$

$k_r = (1.24 \times 10^{+12}$ L/mol·sec$)\exp(-22.22$ kcal/mol$)/[(1.99 \times 10^{-3}$ kcal/mol·K$)(298$ K$)]$

$k_r = 6.62 \times 10^{-5}$ L/mol·sec

4.7   The double bond in propene is another possible nucleophile in the reaction mixture.

$$CH_3CH{=}CH_2 \longrightarrow \underset{\underset{+}{}}{CH_3CH}{-}CH_2{-}\overset{\overset{\displaystyle CH_3}{|}}{CHCH_3}$$

                                                                   a new secondary
                                                                   carbocation

$$\underset{+}{CH_3CHCH_3}$$

Concept Map 4.7   An electrophilic addition reaction to an alkene.

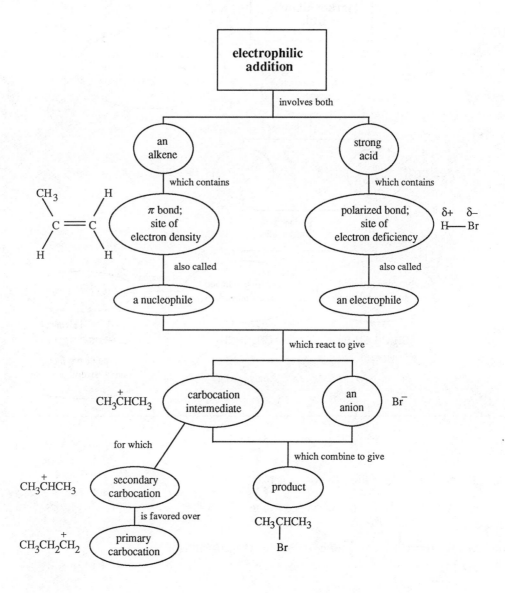

Concept Map 4.8   Markovnikov's Rule.

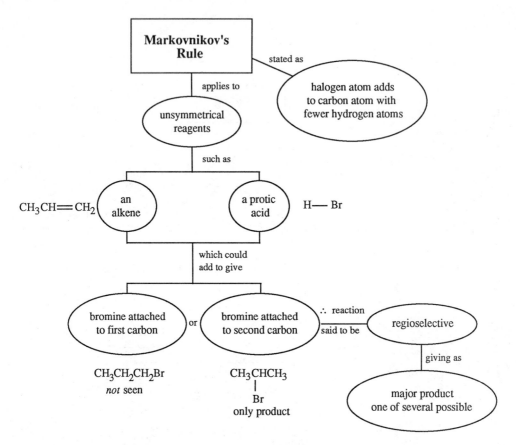

4.8

(a)
$$CH_3\overset{\overset{\displaystyle CH_3}{|}}{C}{=}CH_2 \quad + \quad HBr \quad \longrightarrow \quad CH_3\overset{\overset{\displaystyle CH_3}{|}}{\underset{\underset{\displaystyle Br}{|}}{C}}CH_3$$

(b) $CH_3CH_2CH{=}CH_2 \quad + \quad HBr \quad \longrightarrow \quad CH_3CH_2\underset{\underset{\displaystyle Br}{|}}{C}HCH_3$

(c) $CH_3CH_2CH{=}CHCH_3 \quad + \quad HBr \quad \longrightarrow \quad CH_3CH_2CH_2\underset{\underset{\displaystyle Br}{|}}{C}HCH_3 \quad + \quad CH_3CH_2\underset{\underset{\displaystyle Br}{|}}{C}HCH_2CH_3$

4.9   The methyl cation ($^{+}CH_3$) is less stable than the *n*-propyl cation ($CH_3CH_2CH_2^{+}$) because the methyl cation has no electron-releasing alkyl groups to stabilize the positive charge.

4.10   The *tert*-butyl cation is more stable than the isopropyl cation because the *tert*-butyl cation has three electron-releasing alkyl groups to stabilize the positive charge and the isopropyl group has only two.

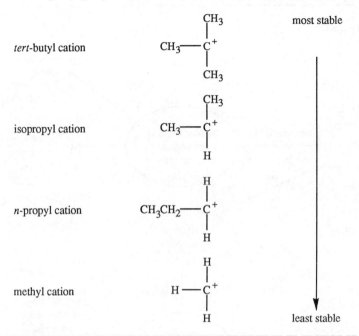

Concept Map 4.9   (see p. 90)

4.11 $\log K_{eq} = -(\Delta G^{\circ}/2.303\ RT)$

At 25 °C (298 K),

for 1-bromopropane

$\log K_{eq} = -(-7.63\ \text{kcal/mol})/[(2.303)(1.99 \times 10^{-3}\ \text{kcal/mol·K})(298\ \text{K})]$

$\log K_{eq} = 5.59$

$K_{eq} = 10^{5.59} = 10^{0.59} \times 10^5 = 3.89 \times 10^5$

for 2-bromopropane

$\log K_{eq} = -(-8.77\ \text{kcal/mol})/[(2.303)(1.99 \times 10^{-3}\ \text{kcal/mol·K})(298\ \text{K})]$

$\log K_{eq} = 6.42$

$K_{eq} = 10^{6.42} = 10^{0.42} \times 10^6 = 2.63 \times 10^6$

4.11 (cont)

Note that the answers that you get may not correspond exactly to the numbers obtained here depending on whether you round off to the correct number of significant figures at each stage of the calculation or carry all figures through and round off at the end of the calculation.

— —

Concept Map 4.9   Carbocations.

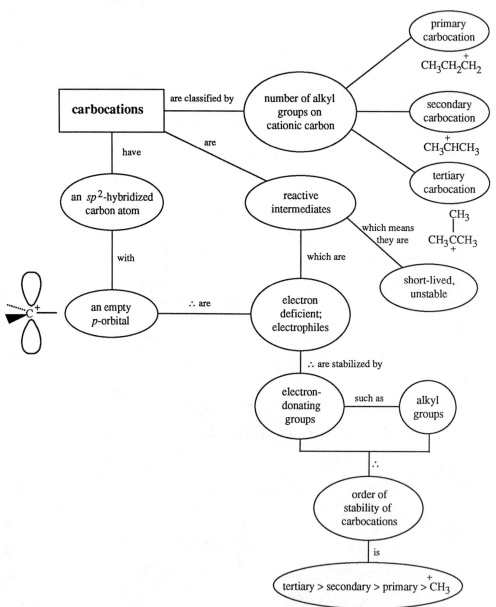

Concept Map 4.10   Energy diagrams.

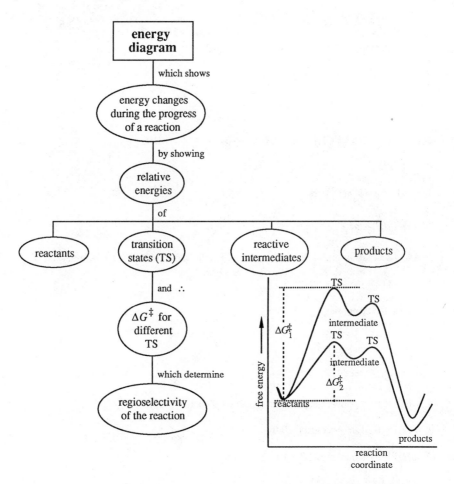

- -

4.12   (a) $CH_3CH{=\!=\!=}CH_2$ (g)   +   HI (g) $\longrightarrow$ $CH_3\underset{|}{C}HCH_3$ (g)

                                                                                    I

(b) Rate = k $[CH_3CH{=}CH_2][HI]$

4.13   Concentration is defined as the number of moles of a substance in a given volume. For an ideal gas, the concentration at constant temperature and volume is proportional to the pressure.

$$P = \frac{n}{V}RT$$

$$P_{propene} = P_{HI} = (45 \text{ mm Hg})/(760 \text{ mm Hg/atm})$$

4.13 (cont)

Rate = $k\,(P_{\text{propene}})(P_{\text{HI}})$

Rate = $(1.66 \times 10^{-6}/\text{atm·sec})(45/760)^2 \text{atm}^2$

Rate = 5.8×10^{-9} atm/sec

4.14   For $CH_3CH{=\!\!=}CH_2$ + HI $\longrightarrow$ $CH_3\underset{\underset{\displaystyle I}{|}}{C}HCH_3$

$k_r = Ae^{-E_a/RT}$

$\ln k_r = (\ln A) - E_a/RT$

$\ln A = E_a/RT - \ln k_r$

$\ln A = (+22.4\ \text{kcal/mol})/[(1.99 \times 10^{-3}\ \text{kcal/mol·K})(490\ \text{K})] + \ln (2.61 \times 10^{-3}\ \text{L/mol·sec})$

$\ln A = (23.0) + (-5.95) = 17.0$

$A = 2.5 \times 10^7$ L/mol·sec

4.15   For $CH_3CH{=\!\!=}CH_2$ + HI $\longrightarrow$ $CH_3CH_2\overset{+}{C}H_2$ + I^-

$k_r = (2.5 \times 10^7\ \text{L/mol·sec})\exp(-38\ \text{kcal/mol})/[(1.99 \times 10^{-3}\ \text{kcal/mol·K})(490\ \text{K})]$

$k_r = (2.5 \times 10^7\ \text{L/mol·sec})\exp(-39.0)$

$k_r = (2.5 \times 10^7\ \text{L/mol·sec})(1.2 \times 10^{-17})$

$k_r = 3.0 \times 10^{-10}$ L/mol·sec

Rate for the formation of the isopropyl cation = $k_{\text{isopropyl}}[CH_3CH{=}CH_2][HI]$

Rate for the formation of the n-propyl cation = $k_{n\text{-propyl}}[CH_3CH{=}CH_2][HI]$

$$\frac{k_{\text{isopropyl}}}{k_{n\text{-propyl}}} = \frac{2.61 \times 10^{-3}\ \text{L/mol·sec}}{3.0 \times 10^{-10}\ \text{L/mol·sec}}$$

$$\frac{k_{\text{isopropyl}}}{k_{n\text{-propyl}}} = 8.7 \times 10^6$$

In the reaction of propene with hydrogen iodide, 8.7 million isopropyl cations are formed for *every* n-propyl cation.

4.16

(a)

$$CH_3CH_2 - \overset{..}{\underset{..}{Br}} : \longrightarrow CH_3CH_2 - \overset{..}{\underset{..}{I}} : \quad + \quad : \overset{..}{\underset{..}{Br}} : ^-$$

electrophile

nucleophile $\longrightarrow$ $: \overset{..}{\underset{..}{I}} : ^-$

(b)

$$CH_3\overset{\overset{CH_3}{|}}{CH} - \overset{..}{\underset{..}{Br}} : \longrightarrow CH_3\overset{\overset{CH_3}{|}}{CH} - C \equiv N : \quad + \quad : \overset{..}{\underset{..}{Br}} : ^-$$

electrophile

$^- : C \equiv N :$

nucleophile

Note that even though both carbon and nitrogen have nonbonding electron pairs, the electrons on the carbon are more available for bonding in nucleophilic substitution reactions because carbon is less electronegative than nitrogen.

(c)

$$CH_3CH_2CH = CH_2 \quad \longrightarrow \quad CH_3CH_2\overset{+}{CH} - \overset{\overset{}{|}}{\underset{H}{CH_2}} \quad \longrightarrow \quad CH_3CH_2CHCH_3$$

nucleophile   electrophile $: \overset{..}{\underset{..}{Cl}} :$

$$H - \overset{..}{\underset{..}{Cl}} :$$

nucleophile $\longrightarrow$ $: \overset{..}{\underset{..}{Cl}} : ^-$

electrophile

(d)

$$CH_3CH_2\overset{\overset{CH_3}{|}}{CH} - \overset{..}{\underset{..}{Br}} : \longrightarrow CH_3CH_2\overset{\overset{CH_3}{|}}{CH} - \overset{..}{\underset{..}{S}} - CH_3 \quad + \quad : \overset{..}{\underset{..}{Br}} : ^-$$

electrophile

$$CH_3 - \overset{..}{\underset{..}{S}} : ^- \longleftarrow \text{nucleophile}$$

(e)

$$CH_3CH = CH_2 \quad + \quad \overset{\delta+}{Cl} - \overset{\delta-}{Cl} \quad \longrightarrow \quad CH_3\overset{}{\underset{Cl}{CH}}CH_2 - Cl$$

nucleophile   electrophile

4.17

reagent	type	reason for choice
(a) $CH_3O^-\ Na^+$	nucleophile	The carbon atom bonded to the iodine atom in ethyl iodide is an electrophilic carbon and will react with a reagent that is a nucleophile. The reaction is a nucleophilic substitution.
(b) H_2SO_4	electrophile	The double bond in 1-butene is a nucleophile and reacts with the electrophilic proton in sulfuric acid. The carbocation produced in the first step of the reaction is an electrophile and will react the the conjugate base of sulfuric acid, a nucleophile, which is also produced in the first step of the reaction. The overall reaction is an electrophilic addition.
(c) Br_2	electrophile	The nucleophilic double bond in 1-butene will react with the electrophile bromine (see Chapter 8 for the mechanism of the addition of bromine to alkenes). The overall reaction is an electrophilic addition.
(d) NH_3	nucleophile	The carbon atom bonded to the bromine atom in *n*-propyl bromide is an electrophilic carbon and will react with a reagent that is a nucleophile. The reaction is a nucleophilic substitution.

4.18

(a)

$$CH_3CH_2CH{=\!=}CH_2 \xrightarrow{\ ?\ } CH_3CH_2\underset{\underset{\displaystyle SH}{|}}{C}HCH_3$$

The elements of hydrogen sulfide, H_2S, have added to the double bond. The double bond is a nucleophile and is protonated by strong Brønsted acids. H_2S, with a pK_a ~10.5, is not sufficiently acidic(hydrogen sulfide is similar to ethanethiol, CH_3CH_2SH, and can be expected to have a similar pK_a). A hydrogen halide such as HBr (pK_a –8) is. In the product that is formed, the bromine atom will be attached to an electrophilic carbon atom and will leave when the hydrogen sulfide anion is used as a nucleophile.

$$CH_3CH_2CH{=\!=}CH_2 \xrightarrow{\ HBr\ } CH_3CH_2\underset{\underset{\displaystyle Br}{|}}{C}HCH_3$$

$$CH_3CH_2\underset{\underset{\displaystyle Br}{|}}{C}HCH_3 \xrightarrow{\ NaSH\ } CH_3CH_2\underset{\underset{\displaystyle SH}{|}}{C}HCH_3 \ + \ NaBr$$

(b)

$$CH_3CH{=\!=}CHCH_3 \xrightarrow{\ ?\ } CH_3\underset{\underset{\displaystyle Cl}{|}}{C}HCH_2CH_3$$

4.18 (cont)

(b) (cont)

The elements of hydrogen chloride, HCl, have added to the double bond. Unlike H_2S, hydrogen chloride is a strong enough acid (pK_a −7) to protonate the double bond. The cation produced by protonation will then react with the conjugate base of hydrogen chloride to give the desired product.

$$CH_3CH\!\!=\!\!CHCH_3 \xrightarrow{\ HCl\ } CH_3CH\!\!-\!\!\overset{+}{C}HCH_3 + Cl^- \longrightarrow CH_3\underset{Cl}{C}HCH_2CH_3$$

$$\underset{H}{}$$

(c) $$CH_2\!\!=\!\!CH_2 \xrightarrow{\ ?\ } CH_3CH_2\overset{+}{N}H_3\ \ I^-$$

The elements of the ammonium ion, $NH_4{}^+$, have added to the double bond. $NH_4{}^+$ is not sufficiently acidic (pK_a 9.4) but HI (pK_a −9) is. Ammonia can then be used as a nucleophile to react with the electrophilic carbon atom in ethyl iodide, the product of the first reaction.

$$CH_2\!\!=\!\!CH_2 \xrightarrow{\ HI\ } CH_3CH_2I \xrightarrow{\ NH_3\ } CH_3CH_2\overset{+}{N}H_3\ \ I^-$$

(d) $$CH_3\underset{Br}{C}HCH_2CH_2CH_3 \xrightarrow{\ ?\ } CH_3\underset{CN}{C}HCH_2CH_2CH_3$$

Cyanide ion has substituted for bromide ion. This is a nucleophilic substitution reaction on 2-bromopentane which requires only one reagent, NaCN.

$$CH_3\underset{Br}{C}HCH_2CH_2CH_3 \xrightarrow{\ NaCN\ } CH_3\underset{CN}{C}HCH_2CH_2CH_3 + NaBr$$

4.19

(a)

4.19 (cont)

(b)

electrophile

$CH_3CH_2CH_2$—Br: $\longrightarrow$ $CH_3CH_2CH_2$—$\overset{+}{P}H_3$: Br: $^-$

:PH$_3$

nucleophile

(c)

electrophile

CH_3CH_2—I: $\longrightarrow$ CH_3CH_2—S—CH_2CH_3 +   : I: $^-$

CH_3CH_2—S: $\longleftarrow$ nucleophile

(d)

nucleophile

CH_3
  \
   C$=$CH$_2$ $\longrightarrow$ electrophile CH_3—$\overset{+}{C}$—CH$_3$ $\longrightarrow$ CH_3—C—CH$_3$
  / | |
CH_3 CH$_3$ CH$_3$
                                                                             : Cl:

H—Cl:

electrophile

nucleophile $\longrightarrow$: Cl: $^-$

(e)

$CH_3CH_2CH$$=$$CHCH_3$ +   Br$_2$ $\longrightarrow$ CH_3CH_2CH—$CHCH_3$
                                                              | |
nucleophile   electrophile   Br   Br

4.20

	reagent	type	reason for choice
(a)	NaSH	nucleophile	The carbon atom bonded to the bromine atom in methyl bromide is an electrophilic carbon and will react with a reagent that is a nucleophile. The reaction is a nucleophilic substitution.
(b)	NaCN	nucleophile	The carbon atom bonded to the bromine atom in ethyl bromide is an electrophilic carbon and will react with a reagent that is a nucleophile. The reaction is a nucleophilic substitution.
(c)	Cl$_2$	electrophile	The nucleophilic double bond in ethylene will react with the electrophile chlorine (see Chapter 8 for the mechanism of the addition of chlorine to alkenes). The reaction is an electrophilic addition.

4.20 (cont)

(d)   CH₃CH₂O⁻ Na⁺ nucleophile   The carbon atom bonded to the bromine atom in *n*-propyl bromide is an electrophilic carbon and will react with a reagent that is a nucleophile. The reaction is a nucleophilic substitution.

(e)   HI   electrophile   The nucleophilic double bond in 1-butene will react with the electrophilic proton of hydrogen iodide and the electrophilic carbocation that is produced will then react with the iodide ion. The reaction is an electrophilic addition.

(f)   NaI   nucleophile   The carbon atom bonded to the bromine atom in isopropyl bromide is an electrophilic carbon and will react with a reagent that is a nucleophile. The reaction is a nucleophilic substitution.

4.21

(a) $CH_2{=}CH_2 \xrightarrow{\ ?\ } CH_3CH_2OCH_3$

The elements of methanol, CH_3OH, have added to the double bond. The double bond is a nucleophile and is protonated by strong Brønsted acids. Methanol is not sufficiently acidic (pK_a 15.5) but a hydrogen halide such as HBr (pK_a −8) is. In the product that is formed, the bromine atom will be attached to an electrophilic carbon atom and will leave when sodium methoxide is used as a nucleophile.

$$CH_2{=}CH_2 \xrightarrow{HBr} CH_3CH_2Br \xrightarrow{CH_3O^-Na^+} CH_3CH_2OCH_3$$

(b) $CH_3CH{=}CH_2 \xrightarrow{\ ?\ } CH_3\underset{I}{C}HCH_3$

The elements of hydrogen iodide, HI, have added to the double bond. Hydrogen iodide is a strong enough acid (pK_a −9) to protonate the double bond. The cation produced by protonation will then react with iodide ion, the conjugate base of hydrogen iodide, to give the desired product.

$$CH_3CH{=}CH_2 \xrightarrow{HI} CH_3\underset{I}{C}HCH_3$$

(c) $CH_3CH_2CH{=}CH_2 \xrightarrow{\ ?\ } CH_3CH_2\underset{OH}{C}HCH_3$

The elements of water, H_2O, have added to the double bond. Water is not sufficiently acidic (pK_a 15.7) but the hydronium ion (pK_a −1.7), formed when a strong acid such as sulfuric acid is added to water, is (See Problem 3.40). The solvent, water, will then act as a nucleophile to react with the carbocation to give the final product.

$$CH_3CH_2CH{=}CH_2 \xrightarrow[H_2SO_4]{H_2O} CH_3CH_2\underset{OH}{C}HCH_3$$

4.21 (cont)

(d)

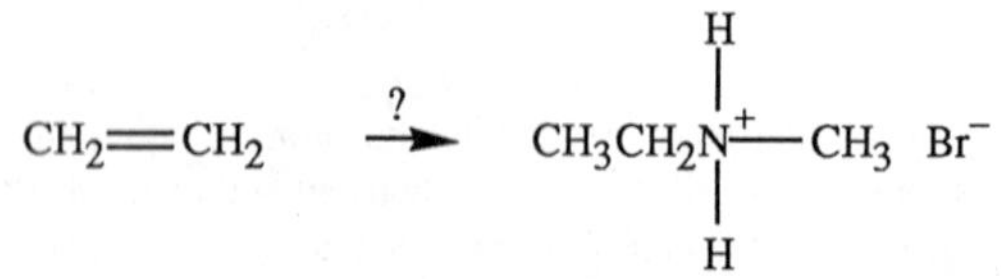

$$CH_2{=}CH_2 \xrightarrow{?} \underset{\underset{H}{|}}{\overset{\overset{H}{|}}{CH_3CH_2\overset{+}{N}}}{-}CH_3 \ Br^-$$

The elements of the methylammonium ion, $CH_3NH_3^+$, have added to the double bond. $CH_3NH_3^+$ is not sufficiently acidic (pK_a 10.6) but HBr (pK_a –8) is. Methylamine can then be used as a nucleophile to react with the electrophilic carbon atom in ethyl bromide, the product of the first reaction.

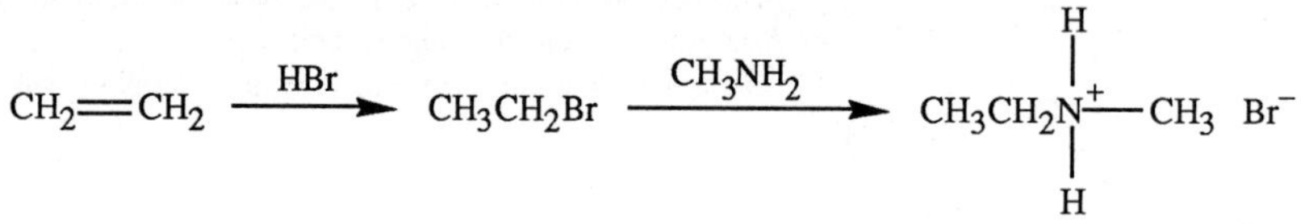

$$CH_2{=}CH_2 \xrightarrow{HBr} CH_3CH_2Br \xrightarrow{CH_3NH_2} \underset{\underset{H}{|}}{\overset{\overset{H}{|}}{CH_3CH_2\overset{+}{N}}}{-}CH_3 \ Br^-$$

4.22

(a)

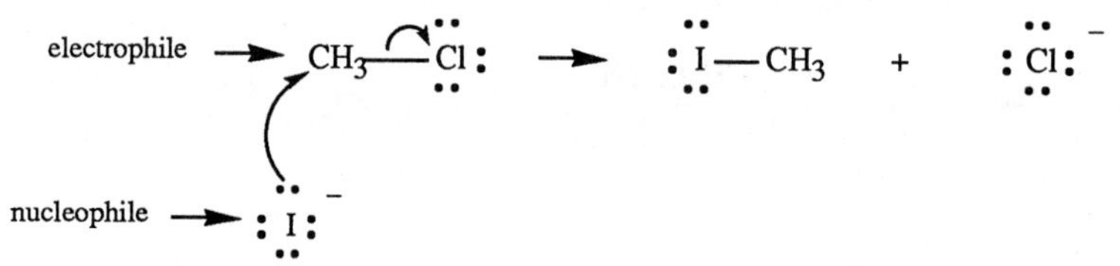

(b)

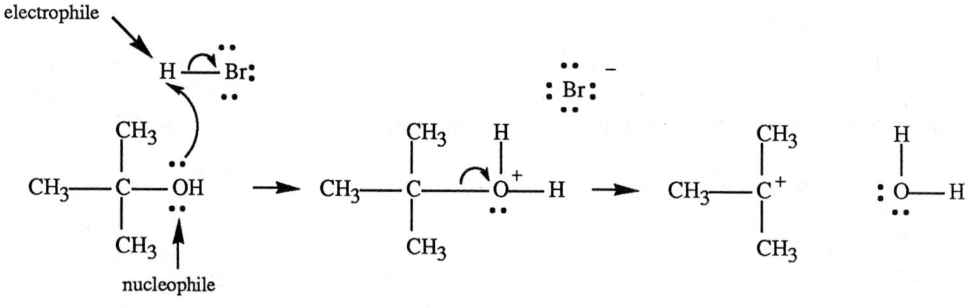

(c)

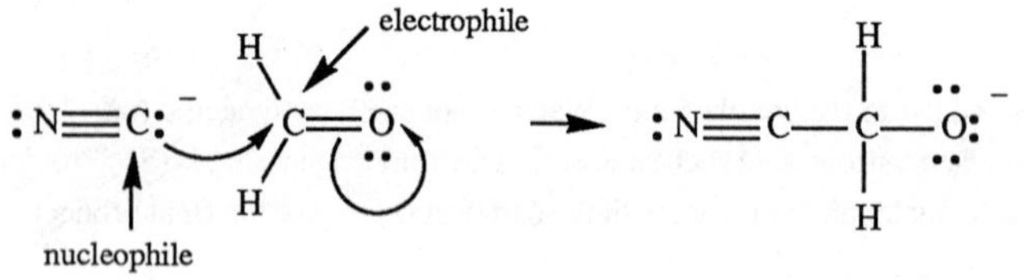

4.22 (cont)

(d)

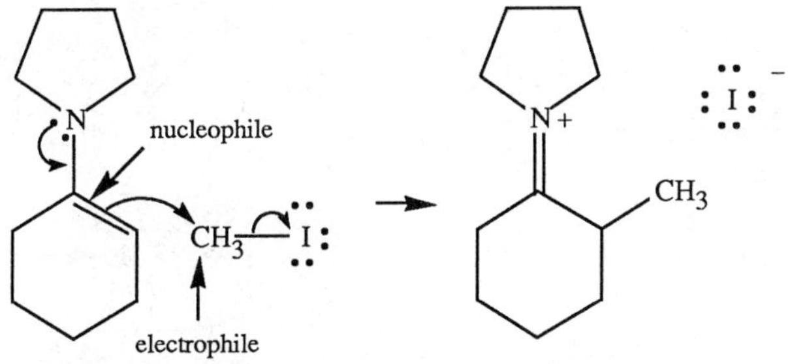

nucleophile

CH$_3$

electrophile

CH$_3$

(e)

acid

base

electrophile

CH$_3$—C—Cl

CH$_3$N—H

CH$_3$ nucleophile

CH$_3$—C—Cl

CH$_3$—N—H

CH$_3$

CH$_3$—C—N—CH$_3$

CH$_3$

H—Cl

CH$_3$—C—N—CH$_3$

CH$_3$

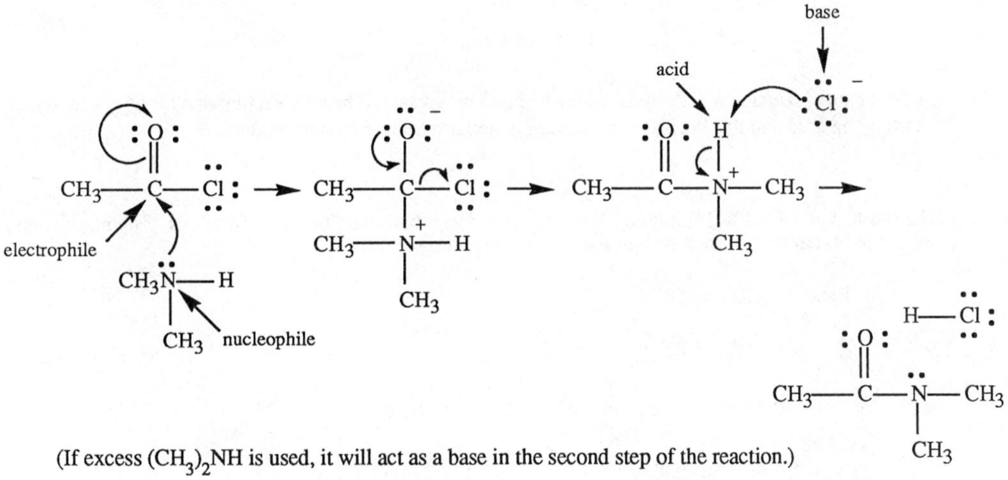

(If excess (CH$_3$)$_2$NH is used, it will act as a base in the second step of the reaction.)

(f)

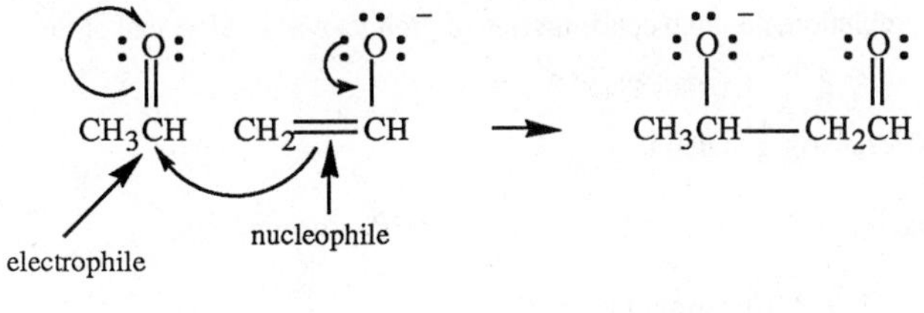

CH$_3$CH   CH$_2$=CH

electrophile

nucleophile

CH$_3$CH—CH$_2$CH

(g)

electrophile

CH$_2$=CH—C—CH$_3$

nucleophile

NH$_3$

CH$_2$—CH=C—CH$_3$

$^+$NH$_3$

CH$_2$—CH—C—CH$_3$

H—N—H ← acid

H

NH$_3$ ← base

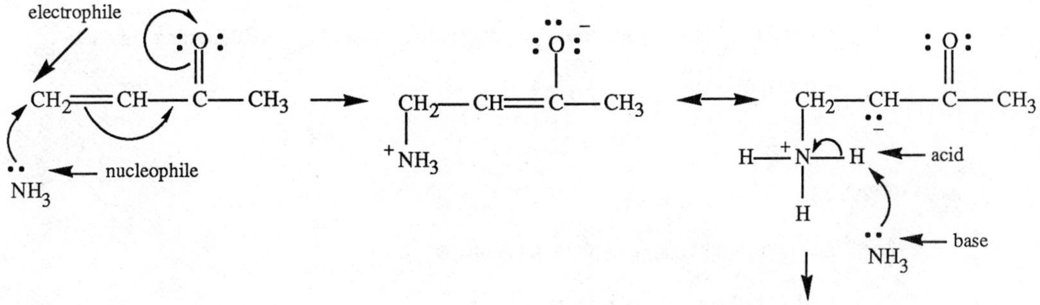

(continued on next page)

4.22 (cont)

(g) (cont)

$$CH_2\!-\!CH\!-\!\overset{\displaystyle :\overset{..}{O}:}{\underset{\displaystyle \|}{C}}\!-\!CH_3 \quad \longleftarrow \quad CH_2\!-\!CH\!-\!\overset{\displaystyle :\overset{..}{O}:}{\underset{\displaystyle \|}{C}}\!-\!CH_3$$

base

$:NH_2$ H $:NH_2$

: NH_3

H   ← acid

$$\underset{\displaystyle H}{\overset{\displaystyle H}{H\!-\!\overset{+}{N}\!-\!H}}$$

4.23   No, the rate of reaction will decrease as the reaction progresses. The rate is a function of the concentration of starting material and this decreases as starting material is converted into product.

4.24   At 308 K, $k = 5.3 \times 10^{-4}$ L/mol·sec. When the reaction is half over, the concentrations of hydroxide ion and of methyl bromide will each be half the initial concentrations.

Rate $= k\,[CH_3Br][OH^-]$

$= k\,(0.5[CH_3Br]_{initial})(0.5[OH^-]_{initial})$

$= 0.25$ (initial rate) $= 0.25(5.3 \times 10^{-9}$ L/mol·sec)

$= 1.3 \times 10^{-9}$ L/mol·sec

4.25   When a reaction has gone to 99% completion, the final concentration of product will be 99% that of the initial starting material concentration.

$[product]_{99\%\ completion} = 0.99[starting\ material]_{initial}$

and the concentration of starting material will be reduced to 1% of the initial concentration.

$[starting\ material]_{99\%\ completion} = 0.01[starting\ material]_{initial}$

The reaction is over when it reaches equilibrium and

$K_{eq} = [product]_{eq}/[starting\ material]_{eq} = 0.99[starting\ material]_{initial}/0.01[starting\ material]_{initial}$

$= 0.99/0.01 = 99$

$\Delta G_r = -RT \ln K_{eq}$

At 25 °C (298 K)

$\Delta G_r = -(1.99 \times 10^{-3}$ kcal/mol·K)(298 K) ln 99

$= -2.7$ kcal/mol

4.26   (a) $CH_3Cl + H_2O \rightarrow CH_3OH + H_3O^+ + Cl^-$

The nucleophile in the reaction is water. The other products of the reaction are the hydronium ion and chloride ion.

(b)

(c)   Rate = k [CH_3Cl][H_2O]

(d) The density of water at room temperature is approximately 1 g/mL. One liter of water contains 1000 g or 55.6 mol (1000 g)/(18 g/mol). The concentration of water in pure water is thus 55.6 M. The concentration of chloromethane that dissolves in one liter of water is only 0.003 moles, which is not enough to significantly change the concentration of water in an aqueous solution of chloromethane. According to the stoichiometry of the reaction, one equivalent of chloromethane reacts with one equivalent of water. At the end of the reaction, assuming all the starting material is converted to product, the final concentration of water will be (55.6 – 0.003) M or, to the correct number of significant figures, 55.6 M. It would not be possible to see a change in the concentration of water as the reaction proceeds. In other words, the concentration of water remains essentially constant as the reaction proceeds, and can be incorporated into the rate constant.

Rate = k'[CH_3Cl] where $k' = k$[H_2O]

This type of reaction, in which the solvent is also the nucleophile, is an example of a **solvolysis reaction.**

4.27

$$CH_3CH{=\!=}CH_2 \;+\; HI \;\longrightarrow\; CH_3\underset{\underset{I}{|}}{C}HCH_3$$

$\Delta G_r^{\circ} = \Delta H_r^{\circ} - T\Delta S_r^{\circ}$

$\Delta H_r^{\circ} = \Delta H_{f,\,\text{isopropyl iodide}}^{\circ} - (\Delta H_{f,\,\text{HI}}^{\circ} + \Delta H_{f,\,\text{propene}}^{\circ})$

$= -19.65$ kcal/mol $- (-1.35$ kcal/mol $+ 2.80$ kcal/mol$) = -21.1$ kcal/mol

$\Delta S_r^{\circ} = \Delta S_{f,\,\text{isopropyl iodide}}^{\circ} - (\Delta S_{f,\,\text{HI}}^{\circ} + \Delta S_{f,\,\text{propene}}^{\circ})$

$= 91.02$ cal/mol·K $- (52.98$ cal/mol·K $+ 73.48$ cal/mol·K$) = -35.44$ cal/mol·K
$= -3.54 \times 10^{-2}$ kcal/mol·K

$\Delta G_r^{\circ} = (-21.1$ kcal/mol$) - (5.00 \times 10^2$ K$)(-3.54 \times 10^{-2}$ kcal/mol·K$)$

$= (-21.1$ kcal/mol$) + (17.7$ kcal/mol$) = -3.40$ kcal/mol

4.27 (cont)

$$\Delta G_r^\circ = -RT \ln K_{eq}$$

$$\ln K_{eq} = -\Delta G_r^\circ / RT$$

$$K_{eq} = \exp(-(-\Delta G_r^\circ / RT)) = \exp[-(-3.40 \text{ kcal/mol})/(1.99 \times 10^{-3} \text{ kcal/mol·K})(5.00 \times 10^2 \text{ K})]$$

$$= 29$$

4.28   (a)

$$\underset{\underset{\text{CH}_3}{|}}{\text{CH}_3\text{C}}\!\!=\!\!\text{CH}_2 \text{ (g)} \quad + \quad \text{H}_2\text{O (l)} \quad \longrightarrow \quad \underset{\underset{\text{OH}}{|}}{\overset{\overset{\text{CH}_3}{|}}{\text{CH}_3\text{CCH}_3}} \text{ (l)}$$

$$\Delta G_r^\circ = \Delta H_r^\circ - T\Delta S_r^\circ$$

$$\Delta H_r^\circ = -85.87 \text{ kcal/mol} - (-4.04 \text{ kcal/mol} - 68.32 \text{ kcal/mol}) = -13.51 \text{ kcal/mol}$$

$$\Delta S_r^\circ = 46.30 \text{ cal/mol·K} - (16.72 \text{ cal/mol·K} + 70.17 \text{ cal/mol·K}) = -40.59 \text{ cal/mol·K}$$

$$= -4.059 \times 10^{-2} \text{ kcal/mol·K}$$

$$\Delta G_r^\circ = (-13.51 \text{ kcal/mol}) - (298 \text{ K})(-4.06 \times 10^{-2} \text{ kcal/mol·K})$$

$$= -1.41 \text{ kcal/mol}$$

$$\Delta G_r^\circ = -RT \ln K_{eq}$$

$$\ln K_{eq} = -\Delta G_r^\circ / RT$$

$$K_{eq} = \exp(-\Delta G_r^\circ / RT) = \exp[-(-1.41 \text{ kcal/mol})/(1.99 \times 10^{-3} \text{ kcal/mol·K})(298 \text{ K})]$$

$$= 10.8$$

(b) When sulfuric acid is added to water, the proton is transferred completely to form the hydronium ion and the hydrogen sulfate anion.

$$\text{H}_2\text{SO}_4 + \text{H}_2\text{O} \rightarrow \text{H}_3\text{O}^+ + \text{HSO}_4^-$$

$$pK_a\, -9 \qquad\qquad pK_a\, -1.7$$

In a dilute solution of H_2SO_4 in water, the electrophile is the hydronium ion, not sulfuric acid. The nucleophile is the double bond of the alkene, which requires a strong acid for protonation. H_2O (pK_a 15.7) is not strong enough, but H_3O^+ is. The hydronium ion is regenerated in the last step of the mechanism, so only a catalytic amount of sulfuric acid is necessary (see Section 8.4 for the mechanism).

4.29   (a) ΔS_r° = (74.27 cal/mol·K + 45.11 cal/mol·K) – (78.32 cal/mol·K) = +41.06 cal/mol·K

$$= +4.106 \times 10^{-2} \text{ kcal/mol·K}$$

The entropy increases because in the reaction two independent molecules, water and cyclohexene, are formed from each molecule of cyclohexanol.

$$\Delta H_r^\circ = (-57.80 \text{ kcal/mol} - 1.28 \text{ kcal/mol}) - (-70.40 \text{ kcal/mol}) = +11.32 \text{ kcal/mol}$$

(b) $\Delta G_r^\circ = \Delta H_r^\circ - T\Delta S_r^\circ$

ΔG_r° = (11.32 kcal/mol) – (298 K)(4.106 $\times$ 10^{-2} kcal/mol·K)

$= -0.916$ kcal/mol

$\Delta G_r^\circ = -RT \ln K_{eq}$

$\ln K_{eq} = -\Delta G_r^\circ / RT$

K_{eq} = exp($-\Delta G_r^\circ / RT$) = exp[$-(-0.916$ kcal/mol)/(1.99 $\times$ 10^{-2} kcal/mol·K)(298 K)]

$= 4.69$, therefore the forward reaction is favored at this temperature.

(c) Because the number of molecules is the same on both sides of the first equation, the change in entropy will be small, and the change in free energy will approximately equal the change in enthalpy. The evolution of heat means that $\Delta H_r^\circ < 0$ (products have a lower enthalpy than starting materials) and therefore that the free energy of the products (protonated alcohol and dihydrogen phosphate ion) is lower than the free energy of the starting material (alcohol and phosphoric acid).

(d) The formation of the very high energy carbocation is the rate-determining step in the addition of an electrophile to an alkene. It is likely that step (2), the formation of the carbocation, is the rate-determining step in the dehydration. Therefore, even though overall energy changes favor the formation of the product at room temperature [part (b)], no product is formed until enough heat is supplied to get over the energy barrier represented by the free energy of activation of the rate-determining step.

5

Alkanes and Cycloalkanes

5.1

(a)

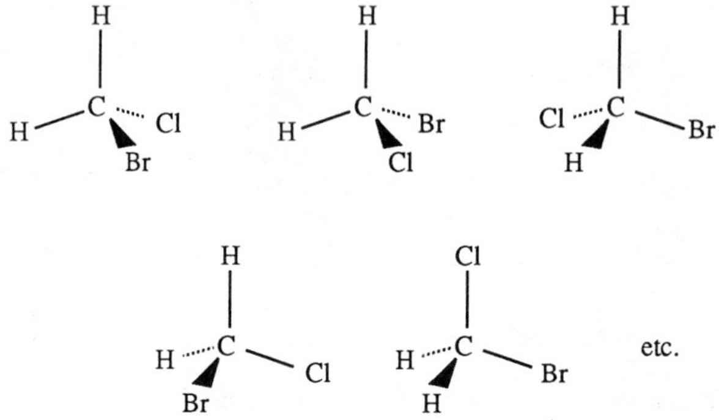

etc.

(b)

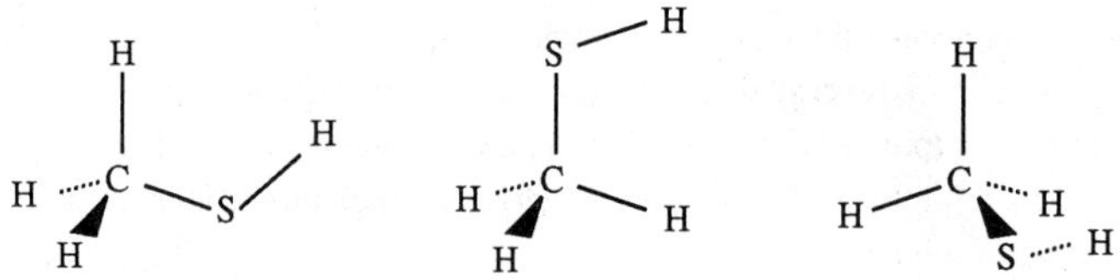

etc.

(c)

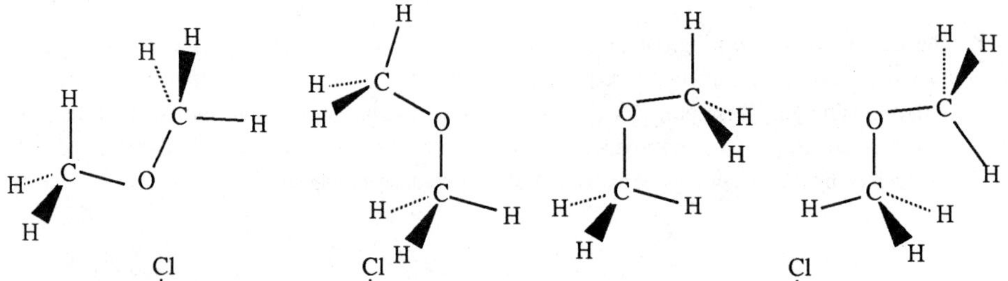

etc.

(d)

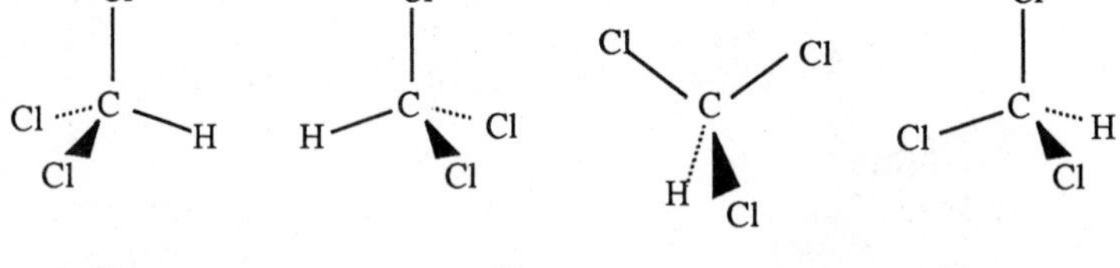

etc.

(e)

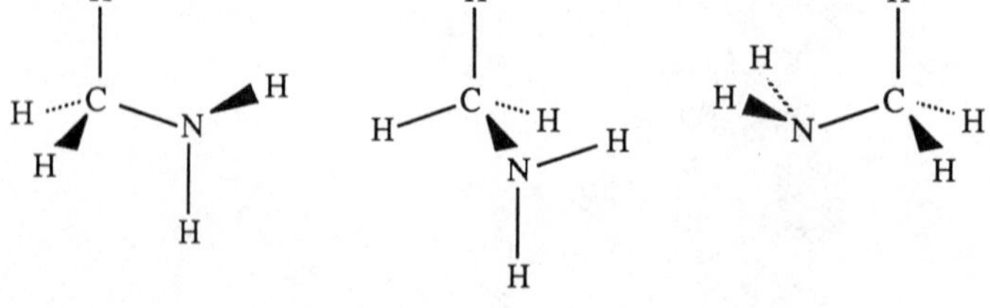

etc.

5.2 $C_2H_3Cl_3$: CH_3CCl_3 $ClCH_2CHCl_2$

Lewis structures

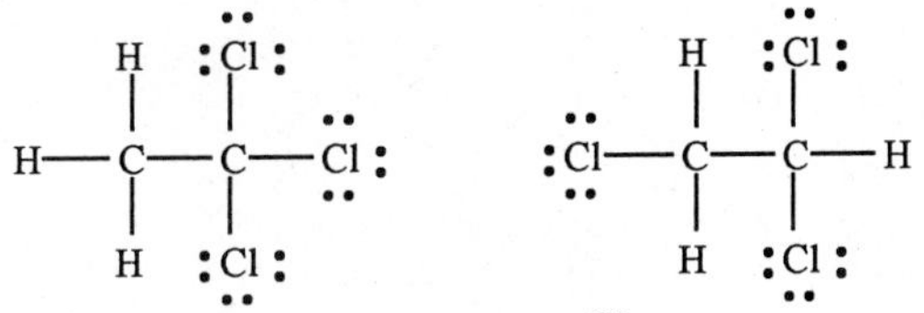

three-dimensional representations

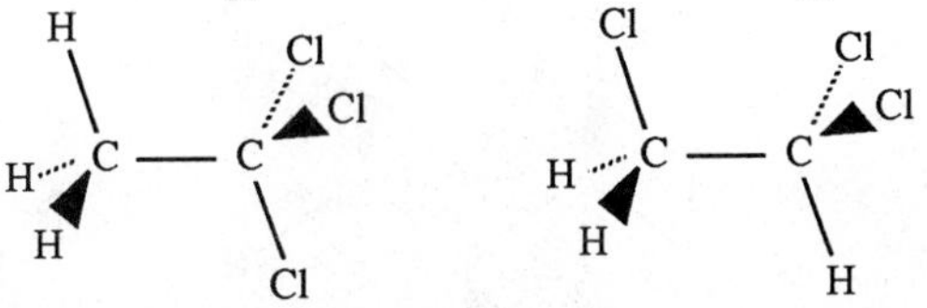

$C_2H_2Cl_4$: $ClCH_2CCl_3$ $Cl_2CHCHCl_2$

Lewis structures

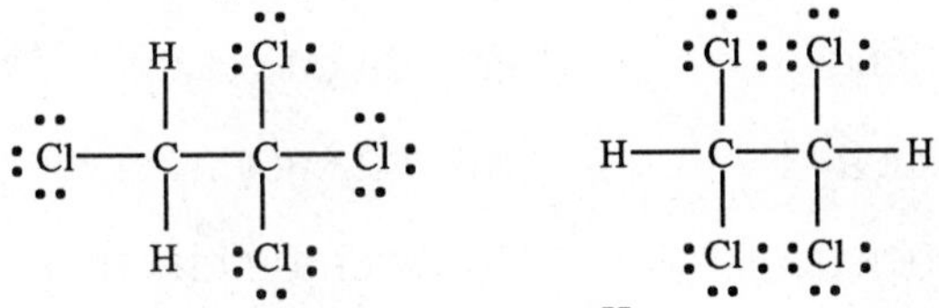

three-dimensional representations

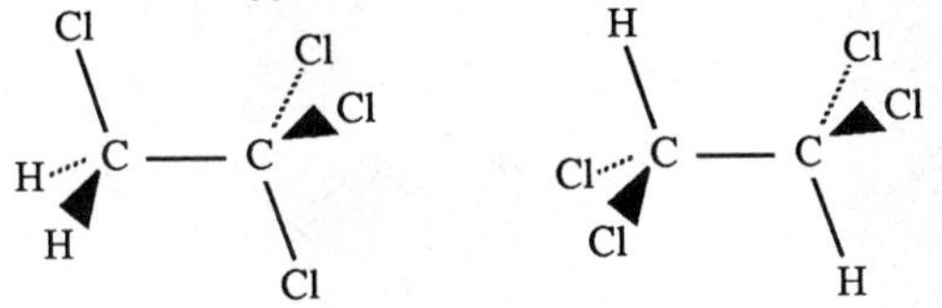

5.3 $C_3H_6Cl_2$:

condensed formulas

(1) $CH_3CH_2CHCl_2$ (2) $CH_3CCl_2CH_3$

(3) $CH_3CHClCH_2Cl$ (4) $ClCH_2CH_2CH_2Cl$

Lewis structures (1)

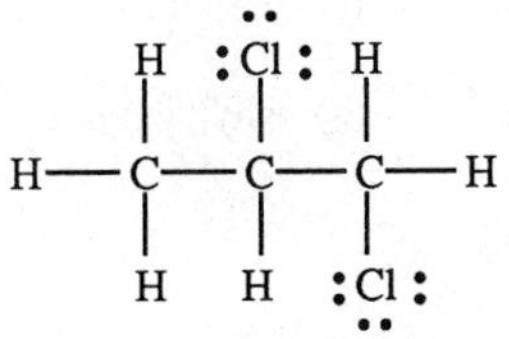

(2)

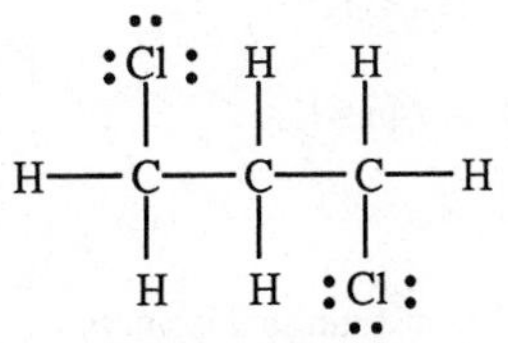

(3)

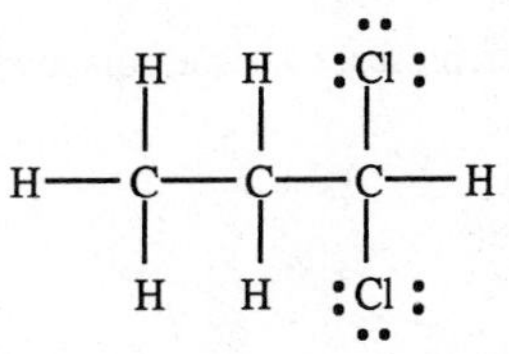

(4)

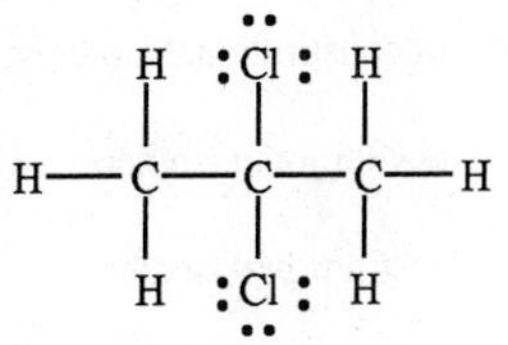

5.3 (cont)

three-dimensional formulas   (1)

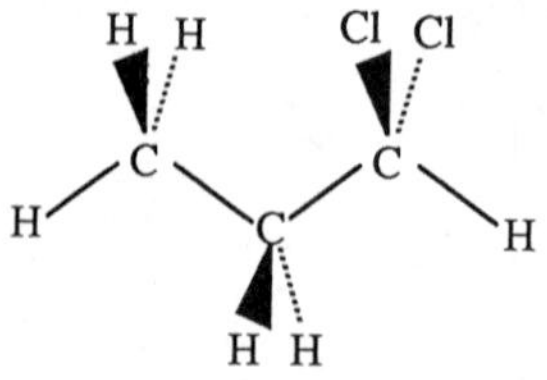

(2)

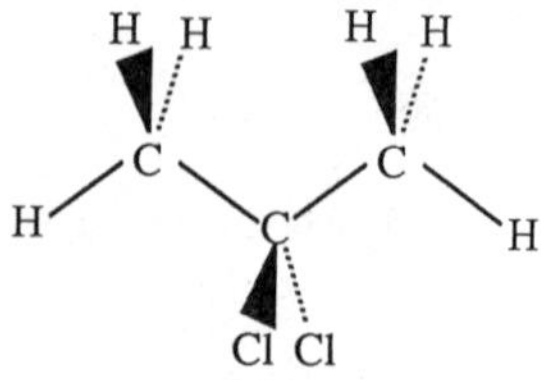

(3)

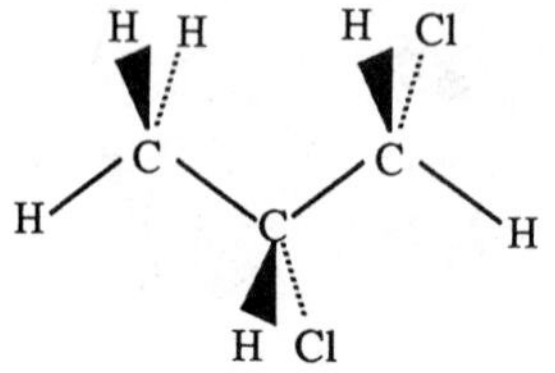

(4)

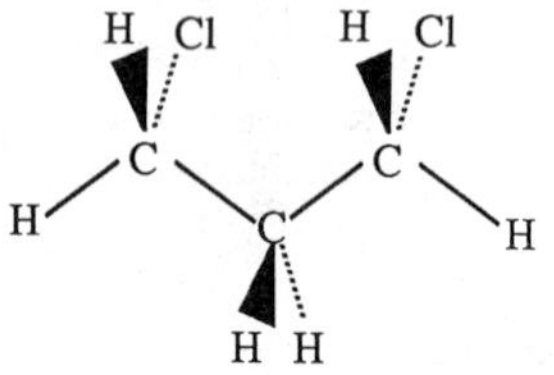

5.4 $C_4H_8Cl_2$

(1) $CH_3CH_2CH_2CHCl_2$

(2) $CH_3CH_2CCl_2CH_3$

(3) $CH_3CH_2CHClCH_2Cl$

(4) $CH_3CHClCH_2CH_2Cl$

(5) $CH_3CHClCHClCH_3$

(6) $ClCH_2CH_2CH_2CH_2Cl$

(7) $CH_3\overset{\overset{\displaystyle CH_3}{|}}{C}HCHCl_2$

(8) $CH_3\overset{\overset{\displaystyle CH_3}{|}}{C}ClCH_2Cl$

(9) $ClCH_2\overset{\overset{\displaystyle CH_3}{|}}{C}HCH_2Cl$

5.5   (a) ethyl

(b) *sec*-butyl

(c) isobutyl

(d) methyl

(e) isopropyl

(f) propyl

(g) *tert*-butyl

(h) ethyl; butyl

5.6   (a) Constitutional isomers

(b) Not constitutional isomers; both formulas represent the same compound

(c) Constitutional isomers

(d) Constitutional isomers

(e) Constitutional isomers

5.6 (cont)

    (f) Constitutional isomers

    (g) Not constitutional isomers; both formulas represent the same compound

5.7

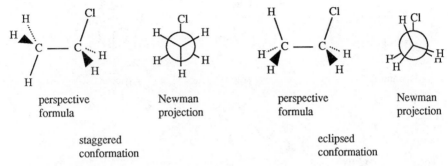

| perspective
formula | Newman
projection | perspective
formula | Newman
projection |

staggered
conformation
                                      eclipsed
conformation

- -

Concept Map 5.1   Conformation.

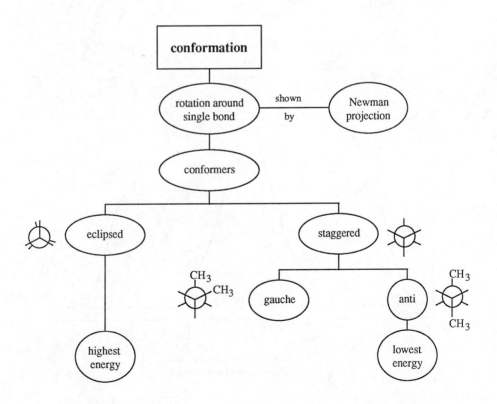

5.8

anti A   eclipsed B   gauche C

eclipsed D   gauche E   eclipsed F

degree of rotation around the bond
between carbon 1 and carbon 2

5.9

no dipole moment
C—Cl bond dipoles cancel

dipole moment
C—Cl bond dipoles do not cancel

At room temperature in the liquid state, a significant portion of 1,2-dichloroethane molecules exist in the gauche conformation.

Even though the negative ends of the C—Cl bond dipoles are close enough in the gauche conformation to repel each other, considerable van der Waals attraction between the chlorine atoms also exists. There is no van der Waals attraction between the chlorine atoms in the anti conformation.

- -

Concept Map 5.2   Representations of organic structures.

5.10   (a) [structure: pent-1-ene]   (b) [structure: chlorocyclopentane]

(c) [structure: cyclobutanol, OH]   (d) [structure: branched alkyne with Br]

(e) [structure: 4-methylpentanoic acid]

5.11   (a)

$$CH_3\text{—}CH_2\text{—}CH_2\text{—}\underset{\underset{CH_3}{|}}{CH}\text{—}\underset{\overset{|}{CH_3}}{CH}\text{—}CH_2\text{—}OH \equiv CH_3CH_2CH_2\underset{\underset{CH_3}{|}}{CH}CHCH_2OH$$

(b)

$$CH_3\text{—}CH_2\text{—}CH_2\text{—}CH_2\text{—}CH_2\text{—}\underset{\overset{|}{Cl}}{CH}\text{—}Cl \equiv CH_3CH_2CH_2CH_2CH_2CHCl_2$$

(c)

$$CH_3\text{—}CH_2\text{—}CH_2\text{—}\underset{\overset{||}{O}}{C}\text{—}\underset{\overset{|}{Br}}{CH}\text{—}CH_3 \equiv CH_3CH_2CH_2CH_2\overset{\overset{O}{||}}{C}CHBrCH_3$$

(d)

$$CH_3\text{—}CH_2\text{—}CH_2\text{—}\underset{\underset{CH_3}{|}}{CH}\text{—}\overset{\overset{O}{||}}{C}\text{—}OCH_3 \equiv CH_3CH_2CH_2\underset{\underset{CH_3}{|}}{CH}COCH_3$$

(e) [structure: cyclopentane ring with CH3 and F substituents]

5.12   (a) 2,2,6,7-tetramethyloctane   (b) 5-ethyl-2,6-dimethyloctane

   (c) 2.3-dimethylpentane   (d) 4-*tert*-butylheptane

   (e) 2.2.5-trimethyl-6-propyldecane   (f) 6-isopropyl-3-methylnonane

5.13

(a)
$$CH_3$$
$$|$$
$$CH_3CCH_2CH_2CH_2CH_2CH_3$$
$$|$$
$$CH_3$$

(b)
$$CH_3 \qquad\qquad CH_2CHCH_3 \qquad CH_3$$
$$CH_3 \qquad\qquad CH_2CHCH_3$$
$$| \qquad\qquad\qquad |$$
$$CH_3CHCH_2CH_2CH_2CHCH_2CH_2CH_2CH_3$$

(c)
$$CH_3 \quad CH_3CHCH_3 \quad CH_3$$
$$| \qquad | \qquad\quad |$$
$$CH_3CHCH_2CH_2CCH_2CH_2CHCH_3$$
$$|$$
$$CH_3CHCH_3$$

(d)
$$CH_2CH_3$$
$$|$$
$$CH_3CH_2CHCHCH_2CH_2CH_2CH_3$$
$$|$$
$$CH_3CCH_3$$
$$|$$
$$CH_3$$

(e)
$$CH_3 \quad CH_3 \quad CH_2CH_3$$
$$| \qquad | \qquad\quad |$$
$$CH_3CCH_2CHCH_2CHCH_2CH_2CH_2CH_2CH_3$$
$$|$$
$$CH_3$$

5.14   Arrows connect equivalent hydrogen atoms on different carbon atoms. All hydrogen atoms on the same carbon
atom are equivalent.

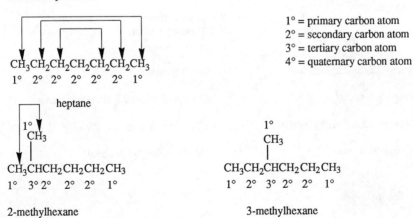

1° = primary carbon atom
2° = secondary carbon atom
3° = tertiary carbon atom
4° = quaternary carbon atom

$$CH_3CH_2CH_2CH_2CH_2CH_2CH_3$$
1°   2°   2°   2°   2°   2°   1°

heptane

1°
$$CH_3$$
$$|$$
$$CH_3CHCH_2CH_2CH_2CH_3$$
1° 3° 2°   2°   2°   1°

2-methylhexane

1°
$$CH_3$$
$$|$$
$$CH_3CH_2CHCH_2CH_2CH_3$$
1°   2°   3° 2°   2°   1°

3-methylhexane

5.14 (cont)

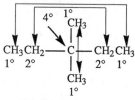

2,2-dimethylpentane

```
1°▼          1°
CH₃         CH₃
 |           |
CH₃CH——CHCH₂CH₃
1°  3°    3° 2° 1°
```

2,3-dimethylpentane

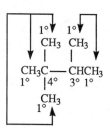

3,3-dimethylpentane

```
   1°      1°
  CH₃     CH₃
   |       |
CH₃CHCH₂CHCH₃
1° 3° 2° 3° 1°
```

2,4-dimethylpentane

all 1° H are equivalent
all 2° H are equivalent
all 3° H are equivalent

```
        1°▼  1°▼
       CH₃  CH₃
        |    |
  CH₃C —— CHCH₃
 1°  |4°  3° 1°
     CH₃
     1°▲
```

2,2,3-trimethylbutane

```
        2° CH₂CH₃  1°
           |
CH₃CH₂CHCH₂CH₃
1°  2°  3° 2°  1°
```

3-ethylpentane

all 1° H are equivalent
all 2° H are equivalent

5.15   (a) 2-bromo-2,3-dimethylpentane

(b) 3-pentanol

(c) 4-chloro-3-hexanol

(d) 4,4-dichloro-2,2-dimethylhexane

(e) 2,3-dibromo-2,3-dimethylbutane

(f) 3-iodononane

(g) 1,1,1-trichlorobutane

(h) 4-*tert*-butyl-3-nonanol

5.16

(a)
```
CH₃CHCH₂CH₂CH₂CH₂CH₂CH₃
   |
   I
```

(b)
```
CH₃CH₂CHCH₂CH₂CH₃
       |
       OH
```

5.16 (cont)

(c) CF_3CH_2OH

(d) CH_3
            |
        $ClCH_2CHCH_2CCH_3$
            | |
            OH CH_3

(e) $HOCH_2CHCH_2OH$
                |
                OH

(f) $CH_3CH_2CH_2CH_2CH_2OH$

(g) $HOCH_2CH_2OH$

(h) CH_2CH_3
            |
        $CH_3CHCCH_2CH_2CH_2CH_3$
            | |
            Cl Cl

(i) CH_3
            |
        CH_3CH_2C——$CHCH_2CH_2CH_3$
            | |
            CH_3 Br

(j) CH_2CH_3
            |
        $CH_3CHCHCH_2CH_2CH_3$
                |
                Br

5.17 $C_6H_{13}Cl$

$CH_3CH_2CH_2CH_2CH_2CH_2Cl$ 1°   1-chlorohexane

$CH_3CH_2CH_2CH_2CHCH_3$ 2°   2-chlorohexane
                          |
                          Cl

$CH_3CH_2CH_2CHCH_2CH_3$ 2°   3-chlorohexane
                      |
                      Cl

CH_3
|
$CH_3CHCH_2CH_2CH_2Cl$ 1°   1-chloro-4-methylpentane

CH_3
|
$CH_3CHCH_2CHCH_3$ 2°   2-chloro-4-methylpentane
              |
              Cl

CH_3
|
$CH_3CHCHCH_2CH_3$ 2°   3-chloro-2-methylpentane
        |
        Cl

CH_3
|
$CH_3CCH_2CH_2CH_3$ 3°   2-chloro-2-methylpentane
|
Cl

5.17 (cont)

$$\underset{\underset{CH_3}{|}}{ClCH_2CHCH_2CH_2CH_3}$$

1°   1-chloro-2-methylpentane

$$\underset{\underset{CH_3}{|}}{CH_3CH_2CHCH_2CH_2Cl}$$

1°   1-chloro-3-methylpentane

$$\underset{\underset{CH_3}{|}}{CH_3CH_2CHCHCH_3}$$
$$\underset{Cl}{|}$$

2°   2-chloro-3-methylpentane

$$\underset{\underset{CH_3}{|}}{CH_3CH_2CCH_2CH_3}$$
$$\underset{Cl}{|}$$

3°   3-chloro-3-methylpentane

$$\underset{\underset{CH_2Cl}{|}}{CH_3CH_2CHCH_2CH_3}$$

1°   3-(chloromethyl)pentane

$$\underset{\underset{CH_3}{|}}{\overset{\overset{CH_3}{|}}{CH_3CCH_2CH_2Cl}}$$

1°   1-chloro-3,3-dimethylbutane

$$\overset{\overset{CH_3}{|}}{CH_3C}—\underset{\underset{CH_3 \quad Cl}{| \quad |}}{CHCH_3}$$

2°   3-chloro-2,2-dimethylbutane

$$\underset{\underset{CH_3}{|}}{\overset{\overset{CH_2Cl}{|}}{CH_3CCH_2CH_3}}$$

1°   1-chloro-2,2-dimethylbutane

$$\underset{\underset{CH_3}{|}}{\overset{\overset{CH_3}{|}}{CH_3CHCHCH_2Cl}}$$

1°   1-chloro-2,3-dimethylbutane

$$\underset{\underset{Cl}{|}}{CH_3CH—\overset{\overset{CH_3}{|}}{C}CH_3}$$

3°   2-chloro-2,3-dimethylbutane

5.18 (1) 1,1-dichlorobutane

(2) 2,2-dichlorobutane

(3) 1,2-dichlorobutane

(4) 1,3-dichlorobutane

(5) 2,3-dichlorobutane

(6) 1,4-dichlorobutane

(7) 1,1-dichloro-2-methylpropane

(8) 1,2-dichloro-2-methylpropane

(9) 1,3-dichloro-2-methylpropane

5.19 (a) ethylbenzene

(b) *sec*-butylbenzene

(c) isobutylbenzene

(d) 1,4-dimethylbenzene (*p*-xylene)

(e) isopropylbenzene

(f) propylbenzene

(g) *tert*-butylbenzene

(h) 1-butyl-4-ethylbenzene

5.20 (a) 1-chloro-3-phenylpropane

(b) 3-iodo-3-phenylhexane

(c) 1,2-dichloro-1-phenylpropane

(d) 2-methyl-2-phenylhexane

(e) 3-phenyl-1-propanol

(f) 3-methyl-1-phenylbutane

5.21 (a)

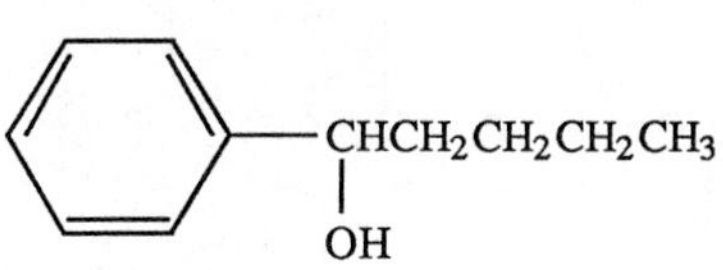

(b)

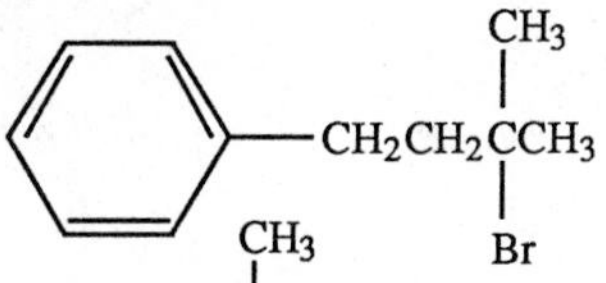

(c)

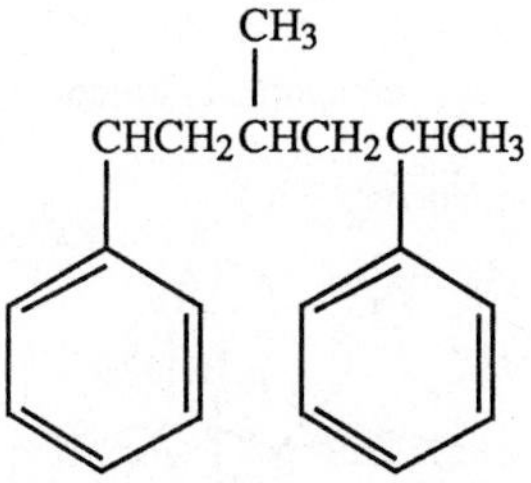

(d)

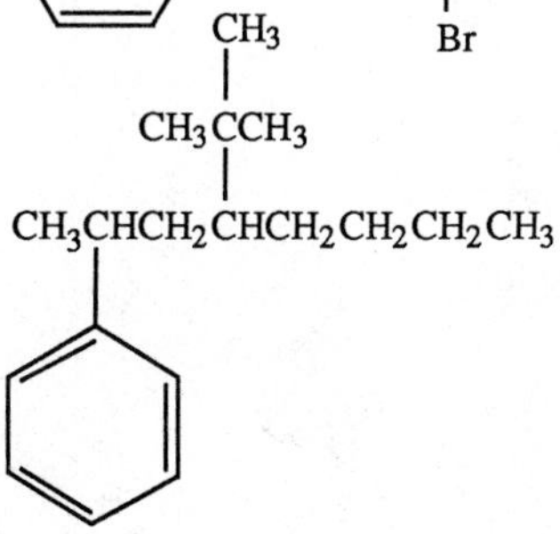

(e)

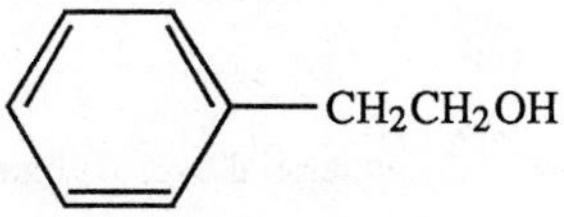

5.22   (a) 1-bromo-1-methylcyclobutane   (b) ethylcyclopropane

(c) 1-ethyl-1-methylcyclopentane   (d) iodocyclopentane

(e) cyclobutanol   (f) bromocyclopropane

5.23   (a)

$$CH_3$$
$$|$$
$$-CCH_3$$
$$|$$
$$CH_3$$

(b) ◁ — OH

(c)

$$CH_2CH_3$$
$$OH$$

(d)

$$CH_3$$
$$|$$
$$-CHCH_3$$

(e) ⬠ — Cl

(f) $CH_3CHCH_2CH_2CH_2CH_2CH_2CH_3$

(g) ⬠ — ⬡

5.24

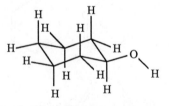

axial chlorine   equatorial chlorine

planar form

chair conformations

axial hydroxyl group   equatorial hydroxyl group

planar form

chair conformations

Concept Map 5.3   Conformation in cyclic compounds.

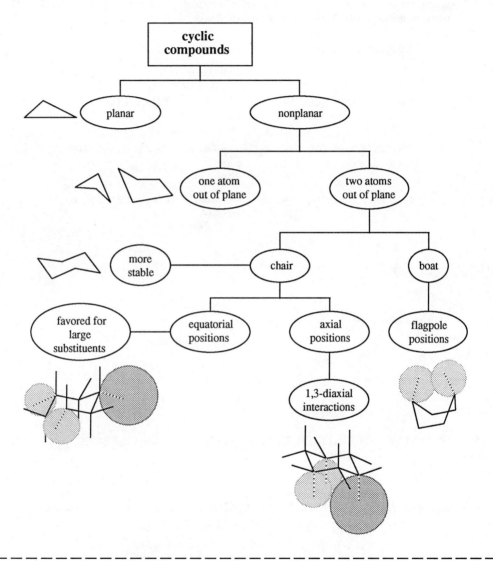

5.25

$$CH_3\overset{\displaystyle CH_3}{\underset{\displaystyle CH_3}{C}}\bullet \quad > \quad CH_3\overset{\displaystyle H}{\underset{\displaystyle CH_3}{C}}\bullet \quad > \quad CH_3\overset{\displaystyle H}{\underset{\displaystyle H}{C}}\bullet \quad > \quad H\overset{\displaystyle H}{\underset{\displaystyle H}{-C}}\bullet$$

5.26   Of the compounds given on p. 174 in the text, only alkenes decolorize a solution of bromine in carbon tetrachloride and react readily with one equivalent of hydrogen in the presence of a platinum catalyst. The compound has to be either 2-buten-1-ol or ethyl vinyl ether.

$$CH_3CH\!=\!CHCH_2OH \;+\; Br_2 \;\xrightarrow[\text{tetrachloride}]{\text{carbon}}\; CH_3CHBrCHBrCH_2OH$$

$$CH_3CH\!=\!CHCH_2OH \;+\; H_2 \;\xrightarrow[\text{catalyst}]{}\; CH_3CH_2CH_2CH_2OH$$

$$CH_3CH_2OCH\!=\!CH_2 \;+\; Br_2 \;\xrightarrow[\text{tetrachloride}]{\text{carbon}}\; CH_3CH_2OCHBrCH_2Br$$

$$CH_3CH_2OCH\!=\!CH_2 \;+\; H_2 \;\xrightarrow[\text{catalyst}]{}\; CH_3CH_2OCH_2CH_3$$

5.27

5.28   triple bond: $CH_3CH_2CH_2CH_2CH_2C\!\equiv\!CH$ $CH_3CHCH_2C\!\equiv\!CCH_3$ with CH_3 substituent

two double bonds: $CH_3CH\!=\!CHCH\!=\!CHCH_2CH_3$ $CH_2\!=\!C\!-\!CHCH\!=\!CH_2$ with two CH_3 substituents

a ring and double bond:

two rings:

5.29   The original compound must have the carbon skeleton present in 2-methylhexane.

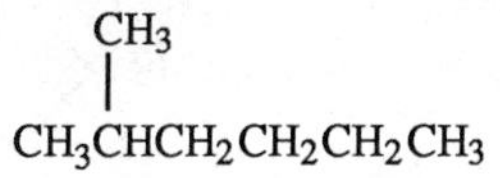

$$\underset{\displaystyle CH_3CHCH_2CH_2CH_2CH_3}{\overset{\displaystyle CH_3}{\vert}}$$

The possible structures are shown below.

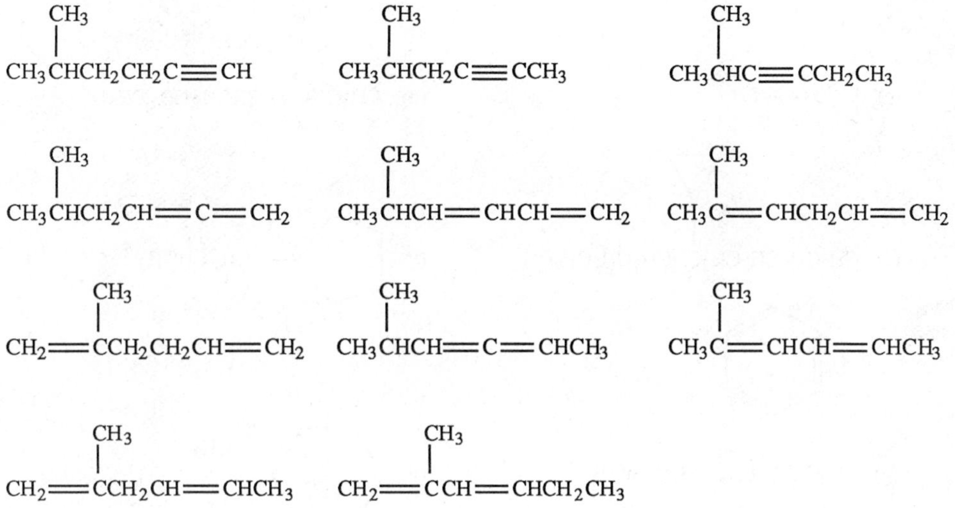

$$\underset{\displaystyle CH_3CHCH_2CH_2C\equiv CH}{\overset{\displaystyle CH_3}{\vert}} \qquad \underset{\displaystyle CH_3CHCH_2C\equiv CCH_3}{\overset{\displaystyle CH_3}{\vert}} \qquad \underset{\displaystyle CH_3CHC\equiv CCH_2CH_3}{\overset{\displaystyle CH_3}{\vert}}$$

$$\underset{\displaystyle CH_3CHCH_2CH=C=CH_2}{\overset{\displaystyle CH_3}{\vert}} \qquad \underset{\displaystyle CH_3CHCH=CHCH=CH_2}{\overset{\displaystyle CH_3}{\vert}} \qquad \underset{\displaystyle CH_3C=CHCH_2CH=CH_2}{\overset{\displaystyle CH_3}{\vert}}$$

$$\underset{\displaystyle CH_2=CCH_2CH_2CH=CH_2}{\overset{\displaystyle CH_3}{\vert}} \qquad \underset{\displaystyle CH_3CHCH=C=CHCH_3}{\overset{\displaystyle CH_3}{\vert}} \qquad \underset{\displaystyle CH_3C=CHCH=CHCH_3}{\overset{\displaystyle CH_3}{\vert}}$$

$$\underset{\displaystyle CH_2=CCH_2CH=CHCH_3}{\overset{\displaystyle CH_3}{\vert}} \qquad \underset{\displaystyle CH_2=CCH=CHCH_2CH_3}{\overset{\displaystyle CH_3}{\vert}}$$

5.30   (a) 1-bromo-1-methylcyclopropane   (b) 2,2,5,5-tetramethylhexane

(c) 1,1-dimethylcyclobutane   (d) 2-chloro-2-methylpentane

(e) isopropylcyclohexane or 2-cyclohexylpropane   (f) 1-ethyl-1-cyclopentanol

(g) 1-chloro-1-ethylcyclohexane   (h) 1-phenyl-1-cyclohexanol

(i) 3-bromo-3-ethyl-1-heptanol   (j) 6-*tert*-butyl-2,3-dichlorooctane or
                                               6,7-dichloro-3-ethyl-2,2-dimethyloctane

5.31

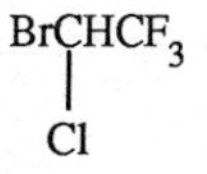

$$\underset{\displaystyle Cl}{\overset{\displaystyle BrCHCF_3}{\vert}}$$

2-bromo-2-chloro-1,1,1-trifluoroethane

5.32   (a) 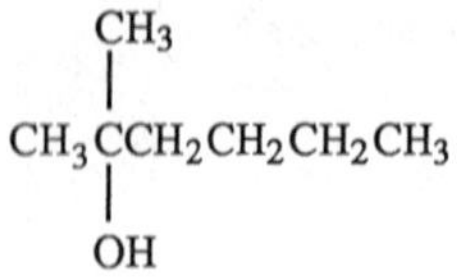
$$\underset{\underset{OH}{|}}{\overset{\overset{CH_3}{|}}{CH_3C}}CH_2CH_2CH_2CH_3$$

(b)

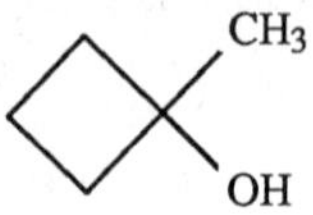

(c) 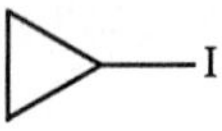
—I

(d) 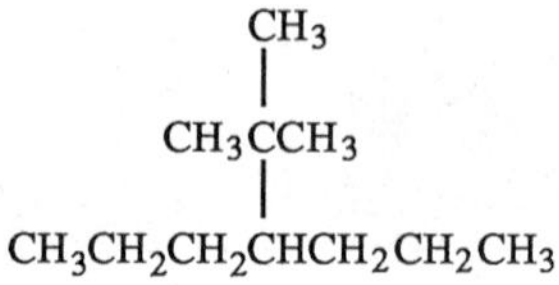
$$CH_3CH_2CH_2\underset{}{CH}CH_2CH_2CH_3$$
with
$$CH_3\overset{\overset{CH_3}{|}}{\underset{|}{C}}CH_3$$
attached

(e) 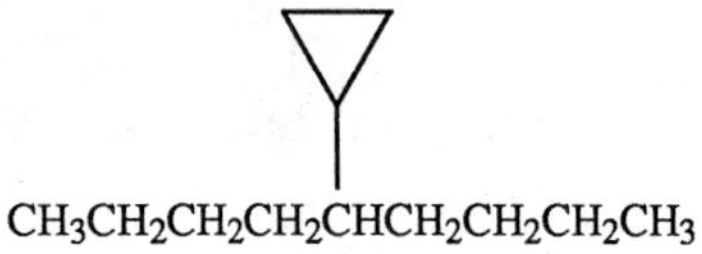 $CH_3CH_2CH_2CH_2CHCH_2CH_2CH_3$

(f) 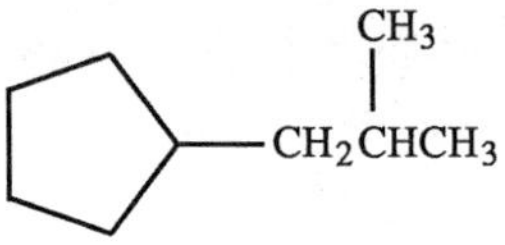
—CH$_2$CHCH$_3$ with CH$_3$

(g) 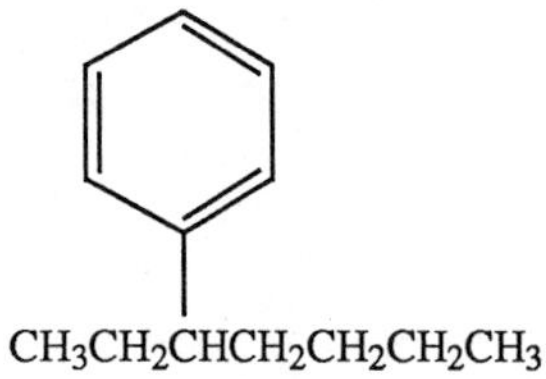 $CH_3CH_2CHCH_2CH_2CH_2CH_3$

(h)

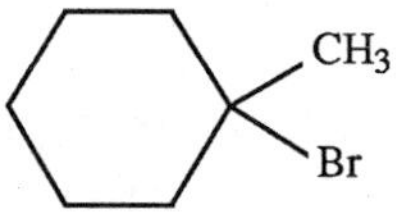

(i) 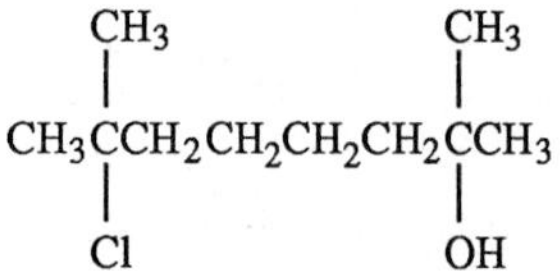
$$\underset{\underset{Cl}{|}}{\overset{\overset{CH_3}{|}}{CH_3C}}CH_2CH_2CH_2CH_2\underset{\underset{OH}{|}}{\overset{\overset{CH_3}{|}}{C}}CH_3$$

5.33   (a) $C_5H_{12}O$

$CH_3CH_2CH_2CH_2CH_2OH$

1-pentanol

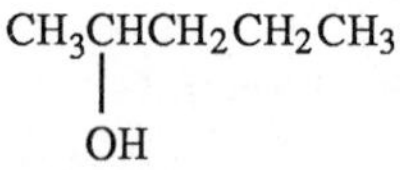

$$CH_3\underset{\underset{OH}{|}}{CH}CH_2CH_2CH_3$$

2-pentanol

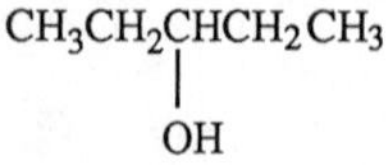

$$CH_3CH_2\underset{\underset{OH}{|}}{CH}CH_2CH_3$$

3-pentanol

$$\underset{\underset{OH}{|}}{\overset{\overset{CH_3}{|}}{CH_3CH}}CH_2CH_2CH_3$$

$$\overset{\overset{CH_3}{|}}{CH_3CH}CH_2CH_2OH$$

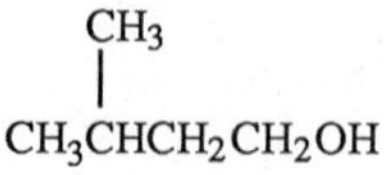

3-methyl-1-butanol

5.33 (a) (cont)

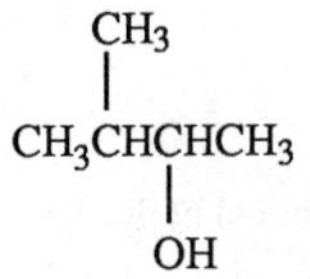

$$CH_3$$
$$|$$
$$CH_3CHCHCH_3$$
$$|$$
$$OH$$

3-methyl-2-butanol

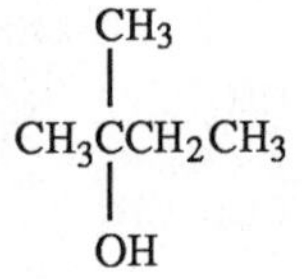

$$CH_3$$
$$|$$
$$CH_3CCH_2CH_3$$
$$|$$
$$OH$$

2-methyl-2-butanol

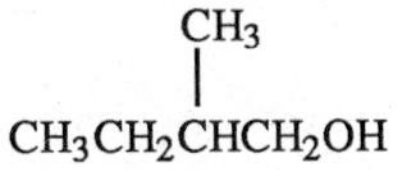

$$CH_3$$
$$|$$
$$CH_3CH_2CHCH_2OH$$

2-methyl-1-butanol

$$CH_3$$
$$|$$
$$CH_3CCH_2OH$$
$$|$$
$$CH_3$$

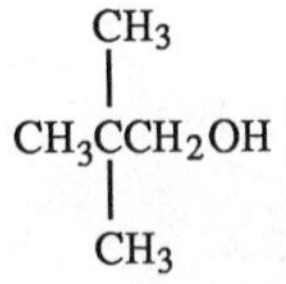

2,2-dimethyl-1-propanol

$$CH_3CH_2CH_2CH_2OCH_3$$

methyl butyl ether

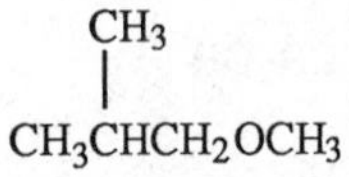

$$CH_3$$
$$|$$
$$CH_3CHCH_2OCH_3$$

methyl isobutyl ether

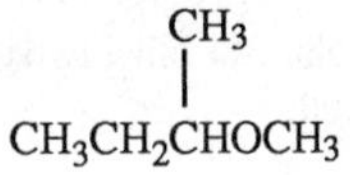

$$CH_3$$
$$|$$
$$CH_3CH_2CHOCH_3$$

methyl *sec*-butyl ether

$$CH_3$$
$$|$$
$$CH_3COCH_3$$
$$|$$
$$CH_3$$

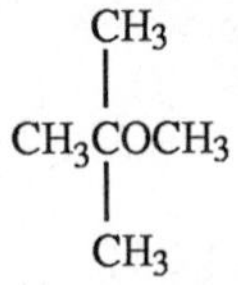

methyl *tert*-butyl ether

$$CH_3CH_2CH_2OCH_2CH_3$$

ethyl propyl ether

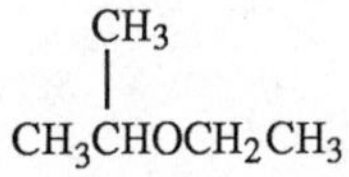

$$CH_3$$
$$|$$
$$CH_3CHOCH_2CH_3$$

ethyl isopropyl ether

(b) $C_3H_5Cl_3$

$$CH_3CH_2CCl_3$$

1,1,1-trichloropropane

$$CH_3CHClCHCl_2$$

1,1,2-trichloropropane

$$ClCH_2CH_2CHCl_2$$

1,1,3-trichloropropane

$$CH_3CCl_2CH_2Cl$$

1,2,2-trichloropropane

$$ClCH_2CHClCH_2Cl$$

1,2,3-trichloropropane

5.34   (a)

$$CH_3CH_2\underset{\underset{CH_3}{|}}{\overset{\overset{CH_3}{|}}{C}}CH_2CH_3$$

most spherical of the three constitutional isomers, least van der Waals attraction between the molecules

(b)

$$CH_3\underset{\underset{OH}{|}}{\overset{\overset{CH_3}{|}}{C}}CH_2CH_3$$

same as (a)

(c)

$$CH_3\underset{\underset{CH_3}{|}}{\overset{\overset{CH_3}{|}}{C}}CH_2OH$$

same as (a)

(d)

$$CH_3\underset{\underset{CH_3}{|}}{\overset{\overset{CH_3}{|}}{C}}OCH_3$$

least polar (the other two are alcohols and form hydrogen bonds

(e)

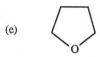

least polar

5.35   (a) C_5H_9Br

Each halogen atom counts as a hydrogen atom in determining the units of unsaturation.

Units of unsaturation = $[(2 \times 5) + 2 - 10]/2 = 1$

Some structural formulas are shown below:

alkenes: $CH_3CH_2CH{=}CHCH_2Br$ $CH_3CH_2CH{=}\underset{\underset{Br}{|}}{C}CH_3$ $CH_3\underset{\underset{Br}{|}}{\overset{\overset{CH_3}{|}}{C}}{=}CCH_3$

cyclic compounds:

[cyclopentane ring]—Br   [bicyclobutane]$\overset{CH_3}{\underset{Br}{}}$ [cyclopropane]—$\underset{\underset{Br}{|}}{C}HCH_3$ [cyclopropane, CH_3]—CH_2Br

(b) $C_5H_{10}O$

Oxygen atoms do not affect the calculation

Units of unsaturation = $[(2 \times 5) + 2 - 10]/2 = 1$

5.35 (cont)

   (b)(cont)

   Some structural formulas are shown below:

aldehydes
and ketones:

$$CH_3CH_2\underset{\underset{CH_3}{|}}{CH}\overset{\overset{O}{\|}}{CH} \qquad CH_3CH_2\overset{\overset{O}{\|}}{C}CH_2CH_3$$

unsaturated
alcohols:

$$CH_2{=}CHCH_2CH_2CH_2OH \qquad CH_3\underset{\underset{OH}{|}}{CH}CH{=}CHCH_3$$

unsaturated
ethers:

$$CH_3OCH_2CH{=}CHCH_3 \qquad CH_2{=}\underset{\underset{CH_3}{|}}{C}CH_2OCH_3 \qquad CH_3\underset{\underset{OCH_3}{|}}{\overset{\overset{HO}{}}{C}}{=}CHCH_3$$

cyclic alcohols:

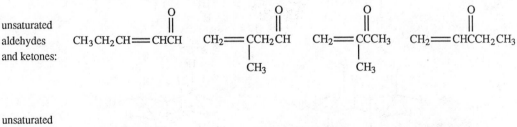

cyclic ethers:

   (c) C_5H_8O

   Units of unsaturation = $[(2 \times 5) + 2 - 8]/2 = 2$

   Some structural formulas are shown below and on the next page:

unsaturated
aldehydes
and ketones:

$$CH_3CH_2CH{=}CH\overset{\overset{O}{\|}}{C}H \qquad CH_2{=}\underset{\underset{CH_3}{|}}{C}CH_2\overset{\overset{O}{\|}}{C}H \qquad CH_2{=}\underset{\underset{CH_3}{|}}{C}\overset{\overset{O}{\|}}{C}CH_3 \qquad CH_2{=}CH\overset{\overset{O}{\|}}{C}CH_2CH_3$$

unsaturated
alcohols and
ethers:

$$CH_3C{\equiv}C\underset{\underset{OH}{|}}{CH}CH_3 \qquad CH_2{=}CHCH{=}CHCH_2OH \qquad CH_2{=}\underset{\underset{OCH_3}{|}}{C}{-}CH{=}CH_2$$

5.35 (cont)

(c) (cont)

cyclic
compounds:

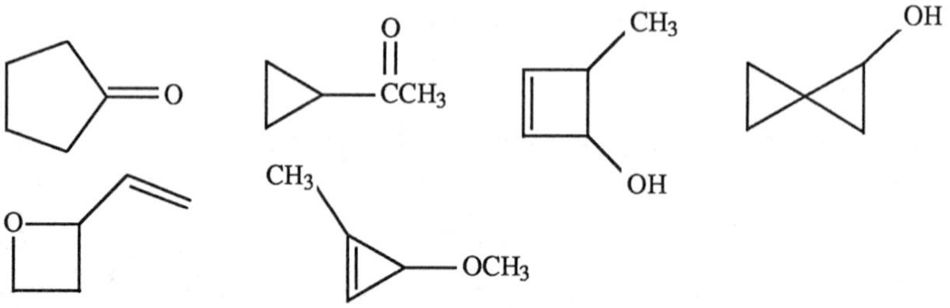

(d)   Cyclic compounds that have rings with four or more carbon atoms will not react easily with bromine in carbon tetrachloride. The ones with the molecular formula of C_5H_9Br are:

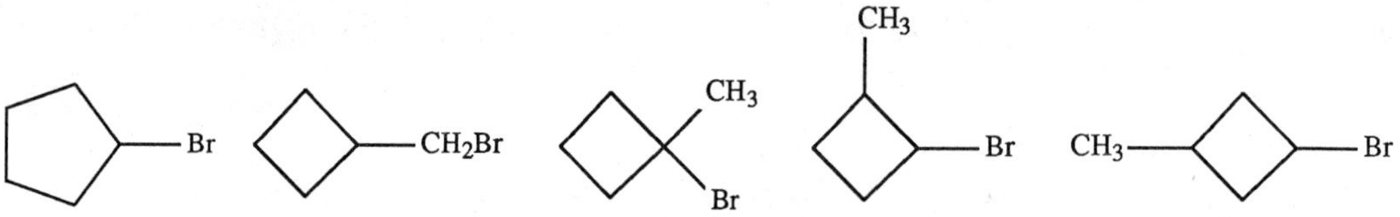

(e)   The hydrogenation data tells us that the original compound was an alkene with the carbon skeleton of 3-methyl-1-butanol.

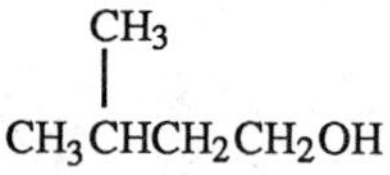

$$\underset{\text{CH}_3\text{CHCH}_2\text{CH}_2\text{OH}}{\overset{\overset{\displaystyle\text{CH}_3}{|}}{}}$$

The following structures are possible.

$$\underset{\text{CH}_3\text{C}=\text{CHCH}_2\text{OH}}{\overset{\overset{\displaystyle\text{CH}_3}{|}}{}} \qquad\qquad \underset{\text{CH}_2=\text{CCH}_2\text{CH}_2\text{OH}}{\overset{\overset{\displaystyle\text{CH}_3}{|}}{}}$$

5.36   For both compounds the conformation in which the subsituent is equatorial is the more stable conformation.

(a)

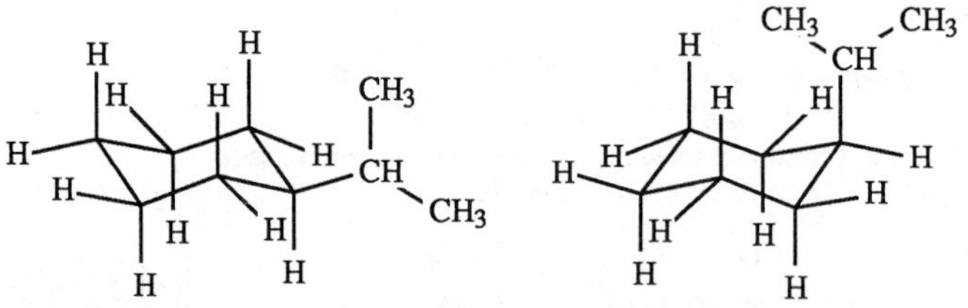

more stable

(b)

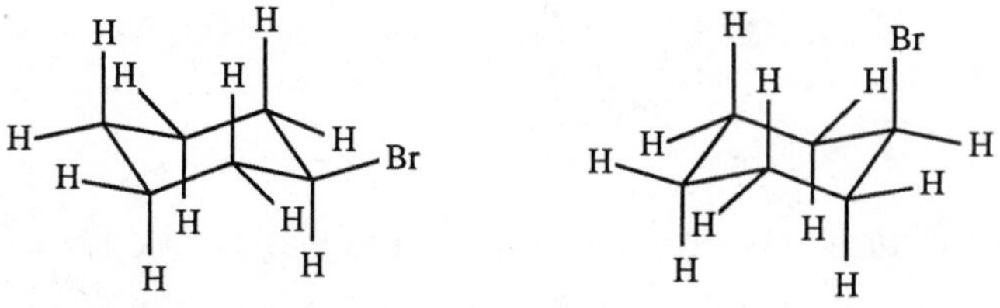

more stable

5.37

(a) $CH_3CH_2NH_2 + HCl \longrightarrow CH_3CH_2NH_3^+ \ Cl^-$

(b) $CH_3CH_2C\equiv CH + NaNH_2 \longrightarrow CH_3CH_2C\equiv C^- Na^+ + NH_3$

(c) $(CH_3CH_2CH_2)_2NH + CH_3CH_2Br \longrightarrow (CH_3CH_2CH_2)_2\overset{+}{N}HCH_2CH_3 \ Br^-$

(d) $CH_3CH_2CH\!=\!CHCH_2CH_3 + HBr \longrightarrow CH_3CH_2CH_2CHBrCH_2CH_3$

(e) $CH_3CH_2CH_2SH + NaOH \longrightarrow CH_3CH_2CH_2S^- Na^+ + H_2O$

(f) $-\overset{+}{N}H_2CH_3 \ Cl^- + NaOH \longrightarrow$ $-NHCH_3 + NaCl + H_2O$

(g) $CH_3CH_2CH\!=\!CHCH_3 + Br_2 \longrightarrow CH_3CH_2CHBrCHBrCH_3$

(h) $-SNa + CH_3CH_2CH_2Br \longrightarrow$ $-SCH_2CH_2CH_3 + NaBr$

(i) $Br-$ $-\overset{O}{\overset{\|}{C}}OH + NaHCO_3 \longrightarrow Br-$ $-\overset{O}{\overset{\|}{C}}O^-Na^+ + H_2O + CO_2\uparrow$

(j) $CH_3CH_2CH_2C\equiv CCH_3 + 2H_2 \xrightarrow[PtO_2]{} CH_3CH_2CH_2CH_2CH_2CH_3$

(k) $CH_3CH_3 + Cl_2 \xrightarrow{h\nu} CH_3CH_2Cl + CH_3CHCl_2 + ClCH_2CH_2Cl + HCl$

5.38

(a) $-SH \xrightarrow[H_2O]{NaOH}$ $-SNa \xrightarrow{\overset{\displaystyle CH_3}{\overset{|}{CH_3CHCH_2CH_2Br}}}$

$-SCH_2CH_2\overset{CH_3}{\overset{|}{C}}HCH_3$

5.38 (cont)

(b) $CH_3CH_2Br \xrightarrow{CH_3NH_2} CH_3CH_2\overset{+}{N}H_2CH_3\ Br^-$

(c) $CH_2{=}CH_2 \xrightarrow{HBr} CH_3CH_2Br \xrightarrow{CH_3ONa} CH_3CH_2OCH_3$

(d) $CH_3CH_2CH_2CH_2NH_2 \xrightarrow{HCl} CH_3CH_2CH_2CH_2\overset{+}{N}H_3\ Cl^-$

(e) $CH_3CH_2C{\equiv}CH \xrightarrow{2\ Br_2} CH_3CH_2CBr_2CHBr_2$

(f) $CH_3CH_2CH_2CH_2OH \xrightarrow{NH_2^-} CH_3CH_2CH_2CH_2O^- \xrightarrow{CH_3Br} CH_3CH_2CH_2CH_2OCH_3$

(g) $CH_3CH{=}CHCH_3 \xrightarrow{HBr} \underset{\underset{Br}{|}}{CH_3CHCH_2CH_3} \xrightarrow{NaCN} \underset{\underset{CN}{|}}{CH_3CHCH_2CH_3}$

(h) $\underset{\underset{OH}{|}}{CH_3CHCH_2CH}{=}CH_2 \xrightarrow[PtO_2]{H_2} \underset{\underset{OH}{|}}{CH_3CHCH_2CH_2CH_3}$

(i)

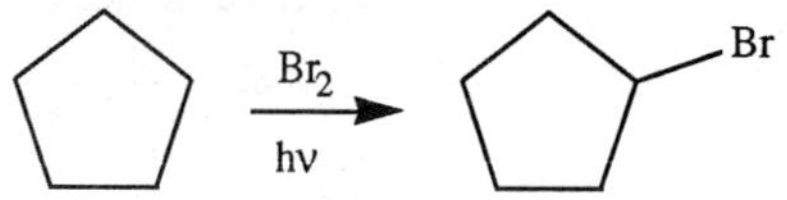

6

Stereochemistry

Concept Map 6.1   Stereochemical relationships.

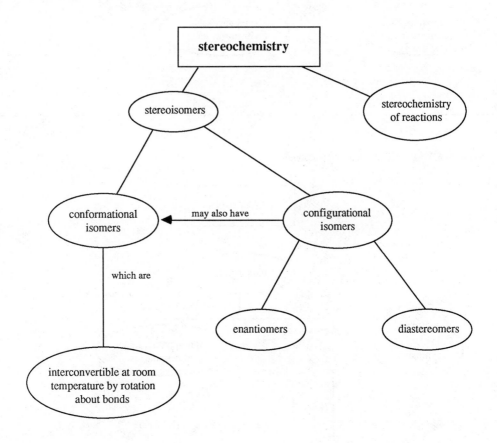

Concept Map 6.2   Chirality.

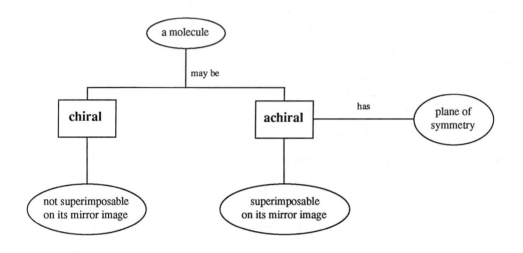

- -

Concept Map 6.3   Definition of enantiomers.

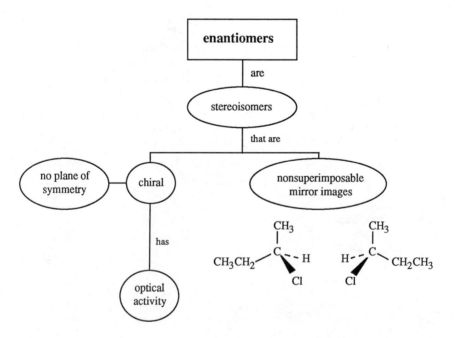

6.1   (a)

If the mug had the words "MOTHER" written on one side, there would no longer be any planes of symmetry. The mug would now be chiral.

If the mug had the words "MOM" instead of "MOTHER" written on one side, it would still be chiral. If the words "MOM" were written opposite the handle, it would be achiral.

(b)

Thread is helically wound around a spool. A helix has no planes of symmetry. A spool with helically wound thread has no plane of symmetry and is chiral.

(c)

An empty spool has many planes of symmetry.

6.2   (a)

helical thread; chiral

(b)

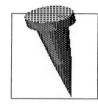

plane of symmetry

achiral

(c)

prongs equal length; achiral
prongs not equal length; chiral

(d)

plane of symmetry

achiral

(e)

plane of symmetry

socks are interchangeable; achiral

(f)

some mittens have a plane of symmetry;
the mittens shown do not. Each mitten
is chiral and is the mirror image isomer of
the other mitten.

(g)

plane of symmetry

achiral

6.2 (cont)

(h)

Notice buttons and buttonholes; chiral

(i)

chiral

6.3   (a)   No

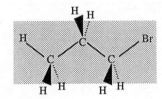

The molecule has a plane of symmetry.

(b)   Yes

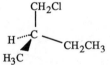

(c)   Yes

(d)   No

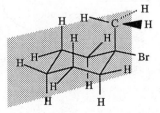

The molecule has a plane of symmetry.

6.3 (cont)

(e)   Yes

$$CH_3CH_2CH_2 \overset{\overset{\displaystyle CH_3}{|}}{\underset{Br}{C}} {\cdots}H \qquad H {\cdots} \overset{\overset{\displaystyle CH_3}{|}}{\underset{Br}{C}} CH_2CH_2CH_3$$

(f)   Yes

$$CH_3\underset{\underset{\displaystyle CH_3}{|}}{CH}CH_2 \overset{\overset{\displaystyle CH_3}{|}}{\underset{OH}{C}} {\cdots}H \qquad \underset{HO}{H {\cdots}} \overset{\overset{\displaystyle CH_3}{|}}{C} CH_2\underset{\underset{\displaystyle CH_3}{|}}{CH}CH_3$$

- -

Concept Map 6.4   Optical activity.

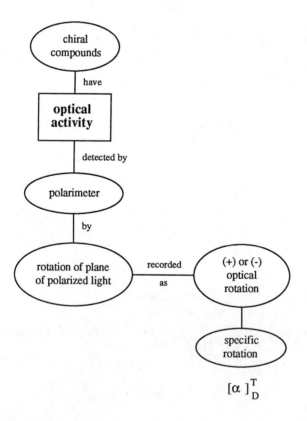

6.4 $\dfrac{\text{optical rotation of unknown solution}}{\text{optical rotation of known concentration}} = \dfrac{+13.3°}{+53.2°} = \dfrac{1}{4}$

∴ the concentration of the unknown solution $= \frac{1}{4} \times$ the concentration of the original solution

$$= \frac{1}{4} \times 20 \text{ g/100 mL} = 5.0 \text{ g/100 mL}$$

6.5 $[\alpha]_D^{20°} = \dfrac{+2.46°}{(1.0 \text{ dm})(5 \text{ g/100 mL})} = 49.2° \ (C \ 0.05, \text{ ethanol})$

6.6   The equation for specific rotation can be rearranged.

$$\alpha = [\alpha]_D^{20°} lc$$

The actual rotation, α, is directly proportional to the concentration. If the solution is diluted, for example, to twice the original volume, the concentration, and, therefore, the optical rotation, will be cut in half. Thus if the original specific rotation is +90°, the actual rotation for the diluted solution will be 45° in a clockwise direction. If the specific rotation is –270°, the actual rotation will be –135°, or 225° in a clockwise direction. In general, decreasing the concentration will cause a positive rotation to decrease in the clockwise direction and a negative rotation to increase in the clockwise direction.

- -

Concept Map 6.5   (see p. 134)

- -

6.7   (a)   enantiomeric excess $= \dfrac{\text{actual rotation}}{\text{rotation of pure enantiomer}} \times 100$

enantiomeric excess $= \dfrac{-14.70°}{-39.5°} \times 100 = 37.2\%$

(b)

(S)-(–)-2-bromobutanoic acid   (R)-(+)-2-bromobutanoic acid
        68.6%   31.4%

The mixture containing a 37.2% enantiomeric excess of the levorotatory compound is 100 – 37.2 or 62.8% racemic. The racemic portion of the mixture contains equal amounts (31.4%) of each enantiomer. Therefore the solution contains 31.4% of the dextrorotatory and (31.4 + 37.2)% or 68.6% of the levorotatory isomer.

Concept Map 6.5   Formation of racemic mixtures in chemical reactions.

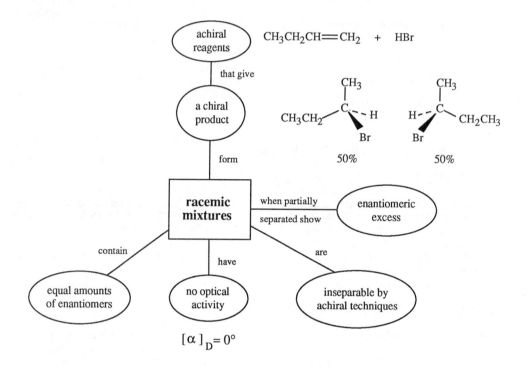

- -

6.8   (a) (*S*)-3-hexanol   (b) (*S*)-2-bromopentane

        (c) (*R*)-3-iodo-3-methylhexane   (d) (*R*)-1-bromo-1-chloropropane

        (e) (*R*)-1,2-dichlorobutane   (f) (*S*)-4-chloro-2-methyl-1-butanol

        (g) (*S*)-1-chloro-3,4-dimethylpentane   (h) (*R*)-4-bromo-2-butanol

6.9   (a)   (b)   (c)

        (d)   (e)   (f)

6.10   (a)

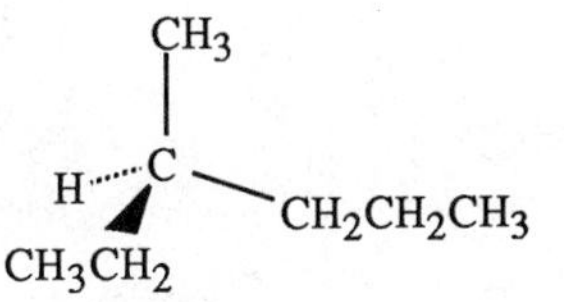

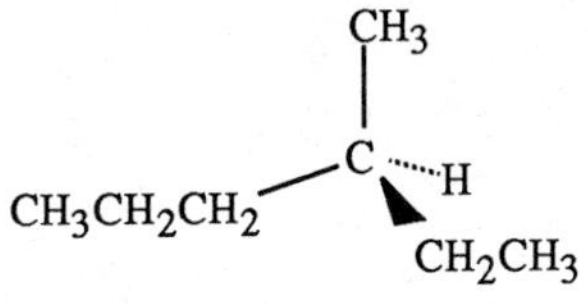

(*R*)-3-methylhexane   (*S*)-3-methylhexane

These are enantiomers

(b)

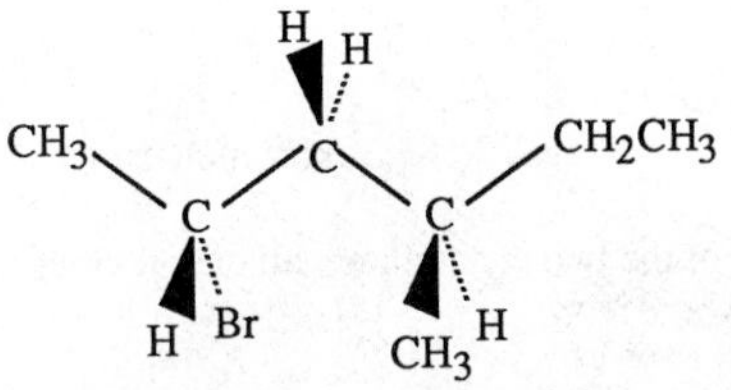

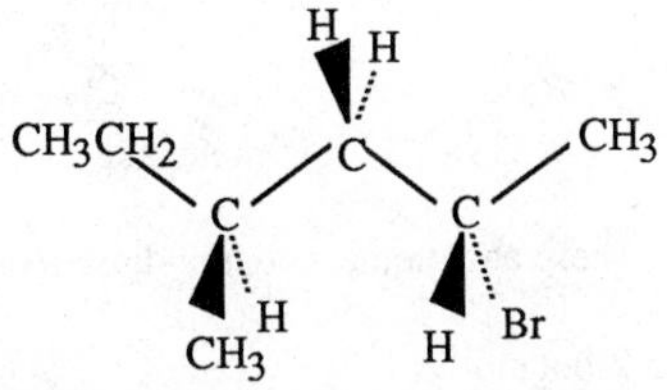

(2*R*, 4*R*)-2-bromo-4-methylhexane   (2*S*, 4*S*)-2-bromo-4-methylhexane

These enantiomers are the diastereomers of the two enantiomers below.

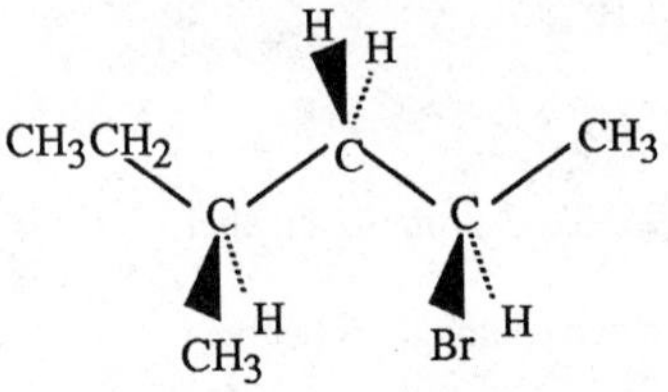

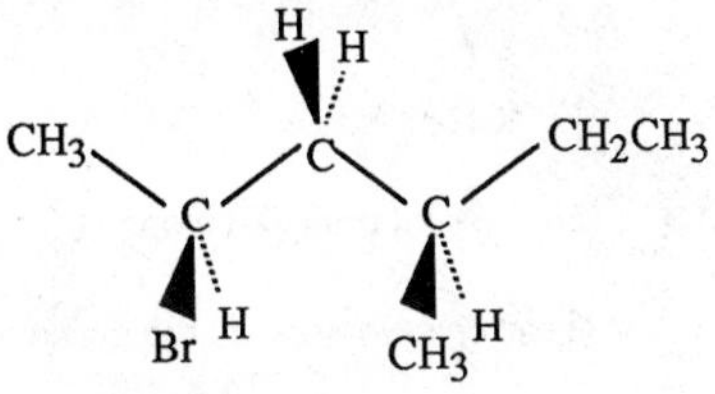

(2*R*, 4*S*)-2-bromo-4-methylhexane   (2*S*, 4*R*)-2-bromo-4-methylhexane

These enantiomers are the diastereomers of the two enantiomers above.

(c)

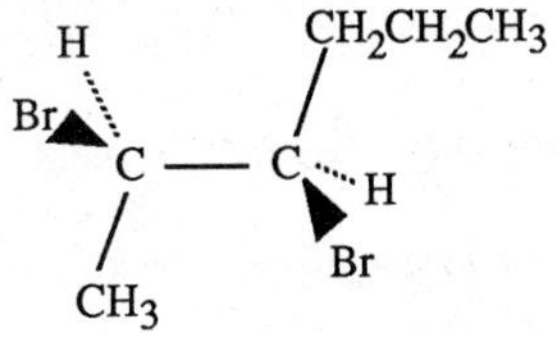

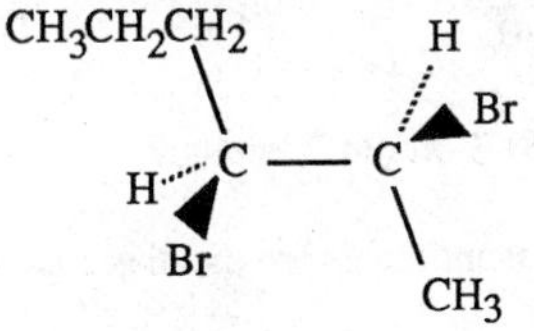

(2*R*, 3*R*)-2,3-dibromohexane   (2*S*, 3*S*)-2,3-dibromohexane

These enantiomers are the diastereomers of the two enantiomers on the next page.

6.10 (cont)

(c) (cont)

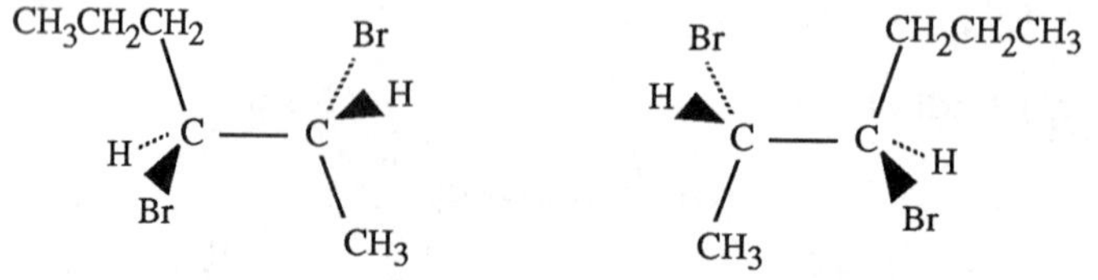

(2R, 3S)-2,3-dibromohexane   (2S, 3R)-2,3-dibromohexane

These enantiomers are the diastereomers of the two enantiomers on the previous page.

6.11   3-bromo-2-butanol

(2R, 3R)-3-bromo-2-butanol   (2S, 3S)-3-bromo-2-butanol

These enantiomers are the diastereomers of the two enantiomers below.

(2R, 3S)-3-bromo-2-butanol   (2S, 3R)-3-bromo-2-butanol

These enantiomers are the diastereomers of the two enantiomers above.

2-chloro-3-methylheptane

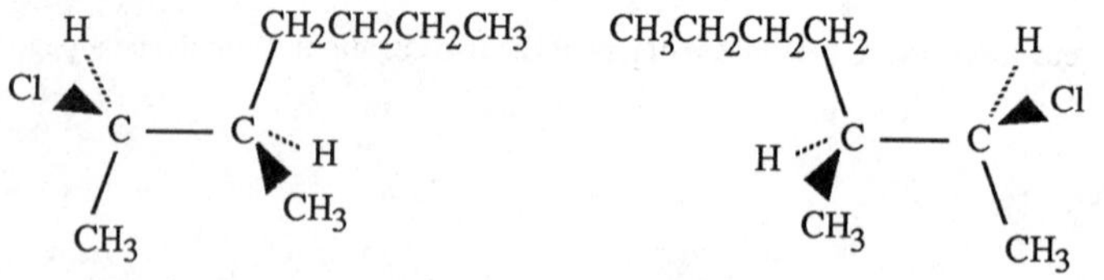

(2R, 3R)-2-chloro-3-methylheptane   (2S, 3S)-2-chloro-3-methylheptane

These enantiomers are the diastereomers of the two enantiomers on the next page.

6.11 (cont)

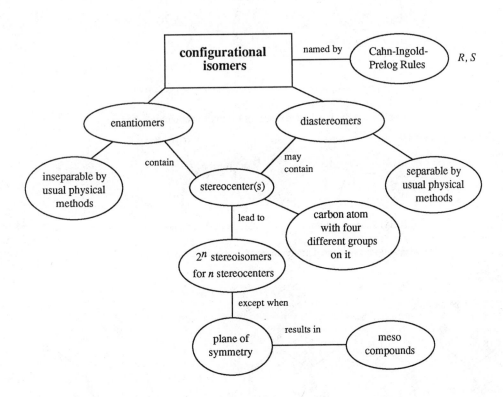

(2*R*, 3*S*)-2-chloro-3-methylheptane   (2*S*, 3*R*)-2-chloro-3-methylheptane

These enantiomers are the diastereomers of the two enantiomers on the previous page.

- -

Concept Map 6.6   Configurational isomers.

6.12   2,3-butanediol

plane of
symmetry

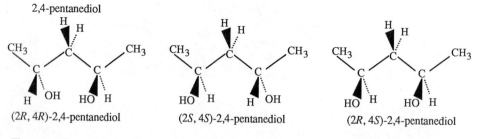

| (2R, 3R)- | (2S, 3S)- | (2R, 3S)- |
| 2,3-butanediol | 2,3-butanediol | 2,3-butanediol |

These enantiomers are the diastereomers of the meso compound   meso compound

2,3-pentanediol

(2R, 3R)-2,3-pentanediol   (2S, 3S)-2,3-pentanediol

These enantiomers are the diastereomers of the two enantiomers below.

(2R, 3S)-2,3-pentanediol   (2S, 3R)-2,3-pentanediol

These enantiomers are the diastereomers of the two enantiomers above.

2,4-pentanediol

(2R, 4R)-2,4-pentanediol   (2S, 4S)-2,4-pentanediol   (2R, 4S)-2,4-pentanediol

These enantiomers are the diastereomers of the meso compound.   meso compound

6.13   2,3-dihydroxybutanedioic acid

(2*R*, 3*R*)-2,3-dihydroxybutanedioic acid   (2*S*, 3*S*)-2,3-dihydroxybutanedioic acid
            (+)-tartaric acid   (–)-tartaric acid

These enantiomers are the diastereomers of the meso compound.

(2*R*, 3*S*)-2,3-dihydroxybutanedioic acid

meso-tartaric acid

6.14   For a compound with *n* stereocenters, there can be a maximum of 2^n stereoisomers. 2,3,4-Tribromohexane has 3 stereocenters and thus a maximum of 8 stereoisomers. The number of different isomers will be reduced if there are meso compounds.

(2*R*,3*R*,4*R*)-2,3,4-tribromohexane   (2*S*,3*S*,4*S*)-2,3,4-tribromohexane

These two compounds are enantiomers and are diastereomers of the other six stereoisomers.

6.14 (cont)

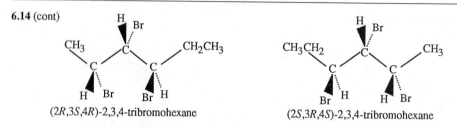

(2R,3S,4R)-2,3,4-tribromohexane   (2S,3R,4S)-2,3,4-tribromohexane

These two compounds are enantiomers and are diastereomers of the other six stereoisomers.

(2R,3R,4S)-2,3,4-tribromohexane   (2S,3S,4R)-2,3,4-tribromohexane

These two compounds are enantiomers and are diastereomers of the other six stereoisomers.

(2S,3R,4R)-2,3,4-tribromohexane   (2R,3S,4S)-2,3,4-tribromohexane

These two compounds are enantiomers and are diastereomers of the other six stereoisomers.

6.15

(1R, 3S)-1,3-dimethyl-   (1R, 3R)-1,3-dimethyl-   (1S, 3S)-1,3-dimethyl-
cyclopentane   cyclopentane   cyclopentane

cis-1,3-dimethylcyclopentane   *trans*-1,3-dimethylcyclopentane

a meso compound   enantiomers

                          diastereomers

Note that in the meso compound the stereocenters have opposite configuration.

6.15 (cont)

   One way of seeing the configuration of the stereocenters of a cyclic compound is to redraw the ring to put the lowest priority group on a bond going back. If the lowest priority group on the stereocenter is one of the ring substituents, the ring should be rotated so the stereocenter is at the back. This is shown below for the meso compound.

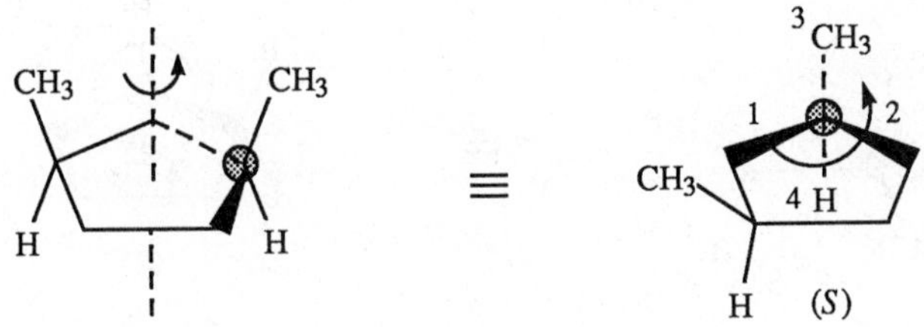

   If the lowest priority group is part of the ring, the ring should be rotated so the stereocenter is at the front. This again puts the lowest priority group on a bond going back. This is shown below for 1,3-dichloro-1,3-dibromocyclopentane.

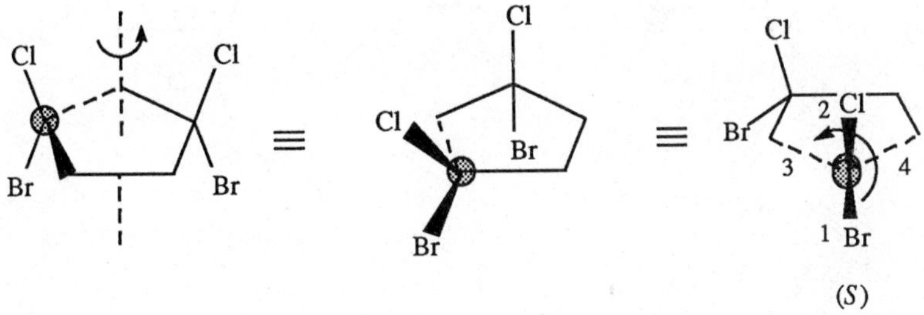

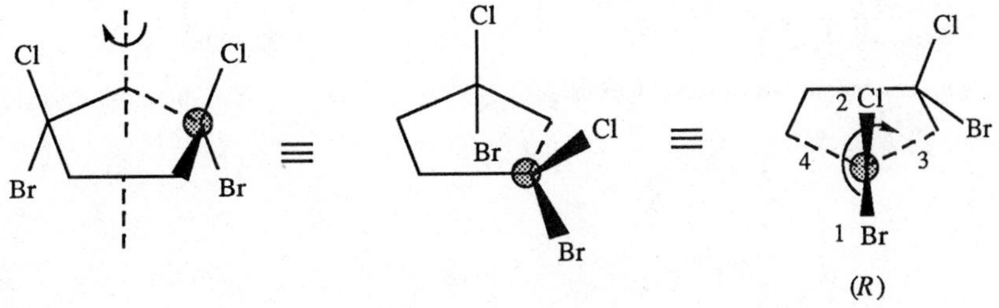

6.16

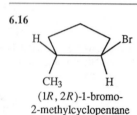

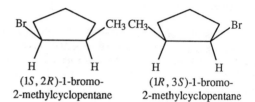

(1*R*, 2*R*)-1-bromo-
2-methylcyclopentane

(1*S*, 2*S*)-1-bromo-
2-methylcyclopentane

(1*S*, 2*R*)-1-bromo-
2-methylcyclopentane

(1*R*, 3*S*)-1-bromo-
2-methylcyclopentane

trans-1-bromo-2--methylcyclopentane   *cis*-1-bromo-2-methylcyclopentane

enantiomers   enantiomers

|_____diastereomers_____|

The assignment of configuration for (1*R*, 2*R*)-1-bromo-2-methylcyclopentane is shown below.

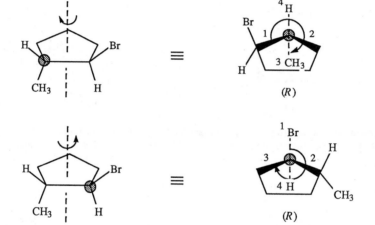

6.17

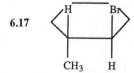

trans-1-bromo-3-methylcyclohexane   *cis*-1-bromo-3-methylcyclohexane

6.17 (cont)

enantiomers   enantiomers

trans-1-bromo-4-methylcyclohexane   *cis*-1-bromo-4-methylcyclohexane

6.18

(a)

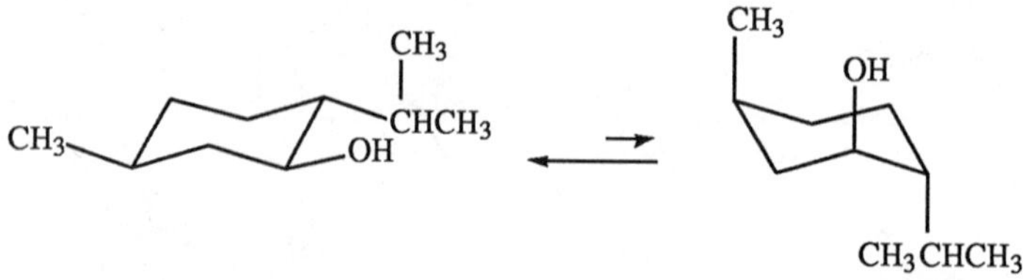

all substituents equatorial;
the most stable isomer of
2-isopropyl-5-methylcyclohexanol

all substituents axial;
the least stable isomer of
2-isopropyl-5-methylcyclohexanol

(b)

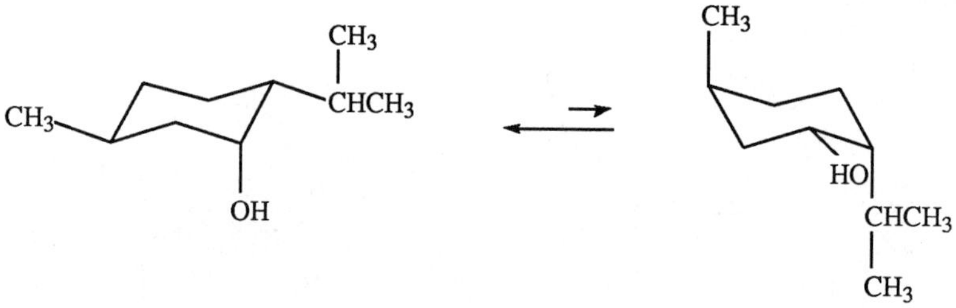

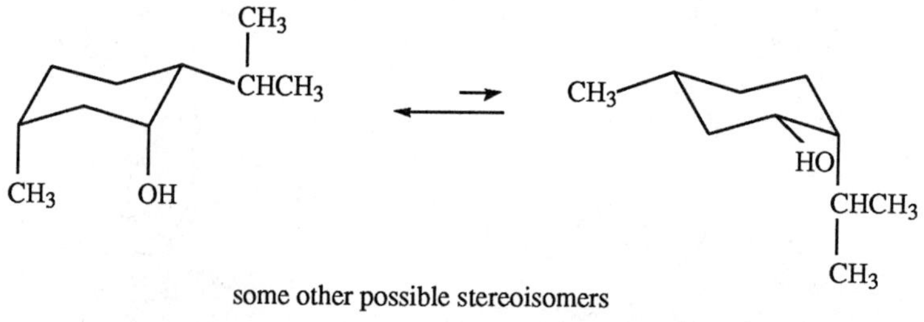

some other possible stereoisomers

Concept Map 6.7   Diastereomers.

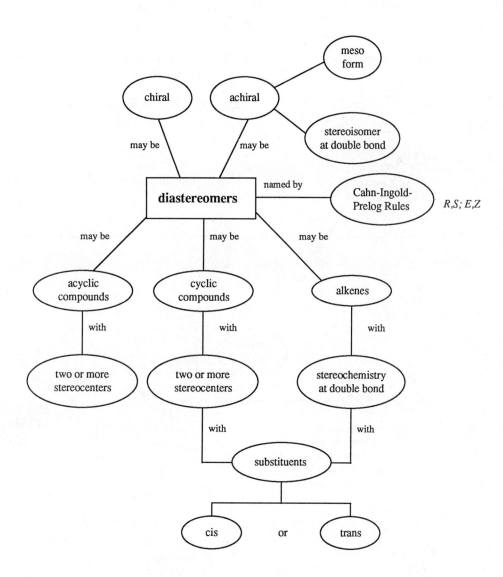

6.19

$$\begin{array}{c} CH_3CH_2 \quad\quad CH_3 \\ \diagdown \quad\quad\diagup \\ C = C \\ \diagup \quad\quad\diagdown \\ H \quad\quad\quad CH_2CH_3 \end{array}$$

(*E*)-3-methyl-3-hexene

$$\begin{array}{c} Br \quad\quad\quad CH_2Br \\ \diagdown \quad\quad\diagup \\ C = C \\ \diagup \quad\quad\diagdown \\ Cl \quad\quad\quad CH_3 \end{array}$$

(*Z*)-1,3-dibromo-1-chloro-2-
methylpropene

$$\begin{array}{c} \quad\quad\quad\quad O \\ \quad\quad\quad\quad \| \\ ClCH_2CH_2 \quad\quad COH \\ \diagdown \quad\quad\diagup \\ C = C \\ \diagup \quad\quad\diagdown \\ CH_3CH_2 \quad\quad H \end{array}$$

(*Z*)-5-chloro-3-ethyl-
2-pentenoic acid

$$\begin{array}{c} CH_3CH_2 \quad\quad C \equiv CH \\ \diagdown \quad\quad\diagup \\ C = C \\ \diagup \quad\quad\diagdown \\ H \quad\quad\quad CH_2CH_3 \end{array}$$

(*Z*)-3-ethyl-3-hexen-1-yne

Concept Map 6.8   The process of resolution.

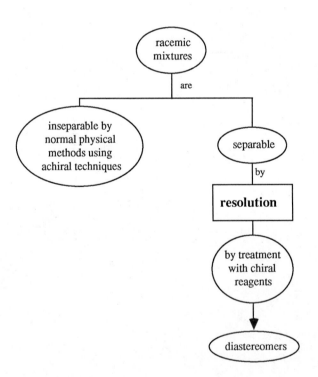

6.20

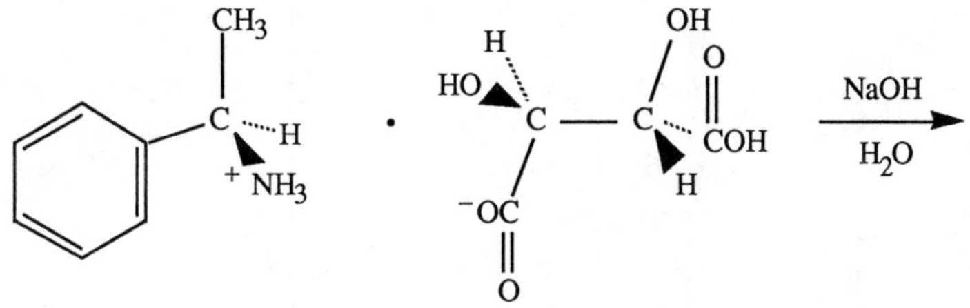

water-soluble salt

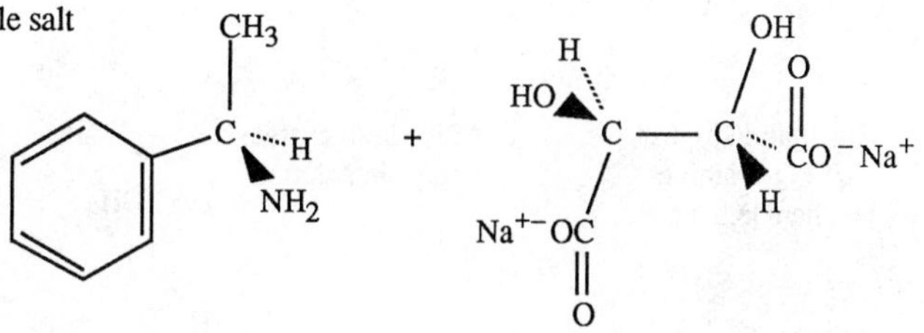

water-insoluble amine   water-soluble salt

6.21

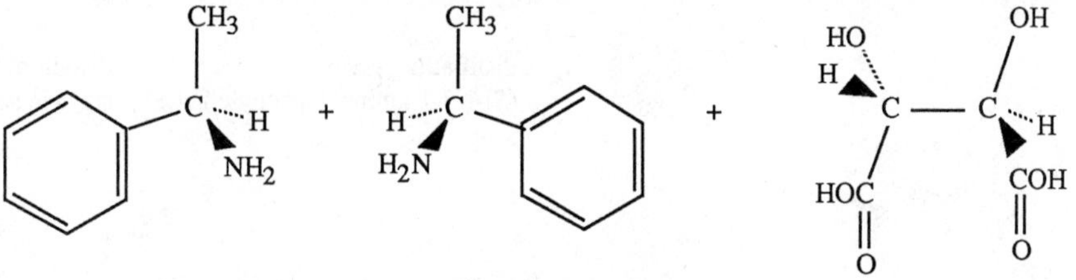

(*R*)-(+)-1-amino-1-phenylethane   (*S*)-(–)-1-amino-1-phenylethane   (2*R*,3*R*)-(+)-tartaric acid

methanol

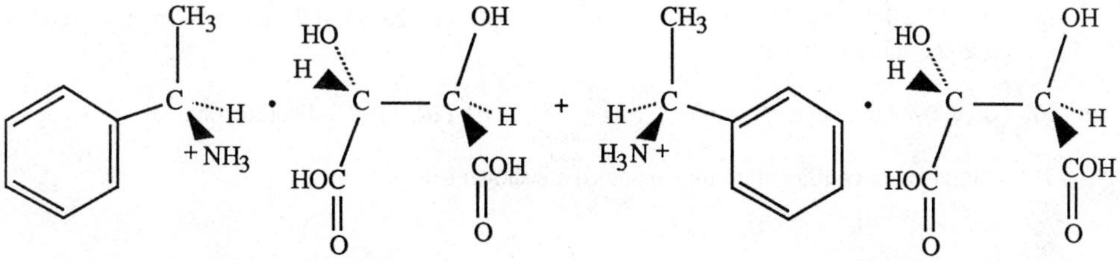

salt of (*R*)-(+)-1-amino-1-phenylethane
with (2*R*,3*R*)-(+)-tartaric acid; more
soluble in methanol; stays in solution

salt of (*S*)-(–)-1-amino-1-phenylethane
with (2*R*,3*R*)-(+)-tartaric acid; less
soluble in methanol; crystallizes out

NaOH
H_2O

NaOH
H_2O

6.21 (cont)

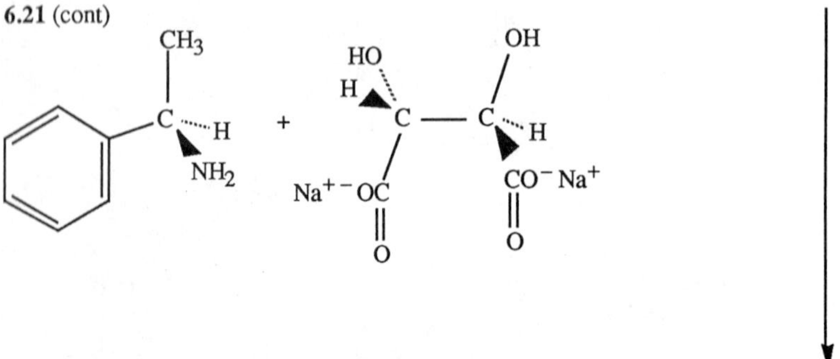

distillation gives
(R)-(+)-1-amino-
1-phenylethane

disodium tartrate
stays in solution

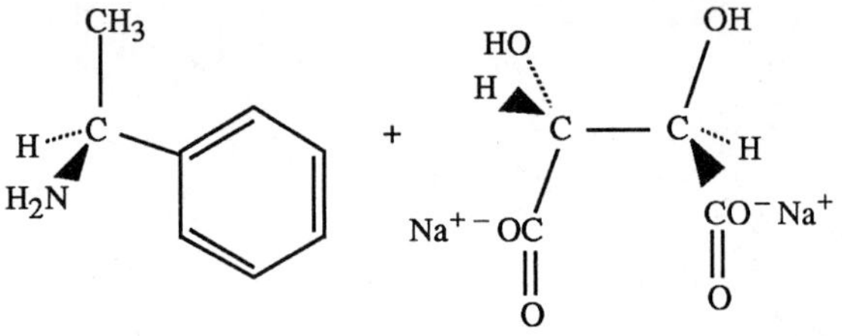

distillation gives
(S)-(–)-1-amino-1-phenylethane

disodium tartrate
stays in solution

6.22   (a) (S)-3-bromo-2-methylpentane

(c) butane

(e) (R)-3-methyl-3-hexanol

(g) (2R, 3S)-2,3-dibromobutane
     meso-2,3-dibromobutane

(i) (1R, 2R)-2-bromocyclopentanol

(b) (S)-1-phenyl-1-butanol

(d) (S)-1,2-dibromopentane

(f) (S)-3-methyl-3-hexanol

(h) (1S, 2S, 4S)-1,2-dibromo-4-methylcyclohexane

(j) (2R, 3S)-2,3-dichloropentane

The assignment of configuration for (h) and (i) are shown below.

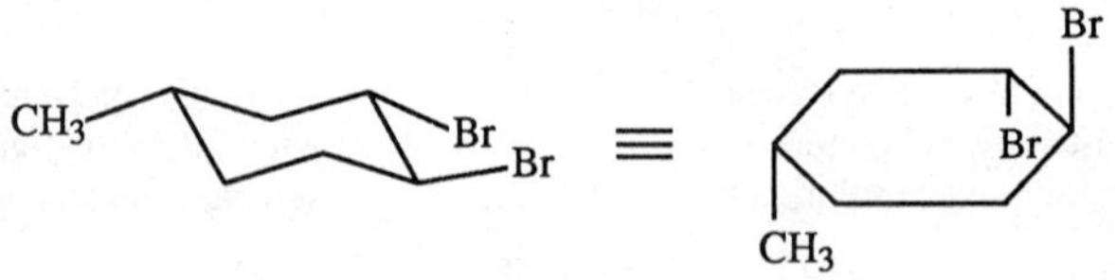

(1S, 2S, 4S)-1,2-dibromo-4-methylcyclohexane

6.22 (cont)

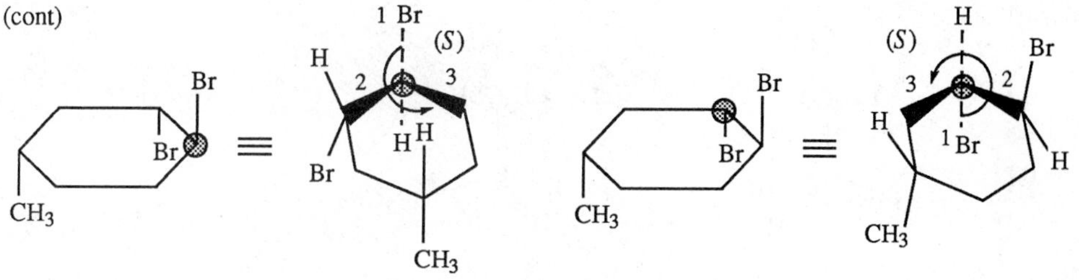

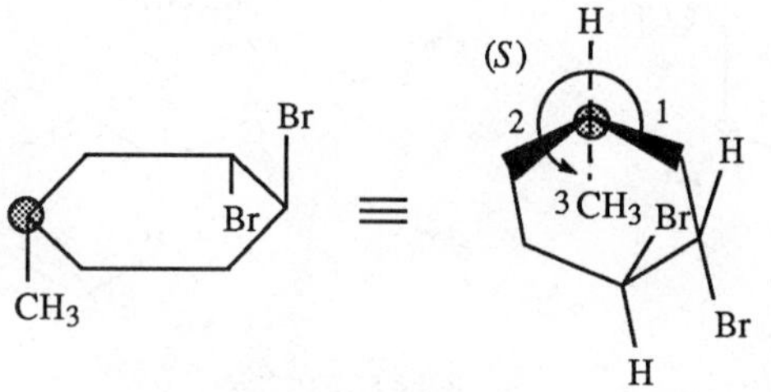

(1*R*, 2*R*)-2-bromocyclopentanol

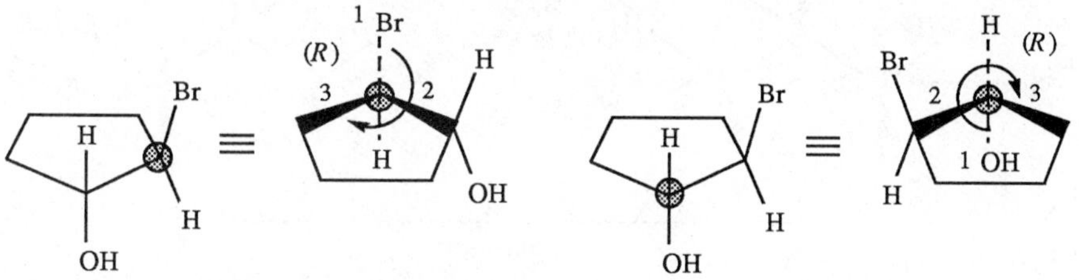

6.23   b, c, e, h, i, and k have stereocenters.

(b)

(c)

(e)

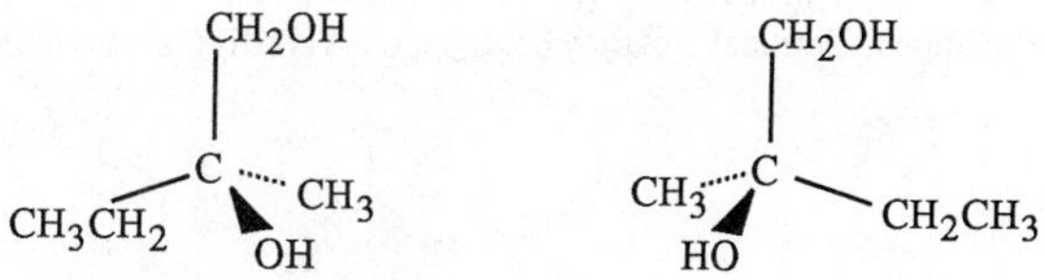

6.23 (cont)

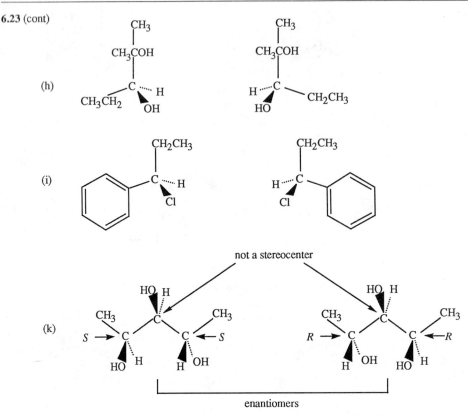

(h)

(i)

not a stereocenter

(k)

enantiomers

The central atom in each of these enantiomers is not a stereocenter because two of the substituents on the carbon are identical (including their configurations).

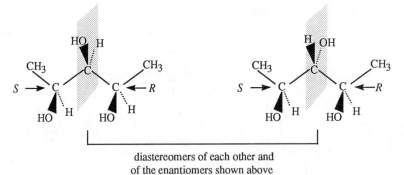

diastereomers of each other and
of the enantiomers shown above

The central carbon atom in each of these meso forms is a stereocenter but because of the symmetry of the molecule, the resulting compounds are not optically active. The two groups containing stereocenters that are attached to the center carbon atom appear identical, but they have opposite configurations and are therefore not identical.

6.24   (a) enantiomers   (The molecule on the left has the *S* configuration, the one on the right, the
                                                R configuration.)

        (b) identical

        (c) identical

        (d) identical

        (e) identical

        (f) enantiomers   (The molecule on the left has the *R* configuration; the other, the *S* configu-
                                            ration.)

        (g) enantiomers   (The molecule on the left has the *R* configuration; the other, the *S* configu-
                                            ration. Note also that the molecule on the right is not the mirror image of
                                            the molecule on the left but is a conformer of the mirror image.)

        (h) constitutional isomers   [(*R*)-3-bromo-1-chlorobutane and (2*R*,3*S*)-2-bromo-3-chlorobutane]

        (i) identical   (Both have the *R* configuration.)

        (j) conformers

        (k) constitutional isomers   [(*R*)-2-bromo-1-chlorobutane and (*S*)-3-bromo-1-chlorobutane]

        (l) conformers   (The chlorine atom is equatorial in the structure on the left and axial in the
                                            one on the right.)

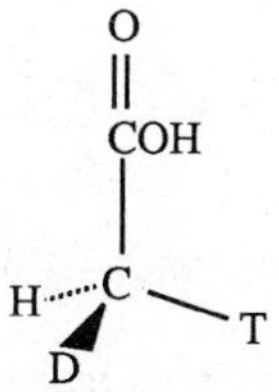

6.25

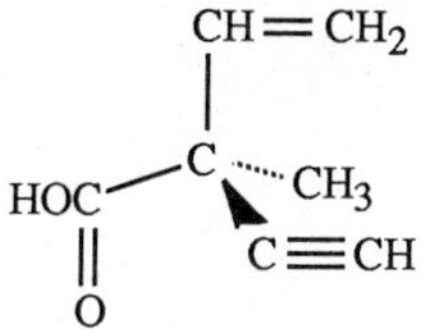

6.26

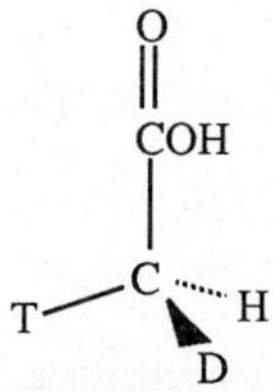

6.27

$$\text{enantiomeric excess} = \frac{\text{actual rotation}}{\text{rotation of pure enantiomer}} \times 100$$

$$88 = \frac{-4.1°}{\text{rotation of pure enantiomer}} \times 100$$

rotation of pure enantiomer $= -4.7$

(*S*)-(–)-citronellol

6.28   (a)

(*R*)-(–)-1-amino-2-propanol
hydrochloride

(*S*)-(+)-1-amino-2-propanol
hydrochloride

(b) The *S*-enantiomer was recovered with higher optical purity (higher absolute specific rotation.)

(c) $\begin{array}{l}\text{enantiomeric excess} \\ \text{for the } R\text{-isomer}\end{array} = \dfrac{-31.5°}{-35°} \times 100 = 90\%$

6.29 $\alpha = [\alpha]_D^{20°} lc$

$\alpha = (-92°)(0.5 \text{ dm})(1 \text{ g}/100 \text{ mL})$

$\alpha = -0.46°$

6.30   (a) $\text{enantiomeric excess} = \dfrac{\text{actual rotation}}{\text{rotation of pure enantiomer}} \times 100$

$[\alpha]_D^{20}$ for pure (+) enantiomer $= +13.90°$, therefore $[\alpha]_D^{20}$ for pure (–) enantiomer $= -13.90°$

$$\text{enantiomeric excess} = \frac{-3.5°}{-13.90°} \times 100 = 25\%$$

The mixture contains a 25% enantiomeric excess of the levorotatory compound and is $100-25$ or 75% racemic. The racemic portion of the mixture contains equal amounts (37.5%) of each enantiomer. Therefore the solution contains 37.5% of (+)-2-butanol and $(37.5 + 25)\%$ or 62.5% of (–)-2-butanol.

6.31

CH₃
|
CH₃NCH₂

$(2S, 3R)$-4-dimethylamino-1,2-
diphenyl-3-methyl-2-butanol

medically useful form

$(2R, 3S)$-4-dimethylamino-1,2-
diphenyl-3-methyl-2-butanol

CH₃
|
CH₃NCH₂

$(2S, 3S)$-4-dimethylamino-1,2-
diphenyl-3-methyl-2-butanol

$(2R, 3R)$-4-dimethylamino-1,2-
diphenyl-3-methyl-2-butanol

6.32

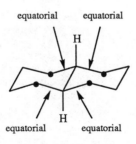

equatorial   equatorial

H

equatorial   equatorial
H

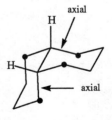

axial
H

H

axial

In *trans*-decalin, the methylene
groups on the ring junction are
both equatorial in each ring;
more stable

In *cis*-decalin, one of the methylene
groups on the ring junction is
axial in each case; less stable

7

Nucleophilic Substitution and Elimination Reactions

7.1

$$\underset{\underset{\text{CH}_3\text{CHCH}_3}{|}}{\overset{\overset{\text{Br}}{|}}{}} \xrightarrow{\text{H}_2\text{O or NaOH/H}_2\text{O}} \underset{\underset{\text{CH}_3\text{CHCH}_3}{|}}{\overset{\overset{\text{OH}}{|}}{}} + \text{CH}_3\text{CH}=\text{CH}_2$$

In H_2O, you would expect a larger proportion of the product to be isopropyl alcohol than when NaOH is added. Hydroxide ion is a stronger base than water and will give more of the elimination product (propene) than water alone. Water is the nucleophile when only water is present; hydroxide ion is the nucleophile when it is present.

7.2 (a) —S⁻ > —O⁻ (b) OH⁻ > $\text{CH}_3\overset{\overset{\displaystyle O}{\|}}{\text{C}}\text{O}^-$

(Sulfur is more polarizable than oxygen.) (The hydroxide ion is more basic than the acetate ion.)

(c) OH⁻ > NO₃⁻ (d) (CH₃)₂NH > NH₃

(The hydroxide ion is more basic than the nitrate ion.) (Dimethylamine is more basic than ammonia.)

(e) CH₃O⁻ > OH⁻

(Methoxide ion is slightly less basic but more polarizable than hydroxide ion.)

Concept Map 7.1   The factors that are important in determining nucleophilicity.

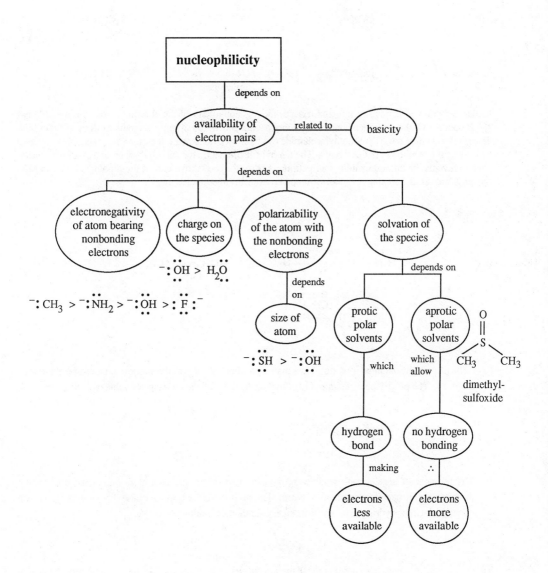

7.3   The increase in the relative rates of the substitution reaction in this series of solvents is related to the decreasing ability of the solvent to hydrogen bond to the nucleophile, chloride ion, and to the steric hindrance of the positive end of the dipole of the carbonyl group.

$$CH_3OH \qquad\qquad HCNH_2 \qquad\qquad HCN(CH_3)_2 \qquad\qquad CH_3CN(CH_3)_2$$

with O double-bonded above the carbonyl carbons of formamide, N,N-dimethylformamide, and N,N-dimethylacetamide

methanol   formamide   N,N-dimethyl-   N,N-dimethyl-
                                     formamide   acetamide

Methanol hydrogen bonds strongly and reduces the nucleophilicity of the chloride anion. Formamide has N—H bonds, so it also hydrogen bonds with chloride anion, but not as strongly as methanol does. Neither N,N-dimethylformamide nor N,N-dimethylacetamide can hydrogen bond at all, therefore the nucleophile behaves as a strong nucleophile in both solvents. The methyl substituent on the carbonyl group of N,N-dimethylacetamide prevents the chloride anion from getting as close to the positive end of the dipole as it can in N,N-dimethylformamide, making the chloride anion an even stronger nucleophile.

7.4   (a)

$$H_2N-OH \xrightarrow{\;HCl\;} H_3\overset{+}{N}-OH \qquad :\overset{..}{\underset{..}{Cl}}:^-$$

$$H_2N-OH \xrightarrow{\;CH_3I\;} CH_3\overset{\overset{\displaystyle H}{|}}{\underset{\underset{\displaystyle H}{|}}{\overset{+}{N}}}-OH \qquad :\overset{..}{\underset{..}{I}}:^-$$

(b)   The pK_as of the oxonium ions are all less than zero while the pK_as of the ammonium ions are all between 4 and 11. The pK_a of the conjugate acid of hydroxylamine falls in the range of values for the ammonium ion.

$$pK_a \quad NH_4^+ \qquad 9.4$$
$$pK_a \quad H_3\overset{+}{N}OH \qquad 5.97$$

The presence of an electron-withdrawing oxygen atom on the nitrogen atom makes the nonbonding electrons on nitrogen less available. Hydroxylamine is therefore a weaker base than ammonia, and the conjugate acid of hydroxylamine is a stronger acid than ammonium ion.

7.5   The hydronium ion comes from protonation of the solvent, water, by the acid, HBr, that was added to the reaction mixture.

$$H_2O \;+\; HBr \longrightarrow H_3O^+ \;+\; Br^-$$

7.6   The other nucleophiles that are present are the starting alcohol and water.

Two additional substitution reactions are possible. The carbocation can react with water to regenerate the starting alcohol or with another alcohol molecule to produce an ether.

Nucleophiles are also bases and can remove a proton from the carbocation to give an alkene, the elimination product.

Concept Map 7.2   (see p. 158)

7.7   The presence of water in the test tube increases the polarity of the solvent, allowing ionization to occur. Therefore some of the reactions observed will be those of carbocations. Under such conditions, the order of reactivity seen for the alkyl halides will not be that expected of bimolecular substitution reactions.

7.8   The initial rate of the reaction will be

rate = k [CH_3CH_2Br] [OH^-]

rate = $(1.7 \times 10^{-3}$ L/mol·s)(0.05 mol/L)(0.07 mol/L)

rate = 6×10^{-6} mol/L·s

If the concentration of bromoethane is doubled (increased from 0.05 mol/L to 0.1 mol/L), the rate will double. The new rate would be ~ 1×10^{-5} mol/L·s.

Concept Map 7.2   The factors that determine whether a substituent is a good leaving group.

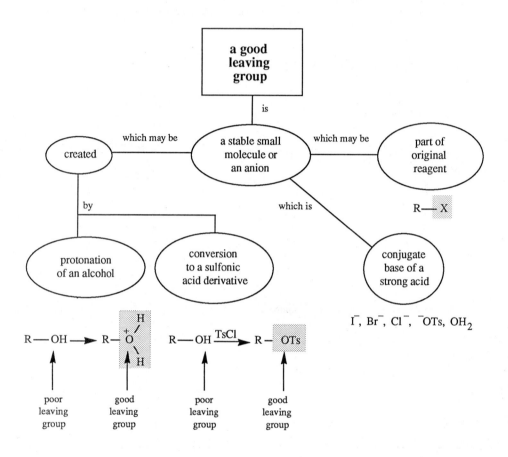

- -

7.9   For the reaction

$$CH_3CH_2CH_2CH_2Br \ + \ K^+ I^- \ \xrightarrow{\text{acetone}} \ CH_3CH_2CH_2CH_2I \ + \ K^+ Br^- \!\downarrow$$

The rate at which *n*-butyl bromide is used up is:

rate $= k\,[CH_3CH_2CH_2CH_2Br]\,[I^-]$

rate $= (1.09 \times 10^{-1}\ \text{L/mol·min})(0.06\ \text{mol/L})(0.02\ \text{mol/L})$

rate $= 1.3 \times 10^{-4}\ \text{mol/L·min}$

7.10   The rate constant for the reaction of *n*-butyl chloride with iodide ion can be calculated using the Arrhenius equation.

$$k = Ae^{-E_a/RT}$$

$$k = (2.24 \times 10^{11}\ \text{L/mol·s})\{\exp(-22.2\ \text{kcal/mol})/[(1.99 \times 10^{-3}\ \text{kcal/mol·K})(333\text{K})]\}$$

$$k = (2.24 \times 10^{11}\ \text{L/mol·s})(\exp-33.5)$$

$$k = (2.24 \times 10^{11}\ \text{L/mol·s})(2.82 \times 10^{-15})$$

$$k = 6.32 \times 10^{-4}\ \text{L/mol·s}$$

7.11

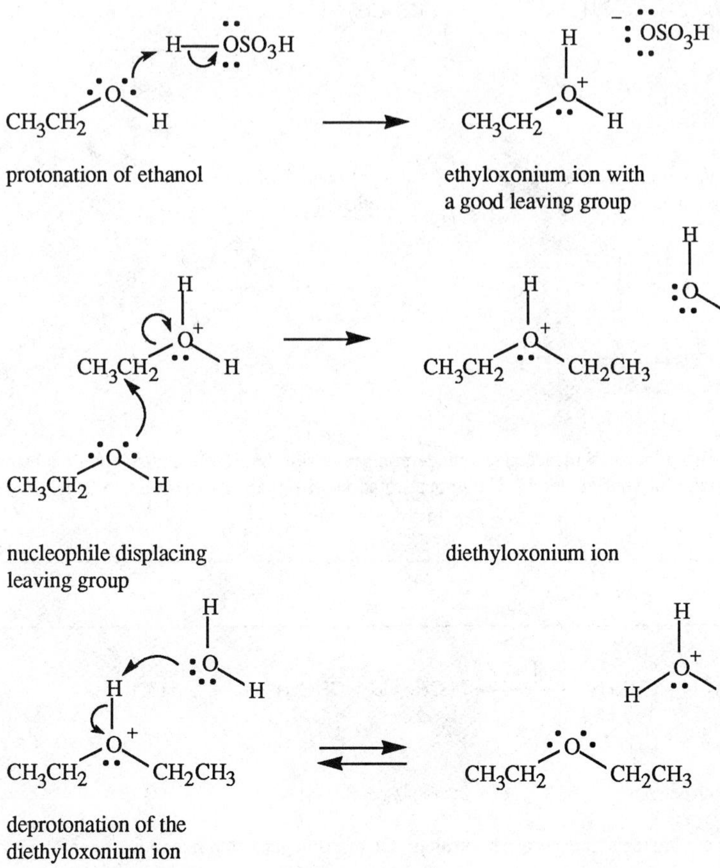

protonation of ethanol

ethyloxonium ion with
a good leaving group

nucleophile displacing
leaving group

diethyloxonium ion

deprotonation of the
diethyloxonium ion

7.12

protonation of
diethyl ether

nucleophile displacing
good leaving group

nucleophile displacing
good leaving group

protonation of
ethanol

7.13   (*S*)-(+)-2-Bromobutane will racemize by undergoing reversible bimolecular nucleophilic substitution reactions
with inversion of configuration. The optical rotation will gradually decrease to 0° as the system reaches
equilibrium.

— —

Concept Map 7.3   (see p. 162)

— —

7.14   (1) $CH_3(CH_2)_5\overset{*}{C}HCH_3$ $\xrightarrow{\text{TsCl}}$ $CH_3(CH_2)_5\overset{*}{C}HCH_3$ + HCl
                    | |
                   OH   OTs

        (+)-2-octanol   (+)-2-octyl tosylate

No bonds are broken at the chiral carbon atom; the reaction goes with retention of configuration.

7.14 (cont)

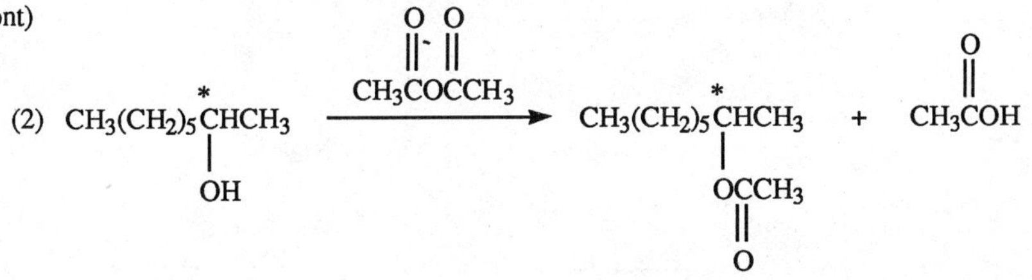

$$\underset{\text{(--)-2-octanol}}{\underset{|}{\overset{*}{\underset{\text{OH}}{\text{CH}_3(\text{CH}_2)_5\text{CHCH}_3}}}} \xrightarrow{\overset{\text{O O}}{\underset{}{\text{CH}_3\overset{||}{\text{C}}\text{O}\overset{||}{\text{C}}\text{CH}_3}}} \underset{\text{(--)-2-octyl acetate}}{\underset{|}{\overset{*}{\underset{\underset{O}{\overset{||}{\text{OCCH}_3}}}{\text{CH}_3(\text{CH}_2)_5\text{CHCH}_3}}}} + \underset{}{\overset{\overset{\text{O}}{||}}{\text{CH}_3\text{COH}}}$$

(2)

The starting alcohol here is of opposite configuration to the alcohol we had in equation 1; we know this because it has the opposite sign of rotation. The reaction is not one that we recognize, and the sign of rotation does not tell us whether the configuration of the acetate is inverted or not. Let us reserve judgement until we look at the next reaction.

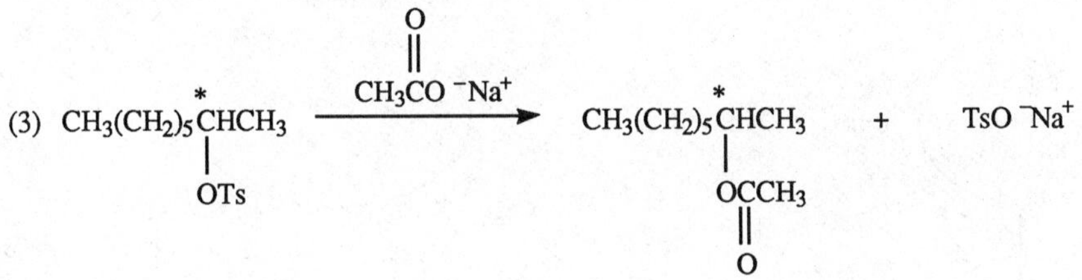

$$\underset{\text{(+)-2-octyl tosylate}}{\underset{|}{\overset{*}{\underset{\text{OTs}}{\text{CH}_3(\text{CH}_2)_5\text{CHCH}_3}}}} \xrightarrow{\overset{\overset{\text{O}}{||}}{\text{CH}_3\text{CO}^-\text{Na}^+}} \underset{\text{(--)-2-octyl acetate}}{\underset{|}{\overset{*}{\underset{\underset{O}{\overset{||}{\text{OCCH}_3}}}{\text{CH}_3(\text{CH}_2)_5\text{CHCH}_3}}}} + \text{TsO}^-\text{Na}^+$$

(3)

We recognize this as an S_N2 reaction, therefore it goes by inversion of configuration. If the (--)-acetate has the opposite configuration from that of the (+)-tosylate, it must also have a configuration opposite to that of the (+)-alcohol because the formation of the tosylate from the (+)-alcohol did not involve inversion of configuration [see part (1)]. Therefore, the configuration of the (--)-acetate must be the same as that of the (--)-alcohol. Reaction 2 must proceed by retention of configuration.

7.15   The fact that the rate of racemization is twice the rate of substitution of the radioactive iodine shows that every substitution occurs with inversion of configuration.

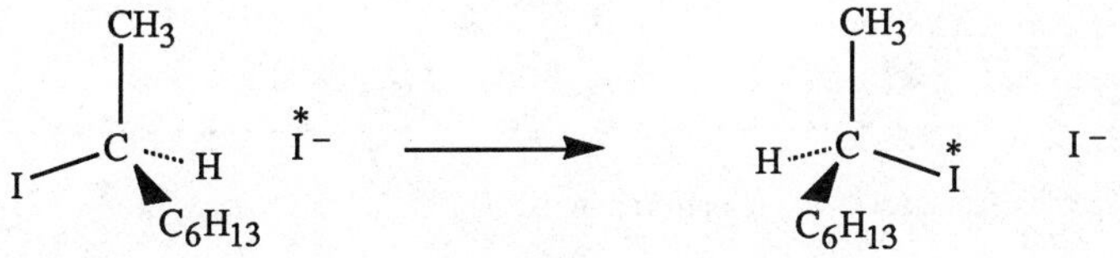

(S)-(+)-2-iodooctane   (R)-(--)-2-iodooctane

inversion of configuration

Each S_N2 reaction incorporates *one* radioactive iodine atom into 2-iodooctane, but results in the loss of optical activity equal to that of two molecules of (S)-(+)-2-iodooctane: the one that had been converted to the (R)-isomer, and the one, the optical activity of which is cancelled by the presence of the (R)-isomer. Therefore the rate of racemization, the rate at which optical activity is lost, is *twice* the rate of the substitution reaction.

Concept Map 7.3   A typical S_N2 reaction.

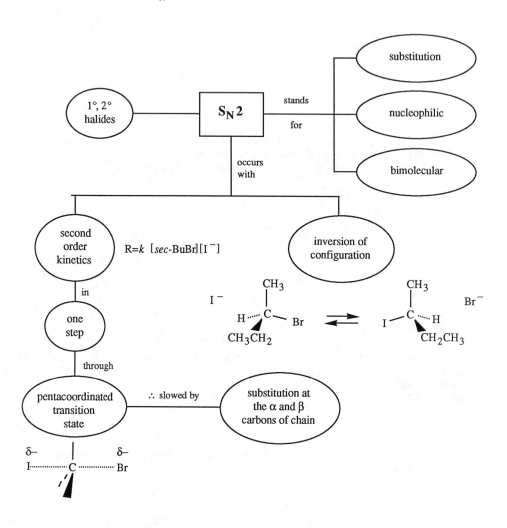

- -

7.16   The following species can act as bases (electron donors).

$$CH_3CH_2\overset{\cdot\cdot}{\underset{\cdot\cdot}{O}}\!-\!H \qquad CH_3\overset{\displaystyle CH_3}{\underset{\displaystyle CH_3}{C}}\!-\!\overset{\cdot\cdot}{\underset{\cdot\cdot}{O}}\!-\!CH_2CH_3 \qquad CH_3\overset{\displaystyle CH_3}{\underset{\displaystyle CH_3}{C}}\!-\!\overset{\cdot\cdot}{\underset{\cdot\cdot}{O}}\!-\!H$$

$$CH_3\overset{\displaystyle CH_3}{C}\!=\!CH_2 \qquad :\!\overset{\cdot\cdot}{\underset{\cdot\cdot}{Cl}}\!:^- \qquad H_2\overset{\cdot\cdot}{\underset{\cdot\cdot}{O}} \qquad {}^-:\!\overset{\cdot\cdot}{\underset{\cdot\cdot}{O}}H$$

7.17   For the reaction

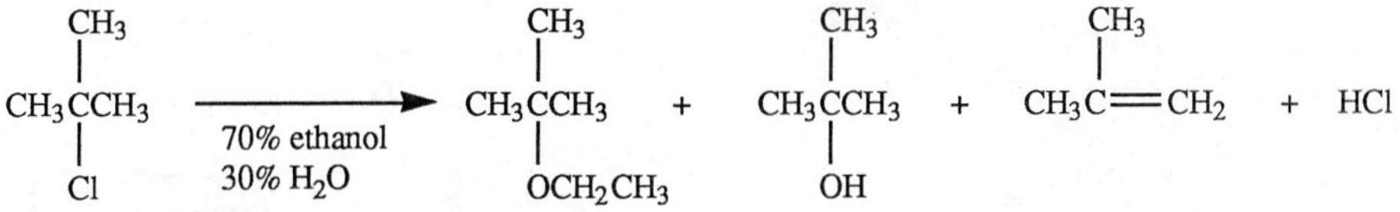

rate   = k [*tert*-butyl chloride]

rate   =   (0.145/h)(0.0824 mol/L)

rate   = 1.19×10^{-2} mol/L·h

The units in which this rate is expressed are mol/L·h. The rate constant will remain the same, but the rate of the reaction will decrease as the *tert*-butyl chloride is used up in the reaction. When half of the *tert*-butyl chloride is gone, the rate of the reaction will be half of what it was initially or 5.95×10^{-3} mol/L·h.

— —

Concept Map 7.4   (see p. 164)

— —

7.18

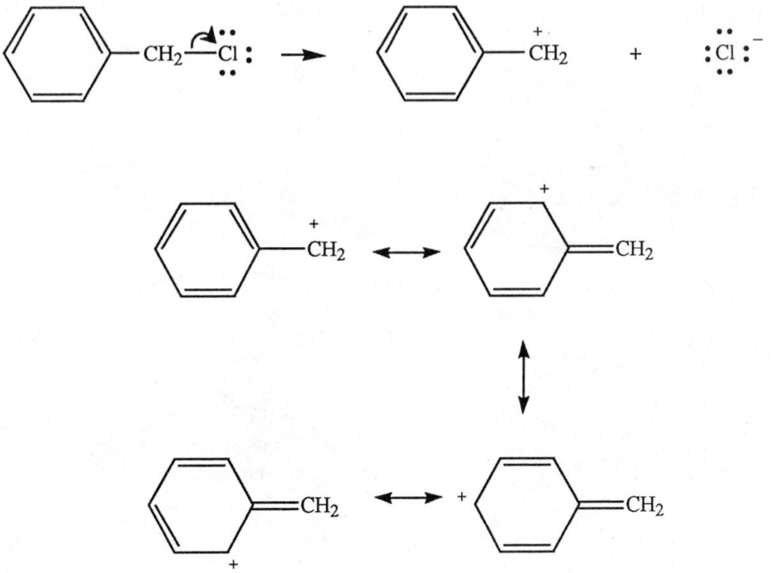

Concept Map 7.4   The S$_N$1 reaction.

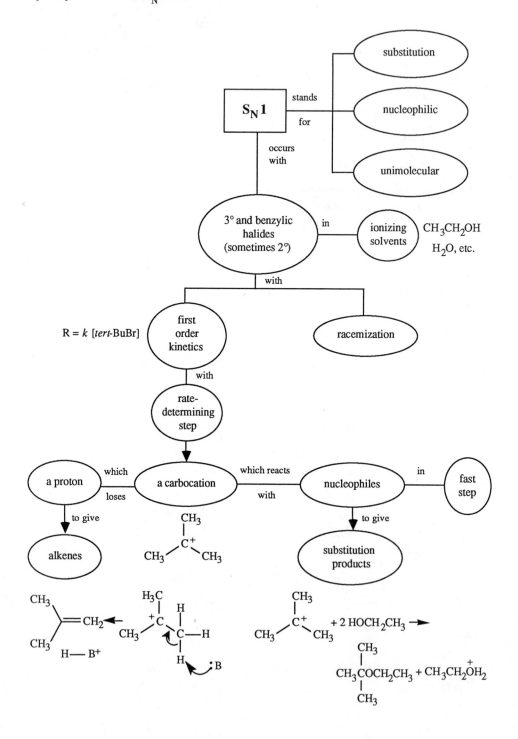

7.19 S_N1 mechanism

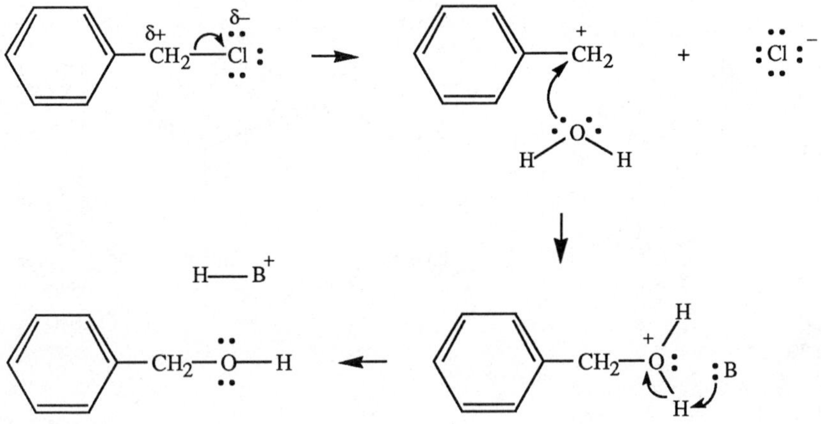

S_N2 mechanism

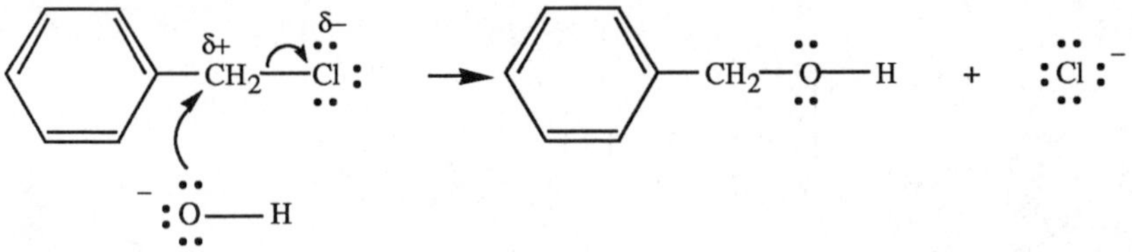

7.20   (a) $CH_3CH_2CH_2CH_2CH_2Br$ $\xrightarrow[\substack{ethanol \\ \Delta}]{KOH}$ $CH_3CH_2CH_2CH=CH_2$ + H_2O + K^+Br^-

(b)

(c)

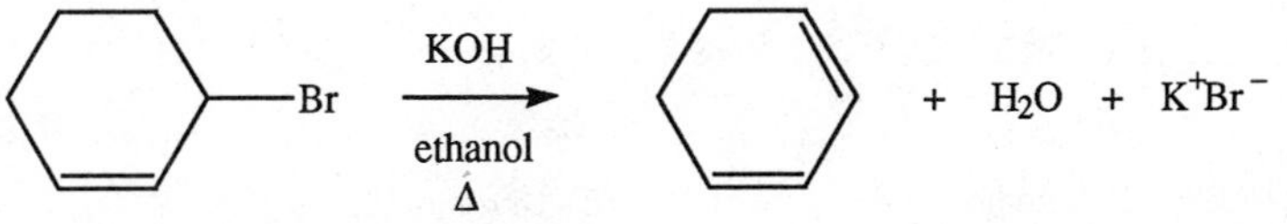

[Note that elimination in the other direction would give an allene-type of structure (Problem 2.37] in a six-membered ring, a highly strained situation.]

7.20 (cont)

(d)

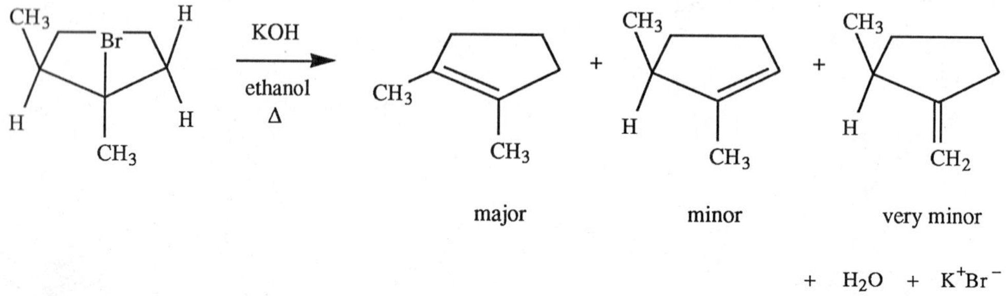

major   minor   very minor

+ H_2O + K^+Br^-

7.21   Yes

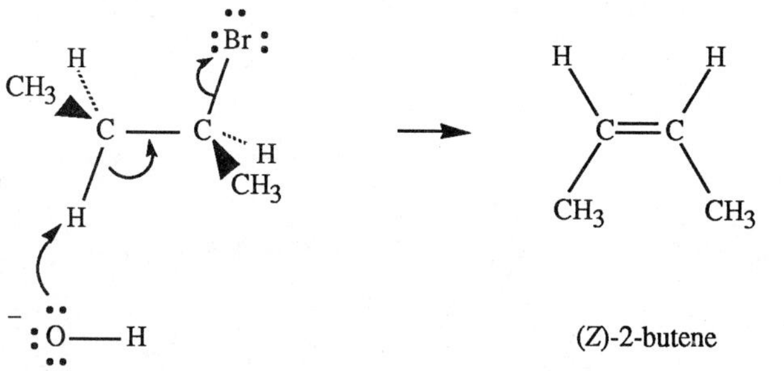

(Z)-2-butene

The transition state for the formation of (E)-2-butene is less crowded, more favorable, and lower in energy than the transition state for the formation of (Z)-2-butene where the two methyl groups are close together.

7.22

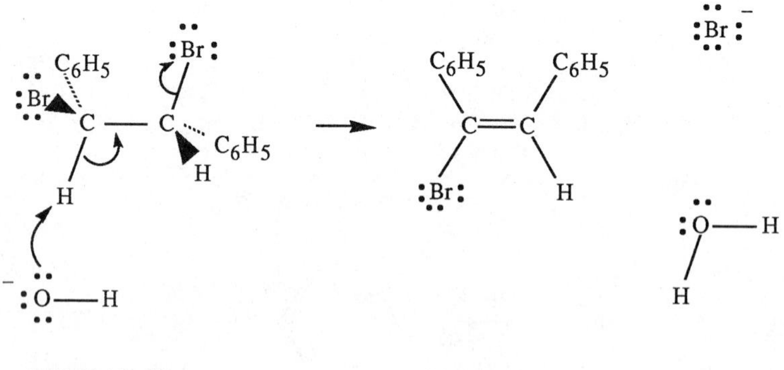

(1R,2S)-1,2-dibromo-   (E)-1-bromo-1,2-
diphenylethane   diphenylethene

7.22 (cont)

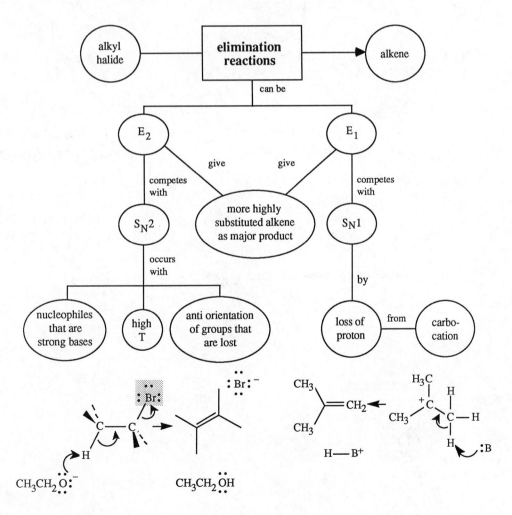

(1*R*,2*R*)-1,2-dibromo-
diphenylethane

(*Z*)-1-bromo-1,2-
diphenylethene

Concept Map 7.5   A comparison of E_1 and E_2 reactions.

7.23   **1.**

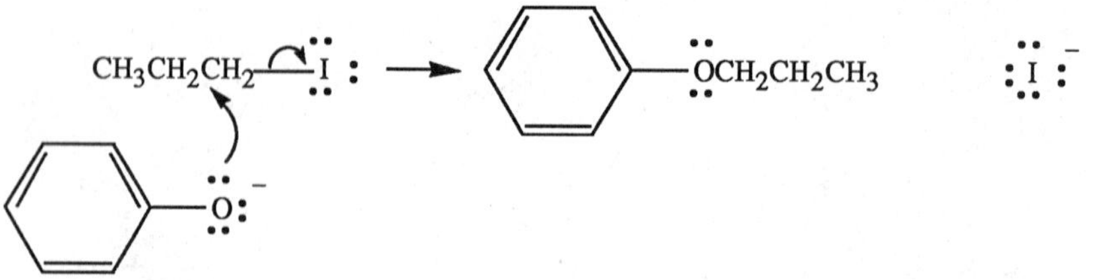

2.

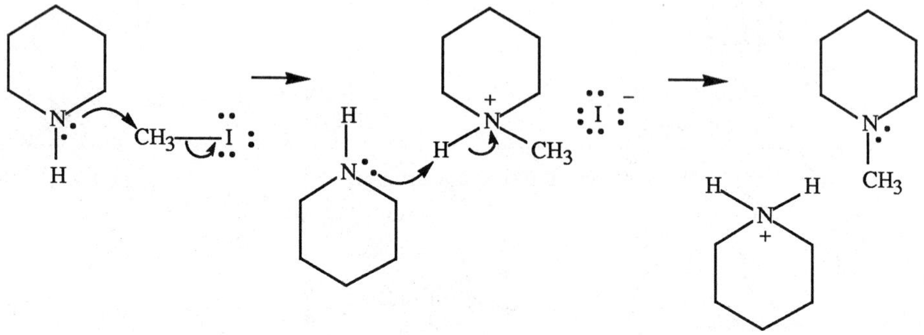

3.

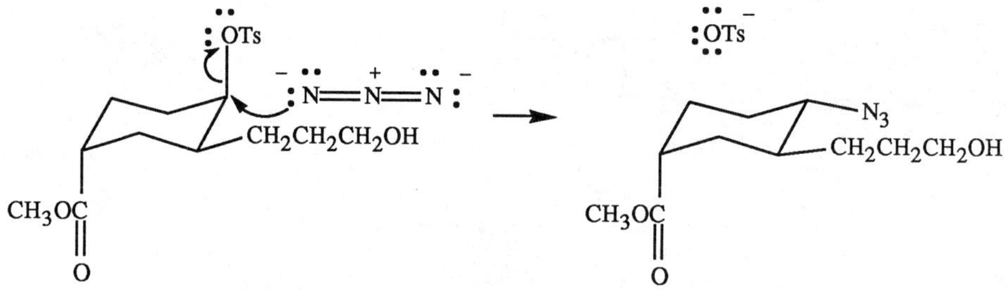

4.

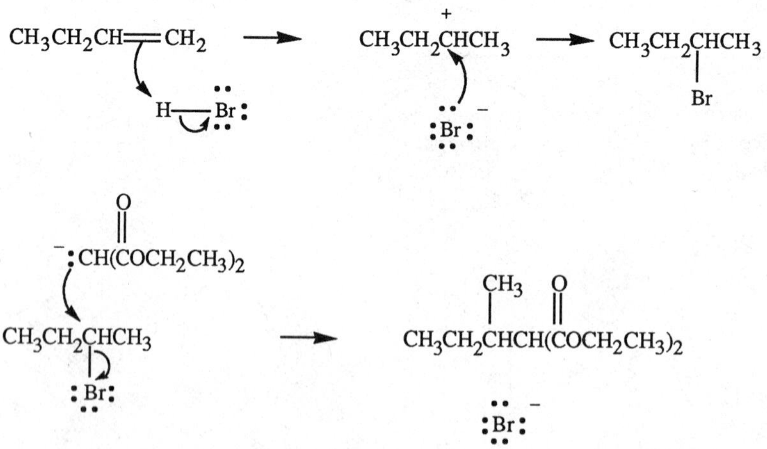

7.24   1.

$$+ \text{ NaOTs}$$

2.

$$+ \text{ NaBr}$$

7.25   1.   What functional groups are present in the starting material and in the product?

The starting material is a terminal alkene; the product contains an azide group.

2.   How do the carbon skeletons of the two compounds compare? How many carbon atoms does each contain? Are there any rings? What are the positions of branches and functional groups on the carbon skeleton?

The carbon skeletons are identical in starting material and products. The azide functional group in the product is on the second carbon atom of the chain, carbon two of the double bond in the starting material.

3.   How have the functional groups changed in going from the starting material to the product? Does the starting material have a good leaving group?

The double bond has disappeared. The product is an azide. There is no good leaving group in the starting material

4.   Is it possible to dissect the structure to see which bonds must have been broken and which formed?

bonds broken   bonds formed

5.   Do we recognize any part of the product molecule as coming from a good nucleophile or an electrophilic addition?

The azide group, —N_3, comes from N_3^-, a good nucleophile.

6.   What type of compound would be a good precursor to the product?

A compound that has a good leaving group where the azide group is in the product.

7.25 (cont)

6. (cont)

$$\text{Ph—CH}_2\text{CH}_2\text{CHCH}_3$$
$$\overset{|}{\text{X}}$$

7. After this last step, do we see how to get from starting material to product? If not, we need to analyze the structure obtained in step six by applying questions 5 and 6 to it.

    Since there is no good leaving group on the starting material, we need to repeat steps 5 and 6.

5'. Do we recognize any parts of the compound we want coming from a good nucleophile or an electrophilic addition?

    A good leaving group such as a halide can come from an electrophilic addition of a hydrogen halide to the double bond.

6'. What type of compound would be a good precursor to the product?

    The starting material is the precursor that we need. We can now complete the problem.

$$\text{Ph—CH}_2\text{CH}_2\text{CHCH}_3 \qquad \text{Ph—CH}_2\text{CH}_2\text{CH}=\text{CH}_2$$
$$\overset{|}{\text{N}_3}$$

$$\nwarrow \overset{\text{NaN}_3}{\text{ethanol}} \qquad\qquad \swarrow \text{HBr}$$

$$\text{Ph—CH}_2\text{CH}_2\text{CHCH}_3$$
$$\overset{|}{\text{Br}}$$

- -

Concept Map 7.6   (see pp. 172-173)

- -

7.26

(a) $CH_3CH_2CH_2CH_2Br \xrightarrow[\Delta]{\text{KOH}} \underset{\text{ethanol}}{} CH_3CH_2CH{=\!\!=}CH_2 + CH_3CH_2CH_2CH_2OH + H_2O + K^+Br^-$

major product with strong
base at high temperature

(b) $CH_3CH_2CH_2CH_2Br \xrightarrow[\substack{50\% \text{ aqueous} \\ \text{ethanol}}]{\text{0.1 M NaOH}} CH_3CH_2CH_2CH_2OH + Na^+Br^-$

(c) $CH_3CH_2CH_2CH_2Br \xrightarrow{NH_3} CH_3CH_2CH_2CH_2\overset{+}{N}H_3 \ Br^- \xrightarrow{NH_3 \text{ (excess)}}$

$CH_3CH_2CH_2CH_2NH_2 + NH_4{}^+Br^-$

(d) $CH_3CH_2CH_2CH_2Br \xrightarrow{Na^+N_3{}^-} CH_3CH_2CH_2CH_2N_3 + Na^+Br^-$

(e) $CH_3CH_2CH_2CH_2Br \xrightarrow{Na^+CN^-} CH_3CH_2CH_2CH_2CN + Na^+Br^-$

(f) $CH_3CH_2CH_2CH_2Br \xrightarrow{CH_3CH_2S^-Na^+} CH_3CH_2CH_2CH_2SCH_2CH_3 + Na^+Br^-$

(g) $CH_3CH_2CH_2CH_2Br \xrightarrow{CH_3\overset{\displaystyle O}{\overset{\|}{C}}O^-Na^+} CH_3CH_2CH_2CH_2O\overset{\displaystyle O}{\overset{\|}{C}}CH_3 + Na^+Br^-$

(h) $CH_3CH_2CH_2CH_2Br \xrightarrow{CH_3CH_2C{\equiv}C^-Na^+} CH_3CH_2CH_2CH_2C{\equiv}CCH_2CH_3 + Na^+Br^-$

(i) $CH_3CH_2CH_2CH_2Br \xrightarrow{Na^+ {}^-CH(\overset{\displaystyle O}{\overset{\|}{C}}OCH_2CH_3)_2} CH_3CH_2CH_2CH_2CH(\overset{\displaystyle O}{\overset{\|}{C}}OCH_2CH_3)_2 + Na^+Br^-$

Concept Map 7.6   Chapter summary.

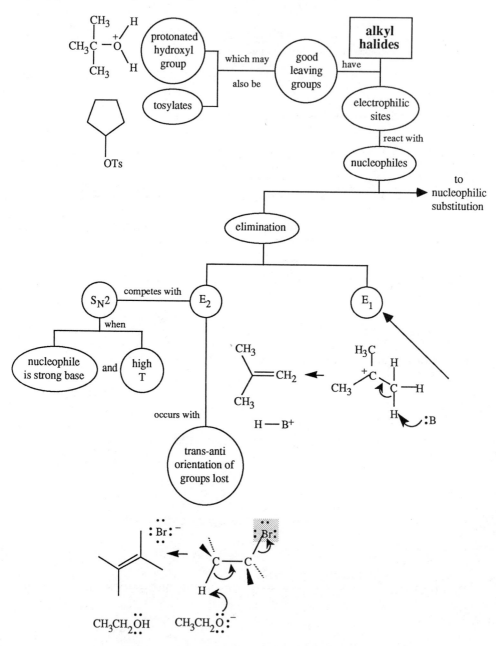

alkyl halides

have

electrophilic sites

good leaving groups

which may

also be

protonated hydroxyl group

tosylates

CH_3 H

$CH_3C\overset{+}{-}O$

CH_3 H

OTs

react with

nucleophiles

to elimination

nucleophilic substitution

3° and benzylic halides

occurs

with

S_N1

occurs

with

1°, 2°, benzylic halides

first order kinetics

two-step reaction

one-step reaction

second order kinetics

$R=k\ [sec\text{-}BuBr][I^-]$

$R=k\ [tert\text{-}BuBr]$

CH_3

$CH_3\overset{\displaystyle CH_3}{\underset{\displaystyle CH_3}{C^+}}$

carbo-cation

penta-coordinated TS

$\delta-$ I $\cdots\cdots$ C $\cdots\cdots$ Br $\delta-$

gives

which

inversion of configuration

to E_1

loses a proton

adds nucleophile

to give

racemized products

CH_3 CH_3

$CH_3\overset{CH_3}{\underset{CH_3}{C^+}}$ + 2 HOCH$_2$CH$_3$ → CH$_3$COCH$_2$CH$_3$ + CH$_3$CH$_2\overset{+}{O}H_2$

CH_3

7.27

(a)

leaving group

⬡—CH$_2$CH$_2$CH$_2$OTs + N$_3$$^-$ ⟶ ⬡—CH$_2$CH$_2$CH$_2$N$_3$ + TsO$^-$

nucleophile

(b)

leaving group

⬡—CH$_2$CH$_2$Cl + NaCN ⟶ ⬡—CH$_2$CH$_2$CN + NaCl

$^-$:C≡N: ⟷ :C≡N:$^-$

more nucleophilic

(c)   CH$_3$CH$_2$CH$_2$CH$_2$Br + CH$_3$CH$_2$S$^-$Na$^+$ ⟶ CH$_3$CH$_2$CH$_2$CH$_2$SCH$_2$CH$_3$ + NaBr

leaving group   nucleophile

(d)   CH$_3$CH$_2$CH$_2$Cl + CH$_3$C≡C$^-$Na$^+$ ⟶ CH$_3$CH$_2$CH$_2$C≡CCH$_3$ + NaCl

leaving group   nucleophile

(e)
CH$_3$

CH$_3$CHCH$_2$CH$_2$Br + NH$_3$ (excess) ⟶

CH$_3$

CH$_3$CHCH$_2$CH$_2$NH$_2$ + NH$_4$Br

leaving group   nucleophile

(f)
O
‖
CH$_3$I + Na$^+$ $^-$CH(COCH$_2$CH$_3$)$_2$ ⟶

O
‖
CH$_3$CH(COCH$_2$CH$_3$)$_2$ + NaI

leaving group   nucleophile

(g)   CH$_3$CH$_2$I + P(CH$_2$CH$_2$CH$_2$CH$_3$)$_3$ ⟶ CH$_3$CH$_2$P$^+$(CH$_2$CH$_2$CH$_2$CH$_3$)$_3$ I$^-$

leaving group   nucleophile

7.27 (cont)

(h) $HOCH_2CH_2Cl$ + CH_3SNa ⟶ $HOCH_2CH_2SCH_3$ + $NaCl$

⬆ leaving group   ⬆ nucleophile

7.28

			AgNO$_3$ in ethanol S$_N$1	NaI in acetone S$_N$2
(a)	$CH_3CH_2CH_2Br$	1° halide	no ppt	ppt
(b)	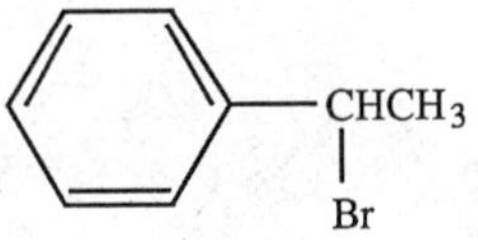 $CHCH_3$ Br	2° and benzylic halide	ppt	ppt
(c)	$CH_3CH_2CHCH_3$ Br	2° halide	ppt	ppt
(d)	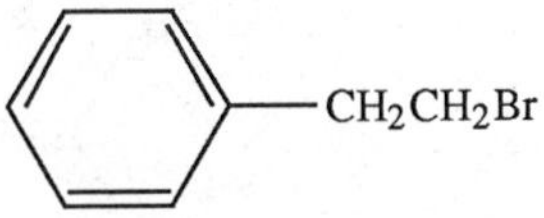 CH_2CH_2Br	1° halide	no ppt	ppt
(e)	CH_3 $CH_3CCH_2CH_3$ Br	3° halide	ppt	no ppt

7.29

(a)

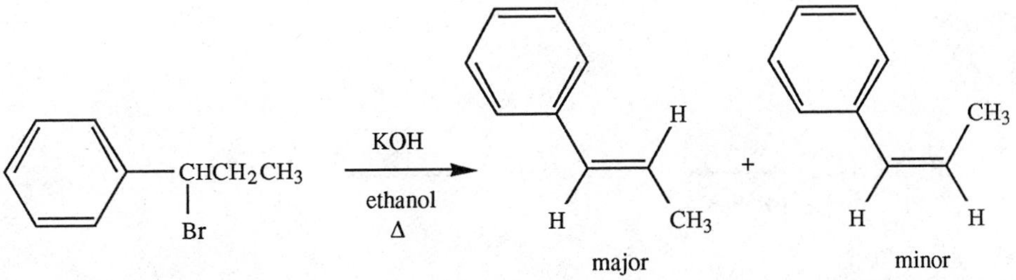

major   +   minor

7.29 (cont)

(b)

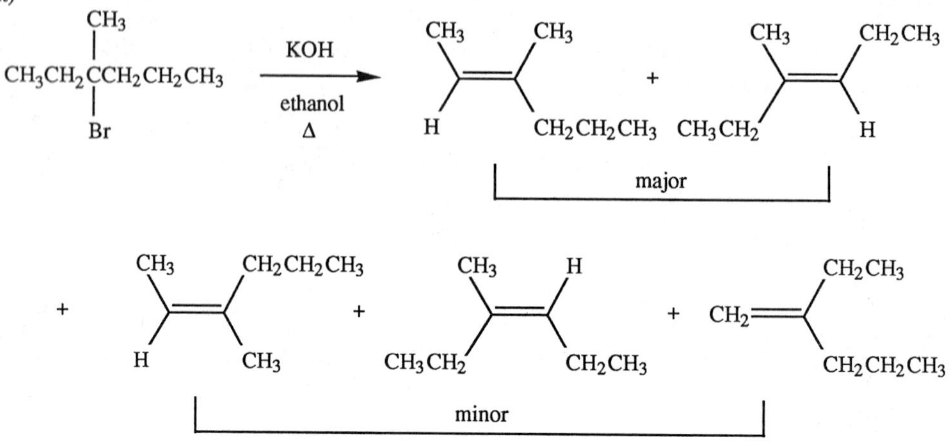

(c)

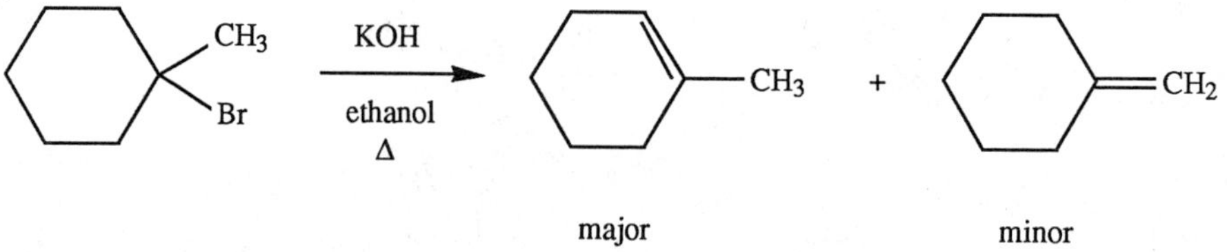

(d)

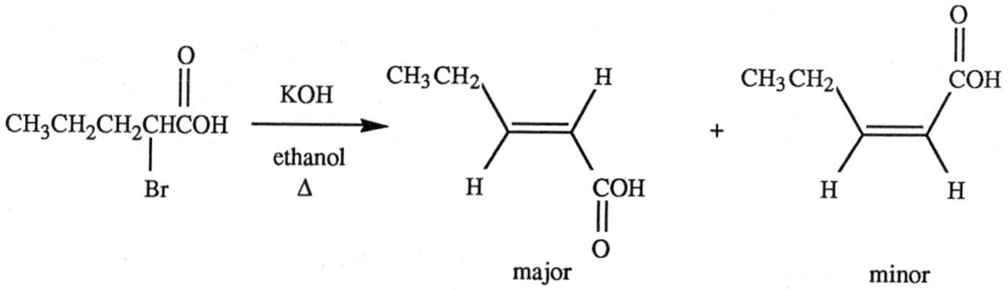

(e)

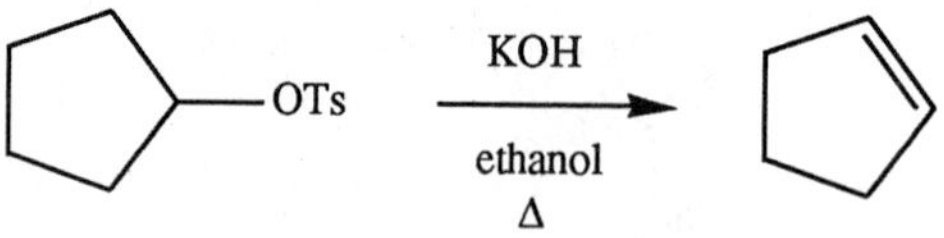

7.30

(a)

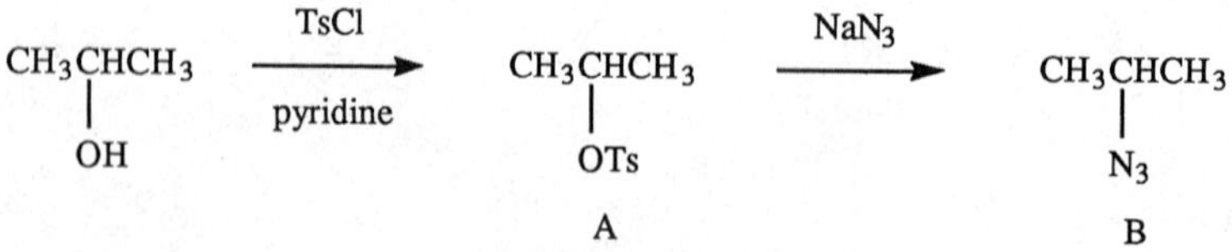

7.29 (cont)

(b) $CH_3CHCH_2CH_2CH_2CH_2Br$ (with CH_3 above) $\xrightarrow[\text{ethanol}]{\text{NaCN}}$ $CH_3CHCH_2CH_2CH_2CH_2CN$ (with CH_3 above)

C

(c) $-CH_2OH$ $\xrightarrow[\text{pyridine}]{\text{TsCl}}$ $-CH_2OTs$ $\xrightarrow[\substack{\text{ethanol} \\ \Delta}]{\text{KOH}}$ $=CH_2$

D   E

(d) $-Br$ $\xrightarrow[\substack{\text{ethanol} \\ \Delta}]{\text{KOH}}$

F

(e) $CH_3CH_2CH_2C=CHCH_2CH_2OH$ (with CH_3 above) $\xrightarrow[\text{pyridine}]{\text{TsCl}}$

$CH_3CH_2CH_2C=CHCH_2CH_2OTs$ (with CH_3 above) $\xrightarrow[\text{acetone}]{\text{NaI}}$ $CH_3CH_2CH_2C=CHCH_2CH_2I$ (with CH_3 above)

G   H

(f) $\xrightarrow[\substack{\text{ethanol} \\ \Delta}]{\text{KOH}}$ (CH₂CH₃) $+$ (CH₂CH₃)

major product   minor product
I

$\downarrow$ HI

J

(g) $\xrightarrow[\text{pyridine}]{\text{TsCl}}$ $\xrightarrow[\substack{\text{dimethyl-} \\ \text{formamide}}]{\text{NaCN}}$

K   L

7.31   The answers to parts (a) and (e) analyze the problems as shown in the Problem Solving Skills section of the chapter. The rest of the problems show only the equations, written so as to emphasize the necessity of reasoning backward from the structure of the desired product to the structure of the starting material.

(a)

1.   What functional groups are present in the starting material and the product?

The starting material is an alcohol and the product is a thioether (a sulfur atom bonded to a carbon atom on each side).

2.   How do the carbon skeletons of the two compounds compare? How many carbon atoms does each contain? Are there any rings? What are the positions of branches and functional groups on the carbon skeletons?

The product has two more carbon atoms than the starting material. These carbon atoms are on the sulfur atom. The thioether group in the product is on the same carbon atom that the hydroxyl group was attached to in the starting material.

3.   How do the functional groups change in going from starting material to product? Does the starting material have a good leaving group?

The hydroxyl group has been replaced by a thioether. The hydroxyl group is not a good leaving group but can be converted into one.

4.   Is it possible to dissect the structures of the starting material and product to see which bonds must be broken and which formed?

bond broken   bond formed

5.   Do we recognize any part of the product molecule as coming from a good nucleophile or an electrophilic addition?

The ethanethiolate anion, $CH_3CH_2S^-$, is a good nucleophile.

6.   What type of compound would be a good precursor to the product?

An alkyl halide or tosylate (a compound with a good leaving group) would be a good precursor.

7.31 (a) (cont)

7. After this last step, do we see how to get from starting material to product? If not, we need to analyze the structure obtained in step 6 by applying questions 5 and 6 to it.

We now can complete the synthesis.

The compound could also have been synthesized using an S_N1 reaction with ethanethiol in an ionizing solvent such as water, but the product of the reaction with the solvent would also form. When possible, it is better to use the S_N2 reaction for synthesis, instead of an S_N1 reaction, because there will be fewer competing side reactions to lower the yield of the desired product.

(b) $CH_3CH_2CH_2CH_2CH_2I \xleftarrow{\text{HI}} CH_3CH_2CH_2CH_2CH_2OH$

(c) $CH_3CH_2CH_2CH_2C{\equiv}CH \xleftarrow{\text{HC}{\equiv}\text{C}^-\text{Na}^+} CH_3CH_2CH_2CH_2Br$

(d)

(e)

1. What functional groups are present in the starting material and the product?

The starting material is an alcohol and the product is an amine.

2. How do the carbon skeletons of the two compounds compare? How many carbon atoms does each contain? Are there any rings? What are the positions of branches and functional groups on the carbon skeletons?

The carbon skeletons of the two compounds are the same. The compounds have opposite stereochemistry.

7.31 (f) (cont)

3. How do the functional groups change in going from starting material to product? Does the starting material have a good leaving group?

    A hydroxyl group has been replaced by an amino group. The hydroxyl group is not a good leaving group but can be converted into one.

4. Is it possible to dissect the structures of the starting material and product to see which bonds must be broken and which formed?

bond broken   bond formed

5. Do we recognize any part of the product molecule as coming from a good nucleophile or an electrophilic addition?

    The amino group comes from ammonia, NH_3, a good nucleophile.

6. What type of compound would be a good precursor to the product?

    A compound with a good leaving group would be a good precursor.

    The precursor must have the same stereochemistry as the alcohol, so that an S_N2 reaction with ammonia gives the amine with the inverted configuration.

7. After this last step, do we see how to get from starting material to product? If not, we need to analyze the structure obtained in step 6 by applying questions 5 and 6 to it.

    We know how to convert a hydroxyl group into a leaving group with retention of configuration. We can now finish the synthesis.

7.31 (cont)

(f)

$$CH_3CHCH_3 \xleftarrow{\quad CH_3CH_2S^-Na^+ \quad} CH_3CHCH_3 \xleftarrow{\quad HBr \quad} CH_3CH{=\!=}CH_2$$

$\quad\quad | \quad\quad\quad\quad\quad\quad\quad\quad\quad\quad\quad\quad | $

$\quad SCH_2CH_3 \quad\quad\quad\quad\quad\quad\quad\quad Br$

(g)

$$CH_3CH_2CH_2CH_2CH_2CH(\overset{\overset{\displaystyle O}{\|}}{C}OCH_2CH_3)_2 \xleftarrow{\quad Na^+ \; ^-CH(\overset{\overset{\displaystyle O}{\|}}{C}OCH_2CH_3)_2 \quad} CH_3CH_2CH_2CH_2CH_2Br$$

7.32 (a) $SH^- > Cl^-$

(b) $(CH_3)_3P > (CH_3)_3B$ (The boron atom has an empty orbital and is an electrophile; the phosphorus atom has nonbonding electrons.)

(c) $CH_3NH^- > CH_3NH_2$

(d) $CH_3CH_2OH > CH_3CH_2Cl$

(e) $CH_3SCH_3 > CH_3OCH_3$

7.33 Acetone is the least ionizing of the solvents listed. The reaction conditions in part (b) will favor S_N2 substitution with inversion of configuration.

7.34 (a) Rate $= k \, [CH_3CH_2CH_2CH_2Br] \, [NaOH]$

(b) The problem does not specify whether the energy diagram should show differences in free energy or enthalpy. The diagram on the next page is drawn showing the experimental energy of activation, E_a, and the enthalpy of reaction, ΔH_r.

7.34 (b) (cont)

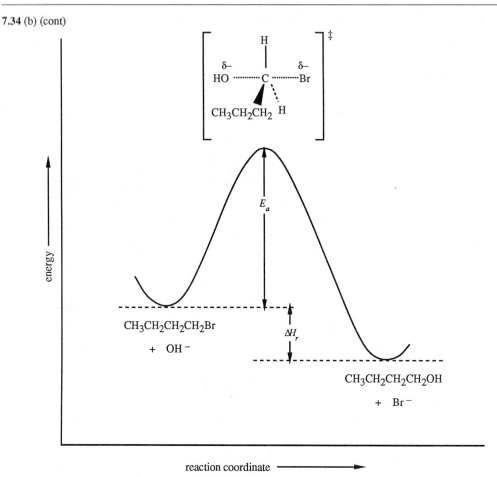

(c) The rate will be doubled.

(d) The rate will be halved.

(e) If some of the ethanol is replaced with water, there will be only a slight effect on the rate since there is only a slight build up of positive charge in the transition state on the carbon atom undergoing the substitution.

7.35   (a) Rate $= k \, [\, C_6H_5\overset{\displaystyle C_6H_5}{\underset{\displaystyle Br}{C}}CH_3 \,]$

7.35 (cont)

    (b) See the note to the energy diagram in Problem 7.34. This diagram is drawn showing free energy of activation and the free energy of reaction.

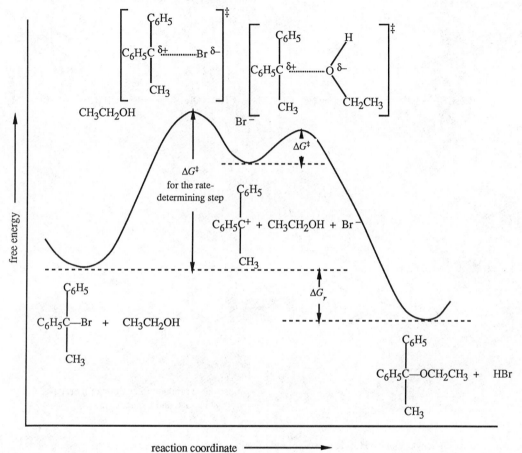

    (c) The rate will double.

    (d) Water is a better ionizing solvent than ethanol. The reaction involves the formation of a carbocation resulting in the separation of charge in the rate-determining step, so an increase in the water content of the solvent will cause an increase in the rate of the reaction by stabilizing the activated complex at the transition state.

7.36

$$CH_3CH{=}CHCH_2 \overset{\frown}{\quad} \ddot{C}l: \longrightarrow [CH_3CH{=}CH{-}\overset{+}{C}H_2 \longleftrightarrow CH_3\overset{+}{C}H{-}CH{=}CH_2] \ :\ddot{C}l:^-$$

resonance stabilized carbocation, an allylic cation, has
partial positive character at carbon 1 and carbon 3

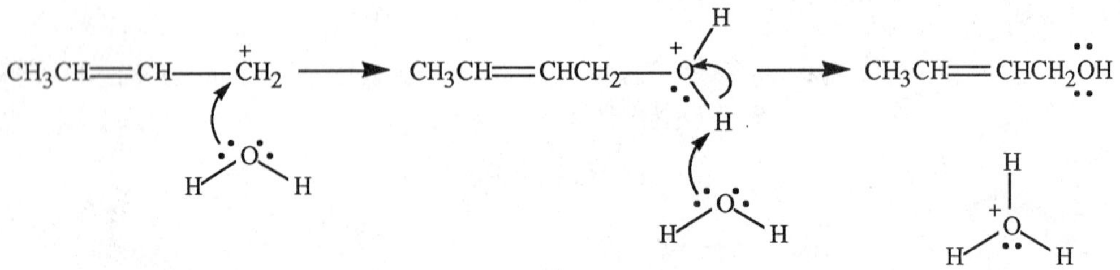

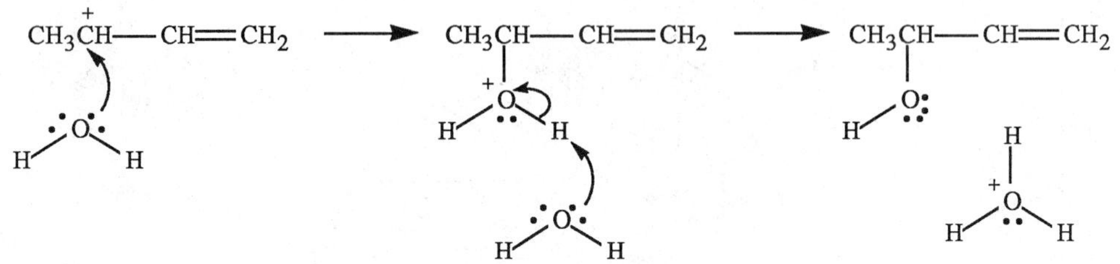

$$CH_3CH=CH-\overset{+}{C}H_2 \longrightarrow CH_3CH=CHCH_2-\overset{+}{O} \longrightarrow CH_3CH=CHCH_2OH$$

$$CH_3\overset{+}{C}H-CH=CH_2 \longrightarrow CH_3CH-CH=CH_2 \longrightarrow CH_3CH-CH=CH_2$$

7.37

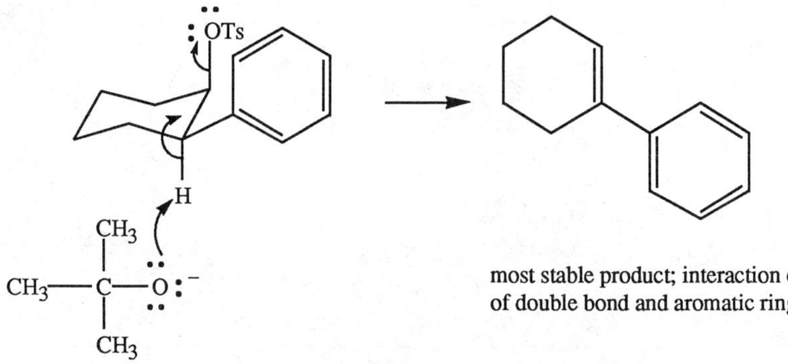

most stable product; interaction of *p* orbitals
of double bond and aromatic ring

In *cis*-2-phenylcyclohexyl tosylate a stable conformation of the cyclohexane ring has the phenyl group in the equatorial position, and the leaving group, the tosyl group, in the axial position required for the E_2 reaction to occur to give the most stable product (see p. 259 in the text).

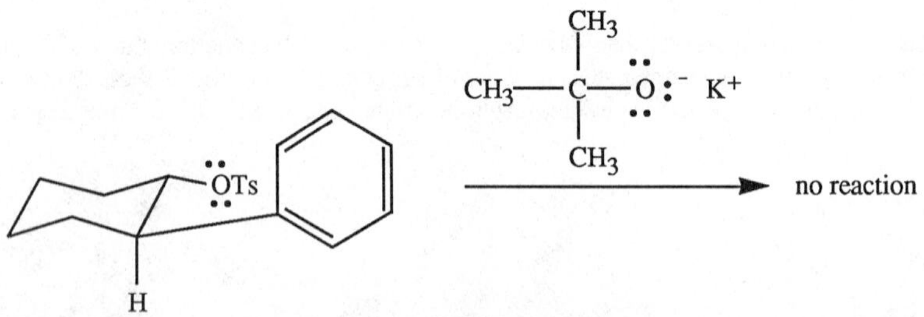

no reaction

In *trans*-2-phenylcyclohexyl tosylate the bulky phenyl ring prevents the molecule from achieving the axial conformation required for elimination, and no reaction takes place.

7.38

(a)

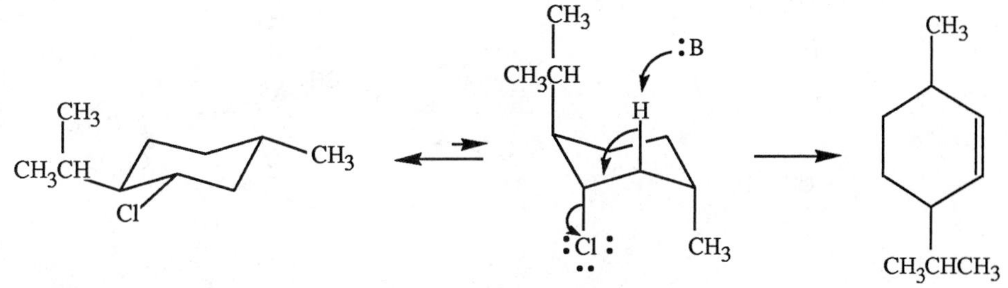

menthyl chloride;
all of the substituents are equatorial
in this chair conformation of the
cyclohexane ring

only one hydrogen is axial
to the axial chlorine

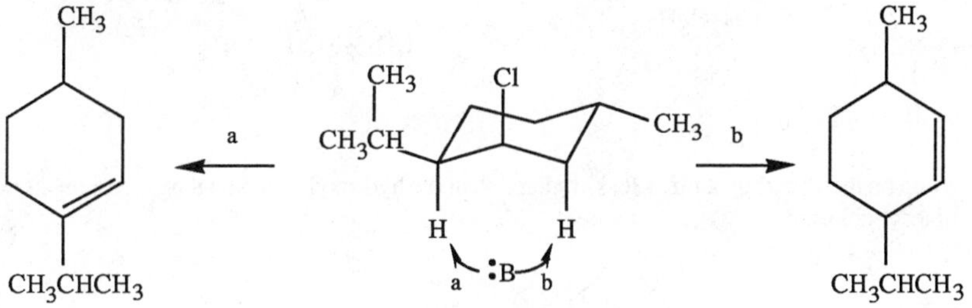

neomenthyl chloride;
two hydrogens axial to
the axial chlorine

(b) When the reaction with menthyl chloride is carried out in aqueous ethanol, a solvent which supports ionization better than pure ethanol, a planar carbocation is formed, and a hydrogen on either side of the cationic center is eliminated to give the two products observed.

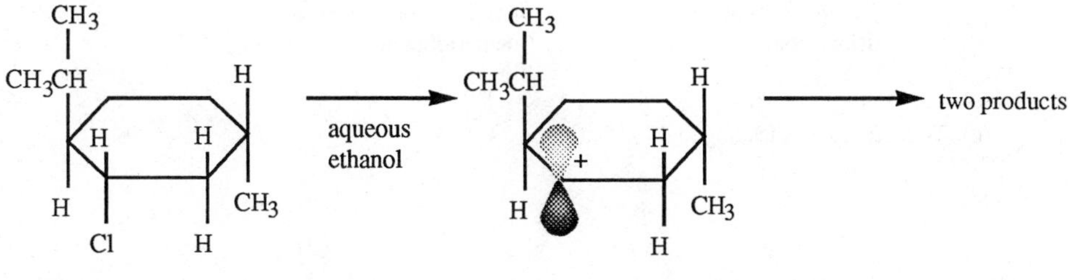

planar carbocation

7.39

(a)

$$\text{CH}_3-\overset{\text{H}}{\underset{\text{CH}_2\text{OH}}{\text{C}}}-\overset{\text{OH}}{\underset{\text{CH}_2\text{C}_6\text{H}_5}{\text{C}}}-\text{C}_6\text{H}_5 \quad \xrightarrow[\text{pyridine}]{\text{TsCl}} \quad \text{CH}_3-\overset{\text{H}}{\underset{\text{CH}_2\text{OTs}}{\text{C}}}-\overset{\text{OH}}{\underset{\text{CH}_2\text{C}_6\text{H}_5}{\text{C}}}-\text{C}_6\text{H}_5$$

$\downarrow$ (CH$_3$)$_2$NH (excess)

$$\text{Cl}^-\ (\text{CH}_3)_2\overset{+}{\underset{\text{H}}{\text{N}}}\text{CH}_2 \ \ \text{CH}_3-\overset{\text{H}}{\text{C}}-\overset{\text{OH}}{\underset{\text{CH}_2\text{C}_6\text{H}_5}{\text{C}}}-\text{C}_6\text{H}_5 \quad \xleftarrow{\text{HCl}} \quad (\text{CH}_3)_2\text{NCH}_2 \ \ \text{CH}_3-\overset{\text{H}}{\text{C}}-\overset{\text{OH}}{\underset{\text{CH}_2\text{C}_6\text{H}_5}{\text{C}}}-\text{C}_6\text{H}_5$$

(b) The hydroxyl group on the 1° carbon atom is less hindered than the hydroxyl group on the 3° carbon atom and will react faster with tosyl chloride.

7.40

(a)

$$\text{D}\overset{\text{H}}{\underset{\text{CH}_3}{\text{C}}}-\overset{\text{Br}}{\underset{\text{CH}_3}{\text{C}}}\text{H} \qquad\qquad \text{H}\overset{\text{D}}{\underset{\text{CH}_3}{\text{C}}}-\overset{\text{Br}}{\underset{\text{CH}_3}{\text{C}}}\text{H}$$

       (2*S*,3*R*)-2-bromo-3-   (2*S*,3*S*)-2-bromo-3-
       deuteriobutane   deuteriobutane

(b) Transition states leading to (*E*)-2-butene:

$$\overset{\delta-}{\text{Br}}$$
$$\text{H}\overset{\text{CH}_3}{\text{C}}=\overset{\text{CH}_3}{\text{C}}\text{H}$$
$$\text{CH}_3\text{CH}_2\overset{\delta-}{\text{O}}\cdots\cdots\text{D}$$

from the (2*S*,3*R*)-isomer, leads to loss of the deuterium atom

$$\overset{\delta-}{\text{Br}}$$
$$\text{D}\overset{\text{CH}_3}{\text{C}}=\overset{\text{CH}_3}{\text{C}}\text{H}$$
$$\text{CH}_3\text{CH}_2\overset{\delta-}{\text{O}}\cdots\cdots\text{H}$$

from the (2*S*,3*S*)-isomer, leads to retention of the deuterium atom

7.40 (b) (cont)

Transition states leading to (Z)-2-butene:

from the (2S,3R)-isomer,
leads to retention of the
deuterium atom

from the (2S,3S)-isomer,
leads to loss of the
deuterium atom

7.41   (a) The similarity of the products from the two reactions suggests that both reactions proceed through the same intermediate, a carbocation.

(b) The fact that they proceed at different rates suggests that the formation of the carbocation is the rate-determining step. The higher rate for the first reaction suggests that chloride ion is a better leaving group than dimethyl sulfide.

7.42

thiocyanate ion

cyanate ion

In the thiocyanate ion, the electrons on the sulfur are more polarizable than those on the nitrogen, and that end of the ion is the better nucleophile. In the cyanate ion, the electrons on the nitrogen are more polarizable than those on the oxygen, and the nitrogen is a better nucleophile.

7.43

1.   What functional groups are present in the starting material and the product?

The starting material has two hydroxyl groups. One hydroxyl group is on a primary carbon atom and the other is on a tertiary carbon atom. The product contains a cyano group on the primary carbon atom and a hydroxyl group on the tertiary carbon atom.

7.43 (cont)

2. How do the carbon skeletons of the two compounds compare? How many carbon atoms does each contain? Are there any rings? What are the positions of branches and functional groups on the carbon skeletons?

   There is one more carbon atom in the product. It has been added to carbon 1 of the starting material and is part of the cyano group.

3. How do the functional groups change in going from starting material to product? Does the starting material have a good leaving group?

   The primary hydroxyl group has been replaced by a cyano group. There is no good leaving group in the starting material but a hydroxyl group can be converted into one.

4. Is it possible to dissect the structure to see which bonds must have been broken and which formed?

5. Do we recognize any parts of the product molecule as coming from a good nucleophile or an electrophilic addition?

   The cyano group, —CN, comes from CN⁻, a good nucleophile.

6. What type of compound would be a good precursor to the product?

   Something with a good leaving group, such as water, a halide, or a tosylate group, where the hydroxyl group is would be a good precursor. We cannot use strong acid to convert the hydroxyl group into a better leaving group because the cyanide anion is a better base than the alcohol and would deprotonate the oxonium ion.

7. Do we now see how to get from starting material to product, or do we need to repeat steps 5 and 6?

   The precursor containing the tosylate group can easily be made from the starting materical and tosyl chloride. A primary alcohol will react faster with tosyl chloride than a tertiary alcohol (see Problem 7.39). We can now complete the problem.

7.43 (cont)

Complete correct answer:

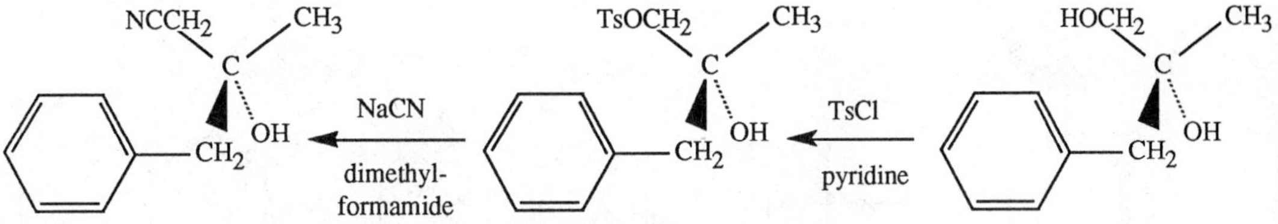

7.44

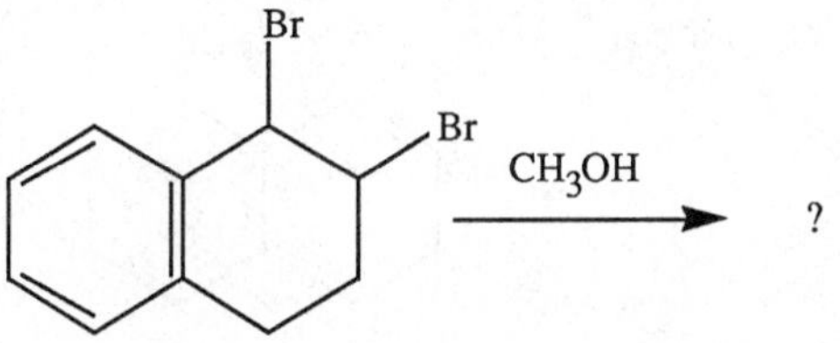

1. To what functional group class does the reactant belong? What is the electronic character of the functional group?

    The reactant contains two bromine atoms. The carbon atoms to which the bromine atoms are attached have partial positive characters and are thus electrophilic. One of the bromine atoms is on a 2° carbon atom, the other bromine atom is on a 2° and benzylic carbon atom.

2. Does the reactant have a good leaving group?

    Yes; the bromine atom will leave as the weakly basic bromide anion.

3. What reagents are present? Are they good acids, bases, nucleophiles, or electrophiles? Is there an ionizing solvent present?

    Methanol is the reagent and is a weak nucleophile. It is also an ionizing solvent.

4. What is the most likely first step for the reaction: protonation or deprotonation, attack by a nucleophile, or attack by an electrophile?

    A benzylic halide ionizes in an ionizing solvent to form a resonance-stabilized benzylic carbocation.

7.44 (cont)

4. (cont)

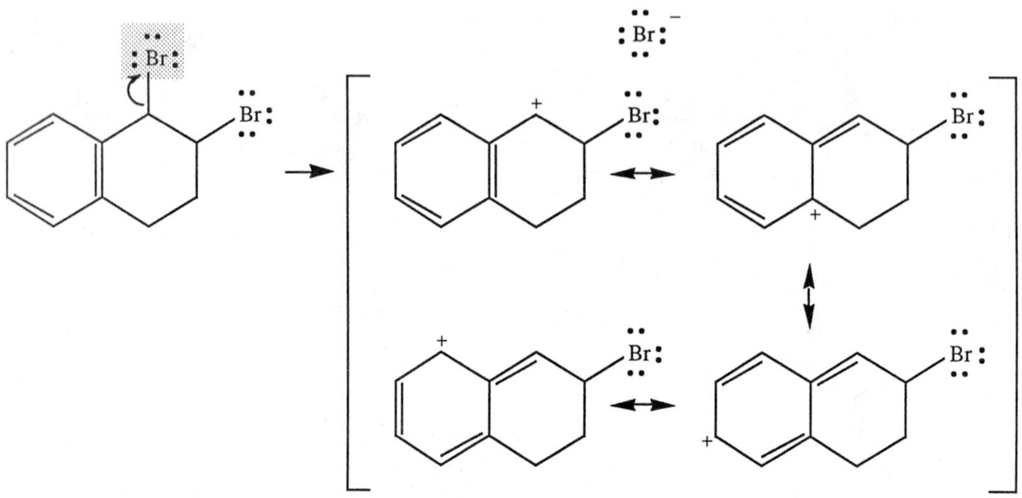

This is the slow step of an S_N1 reaction. The ionization of the benzylic bromide is faster than the reaction of the 2° halide with the weak nucleophile, methanol.

5. What are the properties of the species present in the reaction mixture after the first step? What is likely to happen next?

The carbocation is a strong electrophile and will react with methanol in the second, fast step of an S_N1 reaction.

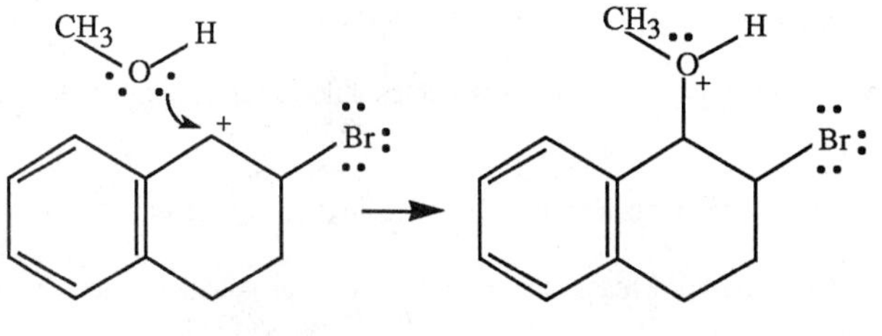

The species created in the second step is an oxonium ion, a strong acid. Proton transfer to another molecule of methanol will take place.

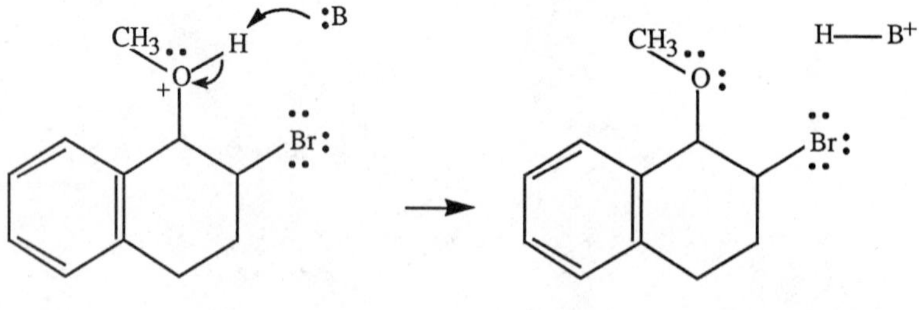

7.44 (5) (cont)

(A side reaction is loss of a proton from the carbocation to give an alkene.)

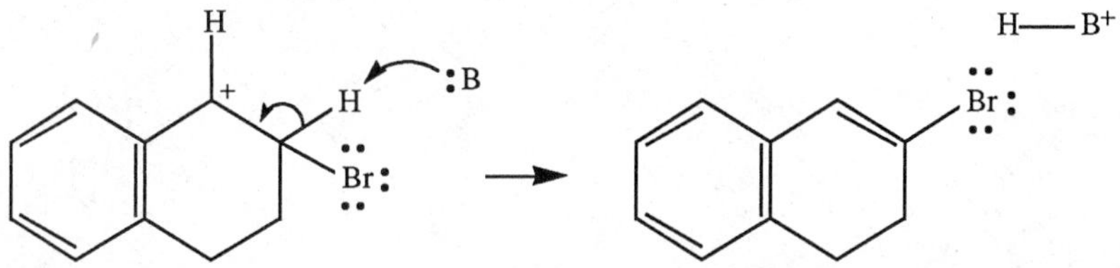

6. What is the stereochemistry of the reaction?

There are two stereocenters in the starting material, therefore it exists as a mixture of four stereoisomers. In an S_N1 reaction, a planar carbocation is formed and the carbon undergoing the ionization loses its chirality. No bonds are broken at the other stereocenter, and the configuration is retained. Assuming that we start with a mixture of stereoisomers, we will get a mixture of stereoisomers in the product. The conversion of just one of the possible stereoisomers into the product is shown below.

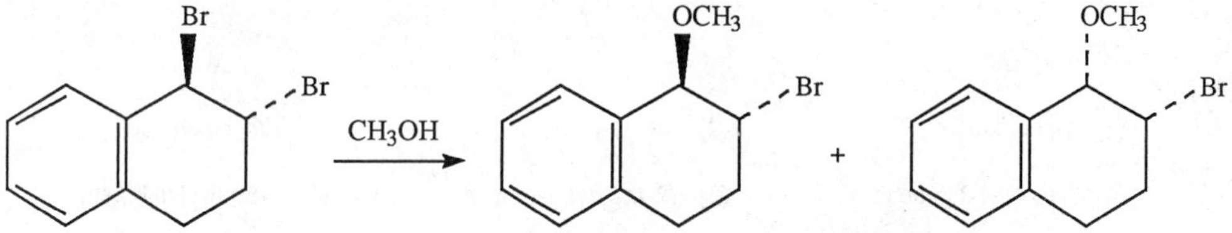

Because stereochemistry is not shown for the starting material, we do not show the stereochemistry of the product either in our answer.

Complete correct answer:

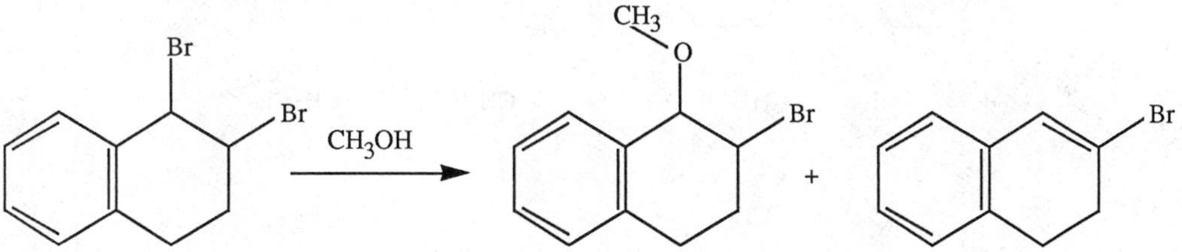

8

Alkenes

8.1 C$_5$H$_{10}$

(1)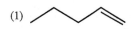

(2)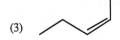

(3)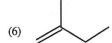

(4)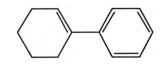

(5)

(6)

8.2

(1) 1-pentene

(2) (*E*)-2-pentene

(3) (*Z*)-2-pentene

(4) 3-methyl-1-butene

(5) 2-methyl-2-butene

(6) 1-methyl-1-butene

8.3 (a)

(b)

CH$_3$

CH$_3$

(c)

CH$_3$ H

C=C

CH$_3$

(d)

CH$_3$CH$_2$ CH$_2$Cl

C=C

Cl   CH$_3$

8.4 No. Each carbon atom of the double bond must have two different substituents for this type of isomerism to be possible.

192

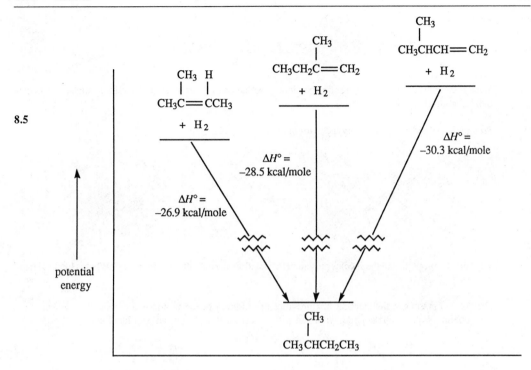

8.5

8.6   The Problem-Solving Skills questions will be used to answer part (a). Only the equations will be shown for the rest of the problem.

(a) $\xrightarrow[\text{H}_2\text{SO}_4]{\text{H}_2\text{O}}$?

1. To what functional group class does the reactant belong? What is the electronic character of the functional group?

   The reactant is an alkene. The double **bond** of the alkene is a nucleophile.

2. Does the reactant have a good leaving group?

   No.

3. What reagents are present? Are they good acids, bases, nucleophiles, or electrophiles? Is there an ionizing solvent present?

   The reagent above the arrow, water, is a weak nucleophile and base. Sulfuric acid, the catalyst under the arrow, is a strong acid. The reactive species in a solution of sulfuric acid in water are actually the hydronium ion, water, and the hydrogen sulfate anion. Water is an ionizing solvent.

8.6 (cont)

(a) (cont) H_2O + H_2SO_4 ⇌ H_3O^+ + HSO_4^-

4. What is the most likely first step for the reaction: protonation or deprotonation, ionization, attack by a nucle-
    ophile, or attack by an electrophile?

    Protonation of the double bond by the hydronium ion is the first step.

5. What are the properties of the species present in the reaction mixture after the first step? What is likely to hap-
    pen next?

    The species present in the reaction mixture after the addition of the proton is a carbocation, which is an
    electrophile. Water, a nucleophile, will react with the carbocation to give an oxonium ion.

    The oxonium ion is a strong acid and will transfer a proton to the excess water.

6. What is the stereochemistry of the reaction?

    The product has no stereocenter and will have no stereochemistry.

    Complete correct answer:

8.6 (cont)

(b)

$$\xrightarrow[\text{H}_2\text{SO}_4]{\text{H}_2\text{O}}$$

(c) $CH_3CH_2CH_2CH_2CH=CH_2$ $\xrightarrow[\text{H}_2\text{SO}_4]{\text{H}_2\text{O}}$ $CH_3CH_2CH_2CH_2CHCH_3$
                                                                                            |
                                                                                           OH

racemic mixture

(d)

$$\underset{\underset{\text{OH}}{|}}{\overset{\overset{\text{CH}_3}{|}}{CH_3CCH_3}} \quad \xrightarrow[\Delta]{\text{H}_3\text{PO}_4} \quad \underset{}{\overset{\overset{\text{CH}_3}{|}}{CH_3C=CH_2}}$$

(e) $\underset{\underset{\text{OH}}{|}}{CH_3CHCH_3}$ $\xrightarrow[\Delta]{\text{H}_3\text{PO}_4}$ $CH_3CH=CH_2$

(f) $CH_2=CH_2$ $\xrightarrow[\text{H}_2\text{SO}_4]{\text{H}_2\text{O}}$ CH_3CH_2OH

(g) $\underset{}{\overset{\overset{\text{CH}_3}{|}}{CH_3C}}=CHCH_2CH_3$ $\xrightarrow[\text{H}_2\text{SO}_4]{\text{H}_2\text{O}}$ $\underset{\underset{\text{OH}}{|}}{\overset{\overset{\text{CH}_3}{|}}{CH_3CHCH_2CH_2CH_3}}$

(h)

$$\xrightarrow[\Delta]{\text{H}_3\text{PO}_4}$$

Concept Map 8.1   Reactions that form carbocations.

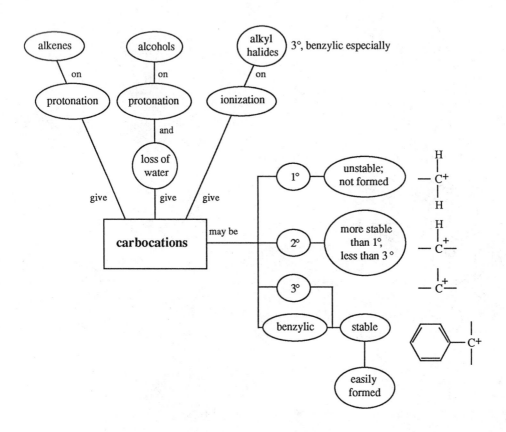

Concept Map 8.2   Reactions of carbocations.

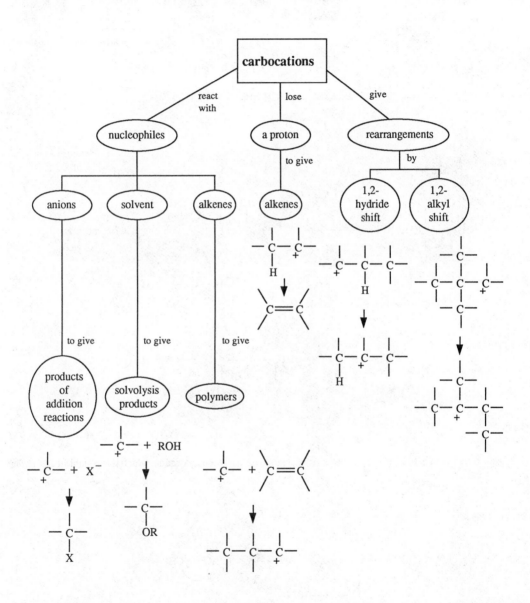

8.7

$$CH_3\overset{\overset{\displaystyle CH_3}{|}}{C}=\overset{\overset{\displaystyle CH_3}{|}}{C}CH_3 \longrightarrow CH_3\overset{\overset{\displaystyle CH_3}{|}}{\underset{+}{C}}-\overset{\overset{\displaystyle CH_3}{|}}{\underset{H}{C}}CH_3$$

$$H-\overset{\cdot\cdot}{\underset{\cdot\cdot}{O}}-PO_3H_2 \qquad\qquad {}^-\overset{\cdot\cdot}{\underset{\cdot\cdot}{O}}-PO_3H_2$$

$$\overset{CH_2}{\underset{H}{|}}-\overset{\overset{\displaystyle CH_3}{|}}{\underset{+}{C}}-\overset{\overset{\displaystyle CH_3}{|}}{\underset{H}{C}}CH_3 \longrightarrow CH_2=\overset{\overset{\displaystyle CH_3}{|}}{C}-\overset{\overset{\displaystyle CH_3}{|}}{\underset{H}{C}}CH_3$$

$${}^-\overset{\cdot\cdot}{\underset{\cdot\cdot}{O}}-PO_3H_2 \qquad\qquad H-\overset{\cdot\cdot}{\underset{\cdot\cdot}{O}}-PO_3H_2$$

$$CH_3\overset{\overset{\displaystyle CH_3}{|}}{\underset{+}{C}}-\overset{\overset{\displaystyle CH_3}{|}}{\underset{H}{C}}CH_3 \longrightarrow CH_3\overset{\overset{\displaystyle CH_3}{|}}{C}=\overset{\overset{\displaystyle CH_3}{|}}{C}CH_3 \quad \text{(starting material regenerated)}$$

$${}^-\overset{\cdot\cdot}{O}-PO_3H_2 \qquad\qquad H-\overset{\cdot\cdot}{\underset{\cdot\cdot}{O}}-PO_3H_2$$

$$CH_3-\overset{\overset{\displaystyle CH_3}{|}}{\underset{\underset{\displaystyle CH_3}{|}}{\underset{+}{C}}}-\overset{\overset{\displaystyle H}{|}}{C}CH_3 \longrightarrow CH_3\overset{\overset{\displaystyle CH_3}{|}}{\underset{\underset{\displaystyle CH_3}{|}}{C}}-\overset{\overset{\displaystyle H}{|}}{\underset{+}{\underset{\underset{\displaystyle H}{|}}{C}}}CH_2 \longrightarrow CH_3\overset{\overset{\displaystyle CH_3}{|}}{\underset{\underset{\displaystyle CH_3}{|}}{C}}-\overset{\overset{\displaystyle H}{|}}{C}=CH_2$$

$${}^-\overset{\cdot\cdot}{\underset{\cdot\cdot}{O}}-PO_3H_2 \qquad\qquad H-\overset{\cdot\cdot}{\underset{\cdot\cdot}{O}}-PO_3H_2$$

Note that very little of this product is seen. It arises from a secondary cation, which is less stable than a tertiary cation.

8.8

$$\overset{\overset{\displaystyle CH_3}{|}}{\underset{\underset{\displaystyle CH_3}{|}}{CH_3CCH_2}}-\overset{\cdot\cdot}{\underset{\cdot\cdot}{O}}\overset{H}{:} \longrightarrow CH_3\overset{\overset{\displaystyle CH_3}{|}}{\underset{\underset{\displaystyle CH_3}{|}}{C}}-CH_2\overset{+}{\underset{}{\overset{H}{O}}}:\overset{}{\underset{\displaystyle H}{}}$$

$$H-\overset{\cdot\cdot}{\underset{\cdot\cdot}{Br}}: \qquad\qquad :\overset{\cdot\cdot}{\underset{\cdot\cdot}{Br}}:^-$$

     ↓ ionization with
     1,2-methyl shift

8.8 (cont)

8.9   Note: H—B$^+$ stands for a Brønsted-Lowry acid and :B stands for the conjugate base of the acid. They are used in mechanisms when more than one acid can transfer a proton and more than one base can accept a proton.

8.9 (cont)

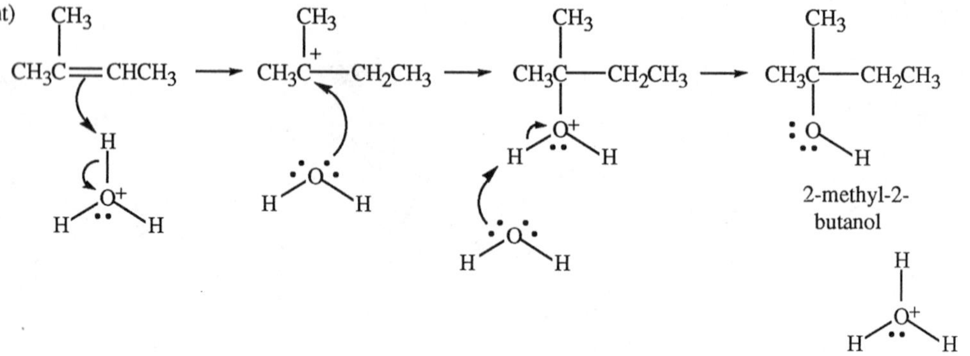

2-methyl-2-butanol

8.10   1.   Attack at carbon 2 of the enantiomeric bromonium ions derived from (Z)-2-pentene.

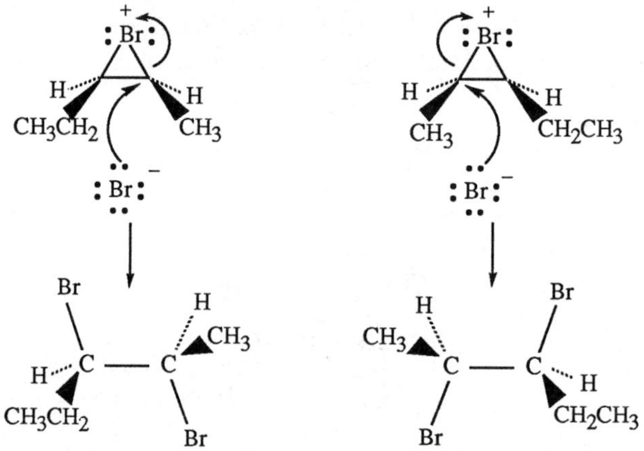

(2R, 3R)-2,3-dibromopentane   (2S, 3S)-2,3-dibromopentane

This is the same mixture as that obtained by attack at carbon 3 shown on p. 293 in the text.

2.   Attack at carbon 2 of the enantiomeric bromonium ions derived from (E)-2-pentene.

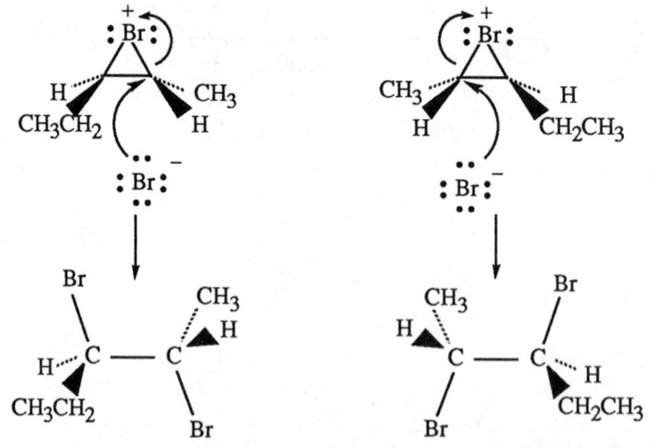

(2S, 3R)-2,3-dibromopentane   (2R, 3S)-2,3-dibromopentane

8.10 (cont)

Attack at carbon 3 of the enantiomeric bromonium ions derived from (*E*)-2-pentene.

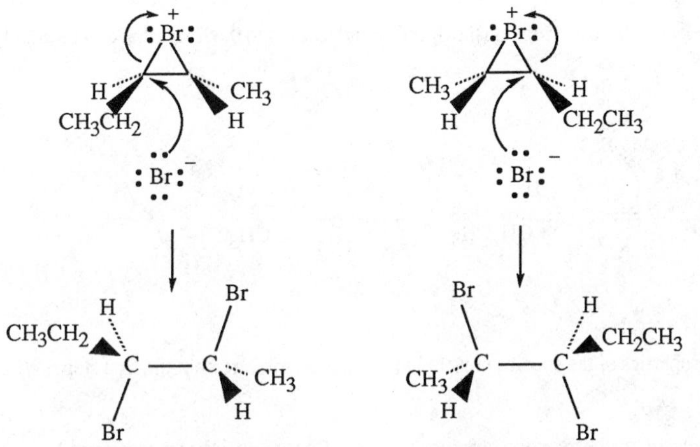

(2*R*, 3*S*)-2,3-dibromopentane   (2*S*,3*R*)-2,3-dibromopentane

This is the same mixture as that obtained by attack at carbon 2.

8.11   Part (a) will be analyzed in detail. Only the equations will be shown for the rest of the problem.

(a)

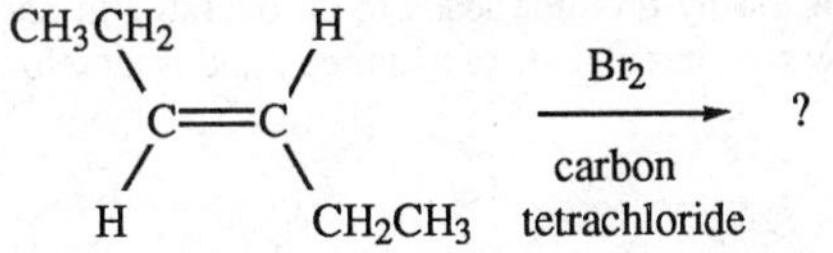

1. To what functional group class does the reactant belong? What is the electronic character of the functional group?

    The reactant is an alkene. The double bond of the alkene is a nucleophile.

2. Does the reactant have a good leaving group?

    No.

3. What reagents are present? Are they good acids, bases, nucleophiles, or electrophiles? Is there an ionizing solvent present?

    Bromine, above the arrow, is an electrophile. The solvent, carbon tetrachloride, is not an ionizing solvent.

8.11 (cont)

4. What is the most likely first step for the reaction: protonation or deprotonation, ionization, attack by a nucleophile, or attack by an electrophile?

Attack by the electrophilic bromine on the double bond leading to the formation of a bromonium ion is the first step.

5. What are the properties of the species present in the reaction mixture after the first step? What is likely to happen next?

The species present in the reaction mixture are the bromonium ion, an electrophile, and bromide ion, a nucleophile. The bromonium ion will react with bromide ion to give the dibromide product.

6. What is the stereochemistry of the reaction?

The opening of the bromonium ion by bromide leads to an overall anti addition and the creation of two stereocenters. The product, however, has a plane of symmetry and is therefore a meso compound.

Complete correct answer:

(3*R*,4*S*)-3,4-dibromohexane
meso-3,4-dibromohexane

8.11 (cont)

(b)

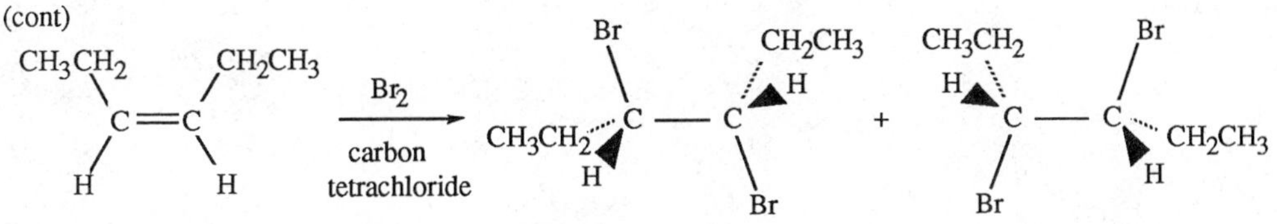

racemic mixture

(c)

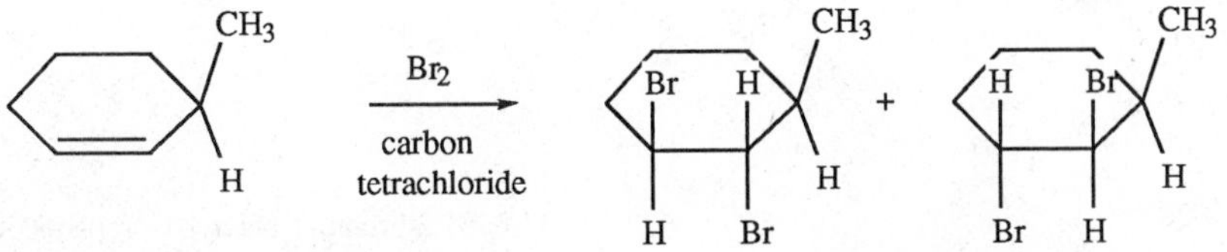

diasteromers

(d)

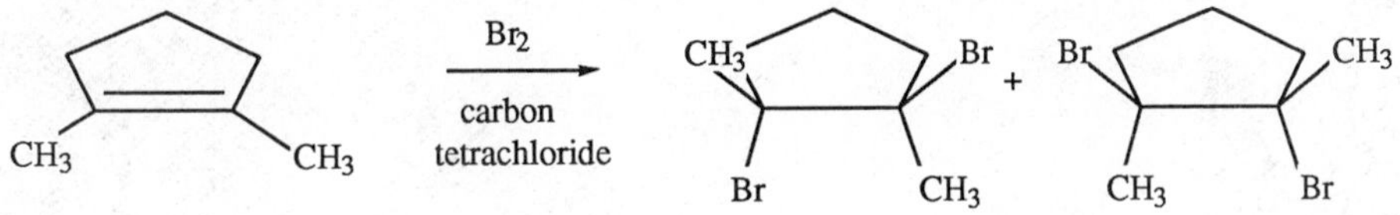

enantiomers

8.12   Formation of product from an unsymmetrical bromonium ion intermediate.

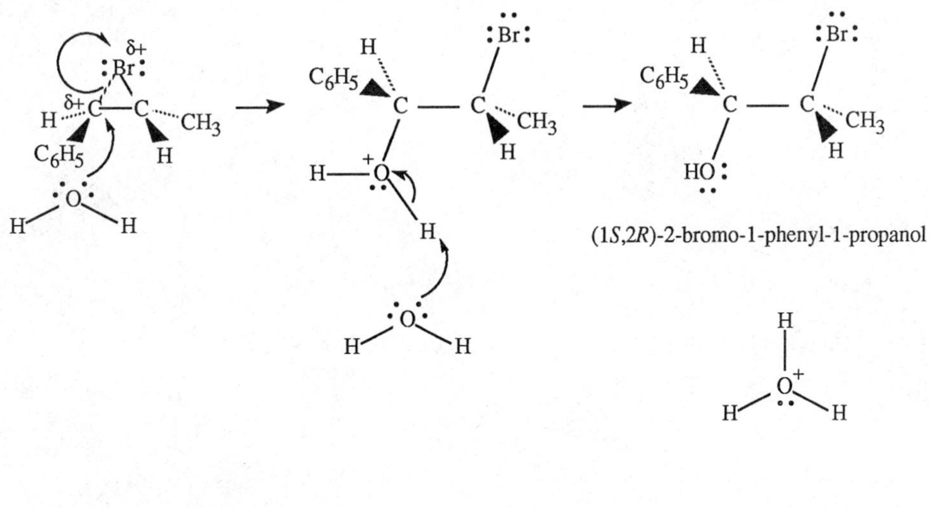

(1*S*,2*R*)-2-bromo-1-phenyl-1-propanol

8.12 (cont)

(1*R*,2*S*)-2-bromo-1-phenyl-1-propanol

Formation of products from a carbocation.

(1*S*, 2*S*)-2-bromo-1-
phenyl-1-propanol

(1*R*, 2*S*)-2-bromo-1-
phenyl-1-propanol

8.12 (cont)

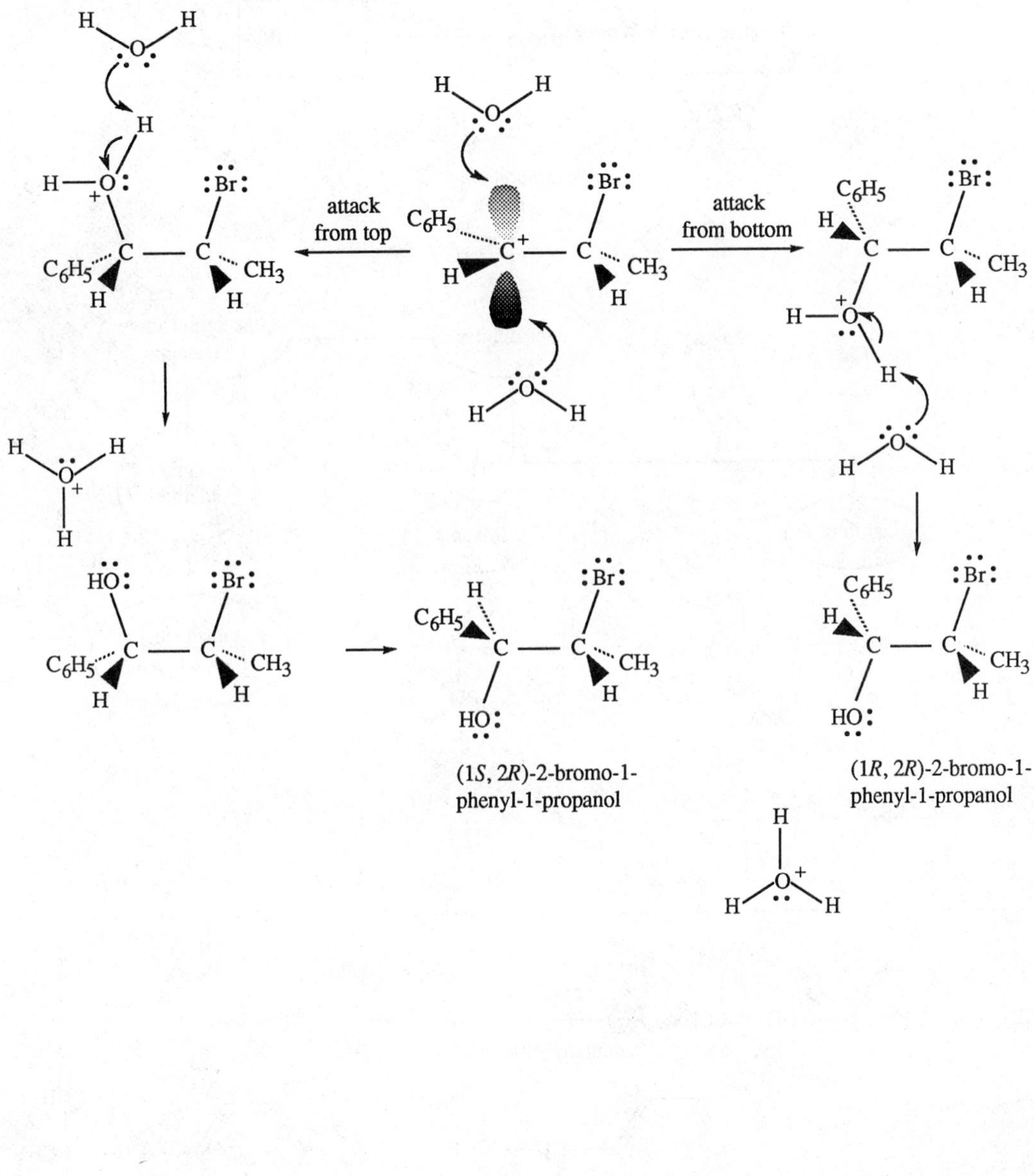

attack
from top

attack
from bottom

(1*S*, 2*R*)-2-bromo-1-
phenyl-1-propanol

(1*R*, 2*R*)-2-bromo-1-
phenyl-1-propanol

Concept Map 8.3   The addition of bromine to alkenes.

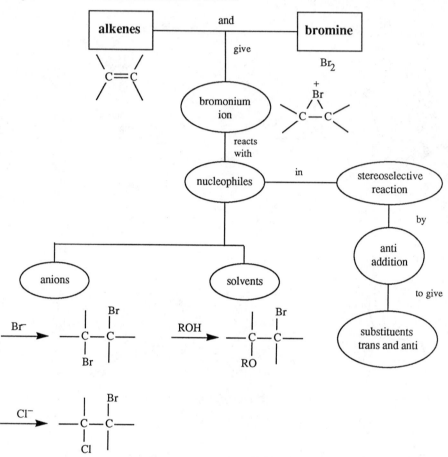

8.13

mixture of enantiomers   mixture of enantiomers

Benzylic and tertiary carbocations are both relatively stable intermediates. Therefore, both the benzylic carbon atom and the tertiary carbon atom have about the same amount of partial charge in the bromonium ion. Attack by the nucleophile can occur at either carbon.

8.14

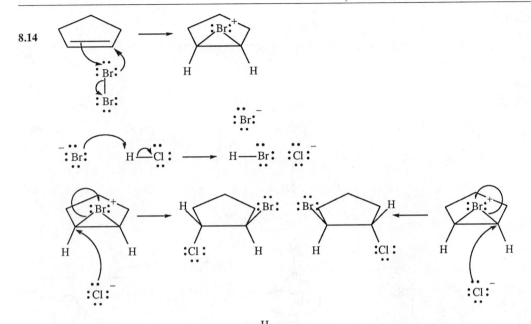

8.15

$$H : B : H$$
with H above B

shape is trigonal planar;
hybridization of the
boron atom is sp^2

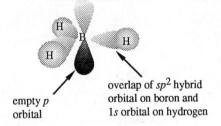

empty *p*
orbital

overlap of sp^2 hybrid
orbital on boron and
1*s* orbital on hydrogen

8.16

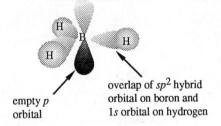

$\xrightarrow{\quad BH_3 \quad}$

8.16 (cont)

$$CH_3(CH_2)_7CH{=}CH_2 \xrightarrow{BH_3}$$

$$\underset{\displaystyle \overset{|}{CH_2CH_2(CH_2)_7CH_3}}{CH_3(CH_2)_7CH_2CH_2BCH_2CH_2(CH_2)_7CH_3}$$

$$\underset{\displaystyle CH_3}{\overset{\displaystyle CH_3}{\underset{|}{CH_3C}}}-CH{=}CH_2 \xrightarrow{BH_3}$$

$$CH_3CCH_2CH_2\!-\!B\!-\!CH_2CH_2CCH_3$$

with the groups CH_3, $CH_2CH_2CCH_3$, CH_3, and CH_3

8.17 (a) $CH_3CH_2CH_2\overset{\displaystyle CH_3}{\underset{|}{C}}{=}CH_2$ $\xrightarrow{\; B\text{-H}\;}$ $CH_3CH_2CH_2\overset{\displaystyle CH_3}{\underset{|}{C}}HCH_2{-}B$

(b) $\langle\!\!\!\!\!\!\!\!\!\!\!\!\!\!\text{phenyl}\rangle{-}CH{=}CH_2$ $\xrightarrow{\; B\text{-H}\;}$ $\langle\text{phenyl}\rangle{-}CH_2CH_2{-}B$

(c) $CH_3\overset{\displaystyle CH_3}{\underset{|}{C}}CH_3$, $\underset{H}{\overset{CH_3\;\;\;\;CH_3}{C{=}C}}\underset{H}{}$ $\xrightarrow{\; B\text{-H}\;}$ $CH_3\overset{\displaystyle CH_3}{\underset{\displaystyle CH_3}{\underset{|}{\overset{|}{C}}}}{-}CH_2{-}\underset{B}{CHCH_3}$

(d) $\underset{CH_3}{\overset{CH_3}{C}}{=}\underset{CH_3}{\overset{CH_3}{C}}$ $\xrightarrow{\; B\text{-H}\;}$ $CH_3CH{-}\underset{B}{\overset{CH_3\;\;\;\;\;CH_3}{\underset{|}{\overset{|}{C}}CH_3}}$

8.17 (cont)

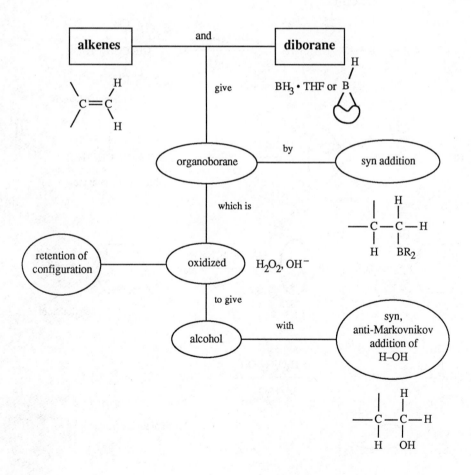

- -

Concept Map 8.4   The hydroboration reaction.

8.18

(a)

$$\text{tricyclopentylborane} \xrightarrow[\text{H}_2\text{O}]{\text{H}_2\text{O}_2,\ \text{NaOH}} \text{cyclopentanol}$$

(b)

CH₂CH₂(CH₂)₇CH₃
|
$$\text{CH}_3(\text{CH}_2)_7\text{CH}_2\text{CH}_2\text{BCH}_2\text{CH}_2(\text{CH}_2)_7\text{CH}_3 \xrightarrow[\text{H}_2\text{O}]{\text{H}_2\text{O}_2,\ \text{NaOH}} \text{CH}_3(\text{CH}_2)_8\text{CH}_2\text{OH}$$

1-decanol

(c)

$$\xrightarrow[\text{H}_2\text{O}]{\text{H}_2\text{O}_2,\ \text{NaOH}}$$

3,3-dimethyl-1-butanol

(a')

CH₃
|
$$\text{CH}_3\text{CH}_2\text{CH}_2\text{CHCH}_2\text{-B} \xrightarrow[\text{H}_2\text{O}]{\text{H}_2\text{O}_2,\ \text{NaOH}}$$

CH₃
|
CH₃CH₂CH₂CHCH₂OH

2-methyl-1-pentanol

(b')

$$\text{C}_6\text{H}_5\text{-CH}_2\text{CH}_2\text{-B} \xrightarrow[\text{H}_2\text{O}]{\text{H}_2\text{O}_2,\ \text{NaOH}} \text{C}_6\text{H}_5\text{-CH}_2\text{CH}_2\text{OH}$$

2-phenylethanol

(c')

CH₃
|
$$\text{CH}_3\text{C-CH}_2\text{-CHCH}_3 \xrightarrow[\text{H}_2\text{O}]{\text{H}_2\text{O}_2,\ \text{NaOH}}$$
|
CH₃

CH₃
|
CH₃C---CH₂---CHCH₃
| |
CH₃   OH

4,4-dimethyl-2-pentanol

8.18 (cont)

(d')

$$CH_3CH-\underset{|}{\overset{|}{C}}CH_3 \quad \xrightarrow[\text{H}_2\text{O}]{\text{H}_2\text{O}_2,\ \text{NaOH}} \quad$$

with CH₃ groups and B bonded below.

$$CH_3CH-\overset{CH_3}{\underset{CH_3}{\overset{|}{C}}}CH_3$$
$$\underset{OH}{|}$$

2,3-dimethyl-2-butanol

(e')

$$CH_3CH_2CH_2CHCH_2CH_3 \quad \xrightarrow[\text{H}_2\text{O}]{\text{H}_2\text{O}_2,\ \text{NaOH}} \quad CH_3CH_2CH_2CHCH_2CH_3$$
$$\underset{OH}{|}$$

3-hexanol

8.19

(a)

$$CH_3\overset{CH_3}{\underset{CH_3}{\overset{|}{C}}}CH=CH_2 \quad \xrightarrow[\text{Ni}]{\text{H}_2} \quad CH_3\overset{CH_3}{\underset{CH_3}{\overset{|}{C}}}CH_2CH_3$$

A

(b)

$$CH_3\overset{CH_3}{\underset{CH_3}{\overset{|}{C}}}=CCH_3 \quad \xrightarrow[\text{Ni}]{\text{H}_2} \quad CH_3\overset{CH_3}{\underset{CH_3}{\overset{|}{C}}}HCHCH_3$$

B

(c)

$$CH_3CH_2\overset{CH_3}{\underset{OH}{\overset{|}{C}}}CH_2CH_3 \quad \xrightarrow[\Delta]{\text{H}_2\text{SO}_4} \quad CH_3CH=\overset{CH_3}{\overset{|}{C}}CH_2CH_3 \quad + \quad CH_3CH_2\overset{CH_2}{\overset{\|}{C}}CH_2CH_3 \quad \xrightarrow[\text{Ni}]{\text{H}_2}$$

C   D

$$CH_3CH_2\overset{CH_3}{\underset{}{\overset{|}{C}}}HCH_2CH_3$$

E

(d)

$$CH_3\overset{CH_3}{\underset{CH_3}{\overset{|}{C}}}HC=CH_2 \quad \xrightarrow[\text{Ni}]{\text{H}_2} \quad CH_3\overset{CH_3}{\underset{CH_3}{\overset{|}{C}}}HCHCH_3$$

F

8.19 (cont)

(e)

$$\underset{\underset{\text{OH}}{|}}{\overset{\underset{\text{CH}_3}{|}}{CH_3CCH_2CH_2CH_3}} \xrightarrow[\Delta]{H_3PO_4} \underset{G}{\overset{\underset{\text{CH}_3}{|}}{CH_3C}\!\!=\!\!CHCH_2CH_3} \; + \; \underset{H}{\overset{\underset{\text{CH}_3}{|}}{CH_2}\!\!=\!\!CCH_2CH_2CH_3} \xrightarrow[\text{Ni}]{H_2}$$

$$\underset{I}{\overset{\underset{\text{CH}_3}{|}}{CH_3CHCH_2CH_2CH_3}}$$

Concept Map 8.5   Catalytic hydrogenation reactions.

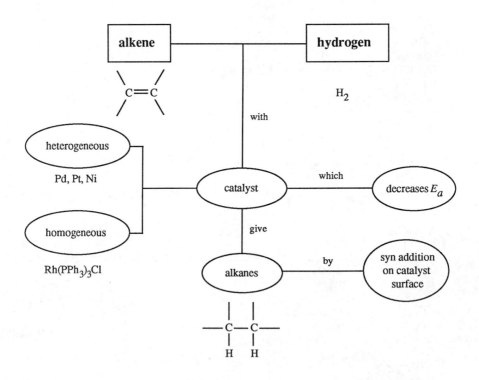

8.20

(a) $CH_3CH_2CH\!\!=\!\!CHCH_2CH_3$ $\xrightarrow[\text{Pt}]{H_2}$ $CH_3CH_2CH_2CH_2CH_2CH_3$

(b) $\underset{\displaystyle\quad}{CH_3CH_2CH_2\overset{\displaystyle CH_3}{\overset{|}{C}}\!\!=\!\!CH_2}$ $\xrightarrow[\substack{RhCl[P(C_6H_5)_3]_3 \\ \text{benzene}}]{H_2}$ $CH_3CH_2CH_2\overset{\displaystyle CH_3}{\overset{|}{C}}HCH_3$

(c) $\xrightarrow[\text{Pd/C}]{H_2}$

(d) $\xrightarrow[\text{Pd/C}]{H_2\ (\text{excess})}$

(e) $CH_3\underset{\displaystyle CH_3}{\overset{\displaystyle CH_3}{\overset{|}{\underset{|}{C}}}}CH\!\!=\!\!CH_2$ $\xrightarrow[\substack{RhCl[P(C_6H_5)_3]_3 \\ \text{benzene}}]{H_2}$ $CH_3\underset{\displaystyle CH_3}{\overset{\displaystyle CH_3}{\overset{|}{\underset{|}{C}}}}CH_2CH_3$

(f) $\xrightarrow[\substack{RhCl[P(C_6H_5)_3]_3 \\ \text{benzene}}]{H_2}$

(g) $CH_3CH_2CH_2CH_2CH_2CH\!\!=\!\!CH_2$ $\xrightarrow[\substack{RhCl[P(C_6H_5)_3]_3 \\ \text{benzene}}]{H_2}$ $CH_3CH_2CH_2CH_2CH_2CH_2CH_3$

8.21

(a) $CH_3\overset{\displaystyle CH_3}{\overset{|}{C}}\!\!=\!\!CHCH_3$ $\xrightarrow[\substack{\text{Zn} \\ H_2O}]{O_3}$ $CH_3\overset{\displaystyle O}{\overset{\|}{C}}CH_3$ + $CH_3\overset{\displaystyle O}{\overset{\|}{C}}H$

(b) $CH_3\overset{\displaystyle CH_3}{\overset{|}{C}}HCH\!\!=\!\!\overset{\displaystyle CH_2CH_3}{\overset{|}{C}}CH_2CH_3$ $\xrightarrow{O_3,\ CH_3OH}$ $\xrightarrow{(CH_3)_2S}$ $CH_3\overset{\displaystyle CH_3}{\overset{|}{C}}H\!\!-\!\!\overset{\displaystyle O}{\overset{\|}{C}}H$ + $CH_3CH_2\overset{\displaystyle O}{\overset{\|}{C}}CH_2CH_3$

8.21 (cont)

(c)
$$\underset{\begin{array}{c}CH_3\\|\end{array}}{}\underset{\begin{array}{c}CH_2CH_3\\|\end{array}}{}$$
$$CH_3CHCH{=}CCH_2CH_3 \xrightarrow[H_2O]{O_3} \xrightarrow{H_2O_2,\,NaOH} \xrightarrow{H_3O^+} CH_3\underset{|}{\overset{CH_3}{C}}H{-}\overset{O}{\overset{||}{C}}OH \;+\; CH_3CH_2\overset{O}{\overset{||}{C}}CH_2CH_3$$

(d)

$$\xrightarrow{O_3,\,CH_3OH} \xrightarrow{(CH_3)_2S} CH_3\overset{O}{\overset{||}{C}}CH_2CH_2CH_2CH_2\overset{O}{\overset{||}{C}}H$$

(e)

$$\xrightarrow[\substack{dichloromethane\\0\,°C}]{O_3} \xrightarrow{H_2O_2,\,NaOH} 2\;Na^+\;{}^-O\overset{O}{\overset{||}{C}}CH_2CH_2CH_2CH_2\overset{O}{\overset{||}{C}}O\;{}^-Na^+$$

(f)
$$\underset{\begin{array}{c}CH_3\\|\end{array}}{}$$
$$CH_3CH_2C{=}CH_2 \xrightarrow{O_3} \xrightarrow[H_2O]{Zn} CH_3CH_2\overset{O}{\overset{||}{C}}CH_3 \;+\; \overset{O}{\overset{||}{H}}CH$$

(g)
$$CH_3(CH_2)_{13}CH{=}CH_2 \xrightarrow[H_2O]{O_3} \xrightarrow{H_2O_2,\,NaOH} \xrightarrow{H_3O^+} CH_3(CH_2)_{13}\overset{O}{\overset{||}{C}}OH \;+\; H\overset{O}{\overset{||}{C}}OH$$

8.22
$$\underset{\begin{array}{c}CH_3CH_2\;\;CH_3\\||\end{array}}{}$$
$$CH_3CH{=}CCH_2\underset{|}{C}CH_3 \;+\; CH_3CH_2\underset{|}{C}{=}\underset{|}{C}HCCH_3 \xrightarrow{O_3} \xrightarrow[H_2O]{Zn}$$
$$\underset{CH_3}{} \qquad\qquad\qquad \underset{CH_3}{}$$

              90%   10%

$$CH_3\overset{O}{\overset{||}{C}}H \;+\; CH_3CH_2\overset{O}{\overset{||}{C}}CH_2\underset{\underset{CH_3}{|}}{\overset{CH_3}{\overset{|}{C}}}CH_3 \;+\; CH_3CH_2\overset{O}{\overset{||}{C}}CH_2CH_3 \;+\; CH_3\underset{\underset{CH_3}{|}}{\overset{CH_3}{\overset{|}{C}}}{-}\overset{O}{\overset{||}{C}}H$$

       ~45%   ~45%   ~5%   ~5%

8.23 $CH_3CH_2CCH_2CH_3$ + CH_3CH $\xleftarrow{\text{ozonolysis}}$ $CH_3CH{=}CCH_2CH_3$ $\xleftarrow{\text{dehydration}}$ $CH_3CH_2CCH_2CH_3$

(with O double bonds on the two carbonyl carbons at left; center alkene bears a CH_2CH_3 substituent; right structure bears CH_2CH_3 and OH substituents)

original alcohol

8.24

(a) $CH_3CHCH{=}CH_2$ (with CH_3 substituent) $\xrightarrow[\substack{H_2O \\ 0\,°C}]{KMnO_4,\ KOH}$ a diol product

CH_3CH— C (bearing CH_2OH, H, OH, and CH_3)   and enantiomer

(b)   cyclopentene $\xrightarrow[\substack{\text{diethyl ether} \\ \text{pyridine}}]{OsO_4}$ osmate ester $\xrightarrow[\substack{\text{mannitol} \\ H_2O}]{KOH}$ cis-diol

(c)   1,2-dimethylcyclohexene $\xrightarrow[\substack{\text{diethyl ether} \\ \text{pyridine}}]{OsO_4}$ osmate ester $\xrightarrow[\substack{\text{mannitol} \\ H_2O}]{KOH}$ diol (HO, OH)

(d) $C{=}C$ (bearing CH_3, H on left carbon; H, CH_3 on right carbon) $\xrightarrow[\substack{\text{diethyl ether} \\ \text{pyridine}}]{OsO_4}$ osmate ester $\xrightarrow[\substack{\text{mannitol} \\ H_2O}]{KOH}$

diol product (CH_3, H, HO, OH)   and enantiomer $\longrightarrow$ diol product (HO, H, CH_3, OH)   and enantiomer

8.24 (cont)

(e)

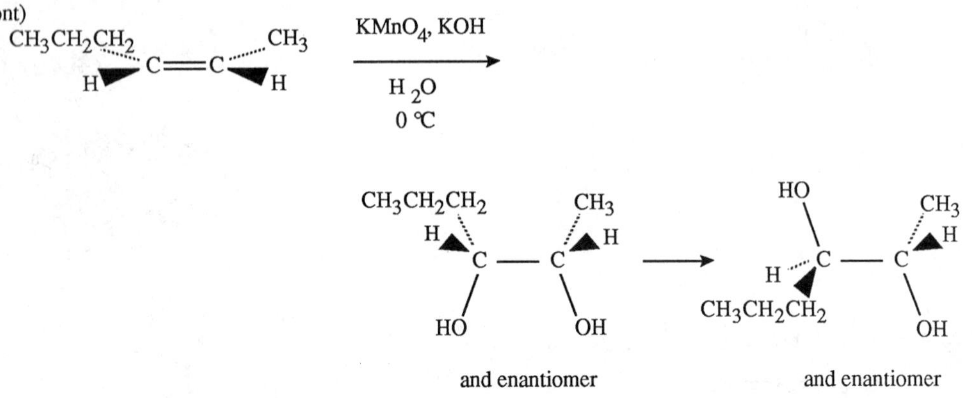

and enantiomer   and enantiomer

8.25

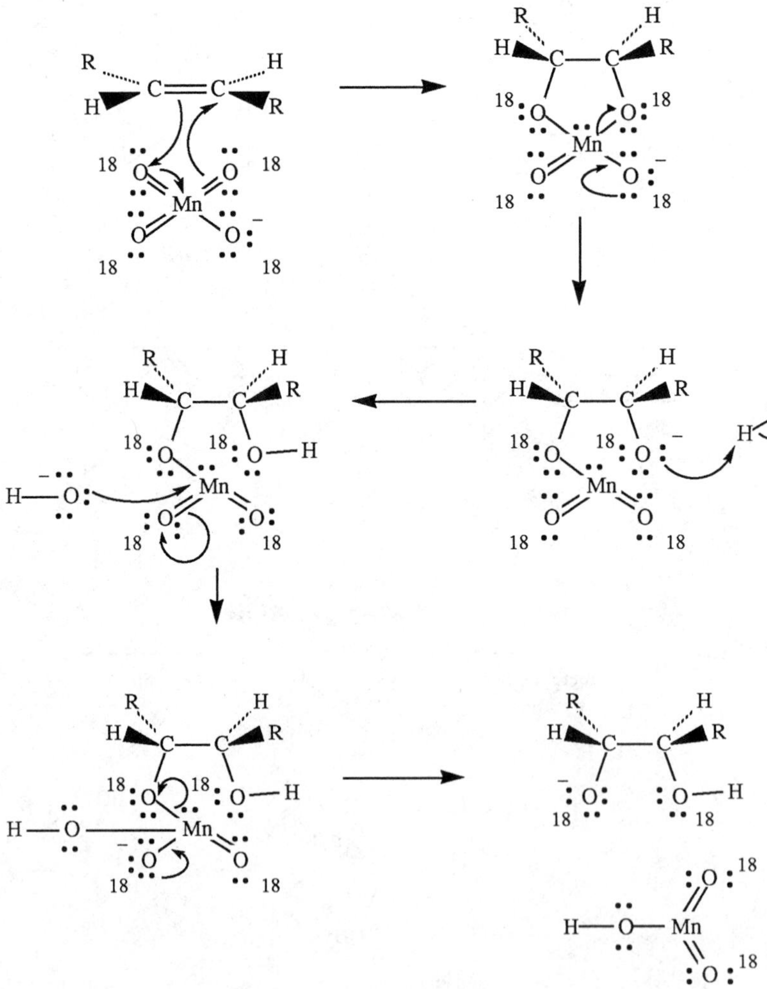

8.25 (cont)

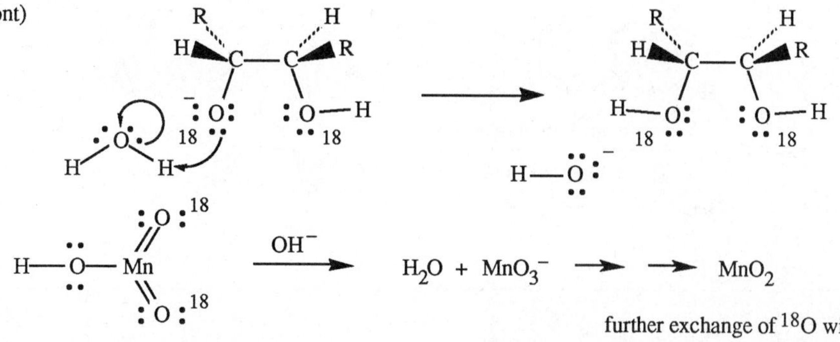

further exchange of ^{18}O with the
solvent will occur for the manganate ion.

Concept Map 8.6   (see p. 218)

8.26

(a)

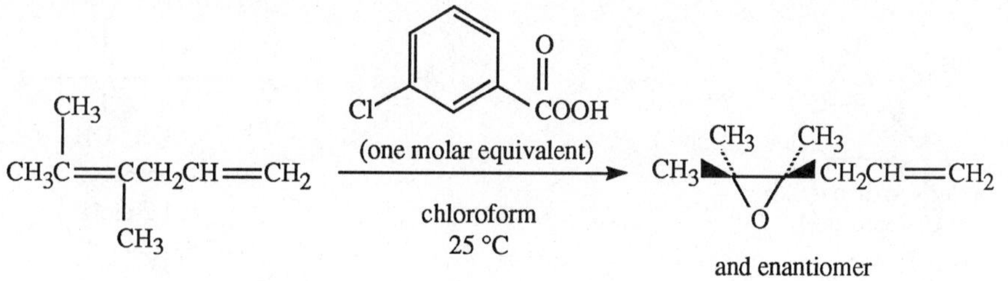

and enantiomer

(b)

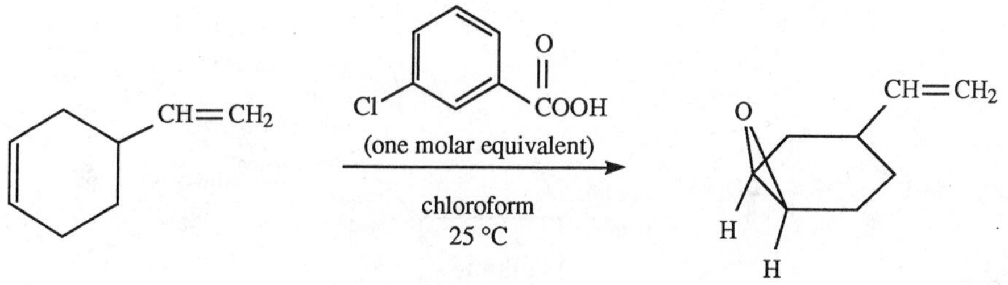

mixture of diastereomers

(c)

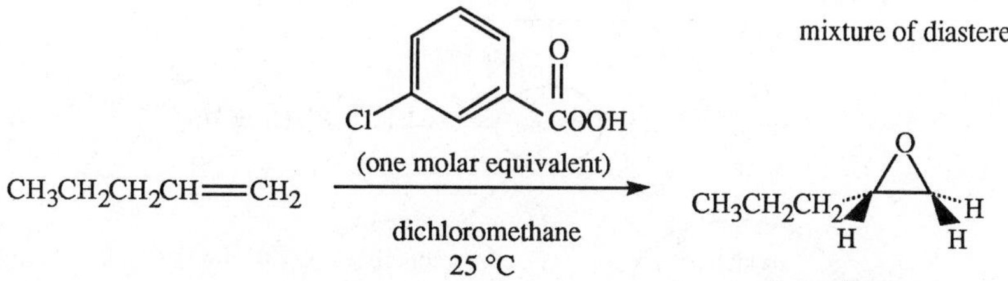

and enantiomer

(d)

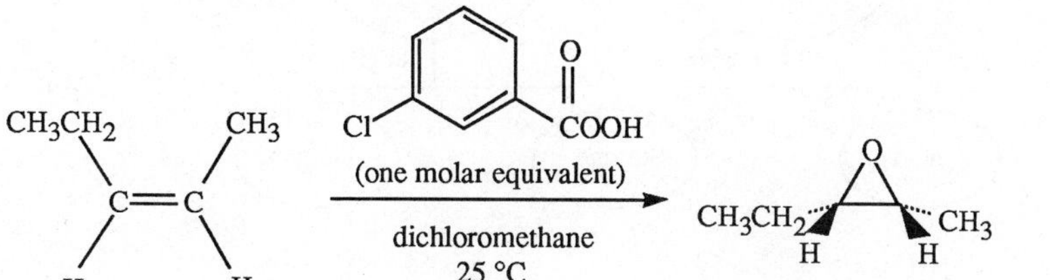

and enantiomer

8.26 (cont)

(e)

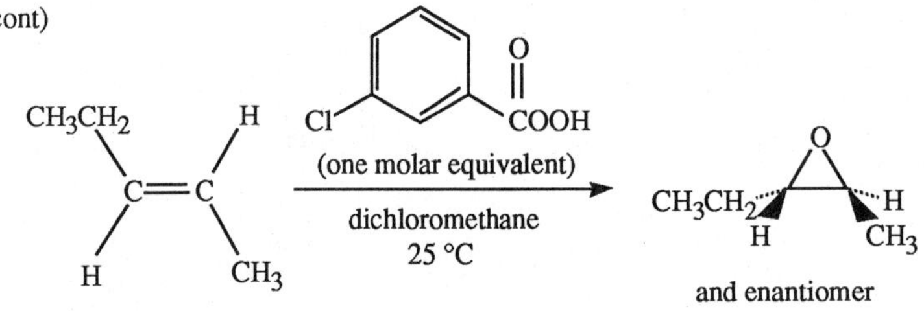

CH3CH2
H
C=C
H
CH3
Cl
O
COOH
(one molar equivalent)
dichloromethane
25 °C
O
CH3CH2
H
H
CH3
and enantiomer

Concept Map 8.6 The oxidation reactions of alkenes.

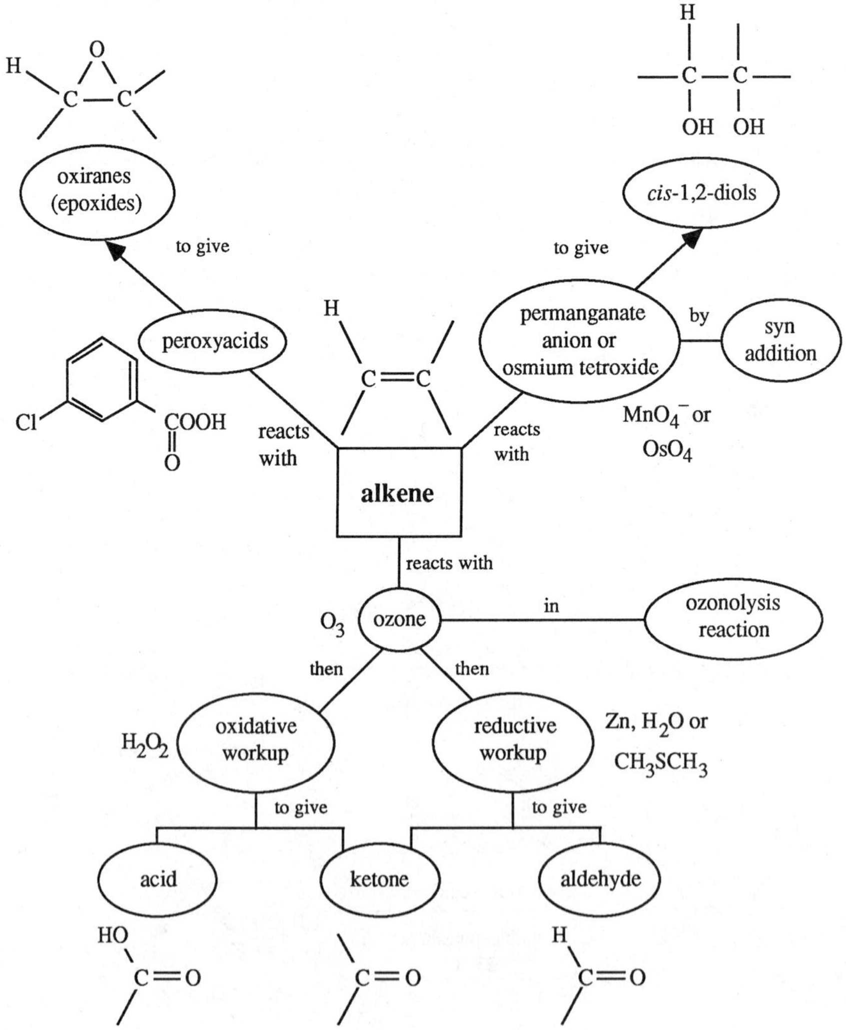

oxiranes (epoxides)
to give
peroxyacids
Cl
COOH
O
reacts with
alkene
reacts with
permanganate anion or osmium tetroxide
MnO_4^- or OsO_4
to give
cis-1,2-diols
OH OH
by
syn addition
reacts with
O_3
ozone
in
ozonolysis reaction
then
then
H_2O_2
oxidative workup
reductive workup
Zn, H_2O or CH_3SCH_3
to give
to give
acid
ketone
aldehyde
HO
C=O
C=O
H
C=O

8.27

1.   What functional groups are present in the starting material and the product?

The starting material is an alkene. It contains a double bond. The product is a diol with the two hydroxyl groups on adjacent carbon atoms. Specific stereochemistry is given for both the starting material and one of the enantiomeric products. The product is a racemic mixture.

(E)-2-butene   (2S, 3S)-2,3-butanediol   more stable anti
                         drawn in the eclipsed   conformation of
                         conformation. No plane   (2S, 3S)-2,3-butanediol
                         of symmetry is present.
                                         one of the enantiomeric products

2.   How do the carbon skeletons of the two compounds compare? How many carbon atoms does each contain? Are there any rings? What are the positions of branches and functional groups on the carbon skeletons?

The carbon skeletons of the two compounds are identical. Each contains four carbon atoms. The two carbon atoms that were doubly bonded have hydroxyl groups on them in the product. The methyl groups in the starting alkene and in the eclipsed conformation of the product diol are on the opposite side of the two reacting carbon atoms.

3.   How do the functional groups change in going from starting material to product? Does the starting material have a good leaving group?

The double bond has disappeared. The two carbon atoms of the double bond have hydroxyl groups on them. There is no leaving group present in the starting material.

4.   Is it possible to dissect the structures of the starting material and product to see which bonds must be broken and which formed?

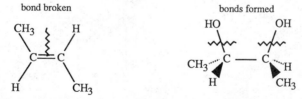

          bond broken   bonds formed

5.   Do we recognize any part of the product molecule as coming from a good nucleophile or an electrophilic addition?

A double bond is converted to a diol by reaction with an oxidizing agent (an electrophile) such as permanganate ion or osmium tetroxide. The reaction of permanganate ion or osmium tetroxide is a syn addition so would give the stereochemistry we need.

6.   What type of compound would be a good precursor to the product?

The starting material can be converted directly to product so no other precursor is needed.

8.27 (cont)

7.   After this last step, do we see how to get from starting material to product? If not, we need to analyze the structure
obtained in step 6 by applying questions 5 and 6 to it.

Complete correct answer:

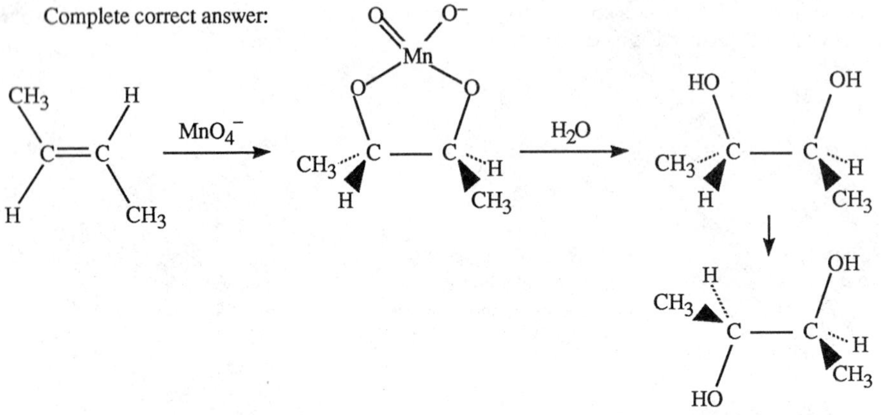

and enantiomer

and enantiomer

8.28   (a) (*E*)-3-methyl-3-heptene   (b) 3-ethyl-2,5-dimethyl-2-hexene

(c) (*E*)-3,4-dichloro-3-heptene   (d) (*Z*)-3,4-dimethyl-3-heptene

(e) 2-methyl-1-hexene   (f) (*Z*)-2-chloro-3-ethyl-3-hexene

(g) 3-bromocyclohexene   (h) (*R*)-1,3-dichlorobutane

(i) (*Z*)-4-bromo-3-ethyl-2-pentene   (j) 3-chloro-2-methyl-2-pentene

(k) *cis*-1,2-dibromocyclobutane   (l) (2*R*, 3*S*)-2-bromo-3-chlorobutane
     meso-1,2-dibromocyclobutane

8.29   (a)   (b) 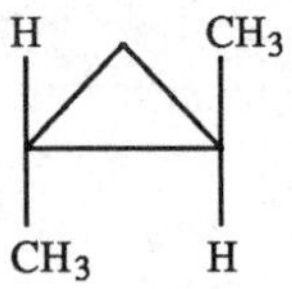   and enantiomer

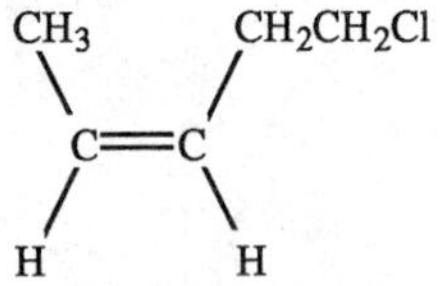

(c)   (d)

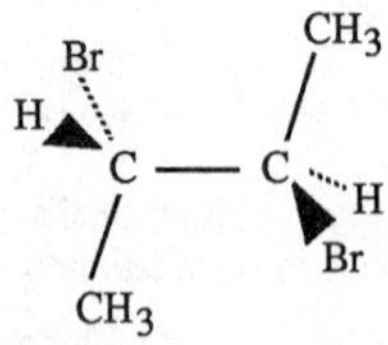

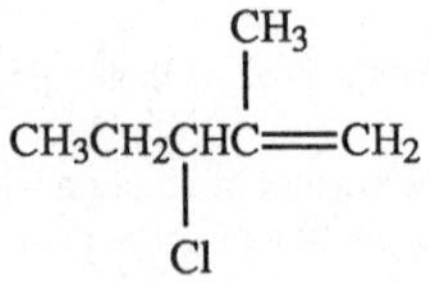

(e) 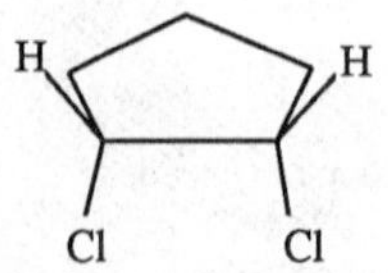   (f)

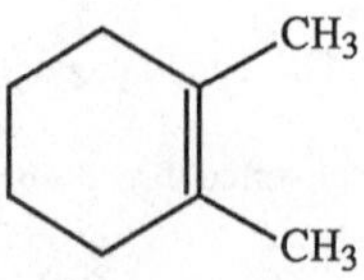

8.30

(a)

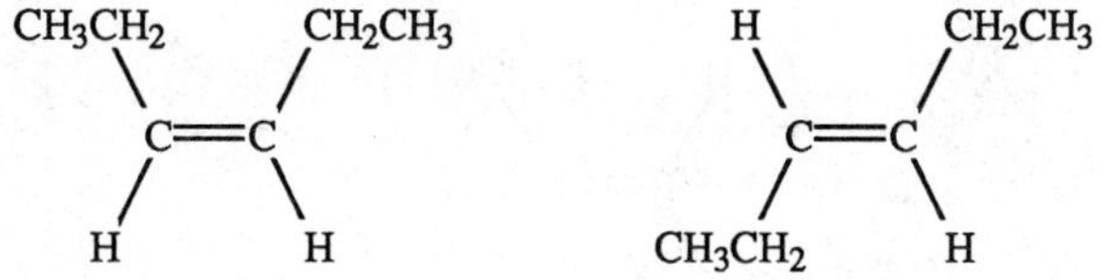

(b)

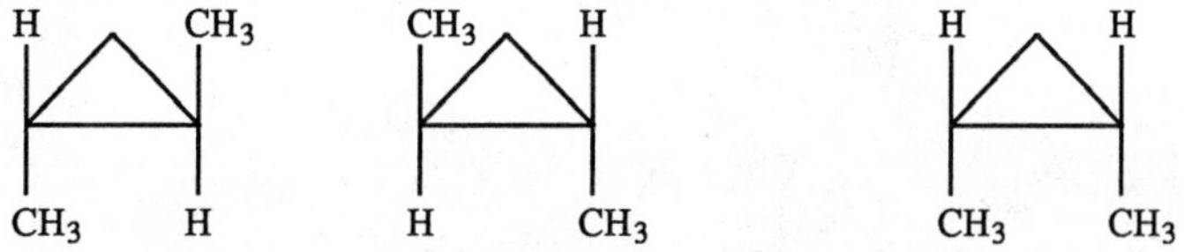

(c)

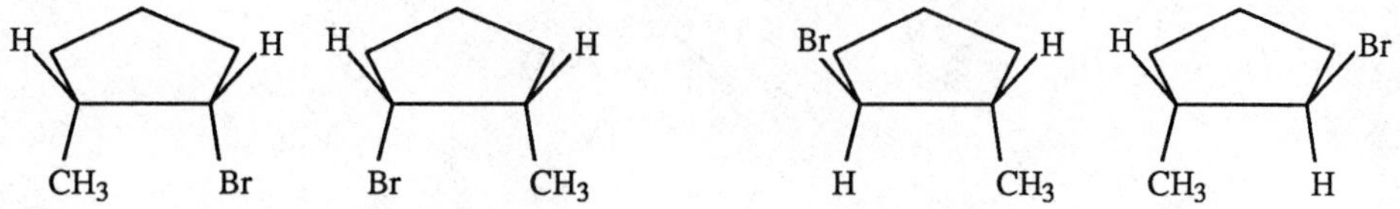

(d)

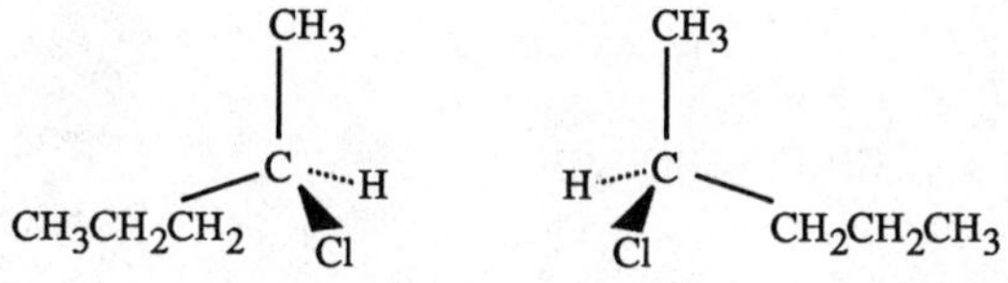

(e)

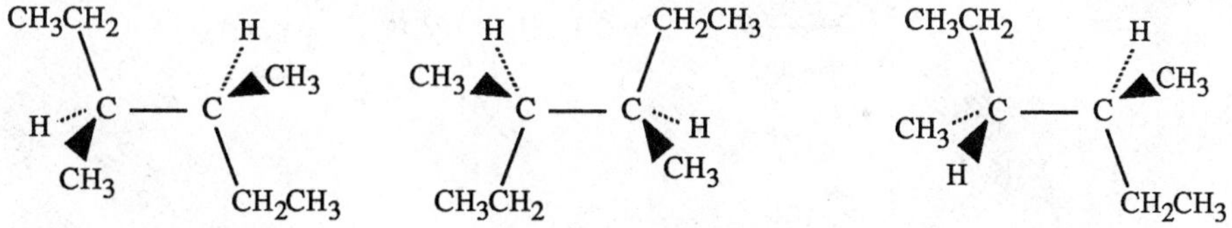

(f)

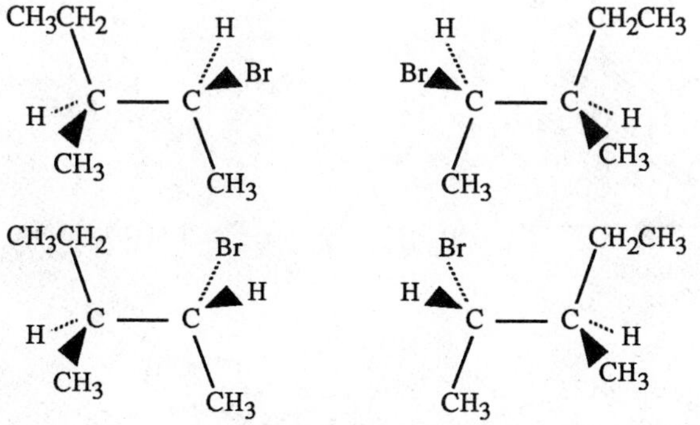

8.30 (cont)

(g)

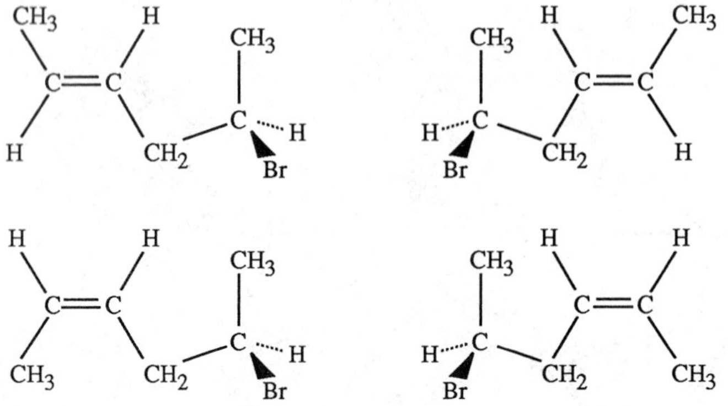

8.31

(a)

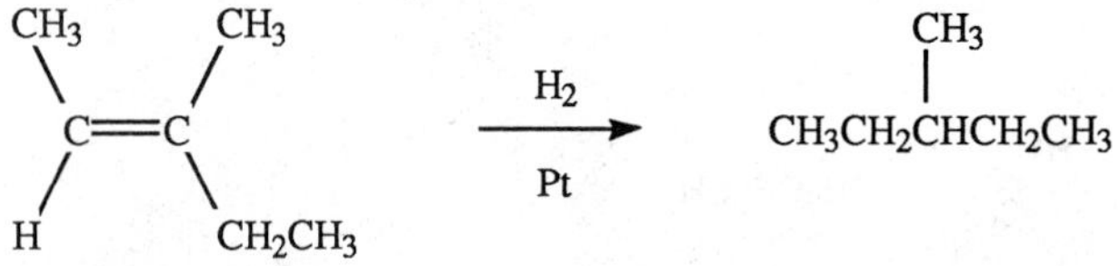

(b)

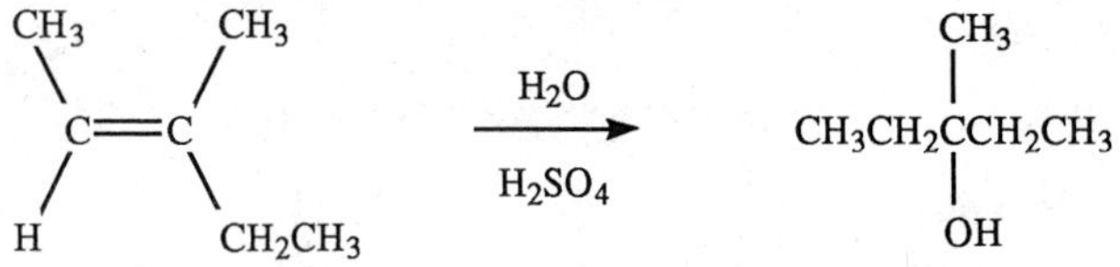

(c)

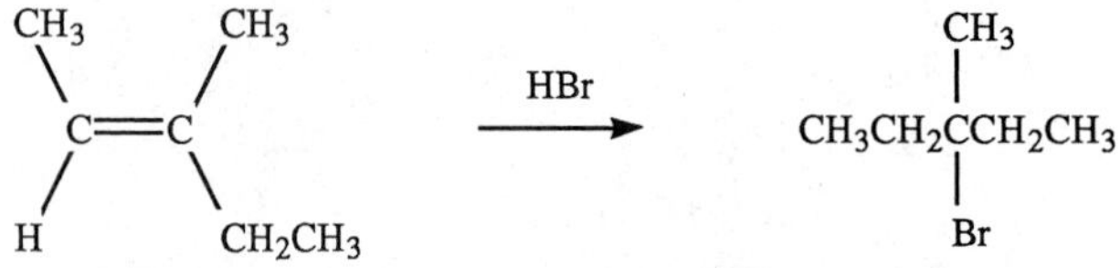

(d)

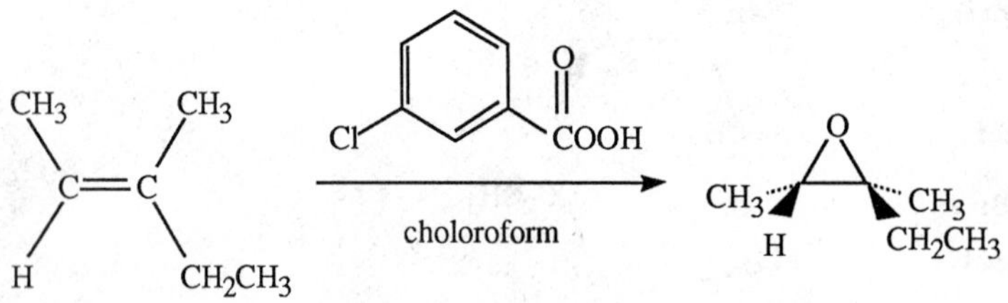

(e)

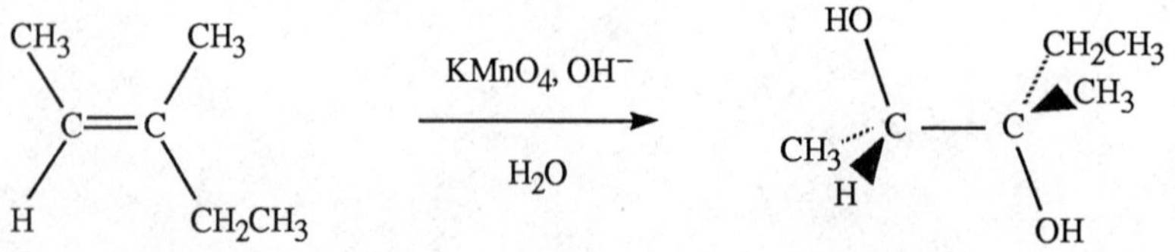

8.31 (cont)

(f)

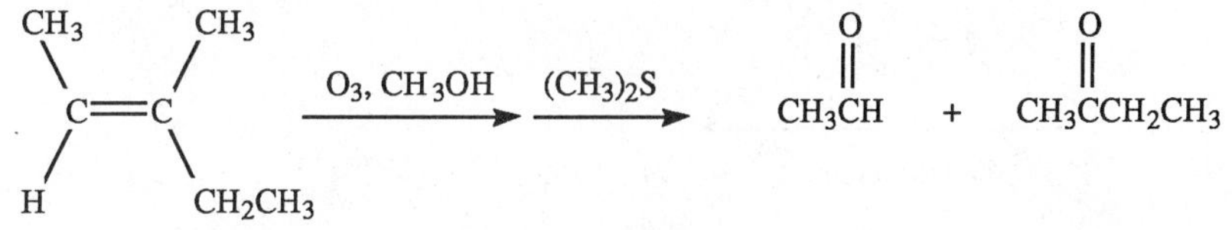

$$CH_3CH \quad + \quad CH_3CCH_2CH_3$$

(g)

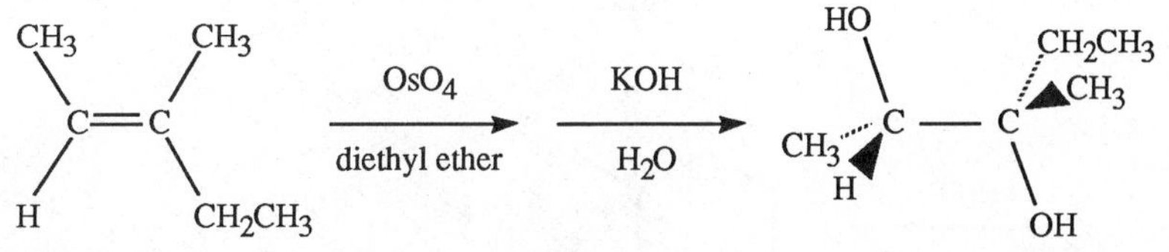

and enantiomer

8.32

(a)

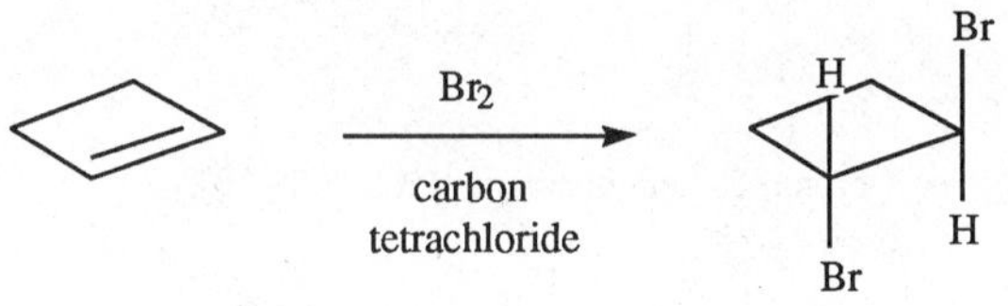

and enantiomer

(b)

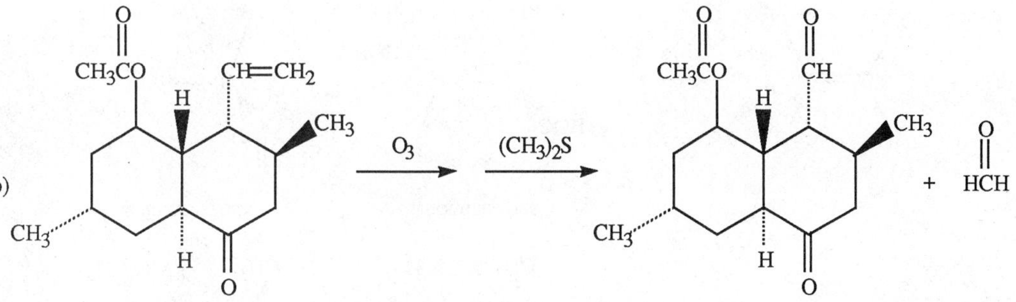

$+ \quad HCH$

(c)

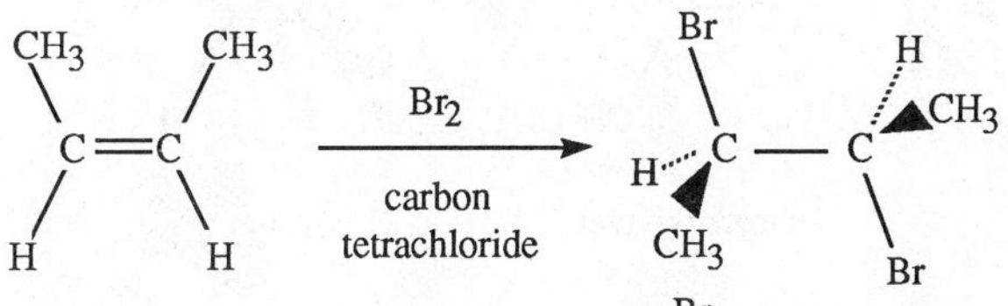

and enantiomer

(d)

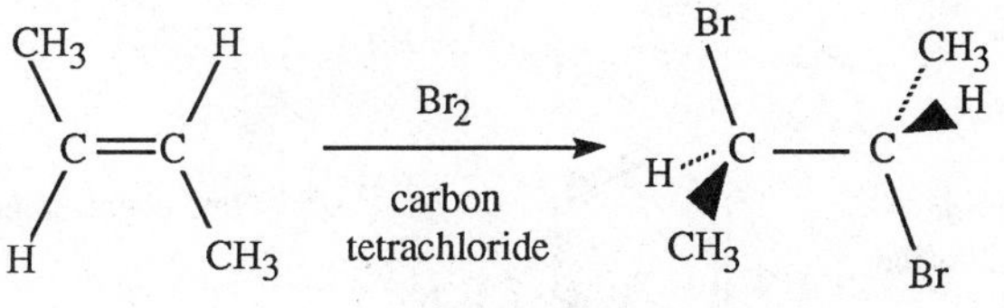

meso compound

8.32 (cont)

(e)

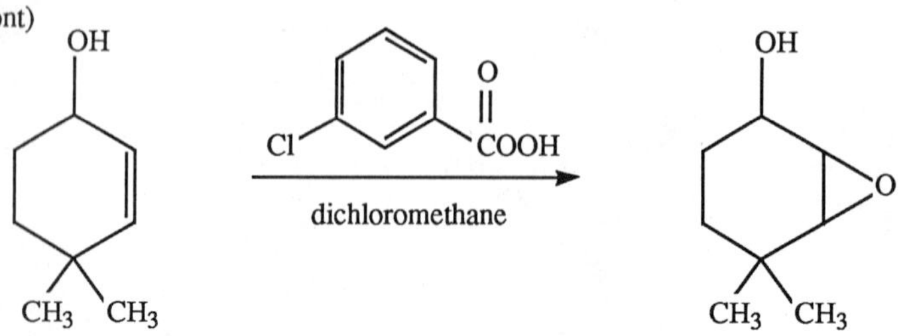
OH
Cl
O
COOH
dichloromethane
O
OH
CH3 CH3
CH3 CH3

(f)

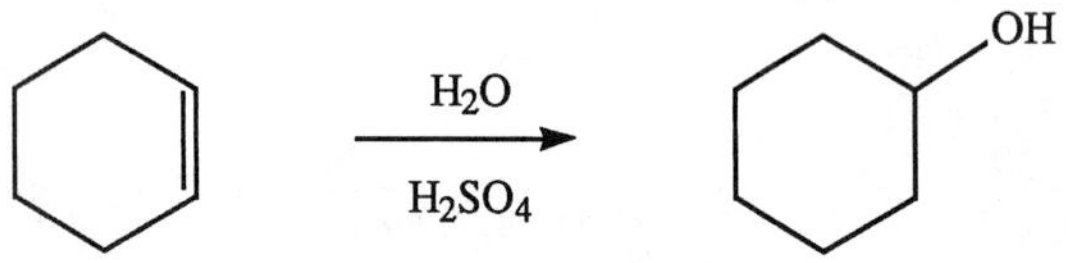
H2O
H2SO4
OH

(g)

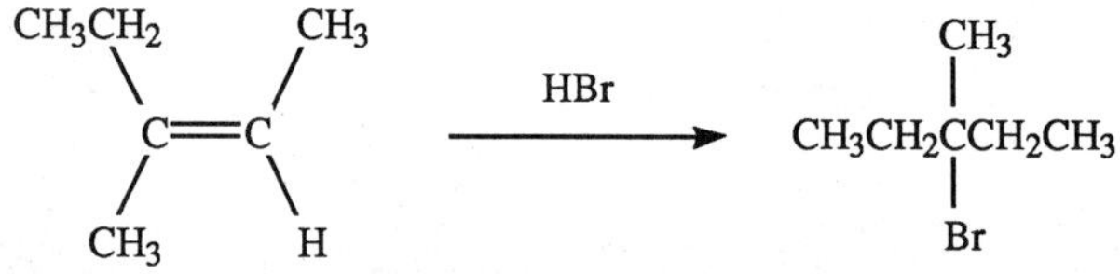
CH3CH2
CH3
C=C
CH3
H
HBr
CH3
CH3CH2CCH2CH3
Br

Note that (*E*)-3-methyl-2-pentene also gives 3-bromo-3-methylpentane [Prob. 8.31 (c)].

(h)

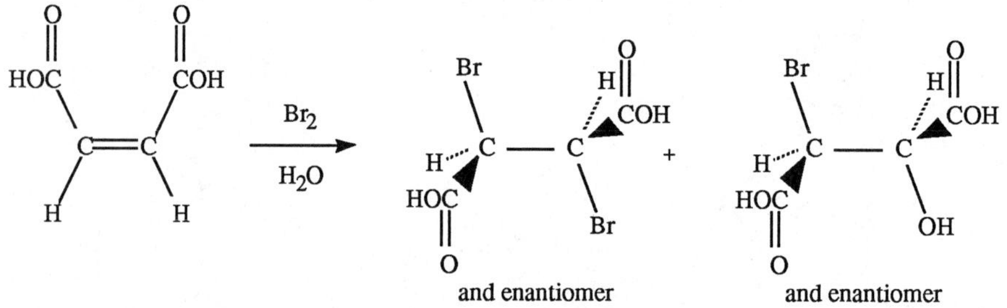
Br2
H2O
and enantiomer
+
and enantiomer

(i)

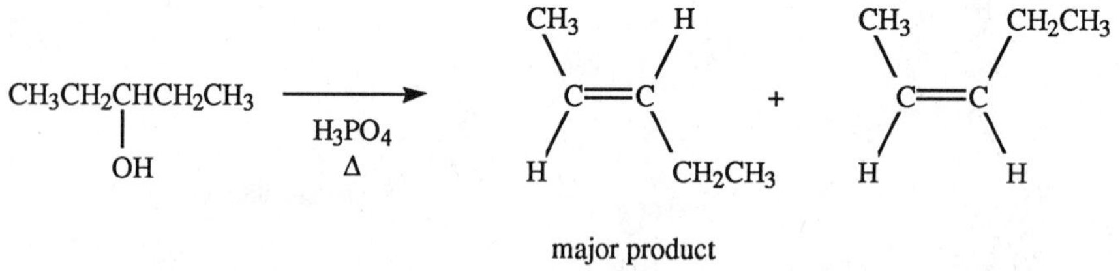
CH3CH2CHCH2CH3
OH
H3PO4
Δ
+
major product

(j)

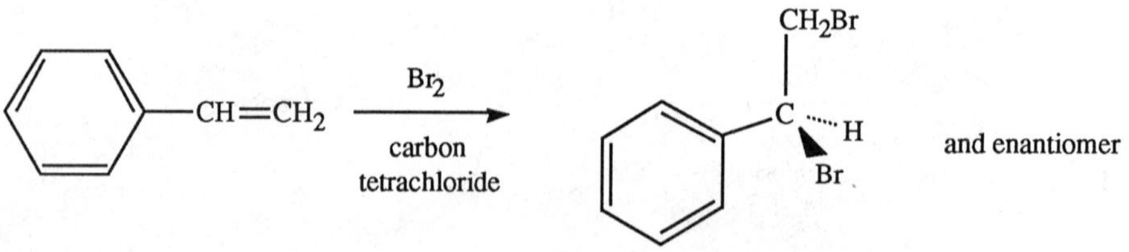
CH=CH2
Br2
carbon
tetrachloride
CH2Br
H
Br
and enantiomer

8.32 (cont)

(k)

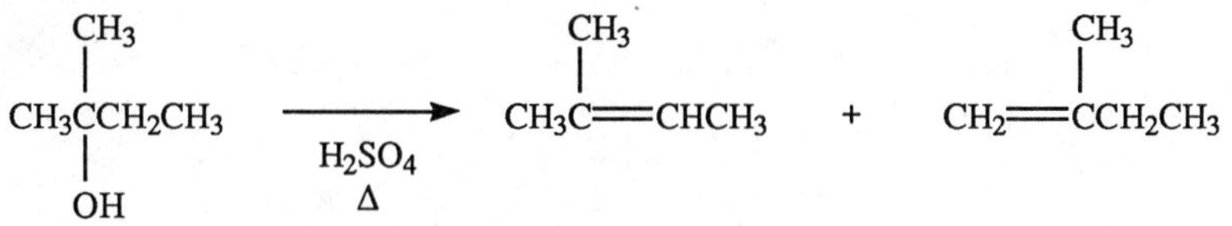

(l)

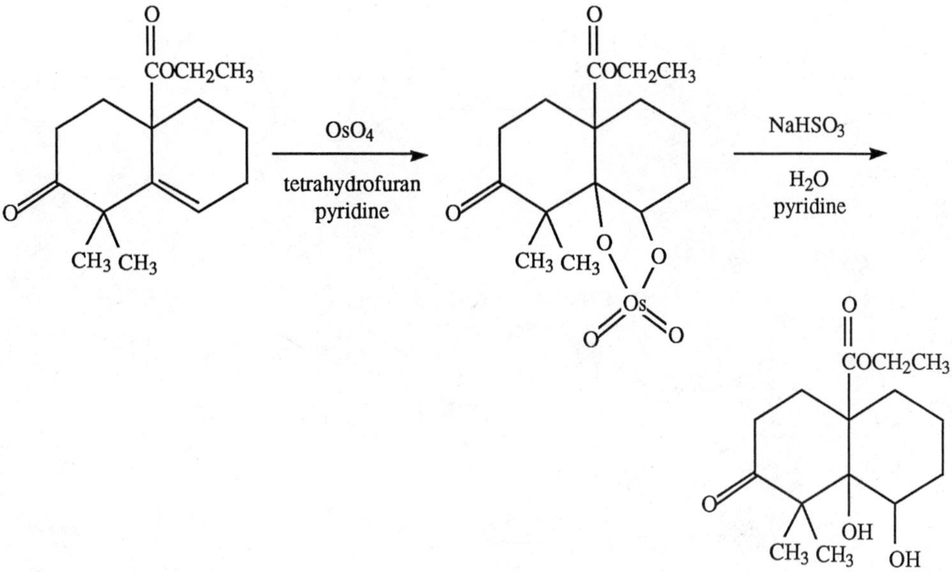

(m)

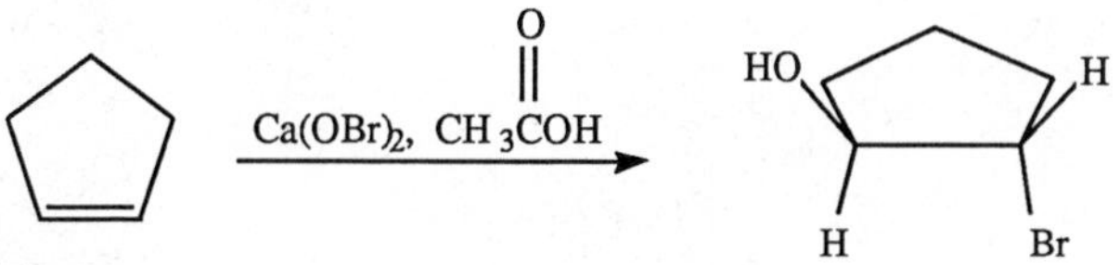

and enantiomer

8.33

(a)

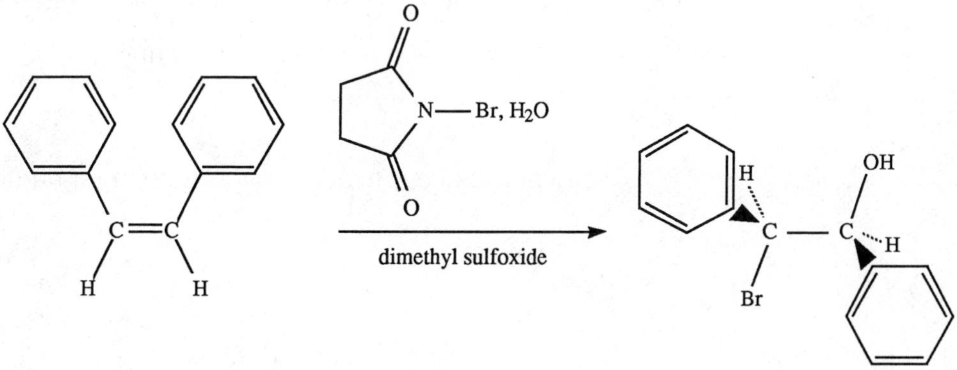

and enantiomer

A

8.33 (cont)

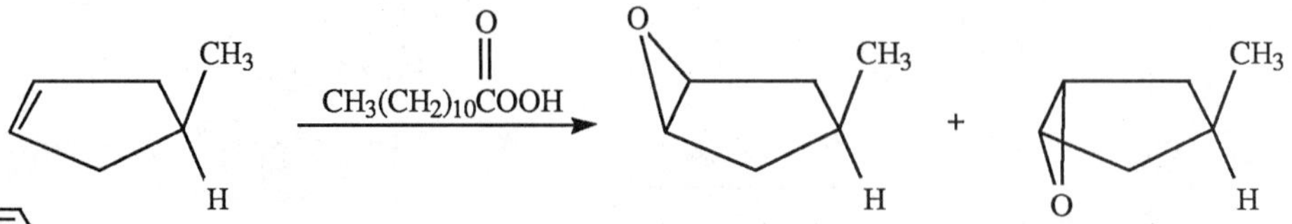

(b)

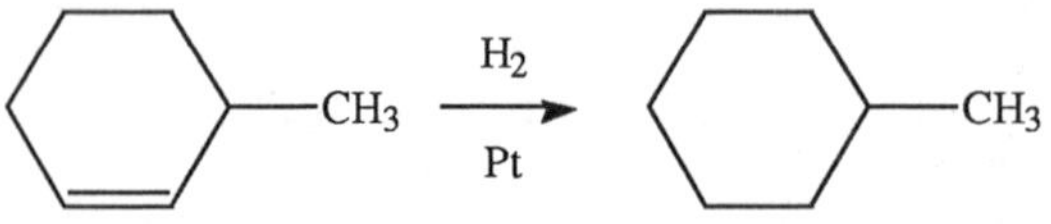

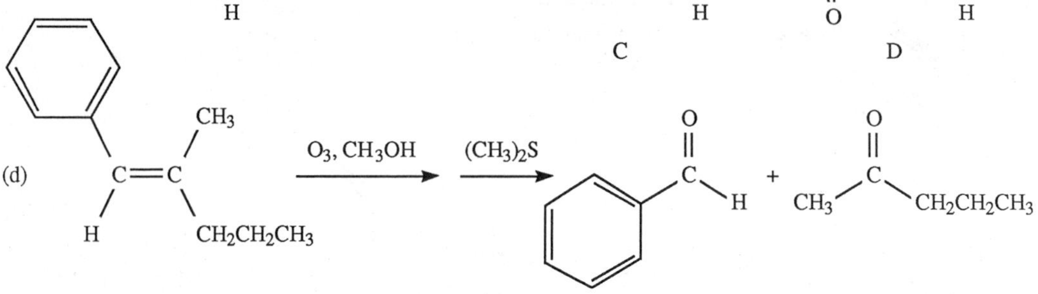

(c)

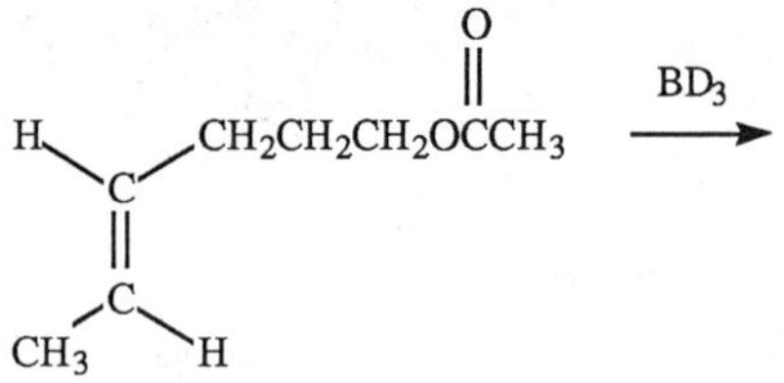

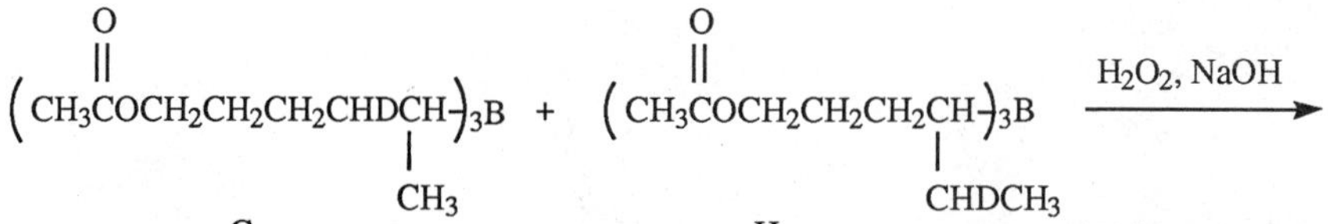

(d)

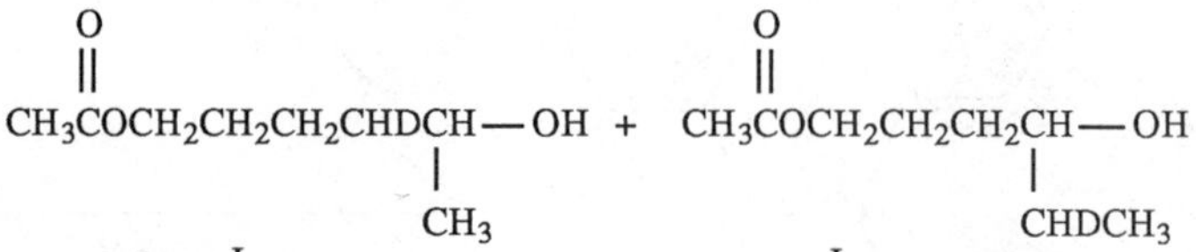

(e)

(f)

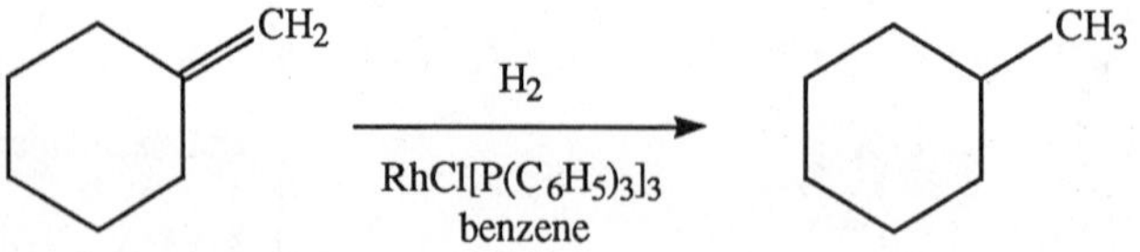

8.33 (cont)

(g)

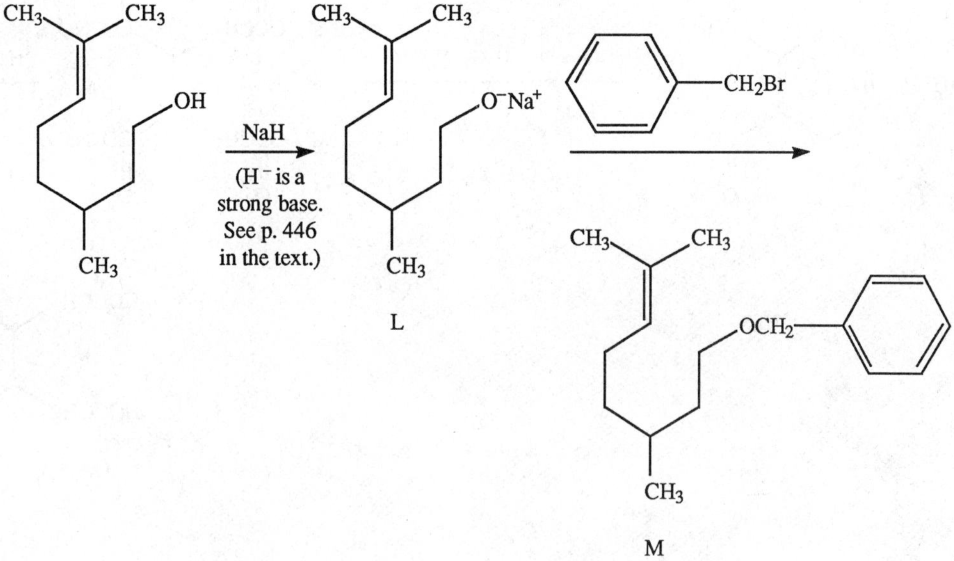
CH3 CH3 OH CH3
NaH
(H⁻ is a strong base. See p. 446 in the text.)
O⁻Na⁺
L
CH2Br
OCH2
M

(h)

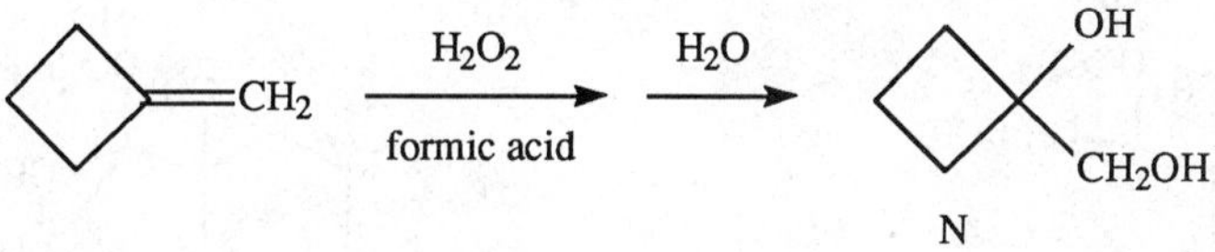
CH2
H2O2
formic acid
H2O
OH
CH2OH
N

(i)

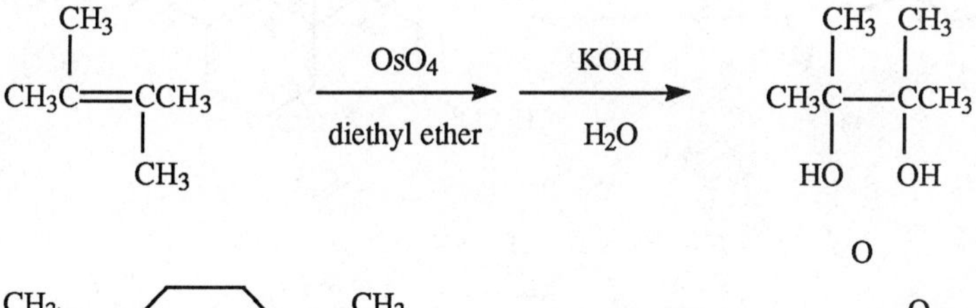
CH3C=CCH3
OsO4
diethyl ether
KOH
H2O
HO OH
O

(j)

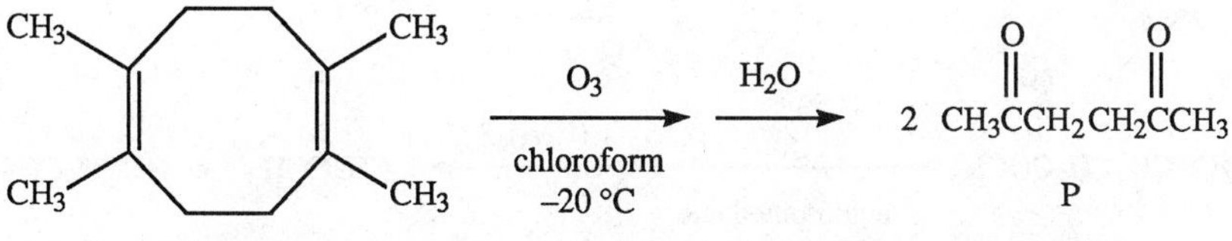
O3
chloroform
–20 °C
H2O
2 CH3CCH2CH2CCH3
P

(k)

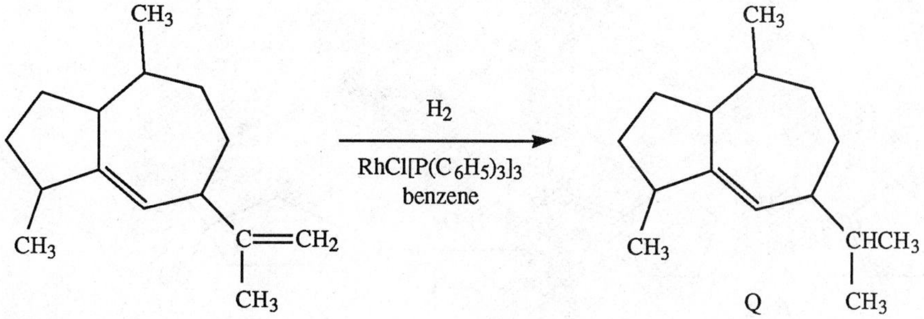
H2
RhCl[P(C6H5)3]3
benzene
CHCH3
Q

8.33 (cont)

(l)

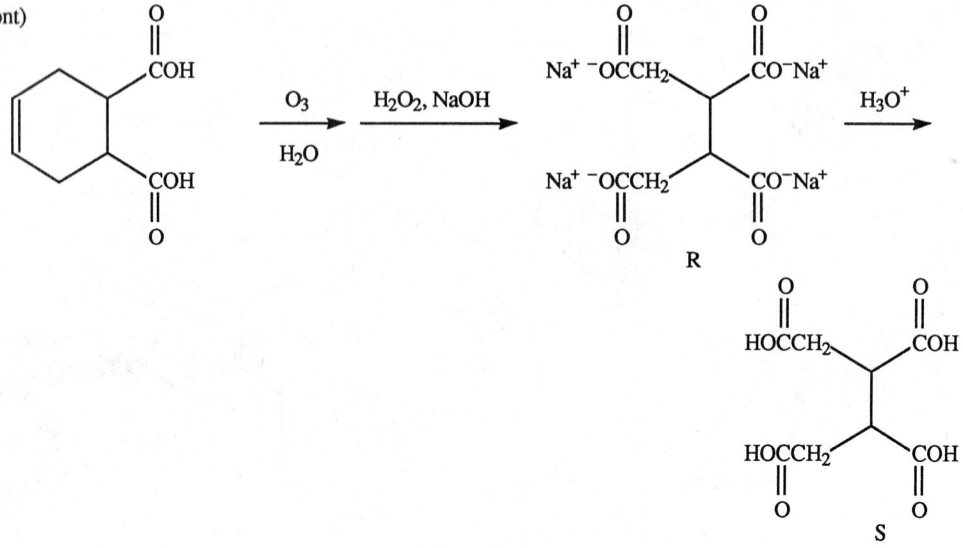

(m)

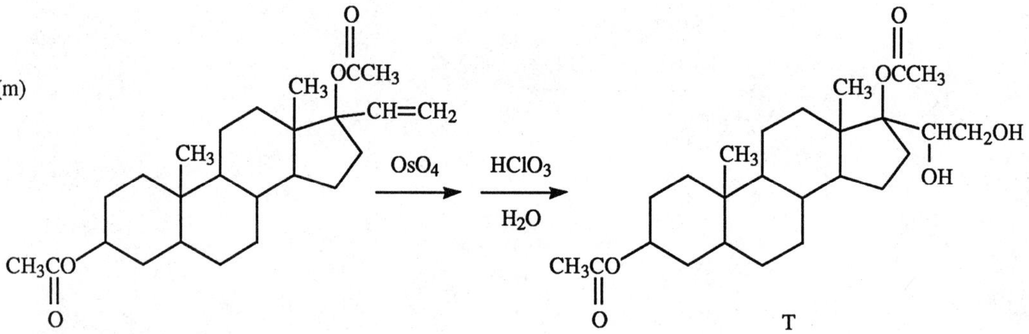

(n)

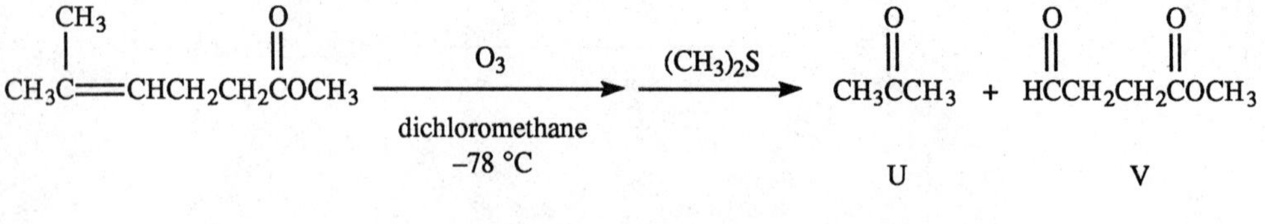

8.34

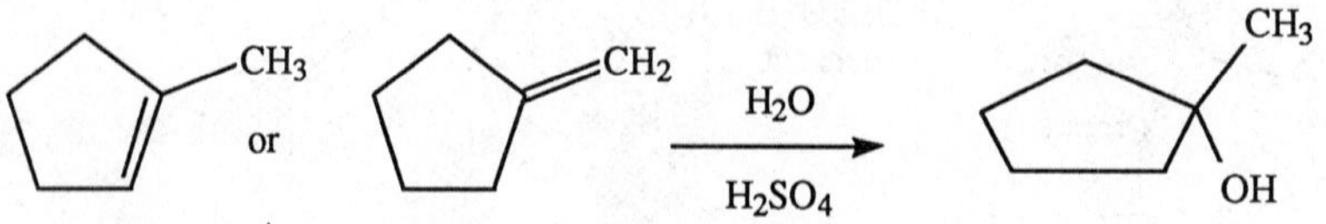

(a)

8.34 (cont)

(b)

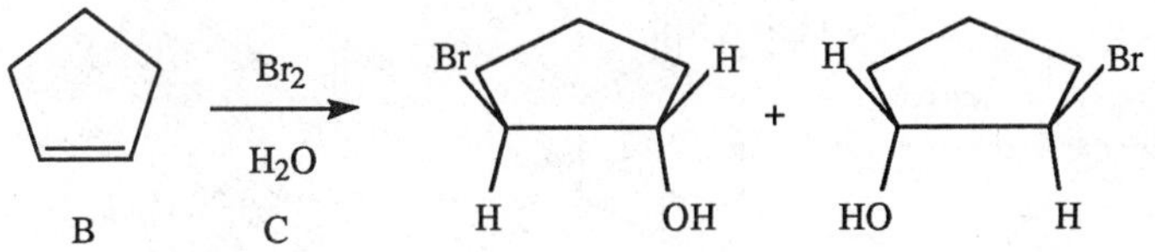
Br2
H2O
B
C
Br
H
H
OH
+
H
Br
HO
H

(c)

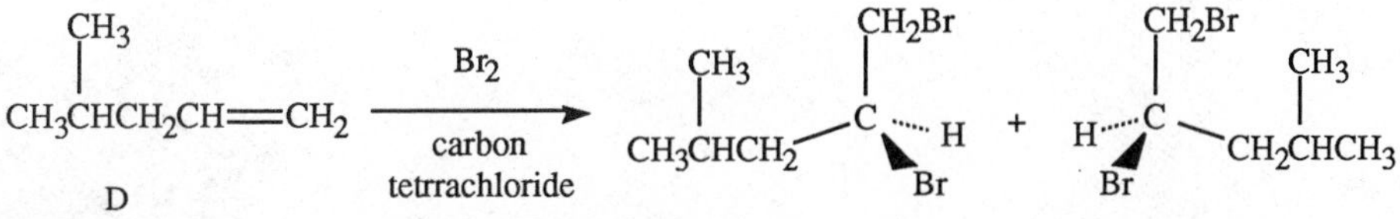
CH3
CH3CHCH2CH═CH2
D
Br2
carbon
tetrrachloride
CH2Br
CH3
CH3CHCH2
C
H
Br
+
CH2Br
CH3
H
C
CH2CHCH3
Br

(d)

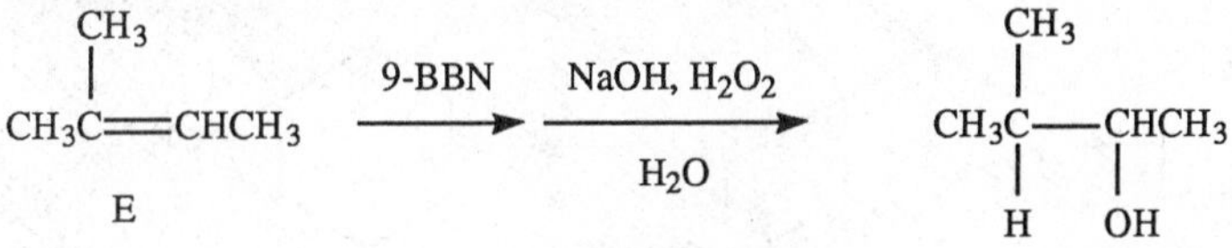
CH3
CH3C═CHCH3
E
9-BBN
NaOH, H2O2
H2O
CH3
CH3C—CHCH3
H
OH

(e)

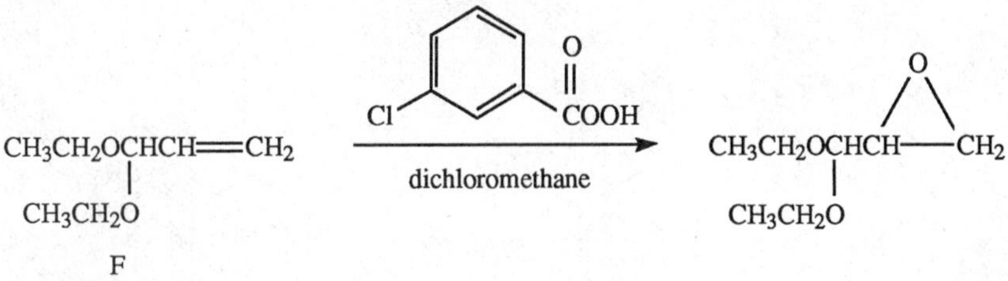
CH3CH2OCHCH═CH2
CH3CH2O
F
Cl
O
COOH
dichloromethane
O
CH3CH2OCHCH—CH2
CH3CH2O

(f)

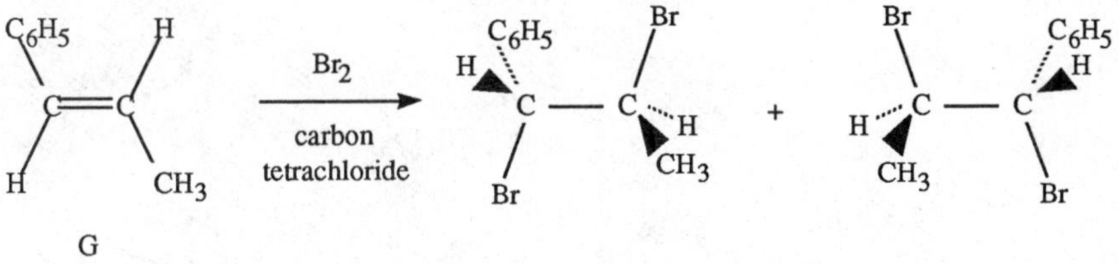
C6H5
H
C═C
H
CH3
G
Br2
carbon
tetrachloride
C6H5
Br
H
C — C
H
CH3
Br
+
Br
C6H5
H
H
C — C
CH3
Br

(g)

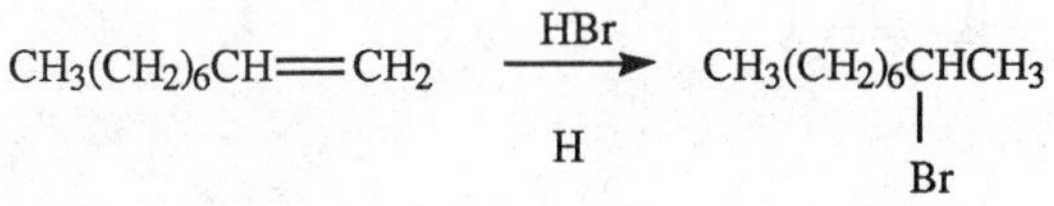
CH3(CH2)6CH═CH2
HBr
H
CH3(CH2)6CHCH3
Br

(h)

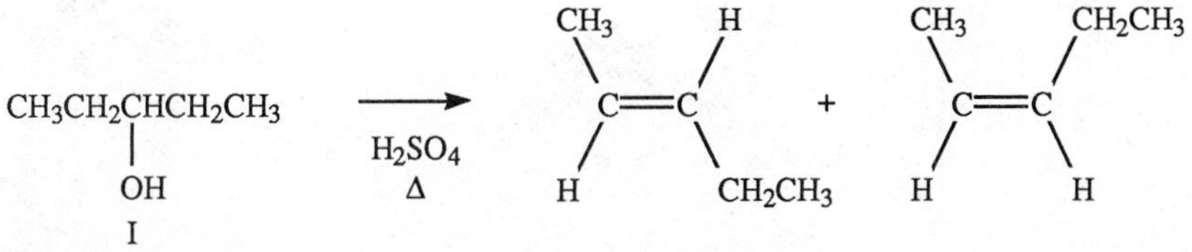
CH3CH2CHCH2CH3
OH
I
H2SO4
Δ
CH3
H
C═C
H
CH2CH3
+
CH3
CH2CH3
C═C
H
H

2-pentanol would give 1-pentene as well as (*E*)- and (*Z*)-2-pentene.

8.34 (cont)

(i)

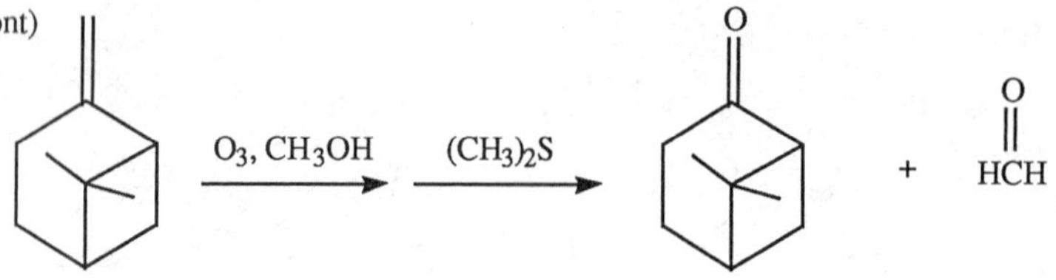

J

8.35

1,2-shift of
sigma bonding
electrons of a
ring carbon;
another type of
alkyl shift

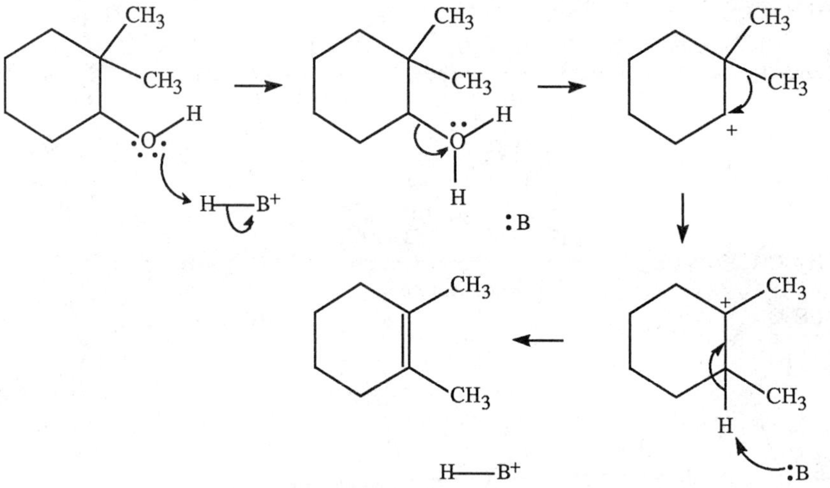

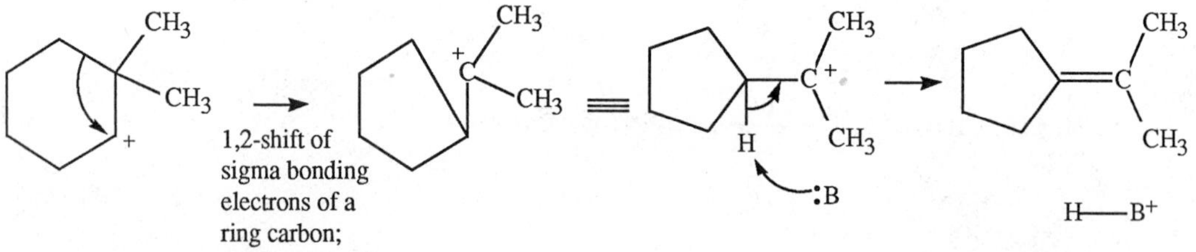

8.36

(a)

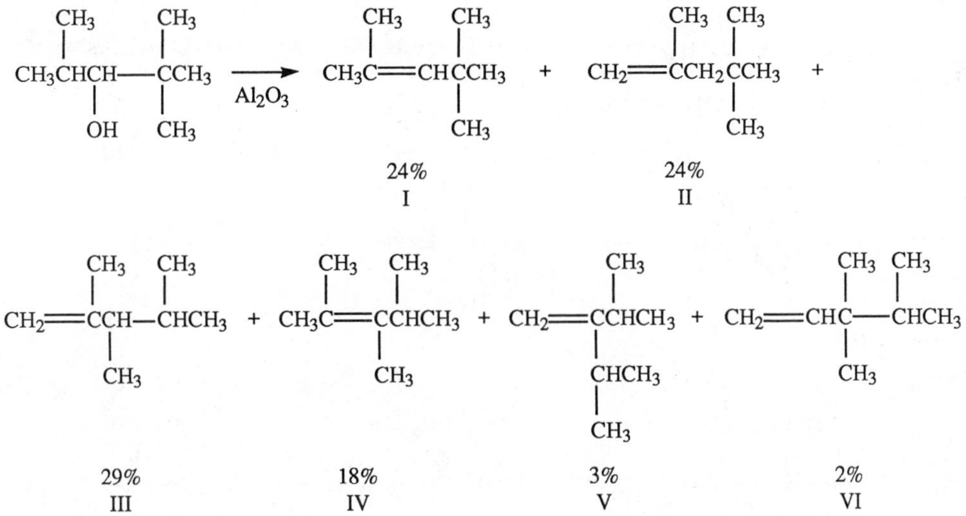

$$CH_3CHCH{-}CCH_3 \xrightarrow{Al_2O_3} CH_3C{=}CHCCH_3 + CH_2{=}CCH_2CCH_3 +$$

24%
I

24%
II

III 29%   IV 18%   V 3%   VI 2%

(b)

$$CH_3CHCH{-}CCH_3 \rightarrow CH_3CHCH{-}CCH_3 \rightarrow CH_3CHCH{-}CCH_3$$

I

1,2-
hydride
shift

II

H{—}B$^+$

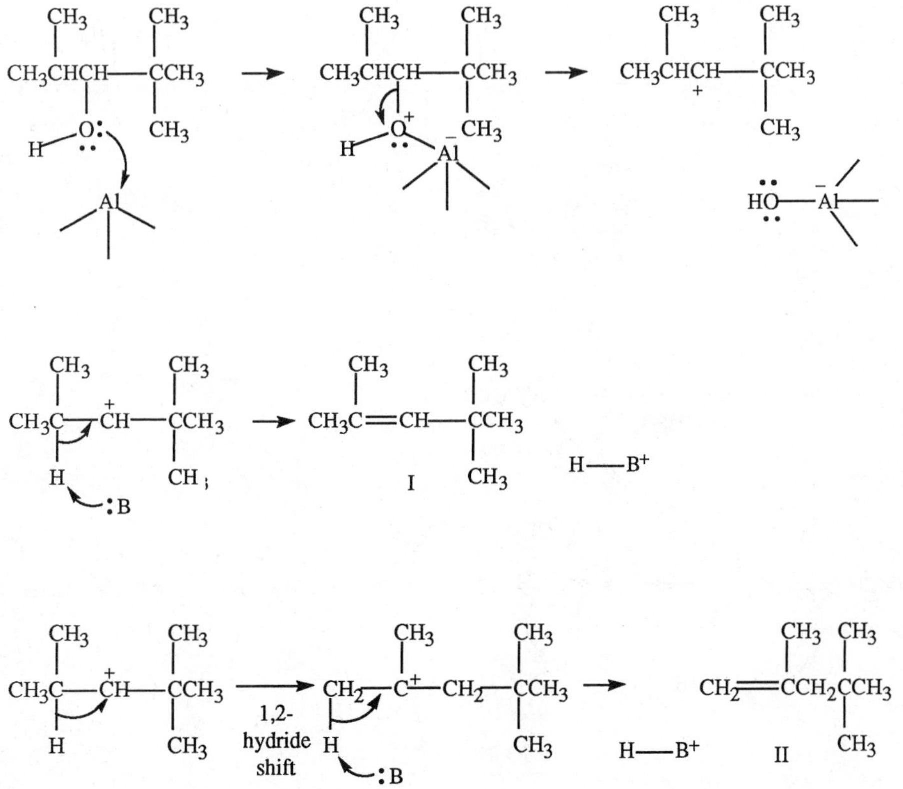

8.36 (cont)

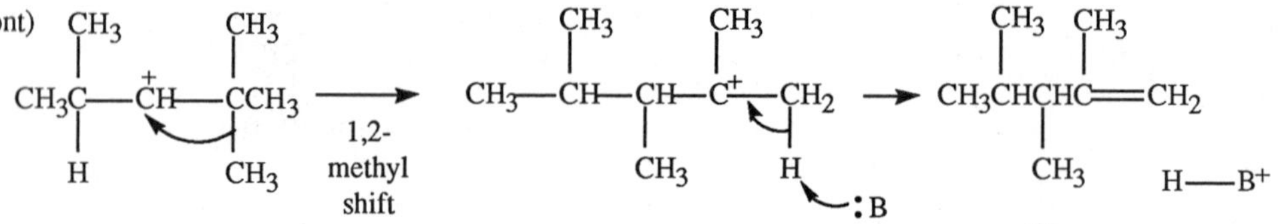

CH$_3$C—CH C—CCH$_3$ $\xrightarrow{\text{1,2-methyl shift}}$ CH$_3$—CH—CH—C$^+$—CH$_2$ $\rightarrow$ CH$_3$CHCHC=CH$_2$

III

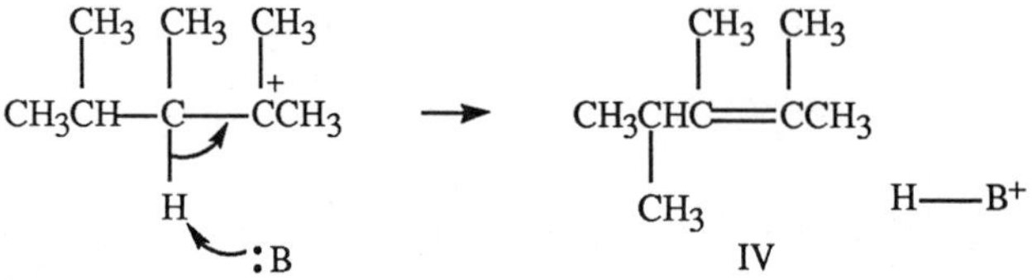

H—B$^+$

CH$_3$CHC=CCH$_3$

IV

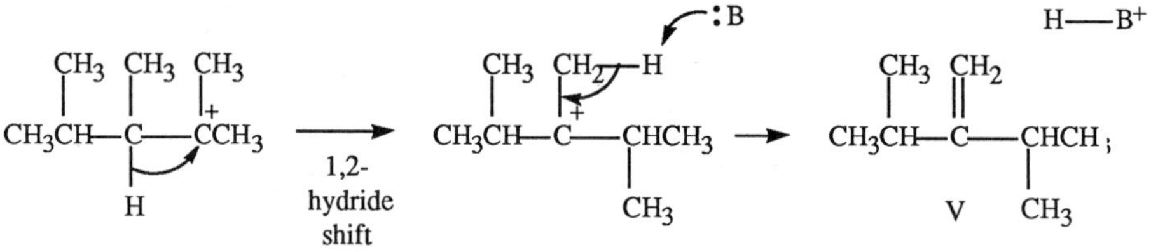

H—B$^+$

$\xrightarrow{\text{1,2-hydride shift}}$ V

CH$_3$CH—CCH=CH$_2$ $\xrightarrow{\text{1,2-methyl shift}}$ CH$_3$CH—CCH—CH$_2$ $\rightarrow$ CH$_3$CH—CCH=CH$_2$

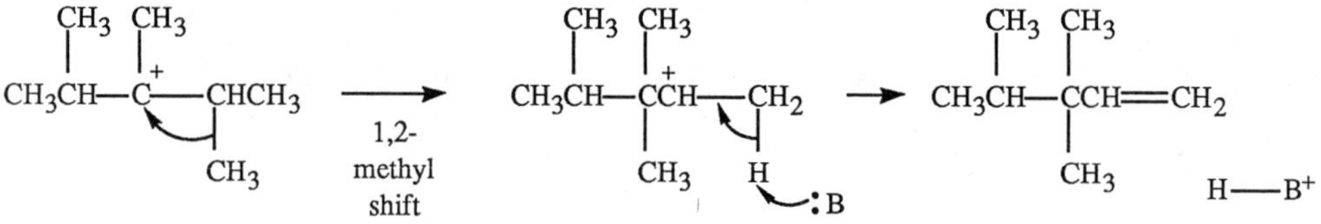

VI

H—B$^+$

8.37   CH$_3$C=CH$_2$ $\xrightarrow[\text{H}_2\text{SO}_4]{\text{CH}_3\text{OH}}$ CH$_3$COCH$_3$

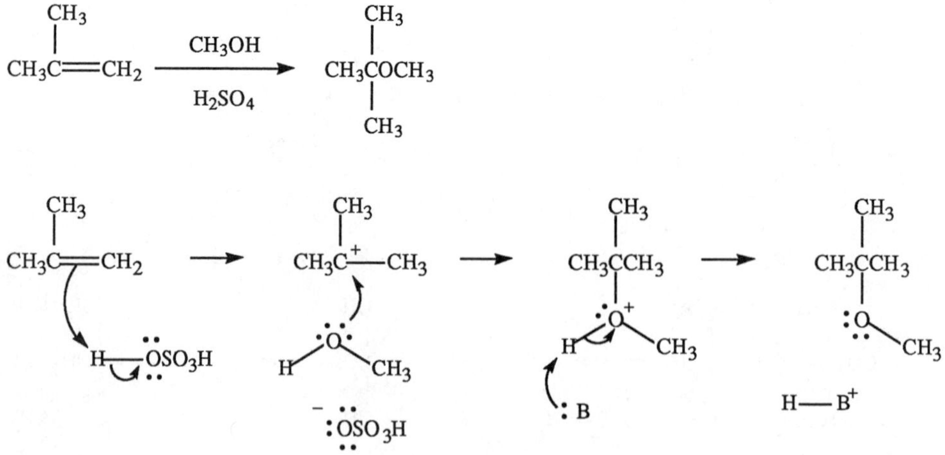

8.38

The most likely first step for the reaction is protonation of the double bond by hydrogen chloride. The most stable carbocation is the one adjacent to the sulfur, which can stabilize the positive charge by delocalization.

The species now present in the reaction mixture is an electrophile. It will react with thiophenol, which is a better nucleophile than chloride ion.

8.38 (cont)

The protonated thioether will now transfer a proton to a base.

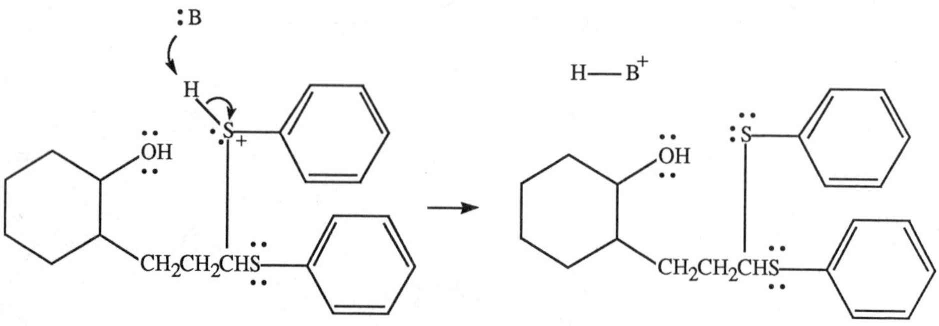

8.39

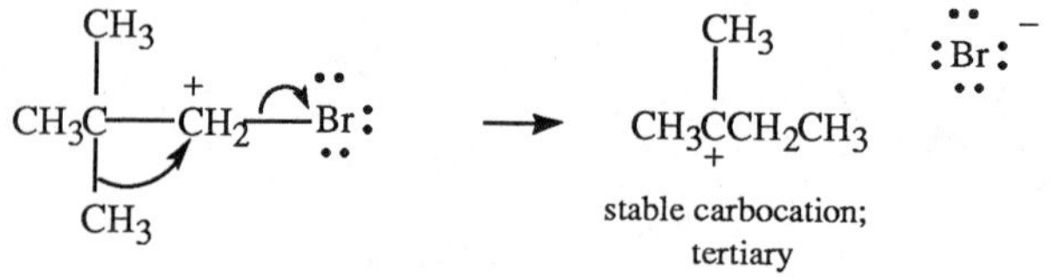

stable carbocation;
tertiary

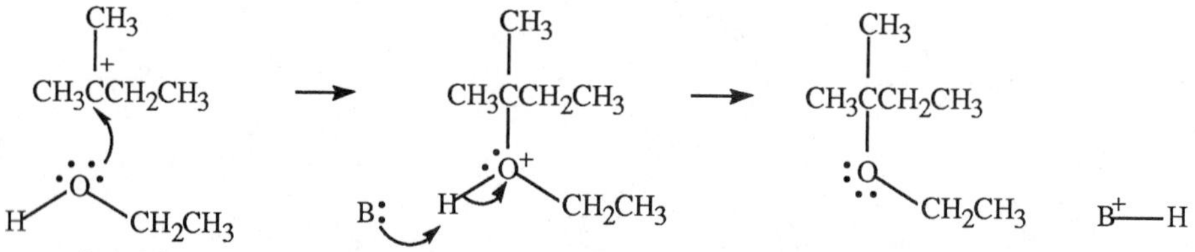

reaction of the
carbocation with
ethanol, a good
nucleophile

deprotonation of
the oxonium ion

2-ethoxy-2-methylbutane

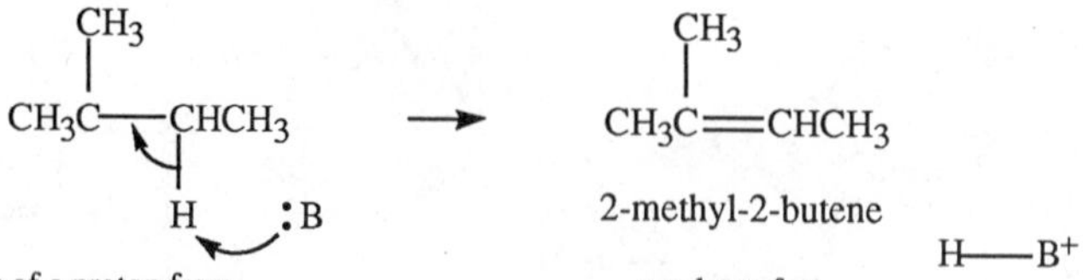

loss of a proton from
the carbocation

2-methyl-2-butene

product of an
elimination reaction

8.40

(a)

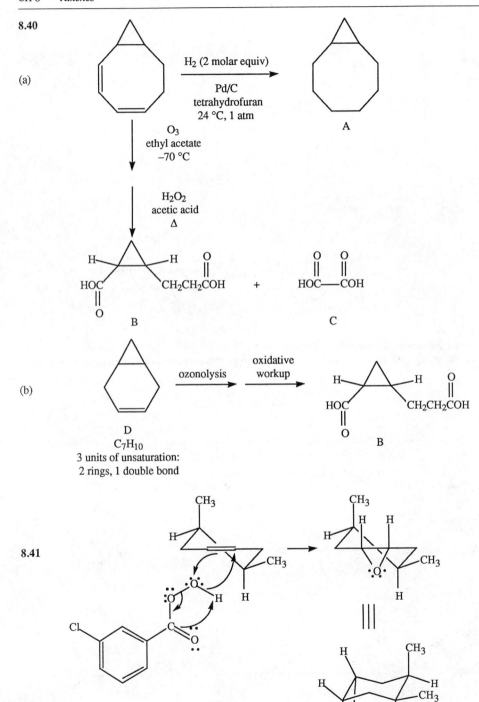

8.41

8.41 (cont)

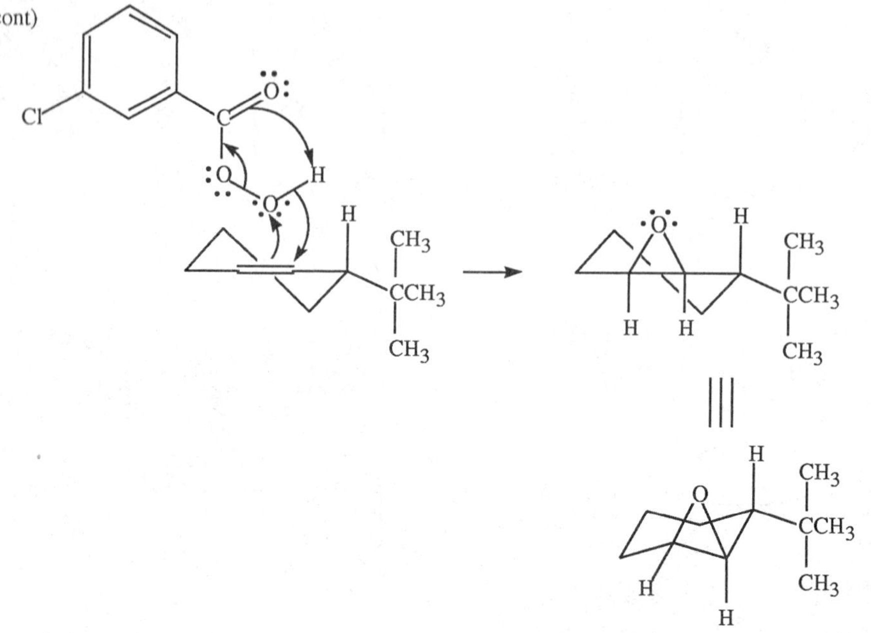

In each case the peroxyacid approaches the double bond from the less hindered side. If the starting material is racemic, the product will also be racemic. None of the diastereomeric product will form in either reaction.

8.42

(a)

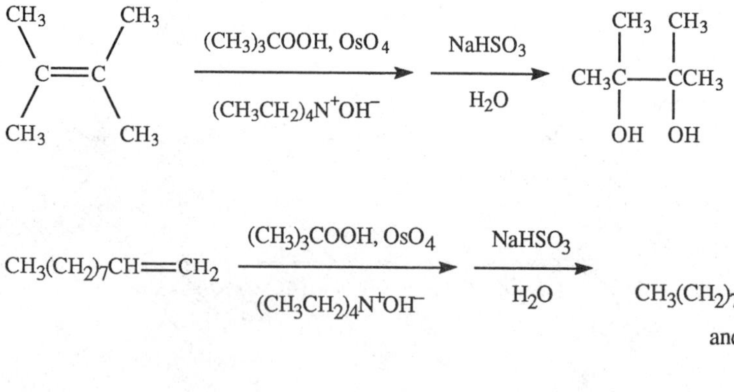

(b)

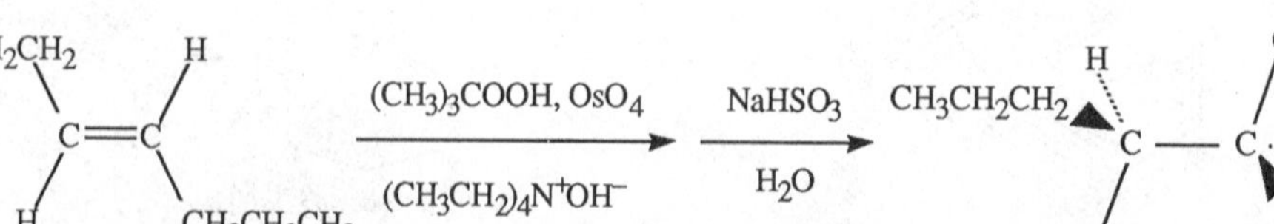

$$CH_3(CH_2)_7CH=CH_2 \xrightarrow[\text{(CH}_3\text{CH}_2)_4\text{N}^+\text{OH}^-]{\text{(CH}_3)_3\text{COOH, OsO}_4} \xrightarrow[\text{H}_2\text{O}]{\text{NaHSO}_3}$$

and enantiomer

(c)

and enantiomer

8.42 (cont)

(d)

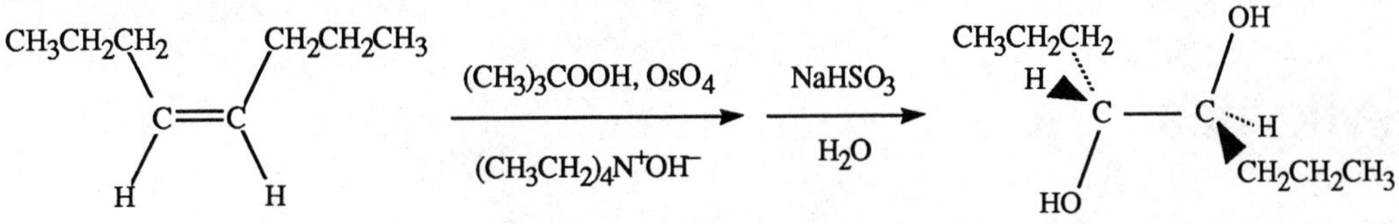

meso compound

8.43

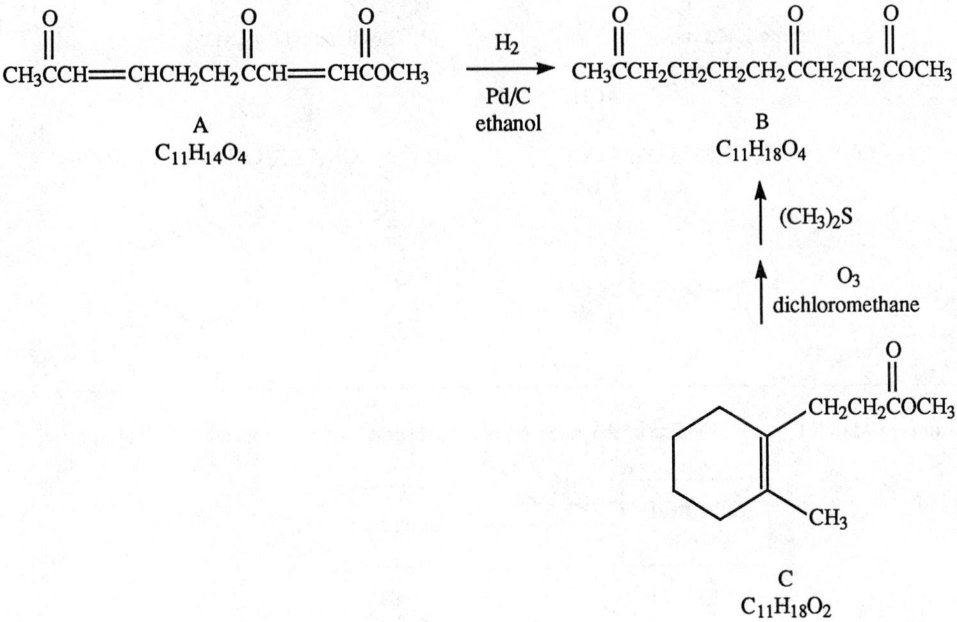

A
$C_{11}H_{14}O_4$

B
$C_{11}H_{18}O_4$

C
$C_{11}H_{18}O_2$

9

Alkynes

9.1 (a) 6,6-dichloro-2-methyl-3-heptyne

(b) 3,3-dimethyl-1-pentyne

(c) 4-bromo-2-hexyne

(d) 2,6-dimethyl-3-octyne

9.2 (a)
$$CH_3CH_2CH_2C\equiv C\overset{\overset{\displaystyle CH_3}{|}}{\underset{\underset{\displaystyle Cl}{|}}{C}}CH_3$$

(b) $CH_3C\equiv CCH_2CHCH_2CH_2CH_3$

(c)
$$CH_3\overset{\overset{\displaystyle CH_3}{|}}{\underset{\underset{\displaystyle CH_3}{|}}{C}}\!-\!C\equiv CCH_2CH_3$$

Concept Map 9.1   Outline of the synthesis of a disubstituted alkyne from a terminal alkyne.

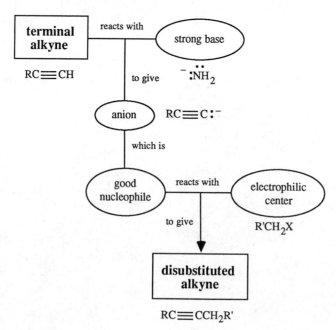

9.3   (a) $CH_3CH_2CH_2C\equiv CH \xrightarrow[\text{NH}_3 \text{ (liq)}]{\text{NaNH}_2} CH_3CH_2CH_2C\equiv C:^-Na^+ \xrightarrow{CH_3CH_2CH_2Br}$

                                                                                    A

$$CH_3CH_2CH_2C\equiv CCH_2CH_2CH_3$$

                                                                                    B

(b) $HC\equiv C:^-Na^+ \xrightarrow[\text{NH}_3 \text{ (liq)}]{\overset{\overset{\displaystyle CH_3}{|}}{CH_3CHCH_2CH_2Br}} \overset{\overset{\displaystyle CH_3}{|}}{CH_3CHCH_2CH_2C\equiv CH}$

        C

                          D

(c) $CH_3CH_2CH_2C\equiv CH \xrightarrow[\text{NH}_3 \text{ (liq)}]{\text{NaNH}_2} CH_3CH_2CH_2C\equiv C:^-Na^+ \xrightarrow{(CH_3)_2SO_4}$

                                                                          F

                     E

$$CH_3CH_2CH_2C\equiv CCH_3$$

                                                                          G

The leaving group in dimethyl sulfate, $(CH_3)_2SO_4$, is the methyl sulfate anion, $^-OSO_3CH_3$.

9.4   (a)

            $\xrightarrow{\text{HBr}}$

(b) $CH_3C\equiv CCH_3 \xrightarrow{\text{HBr (excess)}} \overset{\overset{\displaystyle Br}{|}}{\underset{\underset{\displaystyle Br}{|}}{CH_3CH_2CCH_3}}$

(c) $CH_3CH_2CH_2CH\equiv CH_2 \xrightarrow{\text{HCl}} CH_3CH_2CH_2\underset{\underset{\displaystyle Cl}{|}}{CHCH_3}$

(d)

            $\xrightarrow{\text{HI}}$

(e) $HC\equiv CH \xrightarrow{\text{HBr (excess)}} CH_3CHBr_2$

9.4 (cont)

(f) 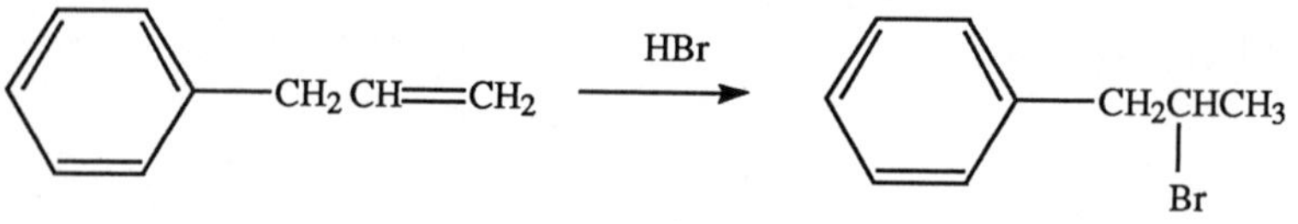

9.5

$$CH_3C{\equiv}C(CH_2)_7\overset{\overset{\large O}{\|}}{C}OH \xrightarrow[H_2SO_4]{80\%} CH_3\overset{\overset{\large O}{\|}}{C}CH_2(CH_2)_7\overset{\overset{\large O}{\|}}{C}OH + CH_3CH_2\overset{\overset{\large O}{\|}}{C}(CH_2)_7\overset{\overset{\large O}{\|}}{C}OH$$

9.6

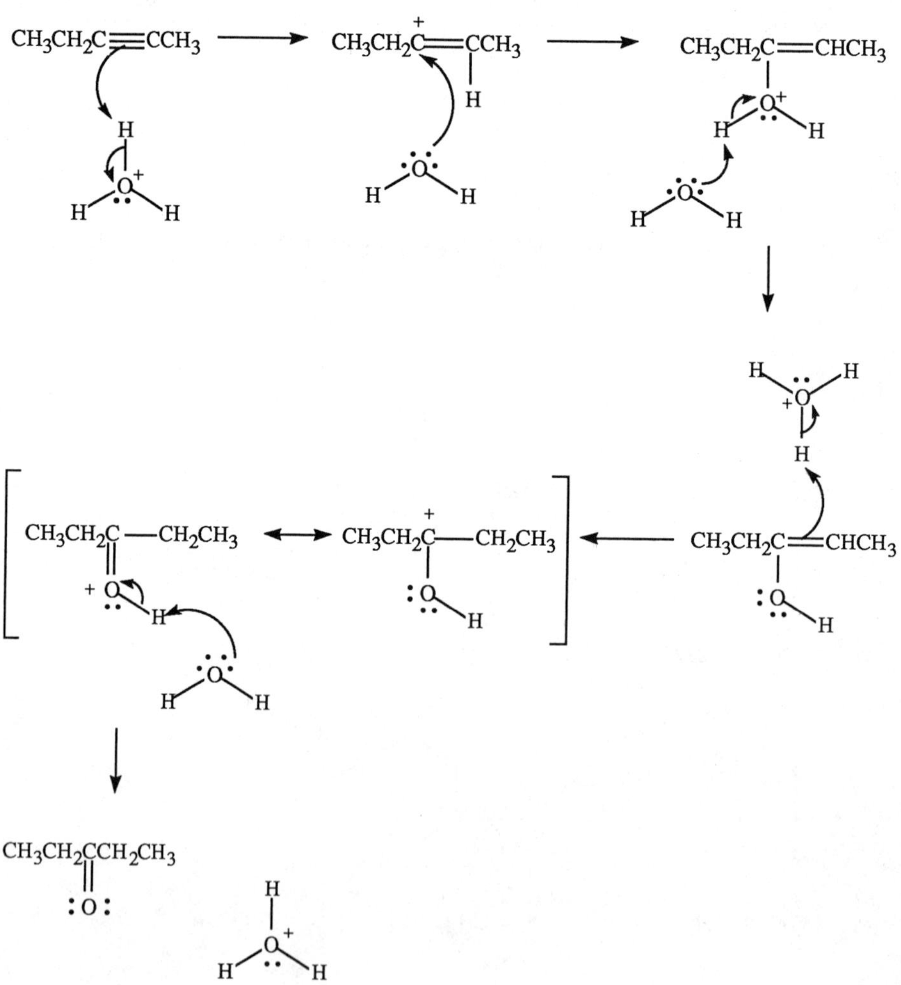

Concept Map 9.2   Electrophilic addition of acids to alkynes.

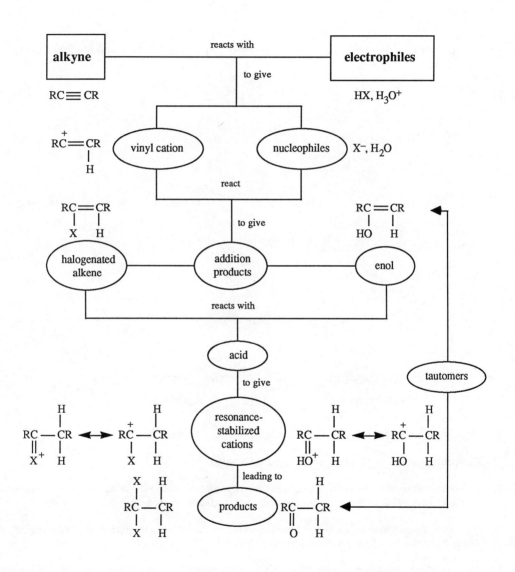

9.7 $HC{\equiv}CH \longrightarrow HC{=}\overset{+}{C}H \longrightarrow CH_2{=}CH \longrightarrow CH_2{=}CH$

$CH_3\overset{}{C}H \longleftarrow CH_3{-}CH \longleftrightarrow CH_3{-}\overset{+}{C}H$

9.8 $CH_3(CH_2)_{11}C{\equiv}CH \xrightarrow{\ ?\ } CH_3(CH_2)_{11}C{\equiv}CCH_2(CH_2)_7CH_3$

                        1-tetradecyne   10-tricosyne

1.   What functional groups are present in the starting material and the product?

The starting material and the product both have a triple bond.

2.   How do the carbon skeletons of the two compounds compare? How many carbon atoms does each contain? Are there any rings? What are the positions of branches and functional groups on the carbon skeletons?

The starting material has a linear chain of 14 carbon atoms; the product has a linear chain of 23 carbon atoms. The starting material is a 1-alkyne and the product is a 10-alkyne.

3.   How do the functional groups change in going from starting material to product? Does the starting material have a good leaving group?

A nonyl group in the product has replaced the terminal hydrogen in the starting material.

4.   Is it possible to dissect the structures of the starting material and product to see which bonds must be broken and which formed?

9.8 (cont)   bond broken   bond formed

(4) (cont) $CH_3(CH_2)_{11}C\equiv C{-}H$ $CH_3(CH_2)_{11}C\equiv C{-}CH_2(CH_2)_7CH_3$

5. Do we recognize any part of the product molecule as coming from a good nucleophile or an electrophilic addition?

    The new carbon-carbon bond must have come from a nucleophilic substitution reaction.

6. What type of compound would be a good precursor to the product?

    The conjugate base of 1-tetradecyne is a good nucleophile for the S_N2 substitution reaction of 1-bromononane.

$$CH_3(CH_2)_{11}C\equiv C{:}^- \xrightarrow{\ CH_3(CH_2)_7CH_2Br\ } CH_3(CH_2)_{11}C\equiv CCH_2(CH_2)_7CH_3$$

7. After this last step, do we see how to get from starting material to product? If not, we need to analyze the structure obtained in step 6 by applying questions 5 and 6 to it.

    Complete correct answer:

$$CH_3(CH_2)_{11}C\equiv CH \xrightarrow[NH_3\ (liq)]{NaNH_2} CH_3(CH_2)_{11}C\equiv C{:}^-Na^+ \xrightarrow{\ CH_3(CH_2)_7CH_2Br\ }$$

$$CH_3(CH_2)_{11}C\equiv CCH_2(CH_2)_7CH_3$$

- -

Concept Map 9.3   Reduction reactions of alkynes.

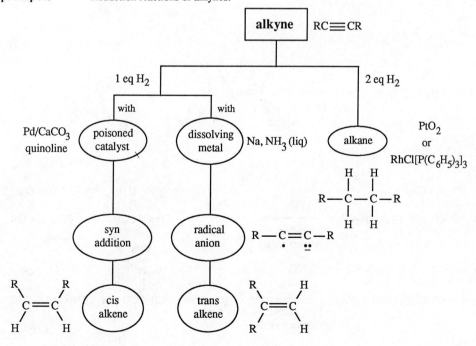

9.9

(a) $CH_3CH_2CH_2CH_2C\equiv CH$ $\xrightarrow[\text{quinoline}]{\text{H}_2 \atop \text{Pd/BaSO}_4}$ $CH_3CH_2CH_2CH_2CH=CH_2$

(b) $CH_3CH_2CH_2CH_2C\equiv CCH_3$ $\xrightarrow[\text{quinoline}]{\text{H}_2 \atop \text{Pd/BaSO}_4}$

$$\underset{H}{\overset{CH_3CH_2CH_2CH_2}{\diagdown}}C=C\underset{H}{\overset{CH_3}{\diagup}}$$

(c) $CH_3CH_2CH_2CH_2C\equiv CCH_3$ $\xrightarrow[\text{NH}_3\text{ (liq)}]{\text{Na}}$

$$\underset{H}{\overset{CH_3CH_2CH_2CH_2}{\diagdown}}C=C\underset{CH_3}{\overset{H}{\diagup}}$$

(d) $CH_3CH_2CH_2C\equiv CCH_3$ $\xrightarrow[\text{RhCl[P(C}_6\text{H}_5)_3]_3]{\text{H}_2}$ $CH_3CH_2CH_2CH_2CH_2CH_3$

(e) $CH_3CH_2CH_2CH_2C\equiv CCH_3$ $\xrightarrow[\text{acetic acid}]{\text{H}_2\text{ (excess)} \atop \text{Pt}}$ $CH_3CH_2CH_2CH_2CH_2CH_2CH_3$

9.10 $CH_3C\equiv CCH_3$ $\xrightarrow[\text{Pd/Al}_2\text{O}_3]{\text{D}_2}$

$$\underset{D}{\overset{CH_3}{\diagup}}C=C\underset{D}{\overset{CH_3}{\diagdown}}$$

9.11

(a)

$$\underset{\underset{\text{}}{|}}{\overset{CH_3}{|}}$$
$CH_3CHC\equiv CCH_2CH_3$ $\xrightarrow[\text{carbon}\atop\text{tetrachloride}]{\text{O}_3}$ $\xrightarrow{\text{H}_2\text{O}}$ $\underset{CH_3}{\overset{\text{O}}{\underset{|}{CH_3CH\overset{\|}{C}OH}}}$ $+$ $CH_3CH_2\overset{\text{O}}{\overset{\|}{C}}OH$

(b)

$$\overset{CH_3}{\underset{|}{}}$$
$CH_3C\equiv CCHCH_2CH_3$ $\xrightarrow[\text{carbon}\atop\text{tetrachloride}]{\text{O}_3}$ $\xrightarrow{\text{H}_2\text{O}}$ $CH_3\overset{\text{O}}{\overset{\|}{C}}OH$ $+$ $\underset{CH_3}{\overset{\text{O}}{\underset{|}{CH_3CH_2CH\overset{\|}{C}OH}}}$

9.12

(a)

$$CH_3\overset{\cdot}{C}H_2CH_2CHCH_3 \xleftarrow{\text{HBr}} CH_3CH_2CH_2CH\!=\!\!CH_2$$
                    |
                    Br

$$\Big\uparrow \begin{array}{c} H_2 \\ Pd/BaSO_4 \\ quinoline \end{array}$$

$$HC\!\equiv\!CH \xrightarrow[\text{NH}_3 \text{ (liq)}]{\text{NaNH}_2} Na^{+\,-}\!:\!C\!\equiv\!CH \xrightarrow{CH_3CH_2CH_2Br} CH_3CH_2CH_2C\!\equiv\!CH$$

(b)

$$\overset{\displaystyle O}{\underset{\displaystyle \|}{}}$$
$$CH_3CH_2CH_2CCH_2CH_3 \xleftarrow[\text{H}_2\text{SO}_4]{\text{H}_2\text{O}} CH_3CH_2C\!\equiv\!CCH_2CH_3 \xleftarrow{CH_3CH_2Br} CH_3CH_2C\!\equiv\!C\!:^-Na^+$$

$$\Big\uparrow \begin{array}{c} NaNH_2 \\ NH_3 \text{ (liq)} \end{array}$$

$$CH_3CH_2C\!\equiv\!CH$$

(c)

$$\xleftarrow[\text{H}_2\text{O}]{\text{KMnO}_4,\ \text{KOH}}$$

$$CH_3CH_2CH_2\diagdown \quad \diagup CH_2CH_3$$
$$\qquad\qquad C\!=\!C$$
$$\qquad\quad H\diagup \qquad \diagdown H$$

and enantiomer

$$\Big\uparrow \begin{array}{c} H_2 \\ Pd/BaSO_4 \\ quinoline \end{array}$$

$$CH_3CH_2CH_2C\!\equiv\!C\!:^-Na^+ \xrightarrow{CH_3CH_2Br} CH_3CH_2CH_2C\!\equiv\!CCH_2CH_3$$

$$\Big\uparrow \begin{array}{c} NaNH_2 \\ NH_3 \text{ (liq)} \end{array}$$

$$CH_3CH_2CH_2C\!\equiv\!CH$$

9.13   (a) (*R*)-2-chloro-4-octyne   (b) (*Z*)-3-methyl-3-heptene

(c) 2-methyl-4-octyn-2-ol   (d) *trans*-1-methyl-2-vinylcyclohexane

9.14   (a)

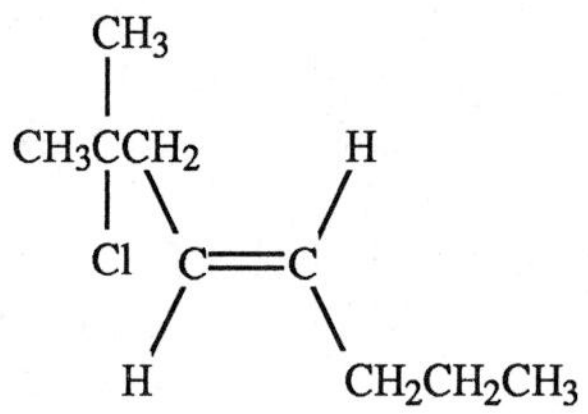

(b)

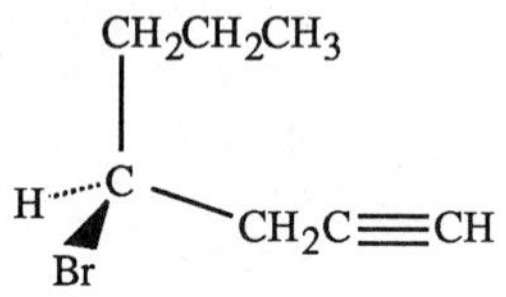

(c)

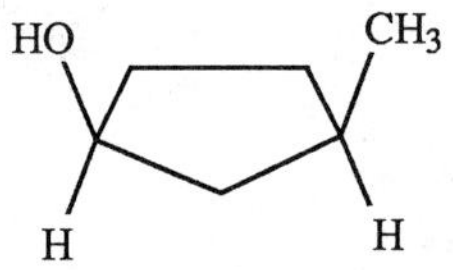

and enantiomer

(d) $CH_3CH_2C \equiv CCH_2CH_2OH$

9.15

(a) $CH_3CH_2C\equiv CCH_3$ $\xrightarrow[\substack{Pd/CaCO_3 \\ quinoline}]{H_2}$

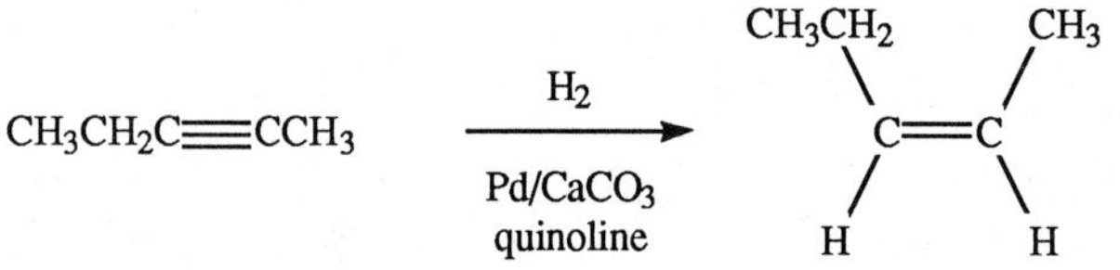

(b) $CH_3CH_2C\equiv CCH_3$ $\xrightarrow[Pt]{H_2 \text{ (excess)}}$ $CH_3CH_2CH_2CH_2CH_3$

(c) $CH_3CH_2C\equiv CCH_3$ $\xrightarrow[NH_3 \text{ (liq)}]{Na}$

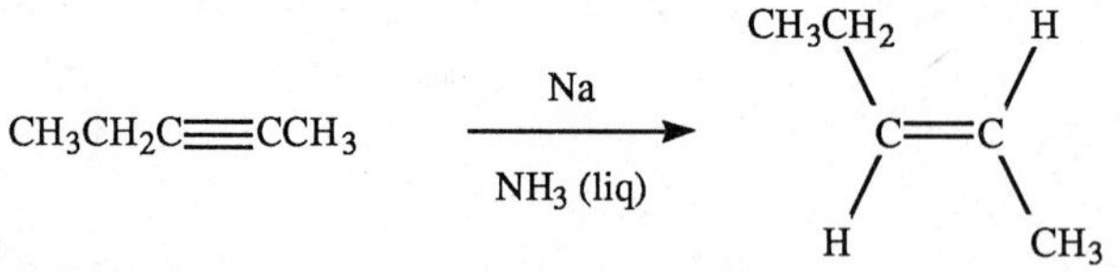

(d)

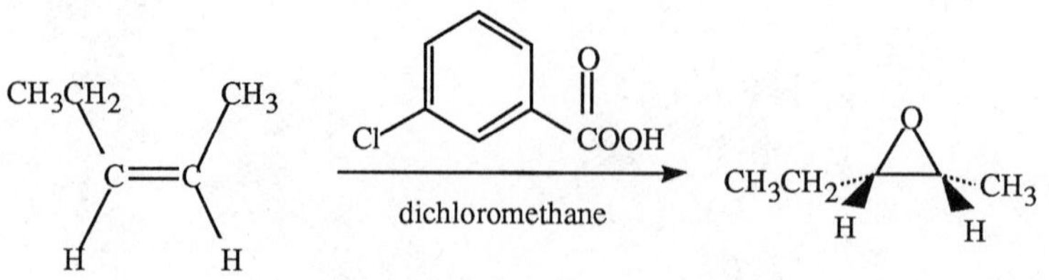

and enantiomer

9.15 (cont)

(e)

CH_3CH_2, H / $C=C$ / H, CH_3
$\xrightarrow{\text{dichloromethane}}$
(3-Cl-benzoic acid, m-CPBA)
$CH_3CH_2\cdots\triangle\cdots H$ (epoxide)   and enantiomer
H CH_3

(f) $CH_3CH_2C\equiv CCH_3$ $\xrightarrow{\text{HBr (1 molar equiv)}}$

CH_3CH_2, Br / $C=C$ / H, CH_3 + CH_3CH_2, H / $C=C$ / Br, CH_3

major products

(g) $CH_3CH_2C\equiv CCH_3$ $\xrightarrow{\text{HBr (2 molar equiv)}}$ $CH_3CH_2CBr_2CH_2CH_3$ + $CH_3CH_2CH_2CBr_2CH_3$

(h) $CH_3CH_2C\equiv CCH_3$ $\xrightarrow{\text{Br}_2\text{ (1 molar equiv)}}$

CH_3CH_2, Br / $C=C$ / Br, CH_3

(i) $CH_3CH_2C\equiv CCH_3$ $\xrightarrow{\text{Br}_2\text{ (2 molar equiv)}}$ $CH_3CH_2CBr_2CBr_2CH_3$

(j) $CH_3CH_2C\equiv CCH_3$ $\xrightarrow[\substack{\text{carbon}\\\text{tetrachloride}}]{O_3}$ $\xrightarrow{H_2O}$

$CH_3CH_2\overset{\overset{\displaystyle O}{\|}}{C}OH$ + $CH_3\overset{\overset{\displaystyle O}{\|}}{C}OH$

(k) $CH_3CH_2C\equiv CCH_3$ $\xrightarrow[\substack{H_2SO_4\\HgSO_4}]{H_2O}$

$CH_3CH_2\overset{\overset{\displaystyle O}{\|}}{C}CH_2CH_3$ + $CH_3CH_2CH_2\overset{\overset{\displaystyle O}{\|}}{C}CH_3$

(l) $CH_3CH_2C\equiv CCH_3$ $\xrightarrow[\text{RhCl[P(C}_6\text{H}_5)_3]_3]{H_2}$ $CH_3CH_2CH_2CH_2CH_3$

9.16

(a)

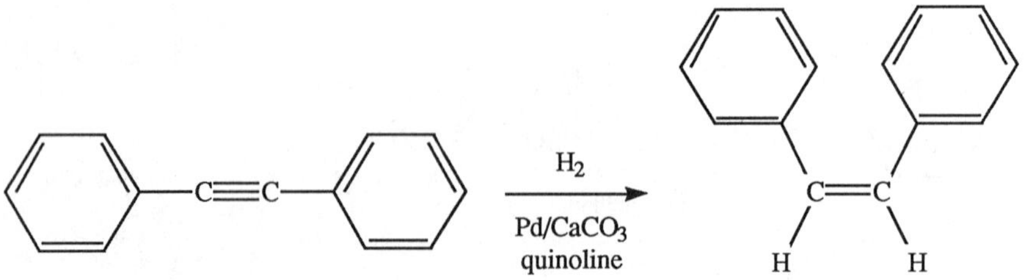

A

(b) $CH_2\!=\!CHBr$ $\xrightarrow{\;HBr\;}$ CH_3CHBr_2

B

(c) $CH_2\!=\!CHCH_2Br$ $\xrightarrow[\substack{\text{carbon}\\\text{tetrachloride}}]{\;Br_2\;}$ $BrCH_2CHBrCH_2Br$

C

(d)

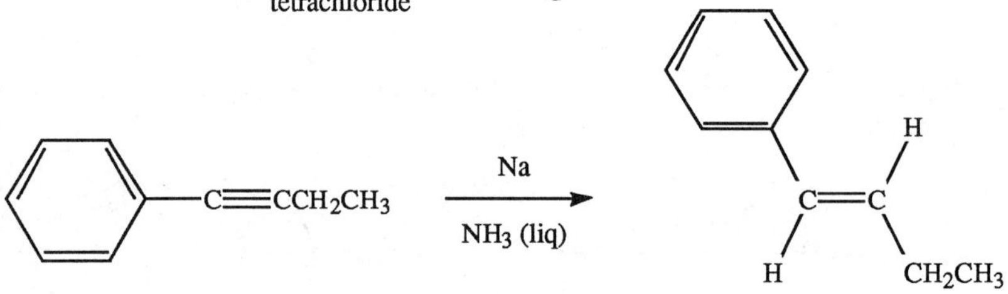

D

(e)

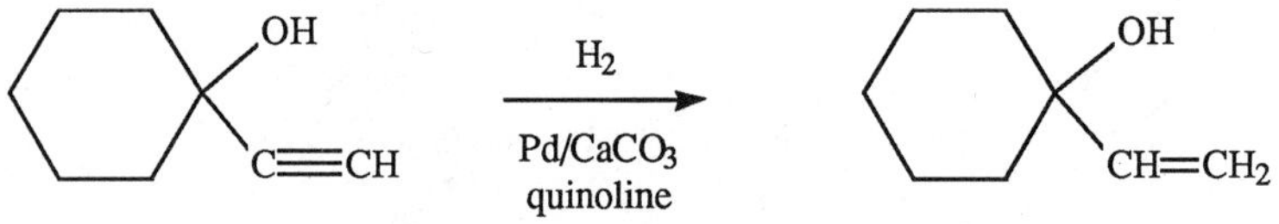

E

(f) $CH_3(CH_2)_7C\!\equiv\!C(CH_2)_7COH$ $\xrightarrow[\substack{\text{carbon}\\\text{tetrachloride}}]{\;O_3\;}$ $\xrightarrow{\;H_2O\;}$ $CH_3(CH_2)_7\overset{\overset{\textstyle O}{\|}}{C}OH$ + $HO\overset{\overset{\textstyle O}{\|}}{C}(CH_2)_7\overset{\overset{\textstyle O}{\|}}{C}OH$

F   G

(g)

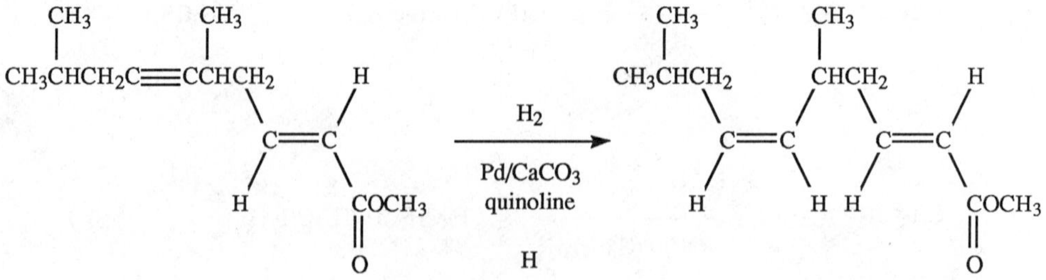

H

9.16 (cont)

(h) $CH_3C{\equiv}CCH_3$ $\xrightarrow[\substack{H_2SO_4 \\ I}]{H_2O}$ $CH_3\overset{\overset{\displaystyle O}{\|}}{C}CH_2CH_3$

(i) $\xrightarrow[\text{NH}_3\text{ (liq)}]{\text{NaNH}_2}$ **J** $\xrightarrow{\text{BrCH}_2\text{CH}_2\text{CH}_2\text{Cl}}$ **K**

$OCH_2C{\equiv}CCH_2CH_2CH_2Cl$ $\xrightarrow{\text{NaCN}}$ $OCH_2C{\equiv}CCH_2CH_2CH_2CN$

L

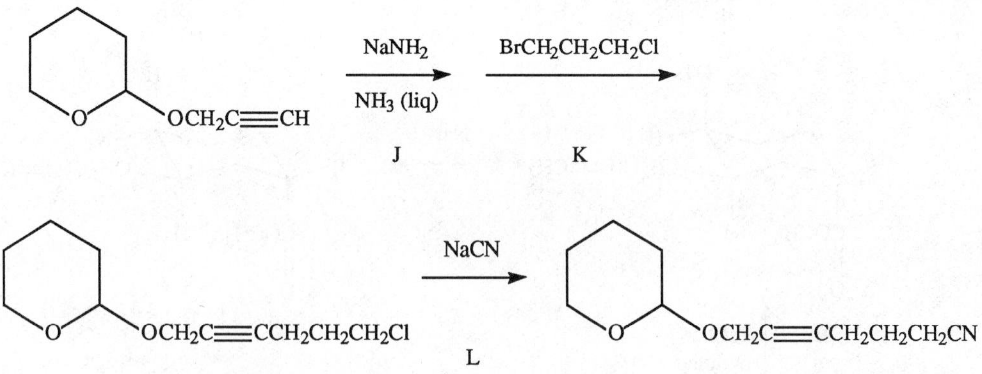

(j) $CH_2{=}CHCH_2CH_2Br$ $\xrightarrow[\text{NH}_3\text{ (liq)}]{HC{\equiv}C^-Na^+}$ $CH_2{=}CHCH_2CH_2C{\equiv}CH$

M

(k) $CH_2{=}CHCHCH_2CH_2OCH_2$ $\xrightarrow{O_3}$ **N** $\xrightarrow{(CH_3)_2S}$ **O**
$\phantom{CH_2{=}CHCH}|$
$\phantom{CH_2{=}CHCH}CH_3$

$O{=}CHCHCH_2CH_2OCH_2$
$\phantom{O{=}CHC}|$
$\phantom{O{=}CHC}CH_3$

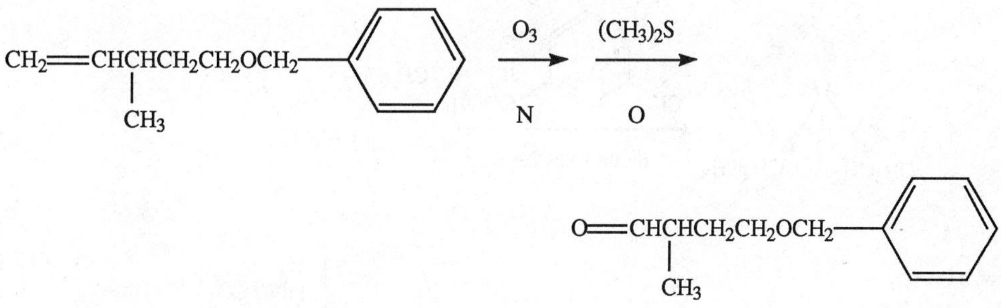

(l) $\overset{\overset{\displaystyle CH_3}{|}}{CH_3CHC}{\equiv}CH$ $\xrightarrow{\text{HBr (excess)}}$ $\overset{\overset{\displaystyle CH_3}{|}}{CH_3CH}CBr_2CH_3$

P

9.17

(a)

(b)

9.17 (cont)

(c)

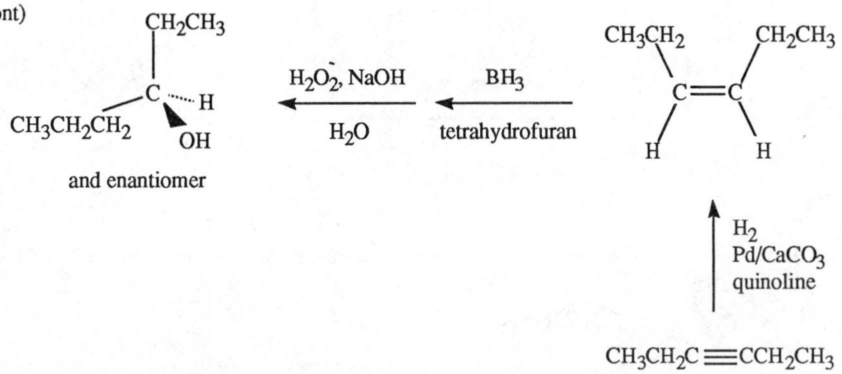

and enantiomer

An alternate synthesis is given below.

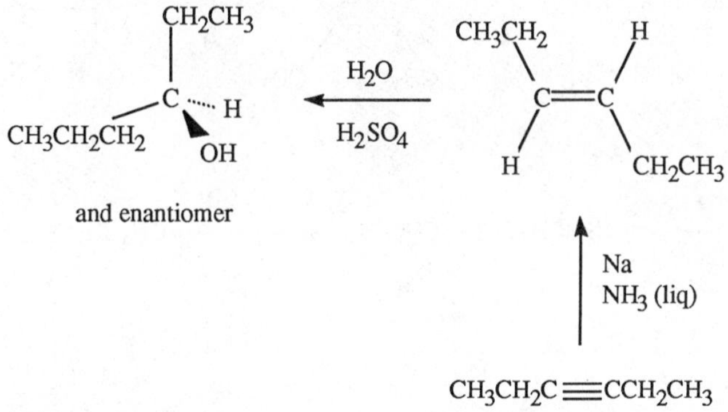

and enantiomer

(d)

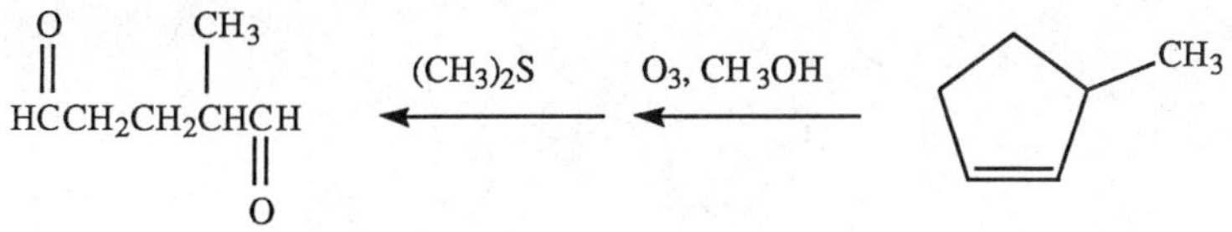

(e)

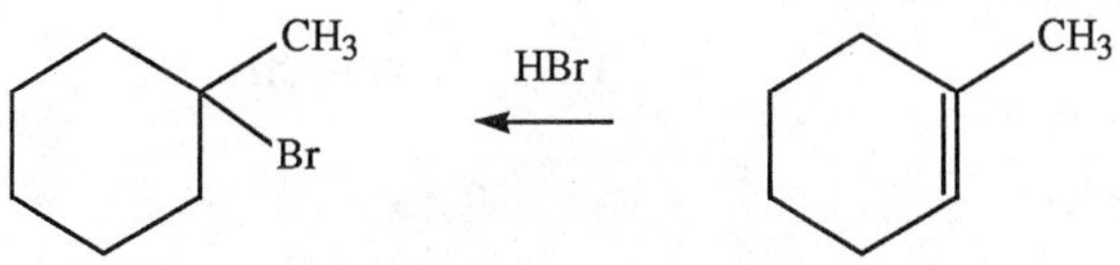

(f)

and enantiomer

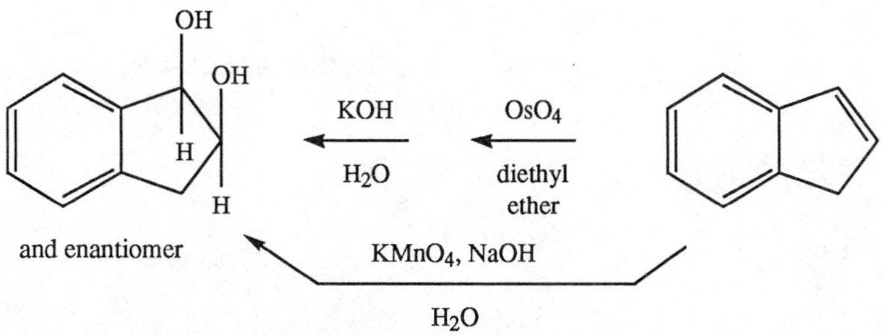

9.17 (cont)

(g)

$$\underset{\substack{H \quad\quad H}}{\overset{\substack{O \quad\quad\quad\quad O \\ \| \quad\quad\quad\quad \| \\ HOCCH_2 \quad\quad COH}}{\triangle}} \xleftarrow{H_3O^+} \xleftarrow[H_2O]{H_2O_2,\ NaOH} \xleftarrow[H_2O]{O_3}$$

(h)

$$\underset{\substack{\| \\ O}}{\overset{\substack{CH_3 \\ | \\ CH_3CHCH}}{}} \xleftarrow[H_2O]{Zn} \xleftarrow{O_3} \overset{\substack{CH_3 \\ | \\ CH_3CHCH=CH_2}}{} \xleftarrow[\substack{ethanol \\ \Delta}]{KOH} \overset{\substack{CH_3 \\ | \\ CH_3CHCH_2CH_2Br}}{}$$

(i)

$$\xleftarrow[dichloromethane]{\overset{O}{\underset{\substack{\| \\ Cl\text{—}\ \ \text{—}COOH}}{}}} \xleftarrow[\substack{H_3PO_4 \\ \Delta}]{} $$

(j)

$$\underset{\substack{H \quad\quad\quad H}}{\overset{\substack{CH_3CH_2 \quad\quad CH_2CH_3 \\ C=C}}{}} \xleftarrow[\substack{Pd/BaSO_4 \\ quinoline}]{H_2} CH_3CH_2C\equiv CCH_2CH_3 \xleftarrow{CH_3CH_2Br} CH_3CH_2C\equiv C:^-Na^+$$

$$\Big\uparrow \substack{NaNH_2 \\ NH_3\ (liq)}$$

$$CH_3CH_2Br \xrightarrow[NH_3\ (liq)]{HC\equiv C:^-Na^+} CH_3CH_2C\equiv CH$$

(k)

$$\underset{\substack{H \quad\quad CH_2CH_3}}{\overset{\substack{O \\ CH_3CH_2\cdots\ \ \ \cdots H}}{}} \xleftarrow[dichloromethane]{\overset{O}{\underset{\substack{\| \\ Cl\text{—}\ \ \text{—}COOH}}{}}} \underset{\substack{H \quad\quad CH_2CH_3}}{\overset{\substack{CH_3CH_2 \quad\quad H \\ C=C}}{}}$$

$$\Big\uparrow \substack{Na \\ NH_3\ (liq)}$$

$$CH_3CH_2C\equiv CCH_2CH_3$$
from (j)

9.18   (a) $C_{13}H_{10}O \xrightarrow[\text{cat}]{H_2} C_{13}H_{14}O$ The corresponding saturated hydrocarbo
would have the formula $C_{13}H_{28}O$.

$C_{13}H_{10}O$ has 9 units of unsaturation $\left(\dfrac{28-10}{2}=9\right)$.

Ater hydrogenation, 7 units of unsaturation remain $\left(\dfrac{28-14}{2}=7\right)$.

Two units of unsaturation $(9-7=2)$ correspond to multiple bonds that are easily hydrogenated.

Possible structures for Carlina-oxide are

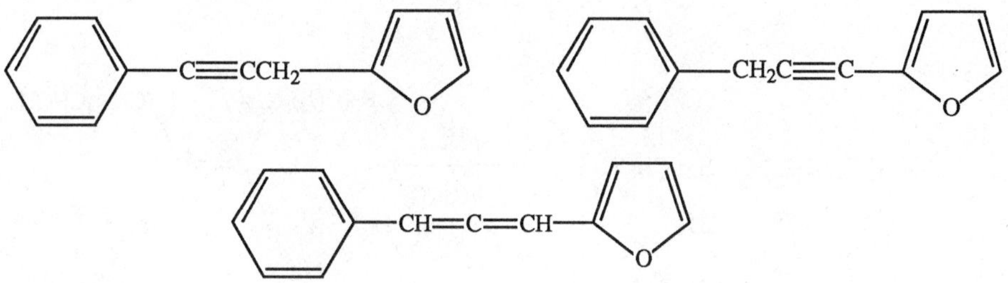

(b)

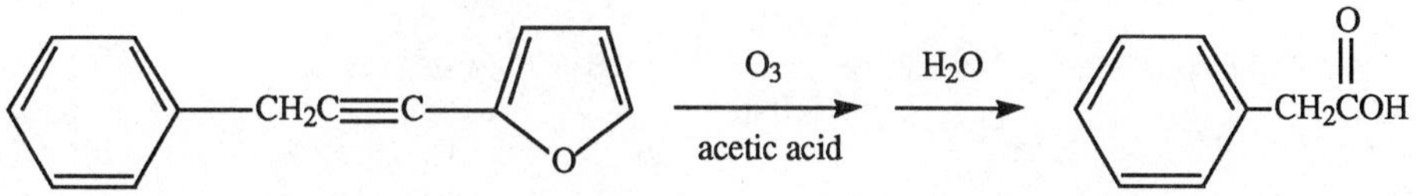

Carlina-oxide   + unisolated compounds

9.19

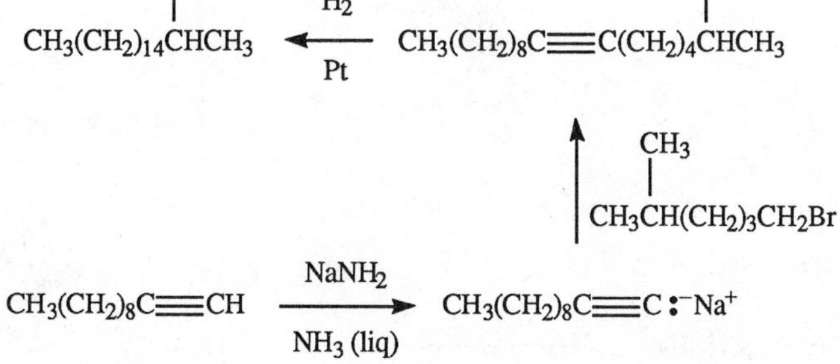

9.20

(a)

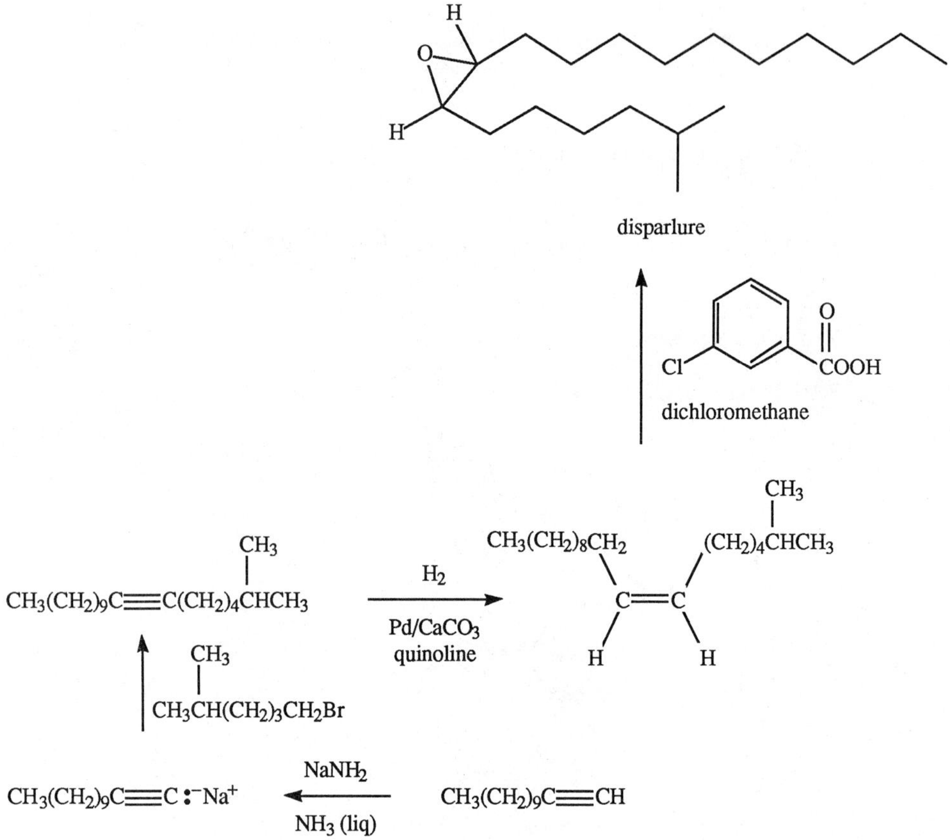

disparlure

3-chlorobenzoic acid (m-chloroperoxybenzoic acid), dichloromethane

$$CH_3(CH_2)_8CH_2 \quad (CH_2)_4CHCH_3 \;|\; CH_3$$
$$C=C$$
$$H \qquad H$$

H_2, Pd/CaCO$_3$, quinoline

$$CH_3(CH_2)_9C\equiv C(CH_2)_4\overset{\overset{\displaystyle CH_3}{|}}{C}HCH_3$$

$$CH_3\overset{\overset{\displaystyle CH_3}{|}}{C}H(CH_2)_3CH_2Br$$

$$CH_3(CH_2)_9C\equiv C:^-Na^+ \quad \xleftarrow[\text{NH}_3\text{ (liq)}]{\text{NaNH}_2} \quad CH_3(CH_2)_9C\equiv CH$$

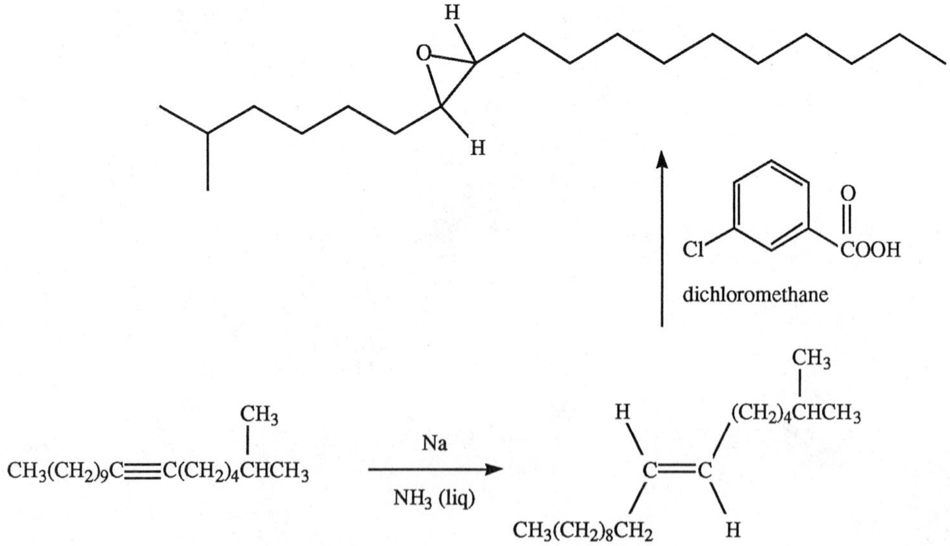

(b)

3-chlorobenzoic acid, dichloromethane

$$H \qquad (CH_2)_4\overset{\overset{\displaystyle CH_3}{|}}{C}HCH_3$$
$$C=C$$
$$CH_3(CH_2)_8CH_2 \qquad H$$

Na, NH$_3$ (liq)

$$CH_3(CH_2)_9C\equiv C(CH_2)_4\overset{\overset{\displaystyle CH_3}{|}}{C}HCH_3$$

9.21

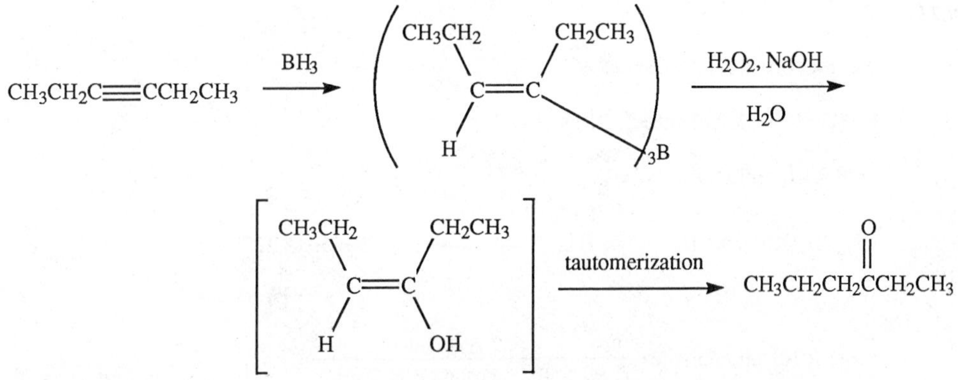
CH3CH2C≡CCH2CH3
BH3
CH3CH2
CH2CH3
C=C
H
3B
H2O2, NaOH
H2O
CH3CH2
CH2CH3
C=C
H
OH
tautomerization
O
CH3CH2CH2CCH2CH3

9.22

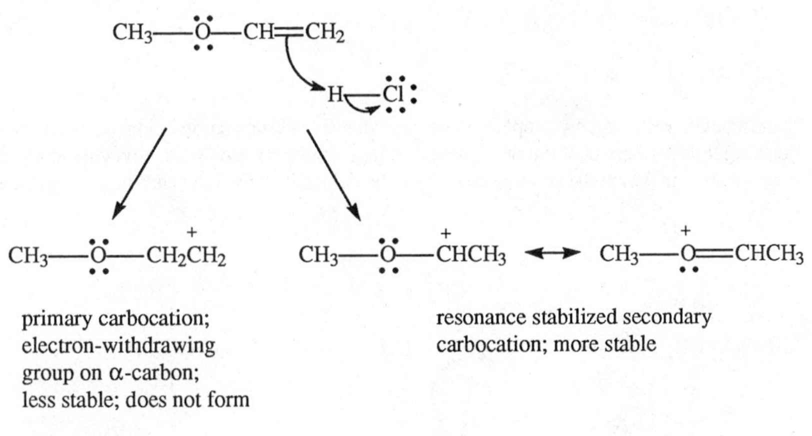
CH3—O—CH=CH2
H—Cl
CH3—O—CH2CH2
CH3—O—CHCH3
CH3—O=CHCH3
primary carbocation;
electron-withdrawing
group on α-carbon;
less stable; does not form
resonance stabilized secondary
carbocation; more stable

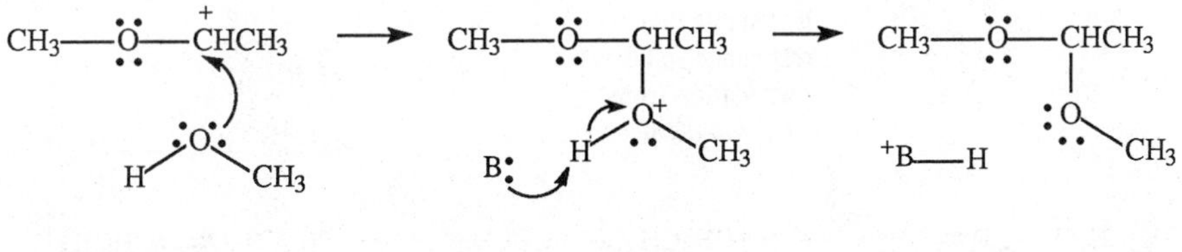
CH3—O—CHCH3
O
H
CH3
CH3—O—CHCH3
B
H
O+
CH3
CH3—O—CHCH3
+B—H
O
CH3

9.23

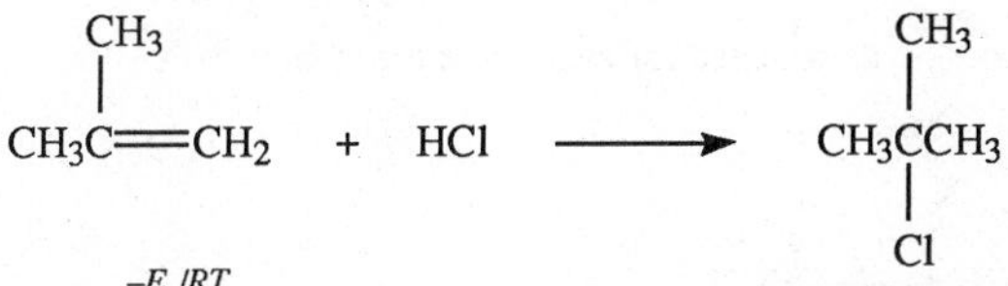
CH3
CH3C=CH2
+
HCl
CH3
CH3CCH3
Cl

$$k = Ae^{-E_a/RT}$$

$$k = (1 \times 10^{11}\ \text{mL/mol•s})(\exp\frac{-28.8\ \text{kcl/mol}}{(1.99 \times 10^{-3}\ \text{kcal/mol•K})(298\ \text{K})})$$

9.23 (cont)

$k = (1 \times 10^{11} \text{ mL/mol·s})(\exp\text{–}48.6)$

$k = (1 \times 10^{11} \text{ mL/mol·s})(8.1 \times 10^{-22})$

$k = 8 \times 10^{-11} \text{ mL/mol·s}$

$$CH_3CH_2OCH{=}CH_2 \quad + \quad HCl \quad \longrightarrow \quad \underset{\displaystyle \underset{Cl}{|}}{CH_3CH_2OCHCH_3}$$

$k = (5 \times 10^8 \text{ mL/mol·s})(\exp\dfrac{-14.7 \text{ kcal/mol}}{(1.99 \times 10^{-3} \text{ kcal/mol·K})(298 \text{ K})})$

$k = (5 \times 10^8 \text{ mL/mol·s})(\exp\text{–}24.8)$

$k = (5 \times 10^8 \text{ mL/mol·s})(1.7 \times 10^{-11})$

$k = 9 \times 10^{-3} \text{ mL/mol·s}$

In the reaction with 2-methylpropene, the intermediate is a tertiary carbocation that is stabilized by the electron-donating inductive effect of the three methyl groups. In the reaction with ethyl vinyl ether, the intermediate is a carbocation in which the positive charge can be delocalized over the carbon atom and the adjacent oxygen atom.

tertiary carbocation
less stable than the
resonance stabilized
carbocation from
ethyl vinyl ether

major contributor

delocalized carbocation; more stable

9.23 (cont)

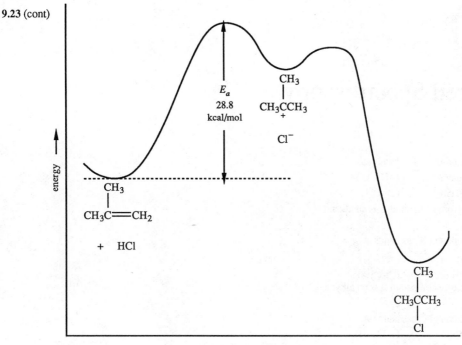

reaction coordinate

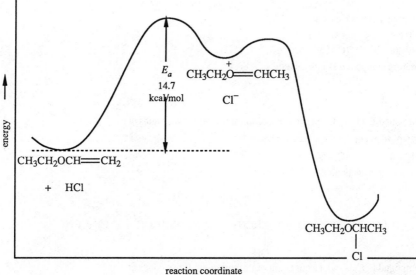

reaction coordinate

The reactive intermediate formed in this reaction is more highly stabilized, and therefore, closer in energy to the energy level of the starting materials than is the case for 2-methylpropene. The E_a for this reaction is therefore lower, and the reaction is faster.

10

Infrared Spectroscopy

10.1 Compound A has bands at

3090 cm^{-1} (=C—H)
1650 cm^{-1} (C=C)

therefore compound A is an alkene.

Compound B has bands at

3300 cm^{-1} (≡C—H)
2100 cm^{-1} (C≡C)
therefore compound B is an alkyne.

Compound C has bands at

3330 cm^{-1} (—O—H)
1060 cm^{-1} (C—O)

therefore compound C is an alcohol.

Compound D has bands at

3080 cm^{-1} (=C—H)
1640 cm^{-1} (C=C)

therefore compound D is an alkene.

10.2 Compound E, $C_4H_{10}O$, has no units of unsaturation. It has bands at

$3325 \text{ (broad) cm}^{-1}$ (—O—H)
1040 cm^{-1} (C—O)

therefore it is an alcohol.
Possible structures are:

$$\begin{array}{ccccc}
 & & & \text{CH}_3 & \text{CH}_3 \\
 & & & | & | \\
\text{CH}_3\text{CH}_2\text{CH}_2\text{CH}_2\text{OH} & \text{CH}_3\text{CH}_2\text{CHCH}_3 & \text{CH}_3\text{CHCH}_2\text{OH} & \text{CH}_3\text{CCH}_3 \\
 & | & & | \\
 & \text{OH} & & \text{OH}
\end{array}$$

Compound F, $C_6H_{12}O$, has one unit of unsaturation. It has bands at

$3325 \text{ (broad) cm}^{-1}$ (—O—H)
1060 cm^{-1} (C—O)

therefore it must be an alcohol. Since there is one unit of unsaturation, and no alkene stretching frequency, it must be a cyclic alcohol.

10.2 (cont)

Possible structures are:

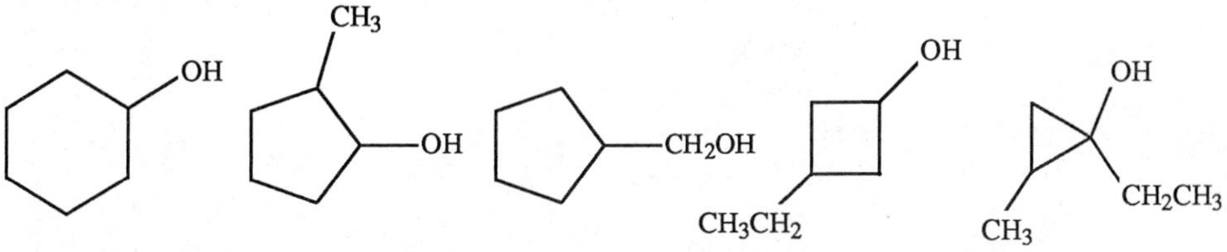

Compound G, $C_6H_{12}O$, has one unit of unsaturation. It has bands at
1700 cm^{-1} (C=O)
The unit of unsaturation corresponds to the carbonyl group, therefore it cannot have a ring.
Possible structures are:

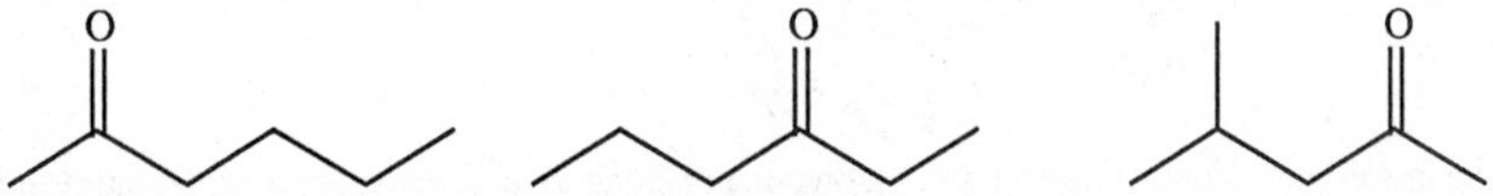

Compound H, C_4H_8O, has one unit of unsaturation. It has bands at
2700 cm^{-1} (O=C—H) (The band at 2900 cm^{-1} is hidden in the alkane band.)
1740 cm^{-1} (C=O)

Possible structures are:

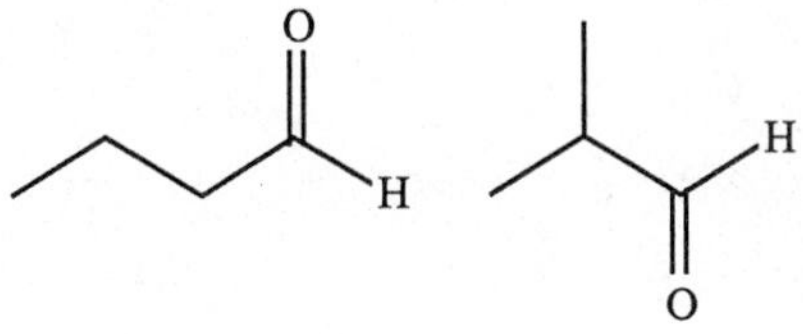

10.3   The best approach to identifying which spectrum belongs to which compound is to first determine what functional groups are present in each compound and then to look for the specific bands in the spectra that correspond to the functional groups.

(a) There is only one alcohol, which should have a broad band at ~3350 cm^{-1} and no band in the carbonyl region. The only spectrum that fits is spectrum 4. Spectrum 4, therefore, must belong to 1-octanol.
     3325 (broad) cm^{-1} (—O—H)
     1060 cm^{-1} (C—O)

(b) Two of the compounds are acids. Acids should have a broad band at ~3300–2400 cm^{-1} and a band in the carbonyl region. Spectra 5 and 6 have the bands that correspond to a carboxylic acid. 3-Butenoic acid is also an alkene, and its spectrum should have a band at ~1600 cm^{-1}. Spectrum 5 has a band at 1645 cm^{-1}, but spectrum 6 does not have a band in this region, therefore spectrum 5 belongs to 3-butenoic acid and spectrum 6 belongs to octanoic acid.

10.3 (cont)

Spectrum 5 (3-butenoic acid).
3100–2500 (broad) cm^{-1} (—O—H)
1715 cm^{-1} (C=O)
1645 cm^{-1} (C=C)
Spectrum 6 (octanoic acid).
3100–2500 (broad) cm^{-1} (—O—H)
1710 cm^{-1} (C=O)

(c)   One of the compounds is an aldehyde, which should have a band in the carbonyl region and a band at 2700 cm^{-1}. Spectrum 2 has these two bands and must, therefore belong to 10-undecenal, an unsaturated aldehyde. Spectrum 2 also has a band at ~1600 cm^{-1}, which confirms the presence of an alkene.
2715 cm^{-1} (O=C—H)
1725 cm^{-1} (C=O)
1640 cm^{-1} (C=C)

(d)   Of the last two spectra to assign (1 and 3), one must belong to an ester (methyl pentanoate) and the other to a ketone (3-methylcyclohexanone). Both have bands in the carbonyl region, but spectrum 1 has broad bands between 1000-1300 cm^{-1}, suggesting that this spectrum belongs to an ester. Spectrum 3 must, therefore, belong to the ketone.
Spectrum 1 (methyl pentanoate)
1740 cm^{-1} (C=O)
1260, 1175 cm^{-1} (C—O)
Spectrum 3 (3-methylcyclohexanone)
1710 cm^{-1} (C=O)

10.4   Compound I has bands at
1740 cm^{-1} (C=O)
1240, 1050 cm^{-1} (C—O)
It is therefore an ester.

Compound J has bands at
3325 (broad) cm^{-1} (—O—H)
1060 cm^{-1} (C—O)
There is no carbonyl absorption; it is therefore an alcohol.

Compound K has bands at
3100 cm^{-1} (=C—H)
1640 cm^{-1} C=C
It is therefore an alkene.

Compound L has bands at
2720 cm^{-1} (O=C—H)
1730 cm^{-1} (C=O)
It is therefore an aldehyde.

10.5   Compound M, C_3H_3Cl, has 2 units of unsaturation. It has bands at

  $3300\ cm^{-1}$ ($\equiv$C—H)

  $2130\ cm^{-1}$ (C$\equiv$C)

The two units of unsaturation correspond to the triple bond. It is a terminal alkyne.
There is only one possible structure.

HC$\equiv$CCH$_2$Cl

Compound N, $C_5H_{10}O$, has one unit of unsaturation. It has bands at

  $1715\ cm^{-1}$ (C=O)

The absorption at $1715\ cm^{-1}$ indicates that it has a carbonyl group. With only one unit of unsaturation, it cannot contain a ring. No absorption at ~$2700\ cm^{-1}$ indicates that it is a ketone and not an aldehyde.
Possible structures are:

Compound O, $C_5H_{10}O_2$, has one unit of unsaturation. It has bands at

  3200–2500 (broad) cm^{-1} (—O—H)

  $1710\ cm^{-1}$ (C=O)

These two bands indicate a carboxylic acid. With one unit of unsaturation in the carbonyl group, the compound cannot contain a ring.
Possible structures are:

Compound P, $C_8H_8O_2$, has five units of unsaturation. It has bands at

  $1725\ cm^{-1}$ (C=O)

  $1600\ cm^{-1}$ (C=C (aromatic))

  1280, 1110 cm^{-1} (C—O)

When a compound has at least four units of unsaturation, an aromatic ring is usually present. This is confirmed by the band at $1600\ cm^{-1}$. The fifth unit of unsaturation is the carbonyl group. The bands at 1280 and $1110\ cm^{-1}$, along with the presence of a carbonyl, indicate an aromatic ester.
There is only one possible structure.

10.6   Compound Q has bands at

3335 (broad) cm^{-1} (—O—H)
3080 cm^{-1} (C=C—H)
1650 cm^{-1} (C=C)
1048 cm^{-1} (C—O)

It is an alcohol and an alkene.

Compound R has bands at

2200 cm^{-1} (C≡C)
1675 cm^{-1} (C=O)

No band is seen at 3300 cm^{-1}, therefore the compound is not a terminal alkyne. The low value for the carbonyl absorption suggests that the carbonyl group is conjugated with the multiple bond. The absence of a band at 2700 cm^{-1} indicates that it is a conjugated ketone and not a conjugated aldehyde.

Compound S has bands at

3460 (broad) cm^{-1} (—O—H)
1710 cm^{-1} (C=O)

There is no band at 2700 cm^{-1}, therefore the compound is a ketone and not an aldehyde. The other functional group in the compound is an alcohol.

Compound T has bands at

3100 cm^{-1} (C=C—H)
1740 cm^{-1} (C=O)
1640 cm^{-1} (C=C)
1240, 1040 cm^{-1} (C—O)

The bands at 1240 and 1040 cm^{-1}, along with the presence of a carbonyl, indicate an ester. The other functional group is an alkene.

10.7   (+)-17-Methyltestosterone

11

Nuclear Magnetic Resonance Spectroscopy

11.1 Arrows point to chemical-shift equivalent hydrogen atoms. Hydrogen atoms on the same methyl group are always chemical-shift equivalent. Hydrogen atoms on the same methylene group are usually, but not always, chemical-shift equivalent.

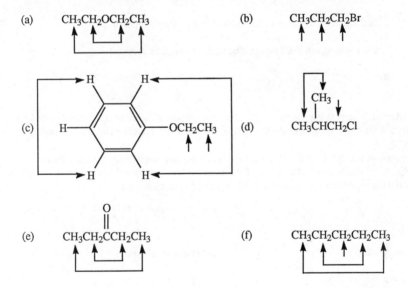

11.2 Compound A is 1,1,2-trichloroethane, $ClCH_2CHCl_2$. This is the only compound of the three for which we expect to see more than one band. The hydrogen atoms attached to the two carbon atoms of 1,1,2-trichloroethane are not chemical-shift equivalent. One is bonded to a carbon atom with two chlorine atoms and absorbs farther from TMS than the other two, which are bonded to a carbon atom with only one chlorine atom.

Acetone and 1,2-dichloroethane will each have only a single band. We can distinguish them from each other by the position of the band in the spectrum.

Compound B is acetone, CH_3CCH_3, with O double bonded. All six hydrogen atoms in acetone are chemical-shift equivalent. A single absorption band is seen in the spectrum at approximately the same place as one of the bands in the spectrum of methyl acetate (Figure 11.1).

263

11.2 (cont)

Compound C is 1,2-dichloroethane, $ClCH_2CH_2Cl$. All four hydrogen atoms in 1,2-dichloroethane are chemical-shift equivalent and give rise to a singlet in the spectrum. The position of the band in the spectrum of 1,2-dichloroethane is similar to that of one of the bands in the spectrum of 1,1,2-trichloroethane.

11.3   (a) $^{28}_{14}Si$ even atomic and mass number; no nuclear magnetic moment

(b) $^{15}_{7}N$ odd mass and atomic number; has nuclear magnetic moment

(c) $^{19}_{9}F$ odd mass and atomic number; has nuclear magnetic moment

(d) $^{31}_{15}P$ odd mass and atomic number; has nuclear magnetic moment

(e) $^{11}_{5}B$ odd mass and atomic number; has nuclear magnetic moment

(f) $^{32}_{16}S$ even atomic and mass number; no nuclear magnetic moment

11.4   The band which moves progressively upfield as the solution of the alcohol is diluted is the band for the hydrogen atom of the hydroxyl group. As the alcohol is diluted, the amount of hydrogen bonding decreases.

Hydrogen bonding further deshields a hydrogen atom because the hydrogen atom involved in hydrogen bonding is partially bonded to another electronegative atom (O, N, or F) and has even less electron density around it than a hydrogen atom which is not involved in hydrogen bonding.

R—O—H   R—O—H - - - - ·O—R
                                                                /
                                                              H

dilute solution   concentrated solution

no hydrogen bonding;   hydrogen bonding; the
the hydrogen atom is less   hydrogen atom is more
deshielded and absorbs   deshielded and absorbs
farther upfield   farther downfield

11.5

The absorption bands between δ 1.0 and 3.0 in the spectrum of cyclohexene fall in the region for hydrogen atoms bonded to sp^3-hybridized carbon atoms. The four hydrogen atoms absorbing farther downfield (δ ~2) are the allylic hydrogen atoms on carbon atoms 3 and 6. These hydrogens are more deshielded than the hydrogens on carbon atoms 4 and 5 because they are closer to the double bond. The four hydrogen atoms farther upfield (δ ~1.5) are those on carbon atoms 4 and 5.

11.6   In this problem, in which we assign the structure of a compound from its molecular formula and spectral data, the proton magnetic resonance data are analyzed in two ways. The chemical shift values for groups of different hydrogen atoms are read off the spectrum, these values are compared with the chemical shift values given in Table 11.1 for different types of hydrogen atoms, and a tentative assignment of partial structure is made. The integration of the spectrum is examined by measuring the relative heights of the steps in the integration curves. The two types of information are summarized in the answers to these problems in the following way:

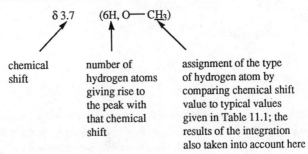

δ 3.7   (6H, O—CH₃)

| chemical shift | number of hydrogen atoms giving rise to the peak with that chemical shift | assignment of the type of hydrogen atom by comparing chemical shift value to typical values given in Table 11.1; the results of the integration also taken into account here |

Compound D, $C_8H_{10}O_2$, has four units of unsaturation. Integration of the spectrum gives a ratio of 1.6/1 or 3/2 for the relative areas under the peaks. The molecular formula tells us that there are 10 hydrogen atoms that absorb, therefore we have six hydrogen atoms that absorb at δ 3.7 and four that absorb at δ 6.8. The results of this reasoning and of our inspection of Table 11.1 are summarized below.

δ 3.7   (6H, —OC$\underline{H}_3$)
δ 6.8   (4H, Ar$\underline{H}$)

The compound is 1,4-dimethyoxybenzene.

Note that the symmetry of the structure of the compound is reflected in the simplicity of the spectrum.

CH₃O—⟨benzene ring⟩—OCH₃

Compound E, C_8H_{10}, has four units of unsaturation. Integration of the spectrum gives a ratio of 1.6/1 or 3/2 (or 6/4 for a total of ten hydrogen atoms in E) for the relative areas under the peaks.

δ 2.3   (6H, ArC$\underline{H}_3$)
δ 6.9   (4H, Ar$\underline{H}$)

As with the spectrum of Compound D, the appearance of the spectrum of Compound E suggests symmetry in the structure. The compound is 1,4-dimethylbenzene (*p*-xylene).

CH₃—⟨benzene ring⟩—CH₃

11.6 (cont)

Compound F, C_7H_7Cl, has four units of unsaturation. (Each halogen atom counts as one hydrogen atom in determining units of unsaturation.) Integration of the spectrum gives a ratio of 2.6/1 or 5/2 (for a total of seven hydrogen atoms in F) for the relative areas under the peaks.

$\delta\,4.4$ (2H, Y—$\underline{CH}_2$—X, where both X and Y deshield the methylene group)

$\delta\,7.3$ (5H, Ar$\underline{H}$)

The compound is benzyl chloride.

Note that in using Table 11.1, we must realize that the values given will not match exactly the chemical shift values that we get from the spectra. We must put all the information together to make assignments of structure. In this problem, once we know that an aryl ring is present, we understand why the hydrogen atoms of the methylene group absorb at lower field than a typical RCH$_2$X group (δ 3.5 in Table 11.1).

Compound G, C_8H_8O, has five units of unsaturation. Integration of the spectrum gives a ratio of 1.7/1 or 5/3 (for a total of eight hydrogen atoms in G) for the relative areas under the peaks.

$\delta\,2.5$ (3H, —C$\underline{CH}_3$)

$\delta\,7\text{–}8$ (5H, Ar$\underline{H}$)

The compound is acetophenone (1-phenylethanone).

The presence of the carbonyl group in the molecule was deduced from the chemical shift of the methyl group, the total number of carbon atoms in the molecule, and the one unit of unsaturation that had to be accounted for once the aromatic ring, with four units of unsaturation, was identified.

11.7   The spectrum of bromoethane has two groups of peaks with an integration giving a ratio of 1.5/1 or 3/2 for the relative areas under the groups of small peaks.

These hydrogen atoms are adjacent to three chemical-shift equivalent hydrogen atoms. They give rise to a quartet centered at δ 3.3.

These hydrogen atoms are adjacent to two chemical-shift equivalent hydrogen atoms. They give rise to a triplet centered at δ 1.8.

11.7 (cont)

There are four possible spin orientations of the two hydrogen atoms of the methylene group. Two of these are indistinguishable. The magnetic field sensed by the adjacent hydrogen atoms varies slightly with a probability of 1:2:1, giving rise to similar relative intensities of the small peaks in the triplet.

(The value of the coupling constant is read off the spectrum.)

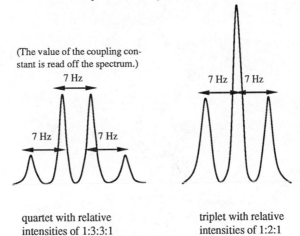

quartet with relative
intensities of 1:3:3:1

triplet with relative
intensities of 1:2:1

The data from the spectrum may be summarized as follows:

δ 1.8   (3H, triplet, J 7 Hz, C$\underline{H}_3$CH$_2$—)
δ 3.3   (2H, quartet, J 7 Hz, CH$_3$C$\underline{H}_2$—)

or as δ 1.8   (3H, t, J 7 Hz, C$\underline{H}_3$CH$_2$—)
        δ 3.3   (2H, q, J 7 Hz, CH$_3$C$\underline{H}_2$—)

11.8   In the problems that follow, a third type of information is added to the two that were used to make structural assignments in Problem 11.6. The splitting patterns are analyzed, coupling constants are measured if necessary, and a determination is made of the number of hydrogen atoms coupling with the ones that give rise to the peak we are analyzing. The information is presented in the way shown at the end of Problem 11.7.

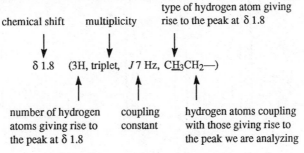

11.8 (cont)

Compound H, C_9H_{12}, has four units of unsaturation. The spectrum has bands at

δ 1.3	(3H, doublet, J 8 Hz, C$\underline{H}_3$—CH—C$\underline{H}_3$)
δ 2.8	(1H, septet, J 8 Hz, CH$_3$—C$\underline{H}$—CH$_3$)
δ 6.8	(5H, singlet, Ar$\underline{H}$)

The compound is isopropylbenzene (cumene).

11.9

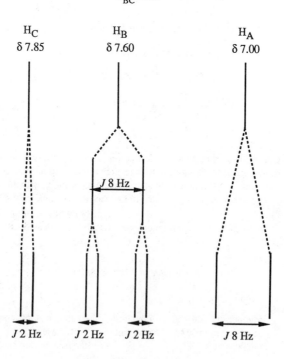

J_{AB} 8 Hz

J_{BC} 2 Hz

11.10

$$CH_3CH_2CH_2Br$$
$$a \quad \underline{b} \quad c$$

$$J_{ab} = J_{bc} = 7 \text{ Hz}$$

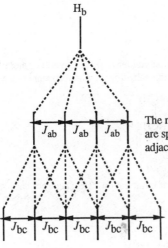

The methylene hydrogen atoms are split into a quartet by the adjacent methyl hydrogen atoms.

Each peak of the quartet is further split into a triplet by the other methylene hydrogen atoms. If the coupling constants for the two types of interactions are the same, the triplets overlap to give a sextet.

11.11   The top spectrum is that of 4-methylbenzonitrile. The symmetrical splitting pattern for the hydrogen atoms on the aromatic ring, similar in appearance to a doublet of doublets, is typical of an aromatic ring that is substituted in the para position with two substituents of differing electronegativity.

The bottom spectrum is that of 3-methylbenzonitrile. An aromatic ring with two substituents of differing electronegativity meta to each other will not give the characteristic doublet of doublets.

11.12   Compound I, $C_5H_{10}O$, has one unit of unsaturation. The spectrum has bands at

δ 1.2   (6H, singlet, $(C\underline{H}_3)_2C$)
δ 3.2   (1H, singlet, —O$\underline{H}$)
δ 4.6–5.0   (2H, multiplet, $C\underline{H}_2$=CH—)
δ 5.2–5.9   (1H, multiplet, CH_2=C$\underline{H}$—)

The compound is 2-methyl-3-buten-2-ol. The assignment of the band at δ 3.2 could not be made until it became clear that the unit of unsaturation is a double bond (bands at δ 4.6–5.9).

$$CH_3$$
$$|$$
$$CH_2\!=\!\!CH\!-\!\!CCH_3$$
$$|$$
$$OH$$

11.12 (cont)

Compound J, C_4H_8O, has one unit of unsaturation. The spectrum has bands at

δ 1.1	(6H, doublet, *J* 8 Hz, $(C\underline{H}_3)_2CH—)$
δ 2.3	(1H, multiplet, *J* 8 Hz, $(CH_3)_2C\underline{H}—)$
δ 9.0	$\overset{\displaystyle O}{\overset{\displaystyle \|}{}}$ (1H, multiplet, $—C\underline{H}$)

The compound is 2-methylpropanal (isobutyraldehyde). The hydrogen atom on the carbonyl group is coupled with a very small coupling constant to the α-hydrogen, which appears as a septet with further fine splitting.

$$\underset{\displaystyle \underset{O}{\|}}{CH_3CH}\overset{\displaystyle \overset{CH_3}{|}}{}CH$$

Compound K, $C_8H_{10}O$, has four units of unsaturation. The spectrum has bands at

δ 2.2	(3H, singlet, $ArC\underline{H}_3$)
δ 3.6	(3H, singlet, $C\underline{H}_3O—)$
δ 6.2–6.8	(4H, multiplet, $Ar\underline{H}$)

The compound is 1-methoxy-4-methylbenzene (*p*-methylanisole).

$$CH_3—\underset{\bigcirc}{}—OCH_3$$

11.13   Compound L, $C_5H_{12}O$, has no units of unsaturation. The infrared spectrum has bands at

3350 cm^{-1}	(—O—H)
3000, 1460 cm^{-1}	(C—H)
1120 cm^{-1}	(C—O)

The proton magnetic resonance spectrum has bands at

δ 0.9	(6H, triplet, *J* 6 Hz, $C\underline{H}_3CH_2$)
δ 1.3	(4H, quintet, *J* 6 Hz, $CH_3C\underline{H}_2CH$)
δ 3.3	(1H, quintet, *J* 6 Hz, $CH_2C\underline{H}CH_2$)
δ 3.6	(1H, singlet, $—O\underline{H}$)

The compound is 3-pentanol. Note how distorted the patterns at δ 0.9 and δ 1.3 are.

$$\underset{\displaystyle \underset{OH}{|}}{CH_3CH_2CH}CH_2CH_3$$

11.14   Compound M, $C_5H_8O_2$, has two units of unsaturation. The infrared spectrum has bands at

3100 cm^{-1}	(=C—H)
1760 cm^{-1}	(C=O)
1675 cm^{-1}	(C=C)
1200 cm^{-1}	(C—O)

The proton magnetic resonance spectrum has bands at

δ 1.8	(3H, multiplet, C$\underline{H}_3$C=C)
δ 1.9	(3H, singlet, C$\underline{H}_3$C=O)
δ 4.3	(2H, multiplet, C$\underline{H}_2$=C)

The compound is 2-propenyl ethanoate (isopropenyl acetate).

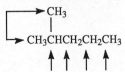

Note that at first glance the two peaks around δ 2.0 appear to be a doublet, but there are no other bands in the spectrum showing the same separation, therefore, the two peaks cannot have arisen because of coupling to another hydrogen atom. We assign the peaks to two nonequivalent methyl groups, one on a double bond and one α to a carbonyl group. One of the peaks for a methyl group is not as high as the other one. The band at δ 4.3 is also thickened. This indicates that the methyl group on the double bond and the vinylic hydrogen atoms are coupled to each other with a small coupling constant.

11.15   Isomer 1 has only two sets of carbon atoms and is 2,3-dimethylbutane.

Isomer 2 has four sets of carbon atoms and is 3-methylpentane.

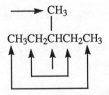

Isomer 3 has five sets of carbon atoms and is 2-methylpentane.

11.16   (a)

Spectrum III has bands for four different types of carbon atoms, one (~50 ppm) for a carbon atom bonded to a nitrogen atom.

(b)

$$CH_3CH_2CHCH_3$$
$$|$$
$$OH$$

Spectrum II has bands for four different types of carbon atoms, one (~70 ppm) for a carbon atom bonded to an oxygen atom.

(c)

$$\overset{O\ \ O}{\underset{\|\ \ \|}{CH_3CH_2COCCH_2CH_3}}$$

Spectrum V has bands for three different types of carbon atoms, one (10 ppm) for an unsubstituted carbon atom, one (~30 ppm) for a carbon α to a carbonyl group, and one (170 ppm) for the carbon atom of the carbonyl group in a carboxylic acid derivative.

(d)

Spectrum IV has bands for two different types of carbon atoms, one (~70 ppm) for a carbon atom bonded to an oxygen atom.

(e)

Spectrum I has bands for eight different types of carbon atoms, four of which (110-150 ppm) are in the alkene region.

11.17   Compound N, C_5H_8O has two units of unsaturation. Three absorption bands in the spectrum for five carbon atoms suggests that the compound has symmetry. One unit of unsaturation is a carbonyl group (219.6 ppm), the other one must be a ring because no bands appear in the region of the spectrum where the carbon atoms of an alkene absorb. The compound is cyclopentanone.

11.18   Compound O, $C_8H_8O_2$, has five units of unsaturation. The spectrum (note the expanded scale) has bands at

δ 3.7	(3H, multiplet, —OC$\underline{H}_3$)
δ 6.8–8	(4H, multiplet, Ar$\underline{H}$)
δ 9.6	(1H, singlet, O=C$\underline{H}$)

The compound is 4-methoxybenzaldehyde.

Compound P, $C_4H_8O_2$, has one unit of unsaturation. The spectrum has bands at

δ 1.2	(3H, triplet, *J* 7 Hz, C$\underline{H}_3$CH$_2$)
δ 2.0	(3H, singlet, C$\underline{H}_3$C=O)
δ 4.1	(2H, quartet, *J* 7 Hz, CH$_3$C$\underline{H}_2$—O)

The compound is ethyl acetate.

$$\overset{O}{\overset{\|}{CH_3COCH_2CH_3}}$$

Compound Q, $C_5H_8O_2$, has two units of unsaturation. The spectrum has bands at

δ 2.2	(3H, singlet, C$\underline{H}_3$—C=C)
δ 2.5	(3H, singlet, C$\underline{H}_3$—C=C)
δ 5.8	(1H, singlet, $\underline{H}$C=C)
δ 12.0	(1H, singlet, —C$\overset{O}{\overset{\|}{}}O\underline{H}$)

The compound is 3-methyl-2-butenoic acid.

Note that at first glance the two peaks around δ 2.0 appear to be a doublet, but there are no other bands in the spectrum showing the same separation, therefore, the two peaks cannot have arisen because of coupling to another hydrogen atom. We assign the peaks to two nonequivalent methyl groups, one cis and one trans to the carboxylic acid group. The band at δ 5.8 shows some evidence of coupling (with a very small coupling constant) with the methyl groups.

11.18 (cont)

Compound R, C_7H_9N, has four units of unsaturation. (When determining the units of unsaturation in a compound containing nitrogen, eliminate the nitrogens from the molecular formula and then subtract one hydrogen for each nitrogen atom to determine the number of hydrogen atoms that corresponds to the saturated formula. The adjusted molecular formula for compound R is C_7H_8.) The spectrum has bands at

δ 2.8	(3H, singlet, C$\underline{\text{H}}_3$N)
δ 3.4	(1H, singlet, N$\underline{\text{H}}$)
δ 6.4–7.4	(5H, singlet, Ar$\underline{\text{H}}$)

The compound is *N*-methylaniline. The broad band at δ 3.4 results from coupling between the nucleus of $^{14}_{7}$N and the hydrogen atom on it.

Compound S, $C_7H_{12}O_3$, has two units of unsaturation. The spectrum has bands at

δ 1.2	(3H, triplet, J 7 Hz, C$\underline{\text{H}}_3$CH$_2$)
δ 2.1	(3H, singlet, C$\underline{\text{H}}_3$C=O)
δ 2.6	(4H, multiplet, —C$\underline{\text{H}}_2$C$\underline{\text{H}}_2$—) The chemical shift value suggests that each of the methylene groups is α to a carbonyl group.
δ 4.1	(2H, quartet, J 7 Hz, CH$_3$C$\underline{\text{H}}_2$—O)

The fragments listed above account for six of the seven carbon atoms, two of the three oxygen atoms, and one of the two units of unsaturation. The remaining carbon and oxygen atoms and one unit of unsaturation make up a carbonyl group. The compound is ethyl 4-oxopentanoate.

$$\underset{\text{CH}_3\text{CCH}_2\text{CH}_2\text{COCH}_2\text{CH}_3}{\overset{\displaystyle\overset{O}{\|}\qquad\overset{O}{\|}}{}}$$

Compound T, $C_6H_{15}N$, has no units of unsaturation. The spectrum has bands at

δ 0.9	(3H, triplet, J 7 Hz, C$\underline{\text{H}}_3$CH$_2$)
δ 2.2	(2H, quartet, J 7 Hz, CH$_3$C$\underline{\text{H}}_2$N)

The fragments listed above account for five hydrogen atoms and two carbon atoms in an ethyl group. The compound has fifteen hydrogen atoms and six carbon atoms, and must be a symmetric compound containing three ethyl groups. The compound is triethylamine.

$$(CH_3CH_2)_3N$$

11.19   Compound U, $C_{10}H_{12}O$, has five units of unsaturation. The infrared spectrum has bands at

   1717 cm^{-1} (C=O)

There are no bands in the infrared at about 2700 and 2900 cm^{-1}, therefore the compound must be a ketone.

The proton magnetic resonance spectrum has bands at

δ 1.8	(3H, singlet, C$\underline{H}_3$C=O)
δ 2.4–2.8	(4H, multiplet, —C$\underline{H}_2$C$\underline{H}_2$—)
δ 6.8	(5H, singlet, Ar$\underline{H}$)

The compound is 4-phenyl-2-butanone.

$$\text{C}_6\text{H}_5\text{—CH}_2\text{CH}_2\overset{\overset{\displaystyle O}{\|}}{\text{C}}\text{CH}_3$$

The peaks around δ 2.5 may be seen as a pair of overlapping triplets.

11.20   Compound V, $C_5H_{12}O$, has no units of unsaturation. The infrared spectrum has bands at

3320 cm^{-1}	(—O—H)
1060 cm^{-1}	(C—O)

The infrared spectrum indicates that V is an alcohol.

The proton magnetic resonance spectrum has bands at

δ 0.8	(6H, doublet, (C$\underline{H}_3$)$_2$CH)
δ 1.0–2.0	(3H, multiplet, —C$\underline{H}$C$\underline{H}_2$—)
δ 2.9	(1H, singlet, —O$\underline{H}$)
δ 3.4	(2H, triplet, —CH$_2$C$\underline{H}_2$—O)

The compound is 3-methyl-1-butanol.

$$\overset{\overset{\displaystyle CH_3}{|}}{\text{CH}_3\text{CHCH}_2\text{CH}_2\text{OH}}$$

11.21

b
CH₃
|
e ⟨benzene ring⟩—CHCH₂CH₃
d   c   a

δ 0.8   (3H, triplet, J_{ac} 7 Hz)
These are the methyl hydrogen atoms labeled a. They are split into a triplet by the two hydrogen atoms on the adjacent methylene group.

δ 1.1   (3H, doublet, J_{bd} 7 Hz)
These are the methyl hydrogen atoms labeled b. They are split into a doublet by the hydrogen atom on the adjacent benzylic carbon atom.

δ 1.4   (2H, multiplet, $J_{ac} = J_{cd} = 7$ Hz)
These are the methylene hydrogen atoms labeled c. They are split by the hydrogen atom on the adjacent benzylic carbon atom and the hydrogen atoms on the adjacent methyl group.

δ 2.4   (1H, sextet, $J_{bd} = J_{cd} = 7$ Hz)
This is the benzylic hydrogen atom labeled d. It is split into a sextet by the hydrogen atoms on the adjacent methylene group and the hydrogen atoms on the adjacent methyl group.

δ 6.8   (5H, singlet)
These are the aromatic hydrogen atoms labeled e. Even though they are not chemical-shift equivalent, the alkyl substituent does not change the magnetic environment enough to cause them to have different chemical shifts. Hydrogen atoms that have the same chemical shift will not couple.

11.22   Compound W has bands in its infrared spectrum at

1685 cm⁻¹   (conjugated C=O)

The infrared spectrum indicates that W is a ketone.

The proton magnetic resonance spectrum has bands at

δ 1.1   (6H, doublet, J 7 Hz, $(CH_3)_2CH$—)
δ 3.3   (1H, septet, $(CH_3)_2CH$—)
δ 7.0–7.7   (5H, multiplet, ArH)

11.22 (cont)

The chemical shift of the methine hydrogen atom and the fact that the aryl hydrogen atoms are split, indicate that the carbonyl group is between the aryl and the isopropyl groups. This is confirmed by the presence a band for a conjugated carbonyl group in the infrared. The compound is 2-methyl-1-phenyl-1-propanone (isobutyrophenone).

11.23

The bands are analyzed below

δ 3.8   (2H, broad multiplet, —N$\underline{H}_2$)
δ 6.3   (1H, doublet, J_{ab} 8 Hz, Ar$\underline{H}_a$)
δ 6.8   (1H, doublet of doublets, J_{ab} 8 Hz, J_{bc} 2 Hz, Ar$\underline{H}_b$)
δ 7.1   (1H, doublet, J_{bc} 2 Hz, Ar$\underline{H}_c$)

H_a is the farthest upfield of the aryl hydrogen atoms because it is next to the amino group, which is strongly electron-donating (see p. 104 in the text), and distant from the electron-withdrawing bromine atoms. It is split into a doublet by H_b with J 8 Hz.

H_b has the next highest chemical shift. It is next to only one bromine atom. A diagram showing the splitting pattern for H_b is shown on the next page.

11.23 (cont)

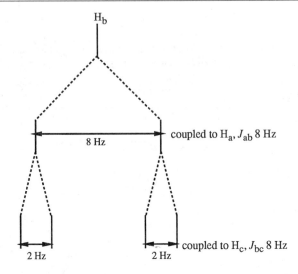

H_c is the most deshielded of the aryl hydrogen atoms because it is between two bromine atoms. It is also split into a doublet by H_b with J 2 Hz.

11.24 The two peaks observed for the methyl groups cannot be due to spin-spin coupling because there are no other bands in the spectrum that also show splitting.

The two separate absorption bands for the two methyl groups tell us that there is partial double bond character between the carbon atom of the carbonyl group and the nitrogen atom, which prevents free rotation. If free rotation were possible, the two methyl groups would become chemically equivalent, and only one chemical shift would be observed for both. Because there is not free rotation, the two methyl groups are in different environments and thus have two different chemical shifts.

Resonance contributors for the amide group showing the existence of partial double bond character for the bond between the carbon atom of the carbonyl group and the nitrogen atom.

11.25 Compound X, $C_4H_8Cl_2$, has no units of unsaturation. We know that there are only two different types of carbon atoms in X because the carbon-13 magnetic resonance spectrum has only two bands. X must be 1,4-dichlorobutane.

11.25 (cont)

$$ClCH_2CH_2CH_2CH_2Cl$$

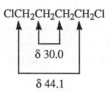

δ 30.0

δ 44.1

2,3-Dichlorobutane also has the symmetry required for this spectrum, but we would expect the methyl groups to absorb at higher field than the chemical shift values observed (Table 11.4).

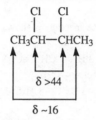

δ >44

δ ~16

Also note that the carbon atoms bearing the chlorine atoms in this compound are even more highly substituted than in 1,4-dichlorobutane. We would expect them to absorb at lower field.

11.26   The infrared spectrum of compound Y shows the presence of an ester (1738 cm^{-1} for the C=O stretching frequency and 1256 and 1173 cm^{-1} for the C—O stretching frequencies).

The proton magnetic resonance spectrum shows the presence of hydrogen atoms bonded to a carbon atom α to a carbonyl group (δ 1.96). The carbon-13 magnetic resonance spectrum shows the presence of at least four different carbon atoms.

δ 22.3	sp^3-carbon atom
δ 28.1	sp^3-carbon atom α to a carbonyl group
δ 79.9	sp^3-carbon atom bonded to an oxygen atom
δ 170.2	ester carbonyl carbon atom

The proton magnetic resonance integration (3:1) has too few hydrogen atoms for the number of carbon atoms indicated by the carbon-13 spectrum. The actual ratio must be a multiple, such as 6:2 or 9:3. We cannot distribute eight hydrogen atoms on a minimum of four carbon atoms in such a way as to give only two chemical shifts. Compound Y must therefore have a total of twelve hydrogen atoms, nine on three methyl groups which are part of a *tert*-butyl group and three on a methyl group which is α to the ester group. Compound Y is *tert*-butyl acetate.

11.27   Compound Z, C_3H_4O, has two units of unsaturation. The infrared spectrum shows the presence of an alcohol (3400 cm^{-1} for the O—H stretching frequency and 1030 cm^{-1} for the C—O stretching frequency) and a terminal alkyne (3300 cm^{-1} for the $\equiv$C—H stretch and 2100 cm^{-1} for the C$\equiv$C stretch). There is only one way in which a triple bond and an alcohol group can be arranged to give a stable compound that has only three carbon atoms.

$HC\equiv CCH_2OH$

This structure is confirmed by both the proton and carbon-13 magnetic resonance spectra.

Analysis of the proton magnetic resonance spectrum is shown below:

δ 2.5	(1H, triplet, *J* 3 Hz, —CH$_2$C$\equiv$C<u>H</u>)
δ 3.1	(1H, broad singlet, —O<u>H</u>)
δ 4.3	(2H, doublet, *J* 3 Hz, O—C<u>H</u>$_2$C$\equiv$CH)

Analysis of the carbon-13 nuclear magnetic resonance spectrum is shown below:

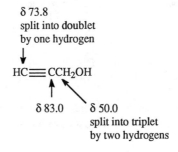

12

Alcohols, Diols, and Ethers

12.1 1,2-Ethanediol (ethylene glycol) is a very polar compound and hydrogen bonds strongly to water. It is infinitely soluble in water. A solution of ethylene glycol in water depresses the freezing point of water so that the mixture does not freeze at temperatures usually seen in the winter time. This keeps pipes and radiators from bursting from the expansion of the water in them as it forms ice. The high boiling point of ethylene glycol ensures that it will not boil out of the radiator of a car with the engine is hot.

12.2

Cholesterol is not very soluble in water even though it has a hydroxyl group because the rest of the molecule is nonpolar and is too large to be brought into solution by the small polar hydroxyl group (Section 1.8B).

12.3 Cholesterol is an alkene as well as an alcohol.

12.3 (cont)

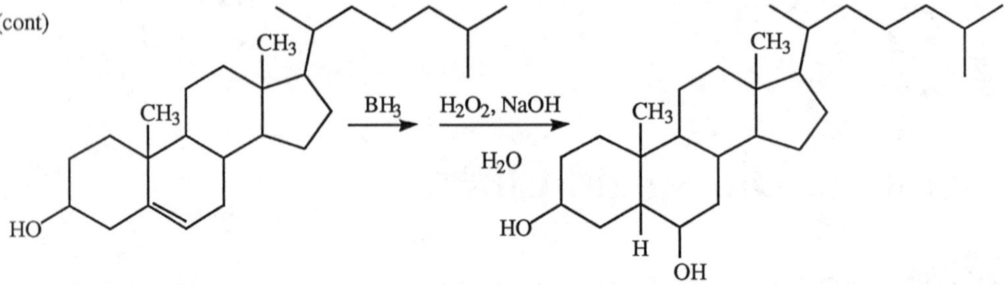

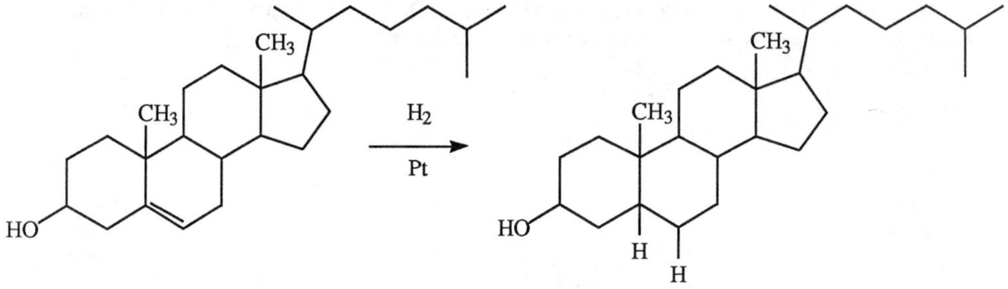

12.4   (a) 3-methyl-1-butanol

(c) 2-pentyn-1-ol

(e) 1-chloro-3-ethoxypropane

(g) cyclohexanol

(i) *cis*-4-*tert*-butylcyclohexanol

(k) 1,3-dimethoxypropane

(b) (*E*)-4-hexen-3-ol

(d) 4,5-dichloro-1-hexanol

(f) diphenyl ether
     (phenoxybenzene)

(h) *trans*-2-methylcyclopentanol

(j) (*R*)-5-hexen-3-ol

(l) 1,2-dimethoxybenzene

12.5   (a)

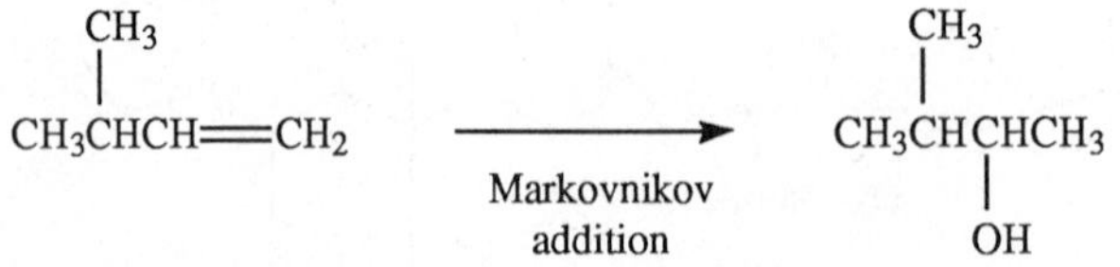

product not observed

12.5 (cont)

(b)

$$CH_3CHCH=CH_2 \longrightarrow CH_3\overset{CH_3}{\underset{H}{\overset{|}{C}}}\overset{+}{-}CHCH_3 \longrightarrow CH_3\overset{CH_3}{\underset{|}{\overset{|}{C}}}\overset{+}{-}CH_2CH_3$$

protonation   1,2-hydride shift   reaction of the more stable
                                                                    tertiary carbocation with the
                                                                    nucleophilic solvent

$$CH_3\overset{CH_3}{\underset{:O:}{\overset{|}{C}}}CH_2CH_3 \longleftarrow CH_3\overset{CH_3}{\underset{|}{\overset{|}{C}}}-CH_2CH_3$$

deprotonation of
the oxonium ion

12.6

$$CH_3\overset{CH_3}{\underset{CH_3}{\overset{|}{C}}}-CH=CH_2 \longrightarrow CH_3\overset{CH_3}{\underset{CH_3}{\overset{|}{C}}}\overset{+}{-}CHCH_3 \longrightarrow CH_3\overset{CH_3}{\underset{|}{\overset{|}{C}}}\overset{+}{-}CHCH_3$$

protonation   1,2-methyl shift   reaction of the more stable
                                                                    tertiary carbocation with the
                                                                    nucleophilic solvent

$$CH_3\overset{CH_3\ CH_3}{\underset{:O:}{\overset{|\quad|}{C}}}-CHCH_3 \longleftarrow CH_3\overset{CH_3}{\underset{|}{\overset{|}{C}}}-CHCH_3$$

deprotonation of the oxonium ion

12.7 $CH_2\!\!=\!\!CHCH_2CH_2CH_2OH$ $\xrightarrow{\ H_3O^+\ }$ $CH_3CHCH_2CH_2CH_2OH$
                                                                                    |
                                                                                    OH

Dehydration would also occur if the temperature were raised.

- -

Concept Map 12.1   Conversion of alkenes to alcohols.

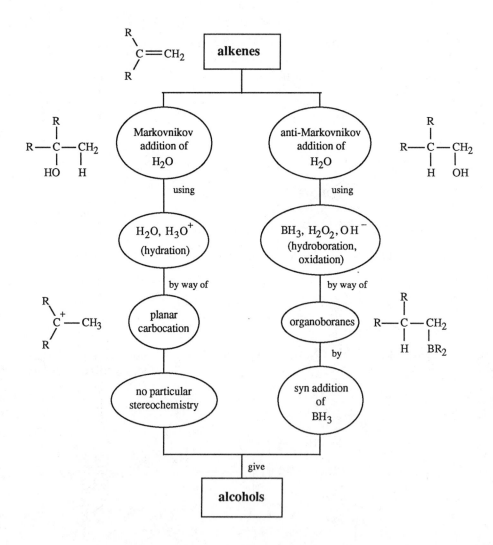

12.8 $CH_3CH_2CHCH_2CH_3$ $\longrightarrow$ $CH_3CH_2CHCH_2CH_3$ $\longrightarrow$ $\overset{+}{CH_3CH_2CHCH_2CH_3}$

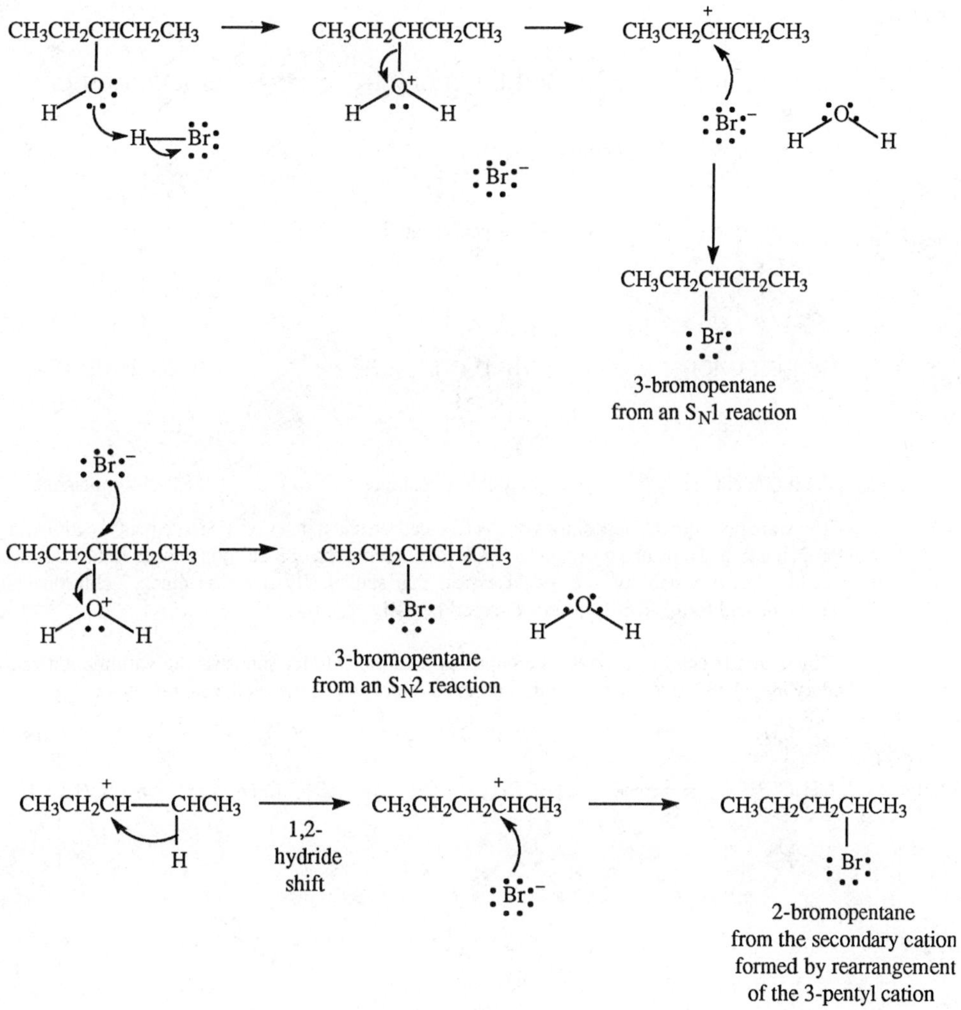

3-bromopentane
from an S_N1 reaction

3-bromopentane
from an S_N2 reaction

$\overset{+}{CH_3CH_2CH}$——$CHCH_3$ $\xrightarrow[\text{shift}]{\text{1,2-}\\\text{hydride}}$ $CH_3CH_2CH_2\overset{+}{C}HCH_3$ $\longrightarrow$ $CH_3CH_2CH_2CHCH_3$

2-bromopentane
from the secondary cation
formed by rearrangement
of the 3-pentyl cation

12.9 $CH_3CH_2CHCH_2CH_3$

ionization assisted by $ZnCl_2$

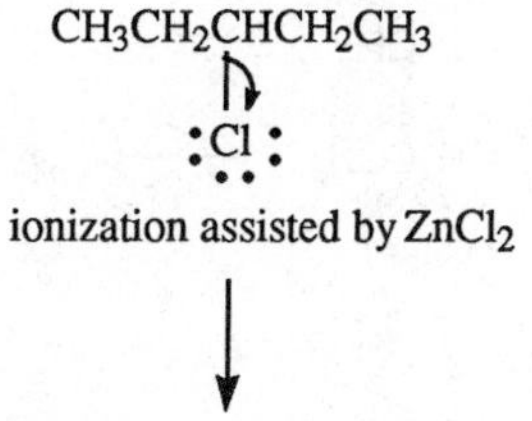

continued on next page

12.9 (cont)

$$CH_3CH_2CH_2\overset{+}{C}HCH_3 \;\;\rightleftarrows\;\; CH_3CH_2\overset{+}{C}HCH_2CH_3 \;\;\rightleftarrows\;\; CH_3\overset{+}{C}HCH_2CH_2CH_3$$

| | 1,2-
hydride
shift | | 1,2-
hydride
shift | |

$:\!\ddot{C}\!l\!:^-$ $\qquad\qquad$ $:\!\ddot{C}\!l\!:^-$ $\qquad\qquad$ $:\!\ddot{C}\!l\!:^-$

2-pentyl cation $\qquad\qquad$ 3-pentyl cation $\qquad\qquad$ 2-pentyl cation

$$CH_3CH_2CH_2CHCH_3 \qquad\qquad CH_3CH_2CHCH_2CH_3 \qquad\qquad CH_3CHCH_2CH_2CH_3$$
$$\overset{|}{:\!\ddot{C}\!l\!:} \qquad\qquad\qquad \overset{|}{:\!\ddot{C}\!l\!:} \qquad\qquad\qquad \overset{|}{:\!\ddot{C}\!l\!:}$$

2-chloropentane $\qquad\qquad$ 3-chloropentane $\qquad\qquad$ 2-chloropentane

There are two equivalent positions on the five-carbon chain that give rise to 2-pentyl cations, and thus to 2-chloropentane, and only one position that leads to the formation of 3-chloropentane. If the two carbocations are roughly equal in stability, we would expect that statistically twice as much 2-chloropentane as 3-chloropentane will form. This is observed experimentally.

Zinc chloride is a Lewis acid. It complexes with the chloride ion assisting with the ionization of the secondary alkyl halides, allowing equilibrium to be established between the cations.

12.10

12.11   (1) With zinc chloride in concentrated hydrochloric acid, we have reaction conditions that allow for ionization of the alkyl halides that are formed as products and for equilibrium among different ionic species. As we saw in Problem 12.9, zinc chloride assists in the ionization of alkyl halides.

12.11   (1) (cont)

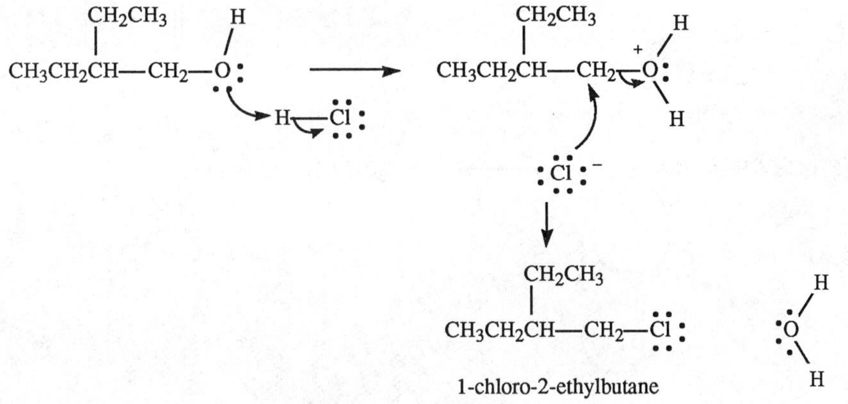

1-chloro-2-ethylbutane

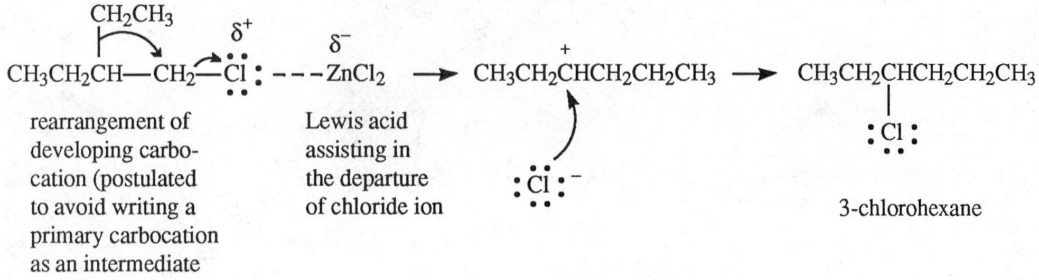

rearrangement of
developing carbo-
cation (postulated
to avoid writing a
primary carbocation
as an intermediate

Lewis acid
assisting in
the departure
of chloride ion

3-chlorohexane

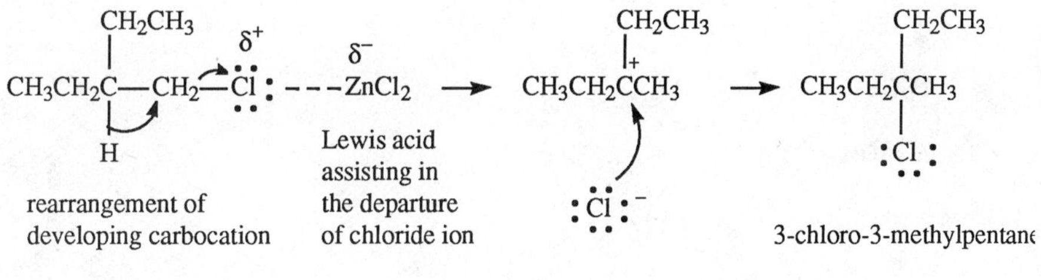

rearrangement of
developing carbocation

Lewis acid
assisting in
the departure
of chloride ion

3-chloro-3-methylpentane

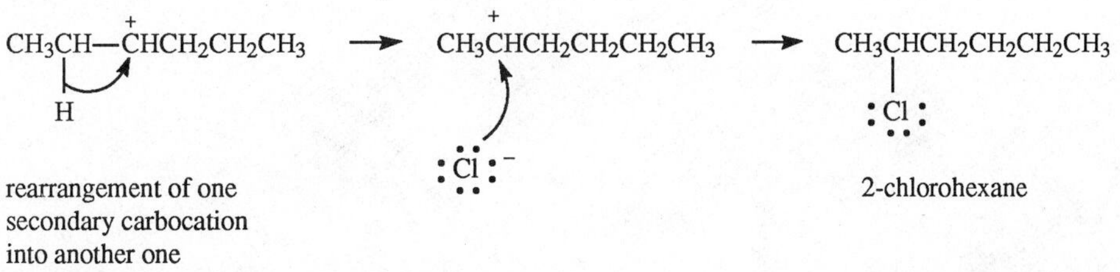

rearrangement of one
secondary carbocation
into another one

2-chlorohexane

(2) With thionyl chloride in pyridine, an S_N2 reaction takes place. The alcohol is converted into the
corresponding halide without rearrangement.

12.11   (2) (cont)

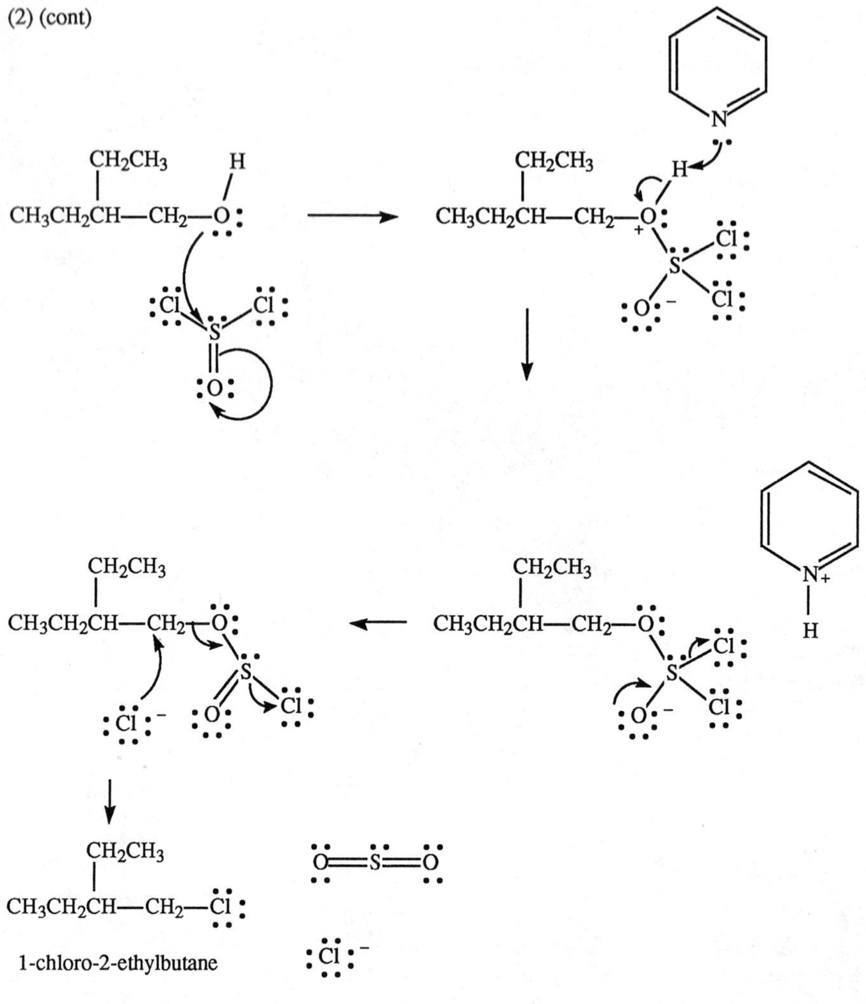

$$CH_3CH_2CH-CH_2-O: \quad \longrightarrow \quad CH_3CH_2CH-CH_2-O: \quad \longrightarrow$$

1-chloro-2-ethylbutane

$$O=S=O$$

$$:Cl:^-$$

Concept Map 12.2   Conversion of alcohols to alkyl halides.

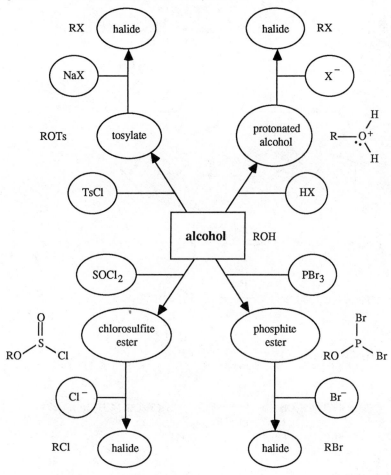

12.12

nucleophilic substitution
on phosphorus

intermediate with a good leaving
group undergoing a nucleophilic
reaction at the chiral carbon atom
with inversion of configuration

12.12 (cont)

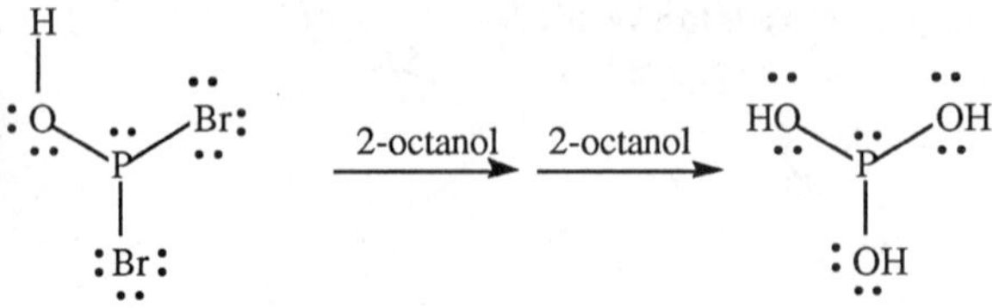

12.13   (a)

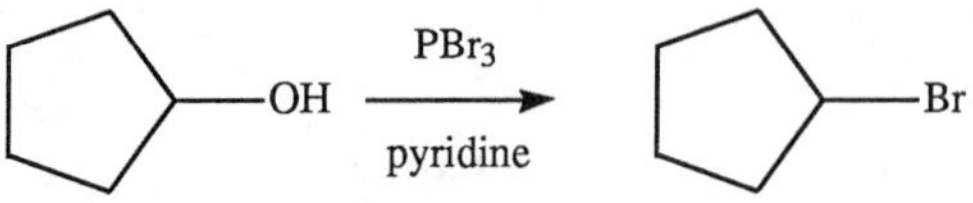

(b)

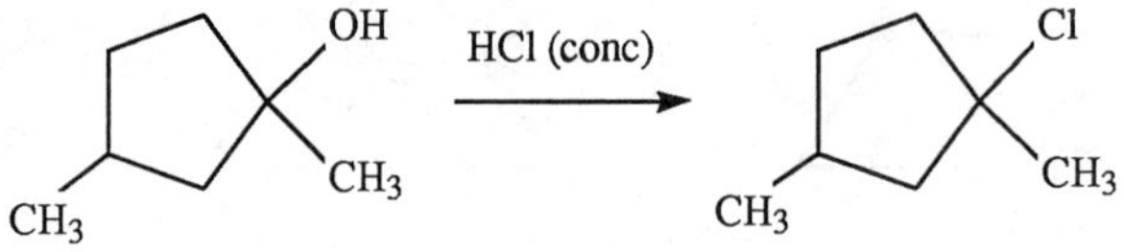

(c) $CH_3(CH_2)_9CH_2OH$ $\xrightarrow[\Delta]{SOCl_2}$ $CH_3(CH_2)_9CH_2Cl$

(d) $CH_3(CH_2)_8CH_2OH$ $\xrightarrow[\Delta]{HBr}$ $CH_3(CH_2)_8CH_2Br$

(e) $CH_3SCH_2CH_2OH$ $\xrightarrow[chloroform]{SOCl_2}$ $CH_3SCH_2CH_2Cl$

(f) $CH_3CH_2OCH_2CH_2OH$ $\xrightarrow[pyridine]{PBr_3}$ $CH_3CH_2OCH_2CH_2Br$

(g)

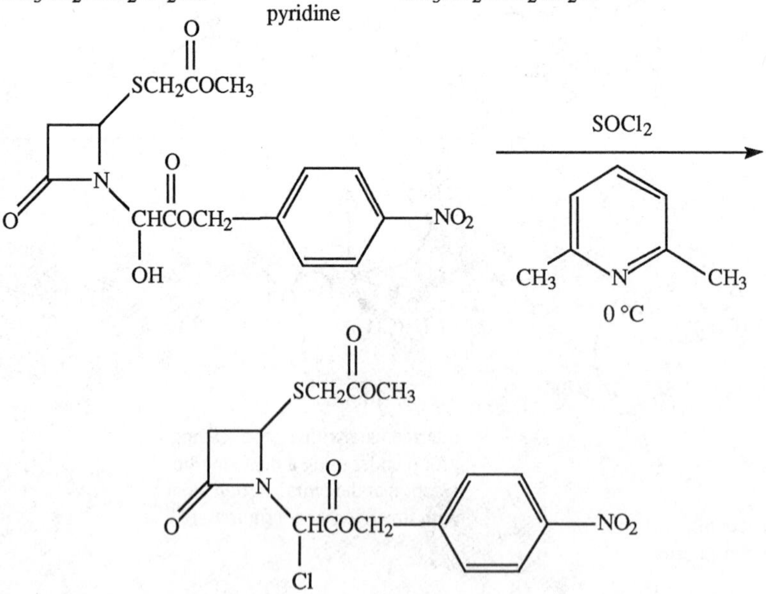

Concept Map 12.3   Oxidation and reduction at carbon atoms.

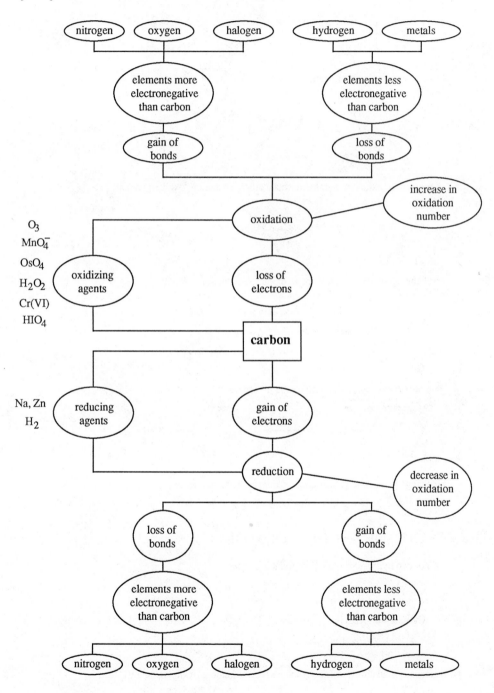

12.14 (a)

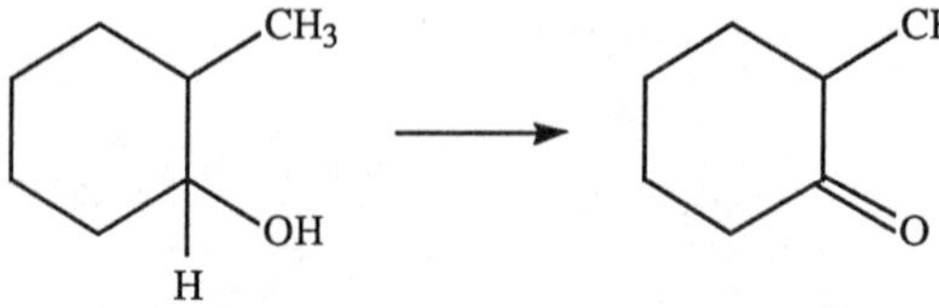
CH3
OH
H
CH3
O

loss of two hydrogens; oxidation; oxidizing agent

(b)

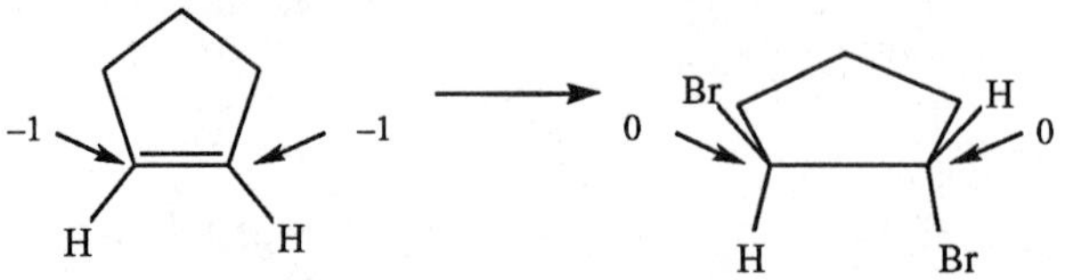
–1
–1
H
H
Br
H
0
0
H
Br

an increase in oxidation number at each carbon; oxidation; oxidizing agent

(c)

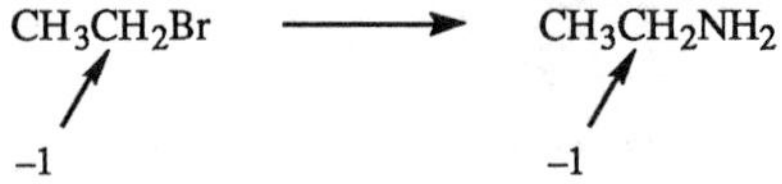
CH3CH2Br
CH3CH2NH2
–1
–1

no change in oxidation state; not an oxidation-reduction reaction

(d)

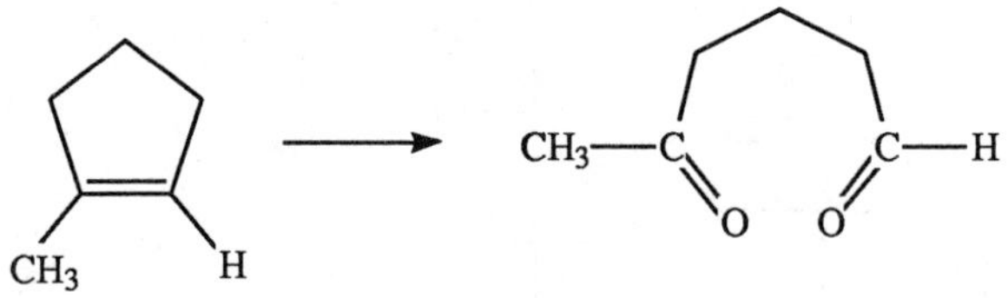
CH3
H
CH3
C
C
H
O
O

gain of two oxygens; oxidation; oxidizing agent

(e)

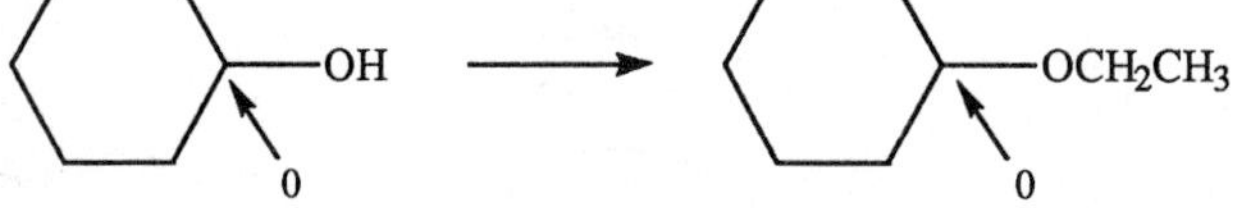
OH
0
OCH2CH3
0

no change in oxidation state; not an oxidation-reduction reaction

(f)

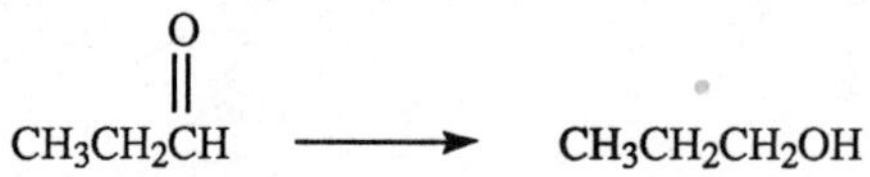
O
CH3CH2CH
CH3CH2CH2OH

gain of two hydrogens; reduction; reducing agent

(g)

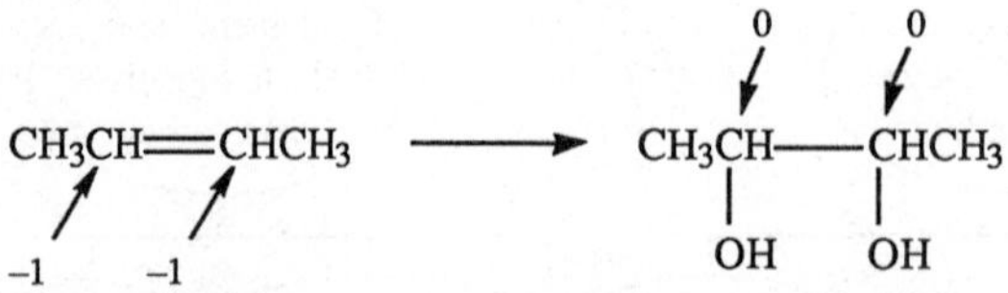
CH3CH=CHCH3
–1
–1
0
0
CH3CH
CHCH3
OH
OH

increase in oxidation number at each carbon; oxidation; oxidizing agent

12.14 (cont)

(h) $CH_3C\equiv CCH_2CH_3 \longrightarrow CH_3CH=CHCH_2CH_3$

gain of two hydrogens; reduction; reducing agent

(i) 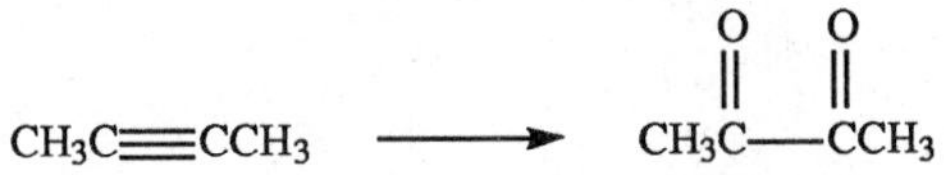 $CH_3C\equiv CCH_3 \longrightarrow$

gain of two oxygens; oxidation; oxidizing agent

(j) 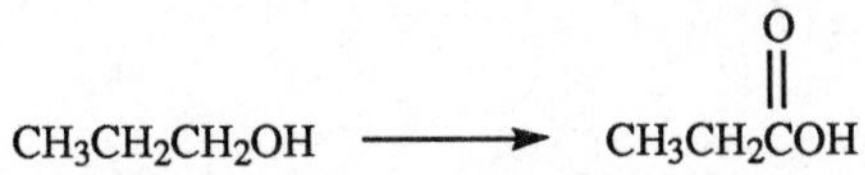 $CH_3CH_2CH_2OH \longrightarrow$

loss of two hydrogens and gain of oxygen; oxidation; oxidizing agent

(k)

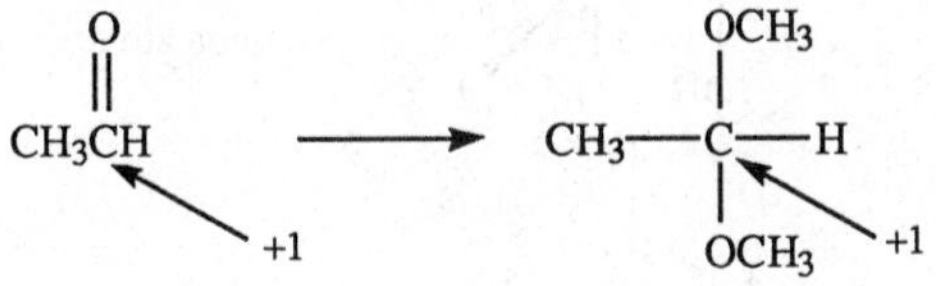

no change in oxidation state; not an oxidation-reduction reaction

12.15

(a)

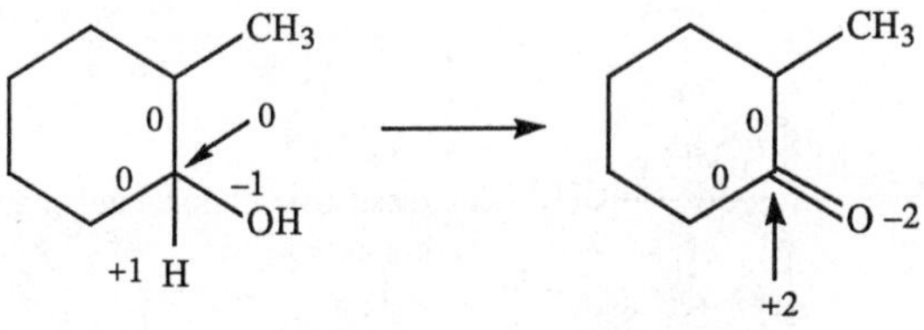

increase in oxidation number

(b)

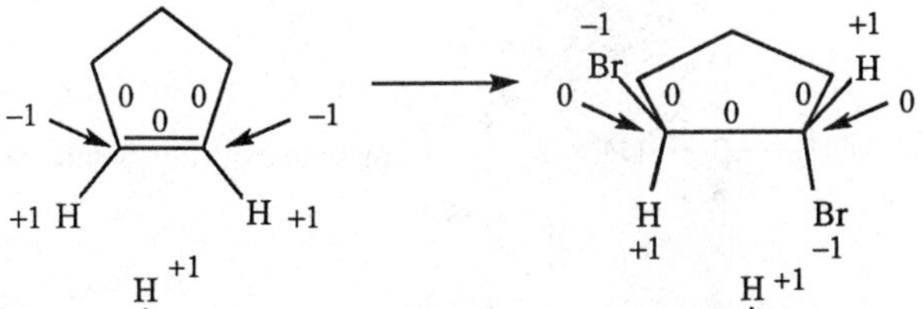

increase in oxidation number at each carbon atom

(c)

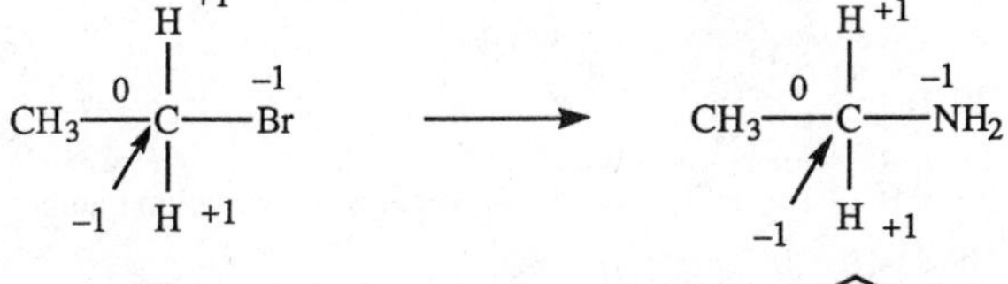

no change in oxidation number

(d)

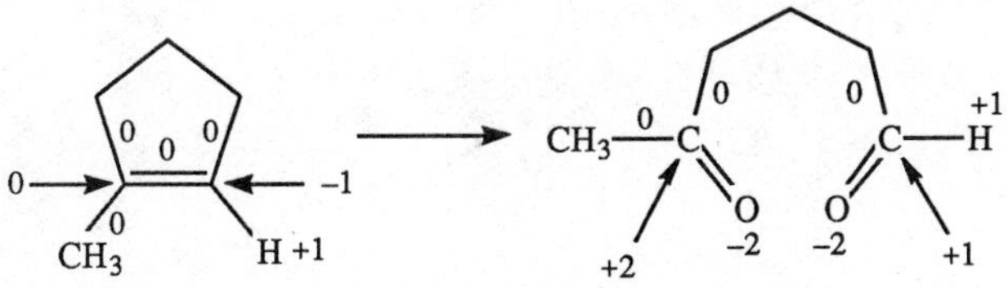

increase in oxidation number at each carbon atom

12.15 (cont)

(e)

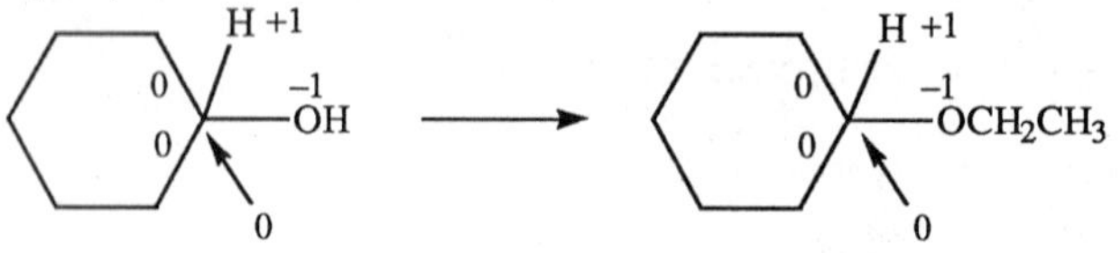

no change in oxidation number

(f) 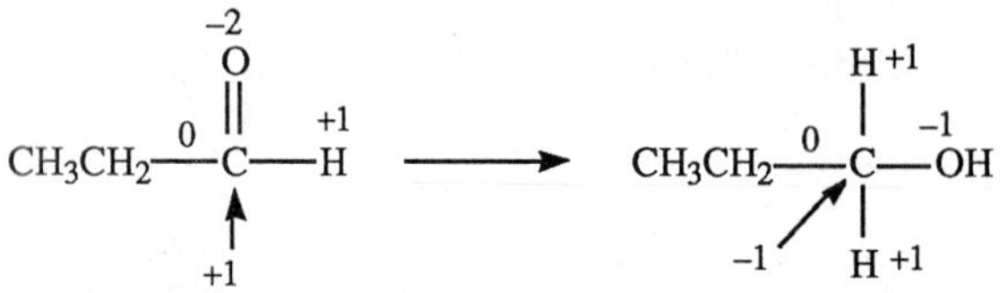 CH_3CH_2—C—H $\longrightarrow$ CH_3CH_2—C—OH

decrease in oxidation number

(g) 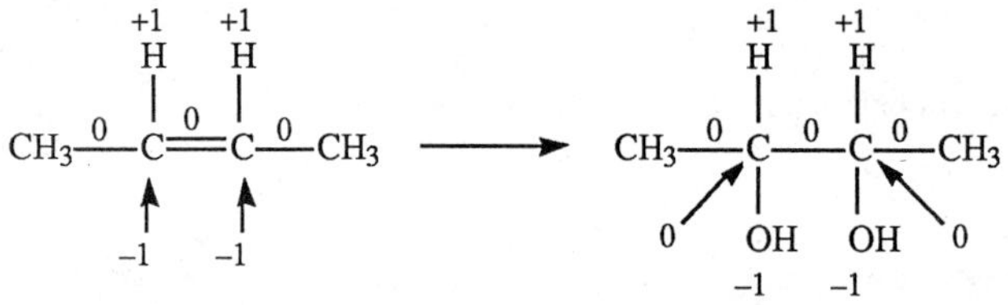 CH_3—C=C—CH_3 $\longrightarrow$ CH_3—C—C—CH_3

increase in oxidation number at each carbon atom

(h) 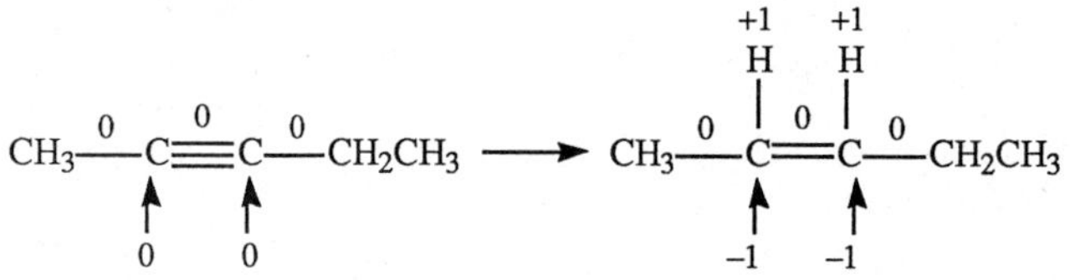 CH_3—C≡C—CH_2CH_3 $\longrightarrow$ CH_3—C=C—CH_2CH_3

decrease in oxidation number at each carbon atom

(i) 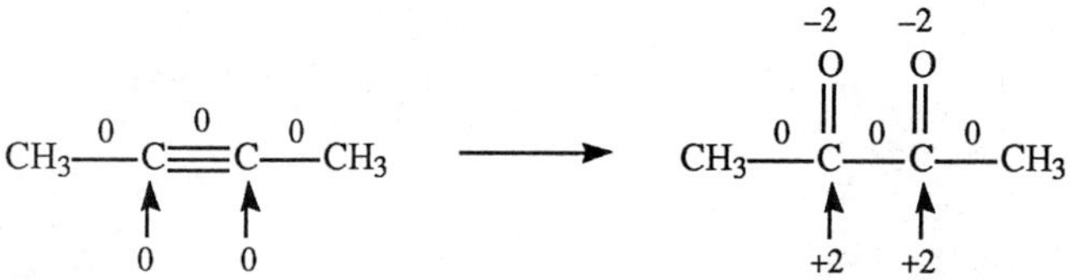 CH_3—C≡C—CH_3 $\longrightarrow$ CH_3—C—C—CH_3

increase in oxidation number at each carbon atom

(j) 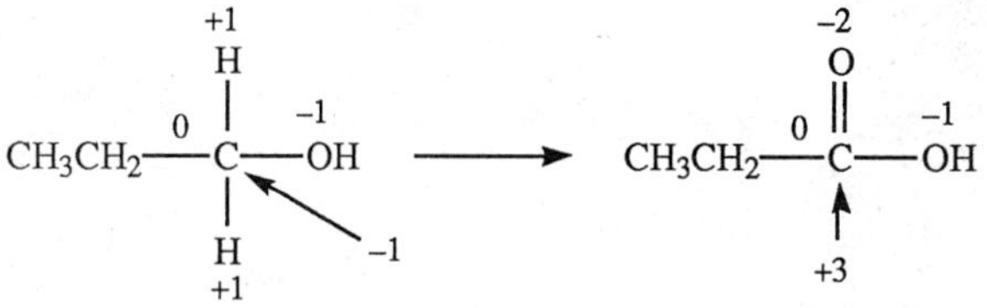 CH_3CH_2—C—OH $\longrightarrow$ CH_3CH_2—C—OH

increase in oxidation number

(k) 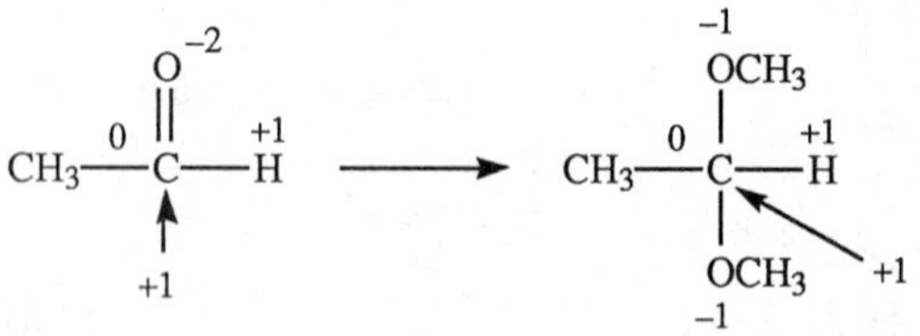 CH_3—C—H $\longrightarrow$ CH_3—C—H

no change in oxidation number

12.16

$$\underset{\text{methane}}{\overset{\overset{+1}{H}}{\underset{\underset{-4}{\overset{|}{H}}\;{+1}}{\overset{+1}{H}-\overset{+1}{\underset{|}{C}}-\overset{+1}{H}}}} \longrightarrow \underset{\substack{\text{chloromethane}\\ \text{methyl chloride}}}{\overset{\overset{+1}{H}}{\underset{\underset{-2}{\overset{|}{H}}\;{+1}}{\overset{+1}{H}-\overset{+1}{\underset{|}{C}}-\overset{-1}{Cl}}}} \longrightarrow \underset{\substack{\text{dichloromethane}\\ \text{methylene chloride}}}{\overset{\overset{-1}{Cl}}{\underset{\underset{0}{\overset{|}{H}}\;{+1}}{\overset{+1}{H}-\overset{+1}{\underset{|}{C}}-\overset{-1}{Cl}}}}$$

$$\underset{\substack{\text{tetrachloromethane}\\ \text{carbon tetrachloride}}}{\overset{\overset{-1}{Cl}}{\underset{\underset{+4}{\overset{|}{Cl}}\;{-1}}{\overset{-1}{Cl}-\overset{-1}{\underset{|}{C}}-\overset{-1}{Cl}}}} \longleftarrow \underset{\substack{\text{trichloromethane}\\ \text{chloroform}}}{\overset{\overset{-1}{Cl}}{\underset{\underset{+2}{\overset{|}{H}}\;{+1}}{\overset{-1}{Cl}-\overset{-1}{\underset{|}{C}}-\overset{-1}{Cl}}}}$$

12.17

(a)

$$\underset{\underset{OH}{\overset{|}{CH_3CHCH_2CHCH_3}}}{\overset{\overset{CH_3}{|}}{}} + \underset{\text{orange}}{CrO_3} \xrightarrow[\text{H}_2\text{SO}_4]{\text{H}_2\text{O}} \underset{\underset{\text{(positive test)}}{\text{blue-green}}}{\overset{\overset{CH_3\quad O}{\overset{|\quad\;\;||}{CH_3CHCH_2CCH_3}}}{}} + Cr^{3+}$$

(b)

(cyclohexane ring with
$$\overset{CH_3}{\underset{CH_3}{\overset{|}{\underset{|}{CCH_3}}}}$$
substituent and OH)

$$\xrightarrow[\text{H}_2\text{SO}_4]{\overset{CrO_3}{\text{H}_2\text{O}}}$$

(cyclohexanone ring with
$$\overset{CH_3}{\underset{CH_3}{\overset{|}{\underset{|}{CCH_3}}}}$$
substituent) $\quad + \quad Cr^{3+}$ (positive test)

(c)

(cholesterol steroid structure, HO–)

$$\xrightarrow[\text{H}_2\text{SO}_4]{\overset{CrO_3}{\text{H}_2\text{O}}}$$

(ketone steroid structure, O=) $\quad + \quad Cr^{3+}$ (positive test)

12.17 (cont)

(d)

$\xrightarrow[\substack{H_2O \\ H_2SO_4}]{CrO_3}$ 3° alcohol, solution remains orange (negative test)

(e) $CH_3CH_2CH_2CH_2CH_2OH$ $\xrightarrow[\substack{H_2O \\ H_2SO_4}]{CrO_3}$ $CH_3CH_2CH_2CH_2\overset{\overset{\displaystyle O}{\|}}{C}OH$ + Cr^{+3}

                                                                                (positive test)

(f)

$$\underset{\substack{| \\ CH_3}}{\overset{\substack{CH_3 \quad CH_3 \\ | \quad\quad |}}{CH_3C\!-\!\!-\!\!CCH_3}}\ \underset{\substack{| \\ OH}}{}$$

$\xrightarrow[\substack{H_2O \\ H_2SO_4}]{CrO_3}$ 3° alcohol, solution remains orange (negative test)

12.18

(1) $\underset{\substack{| \\ CH_3}}{\overset{CH_3}{CH_3C}}\!\!=\!\!CHCH_2CH_2\underset{\substack{| \\ }}{\overset{CH_3}{C}H}CH_2CH_2OH$ $\xrightarrow[\substack{H_2SO_4 \\ H_2O}]{CrO_3}$ $\underset{\substack{| \\ }}{\overset{CH_3}{CH_3C}}\!\!=\!\!CHCH_2CH_2\underset{}{\overset{CH_3}{C}H}CH_2\overset{\overset{\displaystyle O}{\|}}{C}OH$

                                                                                oxidation of a primary alcohol
                                                                                to a carboxylic acid

(2) $\overset{CH_3}{CH_3C}\!\!=\!\!CHCH_2CH_2\overset{CH_3}{C}HCH_2CH_2OH$ $\xrightarrow{\text{strong acid}}$

  $\overset{CH_3}{CH_3C}\!\!=\!\!CHCH_2CH_2\overset{CH_3}{C}HCH\!\!=\!\!CH_2$ + $\overset{CH_3}{CH_3C}\!\!=\!\!CHCH_2CH_2\overset{CH_3}{C}\!\!=\!\!CHCH_3$

                                                dehydration of alcohol

(3) $\overset{CH_3}{CH_3C}\!\!=\!\!CHCH_2CH_2\overset{CH_3}{C}HCH_2CH_2OH$ $\xrightarrow{\text{strong acid}}$ $\underset{\substack{| \\ OH}}{\overset{CH_3}{CH_3C}}CH_2CH_2CH_2\overset{CH_3}{C}HCH_2CH_2OH$

                                                                                hydration of alkene

In addition, polymerization reactions (Section 8.4C) of alkenes occur in the presence of strong acids.

Concept Map 12.4   Reactions of alcohols with Cr(VI) reagents.

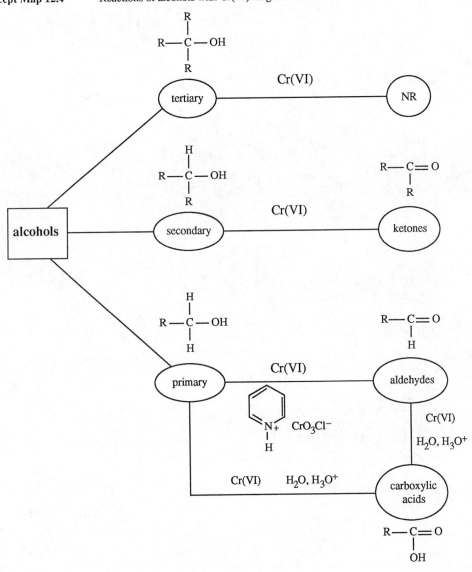

12.19

(a)

12.19 (cont)

(b)

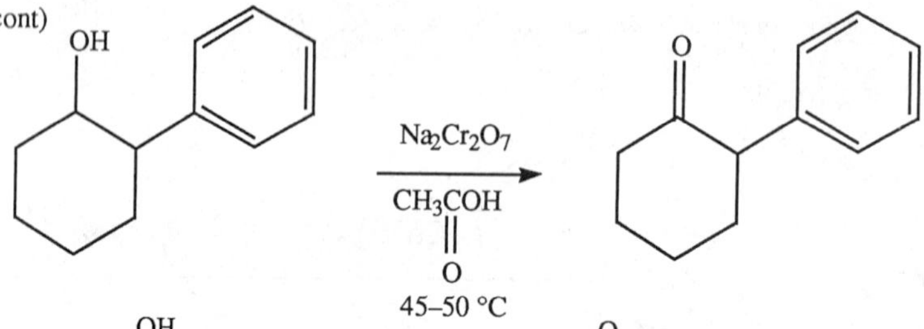

Na$_2$Cr$_2$O$_7$

CH$_3$COH
‖
O
45–50 °C

(c)

CrO$_3$

H$_2$SO$_4$
H$_2$O
acetone

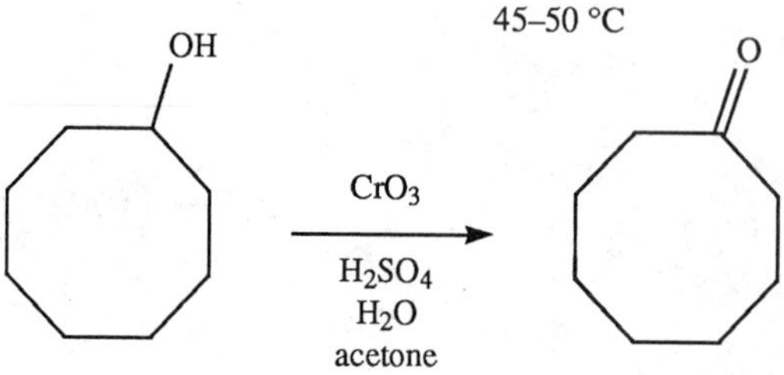

(d)

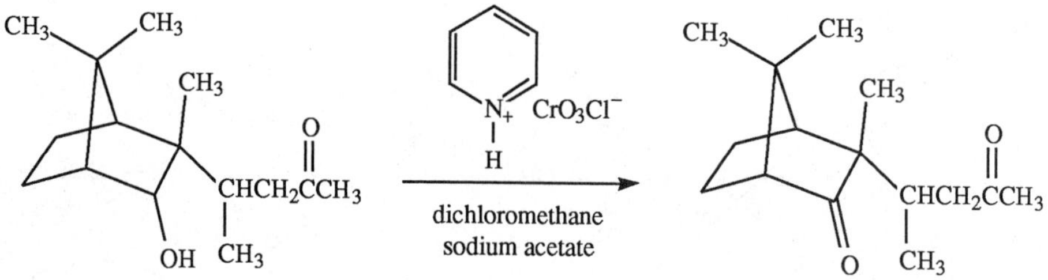

(e)

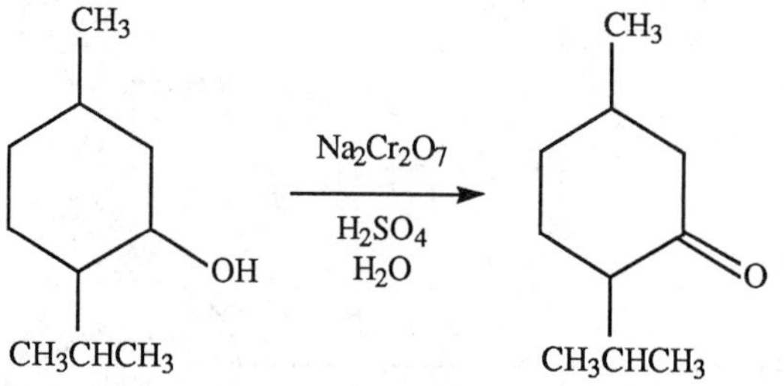

Na$_2$Cr$_2$O$_7$

H$_2$SO$_4$
H$_2$O

12.20

(a)

CrO$_3$

H$_2$SO$_4$
H$_2$O

H$_2$O$_2$, NaOH

H$_2$O

9–BBN

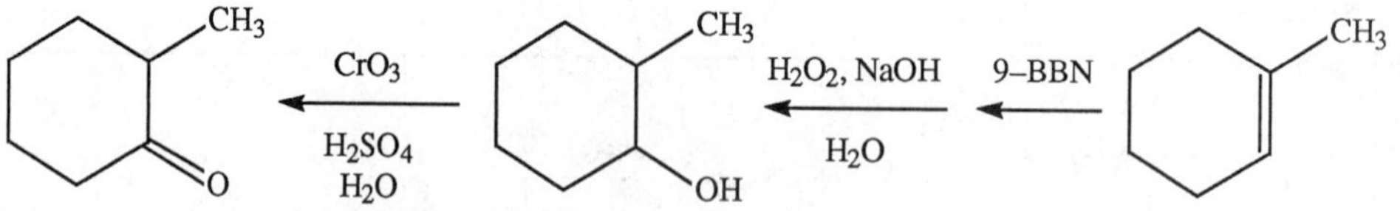

Note that this problem resembles the problem in Problem-Solving Skills Section 12.4D closely. Practice applying the same questions to it.

12.20 (cont)

(b) $CH_3CH_2CHCH_3$ $\xleftarrow{\quad NH_3 \text{ (excess)} \quad}$ $CH_3CH_2CHCH_3$ $\xleftarrow{\quad HBr \quad}$ $CH_3CH_2CH\!\!=\!\!CH_2$
　　　　　　|　　　　　　　　　　　　　　　　　|
　　　　　　NH_2　　　　　　　　　　　　　　　Br

　　　$+\ NH_4{}^+Br^-$

(c)

1. What functional groups are present in the starting material and the product?

    The starting material is an alcohol; the product is an alkyne.

2. How do the carbon skeletons of the two compounds compare? How many carbon atoms does each contain? Are there any rings? What are the positions of branches and functional groups on the carbon skeletons?

    Both compounds contain an aromatic ring. The alcohol group on the benzylic carbon atom has been replaced by a three-carbon chain containing a triple bond (a 1-propynyl group).

3. How do the functional groups change in going from starting material to product? Does the starting material have a good leaving group?

    The benzylic carbon atom is bonded to the first carbon atom of the triple bond. There is no good leaving group present in the starting material.

4. Is it possible to dissect the structures of the starting material and product to see which bonds must be broken and which formed?

    A carbon-oxygen bond was broken. A carbon-carbon bond was formed.

5. Do we recognize any part of the product molecule as coming from a good nucleophile or an electrophilic addition?

    The 1-propynyl group can come from an S_N2 reaction.

12.20 (c) (cont)

6. What type of compound would be a good precursor to the product?

    A compound with a good leaving group, such as an alkyl halide or a tosylate, would be a good precursor to the
    product.

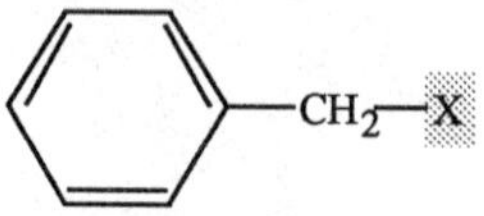

    An alkyl bromide can easily be made from the starting material by reaction with concentrated hydrobromic acid.

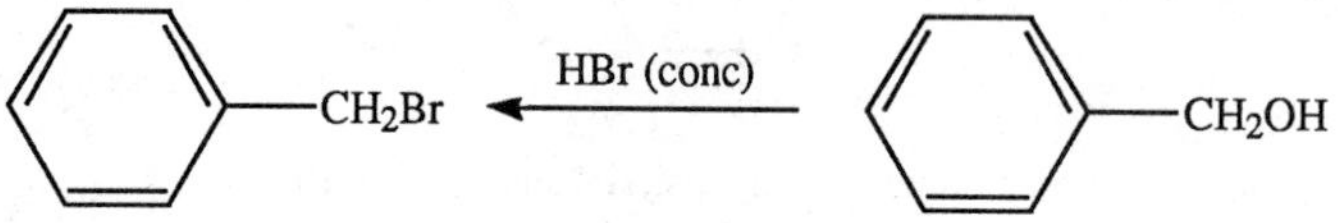

7. After this last step, do we see how to get from starting material to product? If not, we need to analyze the structure
    obtained in step 6 by applying questions 5 and 6 to it.

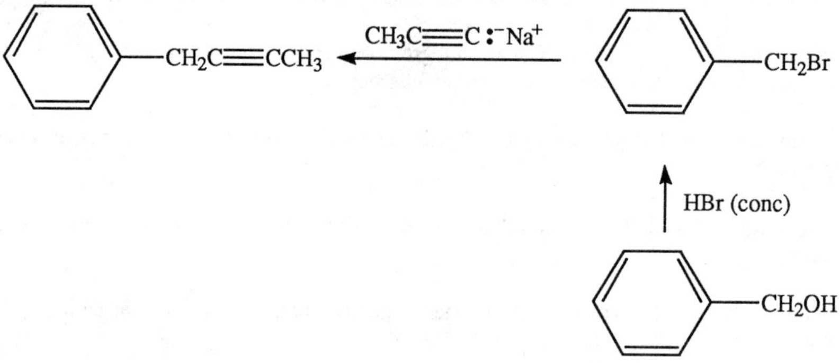

12.21

(a)

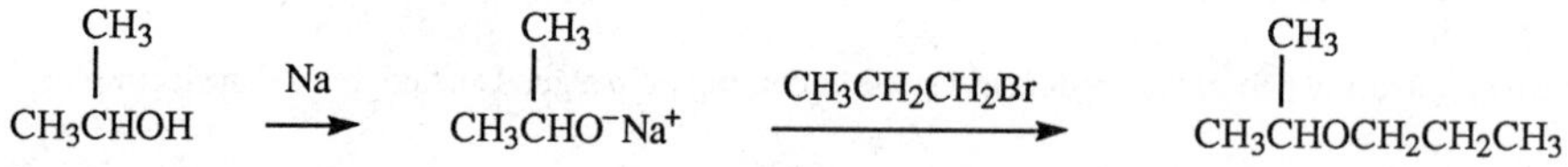

(b)

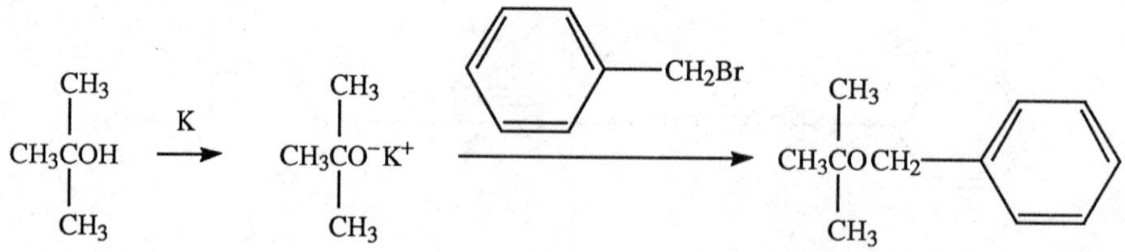

12.21 (cont)

(c) CH_3CH_2OH $\xrightarrow{Na}$ $CH_3CH_2O^-Na^+$ $\xrightarrow{CH_2=CHCH_2Br}$ $CH_3CH_2OCH_2CH=CH_2$

(d) $CH_3CH_2CH_2CH_2OH$ $\xrightarrow{Na}$ $CH_3CH_2CH_2CH_2O^-Na^+$ $\xrightarrow{CH_3I}$ $CH_3CH_2CH_2CH_2OCH_3$

   or CH_3OH $\xrightarrow{Na}$ $CH_3O^-Na^+$ $\xrightarrow{CH_3CH_2CH_2CH_2Br}$ $CH_3CH_2CH_2CH_2OCH_3$

(e) $CH_3CH_2CH_2CH_2CH_2CH_2OH$ $\xrightarrow{Na}$ $CH_3CH_2CH_2CH_2CH_2CH_2O^-Na^+$ $\xrightarrow{CH_3CH_2Br}$

$CH_3CH_2CH_2CH_2CH_2CH_2OCH_2CH_3$

   or CH_3CH_2OH $\xrightarrow{Na}$ $CH_3CH_2O^-\ Na^+$ $\xrightarrow{CH_3CH_2CH_2CH_2CH_2CH_2Br}$

$CH_3CH_2CH_2CH_2CH_2CH_2OCH_2CH_3$

(f)

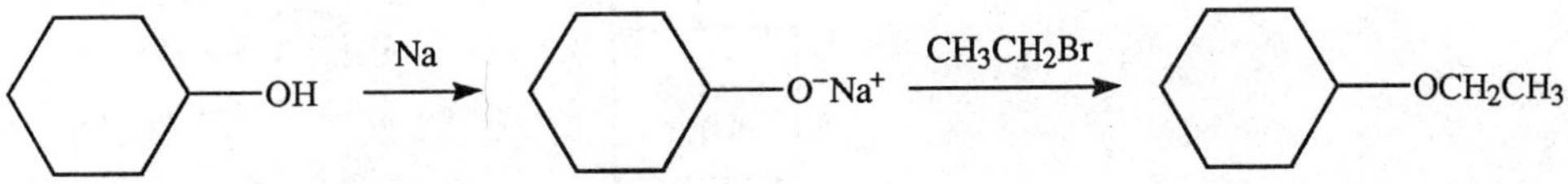

Concept Map 12.5   Preparation of ethers by nucleophilic substitution reactions.

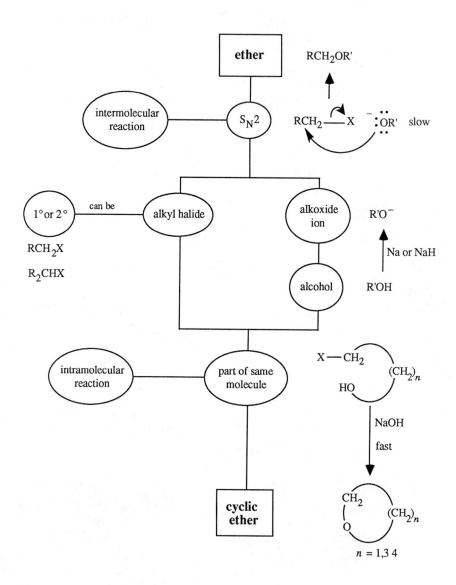

12.22

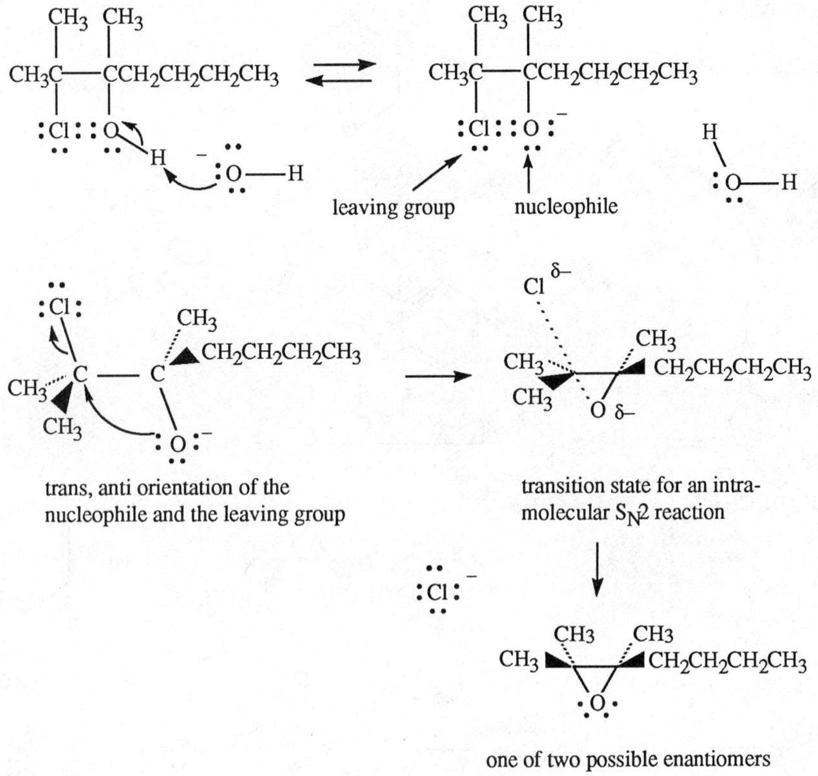

leaving group   nucleophile

trans, anti orientation of the
nucleophile and the leaving group

transition state for an intramolecular S$_N$2 reaction

one of two possible enantiomers

12.23

(a)

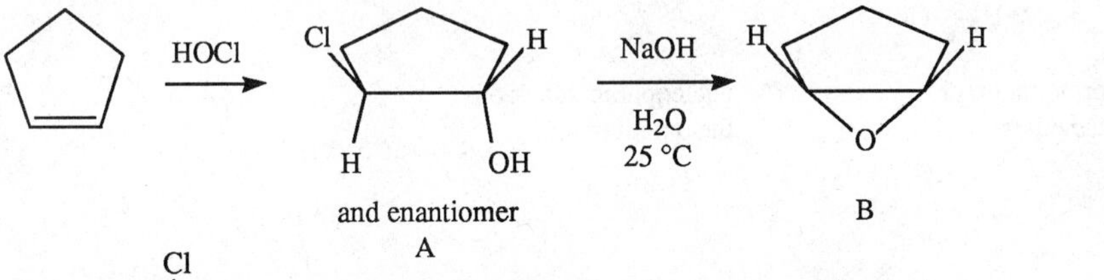

and enantiomer
A

B

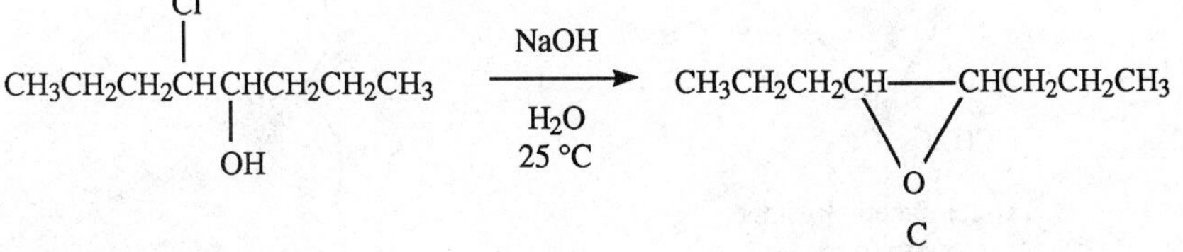

(b) $CH_3CH_2CH_2CHCHCH_2CH_2CH_3$ $\xrightarrow[\substack{H_2O \\ 25\,°C}]{NaOH}$ $CH_3CH_2CH_2CH$——$CHCH_2CH_2CH_3$

C

(c) $ClCH_2CCH_2CH_2CH_2CH_3$ $\xrightarrow[\substack{H_2O \\ 25\,°C}]{NaOH}$ CH_2——$CCH_2CH_2CH_2CH_3$

D

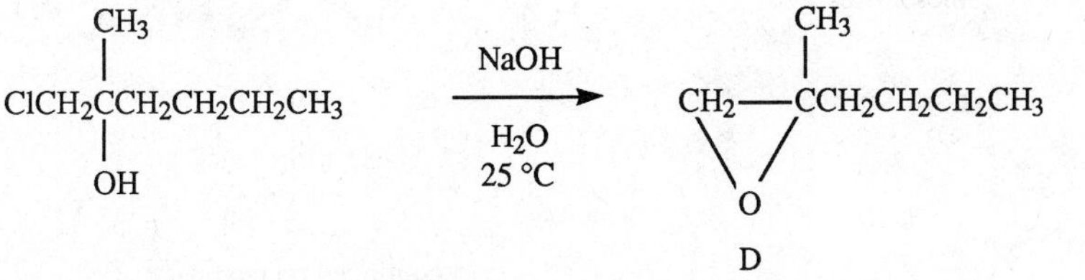

12.23 (cont)

(d)

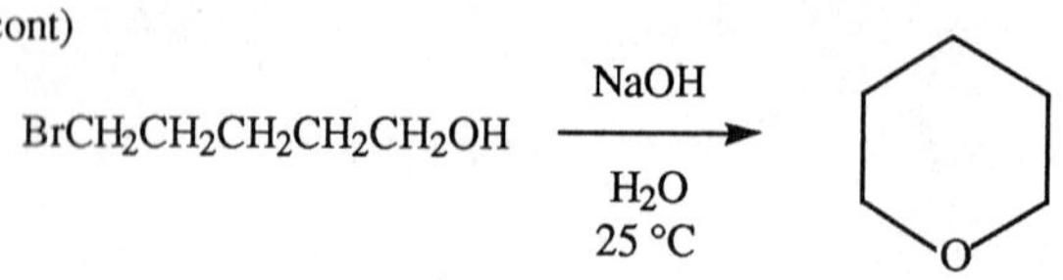
BrCH2CH2CH2CH2CH2OH
NaOH
H2O
25 °C
O

E

(e)

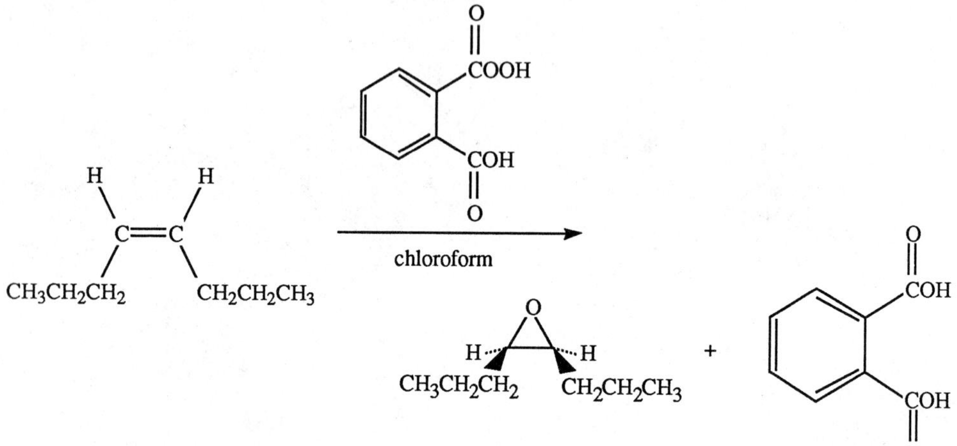
H
H
C=C
CH3CH2CH2
CH2CH2CH3
COOH
COH
chloroform
H
H
CH3CH2CH2
CH2CH2CH3
+
COH
COH

F G

12.24

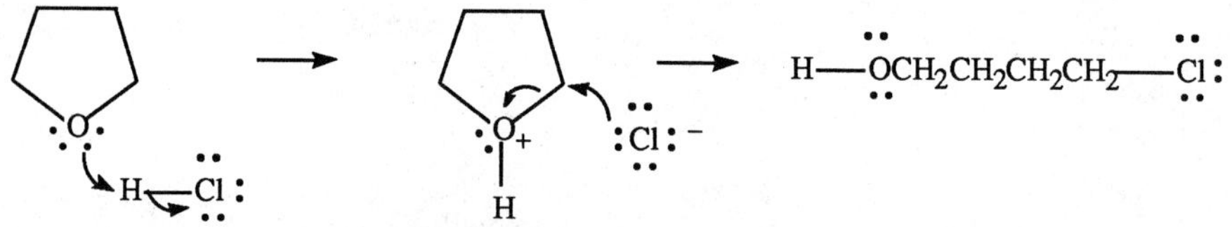
H—Cl
H
Cl −
H—OCH2CH2CH2CH2—Cl

protonation of the ether

nucleophilic attack on the oxonium ion

12.25

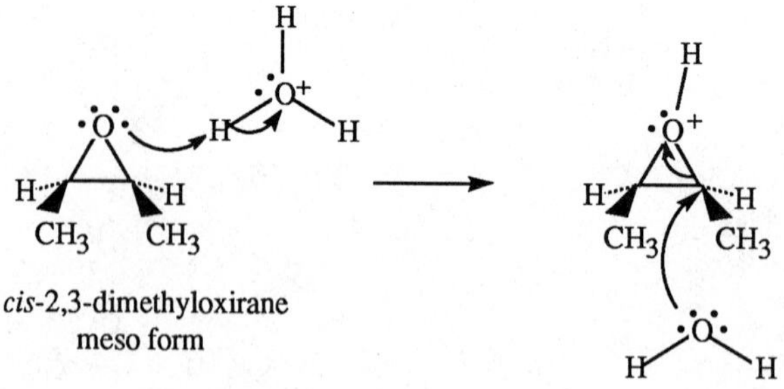
H
H
CH3
CH3
H
H
H
H
H
CH3
CH3
H
H

cis-2,3-dimethyloxirane
meso form

continued on next page

12.25 (cont)

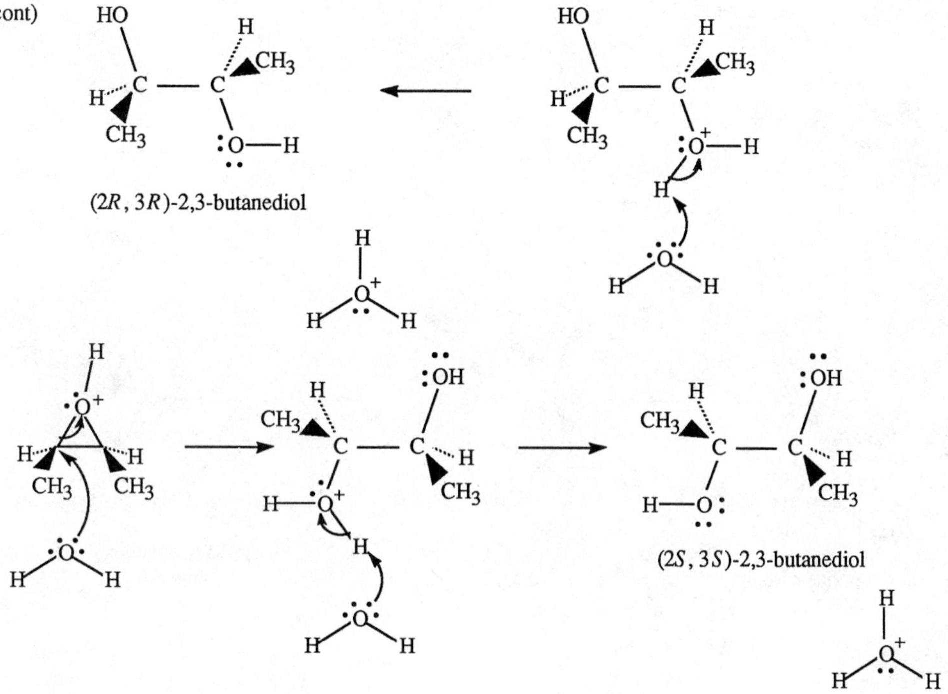

(2R, 3R)-2,3-butanediol

(2S, 3S)-2,3-butanediol

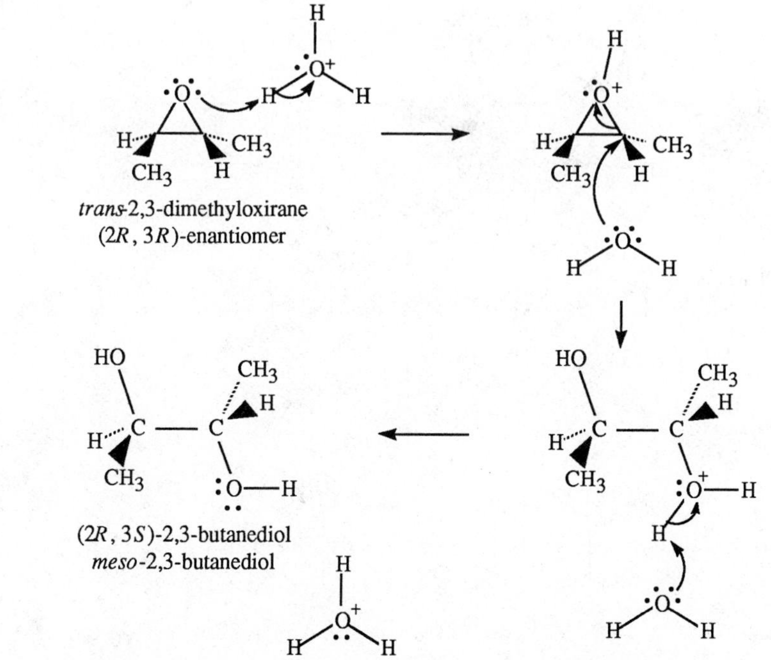

trans-2,3-dimethyloxirane
(2R, 3R)-enantiomer

(2R, 3S)-2,3-butanediol
meso-2,3-butanediol

Note that attack at the other carbon atom also leads to the meso compound. (2S, 3S)-2,3-Dimethyl-oxirane, the enantiomer of the trans isomer shown above, also gives the same compound.

12.26

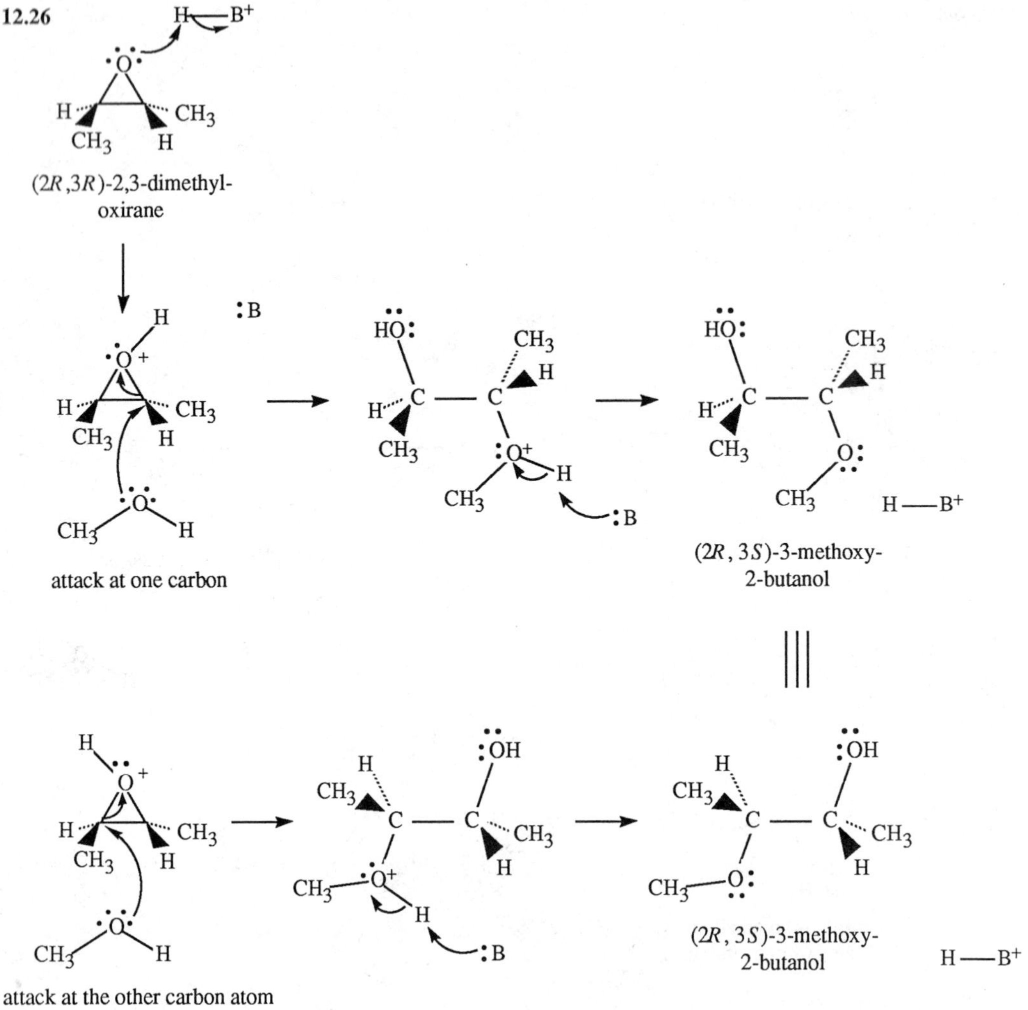

(2R,3R)-2,3-dimethyl-
oxirane

attack at one carbon

(2R, 3S)-3-methoxy-
2-butanol

attack at the other carbon atom

(2R, 3S)-3-methoxy-
2-butanol

12.27

(a) CH_3C——$CHCH_2CH_3$ +   [piperidine] → CH_3C——$CHCH_2CH_3$

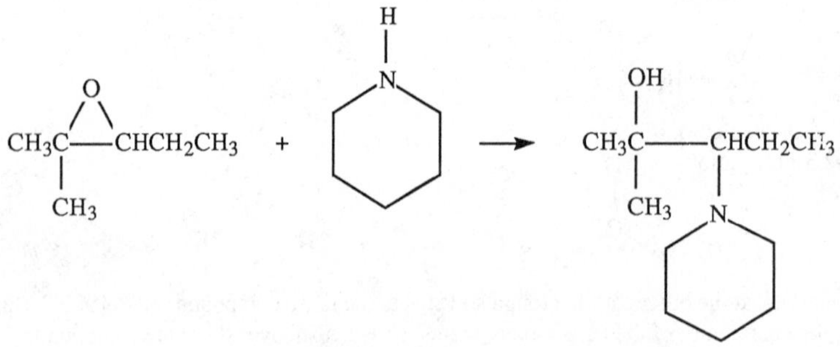

12.27 (cont)

(b) $CH_3CH{-}CHCH_3$ + NaN_3 $\longrightarrow$ $CH_3CHCHCH_3$ (with OH and N_3 substituents)

(c) $CH_3C{-}CH_2$ (with CH_3) + $Na^{18}OH$ $\longrightarrow$ $CH_3CCH_2{}^{18}OH$ (with OH and CH_3)

(d) $CH_3CCH_2CH{-}CH_2$ (with two CH_3) + $CH_3CH_2CH_2S^-Na^+$ $\longrightarrow$ $CH_3CCH_2CHCH_2SCH_2CH_2CH_3$ (with CH_3, OH, CH_3)

(e) $CH_3CH{-}CH_2$ + HBr $\longrightarrow$ CH_3CHCH_2Br (with OH) + CH_3CHCH_2OH (with Br)

- -

Concept Map 12.6   (see p. 308)

- -

12.28 CH_3SH + $CH_3CH_2O^-Na^+$ $\longrightarrow$ $CH_3S^-Na^+$ + CH_3CH_2OH

Concept Map 12.6   Ring-opening reactions of oxiranes.

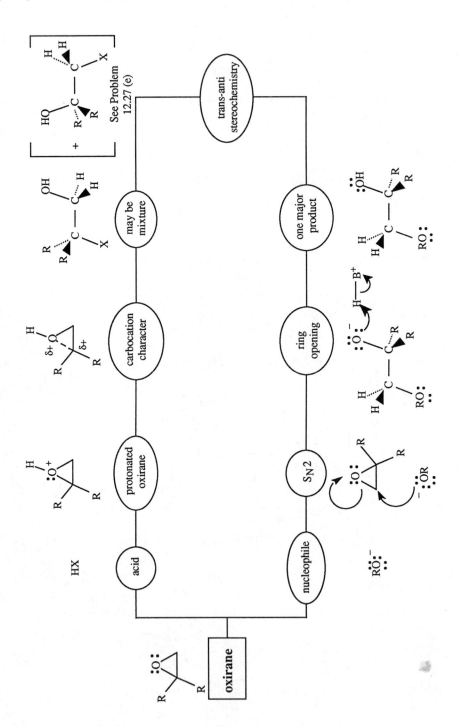

12.29

(a)

and enantiomer   and enantiomer

+

(b)

and enantiomer

1. What functional groups are present in the starting material and the product?

    The starting material is an alkene; the product contains two hydroxyl groups on adjacent carbon atoms.

2. How do the carbon skeletons of the two compounds compare? How many carbon atoms does each contain? Are there any rings? What are the positions of branches and functional groups on the carbon skeletons?

    The carbon skeletons are the same. The alkene has two methyl groups cis to each other on the double bond. The methyl groups are also on the same side of the plane of the page which contains the hydroxyl groups, carbon-2, and carbon-3.

    The two hydroxyl groups are anti to each other.

3. How do the functional groups change in going from starting material to product? Does the starting material have a good leaving group?

    The double bond of the alkene has disappeared. Two hydroxyl groups have been attached to the carbon atoms that were part of the double bond.

12.29 (cont)

4. Is it possible to dissect the structures of the starting material and product to see which bonds must be broken and which formed?

bond broken   bonds formed

A π bond was broken; two carbon-oxygen bonds were formed anti to each other. Carbon is now bonded to elements that are more electronegative that it is; suggesting an oxidation reaction.

5. Do we recognize any part of the product molecule as coming from a good nucleophile or an electrophilic addition?

An alkene can be converted into a compound containing one oxygen by oxidation with the electrophilic reagent, *m*-chloroperoxybenzoic acid. A hydroxyl group can come from water, a nucleophile.

6. What type of compound would be a good precursor to the product?

An oxirane can be converted to the 1,2-diol.

7. After this last step, do we see how to get from starting material to product? If not, we need to analyze the structure obtained in step 6 by applying questions 5 and 6 to it.

12.30

(a)

12.30 (cont)

(b)

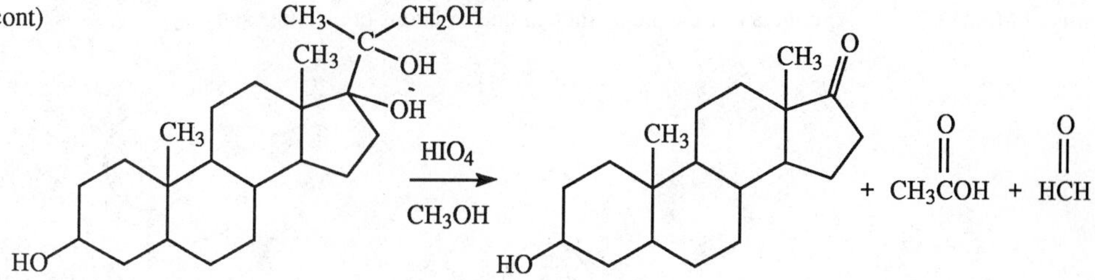

(c)

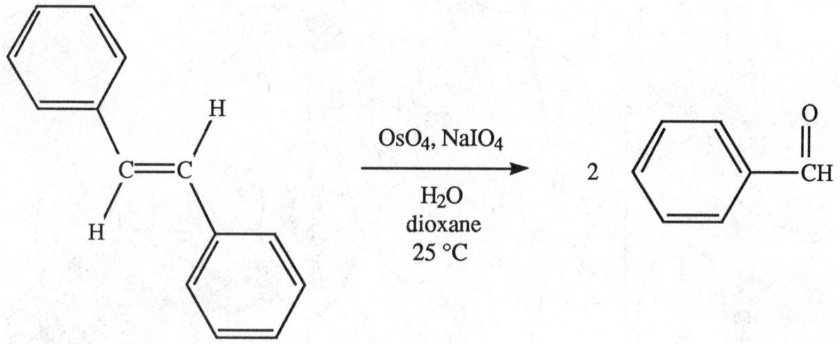

- -

Concept Map 12.7   (see p. 312)

- -

12.31 (a) 1,3-propanediol

(b) 1,2-dimethoxyethane

(c) 3-ethoxy-2,2-dimethyl-1-propanol

(d) 1-bromo-2-methoxyethane

(e) ethoxybenzene
(ethyl phenyl ether)

(f) 2-ethoxyethanol

(g) cyclobutanol

(h) (2*S*)(*E*)-5-methyl-4-hepten-2-ol

(i) (1*R*, 2*R*)-2-bromocyclopentanol

(j) benzyloxyphenylmethane
(dibenzyl ether)

(k) (1*S*, 2*S*)-2-methoxycyclohexanol

(l) 1,1,1-triethoxyethane

Concept Map 12.7   Summary of the preparation of and reactions of alcohols and ethers.

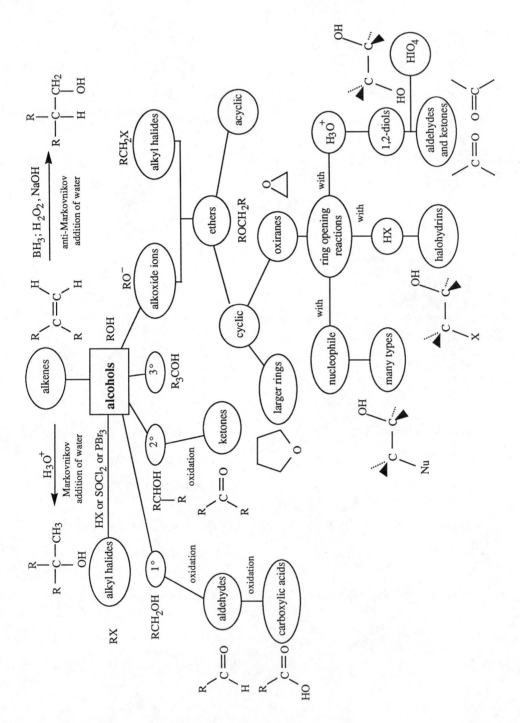

12.32   (a) 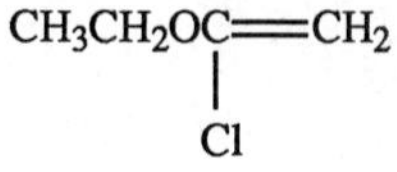 $CH_3CH_2OC{=}CH_2$
                                    $|$
                                    Cl

(b)

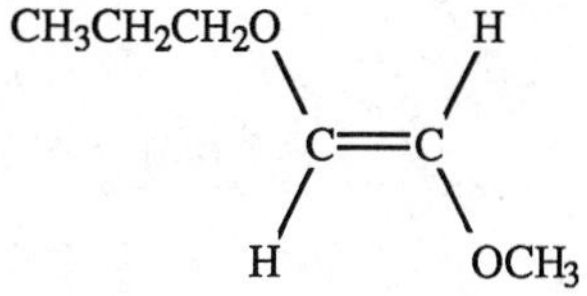

(c) 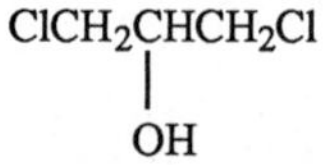 $ClCH_2CHCH_2Cl$
              $|$
              OH

(d) 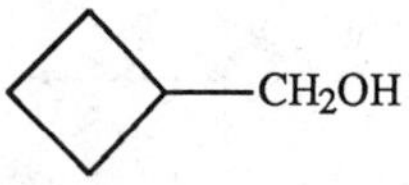 $-CH_2OH$

(e)

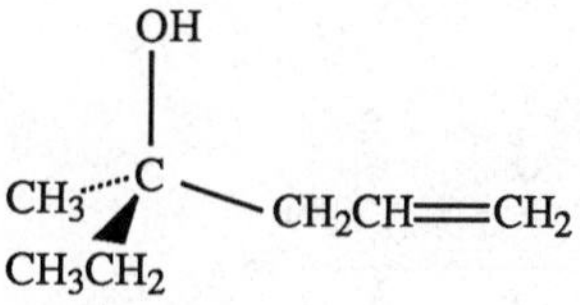

(f)

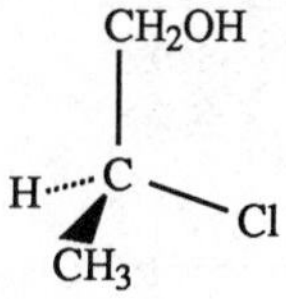

(g)

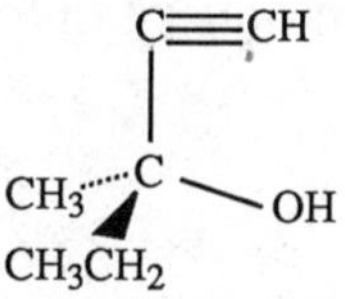

(h) $O_2NCH_2CH_2OH$

(i)

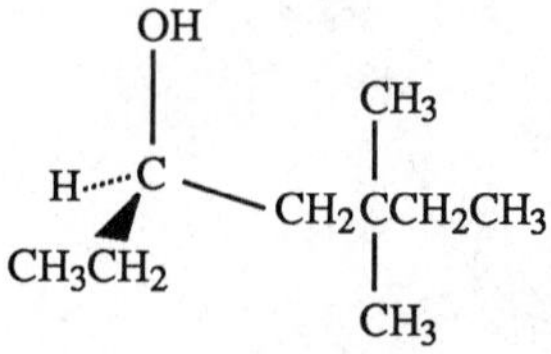

(j)

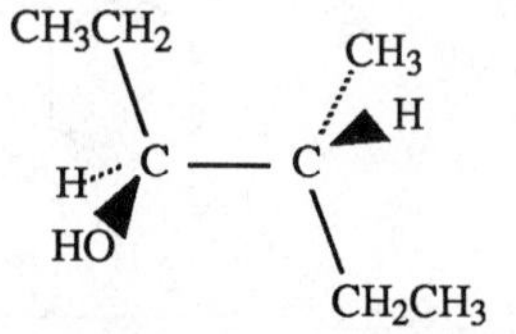

12.33

(a) $CH_3CH_2CH_2OH$ $\xrightarrow[\Delta]{HBr}$ $CH_3CH_2CH_2Br$ $+$ H_2O

(b) $CH_3CH_2CH_2OH$ $\xrightarrow[\Delta]{PBr_3}$ $CH_3CH_2CH_2Br$ $+$ $P(OH)_3$

(c) $CH_3CH_2CH_2OH$ $\xrightarrow[\text{pyridine}]{SOCl_2}$ $CH_3CH_2CH_2Cl$ $+$ $SO_2\uparrow$ $+$

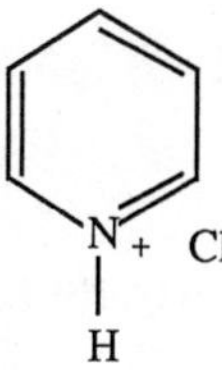

12.33 (cont)

(d) $CH_3CH_2CH_2OH$ $\xrightarrow{NaH}$ $CH_3CH_2CH_2O^-Na^+$ + $H_2\uparrow$

(e) $CH_3CH_2CH_2OH$ $\xrightarrow[\text{cold}]{H_3O^+}$ $CH_3CH_2CH_2\overset{+}{O}H_2$ + H_2O

(f) $CH_3CH_2CH_2OH$ $\xrightarrow[\substack{H_2SO_4 \\ \Delta}]{}$ $CH_3CH{=\!=}CH_2$ + H_2O

(g) $CH_3CH_2CH_2OH$ $\xrightarrow{Na}$ $CH_3CH_2CH_2O^-Na^+$ $\xrightarrow{CH_3CH_2CH_2CH_2Br}$

$CH_3CH_2CH_2OCH_2CH_2CH_2CH_3$ +   NaBr

(h) $CH_3CH_2CH_2OH$ $\xrightarrow[\substack{H_2SO_4 \\ \Delta}]{Na_2Cr_2O_7}$ $CH_3CH_2\overset{\overset{\textstyle O}{\|}}{C}OH$

(i) $CH_3CH_2CH_2OH$ $\xrightarrow[\text{dichloromethane}]{}$ $CH_3CH_2\overset{\overset{\textstyle O}{\|}}{C}H$

12.34

(a) $CH_3CH_2CH_2CH_2\underset{\underset{\textstyle OH}{|}}{C}HCH_3$ $\xrightarrow[\Delta]{HBr}$ $CH_3CH_2CH_2CH_2\underset{\underset{\textstyle Br}{|}}{C}HCH_3$ + $CH_3CH_2CH_2\underset{\underset{\textstyle Br}{|}}{C}HCH_2CH_3$

major

+ H_2O +   some elimination products

(b) $CH_3CH_2CH_2CH_2\underset{\underset{\textstyle OH}{|}}{C}HCH_3$ $\xrightarrow[\Delta]{PBr_3}$ $CH_3CH_2CH_2CH_2\underset{\underset{\textstyle Br}{|}}{C}HCH_3$ + $P(OH)_3$

12.34 (cont)

(c) CH₃CH₂CH₂CH₂CHCH₃ $\xrightarrow[\text{pyridine}]{\text{SOCl}_2}$ CH₃CH₂CH₂CH₂CHCH₃ + SO₂↑ +
     | |
     OH   Cl

(d) CH₃CH₂CH₂CH₂CHCH₃ $\xrightarrow{\text{NaH}}$ CH₃CH₂CH₂CH₂CHCH₃ + H₂↑
     | |
     OH   O⁻Na⁺

(e) CH₃CH₂CH₂CH₂CHCH₃ $\xrightarrow[\text{ZnCl}_2]{\text{HCl}}$ CH₃CH₂CH₂CH₂CHCH₃ + CH₃CH₂CH₂CHCH₂CH₃
     | | |
     OH   Cl   Cl

(f) CH₃CH₂CH₂CH₂CHCH₃ $\xrightarrow[\substack{\text{H}_2\text{SO}_4 \\ \Delta}]{}$ CH₃CH₂CH₂CH=CHCH₃ +
     |
     OH

     CH₃CH₂CH₂CH₂CH=CH₂ + CH₃CH₂CH=CHCH₂CH₃ + H₂O

(g) CH₃CH₂CH₂CH₂CHCH₃ $\xrightarrow[\substack{\text{H}_2\text{SO}_4 \\ \Delta}]{\text{Na}_2\text{Cr}_2\text{O}_7}$ CH₃CH₂CH₂CH₂CCH₃
     | ‖
     OH   O

(h) CH₃CH₂CH₂CH₂CHCH₃ $\xrightarrow[\text{dichloromethane}]{}$ CH₃CH₂CH₂CH₂CCH₃
     | ‖
     OH   O

(i) CH₃CH₂CH₂CH₂CHCH₃ $\xrightarrow[\substack{\text{H}_2\text{SO}_4 \\ \text{H}_2\text{O} \\ \text{acetone}}]{\text{CrO}_3}$ CH₃CH₂CH₂CH₂CCH₃
     | ‖
     OH   O

12.35 (a)   ⬡—OH   H₂SO₄, cold, conc; CrO₃, H₂O, H₂SO₄

12.35 (cont)

(b)

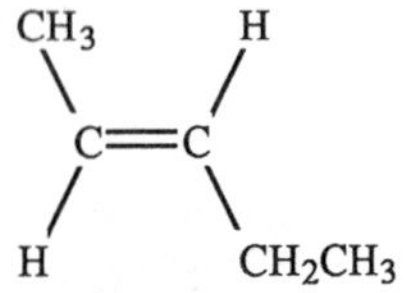

$$
\begin{array}{c}
CH_3 \qquad H \\
C = C \\
H \qquad CH_2CH_3
\end{array}
$$

H_2SO_4, cold, conc; Br_2, carbon tetrachloride

(c)

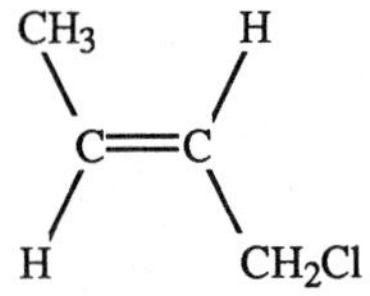

$$
\begin{array}{c}
CH_3 \qquad H \\
C = C \\
H \qquad CH_2Cl
\end{array}
$$

H_2SO_4, cold, conc; Br_2, carbon tetrachloride; $AgNO_3$, CH_3CH_2OH; NaI, acetone

(d)

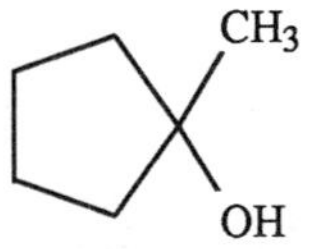

H_2SO_4, cold, conc

(e)

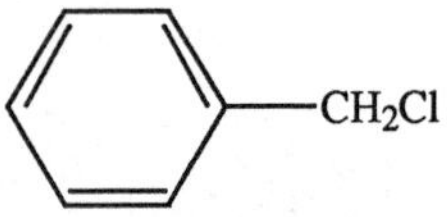

$AgNO_3$, CH_3CH_2OH; NaI, acetone

(f) $CH_3CH_2CH_2CH_2CH_2OH$

H_2SO_4, cold, conc; CrO_3, H_2O, H_2SO_4

(g) $CH_3C\equiv CCH_2CH_3$

H_2SO_4, cold, conc; Br_2, carbon tetrachloride

(h)

(i) $HOCH_2CH_2OH$

H_2O, cold; H_2SO_4, cold, conc; CrO_3, H_2O, H_2SO_4

(j)

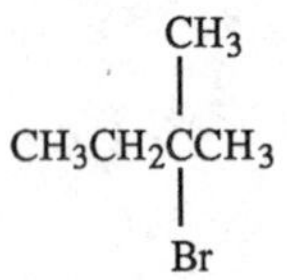

$$
\begin{array}{c}
CH_3 \\
| \\
CH_3CH_2CCH_3 \\
| \\
Br
\end{array}
$$

$AgNO_3$, CH_3CH_2OH

(k)

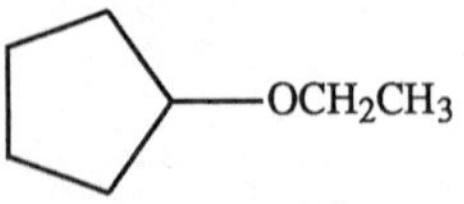

H_2SO_4, cold, conc

(l) $CH_3CH_2CH_2CH_2CH_2CH_2Cl$

NaI, acetone

12.36 (a) The location of the hydroxyl group on the chain affects the solubility in water of the isomeric alcohols. *n*-Butyl alcohol has the polar hydroxyl group at one end of the molecule and a long hydrophobic chain stretching away from it. In *sec*-butyl alcohol all parts of the hydrocarbon portion of the molecule are closer to, and more under the influence of, the polar hydroxyl group. In *tert*-butyl alcohol there is no large hydrophobic chain that is far away from the polar, water-soluble portion of the molecule. The hydroxyl group is able to carry the rest of the molecule into solution with it.

(b) Hydrogen bonding of the alcohol is the most important factor in determining water solubility. The higher molecular weight bromo compound is less soluble than the lower molecular weight, slightly more polar, chloro compound.

(c) The amino alcohol with two hydrogen bonding groups is most soluble in water. The alcohol, which can act as a donor as well as an acceptor of hydrogen bonds, is more soluble in water than the ester, which is only an acceptor of hydrogen bonds.

(d) The more stable the conjugate base (an anion in this example), the stronger is the acid (the alcohol). The *tert*-butoxide anion is destabilized by two effects: the bulk of the methyl groups that interferes with solvation (and therefore stabilization) of the anion, and the electron-donating effect of the methyl groups. The trifluoroethoxide anion, by contrast, is stabilized by the electron-withdrawing effect of the fluorine atoms.

(e) The compound that has the highest boiling point and the highest solubility in water is the one with two hydroxyl groups, which leads to increases hydrogen bonding, both between molecules of the same kind (giving rise to a higher boiling point) and between the compound and water (giving rise to higher solubility in water). Of the compounds with only one hydroxyl group, the one with the higher molecular weight and the longer carbon chain has the higher boiling point (from increased van der Waals attraction) and the lower solubility in water (because the hydrophobic part of the molecule is too large to be brought into solution by the polar hydroxyl group). See Sections 1.8B and 1.8C for review.

(f) The hydroxyl group hydrogen bonds more strongly than the thiol group, and the alcohol is therefore more soluble than the corresponding sulfur compound.

12.37

(a)
$$\underset{\overset{|}{CH_3CHCH_2OH}}{CH_3} \xrightarrow[\text{pyridine}]{SOCl_2} \underset{\overset{|}{CH_3CHCH_2Cl}}{CH_3}$$

(b)
$$\underset{\overset{|}{\underset{\underset{OH}{|}}{CH_3CCH_2CH_3}}}{CH_3} \xrightarrow{HCl\ (conc)} \underset{\overset{|}{\underset{\underset{Cl}{|}}{CH_3CCH_2CH_3}}}{CH_3}$$

(c)
$$\underset{\overset{|}{\underset{\underset{OH}{|}}{CH_3CHCH_2CH_3}}}{} \xrightarrow{PBr_3} \underset{\overset{|}{\underset{\underset{Br}{|}}{CH_3CHCH_2CH_3}}}{}$$

12.37 (cont)

(d)

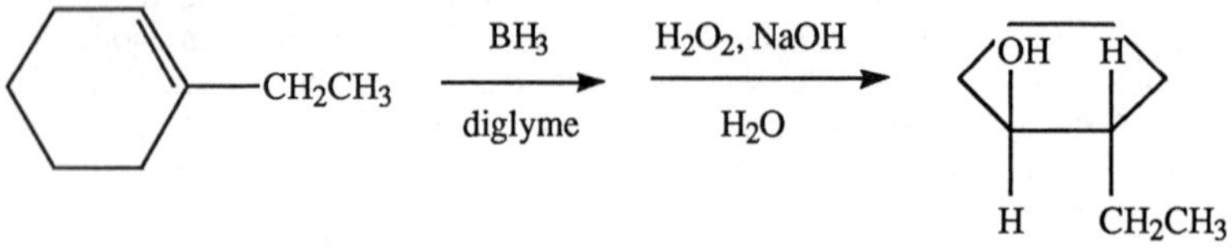

BH$_3$ / diglyme → H$_2$O$_2$, NaOH / H$_2$O →

and enantiomer

(e)   HOCH$_2$(CH$_2$)$_4$CH$_2$OH $\xrightarrow{\text{HBr (g) (excess)}}$ BrCH$_2$(CH$_2$)$_4$CH$_2$Br

(f)

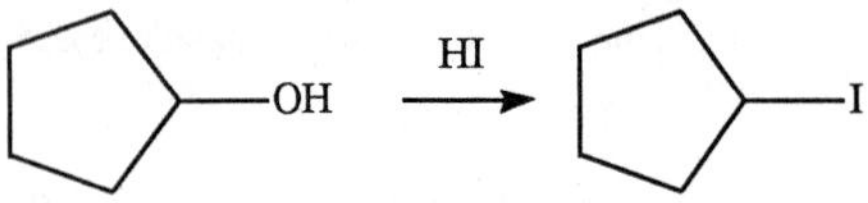

$\xrightarrow{\text{HI}}$

(g)

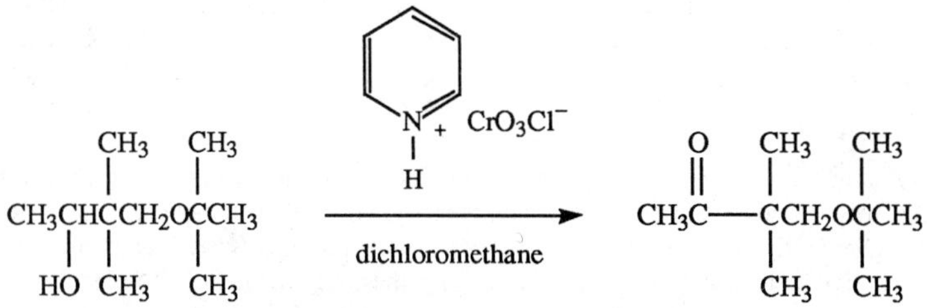

(h)

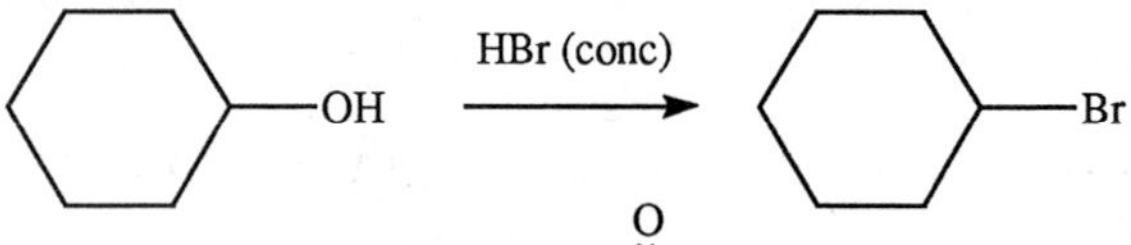

$\xrightarrow{\text{HBr (conc)}}$

(i)

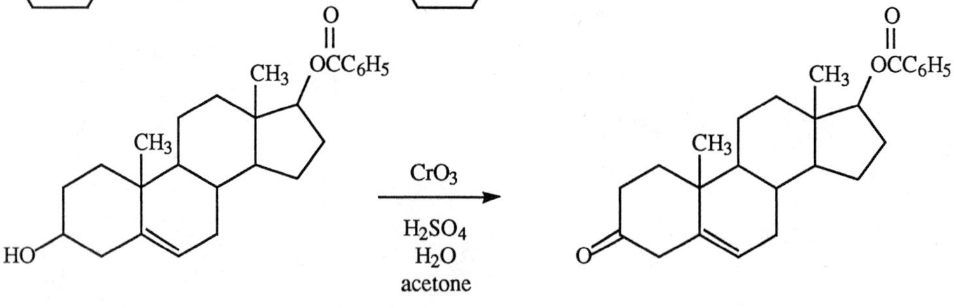

CrO$_3$ / H$_2$SO$_4$ / H$_2$O / acetone →

(j)

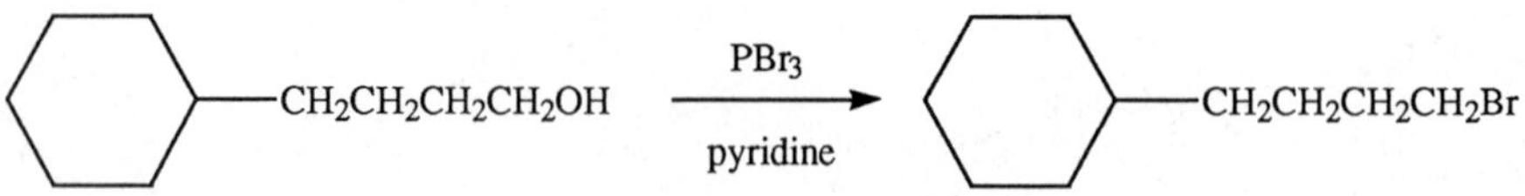

—CH$_2$CH$_2$CH$_2$CH$_2$OH $\xrightarrow[\text{pyridine}]{\text{PBr}_3}$ —CH$_2$CH$_2$CH$_2$CH$_2$Br

(k)

CrO$_3$ / H$_2$SO$_4$ / H$_2$O / acetone →

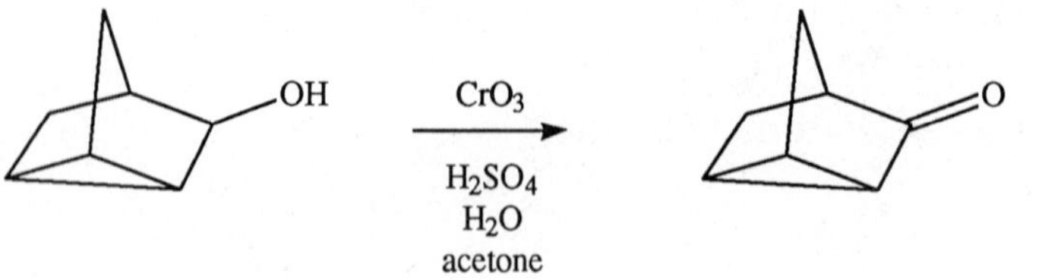

12.37 (cont)

(l)

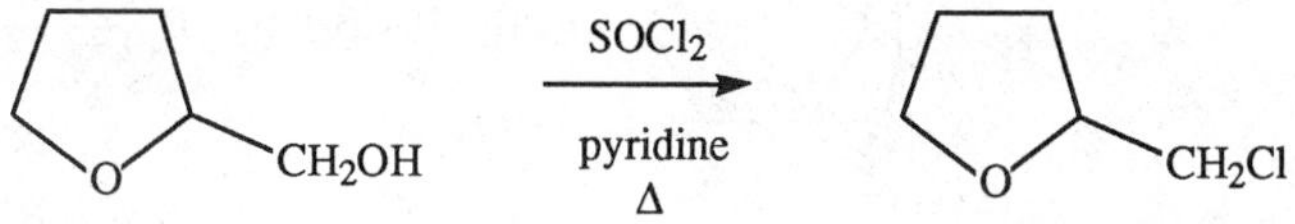

$$\xrightarrow[\substack{\text{pyridine} \\ \Delta}]{\text{SOCl}_2}$$

(m)

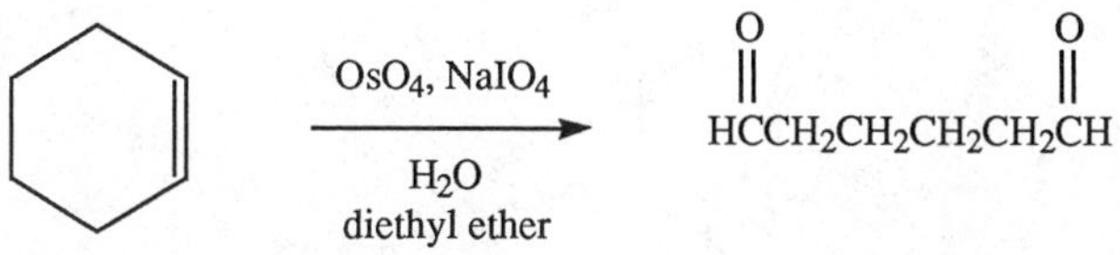

$$\xrightarrow[\substack{\text{H}_2\text{O} \\ \text{diethyl ether}}]{\text{OsO}_4, \text{NaIO}_4}$$

$$\underset{\text{HCCH}_2\text{CH}_2\text{CH}_2\text{CH}_2\text{CH}}{\overset{\text{O}\quad\quad\quad\text{O}}{\|\quad\quad\quad\|}}$$

(n)

$$\xrightarrow[\text{dichloromethane}]{}$$

two diastereomers

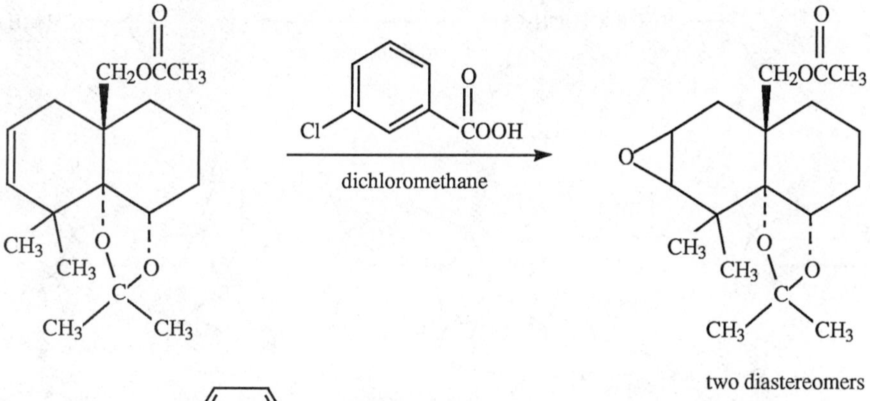

(o)

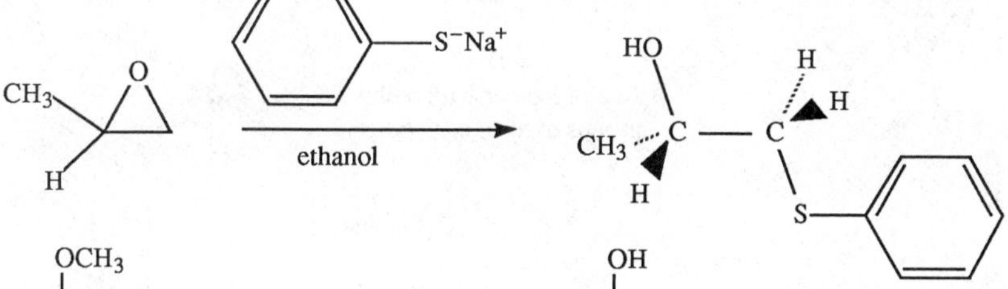

$$\xrightarrow[\text{ethanol}]{}$$

(p)

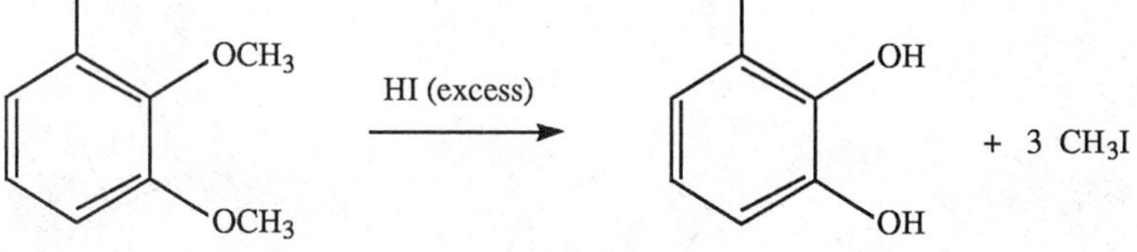

$$\xrightarrow{\text{HI (excess)}}$$

+ 3 CH$_3$I

12.38

(a)

$$\xrightarrow[\text{tetrahydrofuran}]{\text{BH}_3}$$

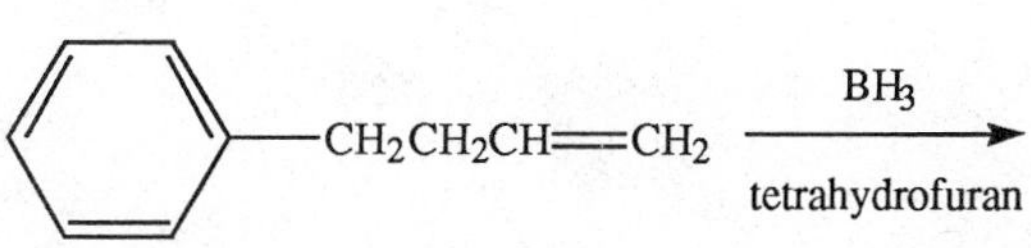

12.38 (cont)
(a) (cont)

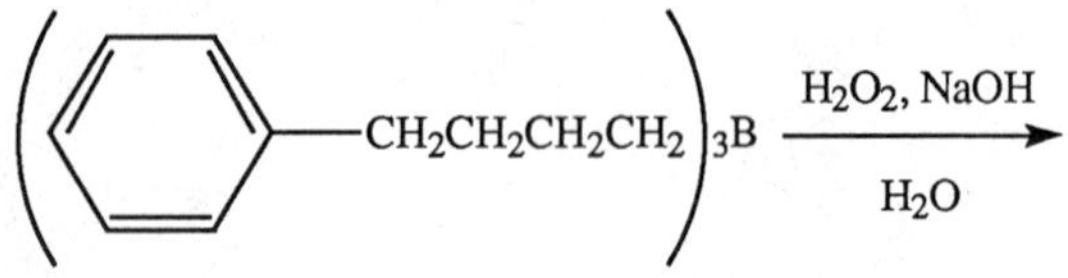

A

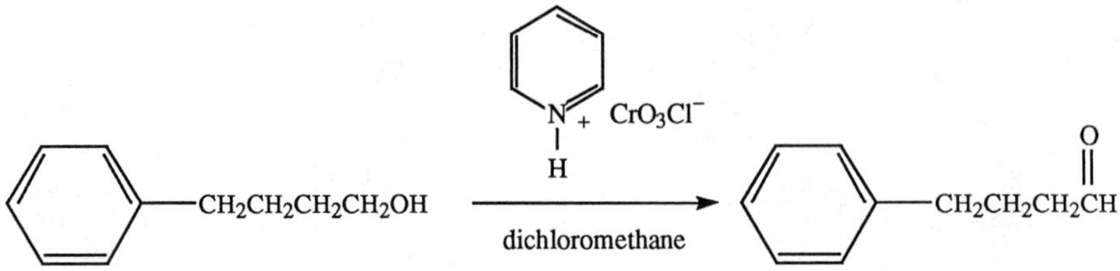

B   C

(b)

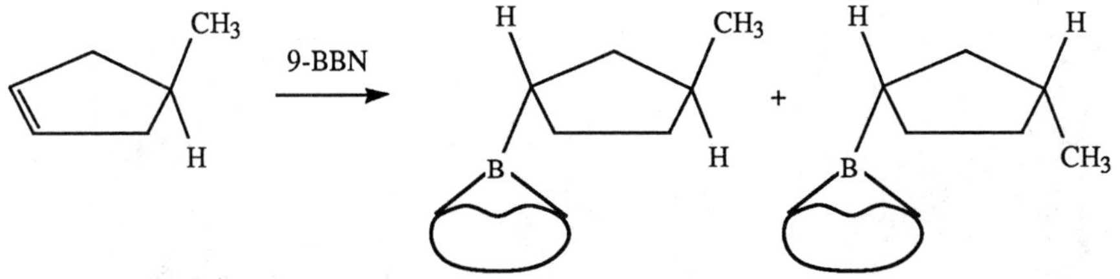

D
(major isomer with bulky
groups trans to each other)

E

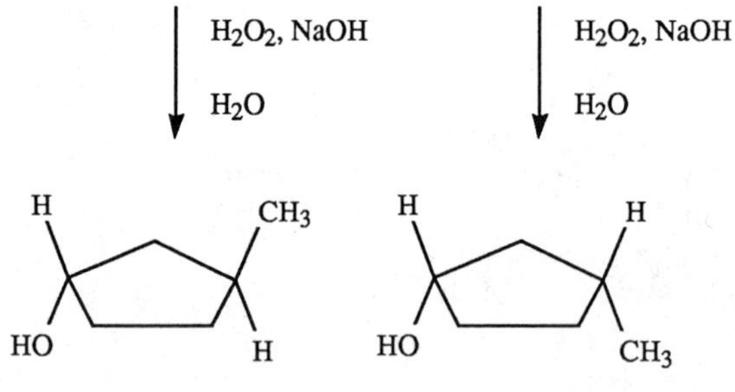

F   G

12.38 (cont)

(c)

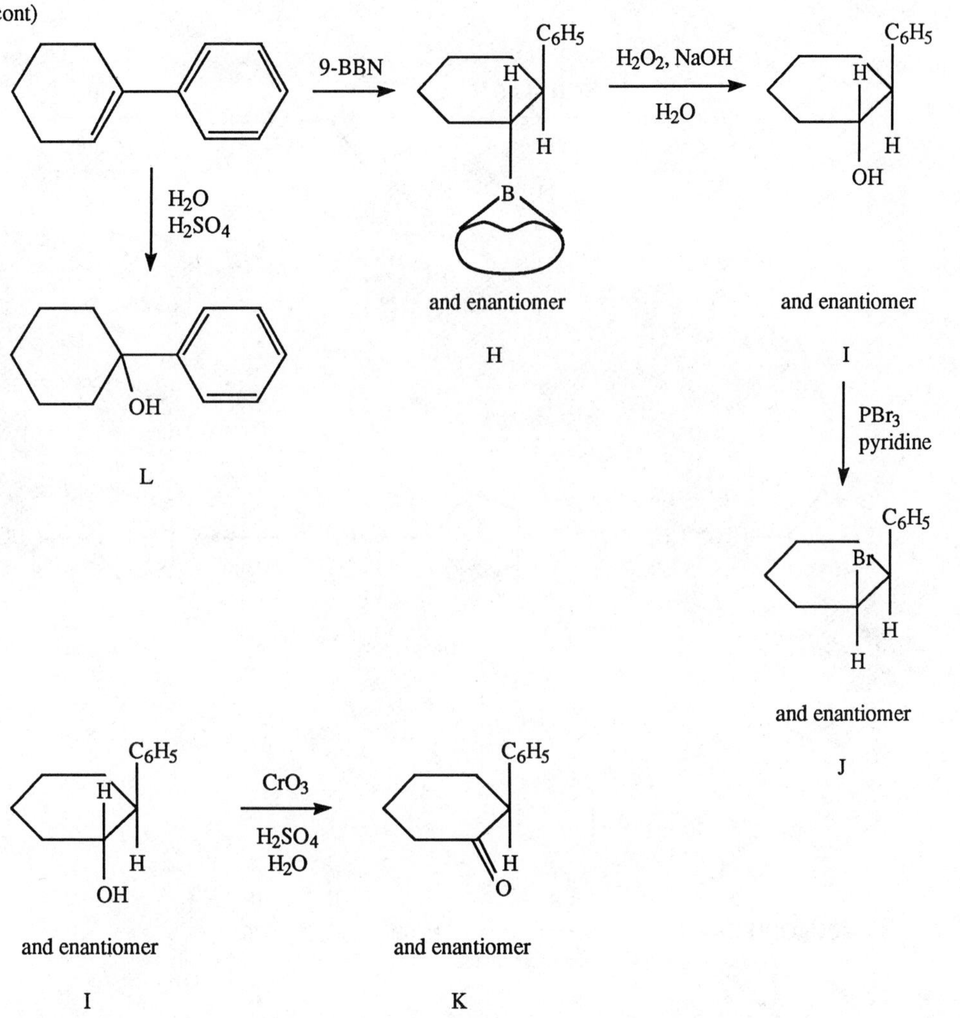

and enantiomer

H

and enantiomer

I

L

PBr$_3$
pyridine

and enantiomer

J

and enantiomer

I

CrO$_3$
H$_2$SO$_4$
H$_2$O

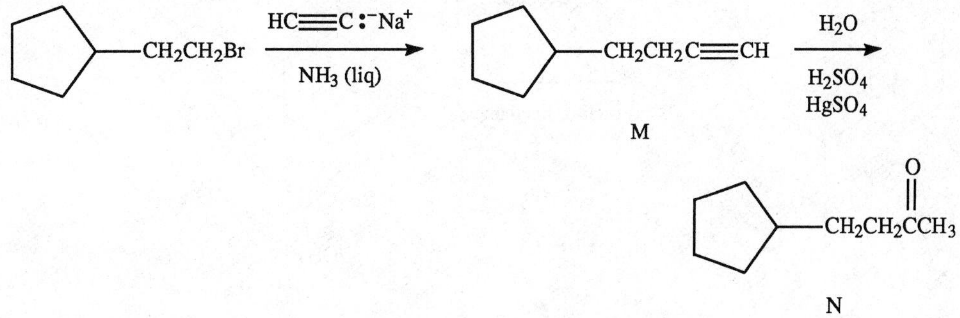

and enantiomer

K

(d)

HC≡C:⁻Na⁺

NH$_3$ (liq)

M

H$_2$O

H$_2$SO$_4$
HgSO$_4$

N

12.38 (cont)

(d) (cont)

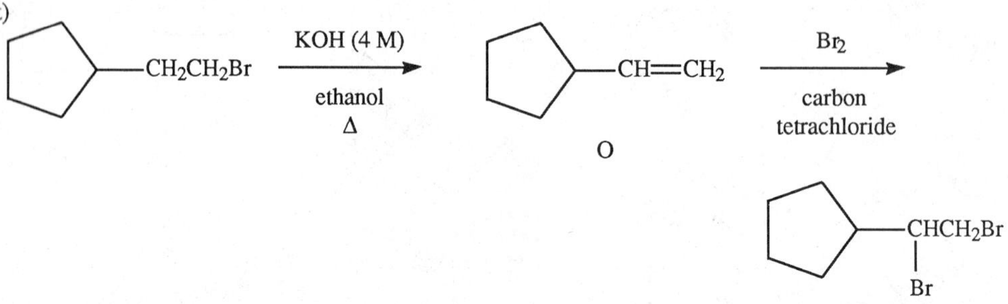

CH2CH2Br
KOH (4 M)
ethanol
Δ
CH=CH2
O
Br2
carbon
tetrachloride
CHCH2Br
Br

racemic mixture

P

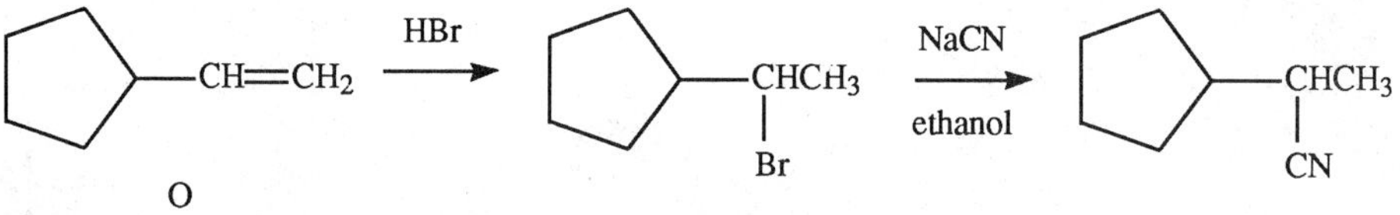

CH=CH2
HBr
CHCH3
Br
NaCN
ethanol
CHCH3
CN

O

racemic mixture

Q

racemic mixture

R

(e) HOCH$_2$(CH$_2$)$_4$CH$_2$OH

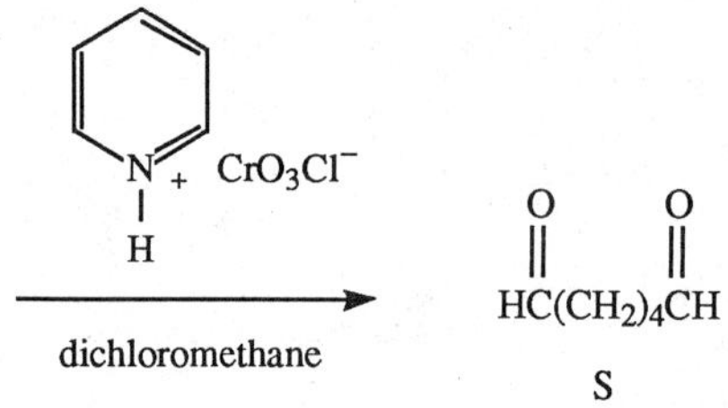

N+ CrO3Cl−
H
dichloromethane
O O
HC(CH2)4CH

S

(f)

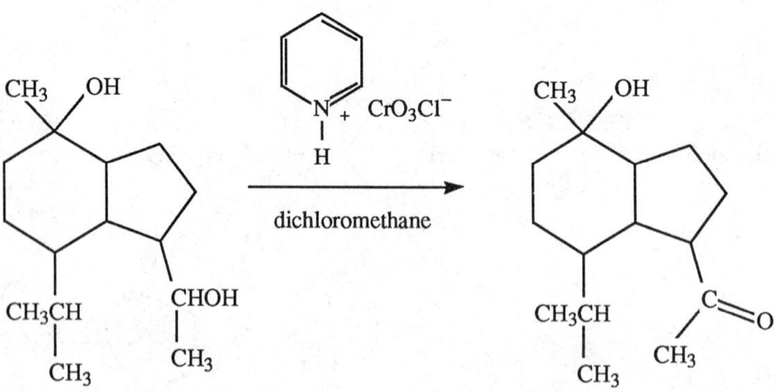

CH3 OH
CHOH
CH3CH
CH3
N+ CrO3Cl−
H
dichloromethane
C=O

T

12.38 (cont)

(g)

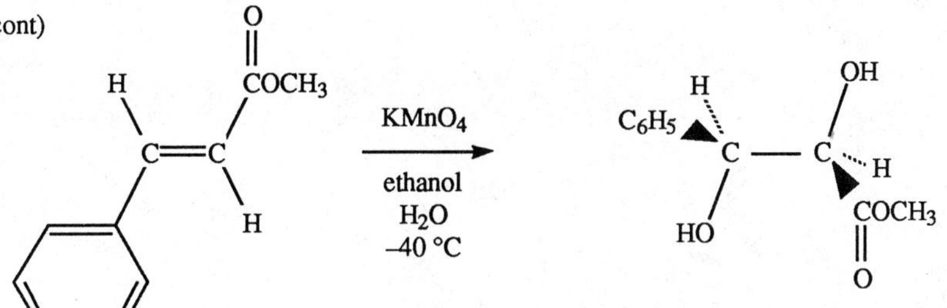

and enantiomer

U

(h)

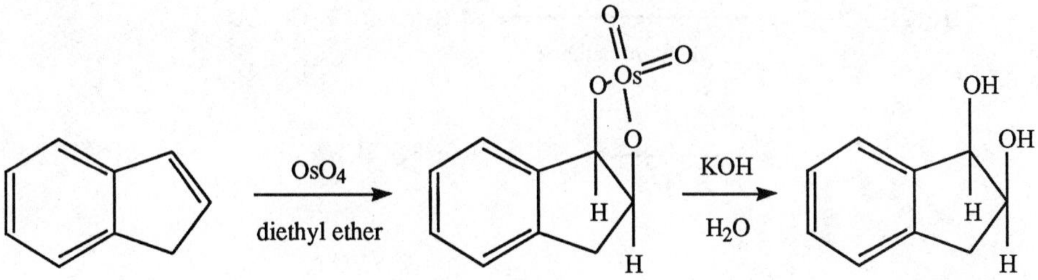

and enantiomer   and enantiomer

V   W

(i)

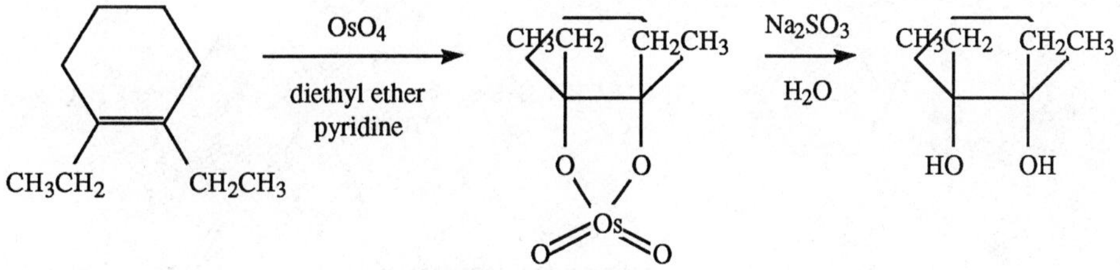

X   Y

(j)

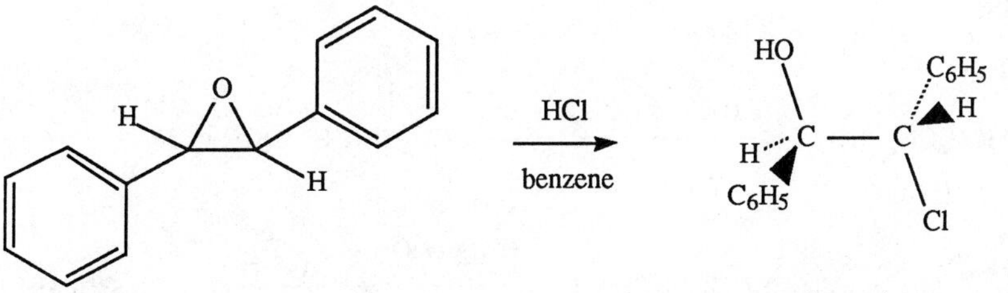

and enantiomer

Z

12.39

(a)

$$\underset{\underset{CH_3}{|}}{CH_3CHCH_2}\overset{\overset{O}{||}}{C}CH_3 \xleftarrow[\substack{H_2SO_4 \\ H_2O}]{CrO_3} \underset{\underset{CH_3}{|}}{CH_3CH}CH_2\underset{\underset{OH}{|}}{CH}CH_3 \xleftarrow{H_3O^+} \underset{\underset{CH_3}{|}}{CH_3CH}CH_2CH=CH_2$$

(b)

$$CH_3(CH_2)_4CH_2\overset{\overset{O}{||}}{C}H \xleftarrow[\text{dichloromethane}]{\text{pyridinium chlorochromate } CrO_3Cl^-} CH_3(CH_2)_4CH_2CH_2OH$$

$$\xleftarrow[H_2O]{H_2O_2,\ NaOH} (CH_3(CH_2)_4CH_2CH_2)_3B \xleftarrow[\text{diglyme}]{BH_3} CH_3(CH_2)_4CH=CH_2$$

(c)

$$C_6H_5-CH_2CH_2CH_2CH_2O-C_6H_5 \xleftarrow{C_6H_5-O^-Na^+}$$

$$C_6H_5-CH_2CH_2CH_2CH_2Br \xleftarrow[\text{pyridine}]{PBr_3} C_6H_5-CH_2CH_2CH_2CH_2OH$$

$$\xleftarrow[H_2O]{H_2O_2,\ NaOH} \xleftarrow{BH_3} C_6H_5-CH_2CH_2CH=CH_2$$

(d)

$$(\underset{\underset{CH_3}{|}}{CH_3CH}CH_2CH_2CH_2)_2O \xleftarrow{\underset{\underset{CH_3}{|}}{CH_3CH}CH_2CH_2CH_2Br} \underset{\underset{CH_3}{|}}{CH_3CH}CH_2CH_2CH_2O^-Na^+ \xleftarrow{Na}$$

$$\underset{\underset{CH_3}{|}}{CH_3CH}CH_2CH_2CH_2OH \xleftarrow[H_2O]{H_2O_2,\ NaOH} \xleftarrow{BH_3} \underset{\underset{CH_3}{|}}{CH_3CH}CH_2CH=CH_2$$

$$\underset{\underset{CH_3}{|}}{CH_3CH}CH_2CH_2CH_2Br \xleftarrow[\text{pyridine}]{PBr_3} \underset{\underset{CH_3}{|}}{CH_3CH}CH_2CH_2CH_2OH$$

12.39 (cont)

(e)

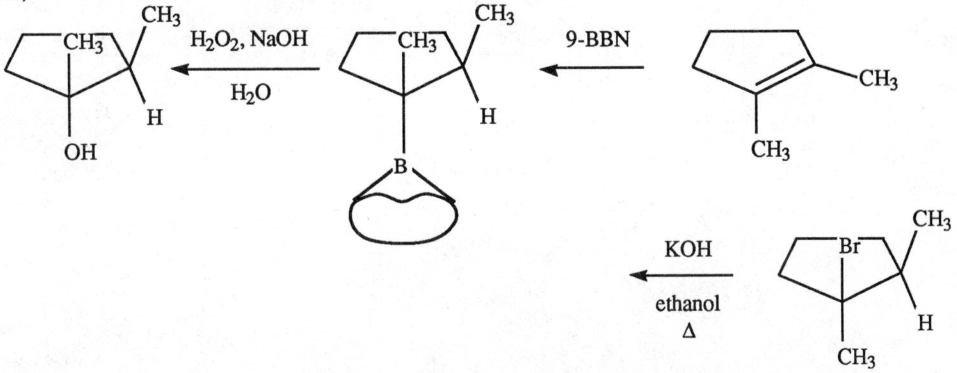
CH3
CH3
H
OH
H2O2, NaOH
H2O
CH3
CH3
H
B
9-BBN
CH3
CH3
KOH
ethanol
Δ
CH3
Br
H
CH3

(f)

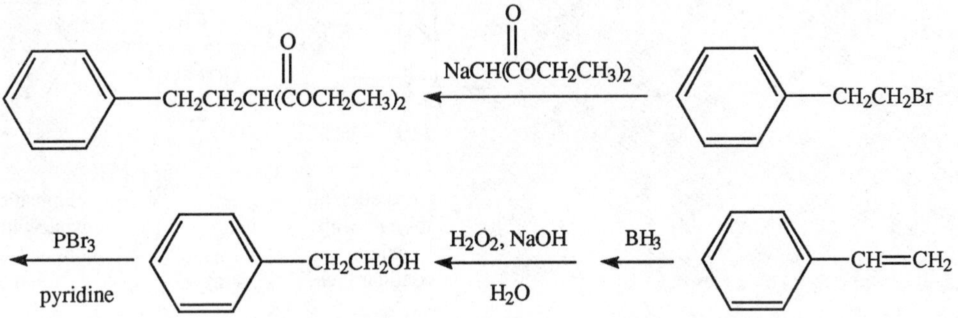
CH2CH2CH(COCH2CH3)2
O
NaCH(COCH2CH3)2
O
CH2CH2Br
PBr3
pyridine
CH2CH2OH
H2O2, NaOH
H2O
BH3
CH=CH2

(g)

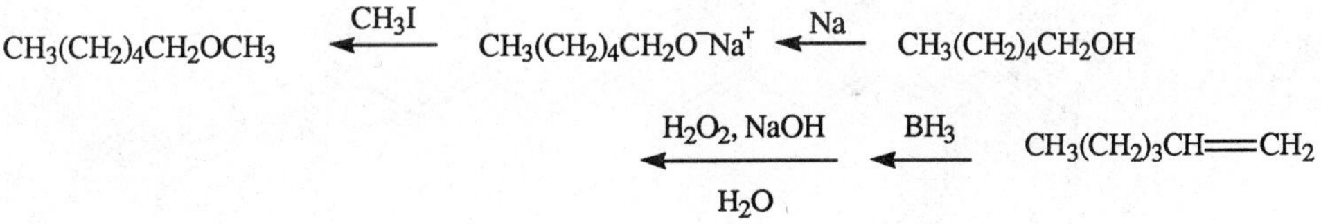
CH3(CH2)4CH2OCH3
CH3I
CH3(CH2)4CH2O⁻Na⁺
Na
CH3(CH2)4CH2OH
H2O2, NaOH
H2O
BH3
CH3(CH2)3CH=CH2

(h)

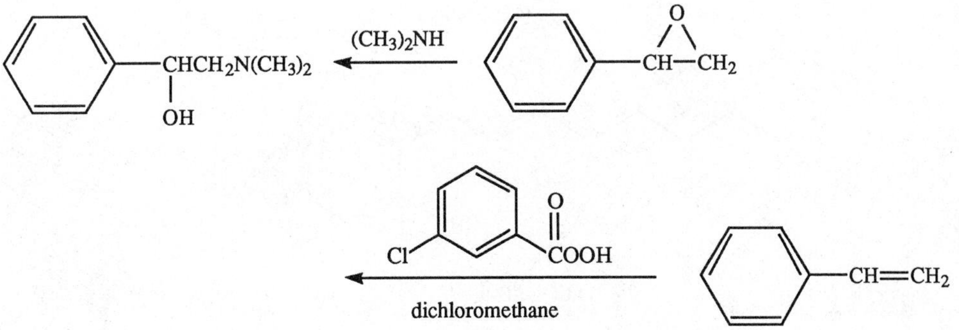
CHCH2N(CH3)2
OH
(CH3)2NH
O
CH—CH2
Cl
O
COOH
dichloromethane
CH=CH2

12.40

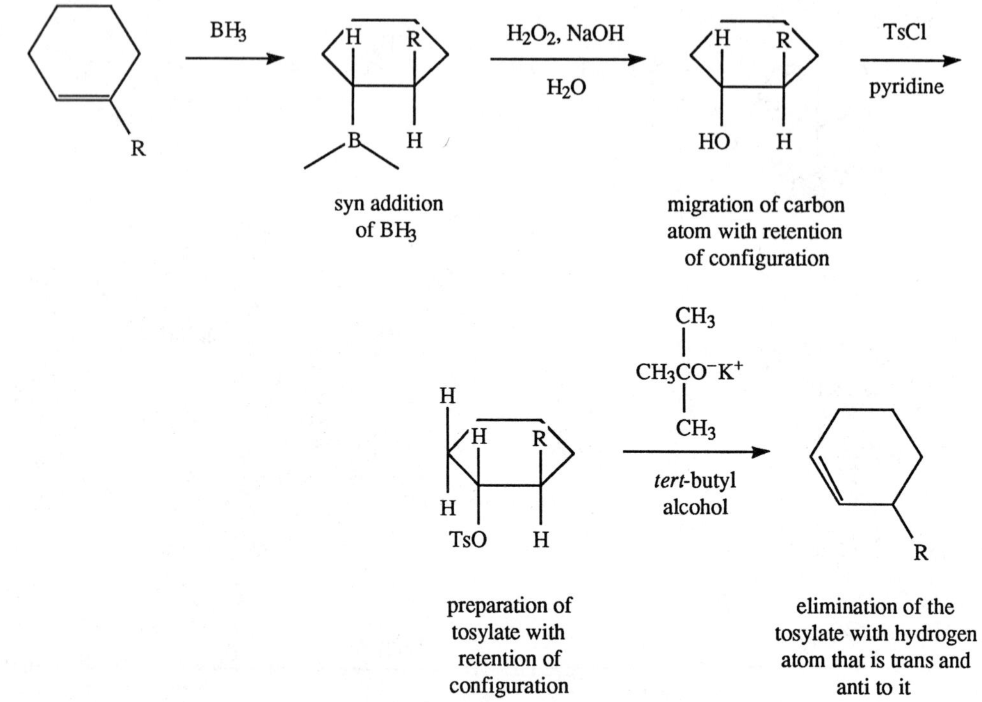
BH3
H R
B H
syn addition of BH3
H2O2, NaOH
H2O
H R
HO H
migration of carbon atom with retention of configuration
TsCl
pyridine
H
H R
H
TsO H
preparation of tosylate with retention of configuration
CH3
CH3CO⁻K⁺
CH3
tert-butyl alcohol
R
elimination of the tosylate with hydrogen atom that is trans and anti to it

12.41 (a)

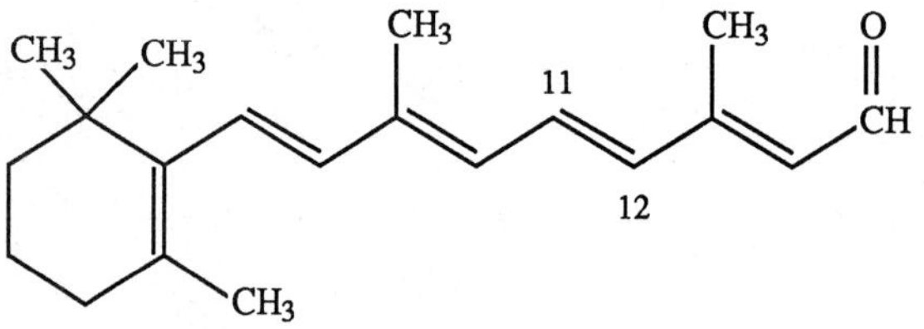
CH3 CH3
CH3
11
12
CH3
O
CH
CH3

(11*E*)-retinal

(b)

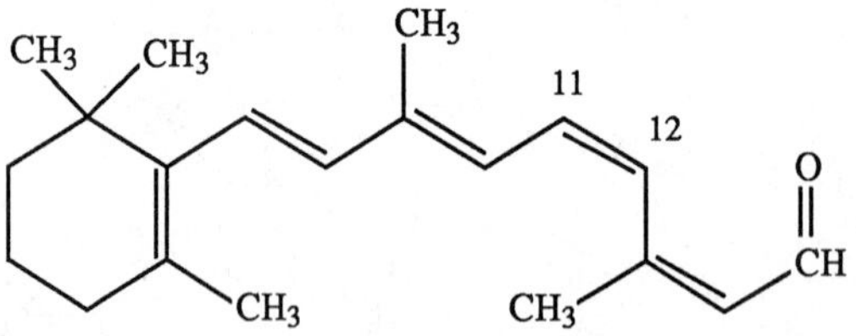
CH3 CH3
CH3
11
12
O
CH
CH3
CH3

(11*Z*)-retinal

(c)

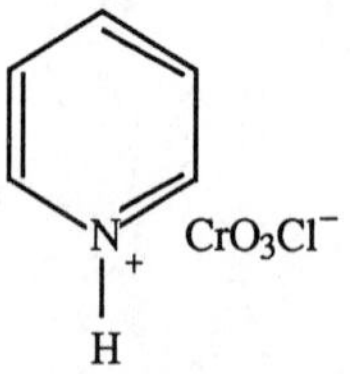
N
+
H
CrO3Cl⁻

in dichloromethane

12.42

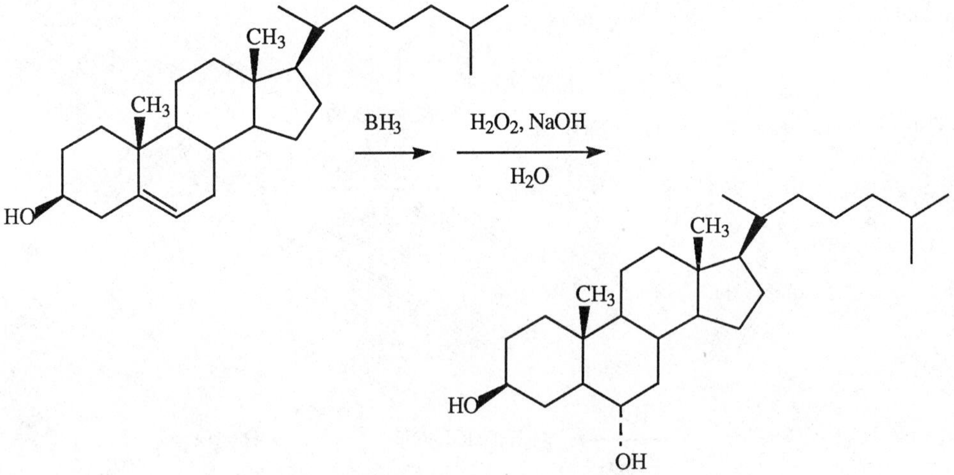

12.43

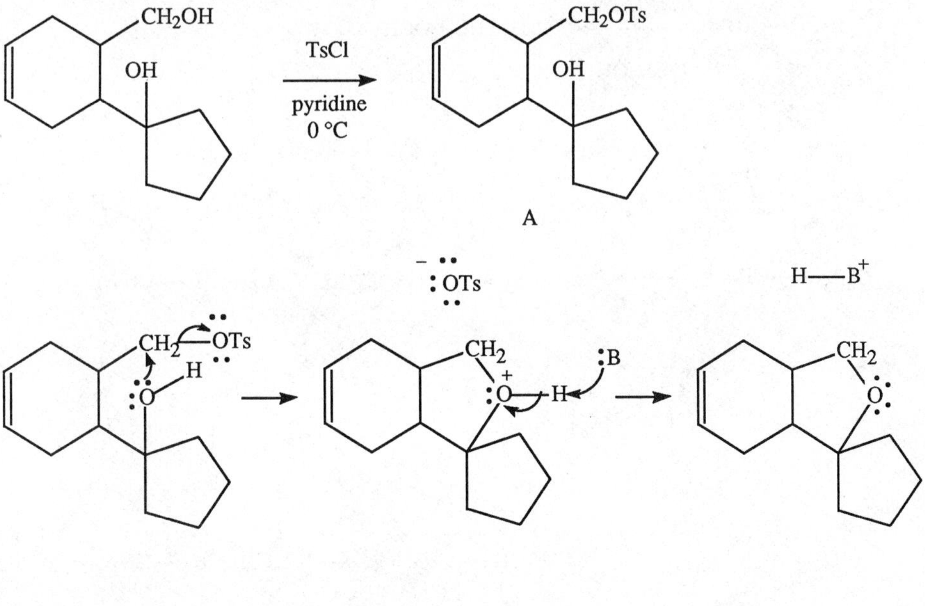

A

12.44 $CH_3C{\equiv}CCH_2C{\equiv}CCH_2CH_2CH_2OH \xrightarrow[\substack{NH_3 \text{ (liq)} \\ \textit{tert}\text{-butyl alcohol} \\ \text{ammonium sulfate}}]{Li}$

12.44 (cont)

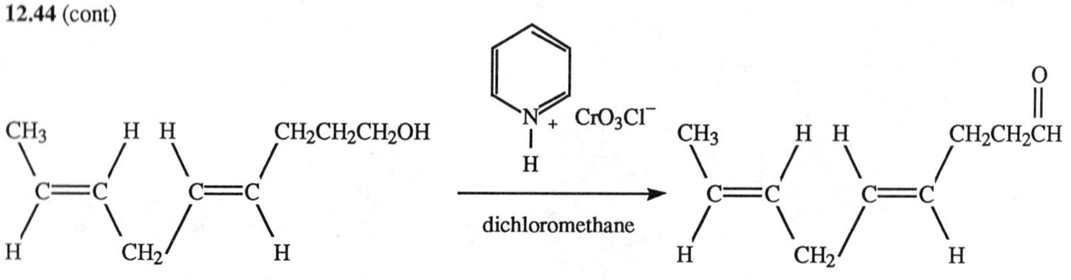

X   dichloromethane   Y

12.45

$$C_6H_5CH\text{—}CH_2 \xrightarrow[H_2SO_4]{CH_3OH} C_6H_5CHCH_2OH$$

with OCH_3

$$C_6H_5CH\text{—}CH_2 \xrightarrow[CH_3OH]{CH_3O^-Na^+} C_6H_5CHCH_2OCH_3$$

with OH

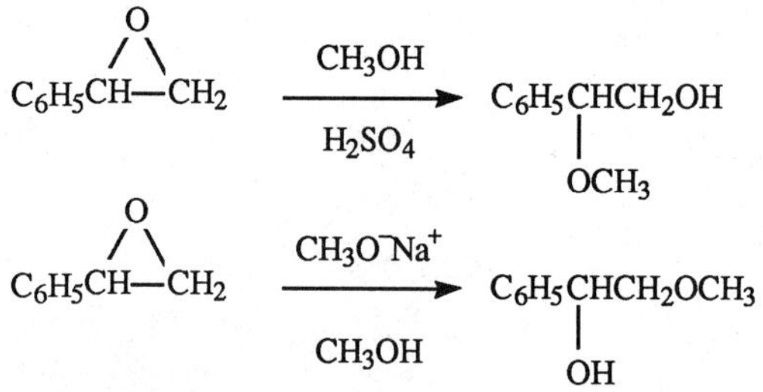

(1)

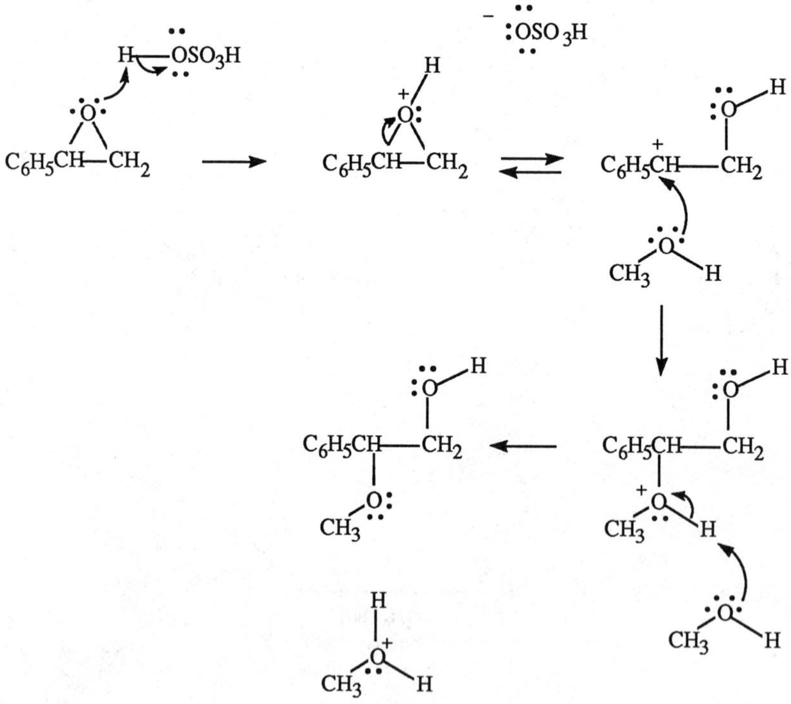

12.45 (cont)

(2)

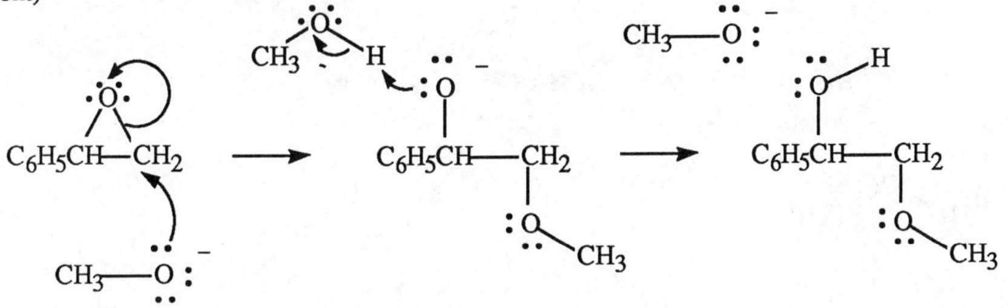

12.46

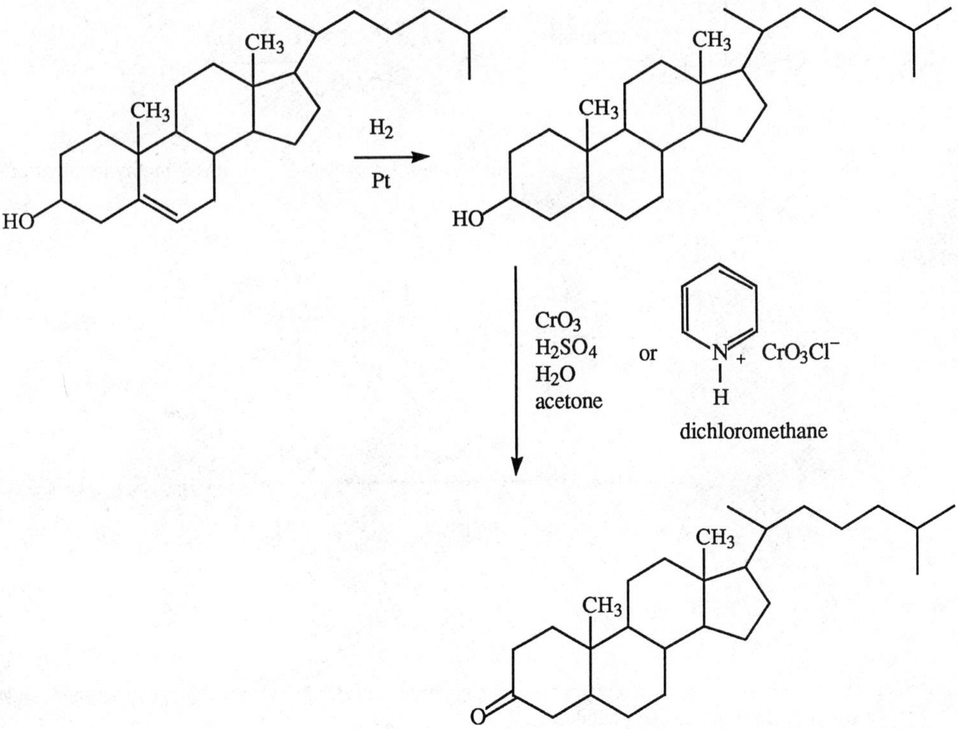

CrO₃
H₂SO₄
H₂O
acetone

or

CrO₃Cl⁻

dichloromethane

12.47

$$CH_3C{=\!=}CH_2 \xrightarrow{\ H^+BF_4^-\ } CH_3\overset{+}{C}CH_3 \quad BF_4^-$$

with CH₃ groups as drawn

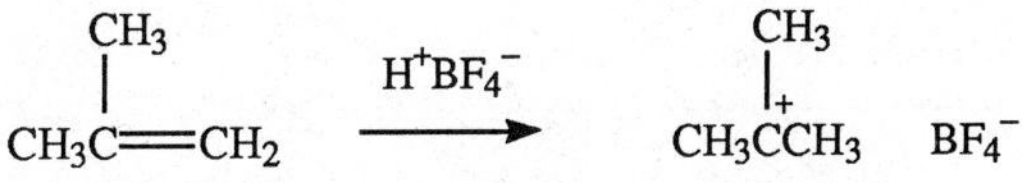

12.47 (cont)

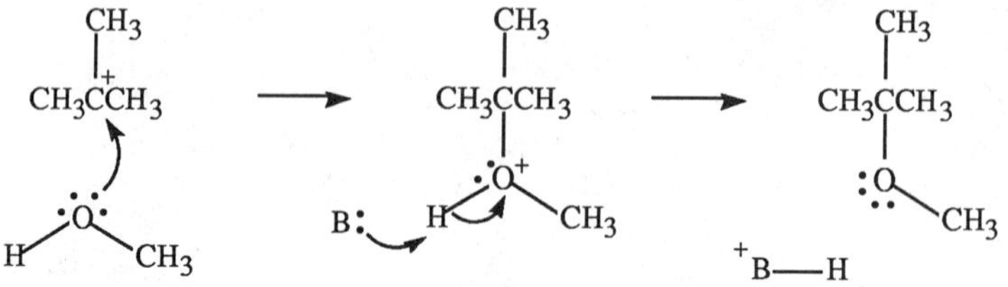

A nucleophilic conjugate base would compete with the alcohol for the carbocation.

12.48 $ClCH_2CH_2CH_2CCH_3$ $\xrightarrow{\text{reduction}}$

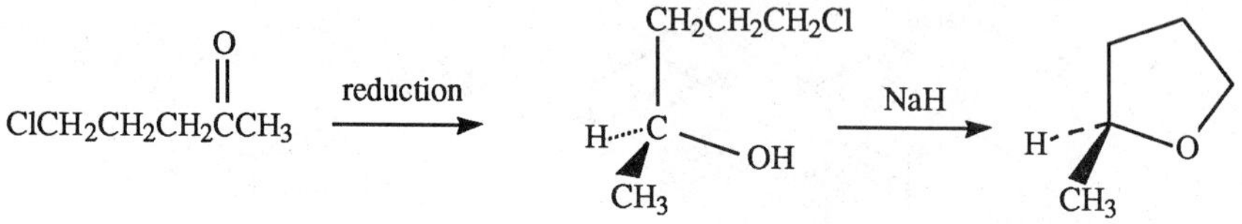

(S)-5-chloro-2-pentanol   (S)-2-methyltetrahydrofuran

12.49

(a) $HC{\equiv}CCH_2OH$ $\xrightarrow[\text{NH}_3 \text{ (liq)}]{2 \text{ LiNH}_2}$ $Li^+{}^-{:}C{\equiv}CCH_2O^-Li^+$

A

$Li^+{}^-{:}C{\equiv}CCH_2O^-Li^+$ $\xrightarrow[\text{H}_2\text{O}]{CH_3(CH_2)_3CH_2Br \quad H_3O^+}$ $CH_3(CH_2)_3CH_2C{\equiv}CCH_2OH$

A   2-octyn-1-ol

(b) $Li^+{}^-{:}C{\equiv}CCH_2OCH_2(CH_2)_3CH_3$

(c) An acetylide anion (charge on carbon) is more nucleophilic than an alkoxide ion (charge on oxygen) and will react faster with the electrophilic alkyl halide.

(d) $HC{\equiv}CCH_2O^-\ Li^+$

12.50

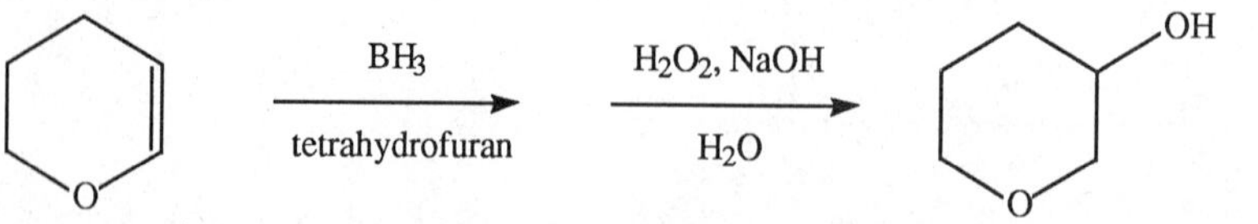

12.50 (cont)

The unsymmetrical alkene can react with diborane by way of two different transition states.

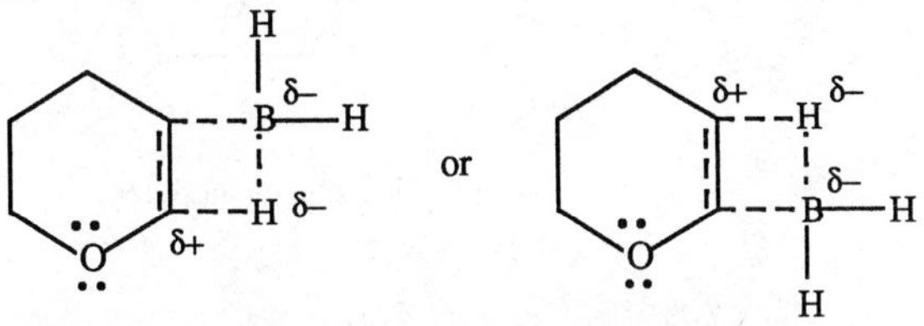

The one with the partial positive charge next to the oxygen is more stable, because the charge can be delocalized to oxygen, shown by resonance contributors written as though an intermediate with a full positive charge forms.

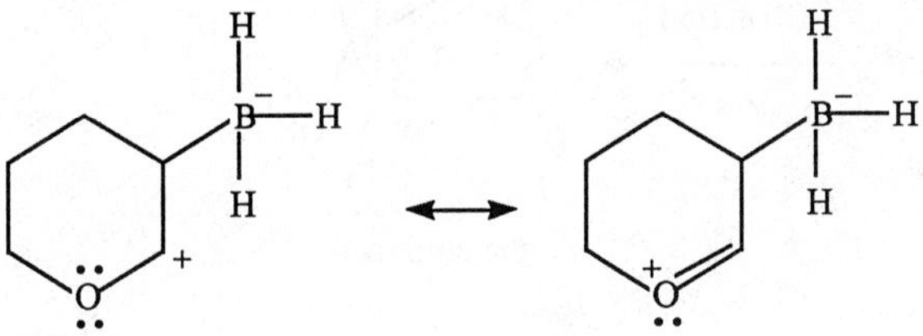

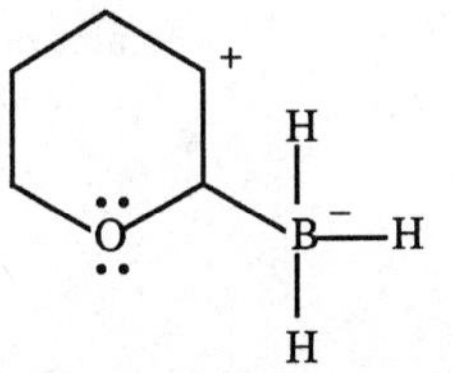

has no additional stabilization

The reaction goes by way of the transition state of lower energy corresponding to the more stable of the two reaction intermediates.

12.51

(a)

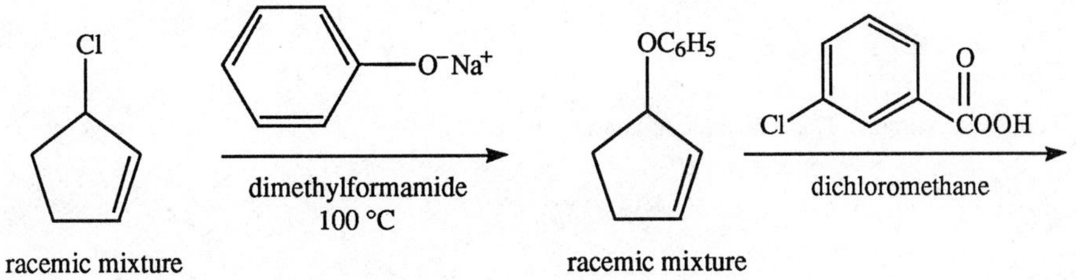
Cl
racemic mixture
O⁻Na⁺
dimethylformamide
100 °C
OC₆H₅
racemic mixture
Cl
O
COOH
dichloromethane

12.51 (cont)

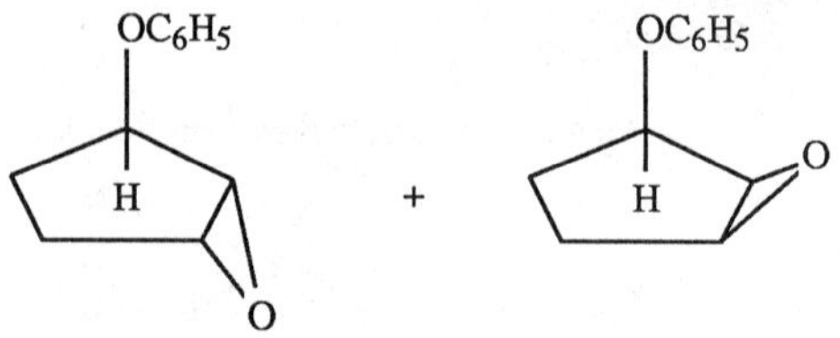

and enantiomer — B

and enantiomer — C

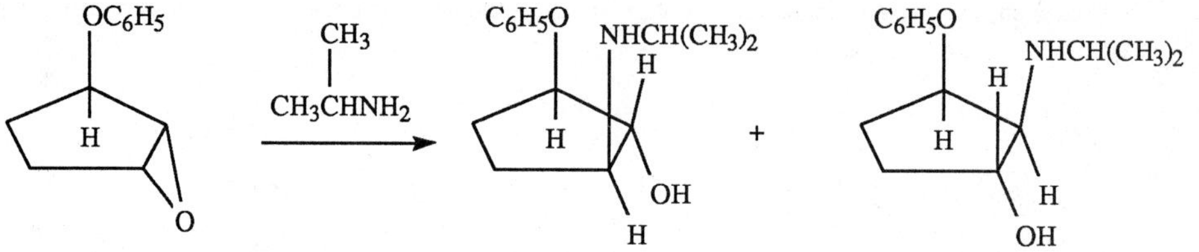

and enantiomer

B

and enantiomer

D

and enantiomer
(very minor product,
bulky phenoxide group
prevents the approach of
amine to the carbon adjacent
to the substituent.)

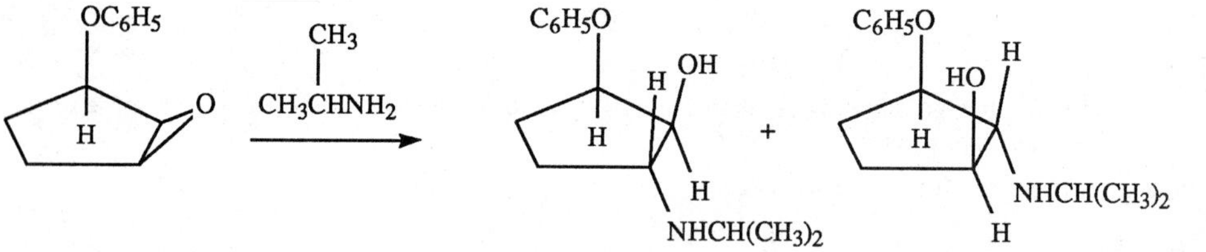

and enantiomer

C

and enantiomer

E

and enantiomer
(minor product)

(b) D and E are diastereomers. D is a racemic mixture.

12.52   The following equations can be used to calculate the free energy of activation from the heat and entropy of activation.

$$\Delta G^{\ddagger} = \Delta H^{\ddagger} - T\Delta S^{\ddagger}$$

For the formation of oxirane

$$\Delta G^{\ddagger} = (23.2 \text{ kcal/mol} - (303 \text{ K})(9.9 \times 10^{-3}\text{kcal/mol·K})$$

$$\Delta G^{\ddagger} = 20.2 \text{ kcal/mol}$$

For the formation of tetrahydrofuran

$$\Delta G^{\ddagger} = (19.8 \text{ kcal/mol} - (303 \text{ K})(-5 \times 10^{-3}\text{kcal/mol·K})$$

$$\Delta G^{\ddagger} = 21.3 \text{ kcal/mol}$$

The formation of oxirane has the smaller free energy of activation and therefore the faster rate. Even though the heat of activation is smaller for the formation of tetrahydrofuran than for oxirane, the reaction giving rise to tetrahydrofuran has a negative (unfavorable) entropy of activation whereas the entropy of activation for

12.53   Compound A, $C_5H_{10}O$, has one unit of unsaturation and Compound B, C_5H_8O, has two units of unsaturation. From the infrared spectrum, A is an alcohol (O—H stretching frequency at 3330 cm^{-1}) and B is a ketone (C=O stretching frequency at 1750 cm^{-1} and no band at 2700 cm^{-1}). The permanganate and bromine tests are negative, therefore neither A nor B has a double bond. They must therefore each contain a ring. The carbonyl stretching frequency is outside of the range given in the table for ketones. This is because the carbonyl stretching frequency for cyclic ketones increases as the size of the ring decreases; 1750 cm^{-1} is typical for five-membered ring ketones while four-membered ring ketones absorb at 1789 cm^{-1}. Compound A must be cyclopentanol and Compound B must be cyclopentanone.

A   B

12.54   Compound C, $C_7H_{16}O$, has no units of unsaturation. The infrared spectrum shows the presence of an alcohol (3300 cm^{-1} for the O—H stretch and 1110 cm^{-1} for the C—O stretch).

Compound D, $C_7H_{14}O$, has one unit of unsaturation. The infrared spectrum shows the presence of an aldehyde (2700 cm^{-1} for the C—H stretch and 1725 cm^{-1} for the C=O stretch).

12.54 (cont)

Analysis of the proton magnetic resonance spectrum of Compound C is shown below:

δ 0.9   (3H, triplet, C$\underline{H}_3$CH$_2$—)
δ 1.3   (10H, multiplet, —(C$\underline{H}_2$)$_5$—)   (A linear chain of methylene groups that have
                                                                 similar chemical shifts and are at least one car-
                                                                 bon away from electron-withdrawing atoms or π
                                                                 bonds will usually appear as a complex multiplet
                                                                 in this region.)
δ 2.2   (1H, singlet, —O$\underline{H}$)
δ 3.6   (2H, triplet, —CH$_2$C$\underline{H}_2$—O—)

The carbon-13 magnetic resonance spectrum shows six alkyl carbon atoms (14.2, 23.1, 26.4, 29.7, 32.4, and 33.2 ppm) suggesting a linear chain. The seventh carbon atom (62.2) falls in the range of a carbon atom bonded to an oxygen atom. Compound C is 1-heptanol, CH$_3$(CH$_2$)$_5$CH$_2$OH. It must be oxidized by pyridinium chlorochromate to heptanal.

Analysis of the proton magnetic resonance spectrum of Compound D confirms that it is heptanal.

δ 0.9   (3H, triplet, C$\underline{H}_3$CH$_2$—)
δ 1.3   (8H, multiplet, —(C$\underline{H}_2$)$_4$—)
δ 2.4   (2H, multiplet, —C$\underline{H}_2$C=O)

$$\overset{\displaystyle O}{\overset{\displaystyle \|}{}}$$

δ 9.8   (1H, triplet, —C$\underline{H}_2$CH)

The carbon-13 magnetic resonance spectrum confirms the presence of an aldehyde functional group (202.2 ppm).

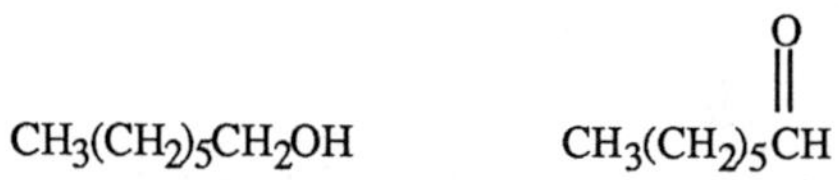

$$\overset{\displaystyle \qquad\qquad\qquad O}{\overset{\displaystyle \qquad\qquad\qquad \|}{CH_3(CH_2)_5CH_2OH \qquad\qquad CH_3(CH_2)_5CH}}$$

                    C   D

12.55   Compound E, C$_5$H$_{12}$O, has no units of unsaturation. The infrared spectrum shows the presence of an alcohol (3400 cm^{-1} for the O—H stretch). Compound E does not react with chromium trioxide so it must be a tertiary alcohol. Only one structural formula is possible for a five-carbon tertiary alcohol. Compound E must be 2-methyl-2-butanol (*tert*-amyl alcohol).

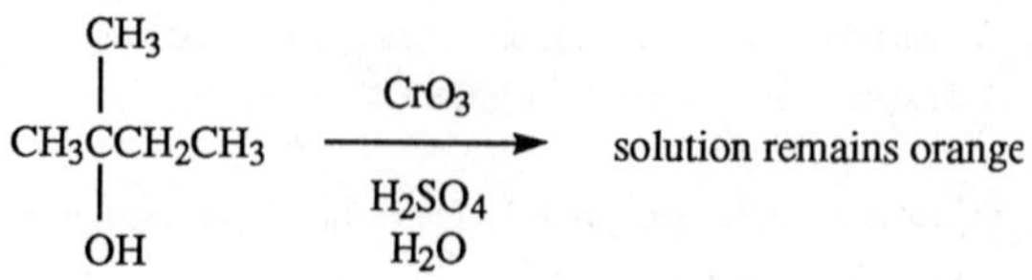

$$\begin{array}{c} CH_3 \\ | \\ CH_3CCH_2CH_3 \\ | \\ OH \\ \\ E \end{array} \quad \xrightarrow[\substack{H_2SO_4 \\ H_2O}]{CrO_3} \quad \text{solution remains orange}$$

12.55 (cont)

Analysis of the magnetic resonance spectra of Compound E confirms the structure. The proton magnetic resonances are given below:

δ 0.9	(multiplet, (C$\underline{H}_3$CH$_2$—)
δ 1.2	(singlet, (C$\underline{H}_3$)$_2$C—)
δ 1.5	(multiplet,—C$\underline{H}_2$CH$_3$)

There are no bands for a hydrogen atom on a carbon bonded to an oxygen atom (~4 ppm), which also confirms that Compound E is a tertiary alcohol.

The carbon-13 magnetic resonance spectrum shows only four types of carbon atoms. Three of the types are alkyl carbon atoms (8.8, 28.9, and 36.8 ppm) and one is a tertiary carbon atom bonded to an oxygen atom (70.6 ppm).

Compound F, $C_5H_{12}O$, also has no units of unsaturation. The infrared spectrum again shows the presence of an alcohol (3400 cm^{-1} for the O—H stretch). A positive reaction with chromium trioxide points to a secondary or primary alcohol. The possible structures are shown below.

Primary alcohols:

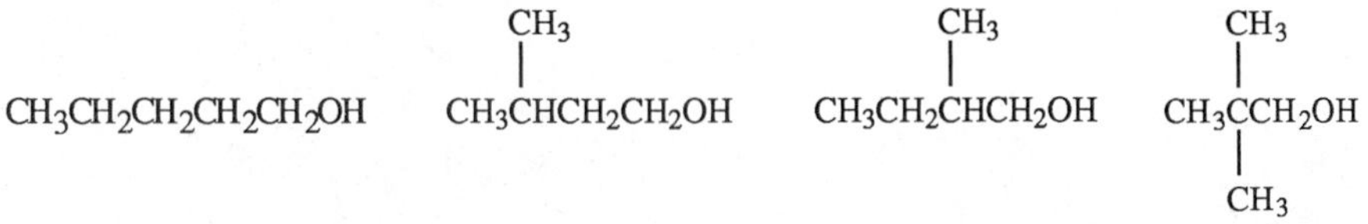

$$CH_3CH_2CH_2CH_2CH_2OH \qquad CH_3\underset{\displaystyle |}{\overset{\displaystyle CH_3}{C}}HCH_2CH_2OH \qquad CH_3CH_2\underset{\displaystyle |}{\overset{\displaystyle CH_3}{C}}HCH_2OH \qquad CH_3\underset{\displaystyle |}{\overset{\displaystyle CH_3}{\underset{\displaystyle CH_3}{C}}}CH_2OH$$

Secondary alcohols:

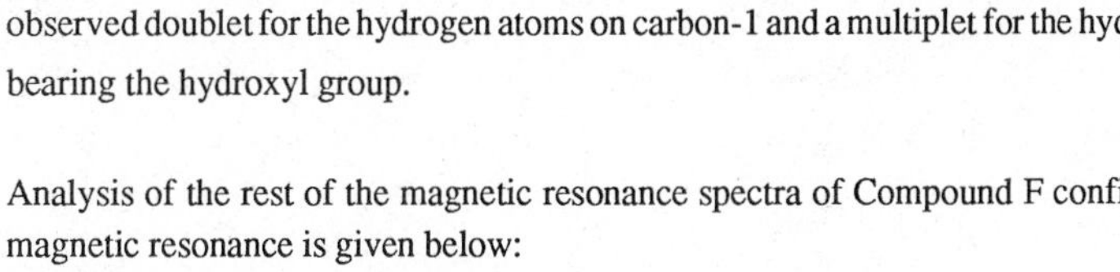

$$CH_3CH_2CH_2\underset{\displaystyle |}{\underset{\displaystyle OH}{C}}HCH_3 \qquad CH_3CH_2\underset{\displaystyle |}{\underset{\displaystyle OH}{C}}HCH_2CH_3 \qquad CH_3\underset{\displaystyle |}{\underset{\displaystyle OH}{C}}H\overset{\displaystyle CH_3}{\overset{\displaystyle |}{C}}HCH_3$$

The carbon-13 magnetic resonance spectrum shows five bands. The three alcohols that have five different types of carbon atoms are 1-pentanol, 2-pentanol, and 2-methyl-1-butanol. Of these, only 2-pentanol will have the observed doublet for the hydrogen atoms on carbon-1 and a multiplet for the hydrogen atom on the carbon atom bearing the hydroxyl group.

Analysis of the rest of the magnetic resonance spectra of Compound F confirms the structure. The proton magnetic resonance is given below:

δ 0.9-1.4	(complex multiplet, C$\underline{H}_3$C$\underline{H}_2$C$\underline{H}_2$—)
δ 1.2	(doublet, C$\underline{H}_3$CH—)
δ 3.7	(multiplet, —CH$_2$C$\underline{H}$CH$_3$)
	O—

12.55 (cont)

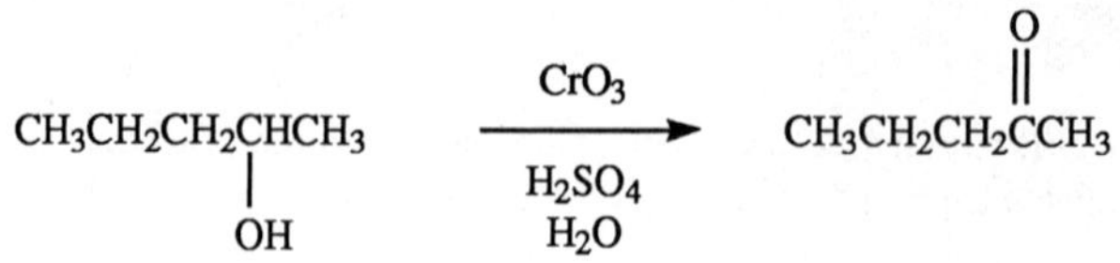

$$CH_3CH_2CH_2CHCH_3 \xrightarrow[\substack{H_2SO_4 \\ H_2O}]{CrO_3} CH_3CH_2CH_2\overset{\displaystyle O}{\overset{\|}{C}}CH_3$$

OH

F

Note that the hydrogen atom on the hydroxyl group is not seen as a separate band in the spectrum of either Compound E or Compound F.

13

Aldehydes and Ketones. Reactions at Electrophilic Carbon Atoms

13.1 (a)

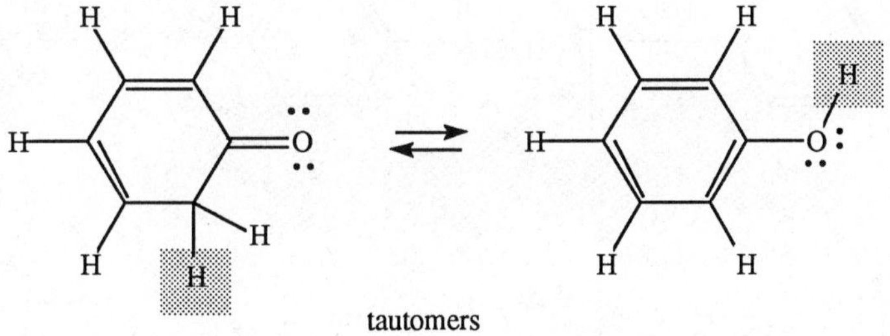

tautomers

(b)

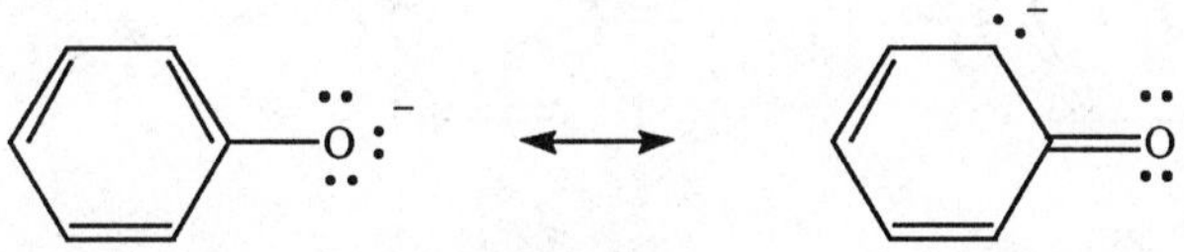

resonance contributors

(c)

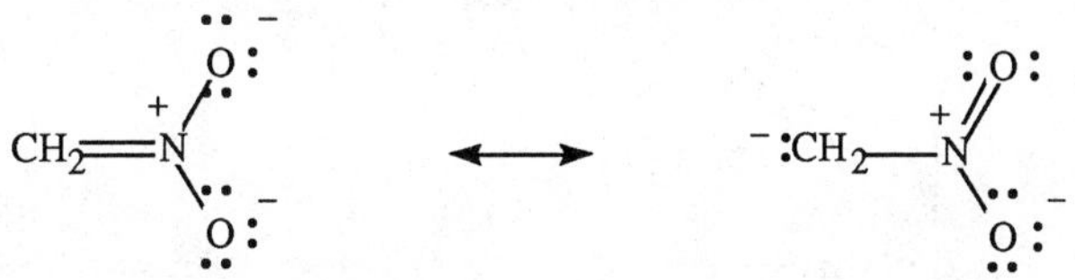

resonance contributors

(d)

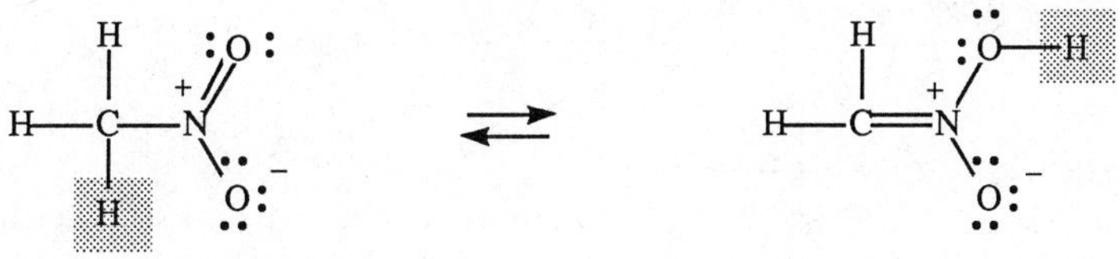

tautomers

13.2

(a)

$$CH_3CH_2CH_2CH \xrightarrow{H_2SO_4} \left[CH_3CH_2CH_2CH \leftrightarrow CH_3CH_2CH_2CH \right] HSO_4^-$$

(with :O: above first structure; :O⁺—H and :O—H resonance forms)

$$\downarrow CH_3CH_2O^-Na^+$$

$$\left[CH_3CH_2\ddot{C}H\!-\!CH \leftrightarrow CH_3CH_2CH\!=\!CH \right] Na^+$$

(with :O: and :O:⁻)

$$HSO_4^-$$

(b)

$$CH_3\!-\!C\!\!-\!\!\text{(phenyl)} \xrightarrow{H_2SO_4} \left[CH_3\!-\!C\!\!-\!\!\text{(phenyl)} \leftrightarrow CH_3\!-\!C\!\!-\!\!\text{(phenyl)} \right.$$

(with :O:, H—O⁺, H—O: resonance forms)

$$\updownarrow$$

$$\left. CH_3\!-\!C\!=\!\text{(phenyl)} \leftrightarrow CH_3\!-\!C\!=\!\text{(phenyl)} \leftrightarrow CH_3\!-\!C\!=\!\text{(phenyl)} \right]$$

(H—O: resonance forms with + on ring)

$$CH_3\!-\!C\!\!-\!\!\text{(phenyl)} \xrightarrow{CH_3CH_2O^-Na^+}$$

(with :O:)

$$\left[^-\!\!:CH_2\!-\!C\!\!-\!\!\text{(phenyl)} \leftrightarrow CH_2\!=\!C\!\!-\!\!\text{(phenyl)} \right]$$

(with :O: and :O:⁻)

(c)

$$\overset{CH_3}{\underset{|}{CH_3C}}\!=\!CHCH_2CH_2CCH_3 \xrightarrow{H_2SO_4}$$

(with :O: above C)

$$\left[\overset{CH_3}{\underset{|}{CH_3C}}\!=\!CHCH_2CH_2CCH_3 \leftrightarrow \overset{CH_3}{\underset{|}{CH_3C}}\!=\!CHCH_2CH_2CCH_3 \right] HSO_4^-$$

(with :O⁺—H and :O—H)

nonbonding electrons become protonated more readily than π electrons do

13.2 (c) (cont)

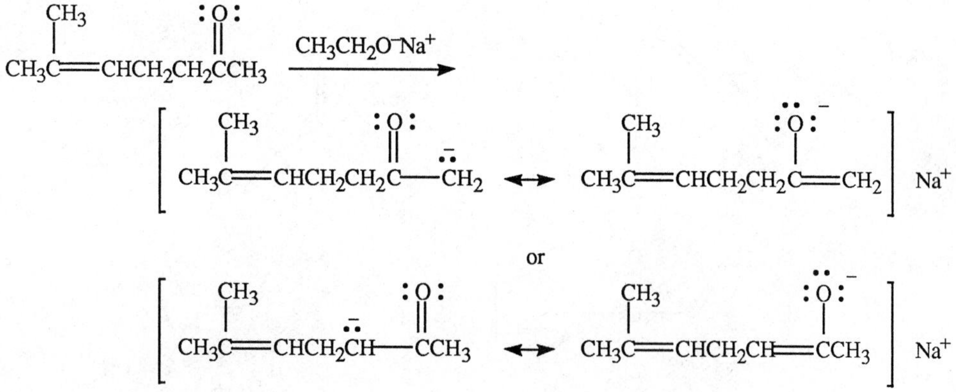

(d)

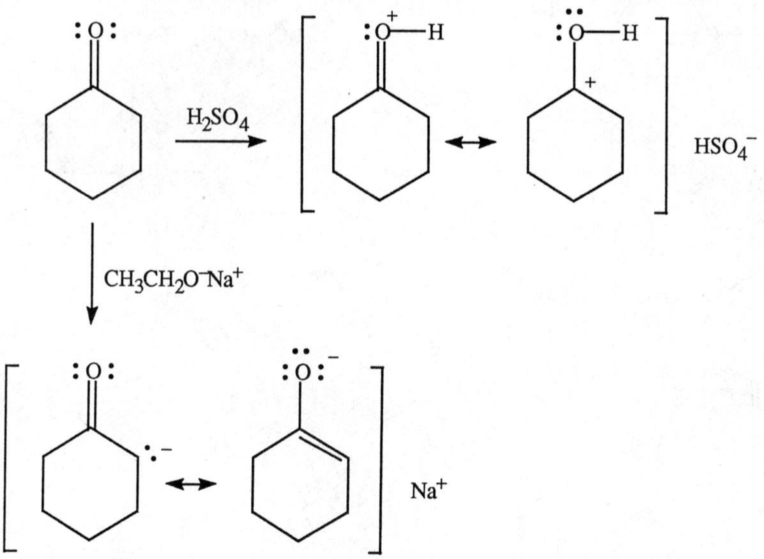

13.3   (a) 3-hydroxy-2-methylpentanal

(c) 4-chlorobenzaldehyde

(e) (*E*)-1,3-diphenyl-2-propen-1-one

(g) 2,2-dibromocyclohexanone

(i) 3,3-dimethylcyclohexanecarbaldehyde

(k) 2-ethyl-2-methylcyclobutanone

(b) 2-pentanone

(d) 4-heptyn-3-one

(f) (*Z*)-6-methyl-5-nonen-2-one

(h) trichloroethanal
     (trichloroacetaldehyde)

(j) 2,4-pentanedione

(l) *cis*-2-propylcyclopentanecarbaldehyde

13.4   (a)

CH$_3$

(b)

CH$_3$ CH$_3$

(c)

(d)

CH$_3$CH$_2$ H

$\overset{\displaystyle C=C}{}$

H   CH$_2$CH$_2\overset{\displaystyle O}{\underset{\displaystyle \|}{C}}$H

(e)   O$_2$N—

—$\overset{\displaystyle O}{\underset{\displaystyle \|}{C}}$H

(f)   CH$_3$CH$_2\overset{\displaystyle O}{\underset{\displaystyle \|}{C}}CH_2CH_2$OH

(g)   H$\overset{\displaystyle O}{\underset{\displaystyle \|}{C}}CH_2CH_2CH_2CH_2\overset{\displaystyle O}{\underset{\displaystyle \|}{C}}$H

(h)

—$\overset{}{\underset{\displaystyle OH}{CH}}\overset{\displaystyle O}{\underset{\displaystyle \|}{C}}$—

(i)

13.5   CH$_3\overset{\displaystyle CH_3}{\underset{}{C}}$=CHCH$_2CH_2CH_2$OH $\xrightarrow[\substack{H_2SO_4 \\ H_2O \\ \Delta}]{Na_2Cr_2O_7}$ CH$_3\overset{\displaystyle CH_3}{\underset{}{C}}$=CHCH$_2CH_2\overset{\displaystyle O}{\underset{\displaystyle \|}{C}}$OH   +

oxidation of the primary
alcohol to a carboxylic acid

CH$_3\overset{\displaystyle CH_3}{\underset{\displaystyle OH}{C}}CH_2CH_2CH_2\overset{\displaystyle O}{\underset{\displaystyle \|}{C}}$OH   +   CH$_3\overset{\displaystyle CH_3}{\underset{}{C}}$=CHCH$_2$CH=CH$_2$

oxidation of the primary alcohol
to a carboxylic acid and
hydration of the double bond

dehydration of primary alcohol

Concept Map 13.1   Some ways to prepare aldehydes and ketones.

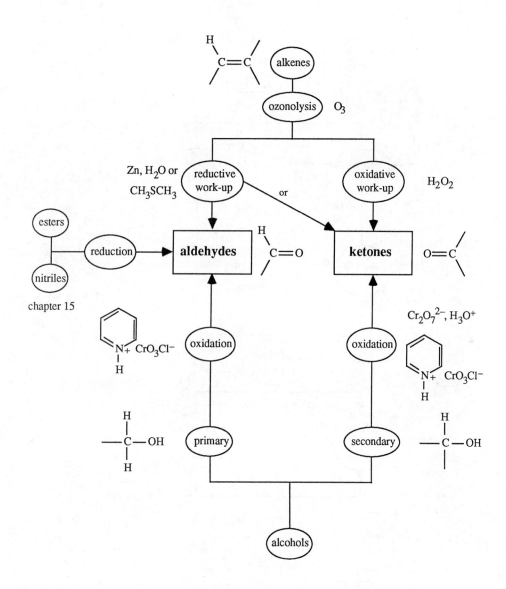

Concept Map 13.2   The relationship between carbonyl compounds, alcohols, and alkyl halides.

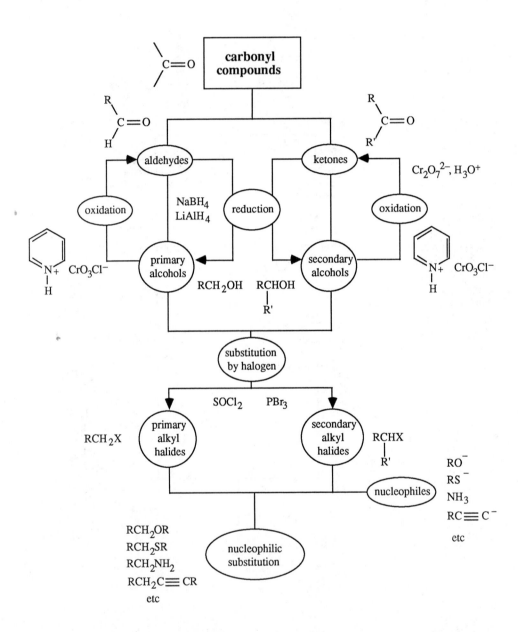

13.6

(a)

$$CH_3O-\text{(ring)}-\overset{\displaystyle O}{\overset{\|}{C}}H \xrightarrow[\text{methanol}]{NaBH_4} CH_3O-\text{(ring)}-CH_2OH$$

(b)

$$CH_3(CH_2)_5\overset{\displaystyle CH_3}{\overset{|}{C}}HCH_2\overset{\displaystyle O}{\overset{\|}{C}}H \xrightarrow[\text{diethyl ether}]{LiAlH_4} \xrightarrow{H_3O^+} CH_3(CH_2)_5\overset{\displaystyle CH_3}{\overset{|}{C}}HCH_2CH_2OH$$

(c)

$$\text{(ring)}-CH_2\overset{\displaystyle O}{\overset{\|}{C}}CH_3 \xrightarrow[\text{diethyl ether}]{LiAlH_4} \xrightarrow{H_3O^+} \text{(ring)}-CH_2\overset{}{\underset{\displaystyle OH}{\overset{|}{C}}}HCH_3$$

(d)

$$\text{(ring)}-\overset{\displaystyle O}{\overset{\|}{C}}H \xrightarrow[\text{diethyl ether}]{LiAlH_4} \xrightarrow{H_3O^+} \text{(ring)}-CH_2OH$$

(e)

$$CH_3\overset{\displaystyle O}{\overset{\|}{C}}CH_2\overset{\displaystyle CH_3}{\overset{|}{C}}HCH_3 \xrightarrow[\text{diethyl ether}]{LiAlH_4} \xrightarrow{H_3O^+} CH_3\overset{}{\underset{\displaystyle OH}{\overset{|}{C}}}HCH_2\overset{\displaystyle CH_3}{\overset{|}{C}}HCH_3$$

(f)

$$CH_3\overset{\displaystyle CH_3}{\overset{|}{C}}=CH\overset{\displaystyle O}{\overset{\|}{C}}CH_3 \xrightarrow[\text{H}_2\text{O}]{NaBH_4} CH_3\overset{\displaystyle CH_3}{\overset{|}{C}}=CH\overset{}{\underset{\displaystyle OH}{\overset{|}{C}}}HCH_3$$

(g)

$$\text{(cyclohexanone)}=O \xrightarrow[\substack{\text{isopropyl}\\\text{alcohol}}]{NaBH_4} \text{(cyclohexane)}-OH$$

(h)

$$\text{(cyclobutanone)}=O \xrightarrow[\text{diethyl ether}]{LiAlH_4} \xrightarrow{H_3O^+} \text{(cyclobutane)}-OH$$

13.7

attack by cyanide ion
can occur at the top
or the bottom of the
carbonyl group

R   *S*

racemic mixture

$\downarrow$ HB$^+$

(*R*)-lactic acid   (*S*)-lactic acid   *R*   *S*

also racemic mixture   racemic mixture

Note that the hydrolysis of each nitrile
proceeds with retention of configuration,
but hydrolysis of the racemic mixture of
nitriles gives rise to a racemic mixture
of hydroxyacids.

13.8

(a) $CH_3CH_2CH_2CH_2Br$ $\xrightarrow[\text{tetrahydro-}\atop\text{furan}]{\text{Li}}$ $CH_3CH_2CH_2CH_2Li$ + Li^+Br^-

(b) $\xrightarrow[\text{diethyl}\atop\text{ether}]{\text{Mg}}$

(c) $CH_2\!\!=\!\!CHCl$ $\xrightarrow[\text{tetrahydro-}\atop\text{furan}\atop\Delta]{\text{Mg}}$ $CH_2\!\!=\!\!CHMgCl$

13.8 (cont)

(d)

$$\text{C}_6\text{H}_5\text{—C}\equiv\text{CH} \xrightarrow[\text{diethyl ether}]{\text{CH}_3\text{CH}_2\text{MgBr}} \text{C}_6\text{H}_5\text{—C}\equiv\text{CMgBr} + \text{CH}_3\text{CH}_3\uparrow$$

(e)

$$\text{C}_6\text{H}_5\text{—MgBr} + \text{CH}_3\overset{\overset{\text{O}}{\|}}{\text{C}}\text{OH} \longrightarrow \text{C}_6\text{H}_5\text{—H} + \text{CH}_3\overset{\overset{\text{O}}{\|}}{\text{C}}\text{O}^-\text{Mg}^{2+}\text{Br}^-$$

(f) $\text{CH}_3\text{CH}_2\text{C}\equiv\text{C}^-\text{Na}^+ + \text{H}_2\text{O} \longrightarrow \text{CH}_3\text{CH}_2\text{C}\equiv\text{CH}\uparrow + \text{Na}^+\text{OH}^-$

(g)

$$\text{C}_6\text{H}_5\text{—CH}_2\text{Br} \xrightarrow[\substack{\text{tetrahydro-}\\\text{furan}}]{\text{Li}} \text{C}_6\text{H}_5\text{—CH}_2\text{Li} + \text{Li}^+\text{Br}^-$$

(h) $\text{CH}_3\text{CH}_2\text{CH}_2\text{Li} + \text{NH}_3 \longrightarrow \text{CH}_3\text{CH}_2\text{CH}_3\uparrow + \text{Li}^+\text{NH}_2^-$

(i) $\text{CH}_3\text{CH}_2\text{MgI} + \text{CH}_3\text{OH} \longrightarrow \text{CH}_3\text{CH}_3\uparrow + \text{CH}_3\text{O}^-\text{Mg}^{2+}\text{I}^-$

(j)

$$\text{C}_6\text{H}_5\text{—C}\equiv\text{CH} \xrightarrow[\text{NH}_3\text{ (liq)}]{\text{Na}^+\text{NH}_2^-} \text{C}_6\text{H}_5\text{—C}\equiv\text{C}^-\text{Na}^+ + \text{NH}_3$$

Concept Map 13.3   Organometallic reagents and their reactions with compounds containing
electrophilic carbon atoms.

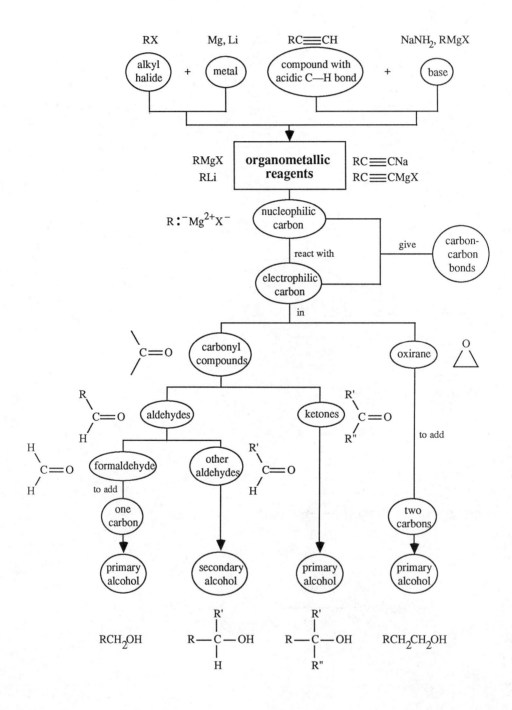

Concept Map 13.4   Some ways to prepare alcohols.

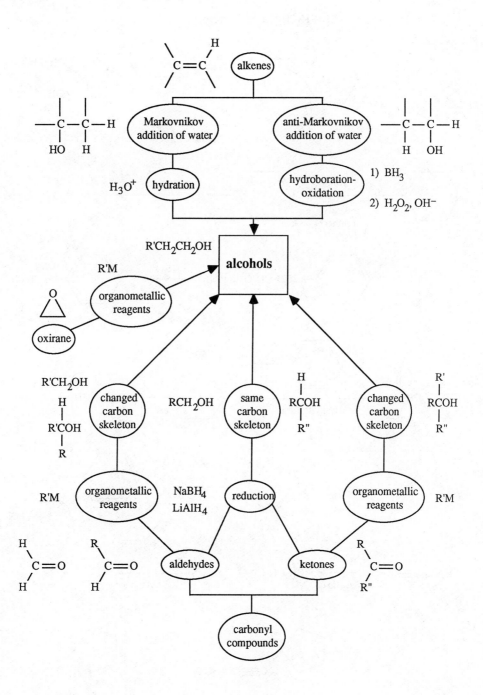

13.9

(a)

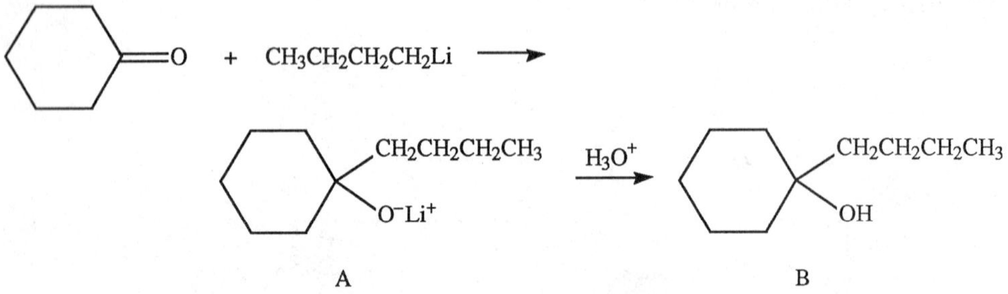

A   B

(b)

C

D   E

(c)

F

G

(d) $CH_3CH_2C\equiv CH$ $\xrightarrow[\text{NH}_3\text{ (liq)}]{\text{NaNH}_2}$ $CH_3CH_2C\equiv C^-Na^+$ $\xrightarrow{\text{diethyl ether}}$

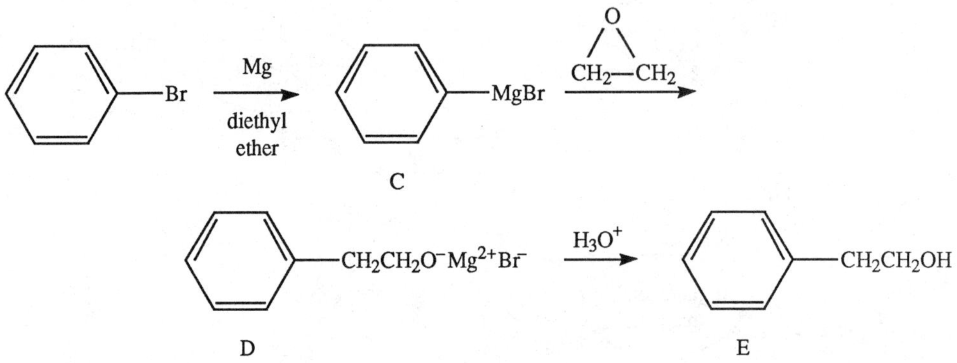

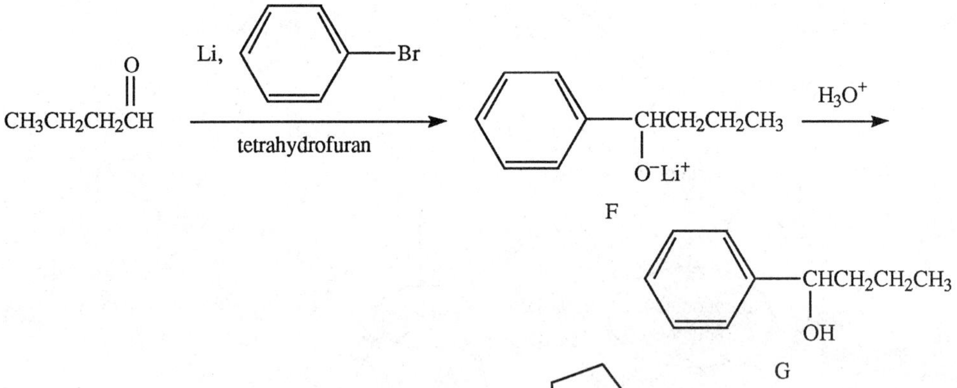

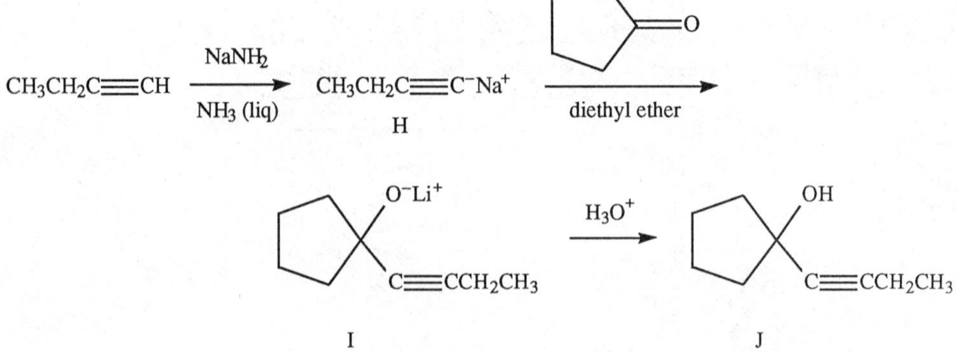

H

I   J

13.9 (cont)

(e)

$$\xrightarrow[\text{diethyl ether}]{\text{CH}_3\text{Li}}$$

K

$$\xrightarrow{\text{H}_3\text{O}^+}$$

L

(f)

$$\underset{\text{CH}_3\text{C}=\text{CHCCH}_3}{\overset{\underset{|}{\text{CH}_3}\quad\overset{\text{O}}{\overset{||}{}}}{}} \xrightarrow[\text{diethyl ether}]{\text{CH}_3\text{CH}_2\text{CH}_2\text{CH}_2\text{Li}}$$

$$\underset{\underset{\text{O}^-\text{Li}^+}{|}}{\underset{\text{CH}_3\text{C}=\text{CHCCH}_3}{\overset{\overset{|}{\text{CH}_3}\quad\overset{|}{\text{CH}_2\text{CH}_2\text{CH}_2\text{CH}_3}}{}}} \xrightarrow{\text{H}_3\text{O}^+} \underset{\underset{\text{OH}}{|}}{\underset{\text{CH}_3\text{C}=\text{CHCCH}_3}{\overset{\overset{|}{\text{CH}_3}\quad\overset{|}{\text{CH}_2\text{CH}_2\text{CH}_2\text{CH}_3}}{}}}$$

M   N

(g)

$$\underset{\text{CH}_3\text{C}=\text{CHCH}_2\text{CH}_2\text{C}=\text{CHCH}_2\text{CH}_2\text{CH}}{\overset{\overset{|}{\text{CH}_3}\qquad\overset{|}{\text{CH}_3}\qquad\overset{\text{O}}{\overset{||}{}}}{}} \xrightarrow[\text{dichloromethane}]{\text{HC}\equiv\text{CMgBr}}$$

$$\underset{\underset{\text{O}^-\text{Mg}^{2+}\text{Br}^-}{|}}{\underset{\text{CH}_3\text{C}=\text{CHCH}_2\text{CH}_2\text{C}=\text{CHCH}_2\text{CH}_2\text{CHC}\equiv\text{CH}}{\overset{\overset{|}{\text{CH}_3}\qquad\qquad\overset{|}{\text{CH}_3}}{}}} \xrightarrow{\text{H}_3\text{O}^+}$$

O

$$\underset{\underset{\text{OH}}{|}}{\underset{\text{CH}_3\text{C}=\text{CHCH}_2\text{CH}_2\text{C}=\text{CHCH}_2\text{CH}_2\text{CHC}\equiv\text{CH}}{\overset{\overset{|}{\text{CH}_3}\qquad\qquad\overset{|}{\text{CH}_3}}{}}}$$

P

13.10

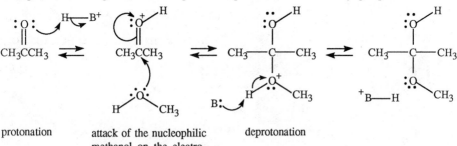

Two stereoisomeric hemiacetals are possible, resulting from the nucleophilic attack of the hydroxyl group at the bottom or the top of the planar carbonyl group. The two hemiacetals are enantiomers of each other.

13.11

(a)

attack of the nucleophilic methanol on the electrophilic carbon atom of the carbonyl group   protonation   deprotonation

(b)   In the presence of strong acid, the first step would be protonation of the carbonyl group.

protonation   attack of the nucleophilic methanol on the electrophilic carbon atom of the carbonyl group   deprotonation

13.12 Essentially all the ^{18}O will wind up in the oxygen of the carbonyl group.

13.13 $CH_2{=}CH{-}\overset{..}{\underset{..}{O}}{-}CH_2CH_3 \longrightarrow \left[CH_3\overset{+}{CH}{-}\overset{..}{\underset{..}{O}}{-}CH_2CH_3 \longleftrightarrow CH_3CH{=}\overset{+}{\underset{..}{O}}{-}CH_2CH_3 \right]$

resonance-stabilized cation
lowers transition state energy
for the hydrolysis reaction

$CH_3CH{=}\overset{..}{\underset{+}{O}}{-}CH_2CH_3 \longrightarrow CH_3CH{-}\overset{..}{\underset{..}{O}}{-}CH_2CH_3 \longrightarrow CH_3CH{-}\overset{..}{\underset{..}{O}}{-}CH_2CH_3$

$CH_3\overset{}{\underset{\|}{C}}H \longleftarrow CH_3\overset{}{\underset{\|}{C}}H + \overset{..}{\underset{H}{O}}{-}CH_2CH_3 \longleftarrow CH_3CH{-}\overset{..}{\underset{+}{O}}{-}CH_2CH_3$

Concept Map 13.5   Hydrates, acetals, ketals.

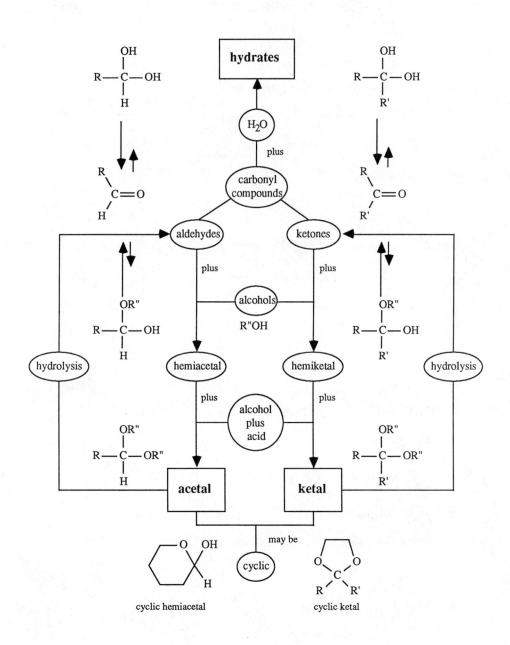

13.14

(a)

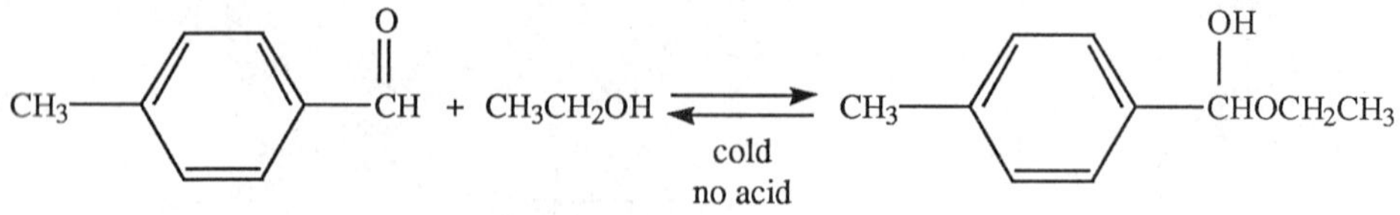

$$CH_3-C_6H_4-CHO + CH_3CH_2OH \underset{\text{no acid}}{\overset{\text{cold}}{\rightleftharpoons}} CH_3-C_6H_4-CHOCH_2CH_3 \; (OH)$$

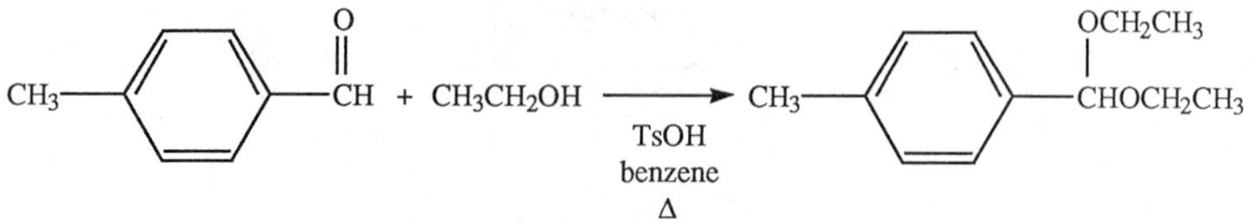

(b)

$$CH_3-C_6H_4-CHO + CH_3CH_2OH \xrightarrow[\Delta]{\substack{\text{TsOH}\\\text{benzene}}} CH_3-C_6H_4-CH(OCH_2CH_3)_2$$

(c)

$$CH_3CH_2CH_2CHO + CH_3OH \xrightarrow[\Delta]{\substack{\text{TsOH}\\\text{benzene}}} CH_3CH_2CH_2CH(OCH_3)_2$$

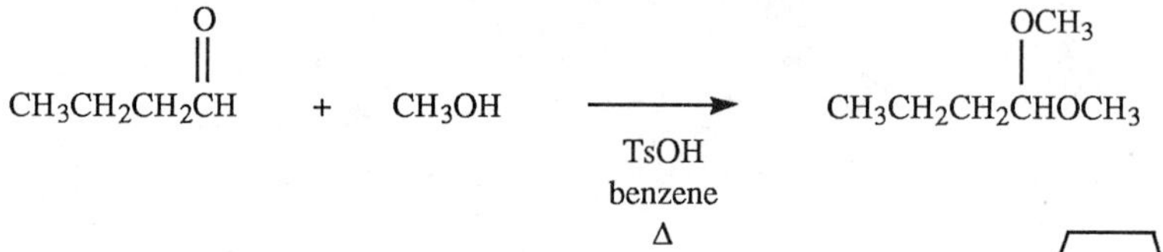

(d)

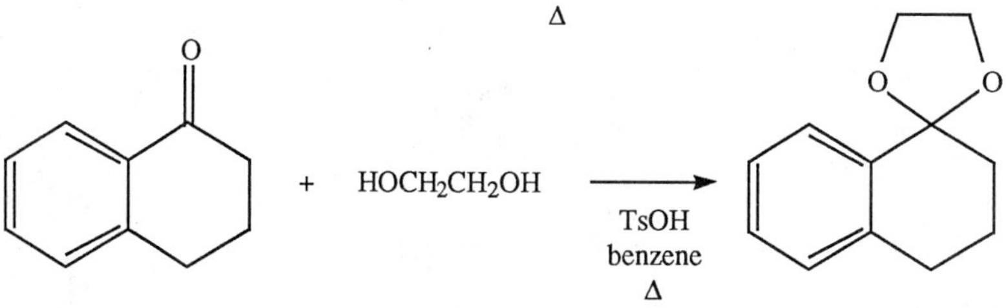

$$\text{(tetralone)} + HOCH_2CH_2OH \xrightarrow[\Delta]{\substack{\text{TsOH}\\\text{benzene}}} \text{(cyclic dioxolane)}$$

13.15

$$\underset{\substack{|\quad|\quad|\\OH\;OH\;OH}}{CH_2-CH-CH_2} + CH_3CCH_3 \longrightarrow C_6H_{12}O_3$$

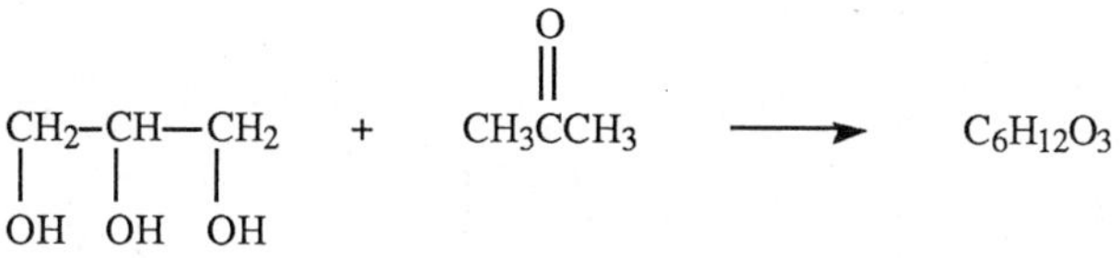

$$C_3H_8O_3 + C_3H_6O = C_6H_{14}O_4$$

$$C_6H_{14}O_4 - C_6H_{12}O_3 = H_2O$$

One molecule of water has been lost in the reaction.

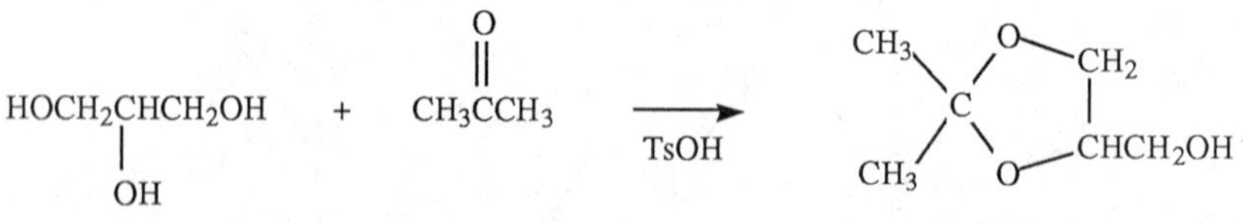

$$HOCH_2CHCH_2OH + CH_3CCH_3 \xrightarrow{\text{TsOH}} \text{a cyclic ketal}$$
$$\underset{OH}{|}$$

a cyclic ketal

13.16

addition of nucleophilic
nitrogen to electrophilic
carbonyl group

deprotonation

protonation

deprotonation

loss of water

protonation

13.17

protonation

nucleophilic attack by
water on electrophilic
carbon atom of
protonated imine

13.17 (cont)

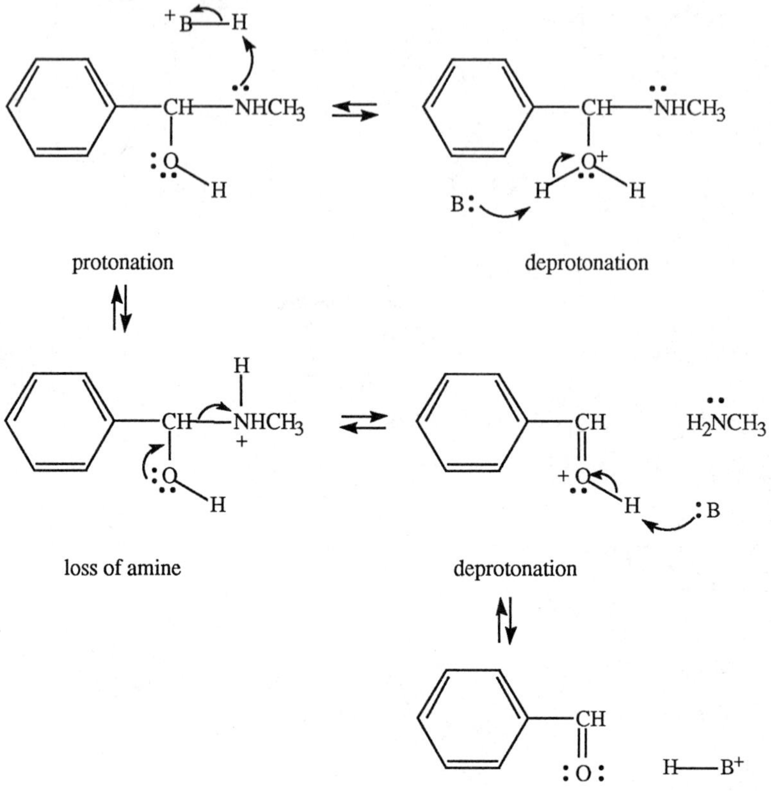

protonation   deprotonation

loss of amine   deprotonation

13.18 CH_3CCH_3 $\rightleftharpoons$ CH_3CCH_3 $\rightleftharpoons$ CH_3CCH_3

protonation

attack of the nucleophilic   deprotonation
amine on the electrophilic
protonated carbonyl group

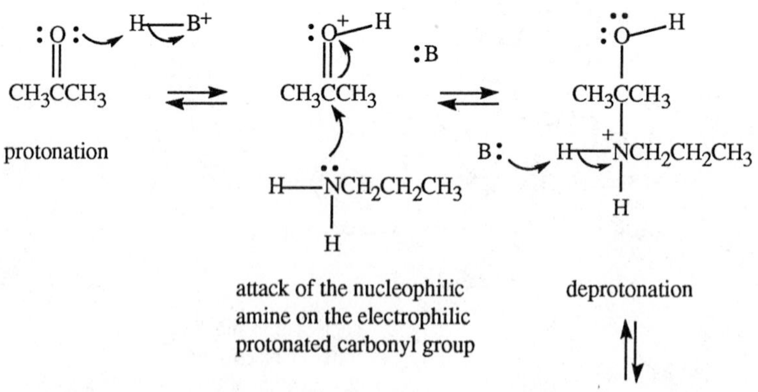

13.18 (cont)

The catalytic amount of hydrochloric acid begins the reaction by protonating the oxygen of the carbonyl group. Sodium hydroxide is added at the end to neutralize the acid in order to prevent catalysis of the reverse reaction, the hydrolysis of the imine (Problem 13.17).

Concept Map 13.6   (see p. 358)

13.19

(a) $CH_3CHCH_2CH_2CH$ (with CH_3 and O) $+$ $HONH_3$ Cl^- $\xrightarrow{CH_3CO^-Na^+}$ $CH_3CHCH_2CH_2CH=NOH$ (with CH_3)

(b) $CH_3CCH_2CH_2CH_3$ (with O) $+$ ⟨benzene ring⟩$-NHNH_2$ $\xrightarrow{acetic\ acid}$ $CH_3CH_2CH_2C=NNH-$⟨benzene ring⟩ (with CH_3)

(c) ⟨cyclopentane ring⟩$=O$ $+$ CH_3-⟨benzene ring⟩$-NH_2$ $\xrightarrow{\Delta}$ ⟨cyclopentane ring⟩$=N-$⟨benzene ring⟩$-CH_3$

(d) ⟨benzene ring⟩$-CH$ (with O) $+$ O_2N-⟨benzene ring with NO_2⟩$-NHNH_2$ $\xrightarrow{H_2SO_4}$

13.19 (cont)

(d) (cont)

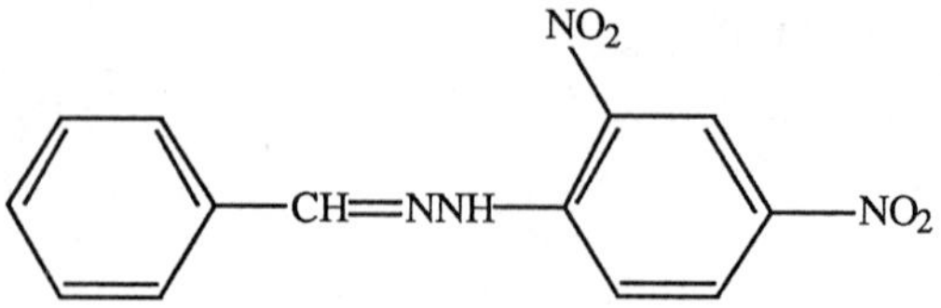

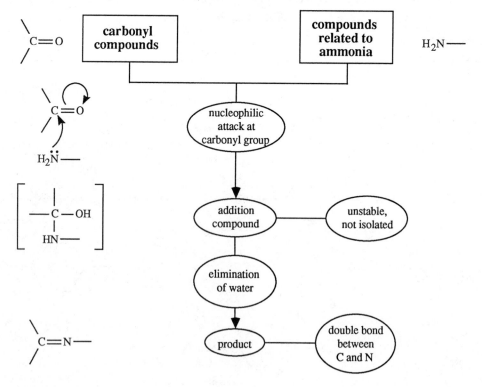

(e) C_6H_5—CCH$_2$CH$_2$CH$_3$ (with C=O) + H$_2$NNH$_2$ $\xrightarrow{\Delta}$ C_6H_5—CCH$_2$CH$_2$CH$_3$ (with C=NNH$_2$)

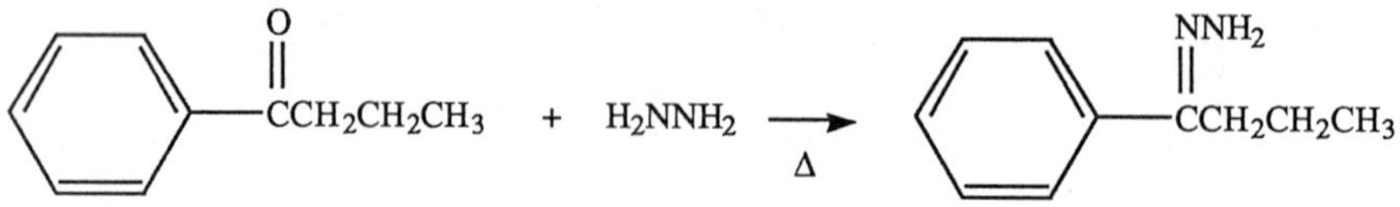

(f) CH$_3$CCH$_2$CH$_3$ (with C=O) + CH$_3$CHCH$_2$NH$_2$ (with CH$_3$) $\xrightarrow[\text{HCl}]{\text{NaOH}}$ CH$_3$CH$_2$C=NCH$_2$CHCH$_3$ (with CH$_3$ and CH$_3$)

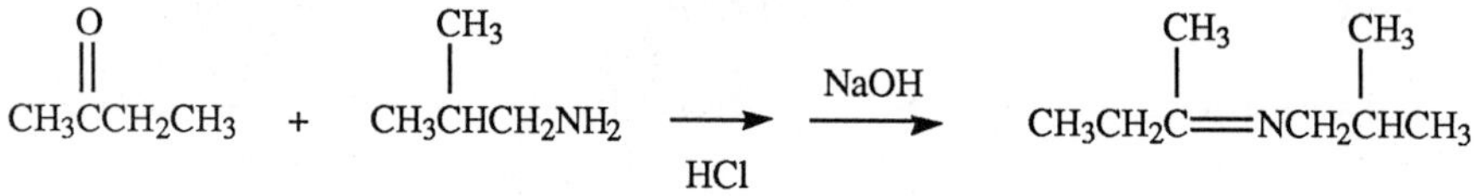

(g) C_6H_5—CH (with C=O) + CH$_3$CH$_2$CH$_2$NH$_2$ $\xrightarrow[\Delta]{\text{benzene}}$ C_6H_5—CH=NCH$_2$CH$_2$CH$_3$

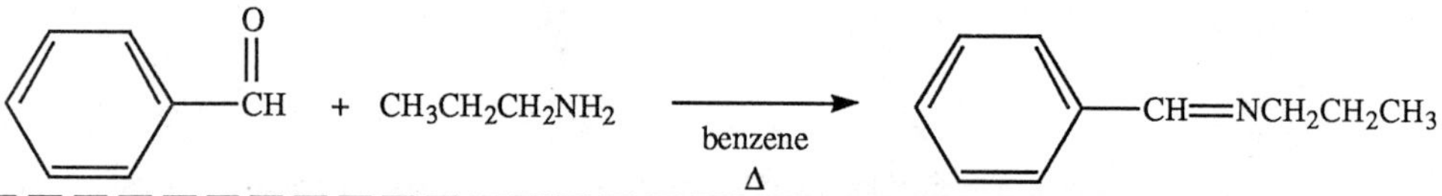

Concept Map 13.6   Reactions of carbonyl compounds with compounds related to ammonia.

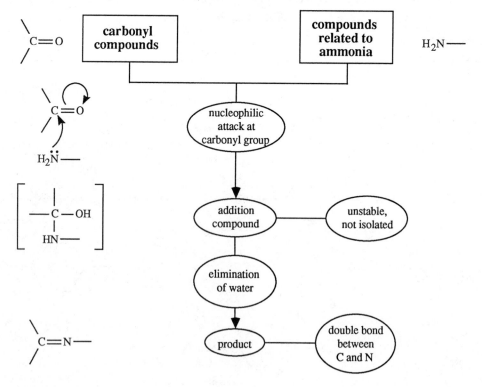

13.20   The major side reaction in the presence of hydrochloric acid would be hydrolysis of the ketal group.

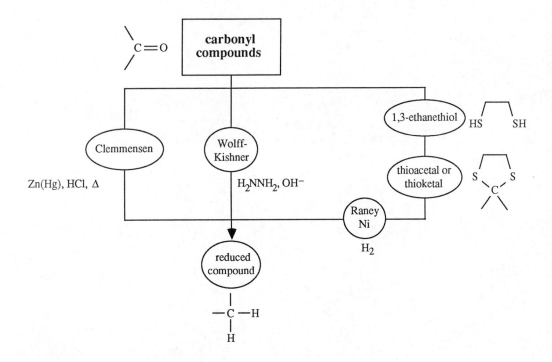

major side reaction

Dehydration of the alcohol function and substitution of a chlorine for the hydroxyl group are also possible side reactions.

Concept Map 13.7   Reduction of carbonyl groups to methylene groups.

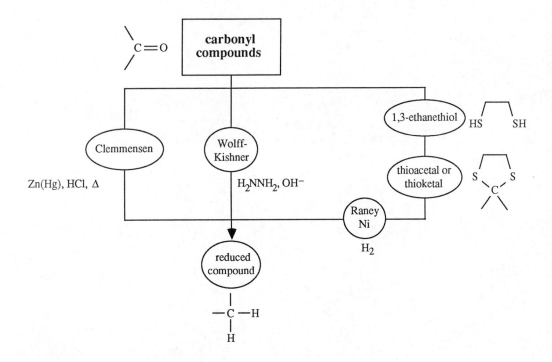

13.21

(a)

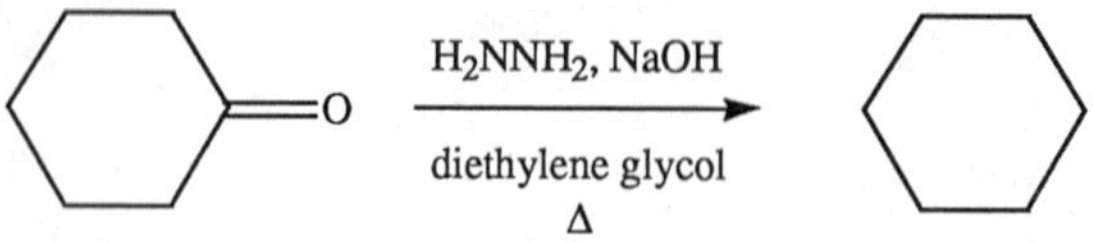

H₂NNH₂, NaOH

diethylene glycol
Δ

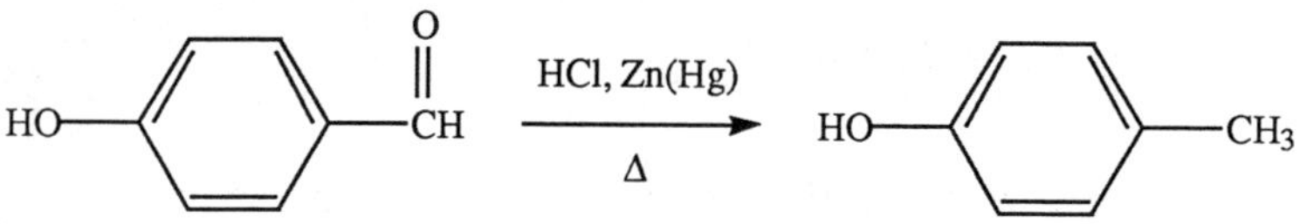

(b)

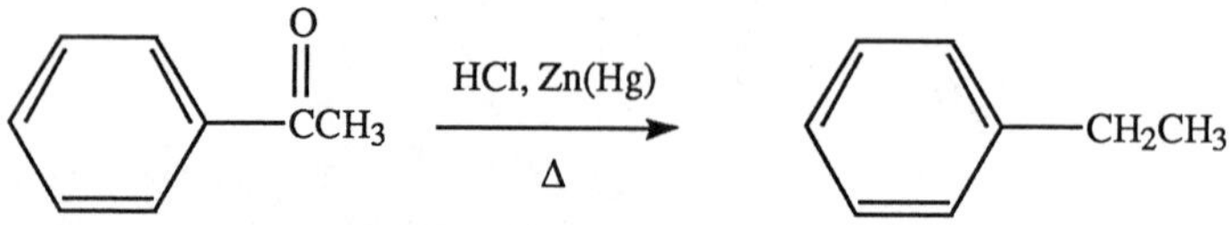

HCl, Zn(Hg)
Δ

(c)

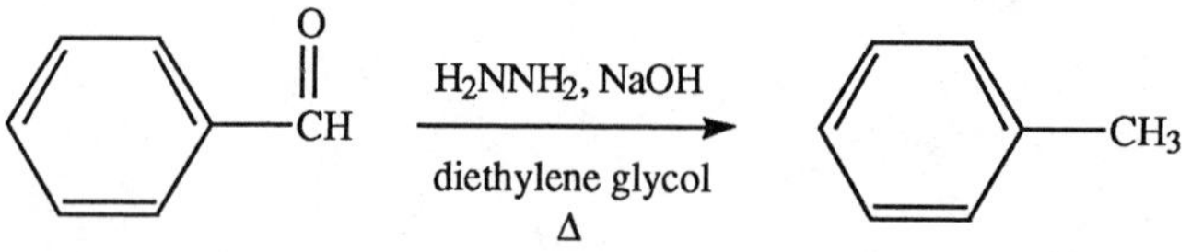

HCl, Zn(Hg)
Δ

(d)

H₂NNH₂, NaOH

diethylene glycol
Δ

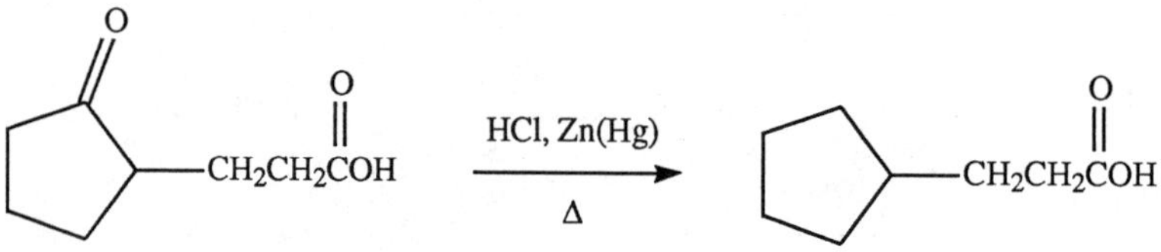

(e)

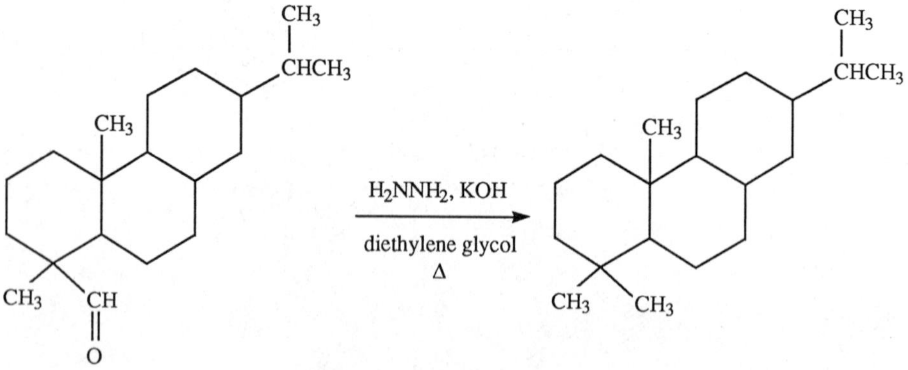

HCl, Zn(Hg)
Δ

(f)

H₂NNH₂, KOH

diethylene glycol
Δ

13.21 (cont)

(g)

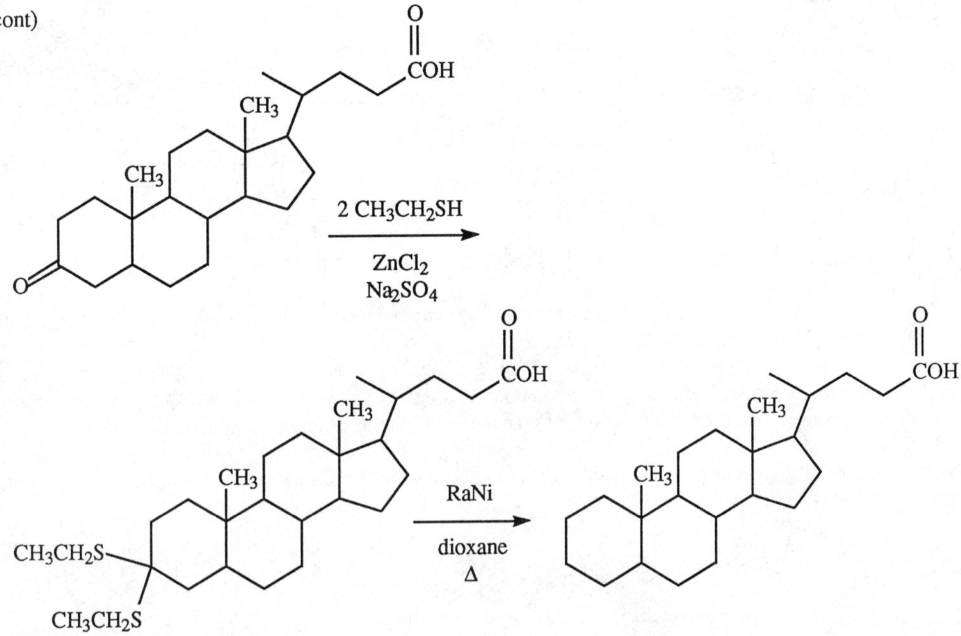

$$2\ CH_3CH_2SH \xrightarrow[Na_2SO_4]{ZnCl_2}$$

$$\xrightarrow[\substack{dioxane \\ \Delta}]{RaNi}$$

(h)

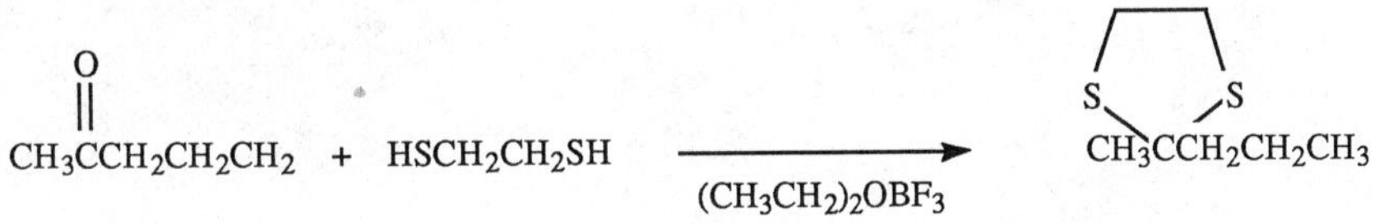

$$\underset{\substack{\| \\ O}}{CH_3CCH_2CH_2CH_2} \quad + \quad HSCH_2CH_2SH \xrightarrow{\quad (CH_3CH_2)_2OBF_3 \quad} \underset{\substack{S\ \ \ \ S}}{CH_3CCH_2CH_2CH_3}$$

(i)

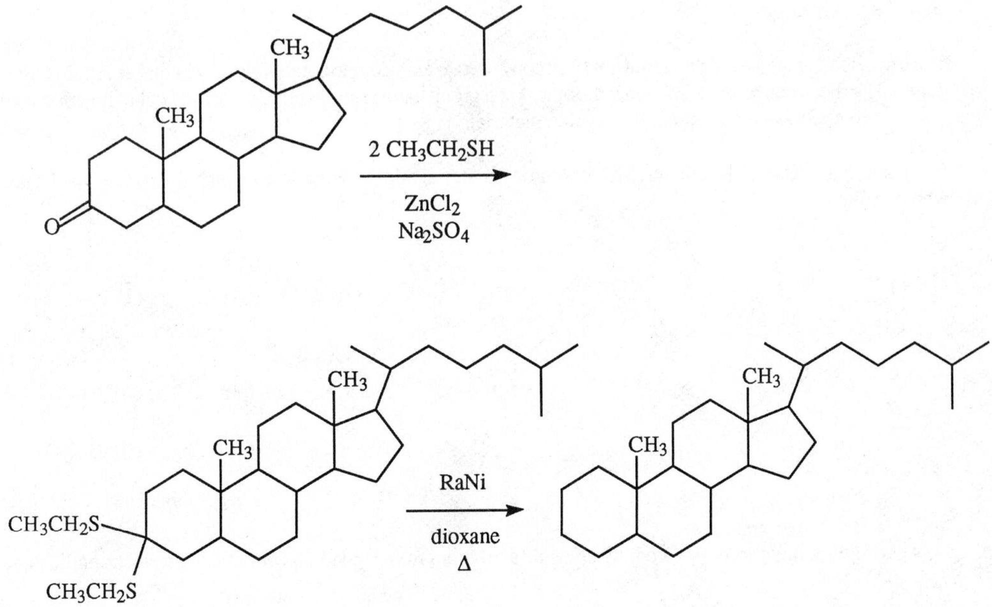

$$2\ CH_3CH_2SH \xrightarrow[Na_2SO_4]{ZnCl_2}$$

$$\xrightarrow[\substack{dioxane \\ \Delta}]{RaNi}$$

13.22

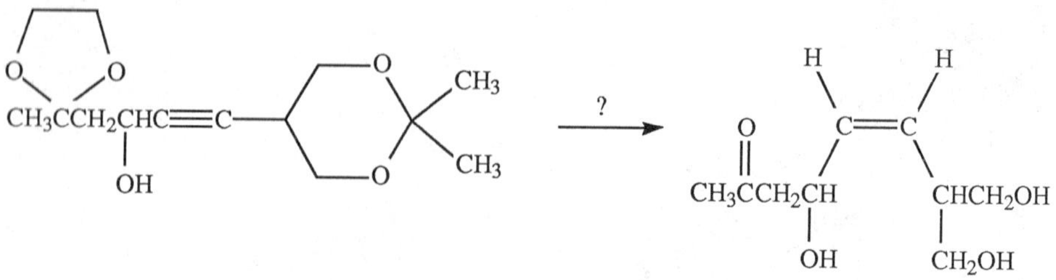

1.   What functional groups are present in the starting material and the product?

The starting material has a triple bond, an alcohol, and two ketal functional groups; the product has a cis double bond, a ketone and three alcohol functional groups.

2.   How do the carbon skeletons of the two compounds compare? How many carbon atoms does each contain? Are there any rings? What are the positions of branches and functional groups on the carbon skeletons?

The atoms of the starting material and product molecules that are the same are shown below. The rings that are present are cyclic ketals.

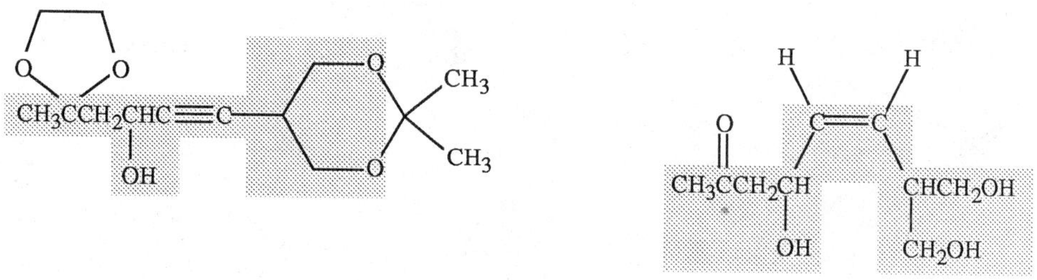

3.   How do the functional groups change in going from starting material to product? Does the starting material have a good leaving group?

The triple bond in the starting material has been converted to a cis double bond. One of the ketal groups in the starting material has been converted to a ketone; the other ketal group has been converted into two alcohol functional groups. There is no good leaving group in the starting material.

4.   Is it possible to dissect the structures of the starting material and product to see which bonds must be broken and which formed?

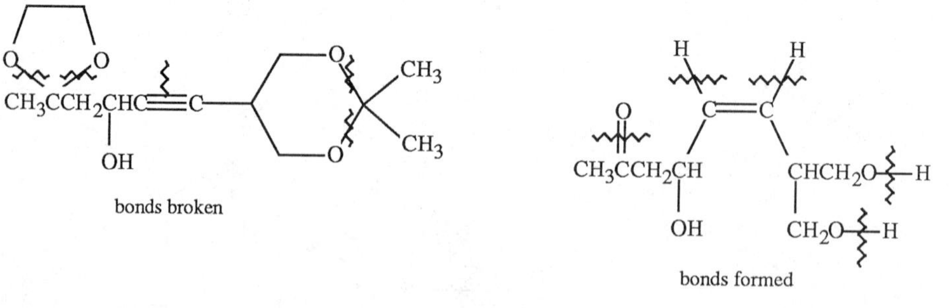

bonds broken

bonds formed

5.   Do we recognize any part of the product molecule as coming from a good nucleophile or an electrophilic addition?

No.

13.22 (cont)

6. What type of compound would be a good precursor to the product?

An alkyne which can be converted to a cis alkene by reduction with hydrogen and the appropriate catalyst.

The ketal groups in the starting material can be removed by acid hydrolysis.

7. After this last step, do we see how to get from starting material to product? If not, we need to analyze the structure obtained in step 6 by applying questions 5 and 6 to it.

The ketal groups are removed before reduction to prevent the acid-catalyzed addition of water to the alkene. Alkynes are less susceptible to hydration by dilute acid than alkenes (Section 9.4B).

Concept Map 13.8   Protecting groups.

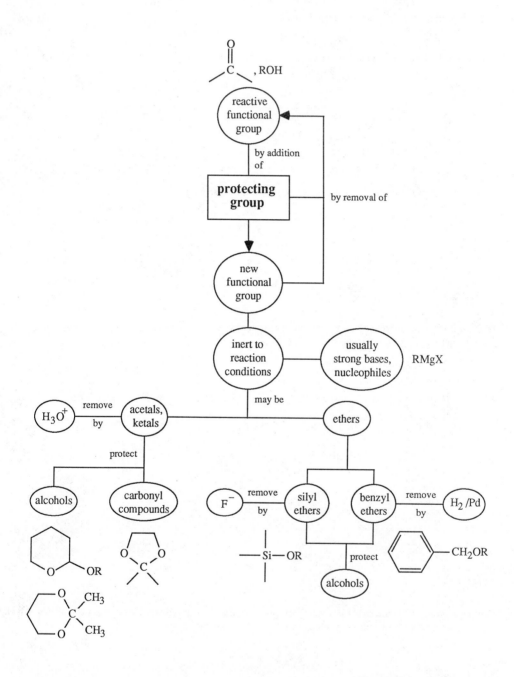

13.23 $CH_2{=\!=}CCH_2CH_2OCH_2$ (with CH_3 branch, attached to benzene ring) $\xrightarrow{?}$ $CH_3CH_2C{-\!-}CHCHCH_2CH_2OCH_2$ (with CH_2 and CH_3 branches, OH group, attached to benzene ring)

1. What functional groups are present in the starting material and the product?

   The starting material has a double bond and an ether functional group. The product has a double bond, and ether and secondary alcohol functional groups.

2. How do the carbon skeletons of the two compounds compare? How many carbon atoms does each contain? Are there any rings? What are the positions of branches and functional groups on the carbon skeletons?

   Carbon 2 of a four carbon segment has added to carbon 1 of the starting material (see structure in part 4).

3. How do the functional groups change in going from starting material to product? Does the starting material have a good leaving group?

   A four carbon segment has added to carbon 1 of the double bond. There is no good leaving group.

4. Is it possible to dissect the structures of the starting material and product to see which bonds must be broken and which formed?

   bonds broken   bonds formed

5. Do we recognize any part of the product molecule as coming from a good nucleophile or an electrophilic addition?

   In order to form a carbon-carbon σ bond, one of the carbon atoms must be a nucleophile and the other an electrophile. A Grignard reagent is a good nucleophile, which can add to the electrophilic carbon atom of an aldehyde to give a secondary alcohol.

6. What type of compound would be a good precursor to the product?

   An aldehyde would be a good precursor to the product we want.

13.23 (cont)

7. After this last step, do we see how to get from starting material to product? If not, we need to analyze the structure obtained in step 6 by applying questions 5 and 6 to it.

   We need to prepare the aldehyde so must apply questions 5 and 6 again.

5'. Do we recognize any part of the precursor molecule as coming from a good nucleophile or an electrophilic addition?

   No.

6'. What type of compound would be a good precursor to the product?

   An aldehyde can be prepared by the oxidation of a primary alcohol.

A primary alcohol can be made by the electrophilic addition of diborane to the terminal double bond in the starting material followed by oxidation of the alkylborane.

7'. After this last step, do we see how to get from starting material to product? If not, we need to analyze the structure obtained in step 6 by applying questions 5 and 6 to it.

13.23 (cont)

7'. (cont)

$$\xleftarrow[H_2O]{H_2O_2,\ NaOH} \xleftarrow{BH_3} \underset{\substack{|\\CH_2=CCH_2CH_2OCH_2-}}{\overset{CH_3}{}}\text{—}\bigcirc$$

13.24 $\underset{\substack{|\\CH_3C}}{\overset{CH_3}{}}=CHCH_2CH=CH_2 \xrightarrow{\text{9-BBN}} \xrightarrow[H_2O]{H_2O_2,\ NaOH}$

$$\underset{\substack{|\\CH_3C}}{\overset{CH_3}{}}=CHCH_2CH_2CH_2OH \xrightarrow[\text{dichloromethane}]{\text{(pyridinium } CrO_3Cl^-)} \underset{\substack{|\\CH_3C}}{\overset{CH_3}{}}=CHCH_2CH_2\overset{O}{\overset{\|}{C}}H \xrightarrow{CH_3MgI}$$

A   B

$$\xrightarrow[H_2O]{NH_4Cl} \underset{\substack{\\CH_3C}}{\overset{CH_3}{}}=CHCH_2CH_2\underset{\substack{|\\OH}}{CHCH_3}$$

13.25 $CH_3CH_2CH=CHCH=CHCHC\equiv CH \xleftarrow[\substack{\\25\ ^\circ C}]{\text{TsOH}} $ (dihydropyran)

$\qquad\qquad\qquad\qquad\quad |$
$\qquad\qquad\qquad\qquad OTHP$

$$\underset{\substack{|\\OH}}{CH_3CH_2CH=CHCH=CHCHC\equiv CH} \xleftarrow[H_2O]{NH_4Cl} \xleftarrow{Na^+\ {}^-C\equiv CH} CH_3CH_2CH=CHCH=CHC\overset{O}{\overset{\|}{}}H$$

13.26

$$\underset{\substack{\\ \\ \underset{H}{}\ C=C\ \underset{H}{}}}{\overset{HC\equiv CCH_2O}{\underset{|}{CH_3CHCH}}\ \ \ \overset{CH_3\ \ CH_3}{\underset{CH_3\ \ CH_3}{CH_2OSi-CCH_3}}} \xleftarrow{HC\equiv CCH_2Cl}$$

13.26 (cont)

$$\underset{\underset{H}{\overset{CH_3}{|}}{\overset{|}{\underset{|}{C}}}=\underset{\underset{H}{}}{\overset{|}{C}}}{\overset{Na^+{}^-O}{\underset{|}{CH_3CHCH}}}\quad \underset{\underset{CH_3\ CH_3}{}}{\overset{CH_3\ CH_3}{CH_2OSi{-}CCH_3}}\xleftarrow{NaH}\quad\underset{\underset{H}{}}{\overset{OH}{CH_3CHCH}}\ \underset{}{C}=C\quad CH_2OSi{-}CCH_3\xleftarrow[\substack{Pd/CaCO_3\\quinoline}]{H_2}$$

(structures drawn with silyl protecting groups)

$$CH_3CHCHC{\equiv}CCH_2OSi{-}CCH_3\xleftarrow[CH_3OH]{NaBH_4}CH_3CHCC{\equiv}CCH_2OSi{-}CCH_3$$

13.27   (a) cyclopentanecarbaldehyde

(b) cyclobutanone

(c) (*E*)-2-hexenal

(d) (*S*)-2-hydroxypropanal

(e) 2-methylcyclohexanone

(f) 5-methyl-2-hexanone

(g) 4-methoxybenzaldehyde
(anisaldehyde)

(h) 3-methyl-2-cyclohexenone

(i) (4*S*, 5*S*)-4-chloro-5-methyl-2-heptanone

(j) 4-phenylbutanal

13.28   (a)

(b)

$$CH_3CH_2CH_2CH_2\overset{\overset{O}{\|}}{C}CH_2Br$$

(c)

(d)

(e)

$$CH_2{=}CHCH{=}CH\overset{\overset{O}{\|}}{C}CH_3$$

(f)

$$\underset{\underset{CH_2CH}{}}{\overset{CH_3}{CH_3CH_2CH_2CHCH}}$$

(g)

(h)

13.29

(a)

$$\underset{\text{CH}_3\text{CH}_2\text{CH}_2\text{CH}}{\overset{\overset{\displaystyle O}{\|}}{}} \xrightarrow[\text{cold}]{\text{H}_2\text{SO}_4 \text{ (conc)}} \underset{\text{CH}_3\text{CH}_2\text{CH}_2\text{CH}}{\overset{\overset{\displaystyle O^+\!\!-\!\!H}{\|}}{}} \quad \text{HSO}_4^-$$

(b)

$$\underset{\text{CH}_3\text{CH}_2\text{CH}_2\text{CH}}{\overset{\overset{\displaystyle O}{\|}}{}} \xrightarrow[\text{H}_2\text{O}]{\text{NaBH}_4} \text{CH}_3\text{CH}_2\text{CH}_2\text{CH}_2\text{OH}$$

(c)

$$\underset{\text{CH}_3\text{CH}_2\text{CH}_2\text{CH}}{\overset{\overset{\displaystyle O}{\|}}{}} \xrightarrow{\underset{\text{H}_2\text{SO}_4}{}} \text{CH}_3\text{CH}_2\text{CH}_2\text{CH}=\text{NNH}—$$

with reagent O$_2$N—（ring with NO$_2$）—NHNH$_2$ and product attached to 2,4-dinitrophenyl ring (NO$_2$, NO$_2$)

(d)

$$\underset{\text{CH}_3\text{CH}_2\text{CH}_2\text{CH}}{\overset{\overset{\displaystyle O}{\|}}{}} \xrightarrow{\text{CH}_3\text{CH}_2\text{CH}_2\text{CH}_2\text{Li}} \xrightarrow{\text{H}_3\text{O}^+} \underset{\overset{\displaystyle |}{\text{OH}}}{\text{CH}_3\text{CH}_2\text{CH}_2\text{CHCH}_2\text{CH}_2\text{CH}_2\text{CH}_3}$$

(e)

$$\underset{\text{CH}_3\text{CH}_2\text{CH}_2\text{CH}}{\overset{\overset{\displaystyle O}{\|}}{}} \xrightarrow[\Delta]{} \text{CH}_3\text{CH}_2\text{CH}_2\text{CH}=\text{N}—$$

with reagent （ring）—NH$_2$ and product phenyl ring

(f)

$$\underset{\text{CH}_3\text{CH}_2\text{CH}_2\text{CH}}{\overset{\overset{\displaystyle O}{\|}}{}} \xrightarrow[\substack{\text{H}_2\text{O} \\ \text{H}_2\text{SO}_4}]{\text{NaCN}} \underset{\overset{\displaystyle |}{\text{OH}}}{\text{CH}_3\text{CH}_2\text{CH}_2\text{CHCN}}$$

(g)

$$\underset{\text{CH}_3\text{CH}_2\text{CH}_2\text{CH}}{\overset{\overset{\displaystyle O}{\|}}{}} \xrightarrow[\substack{\text{H}_3\text{O}^+ \\ \text{H}_2\text{O}}]{\text{Na}_2\text{Cr}_2\text{O}_7} \underset{\text{CH}_3\text{CH}_2\text{CH}_2\text{COH}}{\overset{\overset{\displaystyle O}{\|}}{}} + \text{Cr}^{+3}$$

(h)

$$\underset{\text{CH}_3\text{CH}_2\text{CH}_2\text{CH}}{\overset{\overset{\displaystyle O}{\|}}{}} \xrightarrow[\text{acetone}]{\text{NaI}} \text{no reaction}$$

(i)

$$\underset{\text{CH}_3\text{CH}_2\text{CH}_2\text{CH}}{\overset{\overset{\displaystyle O}{\|}}{}} \xrightarrow[\underset{\overset{\displaystyle \|}{\text{O}}}{\text{CH}_3\text{CO}^-\text{Na}^+}]{\text{HONH}_3^+\text{Cl}^-} \text{CH}_3\text{CH}_2\text{CH}_2\text{CH}=\text{NOH}$$

13.29 (cont)

(j) $CH_3CH_2CH_2\overset{\overset{\displaystyle O}{\|}}{C}H$ $\xrightarrow[\text{diethyl ether}]{CH_3MgI}$ $CH_3CH_2CH_2\underset{\underset{\displaystyle O^-Mg^{2+}I^-}{|}}{C}HCH_3$ $\xrightarrow{H_3O^+}$ $CH_3CH_2CH_2\underset{\underset{\displaystyle OH}{|}}{C}HCH_3$

(k) $CH_3CH_2CH_2\overset{\overset{\displaystyle O}{\|}}{C}H$ $\xrightarrow[\text{H}_2\text{O}]{5\% \text{ NaHCO}_3}$ no reaction

(l) $CH_3CH_2CH_2\overset{\overset{\displaystyle O}{\|}}{C}H$ $\xrightarrow[\substack{\text{diethyl} \\ \text{ether}}]{LiAlH_4 \quad H_3O^+}$ $CH_3CH_2CH_2CH_2OH$

(m) $CH_3CH_2CH_2\overset{\overset{\displaystyle O}{\|}}{C}H$ $\xrightarrow[\substack{\text{diethylene glycol} \\ \Delta}]{H_2NNH_2, \text{ KOH}}$ $CH_3CH_2CH_2CH_3$

(n) $CH_3CH_2CH_2\overset{\overset{\displaystyle O}{\|}}{C}H$ $\xrightarrow[\Delta]{HCl, \text{ Zn(Hg)}}$ $CH_3CH_2CH_2CH_3$

13.30

(a) $CH_3CH_2CH_2\overset{\overset{\displaystyle O}{\|}}{C}CH_3$ $\xrightarrow[\text{cold}]{H_2SO_4 \text{ (cont)}}$ $CH_3CH_2CH_2\overset{\overset{\displaystyle O^+\!\!-H}{\|}}{C}CH_3$ HSO_4^-

(b) $CH_3CH_2CH_2\overset{\overset{\displaystyle O}{\|}}{C}CH_3$ $\xrightarrow[\text{H}_2\text{O}]{NaBH_4}$ $CH_3CH_2CH_2\underset{\underset{\displaystyle OH}{|}}{C}HCH_3$

(c) $CH_3CH_2CH_2\overset{\overset{\displaystyle O}{\|}}{C}CH_3$ $\xrightarrow[\text{H}_2\text{SO}_4]{}$ $CH_3CH_2CH_2\overset{\overset{\displaystyle CH_3}{|}}{C}\!\!=\!\!NNH\!-\!\!\!\bigcirc\!\!-NO_2$
(with O_2N- and NO_2 substituted nitrophenylhydrazine reagent)

(d) $CH_3CH_2CH_2\overset{\overset{\displaystyle O}{\|}}{C}CH_3$ $\xrightarrow[\Delta]{}$ $CH_3CH_2CH_2\overset{\overset{\displaystyle CH_3}{|}}{C}\!\!=\!\!N\!-\!\!\!\bigcirc$
(with aniline $-NH_2$ reagent)

13.30 (cont)

(e) $CH_3CH_2CH_2CCH_3$ (with =O above C) $\xrightarrow[\text{tetrhydrofuran}]{\text{(phenyl)-Li}}$ $CH_3CH_2CH_2C$ (with CH_3 above, phenyl to right, O^-Li^+ below) $\xrightarrow[\text{H}_2\text{O}]{\text{NH}_4\text{Cl}}$

$CH_3CH_2CH_2C$ (with CH_3 above, phenyl to right, OH below)

(f) $CH_3CH_2CH_2CCH_3$ (with =O above) $\xrightarrow[\substack{\text{H}_2\text{O}\\ \text{H}_2\text{SO}_4}]{\text{NaCN}}$ $CH_3CH_2CH_2CCN$ (with CH_3 above, OH below)

(g) $CH_3CH_2CH_2CCH_3$ (with =O above) $\xrightarrow[\substack{\text{H}_3\text{O}^+\\ \text{H}_2\text{O}}]{\text{Na}_2\text{Cr}_2\text{O}_7}$ no reaction

(h) $CH_3CH_2CH_2CCH_3$ (with =O above) $\xrightarrow[\text{ethanol}]{\text{AgNO}_3}$ no reaction

(i) $CH_3CH_2CH_2CCH_3$ (with =O above) $\xrightarrow[\substack{\text{CH}_3\text{CO}^-\text{Na}^+}]{\text{HONH}_3^+\text{Cl}^-}$ (with O below) $CH_3CH_2CH_2C{=}NOH$ (with CH_3 above)

(j) $CH_3CH_2CH_2CCH_3$ (with =O above) $\xrightarrow[\text{diethyl ether}]{\text{CH}_3\text{CH}_2\text{MgI}}$ $CH_3CH_2CH_2CCH_2CH_3$ (with CH_3 above, $O^-Mg^{2+}Br^-$ below) $\xrightarrow{\text{H}_3\text{O}^+}$ $CH_3CH_2CH_2CCH_2CH_3$ (with CH_3 above, OH below)

(k) $CH_3CH_2CH_2CCH_3$ (with =O above) $\xrightarrow[\text{H}_2\text{O}]{\text{5\% NaHCO}_3}$ no reaction

(l) $CH_3CH_2CH_2CCH_3$ (with =O above) $\xrightarrow{\text{CH}_3\text{C}{\equiv}\text{C}^-\text{Na}^+}$ $CH_3CH_2CH_2CC{\equiv}CCH_3$ (with CH_3 above, O^-Na^+ below) $\xrightarrow{\text{NH}_4\text{Cl}}$ $CH_3CH_2CH_2CC{\equiv}CCH_3$ (with CH_3 above, CO^-Na^+ below)

13.30 (cont)

(m) $$CH_3CH_2CH_2\overset{\displaystyle O}{\overset{\|}{C}}CH_3 \xrightarrow[\text{diethylene glycol, }\Delta]{H_2NNH_2,\ KOH} CH_3CH_2CH_2CH_2CH_3$$

(n) $$CH_3CH_2CH_2\overset{\displaystyle O}{\overset{\|}{C}}CH_3 \xrightarrow[\Delta]{HCl,\ Zn(Hg)} CH_3CH_2CH_2CH_2CH_3$$

13.31

(a)

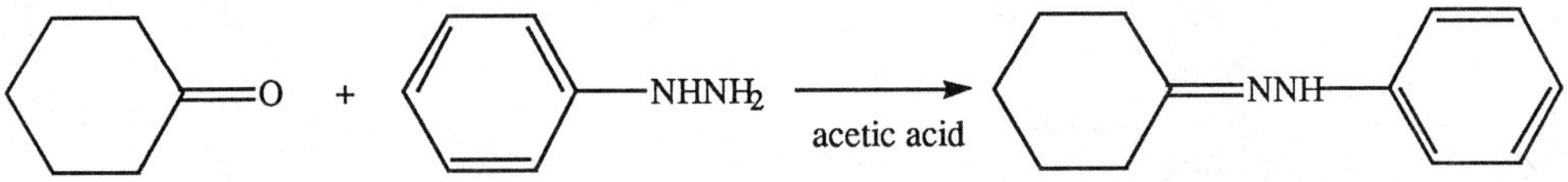

(b)

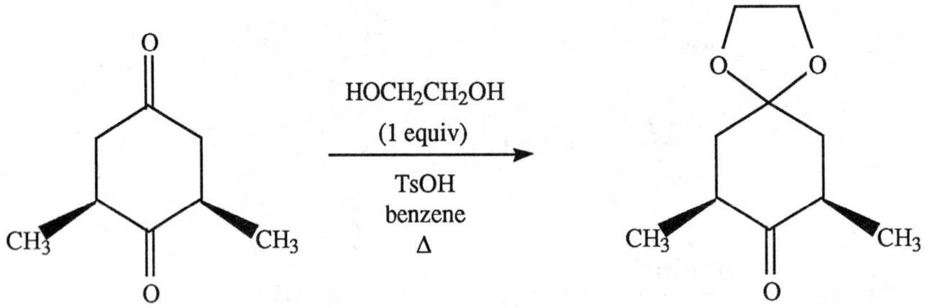

(c)

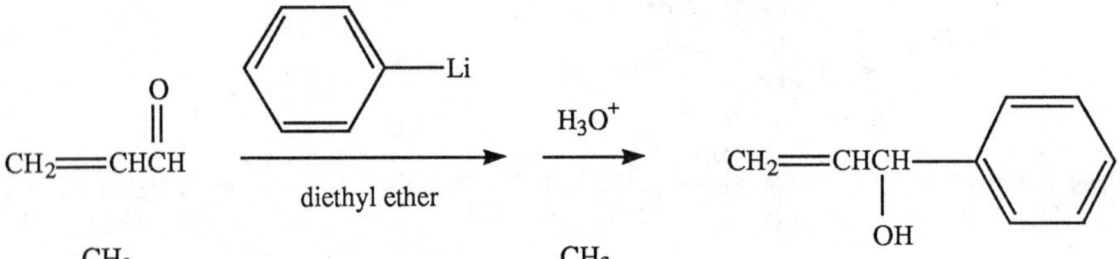

(d)

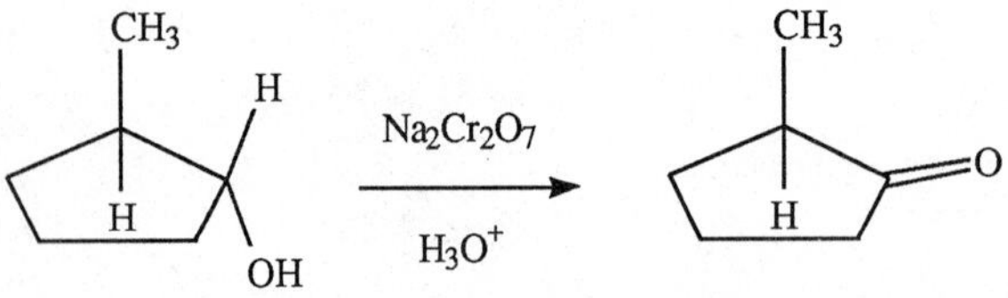

(e)

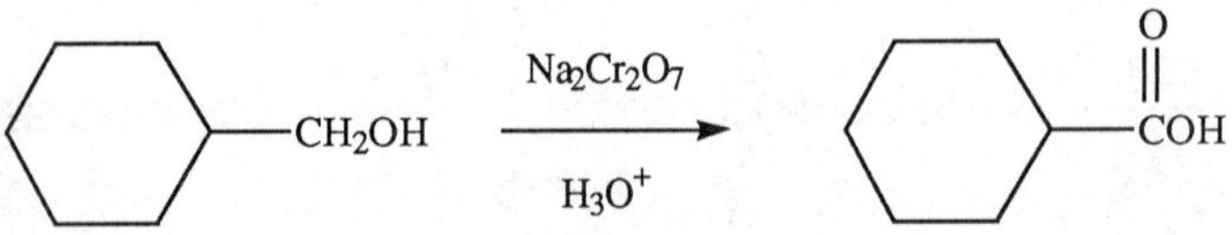

13.31 (cont)

(f)

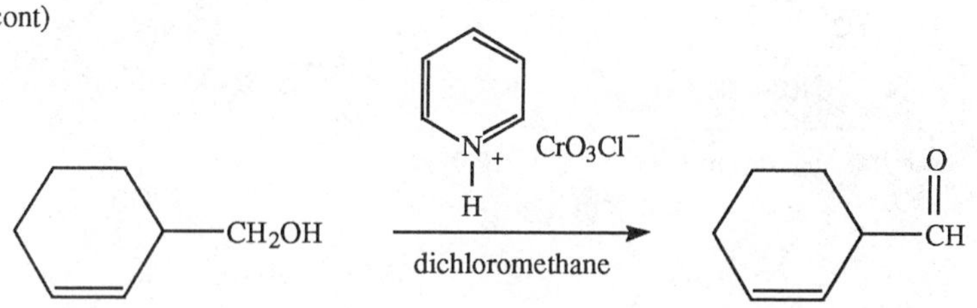

(g)

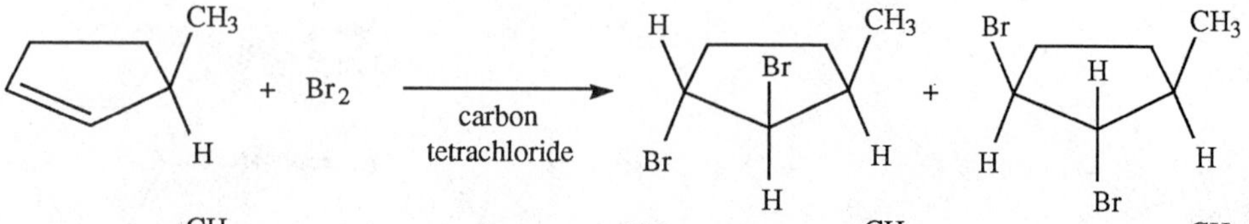

(h)

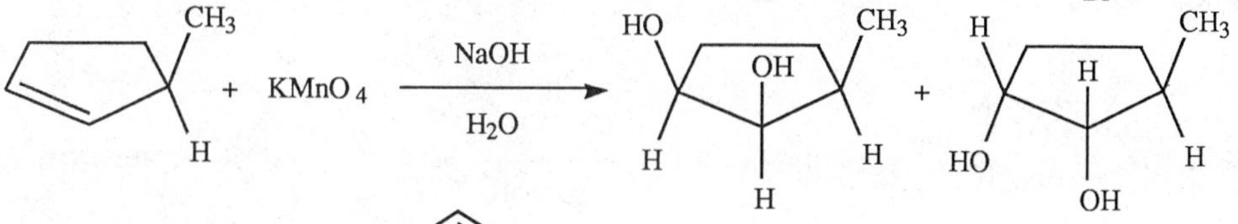

(i)

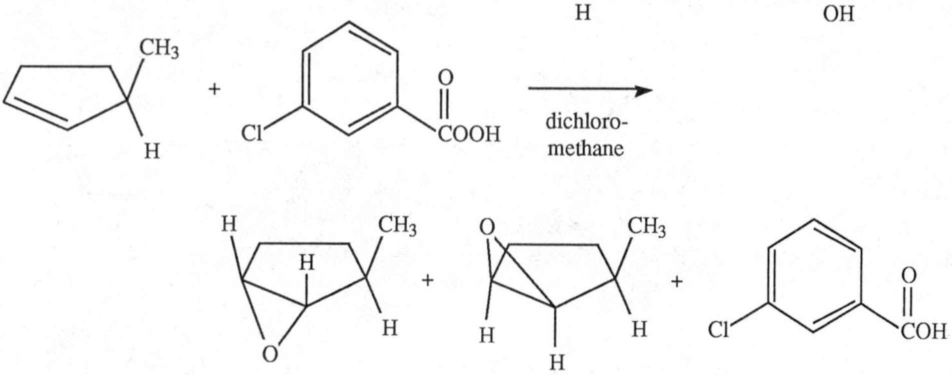

(j)

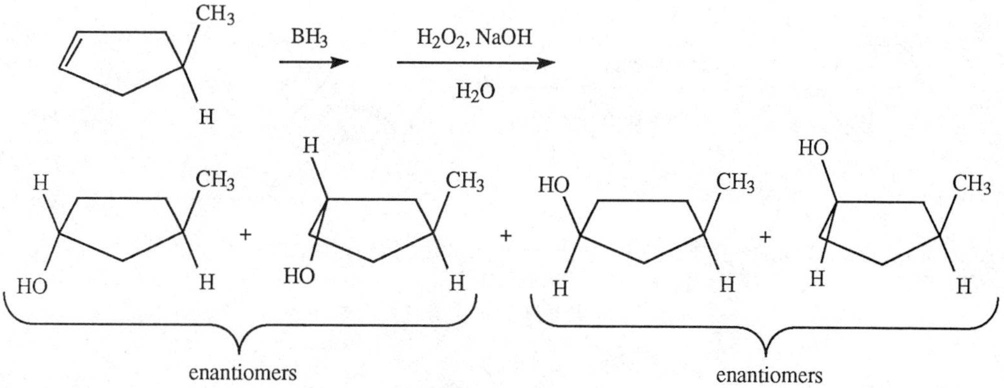

13.31 (cont)

(k)

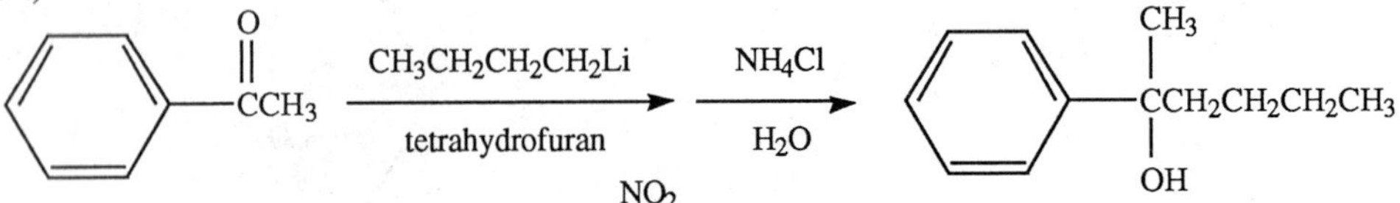

$$\underset{\text{tetrahydrofuran}}{\xrightarrow{CH_3CH_2CH_2CH_2Li}} \qquad \underset{H_2O}{\xrightarrow{NH_4Cl}}$$

(l)

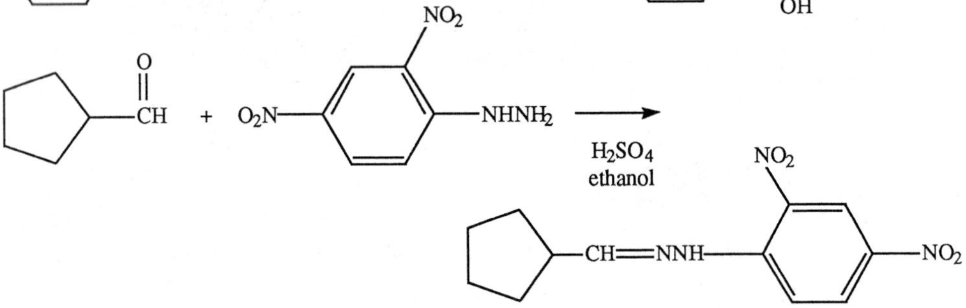

$$\xrightarrow[\text{ethanol}]{H_2SO_4}$$

(m)

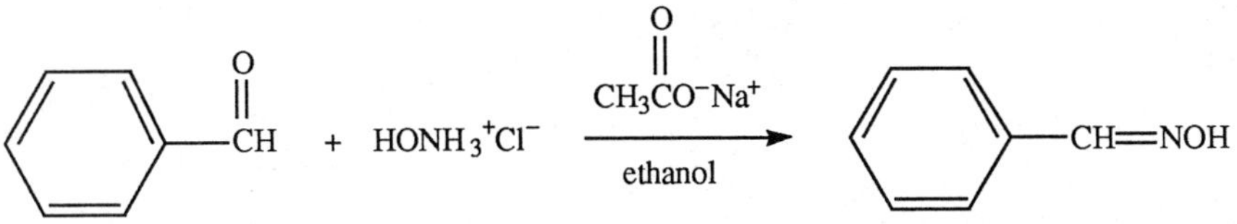

$$\underset{\text{ethanol}}{\xrightarrow{CH_3CO^-Na^+}}$$

(n) CH_3CCH_3 +   ⬡—NH_2 $\xrightarrow{\Delta}$ $CH_3C{=}N$—⬡

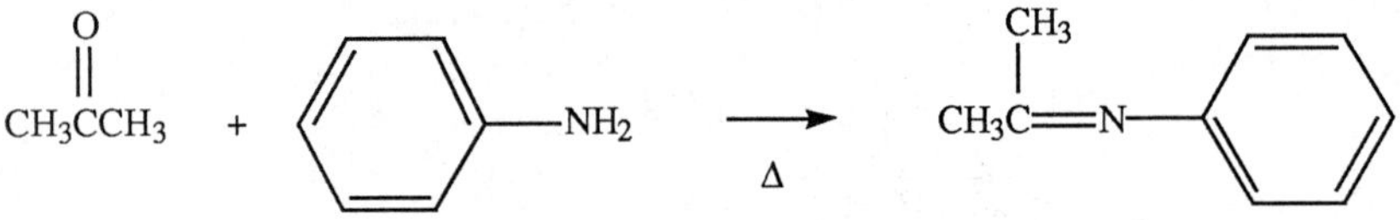

(o) $CH_3CH_2CH_2C{\equiv}CH$ + $NaNH_2$ $\xrightarrow{NH_3 \text{ (liq)}}$ $CH_3CH_2CH_2C{\equiv}C^-Na^+$

(p) $CH_3CH_2CH_2C{\equiv}CH$ + CH_3CH_2MgBr $\xrightarrow{\text{diethyl ether}}$ $CH_3CH_2CH_2C{\equiv}CMgBr$ + $CH_3CH_3{\uparrow}$

(q)

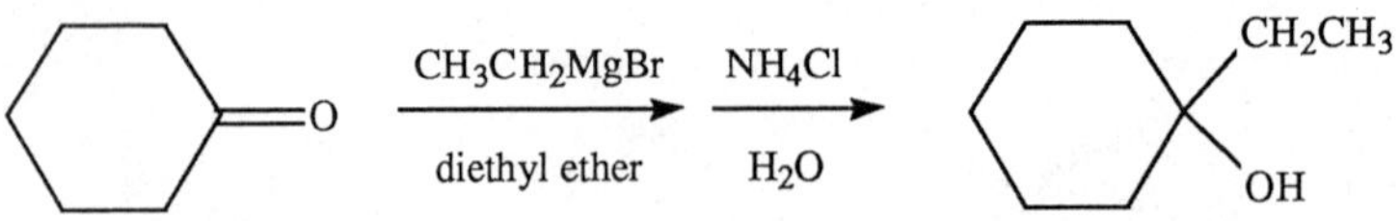

$$\underset{\text{diethyl ether}}{\xrightarrow{CH_3CH_2MgBr}} \qquad \underset{H_2O}{\xrightarrow{NH_4Cl}}$$

(r) $\underset{CH_3CHCH_2CH_2CH}{}$ + CH_3OH $\xrightarrow[\substack{\text{TsOH} \\ \text{benzene} \\ \Delta}]{}$ $CH_3CHCH_2CH_2CHOCH_3$

13.31 (cont)

(s) CH$_3$-⟨benzene ring⟩-CCH$_2$CH$_3$ (with C=O) + HOCH$_2$CH$_2$OH $\xrightarrow[\Delta]{\text{TsOH} \atop \text{benzene}}$

CH$_3$-⟨benzene ring⟩-CCH$_2$CH$_3$ (with dioxolane ring: O, O)

(t) ⟨cyclohexane ring⟩=O + HSCH$_2$CH$_2$OH $\xrightarrow{\text{(CH}_3\text{CH}_2)_2\text{OBF}_3}$ ⟨cyclohexane spiro ring with O and S⟩

(u) $\overset{\text{O}}{\overset{\|}{\text{CH}_3\text{CCH}_2}}\overset{\text{CH}_3}{\underset{|}{\text{CHCH}_3}}$ + $\overset{\text{CH}_3}{\underset{|}{\text{CH}_3\text{CHCH}_2\text{NH}_2}}$ $\xrightarrow[\text{HCl}]{\text{NaOH}}$ $\overset{\text{CH}_3}{\underset{|}{\text{CH}_3\text{CCH}_2\text{CHCH}_3}}$ with NCH$_2$CHCH$_3$ and CH$_3$

(v) HOCH$_2$CH$_2$$\overset{\text{CH}_3}{\underset{\underset{\text{CH}_3}{|}}{\overset{|}{\text{C}}}}CH_2$CH=CHCOCH$_3$ (with O on the CO) $\xrightarrow[\text{dichloromethane}]{\text{pyridinium } \text{CrO}_3\text{Cl}^-}$ HCCH$_2$$\overset{\text{CH}_3}{\underset{\underset{\text{CH}_3}{|}}{\overset{|}{\text{C}}}}CH_2$CH=CHCOCH$_3$ (with two O)

(w) $\overset{\text{O}}{\overset{\|}{\text{CH}_3\text{C}(\text{CH}_2)_8\text{CH}_3}}$ $\xrightarrow[\Delta]{\text{HCl, Zn(Hg)}}$ CH$_3$CH$_2$(CH$_2$)$_8$CH$_3$

(x) $\overset{\text{CH}_3}{\underset{|}{\text{CH}_3\text{CH}}}CH_2$$\overset{\text{O}}{\overset{\|}{\text{C}}}CH_2$$\overset{\text{CH}_3}{\underset{|}{\text{CHCH}_3}}$ $\xrightarrow[\substack{\text{diethylene glycol} \\ \Delta}]{\text{H}_2\text{NNH}_2, \text{NaOH}}$ $\overset{\text{CH}_3}{\underset{|}{\text{CH}_3\text{CH}}}CH_2CH_2CH_2$$\overset{\text{CH}_3}{\underset{|}{\text{CHCH}_3}}$

13.32

(a)

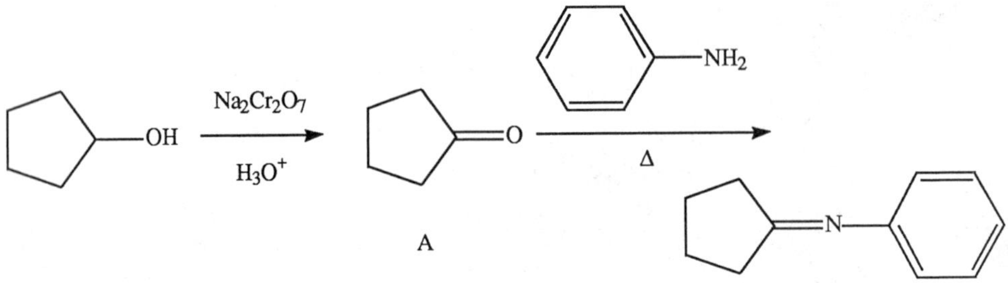

OH
$Na_2Cr_2O_7$
H_3O^+
O
NH_2
Δ
N
A

B

(b)

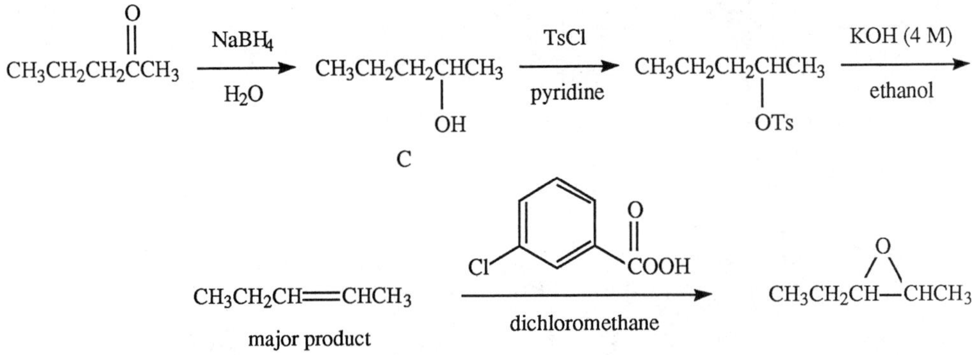

O
$CH_3CH_2CH_2CCH_3$
$NaBH_4$
H_2O
$CH_3CH_2CH_2CHCH_3$
OH
C
TsCl
pyridine
$CH_3CH_2CH_2CHCH_3$
OTs
KOH (4 M)
ethanol
Cl
O
COOH
$CH_3CH_2CH{=}CHCH_3$
major product
dichloromethane
O
$CH_3CH_2CH{-}CHCH_3$

E F

(c)

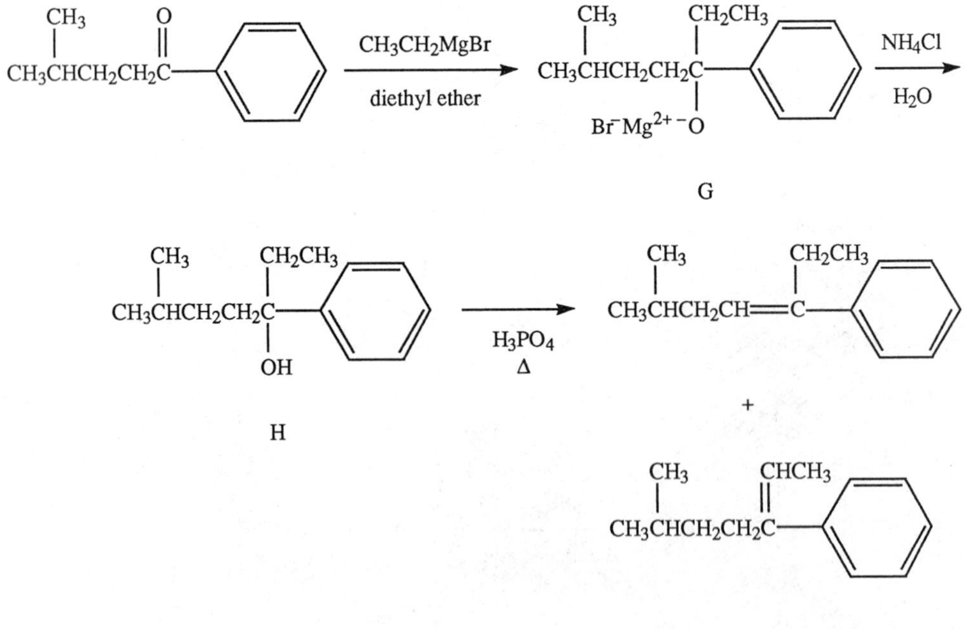

CH_3
O
$CH_3CHCH_2CH_2C$
CH_3CH_2MgBr
diethyl ether
CH_3
CH_2CH_3
$CH_3CHCH_2CH_2C$
$Br^-Mg^{2+}\,^-O$
G
NH_4Cl
H_2O
CH_3
CH_2CH_3
$CH_3CHCH_2CH_2C$
OH
H
H_3PO_4
Δ
CH_3
CH_2CH_3
$CH_3CHCH_2CH{=}C$
+
CH_3
$CHCH_3$
$CH_3CHCH_2CH_2C$

I

13.32 (cont)

(d)

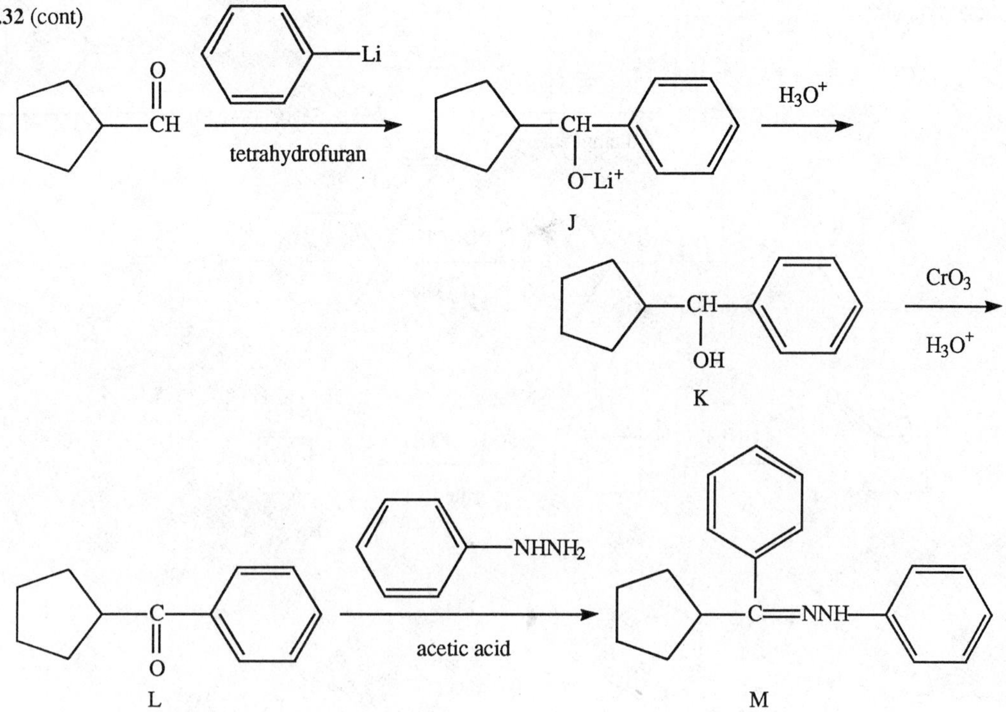

J

K

L

M

(e)

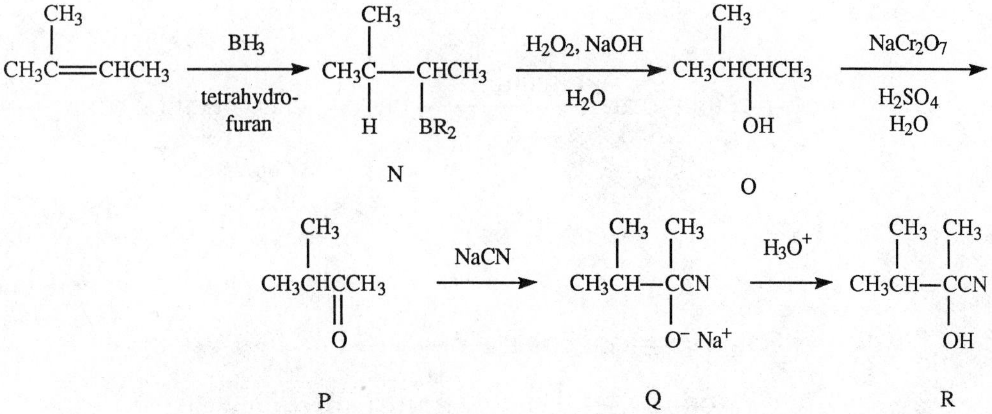

N

O

P

Q

R

(f) $HC\equiv CH$ $\xrightarrow[\text{NH}_3 \text{ (liq)}]{\text{NaNH}_2}$ $HC\equiv C^-Na^+$ $\xrightarrow[\text{diethyl ether}]{}$

S

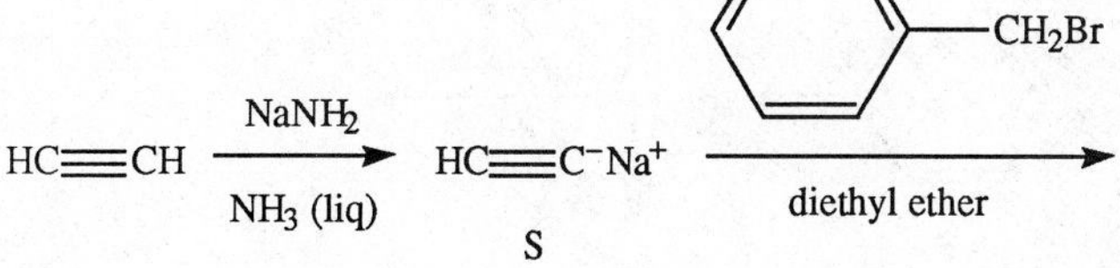

13.32 (f) (cont)

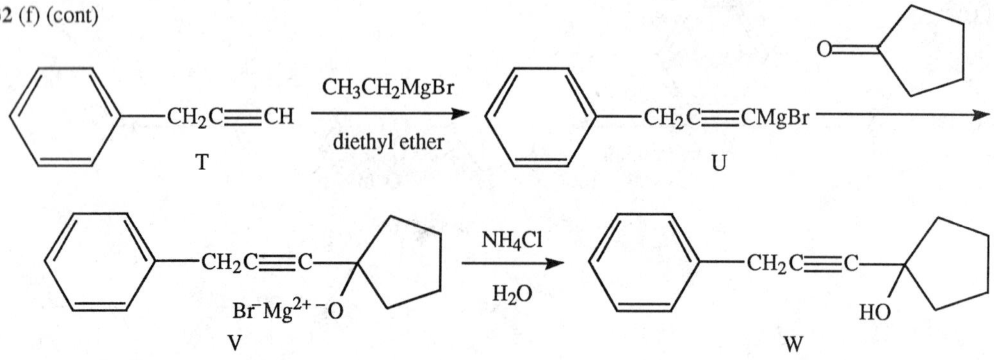
CH2C≡CH
T
CH3CH2MgBr
diethyl ether
CH2C≡CMgBr
U
CH2C≡C
Br⁻Mg2+ ⁻O
V
NH4Cl
H2O
CH2C≡C
HO
W

(g)

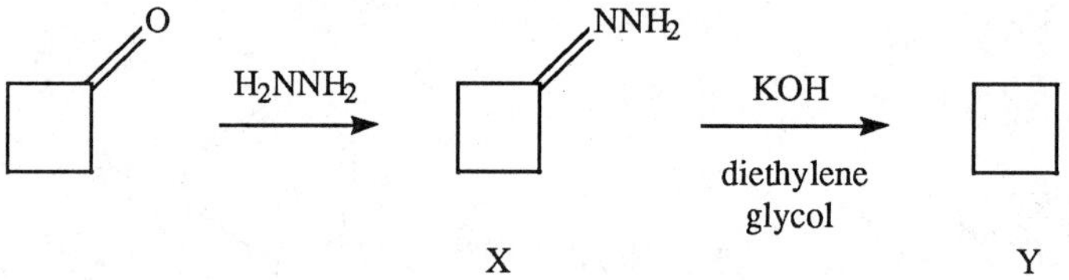
O
H2NNH2
NNH2
X
KOH
diethylene
glycol
Y

13.33

(a)

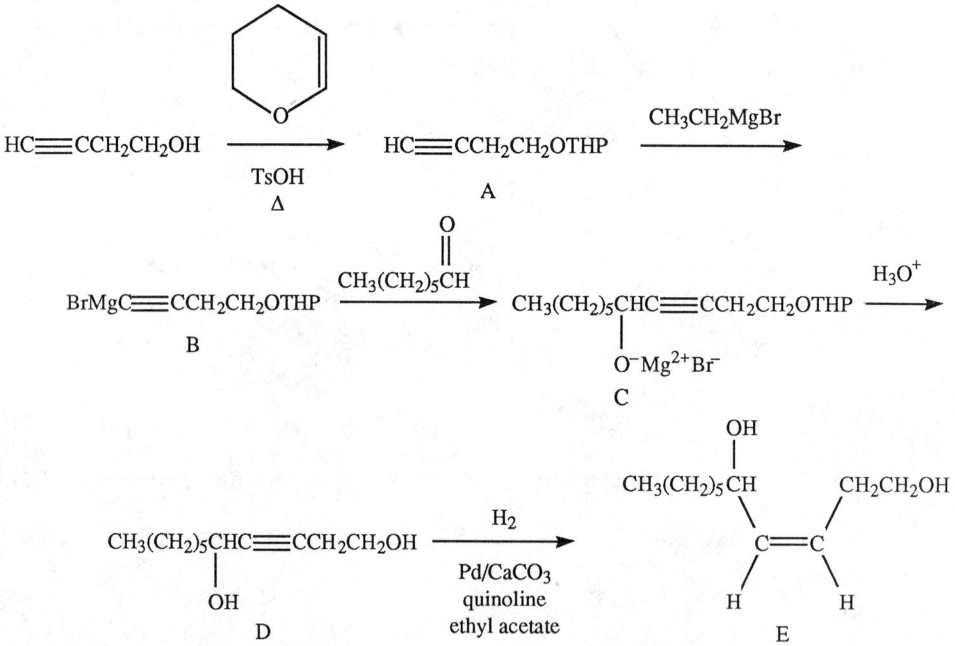
HC≡CCH2CH2OH
O
TsOH
Δ
HC≡CCH2CH2OTHP
A
CH3CH2MgBr
BrMgC≡CCH2CH2OTHP
B
CH3(CH2)5CH
CH3(CH2)5CHC≡CCH2CH2OTHP
O⁻Mg2+Br⁻
C
H3O+
CH3(CH2)5CHC≡CCH2CH2OH
OH
D
H2
Pd/CaCO3
quinoline
ethyl acetate
OH
CH3(CH2)5CH
CH2CH2OH
C=C
H
H
E

(b)

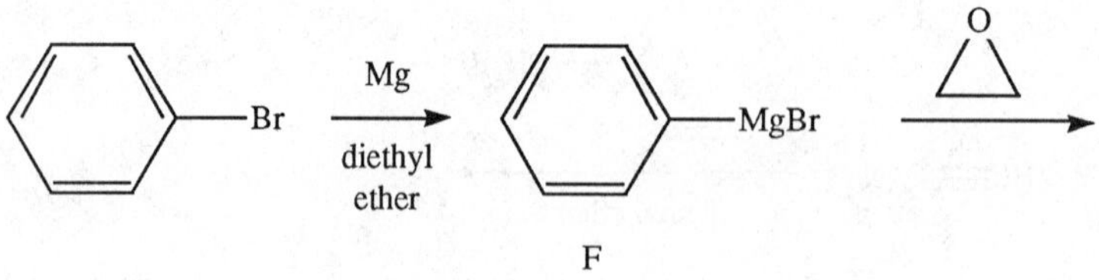
Br
Mg
diethyl
ether
MgBr
F
O

13.33 (cont)

(b) (cont)

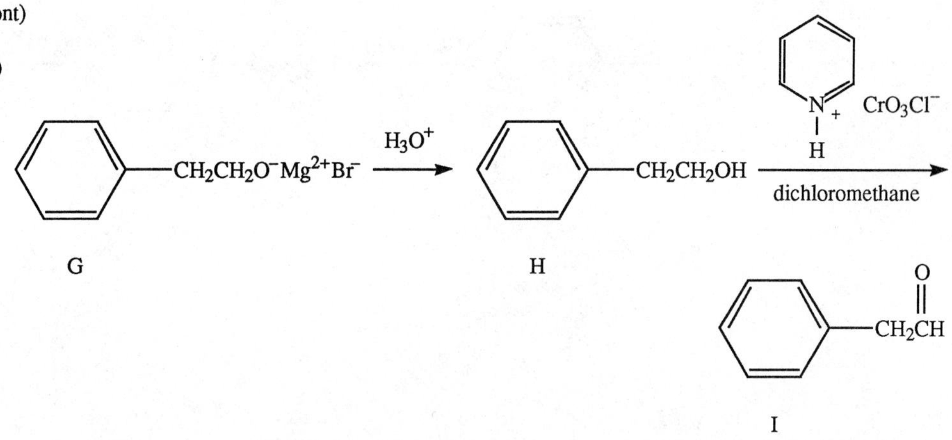

G

H

I

(c)

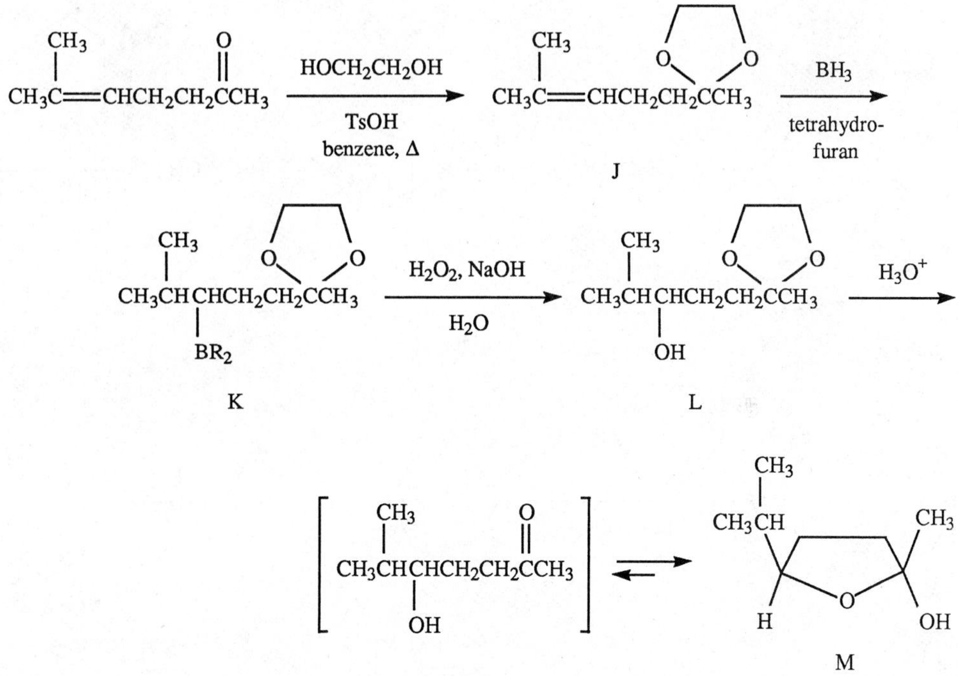

J

K

L

M

(d)

N

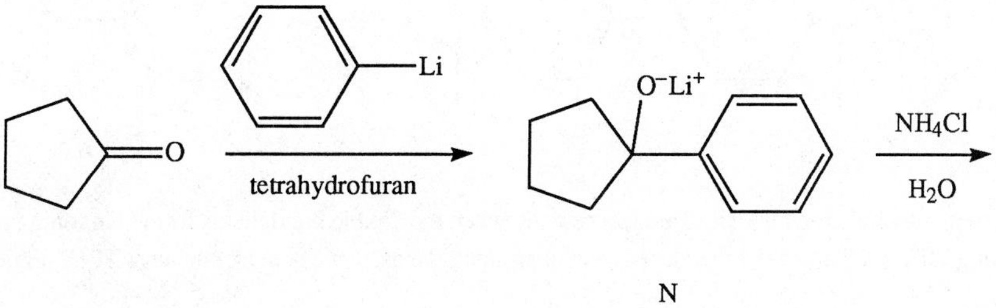

13.33 (cont)

(d) (cont)

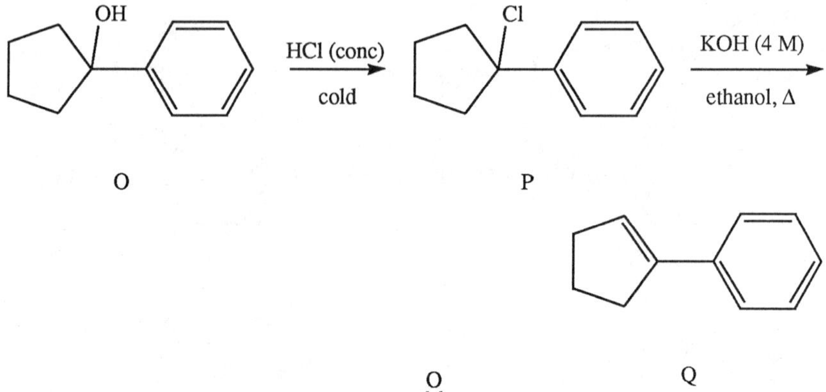

O

P

Q

(e)

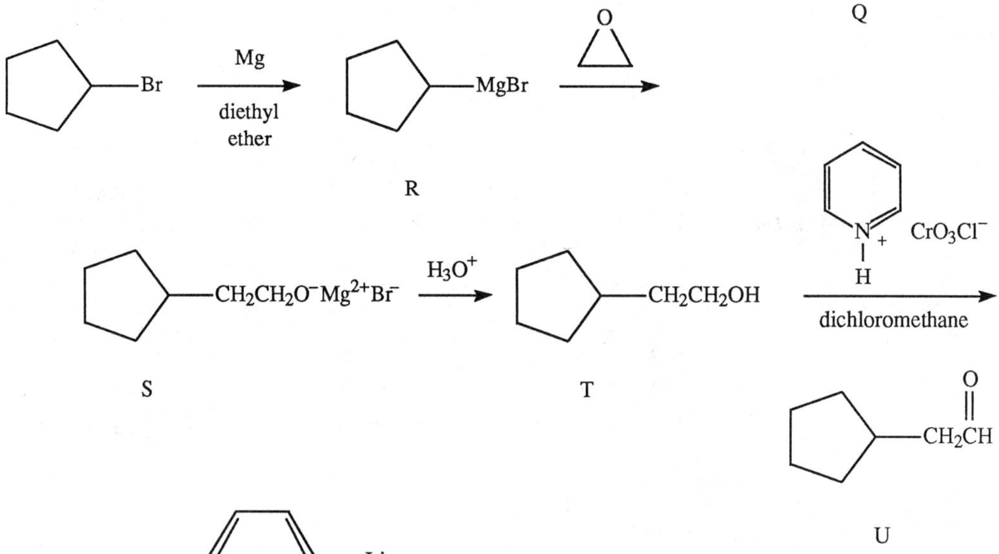

R

S

T

U

(f)

V

W

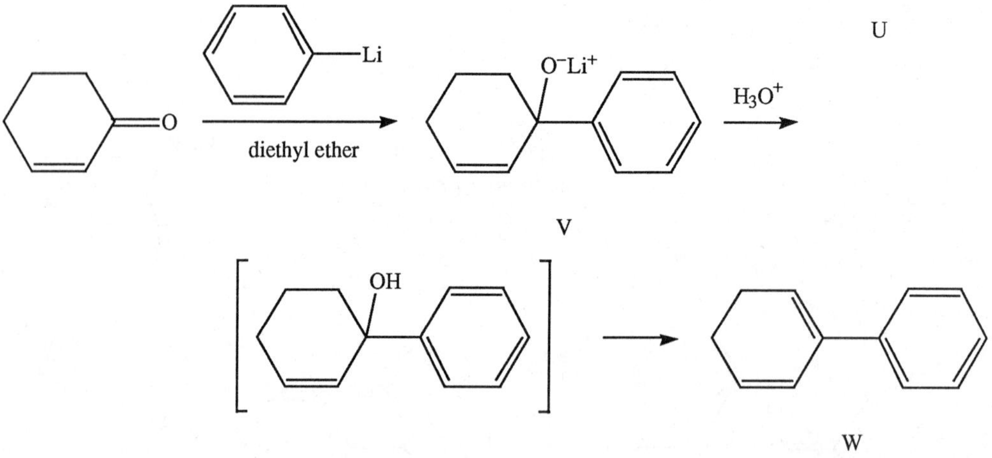

Note that a tertiary alcohol is easily dehydrated, especially when the double bond that is formed is conjugated with an aromatic ring. If we wish to have the alcohol as the product, we use a weak acid, such as NH_4Cl, to protonate the alkoxide.

13.33 (cont)

(g)

$$\xrightarrow[\substack{(CH_3CH_2)_2OBF_3 \\ \text{acetic acid}}]{HSCH_2CH_2SH}$$

X

(h) $CH_2{=}CHCH_2CH_2{-}$ $=O$ $\xrightarrow[\text{chloroform}]{Br_2}$

$BrCH_2\underset{\underset{Br}{|}}{C}HCH_2CH_2{-}$ $=O$ $\xrightarrow[\substack{TsOH \\ \text{benzene, }\Delta}]{HOCH_2CH_2OH}$ $BrCH_2\underset{\underset{Br}{|}}{C}HCH_2CH_2{-}$

Y   Z

(i) $\underset{\underset{CH_3}{|}}{CH_3}CHCH_2CH_2\overset{\overset{O}{\|}}{C}CH_2CH_2\overset{\overset{O}{\|}}{C}OH$ $\xrightarrow[\substack{\text{diethylene glycol} \\ \Delta}]{H_2NNH_2,\ NaOH}$

$\underset{\underset{CH_3}{|}}{CH_3}CHCH_2CH_2CH_2CH_2CH_2\overset{\overset{O}{\|}}{C}O^-Na^+$ $\xrightarrow{H_3O^+}$ $\underset{\underset{CH_3}{|}}{CH_3}CHCH_2CH_2CH_2CH_2CH_2\overset{\overset{O}{\|}}{C}OH$

AA   BB

(j) $CH_3CH_2CH_2CH_2\overset{\overset{O}{\|}}{C}{-}$ $\xrightarrow[\Delta]{HCl,\ Zn(Hg)}$ $CH_3CH_2CH_2CH_2CH_2{-}$

CC

13.33 (cont)

(k)

CH$_2$ ... CH$_3$Li / diethyl ether → ... H$_3$O$^+$ → ...

CH$_3$CHCH$_3$ CH   O

CH$_3$CHCH$_3$ CHCH$_3$ O$^-$Li$^+$

CH$_3$CHCH$_3$ CHCH$_3$ OH

DD   EE

13.34

(a)

$$\underset{O}{\overset{\parallel}{\text{C}_6\text{H}_5\text{—CH}_2\text{CCH}_3}} \xleftarrow[\text{H}_3\text{O}^+]{\text{Na}_2\text{Cr}_2\text{O}_7} \underset{\text{OH}}{\text{C}_6\text{H}_5\text{—CH}_2\text{CHCH}_3} \xleftarrow[\text{H}_2\text{O}]{\text{NH}_4\text{Cl}}$$

C$_6$H$_5$—CH$_2$CHCH$_3$ O$^-$Mg^{2+}Br$^-$ $\xleftarrow{\text{CH}_2\text{—CHCH}_3 \ (O)}$ C$_6$H$_5$—MgBr $\xleftarrow[\text{diethyl ether}]{\text{Mg}}$ C$_6$H$_5$—Br

(b)

$$\underset{\text{H} \quad \text{H}}{\overset{\text{CH}_3\text{CH}_2 \quad \text{CH}_2\text{CH}_2\text{OH}}{\text{C}=\text{C}}} \xleftarrow[\substack{\text{Pd/CaCO}_3 \\ \text{quinoline}}]{\text{H}_2} \text{CH}_3\text{CH}_2\text{C}\equiv\text{CCH}_2\text{CH}_2\text{OH} \xleftarrow{\text{H}_3\text{O}^+}$$

CH$_3$CH$_2$C≡CCH$_2$CH$_2$O$^-$Mg^{2+}Br$^-$ $\xleftarrow{\triangle\text{O}}$ CH$_3$CH$_2$C≡CMgBr $\xleftarrow{\text{CH}_3\text{CH}_2\text{MgBr}}$

CH$_3$CH$_2$C≡CH $\xleftarrow{\text{CH}_3\text{CH}_2\text{Br}}$ Na$^+$ $^-$C≡CH $\xleftarrow[\text{NH}_3 \text{ (liq)}]{\text{NaNH}_2}$ HC≡CH

(c)

$$\underset{O}{\overset{\parallel}{\text{CH}_3\text{CH}_2\text{CH}_2\text{CCH}_2\text{CH}_3}} \xleftarrow[\text{H}_3\text{O}^+]{\text{Na}_2\text{Cr}_2\text{O}_7} \underset{\text{OH}}{\text{CH}_3\text{CH}_2\text{CH}_2\text{CHCH}_2\text{CH}_3} \xleftarrow{\text{H}_3\text{O}^+}$$

CH$_3$CH$_2$CH$_2$CHCH$_2$CH$_3$ O$^-$Mg^{2+}Br$^-$ $\xleftarrow{\overset{O}{\overset{\parallel}{\text{CH}_3\text{CH}_2\text{CH}}}}$ CH$_3$CH$_2$CH$_2$MgBr $\xleftarrow[\substack{\text{diethyl} \\ \text{ether}}]{\text{Mg}}$ CH$_3$CH$_2$CH$_2$Br

13.34 (cont)

(d)

$$CH_3C{=}CHCH_2CH_2CH_3 \xleftarrow[\substack{H_3PO_4 \\ \Delta}]{} CH_3\overset{\overset{CH_3}{|}}{\underset{\underset{OH}{|}}{C}}CH_2CH_2CH_2CH_3 \xleftarrow{H_3O^+}$$

major product

$$CH_3\overset{\overset{CH_3}{|}}{\underset{\underset{O^-Mg^{2+}Br^-}{|}}{C}}CH_2CH_2CH_2CH_3 \xleftarrow{CH_3\overset{O}{\overset{||}{C}}CH_3} CH_3CH_2CH_2CH_2MgBr \xleftarrow[\substack{\text{diethyl} \\ \text{ether}}]{Mg}$$

$$CH_3CH_2CH_2CH_2Br \xleftarrow[\text{pyridine}]{PBr_3} CH_3CH_2CH_2CH_2OH \xleftarrow{H_3O^+}$$

$$CH_3CH_2CH_2CH_2O^-Mg^{2+}Br^- \xleftarrow{\triangle O} CH_3CH_2MgBr \xleftarrow[\substack{\text{diethyl} \\ \text{ether}}]{Mg} CH_3CH_2Br$$

(e)

$$\text{Ph}{-}CH_2CH_2Br \xleftarrow[\text{pyridine}]{PBr_3} \text{Ph}{-}CH_2CH_2OH \xleftarrow{H_3O^+}$$

$$\text{Ph}{-}CH_2CH_2O^-Mg^{2+}Br^- \xleftarrow{\triangle O} \text{Ph}{-}MgBr \xleftarrow[\substack{\text{diethyl} \\ \text{ether}}]{Mg} \text{Ph}{-}Br$$

(f)

$$\begin{array}{c}CH_3CH_2CH_2 \quad\quad Br \\ \underset{\underset{Br}{|}}{\overset{H}{\diagup}}C{-}C\overset{H}{\underset{CH_3}{\diagdown}}\end{array} \xleftarrow[\substack{\text{carbon} \\ \text{tetrachloride}}]{Br_2} \begin{array}{c}CH_3CH_2 \quad\quad H \\ C{=}C \\ H \quad\quad CH_3\end{array} \xleftarrow[\text{NH}_3\text{ (liq)}]{Na}$$

and enantiomer

$$CH_3CH_2CH_2C{\equiv}CCH_3 \xleftarrow{CH_3I} CH_3CH_2CH_2C{\equiv}C^-Na^+ \xleftarrow[\text{NH}_3\text{ (liq)}]{NaNH_2}$$

$$CH_3CH_2CH_2C{\equiv}CH \xleftarrow{CH_3CH_2CH_2Br} Na^{+-}C{\equiv}CH \xleftarrow[\text{NH}_3\text{ (liq)}]{NaNH_2} HC{\equiv}CH$$

13.34 (cont)

(g)

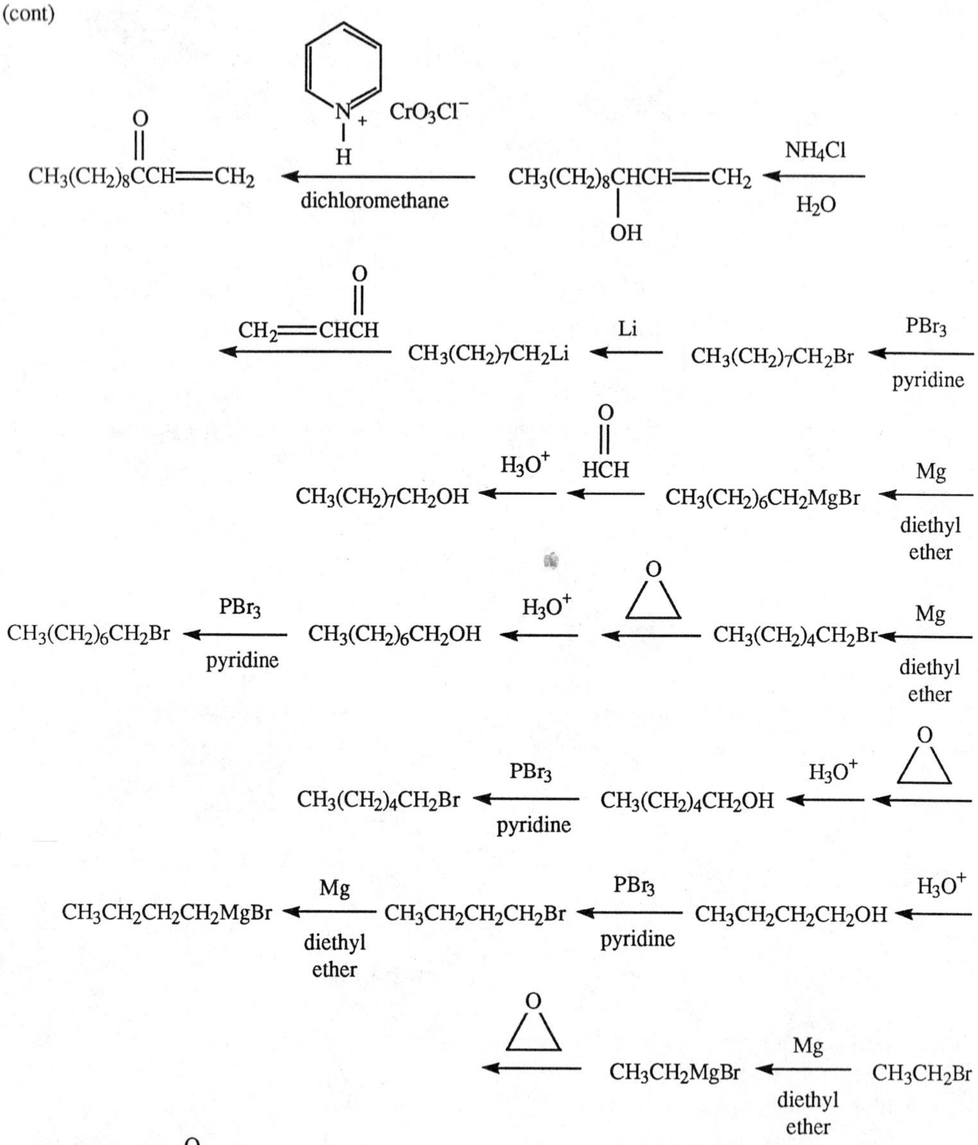
CrO3Cl−
H
CH3(CH2)8CCH═CH2
dichloromethane
CH3(CH2)8CHCH═CH2
OH
NH4Cl
H2O
CH2═CHCH
CH3(CH2)7CH2Li
Li
CH3(CH2)7CH2Br
PBr3
pyridine
H3O+
HCH
CH3(CH2)7CH2OH
CH3(CH2)6CH2MgBr
Mg
diethyl ether
PBr3
CH3(CH2)6CH2Br
pyridine
CH3(CH2)6CH2OH
H3O+
CH3(CH2)4CH2Br
Mg
diethyl ether
PBr3
CH3(CH2)4CH2Br
pyridine
CH3(CH2)4CH2OH
H3O+
Mg
CH3CH2CH2CH2MgBr
diethyl ether
CH3CH2CH2CH2Br
PBr3
pyridine
CH3CH2CH2CH2OH
H3O+
CH3CH2MgBr
Mg
CH3CH2Br
diethyl ether

(h)

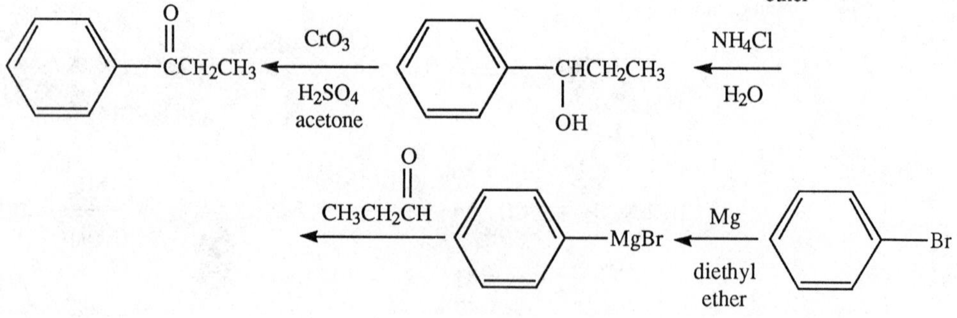
CCH2CH3
CrO3
H2SO4
acetone
CHCH2CH3
OH
NH4Cl
H2O
CH3CH2CH
MgBr
Mg
diethyl ether
Br

13.34 (cont)

(i)

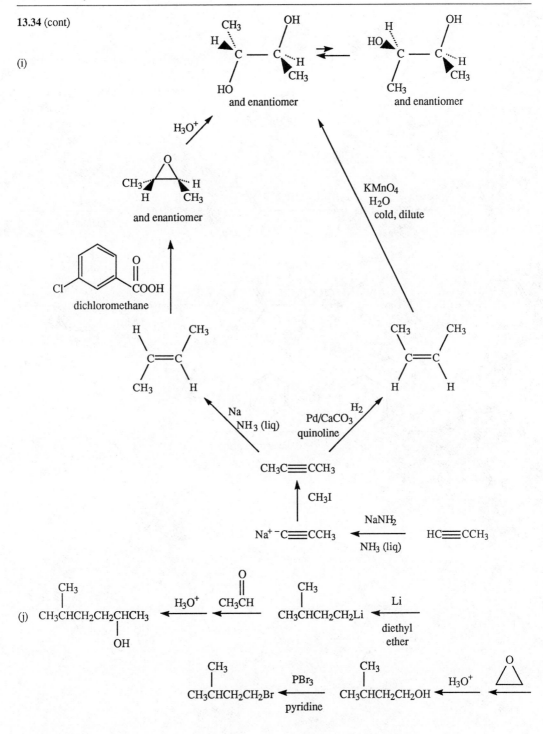

(j)

13.34 (cont)

(j) (cont)

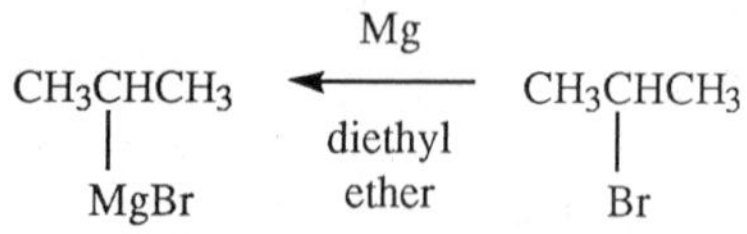

$$\text{CH}_3\text{CHCH}_3 \xleftarrow[\text{diethyl} \atop \text{ether}]{\text{Mg}} \text{CH}_3\text{CHCH}_3$$

with MgBr below the left structure and Br below the right structure.

(k)

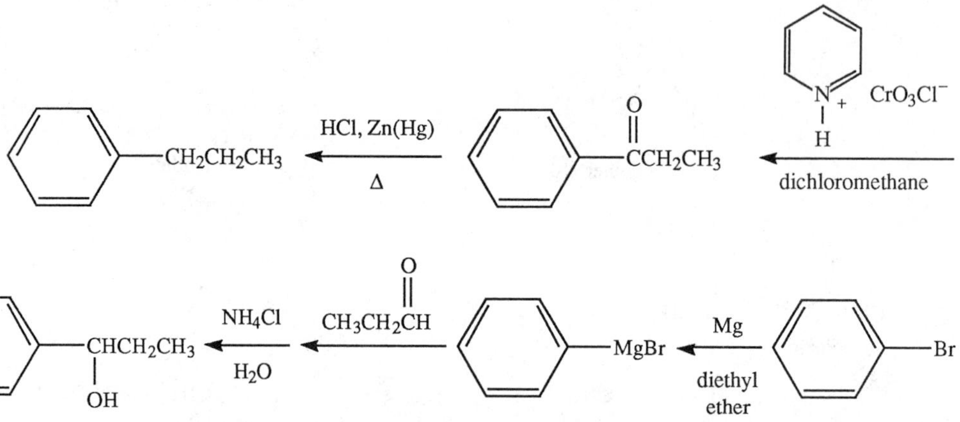

Phenyl—CH₂CH₂CH₃ $\xleftarrow[\Delta]{\text{HCl, Zn(Hg)}}$ Phenyl—C(=O)CH₂CH₃ $\xleftarrow[\text{dichloromethane}]{}$ [pyridinium chlorochromate, PyH⁺ CrO₃Cl⁻]

Phenyl—CHCH₂CH₃ with OH below $\xleftarrow[\text{H}_2\text{O}]{\text{NH}_4\text{Cl}}$ CH₃CH₂CH(=O) + Phenyl—MgBr $\xleftarrow[\text{diethyl} \atop \text{ether}]{\text{Mg}}$ Phenyl—Br

(l)

[epoxide: 1,2-diphenyloxirane with H's] $\xleftarrow[\text{dichloromethane}]{}$ [3-chloroperoxybenzoic acid: Cl-substituted benzene with COOH]

[trans-stilbene, C=C with phenyl and H groups] $\xleftarrow[\Delta]{\text{H}_2\text{SO}_4}$ Phenyl—CH₂CH—Phenyl with OH $\xleftarrow{\text{H}_3\text{O}^+}$

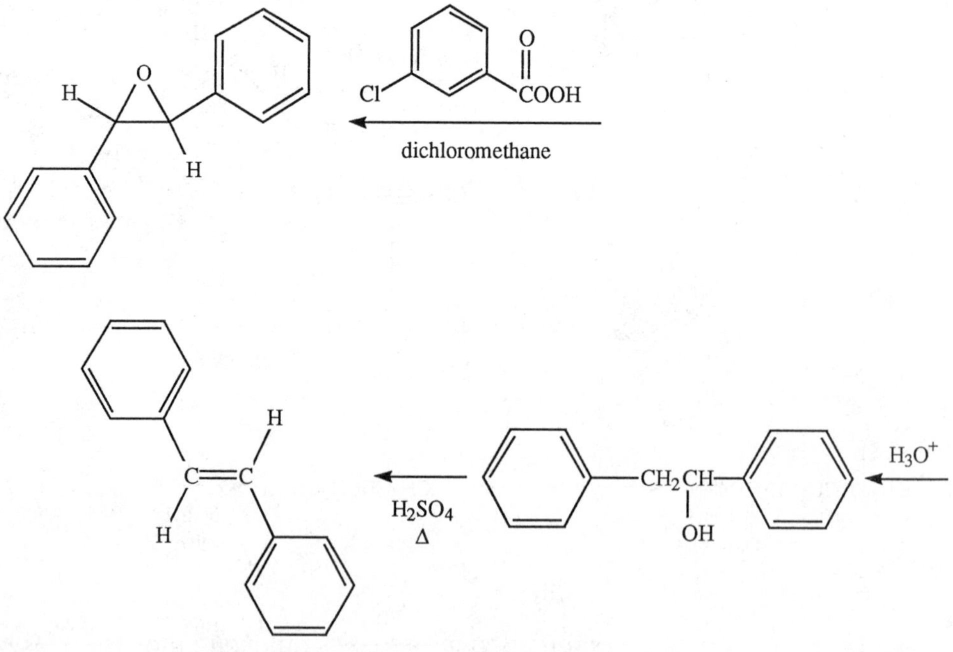

13.34 (cont)

(l) (cont)

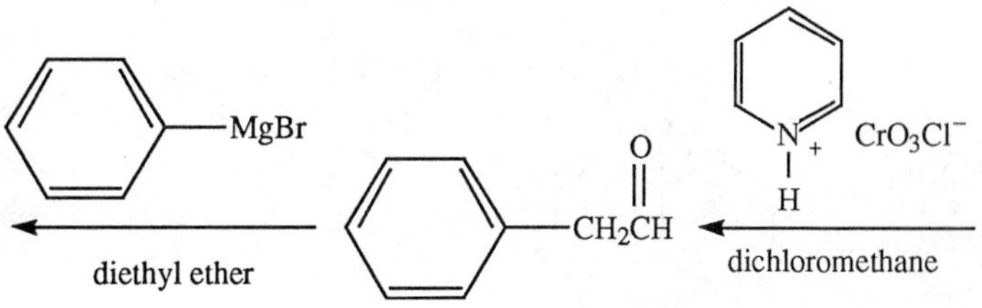

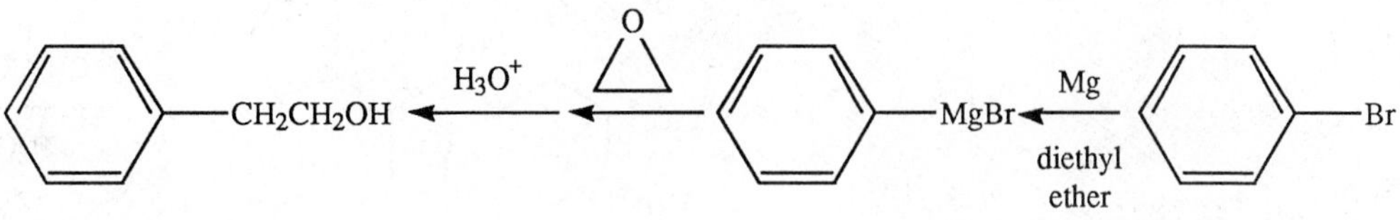

13.35

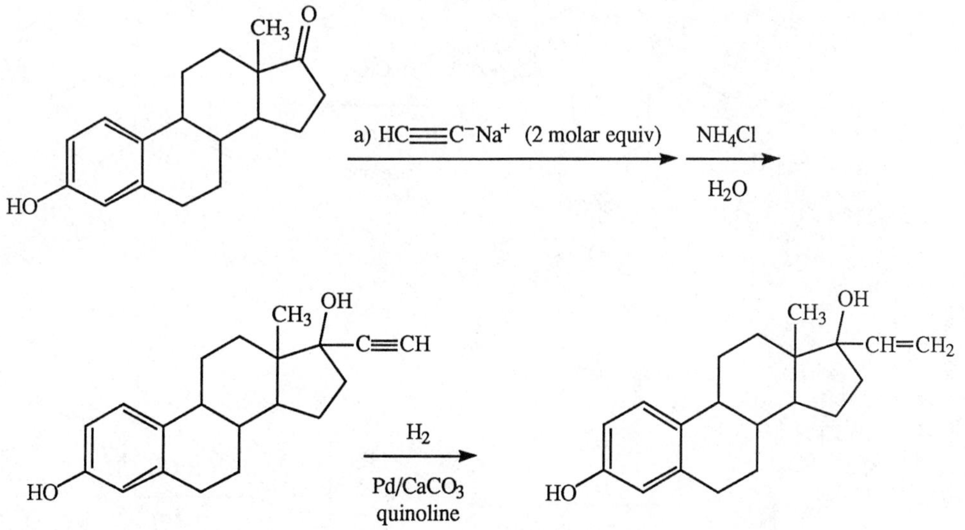

a) HC≡C⁻Na⁺ (2 molar equiv)   NH₄Cl

H₂O

H₂

Pd/CaCO₃
quinoline

13.36

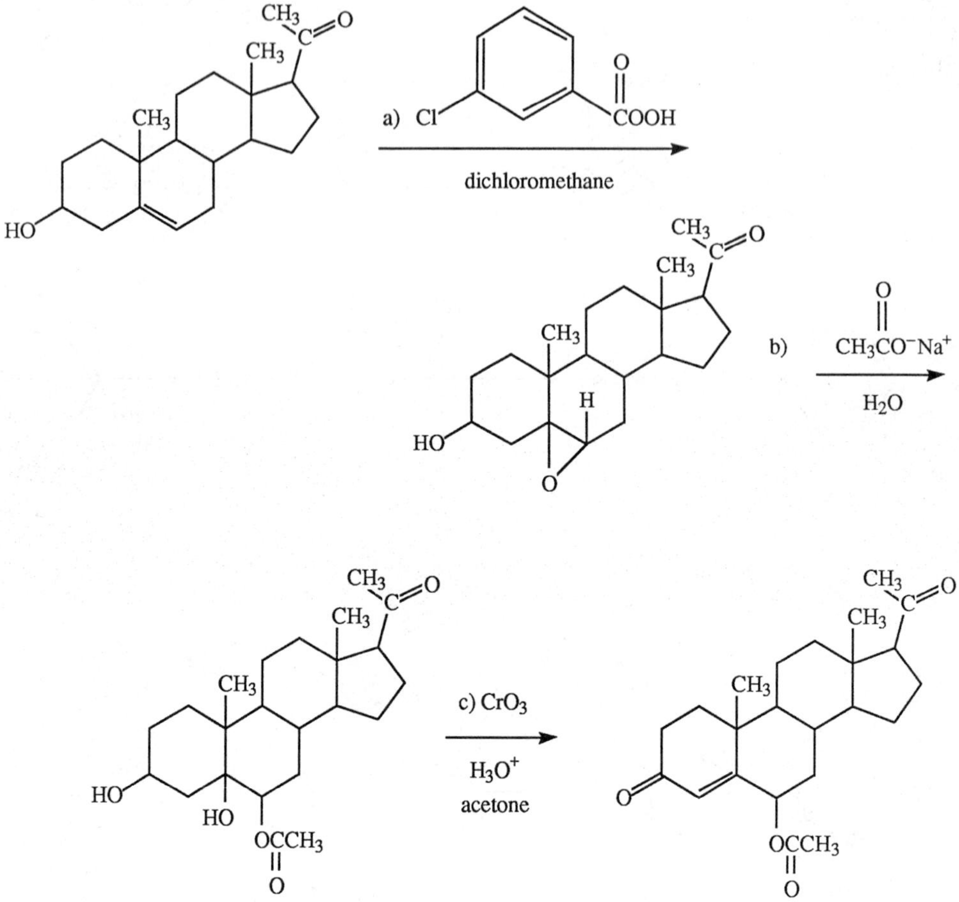

a) Cl — (3-chlorobenzoic acid, COOH)

dichloromethane

b) $CH_3CO^-Na^+$

H_2O

c) CrO_3

H_3O^+

acetone

13.37

pyridinium CrO_3Cl^-

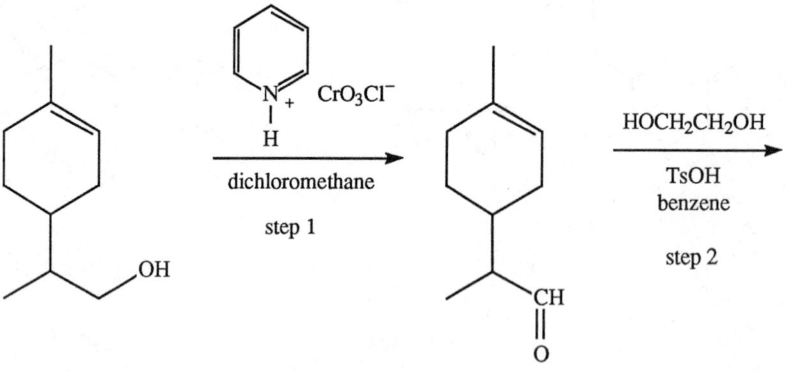

dichloromethane

step 1

$HOCH_2CH_2OH$

TsOH
benzene

step 2

A

13.37 (cont)

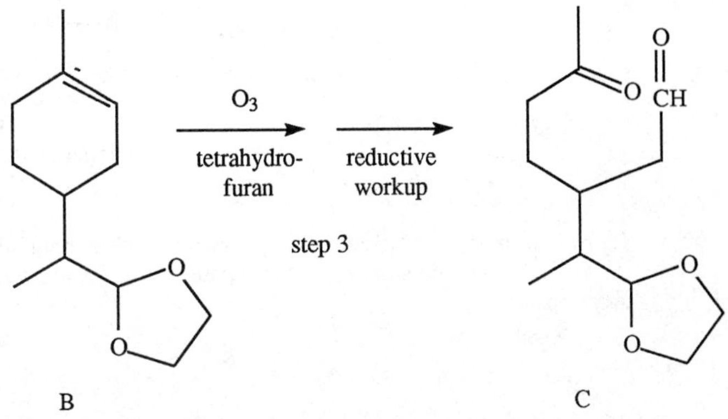

B   C

Step 2 was necessary to protect the aldehyde group from oxidation during ozonolysis.

13.38

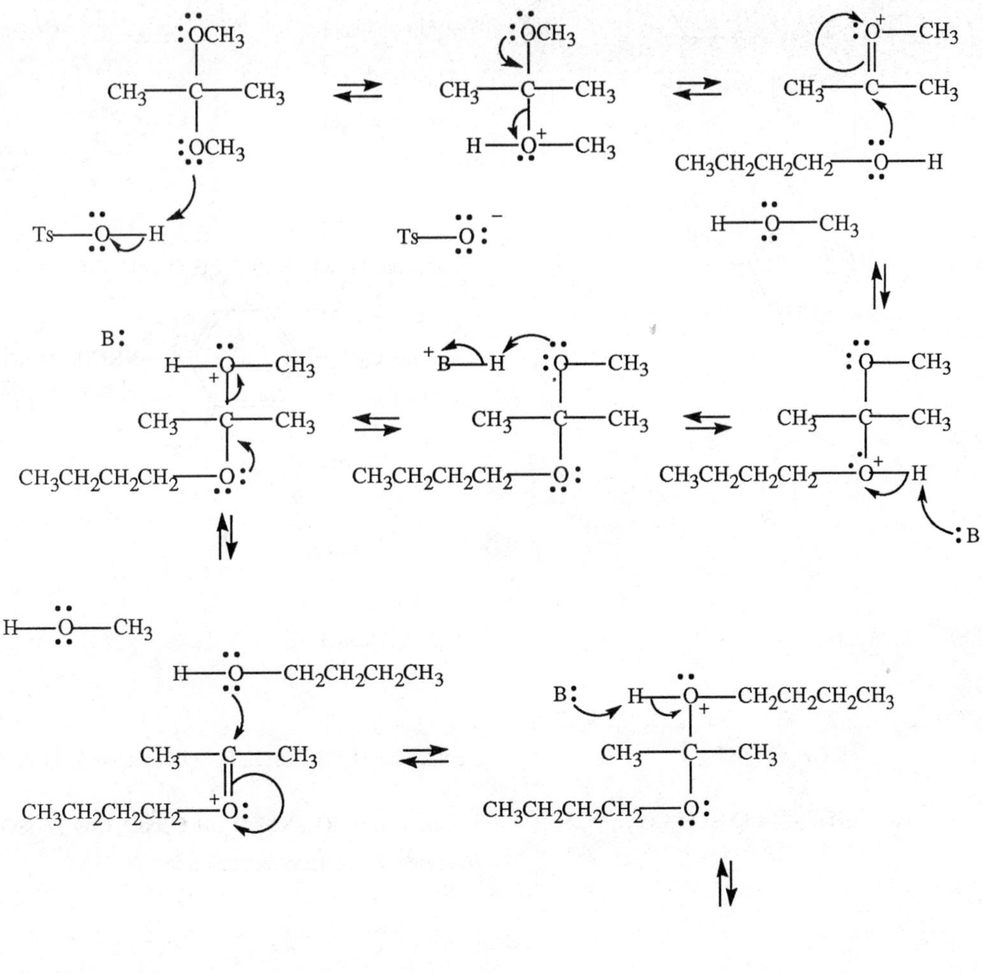

13.38 (cont)

$$^+B\text{---}H$$

$$\ddot{\underset{..}{O}}\text{---}CH_2CH_2CH_2CH_3$$

$$CH_3\text{---}\underset{|}{\overset{|}{C}}\text{---}CH_3$$

$$CH_3CH_2CH_2CH_2\text{---}\ddot{\underset{..}{O}}\!:$$

The reaction proceeds by a series of steps, each of which is an equilibrium. Removal by distillation of methanol, the lowest boiling component of the mixture, as it forms, will push the equilibrium towards the dibutoxypropane.

13.39

(a) $CH_3CH_2CH_2OH$ cold, conc H_2SO_4; CrO_3, H_2SO_4, acetone; H_2O, cold

(b) CH_3-⬡$=O$ cold, conc H_2SO_4; O_2N-⟨benzene⟩$-NHNH_2$,
                                          H_2SO_4, ethanol
                                          with NO_2

(c)   ⬠ with CH_2CH_3 and OH cold, conc H_2SO_4

(d)   ⬡$-CH=O$ cold, conc H_2SO_4; $KMnO_4$, H_2O, cold; CrO_3, H_2SO_4,

                                          acetone; O_2N-⟨benzene⟩$-NHNH_2$, H_2SO_4,
                                          ethanol
                                          with NO_2

(e)   ⬡   none

(f) $CH_3CH_2CH_2\underset{\underset{Br}{|}}{\overset{\overset{CH_3}{|}}{C}}CH_3$ $AgNO_3$, ethanol

(g) $HOCH_2CH_2OH$ cold, conc H_2SO_4; CrO_3, H_2SO_4, acetone; H_2O, cold

(h) $CH_3C\equiv CCH_2CH_2OH$ cold, conc H_2SO_4; $KMnO_4$, H_2O, cold; CrO_3, H_2SO_4,
                                          acetone; Br_2, carbon tetrachloride

13.39 (cont)

(i)

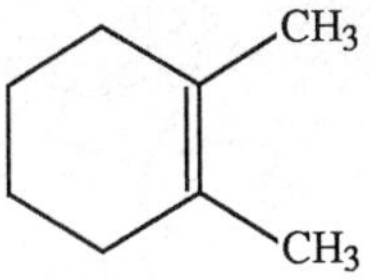

cold, conc H_2SO_4; $KMnO_4$, H_2O, cold; Br_2, carbon tetrachloride

(j)

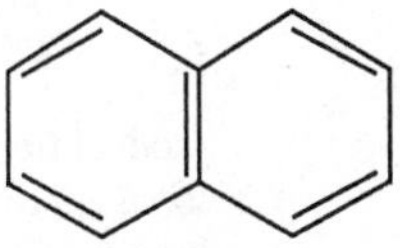

(k)

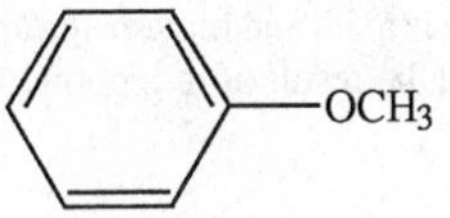

—OCH_3

cold, conc H_2SO_4

(l)

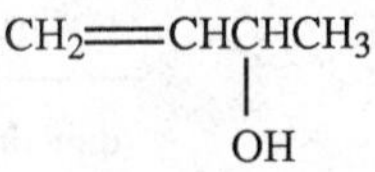

CH_2=$CHCHCH_3$
$\quad\quad\quad |$
$\quad\quad\quad OH$

cold, conc H_2SO_4; $KMnO_4$, H_2O, cold; CrO_3, H_2SO_4, acetone; Br_2, carbon tetrachloride; H_2O, cold

(m)

$CH_3CH_2CH_2Br$

NaI, acetone

(n)

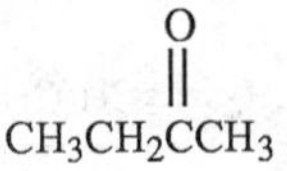

$\quad\quad\quad\; O$
$\quad\quad\quad\; ||$
$CH_3CH_2CCH_3$

cold, conc H_2SO_4; H_2O, cold;

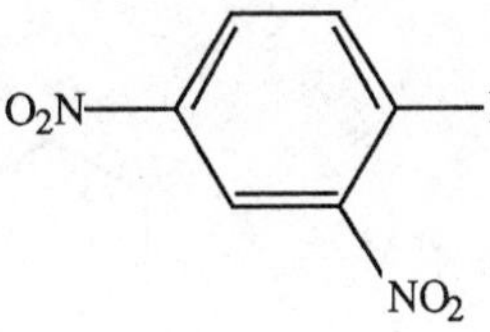

O_2N——$NHNH_2$, H_2SO_4, ethanol
$\quad\quad\quad\quad\quad\quad\quad\quad\; NO_2$

13.40

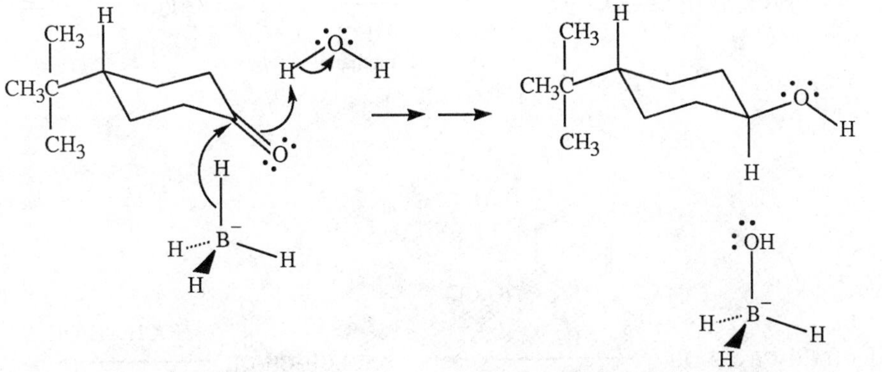

The large *tert*-butyl group occupies an equatorial position on the ring (Section 5.9 C). Reduction takes place so that the major product has the hydroxyl group also in the equatorial position. This means that the hydride is delivered to the carbonyl group from the axial direction.

13.40 (cont)

48%

product from axial
approach of the
borohydride anion

52%

product from equa-
torial approach of the
borohydride anion

In 3,3,5-trimethylcyclohexanone, one of the methyl groups on carbon 3 is axial and hinders the approach of the borohydride anion from the axial direction. The major product is now the result of reduction of the carbonyl group from the less hindered side of the molecule.

13.41 $HC\!\equiv\!CCH_2OH$ +   (dihydropyran) $\xrightarrow{HCl}$ $HC\!\equiv\!CCH_2O$—(tetrahydropyranyl) $\xrightarrow[\text{dimethyl sulfoxide}]{NaH}$

A

$Na^+\ ^-C\!\equiv\!CCH_2O$—(tetrahydropyranyl) $\xrightarrow{CH_2\!=\!CH(CH_2)_8CH_2OTs}$

B

$CH_2\!=\!CH(CH_2)_8CH_2C\!\equiv\!CCH_2O$—(tetrahydropyranyl) $\xrightarrow[\substack{HCl\\ \text{methanol}}]{H_2O}$ $CH_2\!=\!CH(CH_2)_8CH_2C\!\equiv\!CCH_2OH$

C

D

13.42 $BrCH_2CH_2CH_2CH_2OH$ $\xrightarrow[\text{dichloromethane}]{\text{(pyridinium)}\ CrO_3Cl^-}$ $BrCH_2CH_2CH_2\overset{\displaystyle O}{\overset{\|}{C}}H$ $\xrightarrow[\substack{TsOH\\ \text{benzene, }\Delta}]{HOCH_2CH_2OH}$

A

13.42 (cont)

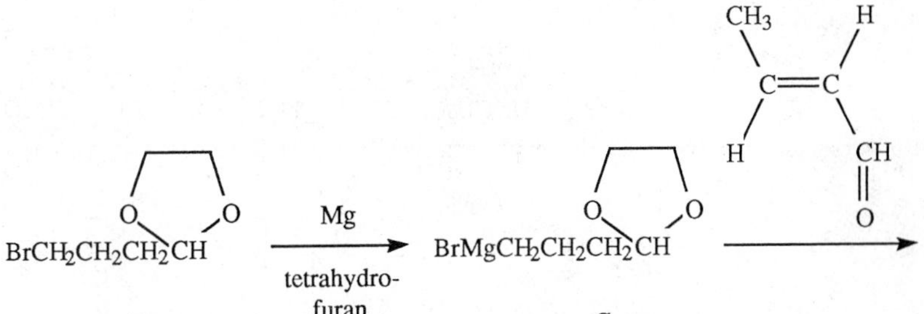

B

C

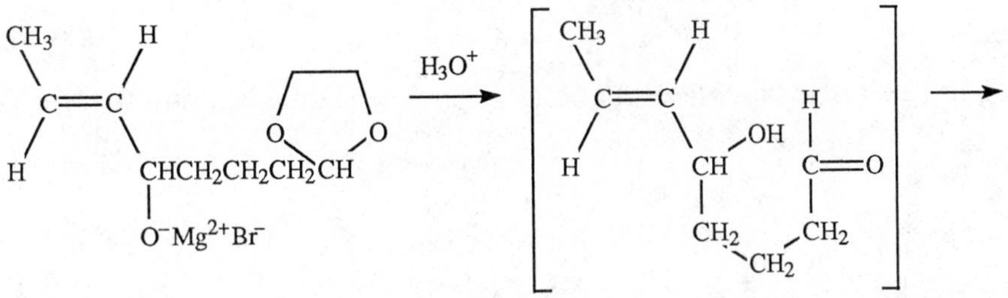

D

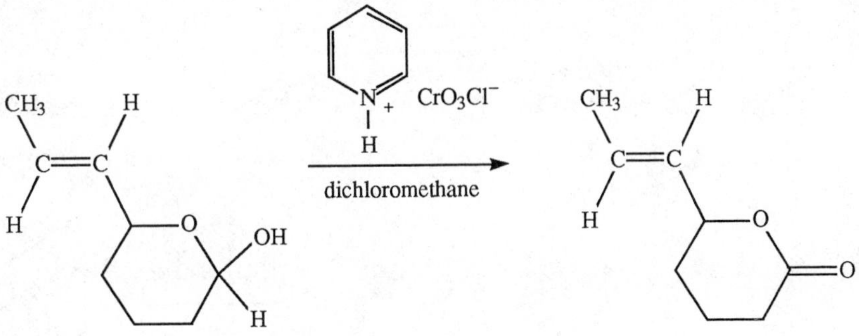

E

$C_8H_{14}O_2$

a stable, cyclic hemiacetal

F

$C_8H_{12}O_2$

three units of unsaturation
(2 double bonds and a ring)

13.43

(a)

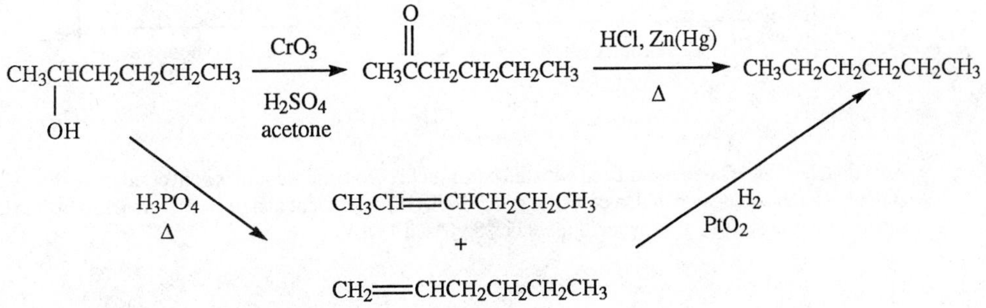

13.43 (cont)

(b)
$$
\underset{\substack{\text{CH}_3 \\ |}}{\text{CH}_3\text{C}}=\text{CHCH}_2\text{CH}_2\overset{\underset{\displaystyle ||}{\text{O}}}{\text{CCH}_3}
\xrightarrow[\substack{\text{TsOH} \\ \text{toluene, } \Delta}]{\text{HOCH}_2\text{CH}_2\text{OH}}
\text{CH}_3\text{C}=\text{CHCH}_2\text{CH}_2\text{C} \quad
\xrightarrow[\substack{\text{H}_3\text{O}^+}]{\underset{\text{Zn}}{\text{O}_3}}
$$

$$
\underset{\displaystyle ||}{\overset{\displaystyle O}{\text{CH}_3\text{CCH}_3}} \quad + \quad \text{HCCH}_2\text{CH}_2\text{CCH}_3
$$

(c)
$$
\text{CH}_2=\text{CHCH}_2\text{CH}_2\overset{\underset{||}{O}}{C}\text{CH}=\text{CHCOCH}_3
\xrightarrow[\substack{\text{TsOH} \\ \text{toluene, } \Delta}]{\text{CH}_3\text{OH}}
\text{CH}_2=\text{CHCH}_2\text{CH}_2\overset{\overset{\text{OCH}_3}{|}}{\underset{\underset{\text{OCH}_3}{|}}{C}}\text{CH}=\text{CHCOCH}_3
$$

(d)
$$
\xrightarrow[\substack{\text{TsOH} \\ \text{toluene, } \Delta}]{\text{HOCH}_2\text{CH}_2\text{OH}} \qquad \xrightarrow[\substack{\text{tetrahydro-} \\ \text{furan}}]{\text{BH}_3}
$$

$$
\xrightarrow[\text{H}_2\text{O}]{\text{H}_2\text{O}_2, \text{NaOH}}
$$

(e)
$$
\xrightarrow[\text{dichloromethane}]{} \qquad \xrightarrow[\text{dichloromethane}]{\text{CrO}_3\text{Cl}^-}
$$

Note that the double bond is oxidized with the peroxyacid before the alcohol is converted to a carbonyl group. A double bond conjugated with a carbonyl group is electron-deficient and not easily attacked by an electron-seeking reagent such as a peroxyacid (Sections 8.9C and 18.1A).

13.43 (cont)

(f)

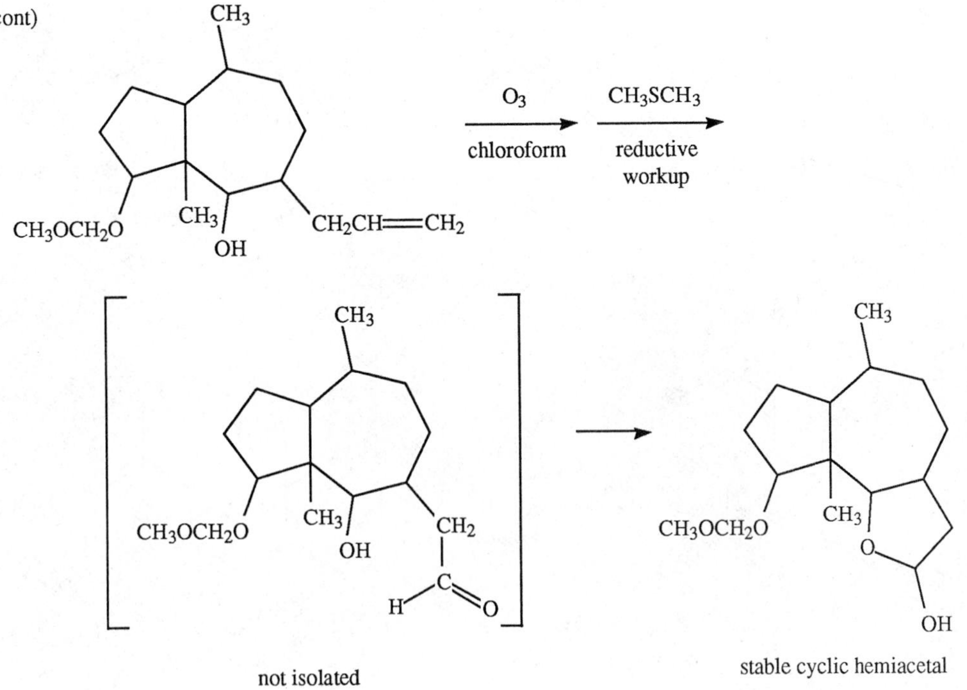

CH3
CH3OCH2O
CH3
OH
CH2CH=CH2
O3
chloroform
CH3SCH3
reductive workup
CH3
CH3OCH2O
CH3
OH
CH2
C
H
O
not isolated
CH3
CH3OCH2O
CH3
O
OH
stable cyclic hemiacetal

(g)

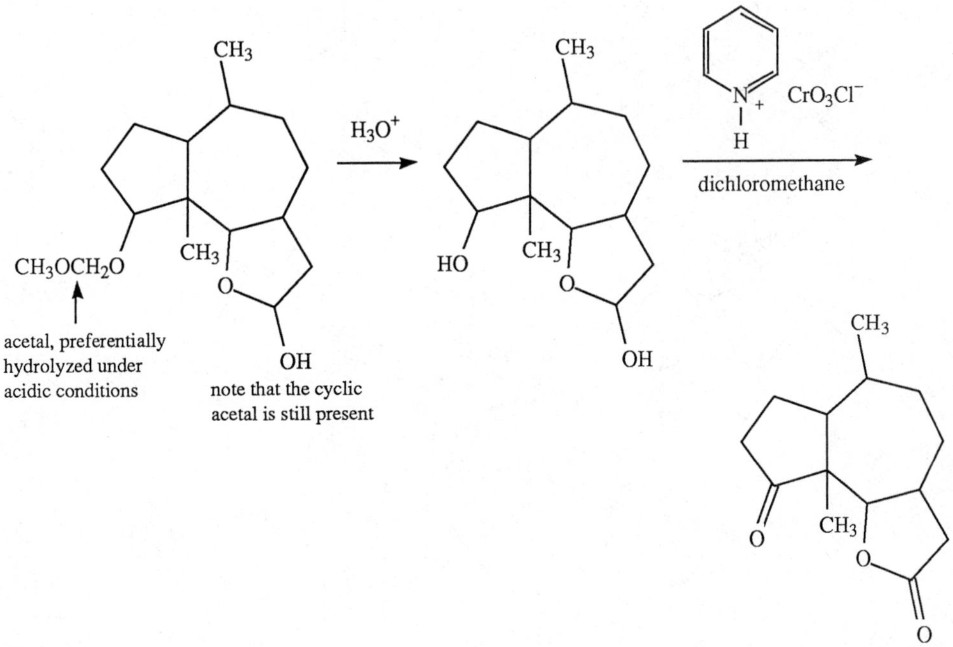

CH3
CH3OCH2O
CH3
O
OH
acetal, preferentially hydrolyzed under acidic conditions
note that the cyclic acetal is still present
H3O+
CH3
HO
CH3
O
OH
N+ CrO3Cl−
H
dichloromethane
CH3
O
CH3
O
O

13.43 (cont)

(h)

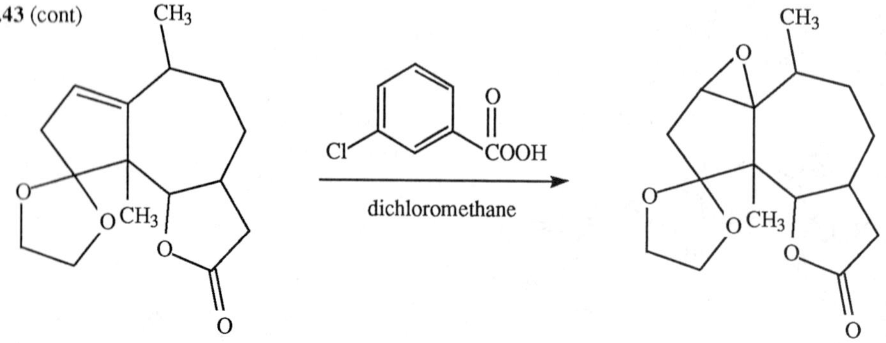

(i)

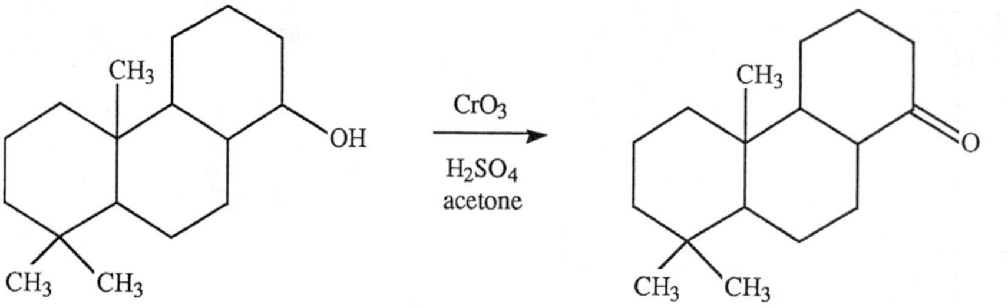

(j)

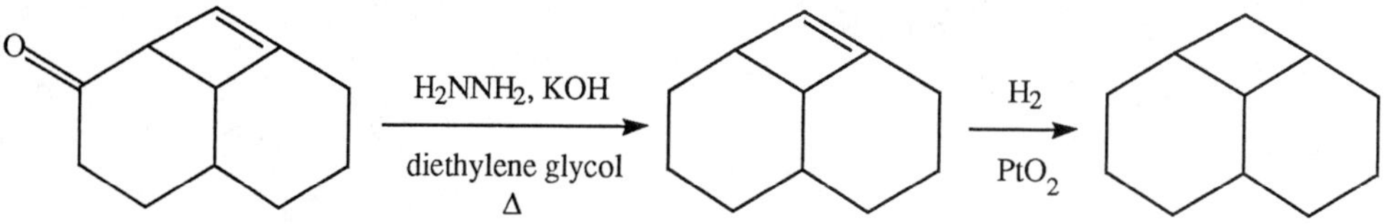

13.44

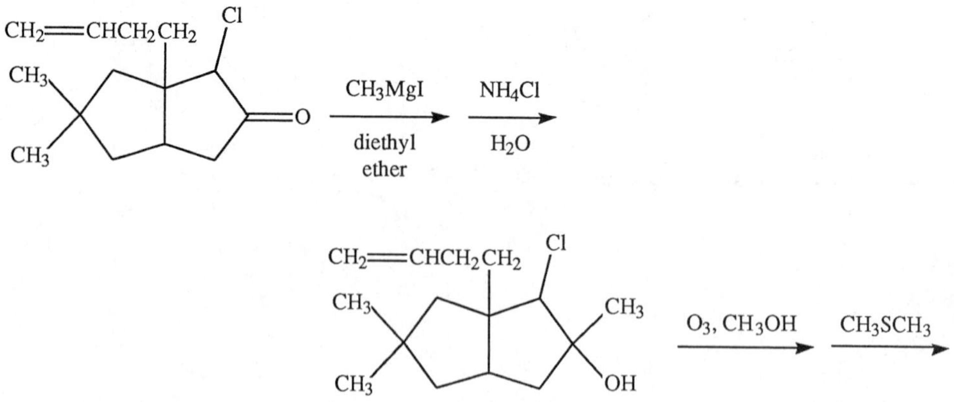

13.44 (cont)

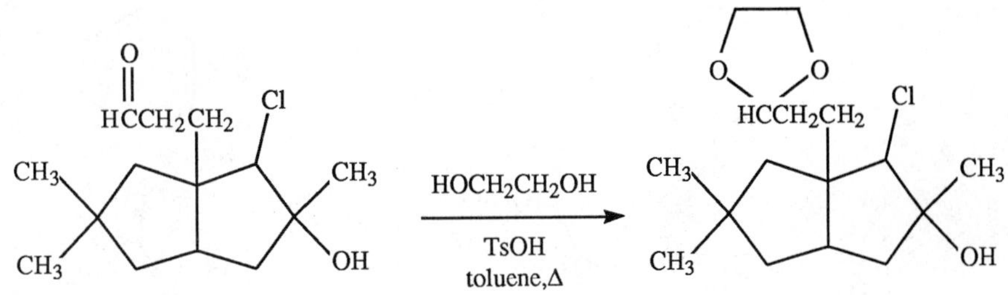

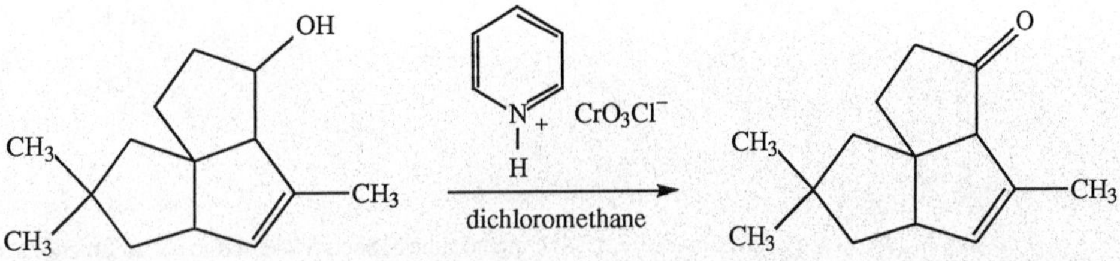

13.45

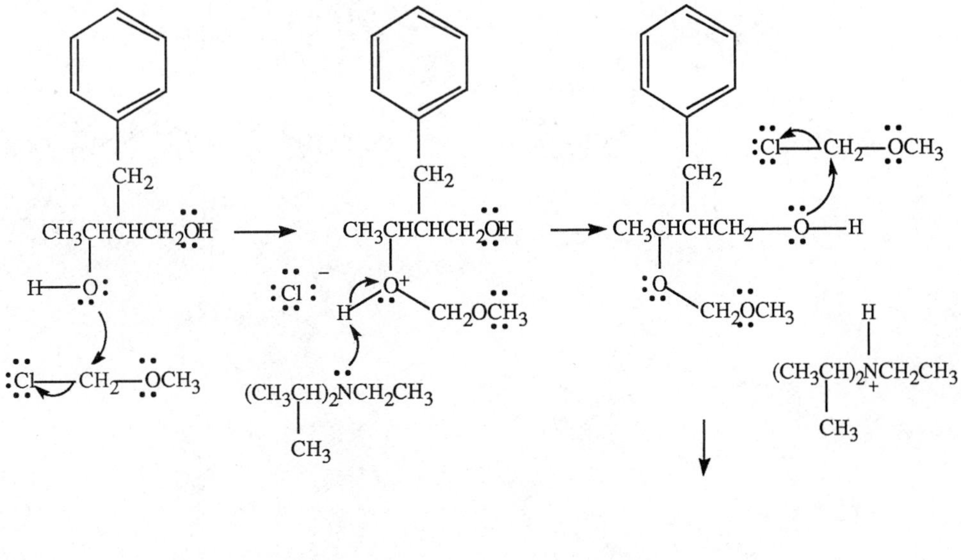

13.45 (cont)

Note that the two hydroxyl groups in the starting material have been converted to acetal functions, acetals of formaldedhye (methanal). The next step is a transacetalization reaction, a conversion of the open-chain acetals to a more stable cyclic acetal of formaldehyde.

13.45 (cont)

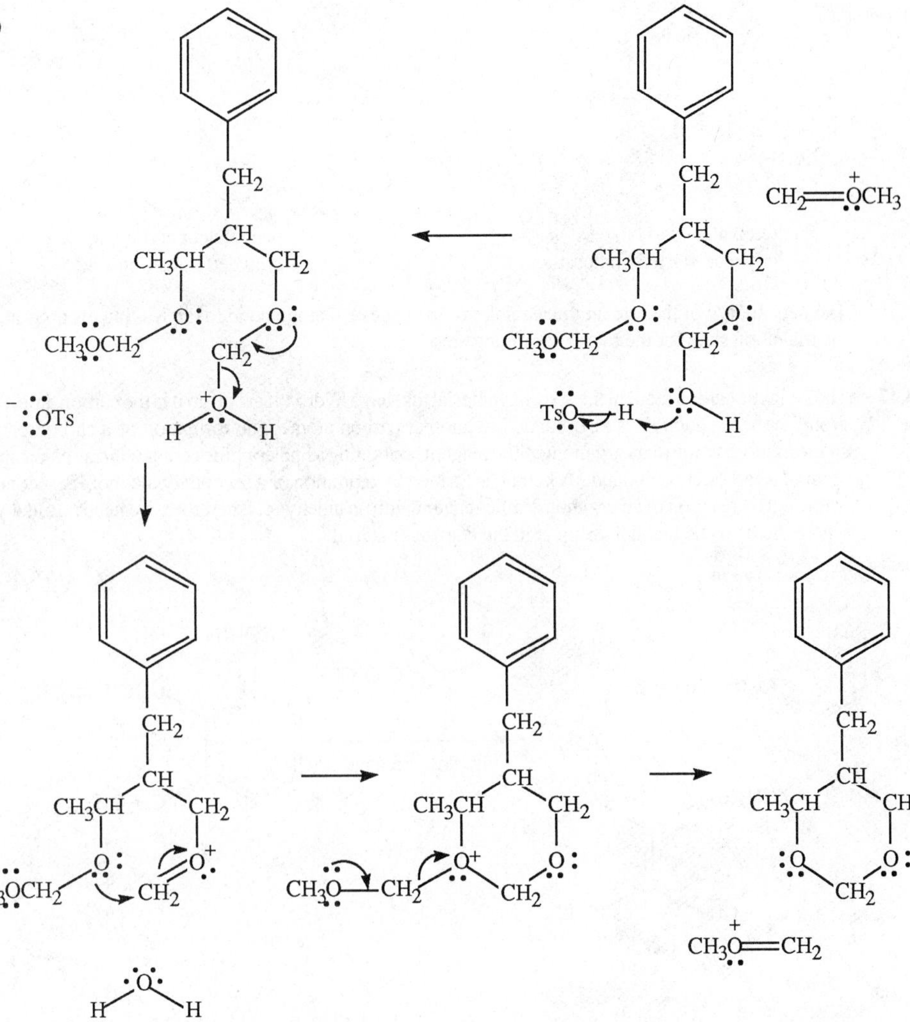

13.46 $CH_3(CH_2)_6CH_2$—$\overset{..}{\underset{..}{Br}}$: $\xrightarrow{\substack{\text{rate-} \\ \text{determining} \\ \text{step}}}$ $CH_3(CH_2)_6\overset{|}{\underset{|}{C}}H_2$:$\overset{..}{\underset{..}{Br}}$:$^-$

          H—$\overset{-}{Al}H_3$ H

                                                                    AlH_3

                                                                       $\downarrow$

                                                        :$\overset{..}{\underset{..}{Br}}$—$AlH_3$

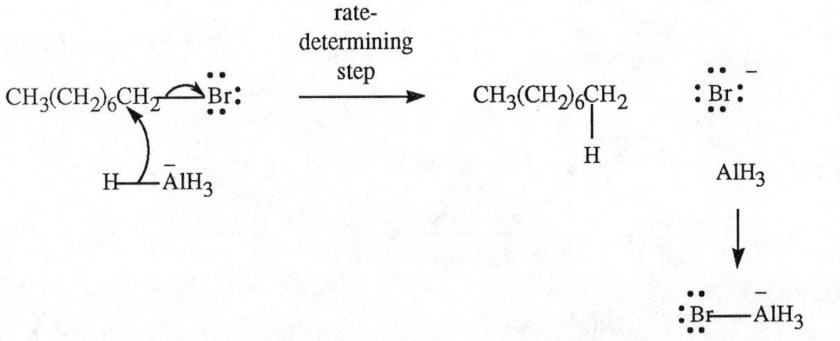

The rate data suggest an S_N2 reaction with hydride ion acting as the nucleophile. The decrease in rate with increased substitution on the β-carbon atom suggests that the transition state for the rate-determining step is more crowded than the starting reagent (Section 7.4C).

13.46 (cont)

tetravalent carbon, less
crowded starting material

pentavalent carbon, more
crowded transition state

The dependence of the rate on the leaving group suggests that the bond to the leaving group is also broken in the transition state for the rate-determining step.

13.47   The benzylic carbon atom in the starting material that has a hydroxyl group on it is the carbon atom of a carbonyl group in the product. It is also bonded to another carbon atom. The formation of a carbon-carbon σ bond suggests the reaction of an organometallic reagent containing a nucleophilic center with the electrophilic carbon atom of a carbonyl compound. A ketone is formed by oxidation of a secondary alcohol. A secondary alcohol is formed by reaction of an organometallic reagent with an aldehyde. Therefore, we need an aldehyde function where the benzylic alcohol group is in the starting material.

13.48

TsOCH₂CH——CH₂ $\xrightarrow{\begin{array}{c}\text{CH}_3\ \text{CH}_3\\ \text{CH}_3\text{C}\!\!-\!\!\text{SiCl}\\ \text{CH}_3\ \text{CH}_3\end{array}}$ TsOCH₂CH——CH₂ $\xrightarrow{\begin{array}{c}\text{O}\\ \|\\ \text{HCCH}\!=\!\!\text{CHO}^-\ ^+\text{N(CH}_2\text{CH}_2\text{CH}_2\text{CH}_3)_4\end{array}}$

with OH, OH below the first structure; TBDMSO, OTBDMS below **A**; imidazole (N–H) shown as reagent.

$$\begin{array}{c}\text{O}\\ \|\\ \text{HCCH}\!=\!\!\text{CHOCH}_2\text{CH}\!\!-\!\!\text{CH}_2\\ \quad\quad\quad\quad\text{TBDMSO}\quad\text{OTBDMS}\\ \textbf{B}\end{array}$$

CH₃CH₂CH=CHCH=CHCHC≡CH $\xrightarrow[\begin{array}{c}\text{tetrahydrofuran}\\ -78\ °\text{C}\end{array}]{\text{R}_2\text{N}^-\text{Li}^+}$ CH₃CH₂CH=CHCH=CHCHC≡C⁻Li⁺

with OTHP below each structure; **C** labels the product.

$$\begin{array}{ccc}\text{O}\\ \|\\ \text{HCCH}\!=\!\!\text{CHOCH}_2\text{CH}\!\!-\!\!\text{CH}_2 & + & \text{CH}_3\text{CH}_2\text{CH}\!=\!\text{CHCH}\!=\!\text{CHCHC}\!\equiv\!\text{C}^-\text{Li}^+ \longrightarrow\\ \quad\quad\text{TBDMSO}\quad\text{OTBDMS} & & \quad\quad\quad\quad\quad\text{OTHP}\\ \textbf{B} & & \textbf{C}\end{array}$$

CH₃CH₂CH=CHCH=CHCHC≡CCHCH=CHOCH₂CH——CH₂

with OTHP, O⁻Li⁺, TBDMSO, OTBDMS below; **D** label.

D $\xrightarrow{\text{several steps}}$ CH₃CH₂(CH=CH)₅OCH₂CHCH₂OTBDMS $\xrightarrow{(\text{CH}_3\text{CH}_2\text{CH}_2\text{CH}_2)_4\text{N}^+\text{F}^-}$

with OTBDMS below the intermediate.

CH₃CH₂(CH=CH)₅OCH₂CHCH₂OH

with OH below.

(±)-fecapentaene

13.49 (a)

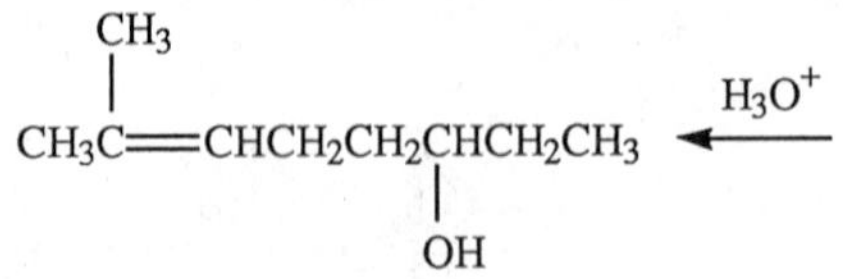

$$CH_3C{=\!=}CHCH_2CH_2CHCH_2CH_3 \xleftarrow{\;H_3O^+\;}$$

with CH3 on the first carbon and OH on the CHCH2CH3 carbon

7-methyl-6-octen-3-ol

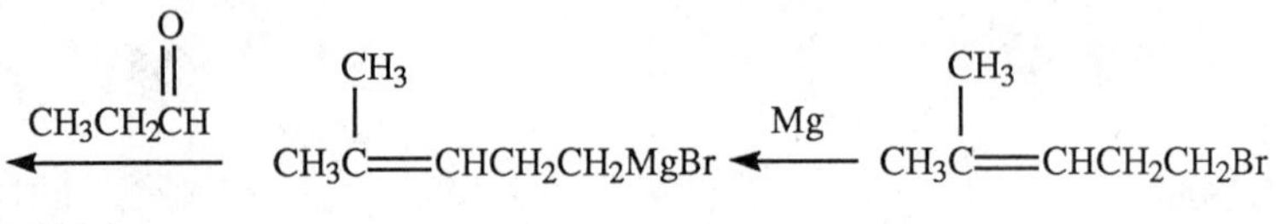

$$CH_3CH_2\overset{\displaystyle O}{\overset{\displaystyle \|}{C}}H \longleftarrow CH_3C{=\!=}CHCH_2CH_2MgBr \xleftarrow{\;Mg\;} CH_3C{=\!=}CHCH_2CH_2Br$$

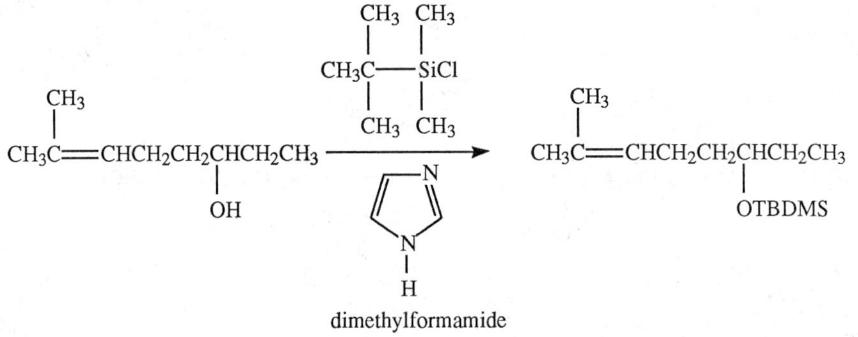

(b)

$$CH_3C{=\!=}CHCH_2CH_2CHCH_2CH_3 \text{ (OH)} \xrightarrow[\text{dimethylformamide}]{} CH_3C{=\!=}CHCH_2CH_2CHCH_2CH_3 \text{ (OTBDMS)}$$

reagent:

CH3C—SiCl with (CH3)2 and (CH3)2 / imidazole (N–H ring)

dimethylformamide

(c)

$$CH_3C{=\!=}CHCH_2CH_2CHCH_2CH_3 \text{ (OTBDMS)} \xrightarrow[\text{dichloromethane}]{}$$

reagent: 3-chloro-benzene-COOH (Cl–C6H4–COOH)

bottom reaction:

epoxide (CH3)2C—O—CH–CH2CH2CHCH2CH3 with H and OTBDMS $\xrightarrow[\substack{\text{tetrahydrofuran}\\0\,°C}]{(CH_3CH_2CH_2CH_2)_4N^+F^-}$ epoxide (CH3)2C—O—CH–CH2CH2CHCH2CH3 with H and OH

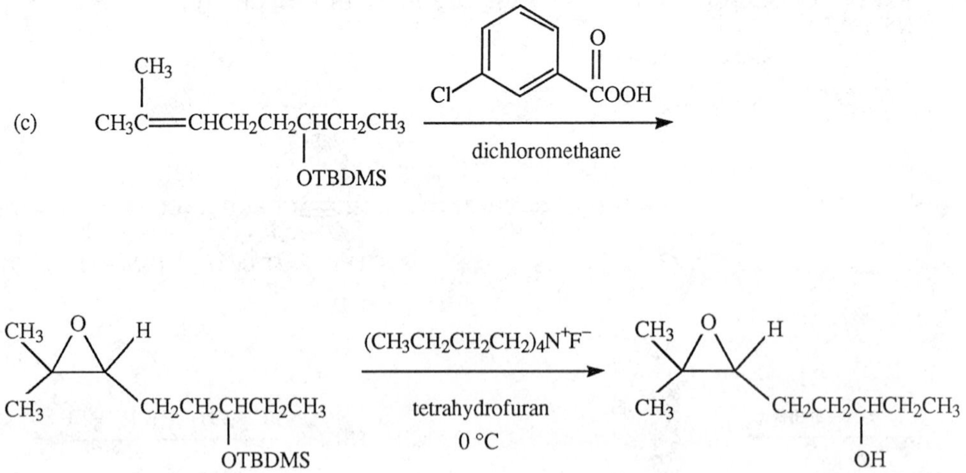

The alcohol functional group is a nucleophile that can open the oxirane ring under acid catalysis (see Section 12.6B).

13.50

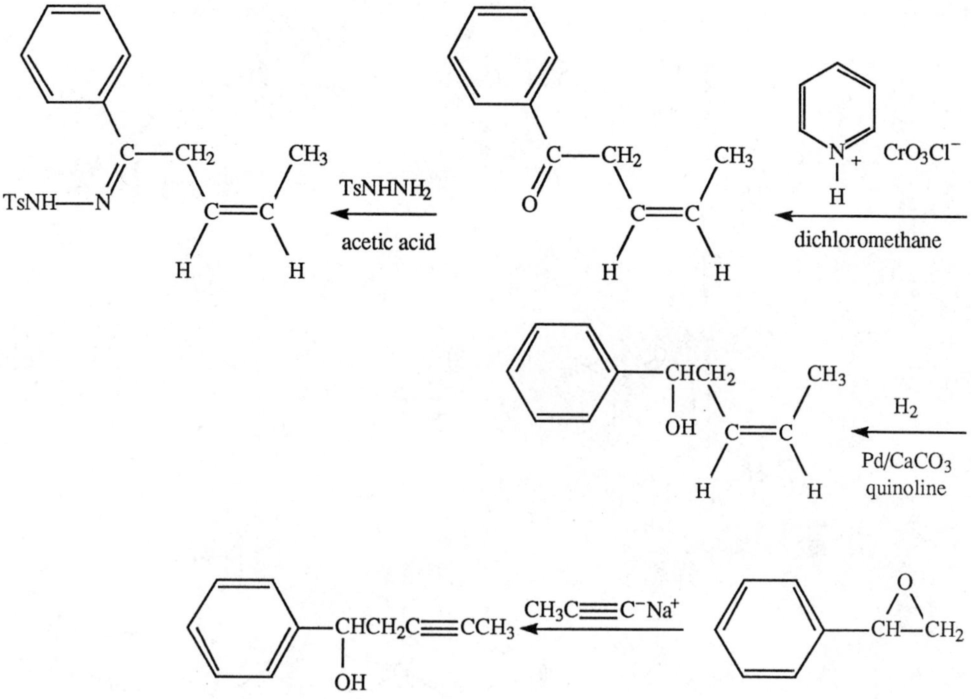

13.51

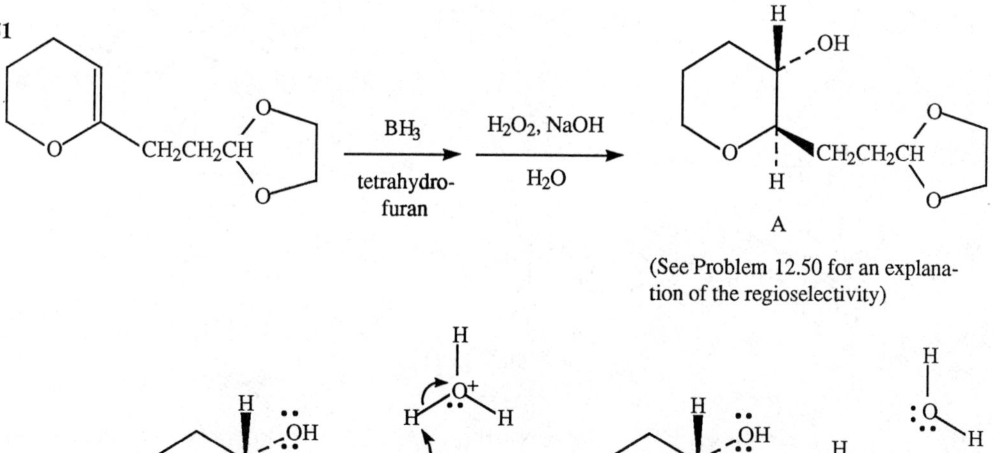

(See Problem 12.50 for an explanation of the regioselectivity)

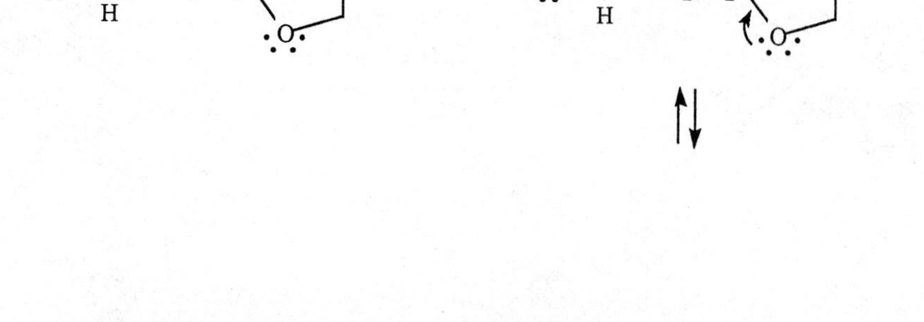

13.51 (cont)

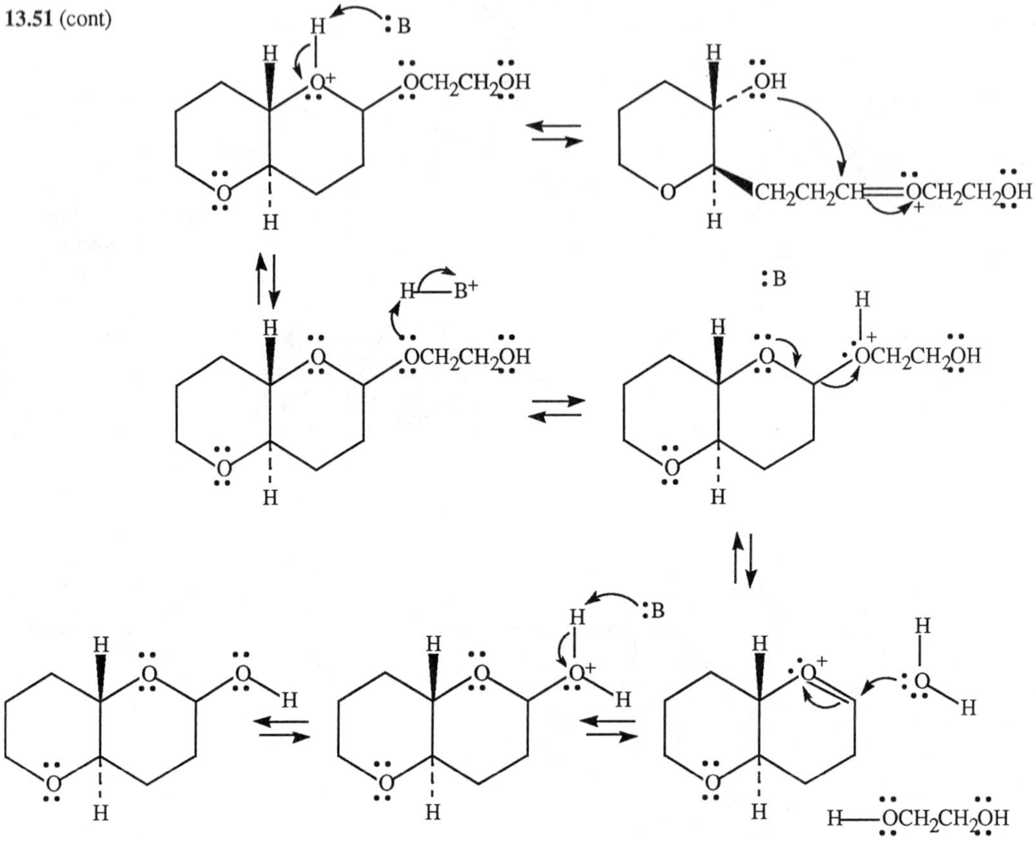

13.52

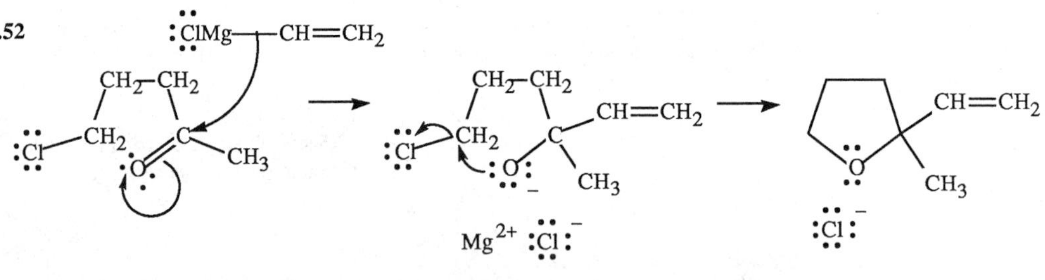

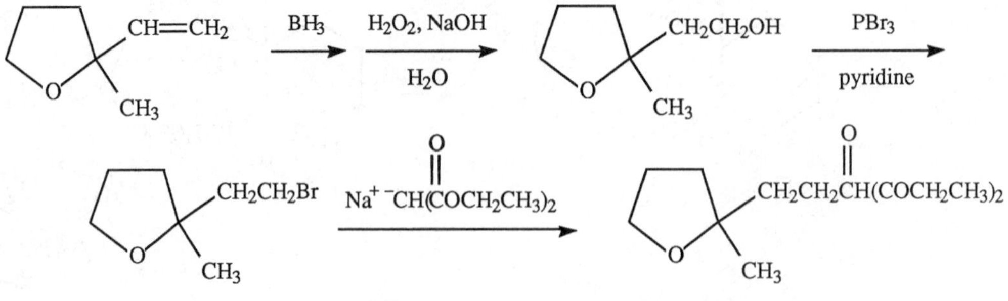

13.53   Compound B, $C_6H_{10}O$, has two units of unsaturation and Compound A, $C_8H_{12}O$, has three units of unsaturation. We can tell from the infrared spectrum that B is a ketone (C=O stretching frequency at $1710\,cm^{-1}$). A carbonyl group only uses up one unit of unsaturation. The other unit of unsaturation must be a ring or a double bond. No absorption bands around $1600\,cm^{-1}$ or above $3000\,cm^{-1}$ appear in the infrared spectrum, so the compound does not contain a double bond. Cyclic ketones smaller than six carbons absorb at frequencies greater than $1725\,cm^{-1}$ (see Problem 12.53), therefore Compound B must be cyclohexanone. Compound A must also contain a cyclohexane ring since it is prepared from Compound B. The infrared spectrum of Compound A tells us that it is an alcohol and a terminal alkyne (broad O—H stretching frequency at about $3400\,cm^{-1}$, $\equiv$C—H at $3308\,cm^{-1}$, and C$\equiv$C at about $2100\,cm^{-1}$). We can determine what has been added to cyclohexanone by subtracting the two molecular formulas.

$$C_8H_{12}O - C_6H_{10}O = C_2H_2 = \text{the elements of acetylene}$$

The elements of acetylene can be added by reaction of sodium acetylide with a ketone followed by acidification. Compound B is 1-ethynylcyclohexanol. The reaction of B to prepare A is shown below:

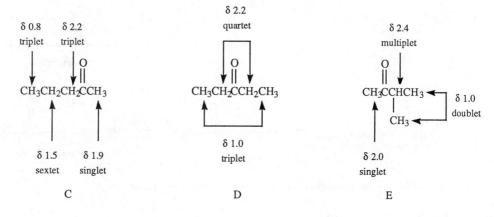

13.54   Compounds C, D, and E, molecular formula $C_5H_{10}O$, all have one unit of unsaturation. The infrared spectra of the three compounds show the presence of a ketone (C=O at $1720\,cm^{-1}$ and no bands at 2900 and $2700\,cm^{-1}$, which tells us that the compounds are not aldehydes). The presence of a ketone is confirmed by the carbon–13 magnetic resonance spectra in which the ketone carbon atom can be seen for C, D, and E (C at 206.6 ppm, D at 209.3 ppm, and E at 211.8 ppm). There are only three ketones that have the molecular formula $C_5H_{10}O$.

The carbon-13 spectra tell us how many types of carbon atoms each compound has. Compound C has five different carbon atoms. In compound D, there are three sets of carbon atoms, which points to symmetry in the molecule. Compound E has four sets of carbon atoms, and therefore must have two which are the same.

The structural formulas are shown below, along with the proton magnetic resonance data for each one.

$\delta\,0.8$ $\delta\,2.2$
triplet   triplet

$CH_3CH_2CH_2CCH_3$

$\delta\,1.5$ $\delta\,1.9$
sextet   singlet

C

$\delta\,2.2$
quartet

$CH_3CH_2CCH_2CH_3$

$\delta\,1.0$
triplet

D

$\delta\,2.4$
multiplet

CH_3CCHCH_3 $\delta\,1.0$
                   doublet
        CH_3

$\delta\,2.0$
singlet

E

14

Carboxylic Acids and Their Derivatives I. Nucleophilic Substitution Reactions at the Carbonyl Group

14.1

(a)

$$CH_3CH_2{-}\overset{\overset{O}{\|}}{C}{-}OH \qquad CH_3CH_2{-}\overset{\overset{O}{\|}}{C}{-}Cl \qquad CH_3CH_2{-}\overset{\overset{O}{\|}}{C}{-}O{-}\overset{\overset{O}{\|}}{C}{-}CH_2CH_3$$

$$CH_3CH_2{-}\overset{\overset{O}{\|}}{C}{-}O{-}CH_2CH_3 \qquad CH_3CH_2{-}\overset{\overset{O}{\|}}{C}{-}NH_2$$

(b)

$$CH_3CH_2{-}\overset{\overset{O}{\|}}{C}{-}\overset{+}{O}H_2 \ HSO_4^- \underset{H_2SO_4}{\rightleftharpoons} CH_3CH_2{-}\overset{\overset{O}{\|}}{C}{-}OH \underset{}{\overset{H_2SO_4}{\rightleftharpoons}}$$

$$CH_3CH_2{-}\overset{\overset{+}{O}{-}H}{\underset{}{\|}}{C}{-}OH \ HSO_4^-$$

major product

$$CH_3CH_2{-}\overset{\overset{O}{\|}}{\underset{\underset{H}{|}}{C}}{-}\overset{+}{O}{-}CH_2CH_3 \underset{H_2SO_4}{\rightleftharpoons} CH_3CH_2{-}\overset{\overset{O}{\|}}{C}{-}O{-}CH_2CH_3 \overset{H_2SO_4}{\rightleftharpoons}$$

$$CH_3CH_2{-}\overset{\overset{+}{O}{-}H \ HSO_4^-}{\underset{}{\|}}{C}{-}O{-}CH_2CH_3$$

major product

$$CH_3CH_2{-}\overset{\overset{O}{\|}}{C}{-}\overset{+}{N}H_3 \underset{H_2SO_4}{\rightleftharpoons} CH_3CH_2{-}\overset{\overset{O}{\|}}{C}{-}NH_2 \overset{H_2SO_4}{\rightleftharpoons} CH_3CH_2{-}\overset{HSO_4^- \ \overset{+}{O}{-}H}{\underset{}{\|}}{C}{-}NH_2$$

major product

(c) The good leaving groups are shaded in the formulas shown in parts (a) and (b).

14.2

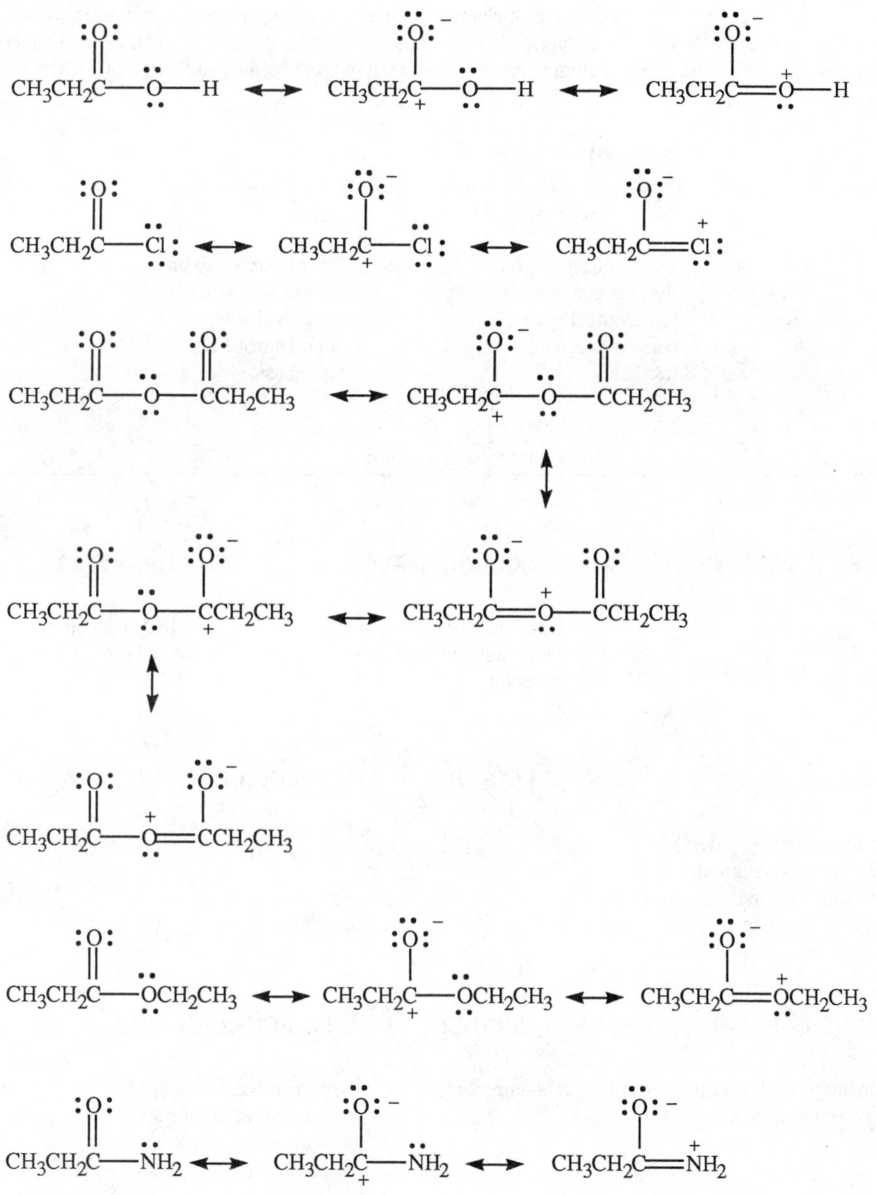

$$CH_3CH_2\overset{\overset{\displaystyle :\!O\!:}{\|}}{C}-\overset{..}{\underset{..}{O}}-H \longleftrightarrow CH_3CH_2\overset{\overset{\displaystyle :\!\ddot O\!:^-}{|}}{\underset{+}{C}}-\overset{..}{O}-H \longleftrightarrow CH_3CH_2\overset{\overset{\displaystyle :\!\ddot O\!:^-}{|}}{C}=\overset{+}{\underset{..}{O}}-H$$

$$CH_3CH_2\overset{\overset{\displaystyle :\!O\!:}{\|}}{C}-\overset{..}{\underset{..}{C}l}\!: \longleftrightarrow CH_3CH_2\overset{\overset{\displaystyle :\!\ddot O\!:^-}{|}}{\underset{+}{C}}-\overset{..}{\underset{..}{C}l}\!: \longleftrightarrow CH_3CH_2\overset{\overset{\displaystyle :\!\ddot O\!:^-}{|}}{C}=\overset{+}{\underset{..}{C}l}\!:$$

$$CH_3CH_2\overset{\overset{\displaystyle :\!O\!:}{\|}}{C}-\overset{..}{\underset{..}{O}}-\overset{\overset{\displaystyle :\!O\!:}{\|}}{C}CH_2CH_3 \longleftrightarrow CH_3CH_2\overset{\overset{\displaystyle :\!\ddot O\!:^-}{|}}{\underset{+}{C}}-\overset{..}{\underset{..}{O}}-\overset{\overset{\displaystyle :\!O\!:}{\|}}{C}CH_2CH_3$$

$$\updownarrow$$

$$CH_3CH_2\overset{\overset{\displaystyle :\!O\!:}{\|}}{C}-\overset{..}{O}-\overset{\overset{\displaystyle :\!\ddot O\!:^-}{|}}{\underset{+}{C}}CH_2CH_3 \longleftrightarrow CH_3CH_2\overset{\overset{\displaystyle :\!\ddot O\!:^-}{|}}{C}=\overset{+}{\underset{..}{O}}-\overset{\overset{\displaystyle :\!O\!:}{\|}}{C}CH_2CH_3$$

$$\updownarrow$$

$$CH_3CH_2\overset{\overset{\displaystyle :\!O\!:}{\|}}{C}-\overset{+}{\underset{..}{O}}=\overset{\overset{\displaystyle :\!\ddot O\!:^-}{|}}{C}CH_2CH_3$$

$$CH_3CH_2\overset{\overset{\displaystyle :\!O\!:}{\|}}{C}-\overset{..}{\underset{..}{O}}CH_2CH_3 \longleftrightarrow CH_3CH_2\overset{\overset{\displaystyle :\!\ddot O\!:^-}{|}}{\underset{+}{C}}-\overset{..}{\underset{..}{O}}CH_2CH_3 \longleftrightarrow CH_3CH_2\overset{\overset{\displaystyle :\!\ddot O\!:^-}{|}}{C}=\overset{+}{\underset{..}{O}}CH_2CH_3$$

$$CH_3CH_2\overset{\overset{\displaystyle :\!O\!:}{\|}}{C}-\overset{..}{N}H_2 \longleftrightarrow CH_3CH_2\overset{\overset{\displaystyle :\!\ddot O\!:^-}{|}}{\underset{+}{C}}-\overset{..}{N}H_2 \longleftrightarrow CH_3CH_2\overset{\overset{\displaystyle :\!\ddot O\!:^-}{|}}{C}=\overset{+}{N}H_2$$

The resonance contributors shown above reduce the positive charge on the carbon atom of the carbonyl group. The carbon atom of the carbonyl group in acid derivatives is thus less electrophilic than the carbon atom of the carbonyl group in aldehydes and ketones.

14.3   Propanamide is much less basic than propylamine because the nitrogen atom in propanamide has a partial positive charge (see the resonance contributors for propanamide in Problem 14.2). This means that there is less electron density on the nitrogen atom in propanamide than in propylamine, and, therefore, lower basicity for the amide.

$$
\underset{\delta+ \quad \delta+}{CH_3CH_2C\overset{\overset{\displaystyle O \; \delta-}{\|}}{\underset{}{}}\!\!\!\overset{\cdot\cdot}{NH_2}}
\qquad\qquad
CH_3CH_2CH_2\!\!-\!\!\overset{\cdot\cdot}{NH_2}
$$

The electron pair on   The electron pair on
the nitrogen atom is   the nitrogen atom is
less available; the   more available; the
amide nitrogen is   amine nitrogen is
less basic   more basic

14.4 ⟶ decreasing solubility ⟶

(a) $CH_3CH_2CH_2CH_2CO^-Na^+$ > $CH_3CH_2CH_2CH_2COH$ (with $\overset{O}{\|}$)   > $CH_3CH_2COCH_2CH_3$ (with $\overset{O}{\|}$)

   ionic   covalent, but hydrogen   only hydrogen bond
                                    bond donor as well as   acceptor
                                    acceptor

(b) $CH_3CH_2CH_2COH$ (with $\overset{O}{\|}$)   > $CH_3CH_2CH_2CH_2OH$ > $CH_3CH_2COCH_2CH_3$ (with $\overset{O}{\|}$)

can hydrogen bond
at the carbonyl group
as well as at the
hydroxyl group

(c) $CH_3CH_2CH_2CNH_2$ (with $\overset{O}{\|}$)   > $CH_3CH_2CH_2COCH_2CH_3$ (with $\overset{O}{\|}$)   > $CH_3CH_2CH_2CO(CH_2)_4CH_3$ (with $\overset{O}{\|}$)

hydrogen bond donor   hydrogen bond acceptor   hydrogen bond acceptor but
as well as acceptor   nonpolar portion of the
                                                          molecule now significantly
                                                          larger than the polar part

14.5

$$
\underset{CH_3CH_3}{\overset{CH_3}{\underset{}{H^{\cdots}C}}}\!\!\diagdown CH_2COH \; (\overset{O}{\|})
\qquad\qquad
(\overset{O}{\|})HOCCH_2\!\!-\!\!\underset{CH_2CH_3}{\overset{CH_3}{C}}^{\cdots}H
$$

   (*R*)-3-methylpentanoic acid   (*S*)-3-methylpentanoic acid

14.5 (cont)

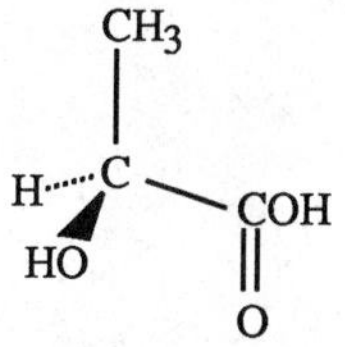

(*S*)-2-hydroxypropanoic acid

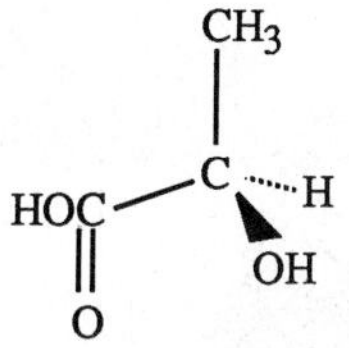

(*R*)-2-hydroxypropanoic acid

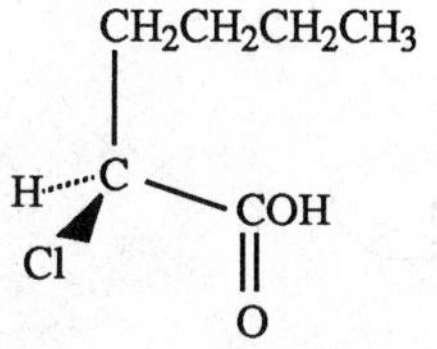

(*S*)-2-chlorohexanoic acid

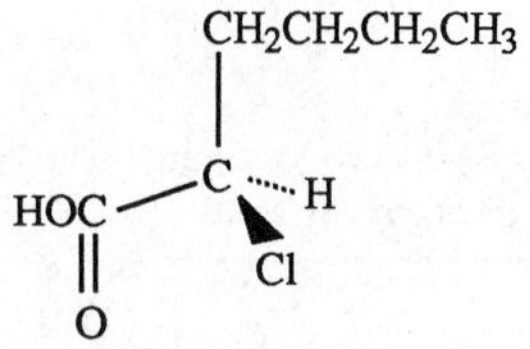

(*R*)-2-chlorohexanoicacid

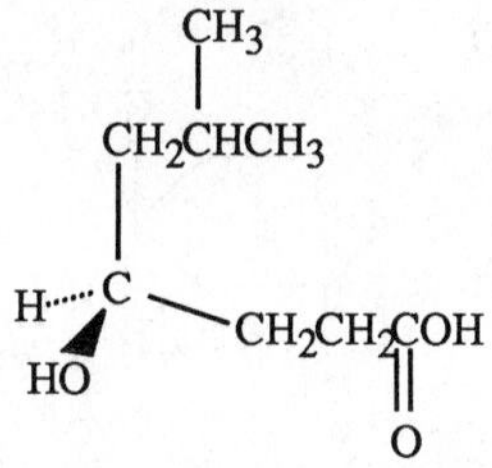

(*R*)-4-hydroxy-6-methylheptanoic acid

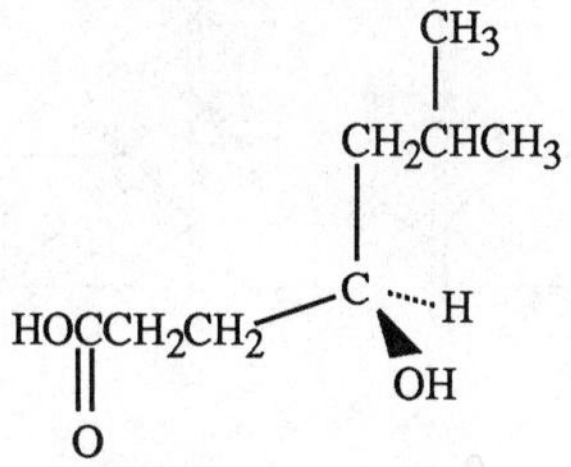

(*S*)-4-hydroxy-6-methylheptanoic acid

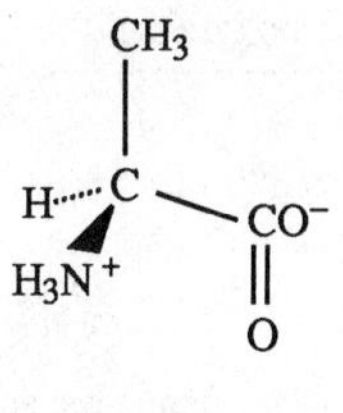

(*S*)-alanine

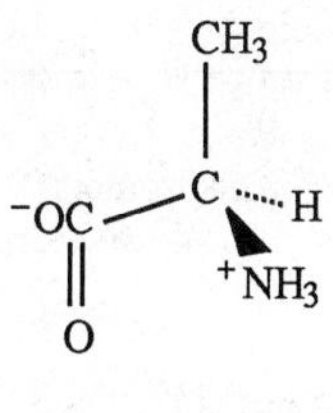

(*R*)-alanine

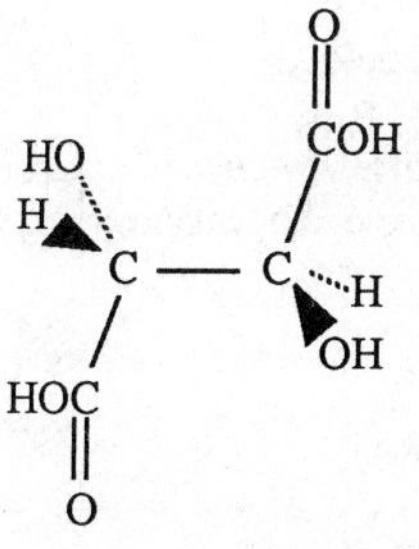

(2*R*, 3*S*)-tartaric acid
meso-tartaric acid

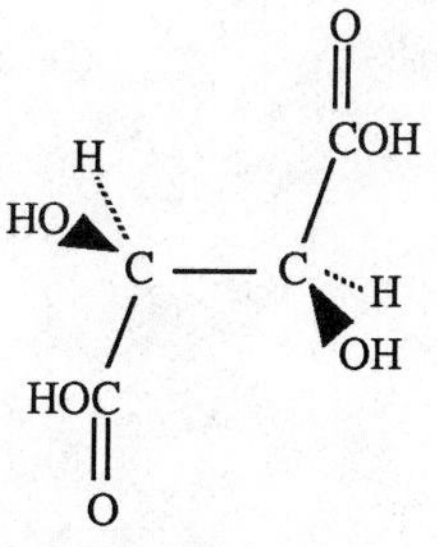

(2*S*, 3*S*)-tartaric acid

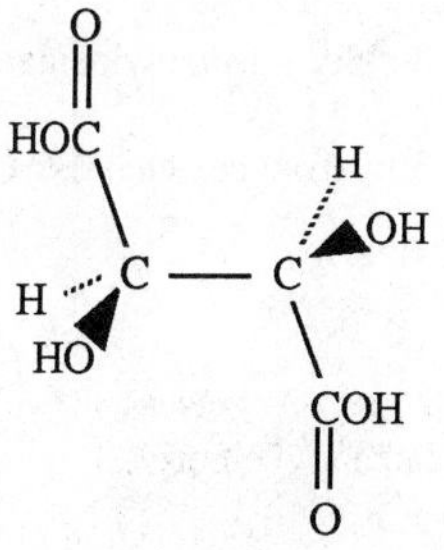

(2*R*, 3*R*)-tartaric acid

14.5 (cont)

(1*R*, 2*R*)-1,2-cyclopentane-
dicarboxylic acid

(1*S*, 2*S*)-1,2-cyclopentane-
dicarboxylic acid

(1*R*, 2*S*)-1,2-cyclopentane-
dicarboxylic acid

trans isomers

cis isomer
meso compound

(1*R*, 3*S*)-3-methyl-
cyclohexane-
carboxylic acid

(1*S*, 3*R*)-3-methyl-
cyclohexane-
carboxylic acid

(1*R*, 3*R*)-3-methyl-
cyclohexane-
carboxylic acid

(1*S*, 3*S*)-3-methyl-
cyclohexane-
carboxylic acid

cis isomers

trans isomers

14.6   (a) undecanoic acid

(c) (*S*)-2-hydroxybutanoic acid

(e) 2-chloro-4-methylpentanoic acid

(g) 3-hydroxyhexanedioic acid

(b) (*E*)-2-pentenoic acid

(d) *m*-bromobenzoic acid

(f) 5-oxohexanoic acid

(h) (1*R*, 2*R*)-2-hydroxycyclopentanecarboxylic acid
(*trans*-2-hydroxycyclopentanecarboxylic acid)

14.7   (a) hexanoyl chloride

(c) methyl (Z)-4-hexenoate

(e) cyclobutanecarbonyl chloride

(b) butanoic anhydride

(d) isobutyl *p*-chlorobenzoate

(f) *p*-chlorobenzoic anhydride

14.8   (a)

$$\begin{array}{c} \text{CH}_3\text{CH}_2 \qquad\quad \overset{\displaystyle O}{\overset{\displaystyle \|}{\text{CH}_2\text{CH}_2\text{COH}}} \\ \diagdown \qquad \diagup \\ \text{C}=\text{C} \\ \diagup \qquad \diagdown \\ \text{H} \qquad\qquad \text{H} \end{array}$$

(b)

(c)

$$\begin{array}{c} \overset{\displaystyle O}{\overset{\displaystyle \|}{\text{COH}}} \\ | \\ \text{H}\cdots\text{C}\!-\!\text{CH}_2\text{CH}_2\text{CH}_3 \\ | \\ \text{Br} \end{array}$$

(d) $\overset{O}{\overset{\|}{\text{HOC}}}(\text{CH}_2)_6\overset{O}{\overset{\|}{\text{COH}}}$

(e) $\text{CH}_3\text{O}\overset{O}{\overset{\|}{\text{C}}}\text{CH}_2\overset{O}{\overset{\|}{\text{C}}}\text{OCH}_3$

(f)

(g)

(h) $\text{Cl}\overset{O}{\overset{\|}{\text{C}}}(\text{CH}_2)_3\overset{O}{\overset{\|}{\text{C}}}\text{Cl}$

(i) $\text{Na}^+{}^-\text{O}\!-\!\overset{O}{\overset{\|}{\text{C}}}\overset{O}{\overset{\|}{\text{C}}}\text{O}^-\text{Na}^+$

(j)

14.9   (a)   propyl benzoate

(b)   ethyl 6-oxooctanoate

(c)   decanoic acid

(d)   *m*-bromobenzoyl chloride

(e)   sodium octadecanoate
(sodium stearate)

(f)   phenylethanoic acid
(phenylacetic acid)

(g)   *N*-methylacetamide

(h)   *p*-methylbenzamide

(i)   (1*R*, 2*R*)-2-bromocyclopentanecarboxylic acid

(j)   (*E*)-2-butenoic acid

(k)   ethyl 3-oxopentanoate

(l)   (*R*)-3-hydroxybutanoic acid

(m)   heptanenitrile

(n)   3-methylbutanoic anhydride

(o)   *N*-phenylbutanamide
(butananilide)

14.10

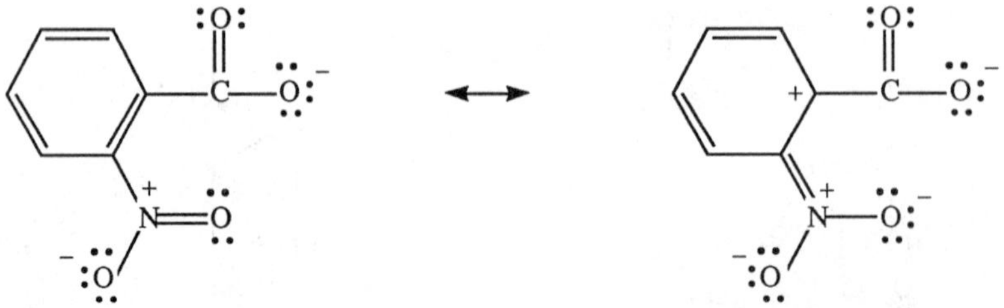

anion of *ortho*-nitro
benzoic acid

positive charge on the carbon
atom that is adjacent to the
carboxylate group;
stabilization of the anion

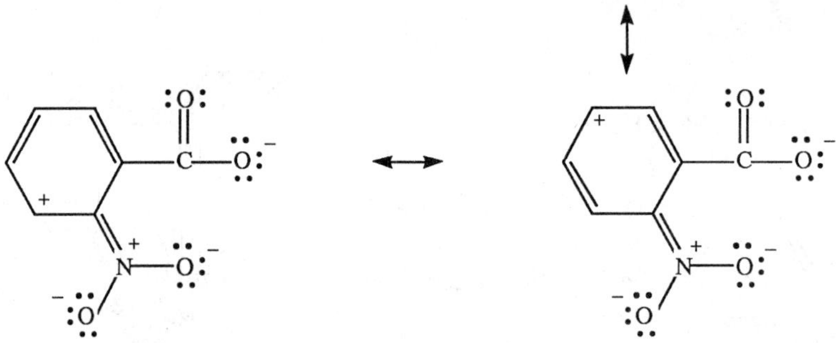

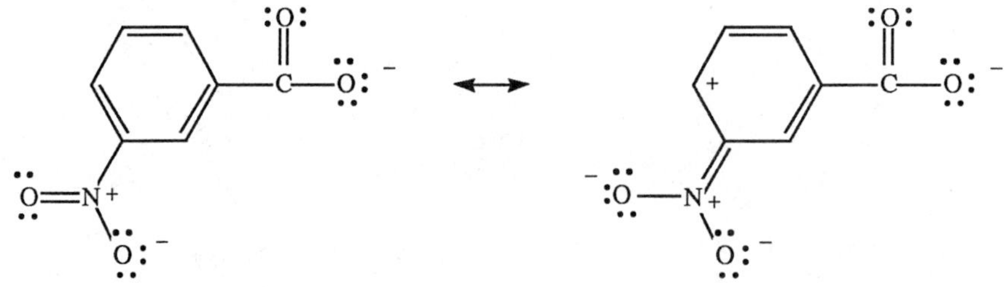

anion of *meta*-nitro
benzoic acid

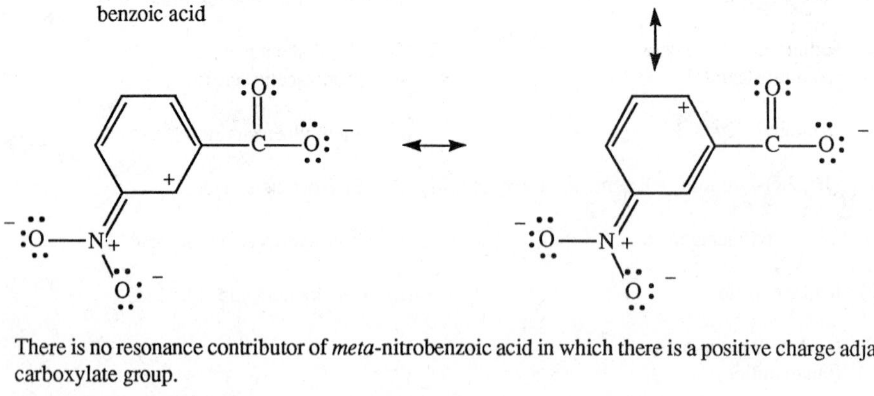

There is no resonance contributor of *meta*-nitrobenzoic acid in which there is a positive charge adjacent to the carboxylate group.

Concept Map 14.1   Preparation of carboxylic acids.

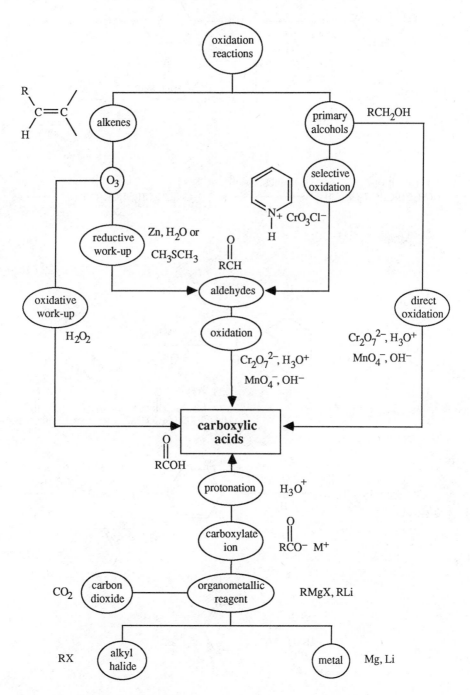

14.11

(a)

$$CH_3-\langle\text{ring}\rangle-Br \xrightarrow[\text{ether}]{\text{Mg} \atop \text{diethyl}} CH_3-\langle\text{ring}\rangle-MgBr \xrightarrow{CO_2}$$

A

$$CH_3-\langle\text{ring}\rangle-CO^-Mg^{2+}Br^- \xrightarrow{H_3O^+} CH_3-\langle\text{ring}\rangle-COH$$

B   **C**

(b)

$$\underset{\text{CH}_3}{\text{CH}_3\text{CH}_2\text{CH}_2\text{CHCH}_2\text{CH}_2\text{OH}} \xrightarrow[\substack{\text{H}_2\text{O} \\ \Delta}]{\text{KMnO}_4,\ \text{NaOH}} \underset{\text{CH}_3\ \ \ \ \text{O}}{\text{CH}_3\text{CH}_2\text{CH}_2\text{CHCH}_2\text{CO}^-\text{Na}^+} \xrightarrow{\text{H}_3\text{O}^+}$$

D

$$\underset{\text{CH}_3\ \ \ \ \text{O}}{\text{CH}_3\text{CH}_2\text{CH}_2\text{CHCH}_2\text{COH}}$$

E

(c)

$$\langle\text{cyclohexene}\rangle \xrightarrow{\substack{\text{O}_3 \\ \text{chloroform}}} \langle\text{F}\rangle \xrightarrow[\substack{\text{H}_2\text{O} \\ \Delta}]{\text{Ag}_2\text{O},\ \text{NaOH}}$$

F

$$Na^+\ ^-OC(CH_2)_4CO^-Na^+ \xrightarrow{H_3O^+} HOC(CH_2)_4COH$$

G   **H**

(d)

$$\langle\text{ring}\rangle-CH_2CH_2CH \xrightarrow[\substack{\text{H}_2\text{O} \\ \Delta}]{\text{Ag}_2\text{O},\ \text{NaOH}} \langle\text{ring}\rangle-CH_2CH_2CO^-Na^+ \xrightarrow{H_3O^+}$$

I

$$\langle\text{ring}\rangle-CH_2CH_2COH$$

J

(e)

$$\langle\text{ring}\rangle-CH_2OH \xrightarrow[\substack{\text{H}_2\text{SO}_4 \\ \text{H}_2\text{O} \\ \Delta}]{\text{CrO}_3} \langle\text{ring}\rangle-COH$$

K

14.11 (cont)

(f) $\underset{\displaystyle CH_3(CH_2)_5CH}{\overset{\displaystyle \overset{O}{\|}}{}} \xrightarrow[\substack{H_2SO_4 \\ H_2O \\ \Delta}]{CrO_3} \underset{\displaystyle CH_3(CH_2)_5COH}{\overset{\displaystyle \overset{O}{\|}}{}}$

L

(g) $CH_3(CH_2)_9CH{=}CH_2 \xrightarrow[\text{chloroform}]{O_3} \xrightarrow{H_2O_2,\ NaOH} \xrightarrow{H_3O^+} \underset{\displaystyle CH_3(CH_2)_9COH}{\overset{\displaystyle \overset{O}{\|}}{}} + \underset{\displaystyle HCOH}{\overset{\displaystyle \overset{O}{\|}}{}}$

M   N

14.12

(a) $C_6H_5{-}CH_2CH_2\overset{\overset{\displaystyle O}{\|}}{C}OH \xrightarrow[\Delta]{SOCl_2} C_6H_5{-}CH_2CH_2\overset{\overset{\displaystyle O}{\|}}{C}Cl$

(b) $CH_3CH_2CH_2\overset{\overset{\displaystyle O}{\|}}{C}OH \xrightarrow{PCl_3} CH_3CH_2CH_2\overset{\overset{\displaystyle O}{\|}}{C}Cl$

(c) $CH_3CH_2CH_2\overset{\overset{\displaystyle O}{\|}}{C}Cl \ + \ Na^+\ ^-OCCH_2CH_2CH_3 \longrightarrow CH_3CH_2CH_2\overset{\overset{\displaystyle O}{\|}}{C}O\overset{\overset{\displaystyle O}{\|}}{C}CH_2CH_2CH_3$

(d) $CH_3(CH_2)_{10}\overset{\overset{\displaystyle O}{\|}}{C}OH \xrightarrow[\Delta]{SOCl_2} CH_3(CH_2)_{10}\overset{\overset{\displaystyle O}{\|}}{C}Cl$

(e) $HOCCH_2CH_2\overset{\overset{\displaystyle O}{\|}}{C}OH \xrightarrow{\overset{\overset{\displaystyle O \ \ O}{\| \ \ \|}}{CH_3COCCH_3}}$

(f) $CH_3CH_2\overset{\overset{\displaystyle O}{\|}}{C}OH \xrightarrow[\substack{\text{clay} \\ \Delta \\ 650\ °C}]{} CH_3CH_2\overset{\overset{\displaystyle O}{\|}}{C}O\overset{\overset{\displaystyle O}{\|}}{C}CH_2CH_3$

14.12 (cont)

(g) 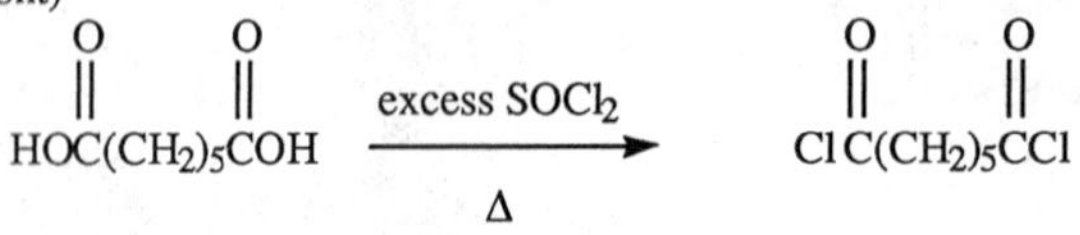 $\overset{O}{\overset{\|}{HOC}}(CH_2)_5\overset{O}{\overset{\|}{COH}}$ $\xrightarrow[\Delta]{\text{excess } SOCl_2}$ $Cl\overset{O}{\overset{\|}{C}}(CH_2)_5\overset{O}{\overset{\|}{C}}Cl$

14.13

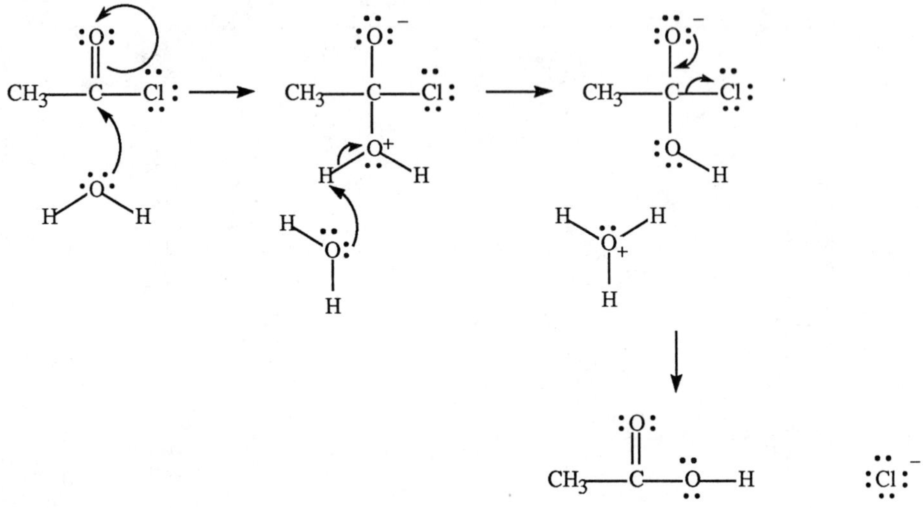

14.14

major   minor   very minor

major   very minor   minor

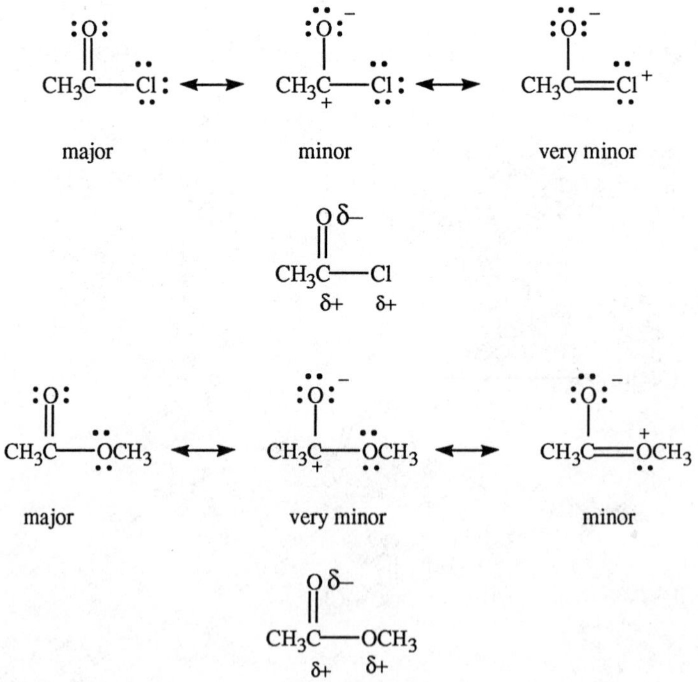

14.14 (cont)

Both chlorine and oxygen have nonbonding electron pairs and can stabilize an adjacent positive charge. The resonance contributor in the ester in which the positive charge is on the oxygen atom makes a greater contribution to the resonance hybrid than the equivalent resonance contributor in the acid chloride in which the positive charge is on the chlorine atom. The carbonyl group in the ester thus has a smaller positive charge than in the chloride, and is less electrophilic.

major

In the anhydride, the electrons of the central oxygen atom are delocalized to two carbonyl groups instead of just to one carbonyl group as in the ester. The carbon atom of each carbonyl group in the anhydride thus has a larger partial positive charge than the carbon atom of the carbonyl group in the ester. Therefore, the electrophilicity of the carbonyl groups in an anhydride is greater than in an ester.

Concept Map 14.2   (see p. 418)

14.15

pK_a 2.8 pK_a 9.3

Chloroacetic acid is a much stronger acid than hydrogen cyanide, therefore the cyanide anion will be converted to hydrogen cyanide and will not be available to act as a nucleophile if chloroacetic acid is not first neutralized with base. In addition, even small amounts of hydrogen cyanide are lethal to humans.

Concept Map 14.2 Hydrolysis reactions of acid derivatives.

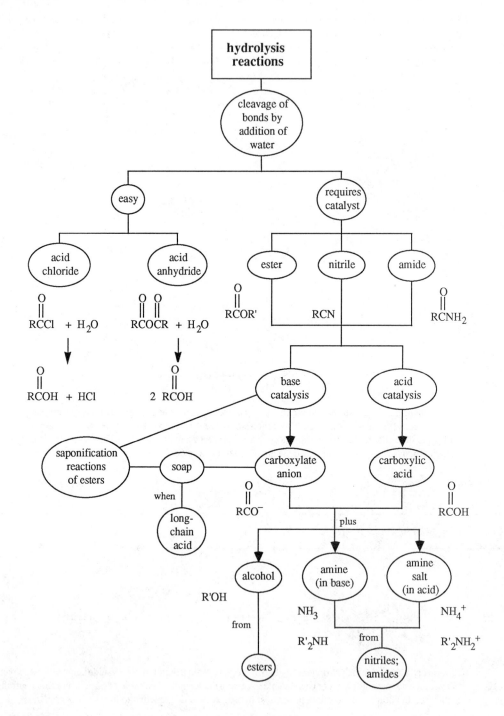

14.16

(a)

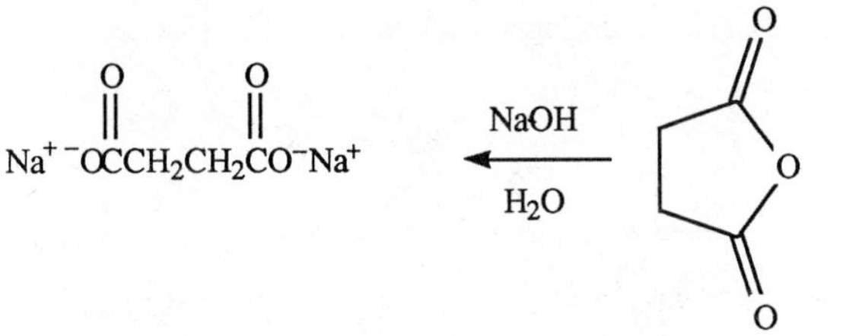

$$Na^+ \ ^-OCCH_2CH_2CO^-Na^+ \xleftarrow[H_2O]{NaOH}$$

(succinic anhydride structure)

(b)

$$CH_3CH_2CH_2\underset{\underset{CH_3}{|}}{CH}COH \xleftarrow{H_3O^+} \xleftarrow{CO_2} CH_3CH_2CH_2\underset{\underset{CH_3}{|}}{CH}MgBr \xleftarrow{Mg}$$

$$CH_3CH_2CH_2\underset{\underset{CH_3}{|}}{CH}Br \xleftarrow[pyridine]{PBr_3} CH_3CH_2CH_2\underset{\underset{CH_3}{|}}{CH}OH \xleftarrow{H_3O^+} \xleftarrow{CH_3MgI} CH_3CH_2CH_2CH$$

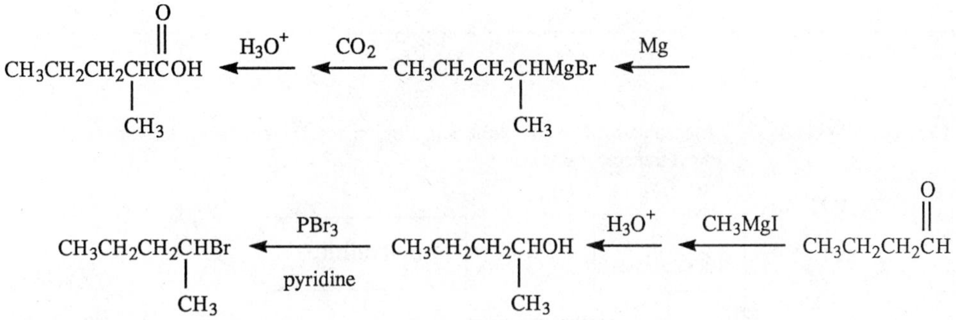

14.17

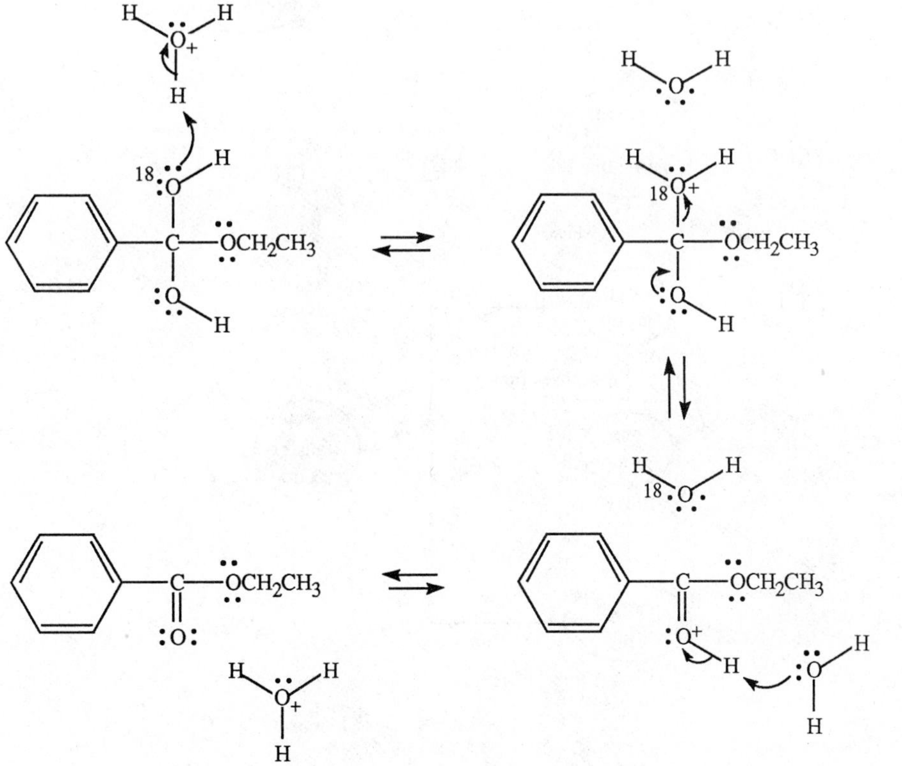

14.18   The two oxygen atoms are equivalent and cannot be distinguished from each other (Section 1.4E).

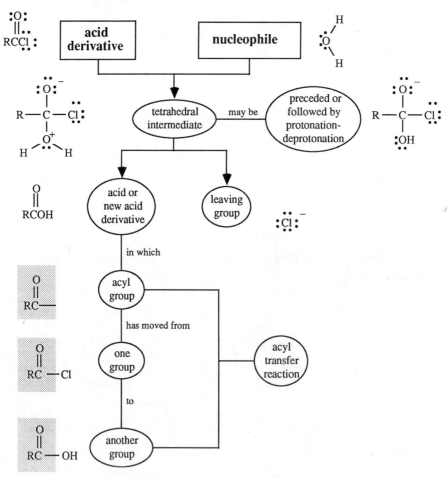

Concept Map 14.3   Mechanism of nucleophilic substitution reactions of acid derivatives.
                           Acyl transfer reactions

14.19   Attack by the hydroxide ion at the carbon atom of the carbonyl group.

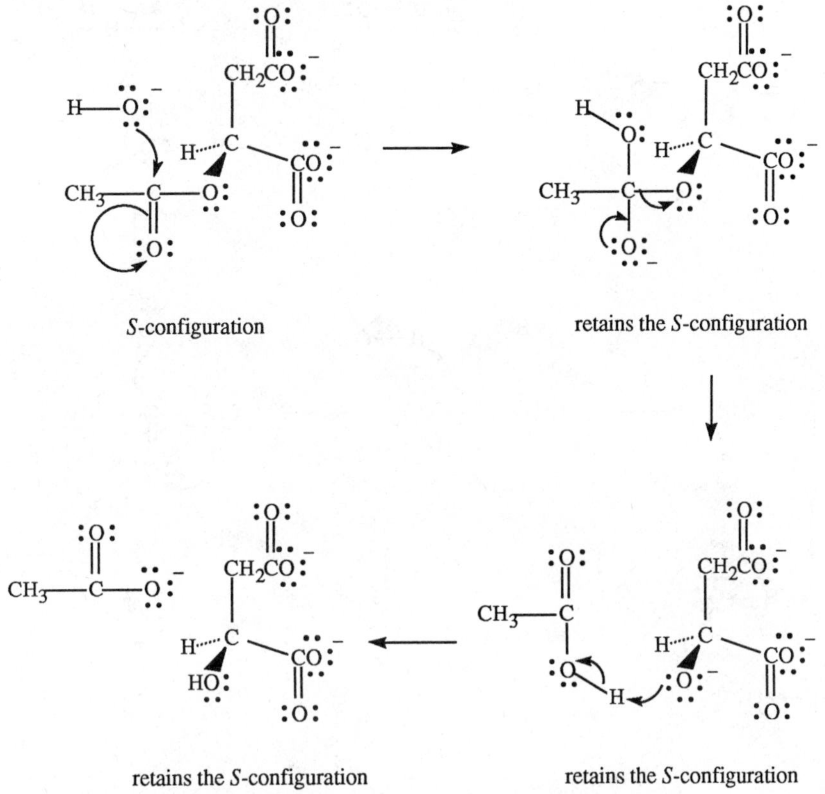

S-configuration

retains the S-configuration

retains the S-configuration

retains the S-configuration

Attack by the hydroxide ion at the stereocenter.

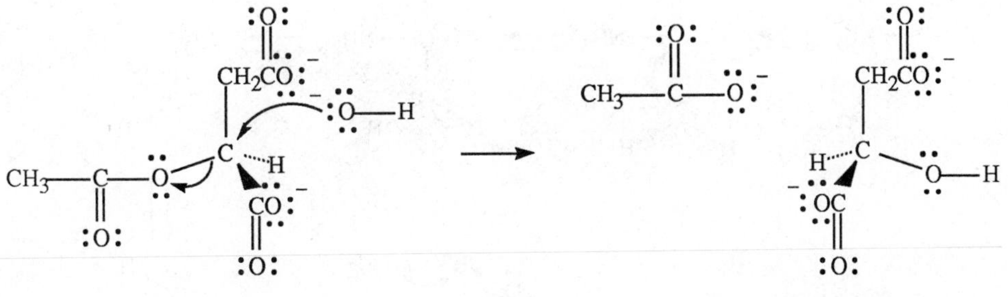

S-configuration

R-configuration

14.20

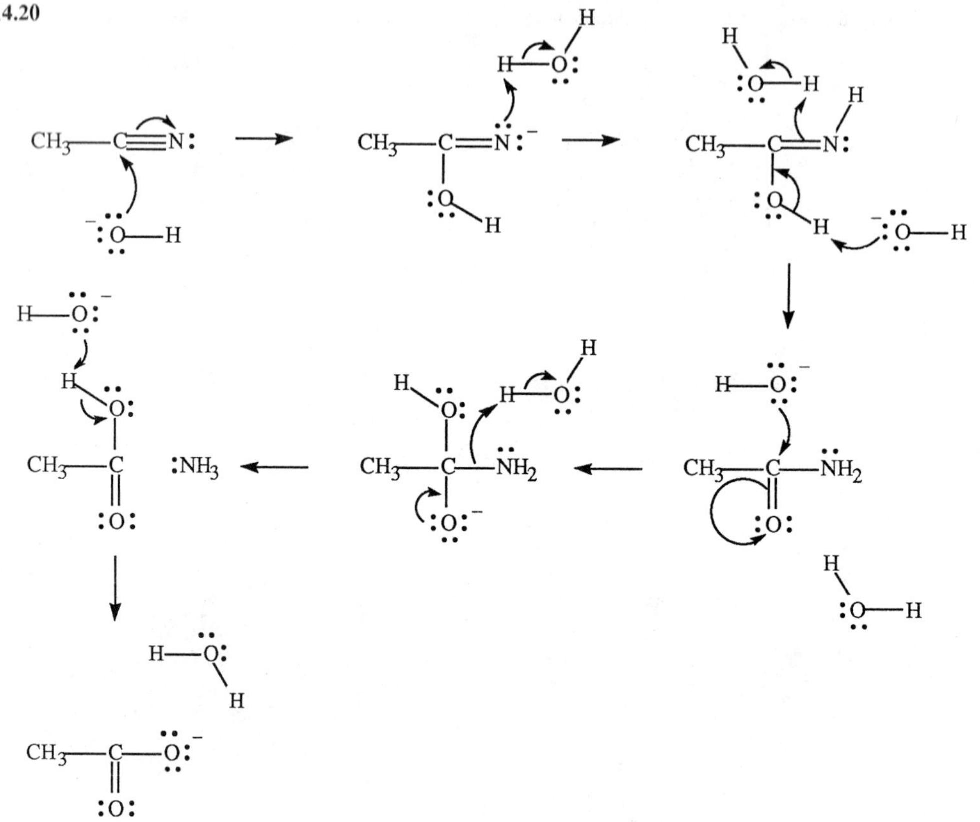

14.21

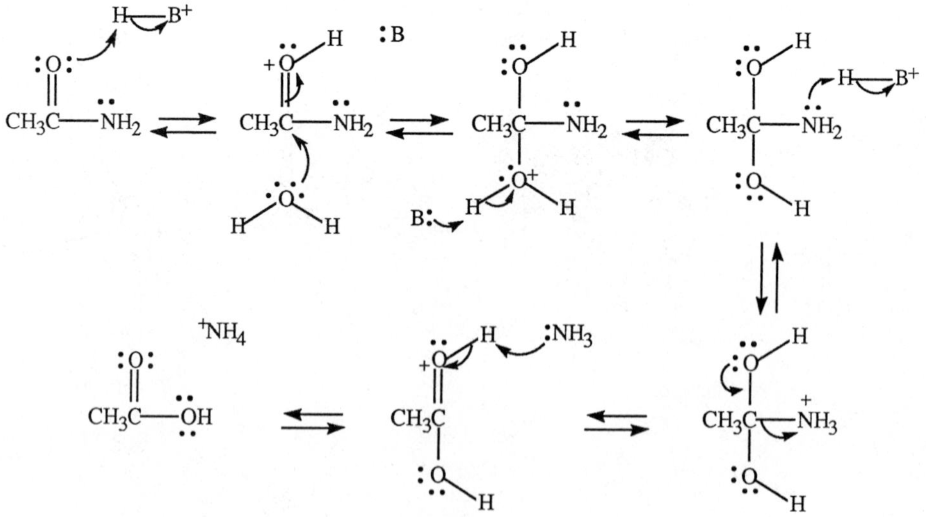

14.22

For the carbon-nitrogen bond to break in the basic hydrolysis of acetamide, the leaving group must be the conjugate base of ammonia, amide anion. In basic solution, it is unlikely that protonation of the nitrogen in the tetrahedral intermediate will take place before the carbon-nitrogen bond begins to break. However, amide anion is such a strong base that it is improbable that it has any real existence in water. The mechanism accounts for this by showing the amide ion taking a proton from water as the tetrahedral intermediate breaks up.

- -

Concept Map 14.4   (see p. 424)

- -

14.23 $CH_3CH_2CH_2CH_2OH$ +   [benzoyl chloride] $\xrightarrow[H_2O]{10\% \text{ NaOH}}$ [phenyl group]$-COCH_2CH_2CH_2CH_3$ with C=O

The other organic product formed would be sodium benzoate.

[benzoyl chloride]$-CCl$ (C=O) $\xrightarrow[H_2O]{10\% \text{ NaOH}}$ [phenyl group]$-CO^-Na^+$ (C=O)

Concept Map 14.4　　　Relative reactivities in nucleophilic substitutions.

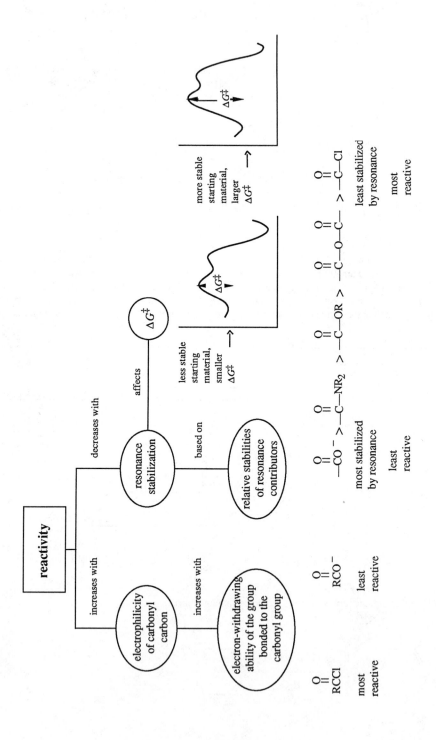

14.24

(a)
$$\underset{\underset{CH_3}{|}}{\overset{\overset{CH_3}{|}}{CH_3}}COH \quad + \quad \underset{}{CH_3\overset{O}{\overset{||}{C}}O\overset{O}{\overset{||}{C}}CH_3} \quad \longrightarrow \quad CH_3\overset{O}{\overset{||}{C}}O\underset{\underset{CH_3}{|}}{\overset{\overset{CH_3}{|}}{C}}CH_3 \quad + \quad CH_3\overset{O}{\overset{||}{C}}OH$$

(b) $CH_3CH_2CH_2CH_2OH \quad + \quad CH_3CH_2CH_2\overset{O}{\overset{||}{C}}Cl \quad \longrightarrow \quad CH_3CH_2CH_2\overset{O}{\overset{||}{C}}OCH_2CH_2CH_2CH_3$

(c)

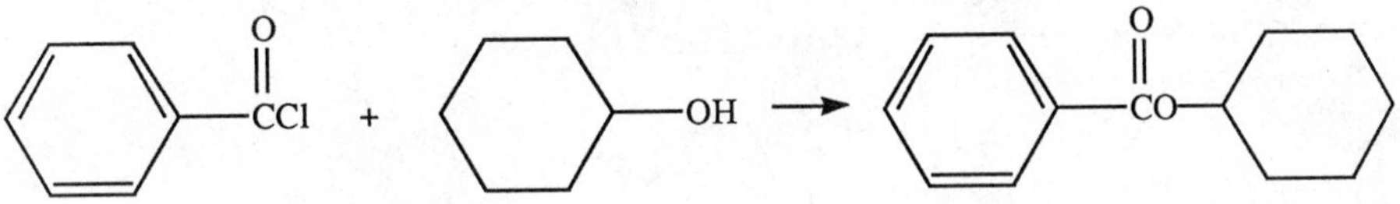

(d)

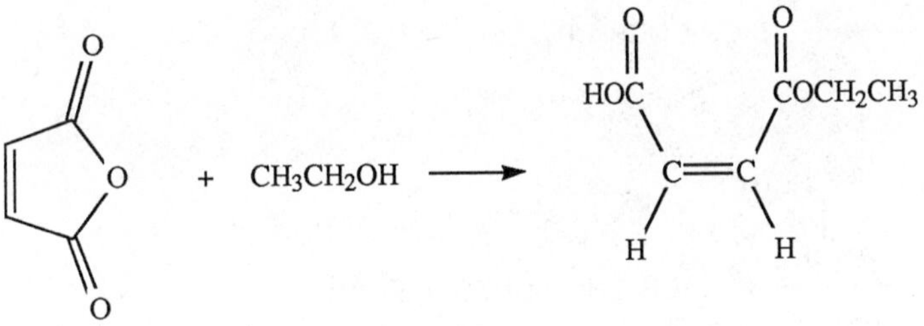

14.25

(a) $\underset{}{HO\overset{O}{\overset{||}{C}}(CH_2)_9\overset{O}{\overset{||}{C}}OH} \quad \xrightarrow[\underset{\Delta}{H_2SO_4}]{CH_3OH \text{ (excess)}} \quad CH_3O\overset{O}{\overset{||}{C}}(CH_2)_9\overset{O}{\overset{||}{C}}OCH_3$

 A

(b)

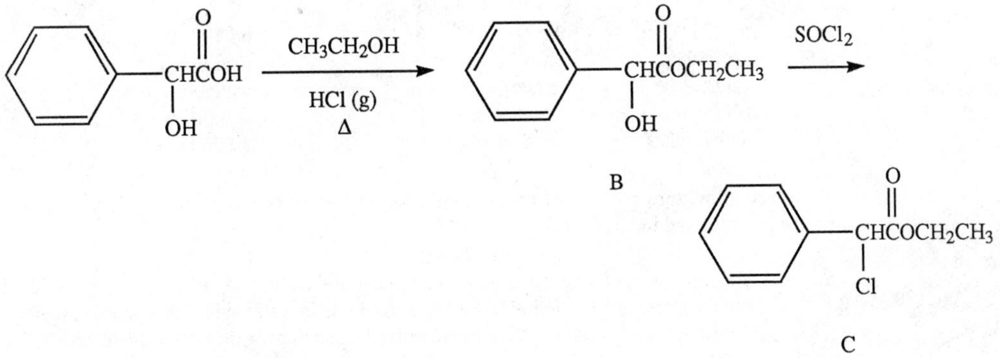

14.25 (cont)

(c) [cyclohexene-fused cyclic anhydride] $\xrightarrow[\substack{\text{TsOH} \\ \Delta}]{\text{CH}_3\text{CH}_2\text{OH (excess)}}$ [cyclohexene with two —$\overset{\text{O}}{\overset{||}{\text{C}}}OCH_2CH_3$ groups]

D

(d) $\text{HOCC}\underset{\overset{||}{\text{O}} \quad \overset{||}{\text{O}}}{\equiv}\text{CCOH}$ $\xrightarrow[\substack{\text{H}_2\text{SO}_4 \\ \Delta}]{\text{CH}_3\text{OH (excess)}}$ $\text{CH}_3\text{OCC}\underset{\overset{||}{\text{O}} \quad \overset{||}{\text{O}}}{\equiv}\text{CCOCH}_3$

E

(e) $\text{BrCH}_2\overset{\overset{\text{O}}{||}}{\text{C}}\text{OH}$ $\xrightarrow[\substack{\text{H}_2\text{SO}_4 \\ \Delta}]{\text{CH}_3\text{CH}_2\text{OH}}$ $\text{BrCH}_2\overset{\overset{\text{O}}{||}}{\text{C}}\text{OCH}_2\text{CH}_3$

F

(f) $\text{CH}_3\text{CH}{=}\text{CHCOH}$ $\xrightarrow[\substack{\text{H}_2\text{SO}_4 \\ \text{benzene} \\ \Delta}]{\overset{\text{CH}_3\text{CH}_2\text{CHCH}_3}{\underset{\text{OH}}{|}}}$ $\text{CH}_3\text{CH}{=}\text{CHCOCCH}_2\text{CH}_3$

G

14.26   (a) [benzene ring]—$\overset{\overset{\text{O}}{||}}{\text{C}}$OH $+$ CH$_3$OH $\underset{\text{H}_2\text{SO}_4}{\rightleftharpoons}$ [benzene ring]—$\overset{\overset{\text{O}}{||}}{\text{C}}OCH_3$ $+$ H$_2$O

(b)

mp 121 °C insoluble in water; soluble in diethyl ether	bp 66 °C miscible with water; soluble in diethyl ether	bp 198 °C insoluble in water; soluble in diethyl ether

The reaction mixture consists of excess methanol, unreacted benzoic acid, sulfuric acid, and the products, water and methyl benzoate.

Step 1: Add more water to the reaction mixture; most of the methanol and the sulfuric acid will go into the water layer. Methyl benzoate is insoluble in water and will appear as an oily layer. Unreacted benzoic acid will remain dissolved in the ester. Add diethyl ether, bp 37 °C, d 0.7, to dissolve the organic compounds. Separate the organic layer from the water layer.

Step 2: Shake the organic layer (diethyl ether, methyl benzoate, benzoic acid) with saturated aqueous sodium bicarbonate.

14.26 (cont) H_2SO_4 + $NaHCO_3$ $\longrightarrow$ $CO_2\uparrow$ + H_2O + SO_4^{2-} + $2\,Na^+$

(traces that   (excess)   soluble in water,
remain in the   insoluble in
methyl   diethyl ether
benzoate)

$\text{C}_6\text{H}_5\text{—COH}$ + $NaHCO_3$ $\longrightarrow$ $\text{C}_6\text{H}_5\text{—CO}^-\text{Na}^+$ + $CO_2\uparrow$ + H_2O

insoluble in water;   soluble in water; insoluble
soluble in diethyl ether   in diethyl ether

The mixture will separate into two layers. The upper layer will be the ether solution containing methyl benzoate, some of the methanol, and small amounts of water. The lower layer will contain sodium sulfate (the salt formed by the reaction of the remaining sulfuric acid with the sodium bicarbonate) sodium benzoate (the salt from the reaction of benzoic acid with sodium bicarbonate) and some methanol.

Step 3:   Separate the ether layer from the water layer. The ether and remaining methanol can be removed by distillation, leaving the higher boiling methyl benzoate behind. This is usually done after the ether layer is dried by adding a solid, anhydrous inorganic salt to it.

Concept Map 14.5   Reactions of acids and acid derivatives with alcohols.

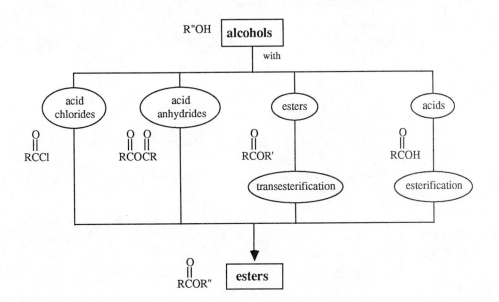

14.27

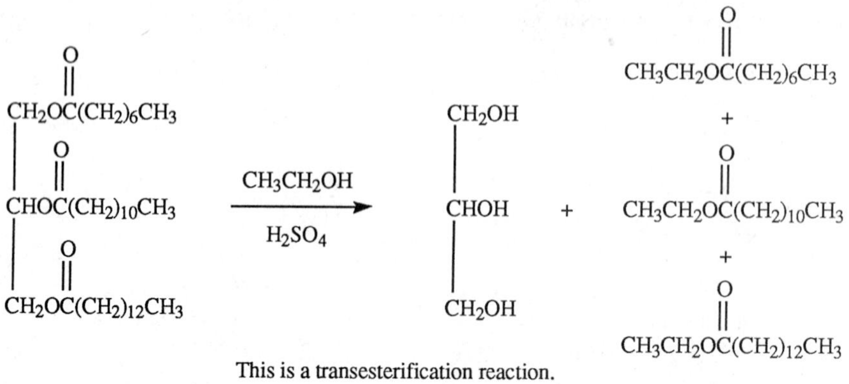

This is a transesterification reaction.

14.28

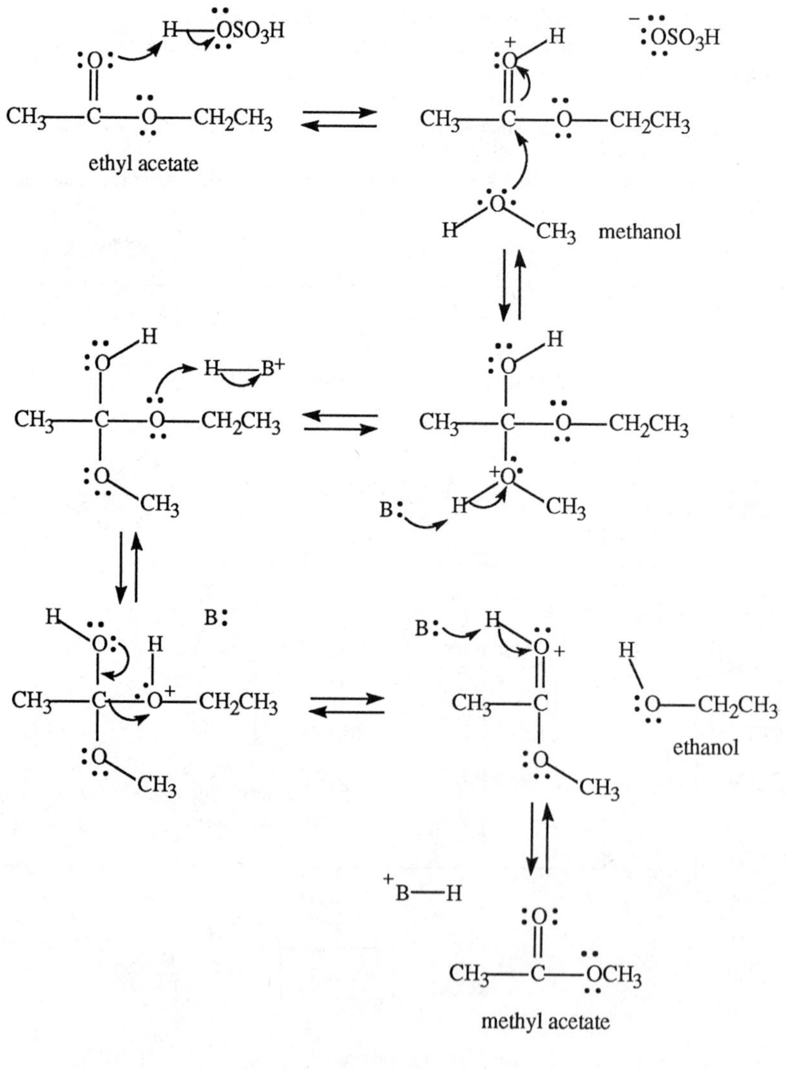

ethyl acetate

methanol

ethanol

methyl acetate

14.29

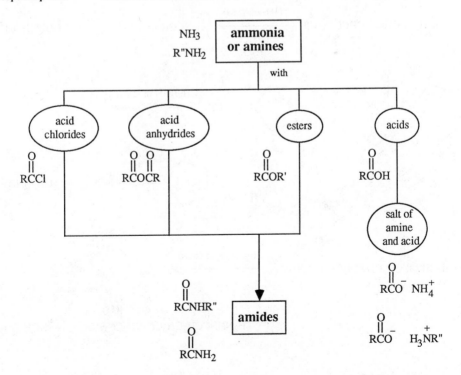

14.30 $\underset{\underset{CH_3\ OH}{|\quad\ \ |}}{HOCH_2C}\overset{\underset{|}{CH_3}\quad\ \overset{O}{\|}\quad\quad\ \overset{O}{\|}}{-CHCNHCH_2CH_2CO^-}\ Ca^{2+}\ ^-\underset{\underset{OH\ \ CH_3}{|\quad\ \ |}}{OCCH_2CH_2NHCCH-CCH_2OH}$

- -

Concept Map 14.6   Reactions of acids and acid derivatives with ammonia or amines.

14.31

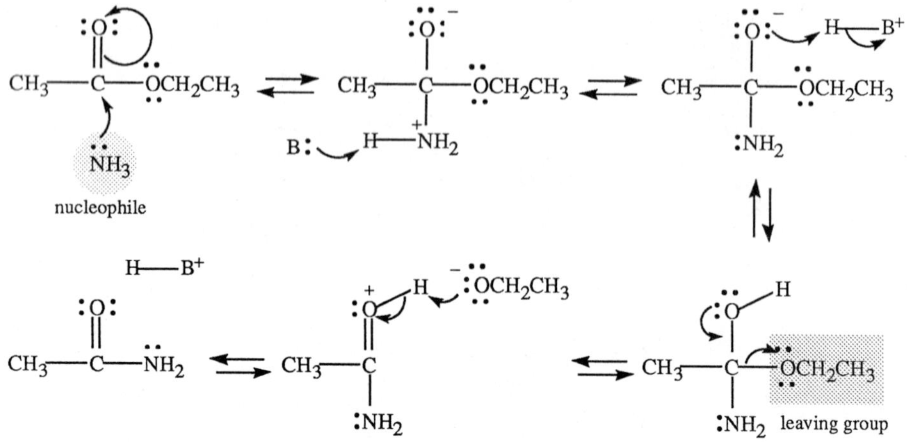

An alkoxide ion is not a good leaving group in an S_N2 reaction, but does serve as a leaving group in acyl transfer reactions. The bond energy of the carbonyl group is so high (Table 2.5, p. 61 in the text) that the reaction forming the carbonyl group occurs even when a good leaving group is not present.

14.32

(a)

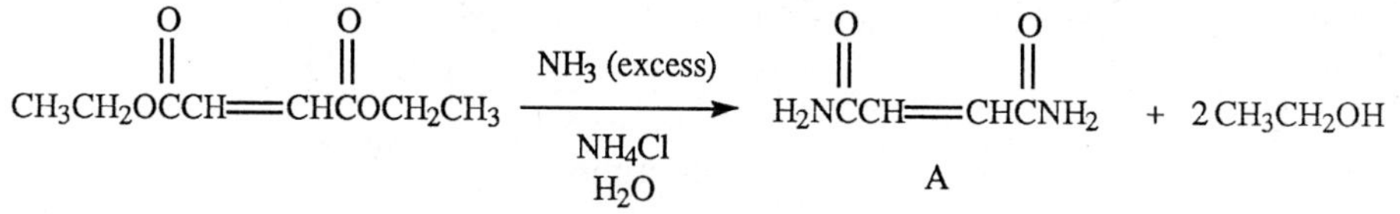

$$CH_3CH_2OCCH{=}CHCOCH_2CH_3 \xrightarrow[\substack{NH_4Cl \\ H_2O}]{NH_3 \text{ (excess)}} H_2NCCH{=}CHCNH_2 + 2\,CH_3CH_2OH$$

A

(b)

B

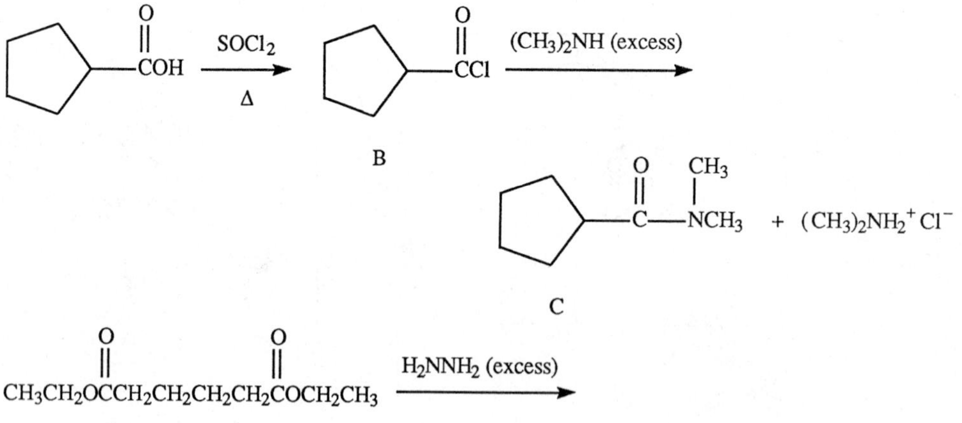

C

(c)

$$CH_3CH_2OCCH_2CH_2CH_2CH_2COCH_2CH_3 \xrightarrow{H_2NNH_2 \text{ (excess)}}$$

$$H_2NNHCCH_2CH_2CH_2CH_2CNHNH_2 + 2\,CH_3CH_2OH$$

D

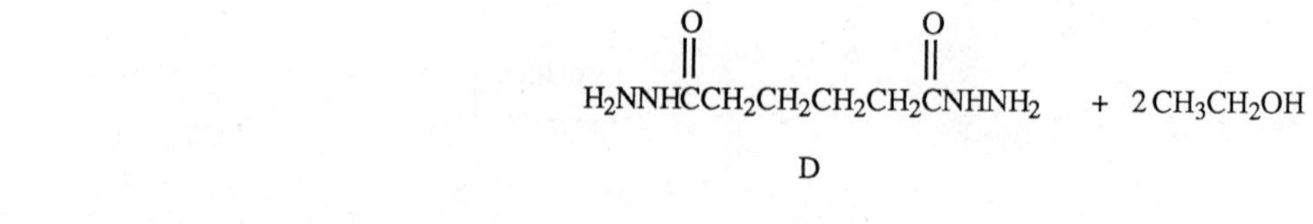

14.32 (cont)

(d)
$$CH_3(CH_2)_4\underset{\underset{CH_2CH_3}{|}}{\overset{\overset{O}{\parallel}}{C}}HCOH \quad \xrightarrow[\Delta]{SOCl_2} \quad CH_3(CH_2)_4\underset{\underset{CH_2CH_3}{|}}{\overset{\overset{O}{\parallel}}{C}}HCCl \quad \xrightarrow[H_2O]{NH_3} \quad CH_3(CH_2)_4\underset{\underset{CH_2CH_3}{|}}{\overset{\overset{O}{\parallel}}{C}}HCNH_2$$

E   F

(e)

NaOH
H₂O

G

(f)
$$N\equiv CCH_2\overset{\overset{O}{\parallel}}{C}OCH_2CH_3 \quad \xrightarrow[\substack{H_2O \\ 0\,°C}]{NH_3} \quad N\equiv CCH_2\overset{\overset{O}{\parallel}}{C}NH_2 \quad + \quad CH_3CH_2OH$$

H

14.33 (a) 2,4-dichlorobenzoic acid   (b) 4-bromohexanoic acid

(c) 2-methylbutanedioic anhydride   (d) 2-methylpentanenitrile
(2-methylsuccinic anhydride)

(e) ethyl (1*R*, 2*R*)-2-ethylcyclohexanecarboxylate (f) *N,N*-dimethylpentanamide

(g) (*R*)-2-hydroxypentanoic acid   (h) 2-chloropentanoyl chloride

(i) 4-ethylbenzanilide   (j) methyl 2-bromo-4-methylpentanoate

(k) ethyl 2-aminobenzoate   (l) 2-methylhexanedioic acid
(2-methyladipic acid)

(m) hexanoic anhydride   (n) (*E*)-4-octenoic acid

(o) (1*R*, 2*R*)-cyclobutanedicarboxylic acid

14.34 (a)

(b)

14.34 (cont)

(c)

Br—⟨benzene ring with Br at position 2⟩—$\overset{\overset{O}{\|}}{C}NH_2$

(d)

CH_3CH_2 and $CH_2\overset{\overset{O}{\|}}{C}OCH_2CH_3$ on a $C=C$ double bond with H, H below

(e) $CH_3CH_2CH_2CH_2CH_2CH_2C\equiv N$

(f) $CH_3CH_2CH_2CH_2CH_2\overset{\overset{O}{\|}}{C}NCH_2CH_3$ with CH_2CH_3 below N

(g) $CH_3CH_2O\overset{\overset{O}{\|}}{C}CH_2CH_2\overset{\overset{CH_3}{|}}{C}\!\!-\!\!\overset{\overset{O}{\|}}{C}OCH_2CH_3$ with CH_3 below

(h) O_2N—⟨benzene ring⟩—$\overset{\overset{O}{\|}}{C}OCH_3$

(i) $CH_3CH_2CH_2CH_2CH_2\overset{\overset{CH_3}{|}}{C}CH_2\overset{|}{C}HCH_2\overset{\overset{O}{\|}}{C}OH$ with CH_3 and OH below

(j)   Br—⟨benzene ring⟩—$\overset{\overset{O}{\|}}{C}\overset{\overset{O}{\|}}{C}$—⟨benzene ring⟩—Br

(k) CH_3O—⟨benzene ring⟩—$\overset{\overset{O}{\|}}{C}Cl$

14.35

(a) CH_3O—⟨benzene ring⟩—NH_2 $\xrightarrow[\substack{\text{acetic acid}\\H_2O\\0\text{-}5\,°C}]{\overset{\overset{O\quad O}{\|\quad\|}}{CH_3COCCH_3}}$ CH_3O—⟨benzene ring⟩—$NH\overset{\overset{O}{\|}}{C}CH_3$ + $CH_3\overset{\overset{O}{\|}}{C}OH$

A

14.35 (cont)

(b) 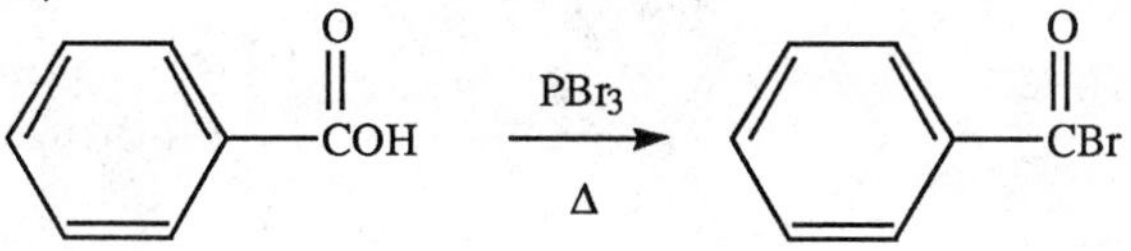

B

(c) 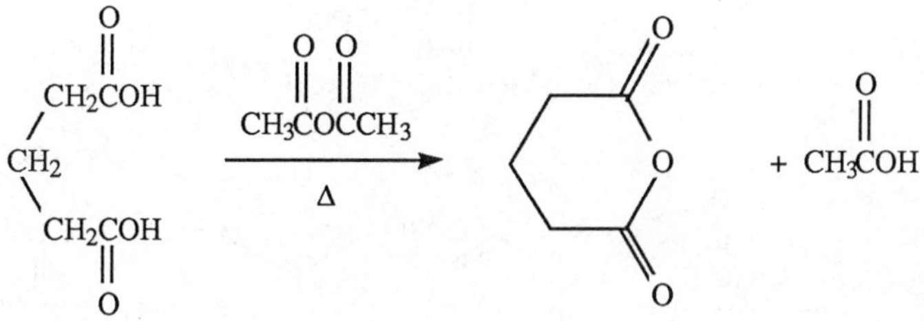

C

(d) 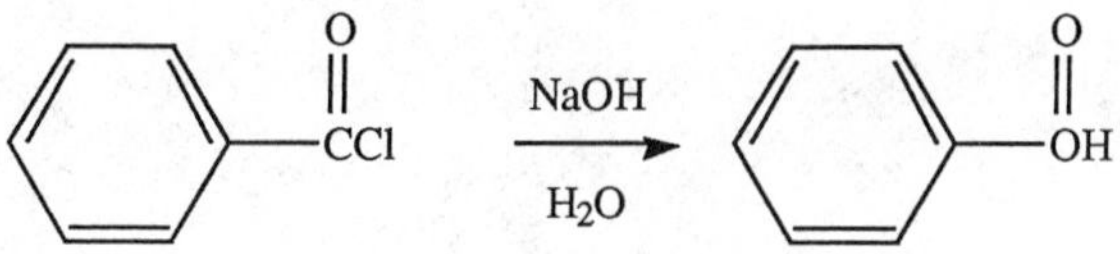

D

(e) 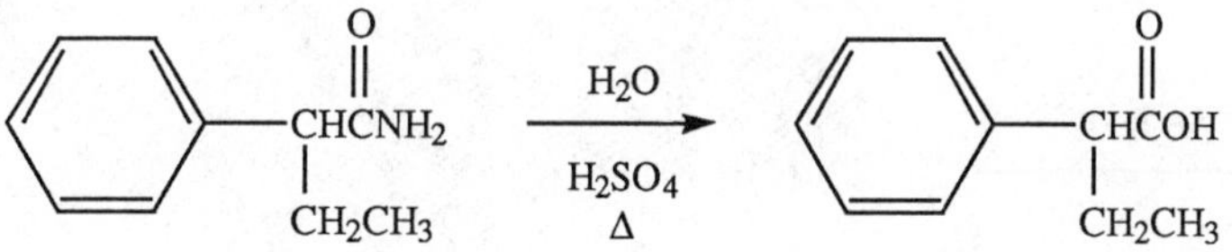

E

(f) 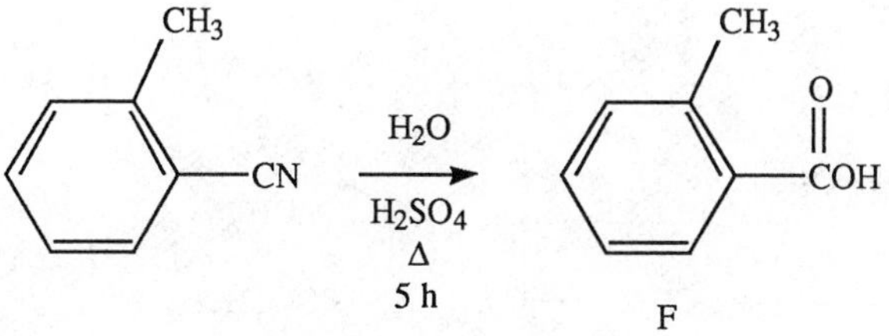

F

(g) $CH_3CH_2\overset{O}{\overset{\|}{C}}Cl \xrightarrow{NH_3\ (excess)} CH_3CH_2\overset{O}{\overset{\|}{C}}NH_2$

G

14.35 (cont)

(h)

$$\text{C}_6\text{H}_5\text{-}\overset{\overset{\displaystyle O}{\|}}{\text{C}}\text{OCH}_2\text{CH}_3 \xrightarrow{\text{H}_2\text{NOH}} \text{C}_6\text{H}_5\text{-}\overset{\overset{\displaystyle O}{\|}}{\text{C}}\text{NHOH} + \text{CH}_3\text{CH}_2\text{OH}$$

H

(i)

$$\text{CH}_3(\text{CH}_2)_5\overset{\overset{\displaystyle O}{\|}}{\text{C}}\text{Cl} \xrightarrow{\text{CH}_3(\text{CH}_2)_5\overset{\overset{\displaystyle O}{\|}}{\text{C}}\text{O}^-\text{Na}^+} \text{CH}_3(\text{CH}_2)_5\overset{\overset{\displaystyle O}{\|}}{\text{C}}\text{O}\overset{\overset{\displaystyle O}{\|}}{\text{C}}(\text{CH}_2)_5\text{CH}_3$$

I

(j)

$$\text{o-C}_6\text{H}_4(\text{CH}_2\text{COOH})(\text{COOH}) \xrightarrow[\Delta]{(\text{CH}_3\text{C})_2\text{O}} J + \text{CH}_3\overset{\overset{\displaystyle O}{\|}}{\text{C}}\text{OH}$$

J

(k)

$$\text{CH}_3(\text{CH}_2)_{10}\overset{\overset{\displaystyle O}{\|}}{\text{C}}\text{O}(\text{CH}_2)_{10}\text{CH}_3 + \text{CH}_3\text{OH} \xrightarrow[\Delta]{\text{H}_2\text{SO}_4} \text{CH}_3(\text{CH}_2)_{10}\overset{\overset{\displaystyle O}{\|}}{\text{C}}\text{OCH}_3 + \text{CH}_3(\text{CH}_2)_{10}\text{OH}$$

K   L

(l)

$$\text{HO}\overset{\overset{\displaystyle O}{\|}}{\text{C}}(\text{CH}_2)_4\overset{\overset{\displaystyle O}{\|}}{\text{C}}\text{OH} \xrightarrow{\text{SOCl}_2 \text{ (excess)}} \text{Cl}\overset{\overset{\displaystyle O}{\|}}{\text{C}}(\text{CH}_2)_4\overset{\overset{\displaystyle O}{\|}}{\text{C}}\text{Cl}$$

M

(m)

$$\text{Cl}\overset{\overset{\displaystyle O}{\|}}{\text{C}}(\text{CH}_2)_4\overset{\overset{\displaystyle O}{\|}}{\text{C}}\text{Cl} \xrightarrow{\substack{(\text{CH}_3)_3\text{COH (excess)} \\ \text{C}_6\text{H}_5\text{N(CH}_3)_2}}$$

M

$$(\text{CH}_3)_3\text{CO}\overset{\overset{\displaystyle O}{\|}}{\text{C}}(\text{CH}_2)_4\overset{\overset{\displaystyle O}{\|}}{\text{C}}\text{OC(CH}_3)_3 +$$

N

$$\text{C}_6\text{H}_5\overset{+}{\text{N}}\text{H(CH}_3)_2 \text{ Cl}^-$$

14.35 (cont)

(n)

O   P

(o)

$$CCl_3CH_2CH_2CH_2CH_2OH \xrightarrow[\substack{H_2O \\ \Delta}]{KMnO_4} CCl_3CH_2CH_2CH_2\overset{\overset{O}{\|}}{C}O^-K^+ \xrightarrow{H_3O^+} CCl_3CH_2CH_2CH_2\overset{\overset{O}{\|}}{C}OH$$

Q   R

14.36

(a)

A
nucleophilic attack at the
less hindered carbon atom

B

(b)

$$\underset{\substack{CH_3 \\ |}}{CH_3CH(CH_2)_{10}CH_2OH} \xrightarrow[\substack{(CH_3CH_2)_3N \\ dichloromethane}]{\overset{\overset{O}{\|}}{CH_3SCl} \; O} \underset{\substack{CH_3 \\ |}}{CH_3CH(CH_2)_{10}CH_2O\overset{\overset{O}{\|}}{\underset{\overset{\|}{O}}{S}}CH_3} \xrightarrow[\substack{tetrahydrofuran \\ dimethyl sulfoxide}]{K^{14}CN}$$

C

$$\underset{\substack{CH_3 \\ |}}{CH_3CH(CH_2)_{10}CH_2{}^{14}CN} \xrightarrow[\substack{ethanol \\ \Delta}]{KOH, H_2O} \underset{\substack{CH_3 \\ |}}{CH_3CH(CH_2)_{10}CH_2{}^{14}\overset{\overset{O}{\|}}{C}O^-K^+} \xrightarrow{H_3O^+} \underset{\substack{CH_3 \\ |}}{CH_3CH(CH_2)_{10}CH_2{}^{14}\overset{\overset{O}{\|}}{C}OH}$$

D   E   F

14.36 (cont)

(c)

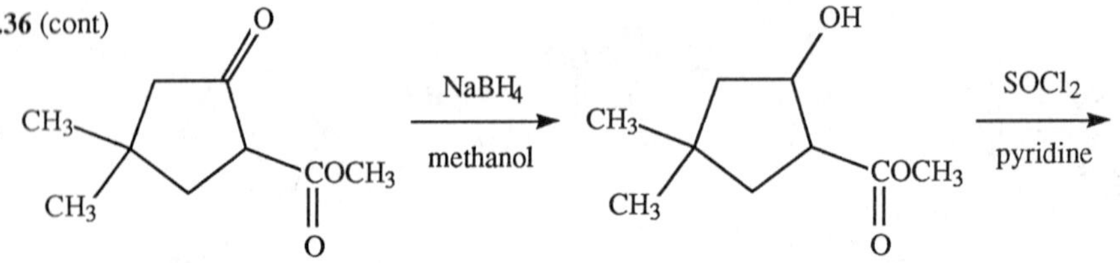

G

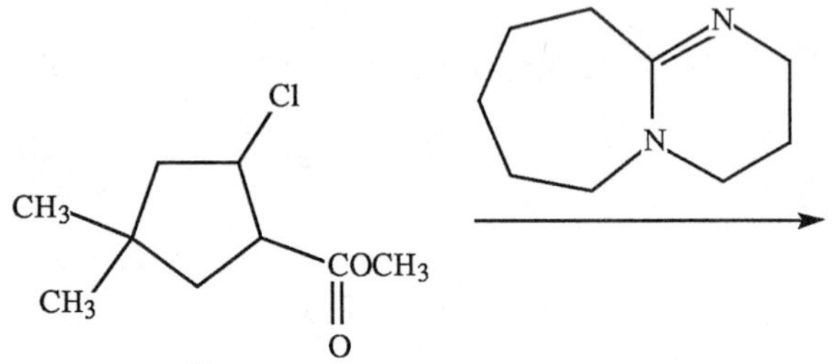

H

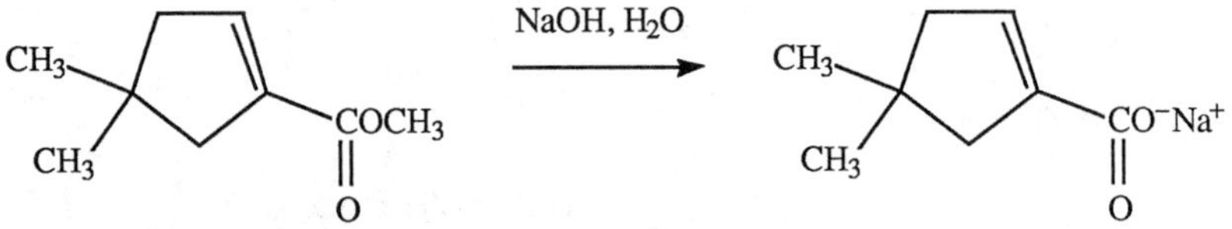

I   J

(A double bond conjugated
with a carbonyl group is more
stable. See Section 16.5A)

14.37 (a)

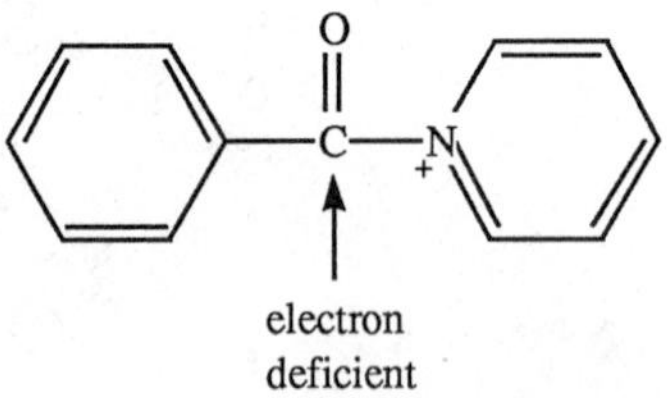

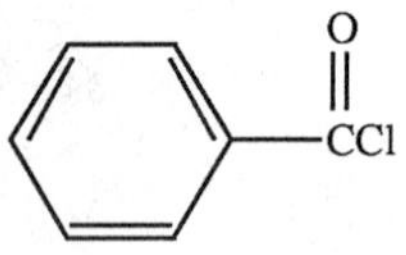

electron
deficient

The intermediate formed when pyridine reacts with benzoyl chloride is an unstable species with a positive
charge adjacent to an electron deficient carbon atom. It has an especially good leaving group because the
positive charge on the nitrogen atom decreases in the transition state, lowering the activation energy.
When chloride ion is the leaving group, as in benzoyl chloride, charge is building up in the transition state,
and the activation energy for that reaction is higher than when pyridine is the leaving group.

14.37 (cont)

(b)

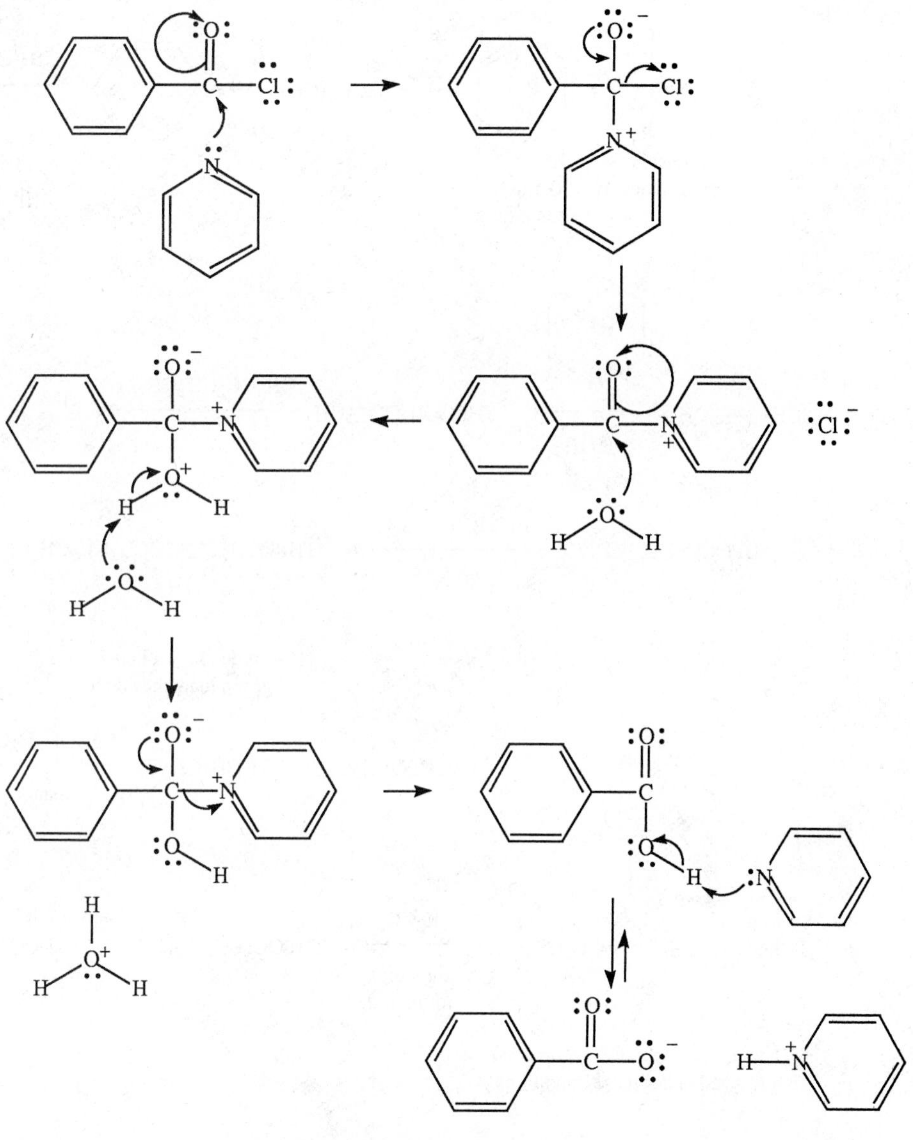

(c)

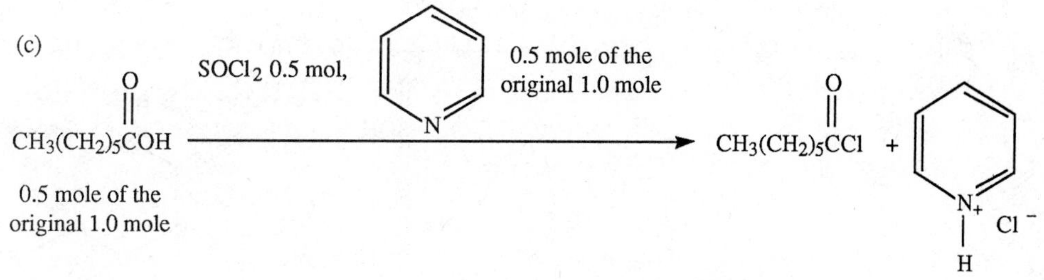
SOCl2 0.5 mol,
0.5 mole of the
original 1.0 mole
CH3(CH2)5COH
0.5 mole of the
original 1.0 mole
CH3(CH2)5CCl
Cl −

14.37 (c) (cont)

$$CH_3(CH_2)_5CCl \ + \ \text{pyridine} \ \rightleftharpoons \ CH_3(CH_2)_5C{-}\overset{+}{N}(\text{pyridine}) \quad Cl^{-} \quad \xrightarrow{\ CH_3(CH_2)_5COH\ }$$

0.5 mole,
product of
previous reaction

0.5 mole
remaining

$$CH_3(CH_2)_5COC(CH_2)_5CH_3$$

14.38　$HOCH_2C{\equiv}CH \ \xrightarrow[\text{TsOH}]{\text{(dihydropyran)}} \ THPOCH_2C{\equiv}CH \ \xrightarrow{\ CH_3CH_2CH_2CH_2Li\ }$

　　　　　　　　　　　　　　　　　　　　　　　　　　A

$$THPOCH_2C{\equiv}C^{-}Li^{+} \ \xrightarrow{\ BrCH_2CH_2CH_2Cl\ } \ THPOCH_2C{\equiv}CCH_2CH_2CH_2Cl \ \xrightarrow[\text{dimethyl sulfoxide}]{\ NaCN\ }$$

　　　　B　　　　　　　　　　　　　　　　　　　　　　C

(Bromide ion is a better leaving
group than choride ion.)

$$THPOCH_2C{\equiv}CCH_2CH_2CH_2CN \ \xrightarrow[\substack{\text{ethanol} \\ \Delta}]{\ NaOH, H_2O\ } \ \xrightarrow[\substack{0\,°C \\ 5\ min}]{\ H_3O^{+}\ }$$

　　　　　　　　　D

$$THPOCH_2C{\equiv}CCH_2CH_2CH_2COH \ \xrightarrow[\substack{\text{dimethyl-} \\ \text{formamide}}]{\ K_2CO_3\ } \ THPOCH_2C{\equiv}CCH_2CH_2CH_2CO^{-}K^{+} \ \xrightarrow{\ CH_3I\ }$$

　　　　　E　　　　　　　　　　　　　　　　　　　　　F

(The tetrahydropyranyl group is not
hydrolyzed under these conditions.)

$$THPOCH_2C{\equiv}CCH_2CH_2CH_2COCH_3 \ \xrightarrow[\substack{\text{(benzenesulfonic acid, } SO_2H) \\ 3\ h}]{\ CH_3OH\ } \ HOCH_2C{\equiv}CCH_2CH_2CH_2COCH_3$$

　　　　　G　　　　　　　　　　　　　　　　　　　　　　　　　H

$$+ \ \text{(2-methoxytetrahydropyran, } OCH_3)$$

14.38 (cont)

The infrared spectrum of Compound H shows the presence of an alcohol ($3450\ cm^{-1}$ for the O—H stretch) and an ester ($1720\ cm^{-1}$ for the C=O stretch). Analysis of the proton magnetic resonance spectrum of Compound H is shown below:

δ 1.6-2.0	(2H, multiplet, —CH$_2$CH$_2$—)
δ 2.2-2.6	(5H, multiplet, —(≡CCH$_2$CH$_2$CH$_2$C=O and —OH)
δ 3.68	(3H, singlet, —OCH$_3$)
δ 4.2-4.3	(2H, multiplet, —OCH$_2$C≡)

Compound H is methyl 7-hydroxy-5-heptynoate

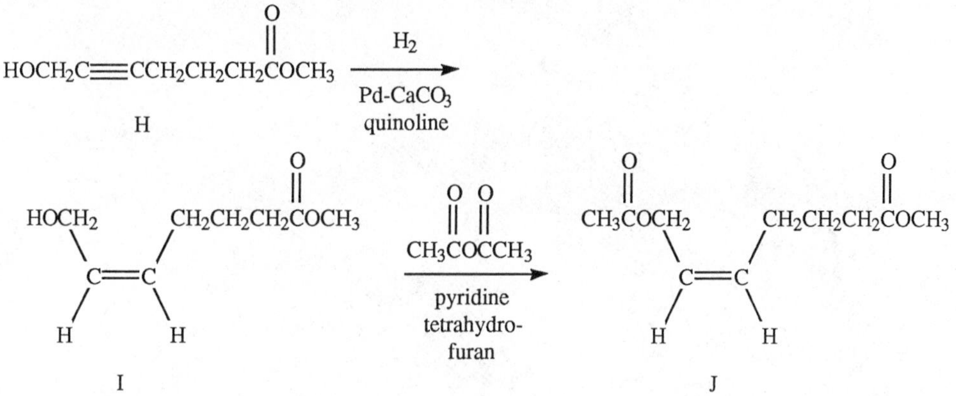

14.39

$$CO_3^{2-} + CH_3OH \rightleftharpoons HCO_3^- + CH_3O^-$$
$$pK_a\ 15 \qquad\qquad pK_a\ 10.2$$

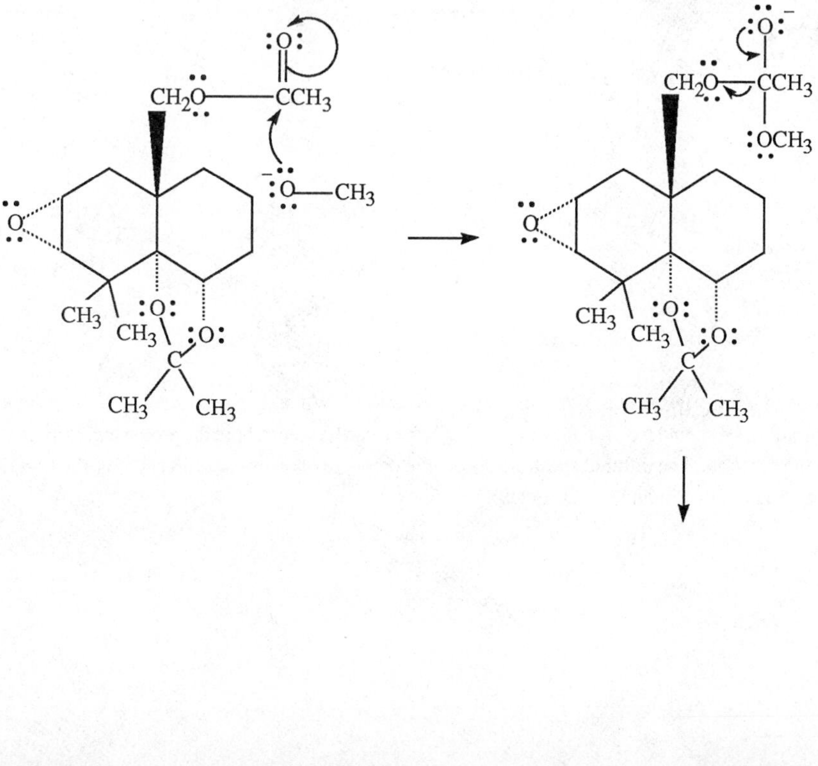

14.39 (cont)

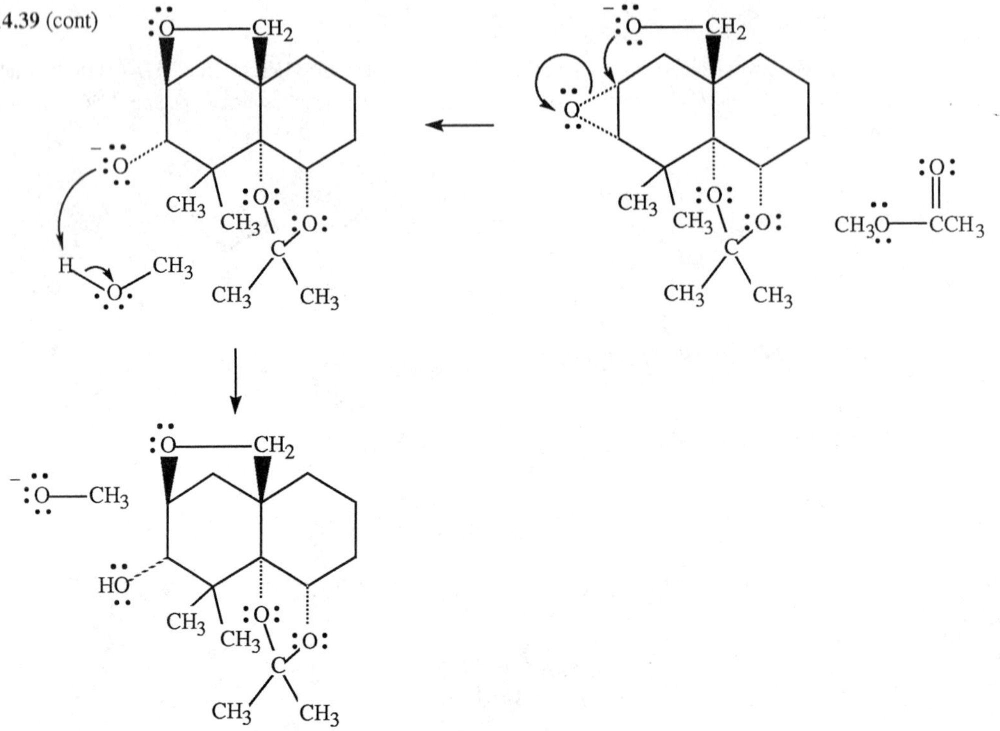

14.40

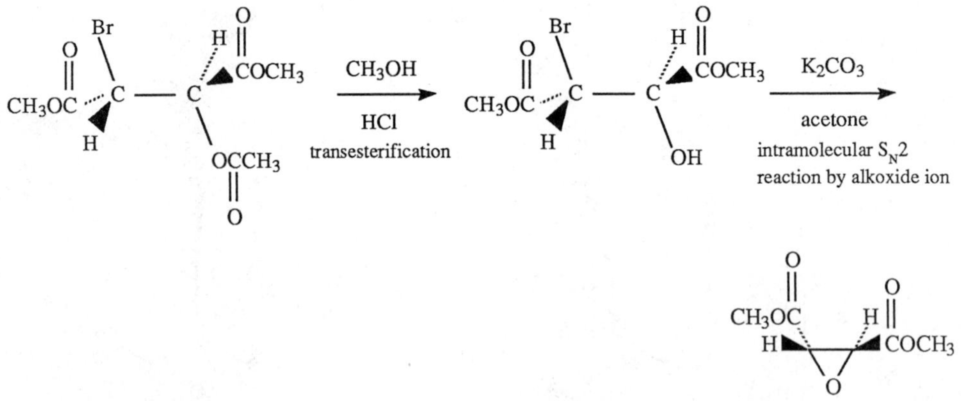

14.41   Compound A, $C_9H_{10}O_2$, has five units of unsaturation. When a compound has four or more units of unsaturation, we should look for an aromatic ring, which can be seen in both the proton and carbon-13 magnetic resonance spectra. The infrared spectrum shows the presence of an ester (1743 cm^{-1} for the C=O stretch and 1229 and 1118 cm^{-1} for the C—O stretch).

14.41 (cont)

Analysis of the proton spectrum of Compound A is shown below:

δ 2.2   (3H, singlet, C<u>H</u>₃C=O

δ 5.2   (2H, singlet, ArC<u>H</u>₂—O—)

δ 7.4   (5H, singlet, Ar<u>H</u>)

The absence of splitting for the aromatic hydrogen atoms tells us that the group bonded to the aromatic ring does not differ significantly from the aryl carbon atoms in electronegativity and in magnetic anisotropy. A tetrahedral carbon atom meets this requirement.

The carbon-13 magnetic resonance spectrum shows two alkyl carbon atoms (20.7 and 66.1 ppm). There are three types of aromatic carbon atoms (128.1, 128.4, and 136.2 ppm), and one carbonyl carbon atom (170.5 ppm). Compound A is benzyl acetate.

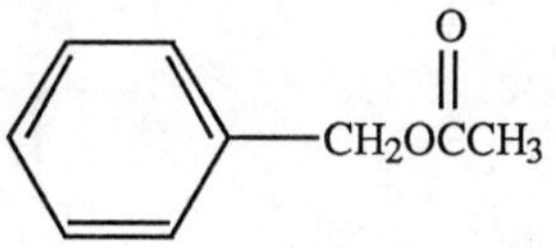

Note that the compound cannot be methyl phenylacetate because the band for the methyl singlet in the proton magnetic resonance spectrum of a methyl ester would be at ~δ 3.7.

15

Carboxylic Acids and Their Derivatives II. Synthetic Transformations and Compounds of Biological Interest

15.1

15.2

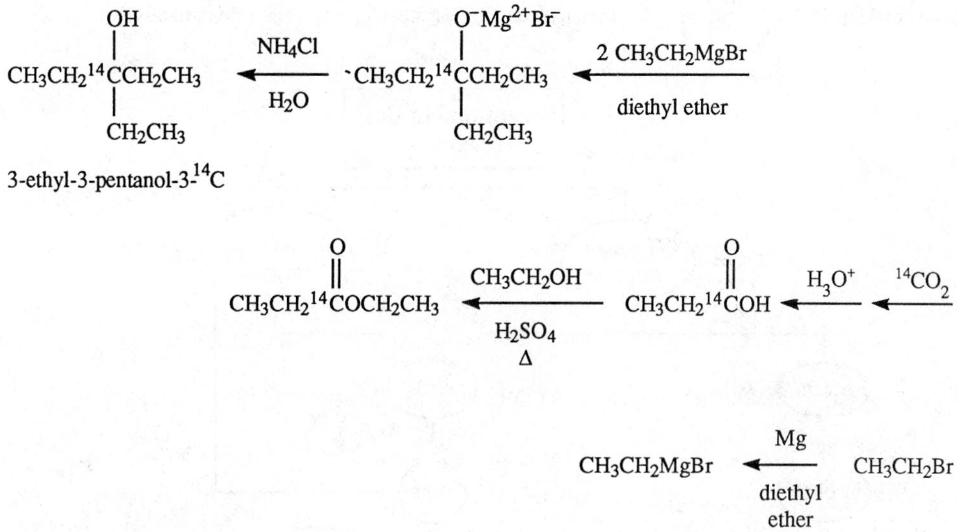

$$\underset{\substack{3\text{-ethyl-3-pentanol-3-}^{14}C}}{\underset{\underset{\underset{CH_2CH_3}{|}}{\overset{\overset{OH}{|}}{CH_3CH_2^{14}CCH_2CH_3}}}{}} \xleftarrow[\text{H}_2\text{O}]{\text{NH}_4\text{Cl}} \underset{\underset{CH_2CH_3}{|}}{\overset{\overset{O^-\text{Mg}^{2+}\text{Br}^-}{|}}{^{\cdot}CH_3CH_2^{14}CCH_2CH_3}} \xleftarrow[\text{diethyl ether}]{2\ CH_3CH_2MgBr}$$

$$\underset{}{\overset{\overset{O}{\|}}{CH_3CH_2^{14}COCH_2CH_3}} \xleftarrow[\substack{\text{H}_2\text{SO}_4 \\ \Delta}]{CH_3CH_2OH} \underset{}{\overset{\overset{O}{\|}}{CH_3CH_2^{14}COH}} \xleftarrow{\text{H}_3\text{O}^+} \xleftarrow{^{14}CO_2}$$

$$CH_3CH_2MgBr \xleftarrow[\substack{\text{diethyl} \\ \text{ether}}]{\text{Mg}} CH_3CH_2Br$$

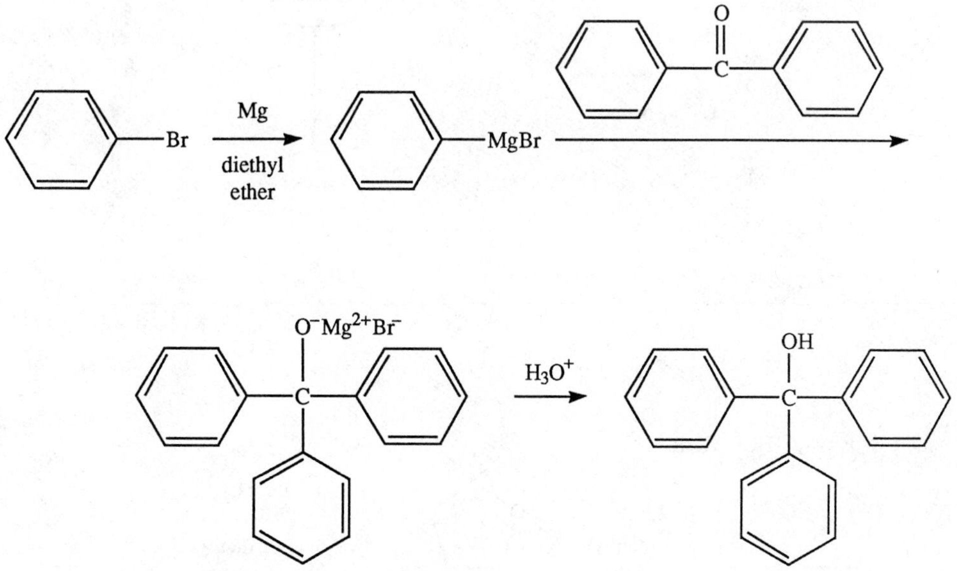

15.3

Concept Map 15.1   Reactions of organometallic reagents with acids and acid derivatives.

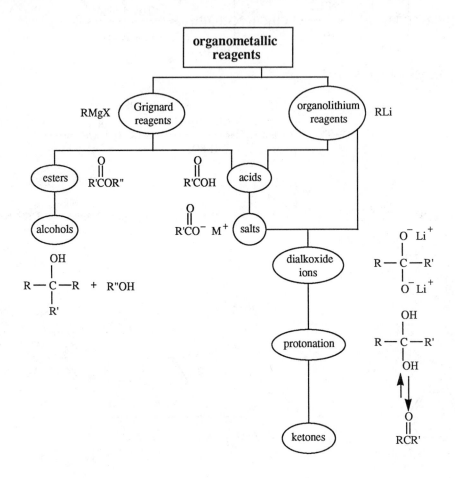

- -

15.4

(a)

15.4 (cont)

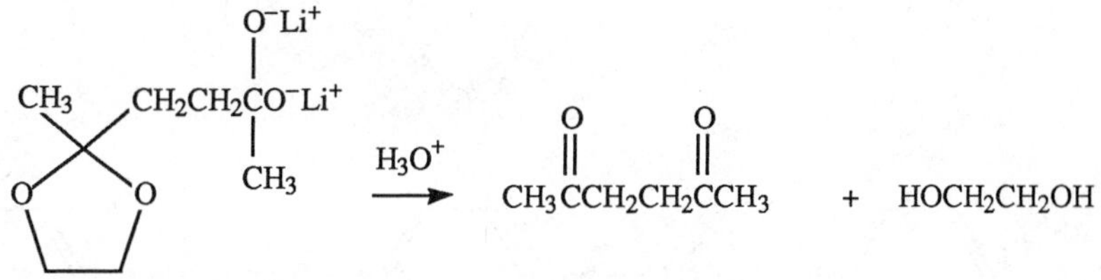

B

C

(b)

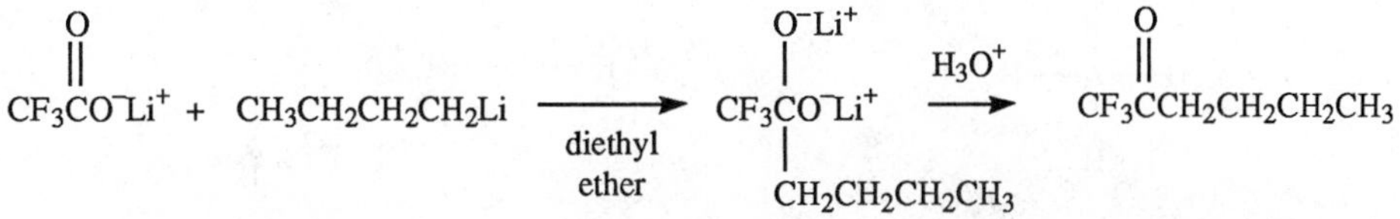

D

E

(c)

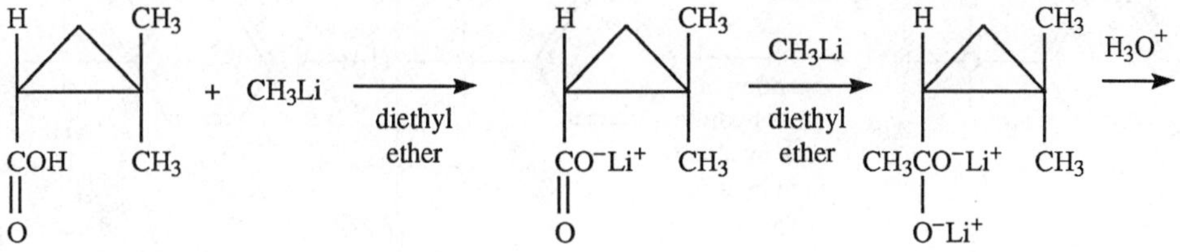

F

G

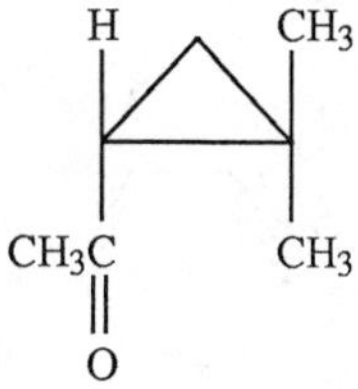

H

(d)

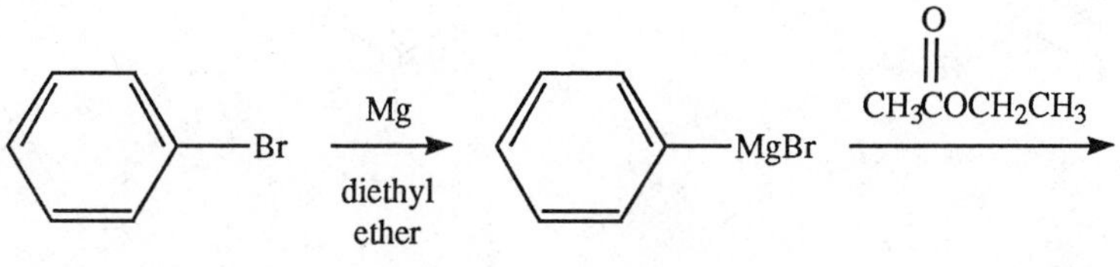

I

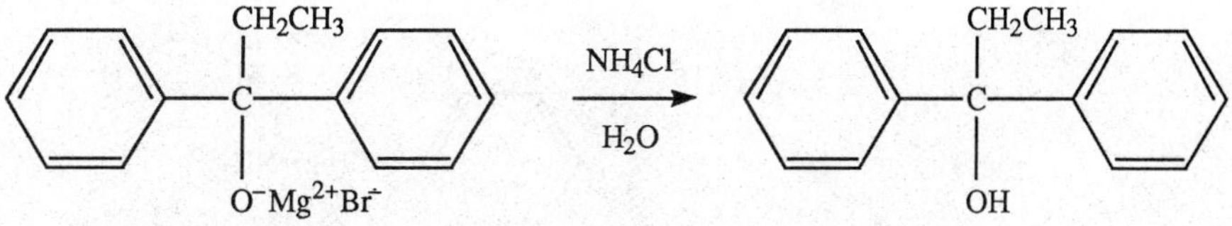

J

K

15.5

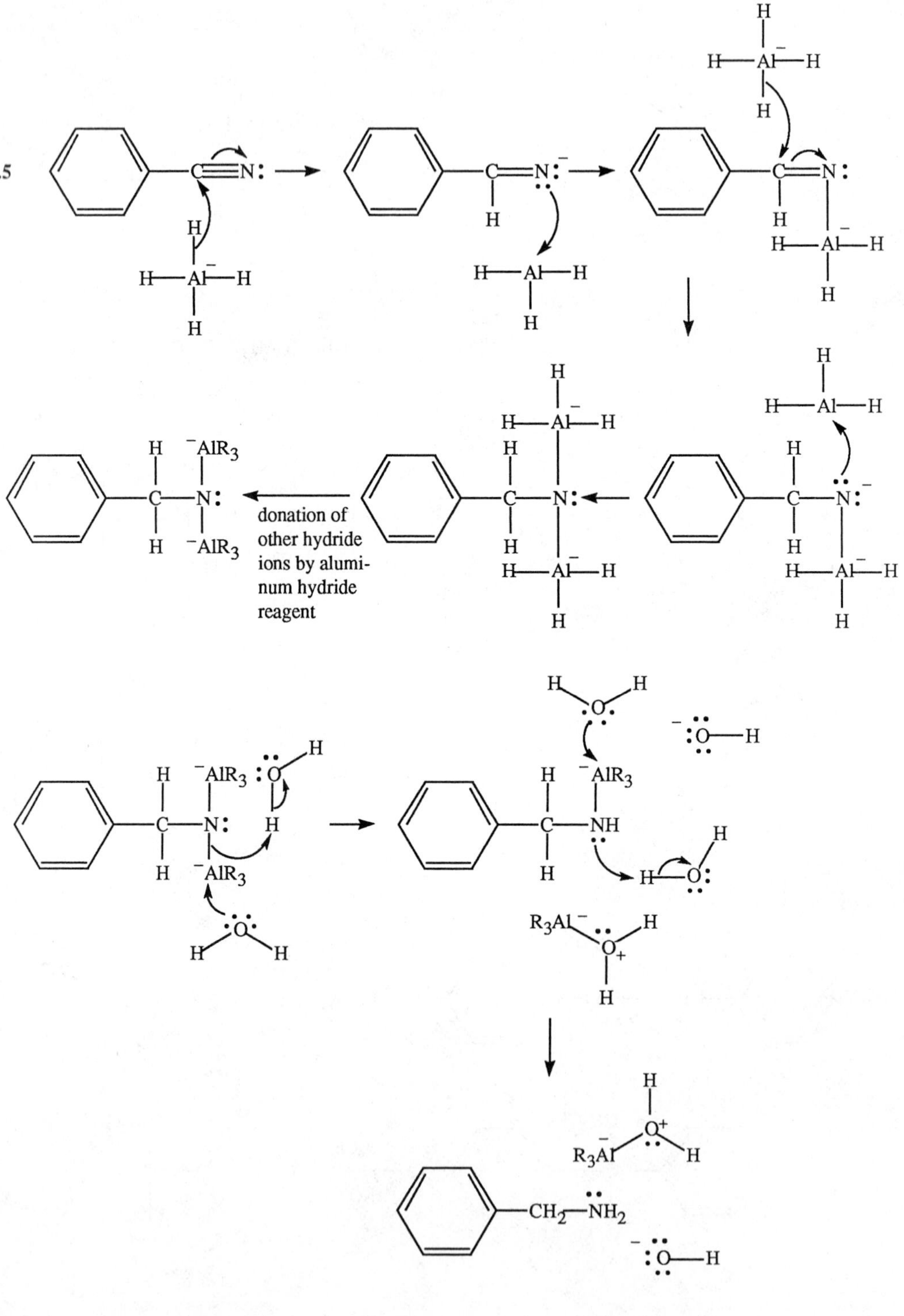

donation of
other hydride
ions by alumi-
num hydride
reagent

15.6

(a)

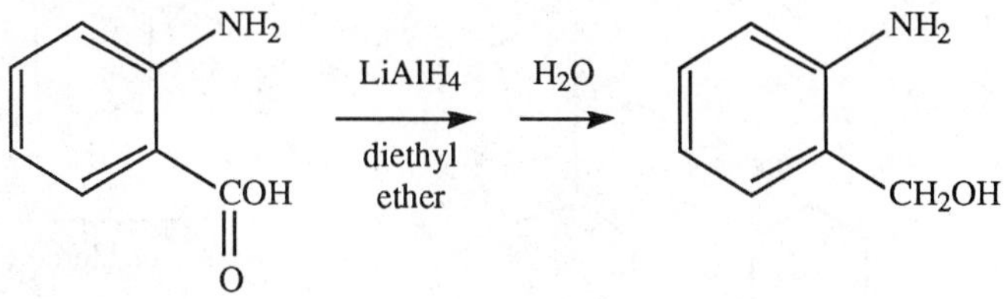

(b)

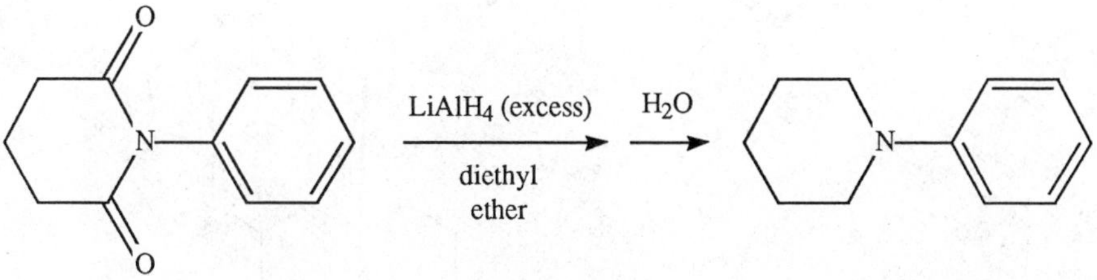

(c)

$$CH_3CH{=}CHCH_2CH_2\overset{\overset{\displaystyle O}{\|}}{C}OCH_3 \quad \xrightarrow[\text{diethyl ether}]{LiAlH_4} \quad \xrightarrow{H_3O^+} \quad CH_3CH{=}CHCH_2CH_2CH_2OH$$

$$+ \quad CH_3OH$$

(d)

$$CH_3CH_2O\overset{\overset{\displaystyle O}{\|}}{C}(CH_2)_4\overset{\overset{\displaystyle O}{\|}}{C}OCH_2CH_3 \quad \xrightarrow[\text{diethyl ether}]{LiAlH_4} \quad \xrightarrow{H_3O^+} \quad HOCH_2(CH_2)_4CH_2OH \quad + \quad CH_3CH_2OH$$

(e)

$$CH_3CH_2\overset{\overset{\displaystyle O}{\|}}{C}NHCH_2CH_3 \quad \xrightarrow[\text{diethyl ether}]{LiAlH_4} \quad \xrightarrow{H_2O} \quad CH_3CH_2CH_2NHCH_2CH_3$$

(f)

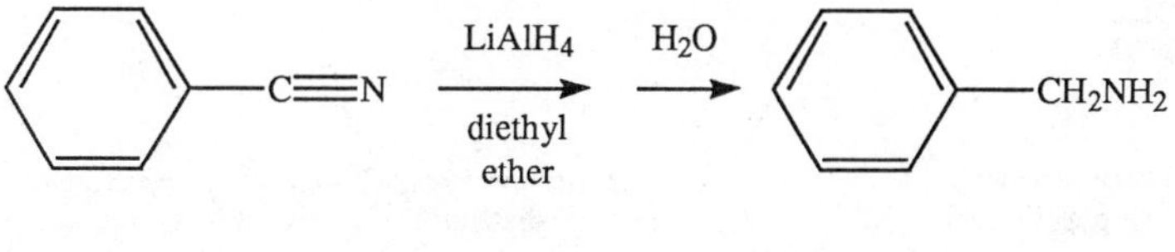

15.7

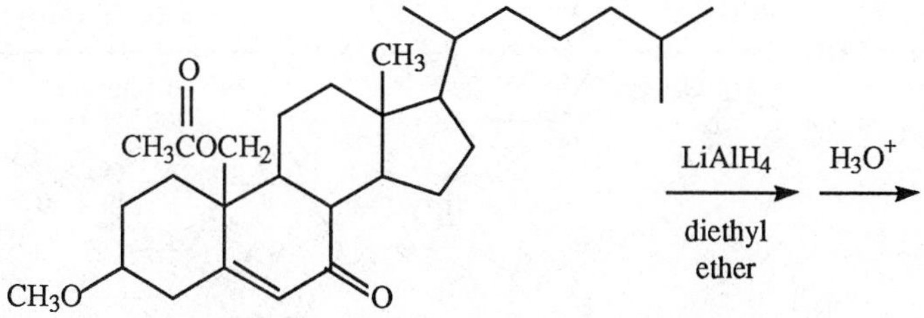

15.7 (cont)

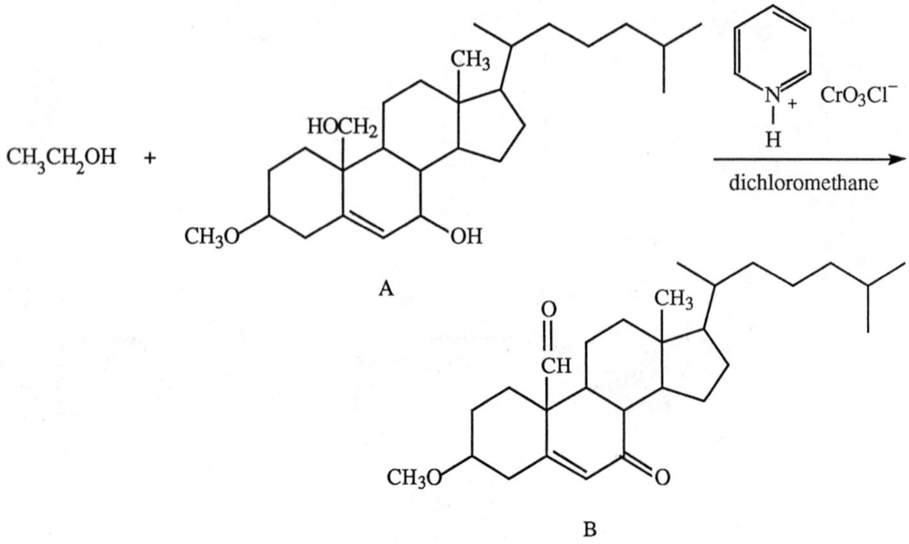

CH_3CH_2OH +

A

B

15.8

(1)

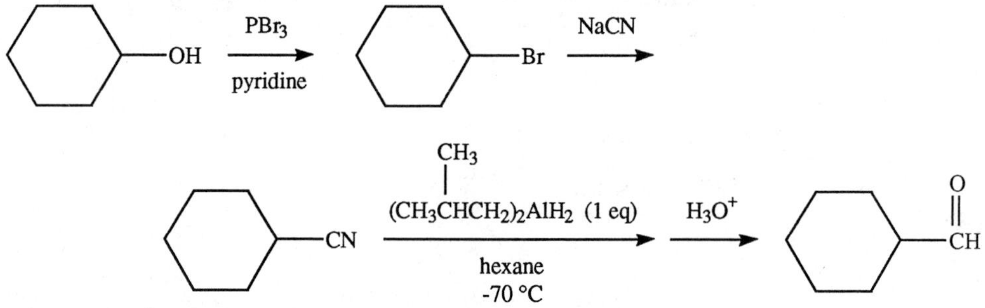

(2)

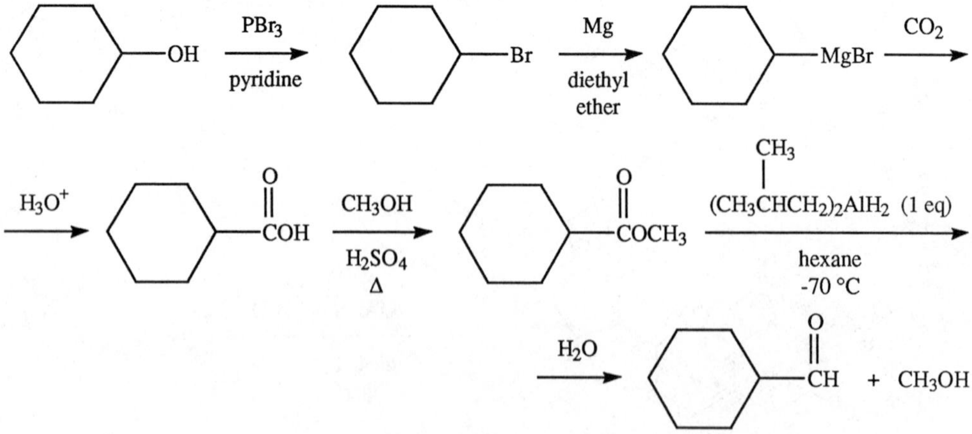

Concept Map 15.2   Reduction of acids and acid derivatives.

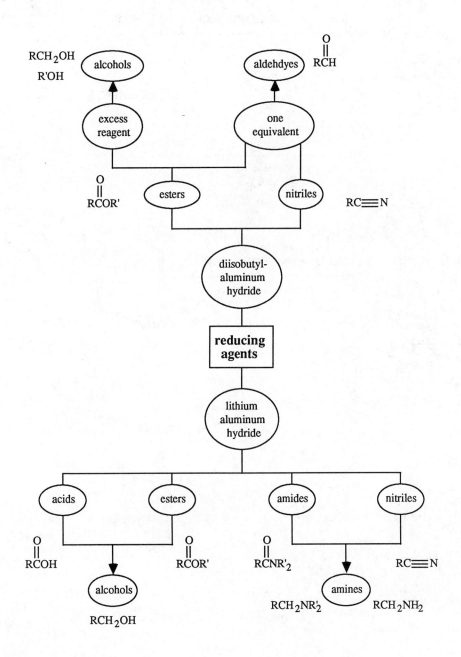

15.9

(a)

$$CH_3CH_2OCH_2(CH_2)_5\overset{\overset{\displaystyle O}{\|}}{C}OCH_2CH_3 \xrightarrow[\substack{hexane \\ -70\ °C}]{\overset{\overset{\displaystyle CH_3}{|}}{(CH_3CHCH_2)_2AlH_2}\ (1\ eq)} \xrightarrow{H_2O}$$

$$CH_3CH_2OCH_2(CH_2)_5\overset{\overset{\displaystyle O}{\|}}{C}H \quad + \quad CH_3CH_2OH$$

(b)

$$\text{(benzoic acid)} \xrightarrow[\substack{benzene \\ 45\ °C,\ 8\ h}]{\overset{\overset{\displaystyle CH_3}{|}}{(CH_3CHCH_2)_2AlH_2}\ (3\ eq)} \xrightarrow{CH_3OH,\ H_2O} \text{(benzyl alcohol) —CH_2OH}$$

(c)

$$\xrightarrow[\substack{benzene \\ 25\ °C}]{\overset{\overset{\displaystyle CH_3}{|}}{(CH_3CHCH_2)_2AlH_2}\ (1\ eq)} \xrightarrow{H_3O^+}$$

$$\quad + \quad \overset{\overset{\displaystyle O}{\|}}{HCH}$$

15.10

(a)

$$\underset{\underset{\displaystyle O}{\|}}{\underset{\displaystyle C\overset{}{H_2}OCCH_3}{\overset{\overset{\displaystyle CH_3}{|}}{CH_3CHCHCH_2CH_2}\overset{\overset{\displaystyle O}{\|}}{O}CCH_3}} \xleftarrow[\substack{pyridine}]{\overset{\overset{\displaystyle O}{\|}}{CH_3CCl}\ (excess)} \underset{\underset{\displaystyle CH_2OH}{|}}{\overset{\overset{\displaystyle CH_3}{|}}{CH_3CHCHCH_2CH_2OH}} \xleftarrow[\substack{diethyl \\ ether}]{H_3O^+\ \ LiAlH_4}$$

$$\underset{\underset{\displaystyle O}{\|}}{\underset{\displaystyle COCH_2CH_3}{\overset{\overset{\displaystyle CH_3}{|}}{CH_3CHCHCH_2}\overset{\overset{\displaystyle O}{\|}}{C}OCH_2CH_3}} \xleftarrow[\substack{Pt}]{H_2} \underset{\underset{\displaystyle O}{\|}}{\underset{\displaystyle COCH_2CH_3}{\overset{\overset{\displaystyle CH_3}{|}}{CH_3C=CCH_2}\overset{\overset{\displaystyle O}{\|}}{C}OCH_2CH_3}}$$

15.10 (cont)

(b)

CH$_2$CH$_2$OH

$\xleftarrow[\text{H}_2\text{O}]{\text{H}_2\text{O}_2,\ \text{NaOH}}$ $\xleftarrow[\text{tetrahydro-furan}]{\text{BH}_3}$

OH

CH=CH$_2$

$\xleftarrow[]{\text{CH}_3\text{CH}_2\text{OH}}$ TsOH

OH

$\xleftarrow[\substack{\text{H}_2\text{O} \\ \Delta}]{\text{NaOH, CH}_3\text{CH}_2\text{OH}}$

CH=CH$_2$

OCCH$_3$
||
O

O
||
+   CH$_3$COCH$_2$CH$_3$

15.11

(a)

O=⟨ ⟩=O (with two O in ring) $\xrightarrow[\substack{\text{H}_2\text{O} \\ \Delta}]{\text{NaOH}}$

O
||
2 HOCH$_2$CO$^-$Na$^+$

(b)

O   O
|| ||
CH$_3$CCH$_2$CH$_2$CH$_2$COH $\xrightarrow[\text{H}_2\text{O}]{\text{NaBH}_4}$

O
||
CH$_3$CHCH$_2$CH$_2$CH$_2$CO$^-$Na$^+$
        |
        OH

$\xrightarrow{\text{H}_3\text{O}^+}$

(lactone with CH$_3$)

(c)

CH$_3$
  \
   O–⟨ ⟩=O
  /
CH$_3$ (lactone)

$\xrightarrow[\substack{\text{H}_2\text{O} \\ \Delta}]{\text{NaOH}}$

CH$_3$ O
  | ||
CH$_3$CCH$_2$CH$_2$CO$^-$Na$^+$
  |
  OH

(d)

OH
|
(tetrahydrofuran ring with OH and H) $\xrightarrow[\text{H}_2\text{SO}_4]{\text{CrO}_3}$ (γ-butyrolactone)

(e)

(cyclobutanone) $\xrightarrow[\text{chloroform}]{\text{O}||\text{C}_6\text{H}_5\text{C–COOH}}$ (γ-butyrolactone)

15.12

$$CH_3OH$$

$$HOCH_2CH_2CH_2CH_2CH_2COCH_3$$

A

CrO₃Cl⁻ / pyridinium chlorochromate

dichloromethane

$$HCCH_2CH_2CH_2CH_2COCH_3$$

B

15.13

?

1. What functional groups are present in the starting material and the product?

    The starting material is a ketone. The product is a ketone and an ester.

2. How do the carbon skeletons of the two compounds compare? How many carbon atoms does each contain? Are there any rings? What are the positions of branches and functional groups on the carbon skeletons?

    The starting material has eight carbon atoms and is a bicyclic compound. The product has nine carbon atoms but only has one ring, a five-membered one, which can also be seen in the bicyclic compound.

3. How do the functional groups change in going from starting material to product? Does the starting material have a good leaving group?

    The ketone of the starting material appears to have been transformed into an ester in the product. A ketone function appears on the five-membered ring in the product. There is no good leaving group in the starting material.

4. Is it possible to dissect the structures of the starting material and product to see which bonds must be broken and which formed?

bond broken   bonds formed

15.13 (cont)

5. Do we recognize any part of the product molecule as coming from a good nucleophile or an electrophilic addition?

    An ester can be made from an alcohol (a nucleophile) and an acid derivative (an electrophile).

6. What type of compound would be a good precursor to the product?

    An alcohol can be oxidized to a ketone.

An alcohol and an ester can be made from a lactone by solvolysis (Problem 15.12).

7. After this last step, do we see how to get from starting material to product? If not, we need to analyze the structure obtained in step 6 by applying questions 5 and 6 to it.

    We need to prepare the lactone.

5'. Do we recognize any part of the product molecule as coming from a good nucleophile or an electrophilic addition?

    The alkyl oxygen of a lactone can come from the nucleophilic oxygen of a peroxyacid.

6'. What type of compound would be a good precursor to the product?

    The Baeyer-Villiger reaction can be used to prepare the lactone from the starting material.

15.13 (cont)

7'. After this last step, do we see how to get from starting material to product? If not, we need to analyze the structure obtained in step 6 by applying questions 5 and 6 to it.

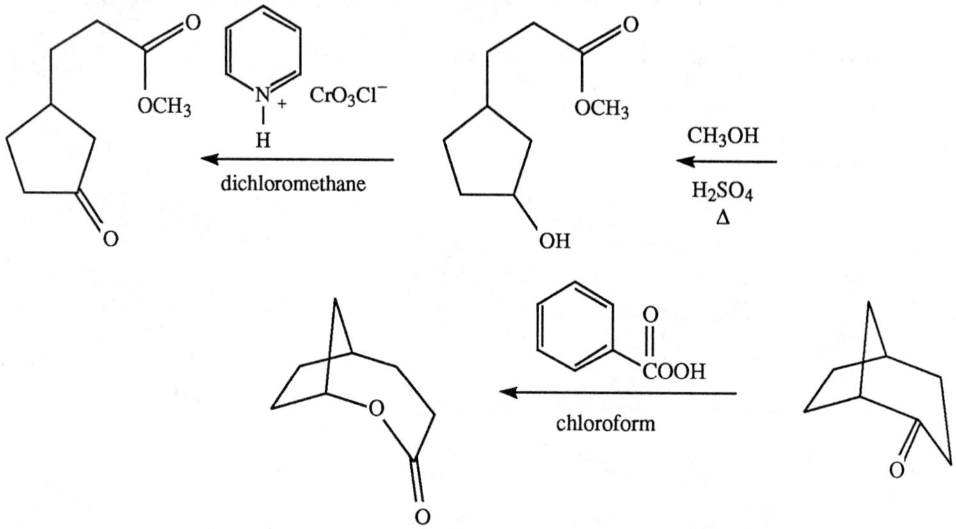

15.14

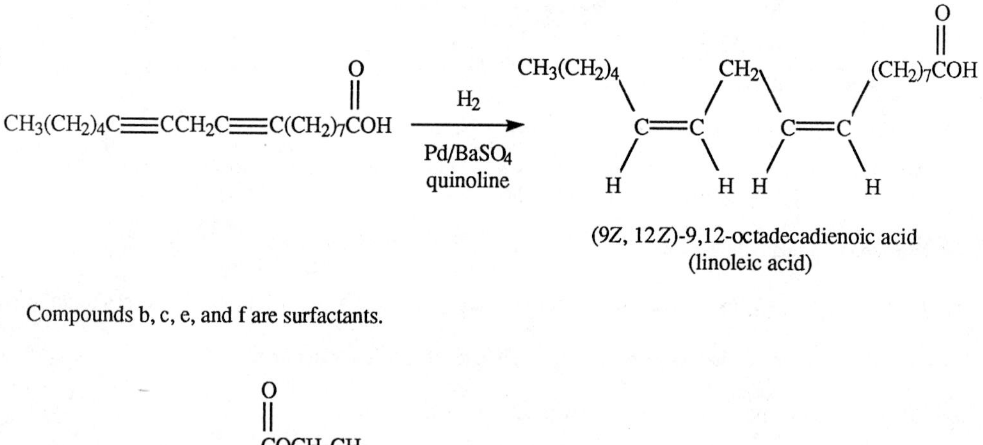

(9Z, 12Z)-9,12-octadecadienoic acid
(linoleic acid)

15.15   Compounds b, c, e, and f are surfactants.

15.16

(a)

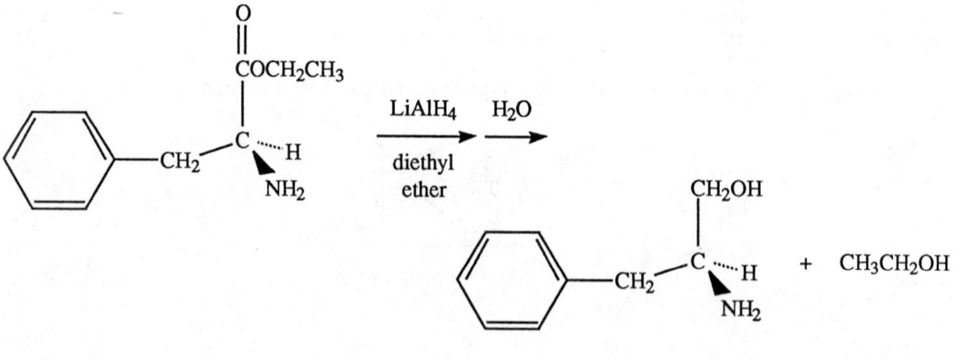

15.16 (cont)

(b)

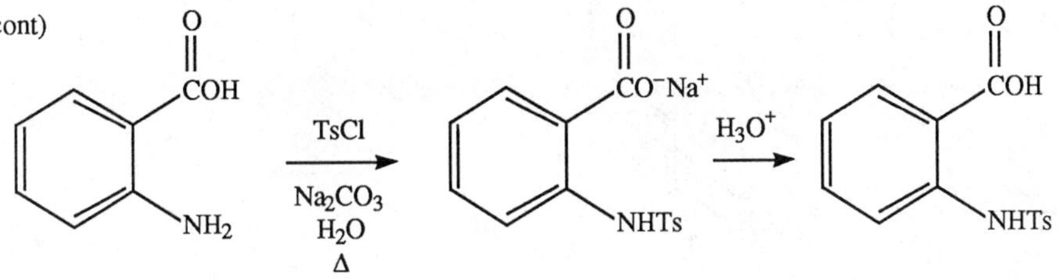

C   D

(c)

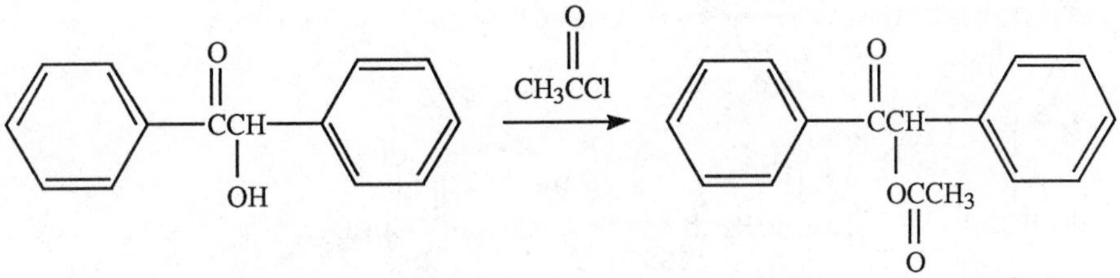

E

(d)

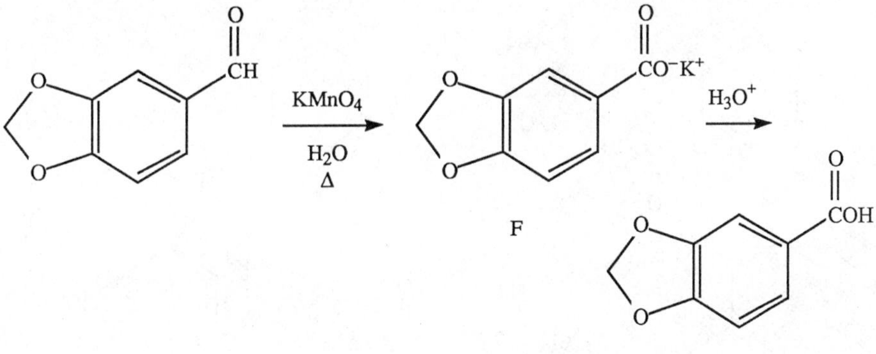

F

G

Note: The cyclic acetal on the aromatic ring is stable and does not open during the protonation of the carboxylate anion.

(e) $CH_3(CH_2)_{14}CH_2OH$ $\xrightarrow[\text{pyridine}]{\text{TsCl}}$ $CH_3(CH_2)_{14}CH_2OTs$

H

(f)

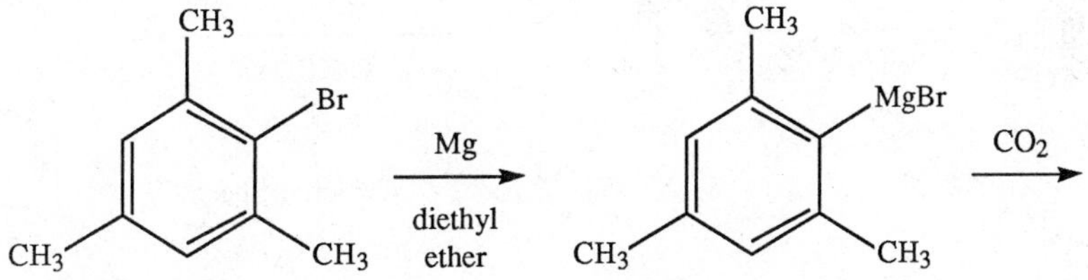

I

15.16 (cont)

(f) (cont)

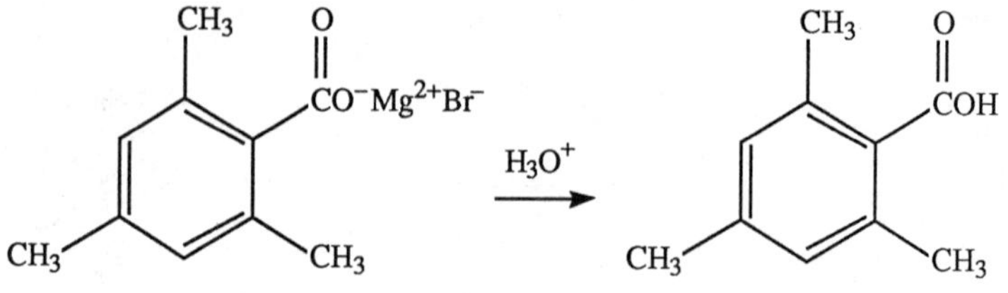

J   K

(g) $BrCH_2(CH_2)_8CH_2OH$ $\xrightarrow[\substack{H_2O \\ \text{acetic acid}}]{CrO_3}$ $BrCH_2(CH_2)_8\overset{\overset{\displaystyle O}{\|}}{C}OH$

L

(h)

M

(i)

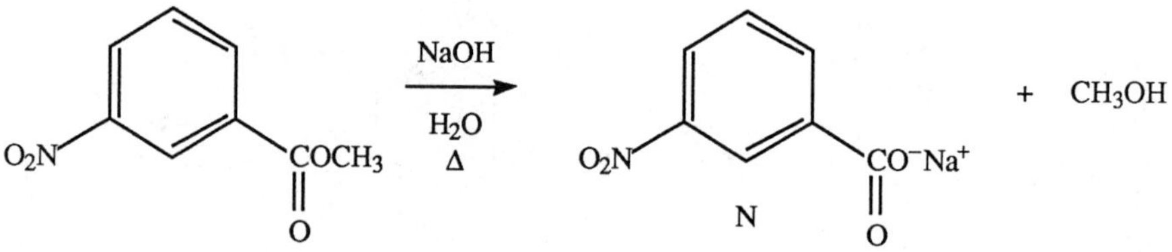

N   + CH_3OH

(j)

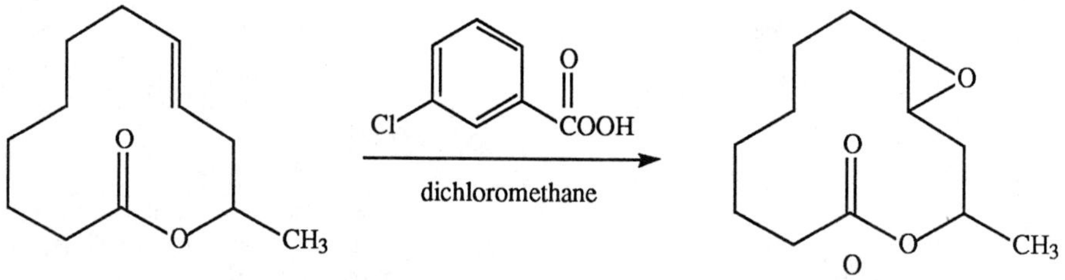

(k)

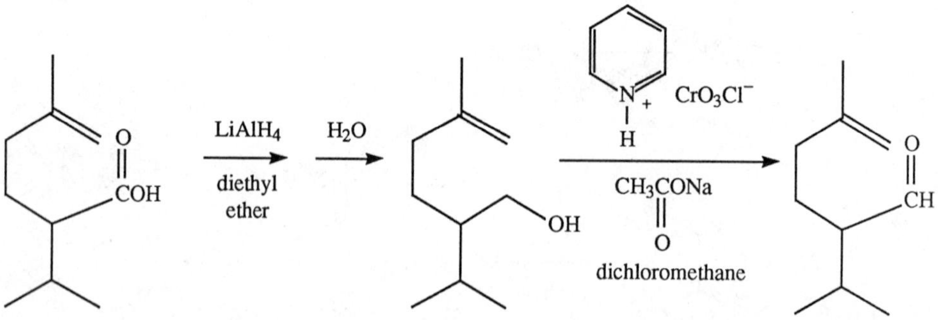

P   Q

13.16 (cont)

(l)

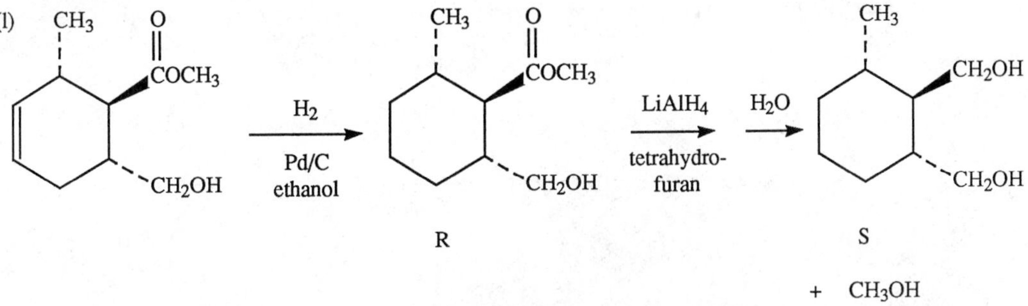

R   S

+   CH$_3$OH

(m)

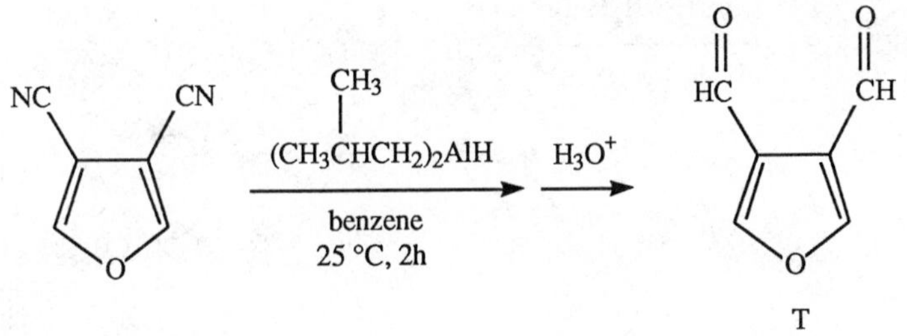

T

(n)

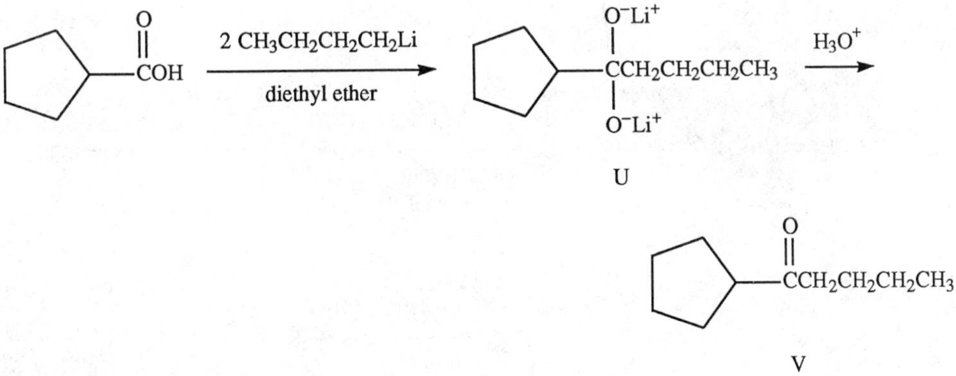

U

V

(o)

W

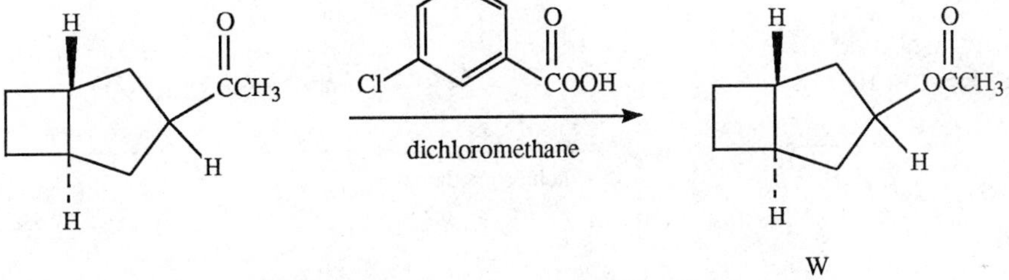

15.17

(a) $CH_2\!\!=\!\!CHCH\!\!=\!\!CHCOH$ (with $\overset{O}{\overset{\|}{}}$ above) $\xrightarrow[\text{diethyl ether}]{\text{LiAlH}_4}$ $\xrightarrow{\text{H}_2\text{O}}$

$CH_2\!\!=\!\!CHCH\!\!=\!\!CHCH_2OH$ $\xrightarrow[\text{dichloromethane}]{}$ $CH_2\!\!=\!\!CHCH\!\!=\!\!CHCH$ (with $\overset{O}{\overset{\|}{}}$)

with pyridinium CrO_3Cl^- reagent over the arrow

A   B

(b) $CH_3CH_2CH_2CH_2COCH_2CH_3$ (with $\overset{O}{\overset{\|}{}}$) $\xrightarrow[\text{diethyl ether}]{2\ CH_3CH_2MgBr}$

$CH_3CH_2CH_2CH_2\underset{\underset{O^-Mg^{2+}Br^-}{|}}{\overset{\overset{CH_2CH_3}{|}}{C}}CH_2CH_3$ $\xrightarrow[\text{H}_2\text{O}]{\text{NH}_4\text{Cl}}$

                                                  C

$CH_3CH_2CH_2CH_2\underset{\underset{OH}{|}}{\overset{\overset{CH_2CH_3}{|}}{C}}CH_2CH_3$

                                                  D

(c) $CH_3CH_2CNHCH_2CH_3$ (with $\overset{O}{\overset{\|}{}}$) $\xrightarrow[\text{diethyl ether}]{\text{LiAlH}_4}$ $\xrightarrow{\text{H}_2\text{O}}$ $CH_3CH_2CH_2NHCH_2CH_3$

                                                                                          E

(d) $CH_3CH\!\!=\!\!CHCH_2C\!\!\equiv\!\!CCH_2CH_2COH$ (with $\overset{O}{\overset{\|}{}}$) $\xrightarrow[\substack{\text{NH}_3\ (\text{liq})\\ \text{ethanol}}]{\text{Li}}$ $\xrightarrow{\text{H}_3\text{O}^+}$

(structure F: cis alkene with $CH_3CH\!\!=\!\!CHCH_2$ and H groups, $C\!\!=\!\!C$, H and CH_2CH_2COH with $\overset{O}{\overset{\|}{}}$)

                                                  F

(e)   (structure F: $CH_3CH\!\!=\!\!CHCH_2$ and H on one carbon, $C\!\!=\!\!C$, H and CH_2CH_2COH with $\overset{O}{\overset{\|}{}}$) $\xrightarrow[\text{diethyl ether}]{\text{LiAlH}_4}$ $\xrightarrow{\text{H}_3\text{O}^+}$

                                                  F

(structure G: $CH_3CH\!\!=\!\!CHCH_2$ and H, $C\!\!=\!\!C$, H and $CH_2CH_2CH_2OH$) $\xrightarrow[\text{dichloromethane}]{}$ (structure H: $CH_3CH\!\!=\!\!CHCH_2$ and H, $C\!\!=\!\!C$, H and CH_2CH_2CH with $\overset{O}{\overset{\|}{}}$)

with pyridinium CrO_3Cl^- reagent over the arrow

                  G   H

15.17 (cont)

(f) $CH_3CH_2NHCCH_2$—⟨benzene ring⟩ $\xrightarrow[\Delta]{\substack{NaOH \\ H_2O}}$ $CH_3CH_2NH_2$ + ⟨benzene ring⟩—$CH_2CO^-Na^+$

with the amide carbonyl $\overset{O}{\overset{\|}{}}$ and the carboxylate carbonyl $\overset{O}{\overset{\|}{}}$

I   J

(g) $CH_3\overset{O}{\overset{\|}{C}}CH_2\overset{O}{\overset{\|}{O}}CH$ + CH_3OH $\xrightarrow[\Delta]{KOH}$ $CH_3\overset{O}{\overset{\|}{C}}CH_2OH$ + $CH_3\overset{O}{\overset{\|}{O}}CH$

K   L

(h) $CH_3\overset{CH_3}{\overset{|}{C}H}CH_2\overset{O}{\overset{\|}{C}}O^-Na^+$ + $CH_3CH_2CH_2\overset{O}{\overset{\|}{C}}Cl$ $\longrightarrow$ $CH_3\overset{CH_3}{\overset{|}{C}H}CH_2\overset{O}{\overset{\|}{C}}\overset{O}{\overset{\|}{O}}CCH_2CH_2CH_3$

M

(i) $CH_3\overset{CH_3}{\underset{CH_3}{\overset{|}{\underset{|}{C}}}}\overset{O}{\overset{\|}{C}}O^-Li^+$ + ⟨benzene ring⟩—Li $\xrightarrow{\substack{diethyl \\ ether}}$

$CH_3\overset{CH_3}{\underset{CH_3}{\overset{|}{\underset{|}{C}}}}\overset{O^-Li^+}{\underset{O^-Li^+}{\overset{|}{\underset{|}{C}}}}$⟨benzene ring⟩ $\xrightarrow{H_3O^+}$ $CH_3\overset{CH_3}{\underset{CH_3}{\overset{|}{\underset{|}{C}}}}\overset{O}{\overset{\|}{C}}$⟨benzene ring⟩

N   O

(j) ⟨benzene ring⟩—$\overset{O}{\overset{\|}{C}}Cl$ + $H_2N(CH_2)_5\overset{O}{\overset{\|}{C}}OH$ $\xrightarrow[H_2O]{NaOH}$ ⟨benzene ring⟩—$\overset{O}{\overset{\|}{C}}NH(CH_2)_5\overset{O}{\overset{\|}{C}}O^-Na^+$ $\xrightarrow{H_3O^+}$

P

⟨benzene ring⟩—$\overset{O}{\overset{\|}{C}}NH(CH_2)_5\overset{O}{\overset{\|}{C}}OH$

Q

15.18

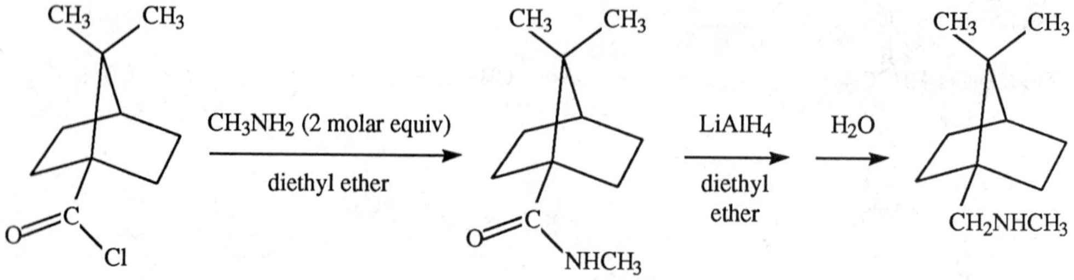

15.19

(a)

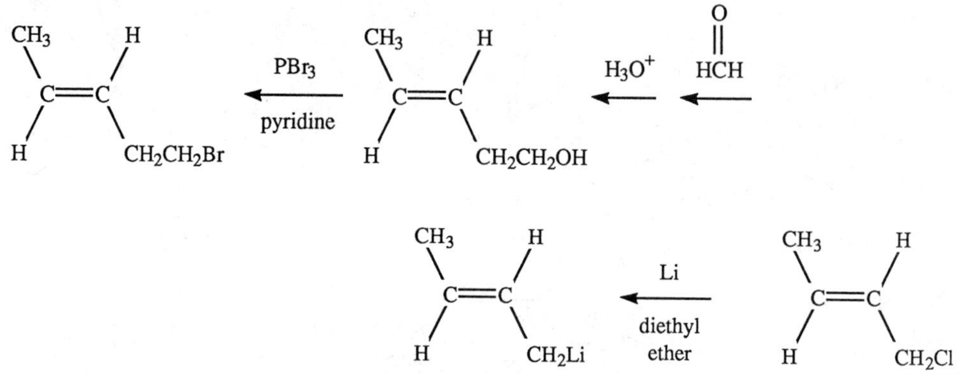

(b)

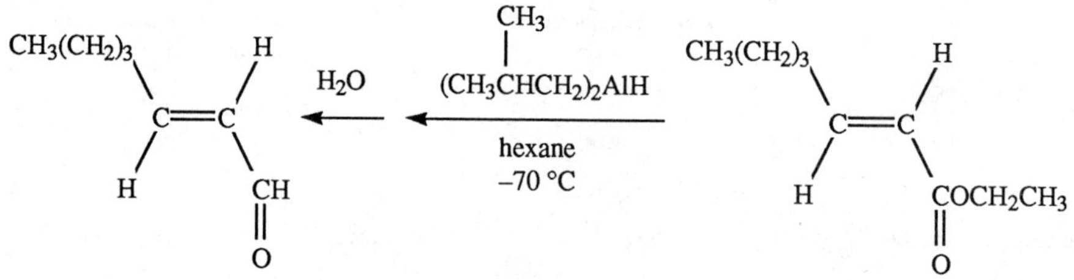

(c)

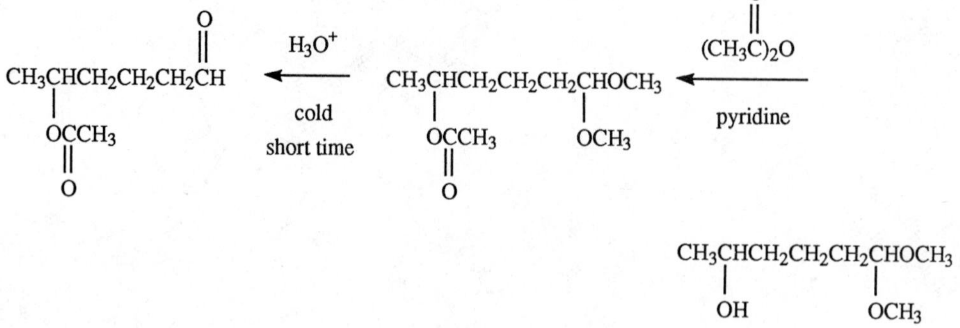

15.19 (cont)

(d)

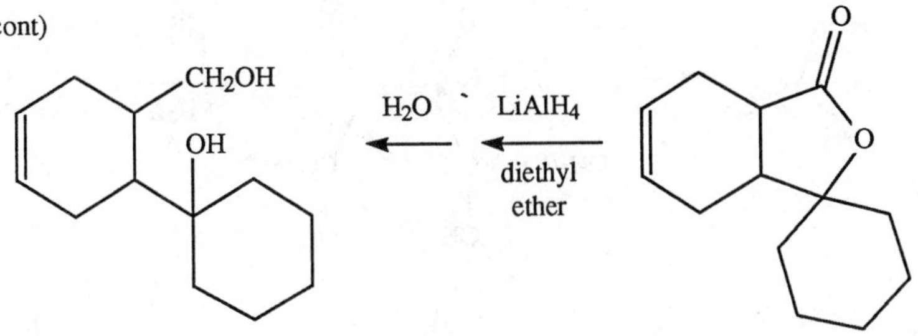

$$\xleftarrow{\text{H}_2\text{O}} \quad \xleftarrow[\substack{\text{diethyl} \\ \text{ether}}]{\text{LiAlH}_4}$$

(e)

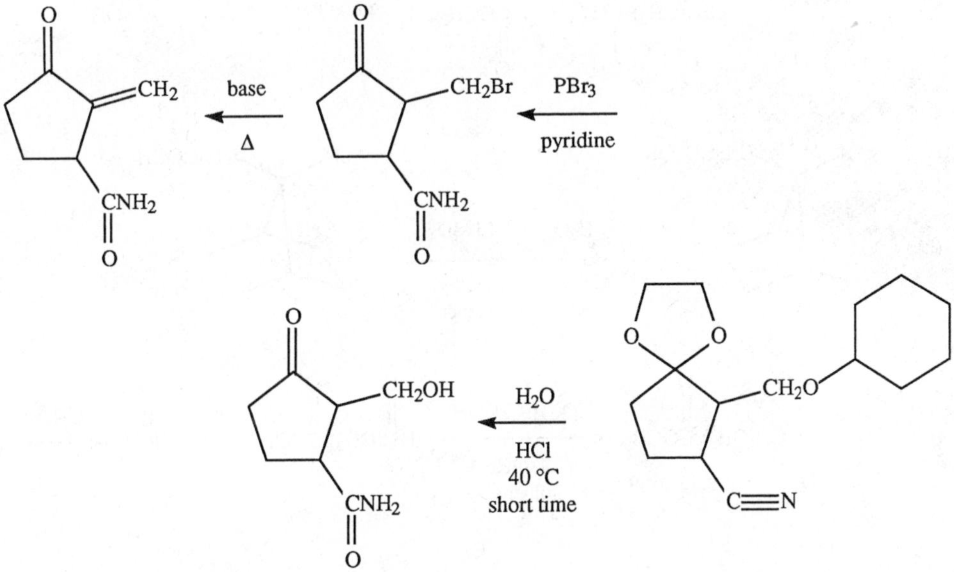

$$\xleftarrow[\Delta]{\text{base}} \qquad \xleftarrow[\text{pyridine}]{\text{PBr}_3}$$

$$\xleftarrow[\substack{\text{HCl} \\ 40\ ^\circ\text{C} \\ \text{short time}}]{\text{H}_2\text{O}}$$

(f) $\text{CH}_3(\text{CH}_2)_7\text{CHC} \equiv \text{CCOCH}_2\text{CH}_3 \xleftarrow[\text{H}_2\text{O}]{\text{NH}_4\text{Cl}} \xleftarrow{\overset{\displaystyle \text{O}}{\overset{\|}{\text{CH}_3(\text{CH}_2)_7\text{CH}}}}$
$\qquad\qquad\qquad\underset{\text{OH}}{|}$

$\text{Na}^+ {}^- \text{C} \equiv \text{CCOCH}_2\text{CH}_3 \xleftarrow[\text{NH}_3\ (\text{liq})]{\text{NaNH}_2} \text{HC} \equiv \text{CCOCH}_2\text{CH}_3$

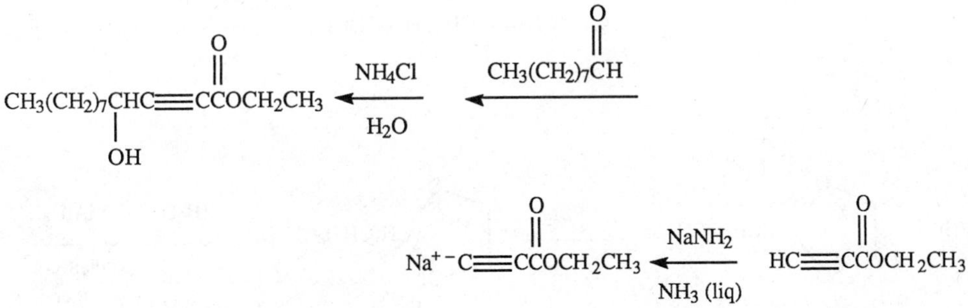

15.19 (cont)

(g)

$$CH_3(CH_2)_7 \quad \xleftarrow{H_3O^+} \quad \overset{H}{\underset{CH_3(CH_2)_7CH}{C}}=\overset{H}{\underset{\underset{O}{\overset{\|}{CO^-Na^+}}}{C}} \quad \xleftarrow[\underset{\Delta}{H_2O}]{NaOH}$$

with $CH_3(CH_2)_7CH$ bearing OH

$$\overset{H}{\underset{CH_3(CH_2)_7CH}{C}}=\overset{H}{\underset{\underset{O}{\overset{\|}{COCH_2CH_3}}}{C}} \quad \xleftarrow[\underset{\text{quinoline}}{Pd/BaSO_4}]{H_2} \quad CH_3(CH_2)_7\underset{OH}{CH}C\equiv C\overset{O}{\overset{\|}{C}}OCH_2CH_3$$

with $CH_3(CH_2)_7CH$ bearing OH

(h)

$$\xleftarrow[\text{diethyl ether}]{H_2O \quad LiAlD_4}$$

bicyclic with CH_2CD_2OH / bicyclic with CH_2COCH_3

(i)

$$\xleftarrow[\underset{\Delta}{\underset{\text{toluene}}{TsOH}}]{HOCH_2CH_2OH} \quad \overset{O}{\overset{\|}{HC}}(CH_2)_8\overset{O}{\overset{\|}{C}}OCH_3 \quad + \quad \overset{O}{\overset{\|}{HC}}H \quad \xleftarrow[]{(CH_3)_2S \quad O_3}$$

dioxolane $CH(CH_2)_8COCH_3$

$$CH_2{=}CH(CH_2)_8\overset{O}{\overset{\|}{C}}OCH_3 \quad \xleftarrow[\underset{\Delta}{TsOH}]{CH_3OH} \quad CH_2{=}CH(CH_2)_8\overset{O}{\overset{\|}{C}}OH$$

(j)

dioxolane $CH(CH_2)_8CH_2Br \quad \xleftarrow[\text{pyridine}]{PBr_3} \quad$ dioxolane $CH(CH_2)_8CH_2OH \quad \xleftarrow[\text{diethyl ether}]{H_2O \quad LiAlH_4}$

dioxolane $CH(CH_2)_8\overset{O}{\overset{\|}{C}}OCH_3$

15.19 (cont)

(k) 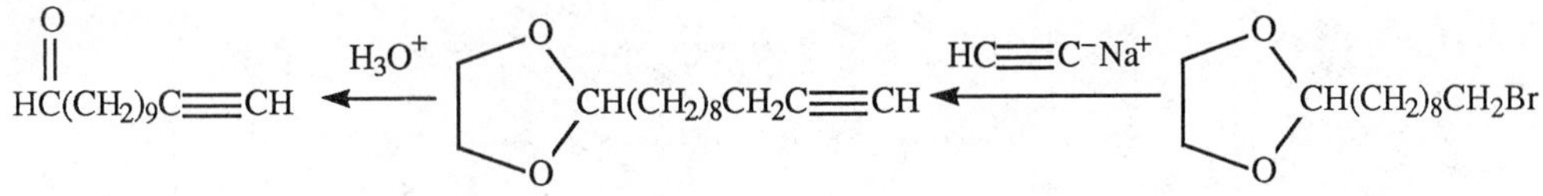 $HC(CH_2)_9C \equiv CH$ $\xleftarrow{H_3O^+}$ [dioxolane]$CH(CH_2)_8CH_2C \equiv CH$ $\xleftarrow{HC \equiv C^- Na^+}$ [dioxolane]$CH(CH_2)_8CH_2Br$

(l)

$\xleftarrow[\text{pyridine}]{CH_3CCl}$

$\xleftarrow[\text{diethyl ether}]{H_2O \quad LiAlH_4}$

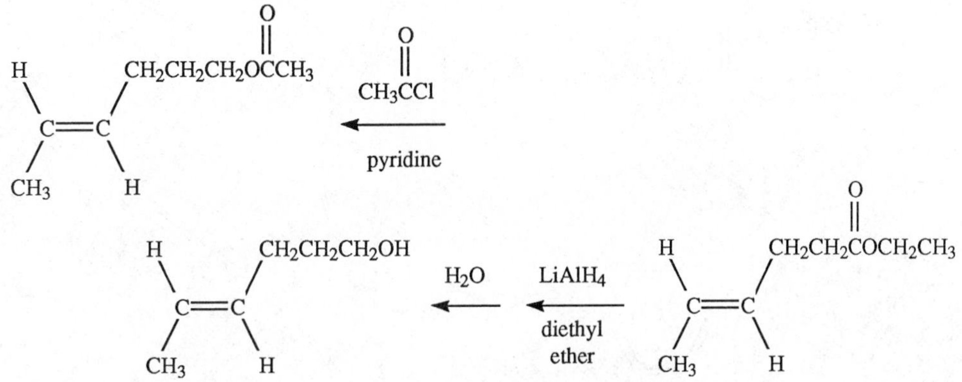

15.20

First reaction:

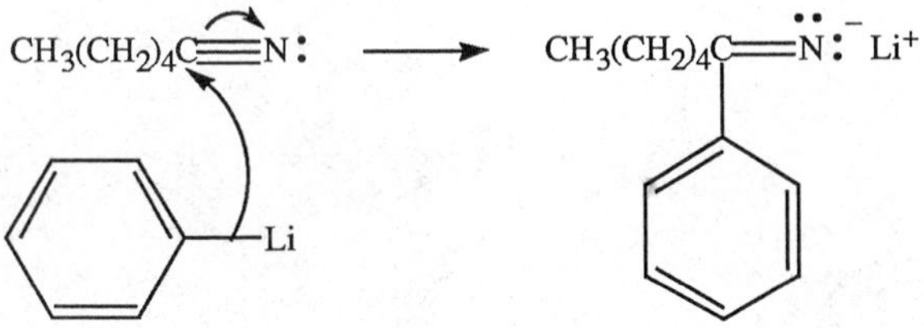

Second reaction:

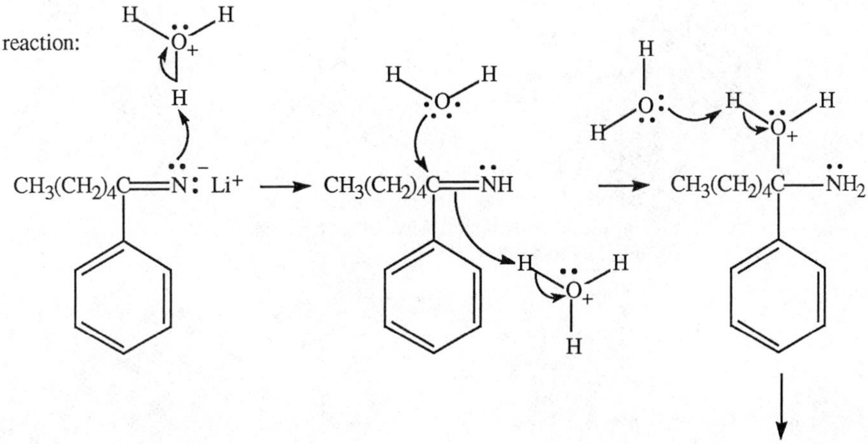

15.20 (cont)

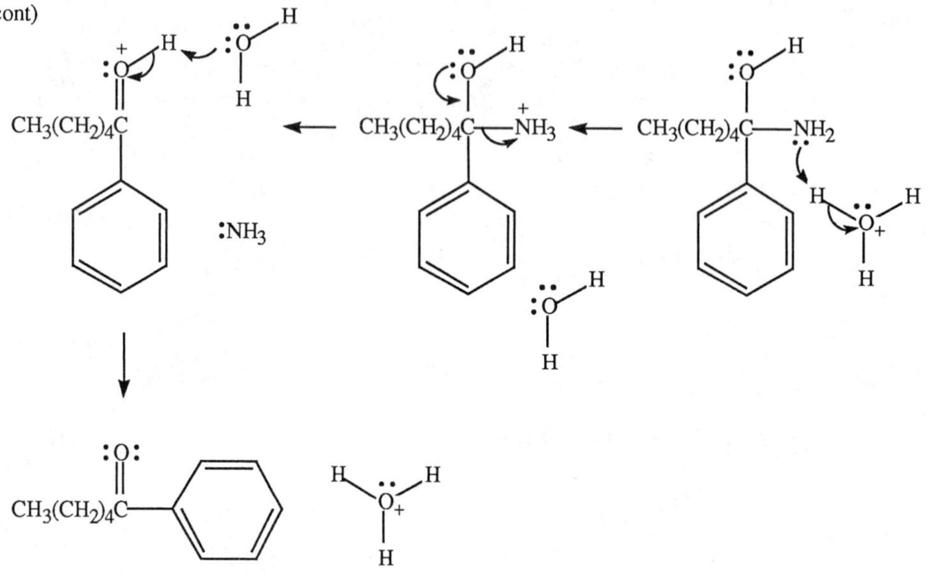

15.21

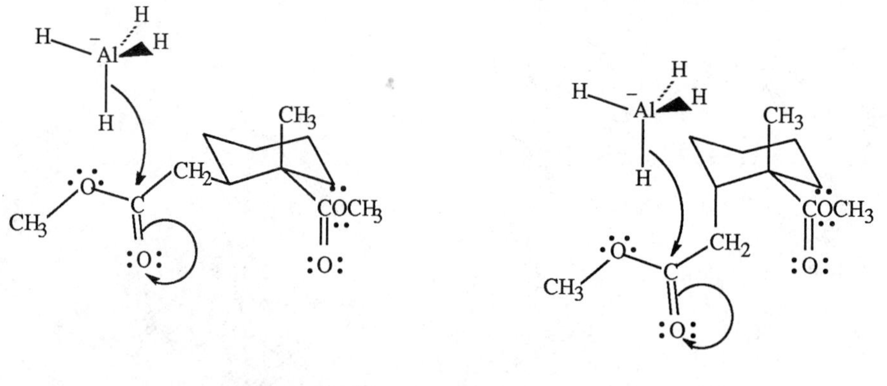

Approach of the aluminum hydride to this
carbonyl is not sterically hindered for either
the cis or trans isomer. Most of the reaction
goes by this pathway.

15.21 (cont)

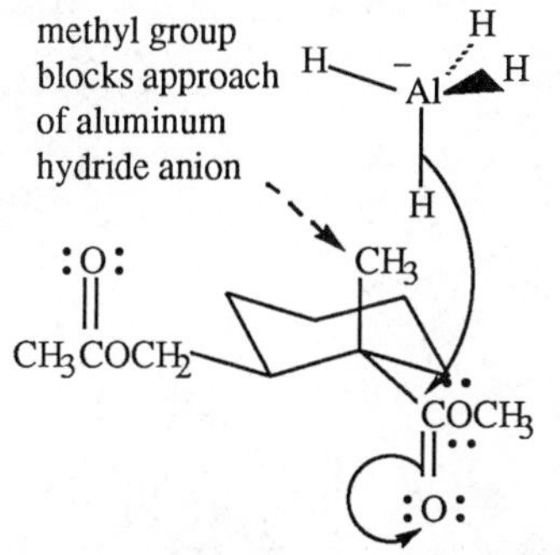

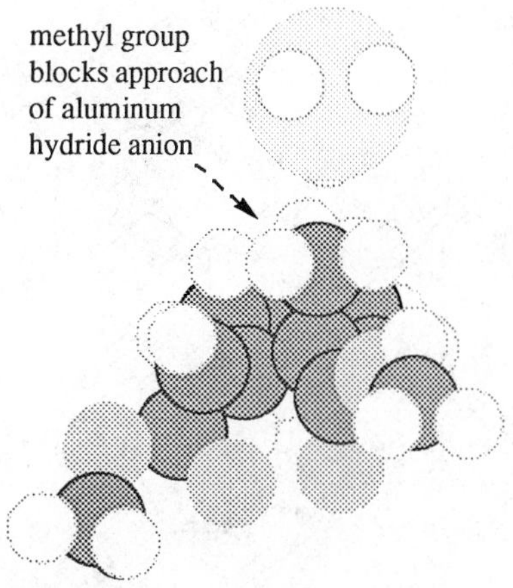

methyl group blocks approach of aluminum hydride anion

methyl group blocks approach of aluminum hydride anion

Approach of the aluminum hydride is sterically hindered for both the cis and trans isomers. This reaction has a higher activation energy than the reaction at the other carbonyl group and, therefore, a slower rate

15.22

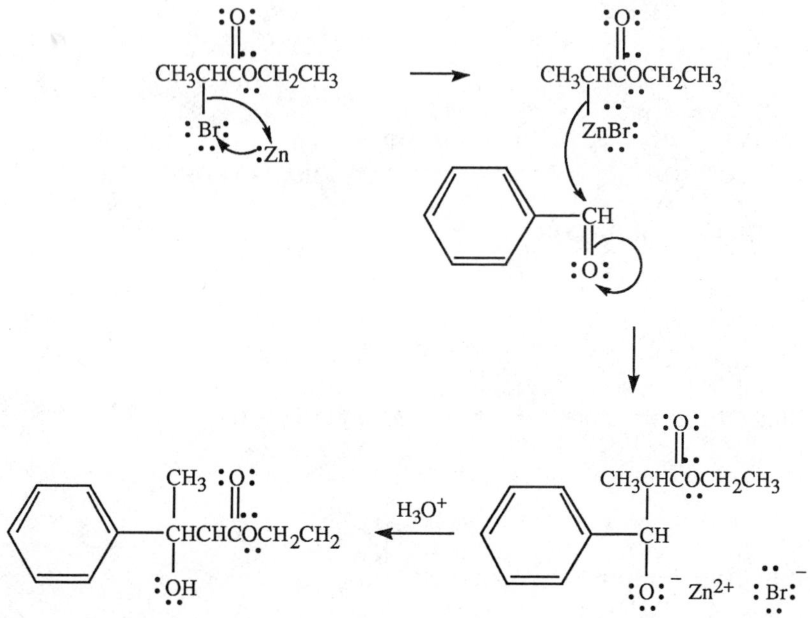

Note the similarity of this reaction to reactions with Grignard and organolithium reagents. The organozinc reagent reacts selectively with the carbonyl group of the aldehyde function in the presence of the ester group.

15.23

(a)

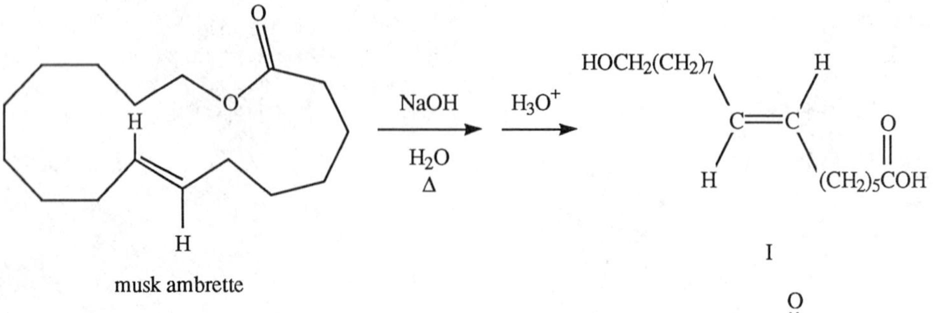

musk ambrette

$$\xrightarrow[\substack{H_2O \\ \Delta}]{NaOH} \quad \xrightarrow{H_3O^+}$$

I

(b)

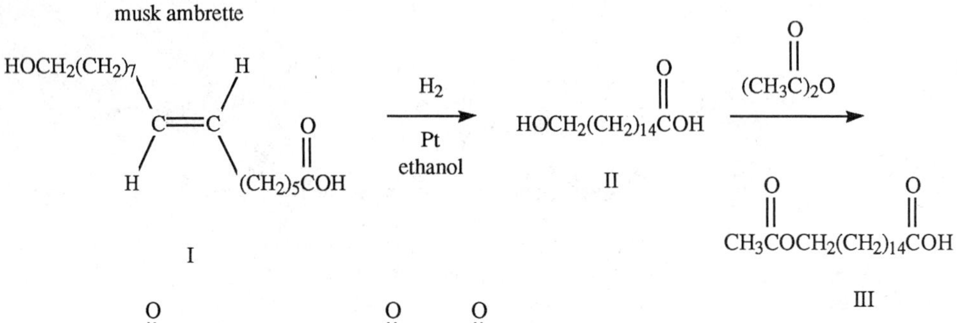

I

$$\xrightarrow[\substack{Pt \\ ethanol}]{H_2} \quad HOCH_2(CH_2)_{14}COH \quad \xrightarrow{(CH_3C)_2O}$$

II

$$CH_3COCH_2(CH_2)_{14}COH$$

III

(c) $\quad HOCH_2(CH_2)_{14}COH \xrightarrow[\substack{H_2SO_4 \\ acetic\ acid}]{CrO_3} HOC(CH_2)_{14}COH$

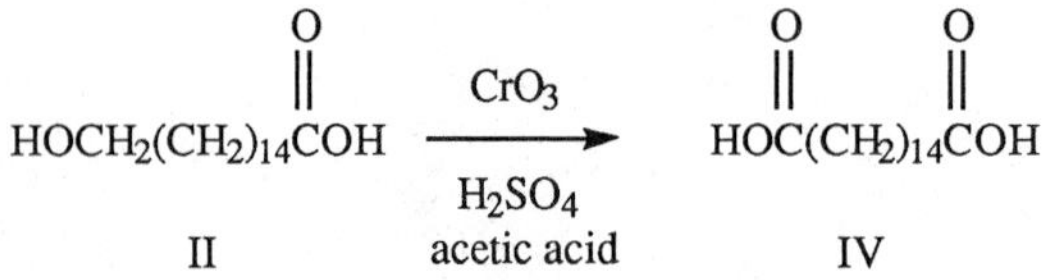

II   IV

(d)

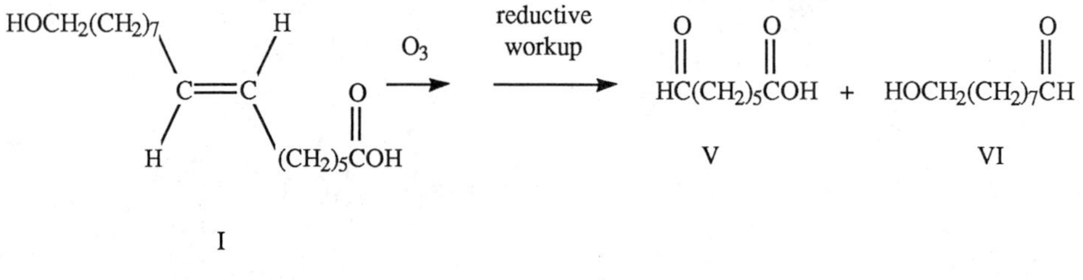

I

$$\xrightarrow{O_3} \xrightarrow{\substack{reductive \\ workup}} HC(CH_2)_5COH \; + \; HOCH_2(CH_2)_7CH$$

V   VI

(e) $\quad HOCH_2(CH_2)_7CH \xrightarrow[\substack{H_2O}]{KMnO_4} \xrightarrow{H_3O^+} HOCH_2(CH_2)_7COH$

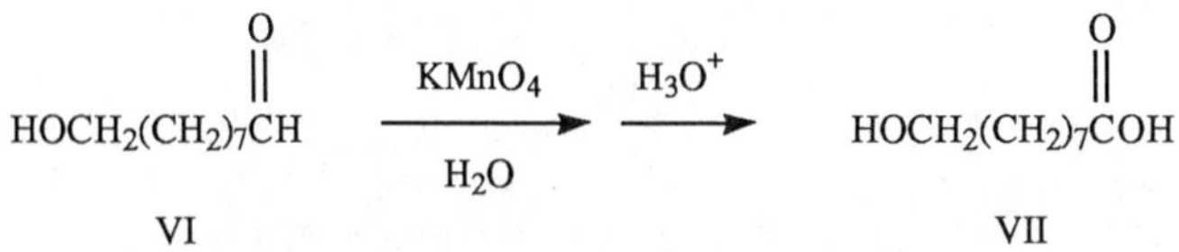

VI   VII

(f) $\quad HOCH_2(CH_2)_7COH \xrightarrow[\substack{H_2SO_4}]{CrO_3} HOC(CH_2)_7COH$

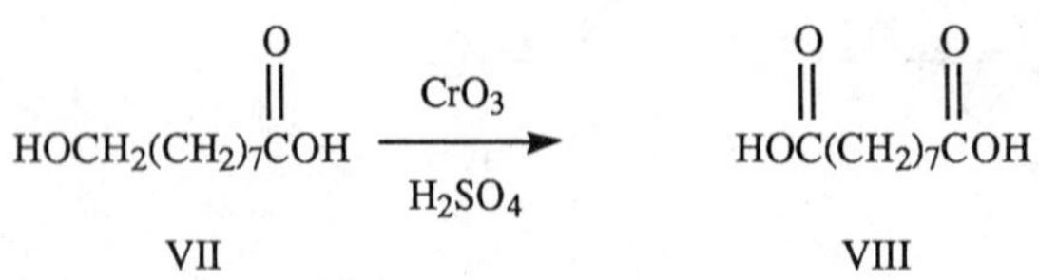

VII   VIII

15.23 (cont)

(g) $\underset{\text{V}}{\overset{\displaystyle\text{O}\quad\text{O}}{\overset{\displaystyle\|\quad\|}{\text{HC(CH}_2)_5\text{COH}}}}$ $\xrightarrow[\substack{\text{HCl}\\\text{Na}_2\text{CO}_3}]{\text{NH}_2\text{OH}}$ $\underset{\text{IX}}{\overset{\displaystyle\overset{\text{O}}{\|}}{\text{HON}\!=\!\!\text{CH(CH}_2)_5\text{COH}}}$

(h) $\underset{\text{IX}}{\text{HON}\!=\!\!\text{CH(CH}_2)_5\overset{\displaystyle\overset{\text{O}}{\|}}{\text{COH}}}$ $\xrightarrow[\substack{\Delta\\\text{dehydrating}\\\text{agent}}]{\overset{\displaystyle\overset{\text{O}}{\|}}{(\text{CH}_3\text{C})_2\text{O}}}$ $\underset{\text{X}}{\text{N}\!\equiv\!\!\text{C(CH}_2)_5\overset{\displaystyle\overset{\text{O}}{\|}}{\text{COH}}}$

(i) $\underset{\text{X}}{\text{N}\!\equiv\!\!\text{C(CH}_2)_5\overset{\displaystyle\overset{\text{O}}{\|}}{\text{COH}}}$ $\xrightarrow[\substack{\text{H}_2\text{O}\\\Delta}]{\text{NaOH}}$ $\xrightarrow{\text{H}_3\text{O}^+}$ $\underset{\text{heptanedioic acid}}{\text{HOC(CH}_2)_5\text{COH}}$ (with two O above)

15.24 $\text{HOCH}_2(\text{CH}_2)_4\overset{\displaystyle\overset{\text{O}}{\|}}{\text{COCH}_2\text{CH}_3}$ $\xrightarrow[\text{dichloromethane}]{\text{pyridinium chlorochromate (CrO}_3\text{Cl}^-)}$ $\underset{\text{A}}{\text{HC(CH}_2)_4\text{COCH}_2\text{CH}_3}$ $\xrightarrow{\text{CH}_2=\text{CHMgBr}}$

$\xrightarrow[\text{H}_2\text{O}]{\text{NH}_4\text{Cl}}$ $\underset{\text{B}}{\underset{\text{OH}}{\text{CH}_2=\text{CHCH(CH}_2)_4\overset{\displaystyle\overset{\text{O}}{\|}}{\text{COCH}_2\text{CH}_3}}}$ $\xrightarrow[\substack{\text{H}_2\text{O}\\\text{ethanol}}]{\text{KOH}}$ $\underset{\text{C}}{\underset{\text{OH}}{\text{CH}_2=\text{CHCH(CH}_2)_4\overset{\displaystyle\overset{\text{O}}{\|}}{\text{CO}^-\text{K}^+}}}$ $\xrightarrow{\text{H}_3\text{O}^+}$

$\underset{\text{D}}{\underset{\text{OH}}{\text{CH}_2=\text{CHCH(CH}_2)_4\overset{\displaystyle\overset{\text{O}}{\|}}{\text{COH}}}}$ $\xrightarrow[\text{pyridine}]{(\text{CH}_3\text{C})_2\text{O}}$ $\underset{\text{E}}{\underset{\underset{\text{O}}{\overset{\|}{\text{OCCH}_3}}}{\text{CH}_2=\text{CHCH(CH}_2)_4\overset{\displaystyle\overset{\text{O}}{\|}}{\text{COH}}}}$

The carbon atom of the carbonyl group of the aldehyde is more electrophilic than the carbon atom of the carbonyl group of the ester and, therefore, reacts more readily with the nucleophilic Grignard reagent. (See Section 14.1A)

15.25

(a)

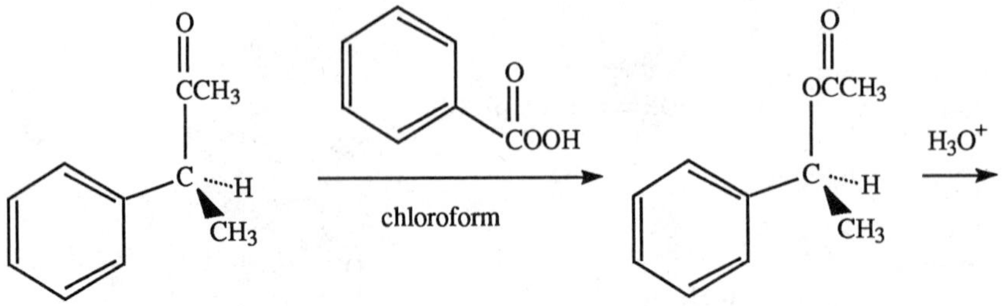

(S)-(+)-3-phenyl-2-butanone

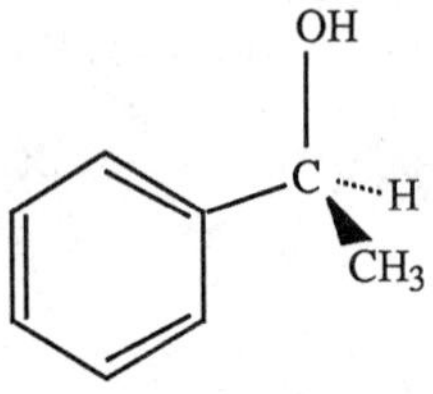

(S)-(−)-1-phenylethanol

The product has the same relative configuration as the starting material. Hydrolysis of an ester proceeds with retention of configuration since the bond to the stereocenter is not broken. Migration of the alkyl group to the oxygen atom in the Baeyer-Villiger reaction must, therefore, also proceed with retention of configuration.

(b)

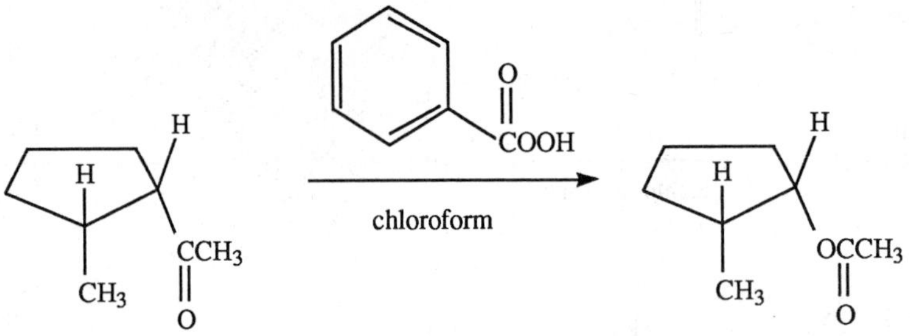

and enantiomer

and enantiomer

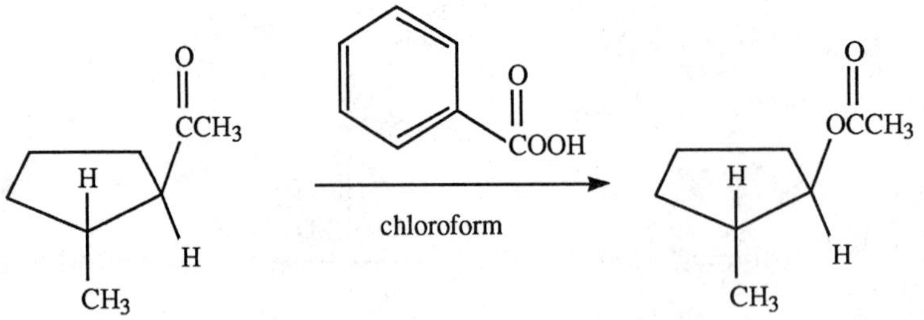

and enantiomer

and enantiomer

15.26

$$CH_3CH_2OCH{=}CH_2$$ / CF_3COH (with $=O$)

(formation of an acetal by reaction with an enol ether; protection of the hydroxyl group)

A

$$\xrightarrow{\text{LiAlH}_4}$$ diethyl ether

$$\xrightarrow{\text{H}_2\text{O}}$$

B

$$\xrightarrow[\text{dichloromethane}]{\text{TsCl, pyridine}}$$

C

$$\xrightarrow[\substack{\text{tetrahydrofuran}\\\text{cold}}]{\text{H}_3\text{O}^+}$$

(hydrolysis of the acetal; removal of the protecting group)

D

$$\xrightarrow{\text{KOH}}$$

(formation of an alkoxide and an intramolecular S_N2 reaction)

E

15.27 $HO(CH_2)_8OH$

$$\xrightarrow[\text{H}_2\text{O}]{\text{HBr (48\%) (1 eq)}}$$ $Br(CH_2)_8OH$ A

$$\xrightarrow[\substack{\text{H}_2\text{O}\\\text{H}_2\text{SO}_4\\\text{acetone}}]{\text{CrO}_3}$$ $Br(CH_2)_7COH$ (with $\overset{O}{\|}$)   B

$$\xrightarrow[\substack{\text{NH}_3 \text{ (liq)}\\\text{tetrahydrofuran}}]{\substack{\text{Li}^+\text{NH}_2^-\\\text{(2 equiv)}}}$$

C

15.27 (cont)

$$\text{Li}^{+-}C\equiv C-\overset{\underset{|}{H}}{C}=\overset{\overset{H}{|}}{C}-CH_2O^-Li^+ \quad + \quad Br(CH_2)_7\overset{\overset{O}{\|}}{C}OH \quad \longrightarrow$$

B

C

$$\underset{HOCH_2}{\overset{H}{\diagdown}}C=C\underset{H}{\overset{C\equiv C(CH_2)_7CO^-Li^+}{\diagup}} \quad \xrightarrow[\text{cold}]{H_3O^+}$$

D

$$\underset{HOCH_2}{\overset{H}{\diagdown}}C=C\underset{H}{\overset{C\equiv C(CH_2)_7\overset{O}{\|}COH}{\diagup}} \quad \xrightarrow[\text{dichloromethane}]{\text{(oxidizing agent similar to)}\; \text{pyridinium } CrO_3Cl^-} \quad \underset{HC\atop\|\atop O}{\overset{H}{\diagdown}}C=C\underset{H}{\overset{C\equiv C(CH_2)_7\overset{O}{\|}COH}{\diagup}}$$

E   F

$$\underset{HC\atop\|\atop O}{\overset{H}{\diagdown}}C=C\underset{H}{\overset{C\equiv C(CH_2)_7\overset{O}{\|}COH}{\diagup}} \quad \xrightarrow[\substack{\text{diethyl ether}\\\text{tetrahydrofuran}}]{\substack{CH_3(CH_2)_3CH_2MgBr\\\text{(2 equivalents)}}} \quad \xrightarrow[H_2O]{NH_4Cl} \quad \underset{CH_3(CH_2)_3CH_2\underset{\overset{|}{OH}}{CH}}{\overset{H}{\diagdown}}C=C\underset{H}{\overset{C\equiv C(CH_2)_7\overset{O}{\|}COH}{\diagup}}$$

F   G

$$\underset{CH_3(CH_2)_3CH_2\underset{\overset{|}{OH}}{CH}}{\overset{H}{\diagdown}}C=C\underset{H}{\overset{C\equiv C(CH_2)_7\overset{O}{\|}COH}{\diagup}} \quad \xrightarrow[\substack{Pd/BaSO_4\\\text{quinoline}}]{H_2} \quad \underset{CH_3(CH_2)_3CH_2\underset{\overset{|}{OH}}{CH}}{\overset{H}{\diagdown}}C=C\underset{H}{\overset{\underset{H}{\diagup}C=C\underset{(CH_2)_7\overset{O}{\|}COH}{\diagdown}}{}}$$

G   coriolic acid

15.28

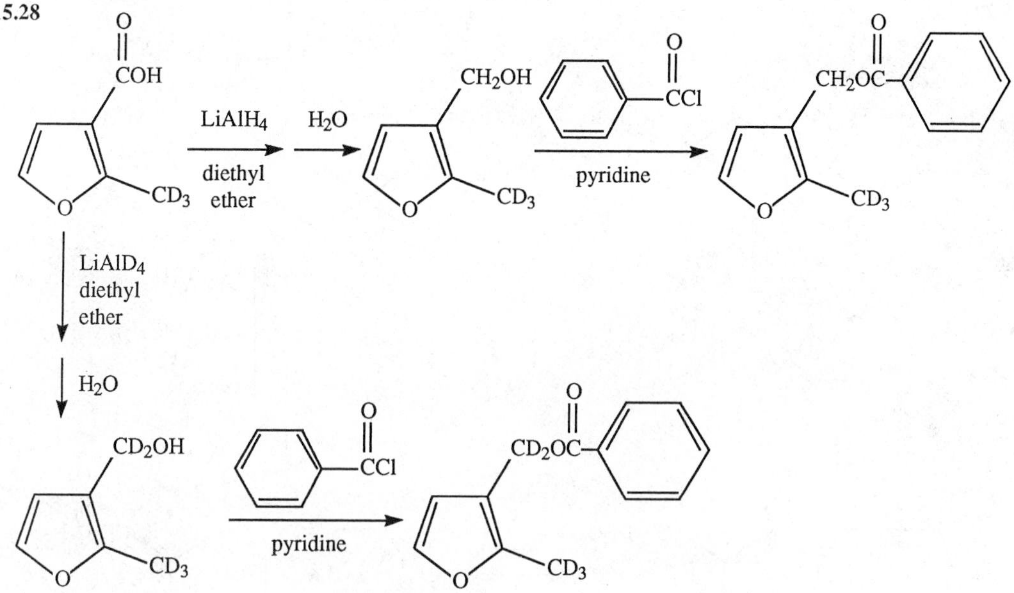
COH
CD3
LiAlH4
diethyl
ether
H2O
CH2OH
CCl
pyridine
CH2OC
LiAlD4
CD2OH
CD2OC

15.29

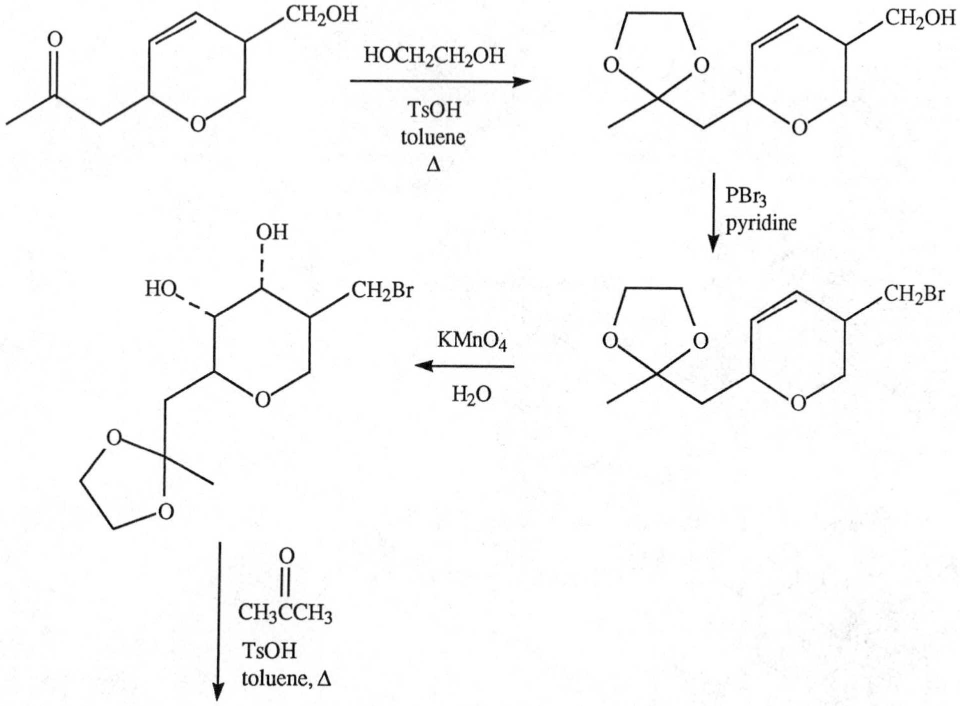
CH2OH
HOCH2CH2OH
TsOH
toluene
Δ
PBr3
pyridine
CH2Br
KMnO4
H2O
OH
HO
CH3CCH3
toluene, Δ

15.29 (cont)

CH_2Br $\xrightarrow[\text{ethanol}]{\text{NaCN}}$ CH_2CN

$$\begin{array}{c} CH_3 \\ | \\ (CH_3CHCH_2)_2AlH \\ (1\ \text{equiv}) \\ \text{benzene} \end{array} \downarrow$$

CH_2CH_2OH $\xleftarrow[\text{CH}_3\text{OH}]{\text{NaBH}_4}$ $CH_2CH \overset{O}{\overset{\|}{}}$

$$\begin{array}{c} CH_3 \ C_6H_5 \\ | \quad\ | \\ CH_3C-SiCl, \\ | \quad\ | \\ CH_3 \ C_6H_5 \end{array} \qquad \underset{\underset{H}{|}}{\text{imidazole}}$$

$CH_2CH_2OSi\overset{\displaystyle C_6H_5}{\underset{\displaystyle C_6H_5}{|}}\overset{\displaystyle CH_3}{\underset{\displaystyle CH_3}{\overset{|}{-}C}}CH_3$

$$\downarrow H_3O^+$$

$CH_2CH_2OSi\overset{\displaystyle C_6H_5}{\underset{\displaystyle C_6H_5}{|}}\overset{\displaystyle CH_3}{\underset{\displaystyle CH_3}{\overset{|}{-}C}}CH_3$

15.30   Compounds A and B, $C_6H_{12}O_2$, have one unit of unsaturation. The carbonyl group of an ester function uses up one unit of unsaturation. Therefore, Compounds A and B cannot have any rings or other π bonds.

Analysis of the proton magnetic spectrum of Compound A is shown below:

δ 0.9, 1.5, and	(8H, consisting of two overlapping triplets, (2) C$\underline{H}_2$CH$_2$—,
δ 1.2-1.8	and a multiplet, —C$\underline{H}_2$—)
δ 2.1	(2H, a distorted triplet, —CH$_2$C$\underline{H}_2$C=O)
δ 4.1	(2H, quartet, CH$_3$C$\underline{H}_2$—O—)

Compound A is ethyl butanoate (ethyl butyrate).

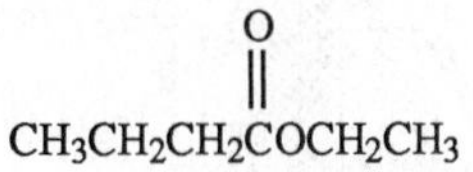

$$\text{CH}_3\text{CH}_2\text{CH}_2\overset{\displaystyle\overset{\text{O}}{\|}}{\text{C}}\text{OCH}_2\text{CH}_3$$

There are only two singlets in the spectrum of Compound B. The band at δ 1.1 is three times as intense as the one at δ 3.5. Since there are twelve hydrogen atoms in the molecular formula, the band at δ 1.1 must represent nine hydrogen atoms and that at δ 3.5 must represent three hydrogen atoms. Analysis of the proton magnetic spectrum of Compound B is shown below:

δ 1.1	(9H, singlet, (C$\underline{H}_3$)$_3$C—)
δ 3.5	(3H, singlet, C$\underline{H}_3$—O—)

Compound B is methyl 2,2-dimethylpropanoate

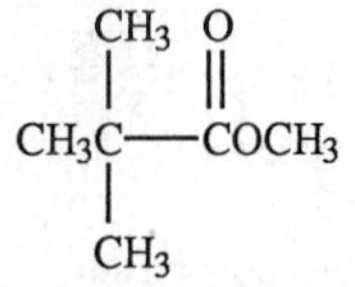

$$\text{CH}_3\overset{\displaystyle\overset{\text{CH}_3}{|}}{\underset{\displaystyle\underset{\text{CH}_3}{|}}{\text{C}}}\overset{\displaystyle\overset{\text{O}}{\|}}{\text{C}}\text{OCH}_3$$

15.31   Compound C, $C_9H_{10}O_3$, has five units of unsaturation. The infrared spectrum shows the presence of a carboxylic acid (3300-2600 cm^{-1} for the O—H and 1719 cm^{-1} for the C=O stretching frequencies).

Analysis of the proton magnetic spectrum of Compound C is shown below:

δ 3.5	(2H, singlet, —C$\underline{H}_2$—)	
δ 3.7	(3H, singlet, C$\underline{H}_3$—O—)	
δ 6.7-7.3	(4H, multiplet, Ar$\underline{H}$)	(This is the splitting pattern for a para substituted benzene ring)
δ 11.6	(1H, singlet, O=CO$\underline{H}$)	

Compound C is (4-methoxyphenyl)acetic acid

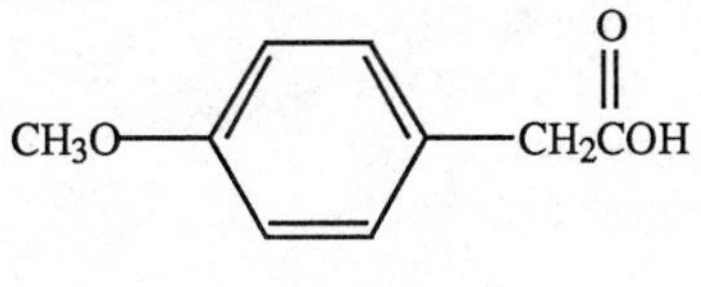

15.32   Compound D has four bands with integrations of 3:2:3:2 going from lower to higher δ values.

Analysis of the proton magnetic spectrum of Compound D is shown below:

δ 1.4	(3H, triplet, J 7 Hz, C$\underline{\text{H}}_3$CH$_2$—)
δ 4.5	(2H, quartet, J 7 Hz, CH$_3$C$\underline{\text{H}}_2$—O—)
δ 7.6-7.9	(3H, multiplet, Ar$\underline{\text{H}}$)
δ 8.1-8.4	(2H, multiplet, Ar$\underline{\text{H}}$)

Compound D is ethyl benzoate.

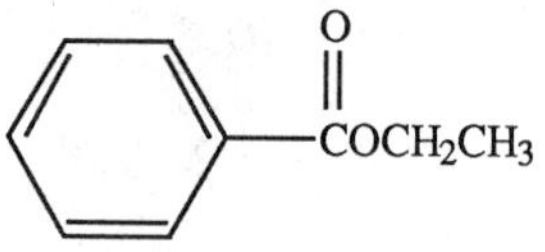

The two hydrogen atoms on carbons 2 and 6 of the aromatic ring are especially deshielded by the carbonyl group of the ester.

Compound E has three bands with integrations of 3:2:3.

Analysis of the proton magnetic spectrum of Compound E is shown below:

δ 1.1	(3H, triplet, J 8 Hz, C$\underline{\text{H}}_3$CH$_2$—)
δ 2.3	(2H, quartet, J 8 Hz, CH$_3$C$\underline{\text{H}}_2$C=O)
δ 3.7	(3H, singlet, C$\underline{\text{H}}_3$O—)

Compound E is methyl propanoate (methyl propionate)

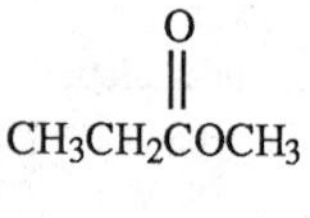

$$CH_3CH_2\overset{\overset{\displaystyle O}{\|}}{C}OCH_3$$

15.33   Compound F, $C_5H_8O_2$, has 2 units of unsaturation. The infrared spectrum shows the presence of an alkene (1635 cm^{-1} for the C=C stretching frequency and 989 cm^{-1} for a C=C—H bending frequency), and an ester (1731 cm^{-1} for the C=O and 1279, 1207, and 1069 cm^{-1} for the C—O stretching frequencies).

Analysis of the proton magnetic spectrum of Compound F is shown below:

δ 1.8	(3H, multiplet, C$\underline{\text{H}}_3$C=CH)
δ 3.5	(3H, singlet, C$\underline{\text{H}}_3$O—)
δ 5.2	(1H, multiplet, C=C$\underline{\text{H}}$)
δ 5.7	(1H, multiplet, C=C$\underline{\text{H}}$)

15.33 (cont)

Compound F is methyl 2-methylpropenoate (methyl methacrylate).

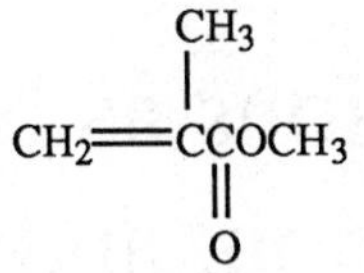

$$CH_2\!\!=\!\!CCOCH_3$$

with CH_3 above the central C, and O double-bonded below the carbonyl carbon.

The allylic and vinyl hydrogen atoms are coupled to each other through the double bond.

15.34 Compound G, $C_8H_{14}O_2$, has 2 units of unsaturation. The infrared spectrum shows the presence of an ester (1736 cm^{-1} for the C=O, and 1160 and 1032 cm^{-1} for the C—O stretching frequencies).

Integration of the proton magnetic resonance spectrum of Compound G gives a 3:2:2 ratio, which adds up to half the number of hydrogen atoms in the molecular formula. This points to symmetry in the molecule. Analysis of the spectrum of is shown below:

δ 1.3   (6H, triplet, C$\underline{H}_3$CH$_2$—)
δ 2.6   (4H, singlet, —C$\underline{H}_2$C=O)
δ 4.3   (4H, quartet, CH$_3$C$\underline{H}_2$O—)

Compound G is diethyl butanedioate (diethyl succinate).

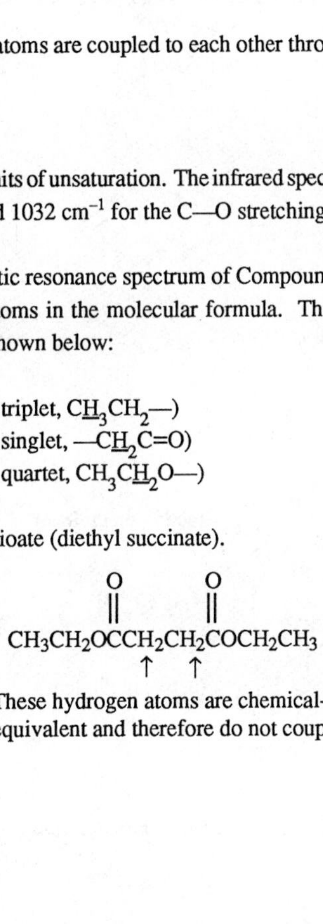

$$CH_3CH_2OCCH_2CH_2COCH_2CH_3$$

with O double bonds above the two carbonyl carbons, and arrows pointing to the two central CH_2 groups.

These hydrogen atoms are chemical-shift
equivalent and therefore do not couple.

16

Enols and Enolate Anions as Nucleophiles I. Halogenation, Alkylation, and Condensation Reactions

16.1 (a)

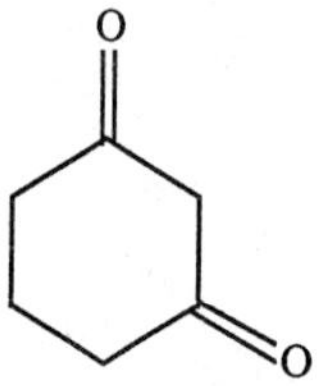

(active methylene)

(b)

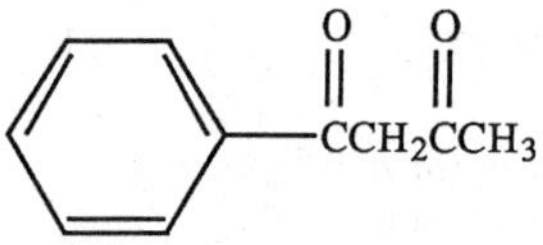

(active methylene)

(c) $CH_3COCH_2CH_3$

(more electrophilic carbonyl carbon atom)

(d) $CH_3OCCH_2COCH_3$

(active methylene)

Concept Map 16.1   Enolization.

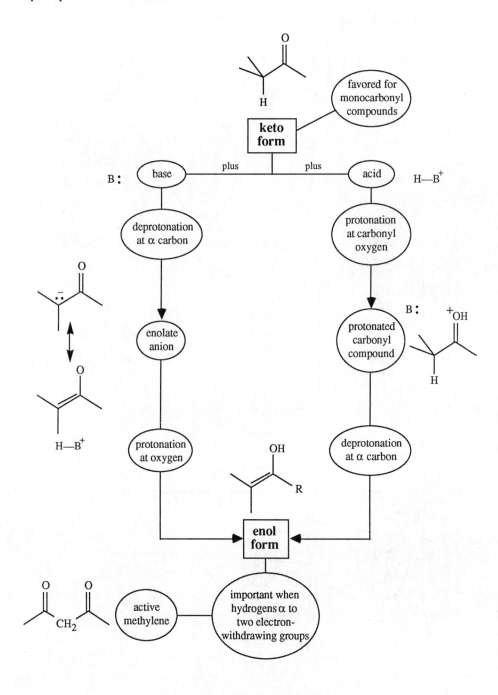

Concept Map 16.2   Thermodynamic and kinetic enolates.

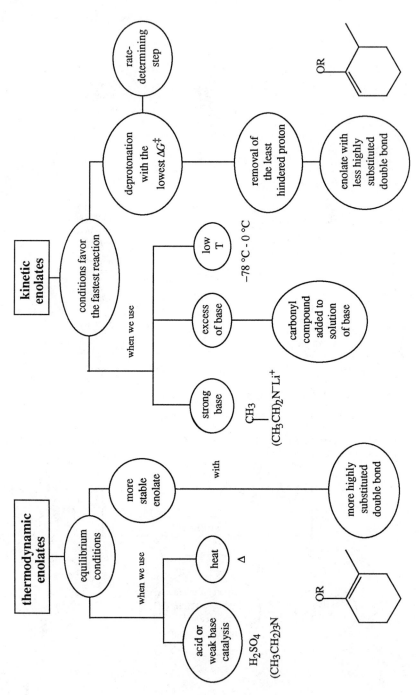

16.2   (a)

(b)

(c) $CH_3COCH_2CH_3$ $\xrightarrow[\text{CH}_3\text{CH}_2\text{OD (excess)}]{\text{CH}_3\text{CH}_2\text{ONa}}$ $CD_3COCH_2CH_3$

(d) $CH_2(COCH_2CH_3)_2$ $\xrightarrow[\text{D}_2\text{O (excess)}]{\text{NaOD}}$ $CD_2(COCH_2CH_3)_2$

16.3

(a) $CH_3CH_2CCH_3$ $\xrightarrow[\text{dimethylformamide} \atop \Delta]{\text{(CH}_3)_3\text{SiCl, (CH}_3\text{CH}_2)_3\text{N}}$ $CH_3CH\!=\!CCH_3$ (with $OSi(CH_3)_3$)

(b) $CH_3CH_2CCH_3$ $\xrightarrow[\text{1,2-dimethyoxyethane}]{\text{(CH}_3\text{CH)}_2\text{N}^-\text{Li}^+}$ $\xrightarrow{\text{(CH}_3)_3\text{SiCl, (CH}_3\text{CH}_2)_3\text{N}}$ $CH_3CH_2C\!=\!CH_2$ (with $OSi(CH_3)_3$)

16.4   (a) $CH_3CCHCCH_3$ (with CH_3) $\xrightarrow[\text{H}_2\text{O}]{\text{Br}_2}$ $CH_3C\!-\!C\!-\!CCH_3$ (with CH_3, Br)

(b)   3-Methyl-2,4-pentanedione will enolize in water. This reaction is an equilibrium reaction, and after a day, the equilibrium concentration of enol is present. The reaction with bromine is fast because the electrophile reacts at the double bond of the enol until the equilibrium concentration of the enol is used up.

16.4 (cont)

Once the enol has been used up, the rate of the overall reaction is determined by the rate at which the enol is formed. That is the slow, rate-determining step of this two-step process.

16.5

For the acid to react with bromine, enolization must occur at the α-carbon atom of the acid. The carboxylic acid is in equilibrium with the ionized form, the carboxylate anion, which bears a negative charge and, therefore, does not enolize readily. Conversion of the acid to the acid bromide makes the carbon atom of the carbonyl group more electrophilic than it is in the carboxylic acid, thus increasing the acidity of the α-hydrogen atom, and promoting enolization. Only a catalytic amount of phosphoric tribromide is used, therefore most of the starting material and the product are present as the carboxylic acid. While the mechanism of the interconversion of the carboxylic acid and acid bromide is not known, it may be seen as taking place by nucleophilic attack of the hydroxyl group of one carboxylic acid on the carbonyl group of the acid bromide giving an intermediate resembling an acid anhydride in structure. Bromide ion could then displace a carboxylic acid, giving rise to a new acid bromide.

Concept Map 16.3   Reactions of enols and enolates with halogen.

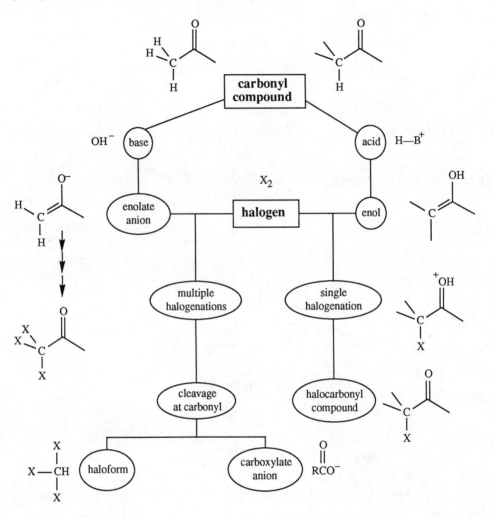

16.6

(a) CH_3CH_2OH + $NaOH$ + Br_2 $\xrightarrow[H_2O]{}$ $CH_3\overset{\overset{\displaystyle O}{\|}}{C}O^-Na^+$ + $CHBr_3$

(b) $CH_3\overset{\overset{\displaystyle CH_3}{|}}{C}=CH\overset{\overset{\displaystyle O}{\|}}{C}CH_3$ + $NaOH$ + I_2 $\xrightarrow[H_2O]{}$ $CH_3\overset{\overset{\displaystyle CH_3}{|}}{C}=CH\overset{\overset{\displaystyle O}{\|}}{C}O^-Na^+$ + CHI_3

16.6 (cont)

(c)

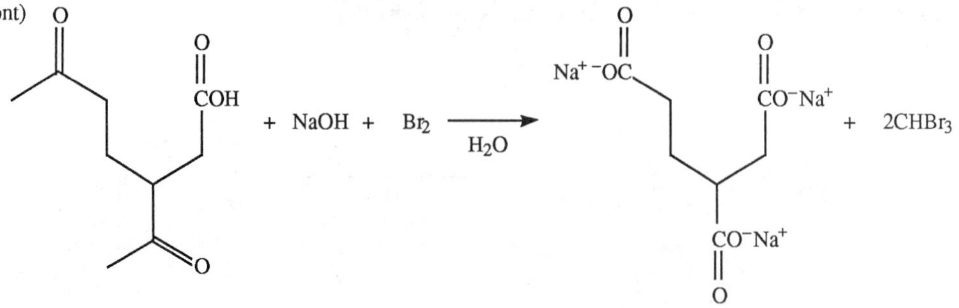

(d)

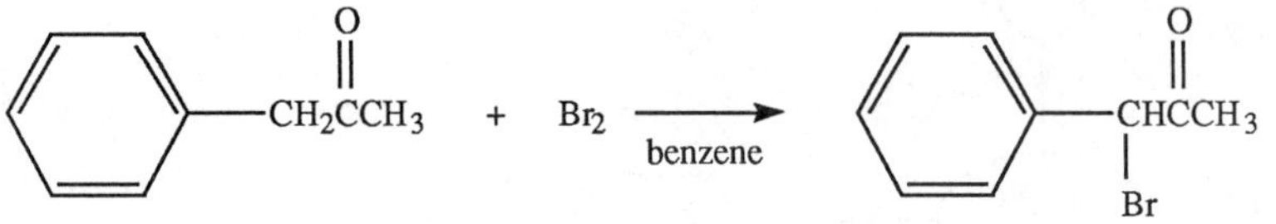

(The more stable enol has
the double bond conjugated
with the aromatic ring.)

(e)

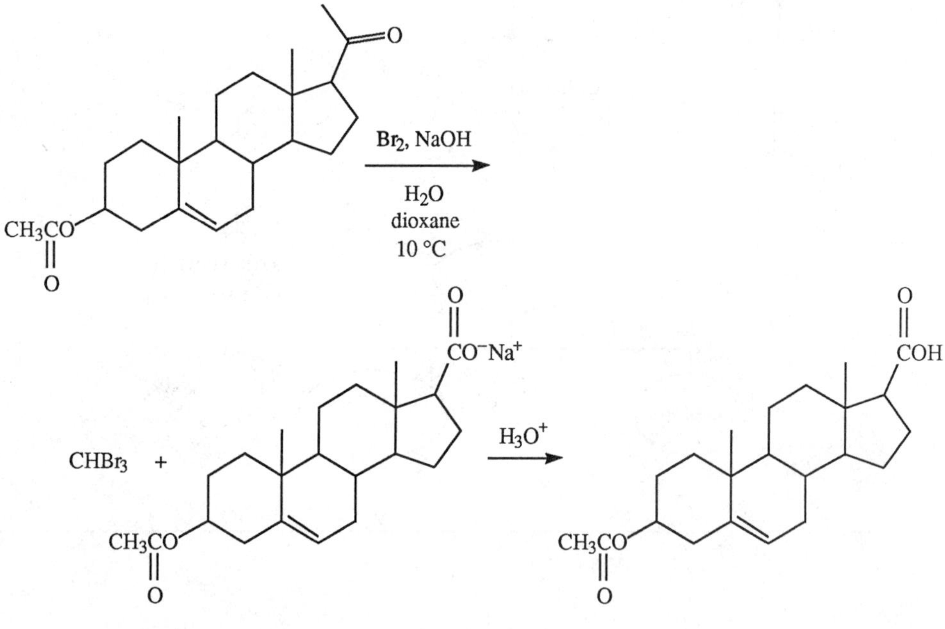

(f)

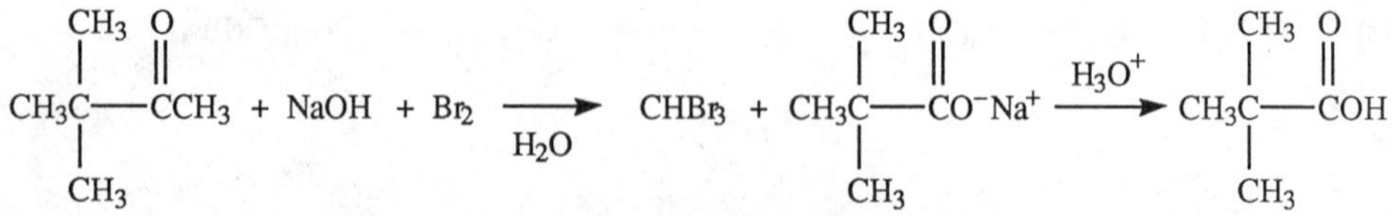

16.6 (cont)

(g) CH_3CCH_3 + NaOH + Cl_2 $\xrightarrow{\ \ H_2O\ \ }$ $CHCl_3$ + $CH_3CO^-Na^+$

16.7

(a)

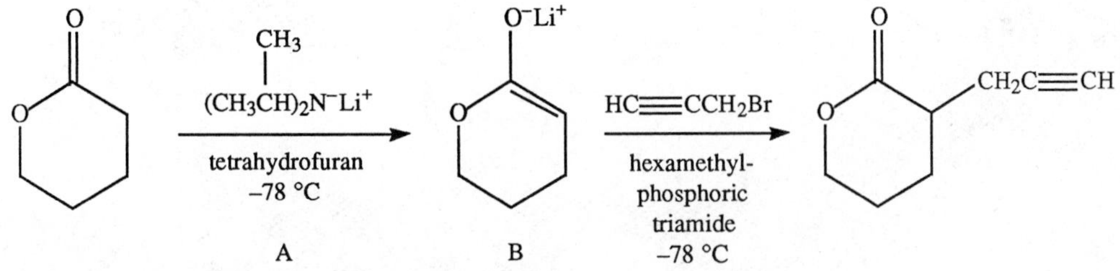

A   B

(b)

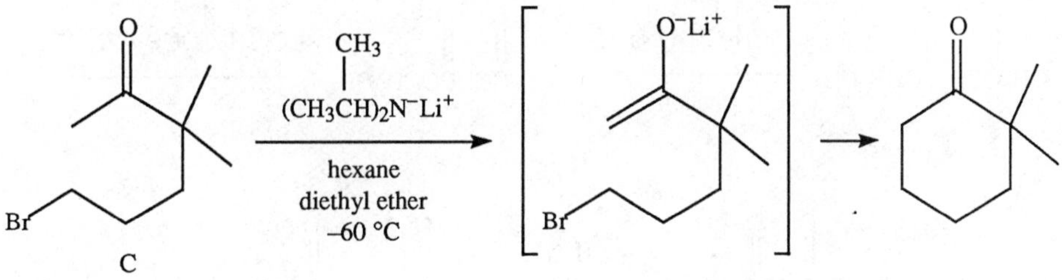

C

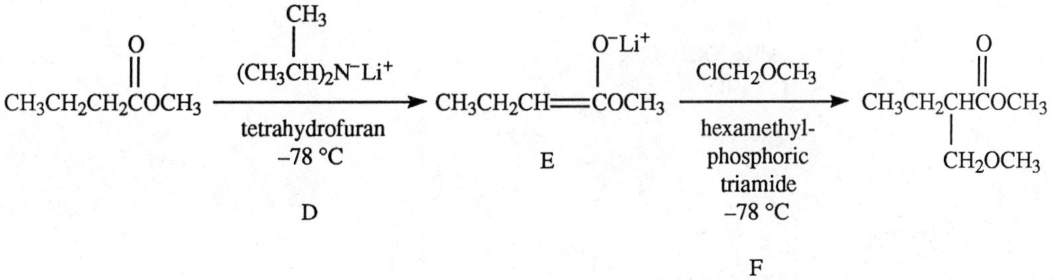

(c) $CH_3CH_2CH_2COCH_3$ $\xrightarrow[\substack{\text{tetrahydrofuran} \\ -78\ ^\circ C}]{\substack{CH_3 \\ | \\ (CH_3CH)_2N^-Li^+}}$ $CH_3CH_2CH=COCH_3$ $\xrightarrow[\substack{\text{hexamethyl-} \\ \text{phosphoric} \\ \text{triamide} \\ -78\ ^\circ C}]{ClCH_2OCH_3}$ $CH_3CH_2CHCOCH_3$

D   E CH_2OCH_3

F

(d)

G   I

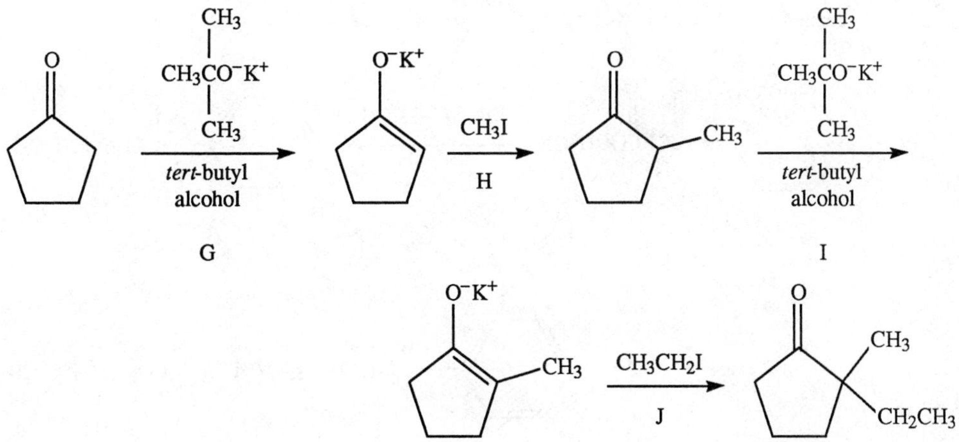

$$\ddot{O}=C=\ddot{O}$$

16.8

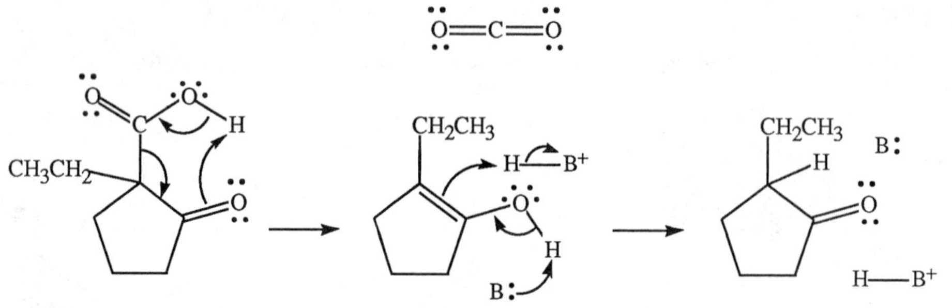

16.9

(a)

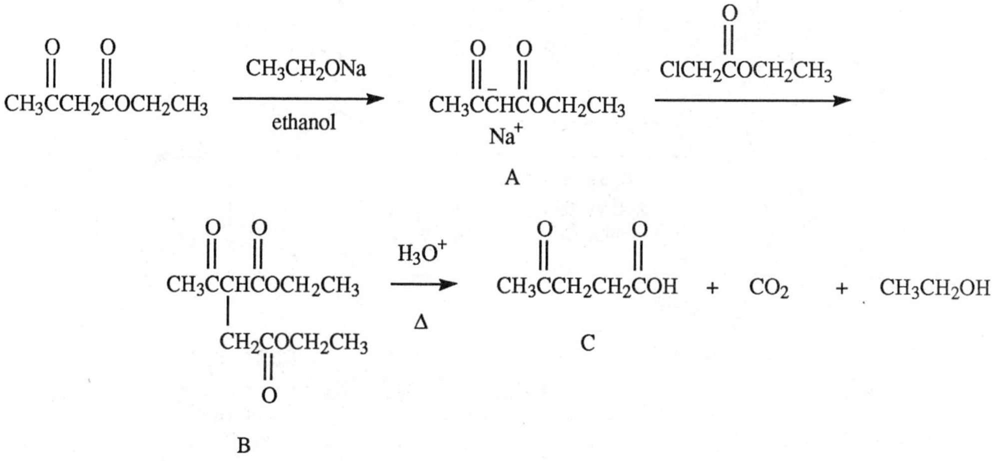

(b)

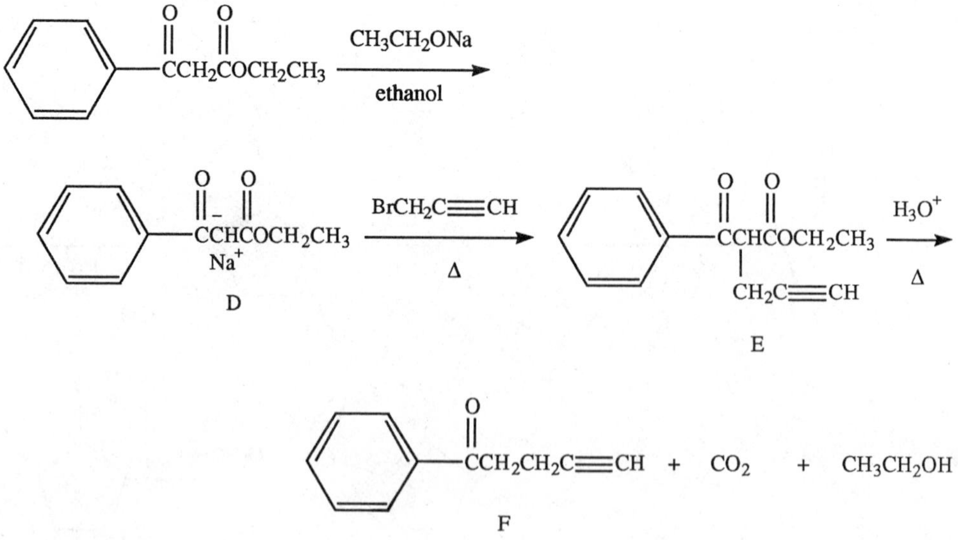

16.9 (cont)

(c)

$$\underset{\text{CH}_3\overset{\text{O}}{\underset{\|}{\text{C}}}\text{CH}_2\overset{\text{O}}{\underset{\|}{\text{C}}}\text{OCH}_2\text{CH}_3}{} \xrightarrow[\text{ethanol}]{\text{CH}_3\text{CH}_2\text{ONa}} \underset{\substack{\text{CH}_3\overset{\text{O}}{\underset{\|}{\text{C}}}\overset{-}{\text{C}}\text{HC}\overset{\text{O}}{\underset{\|}{\text{C}}}\text{OCH}_2\text{CH}_3 \\ \text{Na}^+ \\ \text{G}}}{} \xrightarrow{\text{BrCH}_2\text{CH}_2\overset{\text{O}}{\underset{\|}{\text{C}}}\text{OCH}_2\text{CH}_3}$$

CH₃CCHCOCH₂CH₃
|
CH₂CH₂COCH₂CH₃

H

(d)

$$\text{CH}_3\overset{\text{O}}{\underset{\|}{\text{C}}}\text{CH}_2\overset{\text{O}}{\underset{\|}{\text{C}}}\text{OCH}_2\text{CH}_3 \xrightarrow[\text{ethanol}]{\text{CH}_3\text{CH}_2\text{ONa}} \underset{\substack{\text{Na}^+ \\ \text{I}}}{\text{CH}_3\overset{\text{O}}{\underset{\|}{\text{C}}}\overset{-}{\text{C}}\text{HC}\overset{\text{O}}{\underset{\|}{\text{C}}}\text{OCH}_2\text{CH}_3} \xrightarrow{\text{ClCH}_2\text{CH}_2\text{CH}_2\text{Br}}$$

CH₃CCHCOCH₂CH₃
|
CH₂CH₂CH₂Cl

J

(Br⁻ is a better leaving
group than Cl⁻)

(e)

$$\text{CH}_3\overset{\text{O}}{\underset{\|}{\text{C}}}\text{CH}_2\overset{\text{O}}{\underset{\|}{\text{C}}}\text{OCH}_2\text{CH}_3 \xrightarrow[\text{ethanol}]{\text{CH}_3\text{CH}_2\text{ONa}} \underset{\substack{\text{Na}^+ \\ \text{K}}}{\text{CH}_3\overset{\text{O}}{\underset{\|}{\text{C}}}\overset{-}{\text{C}}\text{HC}\overset{\text{O}}{\underset{\|}{\text{C}}}\text{OCH}_2\text{CH}_3} \xrightarrow{\text{CH}_2{=}\text{CHCH}_2\text{CH}_2\text{Br}}$$

CH₃CCHCOCH₂CH₃
|
CH₂CH₂CH=CH₂

L

$\xrightarrow[\text{toluene}]{\text{Na}}$

$$\text{CH}_3\overset{\text{O}}{\underset{\|}{\text{C}}}{-}\underset{\substack{| \\ \text{CH}_2\text{CH}_2\text{CH}=\text{CH}_2}}{\overset{\text{Na}^+}{\bar{\text{C}}}}{-}\overset{\text{O}}{\underset{\|}{\text{C}}}\text{OCH}_2\text{CH}_3$$

M

$\xrightarrow{\underset{\substack{| \\ \text{CH}_3\text{CHCH}_3}}{\overset{\text{I}}{}}}$

CH₃   O
| ||
CH₃CH   COCH₂CH₃
    \   /
     C
    /   \
CH₃C   CH₂CH₂CH=CH₂
||
O

O

Concept Map 16.4   Alkylation reactions.

16.10

16.11

(a)

$$CH_2(COCH_2CH_3)_2 \xrightarrow[\text{ethanol}]{CH_3CH_2ONa} Na^+ \ ^-CH(COCH_2CH_3)_2 \xrightarrow{CH_3CHCH_3 \ (Br)} CH_3CHCH(COCH_2CH_3)_2 \ (CH_3)$$

A   B

(b)

$$CH_3CH_2CH(COCH_2CH_3)_2 \xrightarrow[\substack{\text{tert-butyl} \\ \text{alcohol}}]{(CH_3)_3CONa} CH_3CH_2\overset{-}{C}(COCH_2CH_3)_2 \ Na^+ \xrightarrow{BrCH_2CH=CH_2}$$

C

$$CH_2=CHCH_2C(COCH_2CH_3)_2$$
$$|$$
$$CH_2CH_3$$

D

(c)

$$CH_2(COCH_2CH_3)_2 \xrightarrow[\text{ethanol}]{CH_3CH_2ONa} Na^+ \ ^-CH(COCH_2CH_3)_2 \xrightarrow{BrCH_2CH_2CH_2Cl}$$

E

$$ClCH_2CH_2CH_2CH(COCH_2CH_3)_2 \xrightarrow[\text{ethanol}]{CH_3CH_2ONa} \left[ClCH_2CH_2CH_2\overset{Na^+}{\overset{-}{C}}(COCH_2CH_3)_2 \right] \longrightarrow$$

F
(Br$^-$ is a better leaving
group than Cl$^-$)

G

H

$$+ \ CO_2 \ + \ 2 \ CH_3CH_2OH$$

(d)

$+ \ Na^+ \ ^-CH(COCH_2CH_3)_2 \longrightarrow Na^+ \ ^-OCH_2CH_2CH(COCH_2CH_3)_2$

I

16.11 (cont)

(e) $CH_2(COCH_2CH_3)_2$ $\xrightarrow[\text{ethanol}]{CH_3CH_2ONa}$ $Na^{+\,-}CH(COCH_2CH_3)_2$ $\xrightarrow{}$

J

K

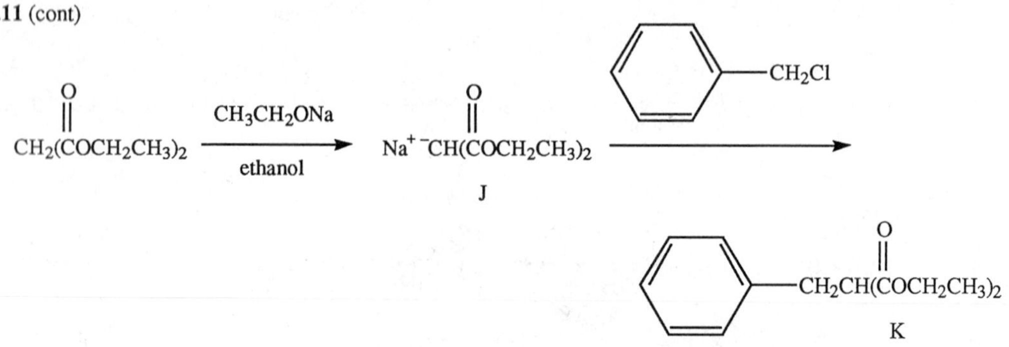

16.12

1. $CH_3CHCH_2CH_2Br$ $\xleftarrow{\text{conc HBr}}$ $CH_3CHCH_2CH_2OH$ $\xleftarrow{H_3O^+}$ $\xleftarrow{}$ $CH_3CHMrBr$ $\xleftarrow[\substack{\text{diethyl}\\\text{ether}}]{Mg}$ CH_3CHBr

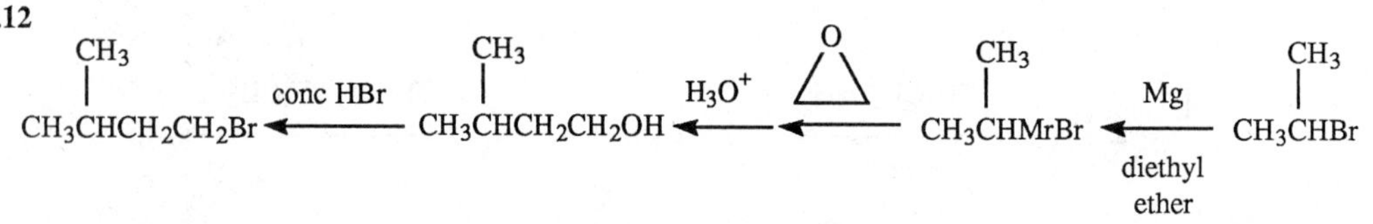

2.

$\xleftarrow{H_3O^+}$

$\xleftarrow[\text{methanol}]{H_2NCNH_2,\ CH_3ONa}$

$CH_3CHCH_2CH_2C(COCH_2CH_3)_2$ (with CH_2CH_3 branch) $\xleftarrow{CH_3CHCH_2CH_2Br}$ $CH_3CH_2\overset{-}{C}(COCH_2CH_3)_2$ K^+ $\xleftarrow[\substack{\textit{tert}\text{-butyl}\\\text{alcohol}}]{(CH_3)_3COK}$

$CH_3CH_2CH(COCH_2CH_3)_2$ $\xleftarrow{CH_3CH_2Br}$ $K^{+\,-}CH(COCH_2CH_3)_2$ $\xleftarrow[\substack{\textit{tert}\text{-butyl}\\\text{alcohol}}]{(CH_3)_3COK}$ $CH_2(COCH_2CH_3)_2$

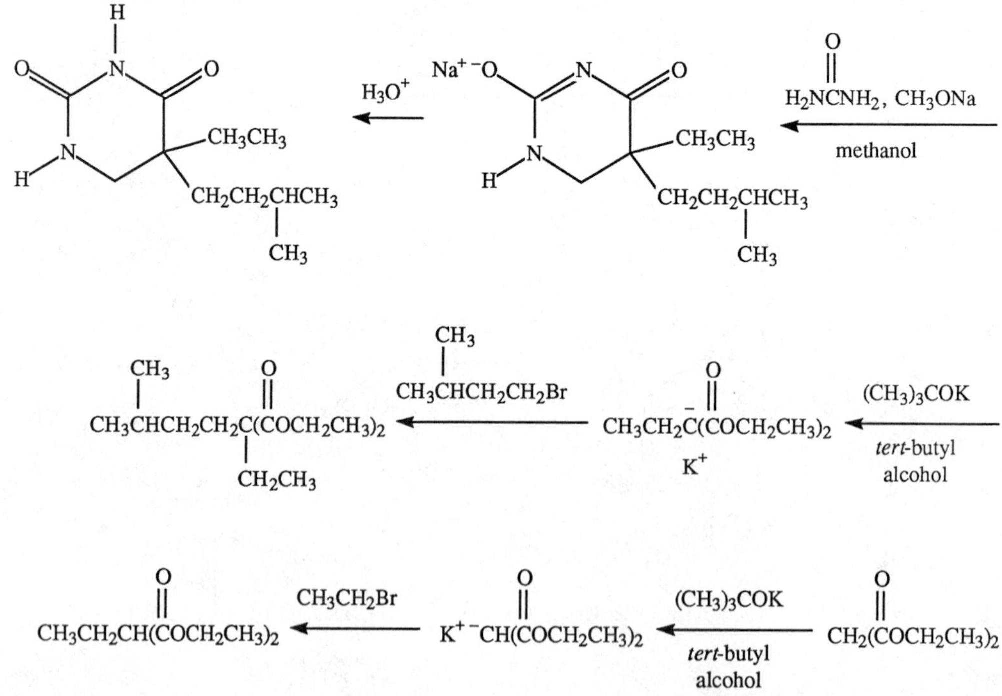

Concept Map 16.5   The aldol condensation.

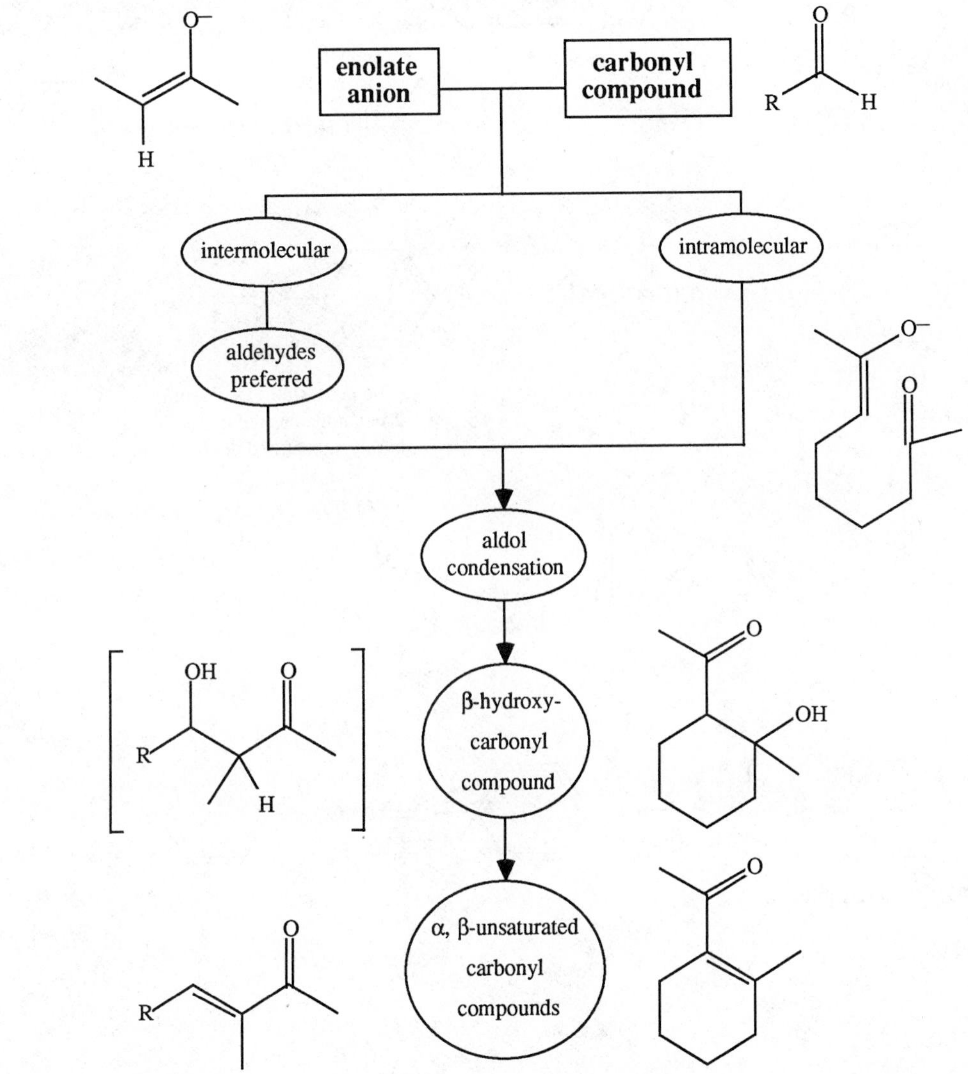

- -

16.13

(a)

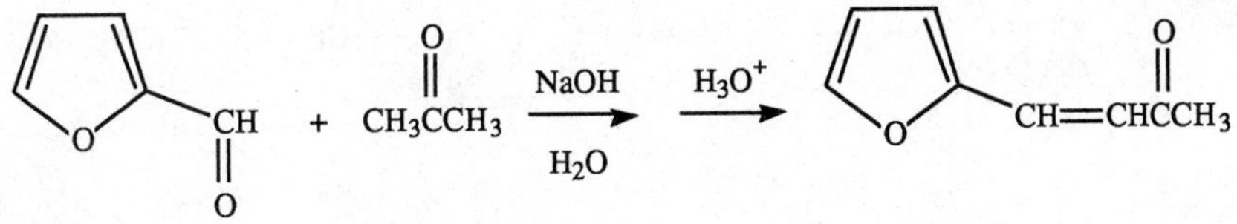

16.13 (cont)

(b)

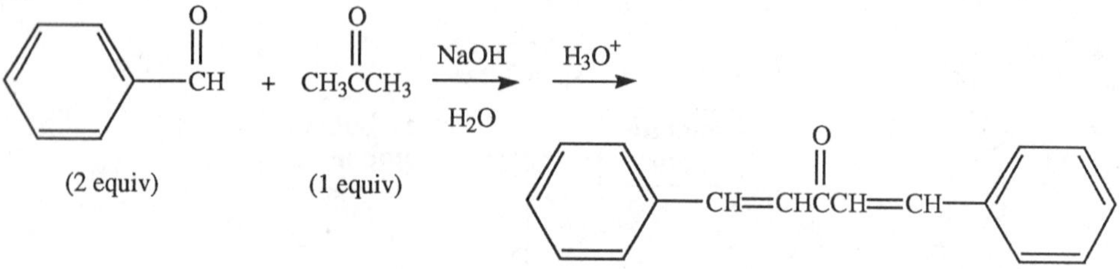

(2 equiv)   (1 equiv)

(c)

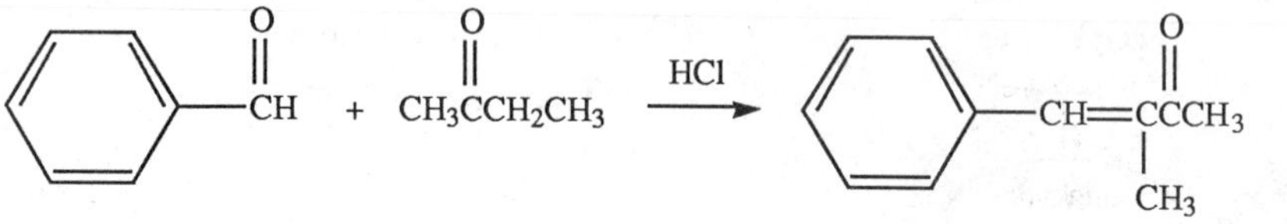

(The more stable enol from
2-butanone has the more
highly substituted double bond)

(d)

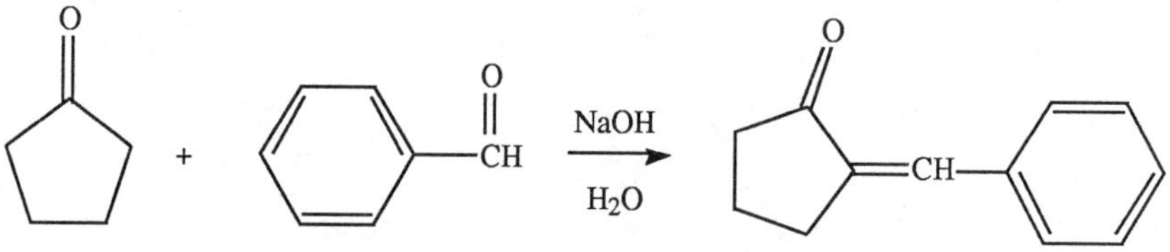

(e)

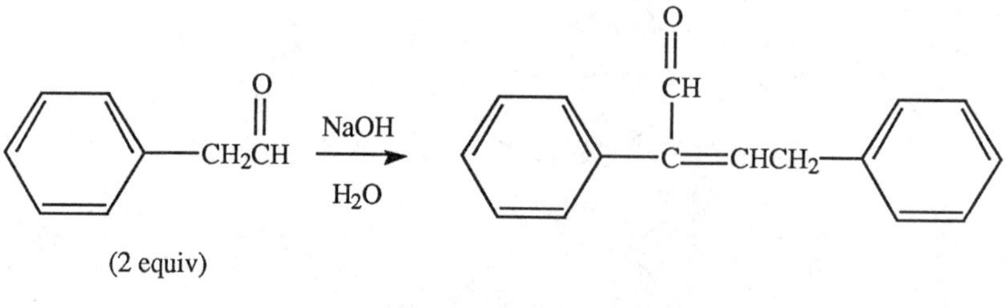

(2 equiv)

16.14

(a)

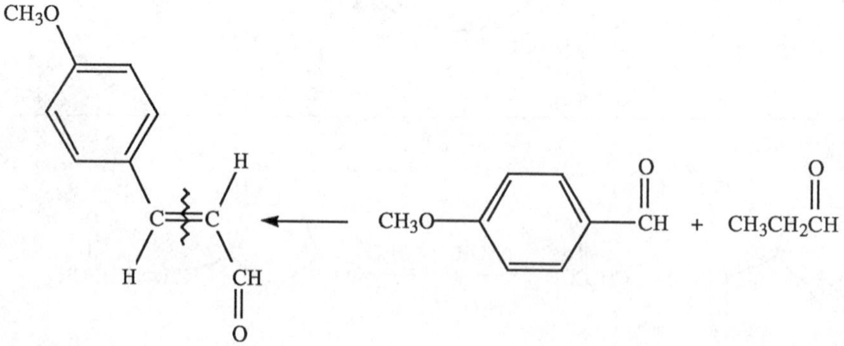

16.14 (cont)

(b)

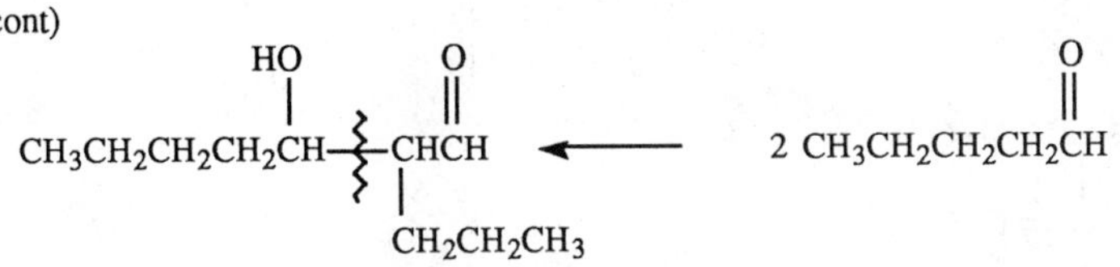

(c)

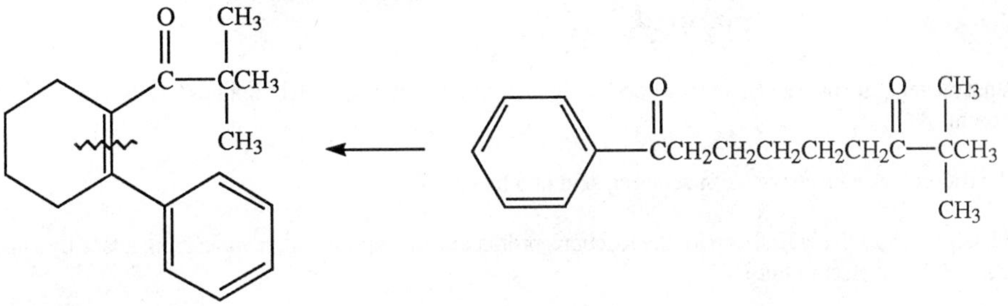

(d)

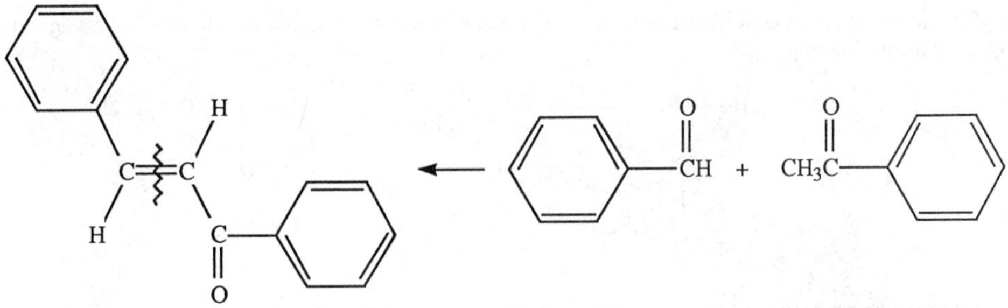

16.15

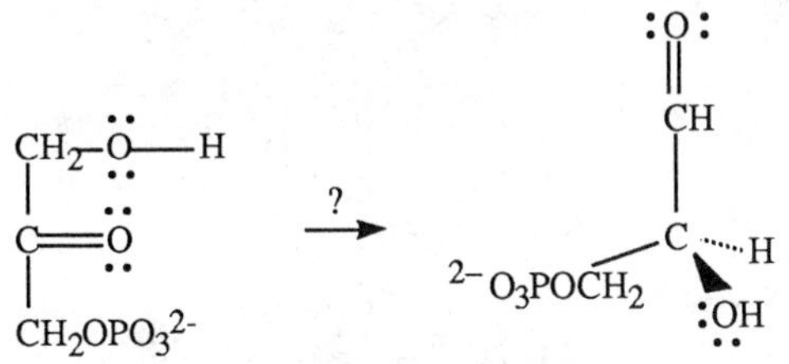

1. To what functional group class does the reactant belong? What is the electronic character of the functional group?

    The reactant is an alcohol, a ketone, and a phosphate ester. The product is an alcohol, an aldehyde and a phosphate ester.

2. How do the structures of the reactants and the desired products compare? How many carbon atoms does each contain? What bonds must be broken and formed to transform starting material into product?

    Both reactant and product have three carbon atoms. There is an alcohol function on carbon 1 and a ketone function on carbon 2 in the reactant. There is an aldehyde function on carbon 1 and an alcohol function on carbon 2 in the product. Both have a phosphate ester on carbon 3. The bonds to be broken and formed are shown on the next page.

16.15 (2) (cont)

3. What reagents are present? Are they good acids, bases, nucleophiles, or electrophiles? Is there an ionizing solvent present?

The triosephosphate enzyme can act as an acid or a base.

4. What is the most likely first step for the reaction: protonation or deprotonation, ionization, attack by a nucleophile, or attack by an electrophile?

Protonation of the carbonyl group of the ester with simultaneous deprotonation of the α hydrogen on carbon 1 is the most likely first step.

5. What are the properties of the species present in the reaction mixture after the first step? What is likely to happen next?

An enediol is formed, which will tautomerize to the product.

6. What is the stereochemistry of the reaction?

The enzyme transforms dihydroxyacetone phosphate stereoselectively into D-glyceraldehyde 3-phosphate.

Concept Map 16.6   Acylation reactions of enolates.

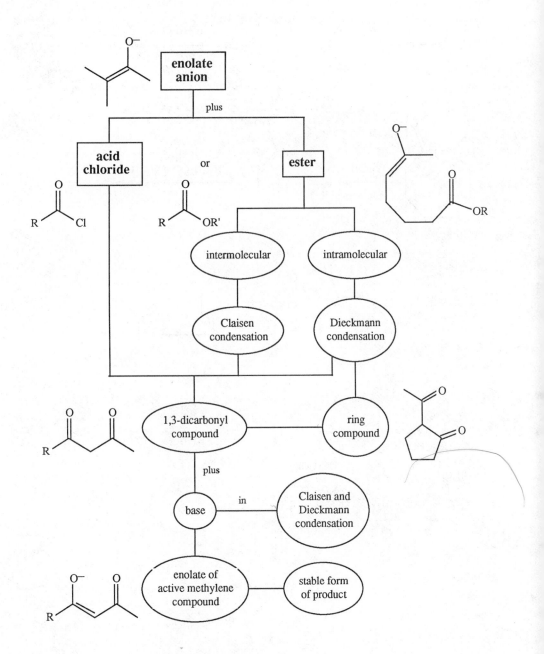

16.16

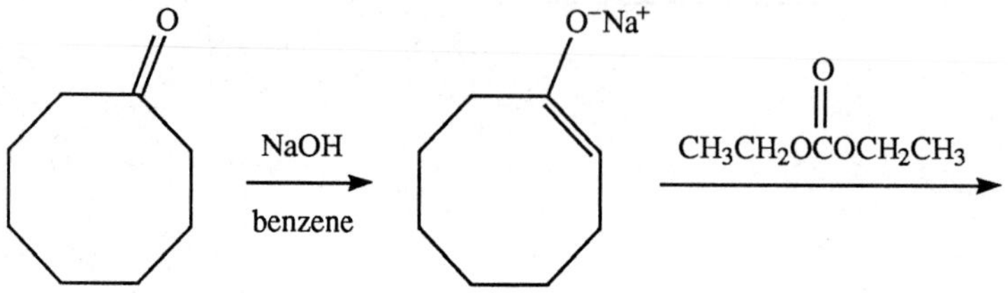

(a)

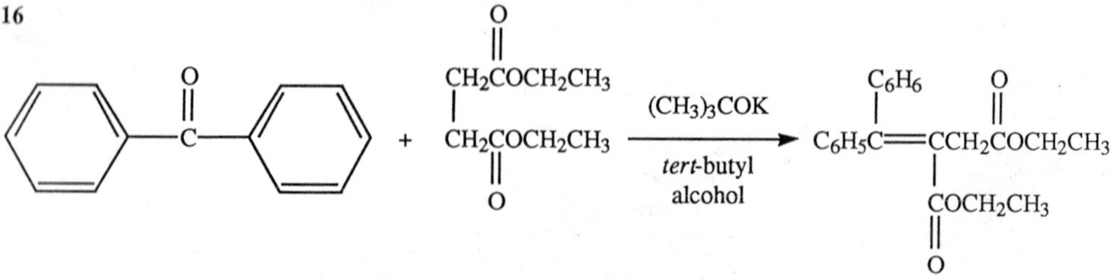

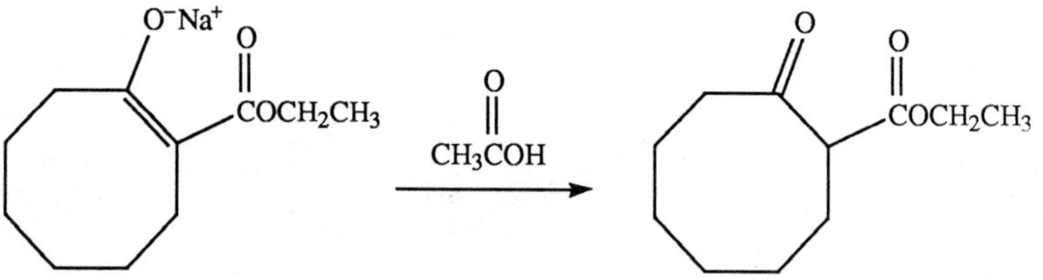

(b)

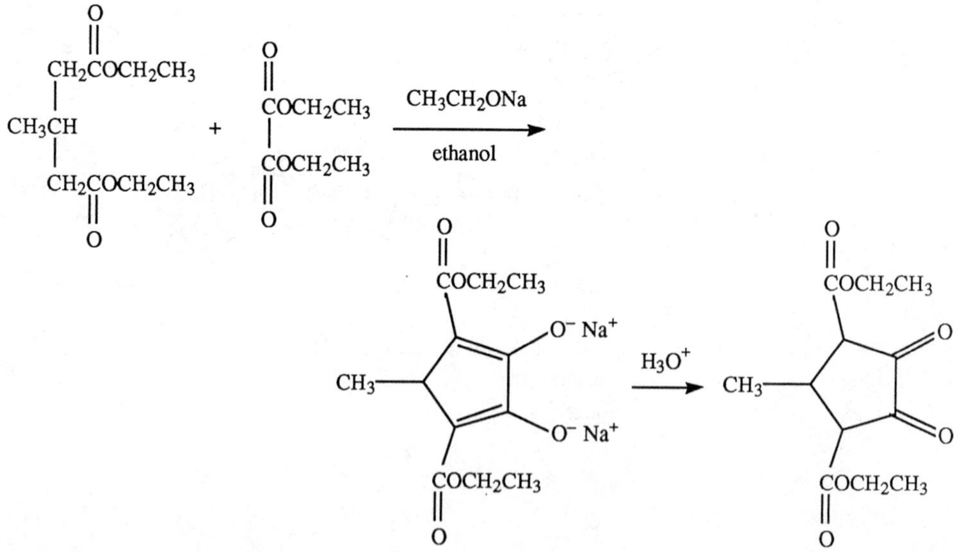

(c)

16.16 (cont)

(d)

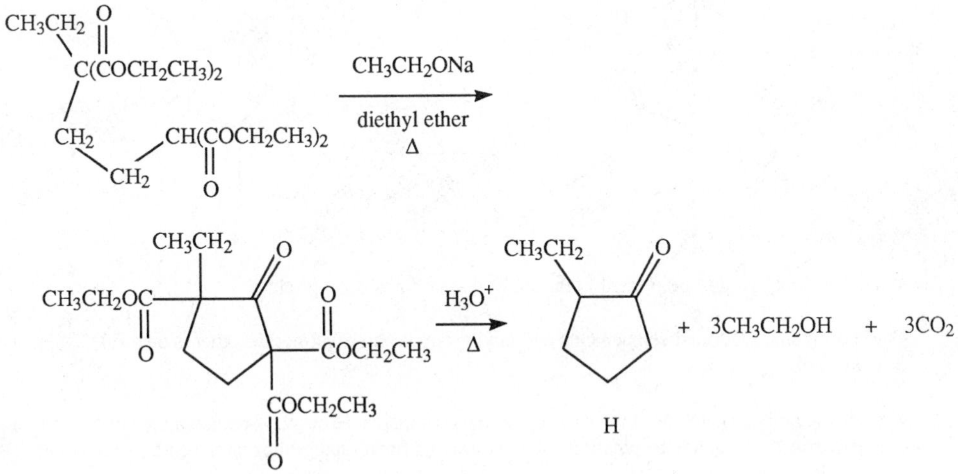

G
(Note that all three ester groups
are β to a carbonyl group)

In part (c) the acidified reaction mixture was not heated, and no hydrolysis and decarboxylation take place. In part (d), hydrolysis of the ester and decarboxylation take place when the acidified reaction mixture is heated.

16.17

(a)

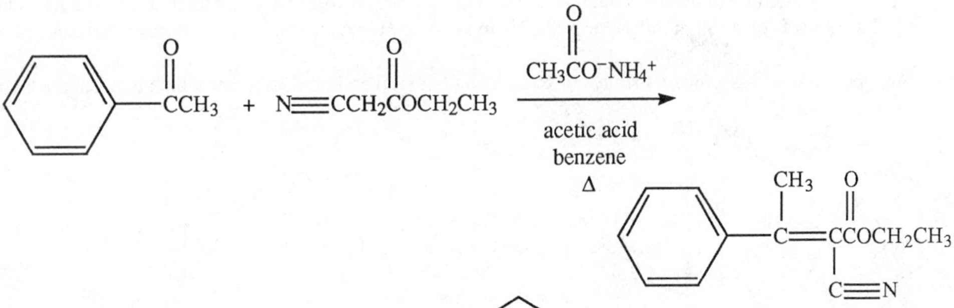

(b)

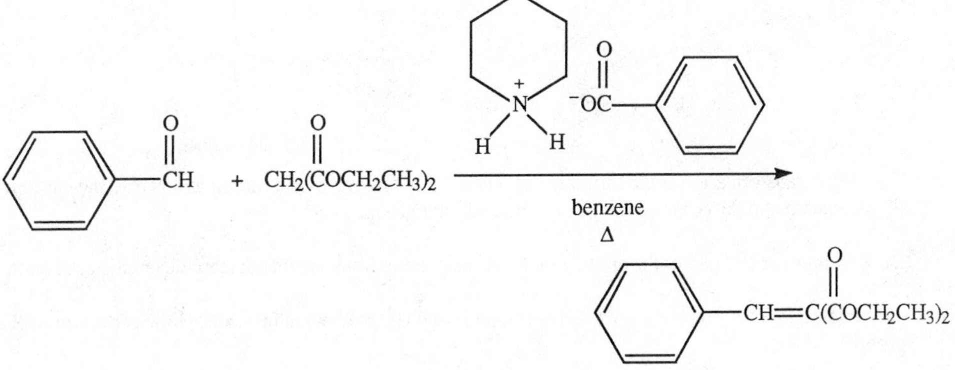

16.18

1. What functional groups are present in the starting material and the product?

    The starting material is a diketone. One carbonyl group is part of a ring; the other is on a side chain. The product is an alcohol.

2. How do the carbon skeletons of the two compounds compare? How many carbon atoms does each contain? Are there any rings? What are the positions of branches and functional groups on the carbon skeletons?

    The product has the same number of carbon atoms as the starting material. There are two five-membered rings in the reactant and three in the product. The carbonyl group of the side chain in the starting material has become an alcohol group in the product and the side chain has been incorporated into the third ring. The new ring in the product is attached at the carbon atom of the carbonyl group in a ring of the starting material.

3. How do the functional groups change in going from starting material to product? Does the starting material have a good leaving group?

    The carbonyl group of the side chain in the starting material has become an alcohol in the product. The carbonyl group that was on a ring has disappeared. There is no good leaving group in the starting material.

4. Is it possible to dissect the structures of the starting material and product to see which bonds must be broken and which formed?

    bonds broken   bonds formed

    The major transformation is the formation of a bond between one of the carbon atoms α to the carbonyl group of the side chain and the carbon atom of the ring carbonyl group.

5. Do we recognize any part of the product molecule as coming from a good nucleophile or an electrophilic addition?

    An enolate anion from a ketone is a good nucleophile and will react with the electrophilic carbon atom of a carbonyl group.

16.18 (4) (cont)

The enolate from the ketone of the side chain will react with the carbonyl group of the ring ketone in an intramolecular aldol condensation. The product will lose water to form an α,β-unsaturated ketone.

6.   What type of compound would be a good precursor to the product?

A ketone that can be reduced to the alcohol.

The ketone that we need for step 6 is formed by the reduction of the α,β-unsaturated ketone we got in step 5.

7.   After this last step, do we see how to get from starting material to product? If not, we need to analyze the structure obtained in step 6 by applying questions 5 and 6 to it.

16.18 (cont) Complete synthesis:

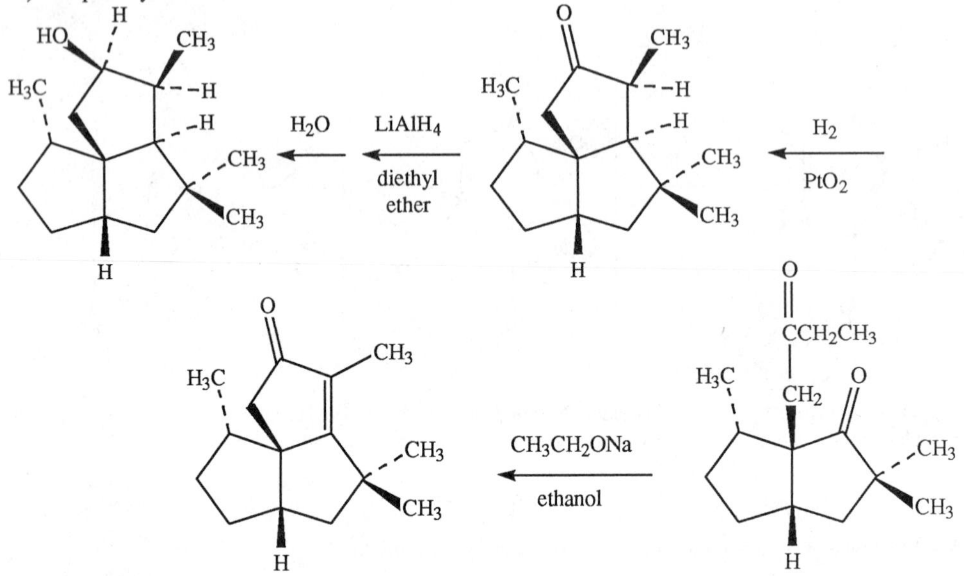
H₂O
LiAlH₄
diethyl ether
H₂
PtO₂
CH₃CH₂ONa
ethanol

16.19

(a)

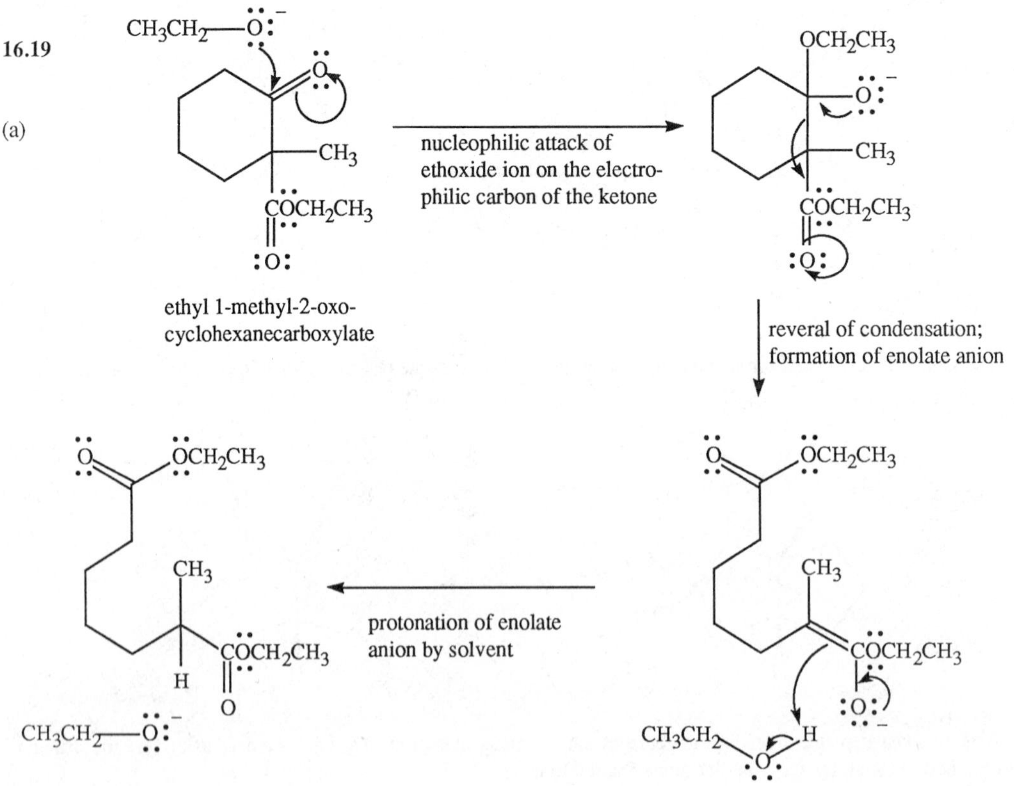
ethyl 1-methyl-2-oxo-cyclohexanecarboxylate
nucleophilic attack of ethoxide ion on the electrophilic carbon of the ketone
reveral of condensation; formation of enolate anion
protonation of enolate anion by solvent

16.19 (cont)

(b)

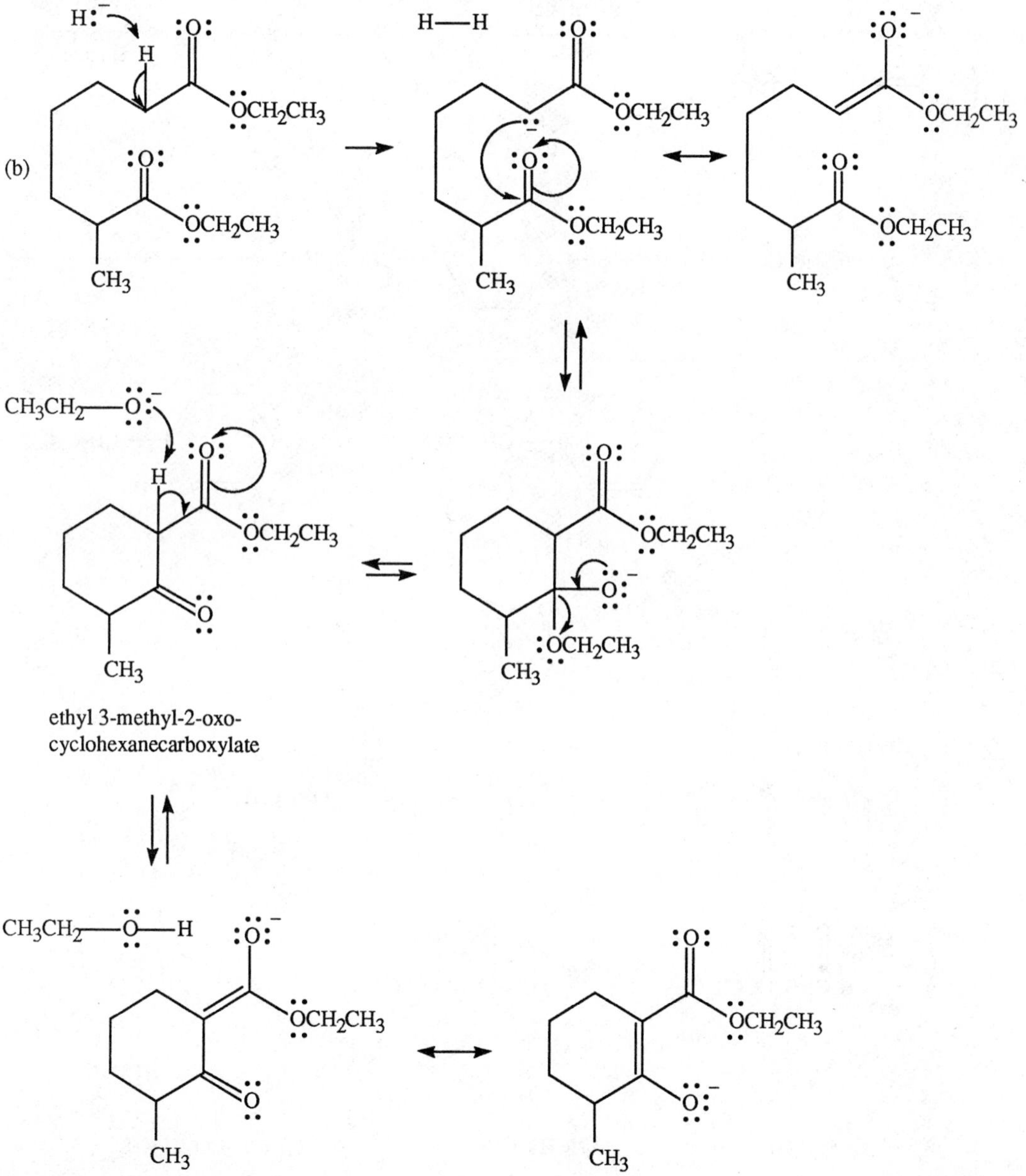

ethyl 3-methyl-2-oxo-
cyclohexanecarboxylate

The product β-ketoesters in Claisen condensations are stabilized by deprotonation to the corresponding enolates. Ethyl 1-methyl-2-oxocyclohexanecarboxylate does not have a hydrogen atom on the carbon atom between the two carbonyl groups, and, cannot, therefore, be stabilized this way. It undergoes reversal of the cyclization reaction. The product resulting from the other cyclization reaction, ethyl 3-methyl-2-oxocyclohexanecarboxylate, does have an active hydrogen. Under the conditions for the reaction, it is the thermodynamically more stable product.

16.20

diethyl ethylphenylmalonate

16.21   (a)

$$\underset{\text{O}}{\overset{\text{O}}{\underset{\|}{\text{CH}_3\text{C}}}}\text{CH}_2\underset{\|}{\overset{\text{O}}{\text{C}}}\text{CF}_3 \quad > \quad \text{CH}_3\underset{\|}{\overset{\text{O}}{\text{C}}}\text{CH}_2\underset{\|}{\overset{\text{O}}{\text{C}}}\text{CH}_3$$

(b)

$\quad > \quad$

(c) $\text{CH}_3\underset{\|}{\overset{\text{O}}{\text{C}}}\text{CH}_2\underset{\|}{\overset{\text{O}}{\text{C}}}\text{OCH}_2\text{CH}_3 \quad > \quad \text{CH}_3\underset{\|}{\overset{\text{O}}{\text{C}}}\text{C}\underset{\underset{\text{CH}_3}{|}}{\text{H}}\underset{\|}{\overset{\text{O}}{\text{C}}}\text{OCH}_2\text{CH}_3$

(d) $\text{O}_2\text{NCH}_2\text{NO}_2 \quad > \quad \text{O}_2\text{NCH}_2\underset{\|}{\overset{\text{O}}{\text{C}}}\text{CH}_3 \quad > \quad \text{O}_2\text{NCH}_2\underset{\|}{\overset{\text{O}}{\text{C}}}\text{OCH}_2\text{CH}_3$

(e) $\text{HC}\underset{\|}{\overset{\text{O}}{}}\text{CH}_2\text{C}\underset{\|}{\overset{\text{O}}{}}\text{H} \quad > \quad \text{CH}_3\text{CH}_2\underset{\|}{\overset{\text{O}}{\text{OC}}}\text{CH}_2\underset{\|}{\overset{\text{O}}{\text{C}}}\text{OCH}_2\text{CH}_3 \quad > \quad \text{CH}_3\text{CH}_2\underset{\|}{\overset{\text{O}}{\text{OC}}}\underset{\underset{\text{CH}_3}{|}}{\text{C}}\text{H}\underset{\|}{\overset{\text{O}}{\text{C}}}\text{OCH}_2\text{CH}_3$

16.22

(a)

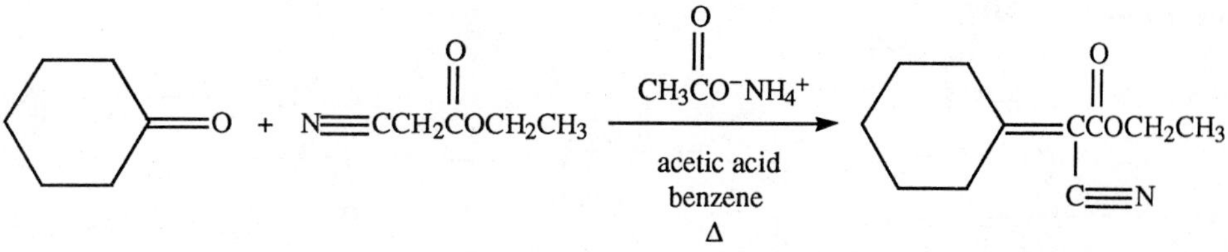

(b)

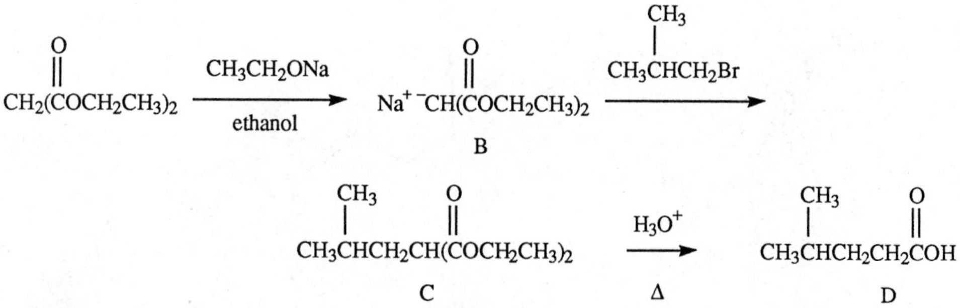

(c)

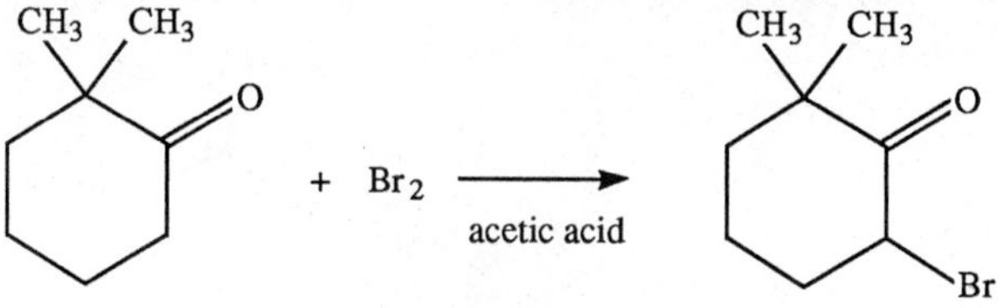

(d)

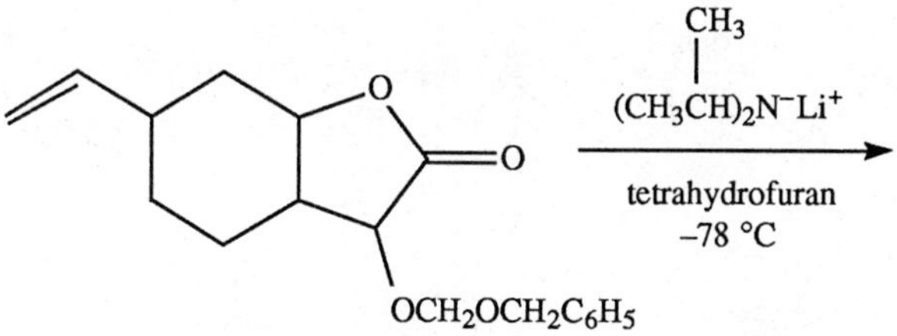

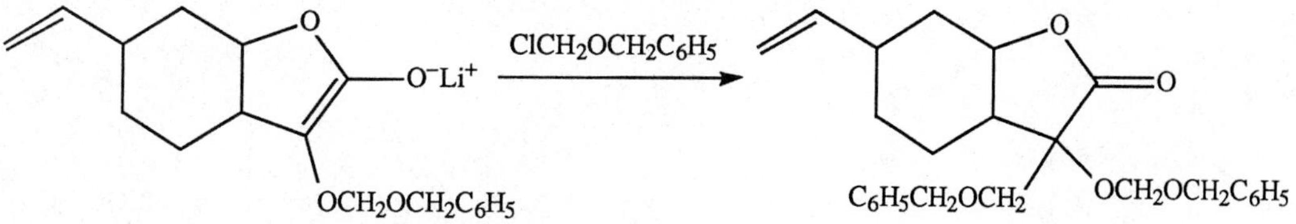

16.22 (cont)

(e) O_2N—⟨benzene ring⟩—$\overset{\overset{O}{\|}}{C}CH_3$ + $CH_3\overset{\overset{O}{\|}}{C}OCH_3$ $\xrightarrow{CH_3ONa}$ O_2N—⟨benzene ring⟩—$\overset{\overset{O}{\|}}{C}CH_2\overset{\overset{O}{\|}}{C}CH_3$

H

(f) $\xrightarrow{CH_3CH_2ONa}$

I

(g) $CH_2(\overset{\overset{O}{\|}}{C}OCH_2CH_3)_2$ $\xrightarrow[\text{ethanol}]{CH_3CH_2ONa}$ $Na^+\ {}^-CH(\overset{\overset{O}{\|}}{C}OCH_2CH_3)_2$

J

K

(h) $CH_3\overset{\overset{O}{\|}}{C}H$ + $CH_2(\overset{\overset{O}{\|}}{C}OCH_2CH_3)_2$ $\xrightarrow[\Delta]{(CH_3\overset{\overset{O}{\|}}{C})_2O}$ $CH_3CH{=}C(\overset{\overset{O}{\|}}{C}OCH_2CH_3)_2$

L

(i) $CH_3\overset{\overset{O}{\|}}{C}CH_2\overset{\overset{O}{\|}}{C}OCH_2CH_3$ $\xrightarrow[H_2O]{NaOH}$ $CH_3\overset{\overset{O}{\|}}{C}\overset{-}{C}H\overset{\overset{O}{\|}}{C}OCH_2CH_3$

Na^+

M

N

16.22 (cont)

(j) $CH_3CCH_2COCH_2CH_3$ + [aromatic aldehyde with CH_3O and OCH_3 substituents] $\xrightarrow[\text{acetic acid}]{\text{piperidine}}$ [product with $CH=CCOCH_2CH_3$ and CCH_3 groups, CH_3O and OCH_3 substituents]

O

(k) CH_3O—[benzene ring]—CCH_3 + I_2 $\xrightarrow[\substack{H_2O \\ \text{dioxane}}]{NaOH}$ CH_3O—[benzene ring]—CO^-Na^+ + CHI_3

                                                                                                P   Q

(l) [benzene ring]—$CH_2C\equiv N$ $\xrightarrow[\text{ethanol}]{CH_3CH_2ONa}$

Na^+

[benzene ring]—$\overset{-}{C}HC\equiv N$ $\xrightarrow{CH_3CH_2OCOCH_2CH_3}$ [benzene ring]—$CHC\equiv N$ with $COCH_2CH_3$ group

                                        R   S

(m) [benzene ring with CH (O) and OCH_3 substituents] + $CH_3CH_2NO_2$ $\xrightarrow[\substack{\text{toluene} \\ \Delta}]{CH_3(CH_2)_3NH_2}$ [benzene ring with $CH=CNO_2$, CH_3, and OCH_3 substituents]

                                                                                                                T

(n) $CH_3(CH_2)_4CH_2CCH_2CH_2CH_2OH$ $\xrightarrow[\text{dichloromethane}]{\text{pyridinium } CrO_3Cl^-}$

$CH_3(CH_2)_4CH_2CCH_2CH_2CH$ $\xrightarrow[\substack{H_2O \\ \text{tetrahydrofuran}}]{KOH}$ [cyclopentenone ring with $(CH_2)_4CH_3$ group]

                            U   V

16.23

Step 1

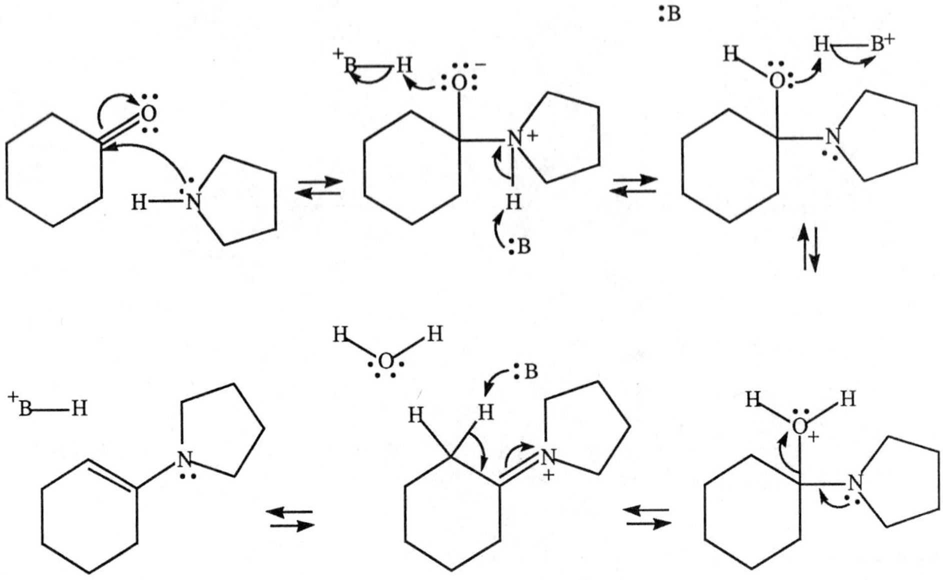

Step 2

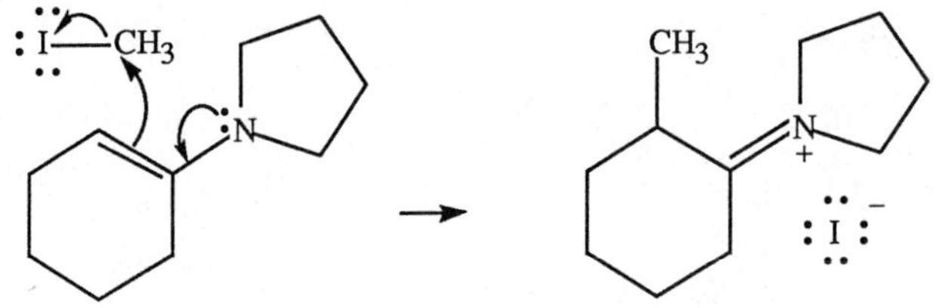

Step 3

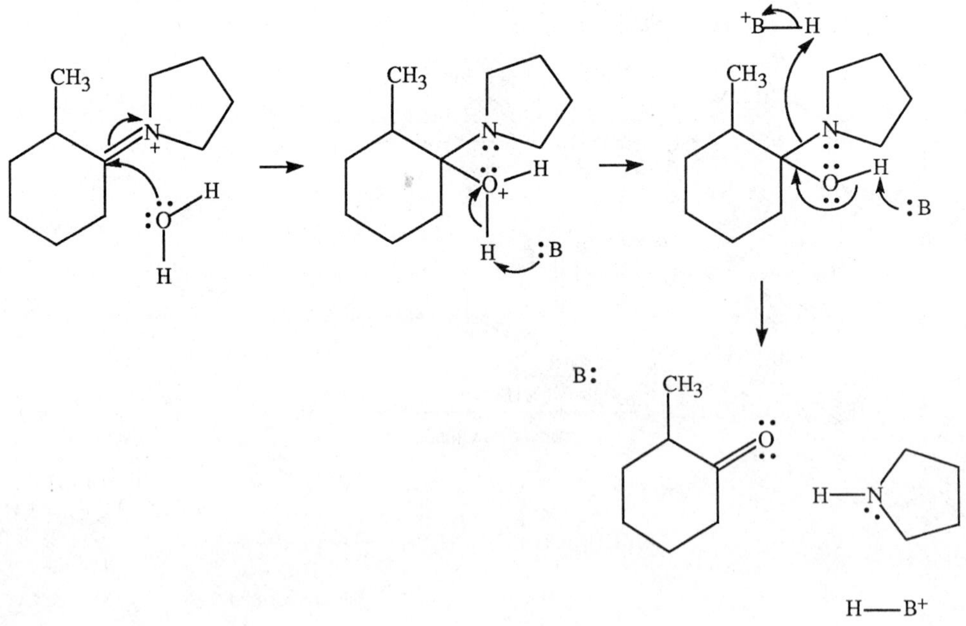

16.24

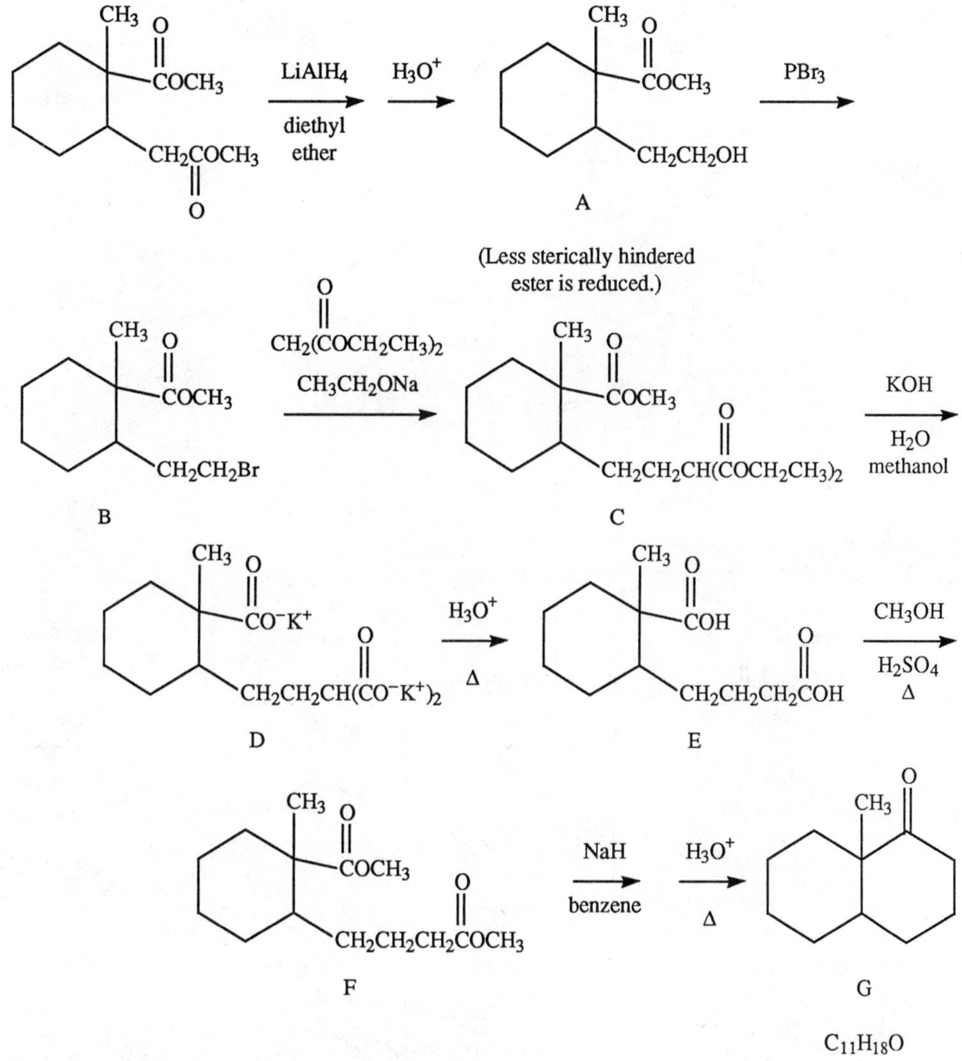

(Less sterically hindered
ester is reduced.)

$C_{11}H_{18}O$

16.25 $CO_3^{2-} + H_2O \rightleftharpoons HCO_3^- + OH^-$

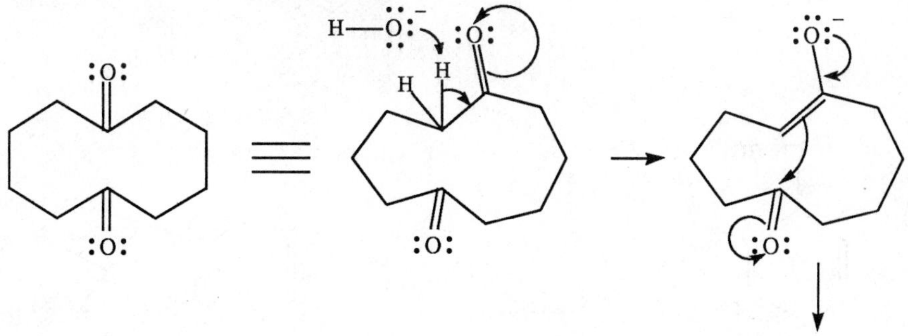

16.25 (cont)

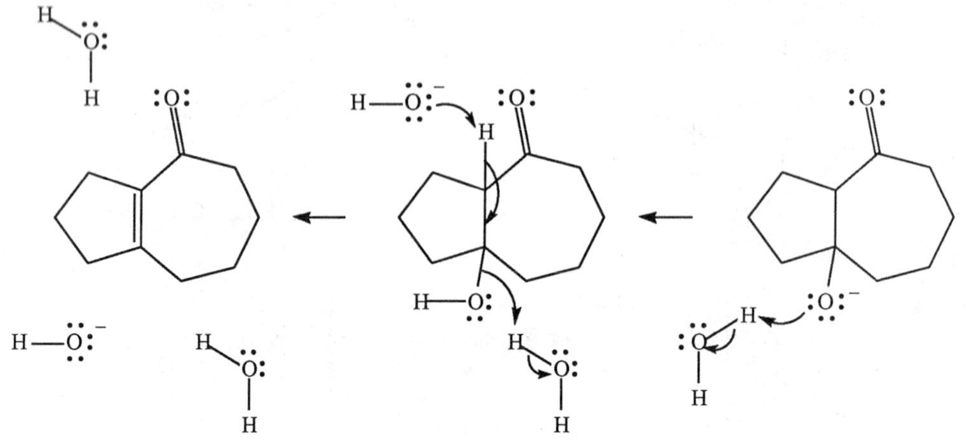

16.26

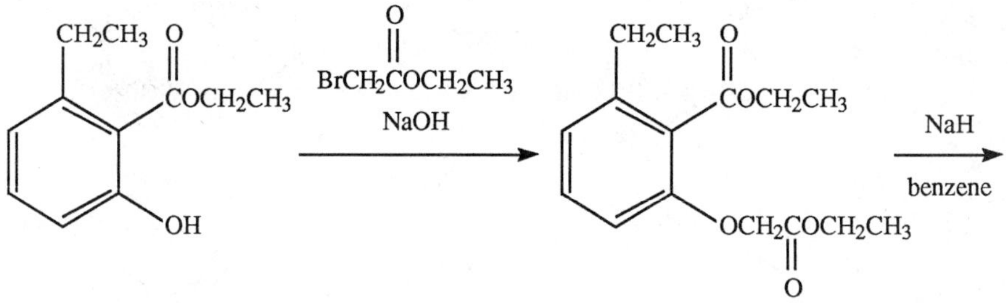

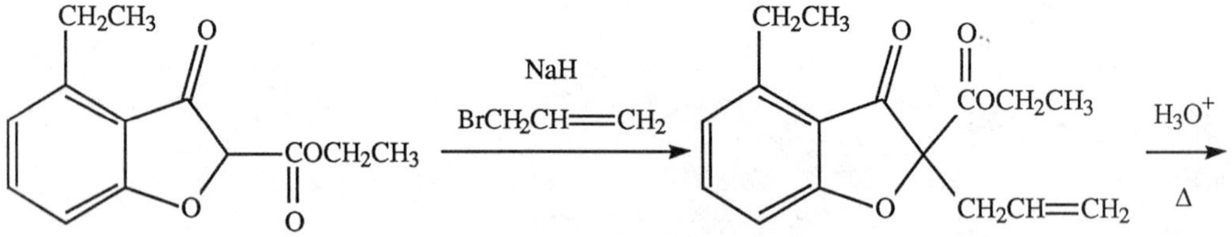

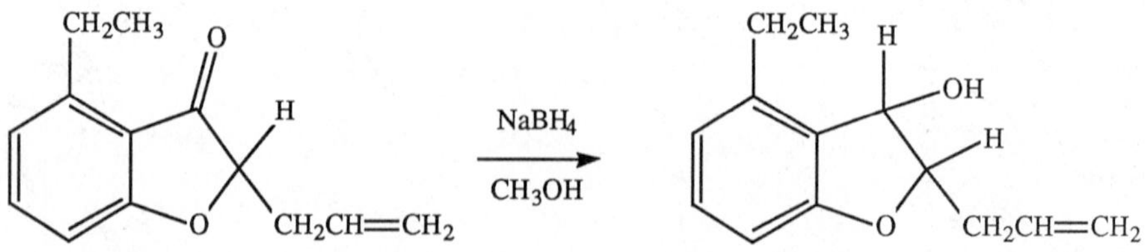

16.27

(a)

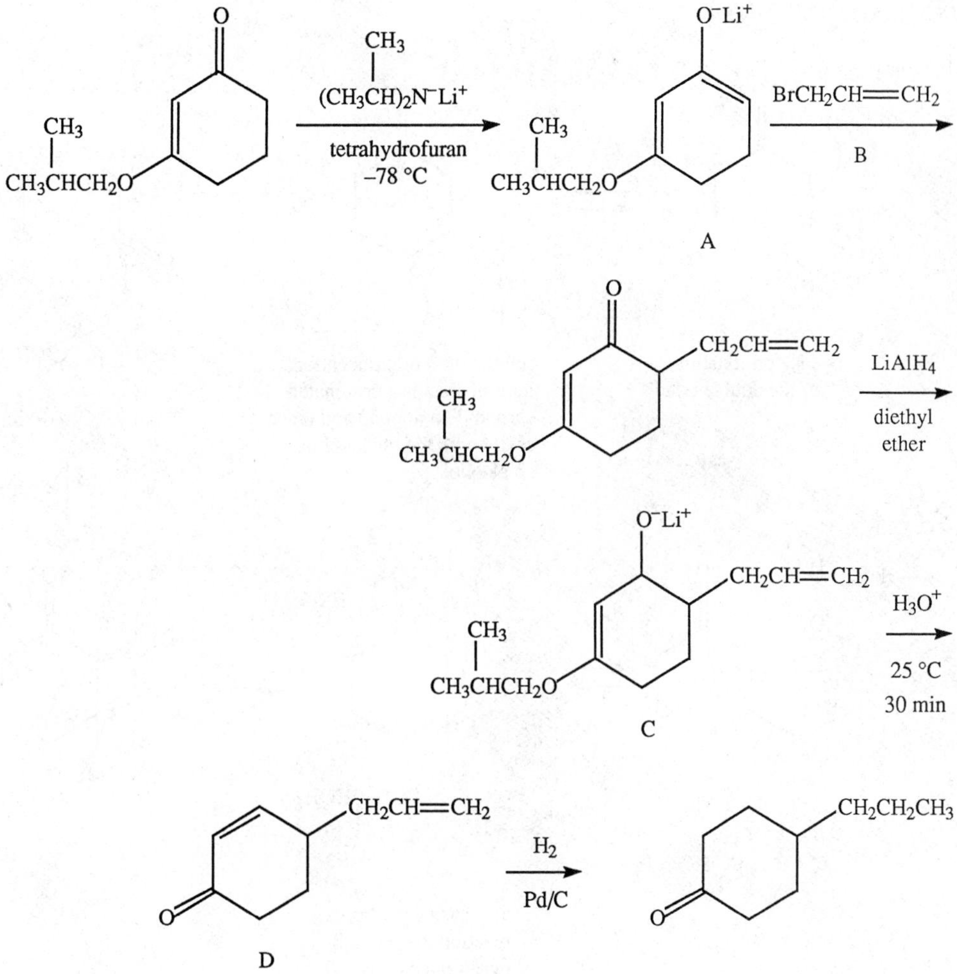
O
CH3
CH3CHCH2O
CH3
(CH3CH)2N⁻Li⁺
tetrahydrofuran
−78 °C
O⁻Li⁺
CH3
CH3CHCH2O
A
BrCH2CH=CH2
B
O
CH2CH=CH2
CH3
CH3CHCH2O
LiAlH4
diethyl ether
O⁻Li⁺
CH2CH=CH2
CH3
CH3CHCH2O
C
H3O⁺
25 °C
30 min
CH2CH=CH2
O
D
H2
Pd/C
CH2CH2CH3
O

(b)

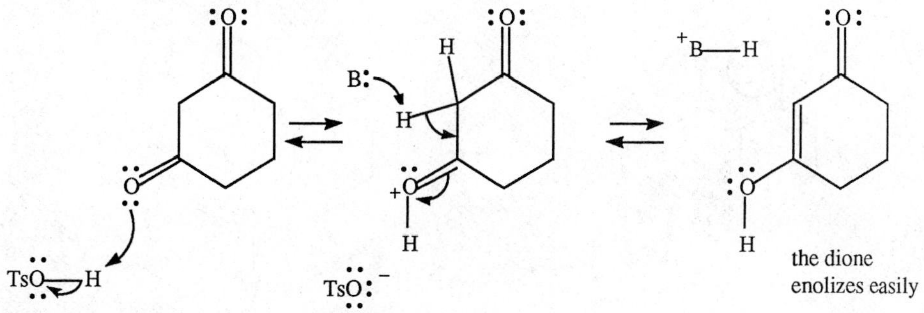
O
O
TsO
H
B:
H
O
H
O
H
TsO:⁻
⁺B—H
O
O
H
the dione
enolizes easily

16.27 (cont)

(b) (cont)

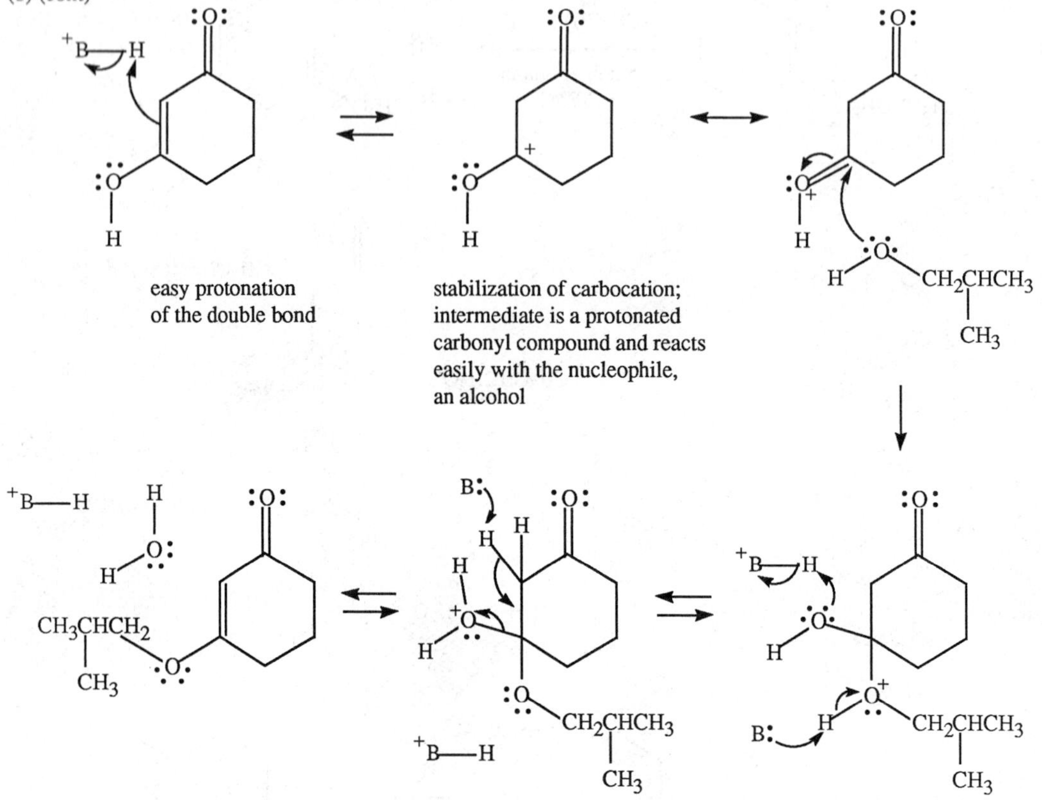

easy protonation
of the double bond

stabilization of carbocation;
intermediate is a protonated
carbonyl compound and reacts
easily with the nucleophile,
an alcohol

also easy dehydration
reaction to give α,β-
unsaturated carbonyl
compound

16.28

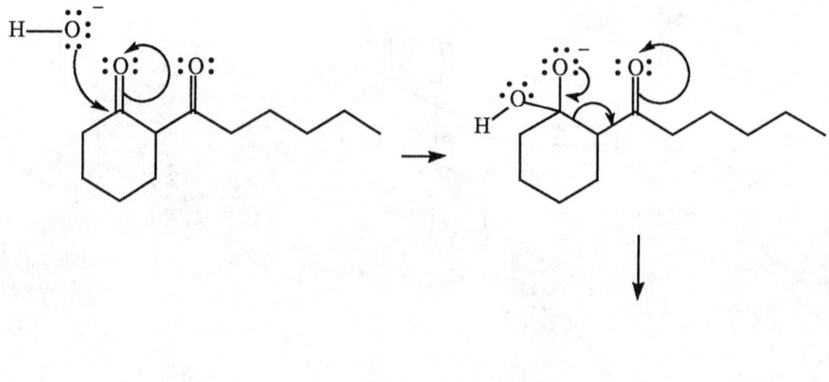

16.28 (cont)

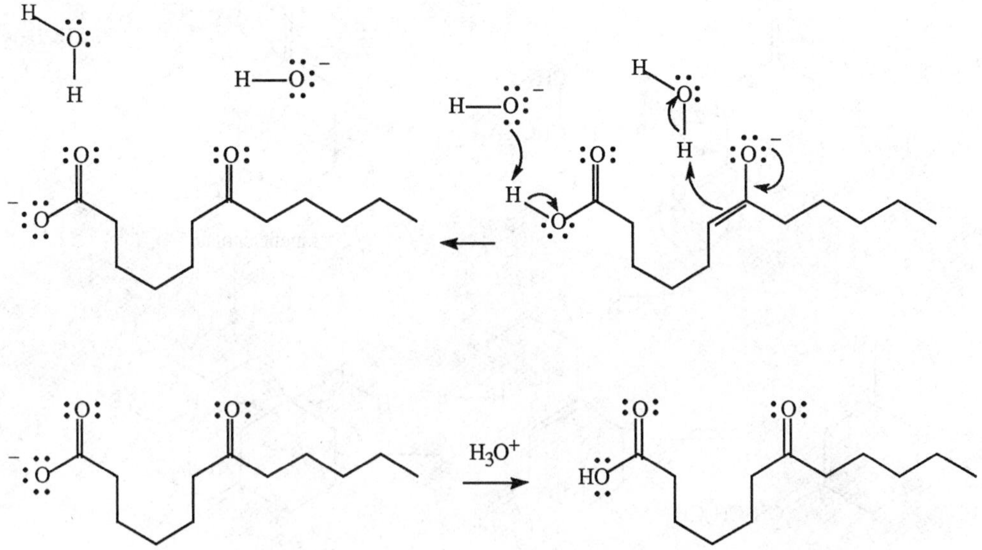

16.29

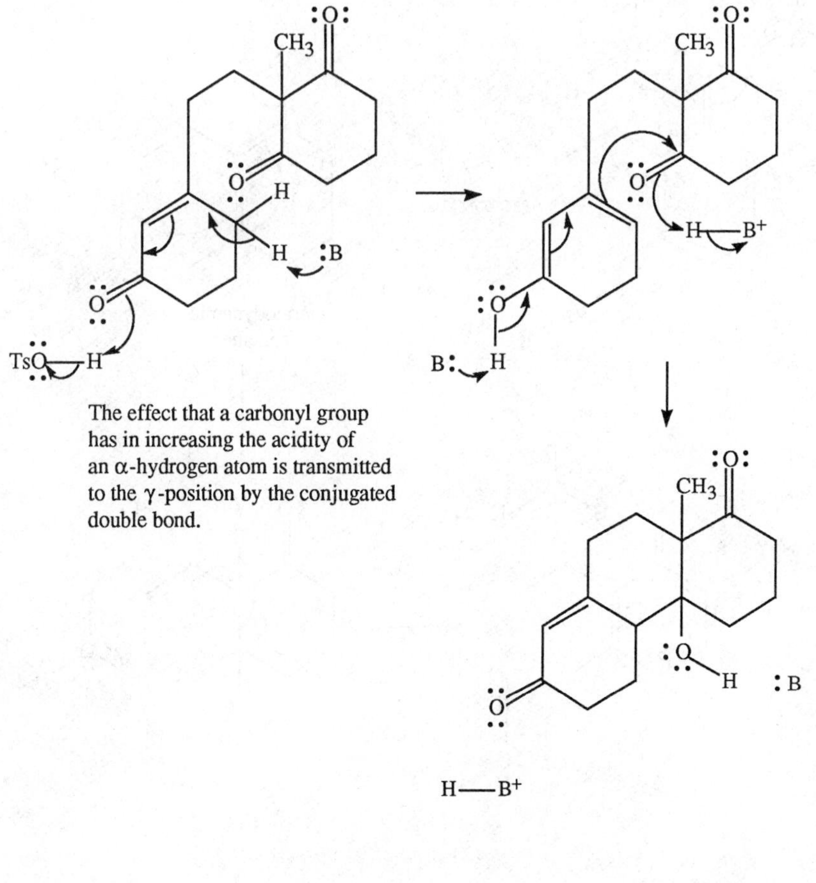

The effect that a carbonyl group
has in increasing the acidity of
an α-hydrogen atom is transmitted
to the γ-position by the conjugated
double bond.

16.30

1.

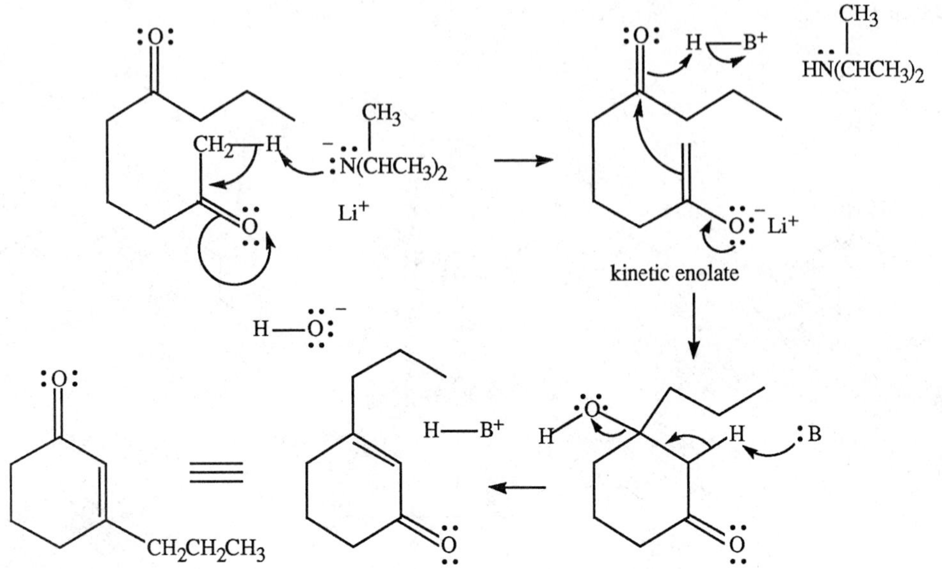

kinetic enolate

H—O:⁻

:O:

≡

CH₂CH₂CH₃

2.

thermodynamic
enolate

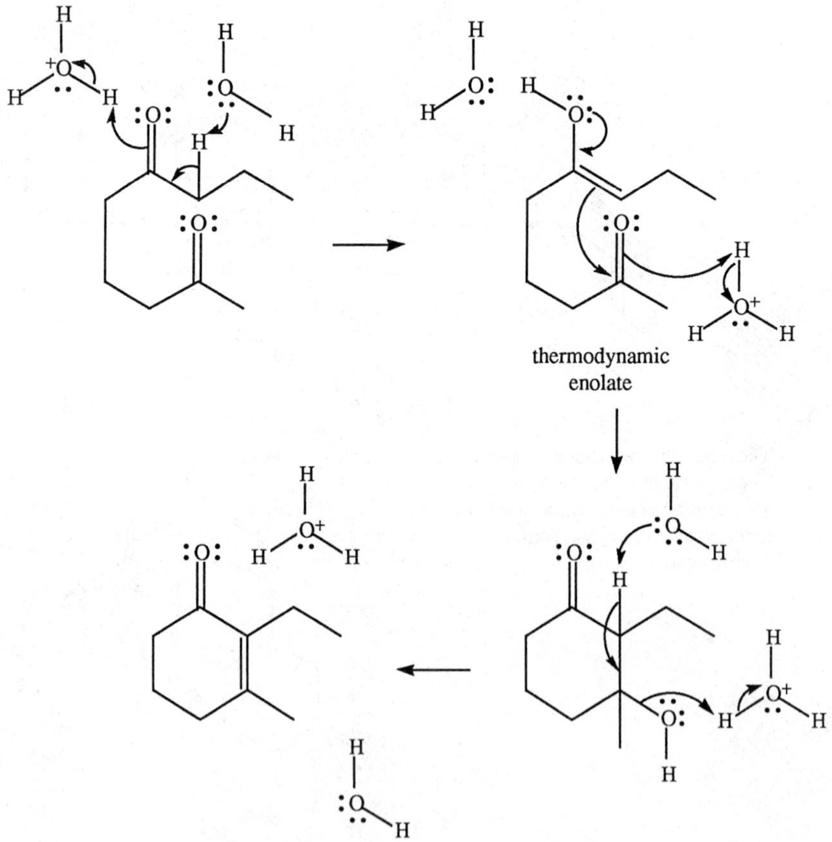

16.31   An examination of the major resonance contributors of the enolate anion from 1-phenyl-2-methyl-1-propanone reveals that there is double bond character to the carbon-carbon bond between carbons 1 and 2 in the structure in which the negative charge is localized on the oxygen atom. To the extent that there is such double bond character, free rotation around that bond is inhibited, and the two methyl groups are no longer chemical-shift equivalent (for example, one is trans and the other one, cis to the oxygen atom), and would be expected to have different chemical shifts.

The experimental observation that there are two chemical shifts for the methyl groups in the region δ 1.0–2.0, suggests that the resonance contributor with the negative charge on oxygen is the more important one. We would expect the positively charged lithium ion to be associated with the oxygen atom rather than with the carbon atom in the enolate anion.

16.32 $2\ CH_3CH_2OH + Na \longrightarrow 2\ CH_3CH_2O^- Na^+ + H_2$

enolate anion,
stable product of
ketolactone in base

16.32 (cont)

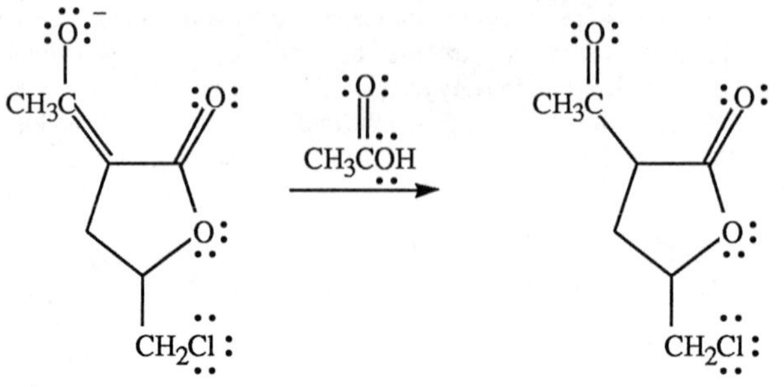

16.33

1.

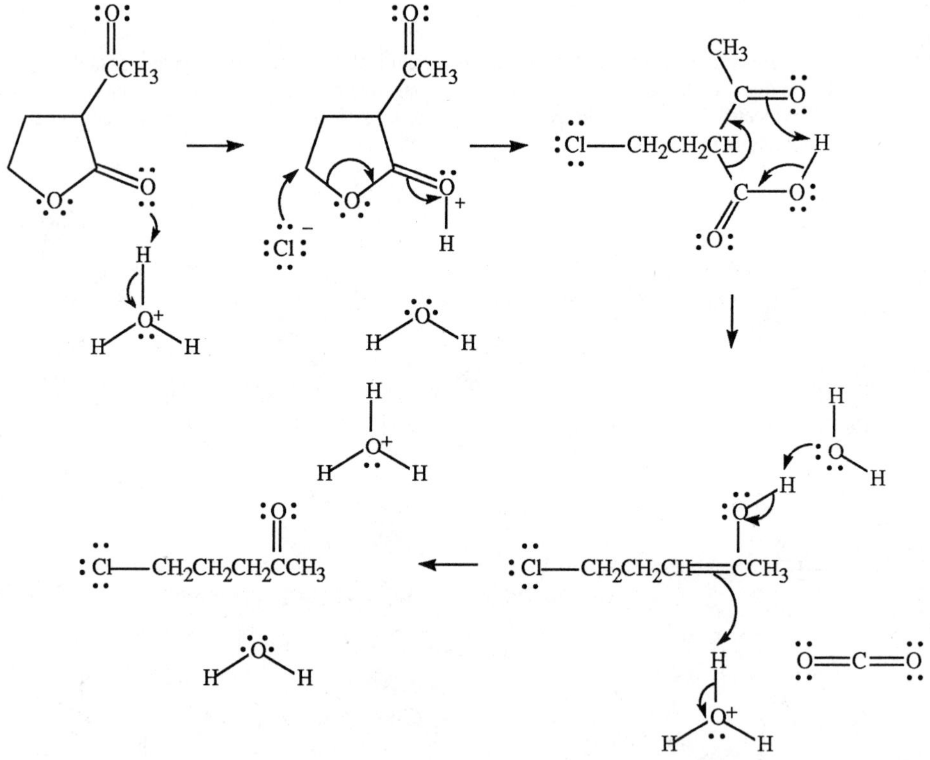

2.

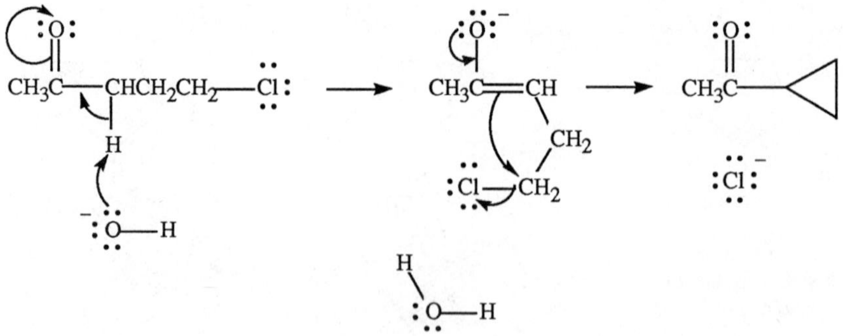

16.33 (cont)

2. (cont)

$$
\begin{array}{c}
\overset{\displaystyle \ddot{\text{O}}:^{-}}{\underset{\displaystyle}{\text{C}}} \\
\text{CH}_3\!\!-\!\!\overset{|}{\text{C}}\!\!=\!\!\text{C}\!\!-\!\!\text{H} \\
\text{H}\!\!-\!\!\text{C}\!\!-\!\!\text{C}\!\!-\!\!\text{H} \\
\overset{|}{:\ddot{\text{Cl}}:}\qquad \text{H}
\end{array}
$$

Note that this reaction is reminiscent of the formation of oxiranes by the intramolecular S_N2 reaction of alkoxide ions derived from halohydrins (Section 12.5C). The thermodynamic enolate of the ketone has the right regiochemistry and stereochemistry to give the three-membered ring.

16.34

$$
\underset{\text{CH}_3\text{CCH}_2\text{COCH}_2\text{CH}_3}{\overset{\text{O}\quad\text{O}}{\|\quad\|}}
\xrightarrow[\text{ethanol}]{\text{CH}_3\text{CH}_2\text{ONa}}
\underset{\underset{\text{Na}^+}{}}{\underset{\text{CH}_3\text{CCHCOCH}_2\text{CH}_3}{\overset{\text{O}\quad\text{O}}{\|\quad\|}}}
\xrightarrow{\text{CH}_3(\text{CH}_2)_4\text{Br}}
$$

A

$$
\underset{\underset{(\text{CH}_2)_4\text{CH}_3}{|}}{\underset{\text{CH}_3\text{CCHCOCH}_2\text{CH}_3}{\overset{\text{O}\quad\text{O}}{\|\quad\|}}}
\xrightarrow[\substack{\text{tetrahydro-}\\\text{furan}}]{\text{NaH}}
\underset{\underset{(\text{CH}_2)_4\text{CH}_3}{|}}{\underset{\text{CH}_3\text{CCCOCH}_2\text{CH}_3}{\overset{\overset{\text{Na}^+}{}}{\overset{\text{O}\quad\text{O}}{\|\quad\|}}}}
\xrightarrow[\text{hexane}]{\text{CH}_3\text{CH}_2\text{CH}_2\text{CH}_2\text{Li}}
$$

B

(the most acidic hydrogen
is the one between the
two carbonyl groups)

C

(the most acidic hydrogens are the
ones α to the carbonyl group)

$$
\underset{\underset{(\text{CH}_2)_4\text{CH}_3}{|}}{\text{Li}\!\!-\!\!\underset{\text{CH}_2\text{CCCOCH}_2\text{CH}_3}{\overset{\overset{\text{Na}^+}{}}{\overset{\text{O}\quad\text{O}}{\|\quad\|}}}}
\xrightarrow{\overset{\text{O}}{\triangle}\!\!-\!\!\text{CH}_3}
\underset{\underset{\text{O}^-\text{Li}^+\quad(\text{CH}_2)_4\text{CH}_3}{|\qquad\qquad|}}{\text{CH}_3\text{CHCH}_2\text{CH}_2\text{CCCOCH}_2\text{CH}_3}^{\overset{\text{Na}^+}{\overset{\text{O}\ \text{O}}{\|\ \|}}}
\xrightarrow[\substack{\text{H}_2\text{O}\\\Delta}]{\text{NaOH}}
$$

D

(the most nucleophilic
carbon is the one α to
only one carbonyl group)

E

16.34 (cont)

$$\text{CH}_3\text{CHCH}_2\text{CH}_2\overset{\text{O}}{\overset{\|}{\text{C}}}\overset{\text{O}}{\overset{\|}{\text{C}}}\text{CO}^-\text{Na}^+ \xrightarrow{\text{H}_2\text{SO}_4} \text{CH}_3\text{CHCH}_2\text{CH}_2\overset{\text{O}}{\overset{\|}{\text{C}}}\text{CH}_2(\text{CH}_2)_4\text{CH}_3 \xrightarrow{\text{CrO}_3}$$

(with Na$^+$ above, O$^-$Li$^+$ and (CH$_2$)$_4$CH$_3$ substituents on F; OH substituent on G)

F

G

+ CO$_2$

$$\text{CH}_3\overset{\text{O}}{\overset{\|}{\text{C}}}\text{CH}_2\text{CH}_2\overset{\text{O}}{\underset{\|}{\text{C}}}\text{CH}_2(\text{CH}_2)_4\text{CH}_3 \xrightarrow[\substack{\text{H}_2\text{O}\\ \text{ethanol}}]{\text{NaOH}}$$

H

dihydrojasmone

$$\text{CH}_3\overset{\text{O}}{\overset{\|}{\text{C}}}\text{CH}_2\text{CH}_2\overset{\text{O}}{\underset{\|}{\text{C}}}\text{CH}_2(\text{CH}_2)_4\text{CH}_3 \xrightarrow[\substack{\text{H}_2\text{O}\\ \text{ethanol}}]{\text{NaOH}} \text{CH}_3\overset{\text{O}}{\overset{\|}{\text{C}}}\text{CH}_2\text{CH}_2\overset{\text{O}^-\text{Na}^+}{\overset{|}{\text{C}}}=\text{CH}(\text{CH}_2)_4\text{CH}_3$$

more stable enolate anion,
containing the more highly
substituted double bond

$$\text{CH}_2=\overset{\text{O}^-\text{Na}^+}{\overset{|}{\text{C}}}\text{CH}_2\text{CH}_2\overset{\text{O}}{\overset{\|}{\text{C}}}\text{CH}_2(\text{CH}_2)_4\text{CH}_3$$

less stable enolate anion,
with the less highly
substituted double bond

dihydrojasmone

this product does not form

16.35 Compound A, $C_{12}H_{20}O_2$, has three units of unsaturation. The infrared spectrum of Compound A shows the presence of an ester functional group (C=O stretching frequency at 1735 cm^{-1} and C—O stretching frequency at 1225 cm^{-1}), and a double bond (=C—H stretching frequency at 2978 cm^{-1}, and =C—H bending frequency at 975 cm^{-1}). The alkene functional group is not likely to be a terminal double bond because the C=C stretching frequency is not seen in the infrared spectrum. The presence of a double bond is confirmed by the proton magnetic resonance spectrum (δ 5.3, 2H). The proton magnetic resonance spectrum also shows the presence of a methyl group bonded to a carbon atom bearing a single hydrogen atom (δ 1.3, 3H, doublet).

$$\text{Compound A} \xrightarrow[\text{Pt}]{\text{H}_2} \text{Compound B, } C_{12}H_{22}O_2$$

Compound B, $C_{12}H_{22}O_2$, still contains two of the three units of unsaturation, one of which is the ester functional group. The other unit of unsaturation must be a ring.

$$\text{Compound B} \xrightarrow[\substack{\text{H}_2\text{O} \\ \Delta}]{\text{NaOH}} \xrightarrow{\text{H}_3\text{O}^+} \text{Compound C, } C_{12}H_{24}O_3$$

Compound C, $C_{12}H_{24}O_3$, has only one unit of unsaturation, which is most likely the carbonyl group of a carboxylic acid since treatment with aqueous base followed by acid will hydrolyze an ester to an alcohol and a carboxylic acid. This reaction suggests that the ester functional group was a lactone.

$$\text{Compound C} \xrightarrow[\substack{\text{acetic} \\ \text{acid}}]{\text{CrO}_3} \text{Compound D, } C_{12}H_{24}O_3$$

$$\text{Compound D} \xrightarrow{\text{I}_2,\ \text{NaOH}} \text{CHI}_3 \ + \ \underset{\text{O}}{\overset{\text{O}}{\text{HOC(CH}_2)_9\text{COH}}}$$

therefore Compound D is $\text{CH}_3\overset{\text{O}}{\underset{}{\text{C}}}\text{(CH}_2)_9\overset{\text{O}}{\underset{}{\text{C}}}\text{OH}$

and Compound C is $\text{CH}_3\underset{\text{OH}}{\text{CH}}\text{(CH}_2)_9\overset{\text{O}}{\underset{}{\text{C}}}\text{OH}$

and Compound B is

16.35 (cont)

The products from the ozonolysis of Compound A allow us to locate the double bond, but does not allow us to determine the stereochemistry of the double bond. If we had more extensive tables correlating infrared absorption frequencies with structure, we would find out that the =C—H bending frequency at 975 cm^{-1} tells us that the double bond is trans.

$$\text{Compound A} \xrightarrow[\substack{\text{ethanol}\\\text{ethyl}\\\text{acetate}}]{\text{O}_3} \xrightarrow[\substack{\text{formic}\\\text{acid}}]{\text{H}_2\text{O}_2} \xrightarrow[\text{ethanol}]{\text{KOH}} \xrightarrow{\text{H}_3\text{O}^+} \underset{\substack{|\\\text{OH}}}{\text{CH}_3\text{CHCH}_2\overset{\text{O}}{\overset{||}{\text{C}}}\text{OH}} + \text{HO}\overset{\text{O}}{\overset{||}{\text{C}}}(\text{CH}_2)_6\overset{\text{O}}{\overset{||}{\text{C}}}\text{OH}$$

Compound A must be

16.36 $\underset{}{\text{CH}_3\overset{\text{O}}{\overset{||}{\text{C}}}\text{CH}_2\text{CH}_2\text{CH}_2\overset{\text{O}}{\overset{||}{\text{C}}}\text{OH}} \xrightarrow[\text{H}_2\text{SO}_4]{\text{D}_2\text{O}} \text{CD}_3\overset{\text{O}}{\overset{||}{\text{C}}}\text{CD}_2\text{CH}_2\text{CH}_2\overset{\text{O}}{\overset{||}{\text{C}}}\text{OH} \xrightarrow[\text{H}_2\text{O}]{\text{NaBH}_4} \xrightarrow{\text{H}_3\text{O}^+}$

$$\left[\underset{\substack{|\\\text{OH}}}{\text{CD}_3\text{CHCD}_2\text{CH}_2\text{CH}_2\overset{\text{O}}{\overset{||}{\text{C}}}\text{OH}} \right] \longrightarrow$$

16.37

16.37 (cont)

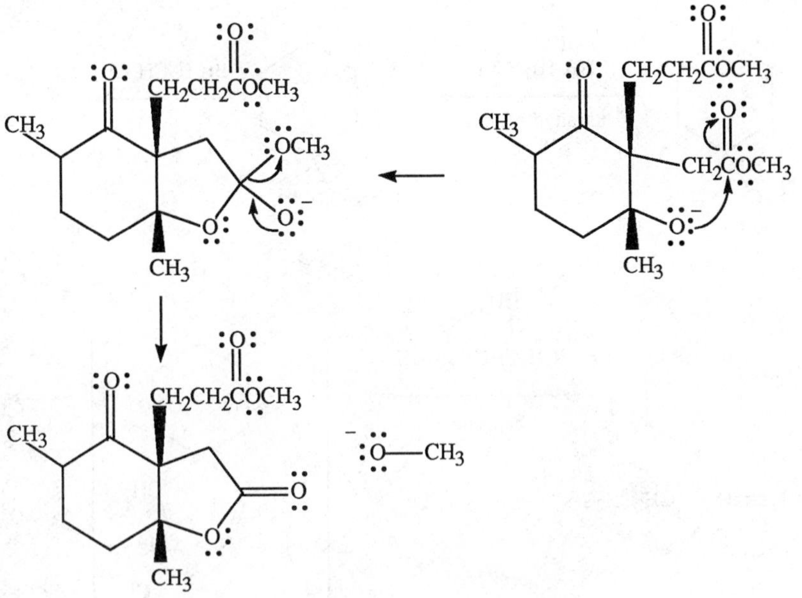

16.38

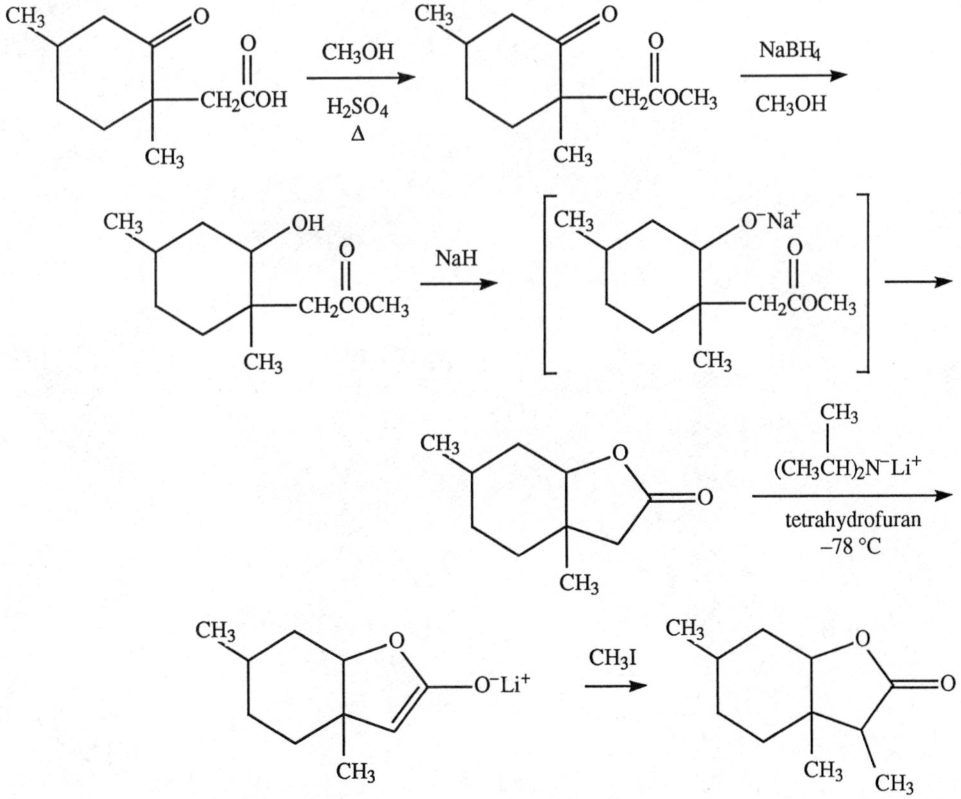

16.39

17

Polyenes

Concept Map 17.1 Different relationships between multiple bonds.

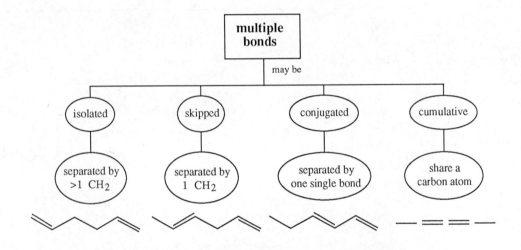

17.1 (a) 1,4-hexadiyne   (b) 2-hexen-4-yne

(c) 1,4-cyclohexadiene   (d) 1,5-cyclooctadiene

(e) (*E*)-1-phenyl-1,4-pentadiene   (f) 1-chloro-2,4-heptadiyne

17.2 (a) $CH_2{=}CHCH_2CH_2\overset{\overset{\displaystyle O}{\|}}{C}CH_3$ (b) $HC{\equiv}CCH_2CH_2C{\equiv}CH$

(c) $CH_2{=}CHCH_2CH{=}CH_2$ (d) $CH_3C{\equiv}CCH_2C{\equiv}CCH_3$

17.2 (cont)

(e)

$$CH_3 \quad H$$
$$\underset{H}{\overset{CH_3}{\diagdown}}C=C\underset{CH_2CCH_2CH=CH_2}{\overset{CH_3}{\diagup}}$$
$$\underset{CH_3}{\overset{|}{CH_3}}$$

(f) $HC\equiv CCH=CHCH=CH_2$

- -

Concept Map 17.2   A conjugated diene.

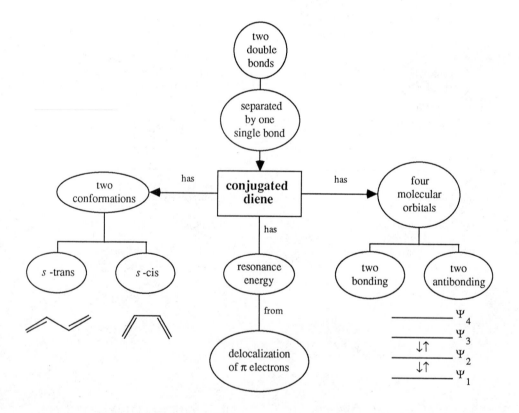

Concept Map 17.3   Addition to dienes.

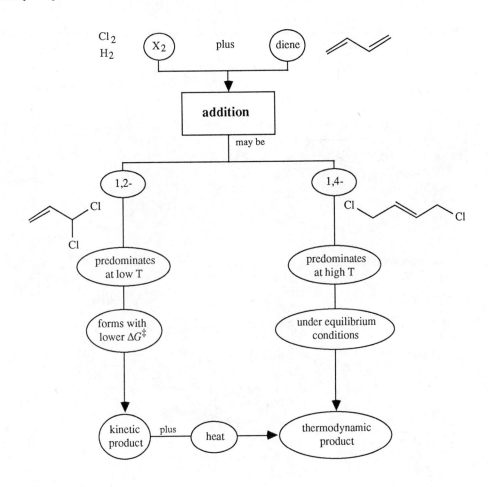

- -

17.3

(a)
$$\text{Br}_2 \text{ (1 molar equiv)}$$
carbon tetrachloride
0 °C

(b) CH_2═$CHCH$═CH_2 $\xrightarrow{\text{HCl (1 molar equiv)}}$ CH_3CHCH═CH_2 + CH_3CH═$CHCH_2Cl$

                                                                          $|$
                                                                          Cl cis and trans

17.3 (cont)

(c)

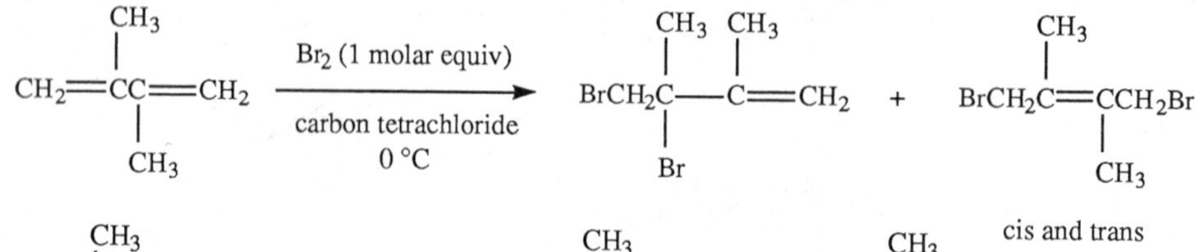

(d)

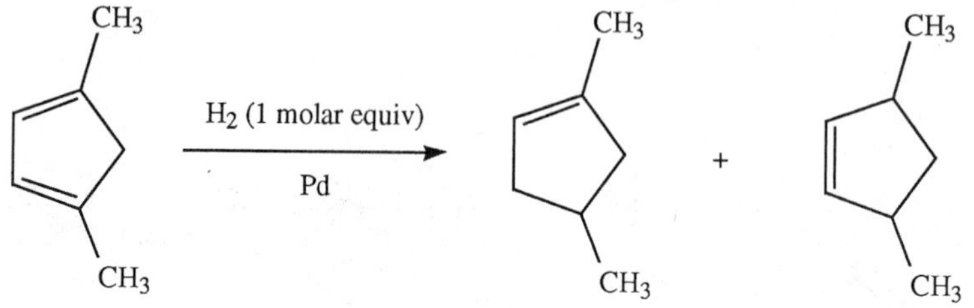

17.4   Formation of 2,3-dimethyl-1,3-butadiene from pinacol

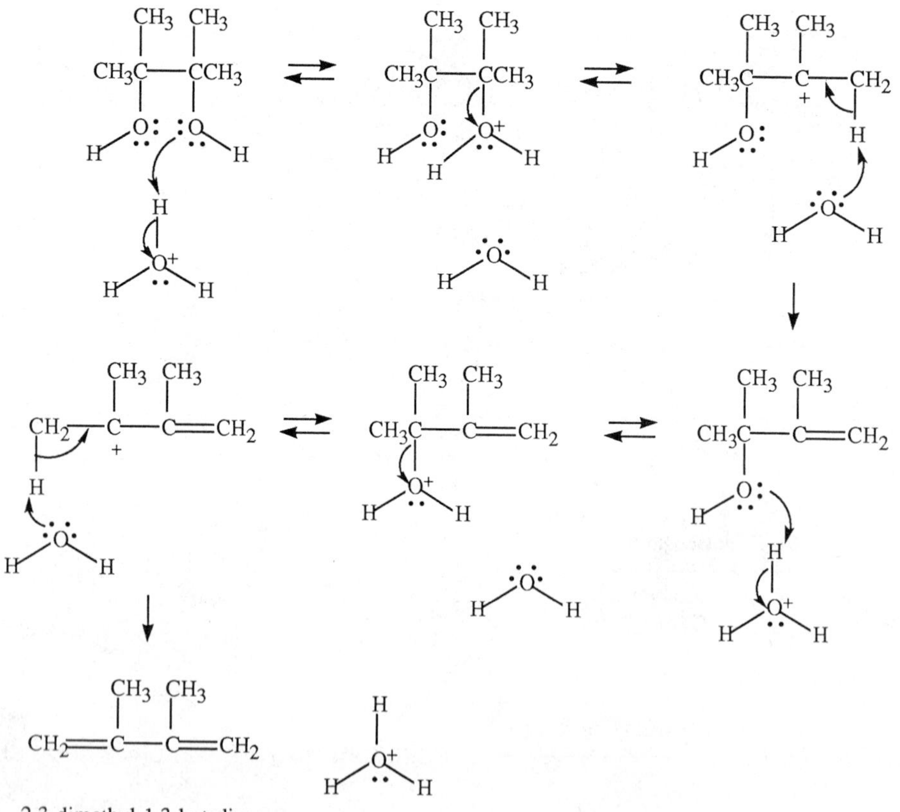

2,3-dimethyl-1,3-butadiene

17.4 (cont)

Formation of pinacolone

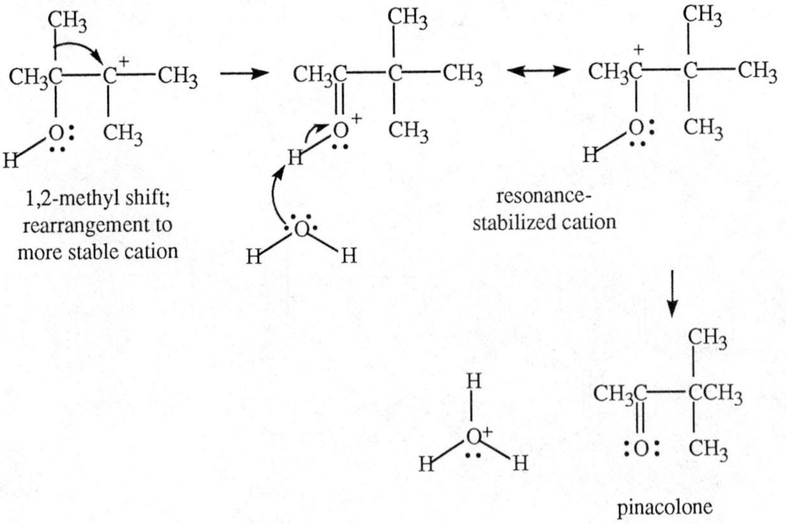

1,2-methyl shift;
rearrangement to
more stable cation

resonance-
stabilized cation

pinacolone

17.5

(a)

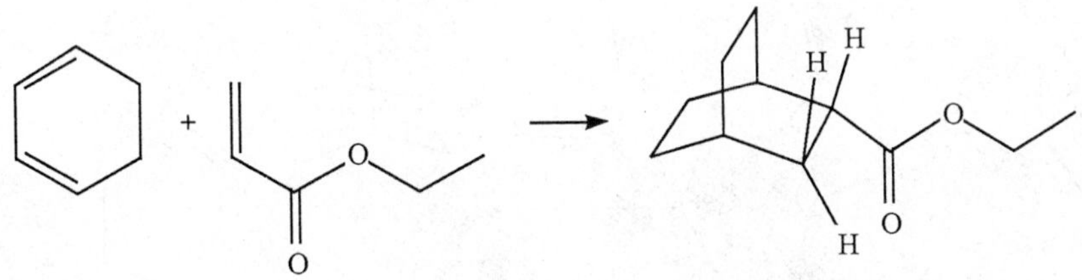

and enantiomer
A

(b)

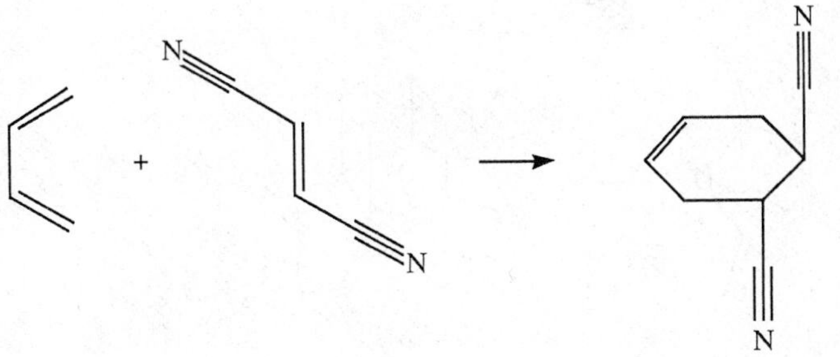

and enantiomer
B

17.5 (cont)

(c)

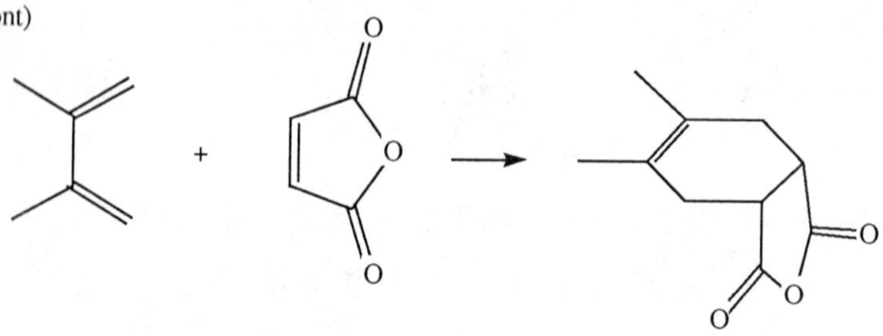

C

(d)

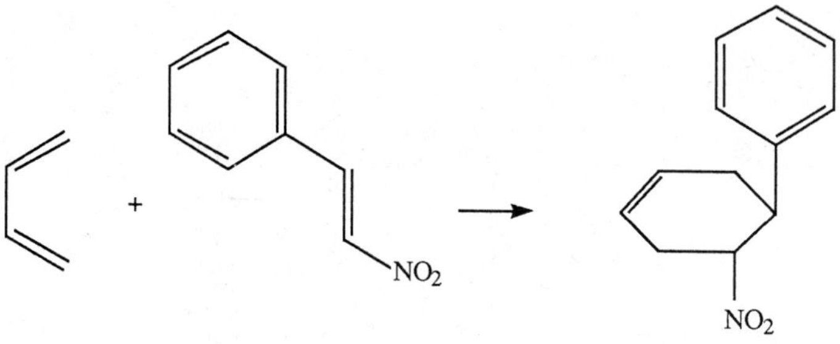

and enantiomer
D

(e)

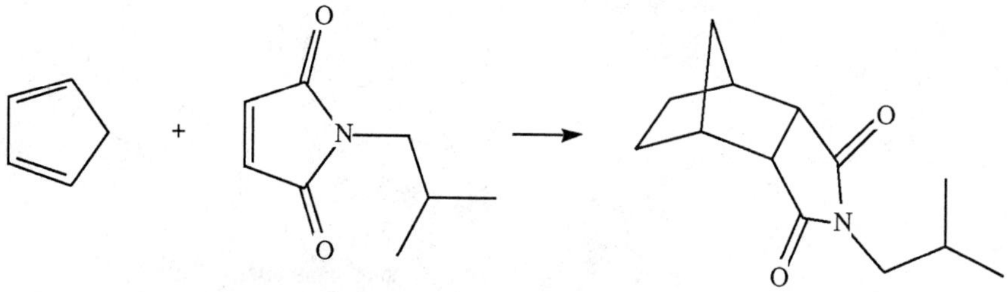

E

(f)

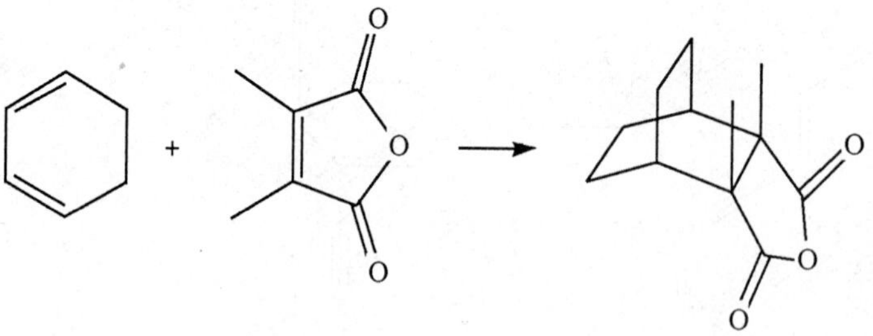

F

17.5 (cont)

(g)

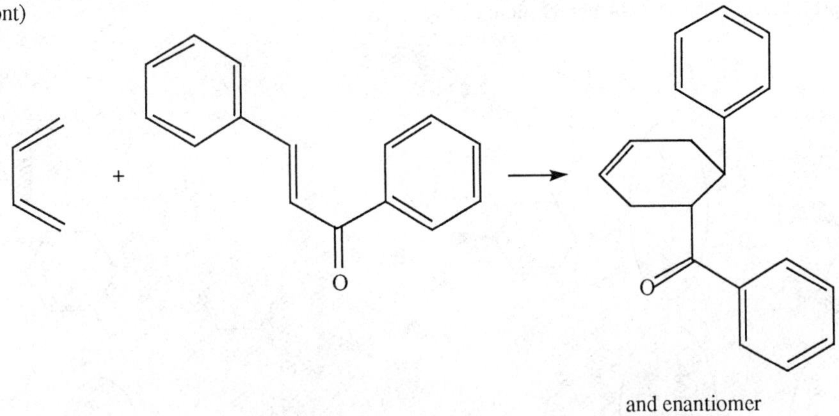

and enantiomer
G

(h)

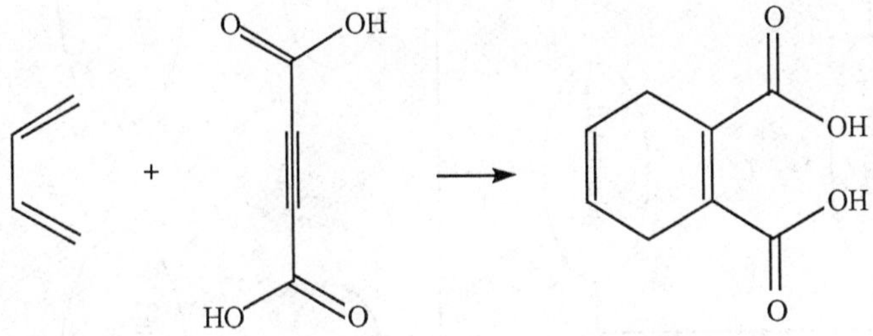

H

17.6

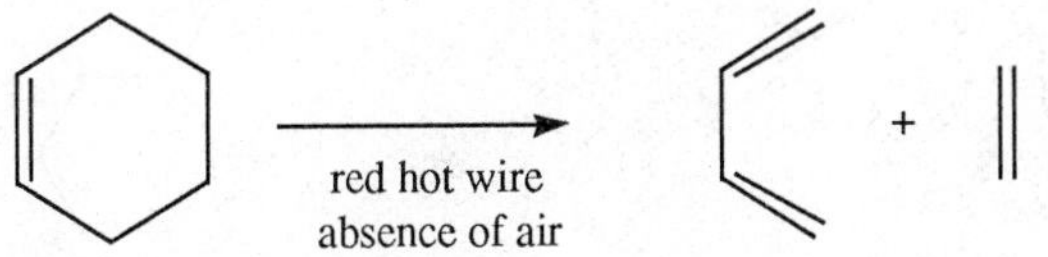

red hot wire
absence of air

Concept Map 17.4   The Diels-Alder reaction.

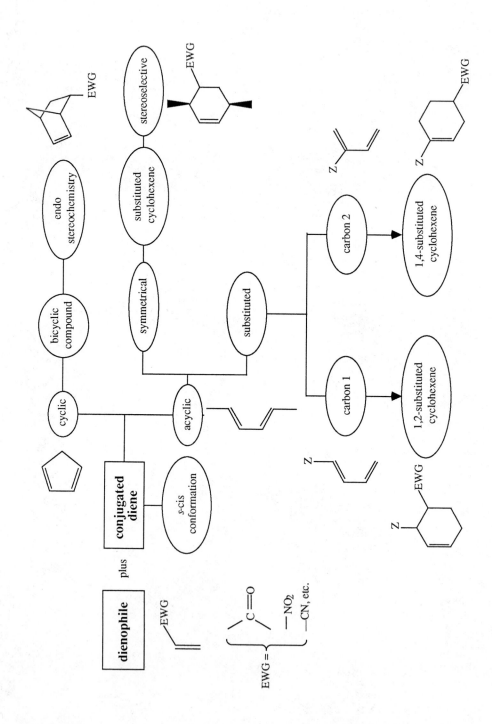

17.7

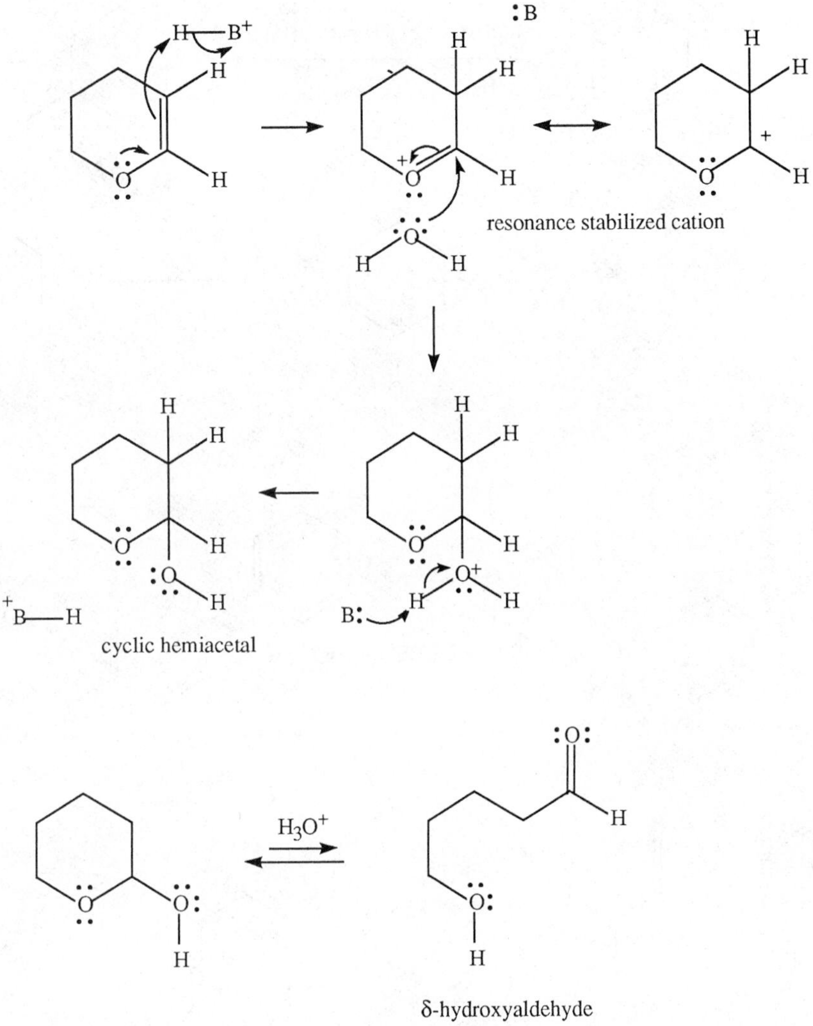
resonance stabilized cation
cyclic hemiacetal
H_3O^+
δ-hydroxyaldehyde

17.8

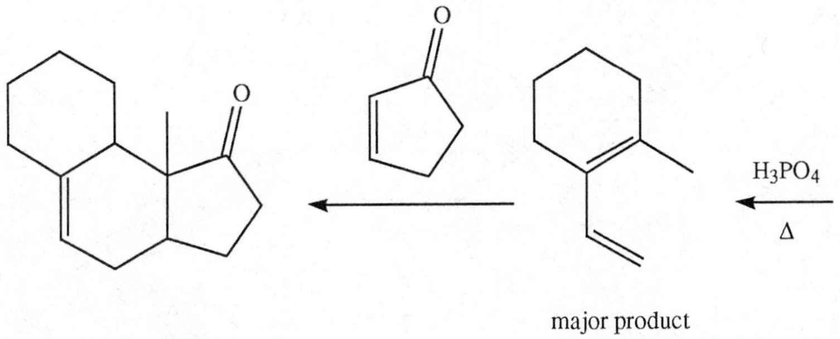
H_3PO_4
Δ
major product

17.8 (cont)

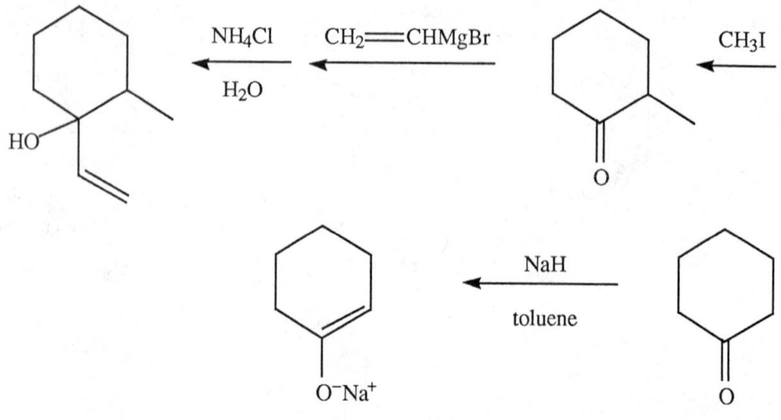

17.9

(a)

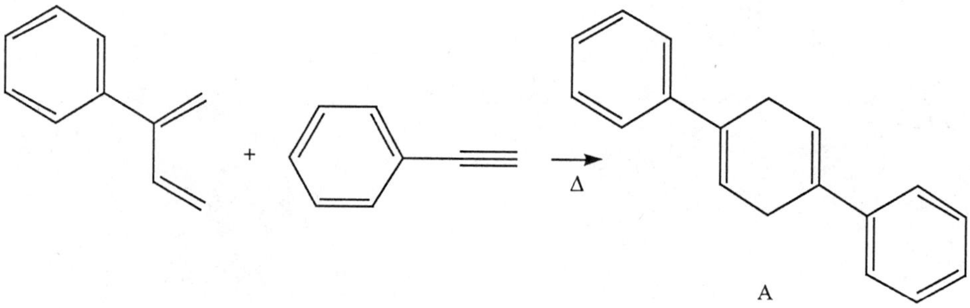

A

(b)

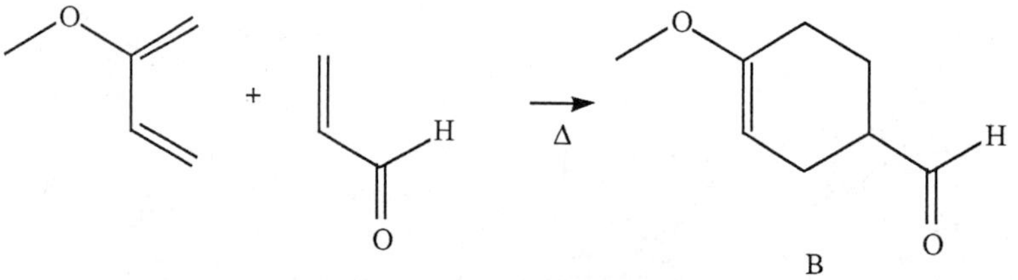

B

(c)

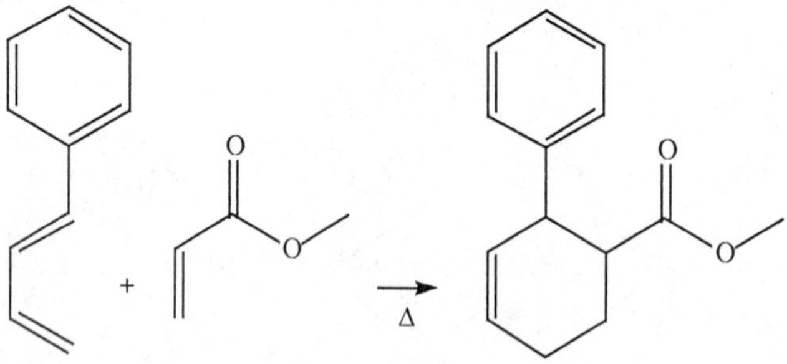

C

17.9 (cont)

(d)

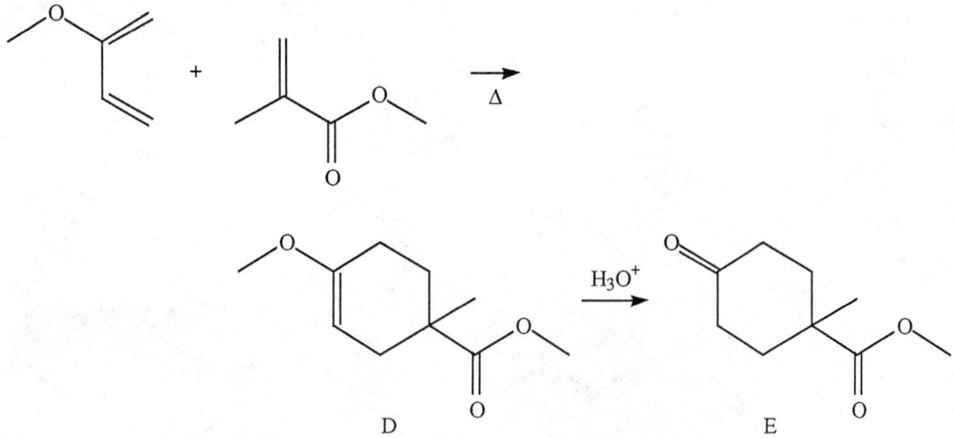

D $\xrightarrow{H_3O^+}$ E

(e)

F

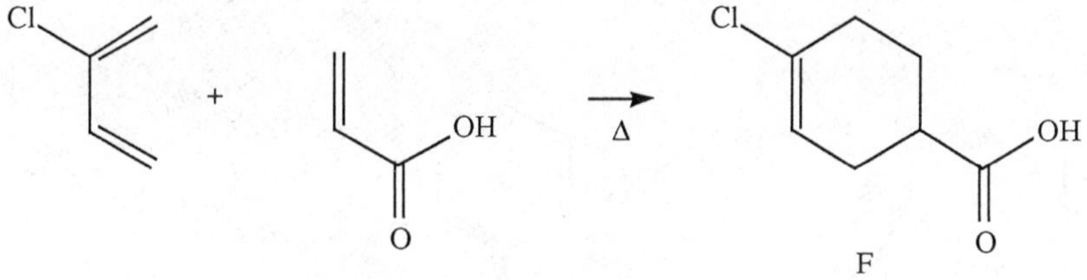

17.10

(a)

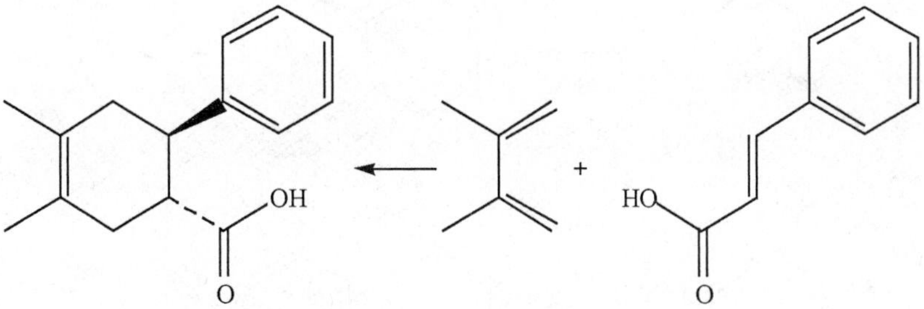

(b)

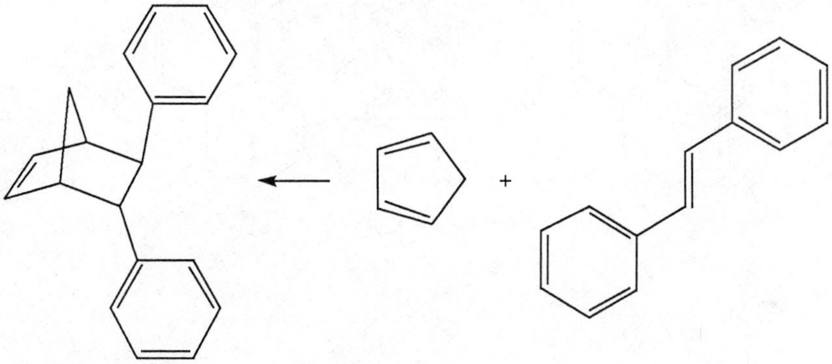

17.10 (cont)

(c)

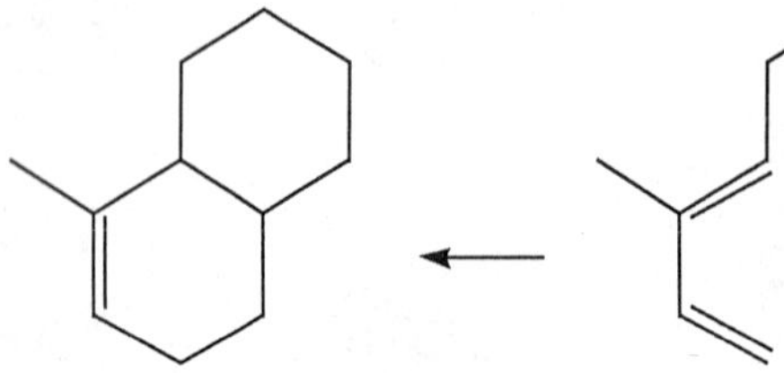

(d)

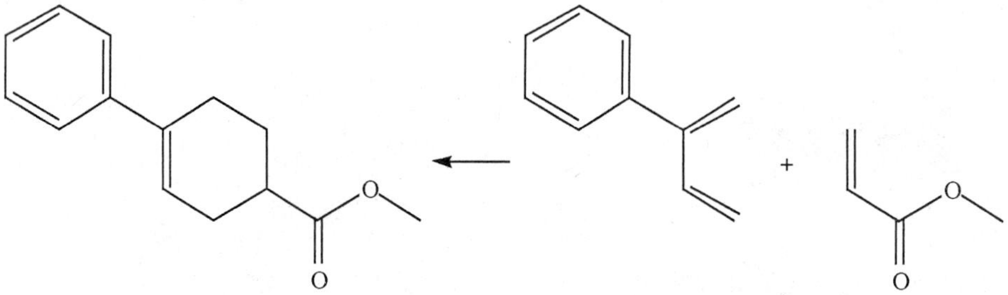

17.11

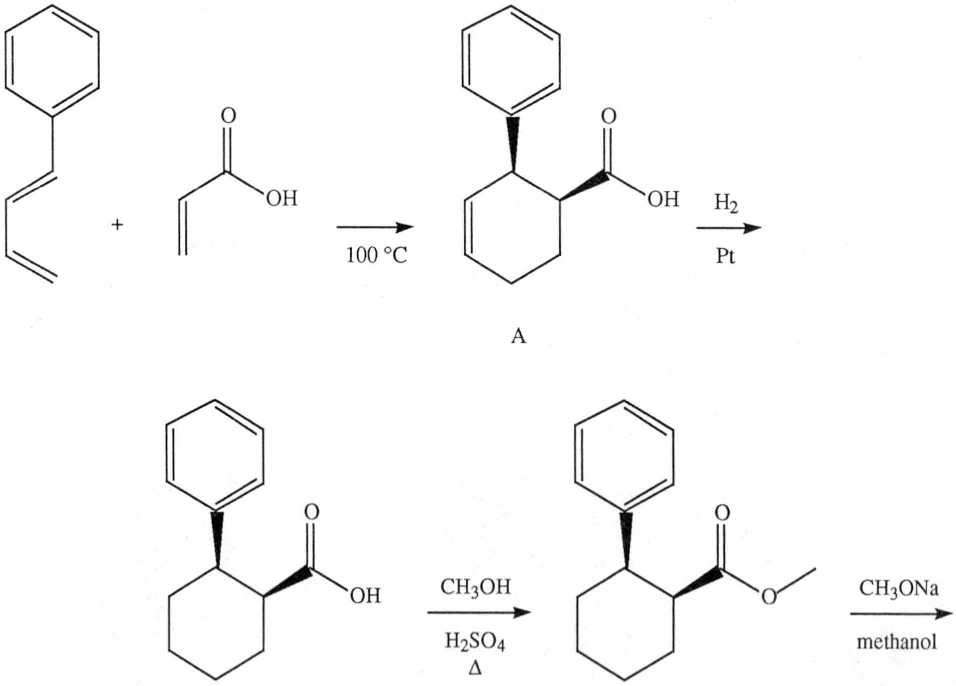

A

B   C

17.11 (cont)

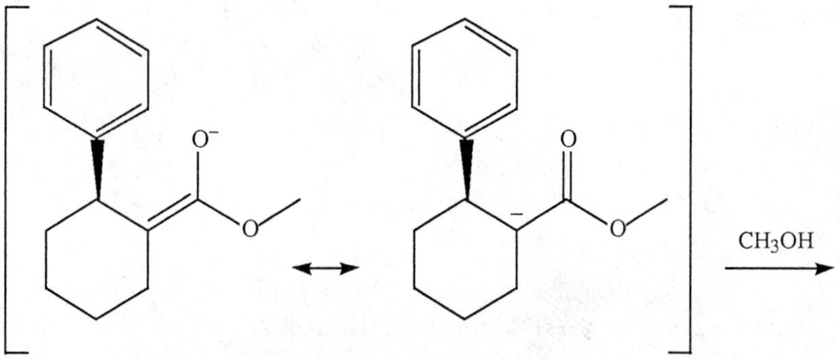

enolate anion of ester (Section 16.1B); loss of
stereochemistry at the α-carbon atom of the ester

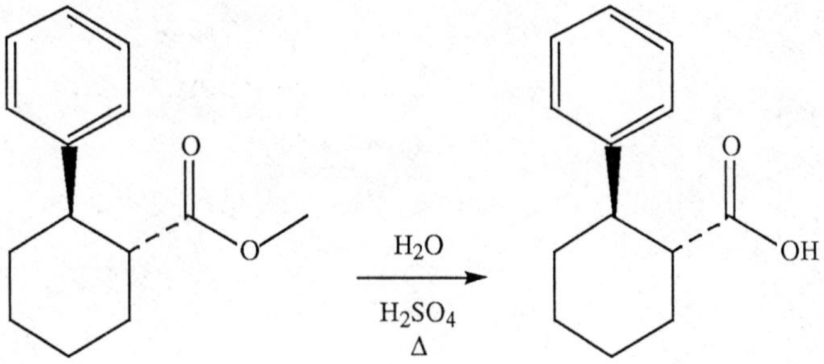

D   E

Generation of the carbanion from the methyl ester followed by reprotonation is an equilibrium reaction and
gives the thermodynamically more stable trans stereoisomer. Compound D is different from Compound C, (and
E from A) therefore the initial Diels-Alder product, Compound A, must have been the cis stereoisomer.

17.12

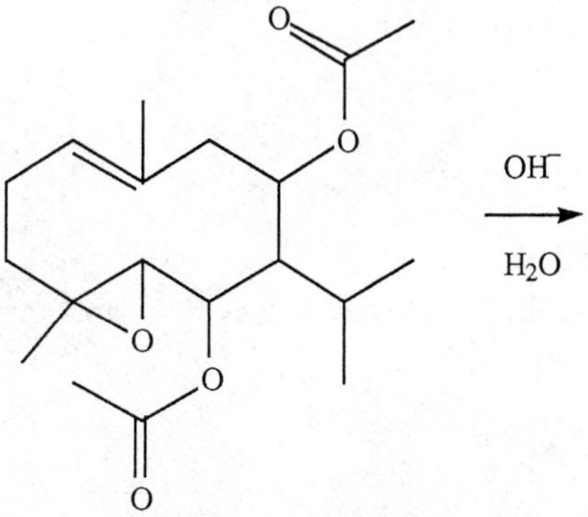

The infrared spectrum of shiromodiol acetate
has the bands typical of an ester (1735 cm^{-1},
C=O of ester; 1240 cm^{-1}, C—O of ester)

17.12 (cont)

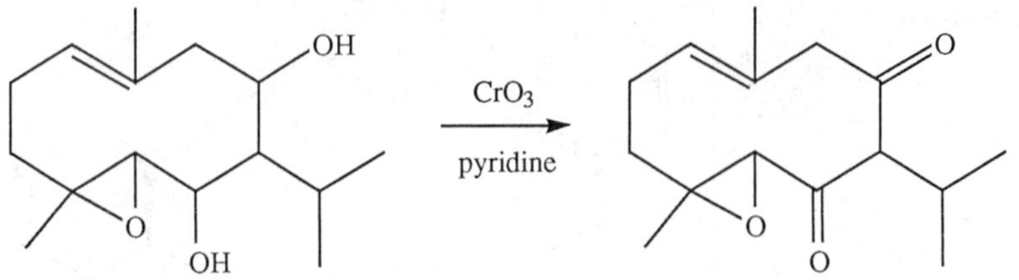

Hydrolysis of the ester functions gives
alcohols (3400 cm^{-1}, O—H of alcohol).
The carbonyl group of the ester is lost.

A

Oxidation converts the alcohol functions
into ketone groups (1720 and 1695 cm^{-1}
for C=O absorption bands).

B

17.13

(a)

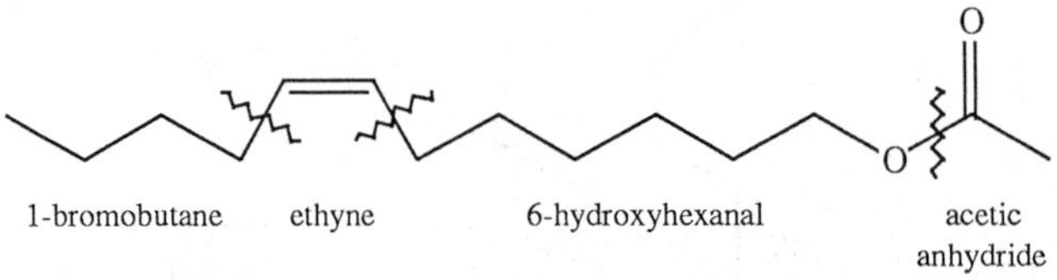

1-bromobutane   ethyne   6-hydroxyhexanal   acetic
anhydride

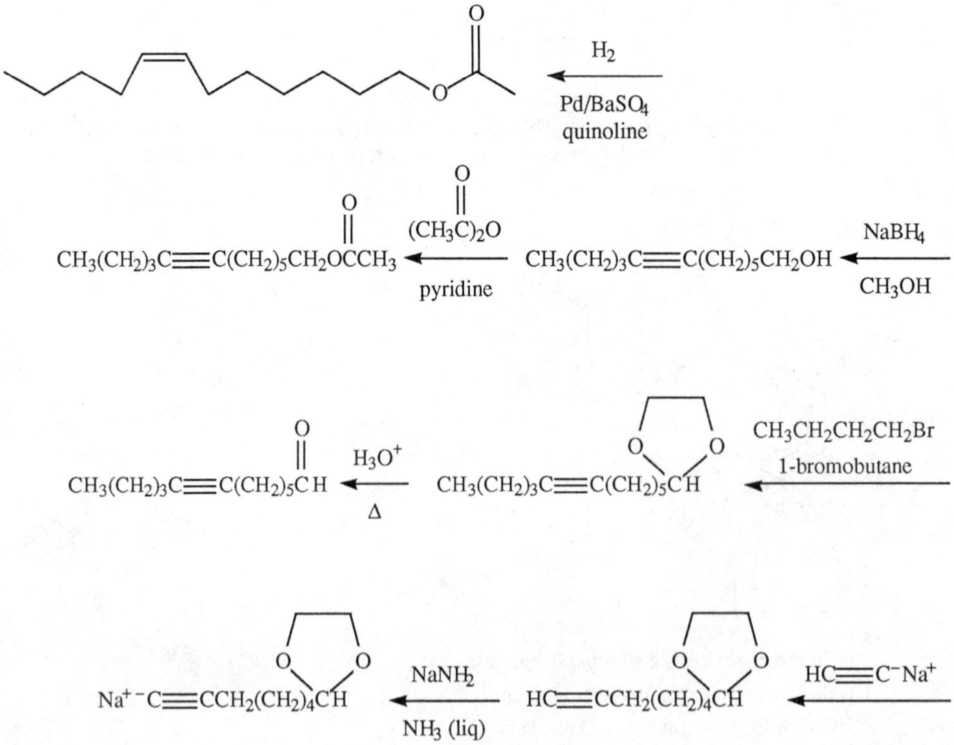

$$\xleftarrow[\substack{\text{Pd/BaSO}_4 \\ \text{quinoline}}]{\text{H}_2}$$

$$CH_3(CH_2)_3C\equiv C(CH_2)_5CH_2OCCH_3 \xleftarrow[\text{pyridine}]{(CH_3C)_2O} CH_3(CH_2)_3C\equiv C(CH_2)_5CH_2OH \xleftarrow[\text{CH}_3\text{OH}]{\text{NaBH}_4}$$

$$CH_3(CH_2)_3C\equiv C(CH_2)_5CH \xleftarrow[\Delta]{H_3O^+} CH_3(CH_2)_3C\equiv C(CH_2)_5CH \xleftarrow[\text{1-bromobutane}]{CH_3CH_2CH_2CH_2Br}$$

$$Na^+{}^-C\equiv CCH_2(CH_2)_4CH \xleftarrow[\text{NH}_3 \text{ (liq)}]{\text{NaNH}_2} HC\equiv CCH_2(CH_2)_4CH \xleftarrow{HC\equiv C^-Na^+}$$

17.13 (a) (cont)

$$\text{BrCH}_2(\text{CH}_2)_4\overset{O}{\underset{}{\text{C}}}\text{H} \quad \xleftarrow[\text{pyridine}]{\text{PBr}_3} \quad \text{HOCH}_2(\text{CH}_2)_4\overset{O}{\underset{}{\text{C}}}\text{H} \quad \xleftarrow[\substack{\text{TsOH}\\\text{benzene}\\\Delta}]{\text{HOCH}_2\text{CH}_2\text{OH}} \quad \text{HOCH}_2(\text{CH}_2)_4\overset{O}{\overset{\|}{\text{C}}}\text{H}$$

6-hydroxyhexanal

$$\text{HC}\equiv\text{C}^-\text{Na}^+ \quad \xleftarrow[\text{NH}_3\ (\text{liq})]{\text{NaNH}_2} \quad \text{HC}\equiv\text{CH}$$
ethyne

Note that the cis double bond, which is the less stable isomer, is introduced as late as possible in the synthesis to prevent isomerization by acidic reagents.

(b)   To get the trans isomer, we would use sodium in liquid ammonia, instead of hydrogen and poisoned catalyst, in the reduction of the triple bond. The reduction in this case would be carried out at the alcohol stage to avoid reaction of the ester with ammonia. Because the trans double bond is the more stable of the two, it is not subject to isomerization as easily as the cis double bond is.

17.14

menthofuran
both are head to tail

bisabolol
both are head to tail

17.14 (cont)

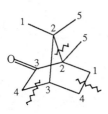

camphor   linaloöl

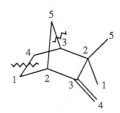

camphene   α-pinene

17.15

(a)

geraniol   nerol

(*E*)-double bond between   (*Z*)-double bond between
carbon 2 and carbon 3   carbon 2 and carbon 3

17.15 (cont)

(a) (cont)

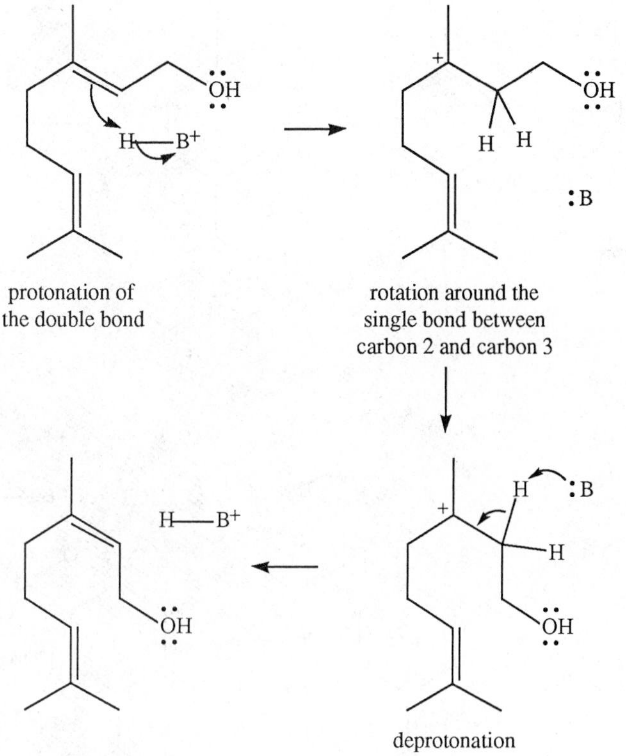
OH
H—B+
H H
:B
protonation of
the double bond
rotation around the
single bond between
carbon 2 and carbon 3
H :B
H
OH
H—B+
OH
deprotonation

(b)

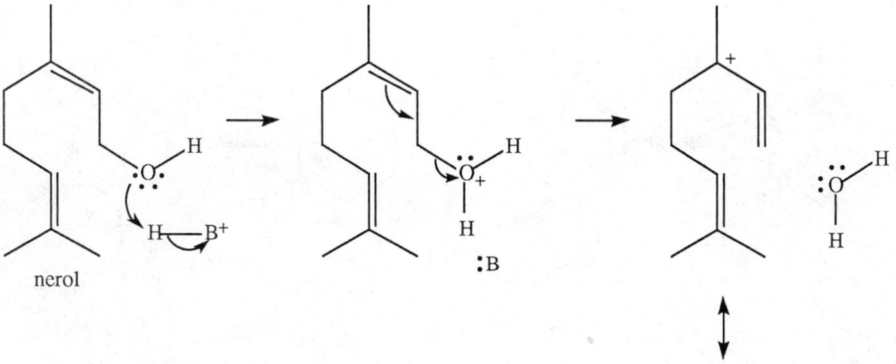
O
H
H—B+
nerol
O+
H
H
:B
+
O
H
H

17.15 (cont)

(b) (cont)

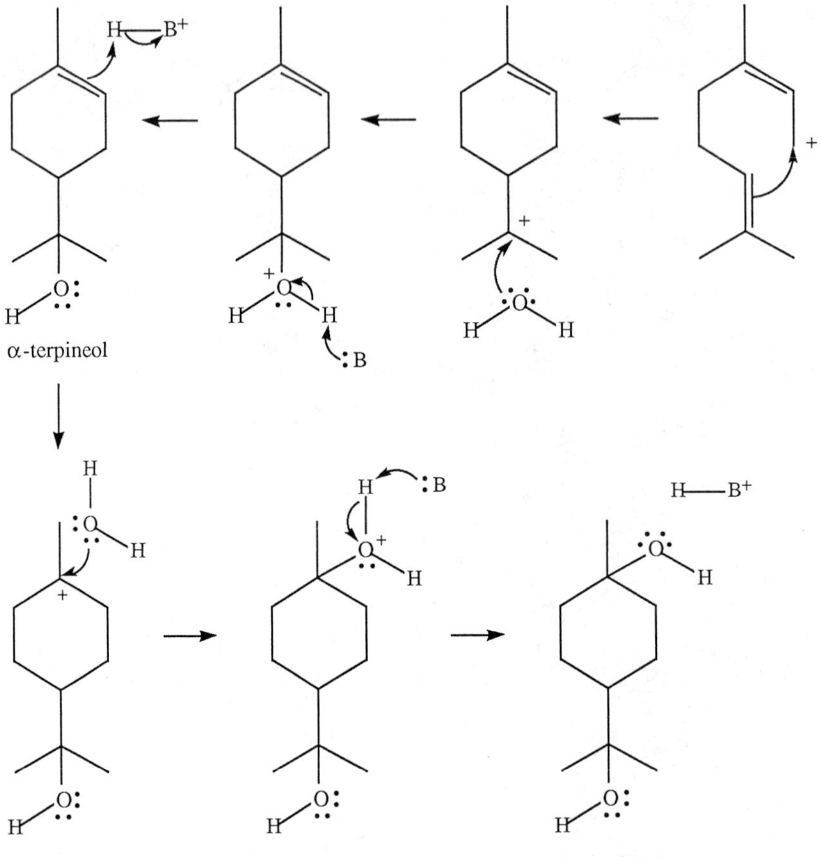

α-terpineol

terpin

(c)

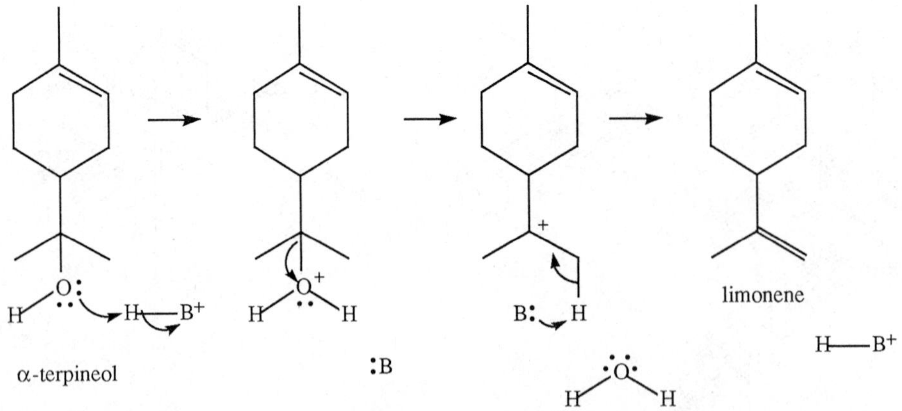

α-terpineol

limonene

17.15 (cont)

(c) (cont)

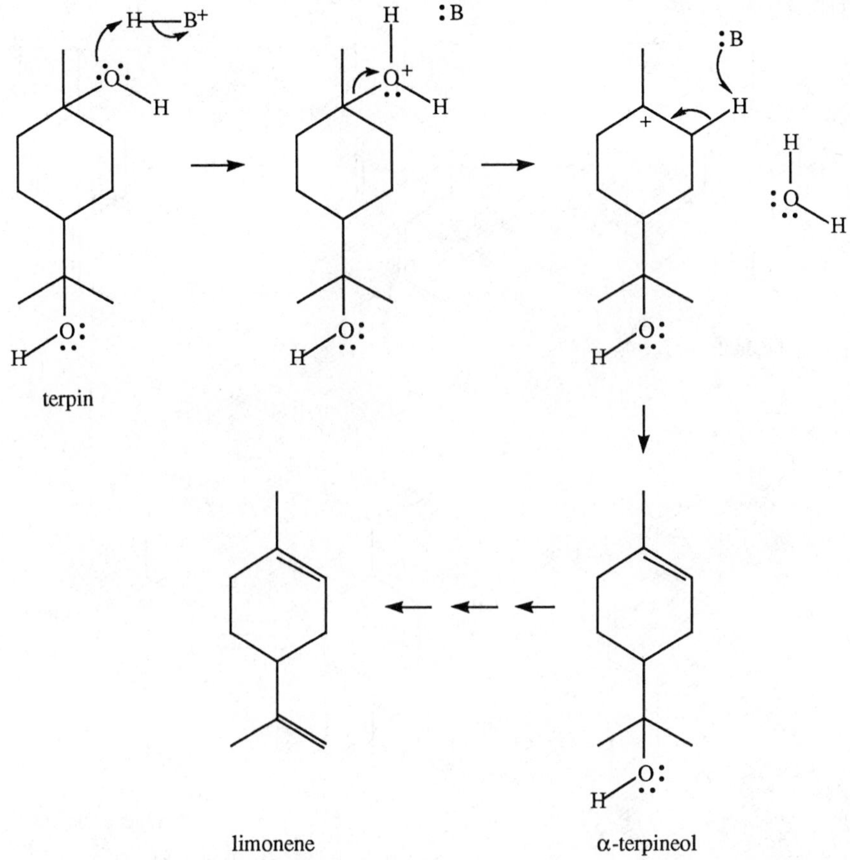

terpin

limonene   α-terpineol

(d)

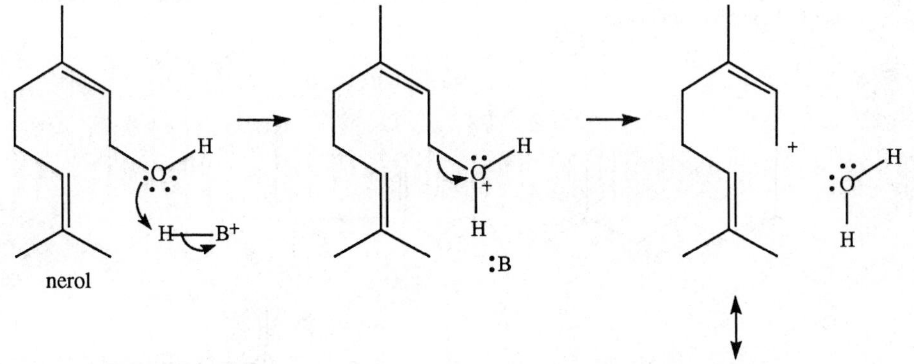

nerol

17.15 (cont)

(d) (cont)

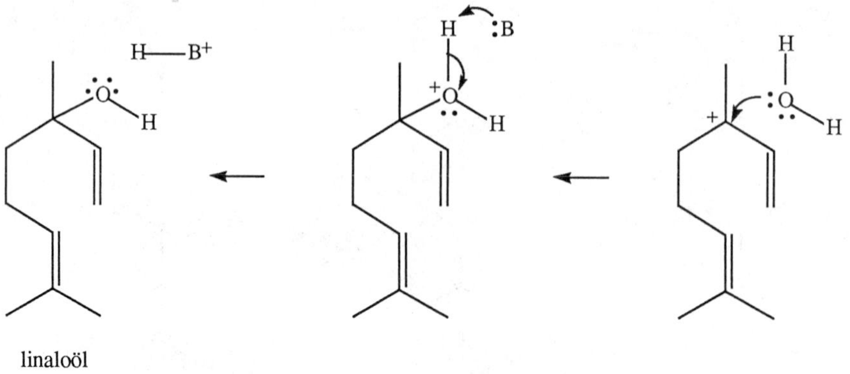

linaloöl

17.16

limonene

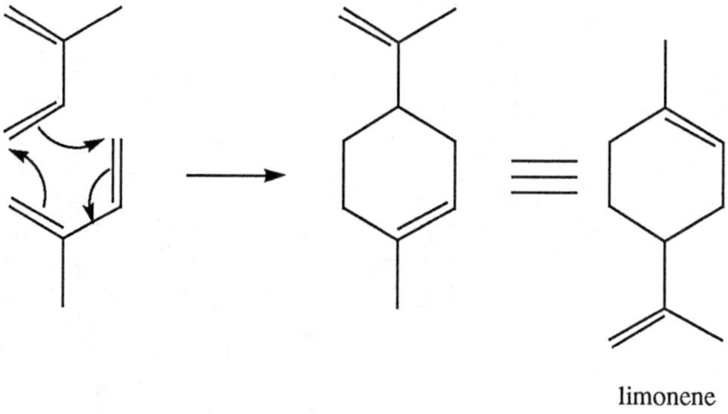

A Diels-Alder reaction between one isoprene unit and the less hindered double bond of the second one in a head-to-tail fashion gives limonene.

17.17

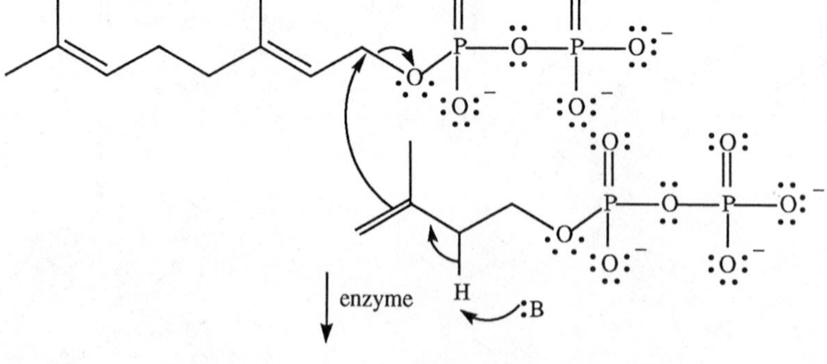

enzyme

17.17 (cont)

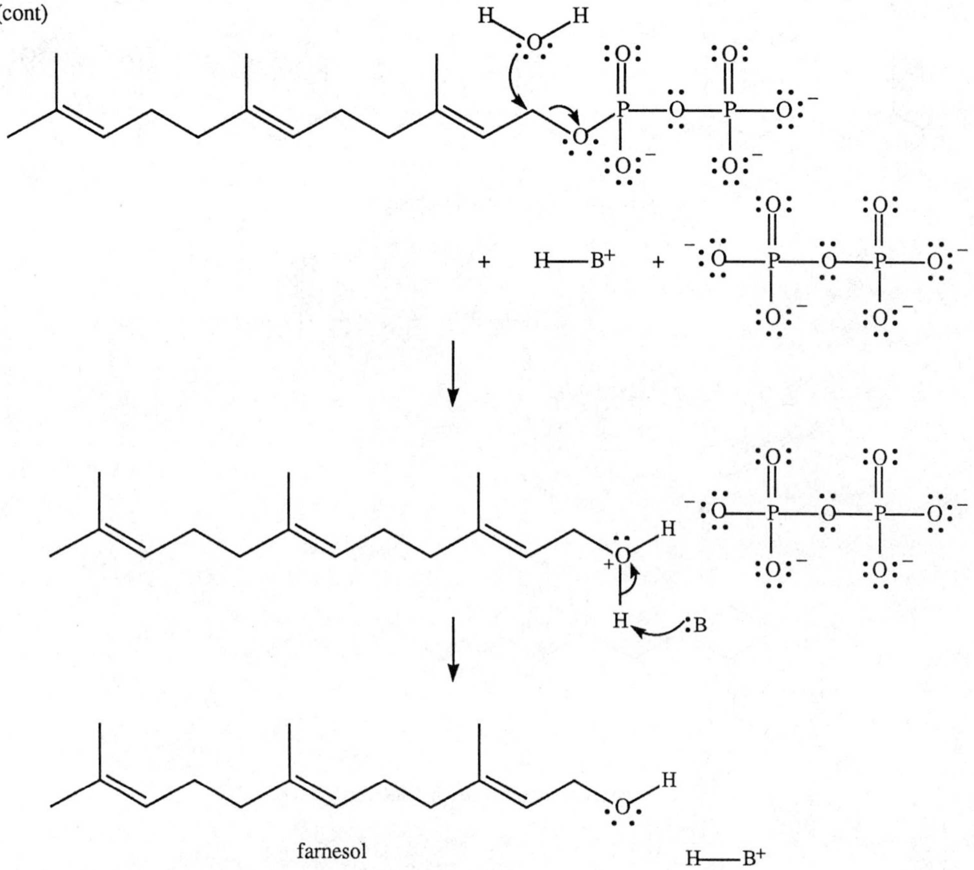

farnesol

H—B⁺

17.18

farnesyl
pyrophosphate

enzyme

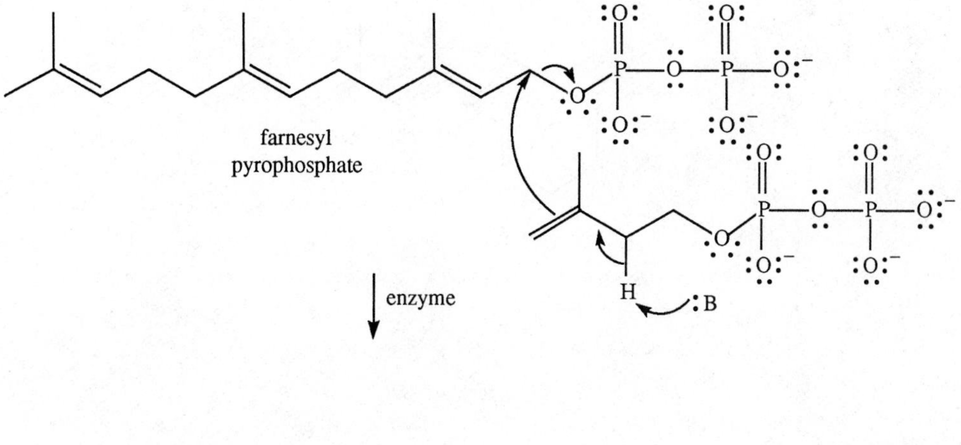

17.18 (cont)

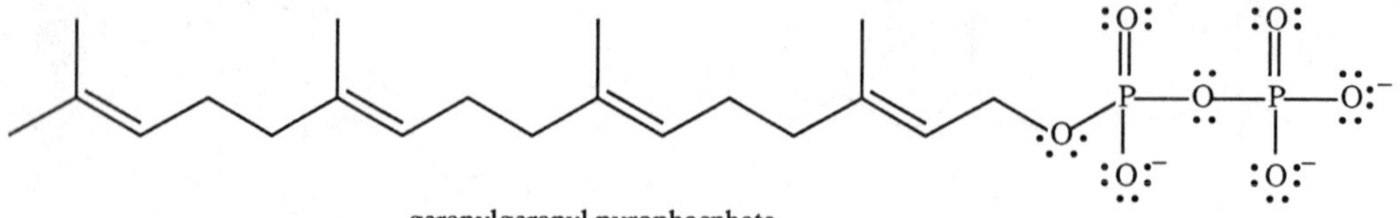

geranylgeranyl pyrophosphate

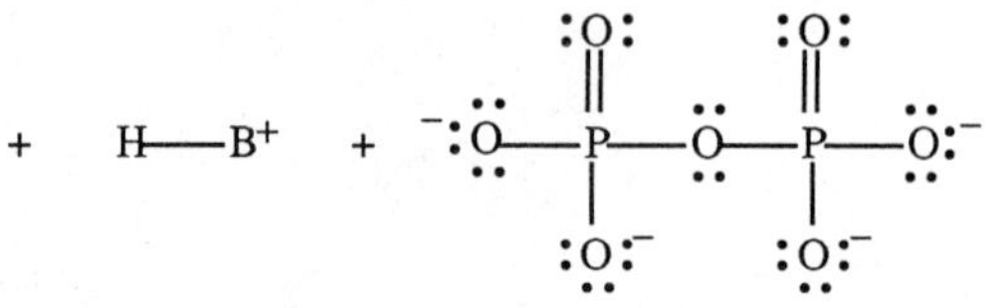

17.19   (a)

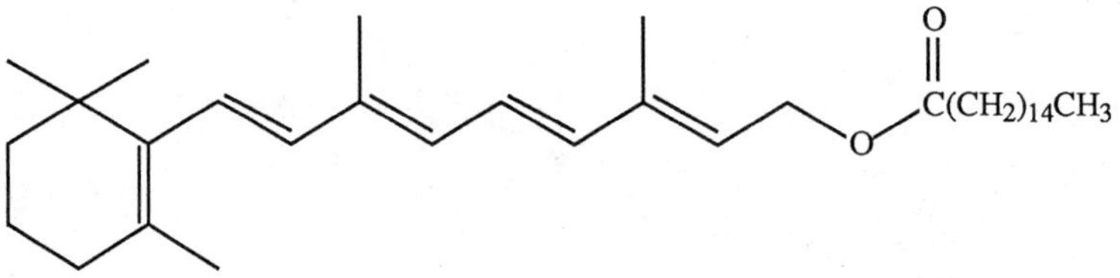

hexadecanoate ester of Vitamin A

(b)

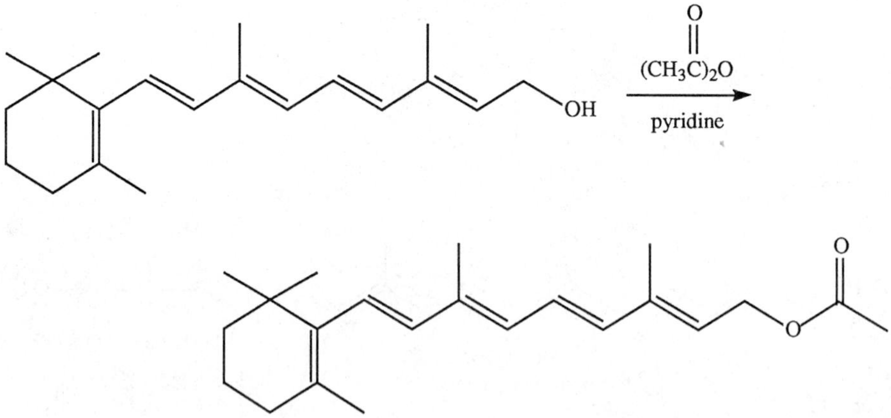

17.20

The most highly substituted double bonds are oxidized preferentially by the peroxyacid.

Concept Map 17.5   Visible and ultraviolet spectroscopy.

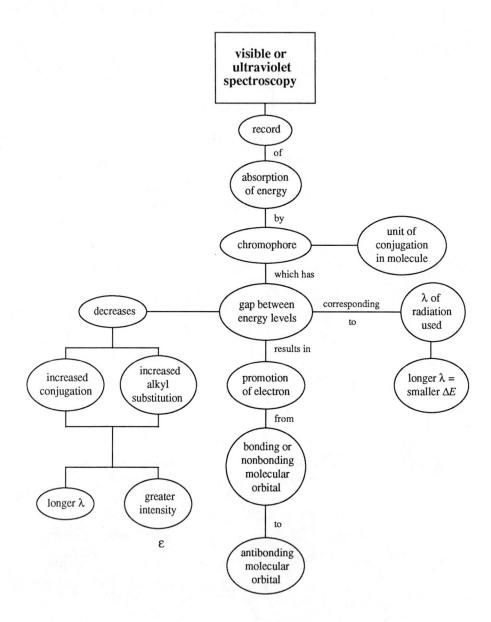

17.21 The compound with the most extended conjugation would be expected to absorb at the longest wavelength.

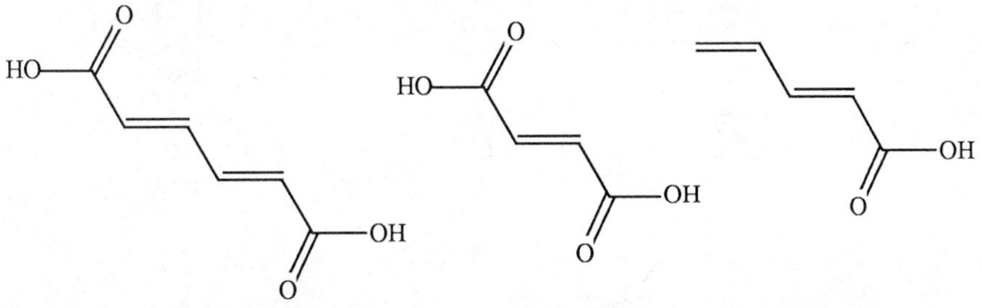

The compound above has the most extensive conjugation, and would absorb at the longest wavelength,

17.22

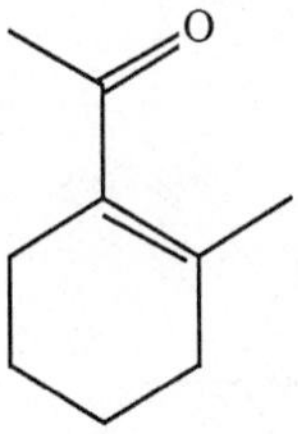

1-Acetyl-2-methyl-1-cyclohexene has two absorption bands, $\lambda_{max}^{ethanol}$ 247 nm (ε 6000) and 305 nm (ε 100). The band at 247 nm corresponds to the $\pi \rightarrow \pi^*$ transition of the conjugated π system and the band at 305 nm to the $n \rightarrow \pi^*$ transition of the carbonyl group.

17.23

(a)

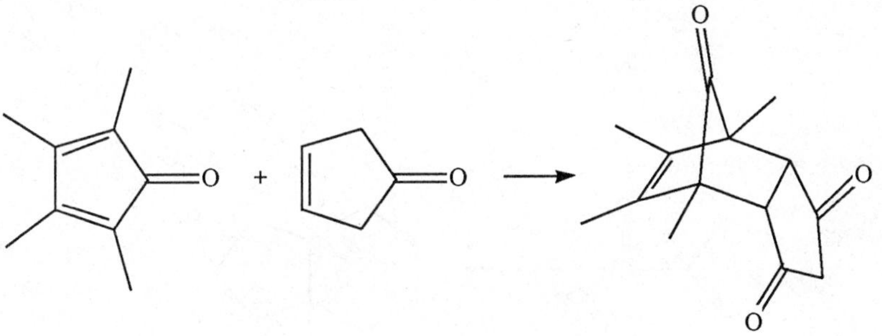

A

(b)

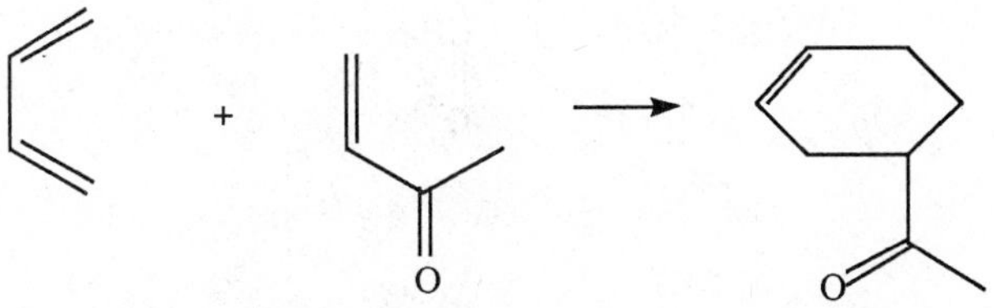

and enantiomer
B

17.23 (cont)

(c)

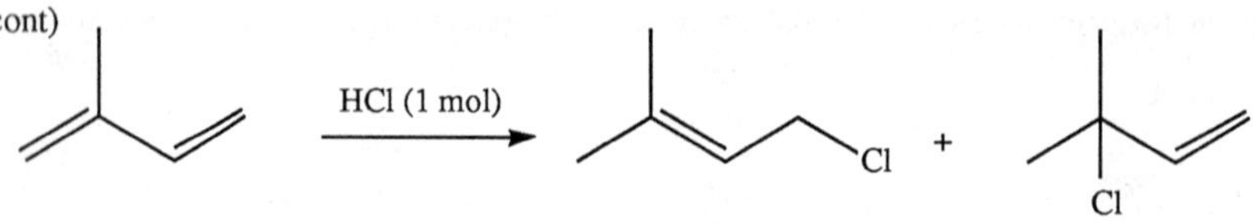

HCl (1 mol)

major
C

D

(d)

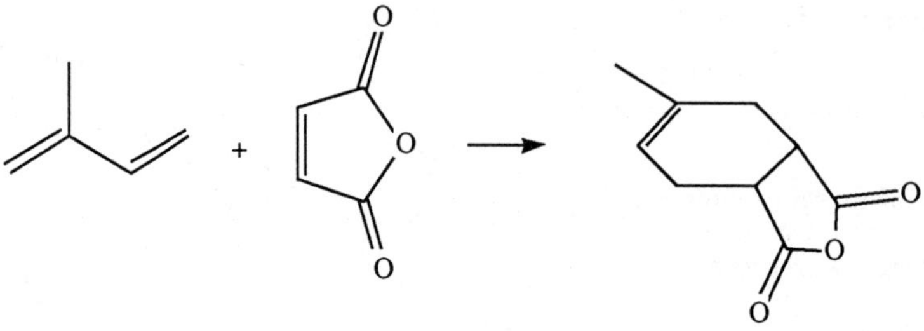

E

(e)

H_2SO_4

Δ

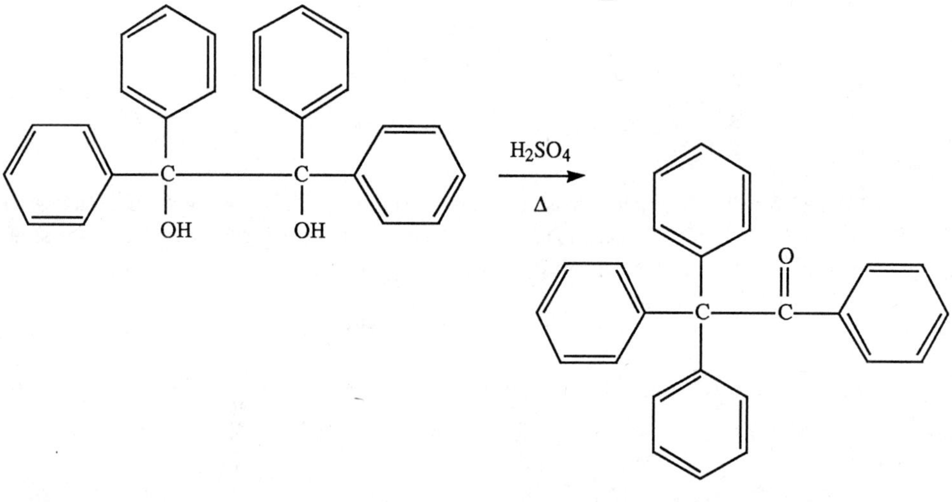

F

(f)

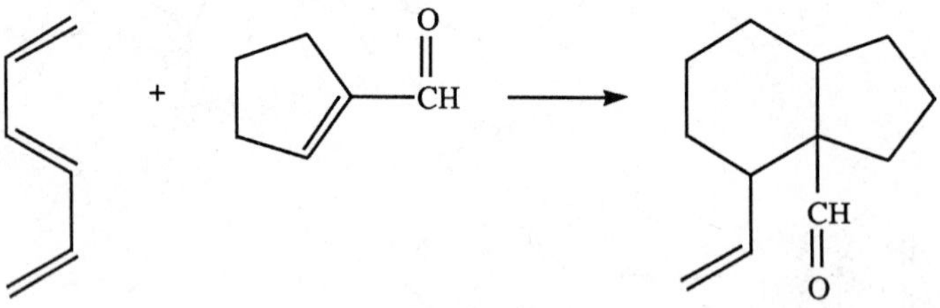

G

17.23 (cont)

(g) $CH_3CH_2CHCH{=}CH_2$
          |
          Br

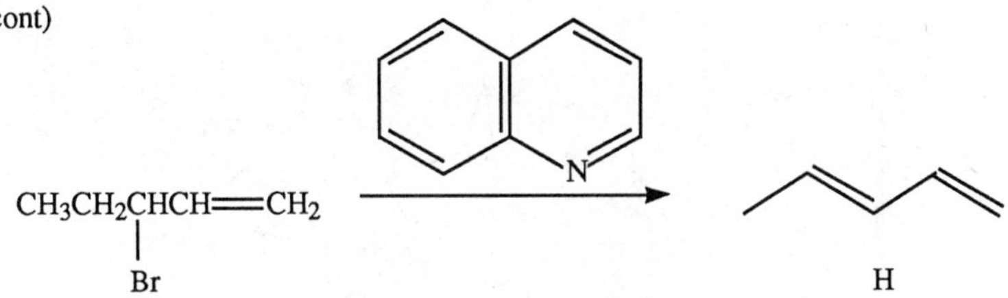

H

(h) $CH_3C{\equiv}CCH_2C{\equiv}CCH_2CH_3$ $\xrightarrow[\text{quinoline}]{\overset{\text{H}_2}{\text{Pd/BaSO}_4}}$

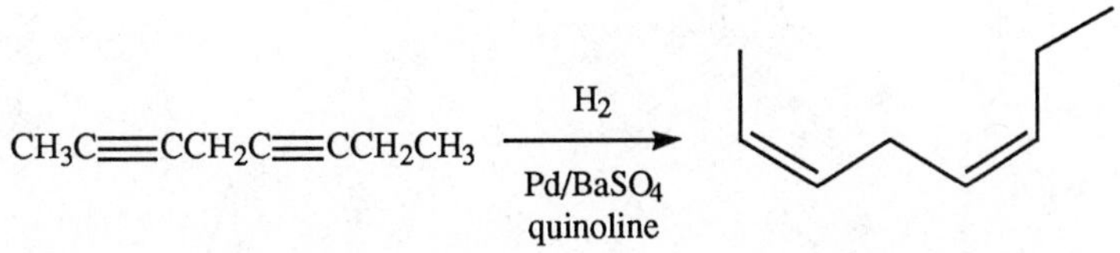

I

(i) $CH_3OCCH_2CH_2CH_2$ CH_2OH $\xrightarrow{CH_3COOH}$

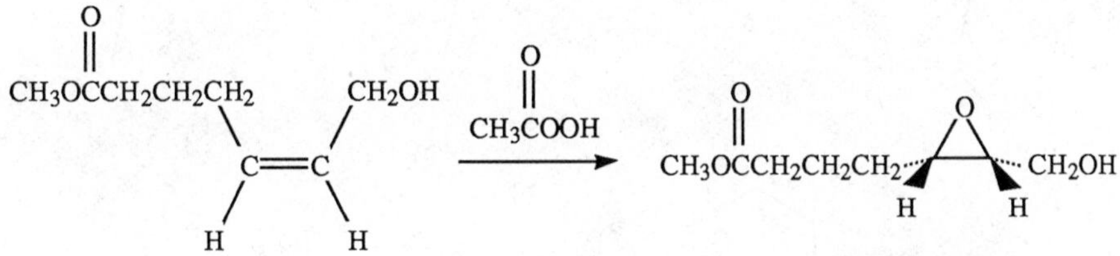

and enantiomer
J

(j) $\xrightarrow{\Delta}$ 2

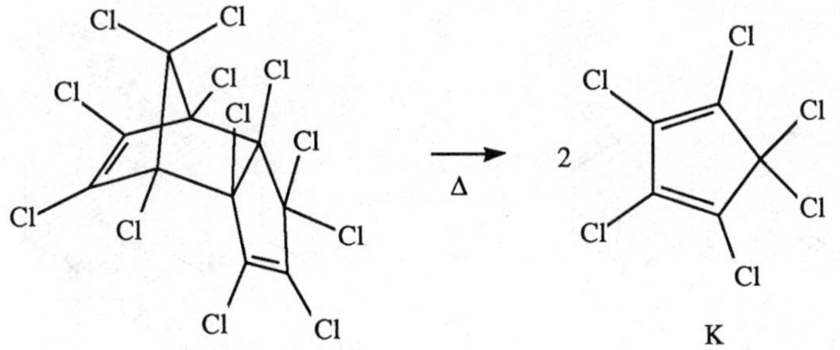

K

(k)

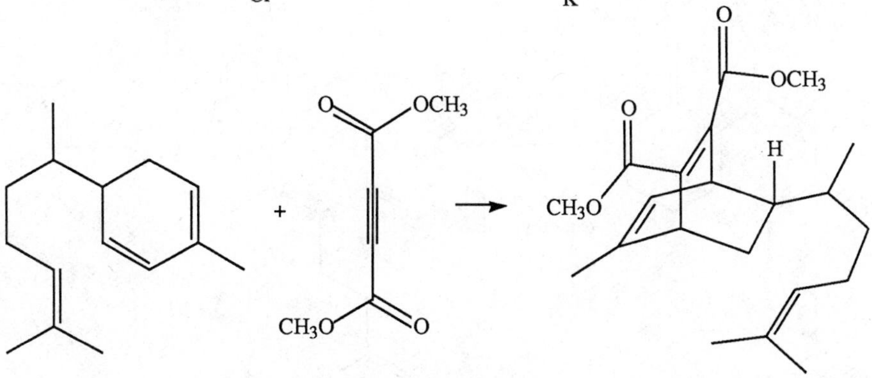

and enantiomer
L

17.23 (cont)

(l)

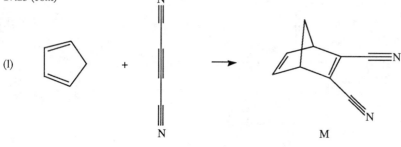

M

(m)

H₂NNH₂, KOH

diethylene glycol
Δ

N

H₂

Pt

O

(n)

BH₃

tetrahydro-
furan

H₂O₂, NaOH

P

Q

(o)

H₂

Pt/C
ethyl acetate
trimethylamine

NaOH

R

17.23 (cont)

(o) (cont)

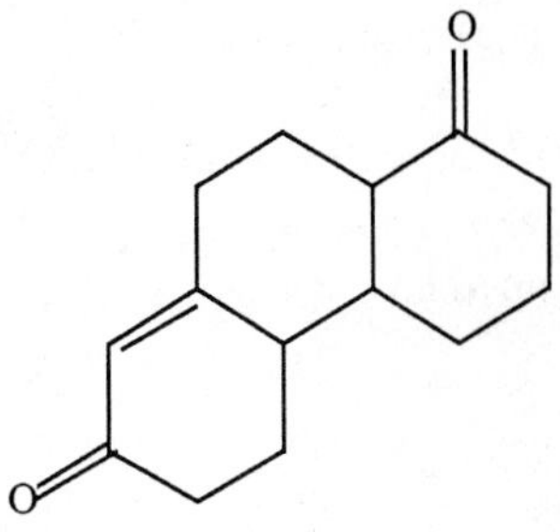

S

(p)

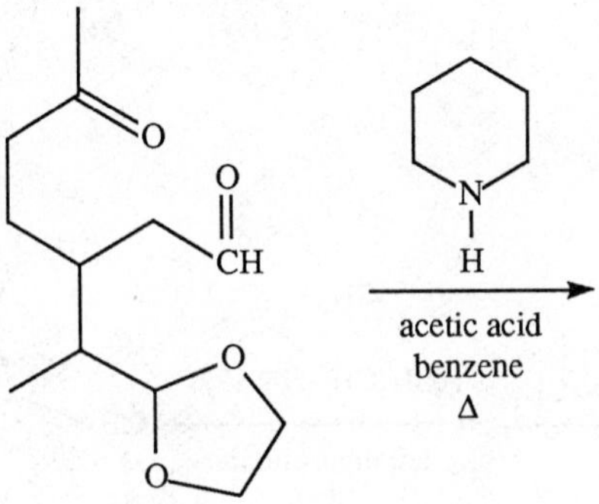

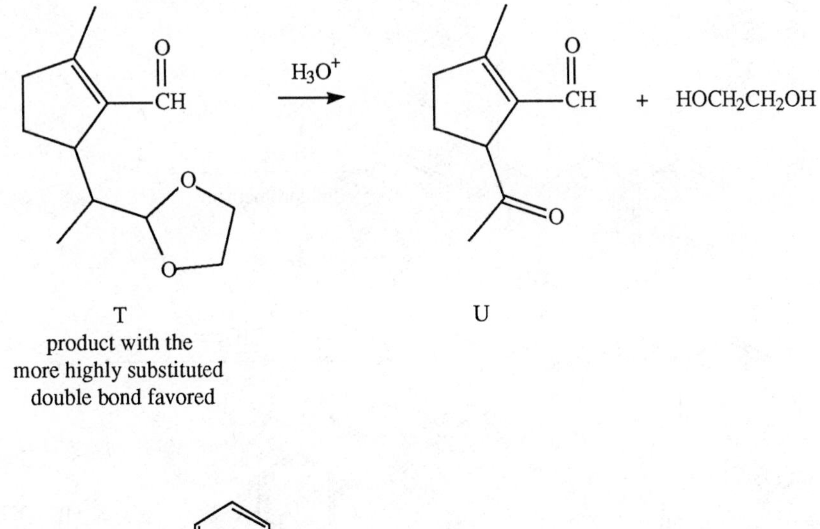

T
product with the
more highly substituted
double bond favored

U

+ HOCH₂CH₂OH

(q)

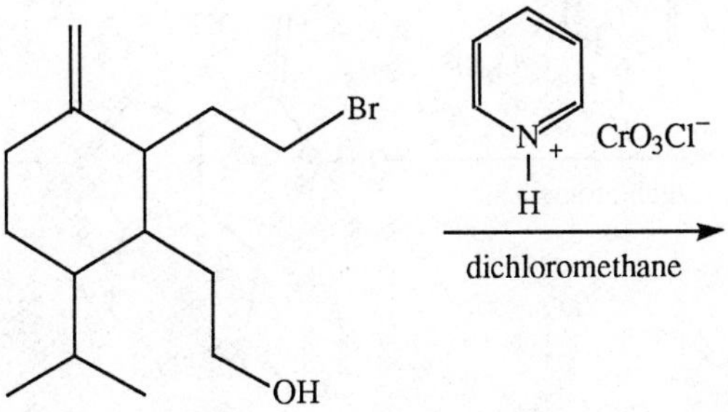

17.23 (cont)

(q) (cont)

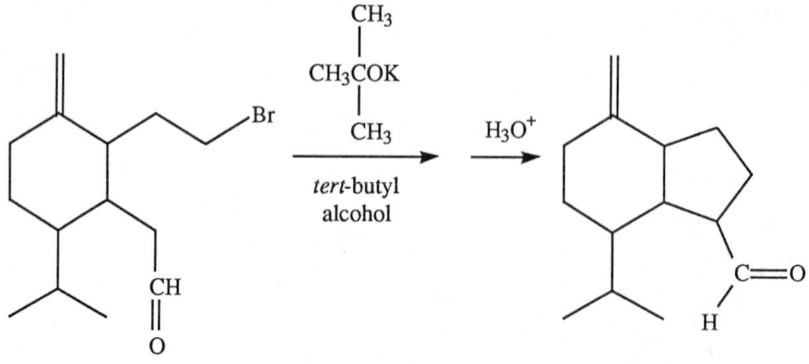

V

W

17.24

(a)

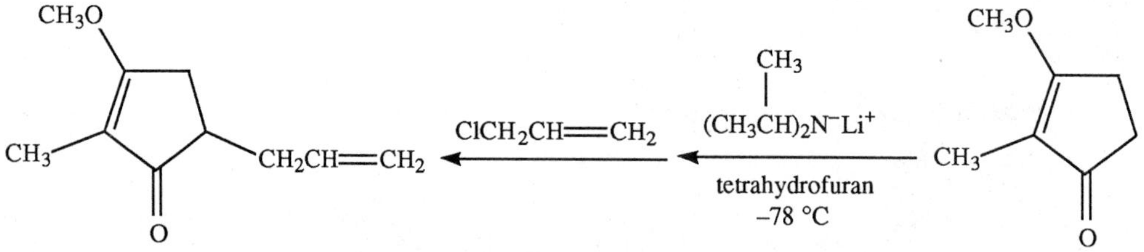

(b)

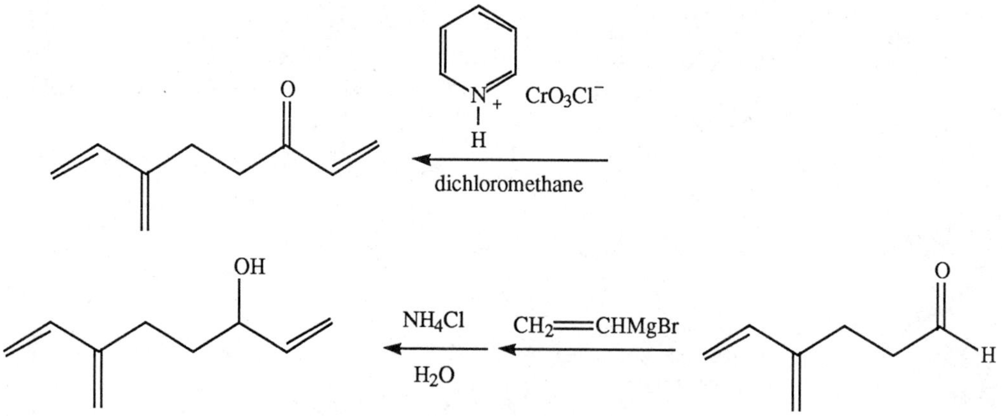

(c)

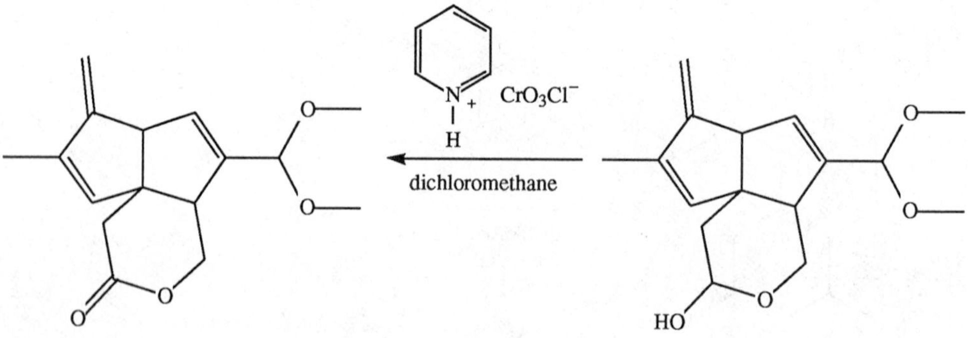

17.24 (cont)

(d)

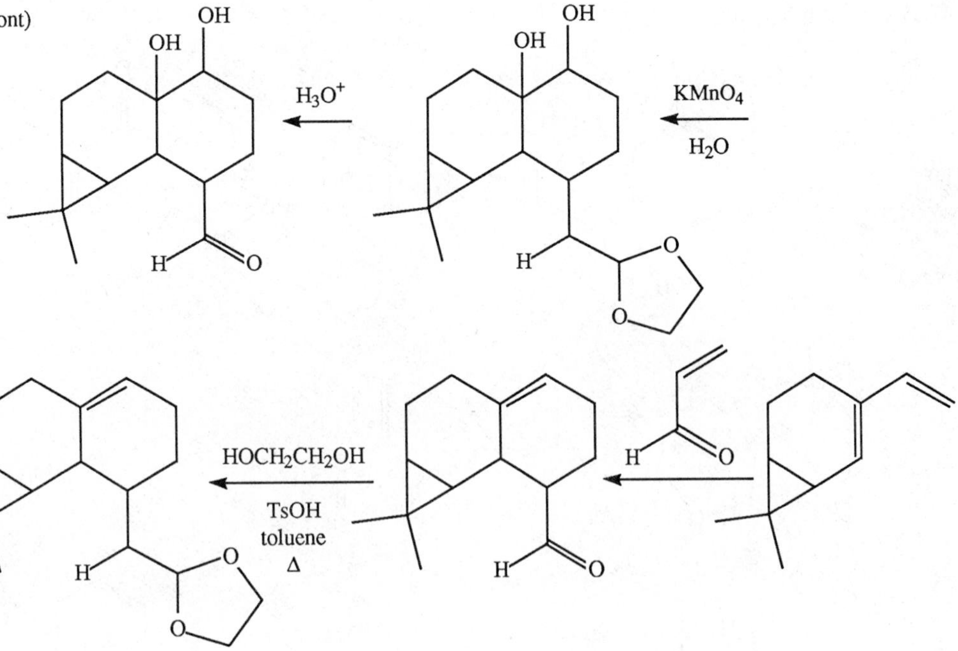
OH
OH
H_3O^+
KMnO$_4$
H_2O
H
O
HOCH$_2$CH$_2$OH
TsOH
toluene
Δ

(e)

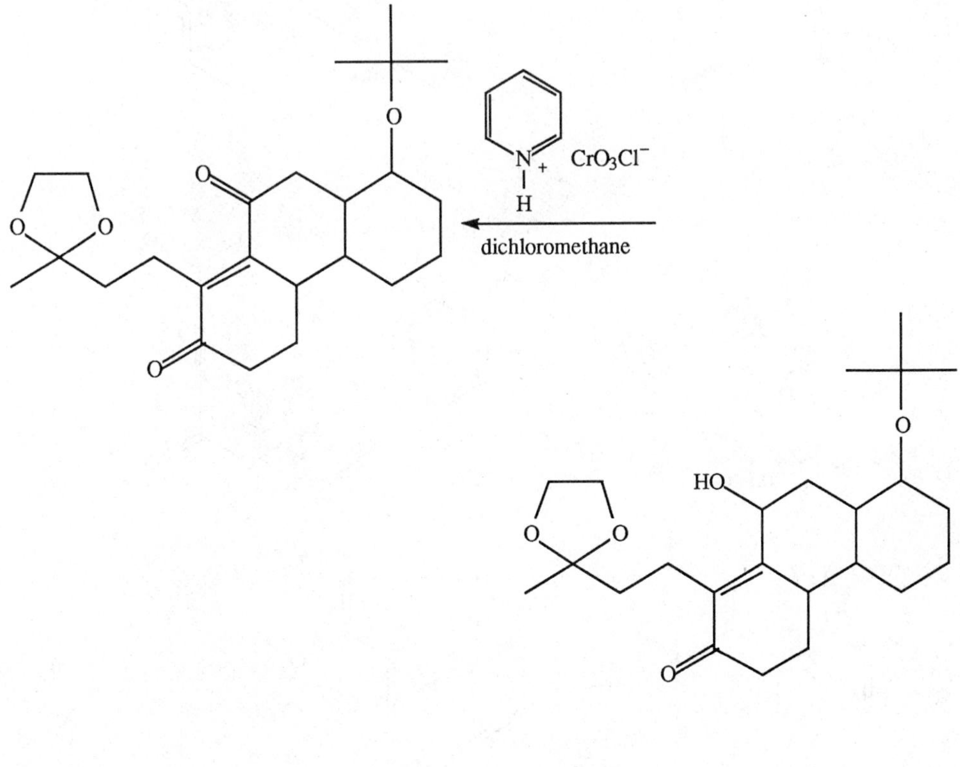
O
N+
H
CrO$_3$Cl$^-$
dichloromethane
HO

17.25   σ bonds formed

(a)

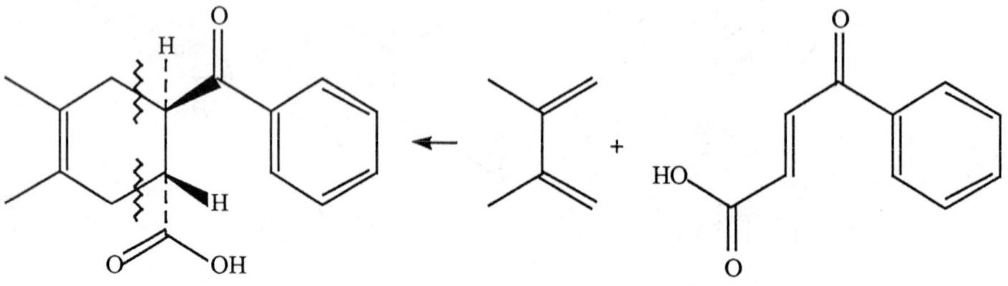

(b)

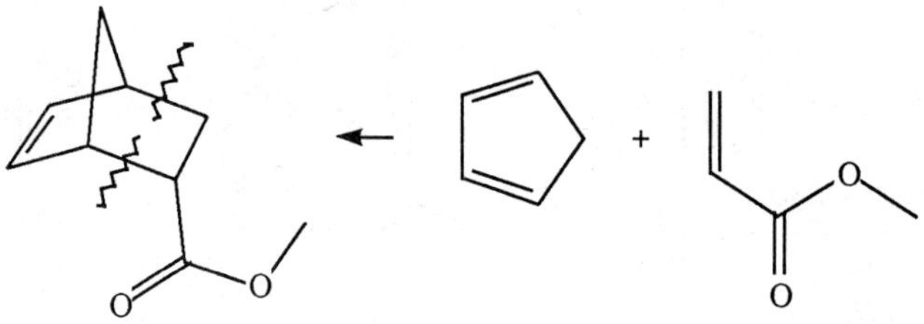

(c)

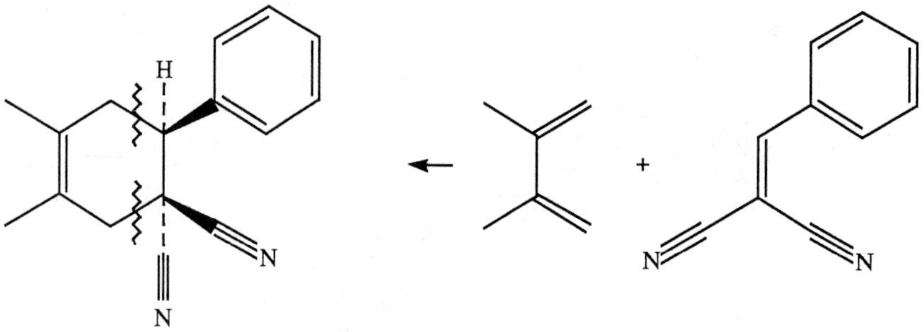

(d)

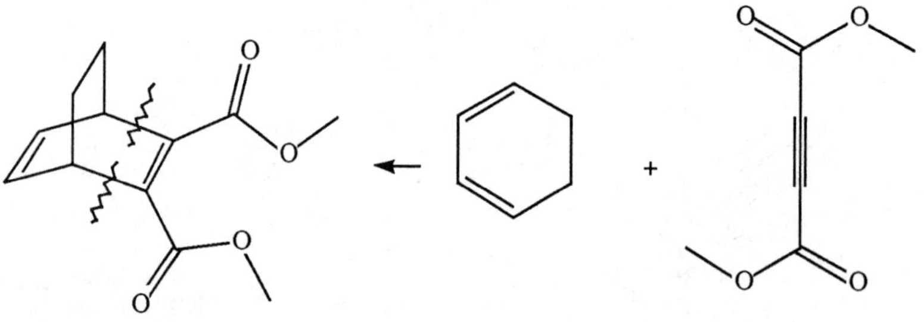

(e)

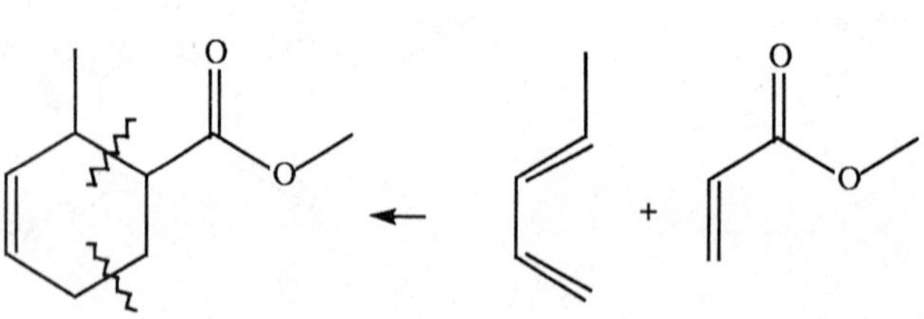

17.25 (cont)

(f)

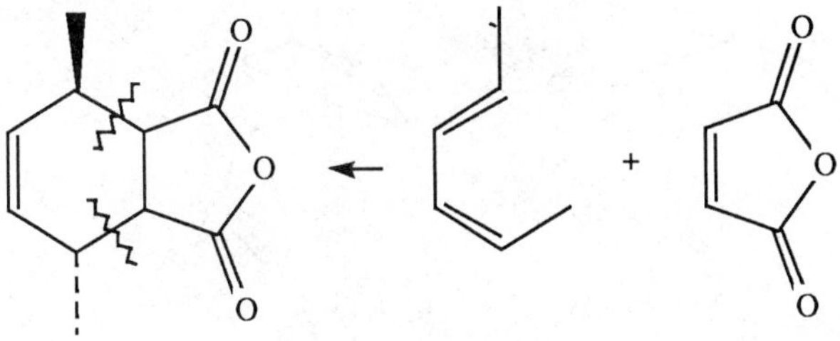

(g)

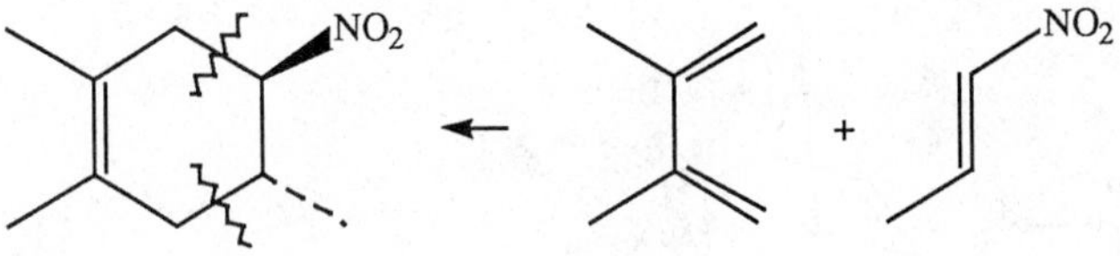

(h)

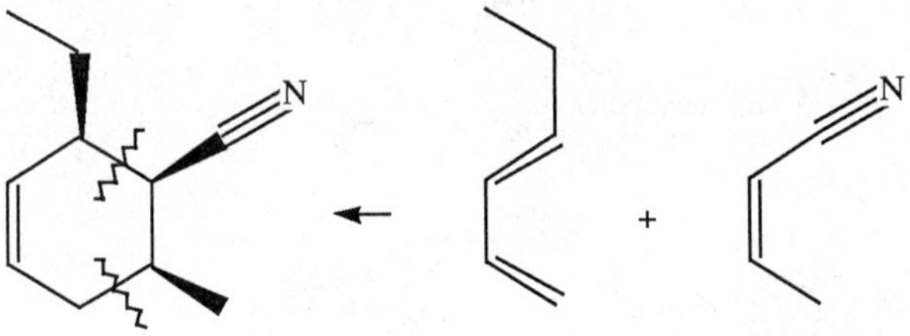

17.26

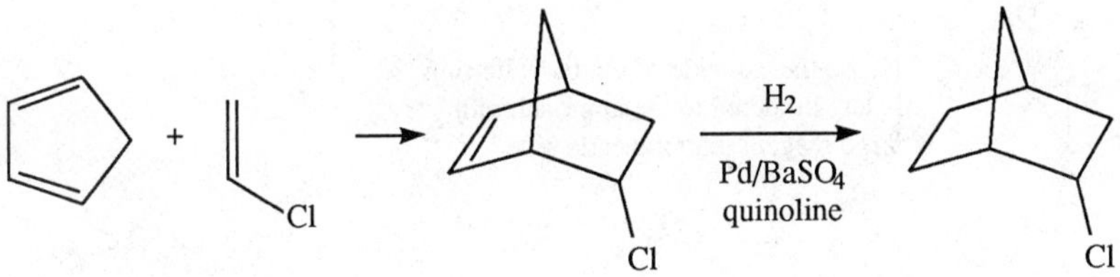

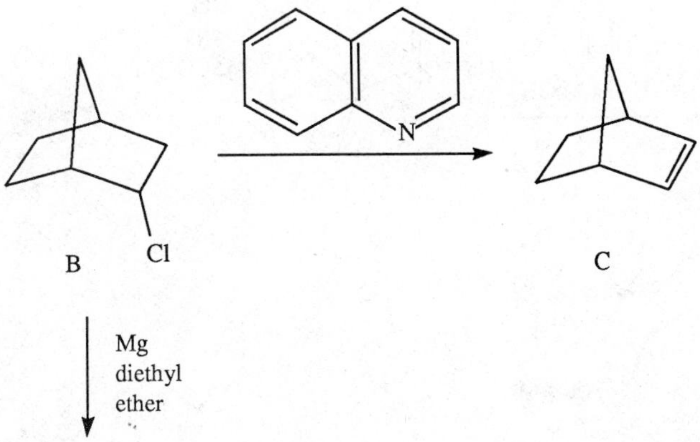

17.26 (cont)

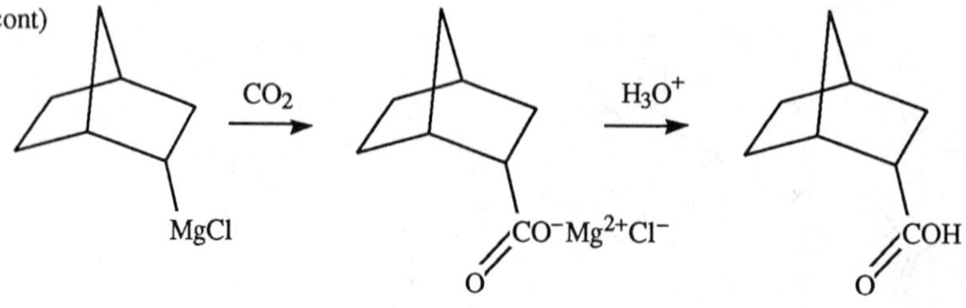
MgCl
CO_2
$\overset{O}{\parallel}$CO⁻Mg²⁺Cl⁻
H_3O^+
COH

D E F

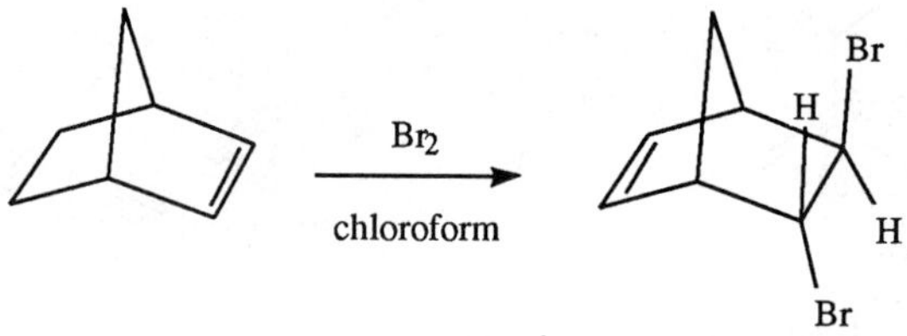
Br_2
chloroform
Br
H
H
Br

and enantiomer

C G

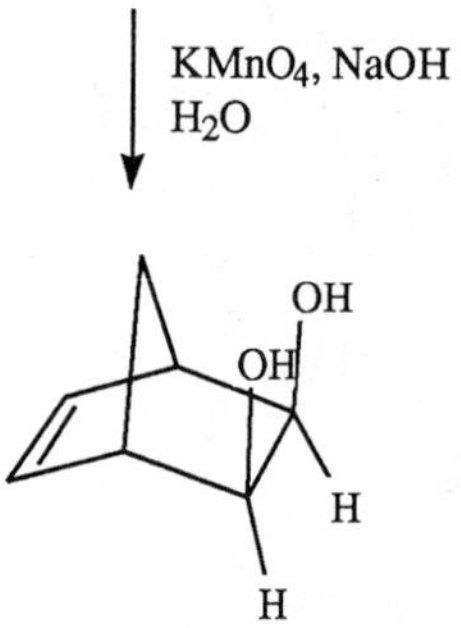
$KMnO_4$, NaOH
H_2O
OH
OH
H
H

Note: the exo side of the bicyclic ring is less hindered to the approach of a large reagent than the endo side.

H

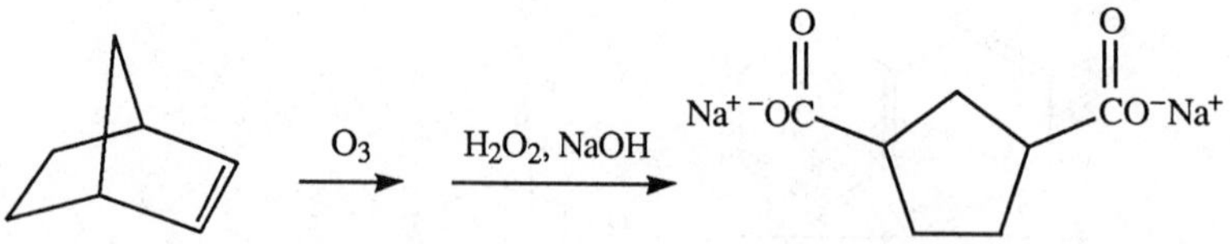
O_3
H_2O_2, NaOH
O
Na^+ $^-$OC
O
CO$^-$$Na^+$

C I

17.27

no steric hindrance;
diene can achieve
s-cis conformation

steric hindrance;
difficult for diene to
achieve *s*-cis conformation

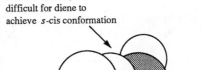

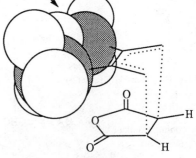

transition state for the
addition of 1,3-butadiene
to maleic anhydride

transition state for the addition of
2,3-di-*tert*-butyl-1,3-butadiene to
maleic anhydride. The bulkiness of
the *tert*-butyl groups prevents the
diene from achieving the *s*-cis conformation needed for the Diels-Alder
reaction. Good overlap of the π system
is also difficult, raising the energy of
the transition state to the point where
the reaction does not proceed.

17.28

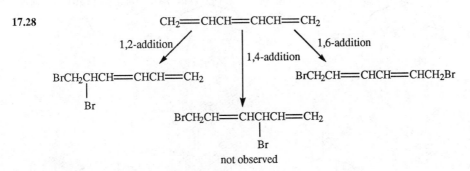

not observed

All three products can be derived from the same intermediate.

17.28 (cont)

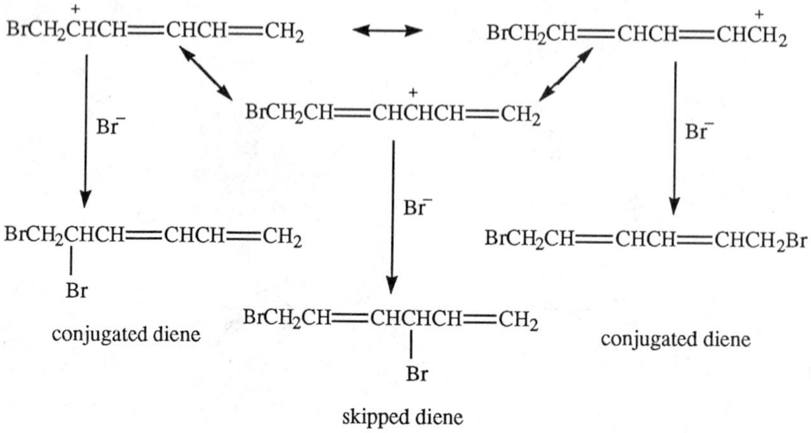

$$\overset{+}{BrCH_2CHCH}=CHCH=CH_2 \quad \longleftrightarrow \quad BrCH_2CH=CHCH=\overset{+}{CHCH_2}$$

$$BrCH_2CH=CH\overset{+}{CH}CH=CH_2$$

Br⁻   Br⁻

Br⁻

$$BrCH_2CHCH=CHCH=CH_2$$
$$\quad\quad | $$
$$\quad\quad Br$$

conjugated diene

$$BrCH_2CH=CHCHCH=CH_2$$
$$\quad\quad\quad | $$
$$\quad\quad\quad Br$$

skipped diene

$$BrCH_2CH=CHCH=CHCH_2Br$$

conjugated diene

The products observed have conjugated double bonds and are thermodynamically more stable than the one with the skipped double bonds.

17.29

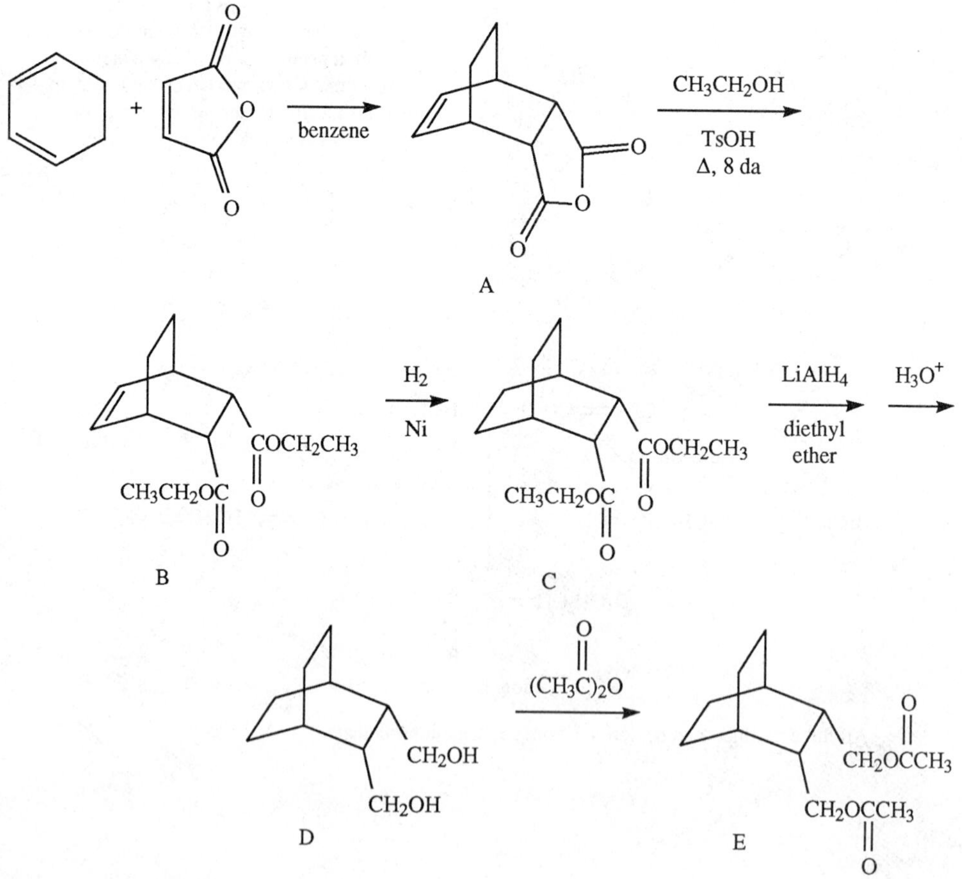

17.30 (a)

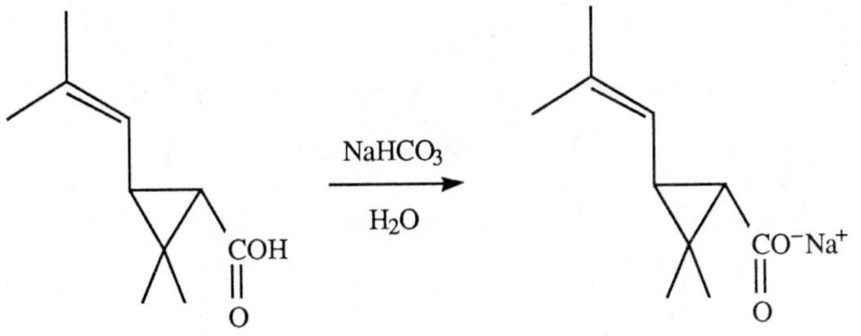
NaHCO3
H2O
COH
O
CO⁻Na⁺
O

(b)

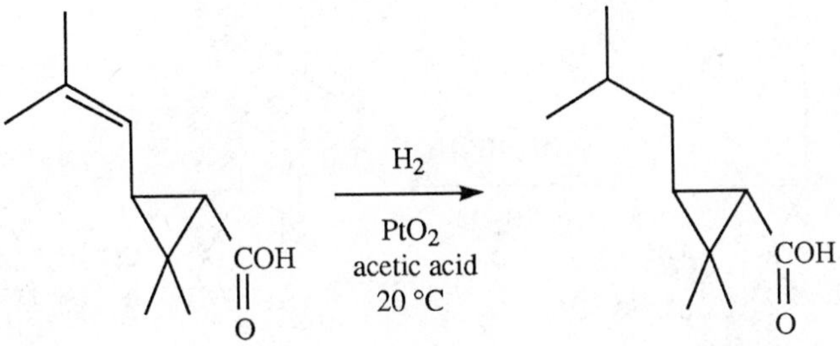
H2
PtO2
acetic acid
20 °C
COH
O
COH
O

(c)

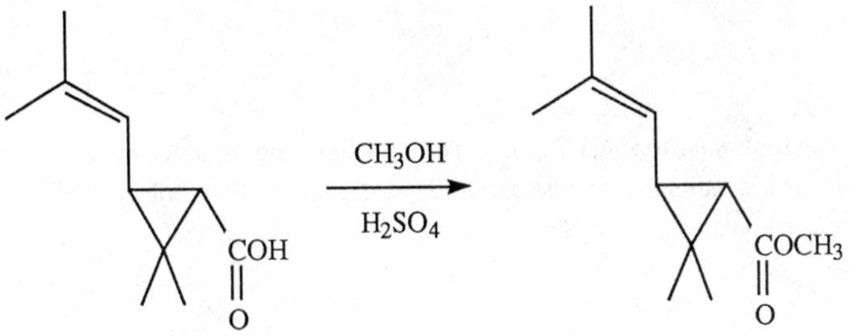
CH3OH
H2SO4
COH
O
COCH3
O

(d)

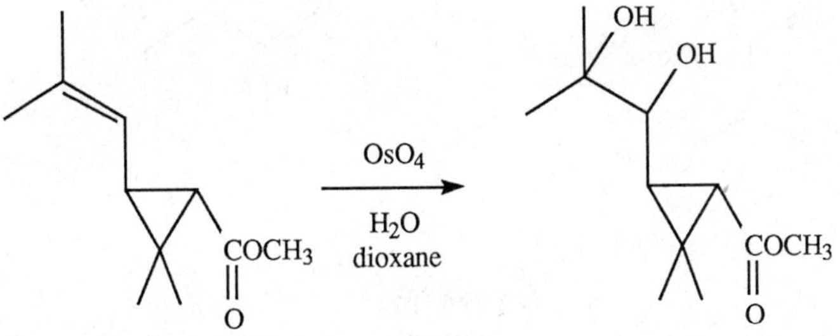
OsO4
H2O
dioxane
COCH3
O
OH
OH
COCH3
O

(e)

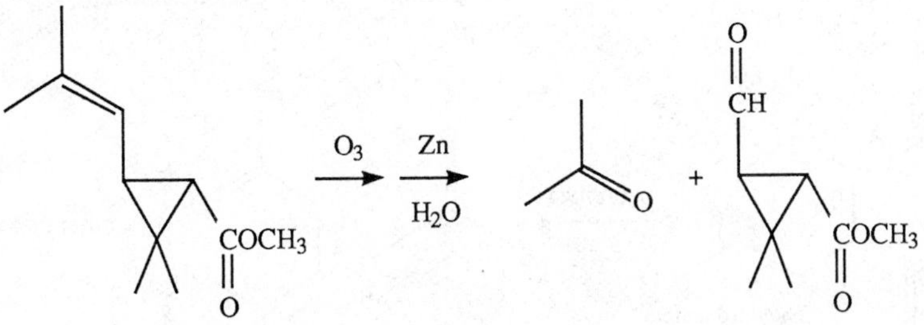
O3
Zn
H2O
COCH3
O
O
+
O
CH
COCH3
O

17.30 (cont)

(f)

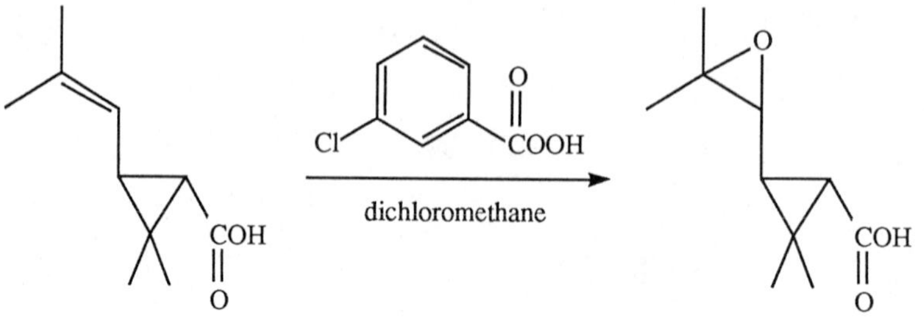
COH
O
Cl
O
COOH
dichloromethane
O
COH
O

(g)

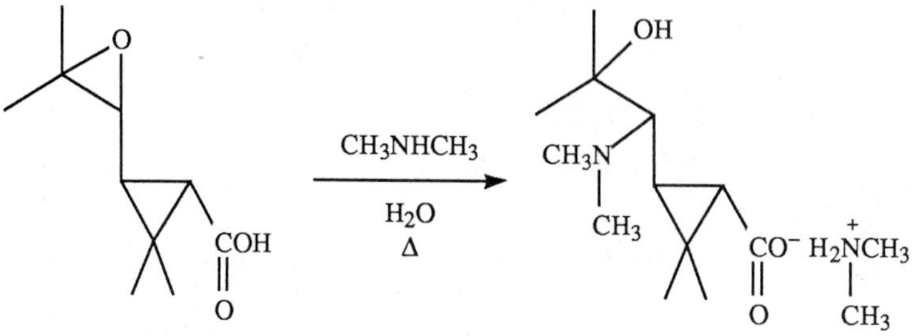
O
COH
O
CH3NHCH3
H2O
Δ
OH
CH3N
CH3
CO⁻ H2NCH3
O
CH3

17.31

(a) Bisabolene has the molecular formula $C_{15}H_{24}$. The corresponding saturated alkane would have the formula $C_{15}H_{32}$. Bisabolene has four units of unsaturation. Hydrogenation of bisabolene gives Compound X, $C_{15}H_{30}$, therefore bisabolene contains one ring and three π bonds.

(b)

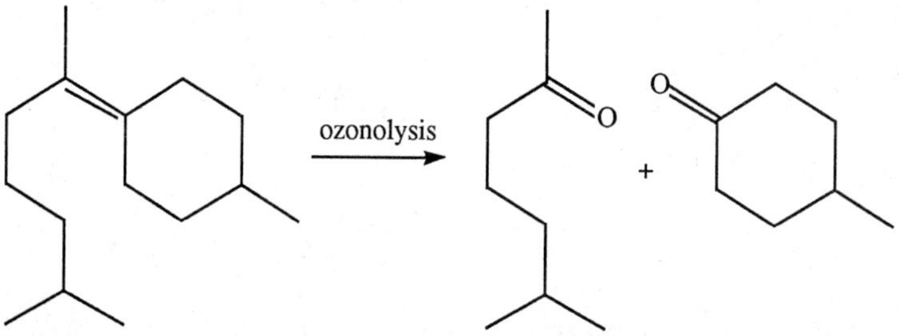
ozonolysis
O
+
O

Compound Y $C_{15}H_{28}$ | 6-methyl-2-heptanone | 4-methyl-cyclohexanone

(c)

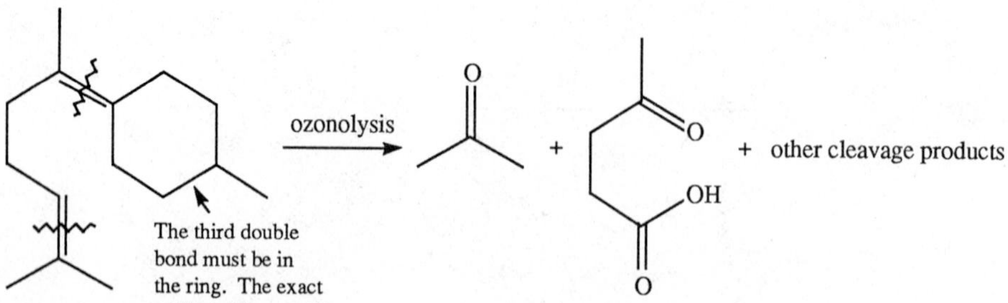
ozonolysis
The third double bond must be in the ring. The exact position is uncertain.
O
+
O
OH
O
+ other cleavage products

acetone 4-oxopentanoic acid

17.31 (cont)

(d)

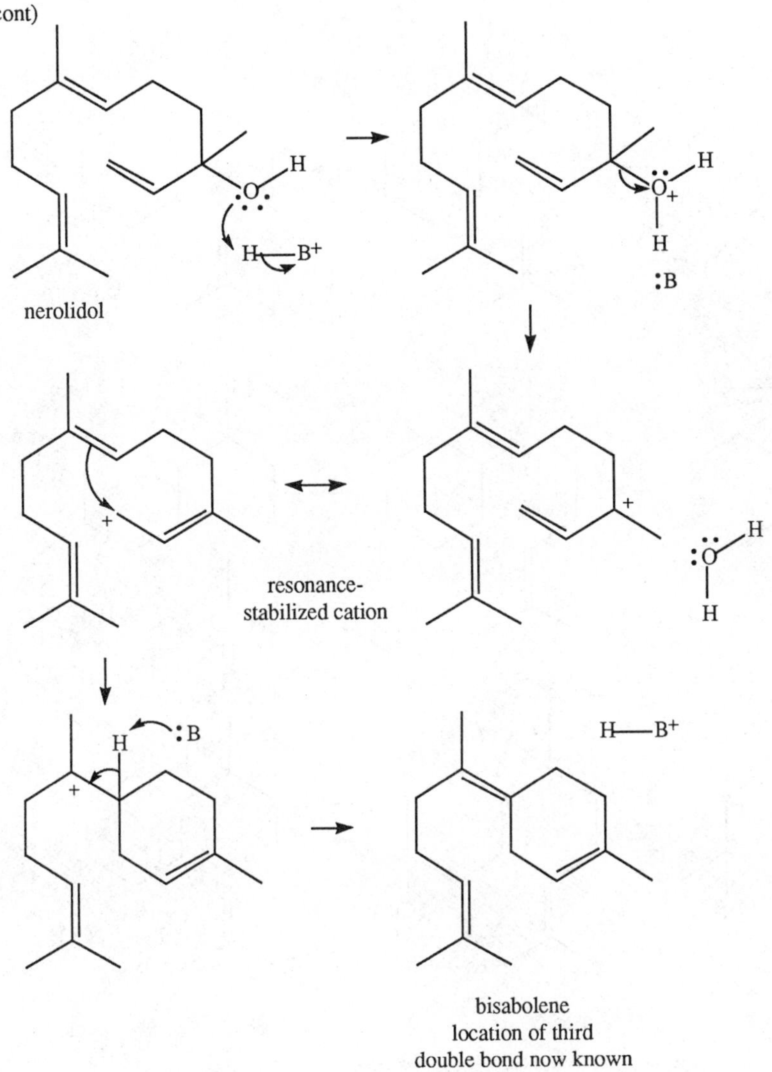

nerolidol

resonance-
stabilized cation

bisabolene
location of third
double bond now known

(e)

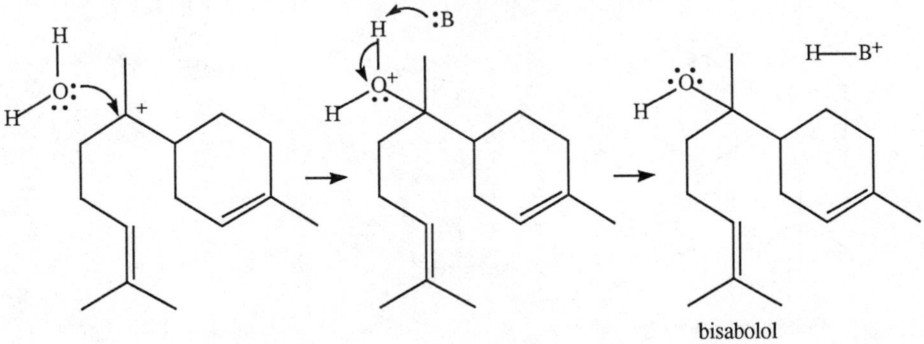

bisabolol

17.31 (cont)

(f)

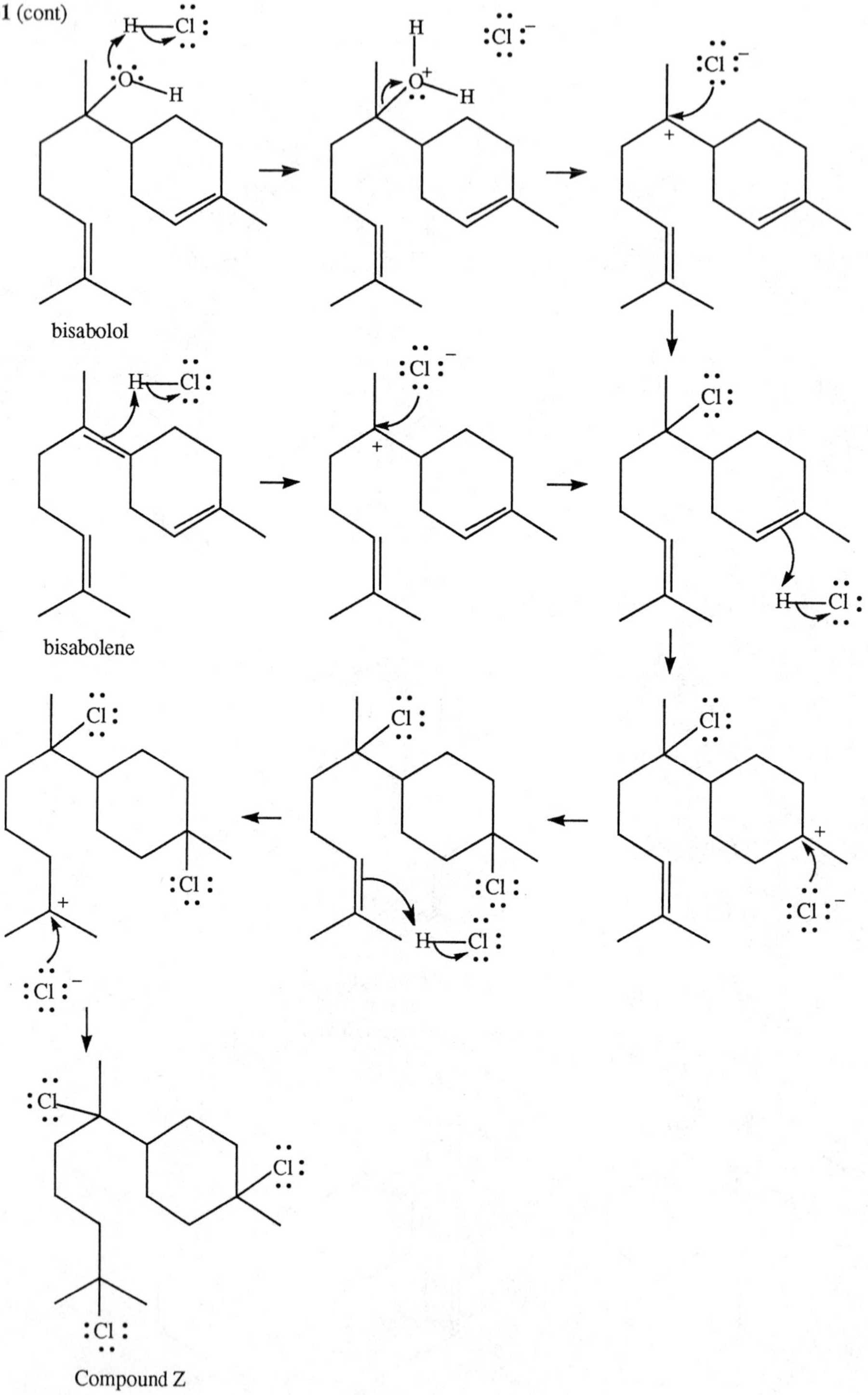

bisabolol

bisabolene

Compound Z

17.32 HC≡CH $\xrightarrow[\text{tetrahydro-}\atop\text{furan}]{\text{Na}}$ HC≡C⁻Na⁺ $\xrightarrow{\text{CH}_3(\text{CH}_2)_9\text{Br}}$
                                                                 A

HC≡C(CH₂)₉CH₃ $\xrightarrow{\text{CH}_3\text{CH}_2\text{CH}_2\text{CH}_2\text{Li}}$ Li⁺ ⁻C≡C(CH₂)₉CH₃ $\xrightarrow[\text{D}]{\overset{\text{CH}_3}{\overset{|}{\text{CH}_3\text{CH(CH}_2)_4\text{Br}}}}$

    B   C

$$\text{CH}_3(\text{CH}_2)_9\text{C}\equiv\text{C(CH}_2)_4\overset{\overset{\text{CH}_3}{|}}{\text{CHCH}_3} \xrightarrow[\substack{\text{Pd/BaSO}_4\\ \text{quinoline}\\ \text{E}}]{\text{H}_2}$$

CH₃(CH₂)₉ ⎯ C=C ⎯ (CH₂)₄CHCH₃ (with CH₃ substituent, H and H on the double bond)
                F

$\xrightarrow[\text{dichloromethane}]{\text{3-chlorobenzoic acid (G)}}$

CH₃(CH₂)₉ — (epoxide) — (CH₂)₄CHCH₃ (with CH₃)

and enantiomer

disparlure

$$\overset{\overset{\text{CH}_3}{|}}{\text{CH}_3\text{CH(CH}_2)_4\text{OH}} \xrightarrow[\substack{\text{pyridine}\\ \text{H}}]{\text{PBr}_3} \overset{\overset{\text{CH}_3}{|}}{\text{CH}_3\text{CH(CH}_2)_4\text{Br}}$$
                         H   D

17.33 (CH₃O-substituted diene) + (nitro-substituted alkene with CH₂CH₂OCH₂CH₃) $\xrightarrow{100\,°\text{C, 6 h}}$ (cyclohexene product)

and enantiomer
A

ν_{max} (cm⁻¹)

2844 C—H (aliphatic)
1658 C=C
1377 —NO₂
1178, 1166, 1117 C—O—C

17.34 (cont)

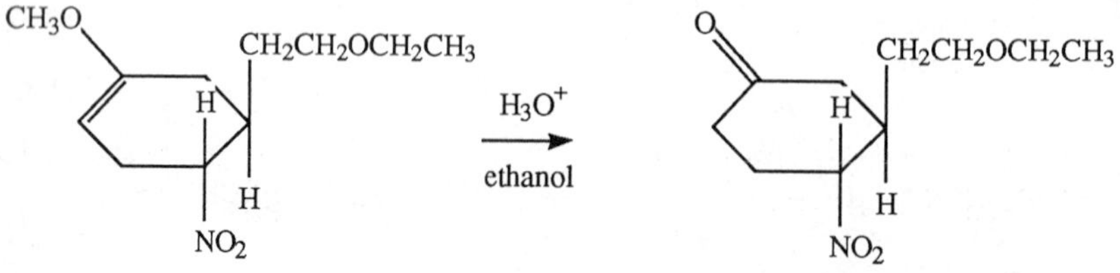

and enantiomer
B

v_{max} (cm^{-1})

2844 C—H (aliphatic)
1722 C=O
1350 —NO$_2$
1117 C—O—C

17.34

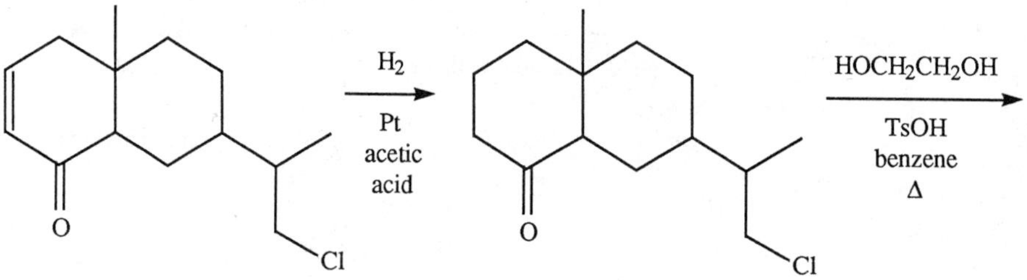

A

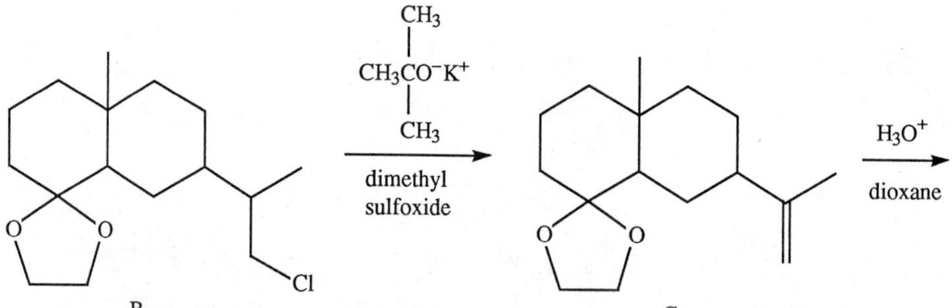

B

C

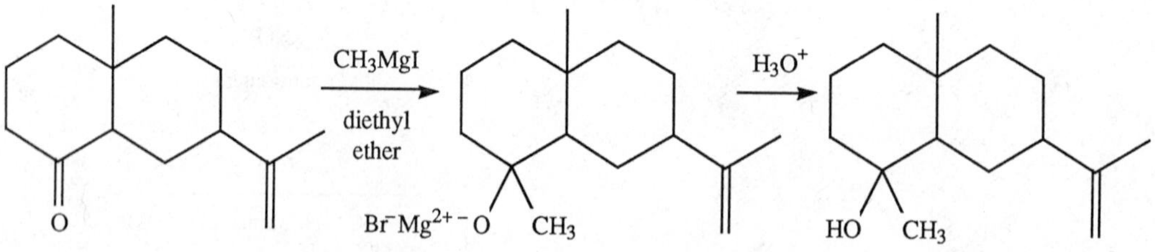

D   E   F
                                                        neointermedeol

17.35

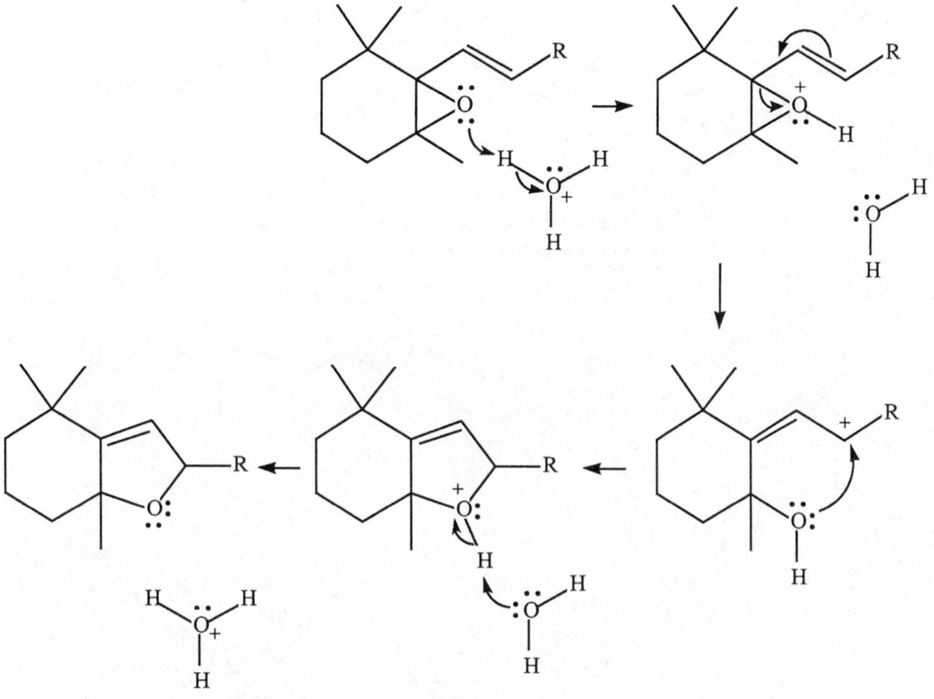

R =

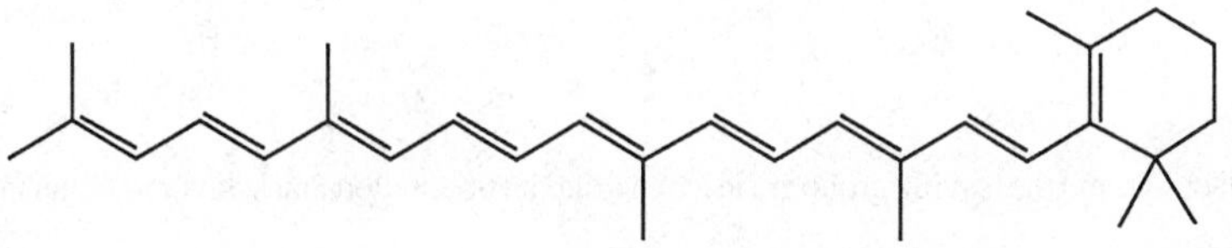

17.36

(a)

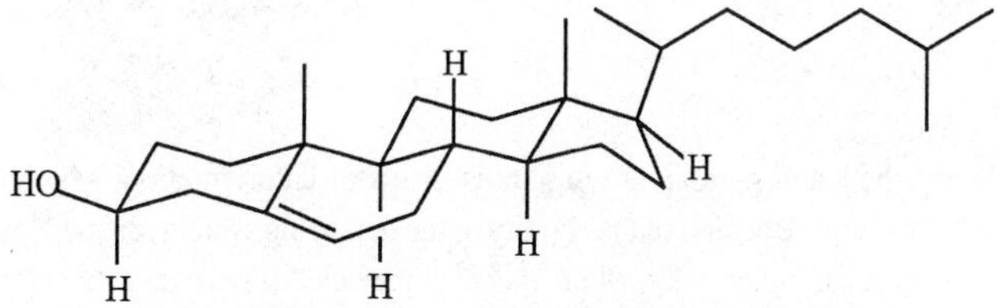

Bromine approaches the molecule from the less
hindered side, away from the methyl group.

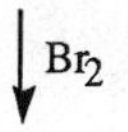

Br$_2$

17.36 (cont)

(a) (cont)

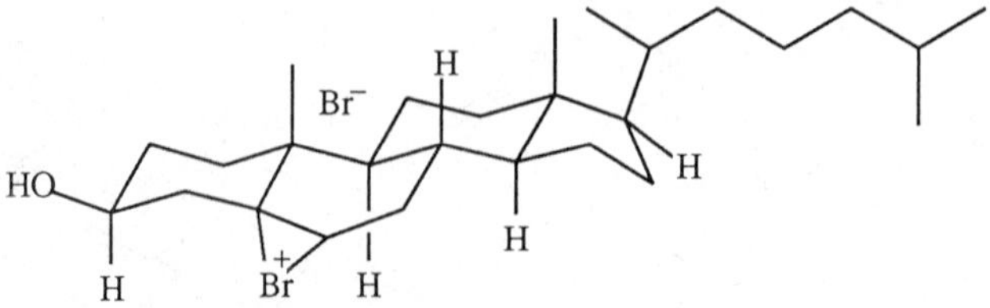

trans diaxial orientation of the bromine atoms
when the bromonium ion opens

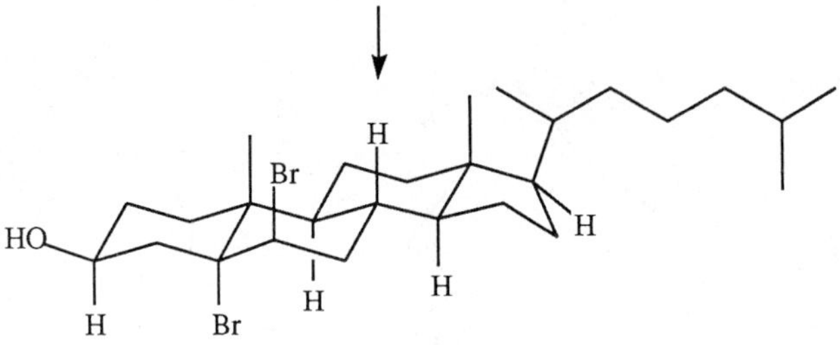

(b)

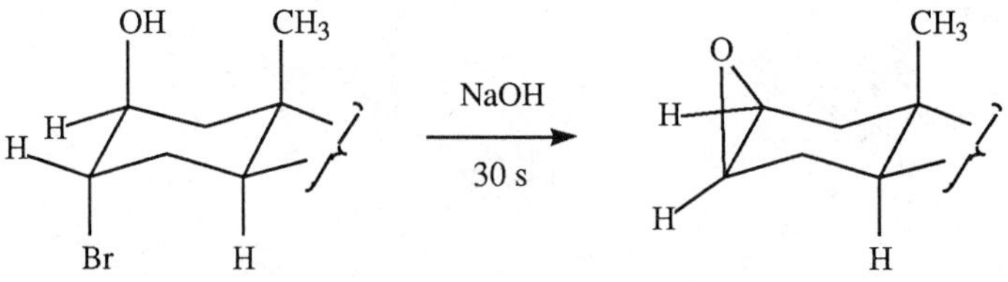

A

The nucleophile and the leaving group are in a trans-diaxial orientation that is favored for an intramolecular S_N2 reaction.

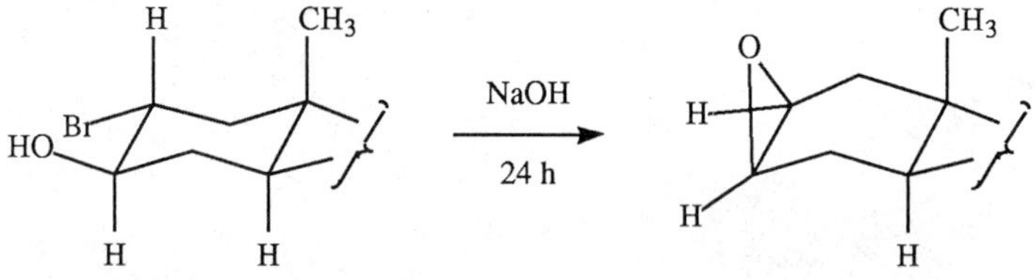

B

The nucleophile and the leaving group are in a trans-diequatorial orientation, which is less favorable for an intramolecular S_N2 reaction. Because of the rigidity of the fused ring system, the two groups can have the trans-diaxial orientation to each other only when the cyclohexane ring is in a boat or twist (Section 5.9C) conformation. Such a conformation is higher in energy than the chair conformation allowed for the reaction of Compound A, so the rate of reaction is much slower for B than for A.

17.37

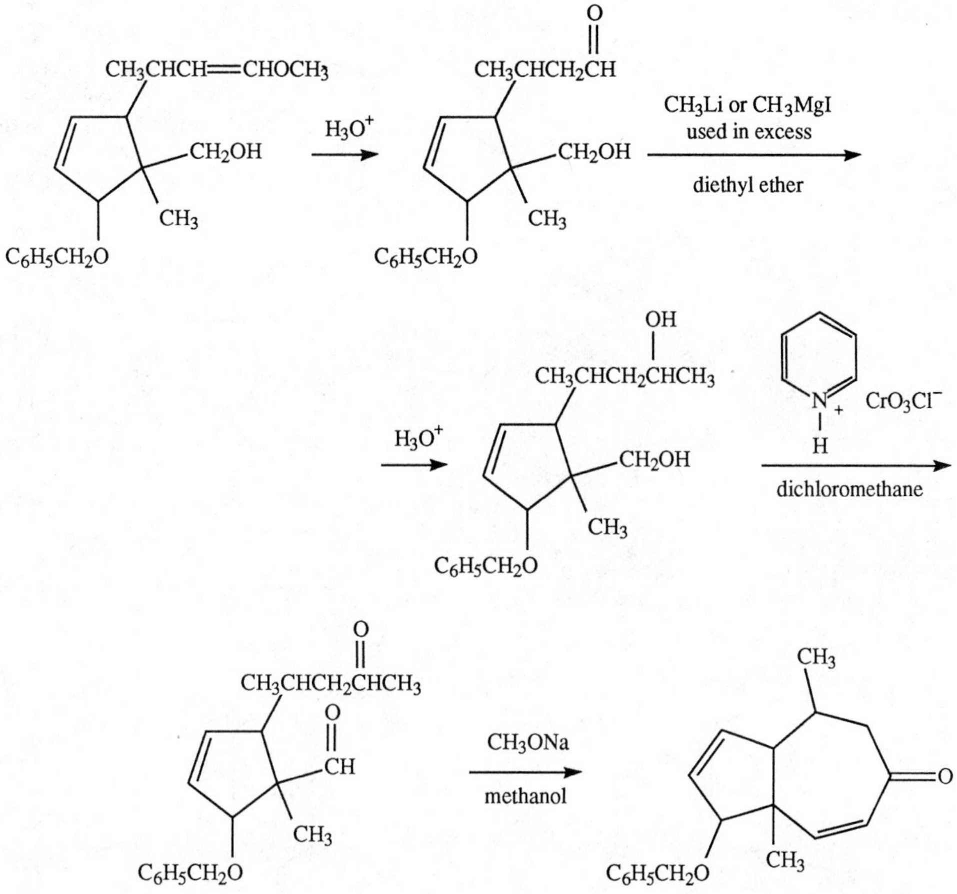

$$\text{CH}_3\text{CHCH}=\text{CHOCH}_3 \qquad \overset{\overset{\displaystyle O}{\parallel}}{\text{CH}_3\text{CHCH}_2\text{CH}}$$

Another product with two fused five-membered rings is theoretically possible. The structure shown for helenalin in the text indicates that the bonds joining the five-membered ring to the seven-membered ring are trans to each other. Such trans stereochemistry at the junction of two fused five-membered rings would create a lot of strain, but is all right at the junction of a five- and seven-membered ring. You may wish to prove this to yourself by building molecular models of both systems.

17.38

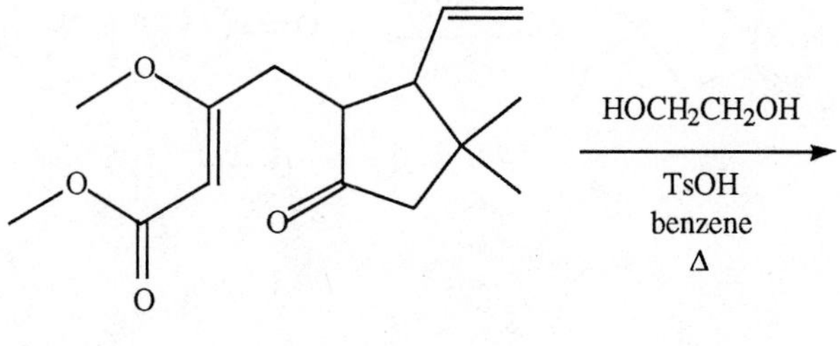

17.38 (cont)

17.39

17.40

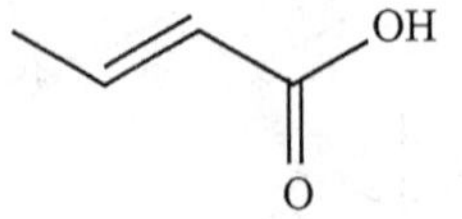

Compound B

λ_{max} 208 nm (ϵ 12,000)

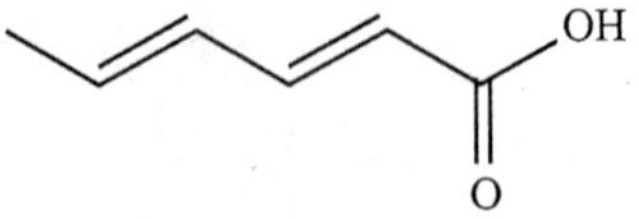

Compound C

λ_{max} 261 nm (ϵ 25,000)

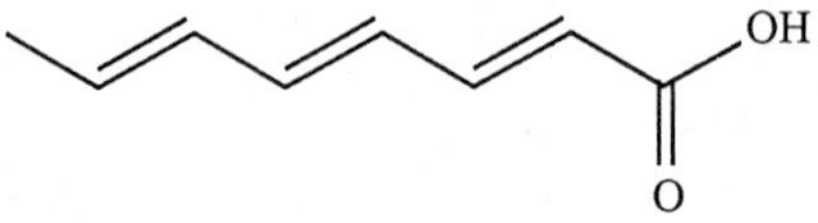

Compound A

λ_{max} 302 nm (ϵ 36,000)

The wavelength of maximum absorption and the intensity of absorption increases with increasing conjugation.

17.41   (a)   A compound with a molecular formula of C_4H_6O has two units of unsaturation.

(b)   The observed λ_{max} 219 nm (ϵ 16,6000) is due to the $\pi \rightarrow \pi^*$ transition of a conjugated system, while λ_{max} 318 nm (ϵ 30) is the $n \rightarrow \pi^*$ transition of a carbonyl group. Only three structures fit both the molecular formula and the ultraviolet spectrum.

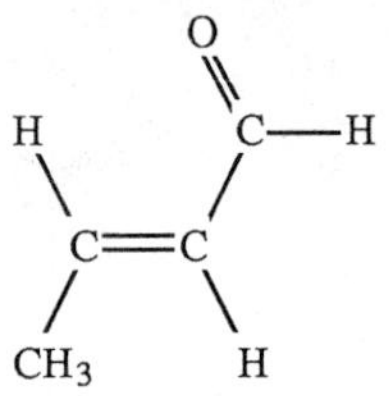

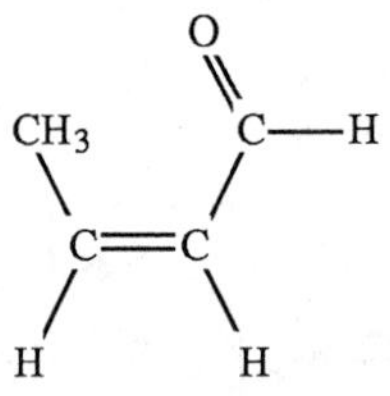

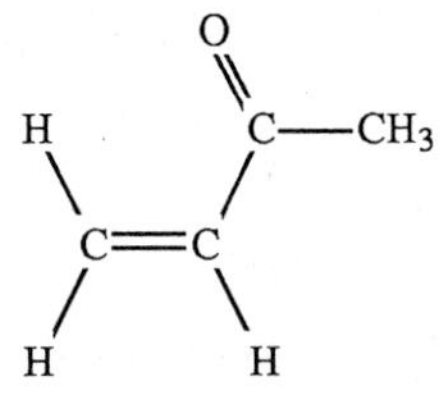

(c)   Some other structures that are not compatible with the data are:

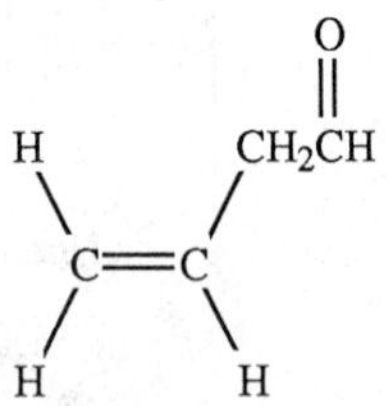

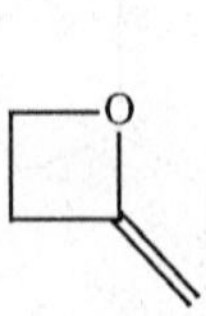

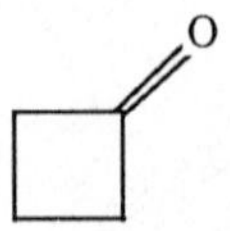

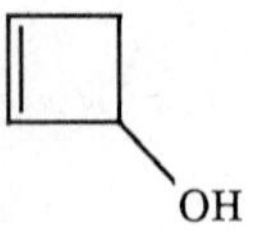

17.42   Of the three structures shown, the one below has the most highly substituted conjugated system and will, therefore, absorb at the highest wavelength.

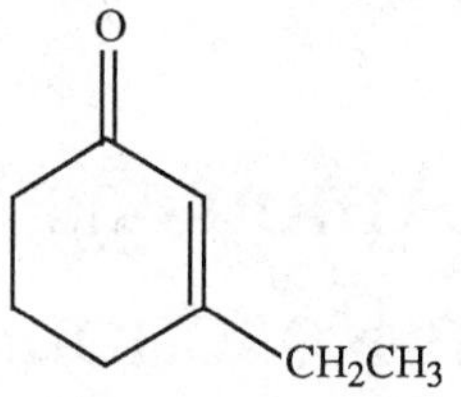

17.43   Of the three irones, β-irone absorbs at the highest wavelength. This suggests that β-irone has the most extended system of conjugation. Of the three structures shown, the one given below has the most extended conjugation and must, therefore, be β-irone.

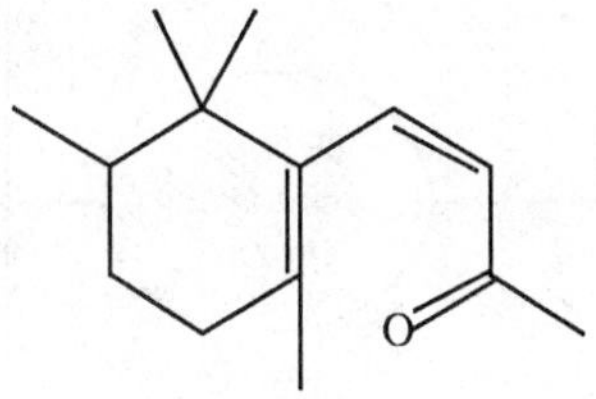

β-irone

17.44

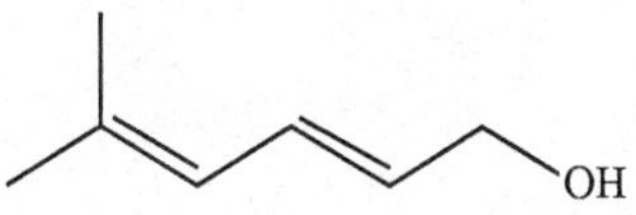

5-Methyl-2,4-hexadien-1-ol has the more highly substituted chromophore and will have the longer wavelength of absorption, 236 nm.

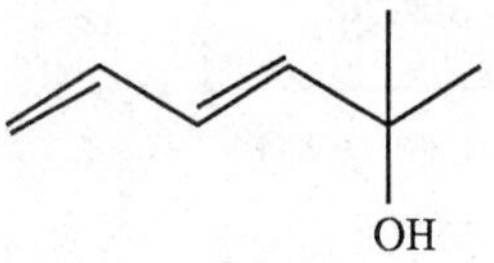

2-Methyl-3,5-hexadien-2-ol will absorb at 223 nm.

18

Enols and Enolate Anions as Nucleophiles II. Conjugate Addition Reactions; Ylides

Concept Map 18.1   Electrophilic alkenes.

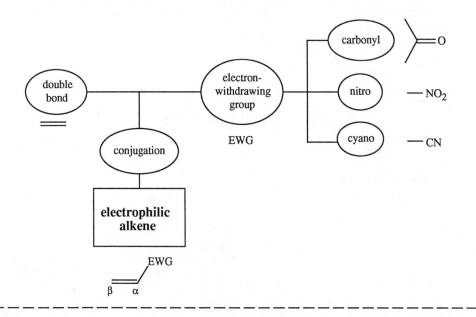

18.1

18.1 (cont)

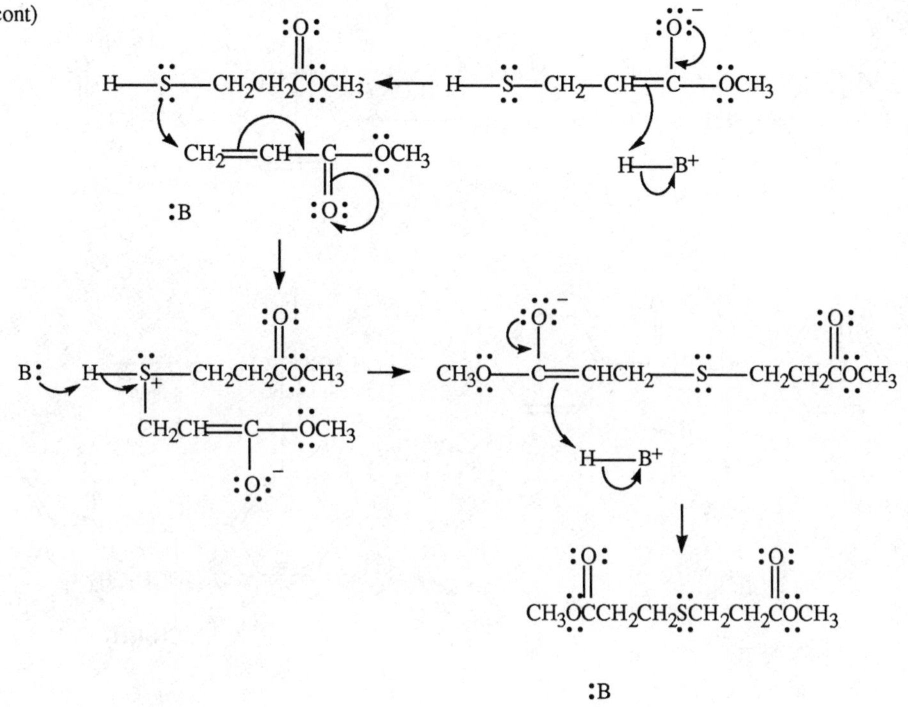

18.2

(a)

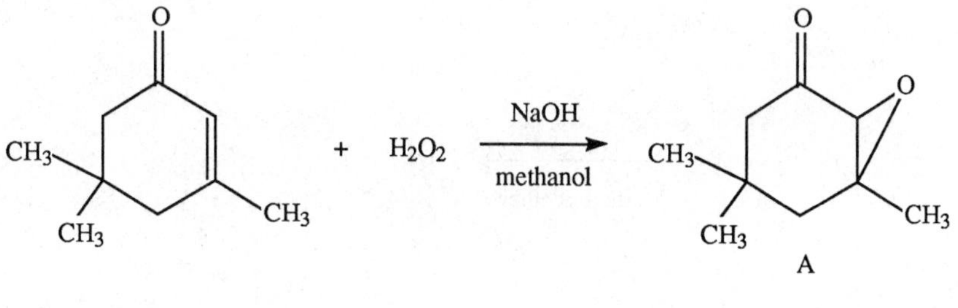

(b)

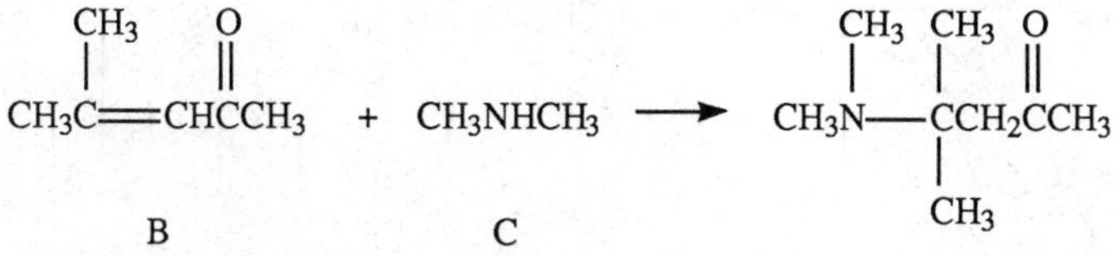

(c) $CH_2\!=\!CHC\!\equiv\!N$ + NH_3 $\xrightarrow{\ H_2O\ }$ $N\!\equiv\!CCH_2CH_2NHCH_2CH_2C\!\equiv\!N$

D

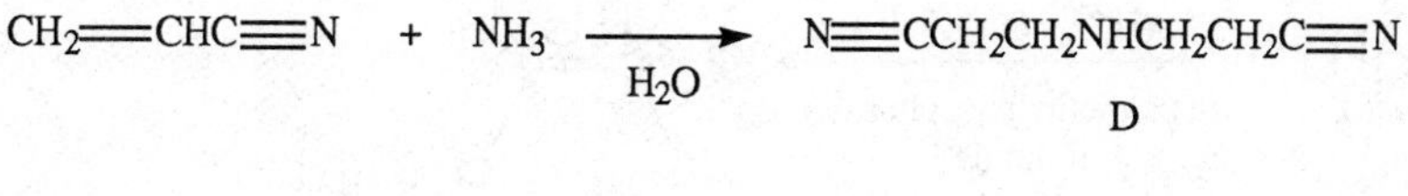

18.2 (cont)

(d)

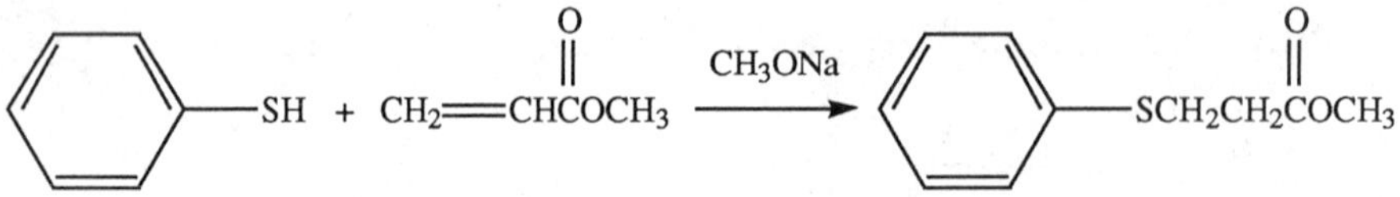

E   F

(e)

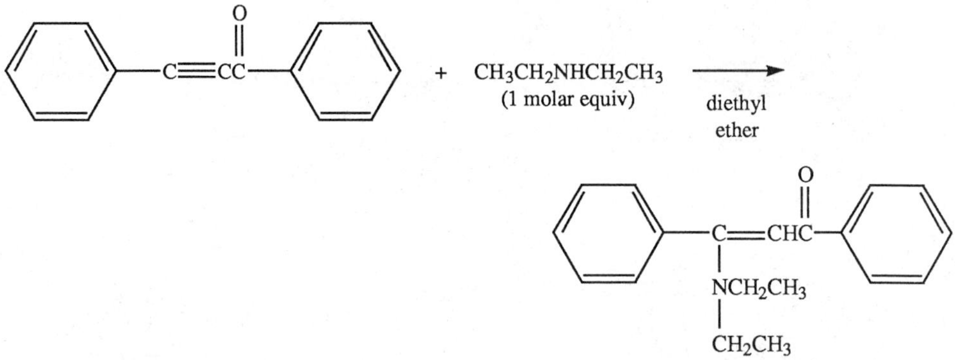

+ CH$_3$CH$_2$NHCH$_2$CH$_3$ ⟶
    (1 molar equiv) diethyl
          ether

G

18.3

(a) CH$_3$CH$_2$CH$_2$CH$_2$Li + CuI ⟶ (CH$_3$CH$_2$CH$_2$CH$_2$)$_2$CuLi
             diethyl ether
              0 °C

(b) CH$_3$CH$_2$CH$_2$CH$_2$CH$_2$I $\xrightarrow[\substack{\text{diethyl ether} \\ 25\,°C}]{\text{(CH}_3\text{)}_2\text{CuLi}}$ CH$_3$CH$_2$CH$_2$CH$_2$CH$_2$CH$_3$

(c) CH$_3$(CH$_2$)$_6$CH$_2$Cl $\xrightarrow[\substack{\text{hexamethylphosphoric} \\ \text{triamide} \\ 25\,°C}]{}$

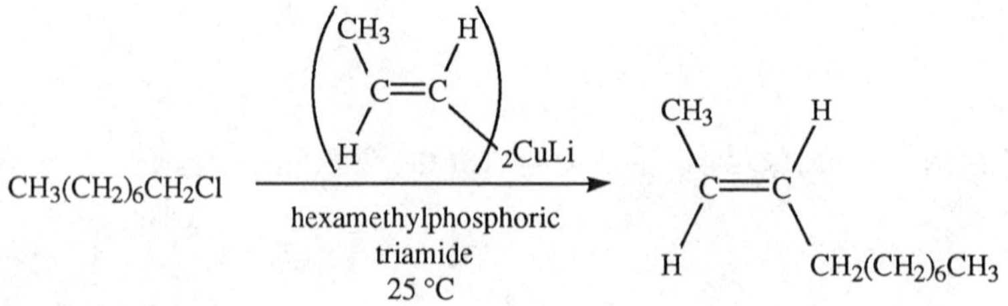

18.3 (cont)

(d)

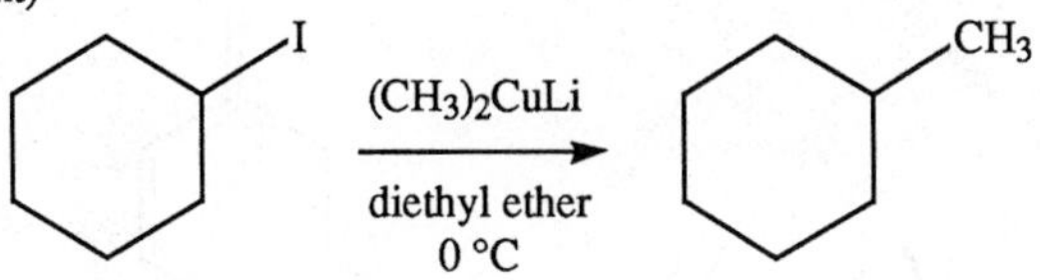

(e)

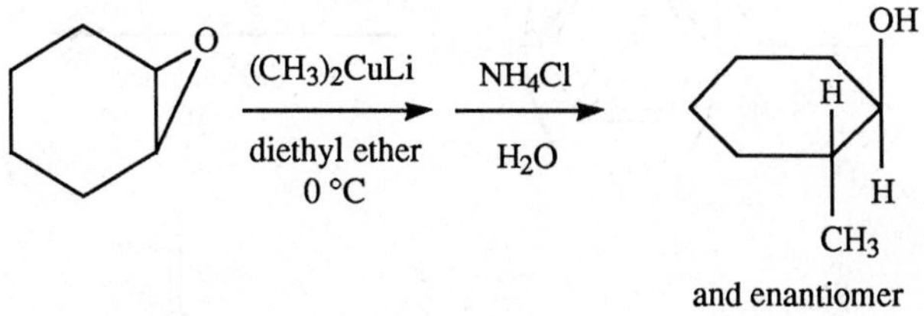

and enantiomer

18.4

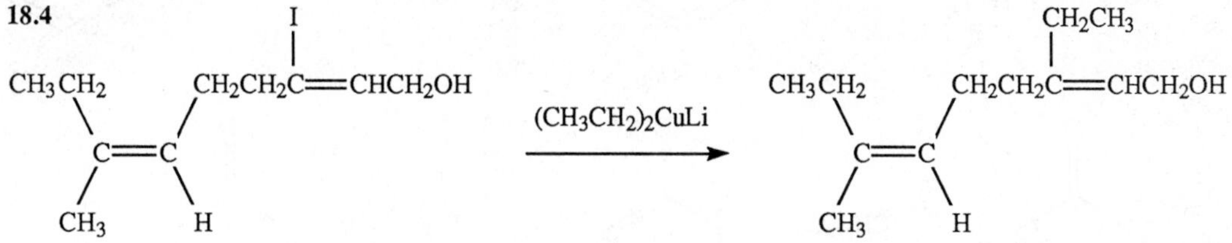

18.5

(a)

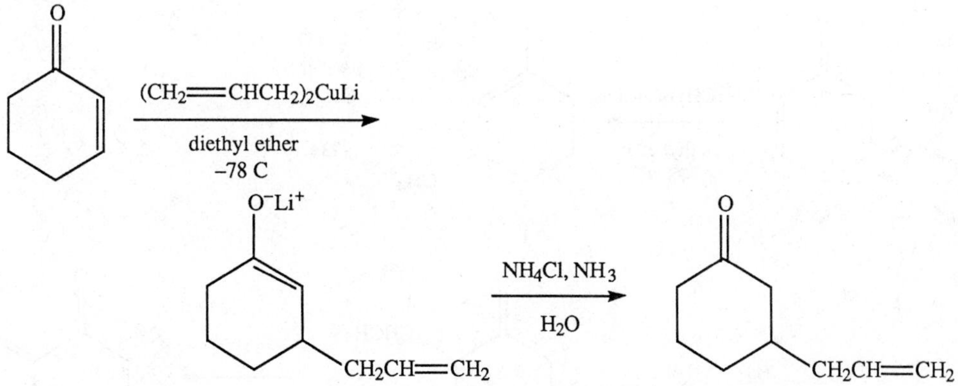

18.5 (cont)

(b)

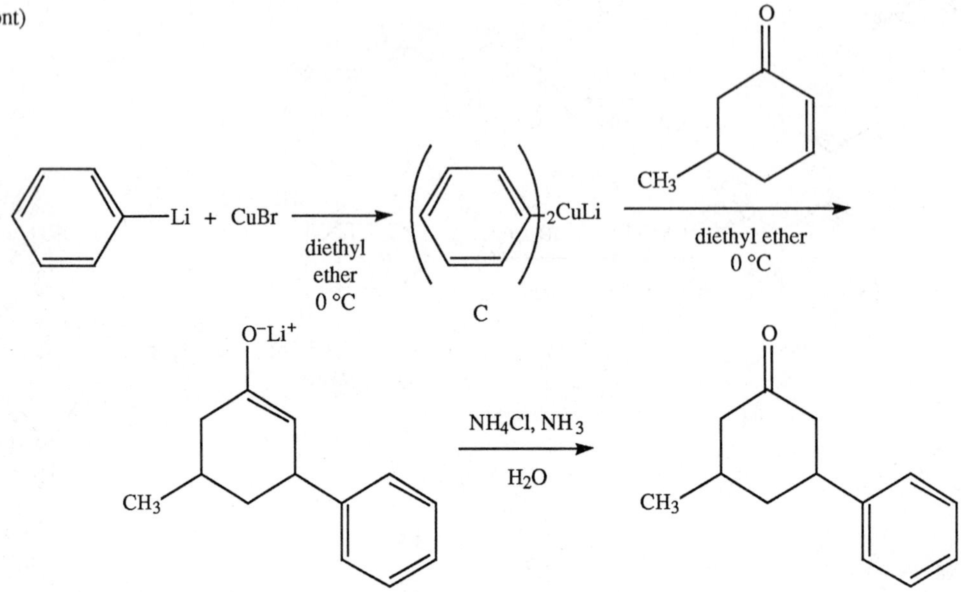

C

D

NH$_4$Cl, NH$_3$

H$_2$O

E

(c)

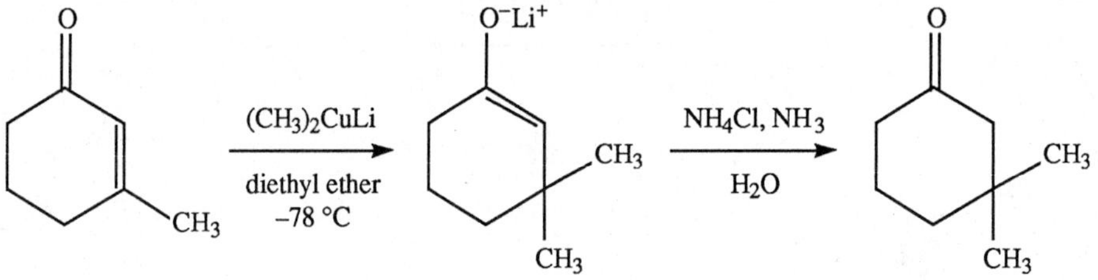

F

(CH$_3$)$_2$CuLi

diethyl ether
−78 °C

G

NH$_4$Cl, NH$_3$

H$_2$O

(d)

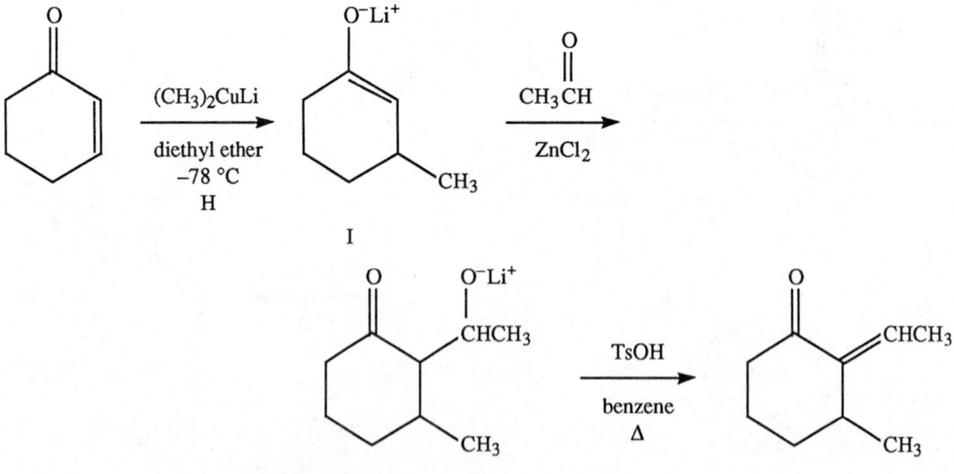

F

(CH$_3$)$_2$CuLi

diethyl ether
−78 °C

H

I

CH$_3$CH

ZnCl$_2$

J

TsOH

benzene
Δ

18.6

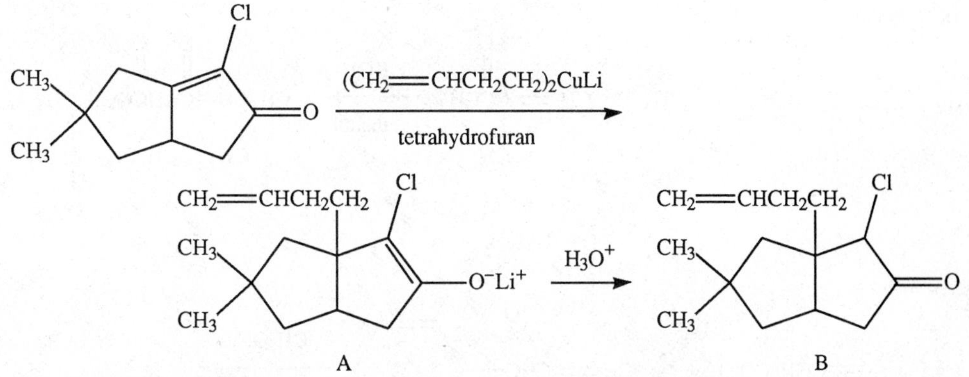

18.7 $RSH + CH_3(CH_2)_8\overset{O}{\overset{\|}{C}}CH{=}CH_2 \longrightarrow CH_3(CH_2)_8\overset{O}{\overset{\|}{C}}CH_2CH_2SR$

18.8

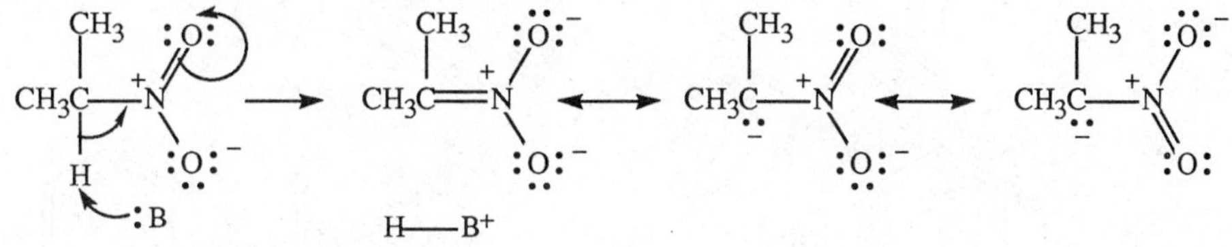

18.9

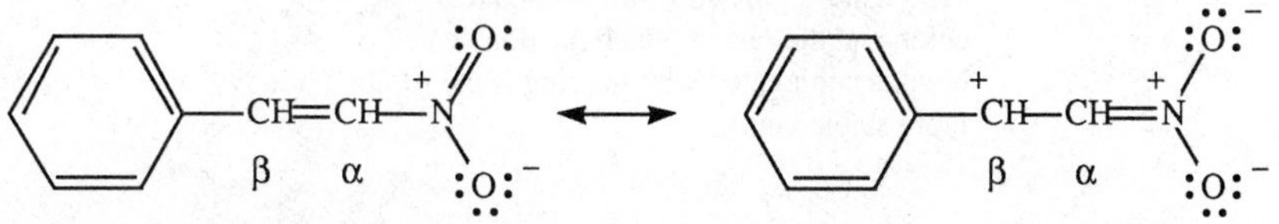

partial positive charge on β-carbon atom

18.10

(a) $CH_2(C{\equiv}N)_2 + 2\ CH_2{=}CH\overset{O}{\overset{\|}{C}}CH_3 \xrightarrow[\text{benzene}]{\text{Na}} CH_3\overset{O}{\overset{\|}{C}}CH_2CH_2C(C{\equiv}N)_2CH_2CH_2\overset{O}{\overset{\|}{C}}CH_3$

18.10 (cont)

(b)

$$\underset{\overset{|}{CH_3}}{CH_3CHC}\overset{\overset{O}{\|}}{C}\overset{\overset{O}{\|}}{CH_2C}OCH_2CH_3 \ + \ CH_2\!\!=\!\!\overset{\overset{O}{\|}}{\underset{\overset{|}{CH_3}}{C}C}CH_3 \ \xrightarrow[\text{ethanol}]{\text{KOH}} \ $$

$$\underset{\overset{|}{CH_3}}{CH_3CH}\!\!-\!\!\overset{\overset{O}{\|}}{C}\underset{\overset{|}{CH_2CH}\!-\!\overset{\overset{O}{\|}}{C}CH_3}{CHC}\overset{\overset{O}{\|}}{O}CH_2CH_3$$
$$\underset{\overset{|}{CH_3}\ \ O}{}$$

(c)

$$N\!\!\equiv\!\!CCH_2\overset{\overset{O}{\|}}{C}OCH_3 \ + \ CH_2\!\!=\!\!CH\overset{\overset{O}{\|}}{C}\!-\!\!\bigcirc \ \xrightarrow[\text{methanol}]{CH_3ONa}$$

$$\bigcirc\!\!-\!\!\overset{\overset{O}{\|}}{C}CH_2CH_2\underset{\overset{|}{\underset{\overset{|}{N}}{\overset{\|}{C}}}}{CH}\overset{\overset{O}{\|}}{C}OCH_3$$

(d)

$$\xrightarrow{NaNH_2} \quad CH_2\!\!=\!\!CHC\!\!\equiv\!\!N \xrightarrow{\quad}$$

(The more highly substituted enolate
anion and the one in which the double
bond is conjugated with the ring is the
more stable one.)

with product: cyclohexanone bearing phenyl and $CH_2CH_2C\!\!\equiv\!\!N$ group.

(e)

$$\bigcirc\!\!-\!\!\underset{\overset{|}{\underset{\overset{|}{N}}{\overset{\|}{C}}}}{CH}\overset{\overset{O}{\|}}{C}OCH_2CH_3 \ + \ CH_2\!\!=\!\!CH\overset{\overset{O}{\|}}{C}CH_3 \ \xrightarrow[\text{benzene}]{Na}$$

$$\bigcirc\!\!-\!\!\underset{\overset{|}{\underset{\overset{|}{O}}{\overset{\|}{C}}OCH_2CH_3}}{\overset{\overset{\overset{N}{\|}}{C}}{C}}CH_2CH_2\overset{\overset{O}{\|}}{C}CH_3$$

18.10 (cont)

(f) 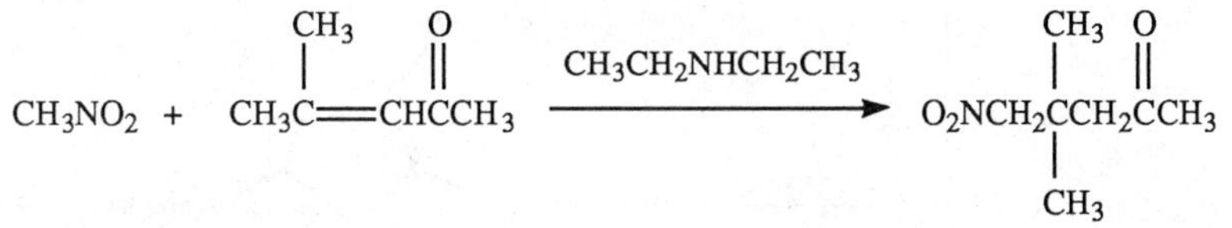   CH$_3$NO$_2$ + CH$_3$C=CHCCH$_3$ $\xrightarrow{\text{CH}_3\text{CH}_2\text{NHCH}_2\text{CH}_3}$ O$_2$NCH$_2$CCH$_2$CCH$_3$

(g)

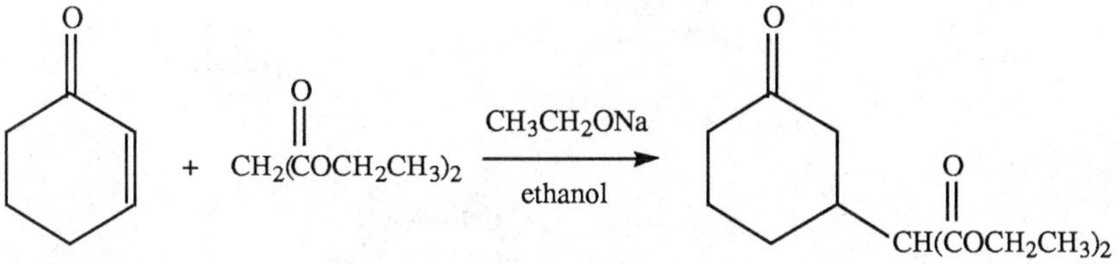

18.11

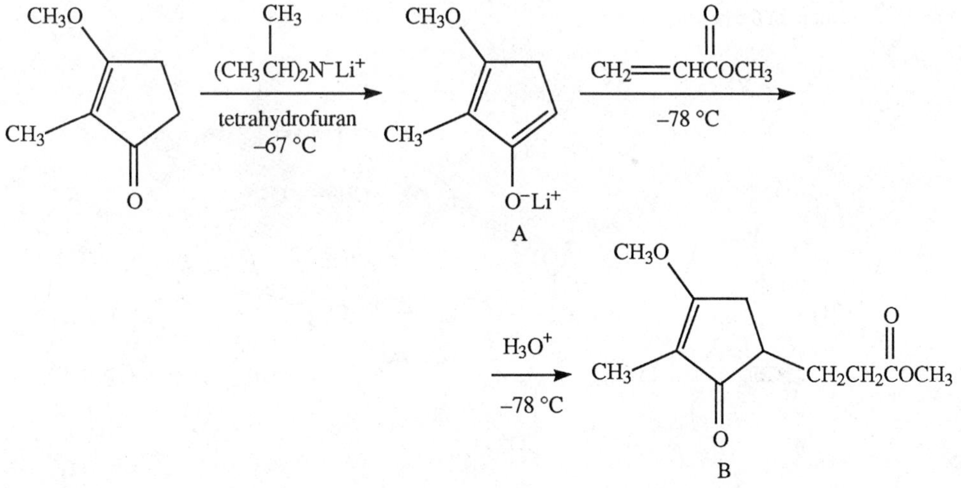

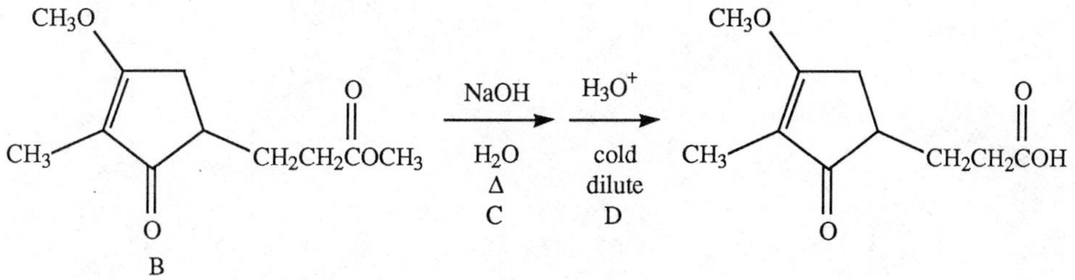

18.12   1.

The carbon atom of the carbonyl group in the carboxylate anion is much less electrophilic than the carbon atom of the ketone carbonyl.

2.

18.12   2. (cont)

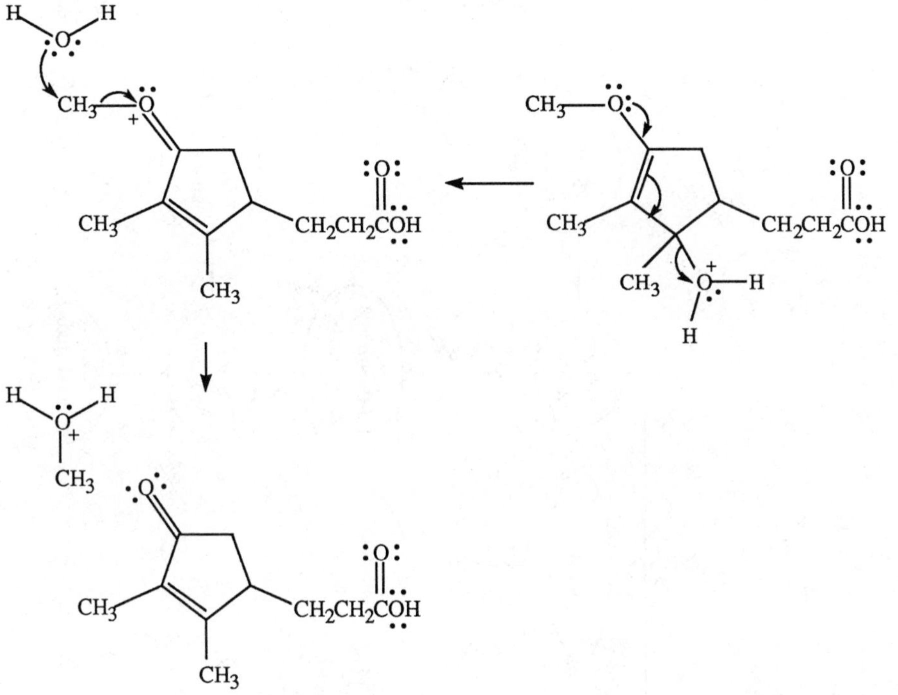

The same result can be obtained by hydrolyzing the enol ether (Problem 13.13), then dehydrating the alcohol.

18.13

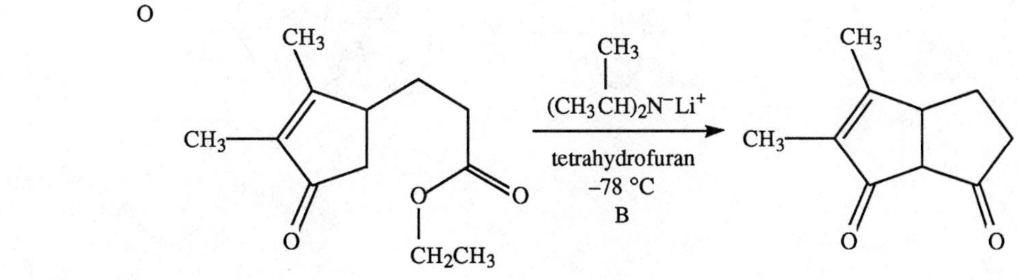

Concept Map 18.2   Reactions of electrophilic alkenes.

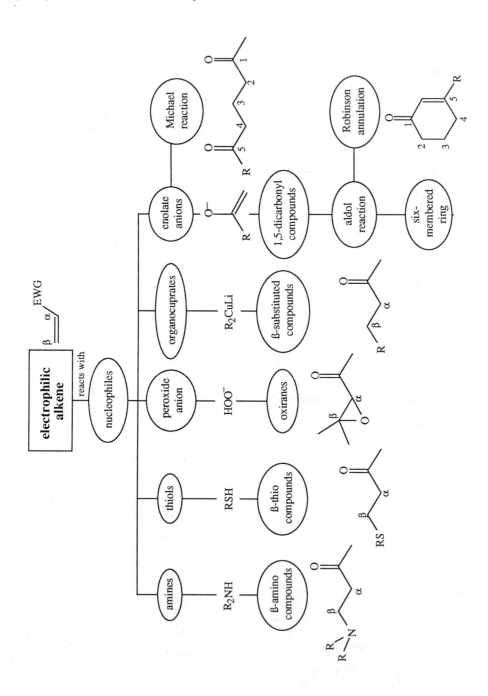

18.14

(a)

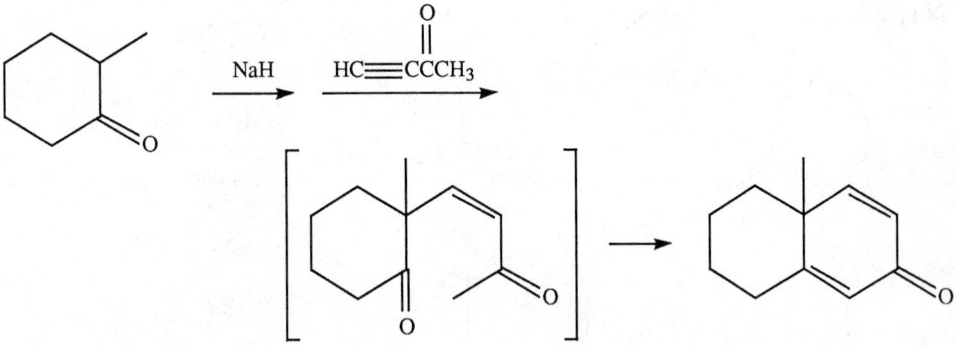
NaH
O
HC≡CCCH3
O
O
O

(b)

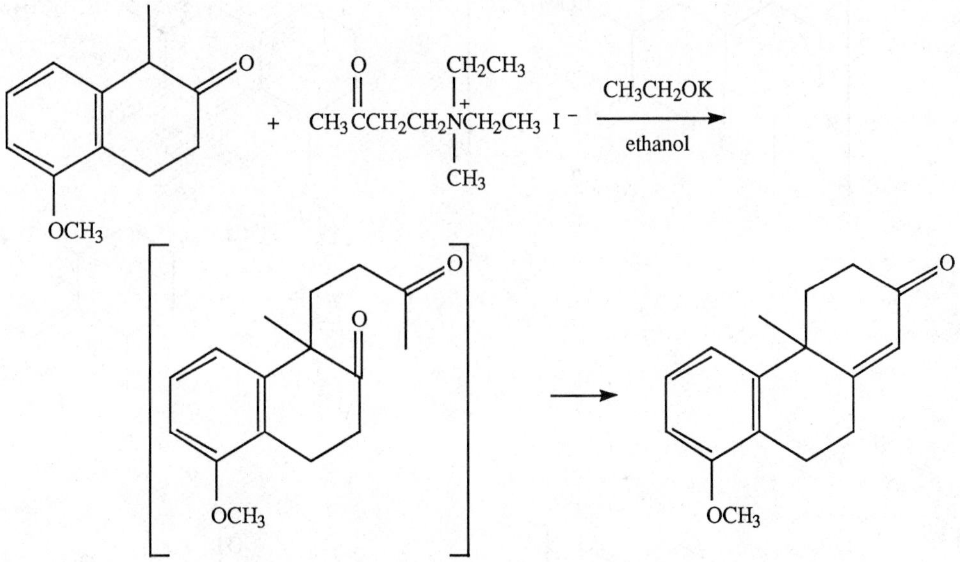
O
OCH3
+
O
CH2CH3
CH3CCH2CH2NCH2CH3 I−
CH3
CH3CH2OK
ethanol
O
O
OCH3
O
OCH3

(c)

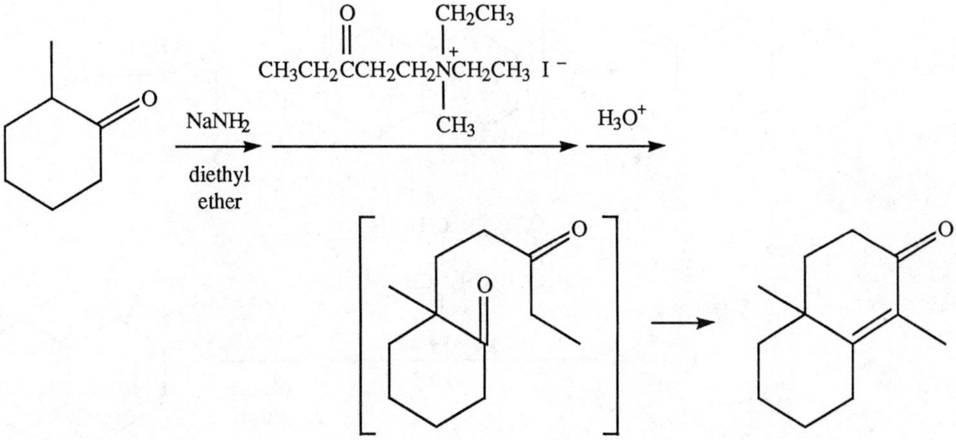
O
NaNH2
diethyl
ether
O
CH2CH3
CH3CH2CCH2CH2NCH2CH3 I−
CH3
H3O+
O
O
O

18.14 (cont)

(d)

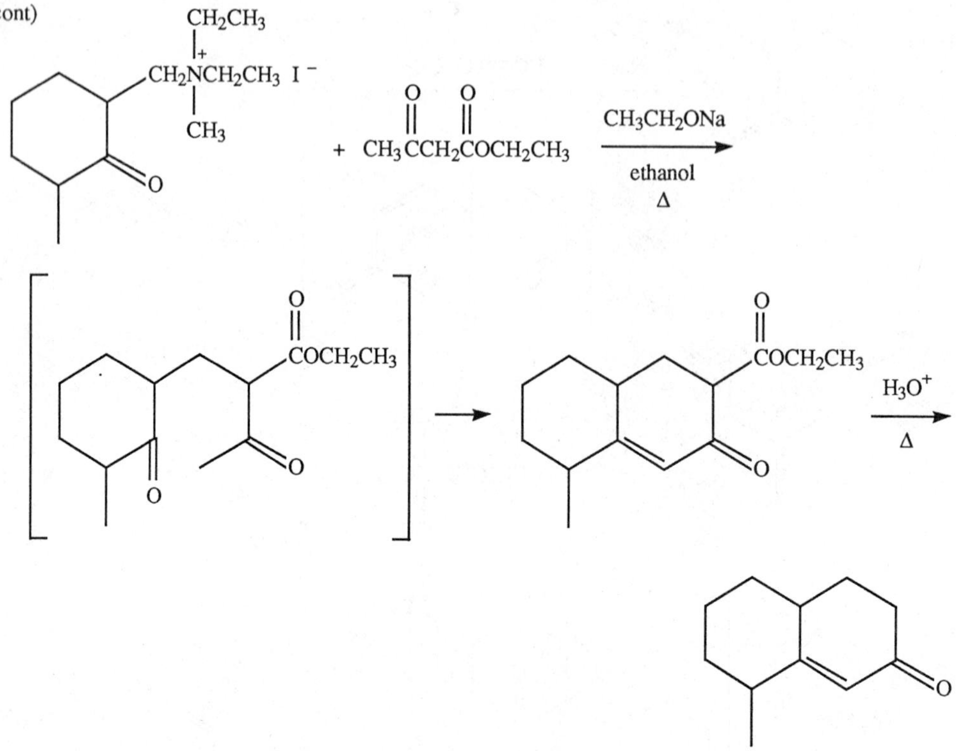

$$+ \quad CH_3\overset{O}{\underset{\|}{C}}CH_2\overset{O}{\underset{\|}{C}}OCH_2CH_3 \quad \xrightarrow[\substack{\text{ethanol} \\ \Delta}]{CH_3CH_2ONa}$$

$\xrightarrow[\Delta]{H_3O^+}$

18.15

$\xleftarrow{H_3O^+}$

$\xleftarrow[\substack{\text{diethyl} \\ \text{ether}}]{NaNH_2}$

$$CH_3\overset{O}{\underset{\|}{C}}CH_2CH_2\overset{\overset{CH_2CH_3}{|+}}{N}CH_2CH_3 \ \ I^-$$
$$\underset{|}{}$$
$$CH_3$$

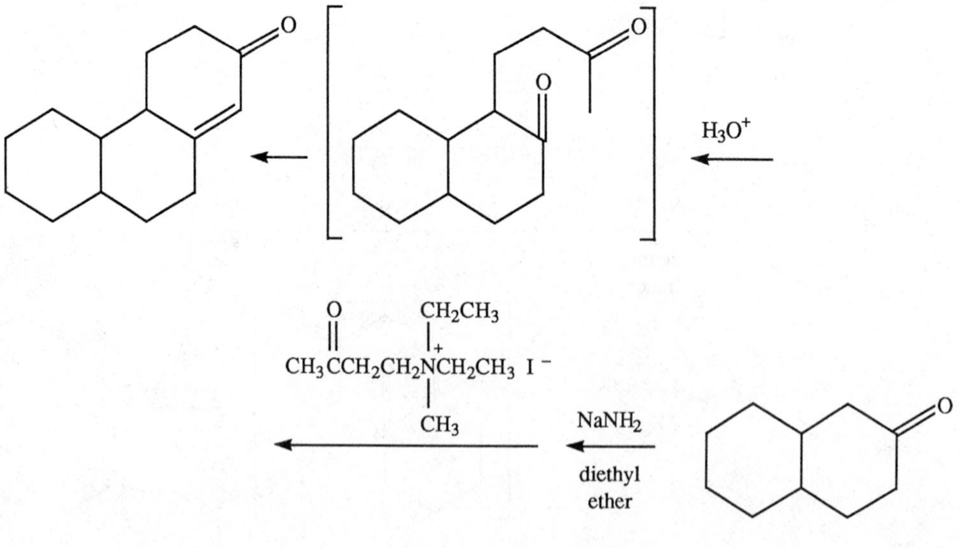

Concept Map 18.3   Ylides.

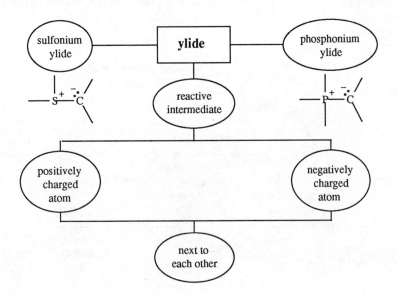

Concept Map 18.4   The Wittig reaction.

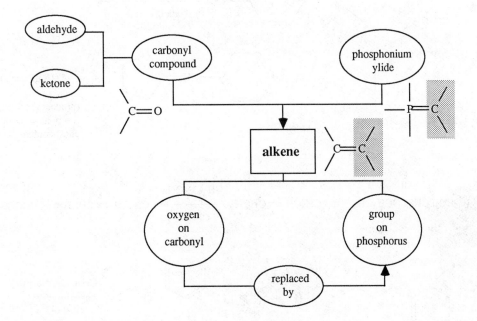

18.16 A carbon atom can be introduced by the reaction of an organometallic reagent such as a Grignard reagent with a carbonyl group (Section 13.6). A double bond can be introduced via an E_2 elimination (Section 7.7B)

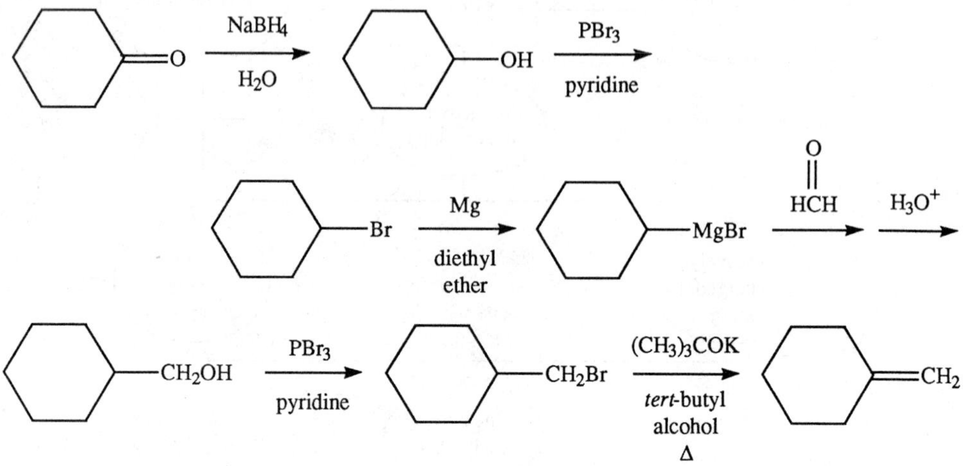
NaBH4
H2O
OH
PBr3
pyridine
Br
Mg
diethyl
ether
MgBr
HCH
H3O+
CH2OH
PBr3
pyridine
CH2Br
(CH3)3COK
tert-butyl
alcohol
Δ
CH2

18.17

(a)

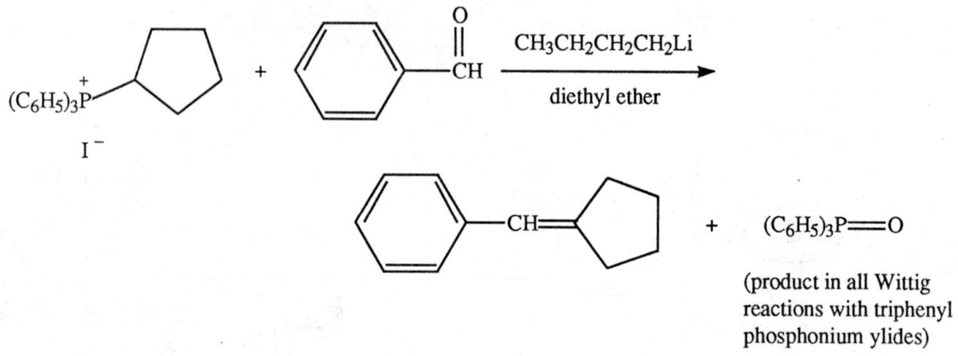
(C6H5)3P
I−
CH
CH3CH2CH2CH2Li
diethyl ether
CH
(C6H5)3P=O
(product in all Wittig
reactions with triphenyl
phosphonium ylides)

(b)

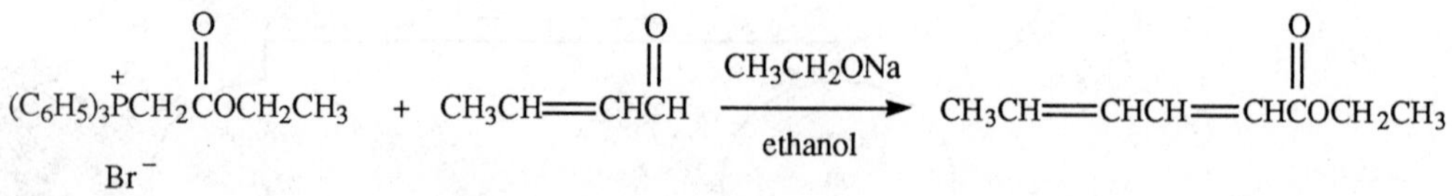
(C6H5)3PCH2COCH2CH3
Br−
CH3CH=CHCH
CH3CH2ONa
ethanol
CH3CH=CHCH=CHCOCH2CH3

(c)

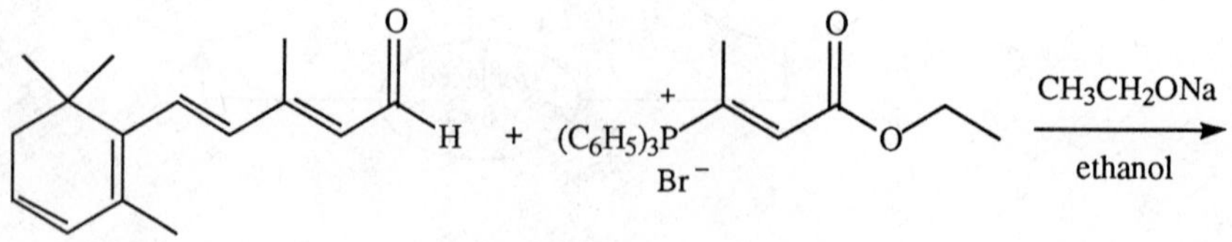
H
(C6H5)3P
Br−
CH3CH2ONa
ethanol

18.17 (cont)

(c) (cont)

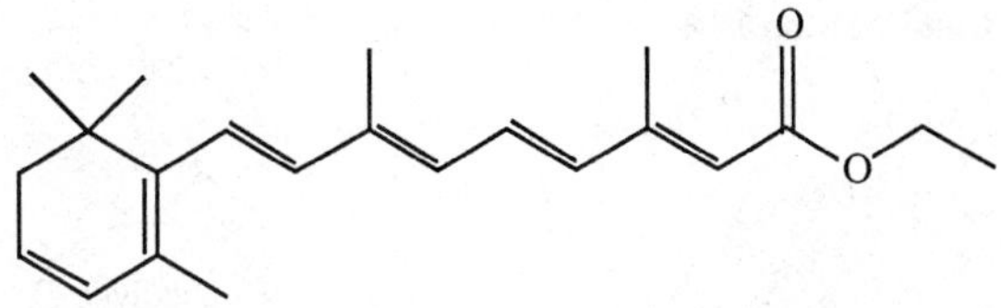

(d)

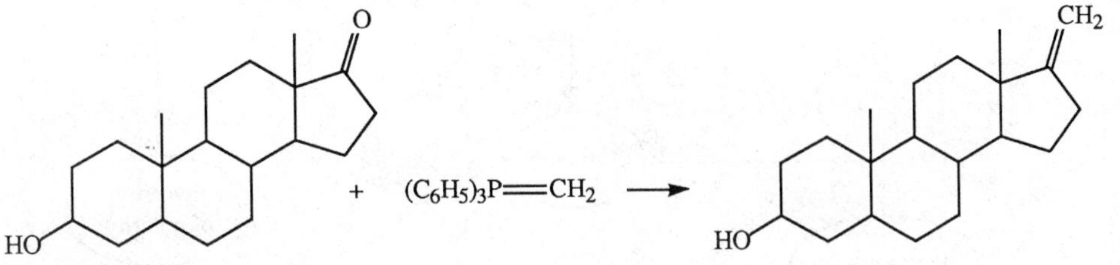

18.18

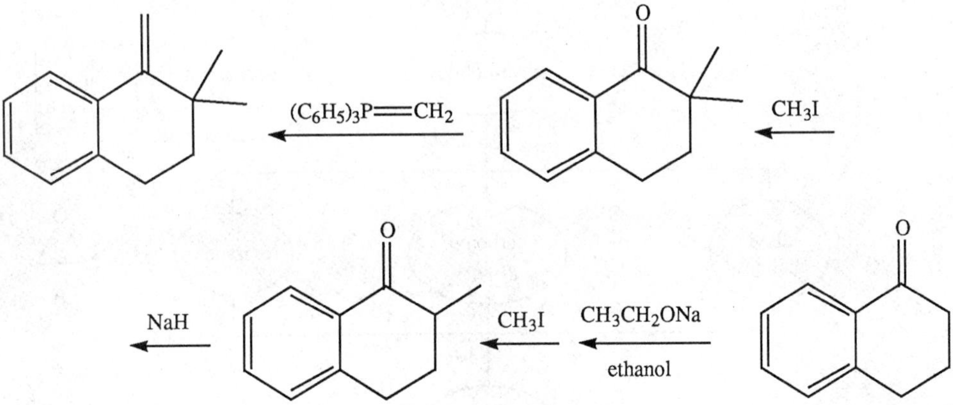

Concept Map 18.5   Dithiane anions.

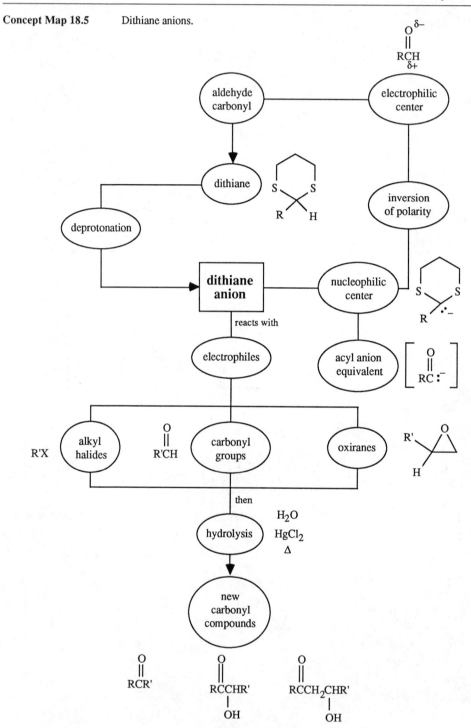

18.19

(a)

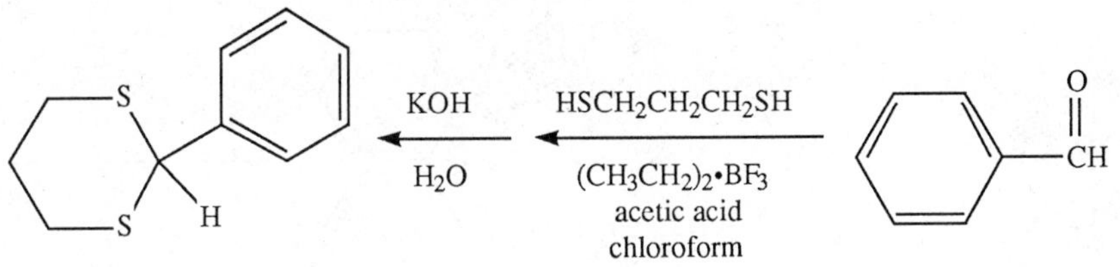

(b)

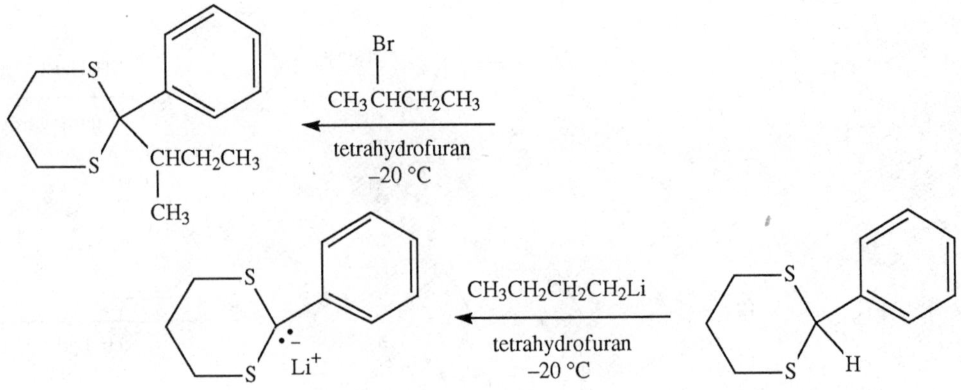

(c)

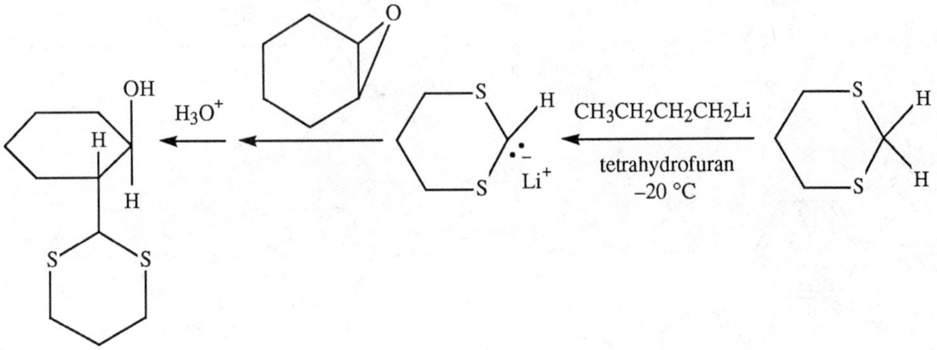

and enantiomer

(d)

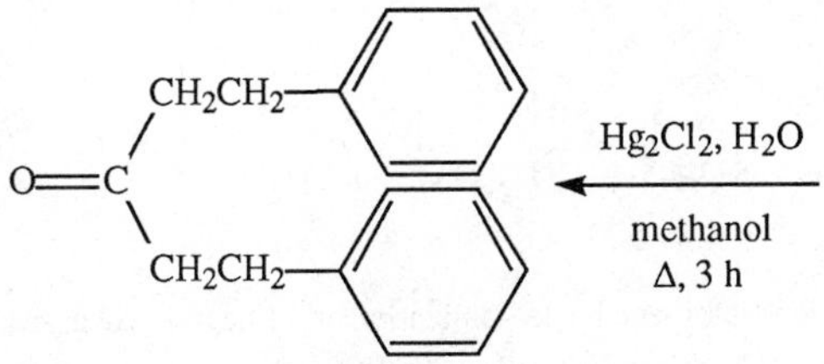

18.19 (cont)

(d) (cont)

$$\xleftarrow[\begin{array}{c}\text{tetrahydrofuran}\\ -20\ ^\circ\text{C}\end{array}]{}$$

$$\xleftarrow[\begin{array}{c}\text{tetrahydrofuran}\\ -20\ ^\circ\text{C}\end{array}]{\text{CH}_3\text{CH}_2\text{CH}_2\text{CH}_2\text{Li}}$$

$$\xleftarrow[\begin{array}{c}\text{tetrahydrofuran}\\ -20\ ^\circ\text{C}\end{array}]{}$$

$$\xleftarrow[\begin{array}{c}\text{tetrahydrofuran}\\ -20\ ^\circ\text{C}\end{array}]{\text{CH}_3\text{CH}_2\text{CH}_2\text{CH}_2\text{Li}} \quad \xleftarrow[\begin{array}{c}\text{H}_2\text{O}\end{array}]{\text{KOH}} \quad \xleftarrow[\begin{array}{c}(\text{CH}_3\text{CH}_2)_2\cdot\text{BF}_3\\ \text{acetic acid}\\ \text{chloroform}\end{array}]{\text{HSCH}_2\text{CH}_2\text{CH}_2\text{SH}} \quad \begin{array}{c}\text{O}\\ \|\\ \text{HCH}\end{array}$$

(e)

$$\xleftarrow[\text{tetrahydrofuran}]{\begin{array}{c}\text{CH}_3\\ |\ +\\ \text{CH}_3\text{S}\!-\!\bar{\text{C}}\text{H}_2\end{array}}$$

18.20   1. $$\text{CH}_3\text{SCH}_3 \ + \ \text{NaH} \ \longrightarrow \ \text{CH}_3\overset{\text{O}}{\overset{\|}{\text{S}}}\!-\!\ddot{\bar{\text{C}}}\text{H}_2\ \text{Na}^+ \ + \ \text{H}_2{\uparrow}$$

    2.   The anion formed from dimethyl sulfoxide is stabilized by delocalization of the negative charge to oxygen.

$$\text{CH}_3\overset{:\ddot{\text{O}}:}{\underset{\ddot{}}{\overset{\|}{\text{S}}}}\!-\!\ddot{\text{C}}\text{H}_2 \quad\longleftrightarrow\quad \text{CH}_3\overset{:\ddot{\text{O}}:^-}{\underset{\ddot{}}{\text{S}}}\!=\!\text{CH}_2$$

18.20 (cont)

3. The anion from dimethyl sulfoxide resembles an enolate anion.

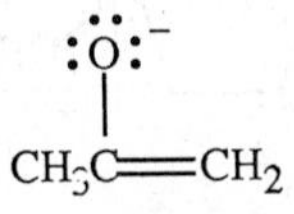

enolate anion
from acetone

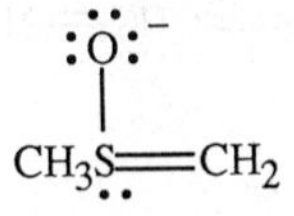

anion from di-
methyl sulfoxide

4.

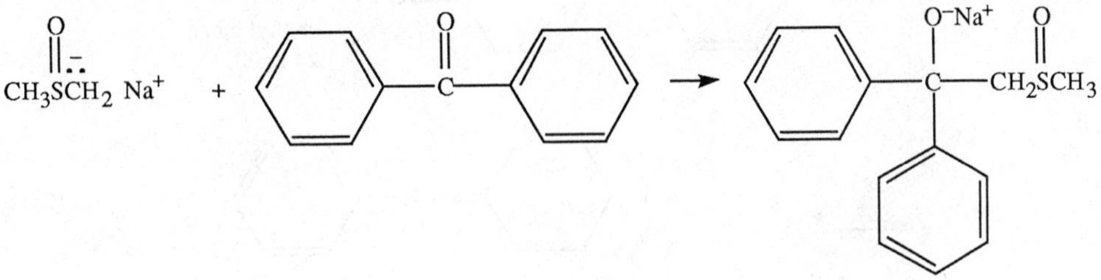

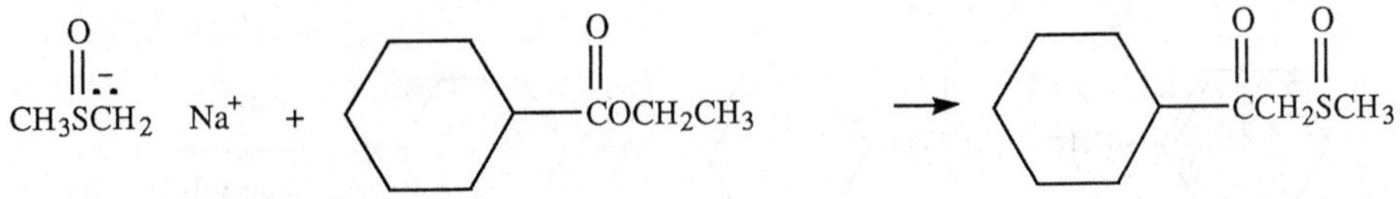

18.21

(a)

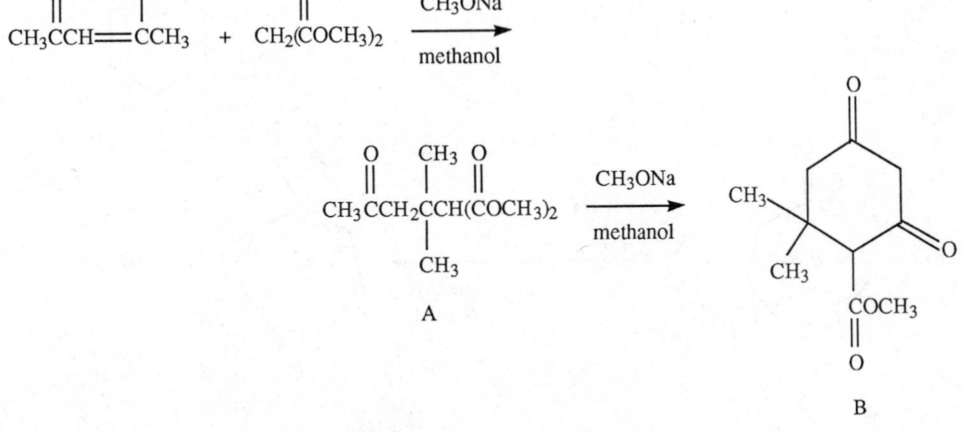

18.21 (cont)

(b)

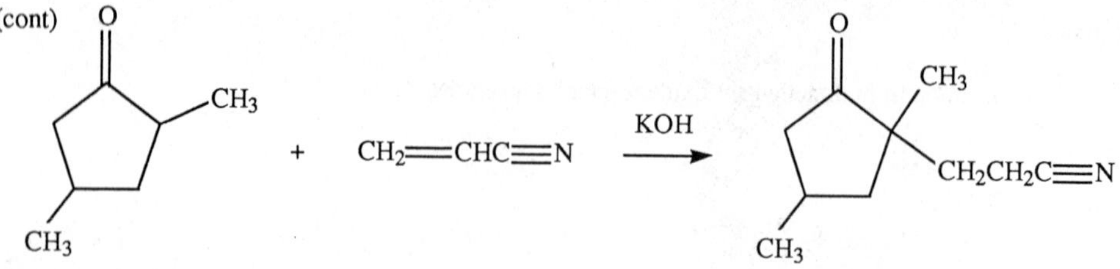

C

(c) $(C_6H_5)_3P$=CH—

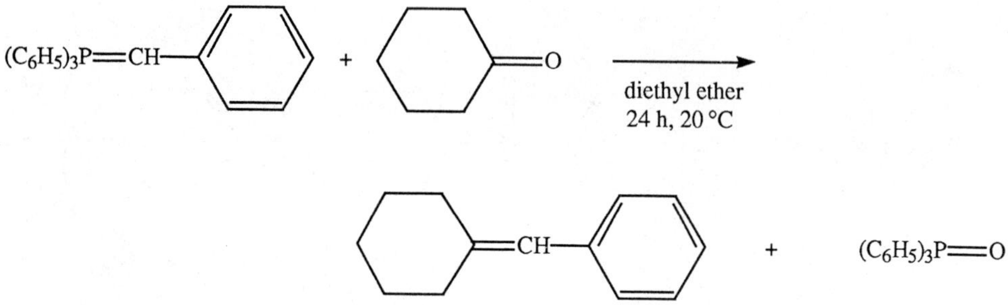

diethyl ether
24 h, 20 °C

D   + $(C_6H_5)_3P$=O

E

(d)

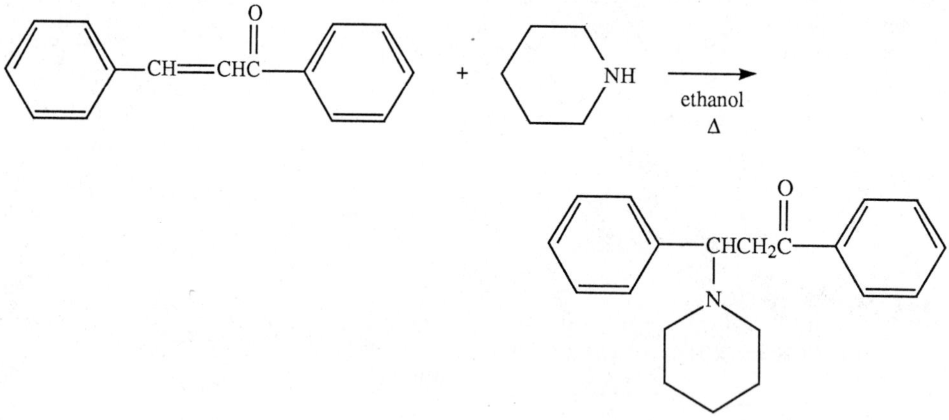

+

ethanol
Δ

F

(e)

+   CH₃S—CH₂

dimethyl
sulfoxide

+ CH_3SCH_3

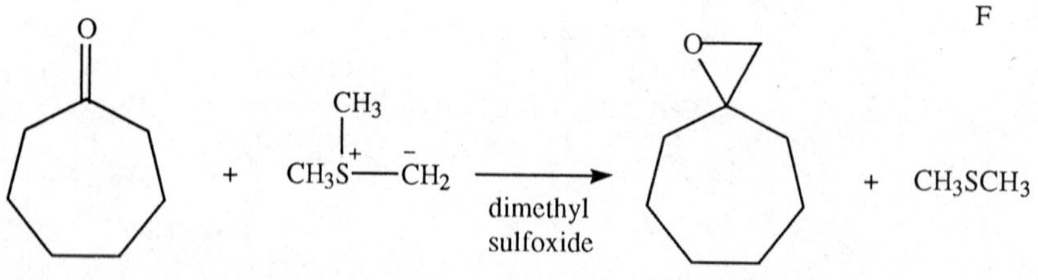

G   H

18.21 (cont)

(f)

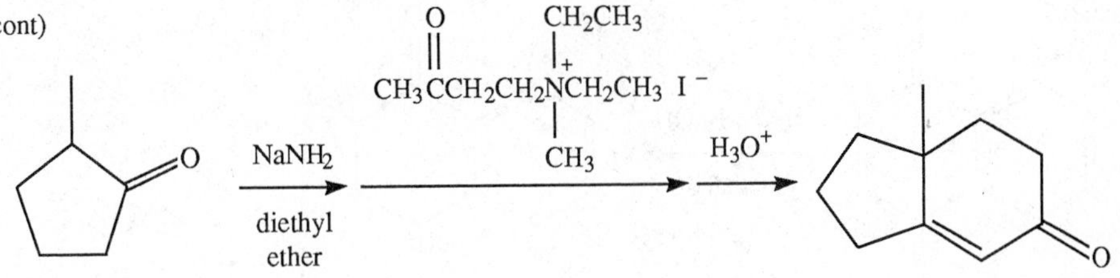

(g)

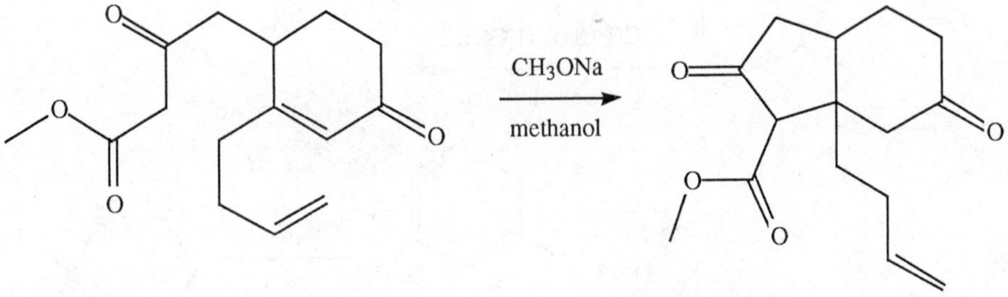

(h)

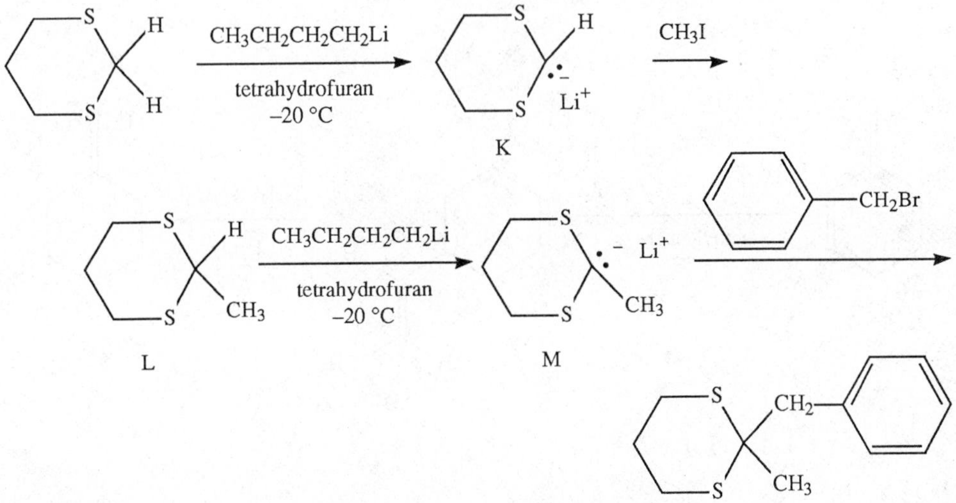

(i) $(C_6H_5)_3P$ + $BrCH_2CH\!=\!CHCOCH_2CH_3$ $\xrightarrow{\text{benzene}}$

$$\underset{O}{\underset{|}{(C_6H_5)_3\overset{+}{P}CH_2CH\!=\!CHCOCH_2CH_3}}\ \overset{Br^-}{}\ \xrightarrow[\text{H}_2\text{O}]{\text{NaOH}}\ (C_6H_5)_3P\!=\!CHCH\!=\!CHCOCH_2CH_3$$

O   P

18.21 (cont)

(j)

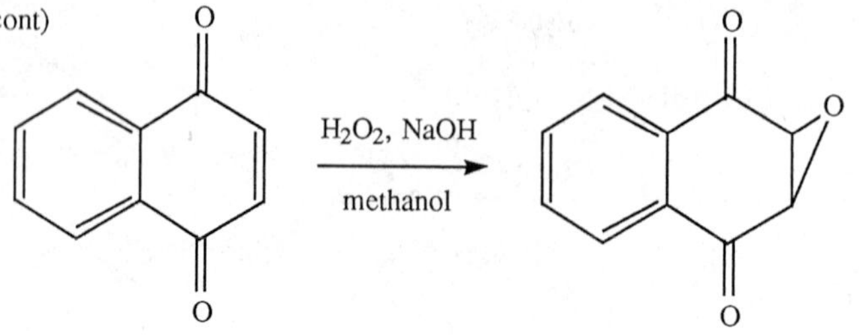

Q

(k)

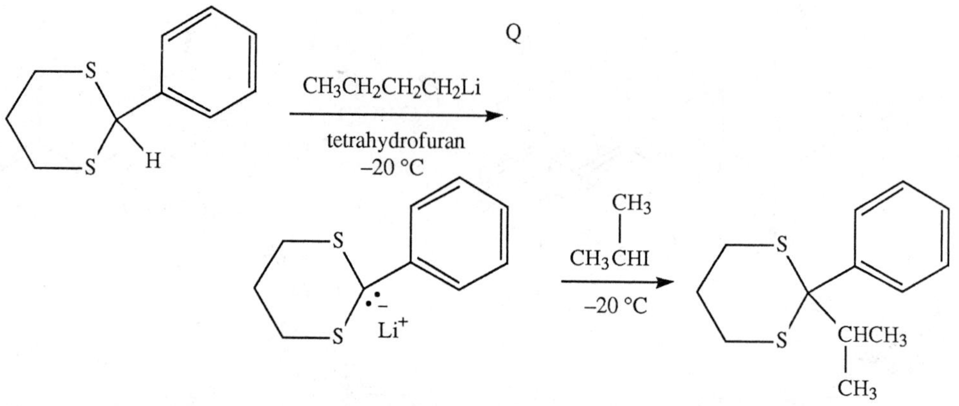

R

S

(l)

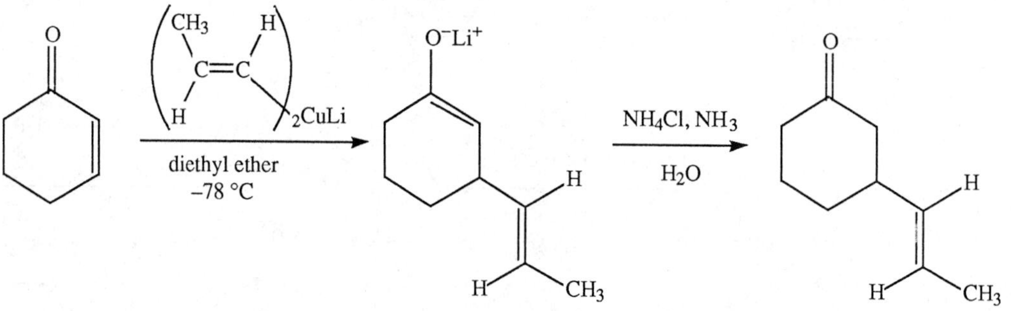

T

U

18.22

(a)

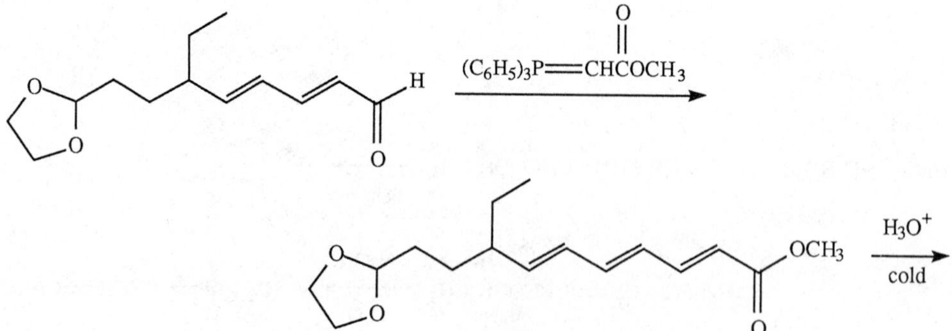

A

18.22 (cont)

(a) (cont)

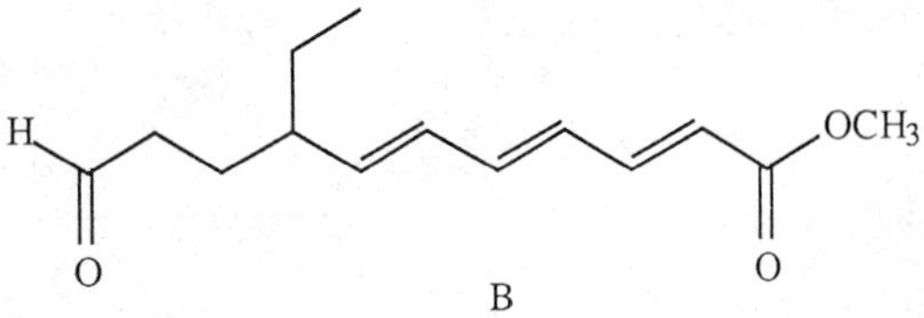

B

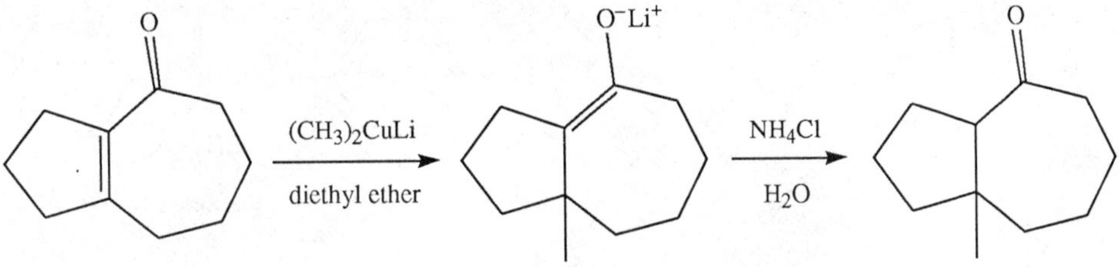

(b)

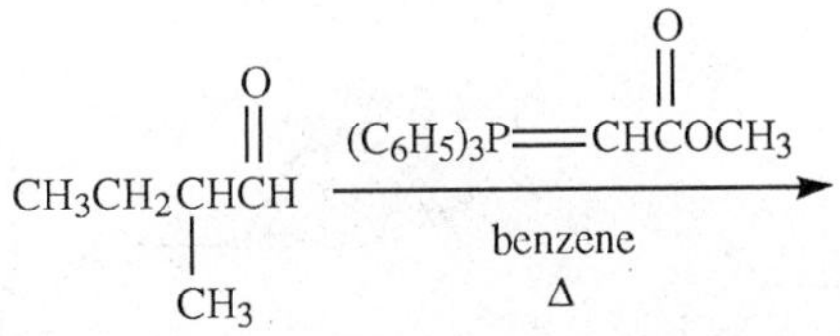

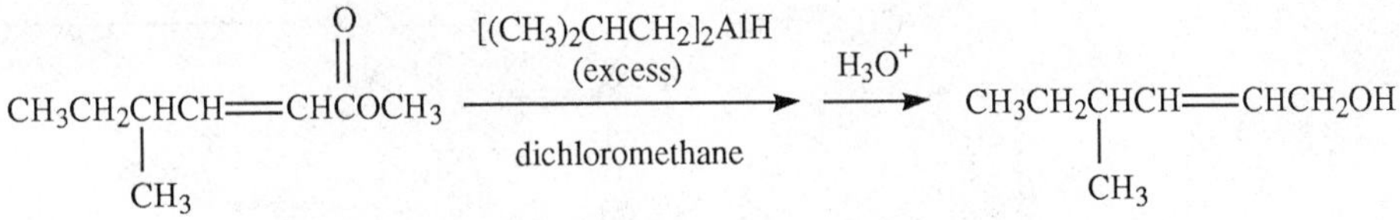

(c)

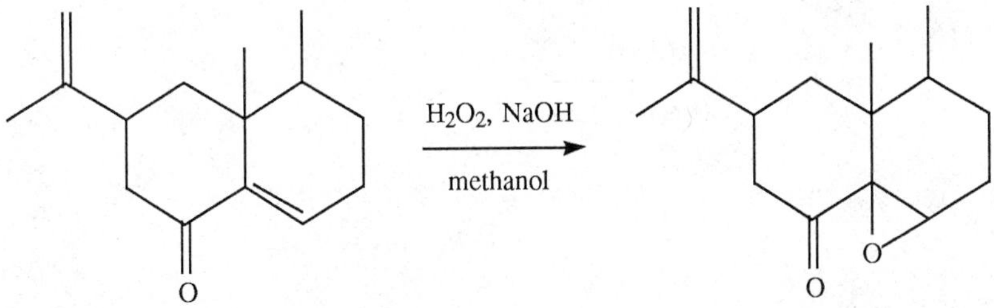

(d)

(e)

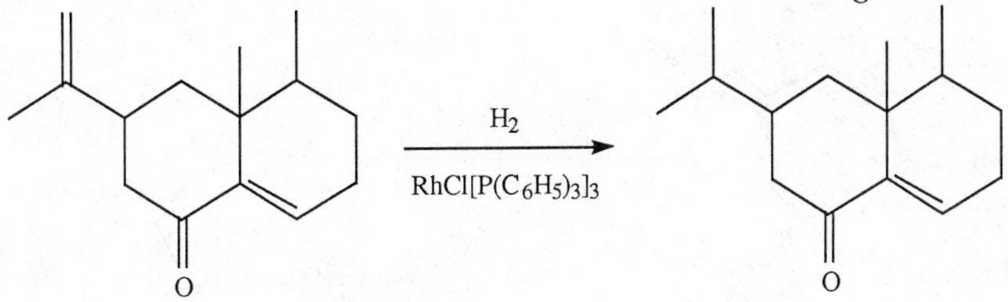

18.22 (cont)

(f)

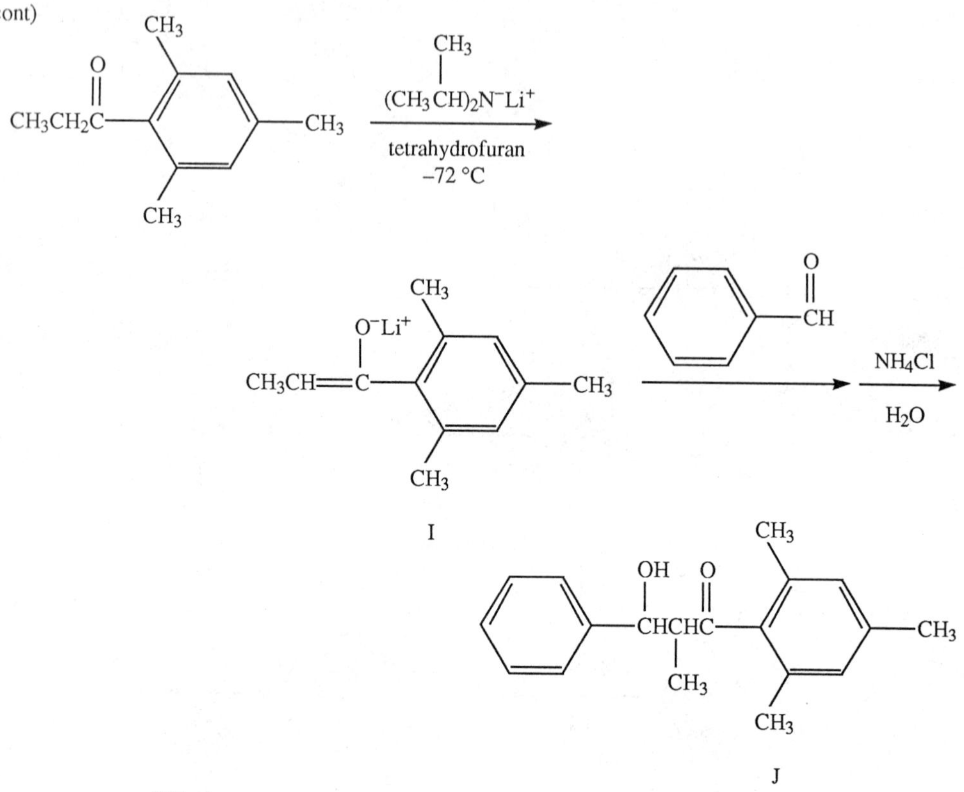
CH3
O
CH3CH2C
CH3
CH3
CH3
(CH3CH)2N⁻Li⁺
tetrahydrofuran
–72 °C
O⁻Li⁺
CH3CH=C
CH3
CH3
CH3
O
CH
NH4Cl
H2O
I
OH O
CHCHC
CH3
CH3
CH3
CH3
J

(g)

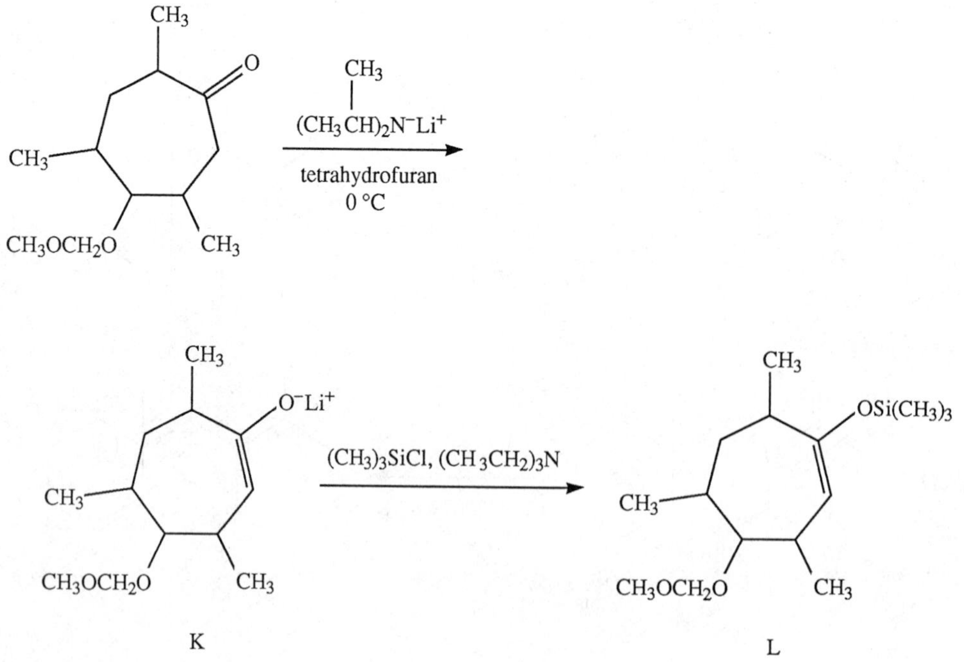
CH3
O
CH3
CH3OCH2O
CH3
CH3
(CH3CH)2N⁻Li⁺
tetrahydrofuran
0 °C
CH3
O⁻Li⁺
CH3
CH3OCH2O
CH3
K
(CH3)3SiCl, (CH3CH2)3N
CH3
OSi(CH3)3
CH3
CH3OCH2O
CH3
L

18.22 (cont)

(h)

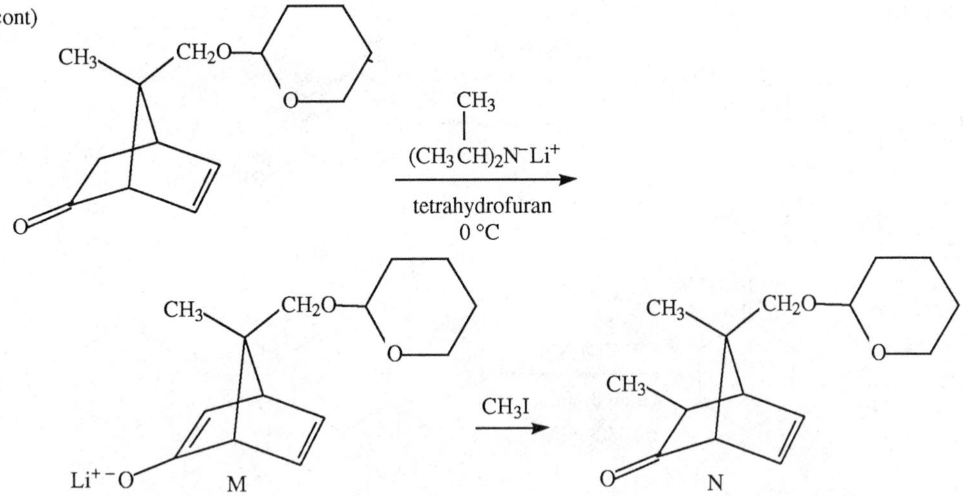

$$CH_3I \longrightarrow N$$

(i)

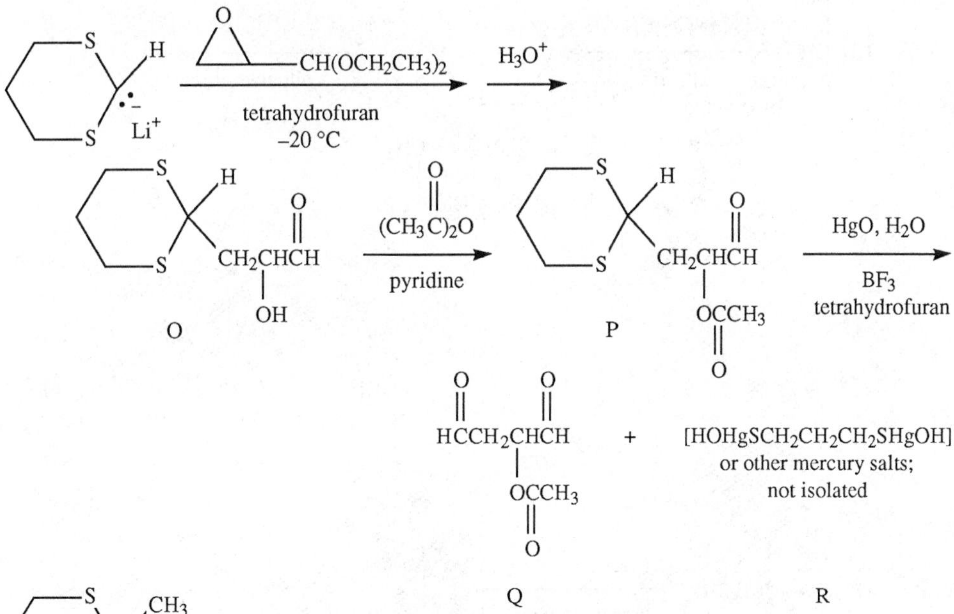

(j)

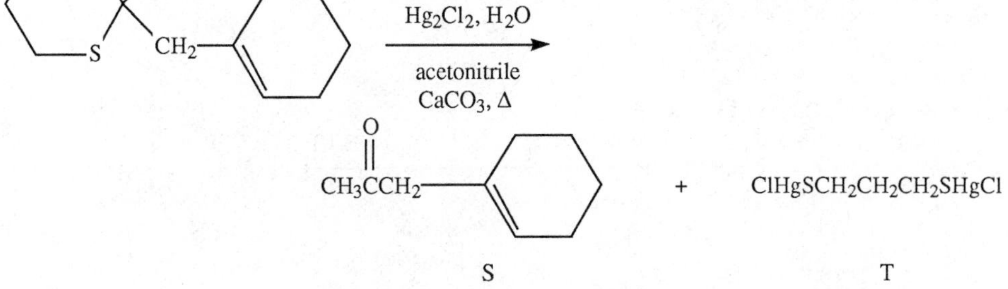

18.22 (cont)

(k)

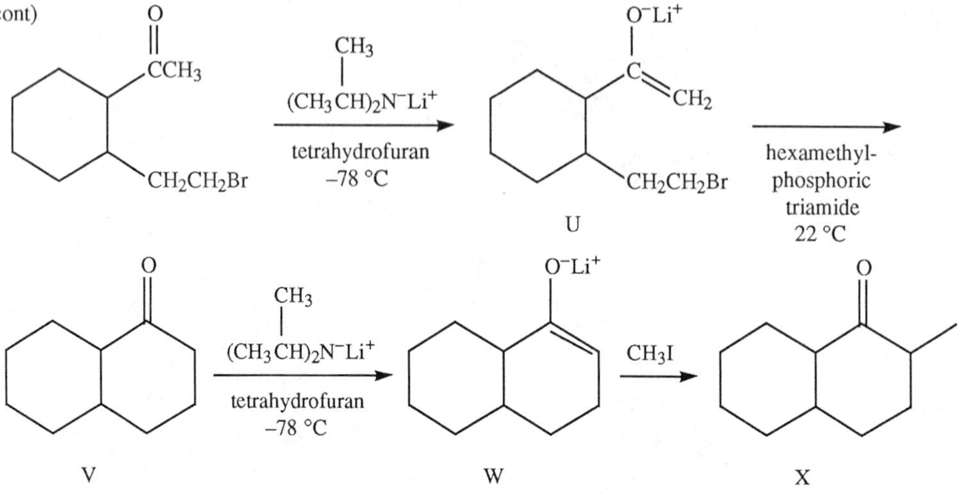

U

hexamethyl-
phosphoric
triamide
22 °C

V

(CH₃CH)₂N⁻Li⁺

tetrahydrofuran
−78 °C

W

CH₃I

X

(l)

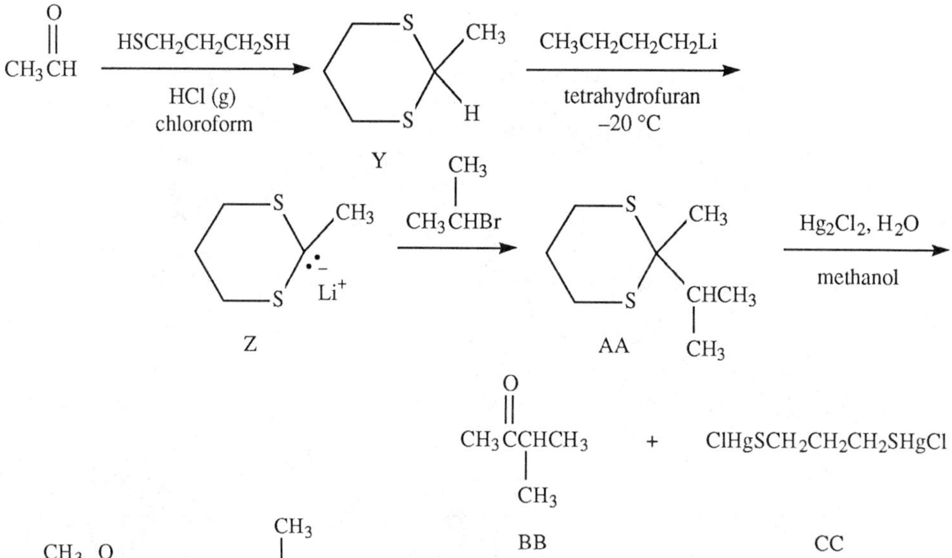

$$CH_3CH$$

O
‖

$$HSCH_2CH_2CH_2SH$$

HCl (g)
chloroform

Y

$$CH_3CH_2CH_2CH_2Li$$

tetrahydrofuran
−20 °C

Z

$$CH_3CHBr$$

AA

$$Hg_2Cl_2, H_2O$$

methanol

BB

CC

(m)

$$CH_3C-CCH_2CH_3$$

(CH₃CH)₂N⁻Li⁺

tetrahydrofuran
−72 °C

DD

NH₄Cl

H₂O

EE

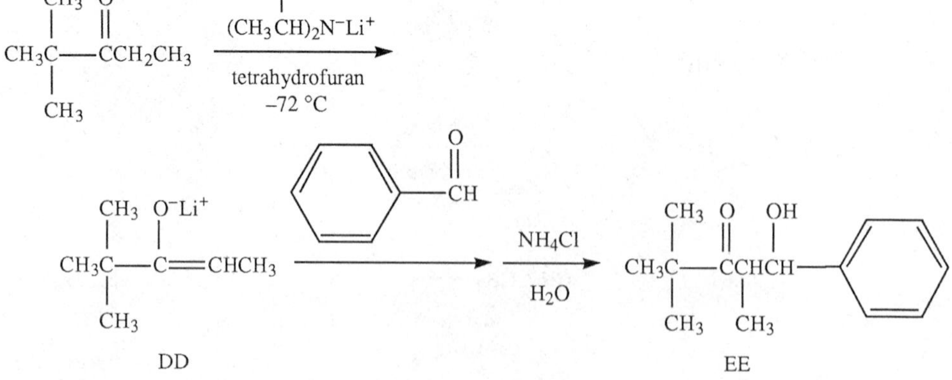

18.23

(a)

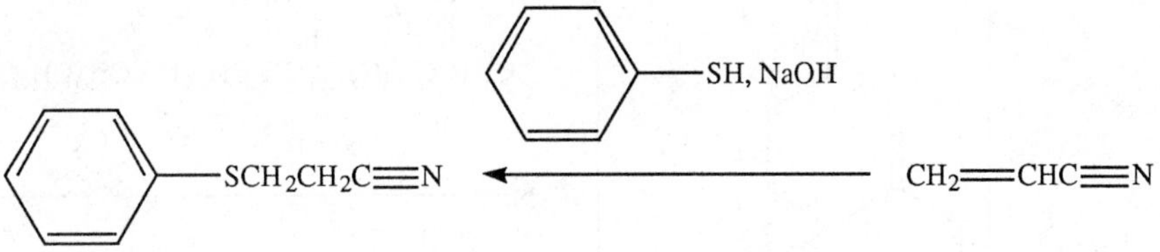

(b)

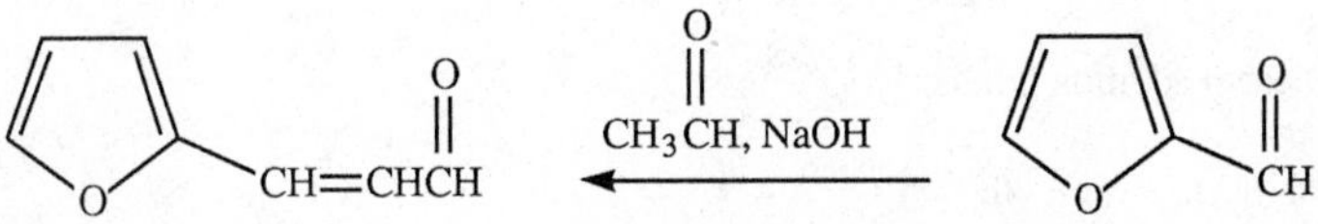

(c)

$$CH_3CH_2CH_2CH_2CH_2CH_2CH=CH_2 \xleftarrow{\quad (CH_2=CHCH_2)_2CuLi \quad} CH_3CH_2CH_2CH_2CH_2I$$

(d)

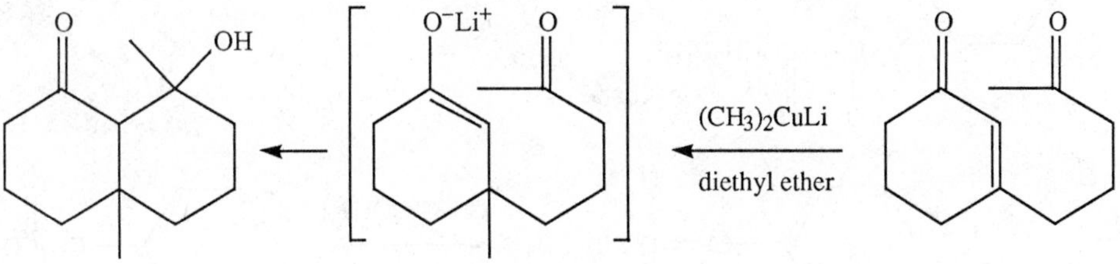

(e)

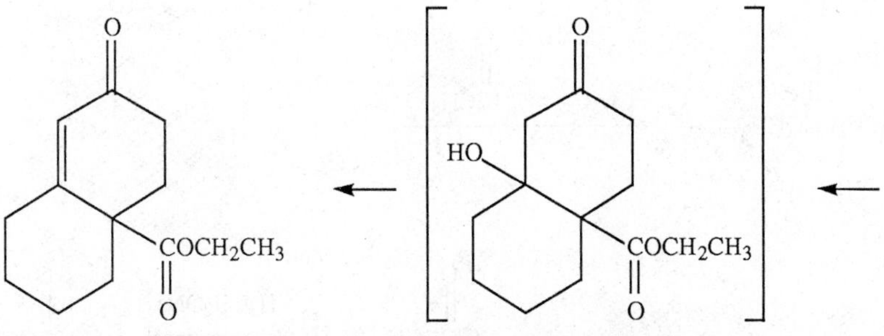

intermediate from
aldol condensation

18.23 (cont)

(e) (cont)

$$\left[\text{Michael addition structure} \right]$$

$$\overset{O}{\underset{\|}{CH_3C}}CH_2CH_2\overset{CH_2CH_3}{\underset{\underset{CH_3}{|}}{\overset{+}{N}}}CH_2CH_3 \ \ I^-, CH_3CH_2O^-Na^+$$

Michael addition product

$$\xleftarrow[\text{dry toluene}]{\text{Na}} \quad CH_3CH_2O\overset{O}{\underset{\|}{C}}(CH_2)_5\overset{O}{\underset{\|}{C}}OCH_2CH_3$$

(f)

$$\xleftarrow{\quad} \quad CH_3CH_2O\overset{O}{\underset{\|}{C}}-\overset{O}{\underset{\|}{C}}OCH_2CH_3$$

$$\xleftarrow[\text{ethanol}]{CH_3CH_2ONa} \quad \text{C}_6H_5-CH_2\overset{O}{\underset{\|}{C}}OCH_2CH_3$$

(g)

$$\overset{O}{\underset{\|}{CH_3C}}CH_2\overset{CH_3}{\underset{\underset{CH_3}{|}}{\overset{|}{C}}}\overset{O}{\underset{\|}{CH}}(\overset{}{C}OCH_2CH_3)_2 \xleftarrow{\quad} \overset{CH_3}{\underset{\|}{CH_3C}}=CH\overset{O}{\underset{\|}{C}}CH_3$$

$$Na^{+ \ -}CH(\overset{O}{\underset{\|}{C}}OCH_2CH_3)_2 \xleftarrow[\text{ethanol}]{CH_3CH_2ONa} CH_2(\overset{O}{\underset{\|}{C}}OCH_2CH_3)_2$$

18.23 (cont)

(h)

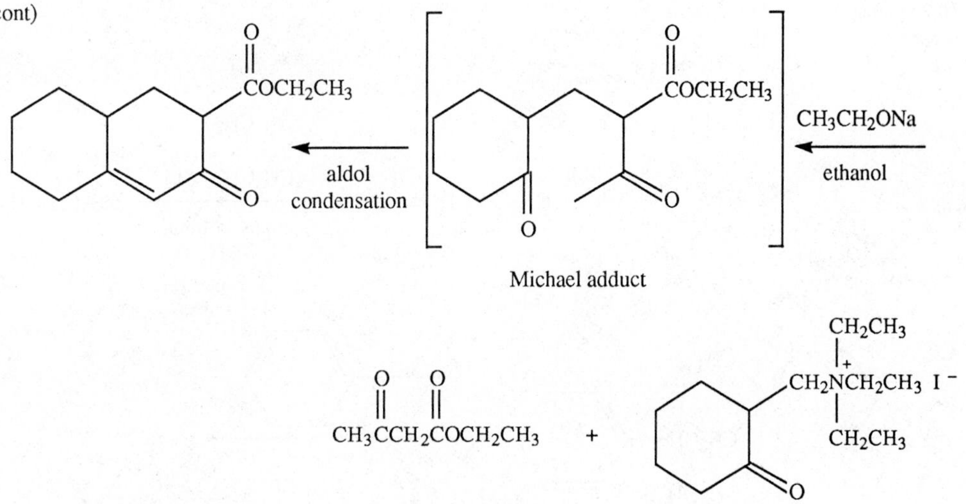

Michael adduct

$$CH_3CCH_2COCH_2CH_3 \quad + \quad$$

(i)

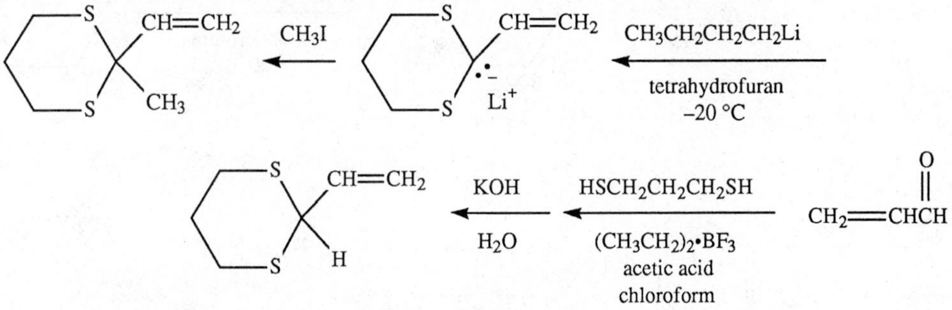

(j)

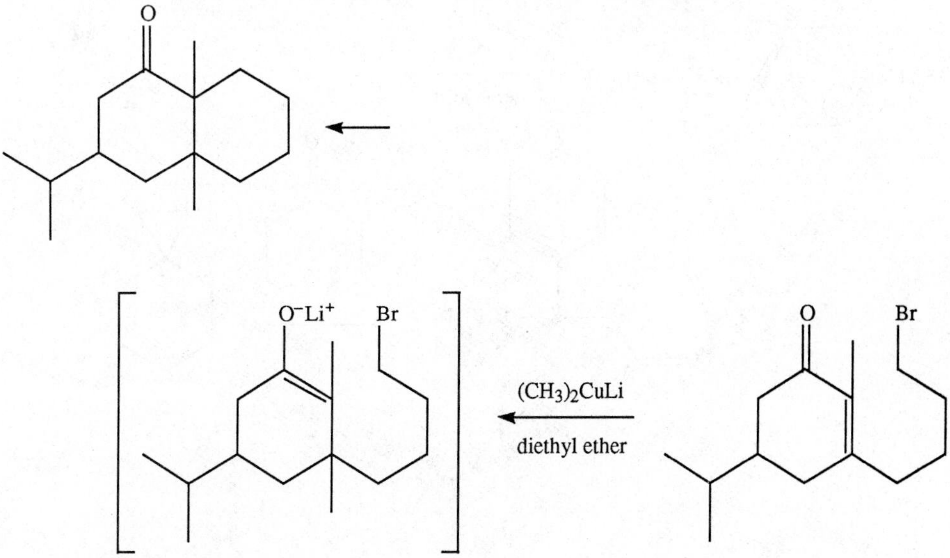

18.23 (cont)

(k)

$$CH_3CH_2CHCOCH_3 \xleftarrow{\quad BrCH_2C\equiv CH \quad}$$

with the $COCH_3$ having O above, and $CH_2C\equiv CH$ below.

$$CH_3CH_2CH=COCH_3 \xleftarrow[\substack{\text{tetrahydrofuran} \\ -78\,°C}]{(CH_3CH)_2N^-Li^+} CH_3CH_2CH_2COCH_3$$

with O^-Li^+ above the $=C$ center, CH_3 above the amide lithium, and O above the right ketone.

(l)

dithiane with CH_3 and CH_2CH—phenyl, OH below $\xleftarrow{\qquad}$ styrene oxide

$$\text{dithiane } CH_3,\ S^{\cdot -}Li^+ \xleftarrow[\substack{\text{tetrahydrofuran} \\ -20\,°C}]{CH_3CH_2CH_2CH_2Li} \text{dithiane } CH_3,\ H$$

18.24

$$(C_6H_5)_2\overset{+}{S}\!\!-\!\!CH_2CH_3 \xrightarrow[\substack{\text{tetrahydro-} \\ \text{furan}}]{\substack{CH_3 \\ | \\ CH_3CLi \\ | \\ CH_3}}$$

$$(C_6H_5)_2\overset{+}{\underset{\cdot\cdot}{S}}\!\!-\!\!\overset{\cdot\cdot}{C}HCH_3 \xrightarrow{\qquad} \quad + \quad (C_6H_5)_2S$$

A   B   C

18.25   (a)

Epoxidation with *m*-chloroperoxybenzoic acid occurs at the double bond with the higher electron density. The other double bond in this molecule is conjugated with a carbonyl group and is, therefore, an "electrophilic alkene."

(b)

18.26

resonance contributors for the carbanion from the sulfone

resonance contributors for the carbanion from the sulfoxide

The negative charge on the sulfone carbanion is delocalized to two oxygen atoms while that on the sulfoxide carbanion is delocalized to only one oxygen atom. The conjugate base of the sulfone is, for this reason, more highly stabilized relative to the acid, and is, therefore, a weaker base than is the conjugate base of the sulfoxide. Another way of rationalizing the difference in basicity is to say that the conjugate base of the sulfone has lower electron density at any one site than the conjugate base of the sulfoxide, and is, therefore, less readily protonated.

18.27

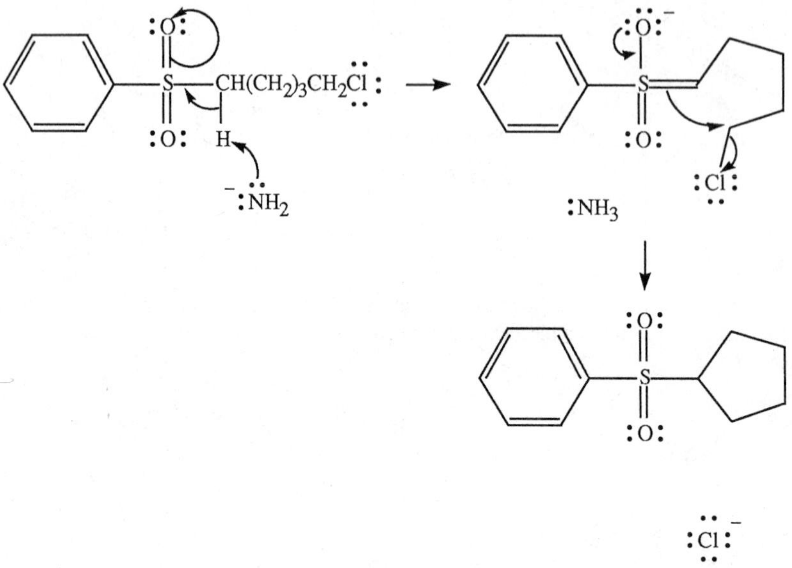
CH(CH2)3CH2Cl
NH2
NH3
Cl

18.28

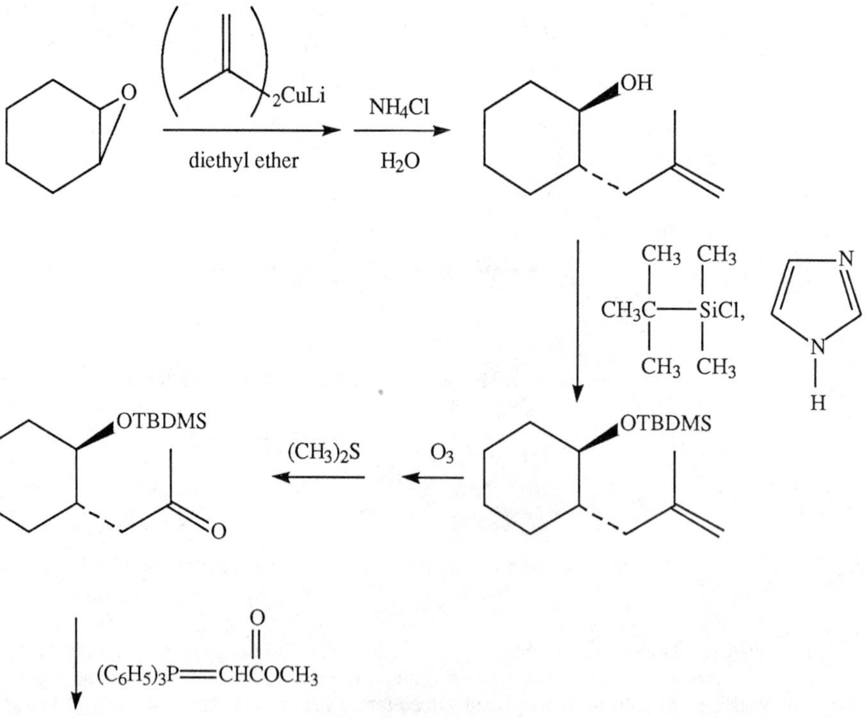
2CuLi
diethyl ether
NH4Cl
H2O
OH
CH3 CH3
CH3C—SiCl,
CH3 CH3
N
H
OTBDMS
O3
(CH3)2S
OTBDMS
O
(C6H5)3P=CHCOCH3

18.28 (cont)

LiAlH$_4$ H$_2$O

diethyl ether

PBr$_3$
pyridine

LiAlH$_4$

diethyl ether

(CH$_3$CH$_2$CH$_2$CH$_2$)N$^+$ F$^-$

18.29

CH$_3$CH$_2$CHCH$_2$ —$\}$— (CH$_2$)$_9$ —$\}$— $\overset{\overset{\displaystyle O}{\|}}{\overset{14}{C}OH}$

|
CH$_3$

2-methyl-   9-bromo-   potassium
butanal   1-nonanol   cyanide

CH$_3$CH$_2$CH(CH$_2$)$_9$CH$_2$$\overset{\overset{\displaystyle O}{\|}}{^{14}C}$OH $\xleftarrow[\Delta]{H_3O^+}$ CH$_3$CH$_2$CH(CH$_2$)$_9$CH$_2$^{14}CN
| |
CH$_3$ CH$_3$

12-methyltetradecanoic acid

$\Big\uparrow$ K^{14}CN
ethanol

CH$_3$CH$_2$CH(CH$_2$)$_9$CH$_2$OH $\xrightarrow[\text{pyridine}]{PBr_3}$ CH$_3$CH$_2$CH(CH$_2$)$_9$CH$_2$Br
| |
CH$_3$ CH$_3$

$\Big\uparrow$ H$_2$O, HClO$_4$
methanol

18.29 (cont)

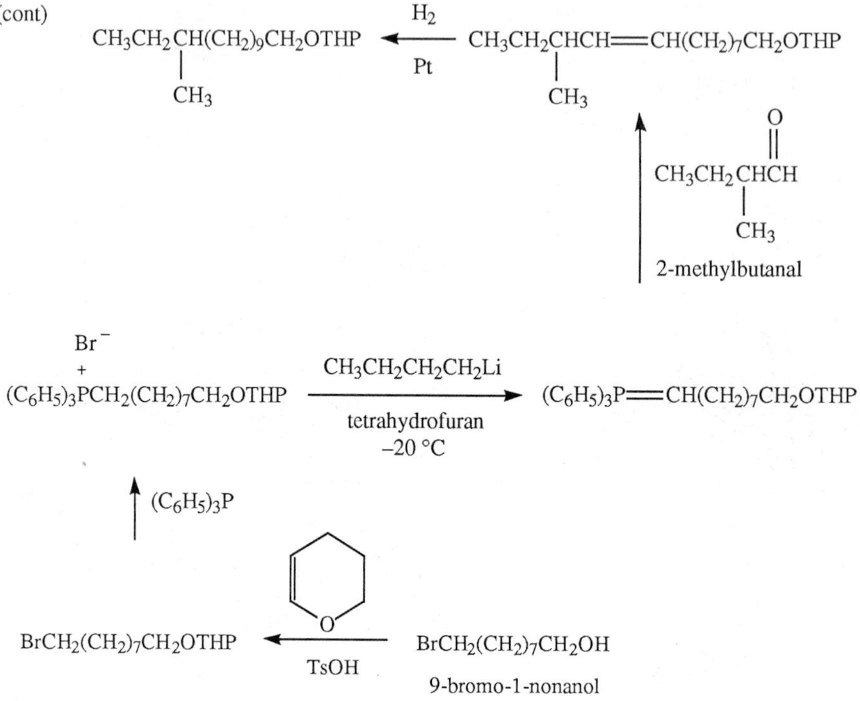

$$CH_3CH_2CH(CH_2)_9CH_2OTHP \xleftarrow[\text{Pt}]{H_2} CH_3CH_2CHCH=CH(CH_2)_7CH_2OTHP$$

$$| \qquad\qquad\qquad\qquad\qquad\qquad |$$
$$CH_3 \qquad\qquad\qquad\qquad\qquad\qquad CH_3$$

$$\begin{array}{c} O \\ \| \\ CH_3CH_2CHCH \\ | \\ CH_3 \end{array}$$

2-methylbutanal

$$\begin{array}{c} Br^- \\ + \\ (C_6H_5)_3PCH_2(CH_2)_7CH_2OTHP \end{array} \xrightarrow[\substack{\text{tetrahydrofuran} \\ -20\ ^\circ C}]{CH_3CH_2CH_2CH_2Li} (C_6H_5)_3P=CH(CH_2)_7CH_2OTHP$$

$(C_6H_5)_3P$

$$BrCH_2(CH_2)_7CH_2OTHP \xleftarrow[\text{TsOH}]{} BrCH_2(CH_2)_7CH_2OH$$

9-bromo-1-nonanol

18.30

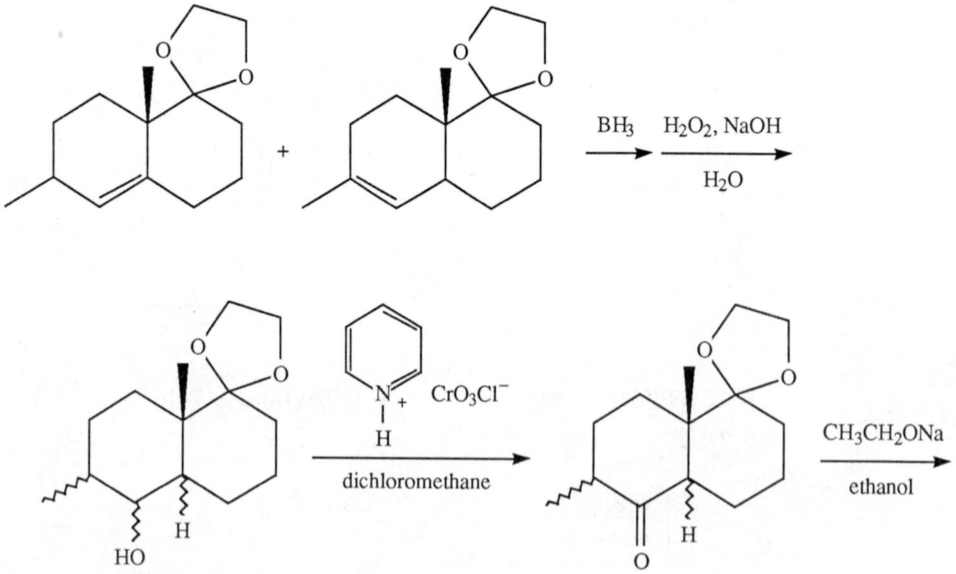

$$\xrightarrow{BH_3} \xrightarrow[H_2O]{H_2O_2,\ NaOH}$$

$$\xrightarrow[\text{dichloromethane}]{} $$

$$\xrightarrow[\text{ethanol}]{CH_3CH_2ONa}$$

18.30 (cont)

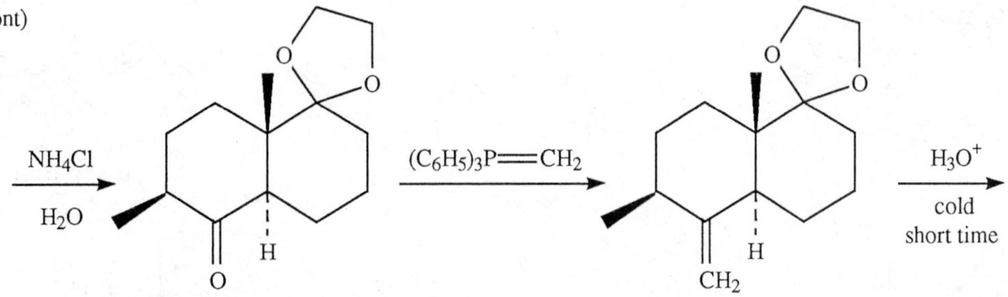

(This is an example of equilibration. Enolization occurs at both sides of the carbonyl group. Protonation of the enolate anion gives the form that is thermodynamically more stable.)

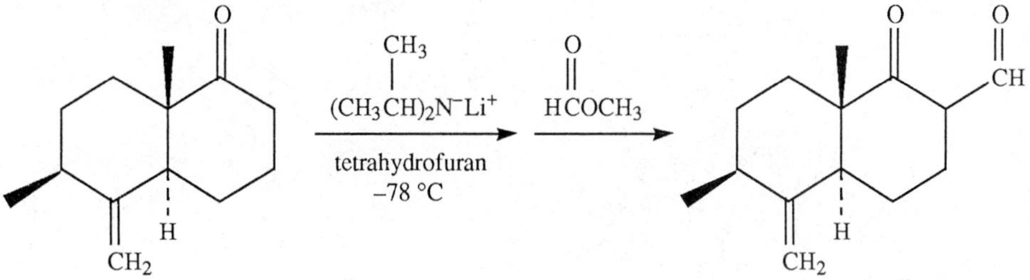

18.31

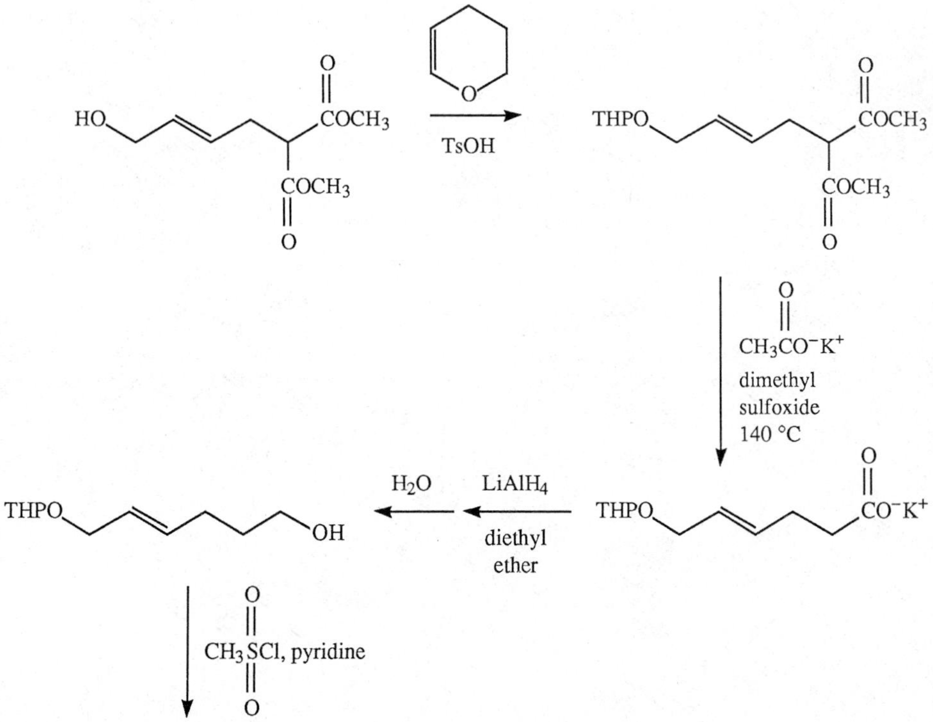

18.31 (cont)

Mechanism of Step 2:

18.31 (cont)

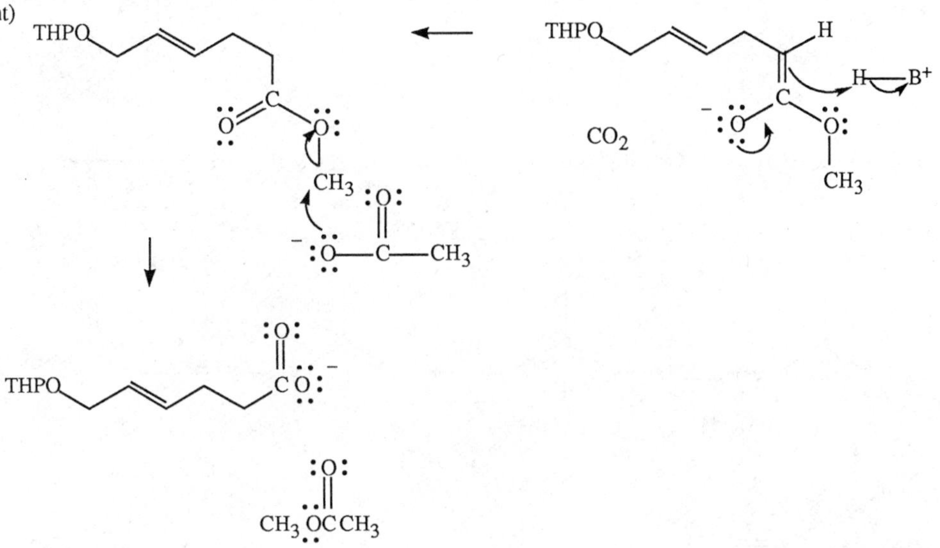

18.32

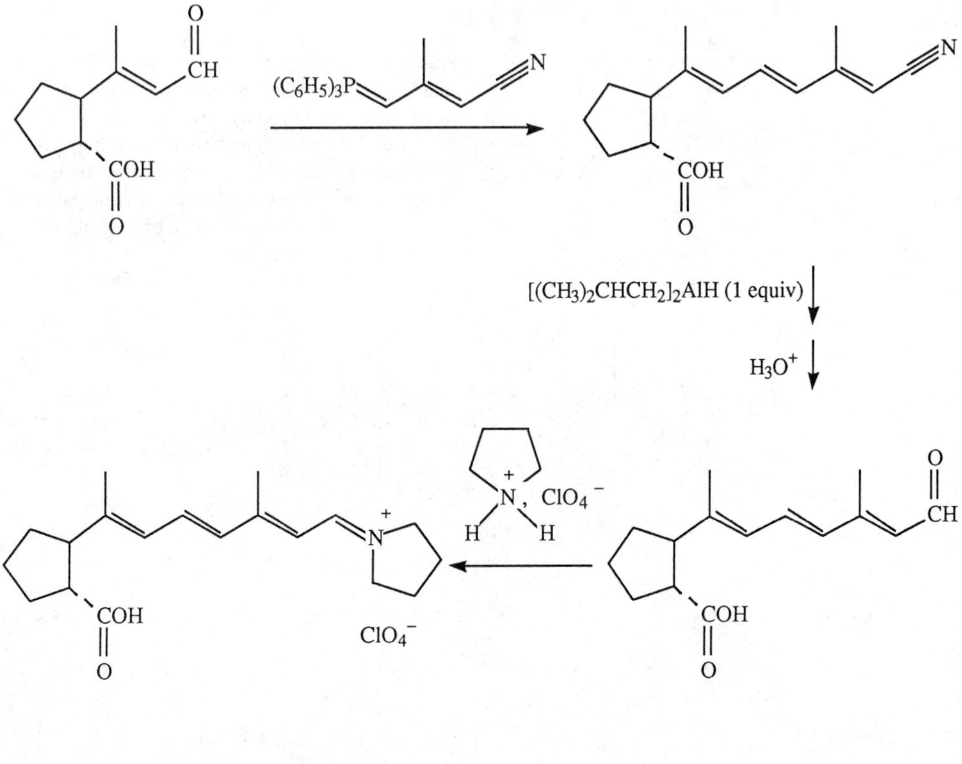

18.33 $HC\equiv CH$ + 2 CH_3CH_2MgBr $\longrightarrow$ $BrMgC\equiv CMgBr$ $\longrightarrow$

diethyl
ether
toluene

A

B

$\xrightarrow{H_3O^+}$

C

$\xrightarrow[\text{quinoline}]{\begin{array}{c}H_2\\ Pd/BaSO_4\end{array}}$

D $\xrightarrow[-10\,°C]{HBr\ (48\%)}$ E $\xrightarrow[\text{benzene}]{2(C_6H_5)_3P}$

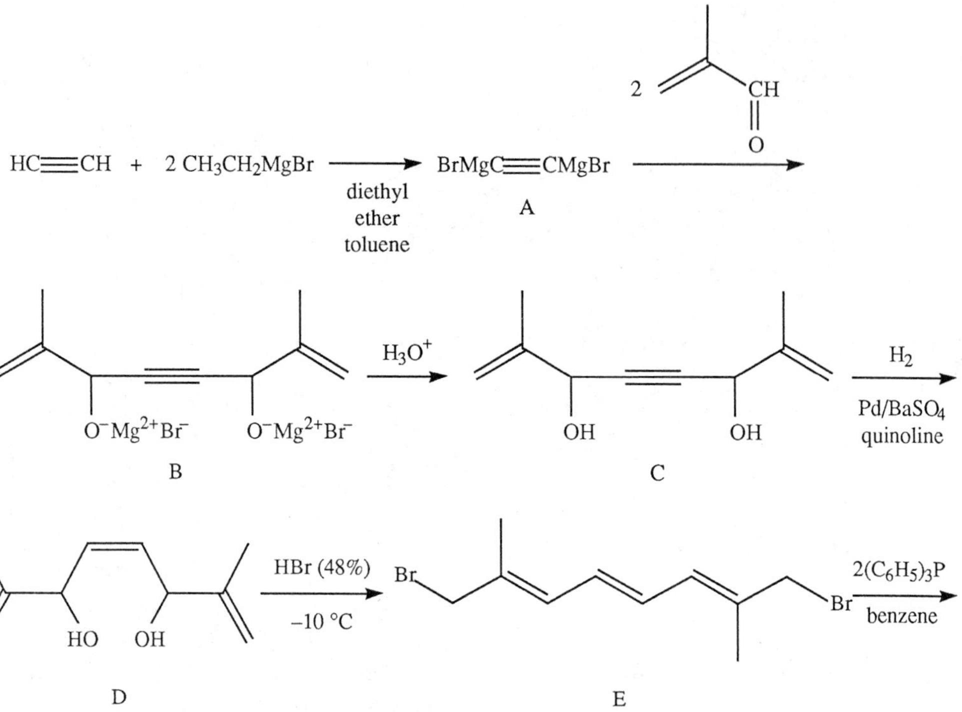

Allylic carbocations are the intermediates
in this reaction. The carbocations, stabilized
by delocalization of charge by resonance,
react with bromide ion to give the most highly
conjugated system. The cis stereochemistry of
the central double bond is lost in the conjugated
triene, which has the most stable trans configuration at each double bond.

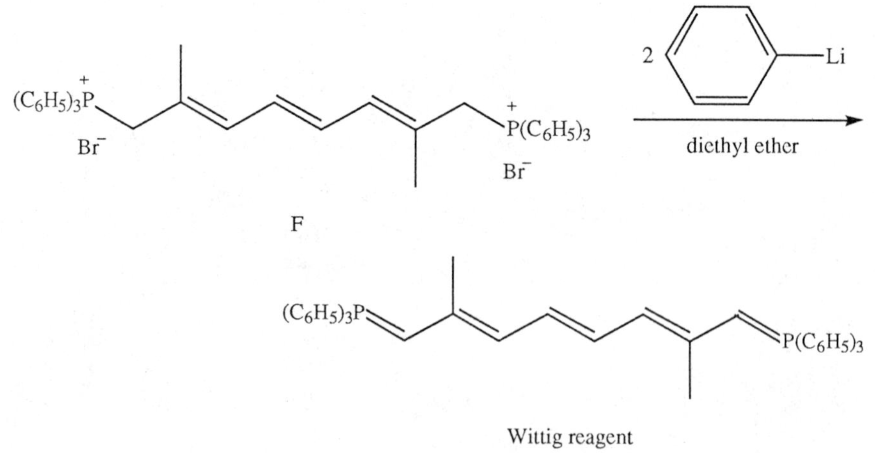

$\xrightarrow[\text{diethyl ether}]{2\ C_6H_5-Li}$

F

Wittig reagent

18.34 $CH_3CHCH_2CH_2MgBr$ $\xrightarrow[\text{benzene}]{}$ $CH_3CHCH_2CH_2CH_2CH_2CH_2O^-Mg^{2+}Br^-$ $\xrightarrow{H_3O^+}$

with CH_3 substituent on first carbon; with CH_3 substituent on product

A

$CH_3CHCH_2CH_2CH_2CH_2CH_2OH$ $\xrightarrow{HBr}$ $CH_3CHCH_2CH_2CH_2CH_2CH_2Br$ $\xrightarrow[\substack{\text{dimethyl-}\\\text{formamide}}]{(C_6H_5)_3P}$

with CH_3 substituent

B   C

$CH_3CHCH_2CH_2CH_2CH_2CH_2P(C_6H_5)_3$ Br^- (+) $\xrightarrow[\substack{\text{hexamethyl-}\\\text{phosphoric}\\\text{triamde}}]{\text{base}}$

with CH_3 substituent

D

$CH_3CHCH_2CH_2CH_2CH_2CH=P(C_6H_5)_3$ $\xrightarrow{CH_3(CH_2)_9\overset{\displaystyle O}{\overset{\|}{C}}H}$

with CH_3 substituent

E

$CH_3CH(CH_2)_4$ and $(CH_2)_9CH_3$ with CH_3 substituent; C=C double bond with H, H $\xrightarrow[\text{dichloromethane}]{}$ 3-Cl-C_6H_4-COOH $CH_3CH(CH_2)_4$—epoxide—$(CH_2)_9CH_3$ with H, H

F   and enantiomer

18.35   1. $CH_3\overset{\displaystyle O}{\overset{\|}{C}}CH_2\overset{\displaystyle O}{\overset{\|}{C}}OCH_2CH_3$ $\xrightarrow[\text{ethanol}]{CH_3CH_2ONa}$ $CH_3\overset{\displaystyle O}{\overset{\|}{C}}\overset{-}{C}HCOCH_2CH_3$ $\xrightarrow[\text{(1 molar equiv)}]{Br(CH_2)_3Br}$

Na^+

A

$CH_3\overset{\displaystyle O}{\overset{\|}{C}}CHCOCH_2CH_3$ $\xrightarrow[\Delta]{\substack{HBr\\(48\%)}}$ $CH_3\overset{\displaystyle O}{\overset{\|}{C}}(CH_2)_3CH_2Br$ $\xrightarrow[\substack{TsOH\\\Delta\\\text{benzene}}]{HOCH_2CH_2OH}$

$(CH_2)_2CH_2Br$

B   C

18.35   1. (cont)

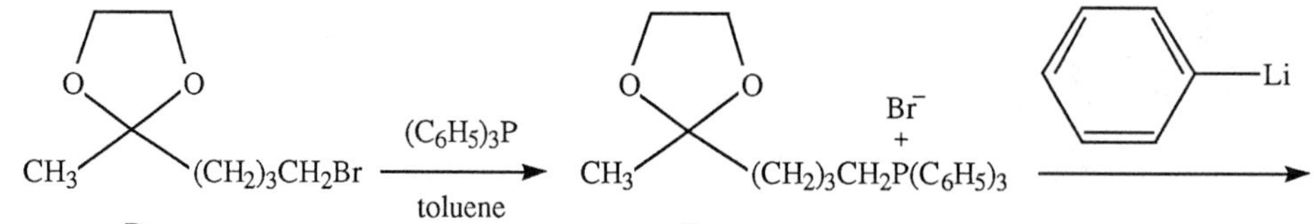

D $\xrightarrow[\text{toluene}]{(C_6H_5)_3P}$ E

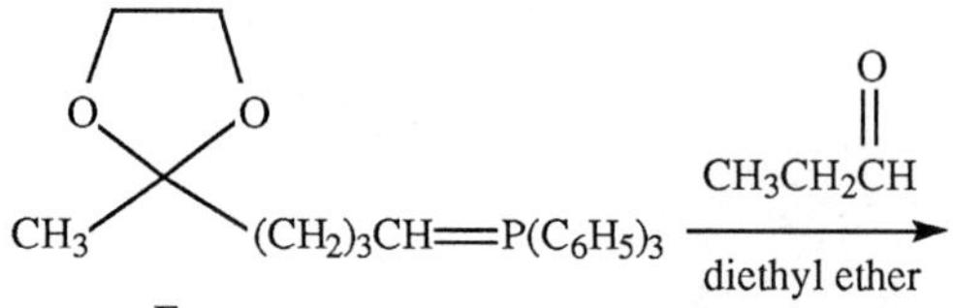

E

F $\xrightarrow[\text{diethyl ether}]{\underset{\displaystyle CH_3CH_2CH}{\overset{\displaystyle O}{\|}}}$

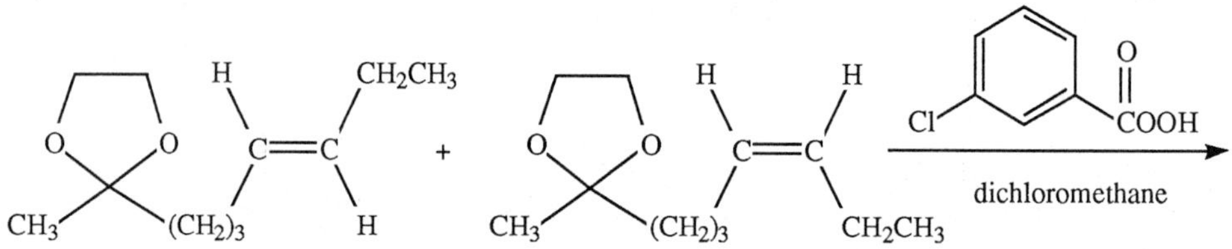

G

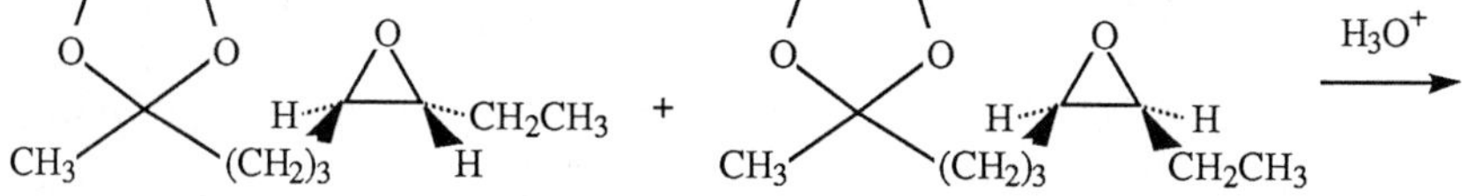

and enantiomer   and enantiomer

H (four stereoisomers)

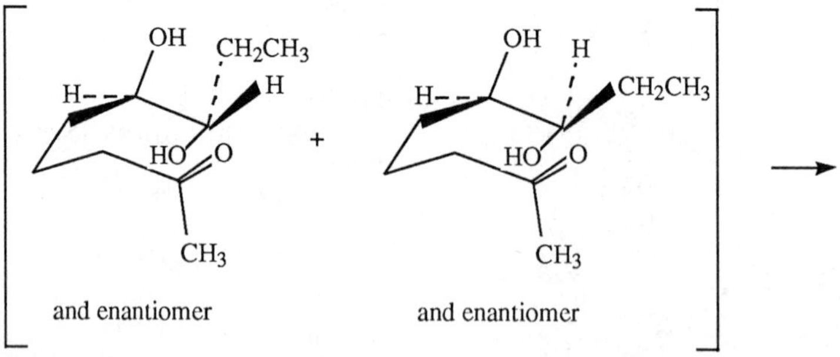

and enantiomer   and enantiomer

I

Removal of the protecting group from the carbonyl group and opening of the oxirane ring in aqueous acid gives I, which, under acidic conditions, forms the stable cyclic ketal of the resulting ketone-diol.

18.35 1. (cont)

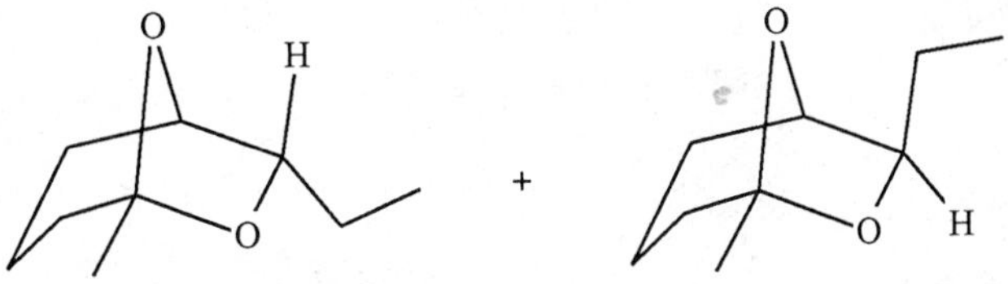

O
H
O
+
O
O
H

and enantiomer — *endo*-brevicomin

and enantiomer — *exo*-brevicomin

2.

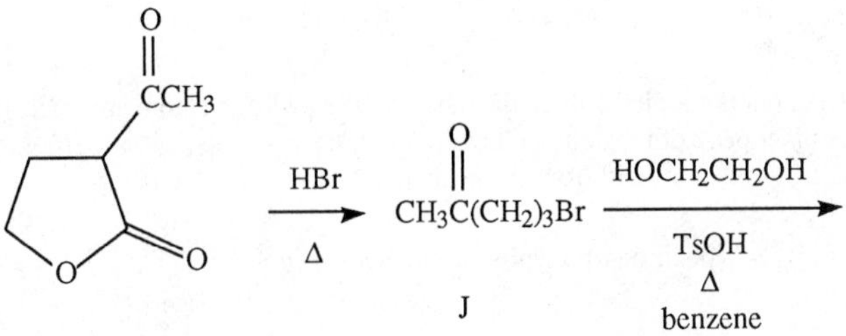

O
CCH3
O
O
HBr
Δ
O
CH3C(CH2)3Br
J
HOCH2CH2OH
TsOH
Δ
benzene

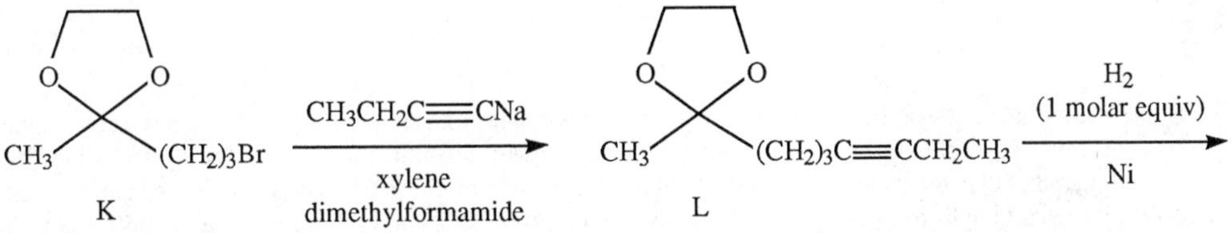

O
O
CH3
(CH2)3Br
K
CH3CH2C≡CNa
xylene
dimethylformamide
O
O
CH3
(CH2)3C≡CCH2CH3
L
H2
(1 molar equiv)
Ni

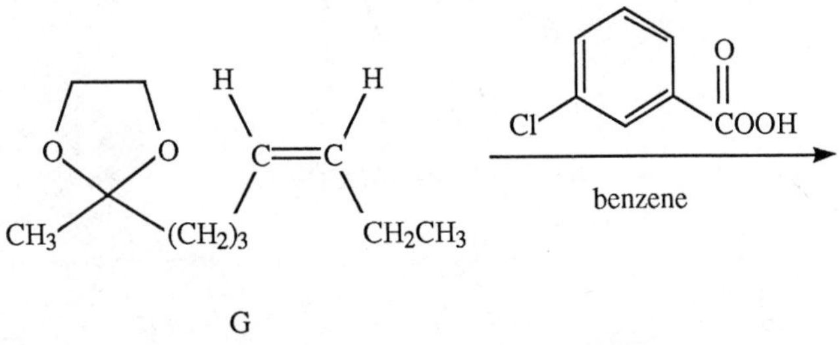

O
O
H
H
C=C
CH3
(CH2)3
CH2CH3
G
Cl
O
COOH
benzene

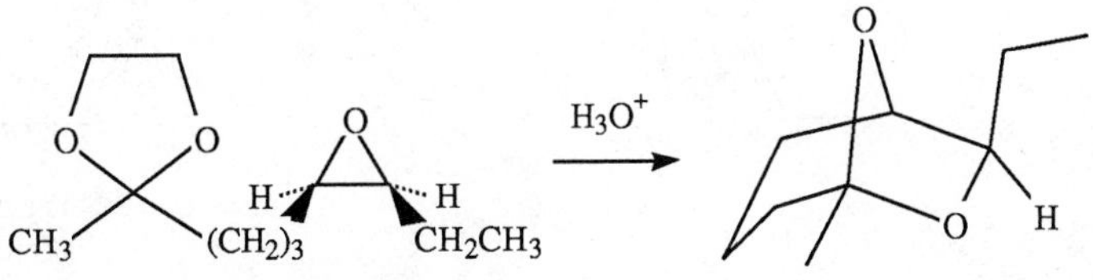

O
O
O
H
H
CH3
(CH2)3
CH2CH3
H3O+
O
O
H

and enantiomer

H (two stereoisomers

and enantiomer

exo-brevicomin

18.36

δ 7.2 (1H, doublet, *J* 16 Hz)

δ 4.0 (2H, quartet, *J* 8 Hz)

δ 1.2 (3H, triplet, *J* 8 Hz)

δ 6.9 - 7.2 (5H, multiplet)

δ 6.0 (1H, doublet, *J* 16 Hz)

Each vinylic hydrogen atom is coupled only to the other vinylic hydrogen atom and each appears as a doublet with *J* ~ 16 Hz. (The inner peak of the doublet downfield of the aromatic peak is part of the multiplet for the aromatic hydrogens. It is 16 Hz upfield from the peak at δ 7.4, which is the outer peak of the doublet.)

position of unsplit vinylic hydrogen

←— 16 Hz —→

The normal range for the chemical shift of vinylic hydrogen atoms is about δ 5–6. Both vinylic hydrogen atoms in ethyl (*E*)-3-phenyl propenoate are shifted farther downfield, indicating they are being deshielded by something other than the carbon-carbon double bond. The higher field (δ 6.0) vinylic hydrogen atom is deshielded by the carbonyl group of the ester function. The lower field vinylic hydrogen atom (δ 7.2) is even more deshielded because it is β to the carbonyl group and adjacent to the phenyl ring.

The chemical shift of the hydrogen atom on the β-carbon atom reflects the lower electron density at this carbon, symbolized by the positive charge shown in the resonance contributor above.

18.37

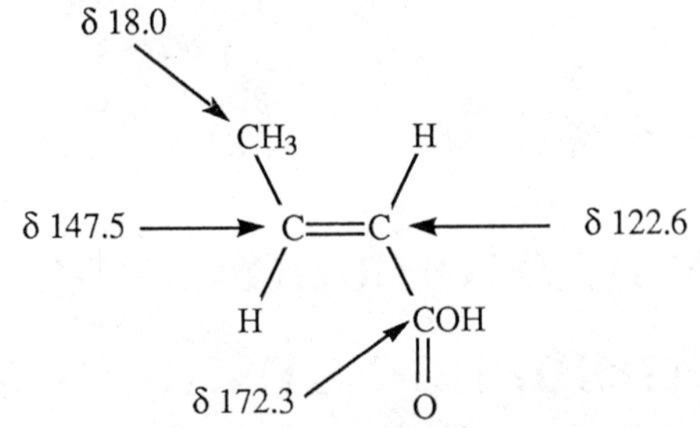

Chemical shifts in C-13 magnetic resonance spectra have essentially the same kind of dependence on electron density as proton chemical shifts. A carbon atom attached to an electron-withdrawing group or atom will be shifted downfield (deshielded) relative to a carbon atom attached to an electron-donating group or atom. The carbon atom on the carboxylic acid group, attached to two oxygens, absorbs farthest downfield and the methyl group farthest upfield. Of the two carbon atoms of the double bond, carbon 3 absorbs farther downfield than carbon 2 because it is β to the carbonyl group.

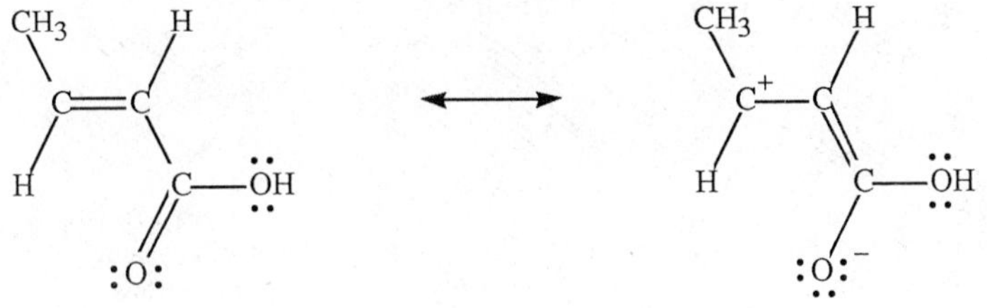

19

The Chemistry of Aromatic Compounds I.
Electrophilic Aromatic Substitution

19.1

yes
planar π system
6 π electrons

no
π system not
conjugated

yes
planar π system
6 π electrons

yes
planar π system
6 π electrons

yes
planar π system
6 π electrons

yes
planar π system
6 π electrons

19.2 The hydrogen atoms on the sp^2-hybridized carbon atoms in thiophene absorb at a much lower field than those bound to the sp^2-hybridized carbon atoms in methyl vinyl sulfide. The hydrogen atoms in thiophene are thus much more deshielded than the vinyl hydrogens in methyl vinyl sulfide. The chemical shifts of the thiophene hydrogen atoms fall in the region where hydrogen atoms on aromatic rings absorb, pointing to the presence of a ring current, which is one of the criteria for aromaticity.

19.2 (cont)

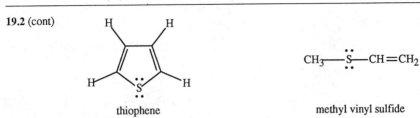

thiophene   methyl vinyl sulfide

(4n + 2) π electrons
with n = 1, therefore
an aromatic system
with delocalization of
electrons over the ring

19.3

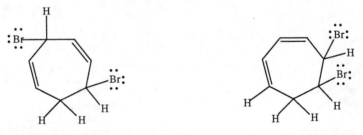

19.4   The dibromide in Problem 19.3 is the 1,6-addition product. Two other possibilities are 1,2- and 1,4-addition products. Either one can lose hydrogen bromide and then ionize to form tropylium bromide.

product from 1,4-addition   product from 1,2-addition
not formed
skipped double bonds

19.5

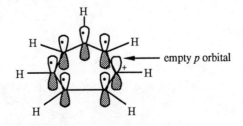

σ bonds between the carbon atoms and
between the carbon and hydrogen atoms

Concept Map 19.1   Aromaticity.

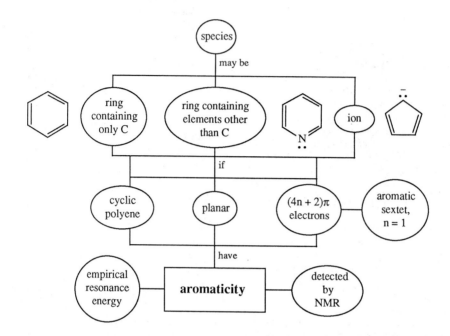

19.6   (a) 1,3-dinitrobenzene
                m-dinitrobenzene

(b) 2-methylnaphthalene
     β-methylnaphthalene

(c) 3-ethylnitrobenzene
     m-ethylnitrobenzene

(d) 4-chlorophenol
     p-chlorophenol

(e) 2,4-dichlorotoluene

(f) 4-*tert*-butylnitrobenzene
     p-*tert*-butylnitrobenzene

19.7   (a)   (b)   (c)

19.7 (cont)

(d)

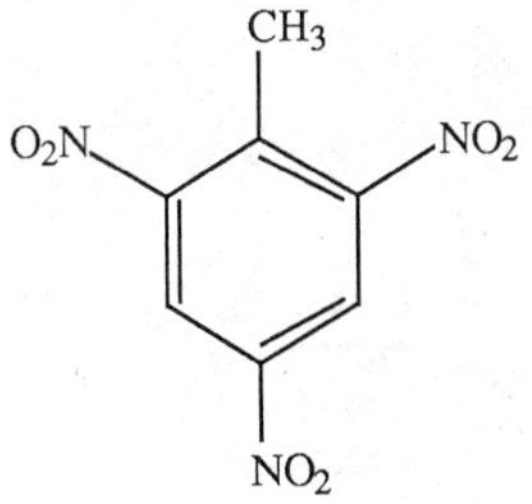

(e)

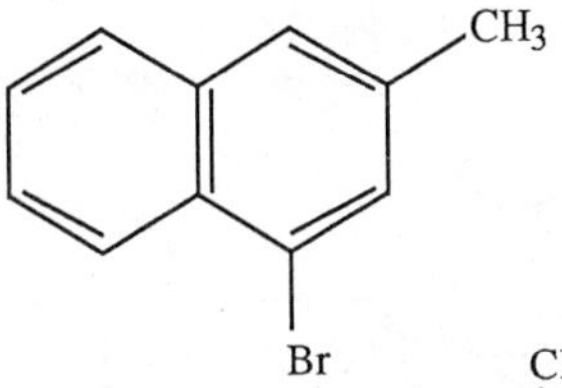

(f)

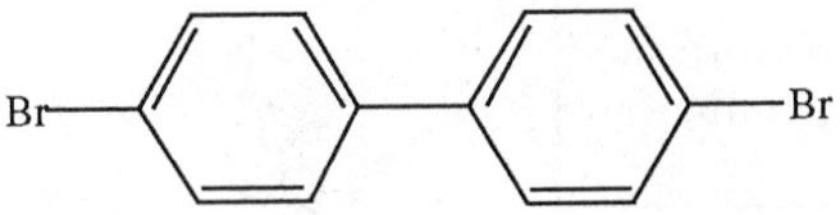

(g)

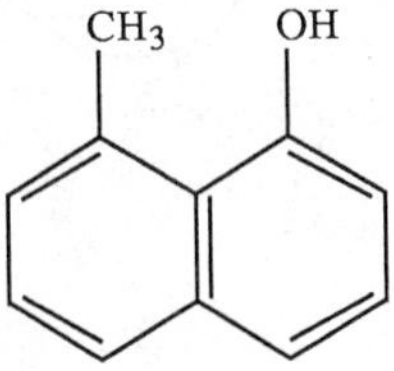

(h)

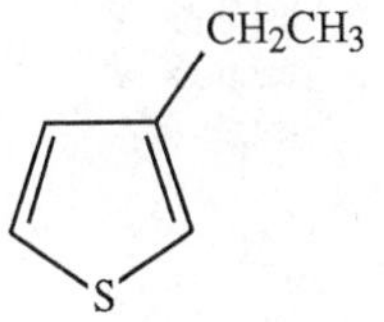

(i)

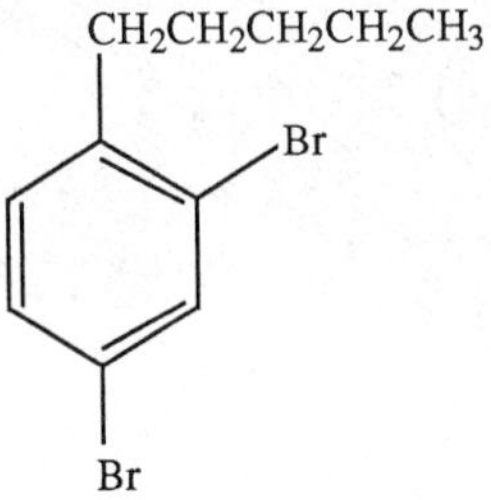

19.8

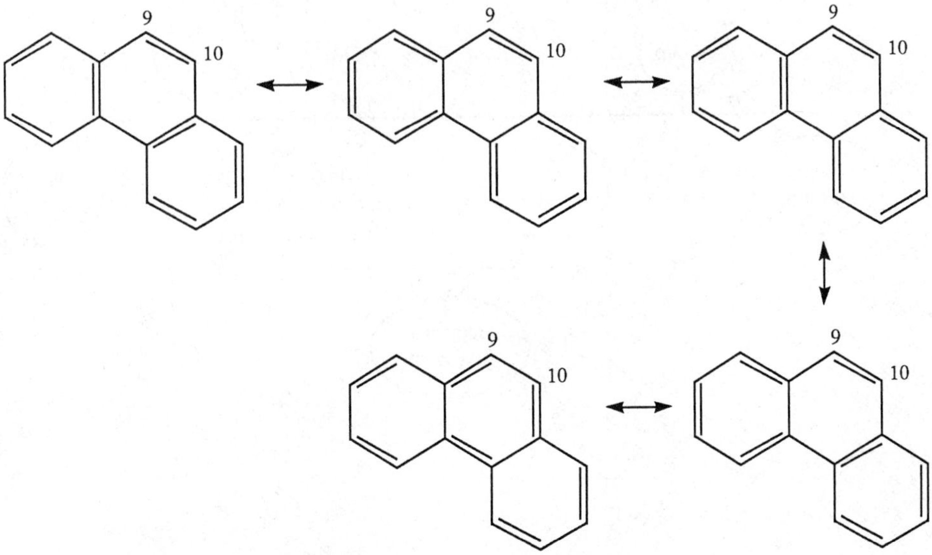

Four of the five resonance contributors of phenanthrene have a double bond at the 9,10 position. That bond, therefore, has more localized double bond character that the other bonds in the molecule.

Concept Map 19.2   Electrophilic aromatic substitution.

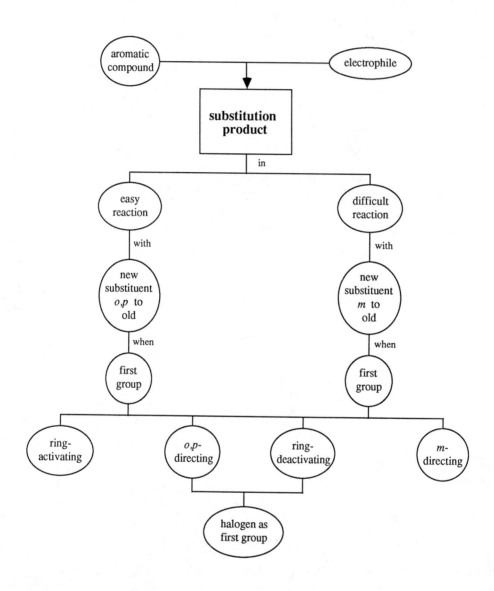

19.9 (a)

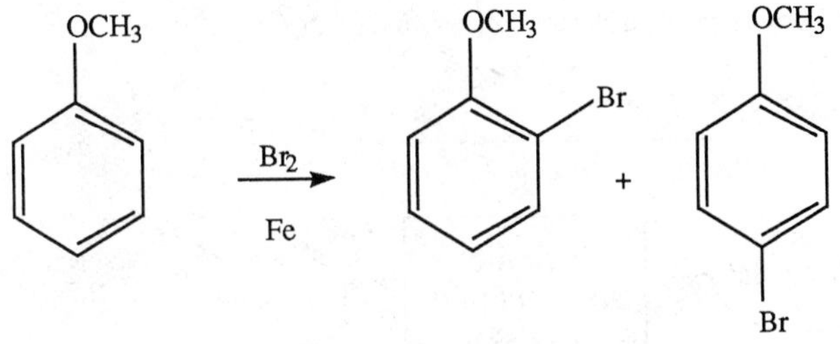
OCH3
Br2
Fe
OCH3
Br
+
OCH3
Br

(b)

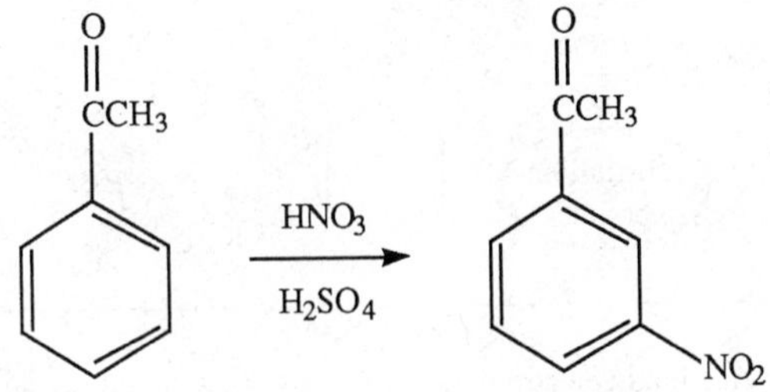
O
CCH3
HNO3
H2SO4
O
CCH3
NO2

(c)

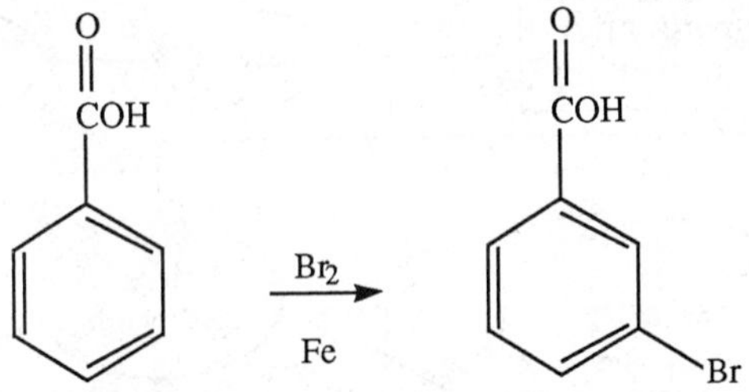
O
COH
Br2
Fe
O
COH
Br

(d)

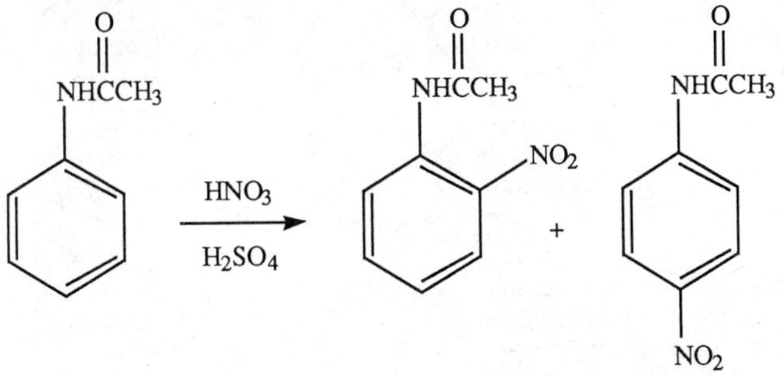
O
NHCCH3
HNO3
H2SO4
O
NHCCH3
NO2
+
O
NHCCH3
NO2

(e)

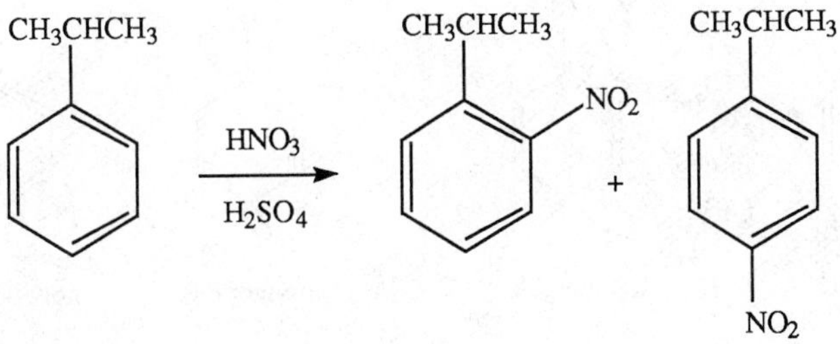
CH3CHCH3
HNO3
H2SO4
CH3CHCH3
NO2
+
CH3CHCH3
NO2

Concept Map 19.3 Essential steps of an electrophilic aromatic substitution.

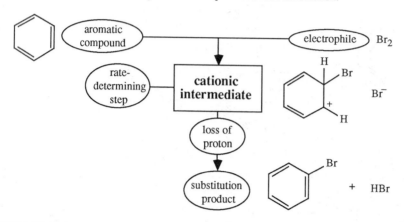

- -

Concept Map 19.4 Reactivity and orientation in electrophilic aromatic substitution.

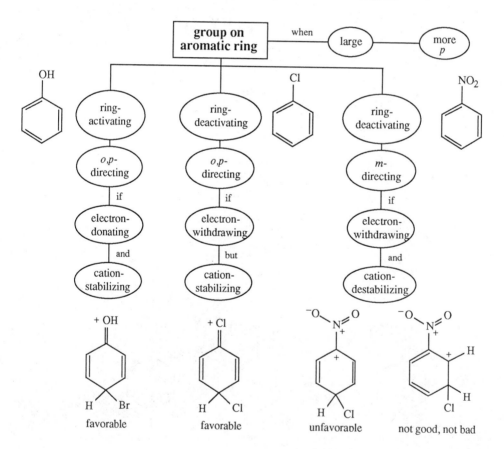

19.10   If electrophilic substitution occurs at the para position of the ring, the intermediate will have the following resonance contributors.

In this resonance contributor
the positive charge on the ring
is next to the positively charged
nitrogen atom; unfavorable

A similar set of resonance contributors can be drawn for ortho substitution. If electrophilic substitution occurs at the meta position, the positive charge on the ring is separated from the positive charge on the nitrogen atom by at least one carbon atom in all resonance contributors.

The intermediate formed on electrophilic attack at the meta position is, therefore, of lower energy than the intermediate formed when reaction occurs at the ortho or para positions.

19.11

electron-withdrawing substituent;
                            meta directing

positive charge on atom directly bonded to ring;
                            meta directing

electron-withdrawing substituent (positively charged
                            nitrogen atom closest to ring);
                            meta directing

positive charge on atom directly bonded to ring;
                            meta directing

19.11 (cont)

will stabilize a positive charge by resonance; ortho,para directing

$CH_3CH=CH—$ will stabilize a positive charge by resonance; ortho,para directing

19.12

Amines are good Brønsted bases. Strong acid converts the amine into a substituted ammonium ion, which now has a positively charged nitrogen atom directly attached to the ring, and is meta directing. The last step of the reaction is the deprotonation of the aromatic ammonium ion by ammonia to generate the free amine.

19.13

The hydroxyl group hydrogen bonds to water and, therefore, undergoes rapid hydrogen-deuterium exchange. Exchange of the hydrogen atoms on the aromatic ring has a high energy of activation and requires strong acid catalysis because, for exchange to occur, the aromaticity of the ring must be disrupted by the formation of a cationic intermediate. In the absence of strong acid, such exchange is negligibly slow.

19.14

(a)

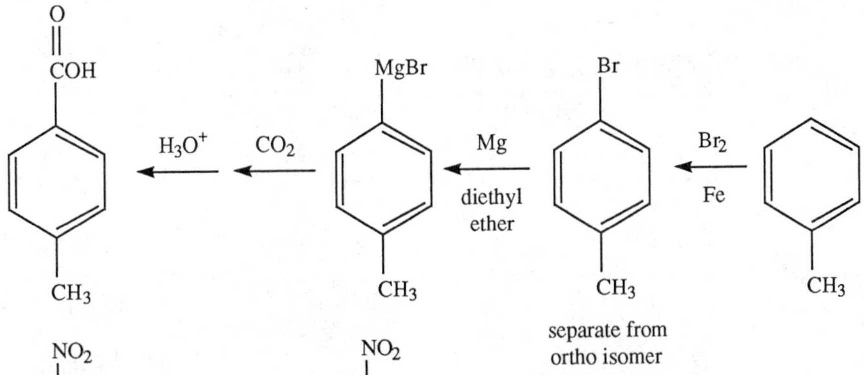

(b)

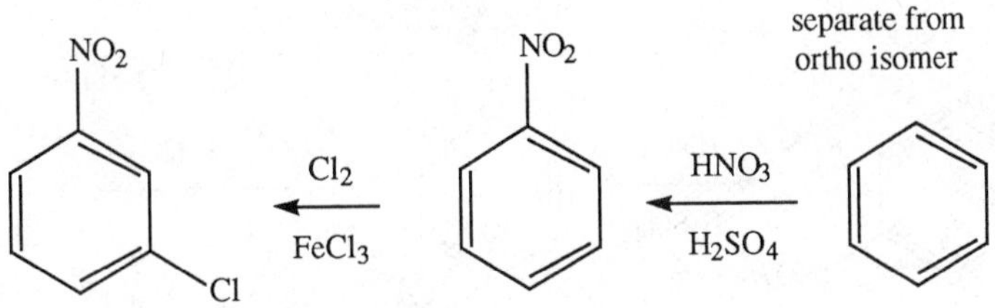

(Note: reversing the order in which these reagents are used would give ortho and para substituted products.)

(c)

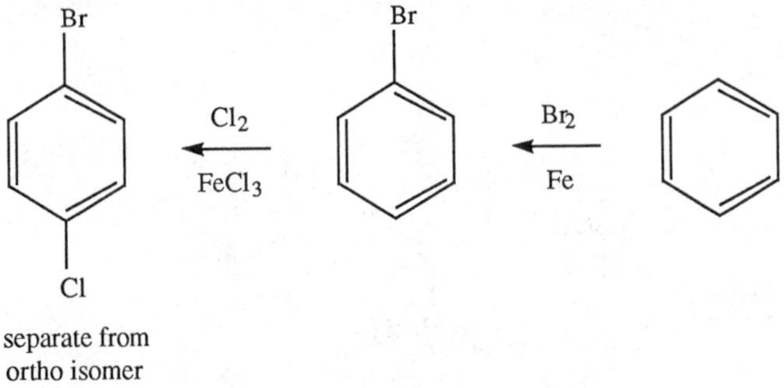

separate from
ortho isomer

(d)

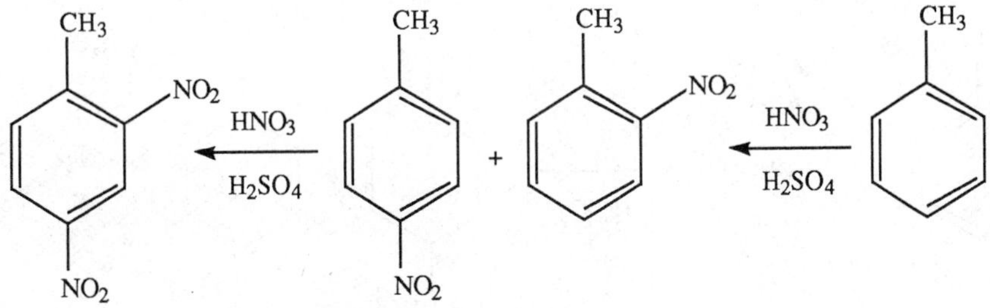

(e)

[from (c)]

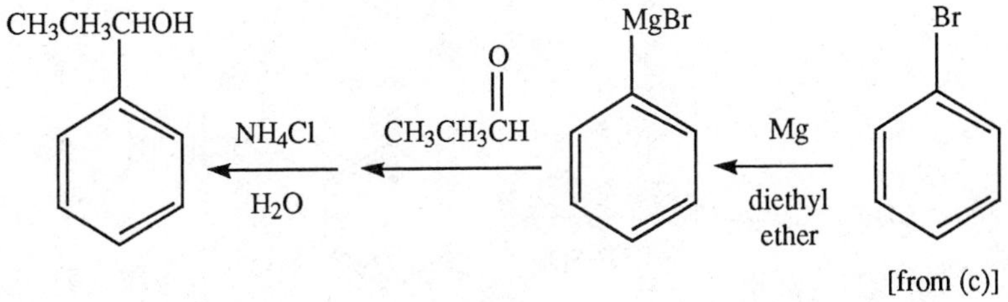

19.14 (cont)

(f)

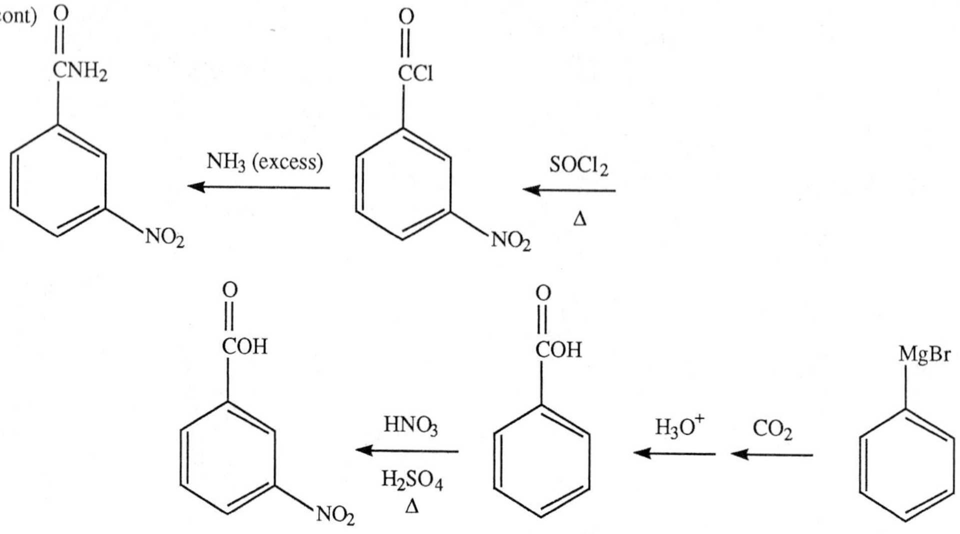

[from (e)]

19.15

(a)

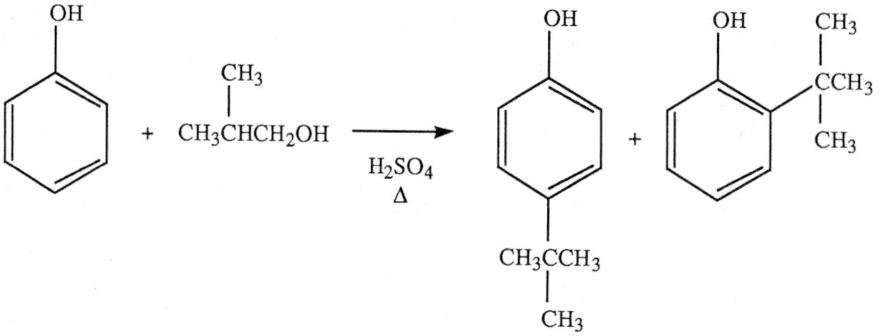

(b)

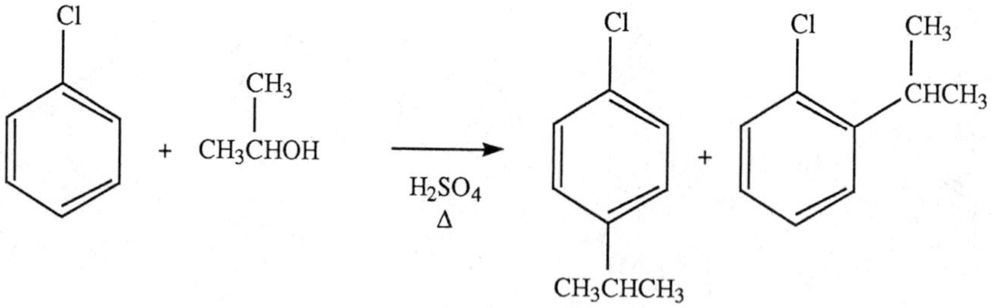

(c)

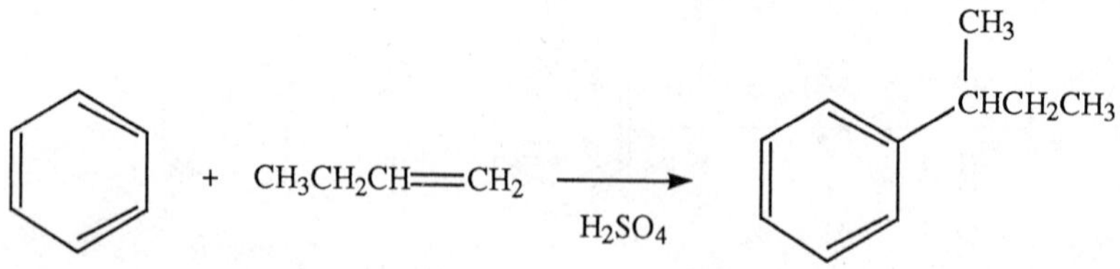

19.15 (cont)

(d)

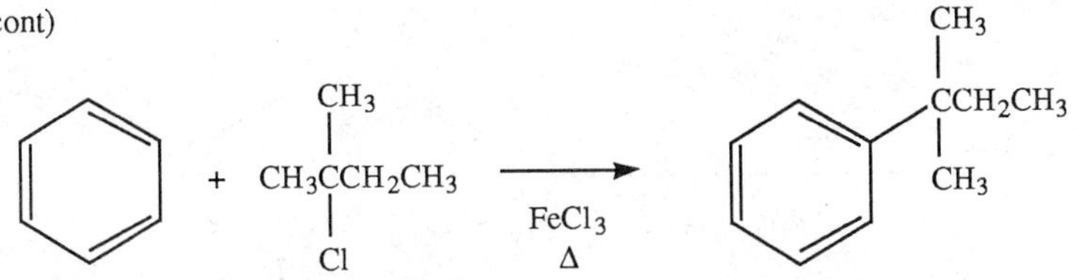

(e)

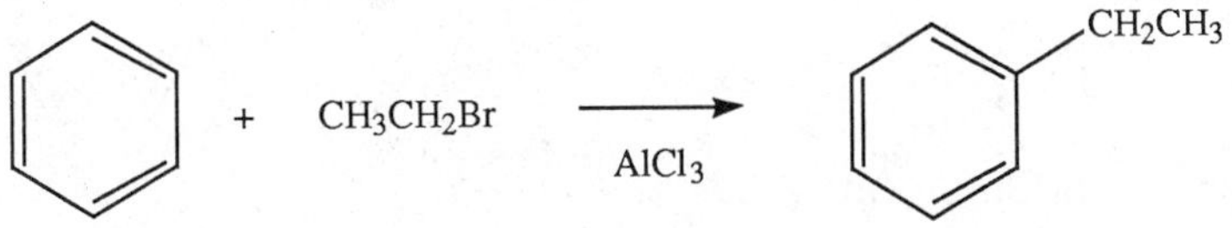

19.16

(a)

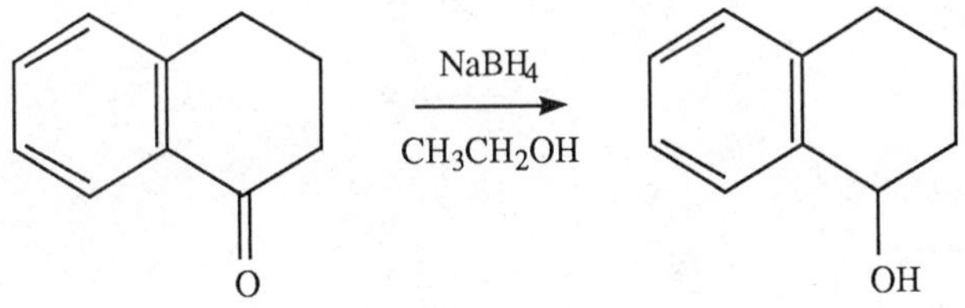

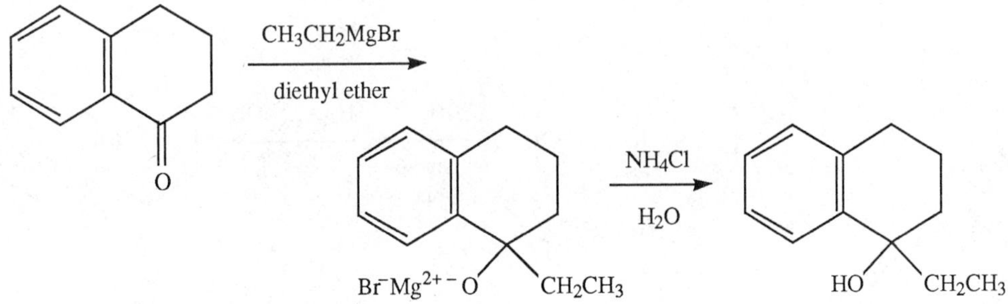

(b)

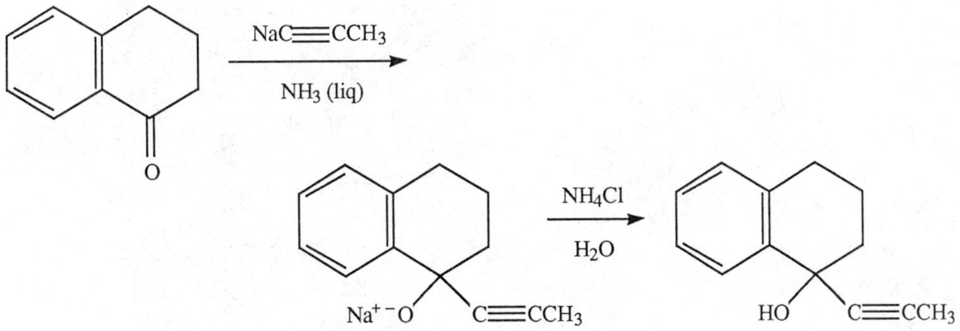

(c)

19.16 (cont)

(d)

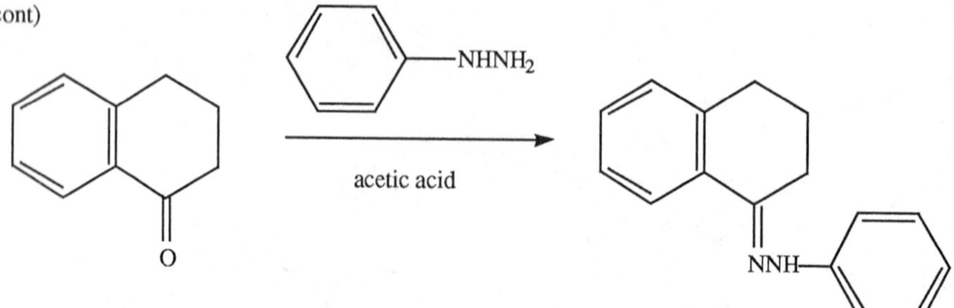

F

(e)

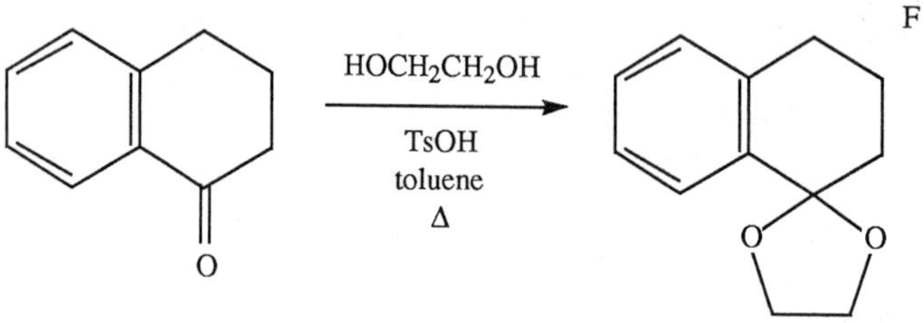

G

(f)

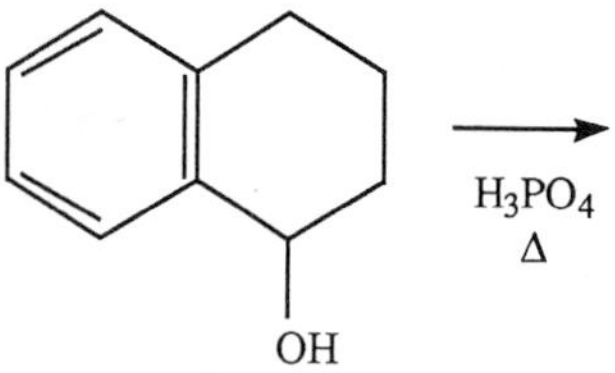

A

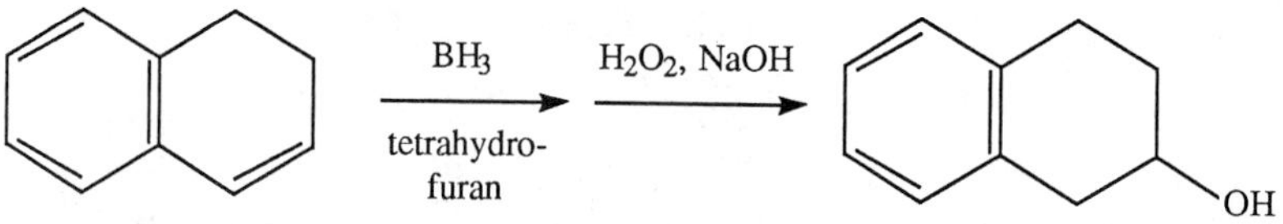

H   I

(g)

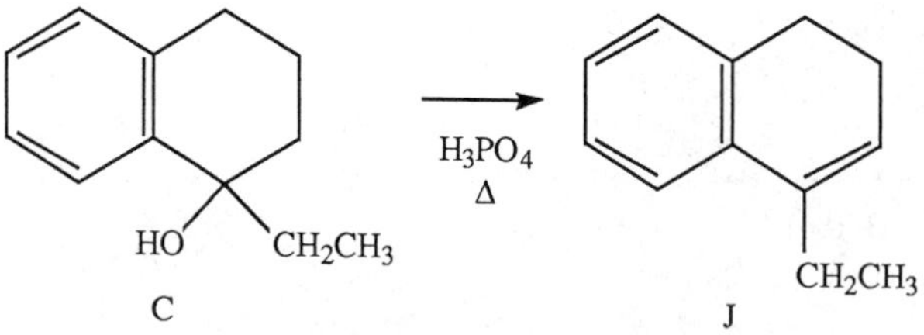

C   J
                                          (major product)

19.16 (cont)

(h)

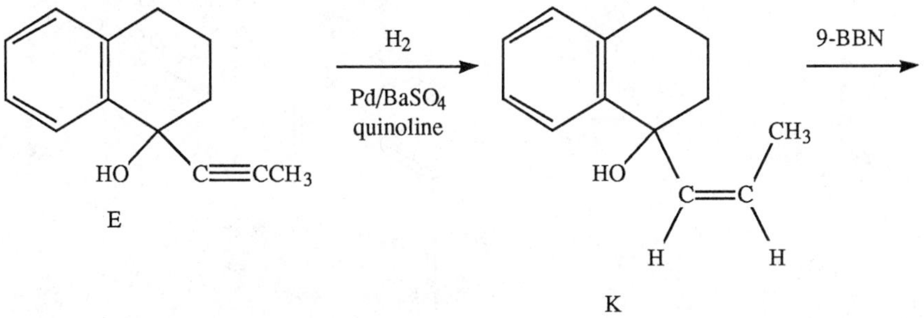

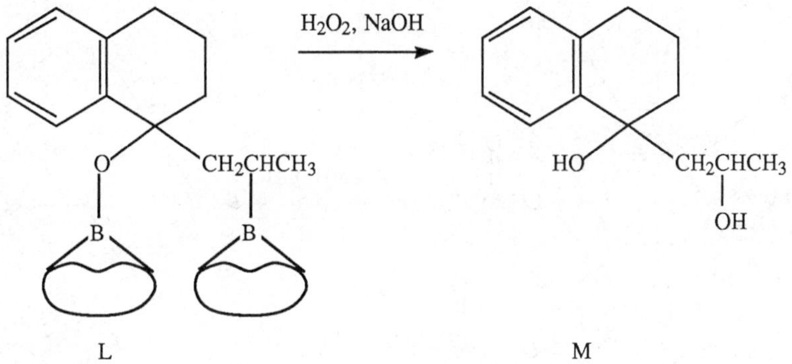

19.17

(a)

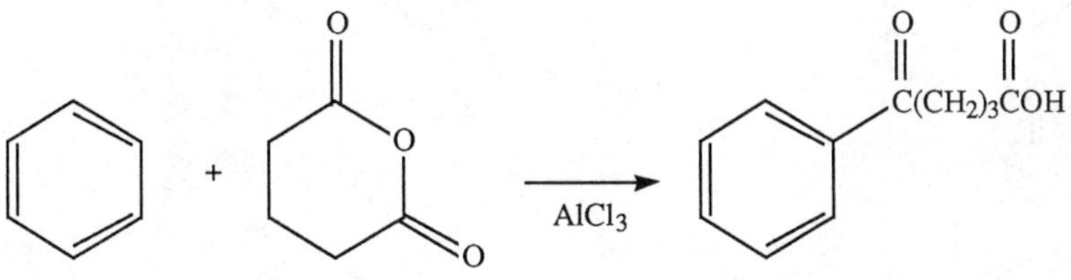

(b)

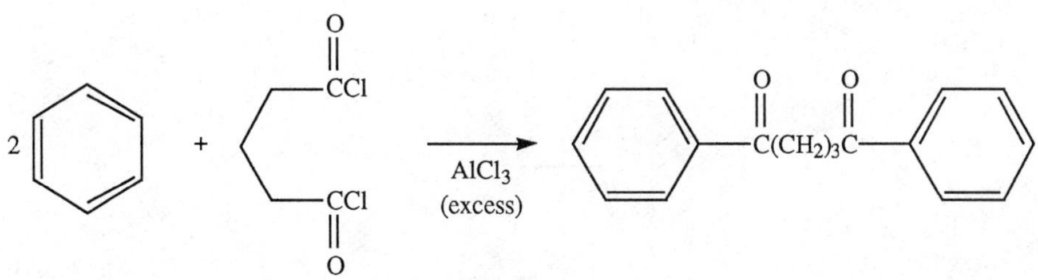

(c)

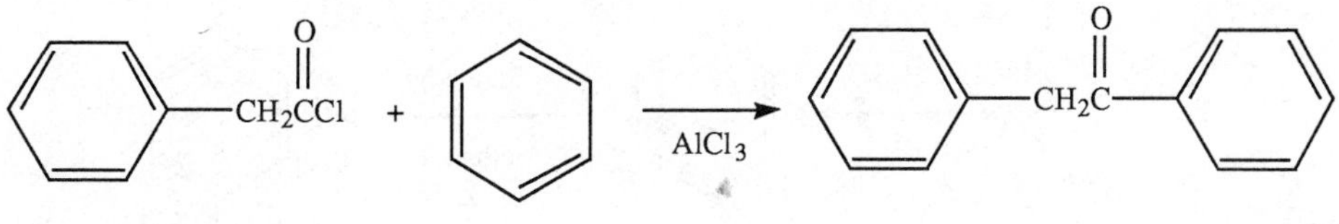

19.17 (cont)

(d)

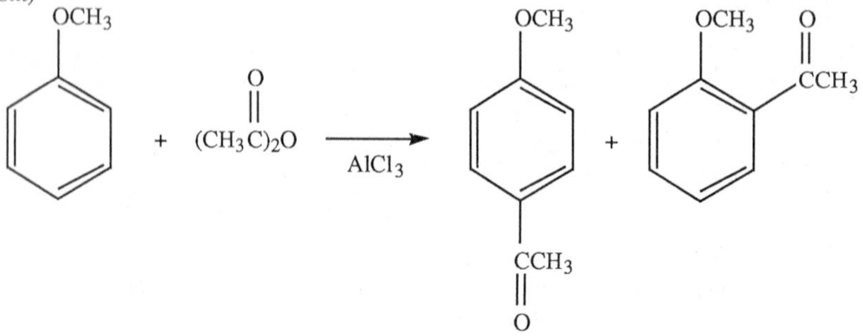

19.18

(a)

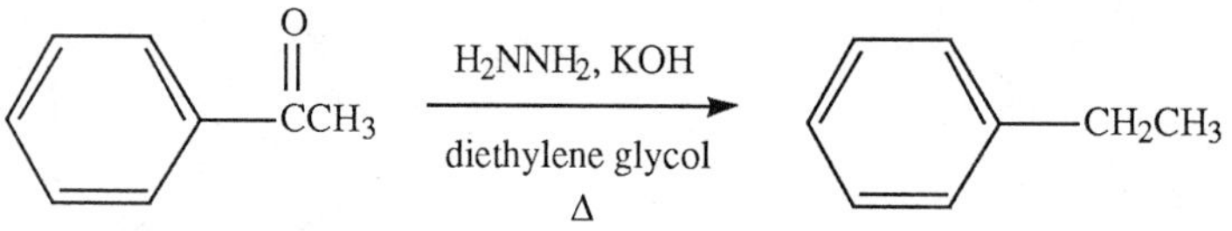

(b)

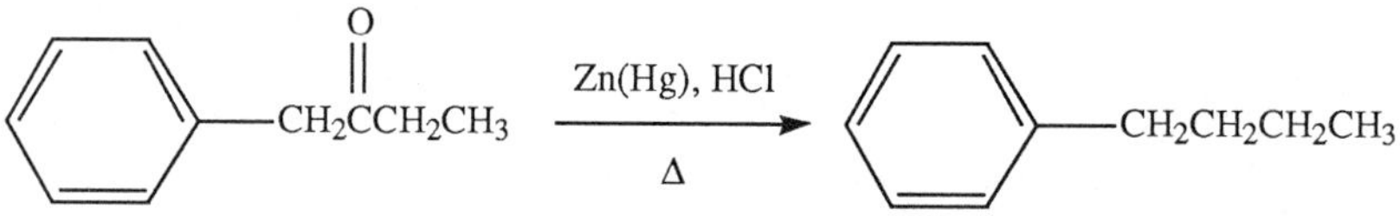

(c)

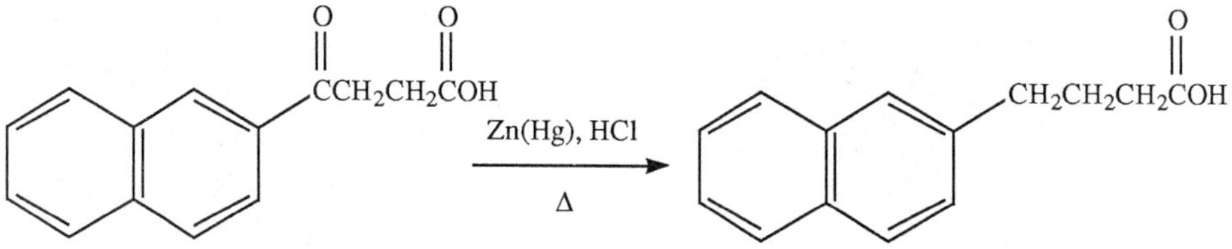

(d)

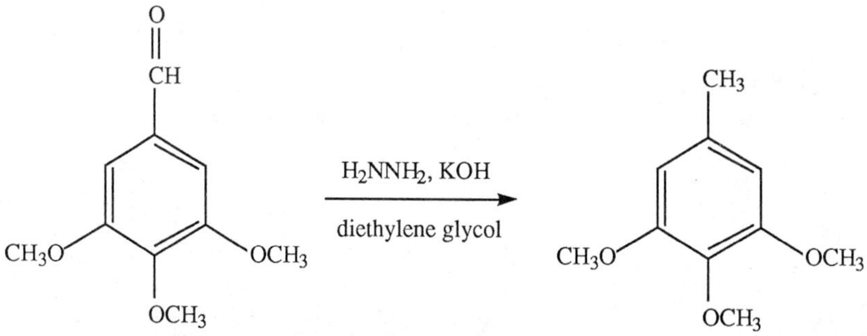

(e)

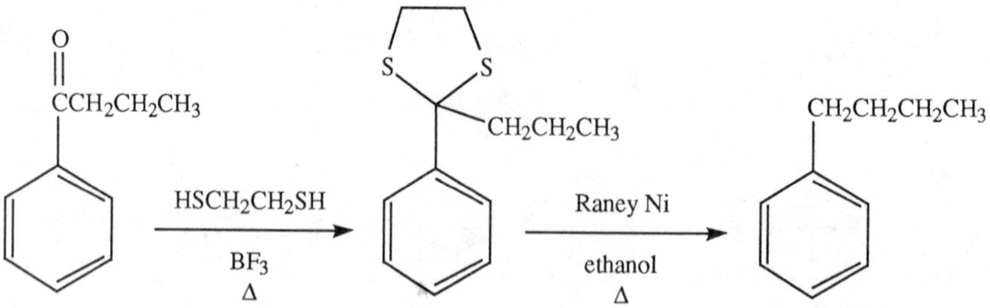

Concept Map 19.5 Electrophiles in aromatic substitution reactions.

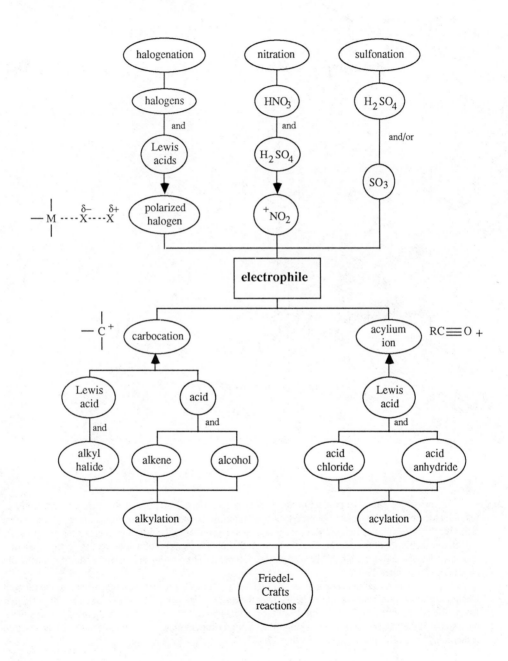

19.19

(a)

CH₃CHCH₃ ⟶ CH₃CHCH₃ + CH₃CHCH₃

(b)

(c)

19.20   At low temperatures the product mixture reflects the relative stabilities of the different carbocationic intermediates resulting from electrophilic attack at the ortho, meta, and para positions on the ring. The transition states for the different reaction pathways are assumed to be similar to the different intermediates. Therefore the relative stabilities of the intermediates determine the relative energies of activation for the three reaction pathways, and the rates at which the three products are formed. The reaction at low temperatures is kinetically controlled (see Section 17.3A).

        The sulfonation reaction is reversible. At high temperatures the reverse reaction as well as the forward reaction occurs. Equilibrium is established among the different isomers. The ratio of products at the higher temperature represents the relative stabilities of the products. *p*-Toluenesulfonic acid is thermodynamically the most stable of the products and is the major product at high temperatures. Under these conditions the reaction is thermodynamically controlled (see Section 17.3A).

19.21

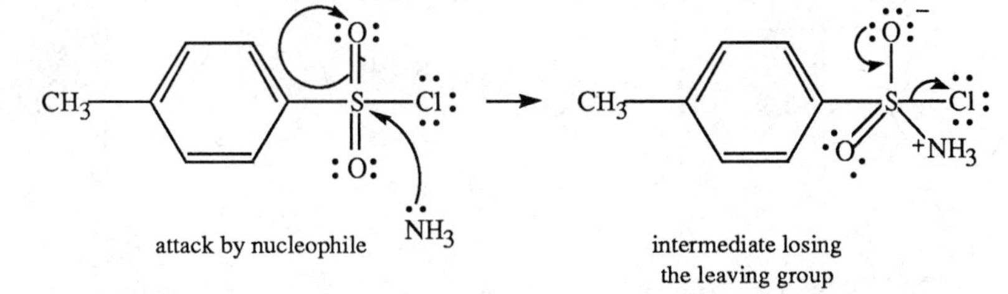

attack by nucleophile   NH₃

four groups around the sulfur atom
in the starting material (note that this
contrasts with three groups around the
carbon atom of the carbonyl group
in acid derivatives) (see Chapter 14,
A Look Ahead)

intermediate losing
the leaving group

five groups around the sulfur
atom in the intermediate

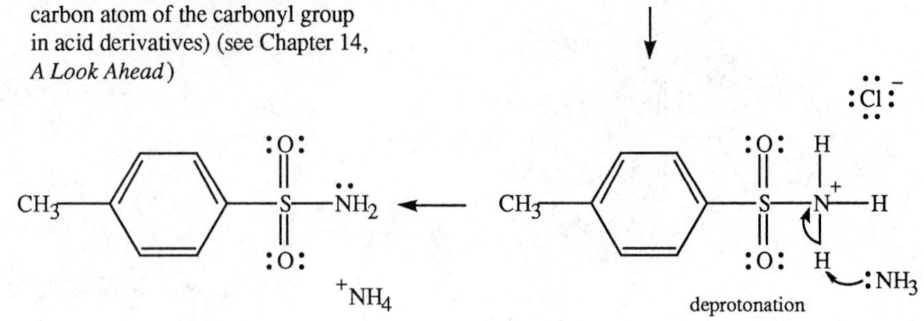

deprotonation

19.22

(a)

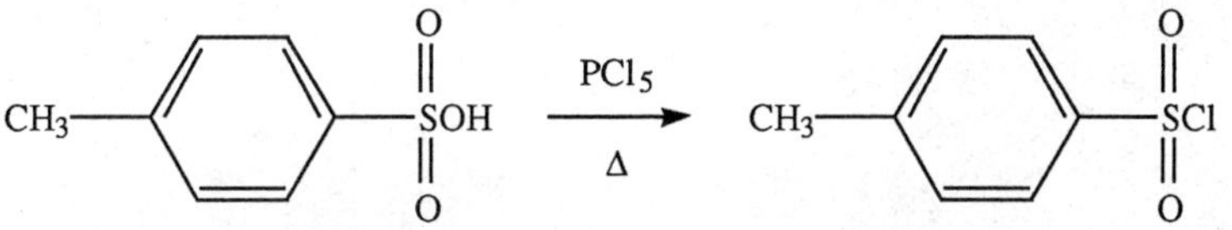

(b)

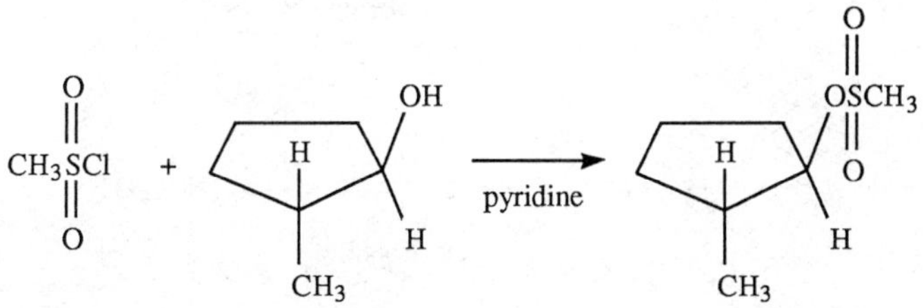

(c)

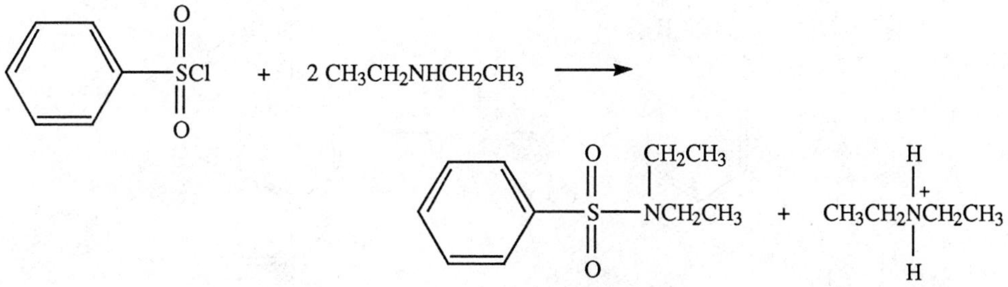

19.22 (cont)

(d)

$$\text{C}_6\text{H}_5\text{-}\underset{\underset{O}{\|}}{\overset{\overset{O}{\|}}{S}}\text{Cl} + \text{CH}_3\text{CH}_2\text{CH}_2\text{CH}_2\text{NH}_2 \xrightarrow[\text{H}_2\text{O}]{\text{NaOH}} \text{C}_6\text{H}_5\text{-}\underset{\underset{O}{\|}}{\overset{\overset{O}{\|}}{S}}\overset{}{N}\text{CH}_2\text{CH}_2\text{CH}_2\text{CH}_3 \quad \text{Na}^+$$

(e)

$$\text{C}_6\text{H}_5\text{-}\underset{\underset{O}{\|}}{\overset{\overset{O \quad \text{Na}^+}{\|}}{S}}\overset{-}{N}\text{CH}_2\text{CH}_2\text{CH}_2\text{CH}_3 \xrightarrow{\text{H}_3\text{O}^+} \text{C}_6\text{H}_5\text{-}\underset{\underset{O}{\|}}{\overset{\overset{O}{\|}}{S}}\text{NHCH}_2\text{CH}_2\text{CH}_2\text{CH}_3$$

(f)

$$\text{C}_6\text{H}_5\text{-}\underset{\underset{O}{\|}}{\overset{\overset{O}{\|}}{S}}\text{Cl} + \text{CH}_3\text{CH}_2\overset{\overset{\text{CH}_2\text{CH}_3}{|}}{N}\text{CH}_2\text{CH}_3 \xrightarrow[\text{H}_2\text{O}]{\text{NaOH}}$$

$$\text{C}_6\text{H}_5\text{-}\underset{\underset{O}{\|}}{\overset{\overset{O}{\|}}{S}}\text{O}^-\text{Na}^+ + \text{CH}_3\text{CH}_2\overset{\overset{\text{CH}_2\text{CH}_3}{|}}{N}\text{CH}_2\text{CH}_3 + \text{Na}^+\text{Cl}^-$$

19.23

$$\text{H}-\ddot{\text{O}}\diagdown_{\text{H}}$$

(a)

$$\text{H}-\ddot{\text{O}}-\ddot{\text{N}}=\ddot{\text{O}} \longrightarrow \text{H}-\overset{+}{\underset{|}{\ddot{\text{O}}}}-\ddot{\text{N}}=\ddot{\text{O}} \longrightarrow \overset{+}{:}\text{N}=\ddot{\text{O}}: \longleftrightarrow :\text{N}=\overset{+}{\ddot{\text{O}}}:$$

$$\text{H}\diagdown\text{B}^+ \qquad\qquad \text{H} \quad :\text{B} \qquad\qquad \text{electrophile}$$

$$\text{C}_6\text{H}_5\text{-}\ddot{\text{N}}(\text{CH}_3)\text{CH}_3 + :\overset{+}{\text{N}}=\ddot{\text{O}}: \longrightarrow \text{CH}_3\ddot{\text{N}}\text{CH}_3\text{-}\text{C}_6\text{H}_4\text{-}\ddot{\text{N}}=\ddot{\text{O}}:$$

19.23 (cont)

(b)   H—O—Cl

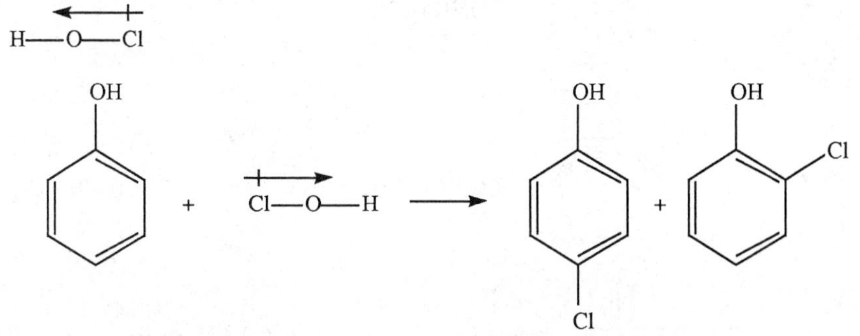

19.24

(a)

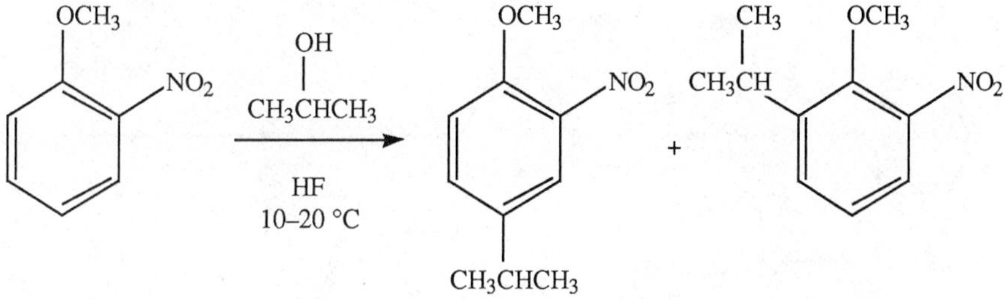

(b)

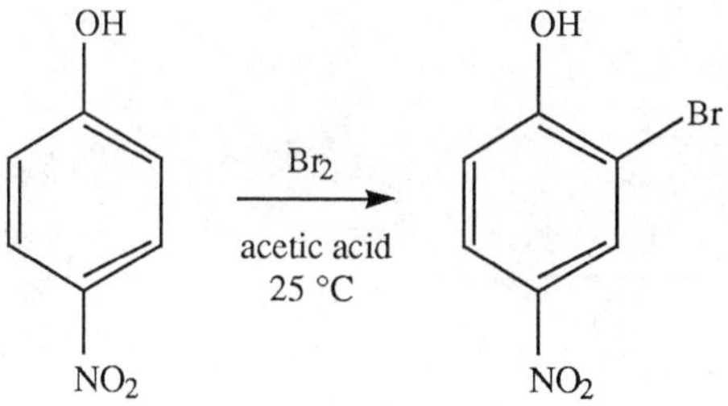

(c)

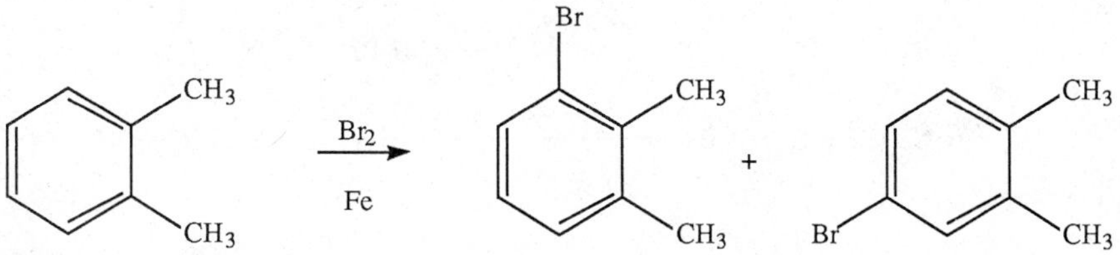

19.25

(a)

$$\text{C}_6\text{H}_5\text{-CH}_2\text{CH}_2\text{CH}_2\text{CH}_3 \xleftarrow{\underset{\Delta}{\text{Zn(Hg), HCl}}}$$

$$\text{C}_6\text{H}_5\text{-CCH}_2\text{CH}_2\text{CH}_3 \;(\text{O}) \xleftarrow[\text{AlCl}_3]{} \text{CH}_3\text{CH}_2\text{CH}_2\text{CCl} \;(\text{O}) \xleftarrow{\text{SOCl}_2}$$

$$\text{CH}_3\text{CH}_2\text{CH}_2\text{COH} \;(\text{O}) \xleftarrow[\substack{\text{H}_2\text{O} \\ \Delta}]{\text{H}_3\text{O}^+ \quad \text{NaOH}} \text{CH}_3\text{CH}_2\text{CH}_2\text{CN} \xleftarrow[\text{ethanol}]{\text{NaCN}} \text{CH}_3\text{CH}_2\text{CH}_2\text{Br}$$

(b)   (1)

$$\text{C}_6\text{H}_5\text{-CH}_2\text{Br} \xleftarrow[\text{pyridine}]{\text{PBr}_3} \text{C}_6\text{H}_5\text{-CH}_2\text{OH} \xleftarrow{\text{H}_3\text{O}^+ \quad \overset{\text{O}}{\overset{\|}{\text{HCH}}}}$$

$$\text{C}_6\text{H}_5\text{-MgBr} \xleftarrow[\substack{\text{diethyl} \\ \text{ether}}]{\text{Mg}} \text{C}_6\text{H}_5\text{-Br} \xleftarrow[\text{Fe}]{\text{Br}_2} \text{C}_6\text{H}_6$$

(2)

$$\underset{\substack{\text{C}_6\text{H}_5\text{-CH}_2 \qquad \text{CH}_2\text{CH}_3 \\ \diagdown \qquad \diagup \\ \text{C}=\text{C} \\ \diagup \qquad \diagdown \\ \text{H} \qquad\qquad \text{H}}}{} \xleftarrow[\substack{\text{Pd/BaSO}_4 \\ \text{quinoline}}]{\text{H}_2}$$

$$\text{C}_6\text{H}_5\text{-CH}_2\text{C}\!\equiv\!\text{CCH}_2\text{CH}_3 \xleftarrow{\text{C}_6\text{H}_5\text{-CH}_2\text{Br}} \text{NaC}\!\equiv\!\text{CCH}_2\text{CH}_3 \xleftarrow[\text{NH}_3 \text{ (liq)}]{\text{NaNH}_2}$$

$$\text{HC}\!\equiv\!\text{CCH}_2\text{CH}_3 \xleftarrow{\text{CH}_3\text{CH}_2\text{Br}} \text{HC}\!\equiv\!\text{CNa} \xleftarrow[\text{NH}_3 \text{ (liq)}]{\text{NaNH}_2} \text{HC}\!\equiv\!\text{CH}$$

19.25 (cont)

(c) (1)   [benzene ring]—CH$_3$ ⟵ $\dfrac{CH_3I}{AlCl_3}$ [benzene ring]

(c) (2)   CH$_3$—[bicyclic ketone structure] ⟵ $\dfrac{}{AlCl_3}$ CH$_3$—[benzene]—CH$_2$CH$_2$—C(=O)Cl ⟵ $\dfrac{SOCl_2}{\Delta}$

CH$_3$—[benzene]—CH$_2$CH$_2$—C(=O)OH ⟵ $\dfrac{Zn(Hg),\ HCl}{\Delta}$ CH$_3$—[benzene]—C(=O)—CH$_2$CH$_2$—C(=O)OH ⟵ $\dfrac{}{AlCl_3}$ [benzene]—CH$_3$

[succinic anhydride structure] ⟵ $\dfrac{(CH_3C)_2O}{\Delta}$ HOCCH$_2$CH$_2$COH ⟵ $\dfrac{H_3O^+}{\begin{array}{c}NaOH\\ H_2O\\ \Delta\end{array}}$

NCCH$_2$CH$_2$CN ⟵ $\dfrac{NaCN\ (excess)}{ethanol}$ BrCH$_2$CH$_2$Br

(d)   [benzene]—C(CH$_3$)(OH)CH$_2$CH$_3$ ⟵ $\dfrac{NH_4Cl}{H_2O}$ $\dfrac{CH_3MgI}{\begin{array}{c}diethyl\\ ether\end{array}}$ [benzene]—C(=O)CH$_2$CH$_3$ ⟵ $\dfrac{CH_3CH_2CCl}{AlCl_3}$ [benzene]

19.26   (a) 3,4-dimethylnitrobenzene

(b) 4-methyl-1-phenylpentane

(c) 2-chloro-6-ethylnaphthalene

(d) 2,4-dinitrotoluene

(e) *m*-bromobenzoic acid

(f) methyl 3,5-dimethylbenzoate

(g) 1-(4-methylphenyl)-1-butanone

(h) *p*-bromobenzaldehyde
4-bromobenzaldehyde

19.27   (a)

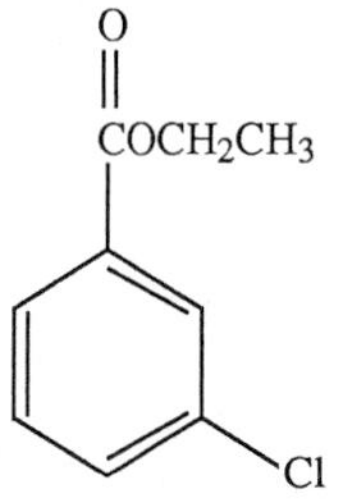

(b)

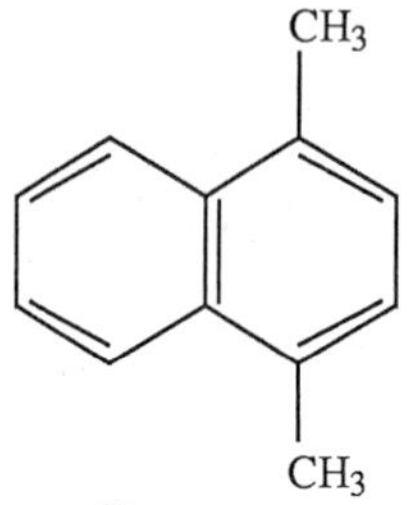

(c)

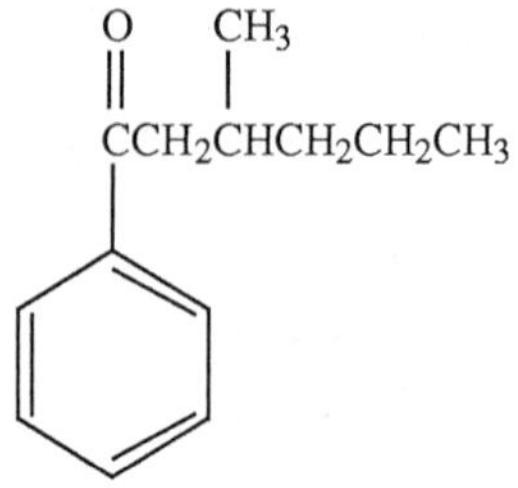

(c)

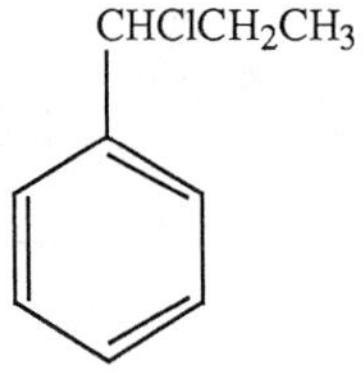

(d)

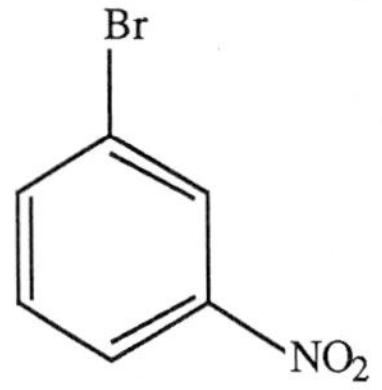

(e)

19.28

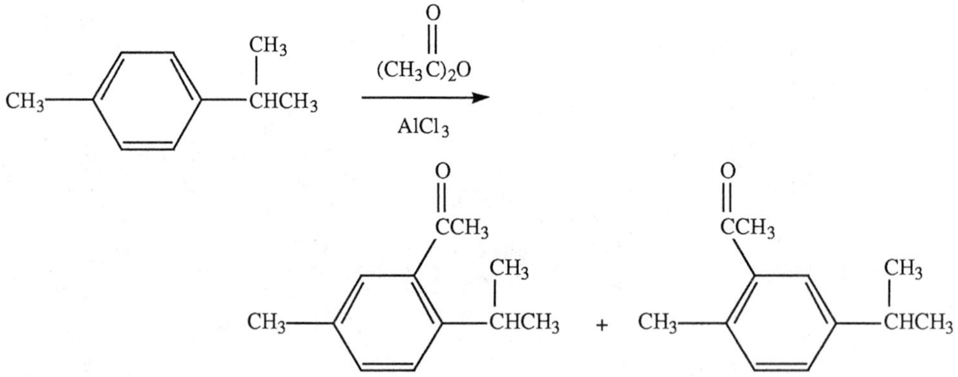

(a)

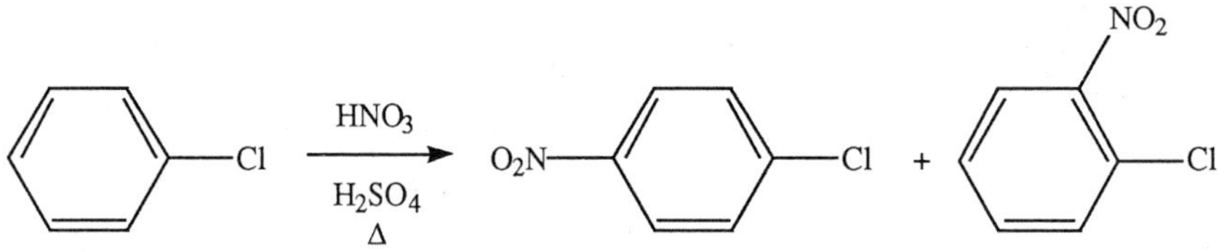

(b)

(c)

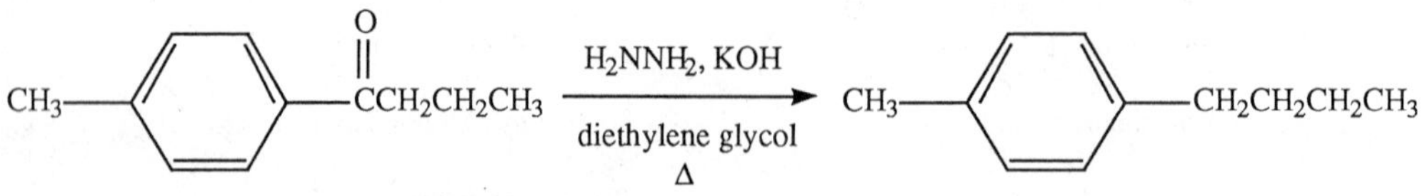

19.28 (cont)

(d)

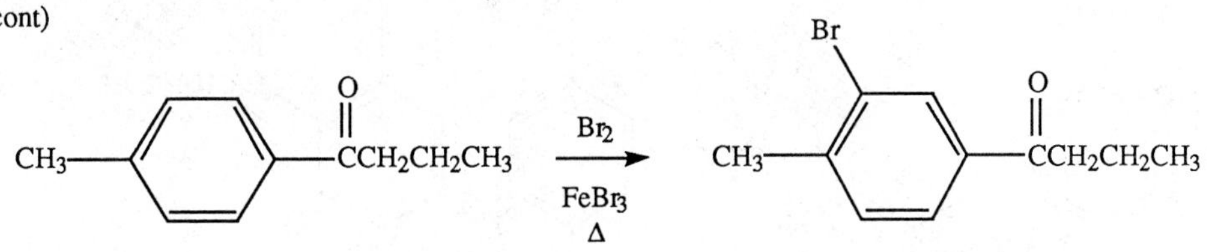

(e)

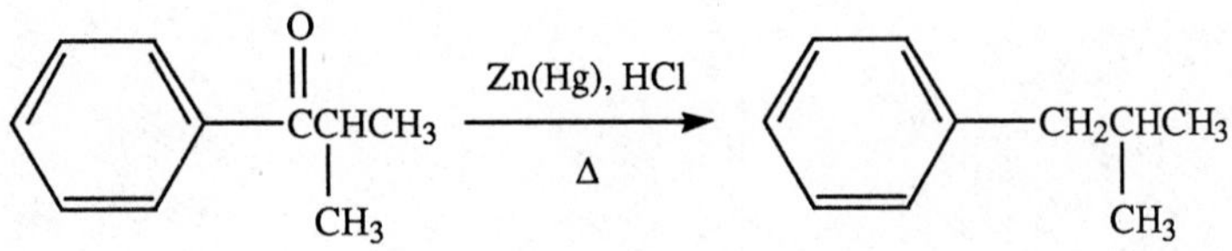

(f)

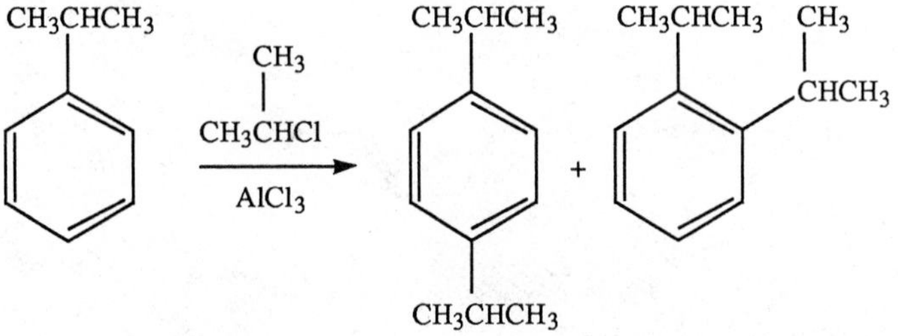

(g)

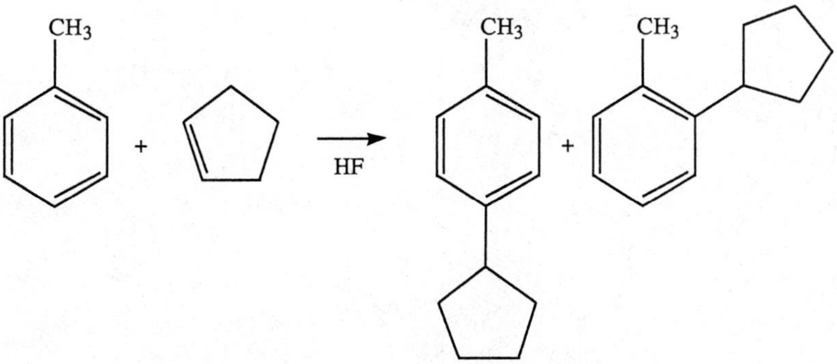

19.28 (cont)

(h)

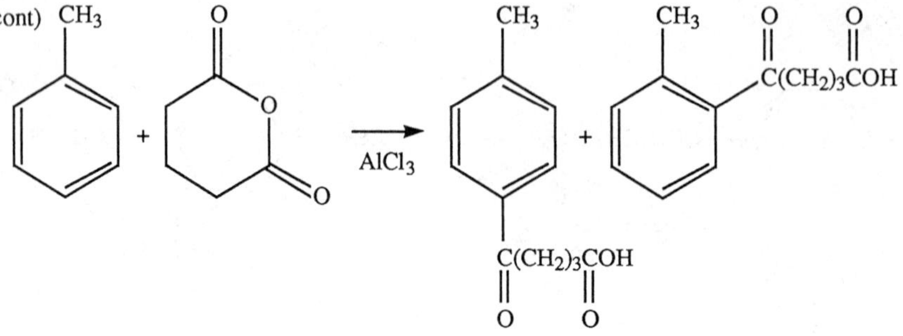

L   M

(i)

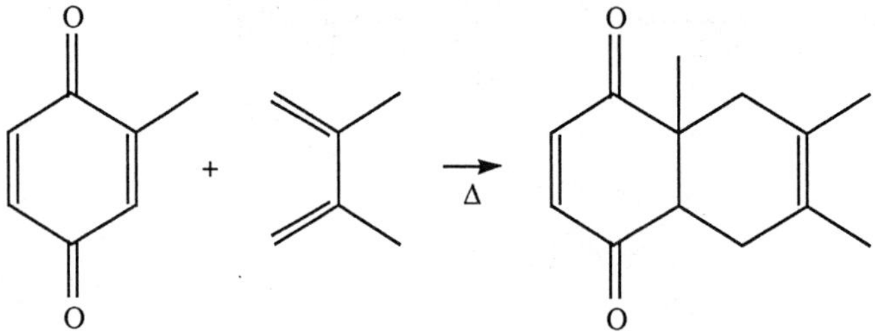

N

(j)

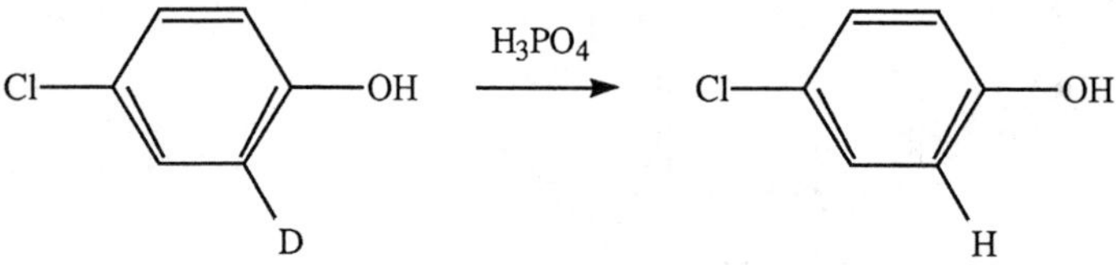

D   H

O

(k)

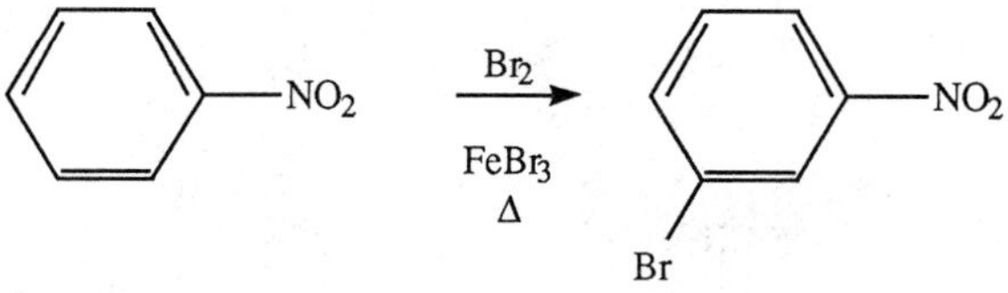

P

(l)

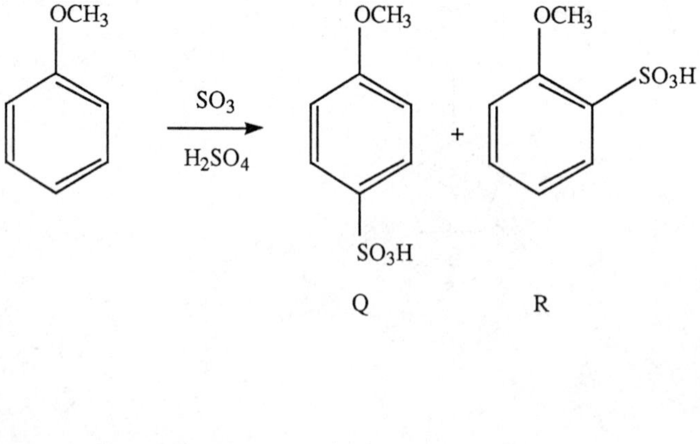

Q   R

19.28 (cont)

(m)

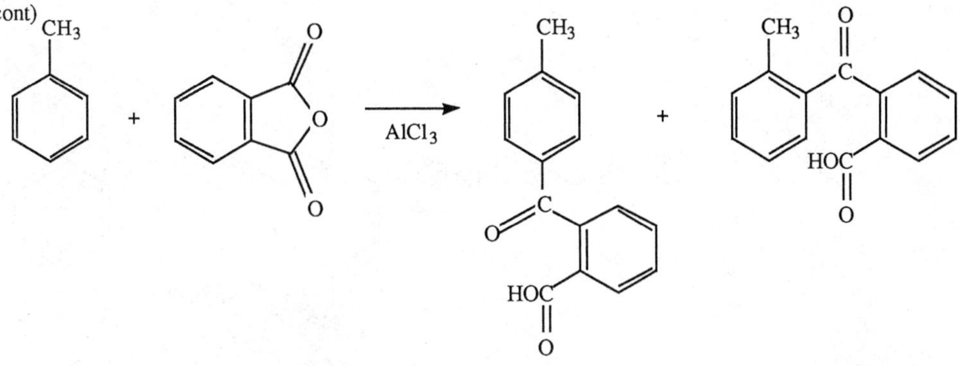

S   T

(n)

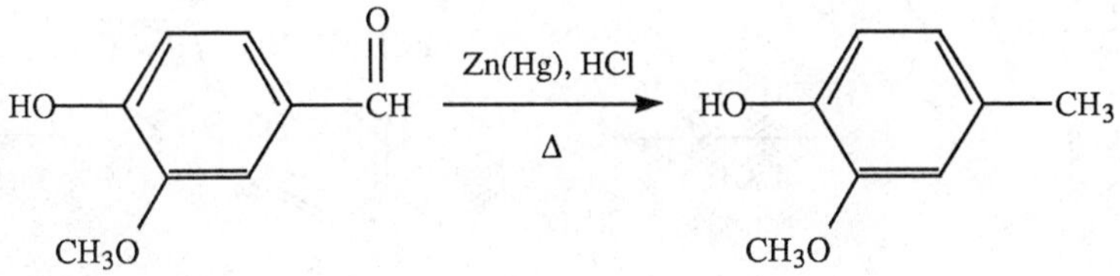

U

(o)

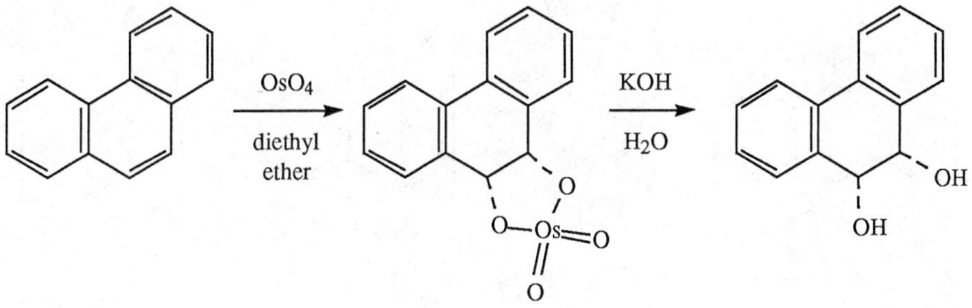

V   W

(p)

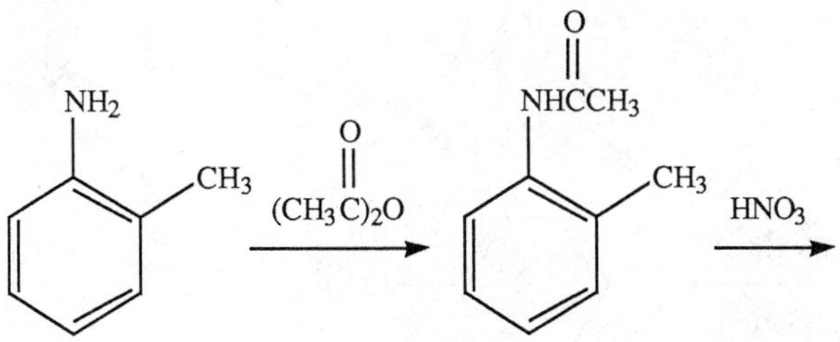

X

19.28 (p) (cont)

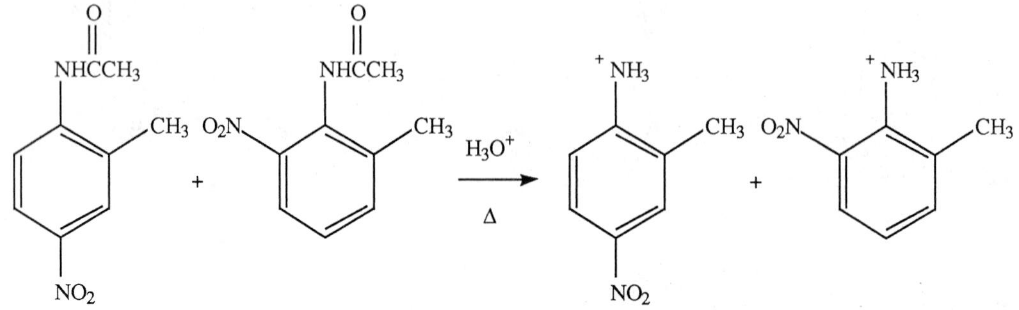

Y   Z   AA   BB

(q)

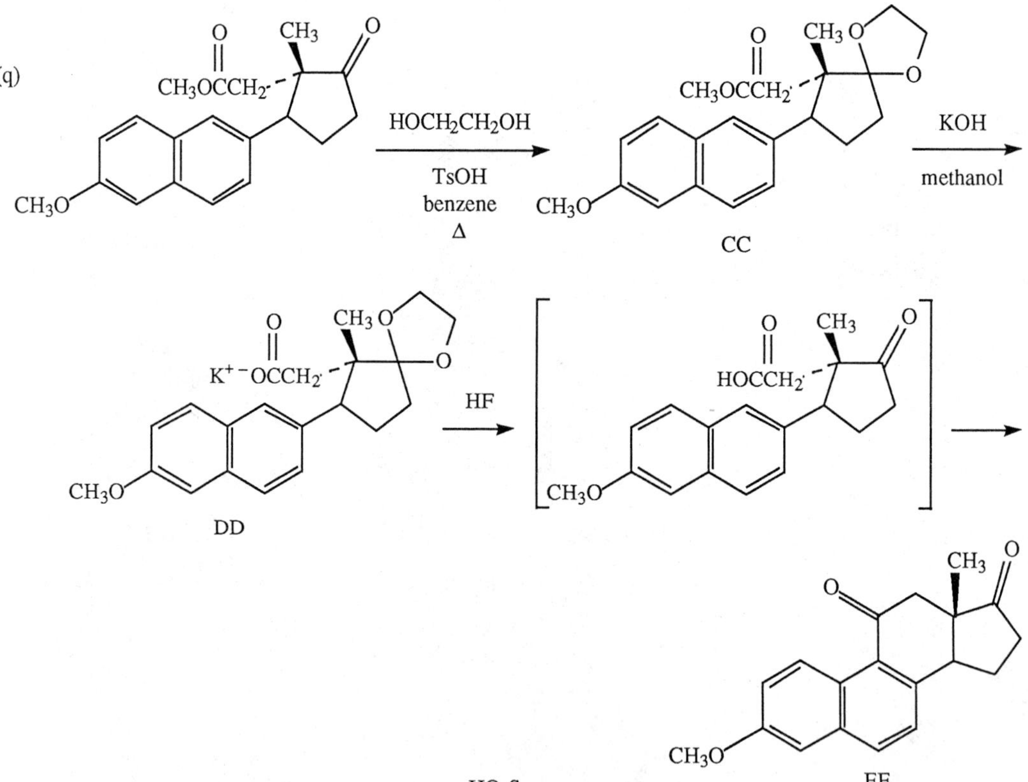

CC

DD

EE

(r)

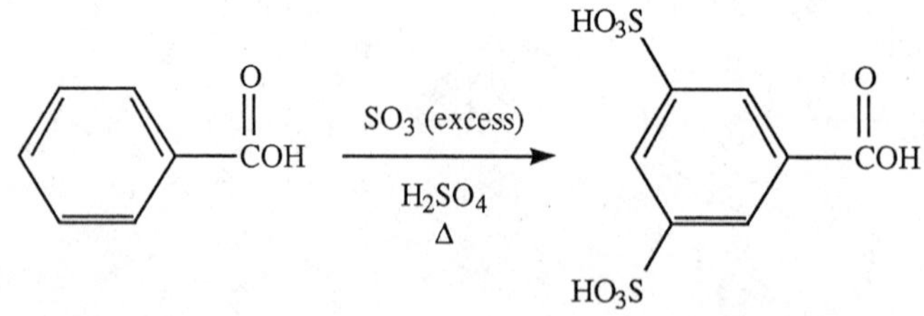

FF

19.29

Mechanism:

S_N1 or S_N2
possible at this stage

19.30   (a)

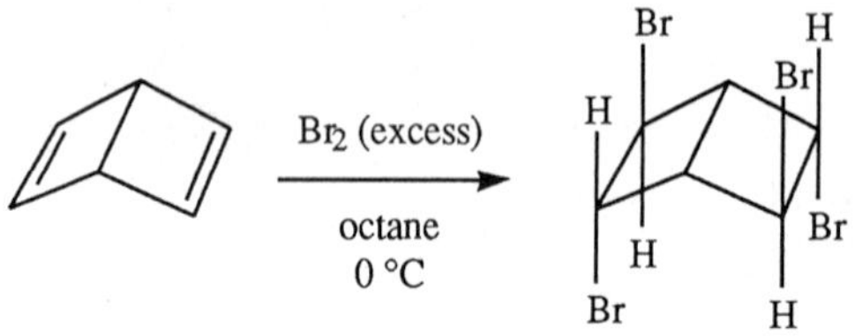

A

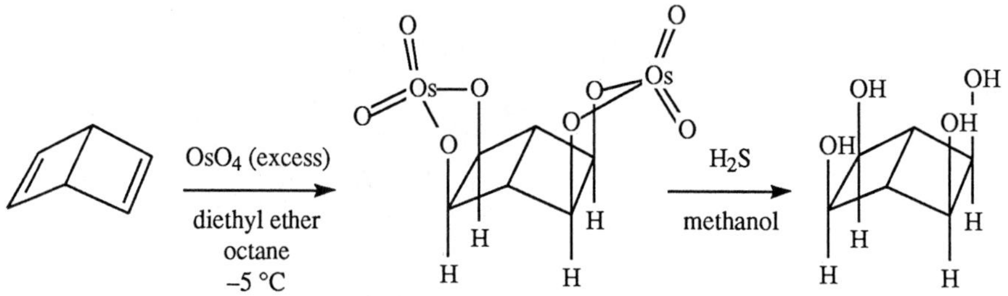

B   C

Only one possible isomer is shown for the reactions above.

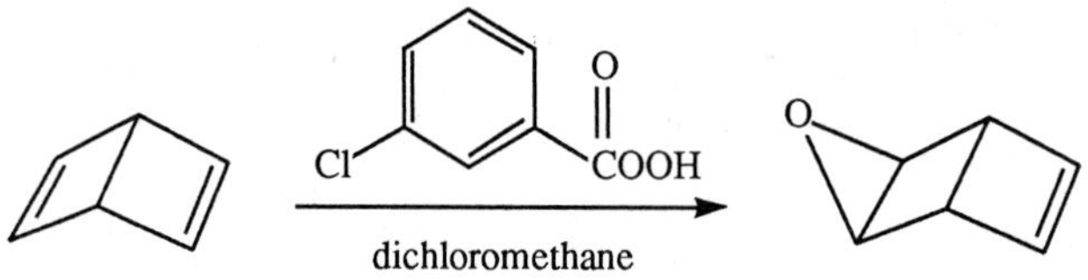

D

(b)

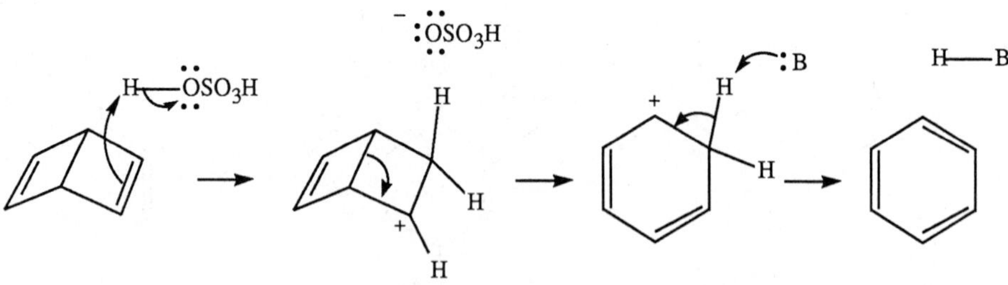

19.31

(a)

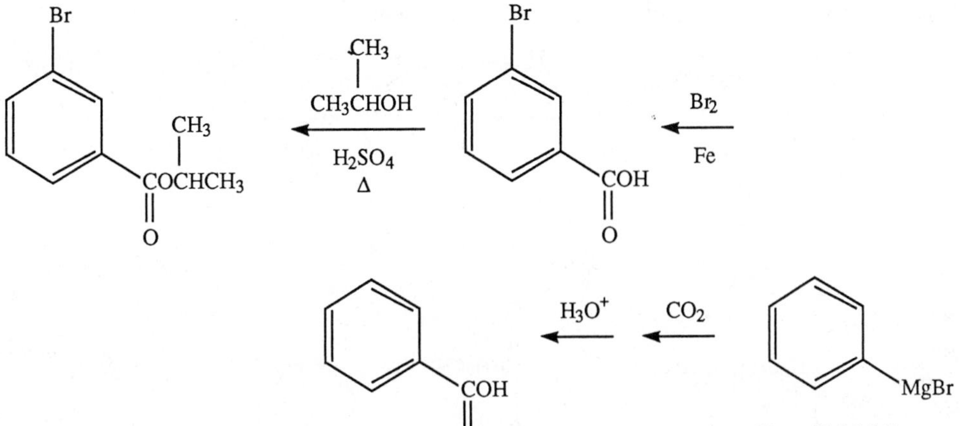

[from 19.25 (b)]

(b)

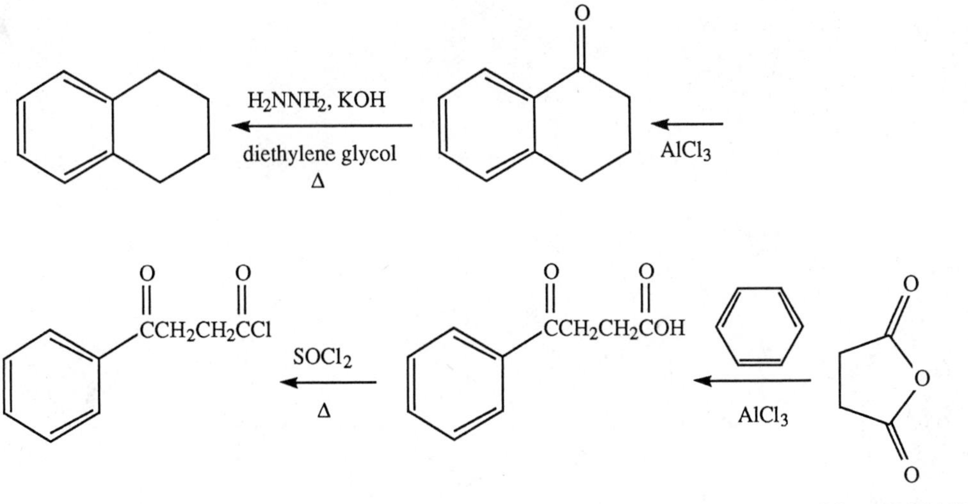

[from 19.25 (c)]

(c)

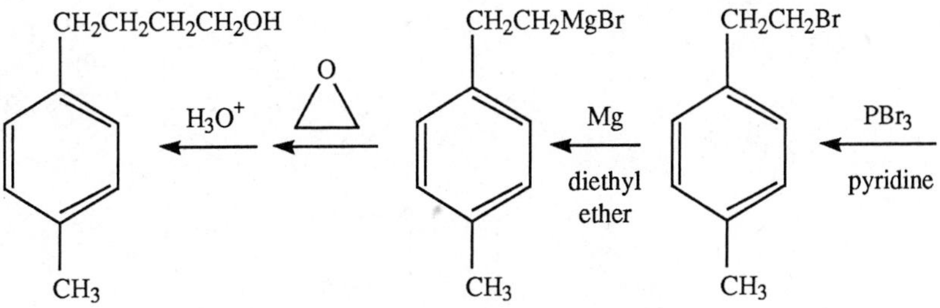

19.31 (c) (cont)

CH$_2$CH$_2$OH

$\xleftarrow{\text{H}_3\text{O}^+}$ ⬡(epoxide) $\xleftarrow{}$ MgBr $\xleftarrow[\text{diethyl ether}]{\text{Mg}}$ Br $\xleftarrow[\text{Fe}]{\text{Br}_2}$ ⬡

(with CH$_3$ substituents on each ring)

(d)

CH$_2$CH$_2$CH$_2$CH$_3$

⬡ $\xleftarrow[\Delta]{\text{Zn(Hg), HCl}}$

O‖CCH$_2$CH$_2$CH$_3$ on ⬡ $\xleftarrow[\text{AlCl}_3]{}$ ⬡ $\xleftarrow{}$ CH$_3$CH$_2$CH$_2$CCl (C=O) $\xleftarrow[\Delta]{\text{SOCl}_2}$

O‖
CH$_3$CH$_2$CH$_2$COH $\xleftarrow[\Delta]{\text{H}_3\text{O}^+}$ CH$_3$CH$_2$CH$_2$C≡N $\xleftarrow[\text{ethanol}]{\text{NaCN}}$ CH$_3$CH$_2$CH$_2$Br

(e)

(1)

O‖CH on ⬡ with CH$_3$ $\xleftarrow[\text{dichloromethane}]{\text{(pyridinium CrO}_3\text{Cl}^-)}$ CH$_2$OH on ⬡ with CH$_3$ $\xleftarrow[]{\text{H}_3\text{O}^+}$ HCH (O=) $\xleftarrow[\text{diethyl ether}]{\text{Mg}}$

Br on ⬡ with CH$_3$ $\xleftarrow[\text{FeBr}_3]{\text{Br}_2}$ ⬡ with CH$_3$

19.31 (e) (cont)

(2)

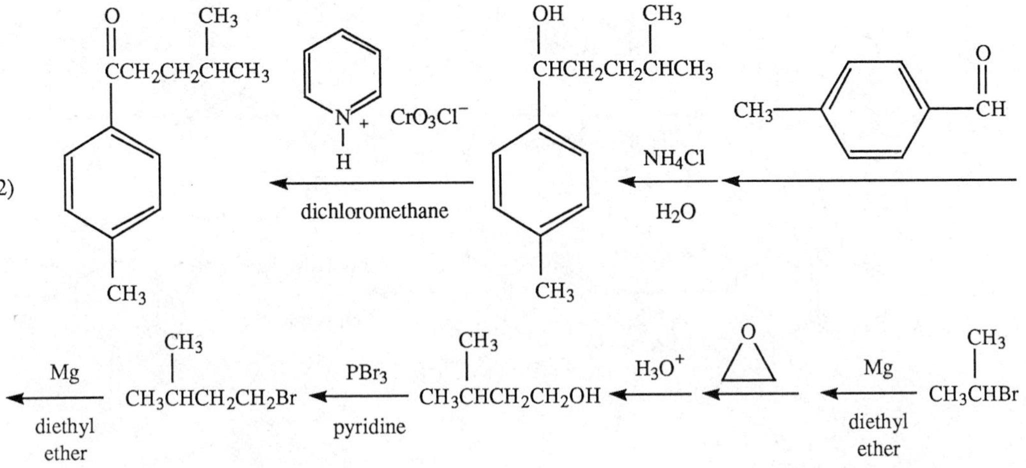

(f)

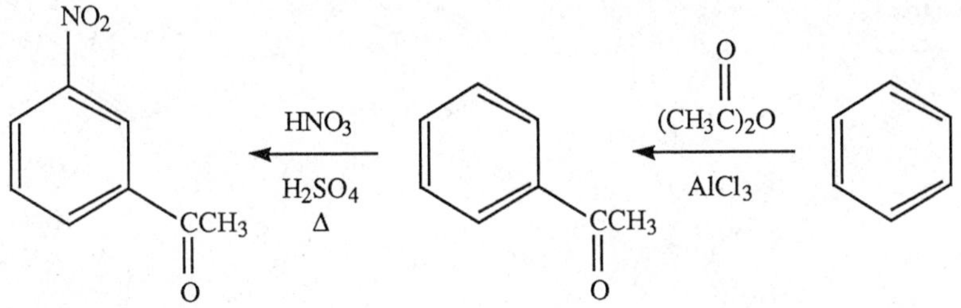

(g)

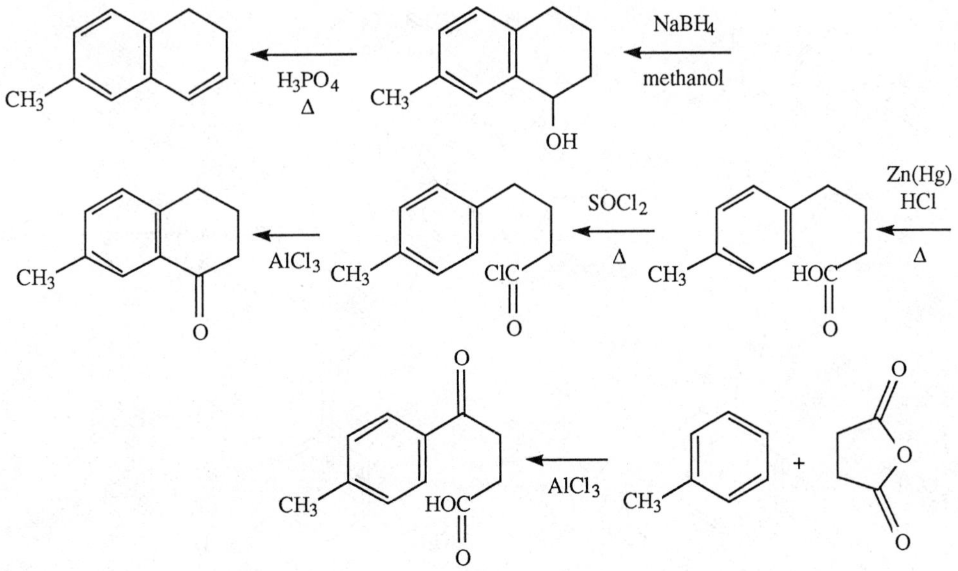

[from 19.25 (c)]

19.31 (cont)

(h)

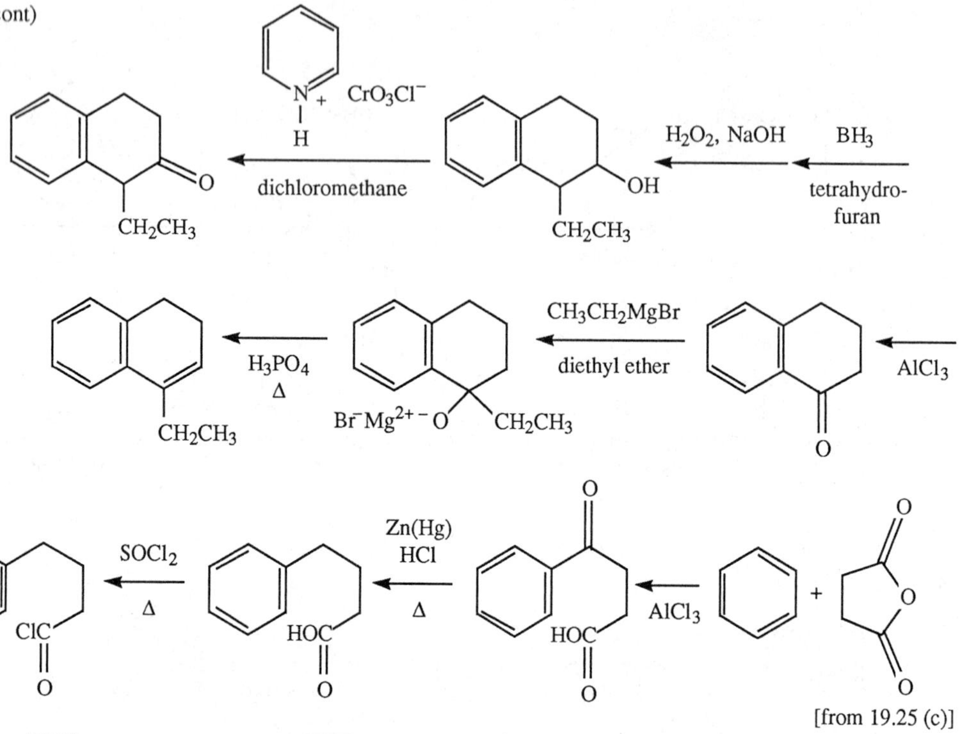

[from 19.25 (c)]

(i)

[from 19.25 (b)]

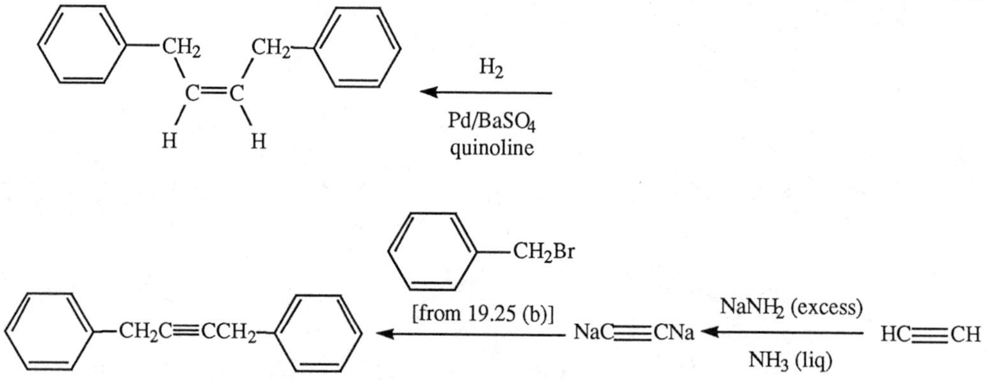

19.32

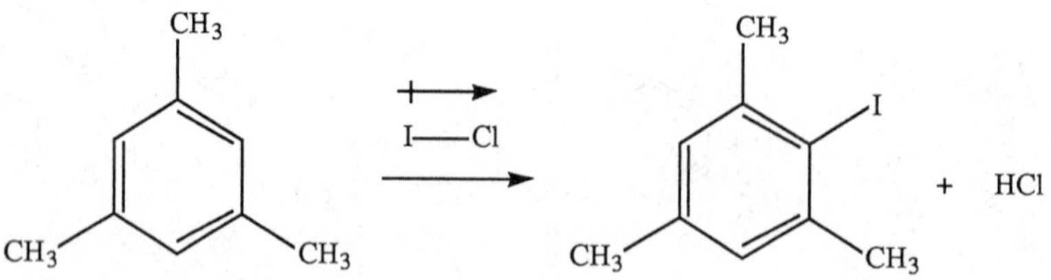

Chlorine is more electronegative than iodine; iodine is, therefore, the electrophile.

19.33

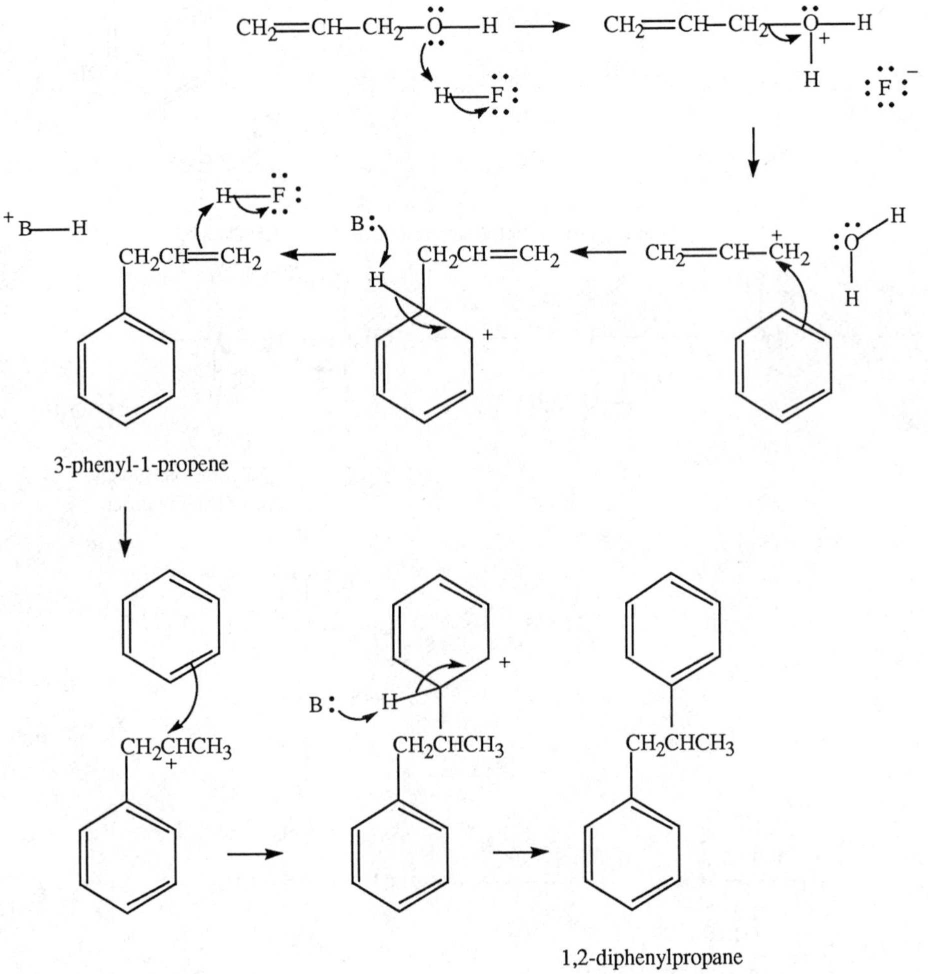

$$CH_2{=}CH{-}CH_2{-}\overset{\cdot\cdot}{\underset{\cdot\cdot}{O}}{-}H \longrightarrow CH_2{=}CH{-}CH_2\overset{+}{\underset{\underset{H}{|}}{\overset{\cdot\cdot}{O}}}{-}H$$

$$H{-}\overset{\cdot\cdot}{\underset{\cdot\cdot}{F}}{:} \qquad\qquad :\overset{\cdot\cdot}{\underset{\cdot\cdot}{F}}:^{-}$$

3-phenyl-1-propene

1,2-diphenylpropane

19.34

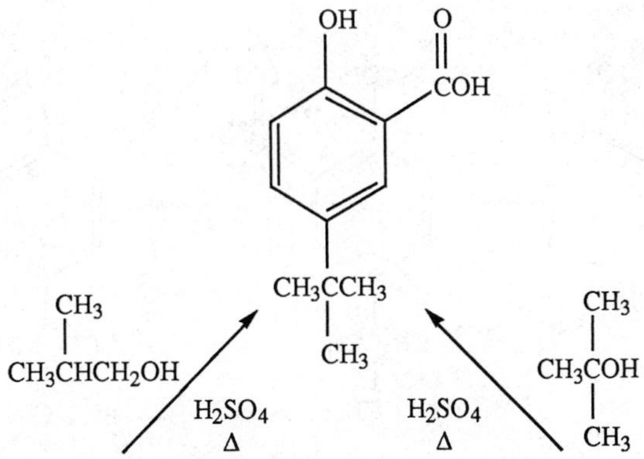

19.34 (cont)

OH O
‖
.COH

OH O
‖
.COH

Both alcohols form the same carbocationic intermediate.

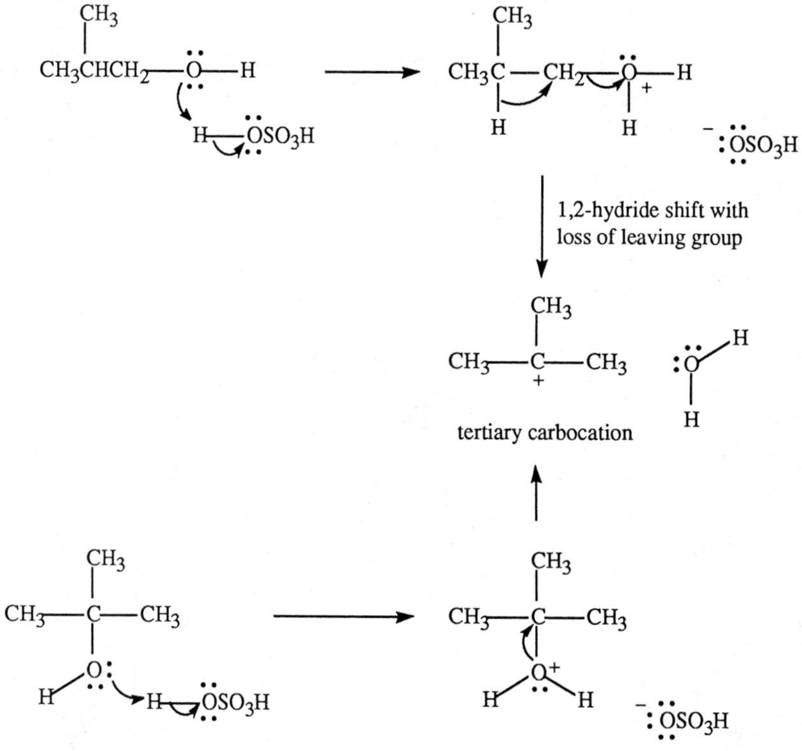

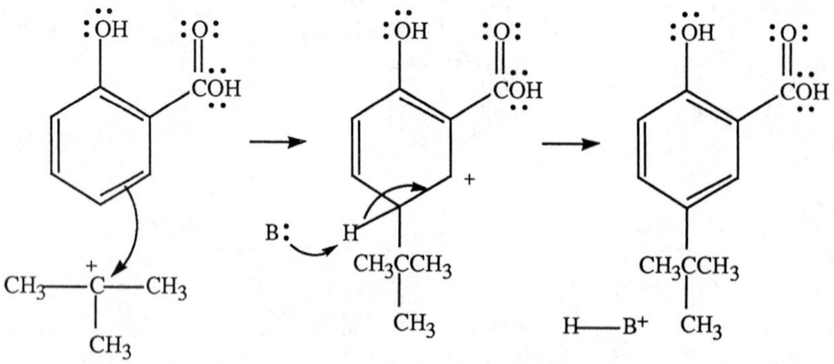

CH₃
|
CH₃CHCH₂—O̤—H ⟶ CH₃C—CH₂—O̤⁺—H
        H—O̤SO₃H   H   H   ⁻:O̤SO₃H

1,2-hydride shift with
loss of leaving group

↓

CH₃
|
CH₃—C—CH₃   :O⟨H
     + |
                    H

tertiary carbocation

↑

CH₃   CH₃
| |
CH₃—C—CH₃   ⟶   CH₃—C—CH₃
     O̤:   O⁺
H   H—O̤SO₃H   H   H
                              ⁻:O̤SO₃H

:O̤H :O̤:   :O̤H :O̤:   :O̤H :O̤:
      ‖   ‖   ‖
      .CO̤H   .CO̤H   .CO̤H

         ⟶   ⟶

                    B: H   +

CH₃—C⁺—CH₃   CH₃CCH₃   CH₃CCH₃
     | | |
     CH₃   CH₃   H—B⁺ CH₃

19.35

Azulene has a planar π system with 10 π electrons, so we would expect the compound to have aromaticity. It has $4n + 2$ electrons with $n = 2$. One particularly interesting resonance contributor for azulene may be regarded as the juxtaposition of an aromatic tropylium ion with an aromatic cyclopentadienyl anion.

tropylium ion $\longrightarrow$ $\longleftarrow$ cyclopentadienyl anion

19.36   The basicity of an amine is determined by the availability of the pair of nonbonding electrons. The nonbonding electrons in ammonia are localized on the nitrogen atom and are readily available for protonation. The nonbonding electrons in pyrrole and indole are part of an aromatic sextet. Protonation of the nitrogen atom in either one of these compounds leads to loss of aromaticity. The conjugate acids of pyrrole and indole are, therefore, of higher energy relative to the bases than the ammonium ion is relative to ammonia.

The acidity of a compound is determined by the relative stabilities of the acid and its conjugate base. When ammonia loses a proton, the amide anion that is formed has the negative charge concentrated on the nitrogen atom. Deprotonation of pyrrole or indole gives anions that are stabilized by resonance. The negative charge is delocalized to other atoms in the ring(s). The conjugate bases of pyrrole and indole are, therefore, weaker bases than the amide anion and pyrrole and indole are thus stronger acids than ammonia.

$:NH_3$ $\longrightarrow$ $^-:NH_2$ anion with localized charge

pyrrole   anion with delocalization of charge   the charge can be delocalized to all four carbons

indole   anion with delocalization of charge

19.37 $CH_3 \overset{\overset{\displaystyle :O:}{\parallel}}{\underset{\bullet\bullet}{S}}CH_3$ +   NaH $\longrightarrow$ Na^+ $^-:CH_2 \overset{\overset{\displaystyle :O:}{\parallel}}{\underset{\bullet\bullet}{S}}CH_3$ + $H_2\uparrow$

cyclononatetraene $4n + 2 = 10$

$4n + 2 = 10$

Both anions have $4n + 2 = 10$ π electrons delocalized in planar rings and are, therefore, aromatic species. They have aromatic stability just as the tropylium ion and the cyclopentadienyl anion do (Section 19.1D)

19.38

Resonance contributors for heptafulvene have the positive charge delocalized over the seven-membered ring in what is a tropylium ion, but the negative charge localized on a carbon atom. In the dicyano compound, the negative charge can be delocalized onto the two nitrogen atoms. The two electron-withdrawing cyano groups thus stabilize the negative character at the carbon atom outside the ring.

19.38 (cont)

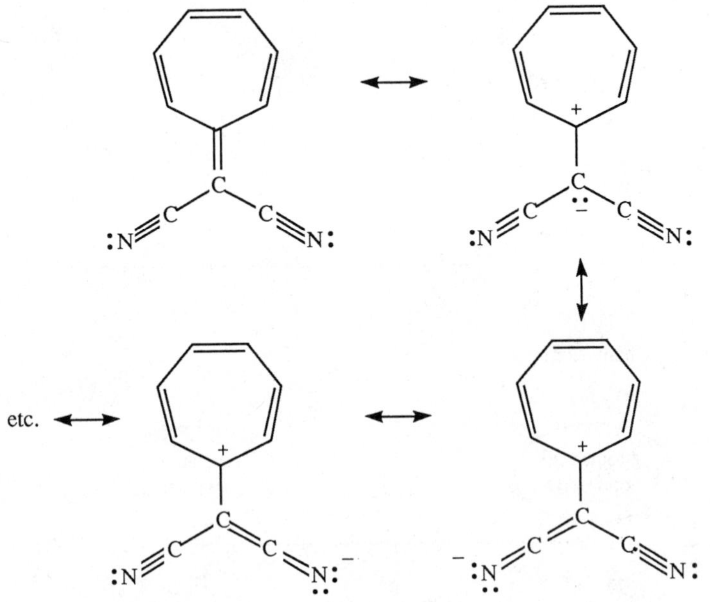

19.39

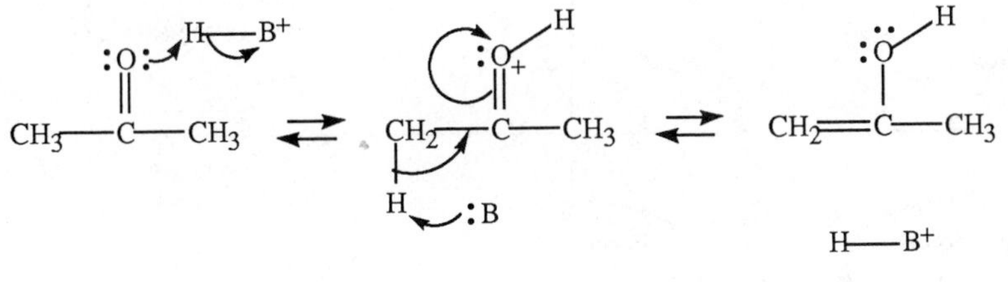

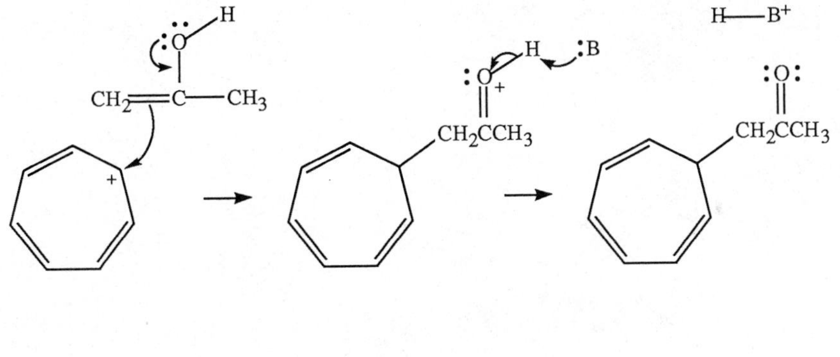

20

Free Radicals

Concept Map 20.1 Chain reactions.

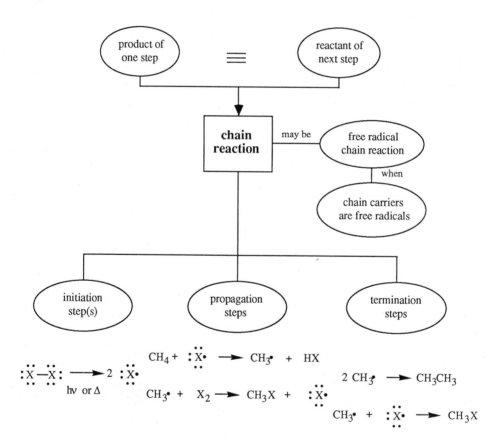

20.1 Propane has six primary hydrogen atoms and two secondary hydrogen atoms for a total of eight hydrogen atoms. If there were no difference in reactivity between primary and secondary hydrogen atoms, for every six primary hydrogen atoms that react, two secondary hydrogen atoms would react, and the product mixture would consist of 75% 1-chloropropane and 25% 2-chloropropane. When the difference in reactivity is considered, for every six primary hydrogen atoms that react, 2×2.5 or five secondary hydrogen atoms would react. The product mixture should thus consist of 55% 1-chloropropane and 45% 2-chloropropane.

$$\frac{6}{6+5} \times 100 = 55\%$$

$$\frac{5}{6+5} \times 100 = 45\%$$

2-Methylpropane has nine primary hydrogen atoms and one tertiary hydrogen atom. If there were no difference in reactivity between primary and tertiary hydrogen atoms, the product mixture would consist of 90% 1-chloro-2-methylpropane (isobutyl chloride) and 10% 2-chloro-2-methylpropane (*tert*-butyl chloride). When the difference in reactivity is considered, for every nine primary hydrogen atoms that react, 1×4 or four tertiary hydrogen atoms would react. The product mixture should thus consist of 69% 1-chloro-2-methylpropane and 31% 2-chloro-2-methylpropane.

$$\frac{9}{9+4} \times 100 = 69\%$$

$$\frac{4}{9+4} \times 100 = 31\%$$

20.2 The selectivity observed at 300 °C depends on small differences between energies of activation for the abstraction of primary, secondary, and tertiary hydrogen atoms. As the temperature increases, the average kinetic energy of chlorine atoms and of alkane molecules increases, and the number of collisions that have energies in excess of the energy of activation for the abstraction of the different kinds of hydrogen atoms also increases. Differences in energies of activation for different reactions become less and less relevant. We see a leveling off of the selectivity of the reactions. If the temperature is high enough, almost all molecules will have energy equal to or greater than the activation energies for any hydrogen to be abstracted, and the rates of reaction would be the same.

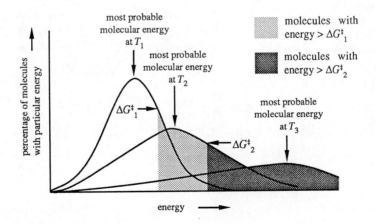

Concept Map 20.2   Halogenation of alkanes.

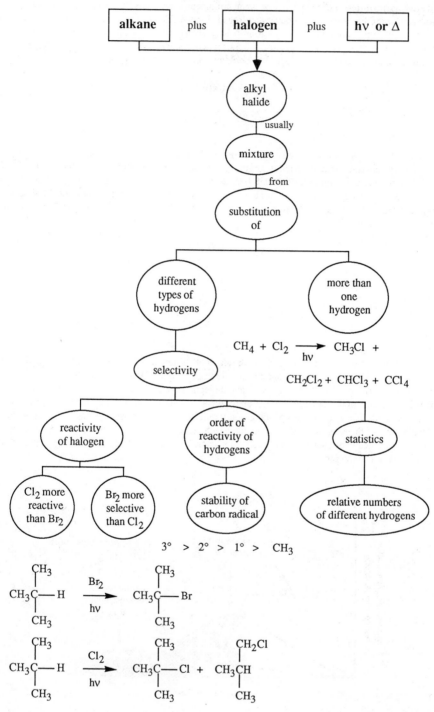

20.3   Neopentane (2,2-dimethylpropane) has only primary hydrogen atoms. The enthalpy for the abstraction of a primary hydrogen atom by a chlorine atom is –5 kcal/mol. The activation energy for this exothermic step is small (~1 kcal/mol, p. 826 in text), and the rate correspondingly fast. The enthalpy for the abstraction of a primary hydrogen atom by a bromine atom, on the other hand, is +12 kcal/mol. The activation energy must be at least as large as the enthalpy, and the rate correspondingly slow.

20.4

(a)

(b)

$$CH_3CHCH_2CH{=}CHCH_3 \xrightarrow[\text{carbon tetrachloride}]{\text{NBS}} CH_3CHCHCH{=}CHCH_3$$

(c)

(d)

This product is reported to form in 58% yield. Another product is also possible, the one in which the allylic hydrogen atom adjacent to the oxygen atom is replaced.

Usually an ether oxygen stabilizes a radical at an adjacent carbon atom. The following resonance contributors are written to explain the effect.

In this case, the electron-withdrawing carbonyl group of the oxygen atom would make the second resonance contributor unfavorable. The product, with two good leaving groups on the same allylic carbon atom, might also be too unstable to isolate.

20.5

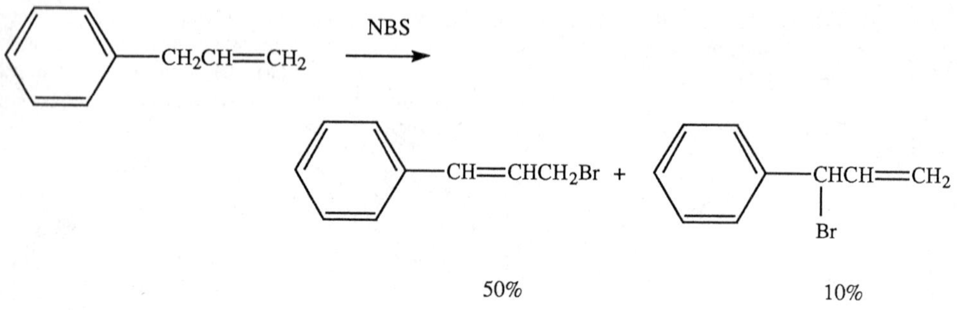

$CH_2CH{=}CH_2$
NBS
$CH{=}CHCH_2Br$ +
$CHCH{=}CH_2$
Br
50%
10%

Initiation step:

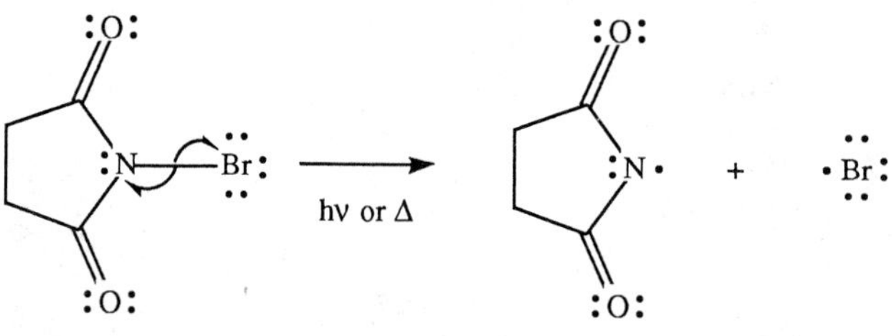

hν or Δ

The role of benzoyl peroxide is not known.

Propagation steps:

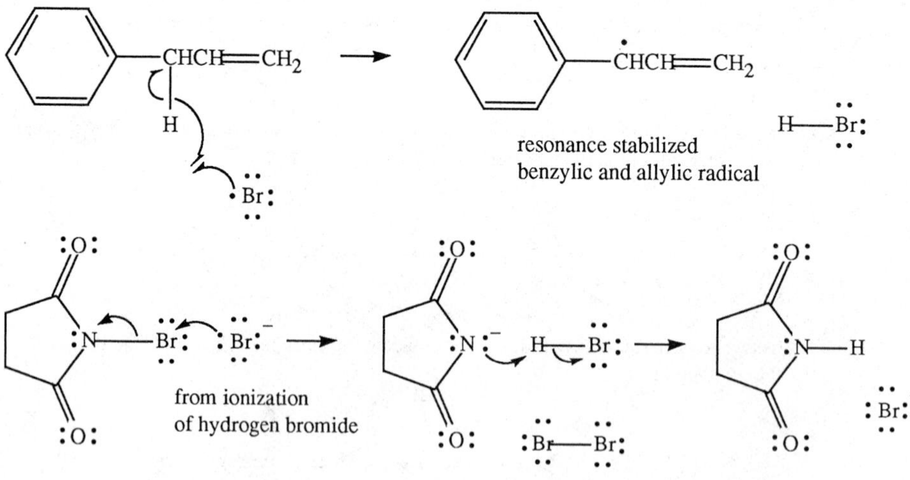

resonance stabilized
benzylic and allylic radical
from ionization
of hydrogen bromide

20.5 (cont)

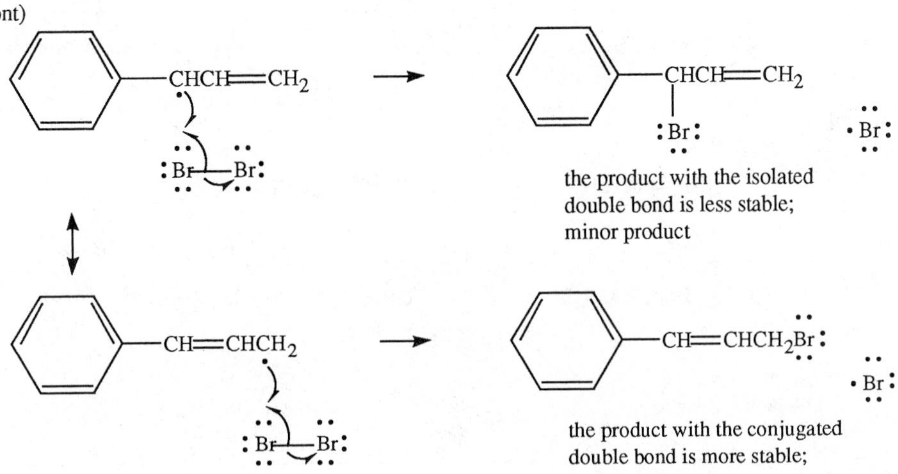

the product with the isolated
double bond is less stable;
minor product

the product with the conjugated
double bond is more stable;
major product

20.6 $CH_3CH_2CH_2CH_2CH\!=\!CHCH_3$ $\xrightarrow{\text{NBS}}$ $CH_3CH_2CH_2CHCH\!=\!CHCH_3$
$\underset{\displaystyle Br}{|}$

60%

$CH_3CH_2CH_2CHCH\!=\!CHCH_3$ $\longleftrightarrow$ $CH_3CH_2CH_2CH\!=\!CHCHCH_3$

$\downarrow Br_2$ $\qquad\qquad\qquad\qquad\qquad\qquad\downarrow Br_2$

$CH_3CH_2CH_2CHCH\!=\!CHCH_3$ $\qquad\qquad$ $CH_3CH_2CH_2CH\!=\!CHCHCH_3$
$\underset{\displaystyle Br}{|}$ $\qquad\qquad\qquad\qquad\qquad\quad$ $\underset{\displaystyle Br}{|}$

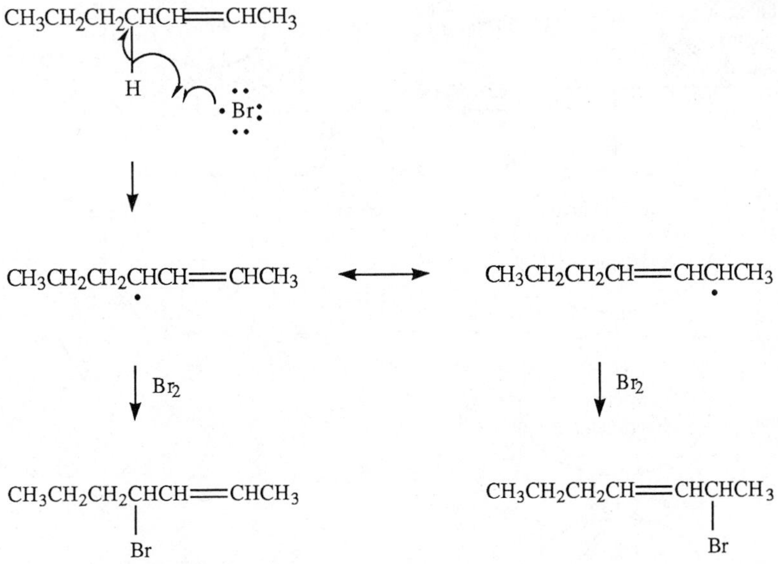

4-bromo-2-heptene $\qquad\qquad\qquad\qquad$ 2-bromo-3-heptene

20.6 (cont)

$$CH_3CH_2CH_2CH_2CH\!=\!CHCH_2$$

H $\cdot\,\ddot{\underset{\cdot\cdot}{Br}}:$

$$CH_3CH_2CH_2CH_2CH\!=\!CHCH_2 \quad\longleftrightarrow\quad CH_3CH_2CH_2CH_2CHCH\!=\!CH_2$$

$\downarrow Br_2$ $\downarrow Br_2$

$$CH_3CH_2CH_2CH_2CH\!=\!CHCH_2Br \qquad\qquad CH_3CH_2CH_2CH_2CHCH\!=\!CH_2$$
$$\underset{Br}{|}$$

1-bromo-2-heptene   3-bromo-1-heptene

- -

Concept Map 20.3   Selective free radical halogenations.

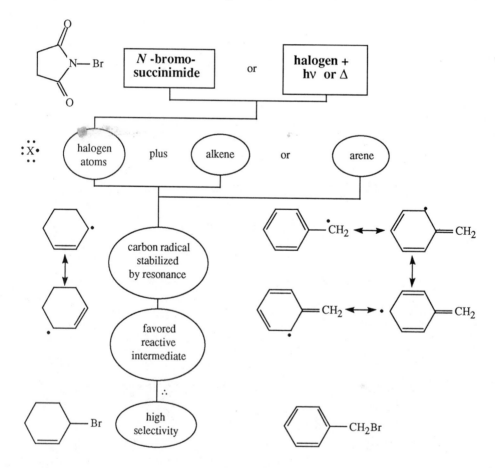

20.7

(a)

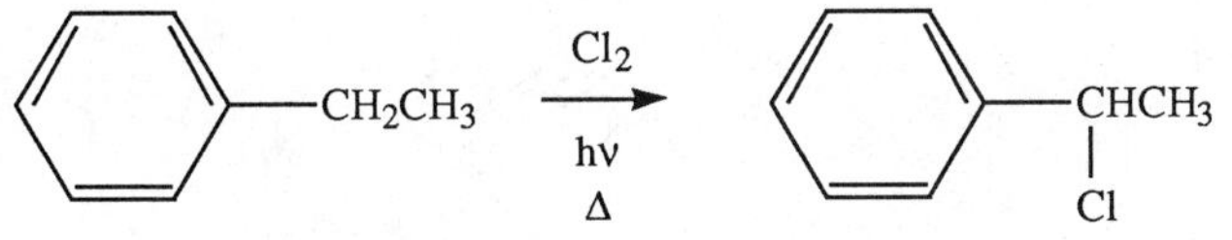

major product

(b)

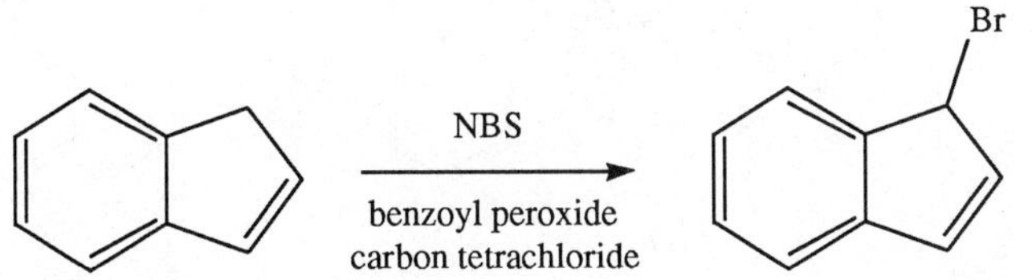

(c)

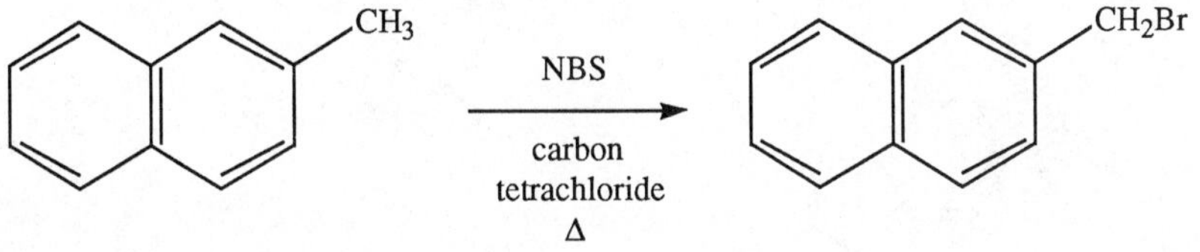

(d)

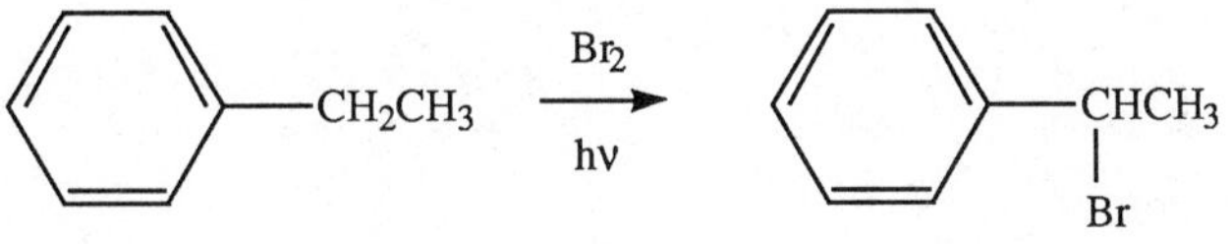

major product

20.8   The unpaired electron can be delocalized over three rings.

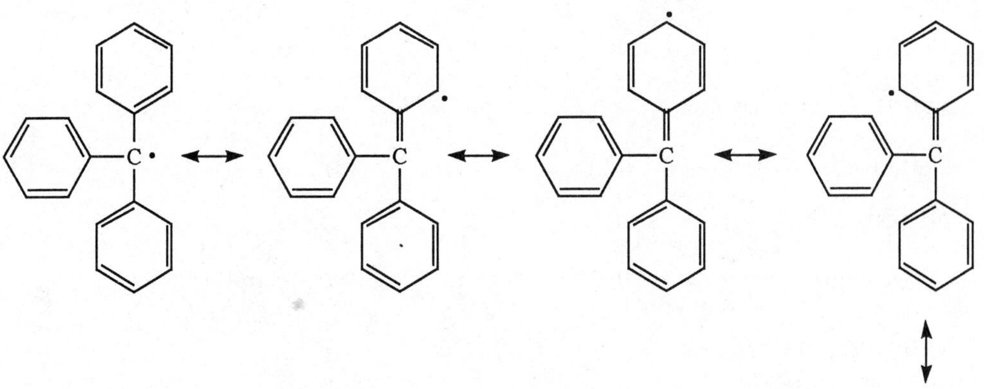

20.8 (cont)

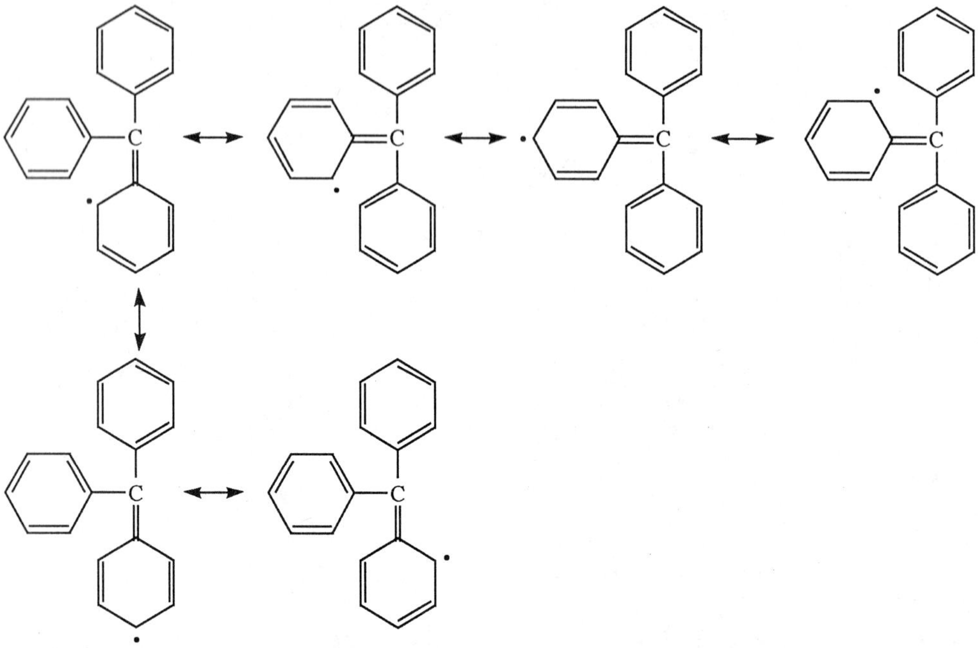

20.9

$$CH_3CHCH_2-H \longrightarrow CH_3CHCH_2 + H\cdot \qquad DH° \quad 98 \text{ kcal/mol}$$

(with CH_3 substituent on the central carbon)

$$CH_3C-H \longrightarrow CH_3C\cdot + H\cdot \qquad DH° \quad 91 \text{ kcal/mol}$$

(with two CH_3 substituents)

$\Delta(DH°) = 98 - 91 = 7 \text{ kcal/mol}$

$$CH_3CH_2CH_2-H \longrightarrow CH_3CH_2CH_2 + H\cdot \qquad DH° \quad 98 \text{ kcal/mol}$$

$$CH_3CH-H \longrightarrow CH_3CH + H\cdot \qquad DH° \quad 95 \text{ kcal/mol}$$

(with CH_3 substituent)

$\Delta(DH°) = 98 - 95 = 3 \text{ kcal/mol}$

20.10   The change in mechanism does not change the relative energy levels for propene, hydrogen bromide, *n*-propyl bromide, and isopropyl bromide because the energies of starting material and products are dependent only on the states (i.e., temperature, pressure, physical state, etc.) and not on how the starting material is converted to product. As long as the initial and final states remain the same in the two reactions, the relative energy levels remain the same. What does change is the nature of the intermediates and their relative energies, as well as the activation energies for each step.

20.11

1.

$$CH_3CO\!-\!OCCH_3 \xrightarrow{\ h\nu\ } 2\ CH_3CO\cdot$$

$$CH_3CO\cdot \quad :Br\!-\!CCl_3 \longrightarrow \ \cdot CCl_3 \ + \ CH_3COBr$$

$$CH_2\!=\!CH(CH_2)_5CH_3$$

$$Cl_3CCH_2CH(CH_2)_5CH_3 \longleftarrow Cl_3CCH_2CH(CH_2)_5CH_3$$
$$\underset{\displaystyle :Br:}{|} \qquad\qquad\qquad\qquad :Br\!-\!CCl_3$$

$$\cdot CCl_3$$

2.

$$CH_3CH_2S\!-\!H \xrightarrow{\ \Delta\ } CH_3CH_2S\cdot \ + \ \cdot H$$

$$CH_2\!=\!CCH_3$$
$$\underset{\displaystyle CH_3}{|}$$

$$\underset{\displaystyle CH_3}{\overset{\displaystyle CH_3}{|}}$$
$$CH_3CH_2SCH_2CHCH_3 \longleftarrow CH_3CH_2SCH_2CCH_3$$

$$CH_3CH_2S\cdot \qquad\qquad\qquad H\!-\!SCH_2CH_3$$

20.11 (cont)

3.

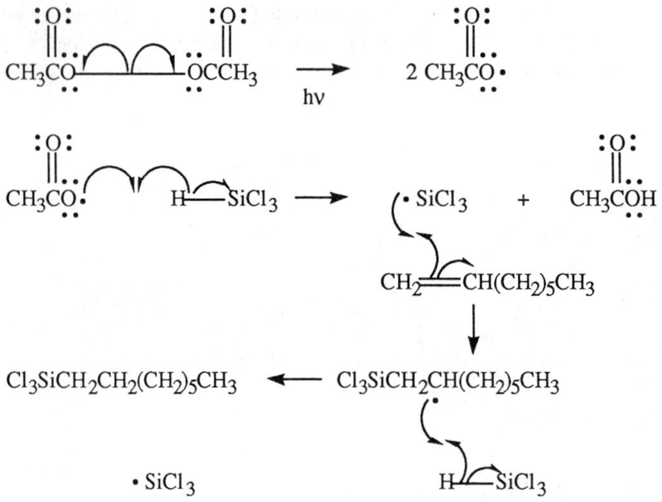

$$CH_3\overset{:O:}{\underset{..}{C}}O\frown\frown O\overset{:O:}{\underset{..}{C}}CH_3 \xrightarrow{h\nu} 2\ CH_3\overset{:O:}{\underset{..}{C}}O\cdot$$

20.12

(a)

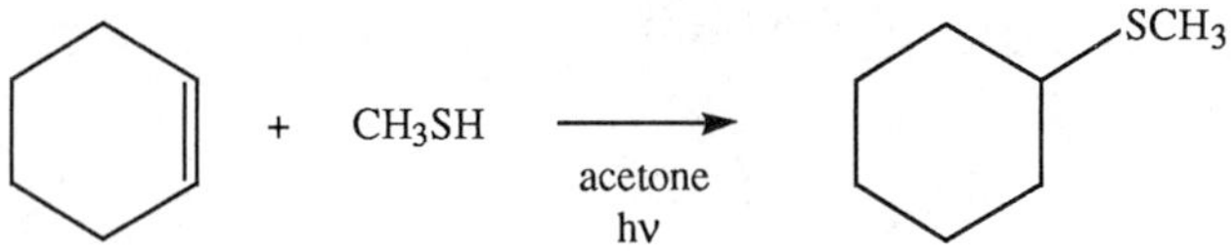

(b) $$CH_3(CH_2)_5CH=CH_2 \xrightarrow[\text{peroxides}]{HBr} CH_3(CH_2)_7Br$$

(c)

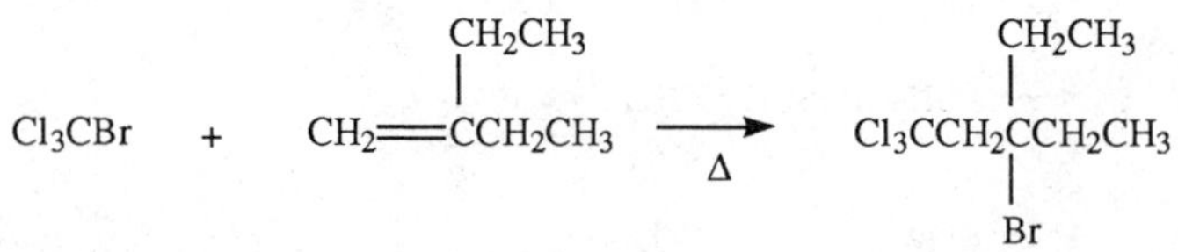

(d) $$CH_3(CH_2)_4SiH_3 + CH_3(CH_2)_5CH=CH_2 \xrightarrow{\text{peroxides}} CH_3(CH_2)_4SiH_2(CH_2)_7CH_3$$

Concept Map 20.4   Free radical addition reactions of alkenes.

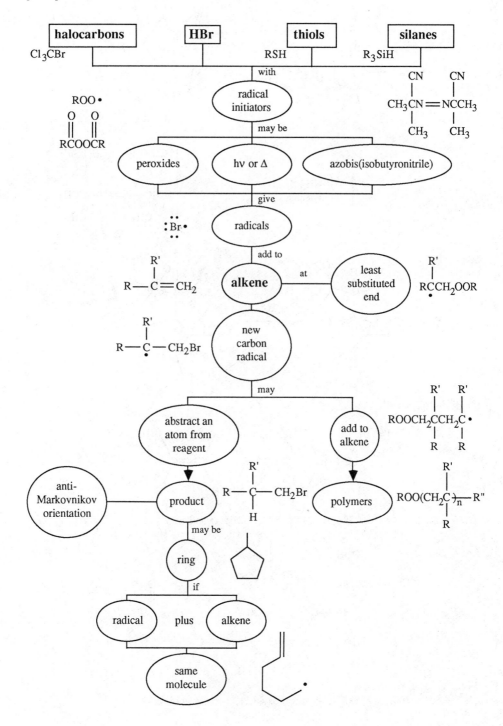

20.13

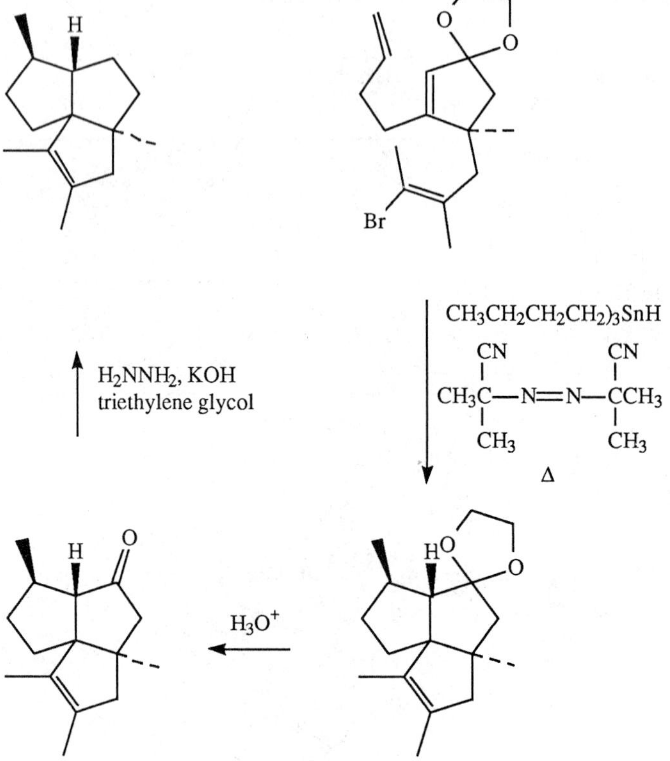
H
O
O
Br
$(CH_3CH_2CH_2CH_2)_3SnH$
CN CN
$CH_3C—N=N—CCH_3$
CH_3 CH_3
Δ
H_2NNH_2, KOH
triethylene glycol
H O
H O O
H_3O^+

20.14

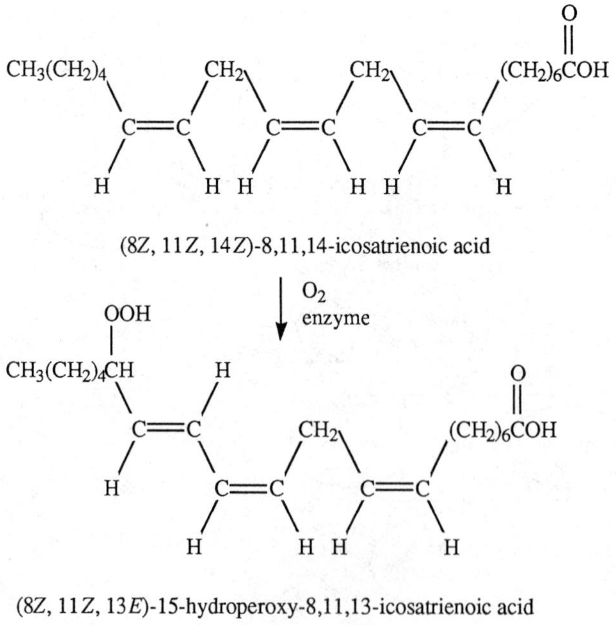
$CH_3(CH_2)_4$
CH_2
CH_2
O
$(CH_2)_6COH$
C=C C=C C=C
H H H H H H
(8Z, 11*Z*, 14*Z*)-8,11,14-icosatrienoic acid
O_2
enzyme
OOH
$CH_3(CH_2)_4CH$
H
O
$(CH_2)_6COH$
C=C
H
CH_2
C=C C=C
H H H H
(8Z, 11*Z*, 13*E*)-15-hydroperoxy-8,11,13-icosatrienoic acid

20.14 (cont)

20.15

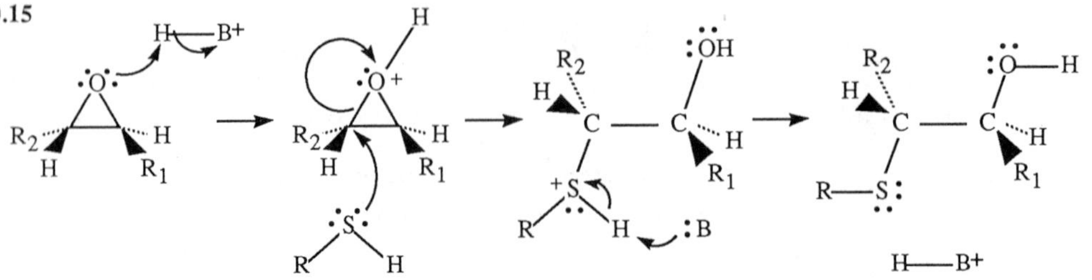

Note that while the opening of the ring in an unsymmetrical oxirane usually gives rise to a mixture of isomers, a reaction in biological systems gives rise to a single product. The mechanism shown above supposes that the reaction is taking place with a preferred orientation at the active site of an enzyme (see Section 26.6 for an example) and that, therefore, only one product is formed.

20.16

(a)

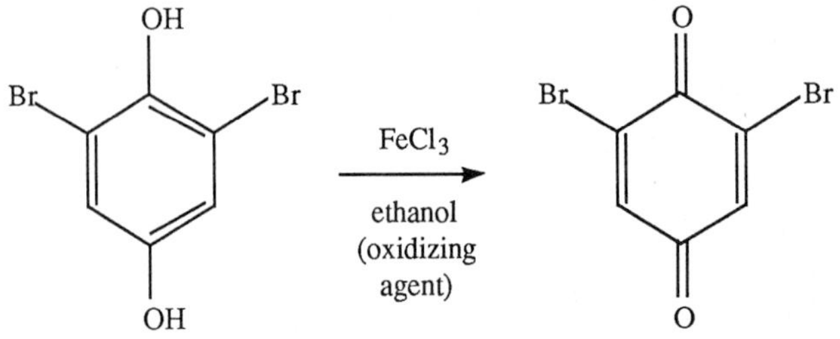

(b)

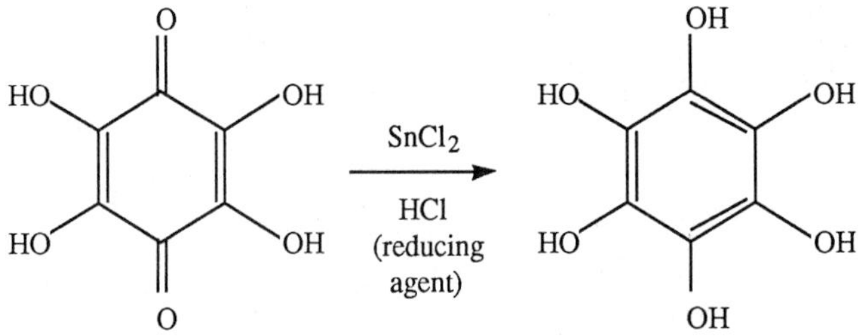

(c)

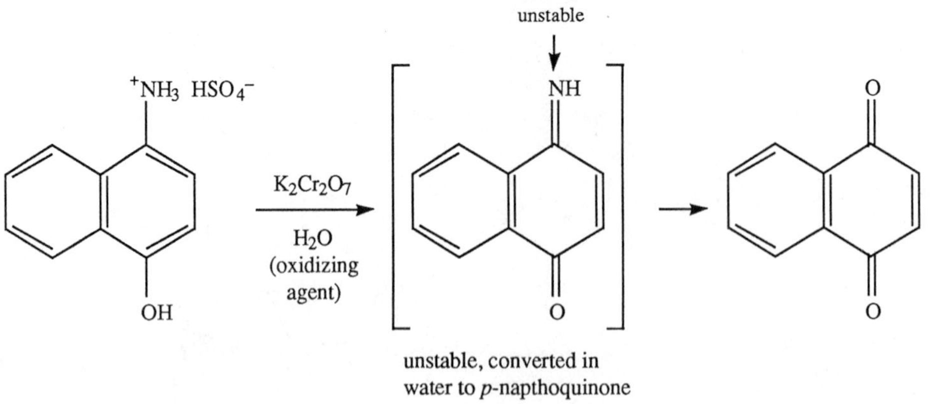

imine; unstable

unstable, converted in water to *p*-napthoquinone

20.16 (cont)

(d)

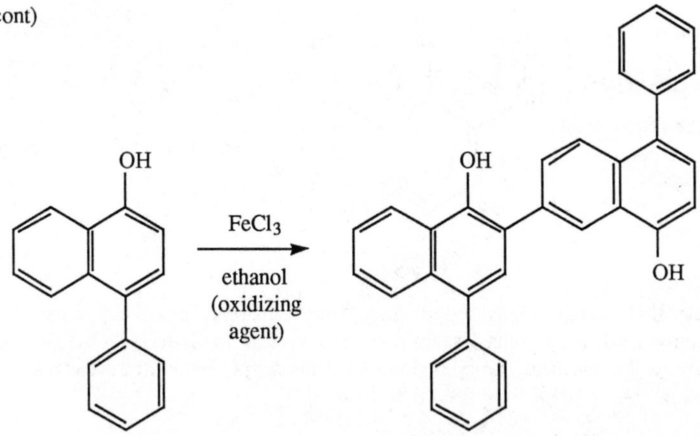
OH
FeCl3
ethanol
(oxidizing
agent)
OH
OH

(one of several possible products)

(e)

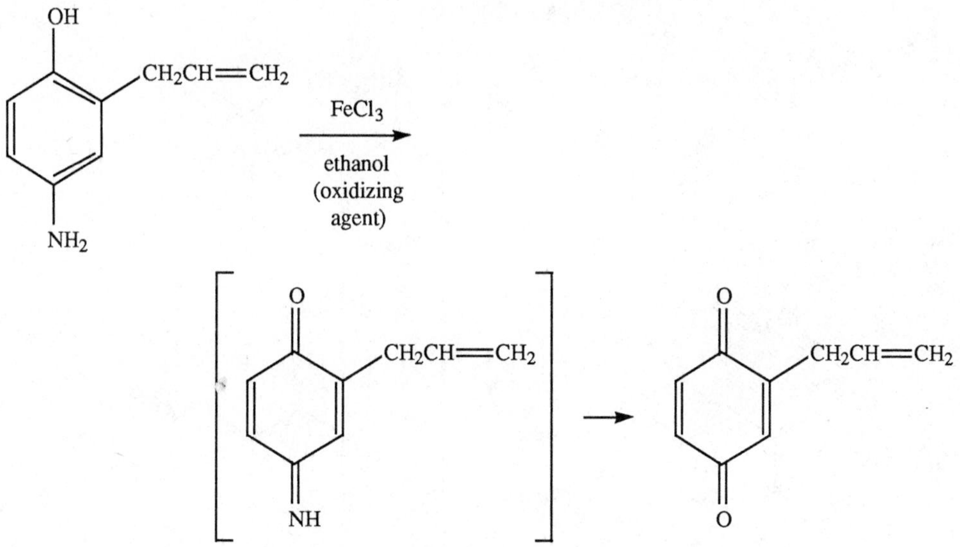
OH
CH2CH=CH2
NH2
FeCl3
ethanol
(oxidizing
agent)
O
CH2CH=CH2
NH
O
CH2CH=CH2
O

20.17

(a)

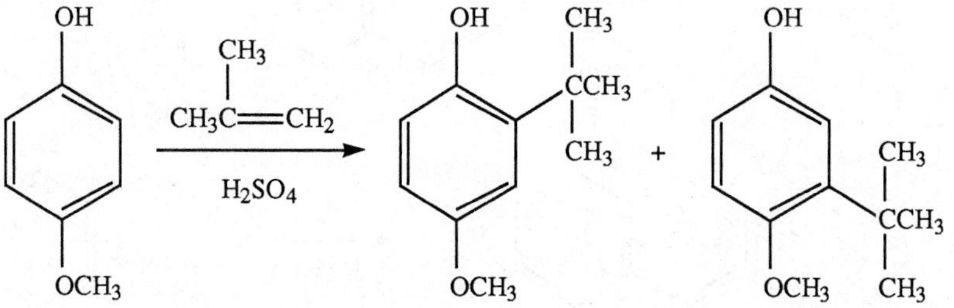
OH
OCH3
CH3
CH3C=CH2
H2SO4
OH
CH3
CCH3
CH3
OCH3
+
OH
CH3
CCH3
OCH3
CH3

20.17 (cont)

(b)

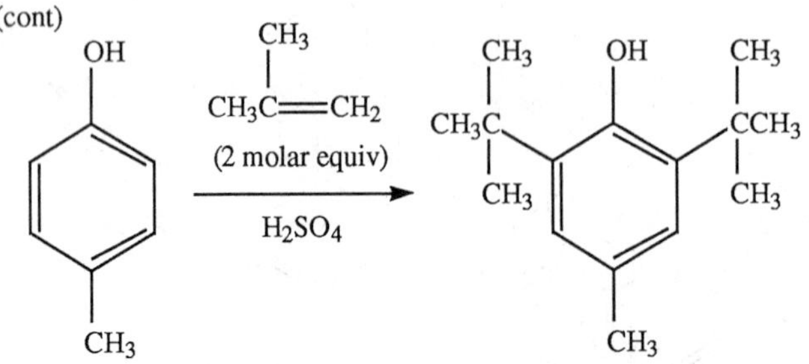

Both syntheses are Friedel-Crafts alkylation reactions. In (a) both the hydroxyl group and the methoxyl group are strong ortho,para directors, leading to attack at positions ortho to each. In (b) the hydroxyl group is a much stronger ortho,para director than the methyl group, and attack is at the two positions ortho to the hydroxyl group.

20.18

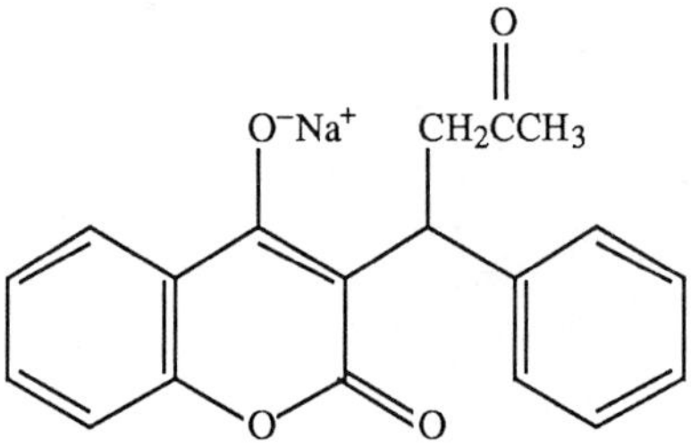

20.19

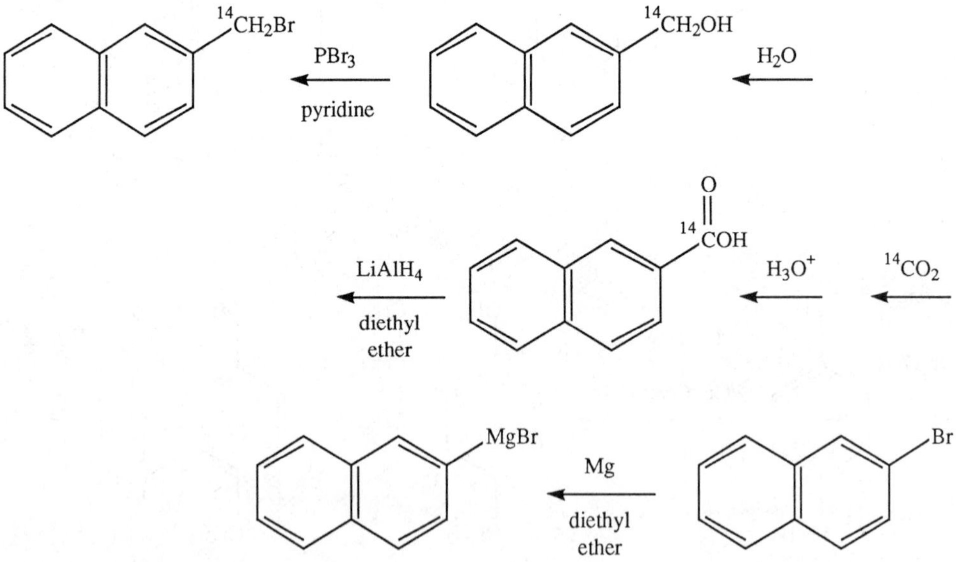

Concept Map 20.5   Oxidation reactions as free radical reactions.

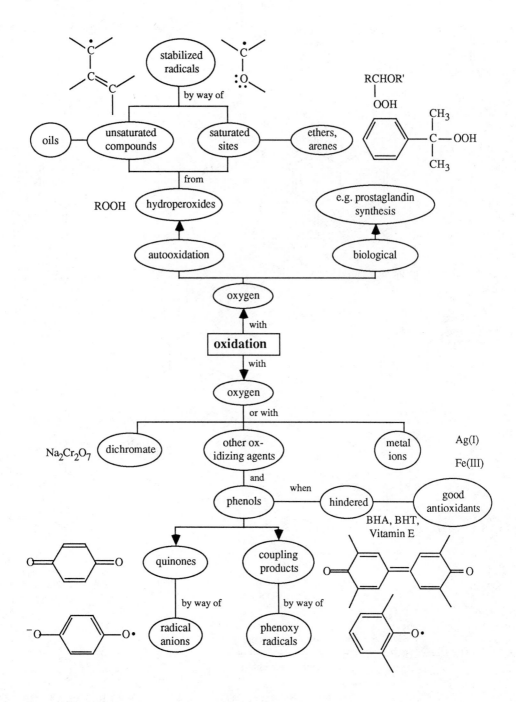

20.20

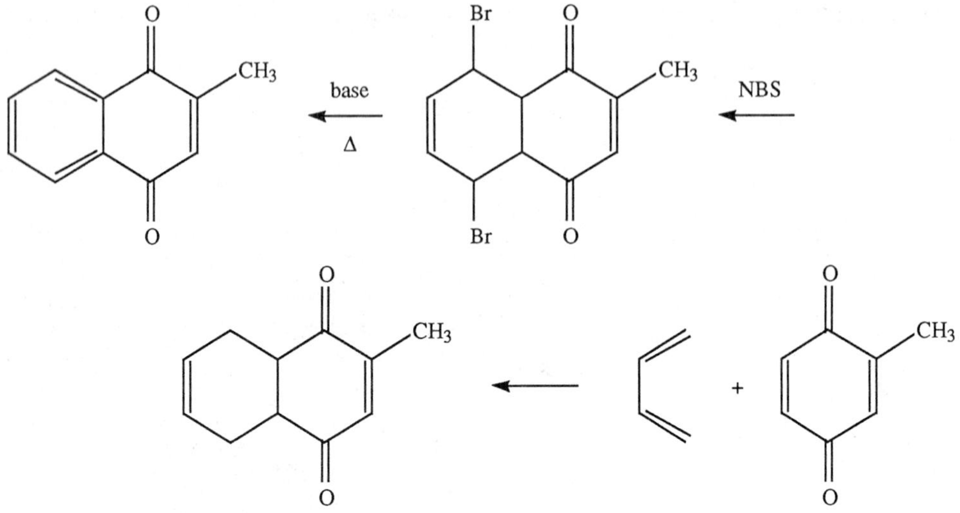

An alternate synthesis is shown below:

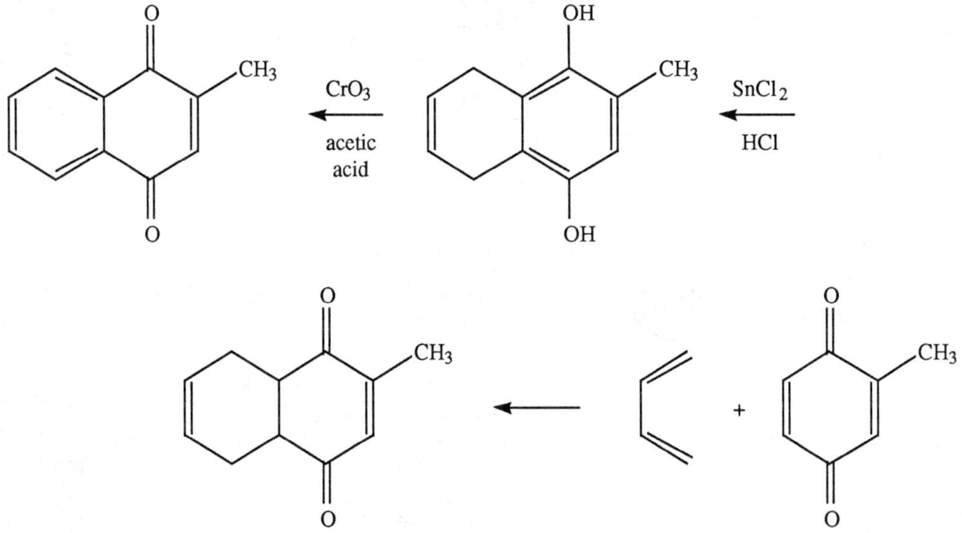

20.21

(a) $CH_3CH\!\!=\!\!CHCH\!\!=\!\!CHCH_3$ $\xrightarrow[\text{carbon tetrachloride}]{\text{Br}_2 \text{ (1 molar equiv)}}$

$CH_3CHBrCHBrCH\!\!=\!\!CHCH_3$ + $CH_3CHBrCH\!\!=\!\!CHCHBrCH_3$

20.21 (cont)

(b)

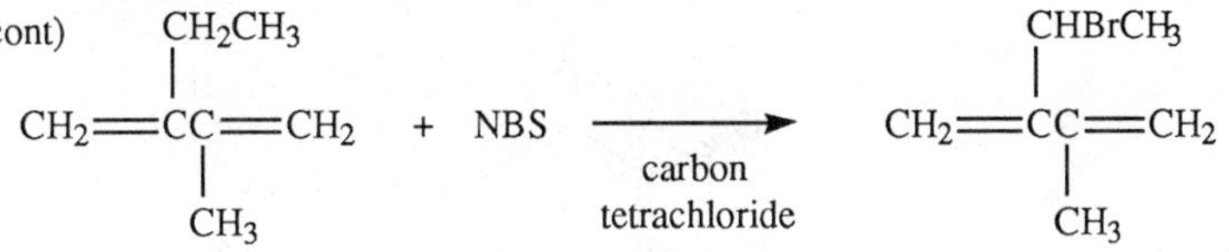

$$CH_2=CC=CH_2 \quad + \quad NBS \quad \xrightarrow{\text{carbon tetrachloride}} \quad CH_2=CC=CH_2$$

with substituents CH_2CH_3 and CH_3 on the left, and $CHBrCH_3$ and CH_3 on the right.

(c)

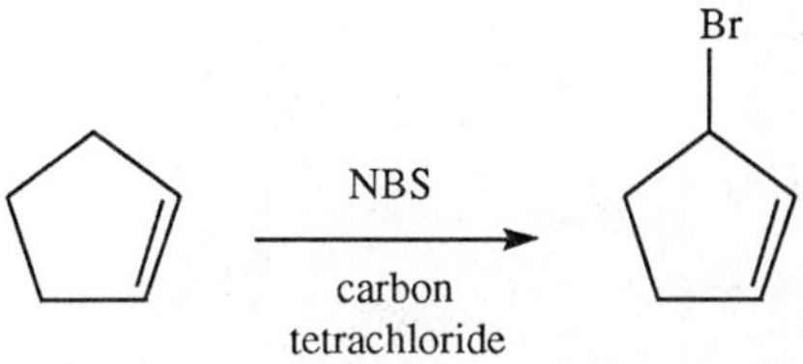

NBS, carbon tetrachloride

Cyclopentene → 3-bromocyclopentene (Br substituent)

(d)

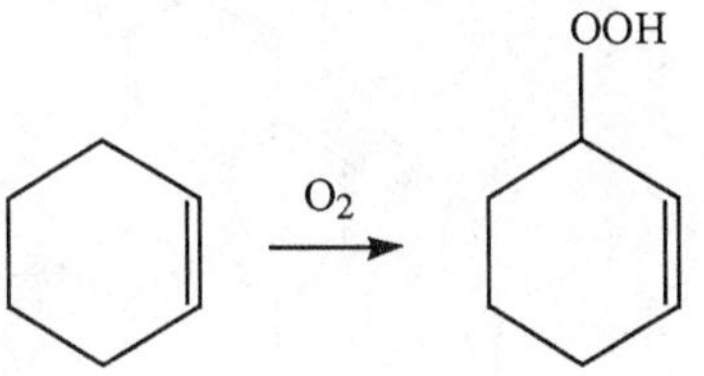

O_2

Cyclohexene → cyclohexenyl hydroperoxide (OOH substituent)

(e)

$$CH_3CH_2CH_3 \quad \xrightarrow[\text{hv}]{\text{Cl}_2} \quad CH_3CH_2CH_2Cl \quad + \quad CH_3CHClCH_3 \quad +$$

major products

$$CH_3CH_2CHCl_2 \quad + \quad CH_3CHClCH_2Cl$$

two of the possible minor products

(f)

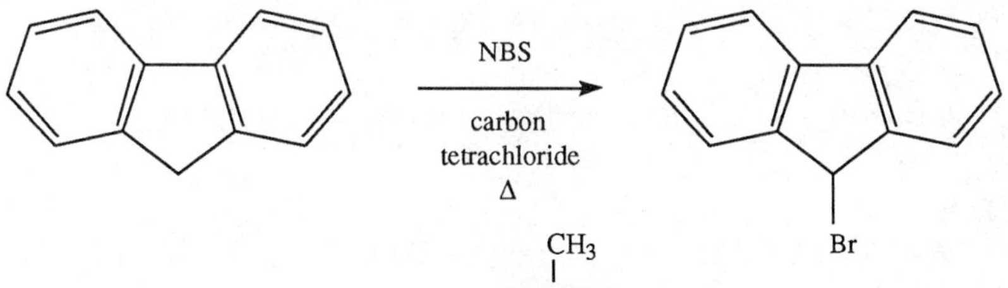

$$\xrightarrow[\substack{\text{carbon tetrachloride} \\ \Delta}]{NBS}$$

Fluorene → 9-bromofluorene (Br substituent)

(g)

$$\xrightarrow[\Delta]{O_2}$$

4-tert-butylphenol (OH, CH_3CCH_3 with CH_3) → biphenyl diol product with CH_3CCH_3 / CH_3 substituents, two OH groups

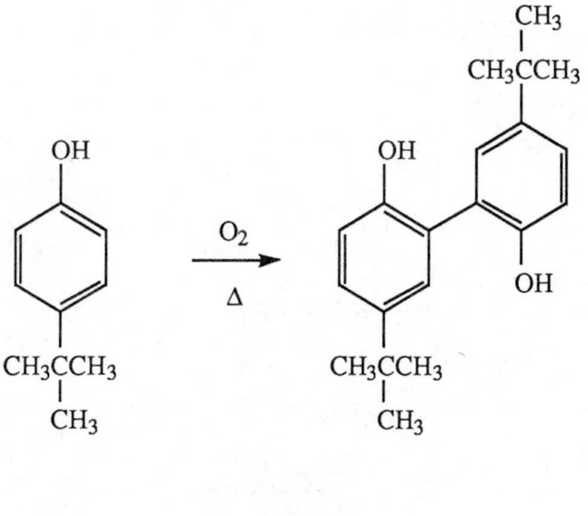

20.21 (cont)

(h)

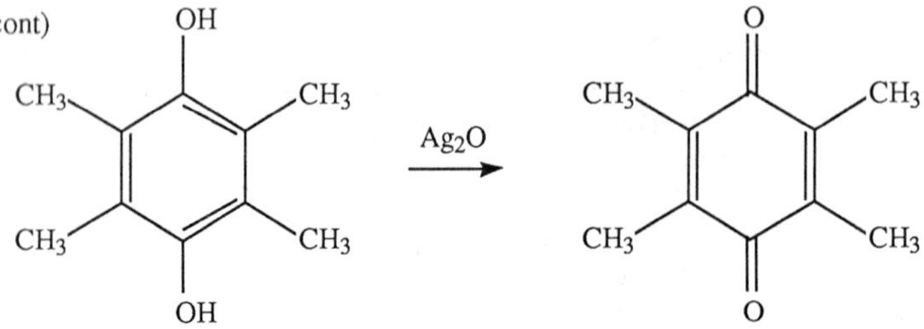

(i)

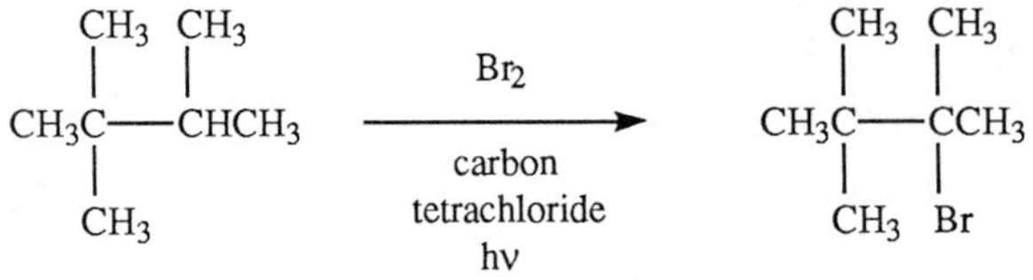

(j)

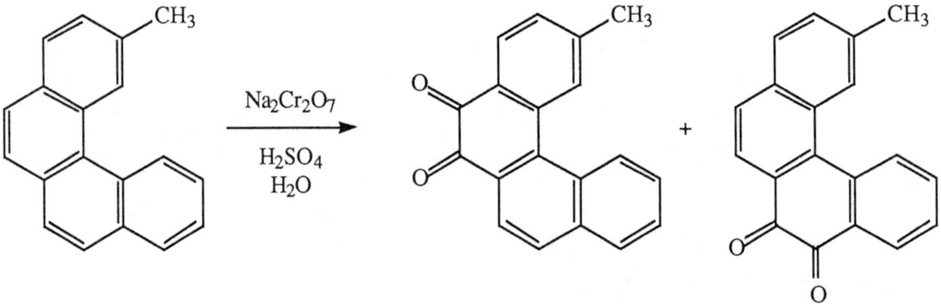

(k)

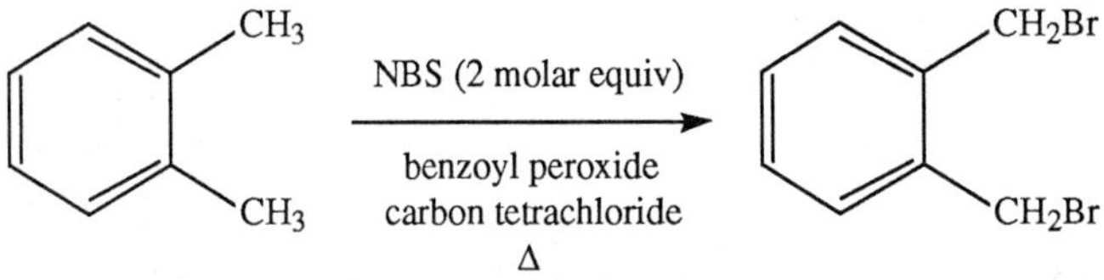

(l)

$$CH_3CH_2CH_2CH{=}CH_2 \ + \ Cl_3SiH \ \xrightarrow{\ 280\ °C\ } \ CH_3CH_2CH_2CH_2CH_2SiCl_3$$

(m)

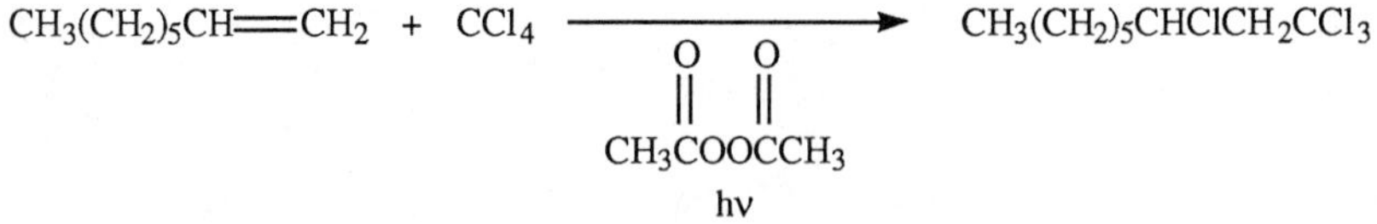

$$CH_3(CH_2)_5CH{=}CH_2 \ + \ CCl_4 \ \xrightarrow[\substack{CH_3COOCCH_3 \\ h\nu}]{} \ CH_3(CH_2)_5CHClCH_2CCl_3$$

(n)

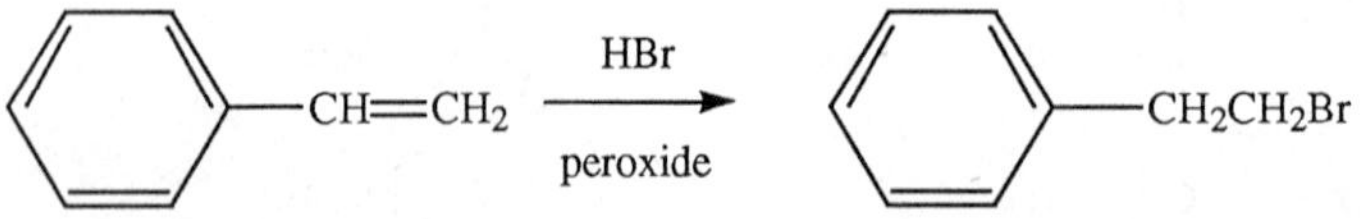

20.21 (cont)

(o)

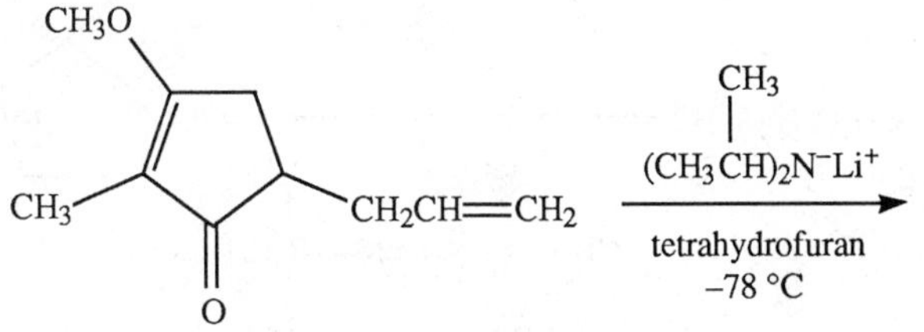

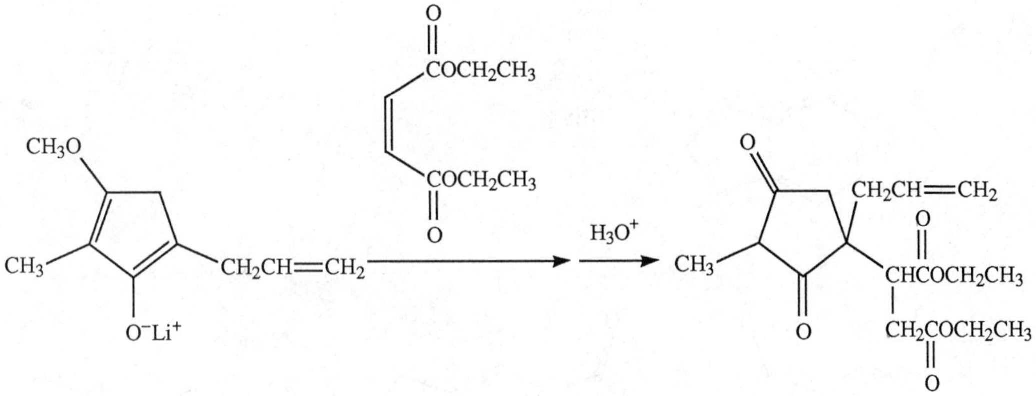

(p) $CH_3CH_2CH_2CH_3$ $\xrightarrow[h\nu]{Cl_2}$ $CH_3CH_2CHClCH_3$ + $CH_3CH_2CH_2CH_2Cl$

+ some polychlorinated products

(q) $\underset{\underset{CH_3CHCH=CH_2}{|}}{CH_3}$ $\xrightarrow[\text{peroxide}]{HBr}$ $\underset{\underset{CH_3CHCH_2CH_2Br}{|}}{CH_3}$

(r) CH_3CH_2SH + $CH_2=CHOCH_2CH_3$ $\xrightarrow{\Delta}$ $CH_3CH_2SCH_2CH_2OCH_2CH_3$

(s) $BrCH_2CH=CH_2$ $\xrightarrow[O_2]{HBr}$ $BrCH_2CH_2CH_2Br$

(t)

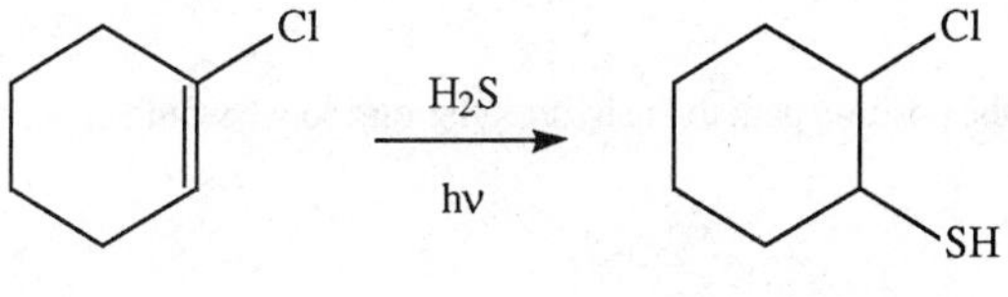

20.21 (cont)

(u)

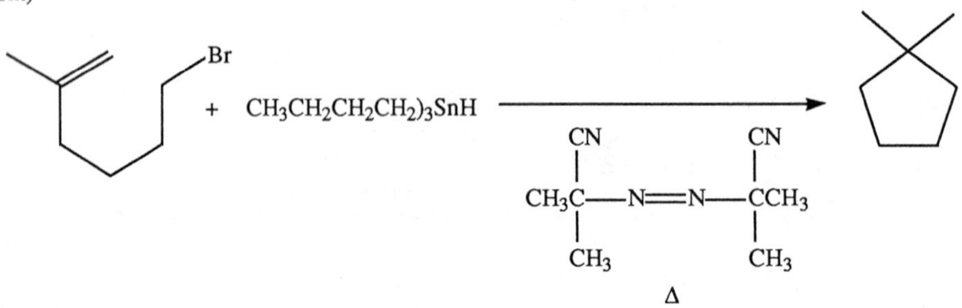

(v)

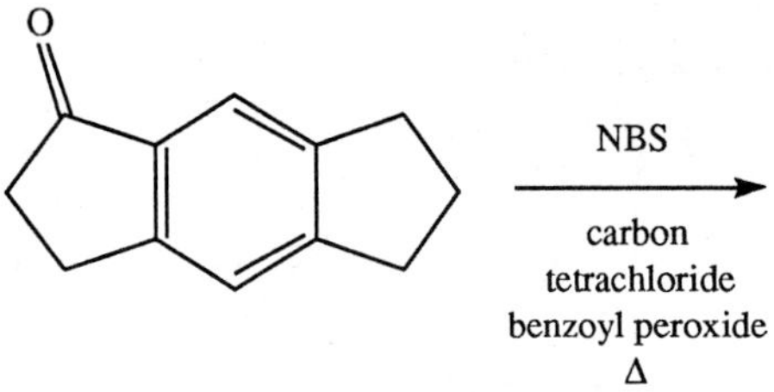

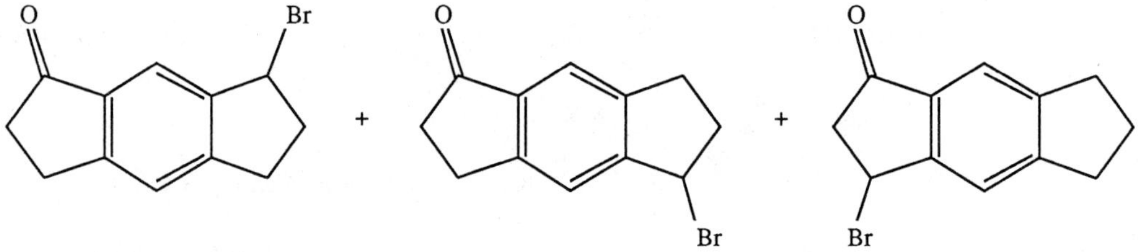

20.22

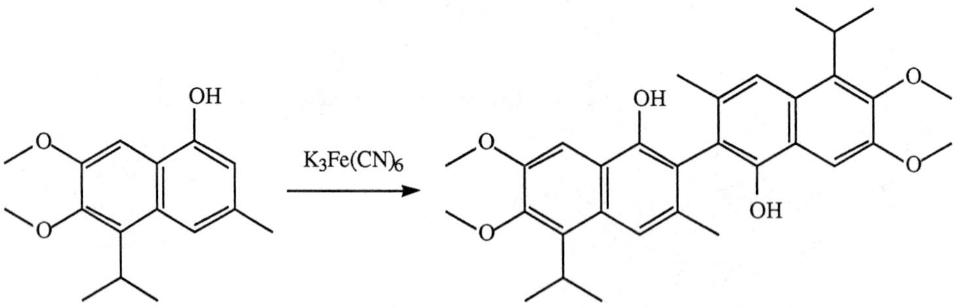

The isopropyl group hinders approach to the position para to the hydroxyl group so we would expect coupling to occur mainly at the ortho position.

20.23

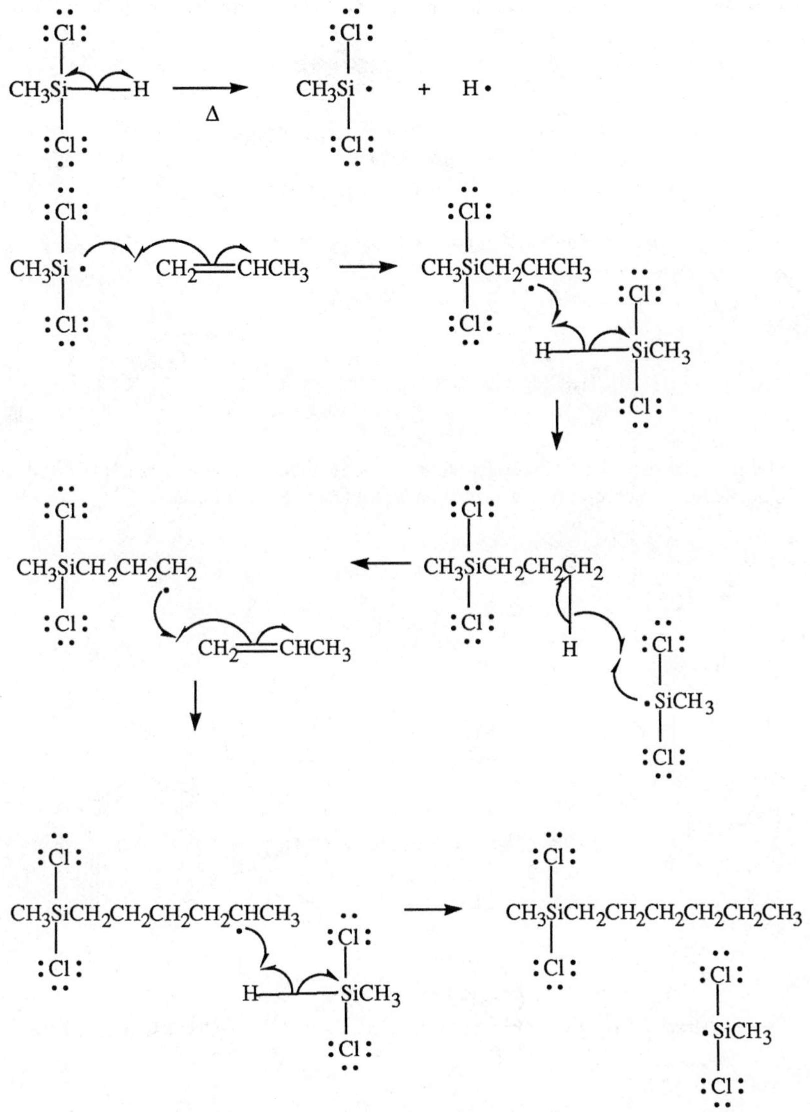

$$CH_3Si(Cl)(Cl)-H \xrightarrow{\Delta} CH_3Si(Cl)(Cl)\cdot \; + \; H\cdot$$

$$CH_3Si(Cl)(Cl)\cdot \; + \; CH_2=CHCH_3 \longrightarrow CH_3SiCH_2\overset{\cdot}{C}HCH_3$$

$$CH_3SiCH_2CH_2CH_2\cdot \; \longleftarrow \; CH_3SiCH_2CH_2CH_2$$

$$CH_2=CHCH_3$$

$$CH_3SiCH_2CH_2CH_2CH_2\overset{\cdot}{C}HCH_3 \longrightarrow CH_3SiCH_2CH_2CH_2CH_2CH_2CH_3$$

$$H-SiCH_3$$

$$\cdot SiCH_3$$

20.24 1. Conditions for the formation of free radicals, such as heat, light, or peroxides, would be recommended for these reactions.

$$CH_3CH{=}O \quad + \quad CH_3(CH_2)_5CH{=}CH_2 \xrightarrow{\text{peroxides}} CH_3C(CH_2)_7CH_3$$
(with carbonyl O above the acetaldehyde and product carbonyl)

cyclohexane ring with OH and H substituents $+ \quad CH_3(CH_2)_5CH{=}CH_2 \xrightarrow{\text{peroxides}}$ cyclohexane ring with OH and $(CH_2)_7CH_3$

cyclohexane ring with NH_2 and H substituents $+ \quad CH_3(CH_2)_5CH{=}CH_2 \xrightarrow{\text{peroxides}}$ cyclohexane ring with NH_2 and $(CH_2)_7CH_3$

2. *tert*-Butyl alcohol does not have a hydrogen atom attached to the carbon atom bonded to the hydroxyl group and would, therefore, not react with 1-octene under free radical conditions.

20.25

(a) $\xleftarrow{\quad CH_3CH_2CH_2CH_2)_3SnH \quad}$

with reagent below arrow:

$$CH_3\overset{\displaystyle CN}{\underset{\displaystyle CH_3}{C}}{-}N{=}N{-}\overset{\displaystyle CN}{\underset{\displaystyle CH_3}{C}}CH_3$$

Δ

(b) phenyl–$CH{=}CHCH_2CH_3 \xleftarrow{\quad (CH_3)_2CuLi \quad}$ phenyl–$CH{=}CHCH_2OTs$

(c) $\xleftarrow{\quad H_3O^+ \quad} \xleftarrow{\quad (CH_3)_2CuLi \quad}$

20.25 (cont)

(d)

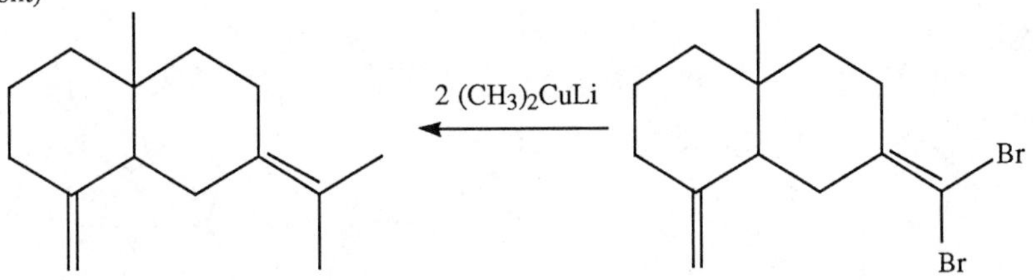

2 (CH₃)₂CuLi

(e)

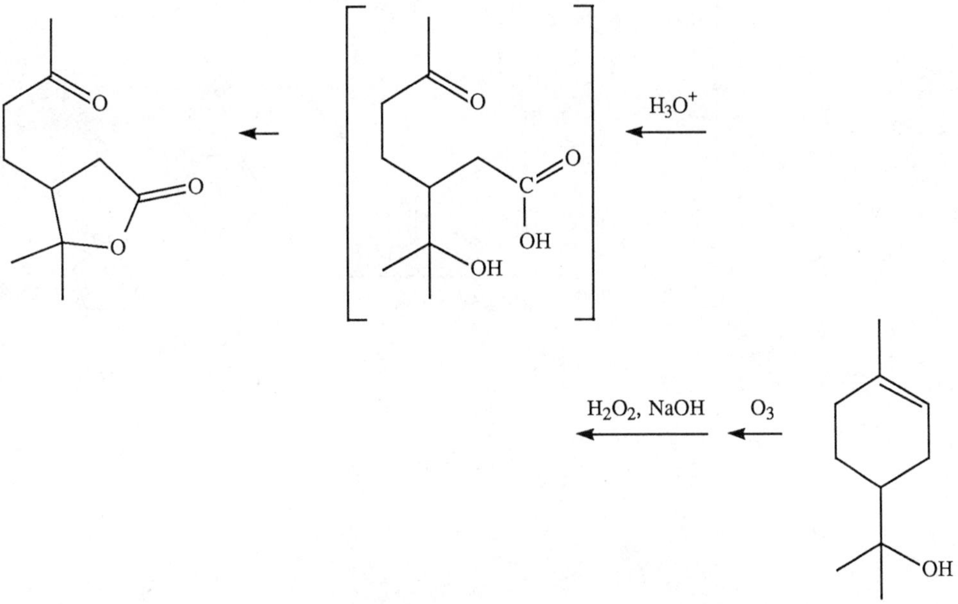

H₃O⁺

H₂O₂, NaOH   O₃

20.26

HOCH₂CH₂OH

TsOH
benzene
Δ

(CH₃)₂S

O₃
methanol

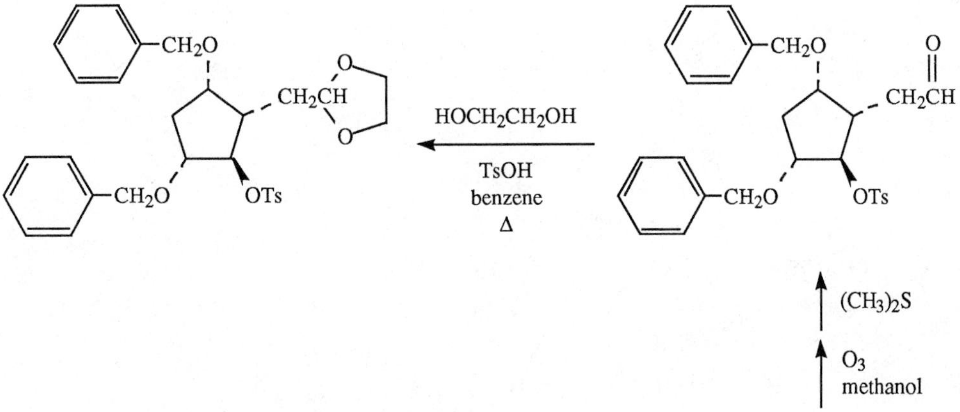

20.26 (cont)

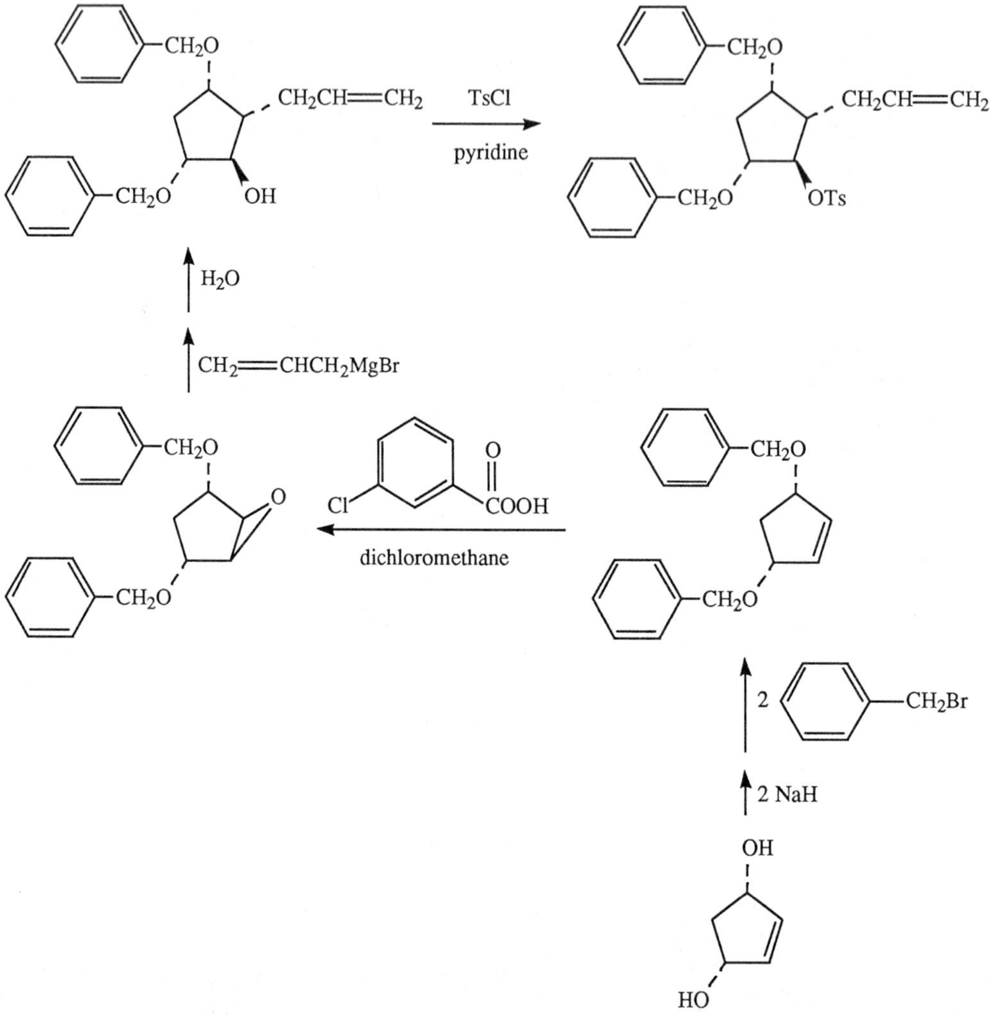

20.27

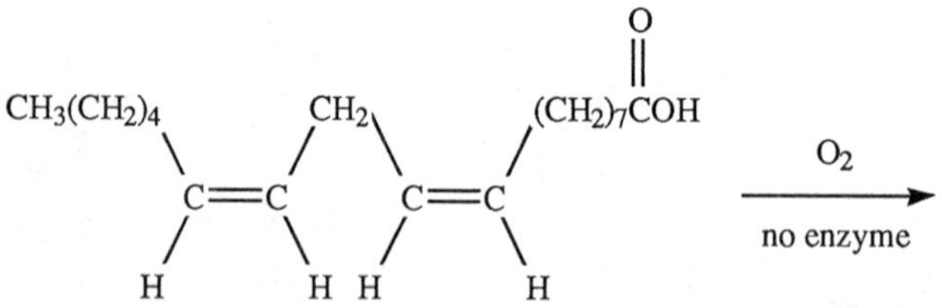

20.27 (cont)

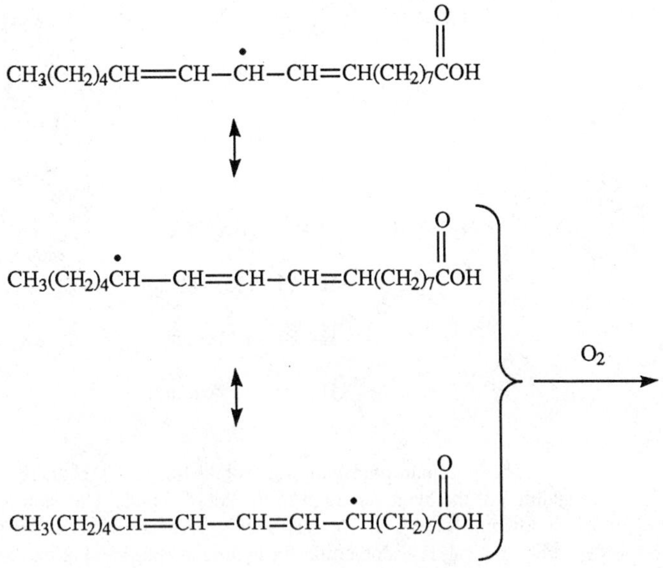

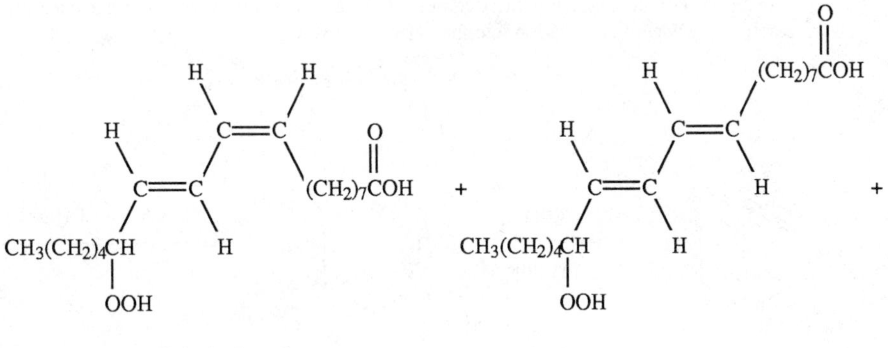

$$CH_3(CH_2)_4CH=CH-\overset{\displaystyle\centerdot}{C}H-CH=CH(CH_2)_7\overset{\displaystyle\overset{O}{\|}}{C}OH$$

$$\updownarrow$$

$$CH_3(CH_2)_4\overset{\displaystyle\centerdot}{C}H-CH=CH-CH=CH(CH_2)_7\overset{\displaystyle\overset{O}{\|}}{C}OH$$

$$\updownarrow$$

$$CH_3(CH_2)_4CH=CH-CH=CH-\overset{\displaystyle\centerdot}{C}H(CH_2)_7\overset{\displaystyle\overset{O}{\|}}{C}OH \Bigg\}\ \xrightarrow{\ O_2\ }$$

biological product

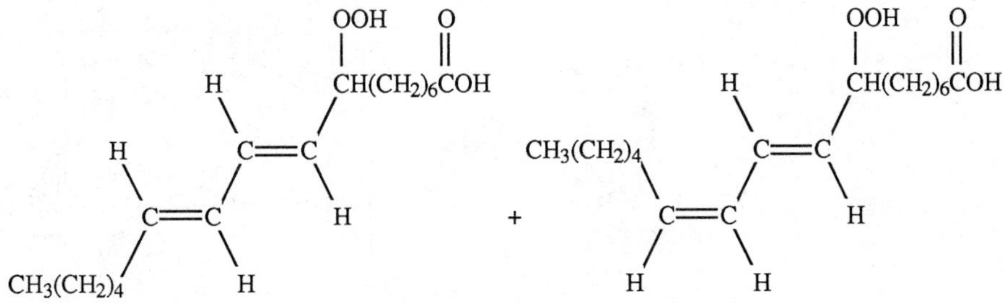

20.28   (1) CH_3CH=CH_2 +   X• $\longrightarrow$ $CH_3\overset{\bullet}{C}HCH_2X$ ΔH_r (kcal/mol)

C=C (π)   ~59 kcal/mol   C—Cl   ~81 kcal/mol   −22

C—Br   ~68 kcal/mol   −9

C—I   ~52 kcal/mol   +7

(2) $CH_3\overset{\bullet}{C}HCH_2X$ +   HX $\longrightarrow$ $CH_3CH_2CH_2X$ ΔH_r (kcal/mol)

H—Cl   103 kcal/mol   C—H (2°)   95 kcal/mol   +8

H—Br   87 kcal/mol   −8

H—I   71 kcal/mol   −24

The first chain-propagating step of the free radical chain reaction of hydrogen halides with propene is the addition of the hydrogen atom to the double bond. This step is exothermic for chlorine and bromine, but endothermic for iodine. The second step, the abstraction of a hydrogen atom from the hydrogen halide by the carbon radical, is endothermic for hydrogen chloride, but exothermic for hydrogen bromide and hydrogen iodide. Therefore, only with hydrogen bromide are both steps of the reaction exothermic and expected to have small energies of activation. For the other two hydrogen halides, one or the other step has a high energy of activation. For such systems, the free radical chain reaction is slow, and other reaction pathways, such as reactions by way of carbocation intermediates, are favored.

20.29

1.

2.

20.29 (cont)

3.

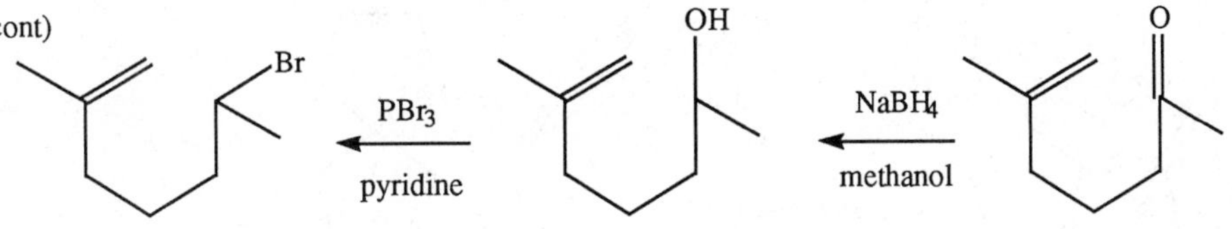

4.

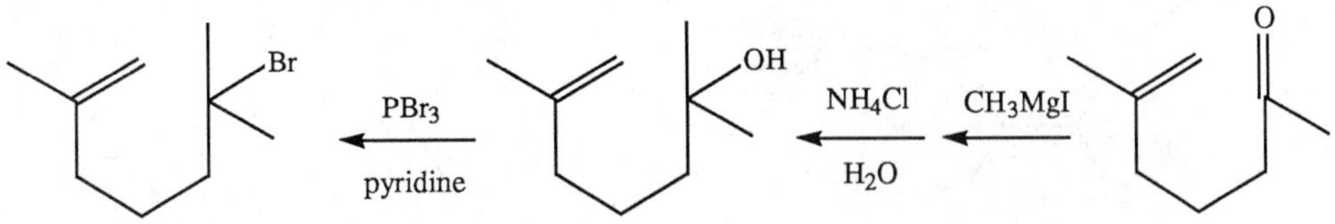

20.30

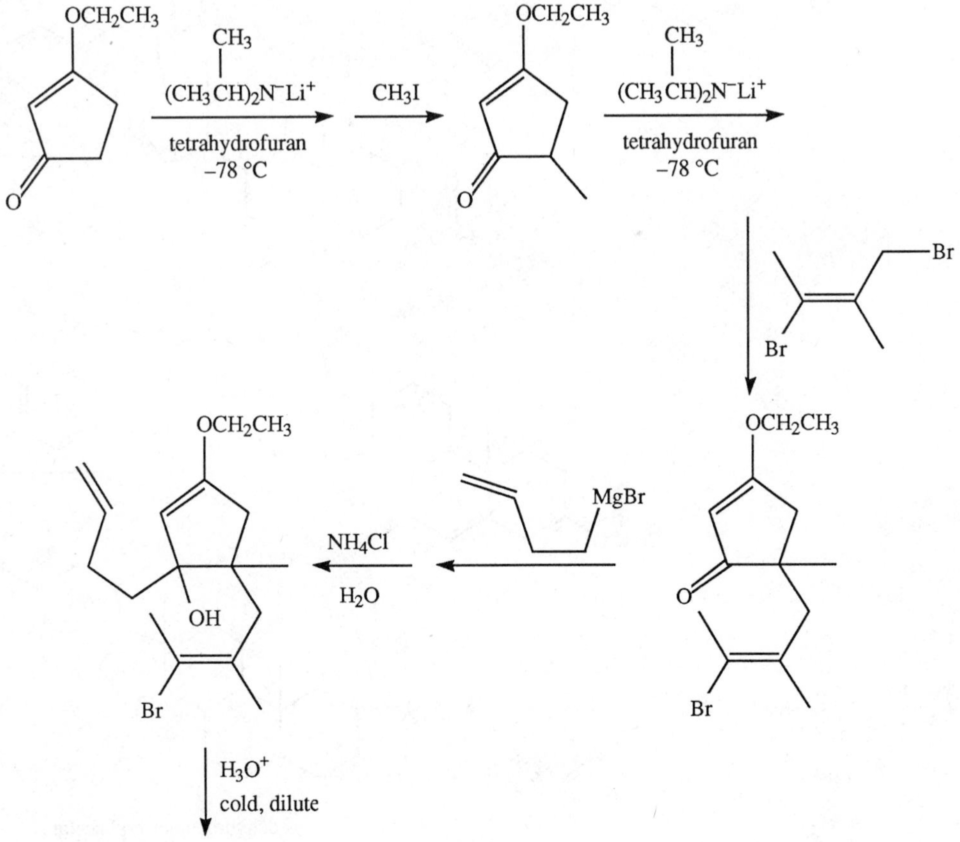

20.30 (cont)

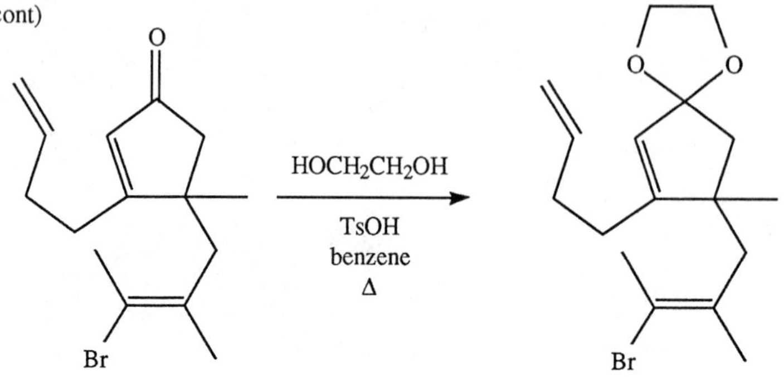

20.31

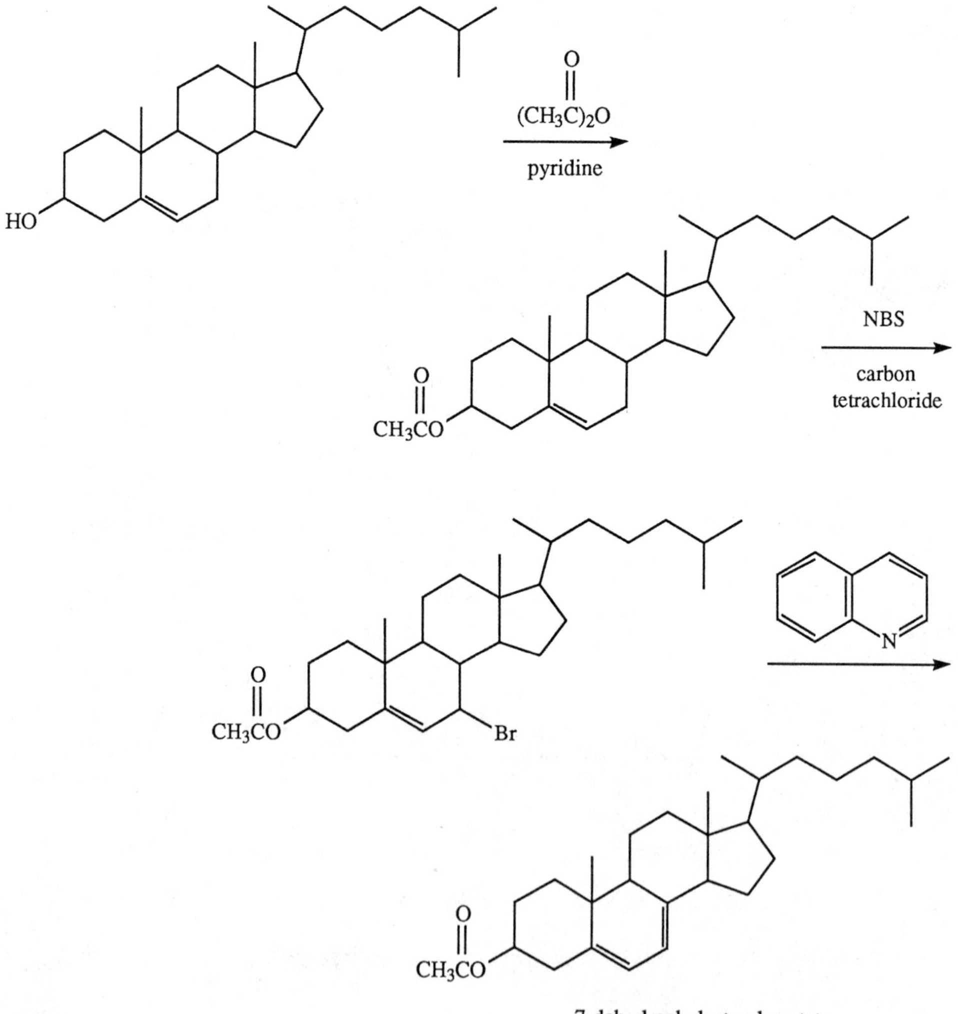

7-dehydrocholesteryl acetate

21

Mass Spectrometry

21.1 If we subtract the m/z value for the base peak from that for the peak for the molecular ion, we can determine the mass of the neutral fragment which is lost.

$$58 \quad - \quad 29 \quad = \quad 29$$

molecular	base	neutral
ion	peak	fragment

The fragment is most likely an ethyl radical, $CH_3\overset{\bullet}{C}H_2$, formed by homolytic cleavage of the bond between the carbon atom of the carbonyl group and the α-carbon atom. The fragment that is charged, with an m/e of 29, is a resonance-stabilized acylium ion, $HC{\equiv}O^+$. Putting these two fragments together gives us the identity of the unknown, propanal.

propanal   molecular ion
                                                m/z 58

M $\overset{+}{\bullet}$	ethyl radical	an acylium ion
m/z 58	not seen in mass spectrum	base peak m/z 29

21.2 A hydrogen atom is lost in going from the molecular ion at m/z 46 to the peak at m/z 45. The lose of a hydroxyl radical (m/z 17) from the molecular ion leads to the base peak at m/z 29. Both of these fragments come from homolytic cleavage of the two bonds to the carbonyl group. Putting these fragments together gives us the structure of the unknown acid, which is formic acid.

formic acid   molecular ion
                                                m/z 46

21.2 (cont)

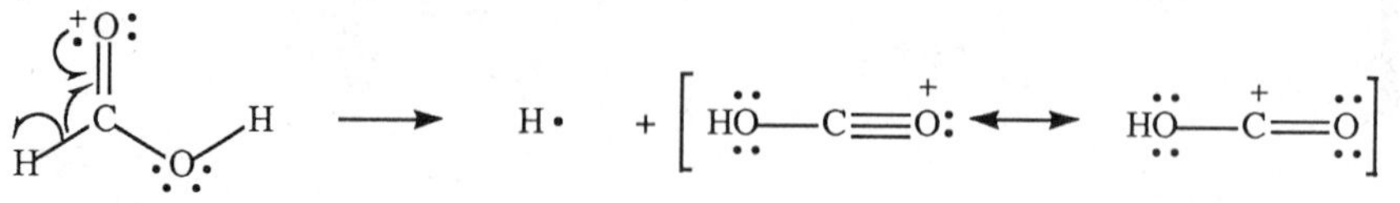

M$^{+\cdot}$ m/z 46 — hydrogen atom — an acylium ion m/z 45

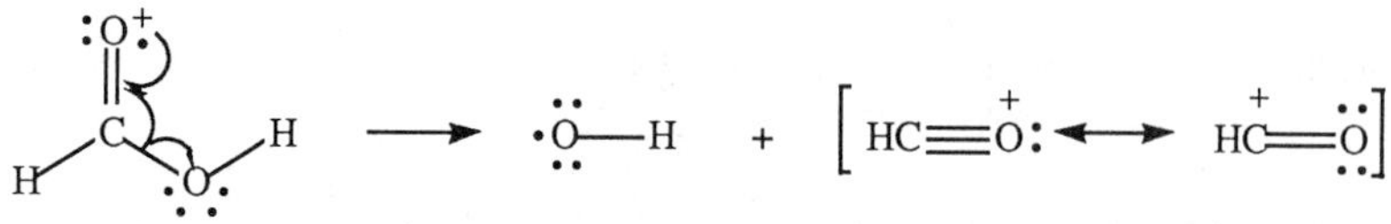

M$^{+\cdot}$ m/z 46 — hydroxyl radical — another acylium ion base peak m/z 29

21.3 Two possible cleavages of the molecular ion of 3-pentanone at the bonds to the α-carbon atom are shown below.

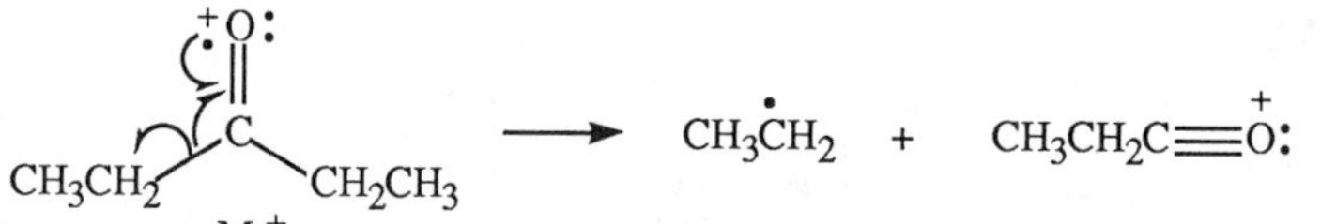

M$^{+\cdot}$ m/z 86 — m/z 57

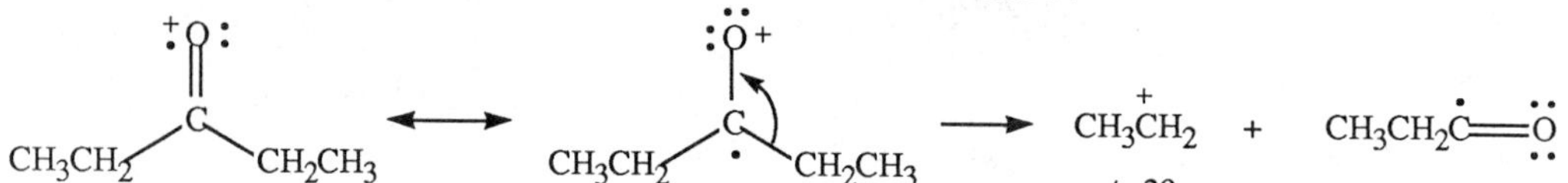

M$^{+\cdot}$ m/z 86 — m/z 29

The expected cleavage of the molecular ion of 2-pentanone at the bonds to the two different α-carbon atoms are shown below.

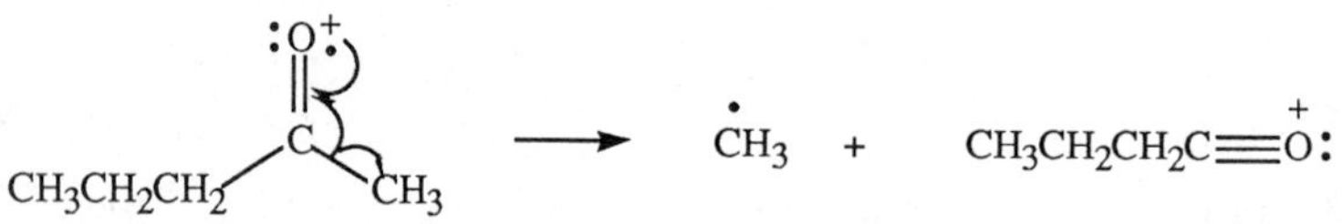

M$^{+\cdot}$ m/z 86 — m/z 71

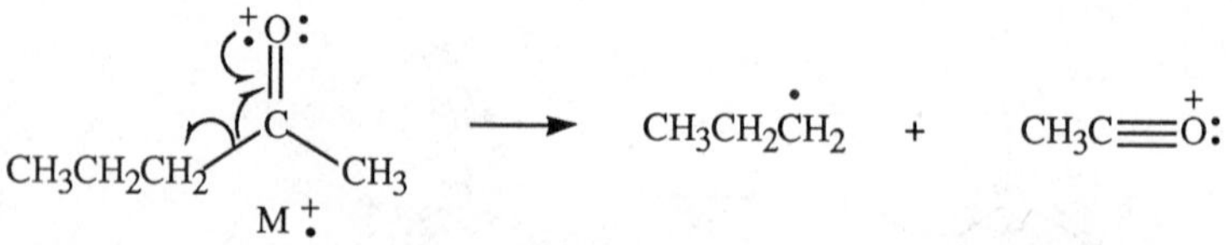

M$^{+\cdot}$ m/z 86 — m/z 43

The mass spectrum with base peaks at m/z 29 and m/z 57 is the spectrum of 3-pentanone. The mass spectrum with the base peak at m/z 43 and a small peak at m/z 71 is the spectrum of 2-pentanone.

21.4 The formation of the base peak in the spectrum of 2,2-dimethylpropanal from the molecular ion is shown below.

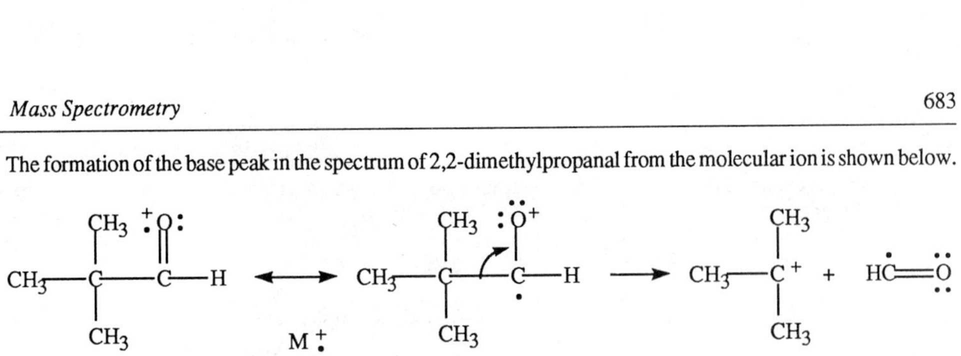

M$^{+\cdot}$ *m/z* 86

tert-butyl cation *m/z* 57

The equivalent fragmentation from the molecular ion of acetaldehyde is shown below.

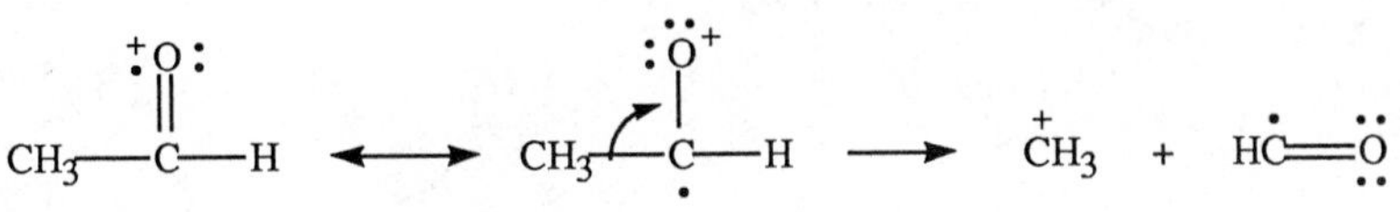

M$^{+\cdot}$ *m/z* 44

methyl cation *m/z* 15

The base peak in the fragmentation of the molecular ion of 2,2-dimethylpropanal is the stable tertiary cation, the *tert*-butyl cation. The comparable species in the fragmentation of the molecular ion of acetaldehyde would be the unstable methyl cation, *m/z* 15, which does not appear in the spectrum of acetaldehyde.

21.5

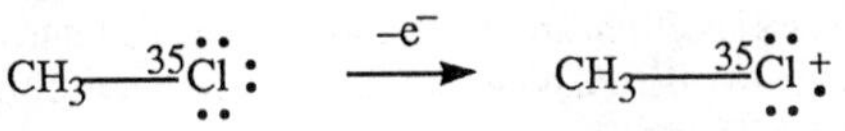

M$^{+\cdot}$ *m/z* 50

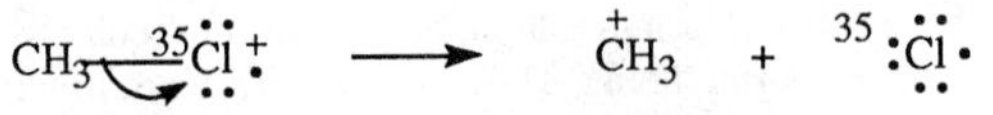

m/z 15

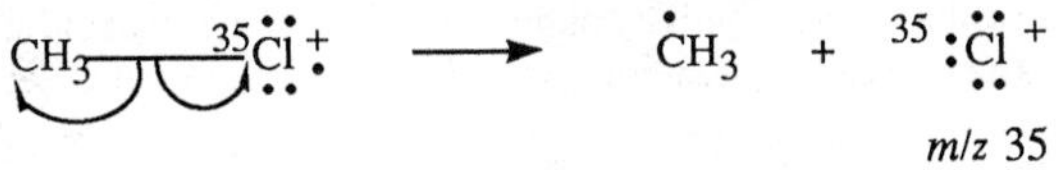

m/z 35 not evident in the spectrum

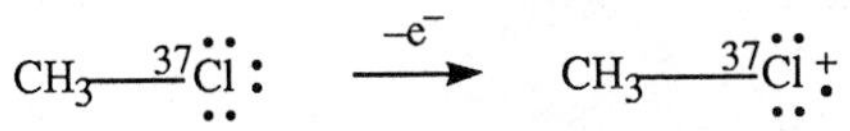

M$^{+\cdot}$ *m/z* 52

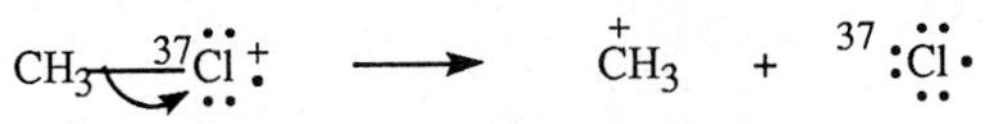

m/z 15

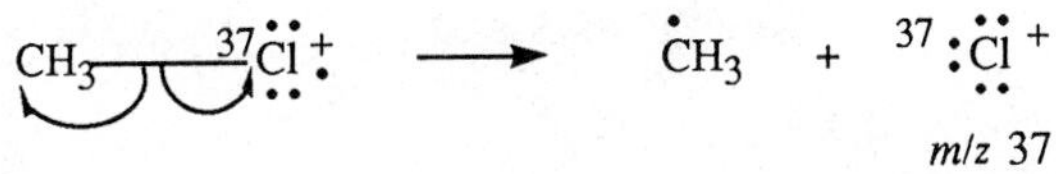

m/z 37 not evident in the spectrum

21.6   The two ions with the highest *m/e* values appear in this spectrum at *m/z* 64 and *m/z* 66. The one at the higher mass is one-third the intensity of the other, suggesting that these are molecular ions for a compound containing chlorine.

$$M \overset{+}{\cdot} \qquad m/z\ 64 \quad (^{35}\text{Cl})$$

$$M \overset{+}{\cdot} \qquad m/z\ 66 \quad (^{37}\text{Cl})$$

An important fragmentation of the molecular ion gives rise to the peak at *m/z* 29. This is the ethyl cation and results from loss of chlorine from the molecular ion. Putting these fragments together, we identify the haloalkane as ethyl chloride (chloroethane).

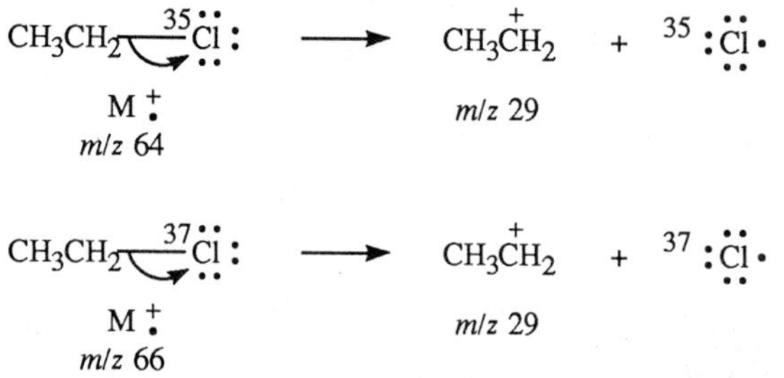

$$\text{CH}_3\text{CH}_2 \overset{35}{\underset{}{\text{Cl}}} : \quad \longrightarrow \quad \text{CH}_3\overset{+}{\text{CH}}_2 \quad + \quad {}^{35} : \overset{..}{\underset{..}{\text{Cl}}} \cdot$$

$$\begin{array}{c} M \overset{+}{\cdot} \\ m/z\ 64 \end{array} \qquad\qquad\qquad m/z\ 29$$

$$\text{CH}_3\text{CH}_2 \overset{37}{\underset{}{\text{Cl}}} : \quad \longrightarrow \quad \text{CH}_3\overset{+}{\text{CH}}_2 \quad + \quad {}^{37} : \overset{..}{\underset{..}{\text{Cl}}} \cdot$$

$$\begin{array}{c} M \overset{+}{\cdot} \\ m/z\ 66 \end{array} \qquad\qquad\qquad m/z\ 29$$

21.7   The two ions at the highest *m/z* values in this mass spectrum are at *m/z* 148 and, equal in intensity to the first, at *m/z* 150. This suggests that one of the two halogens in the unknown is bromine. The presence of bromine is confirmed by the two peaks of equal intensity at *m/z* 79 and 81.

         The first fragmentation, *m/z* 148 to *m/z* 129, or (*m/z* 150 to *m/z* 131) is the loss of a species with a mass of 19, which corresponds to the atomic weight of fluorine. The base peak, which is at *m/z* 69, results from the lose of a bromine atom. The compound can have only one bromine atom, because the molecular weight is too low for more than one, therefore there must be three fluorine atoms. The compound is bromotrifluoromethane, CF_3Br.

         The fragmentations are shown below. The molecular ion is shown as being formed by loss of a nonbonding electron from bromine because the electronegativity of bromine is lower than that of fluorine, and the bromine atom is much larger than a fluorine atom, therefore it would be easier to lose the electron from bromine than from fluorine.

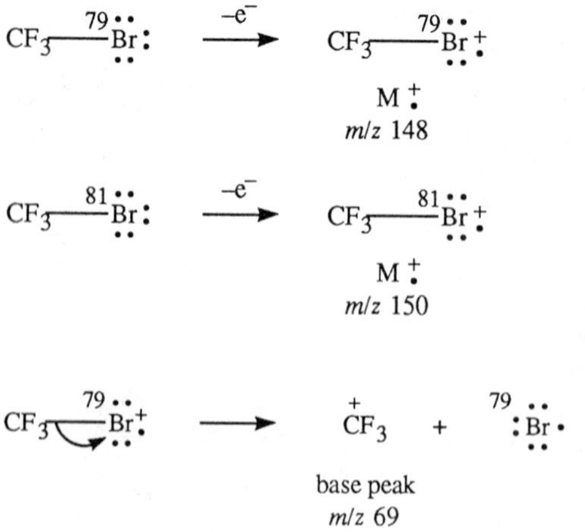

$$\text{CF}_3 \overset{79}{\underset{}{\text{Br}}} : \quad \overset{-e^-}{\longrightarrow} \quad \text{CF}_3 \overset{79}{\underset{..}{\text{Br}}} \overset{+}{\cdot}$$

$$\begin{array}{c} M \overset{+}{\cdot} \\ m/z\ 148 \end{array}$$

$$\text{CF}_3 \overset{81}{\underset{}{\text{Br}}} : \quad \overset{-e^-}{\longrightarrow} \quad \text{CF}_3 \overset{81}{\underset{..}{\text{Br}}} \overset{+}{\cdot}$$

$$\begin{array}{c} M \overset{+}{\cdot} \\ m/z\ 150 \end{array}$$

$$\text{CF}_3 \overset{79}{\underset{}{\text{Br}}} \overset{+}{:} \quad \longrightarrow \quad \overset{+}{\text{CF}}_3 \quad + \quad {}^{79} : \overset{..}{\underset{..}{\text{Br}}} \cdot$$

$$\begin{array}{c} \text{base peak} \\ m/z\ 69 \end{array}$$

21.7 (cont)

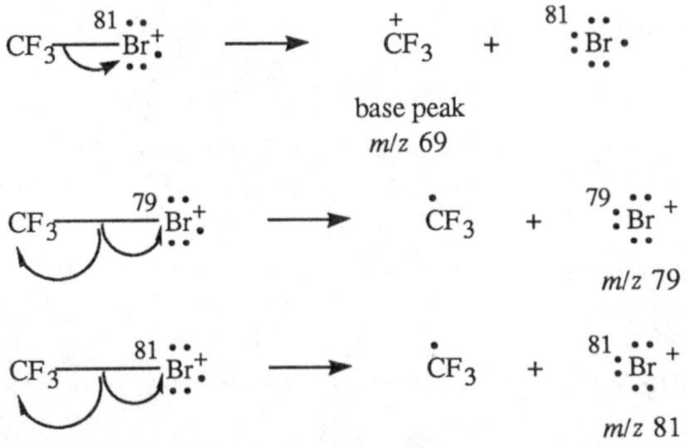

$$CF_3 \overset{81\ \cdot\cdot}{\underset{\cdot\cdot}{Br^+}} \longrightarrow \overset{+}{CF_3} \ + \ \overset{81\ \cdot\cdot}{:\overset{\cdot\cdot}{Br}} \cdot$$

base peak
m/z 69

$$CF_3 \overset{79\ \cdot\cdot}{\underset{\cdot\cdot}{Br^+}} \longrightarrow \overset{\cdot}{CF_3} \ + \ \overset{79\ \cdot\cdot}{:\overset{\cdot\cdot}{Br}} {}^+$$

m/z 79

$$CF_3 \overset{81\ \cdot\cdot}{\underset{\cdot\cdot}{Br^+}} \longrightarrow \overset{\cdot}{CF_3} \ + \ \overset{81\ \cdot\cdot}{:\overset{\cdot\cdot}{Br}} {}^+$$

m/z 81

Even though the majority of the molecular ions result from loss of an electron from the bromine atom, a certain number of the molecular ions will result from loss of an electron from a fluorine atom. A better way of thinking of this, of course, is that all of the electrons belong to the whole molecule. When an electron is lost, the fragments that result from this molecular ion may be rationalized by picturing the deficiency of the electron as being localized at one site or another. The relative intensities of the bands at *m/z* 69 and at *m/z* 129 and 131 in the spectrum are a measure of the ease with which bromine and fluorine accommodate the deficiency of an electron. Localization of the deficiency of an electron on fluorine leads to the following fragmentations.

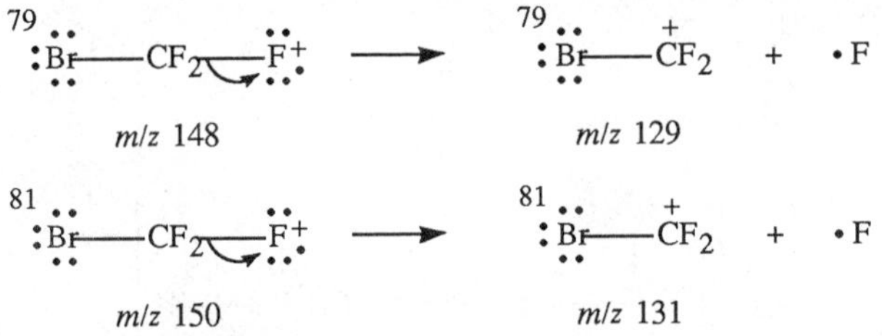

$$\overset{79\ \cdot\cdot}{:\overset{\cdot\cdot}{Br}} {\text{---}} CF_2 {\text{---}} \overset{\cdot\cdot}{\underset{\cdot\cdot}{F^+}} \longrightarrow \overset{79\ \cdot\cdot}{:\overset{\cdot\cdot}{Br}} {\text{---}} \overset{+}{CF_2} \ + \ \cdot F$$

m/z 148   *m/z* 129

$$\overset{81\ \cdot\cdot}{:\overset{\cdot\cdot}{Br}} {\text{---}} CF_2 {\text{---}} \overset{\cdot\cdot}{\underset{\cdot\cdot}{F^+}} \longrightarrow \overset{81\ \cdot\cdot}{:\overset{\cdot\cdot}{Br}} {\text{---}} \overset{+}{CF_2} \ + \ \cdot F$$

m/z 150   *m/z* 131

21.8   The formation of the molecular ion for isobutylamine and the pathway for its fragmentation to the base peak are shown below:

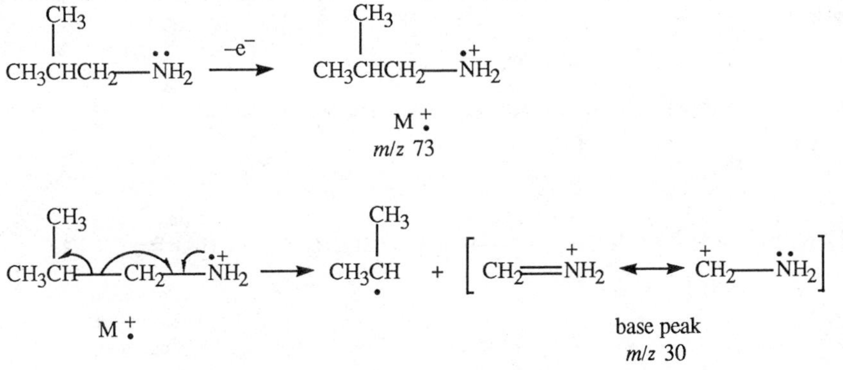

$$\begin{array}{c} CH_3 \\ | \\ CH_3CHCH_2\text{---}\overset{\cdot\cdot}{NH_2} \end{array} \xrightarrow{-e^-} \begin{array}{c} CH_3 \\ | \\ CH_3CHCH_2\text{---}\overset{\cdot+}{NH_2} \end{array}$$

$$M \overset{+}{\cdot}$$
m/z 73

$$\begin{array}{c} CH_3 \\ | \\ CH_3CH\text{---}CH_2\text{---}\overset{\cdot+}{NH_2} \end{array} \longrightarrow \begin{array}{c} CH_3 \\ | \\ CH_3\overset{\cdot}{CH} \end{array} \ + \ \left[CH_2 {=} \overset{+}{NH_2} \longleftrightarrow \overset{+}{CH_2}\text{---}\overset{\cdot\cdot}{NH_2} \right]$$

$$M \overset{+}{\cdot}$$
base peak
m/z 30

21.8 (cont)

The corresponding reaction for *sec*-butylamine is:

$$CH_3CH_2\overset{\overset{\displaystyle CH_3}{|}}{CH}\!\!-\!\!\overset{..}{N}H_2 \quad \xrightarrow{-e^-} \quad CH_3CH_2\overset{\overset{\displaystyle CH_3}{|}}{CH}\!\!-\!\!\overset{\bullet +}{N}H_2$$

M $\overset{+}{\underset{\bullet}{}}$
m/z 73

$$CH_3CH_2\!-\!\overset{\overset{\displaystyle CH_3}{|}}{CH}\!-\!\overset{\bullet +}{N}H_2 \longrightarrow CH_3\overset{\bullet}{C}H_2 \;+\; \left[CH_3CH\!\!=\!\!\overset{+}{N}H_2 \longleftrightarrow CH_3\overset{+}{C}H\!\!-\!\!\overset{..}{N}H_2 \right]$$

M $\overset{+}{\underset{\bullet}{}}$ base peak
                                                                m/z 44

Similarly for *tert*-butylamine:

$$CH_3\overset{\overset{\displaystyle CH_3}{|}}{\underset{\underset{\displaystyle CH_3}{|}}{C}}\!\!-\!\!\overset{..}{N}H_2 \quad \xrightarrow{-e^-} \quad CH_3\overset{\overset{\displaystyle CH_3}{|}}{\underset{\underset{\displaystyle CH_3}{|}}{C}}\!\!-\!\!\overset{\bullet +}{N}H_2$$

M $\overset{+}{\underset{\bullet}{}}$
m/z 73

$$CH_3\!-\!\overset{\overset{\displaystyle CH_3}{|}}{\underset{\underset{\displaystyle CH_3}{|}}{C}}\!-\!\overset{\bullet +}{N}H_2 \longrightarrow \overset{\bullet}{C}H_3 \;+\; \left[CH_3\overset{\overset{\displaystyle CH_3}{|}}{C}\!\!=\!\!\overset{+}{N}H_2 \longleftrightarrow CH_3\overset{\overset{\displaystyle CH_3}{|}}{\underset{+}{C}}\!\!-\!\!\overset{..}{N}H_2 \right]$$

M $\overset{+}{\underset{\bullet}{}}$ base peak
                                                                        m/z 58

The spectrum with the small molecular ion at *m/z* 73 and the base peak at *m/z* 30 is that of isobutylamine. The spectrum with the barely visible molecular ion at *m/z* 73 and the base peak at *m/z* 44 is that of *sec*-butylamine. The spectrum of *tert*-butylamine has a base peak of *m/z* 58. No molecular ion is seen in the spectrum of *tert*-butylamine.

21.9 $CH_3CH_2CH_2CH_2CH\!\!=\!\!CH_2 \quad \longrightarrow \quad CH_3CH_2CH_2CH_2CH\!\!-\!\!\overset{+}{\cdot}CH_2$

1-hexene   M $\overset{+}{\underset{\bullet}{}}$
                                                                m/z 84

$CH_3CH_2CH_2\!-\!CH_2\!-\!CH\!-\!\overset{+}{\cdot}CH_2 \quad \longrightarrow \quad CH_3CH_2\overset{\bullet}{C}H_2 \;+\; CH_2\!\!=\!\!CH\overset{+}{C}H_2$

M $\overset{+}{\underset{\bullet}{}}$ allyl cation
m/z 84   base peak
                                                                    m/z 41

21.10   The expected fragmentations of the molecular ion from 2,2-dimethyl-1-phenylpropane are shown below:

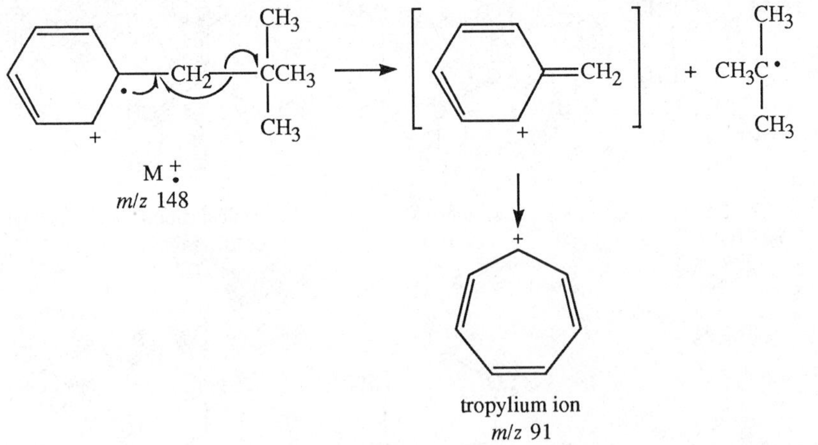

M $\overset{+}{\bullet}$
m/z 148

tropylium ion
m/z 91

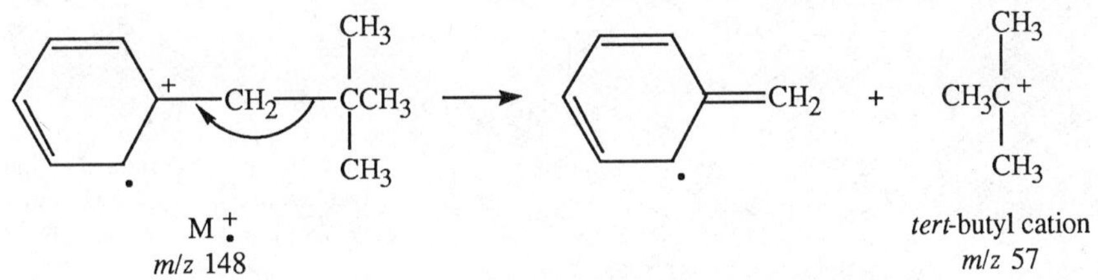

M $\overset{+}{\bullet}$
m/z 148

tert-butyl cation
m/z 57

The expected fragmentation of the molecular ion from 2-methyl-3-phenylbutane is shown below:

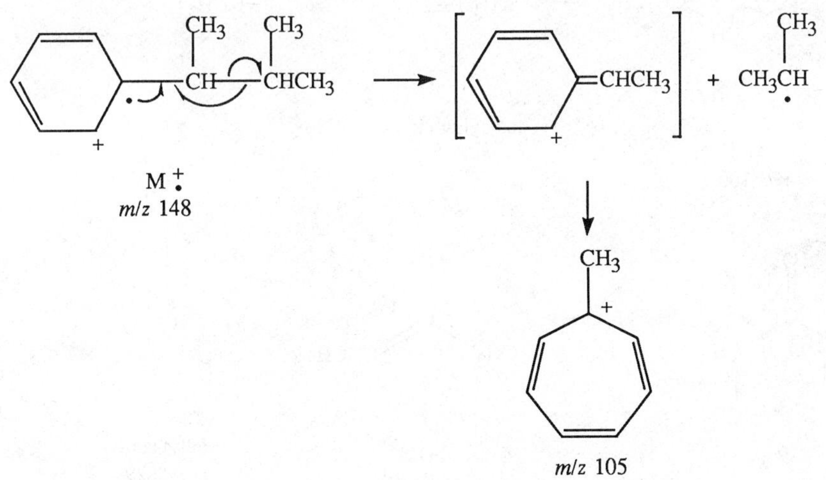

M $\overset{+}{\bullet}$
m/z 148

m/z 105

The spectrum with a peak at *m/z* 91 and the base peak at *m/z* 57 is that of 2,2-dimethyl-1-phenylpropane. The spectrum with the base peak at *m/z* 105 is that of 2-methyl-3-phenylbutane.

21.11

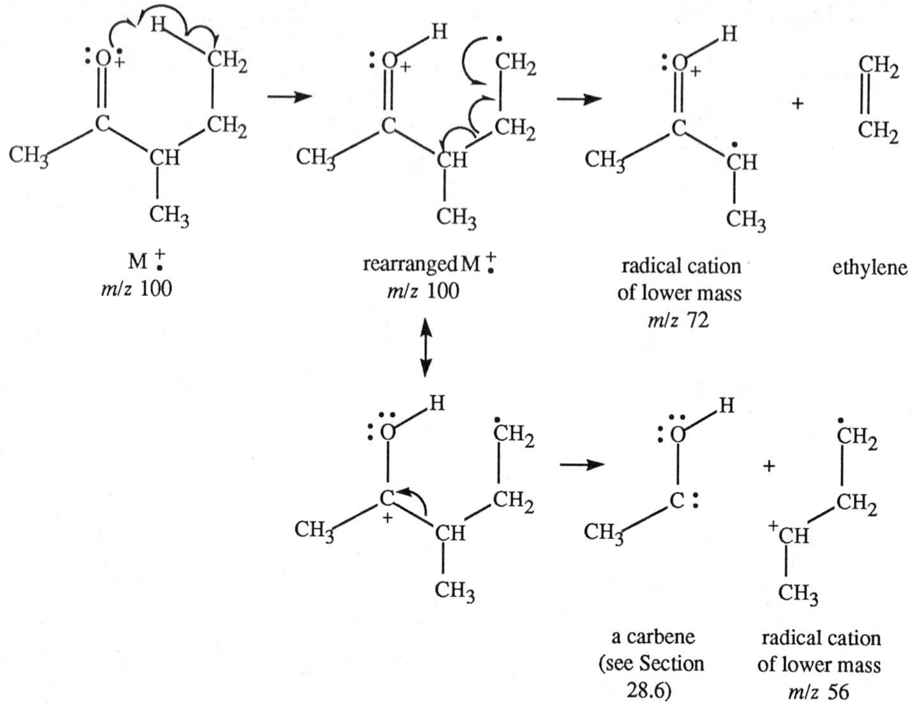

M $\overset{+}{\cdot}$
m/z 100

rearranged M $\overset{+}{\cdot}$
m/z 100

radical cation
of lower mass
m/z 72

ethylene

a carbene
(see Section
28.6)

radical cation
of lower mass
m/z 56

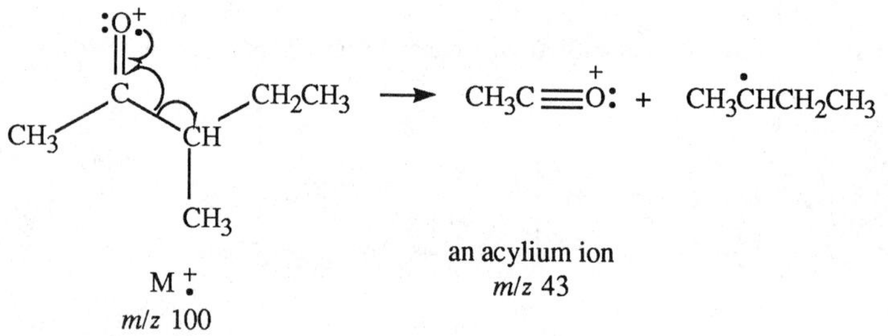

$$CH_3C \equiv \overset{+}{O}: \; + \; CH_3\overset{\cdot}{C}HCH_2CH_3$$

M $\overset{+}{\cdot}$
m/z 100

an acylium ion
m/z 43

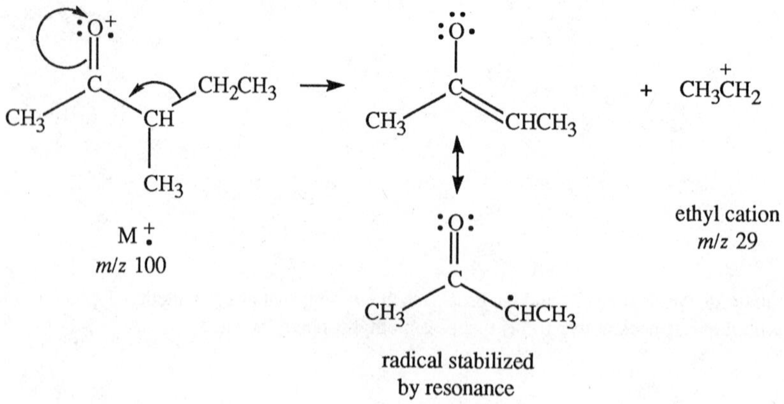

$+ \; CH_3\overset{+}{C}H_2$

ethyl cation
m/z 29

M $\overset{+}{\cdot}$
m/z 100

radical stabilized
by resonance

21.12   One possible rearrangement and fragmentation of the molecular ion from 1-hexene is shown below.

| | | a radical cation of lower mass *m/z* 42 |
| M $\overset{+}{\cdot}$ *m/z* 84 | a rearranged M $\overset{+}{\cdot}$ *m/z* 84 | + CH$_3$CH=CH$_2$ propene |

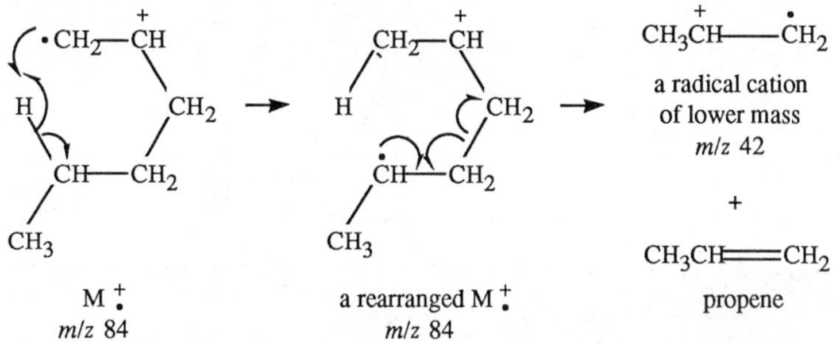

Another possible rearrangement and fragmentation of the molecular ion from 1-hexene is shown below.

| M $\overset{+}{\cdot}$ *m/z* 84 | a rearranged M $\overset{+}{\cdot}$ *m/z* 84 | another radical cation of lower mass *m/z* 56 + CH$_2$=CH$_2$ ethylene |

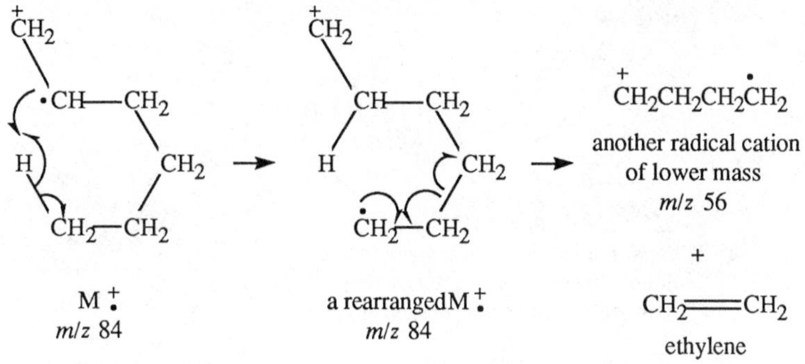

21.13   The infrared spectrum indicates the presence of an alcohol (3338 cm^{-1} is the O—H stretching frequency and 1031 cm^{-1} is the C—O stretching frequency). The proton magnetic resonance spectrum is analyzed below.

δ 1.7	(2H, multiplet, J ~6 Hz, —CH$_2$C$\underline{H}_2$CH$_2$—)
δ 2.5	(2H, triplet, J 6 Hz, ArC$\underline{H}_2$CH$_2$—)
δ 3.3	(2H, triplet, J 6 Hz, —CH$_2$C$\underline{H}_2$O—)
δ 4.0	(1H, singlet, —O$\underline{H}$)
δ 6.7	(5H, singlet, J 6 Hz, Ar$\underline{H}$)

Analysis of subunits in the proton magnetic resonance spectrum gives us a molecular formula of C$_9$H$_{12}$O, which corresponds to a molecular weight of 136. The molecular ion in the mass spectrum is at *m/z* 136. We therefore have all the atoms. The base peak (*m/z* 91) is the tropylium ion, which results from loss of a C$_2$H$_5$O unit and is the result of the following fragmentation.

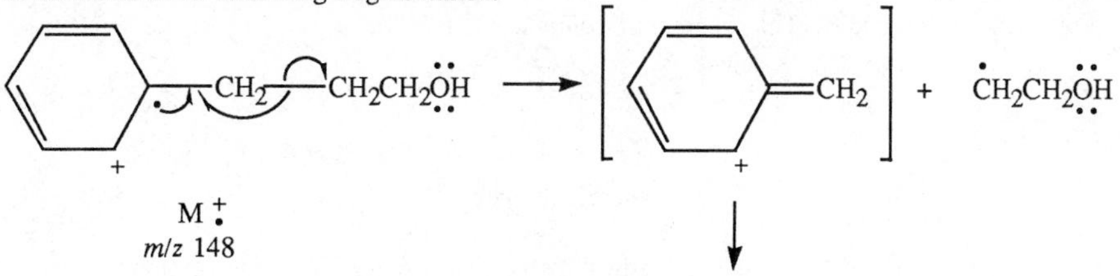

M $\overset{+}{\cdot}$ *m/z* 148

21.13 (cont)

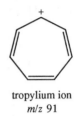

tropylium ion
m/z 91

The compound is 3-phenyl-1-propanol. The C-13 nuclear magnetic resonance data are analyzed below.

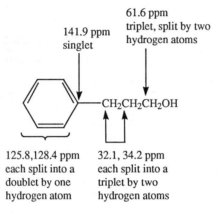

21.14 The mass spectrum of Compound B has two molecular ions at *m/z* 60 and 62 having a ratio of intensities of 3:1. This ratio tells us that the halogen in Compound B is chlorine. If we subtract 35 from 60, we get a mass of 25, which corresponds to the molecular formula C_2H, and the structure $HC \equiv C-$. Compound B is chloroethyne, $HC \equiv CCl$.

21.15 Compound C is an ester and must, therefore, contain the structural unit $-CO-$ (with $\stackrel{O}{\|}$ above CO), which accounts for 44 mass units. The molecular ion appears at *m/z* 60. Subtraction of 44 from 60 leaves 16 mass units, which corresponds to CH_3 and H. The structure of Compound C must be:

methyl formate

The presence of a methoxyl group is confirmed by the presence of a band in the proton magnetic resonance spectrum at δ 3.69. The fragmentations of the molecular ion leading to the ions with *m/z* 31 and 29 are shown on the next page.

21.15 (cont)

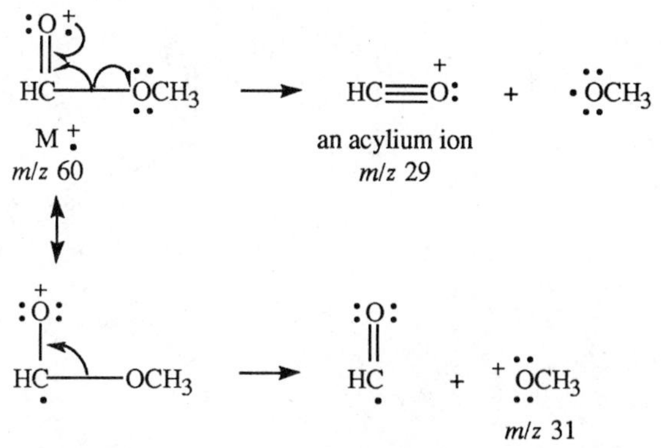

$$
\underset{\substack{\text{M} \overset{+}{\cdot} \\ m/z\ 60}}{\text{HC} \!-\! \text{OCH}_3} \longrightarrow \underset{\substack{\text{an acylium ion} \\ m/z\ 29}}{\text{HC} \!\equiv\! \overset{+}{\text{O}} :} \ + \ \cdot \ddot{\text{O}} \text{CH}_3
$$

$$
\underset{\substack{\text{M} \overset{+}{\cdot} \\ m/z\ 60}}{\text{HC} \!-\! \text{OCH}_3} \longrightarrow \text{HC} \ + \ \overset{+}{} \ddot{\text{O}} \text{CH}_3
$$

$$
m/z\ 31
$$

21.16   The two peaks of equal size in the mass spectrum of Compound D at m/z 79 and 81 tell us that bromine is present. The fact that the molecular ion fragments before it can be recorded suggests that the compound gives rise to a very stable cation, m/z 43. An acylium ion , $CH_3C\equiv O^+$, fits this description. Compound D is acetyl bromide.

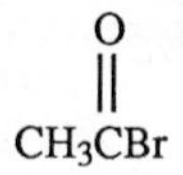

$$
\underset{\text{CH}_3\text{CBr}}{\overset{\displaystyle\text{O}}{\overset{\displaystyle\|}{}}}
$$

(Other possibilities that may come to mind are isopropyl or *n*-propyl bromide. Neither alkyl cation is as stable as the resonance-stabilized acylium ion, and we would therefore expect to see the molecular ion for either of these compounds.)

21.17   The molecular ion of Compound E has m/z 60. There is a difference of 29 mass units between the molecular ion and the base peak at m/z 31, which corresponds to the loss of an ethyl radical and the formation of the oxonium ion, $\overset{+}{\text{CH}_2}\!=\!\text{OH}$, which is usually the base peak in a primary alcohol. Compound E is *n*-propyl alcohol.

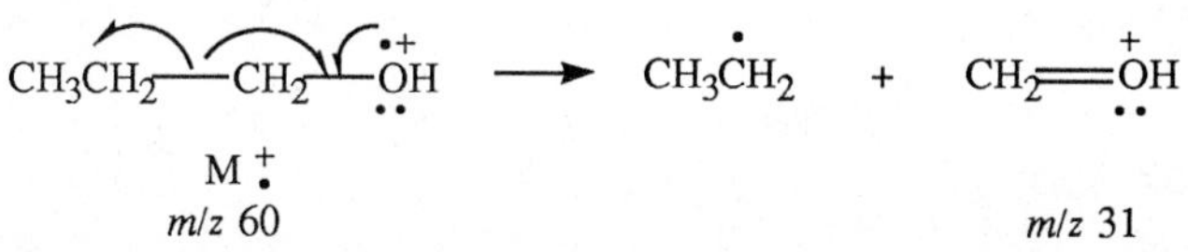

$$
\underset{\substack{\text{M} \overset{+}{\cdot} \\ m/z\ 60}}{\text{CH}_3\text{CH}_2 \!-\! \text{CH}_2 \!-\! \overset{\cdot+}{\ddot{\text{O}}\text{H}}} \longrightarrow \text{CH}_3\dot{\text{C}}\text{H}_2 \ + \ \underset{m/z\ 31}{\text{CH}_2 \!=\! \overset{+}{\ddot{\text{O}}}\text{H}}
$$

Compound F is isomeric with *n*-propyl alcohol, therefore it must be isopropyl alcohol. The fragmentation leading to the base peak is shown below.

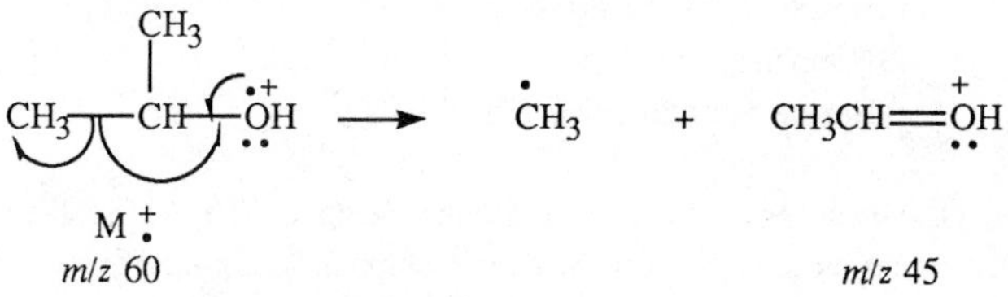

$$
\underset{\substack{\text{M} \overset{+}{\cdot} \\ m/z\ 60}}{\overset{\displaystyle\text{CH}_3}{\underset{\displaystyle|}{\text{CH}_3 \!-\! \text{CH} \!-\! \overset{\cdot+}{\ddot{\text{O}}\text{H}}}}} \longrightarrow \dot{\text{C}}\text{H}_3 \ + \ \underset{m/z\ 45}{\text{CH}_3\text{CH} \!=\! \overset{+}{\ddot{\text{O}}}\text{H}}
$$

21.18

M $\overset{+}{\underset{\bullet}{}}$	rearranged M $\overset{+}{\underset{\bullet}{}}$	base peak	propene
m/z 86		*m/z* 44	
not seen in spectrum			

21.19   The mass spectrum of Compound G has two molecular ions of equal intensity at *m/z* 120 and 122. The presence of these two peaks tells us that Compound G contains bromine. The presence of bromine is confirmed by the two very weak peaks of equal intensity at *m/z* 79 and 81. The base peak at *m/z* 41 is the allyl cation, which results from loss of a bromine atom from the molecular ion. Compound G is 3-bromo-1-propene (allyl bromide). The formation of the base peak in the mass spectrum results from loss of a bromine atom.

M $\overset{+}{\underset{\bullet}{}}$		base peak
m/z 120		*m/z* 41

The proton and C-13 magnetic resonance spectra are analyzed below:

δ 3.7	(2H, doublet, BrC$\underline{H}_2$CH—)
δ 4.9 and 5.1	(2H, two doublets of doublets, —CH=C$\underline{H}_2$)
δ 5.4 – 6.1	(1H, multiplet, —CH$_2$—C$\underline{H}$=CH$_2$)

21.20   The infrared spectrum tells us that Compound H is a conjugated ketone (1676 cm^{-1} is the C=O stretching frequency for a conjugated carbonyl group). The proton magnetic resonance spectrum is analyzed below.

δ 2.3	(3H, singlet, C$\underline{H}_3$C=O)
δ 3.6	(3H, singlet, —OC$\underline{H}_3$)
δ 6.4 – 7.5	(4H, para substitution pattern, Ar$\underline{H}$)

The fragments add up to $C_9H_{10}O_2$, which corresponds to a molecular weight of 150. This is the *m/z* of the molecular ion of Compound H, therefore all atoms are accounted for. Compound H is *p*-methoxyacetophenone.

21.20 (cont)

The base peak, *m/z* 135, results from loss of a methyl radical from the molecular ion, leaving a resonance-stabilized acylium ion with *m/z* 135.

M $\overset{+}{\cdot}$	
m/z 150	*m/z* 135

21.21 The infrared spectrum tells us that Compound I is an aldehyde (1703 cm^{-1} is the C=O stretching frequency and 2730 cm^{-1} distinguishes an aldehyde group from a ketone group).

δ 2.4	(3H, singlet, ArC$\underline{H}_3$)
δ 7.1 – 7.9	(4H, para substitution pattern, Ar$\underline{H}$)
δ 9.9	(1H, singlet, —C$\underline{H}$=O)

These fragments add up to C$_8$H$_8$O corresponding to a molecular weight of 120, which is the *m/z* of the molecular ion of Compound I. Putting the subunits from the proton magnetic resonance together gives us *p*-tolualdehyde. The first base peak, *m/z* 119, results from the loss of a hydrogen atom from the molecular ion, giving an acylium ion. The second base peak, *m/z* 91, is the tropylium ion, which most likely comes from a rearrangement of the methylphenyl cation formed by loss of carbon monoxide from the acylium ion.

M $\overset{+}{\cdot}$	an acylium ion
m/z 120	*m/z* 119

21.21 (cont)

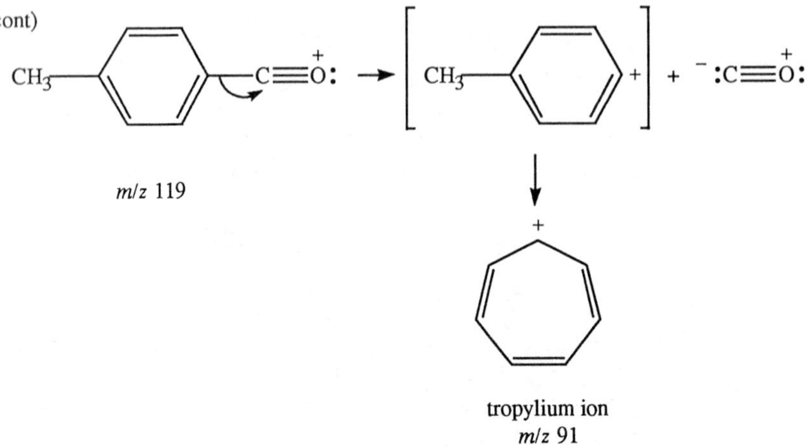

m/z 119

tropylium ion
m/z 91

The C-13 nuclear magnetic resonance is analyzed below.

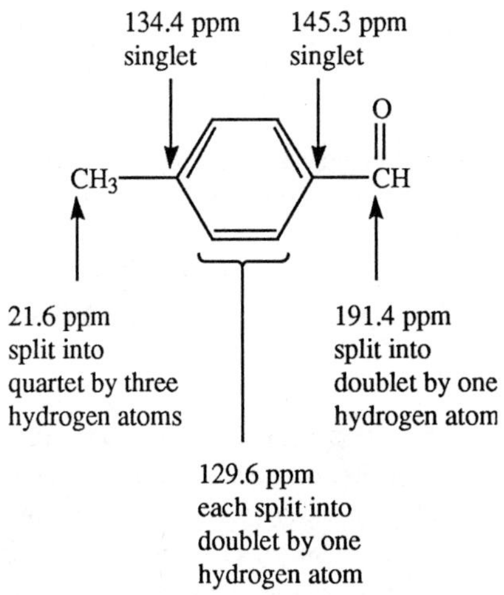

134.4 ppm
singlet

145.3 ppm
singlet

21.6 ppm
split into
quartet by three
hydrogen atoms

191.4 ppm
split into
doublet by one
hydrogen atom

129.6 ppm
each split into
doublet by one
hydrogen atom

22

The Chemistry of Amines

22.1 (a) 2-methyl-1-propanamine
isobutylamine

(b) *cis*-2-ethyl-1-cyclopentanamine
cis-1-amino-2-ethylcyclopentane

(c) *N,N*-diethylbutanamine

(d) 3-cyclopropyl-1-propanamine
3-cyclopropyl-1-propylamine

(e) 3-nitroaniline
m-nitroaniline

(f) 2,4-dibromoaniline

22.2 (a)

$$K_{diss} = \frac{[R_3N \colon][BR'_3]}{[R_3N^{+} \colon^{-} BR'_3]}$$

(b) The dissociation constants for the amine-borane complexes become smaller as the number of substituents on the nitrogen atom increases from none (for ammonia) to two for dimethylamine. When there are three substituents on the nitrogen atom, the dissociation constant is larger than it is for the other two complexes. The dissociation constant measures the strength of the bond between the atoms of the Lewis base, the amine, and the Lewis acid, the borane. The bond strength is influenced by two factors: (1) electronic factors, mainly the availability of the nonbonding electron pair on the nitrogen atom; and (2) steric factors. The values observed for the dissociation constants may be interpreted to mean that alkyl substitution on the nitrogen atom increases the availability of the nonbonding electrons on the nitrogen atom, but that the presence of three substituents on nitrogen leads to steric hindrance that interferes with bonding in the complex between trimethyamine and trimethylborane. The steric factor is more important in this case than in the reaction of the amine with a Brønsted-Lowry acid because trimethylborane is larger than a proton.

22.3 The acidity of a conjugate acid is related to the availability of the nonbonding electron pair on the base. The observed trend for the conjugate acids of the given amines is due to the decreasing availability of the electron pair on the nitrogen. This decreasing availability is caused by the electron-withdrawing ether and cyano substituents which pull electron density away by induction. The inductive effect is greatest when the electron-withdrawing group is closest to the amino group. The cyano group is more electron-withdrawing than the oxygen because of resonance.

22.4

(a)   All of the cyanoanilines are weaker bases than aniline itself because the cyano group is an electron-withdrawing group. The effect of the cyano group is greatest when it is ortho, and smallest when it is meta, to the amino group. Resonance effects are important for the ortho- and para-substituted compounds, but only the inductive effect is important in *m*-cyanoaniline.

nonbonding electrons of the amino group are delocalized to
the ring and to the cyano group

no resonance delocalization of the nonbonding electrons of the
amino group to the cyano group; only the inductive effect

22.4 (cont)

(a) (cont)

delocalization of the nonbonding electrons of the amino group to the
ring and to the cyano group; inductive effect of the cyano group in the
para position weaker than when it is in the ortho position

(b) *p*-Toluidine is a slightly stronger base than aniline because of the electron-donating effect of the methyl group. 4-Aminobenzophenone is a weaker base than aniline because the carbonyl group is an electron-withdrawing group.

The carbonyl group decreases the availability of the nonbonding electrons on the amino group by the inductive effect and by resonance.

The inductive effect of the methyl group increases the availability of the nonbonding electrons on the amino group. In both series of compounds the trends can be rationalized by looking at the availability of the nonbonding electrons on the amino group in the conjugate base. The more available the electrons, the stronger the base, the weaker is the conjugate acid, and the larger the pK_a.

22.5

(a)
$$CH_3CHCO^- + NH_4{}^+Br^- \xleftarrow{\;NH_3 \text{ (excess)}\;} CH_3CHCOH$$
$$\quad\;\; |\hspace{5.3cm}|$$
$$\quad\;\; {}^+NH_3\hspace{4.5cm}Br$$

with carbonyl (C=O) groups above both CH₃CHCO⁻ and CH₃CHCOH.

(b)

⬡—NHCH₃ + ⬡—NH₃⁺I⁻ $\xleftarrow{\begin{array}{c}CH_3I\\(1\text{ equiv})\end{array}}$ ⬡—NH₂

2 equiv

(c)
⬡—CH₂CH₂NH₂ $\xleftarrow{\;NH_3 \text{ (excess)}\;}$ ⬡—CH₂CH₂Br $\xleftarrow[\text{pyridine}]{PBr_3}$

⬡—CH₂CH₂OH $\xleftarrow{H_3O^+}$ $\xleftarrow{\triangle\,O}$ ⬡—MgBr $\xleftarrow[\substack{\text{diethyl}\\ \text{ether}}]{Mg}$

⬡—Br $\xleftarrow[\text{Fe}]{Br_2}$ ⬡

22.6   4-Chlorobutanoic acid is a strong enough acid to protonate the phthalimidate anion, converting it to phthalimide. The nitrogen atom in phthalimide is not nucleophilic because the nonbonding electrons on nitrogen are delocalized to two carbonyl groups.

phthalimidate—N⁻K⁺ + Cl(CH₂)₃COH ⟶ phthalimide—NH + Cl(CH₂)₃CO⁻K⁺

$pK_a\,4.5$

$pK_a\,7.4$

22.7   ClCH₂CH₂CH₂C≡N $\xleftarrow[\text{pyridine}]{SOCl_2}$ HOCH₂CH₂CH₂C≡N $\xleftarrow[\text{ethanol}]{NaCN}$

4-chlorobutanenitrile

HOCH₂CH₂CH₂Cl $\xleftarrow[\text{pyridine}]{SOCl_2\,(1\text{ molar equiv})}$ HOCH₂CH₂CH₂OH

22.8

(a)

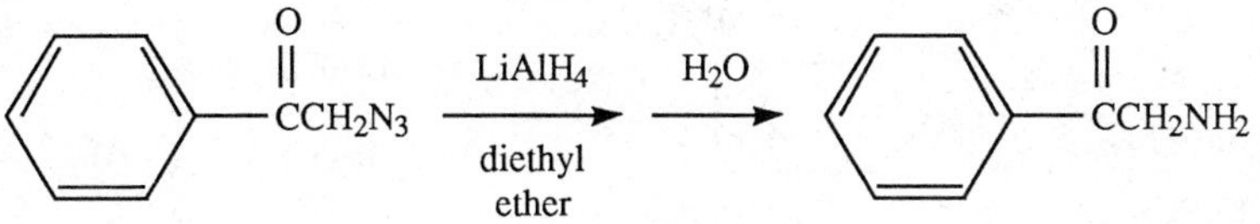

(b) $CH_3(CH_2)_4C \equiv C(CH_2)_8N_3$ $\xrightarrow[\text{quinoline}]{\overset{H_2}{\text{Pd/CaCO}_3}}$

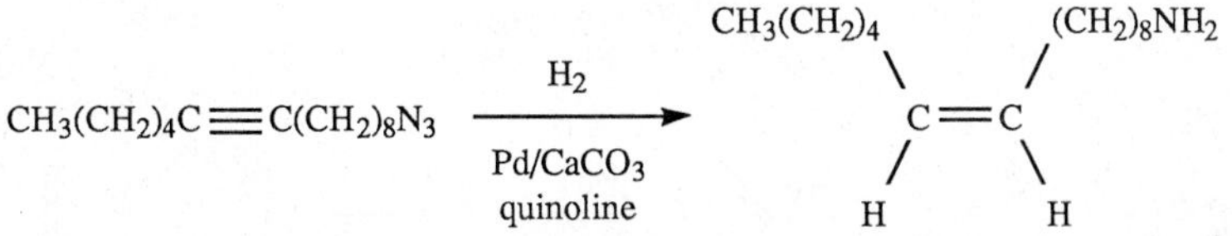

22.9

(a)

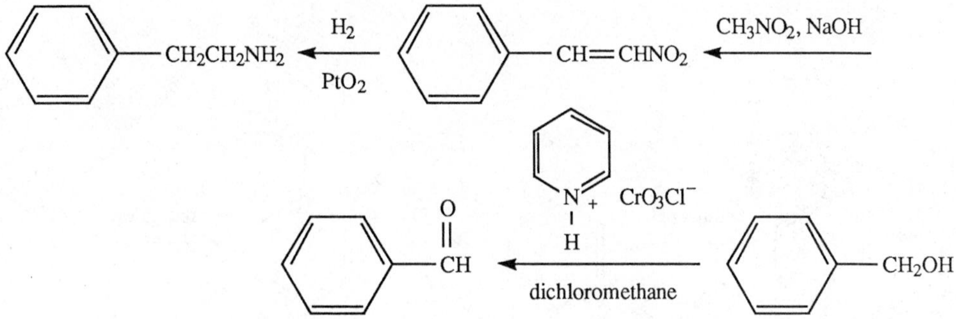

(b)

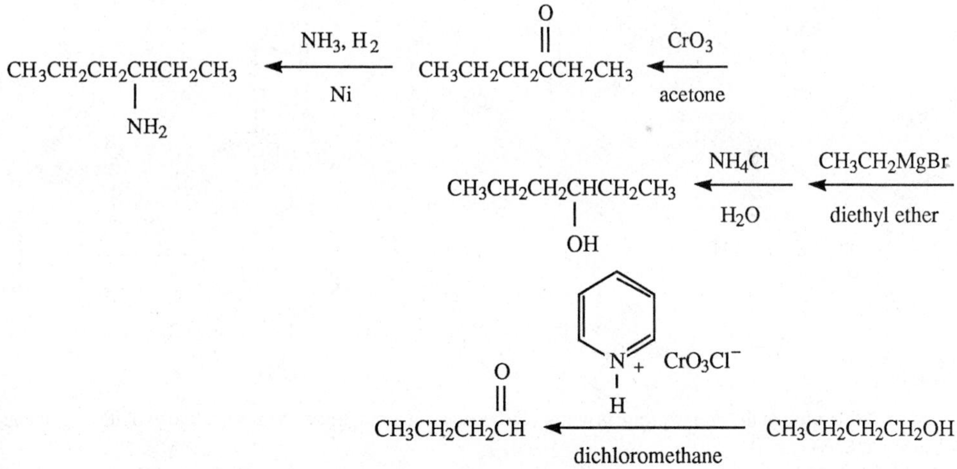

(c)

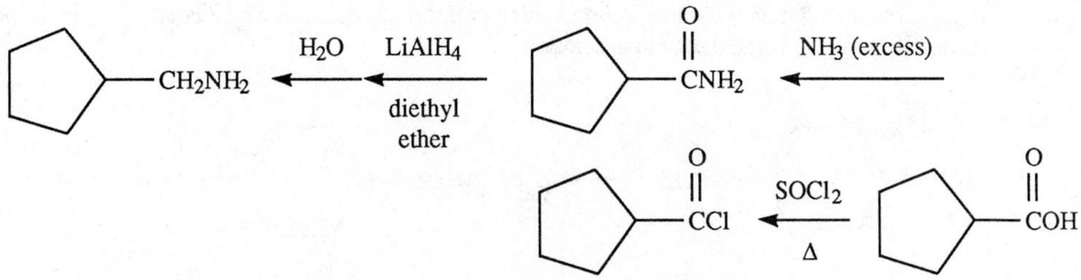

22.9 (cont)

(d)

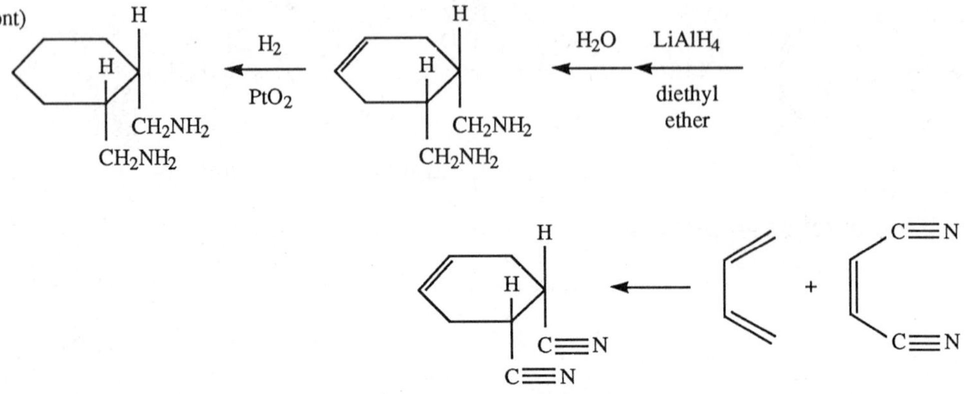

22.10

(a)

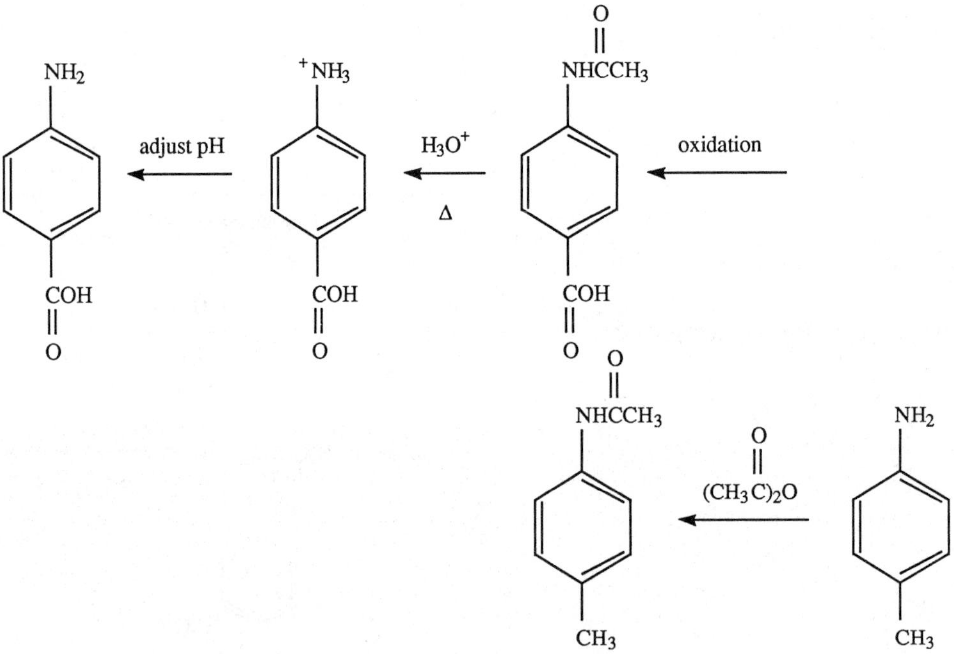

Protection of the amino group is necessary because an aromatic amine is sensitive to oxidizing agents (Section 20.7A).

(b)   If the pH of the solution is too low, deprotonation of the amine function would not occur, and the protonated *p*-aminobenzoic acid would remain in solution.

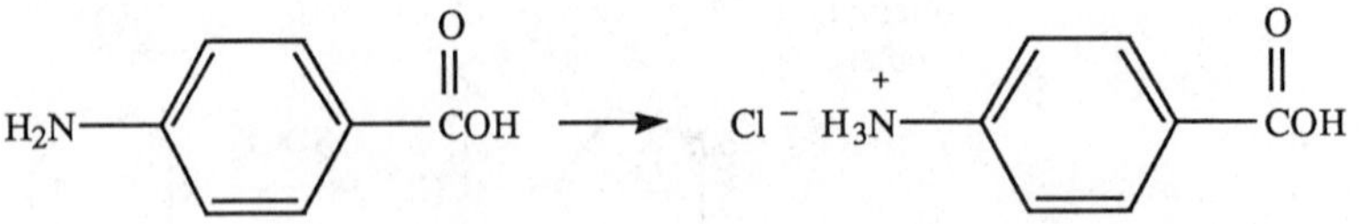

22.10 (cont)

(b) (cont)

If the pH is too high, deprotonation of the carboxylic acid group would occur, and the *p*-aminobenzoate salt would remain in solution.

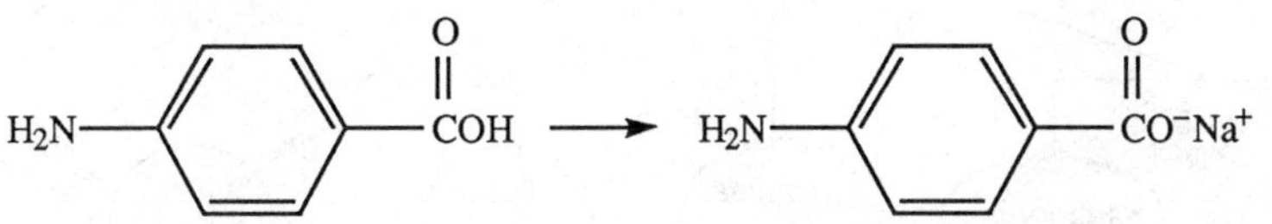

22.11

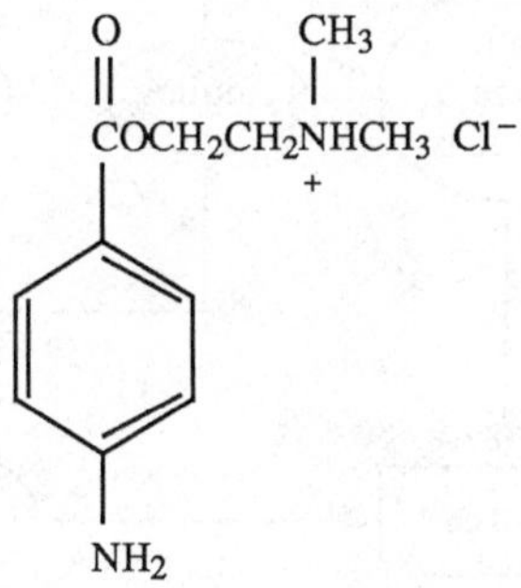

Procaine has two amine functions, an amino group on the aromatic ring, and a tertiary alkyl amine function. The alkyl amine is much more basic (pK_a of a substituted ammonium ion, ~10) than an aryl amine (pK_a of the anilinium ion, 4.6), therefore procaine hydrochloride has the structure shown above.

Concept Map 22.1   Preparation of amines.

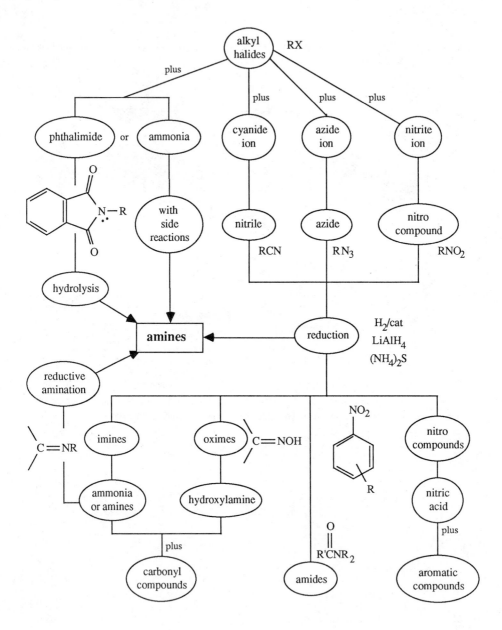

22.12

(a)

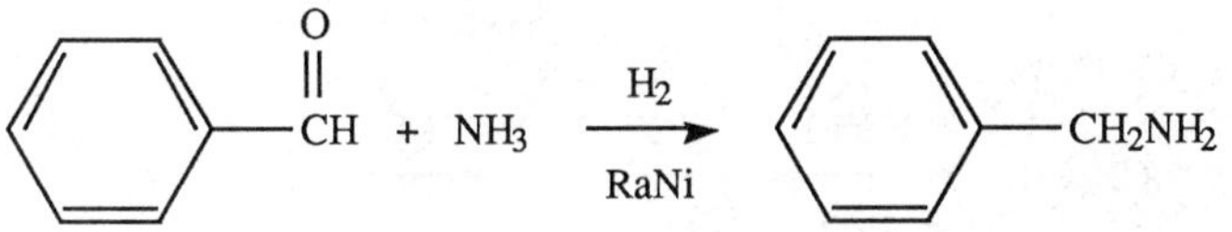

(b)

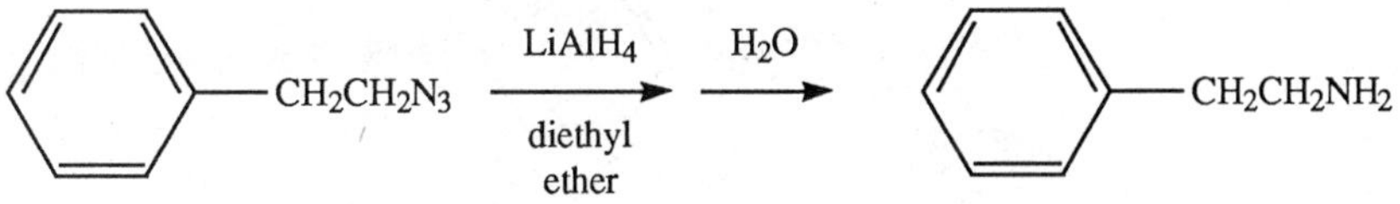

(c)

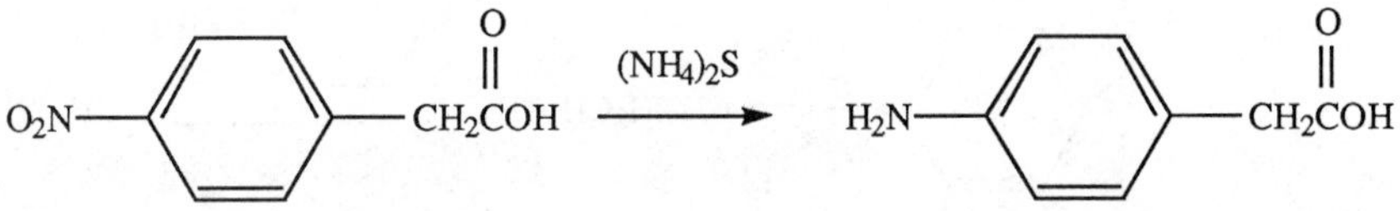

(d)

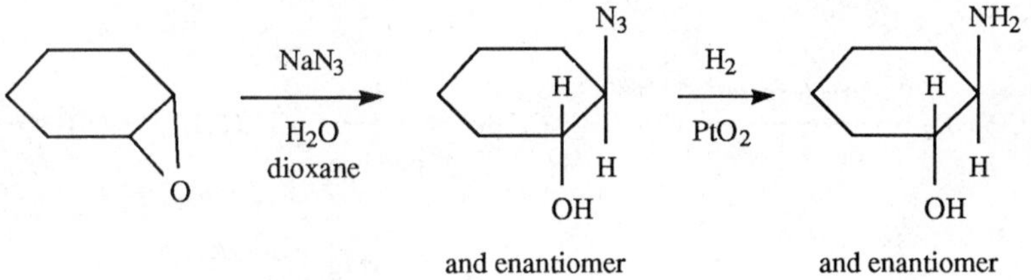

and enantiomer   and enantiomer

(e)

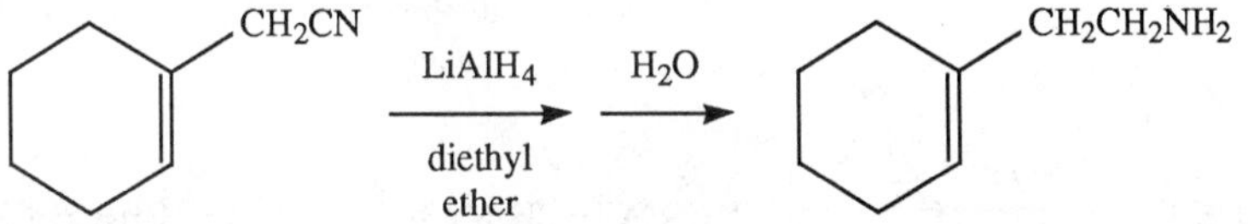

22.13

(a)

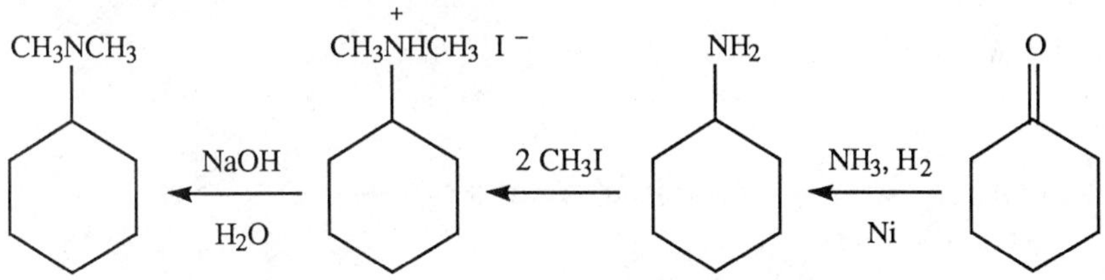

(b)

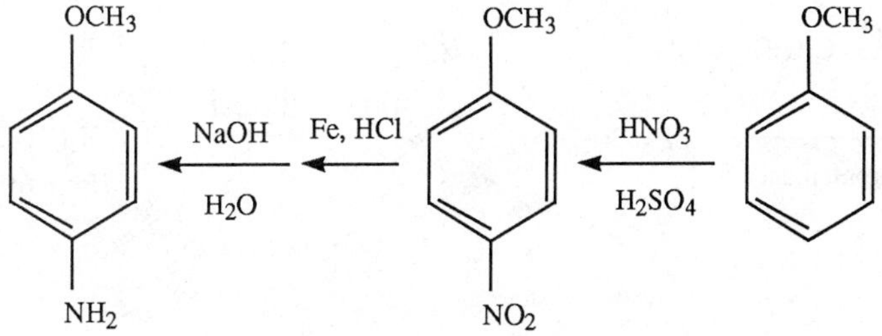

22.13 (cont)

(c)

$$\underset{\overset{|}{\text{NCH}_2\text{CH}_2\text{CH}_2\text{CH}_3}}{\overset{\text{CH}_2\text{CH}_2\text{CH}_2\text{CH}_3}{}}$$

$$\xleftarrow[\text{H}_2\text{O}]{\text{NaOH}} \xleftarrow{\text{CH}_3\text{CH}_2\text{CH}_2\text{CH}_2\text{Br}}$$

$$\text{—NHCH}_2\text{CH}_2\text{CH}_2\text{CH}_3 \xleftarrow[\text{H}_2\text{O}]{\text{NaBH}_4}$$

$$\text{—N}=\text{CHCH}_2\text{CH}_2\text{CH}_3 \xleftarrow{\quad} \overset{\text{—NH}_2}{\quad} \quad \overset{\overset{\text{O}}{\|}}{\text{CH}_3\text{CH}_2\text{CH}_2\text{CH}}$$

(d)

(1) $$\text{CH}_3(\text{CH}_2)_6\text{CH}_2\text{NH}_2 \xleftarrow{\text{NH}_3 \text{ (excess)}} \text{CH}_3(\text{CH}_2)_6\text{CH}_2\text{Br} \xleftarrow[\text{pyridine}]{\text{PBr}_3} \text{CH}_3(\text{CH}_2)_6\text{CH}_2\text{OH} \xleftarrow{\text{H}_3\text{O}^+}$$

$$\xleftarrow{\quad} \text{CH}_3(\text{CH}_2)_4\text{CH}_2\text{MgBr} \xleftarrow[\substack{\text{diethyl} \\ \text{ether}}]{\text{Mg}} \text{CH}_3(\text{CH}_2)_4\text{CH}_2\text{Br} \xleftarrow[\text{pyridine}]{\text{PBr}_3}$$

$$\text{CH}_3(\text{CH}_2)_4\text{CH}_2\text{OH} \xleftarrow{\quad} \xleftarrow{\text{H}_3\text{O}^+} \text{CH}_3\text{CH}_2\text{CH}_2\text{CH}_2\text{MgBr} \xleftarrow[\substack{\text{diethyl} \\ \text{ether}}]{\text{Mg}} \text{CH}_3\text{CH}_2\text{CH}_2\text{CH}_2\text{Br}$$

(2) $$\text{CH}_3(\text{CH}_2)_6\text{CH}_2\text{NH}_2 \xleftarrow[\substack{\text{Pd} \\ \text{acetic acid}}]{\text{H}_2} \text{CH}_3(\text{CH}_2)_5\text{CH}=\text{CHNO}_2 \xleftarrow{\text{CH}_3\text{NO}_2, \text{ NaOH}}$$

$$\underset{\text{CH}_3(\text{CH}_2)_5\text{CH}}{\overset{\overset{\text{O}}{\|}}{}} \xleftarrow[\text{dichloromethane}]{\text{N}^+\text{H } \text{CrO}_3\text{Cl}^-} \text{CH}_3(\text{CH}_2)_5\text{CH}_2\text{OH} \xleftarrow{\text{H}_3\text{O}^+} \xleftarrow{\quad} \text{CH}_3\text{CH}_2\text{CH}_2\text{CH}_2\text{MgBr}$$
$$\text{[from (d) 1.]}$$

22.13 (cont)

(d) (cont)

(3) $CH_3(CH_2)_6CH_2NH_2$ ←$\dfrac{NH_3,\ H_2}{Ni}$ $CH_3(CH_2)_6CH$ (with =O) ←$\dfrac{}{\text{dichloromethane}}$ [pyridine·H$^+$ CrO$_3$Cl$^-$] $CH_3(CH_2)_6CH_2OH$ [from (d) 1.]

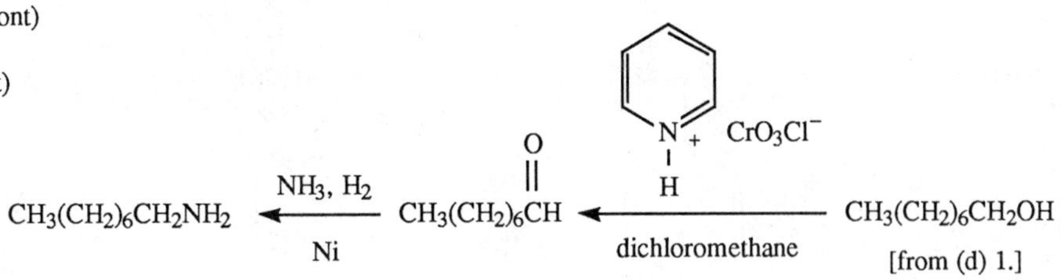

22.14

(a) CH_3NCH_3 (on benzene ring) ←$\dfrac{NaOH}{H_2O}$ $CH_3\overset{+}{N}HCH_3$ I$^-$ (on benzene ring) ←$\dfrac{}{2\ CH_3I}$ NH_2 (on benzene ring) ←$\dfrac{}{NaOH}$ ←$\dfrac{}{Sn,\ HCl}$ NO_2 (on benzene ring) ←$\dfrac{HNO_3}{H_2SO_4}$ (benzene)

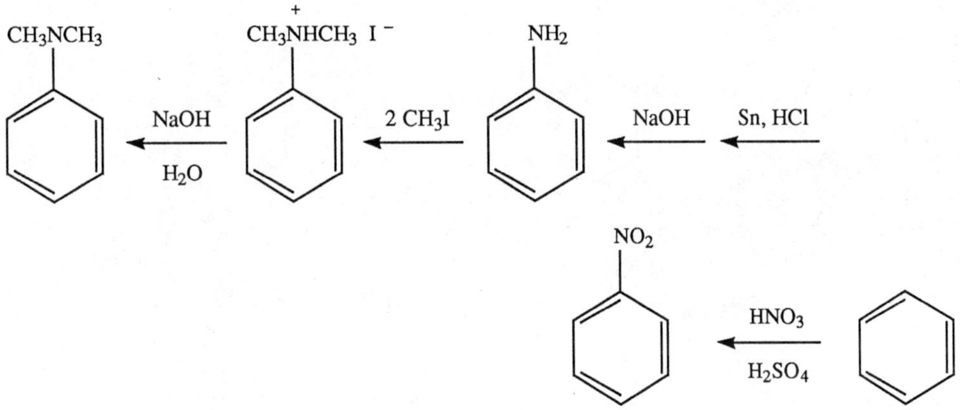

(b) SO_2Cl (on benzene ring) ←$\dfrac{}{PCl_5}$ SO_3H (on benzene ring) ←$\dfrac{SO_3}{H_2SO_4}$ (benzene)

H_2N— (ring) —NHS(=O)(=O)— (ring) ←$\dfrac{H_2}{PtO_2\ \text{ethanol}}$ O_2N— (ring) —NHS(=O)(=O)— (ring) ← SO_2Cl (on benzene ring)

NH_2 / O_2N (on benzene ring) ←$\dfrac{}{(NH_4)_2S}$ NO_2 / O_2N (on benzene ring) ←$\dfrac{HNO_3}{H_2SO_4}$ (benzene)

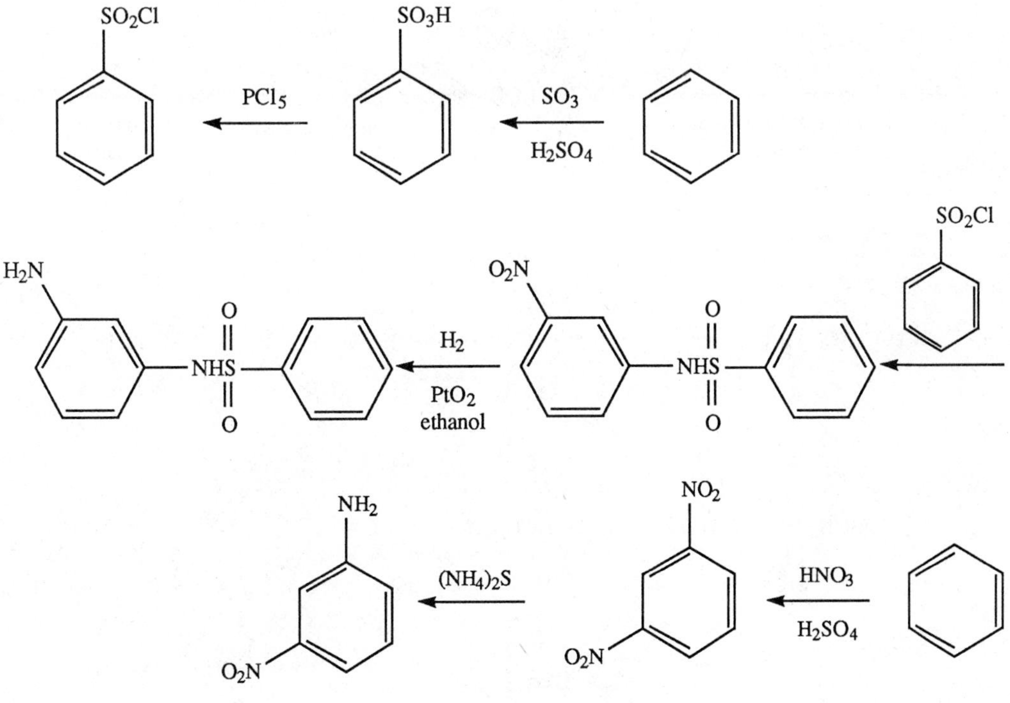

22.14 (cont)

(c)

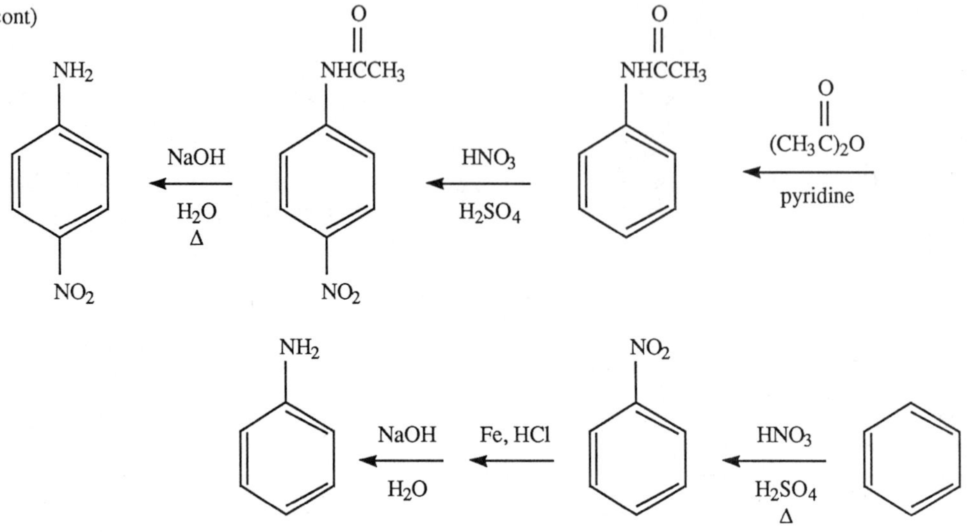

(d)

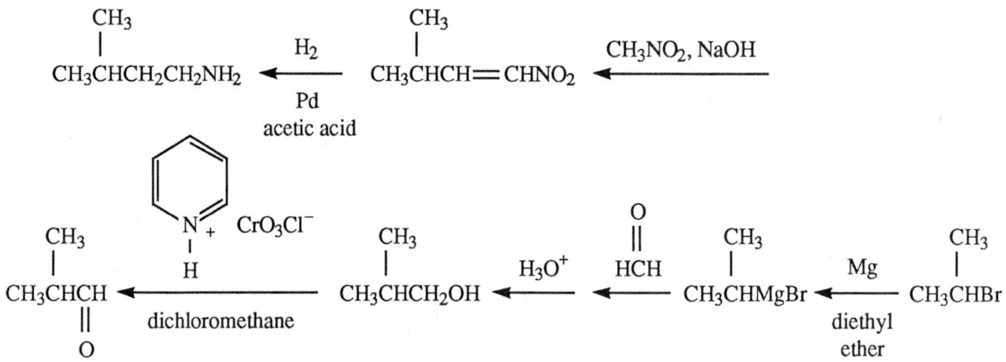

$$CH_3CHCH_2CH_2NH_2 \xleftarrow[\substack{Pd \\ acetic\ acid}]{H_2} CH_3CHCH=CHNO_2 \xleftarrow{CH_3NO_2,\ NaOH}$$
with CH_3 substituents

22.15

(a)

$$CH_3CH_2CH_2CHCH_3 \xrightarrow[\substack{H_2O \\ 0\,°C}]{NaNO_2,\ HCl} CH_3CH_2CH_2CHCH_3 + CH_3CH_2CH_2CHCH_3$$
with NH_2 → Cl / OH

$$CH_3CH_2CHCH_2CH_3 + CH_3CH_2CHCH_2CH_3 + \ \ C=C \ \ + \ \ C=C$$
with Cl / OH and alkene isomers

$$+ \ \ CH_3CH_2CH_2CH=CH_2 \ \ + \ \ N_2\uparrow$$

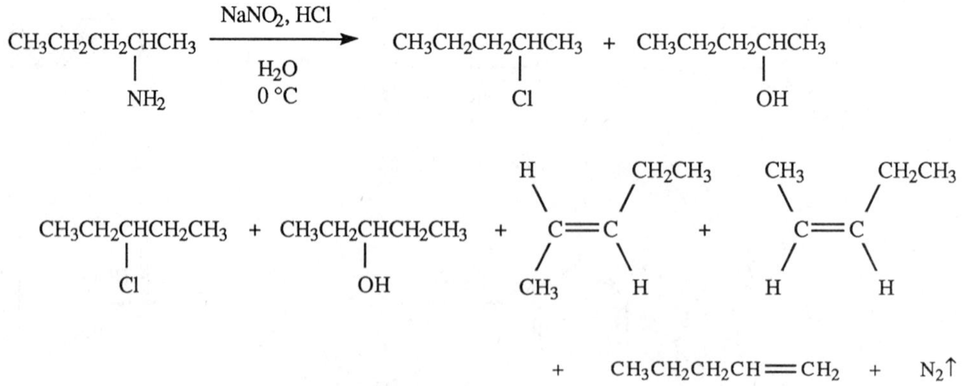

22.15 (cont)

(b)

(c)

$$CH_3CH_2CH_2NCH_3 \xrightarrow[\substack{O \\ \| \\ CH_3CO^-Na^+ \\ \Delta}]{\substack{O \\ \| \\ NaNO_2, \ CH_3COH}}$$

with a CH_3 branch on the N

$$\underset{\substack{O \\ \|}}{CH_3CH_2CH} \ + \ \underset{\substack{CH_3 \\ |}}{CH_3CH_2CH_2N-N=O} \ + \ \underset{\substack{O \\ \|}}{HCH} \ + \ \underset{\substack{CH_3 \\ |}}{CH_3N-N=O}$$

(d)

$$\underset{\substack{O \\ \|}}{CH_3NHCNHCH_3} \xrightarrow[\substack{H_2O \\ 0\,°C}]{NaNO_2, \ HCl} \ \underset{\substack{O \\ \| \\ N=O}}{CH_3NHCNCH_3} \ + \ \underset{\substack{O \\ \| \\ N=O \quad N=O}}{CH_3N-C-NCH_3}$$

(e)

$$\underset{\substack{CH_3 \\ | \\ CH_3}}{CH_3CNH_2} \xrightarrow[\substack{H_2O \\ 0\,°C}]{NaNO_2, \ HCl} \ \underset{\substack{CH_3 \\ | \\ Cl}}{CH_3CCH_3} \ + \ \underset{\substack{CH_3 \\ | \\ OH}}{CH_3CCH_3} \ + \ \underset{\substack{CH_3 \\ |}}{CH_3C=CH_2} \ + \ N_2\uparrow$$

22.16 $R_2NH \ + \ HCl \ \longrightarrow \ \overset{+}{R_2NH_2} \ Cl^-$
water soluble

$R_2NH \ + \ HNO_2 \ \xrightarrow{HCl} \ R_2\ddot{N}-N=\ddot{O} \ \longleftrightarrow \ R_2\overset{+}{N}=N-\ddot{O}{:}^-$

The nonbonding electrons on the amine nitrogen atom are delocalized to the nitroso group. The nitrosoamine is therefore much less basic than the original amine and will not form a salt. A nitros-amine resembles an amide in basicity.

Concept Map 22.2   Nitrosation reactions.

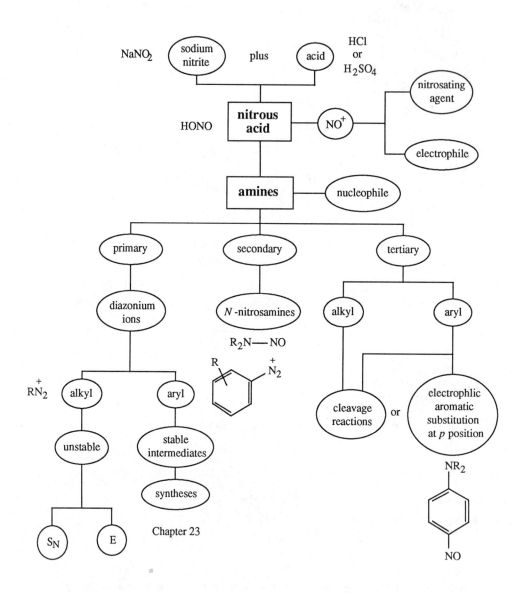

22.17

(a)

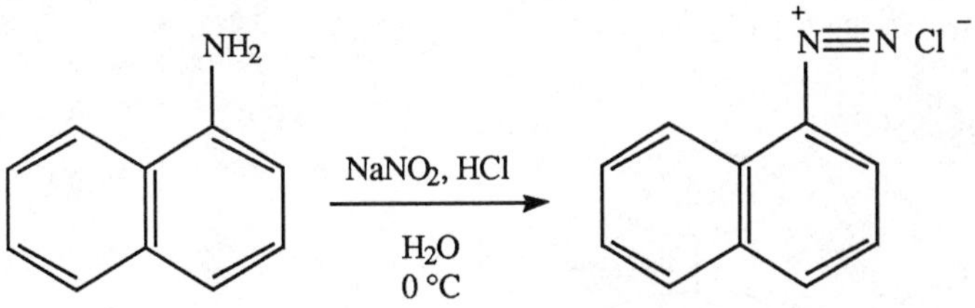

(b)

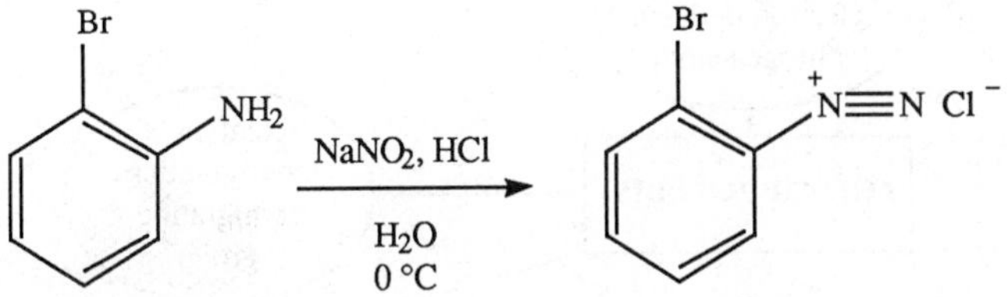

(c)

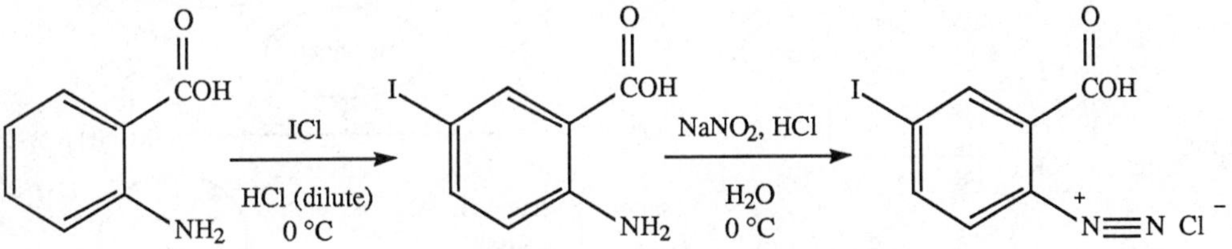

(d)

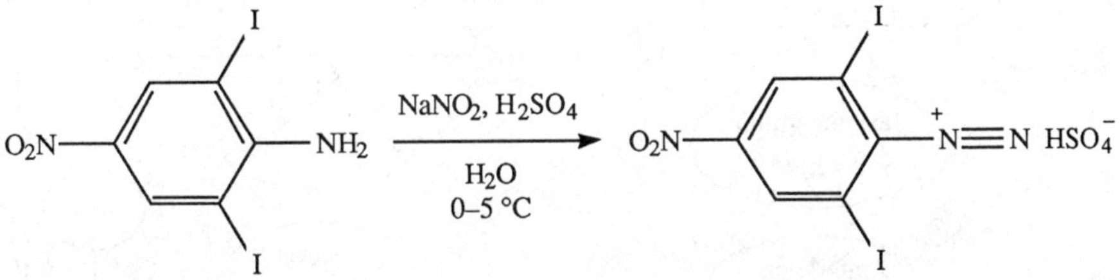

22.18

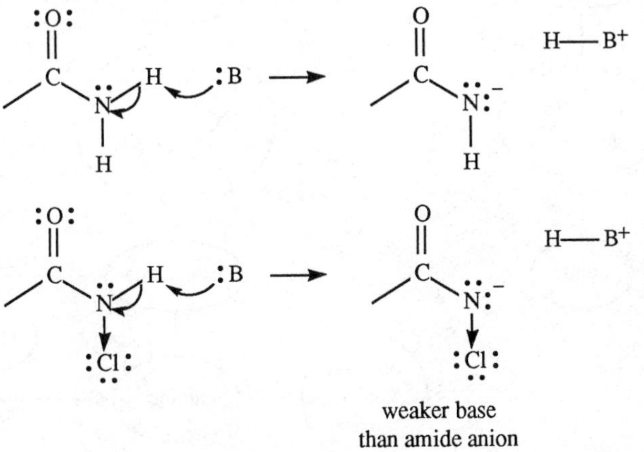

weaker base
than amide anion

Chlorine is more electronegative than hydrogen, therefore it can stabilize a negative charge more effectively. The anion from the *N*-chloroamide is thus a weaker base than the anion from the amide, and its conjugate acid is a stronger acid than the original unsubstituted amide.

Concept Map 22.3   Rearrangements to nitrogen atoms.

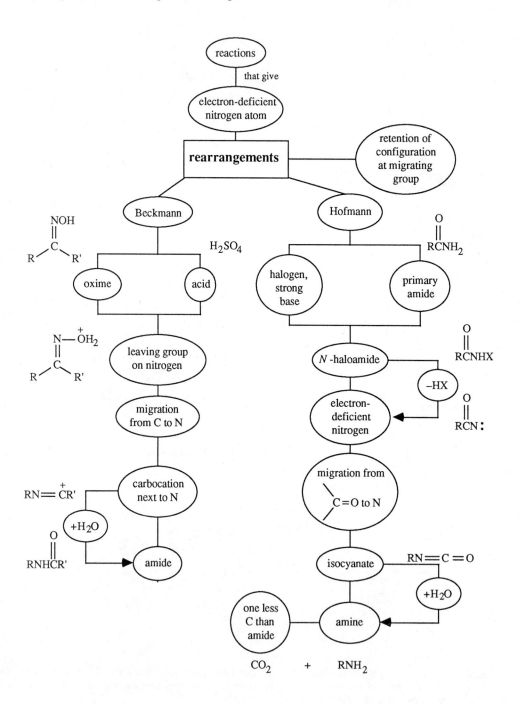

22.19

If an intramolecular reaction with water had occurred, the amide would contain ^{16}O instead of ^{18}O.

22.20

(a)

$$\underset{\text{ClCCl}}{\overset{\text{O}}{\|}} \quad + \quad NH_3 \text{ (excess)} \quad \longrightarrow \quad \underset{H_2NCNH_2}{\overset{\text{O}}{\|}}$$

(b)

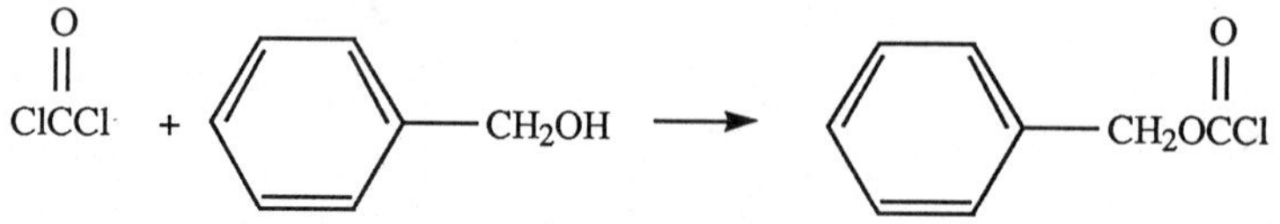

$$\underset{\text{ClCCl}}{\overset{\text{O}}{\|}} \quad + \quad \text{(benzyl alcohol)} - CH_2OH \quad \longrightarrow \quad \text{(benzyl)} - CH_2OCCl$$

(1 molar equiv)

(c)

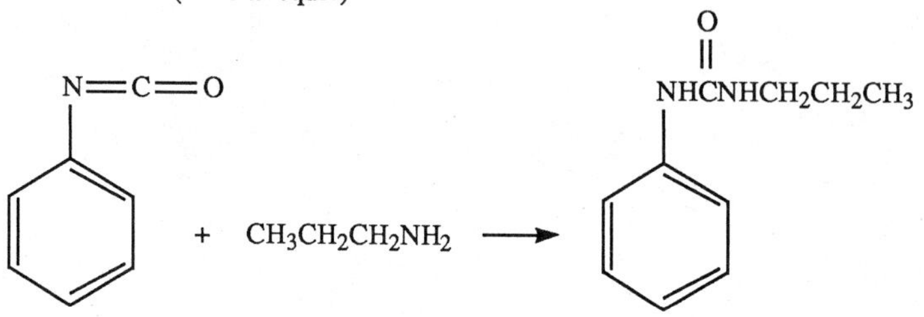

phenyl isocyanate N=C=O   + $CH_3CH_2CH_2NH_2$ $\longrightarrow$ product with NHCNHCH$_2$CH$_2$CH$_3$

(d)

$$\underset{CH_3CH_2OCNH_2}{\overset{\text{O}}{\|}} \quad + \quad H_2O \quad \xrightarrow[\underset{\Delta}{H_3O^+}]{} \quad CH_3CH_2OH \quad + \quad CO_2\uparrow \quad + \quad NH_4^+$$

(e)

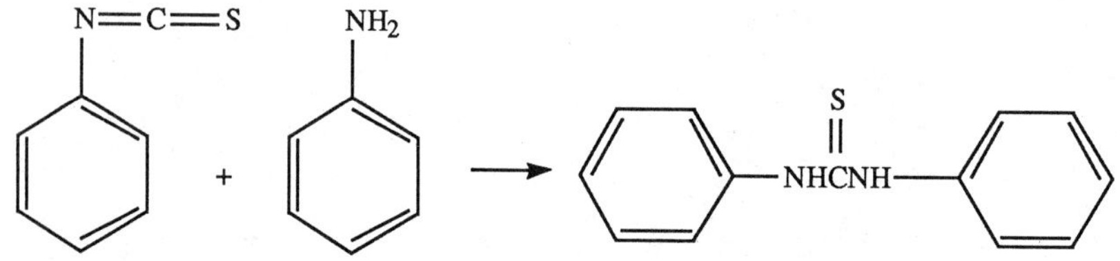

$$\text{phenyl isothiocyanate (N=C=S)} \quad + \quad \text{aniline (NH}_2\text{)} \quad \longrightarrow \quad \text{—NHCNH— (S)}$$

phenyl isothiocyanate   a substituted thiourea

22.21   1.   Degradation in acid

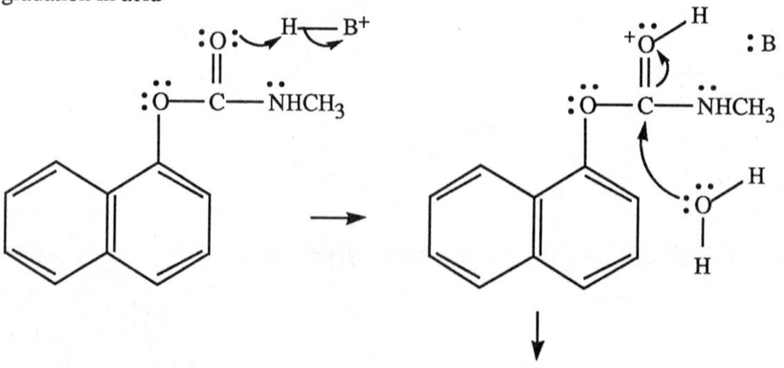

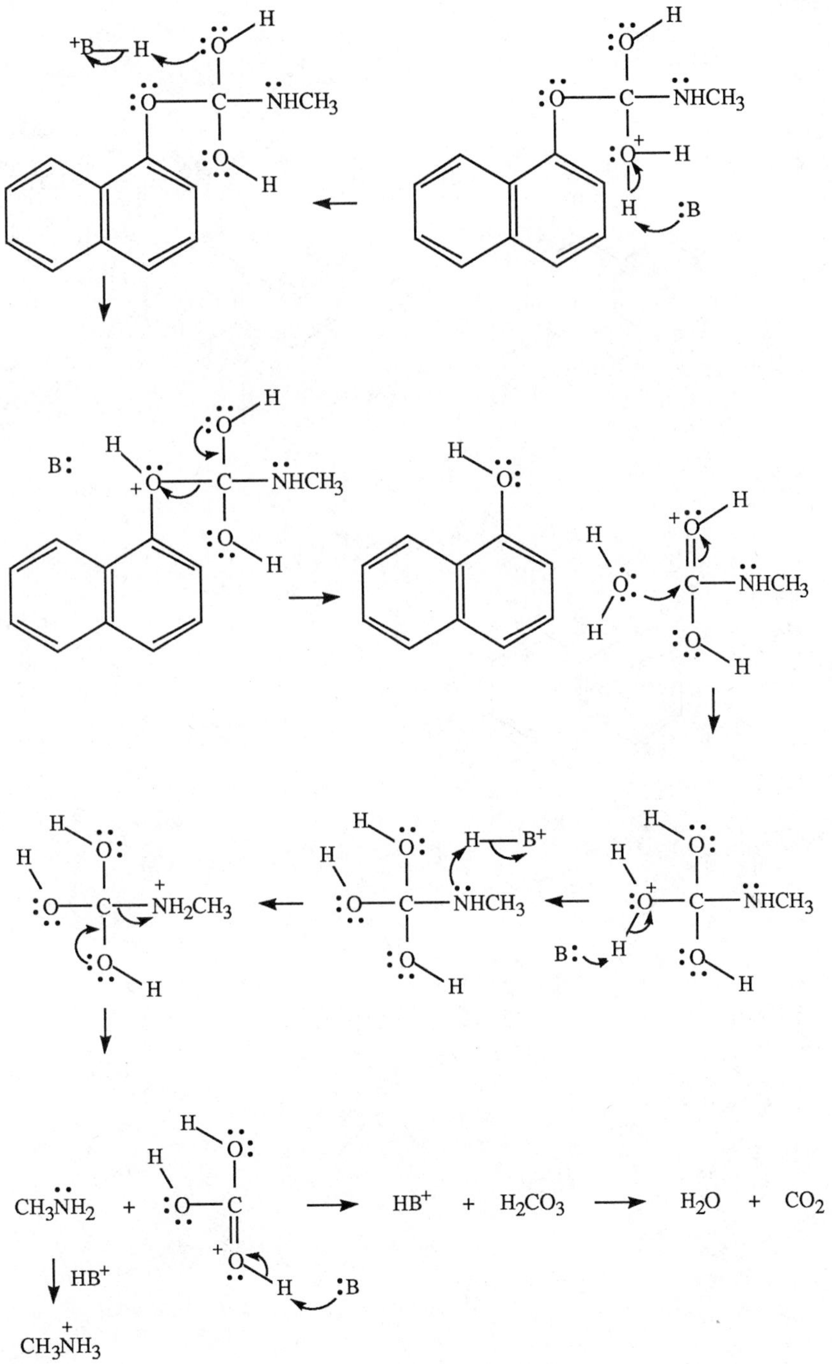

22.21 (cont)

2.   Degradation in base

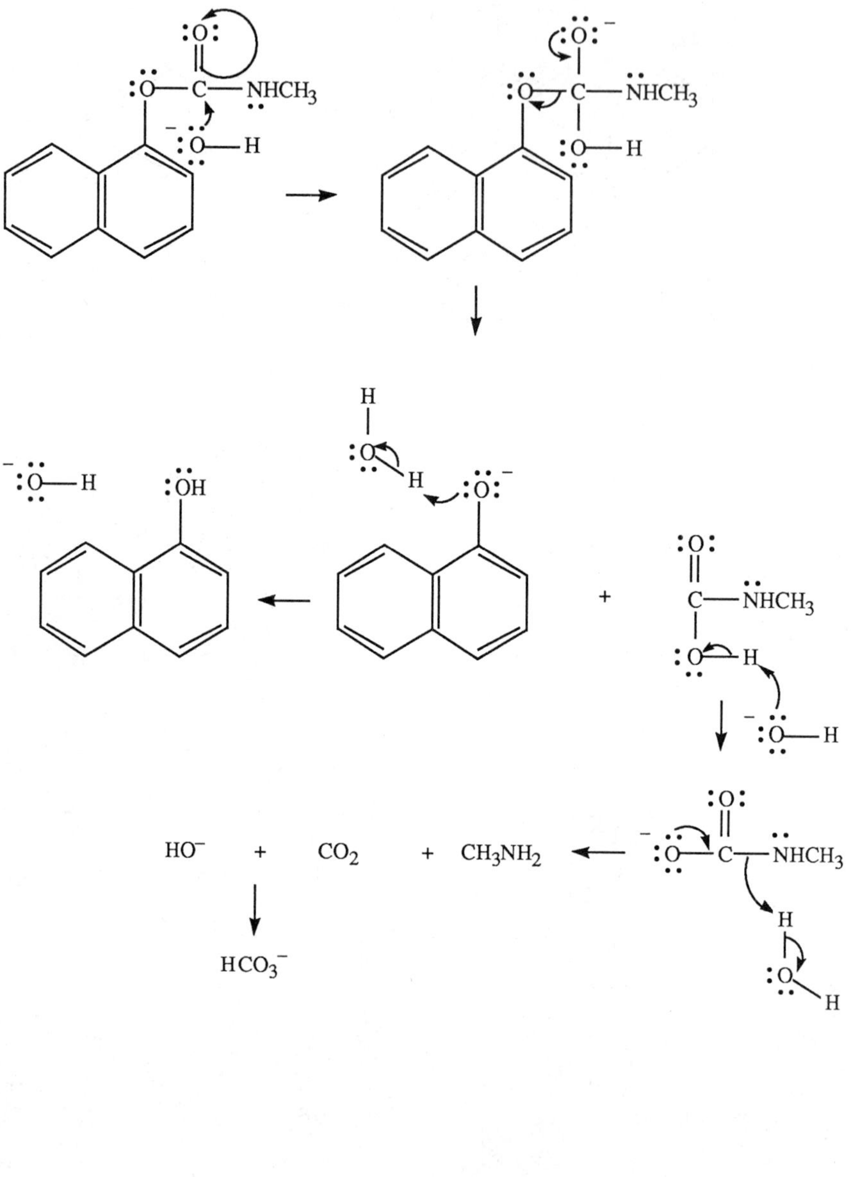

HO⁻   +   CO₂   +   CH₃NH₂

HCO₃⁻

Concept Map 22.4   Reactions of quaternary nitrogen compounds.

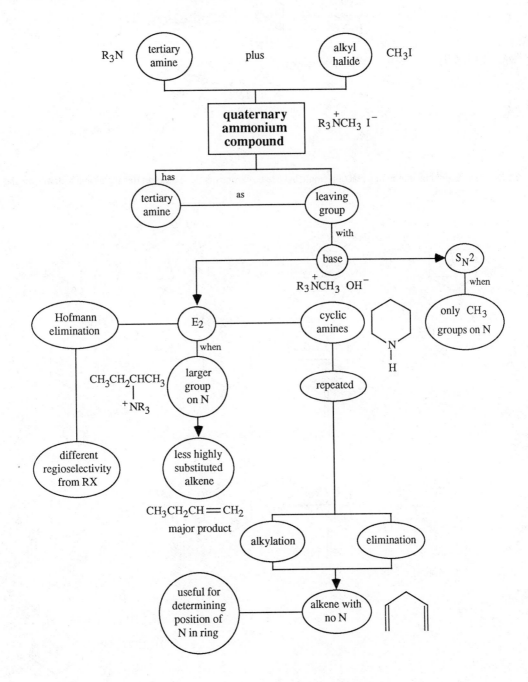

22.22

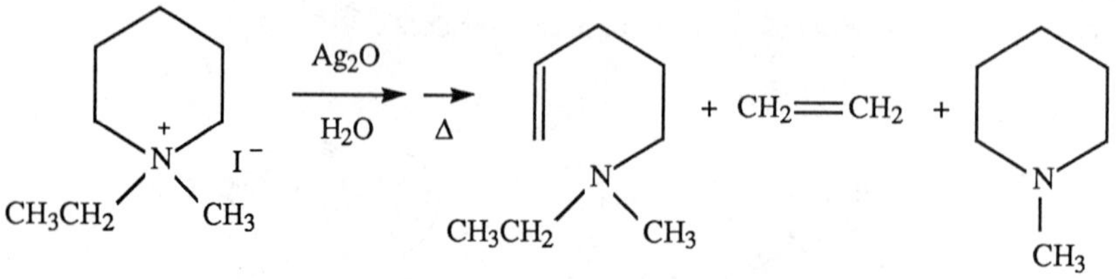

1-(*N*-ethyl-*N*-
methylamino)-
4-pentene

N-methyl-
piperidine

22.23 The Hofmann elimination reaction would have to be carried out three times to free the nitrogen atom from the compound.

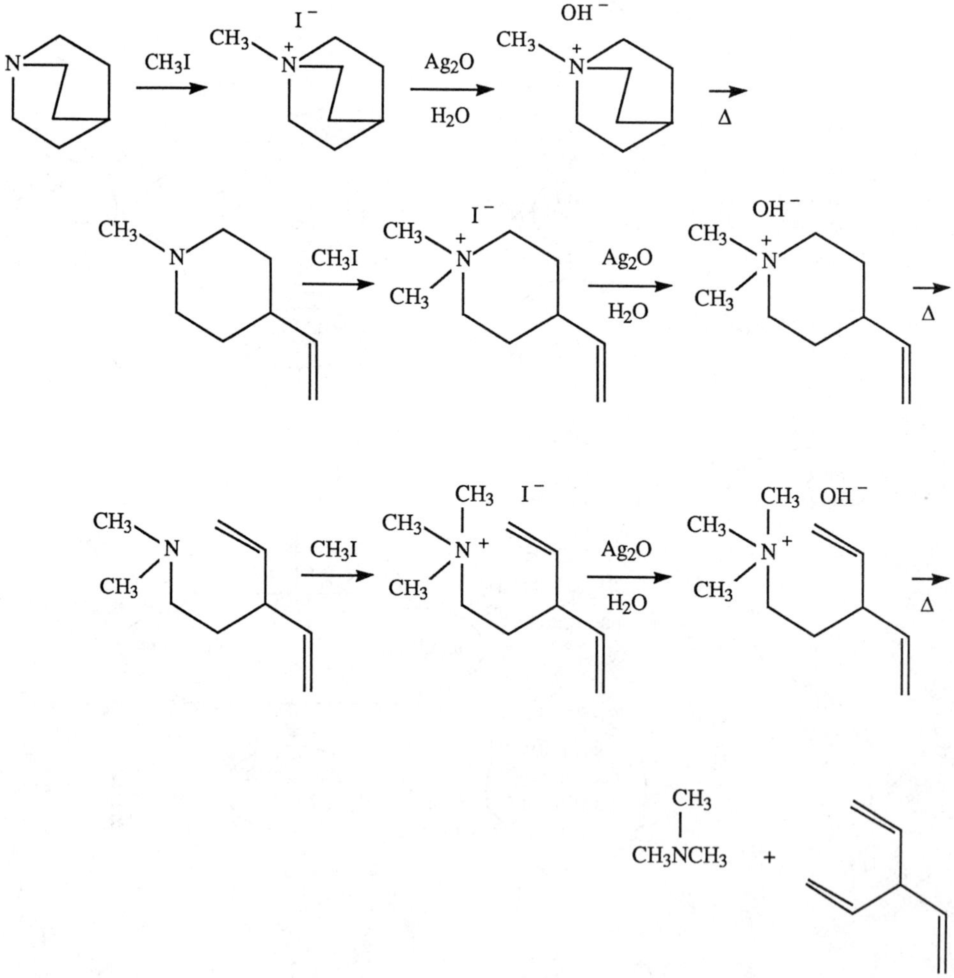

$$CH_3NCH_3$$

with CH_3 on N

+

22.24

(a)

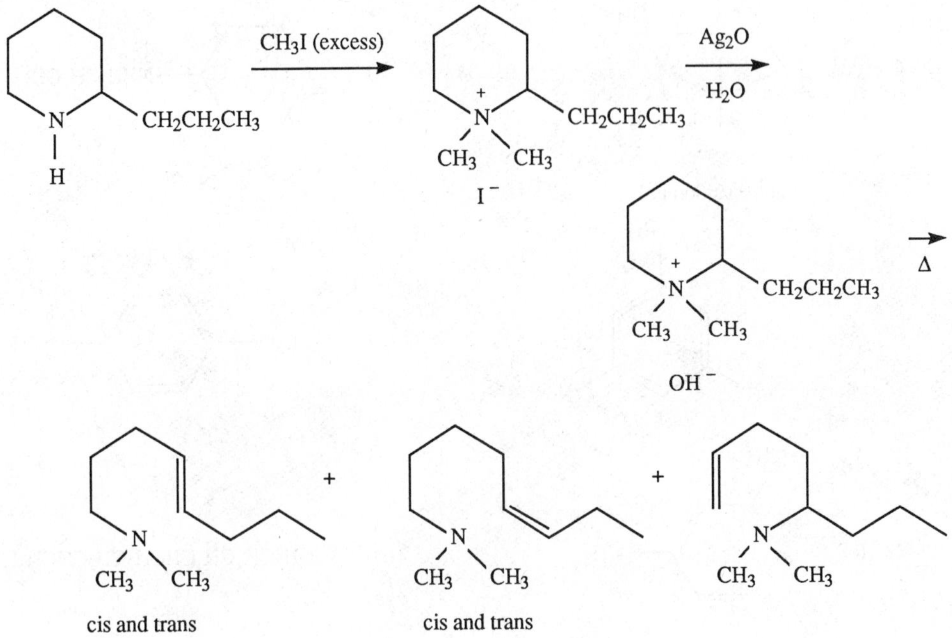

cis and trans   cis and trans

(b)

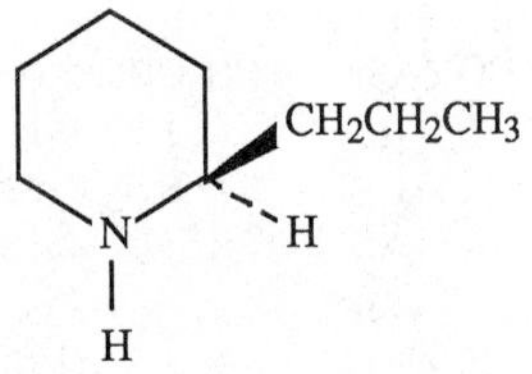

22.25   (a) *trans*-2-ethyl-1-(*N*-methylamino)cyclobutane   (b) 2-methylpyridine

(c) triethylamine   (d) 2-amino-4-methylbenzoic acid

(e) 4-aminobutanoic acid   (f) 2,4,6-trichloroaniline

(g) 5-methyl-2-hexanamine   (h) *cis*-2-aminocyclohexanol
     2-amino-5-methylhexane

(i) *N,N*-diethyl-4-nitroaniline
     N,N-diethyl-*p*-nitroaniline

22.26 (a)

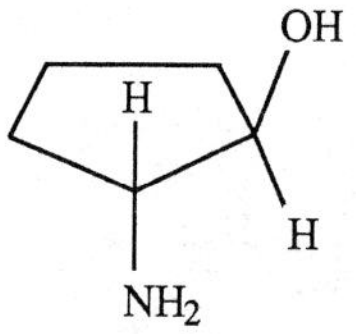
OH
H
H
NH2

and enantiomer

(b)

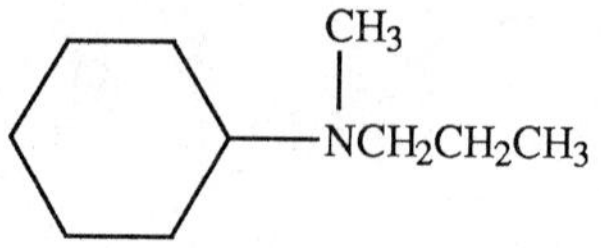
CH3
NCH2CH2CH3

(c)

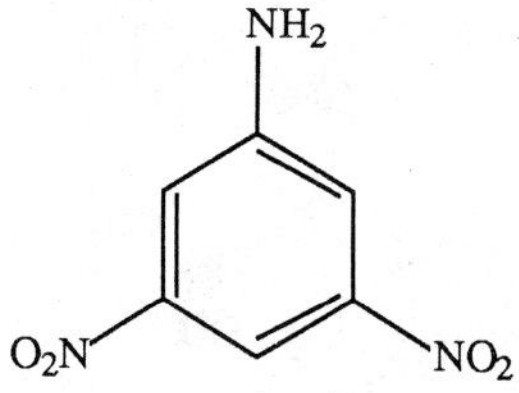
NH2
O2N
NO2

(d)

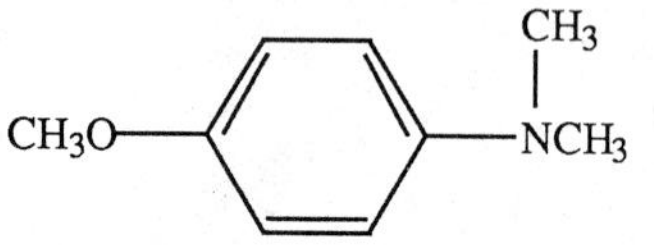
CH3O
CH3
NCH3

(e)

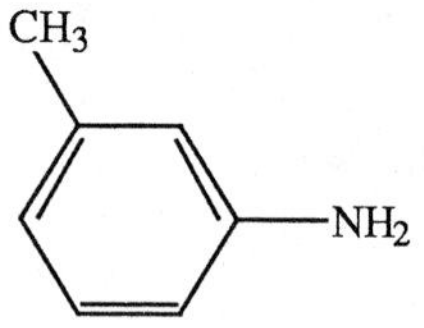
CH3
NH2

(f)

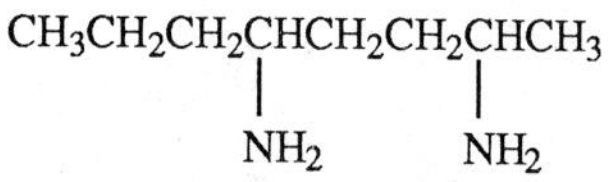
CH3CH2CH2CHCH2CH2CHCH3
NH2
NH2

(g)

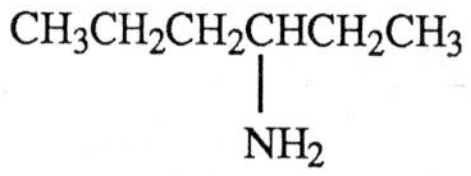
CH3CH2CH2CHCH2CH3
NH2

(h)

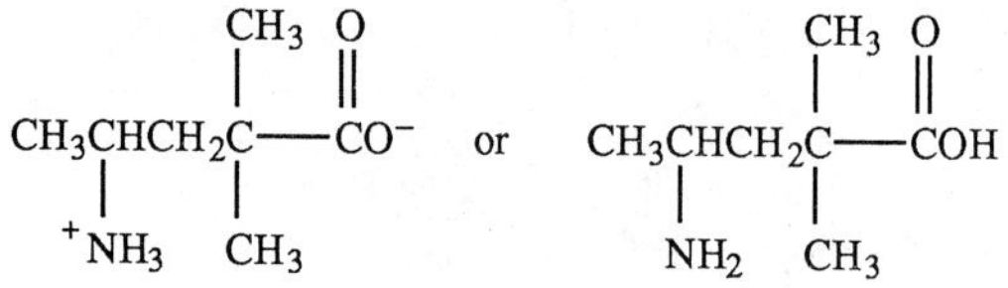
CH3 O
CH3CHCH2C—CO−
+NH3 CH3
or
CH3 O
CH3CHCH2C—COH
NH2 CH3

22.27

(a)

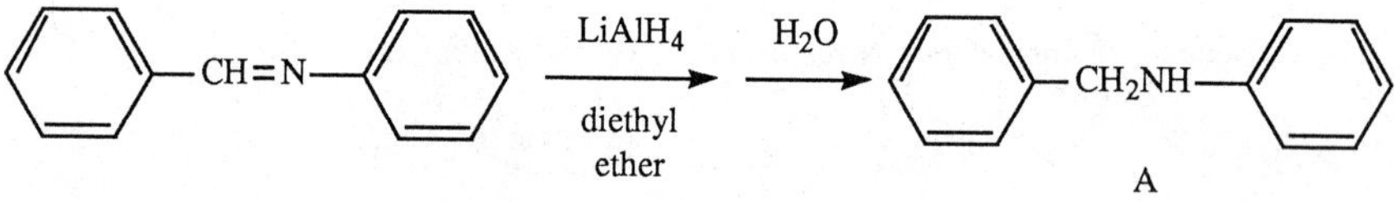
CH=N
LiAlH4
diethyl
ether
H2O
CH2NH
A

(b)

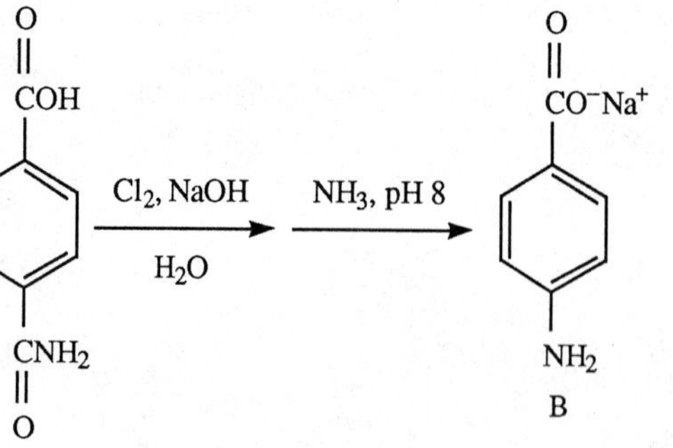
O
COH
CNH2
O
Cl2, NaOH
H2O
NH3, pH 8
O
CO−Na+
NH2
B

22.27 (cont)

(c)

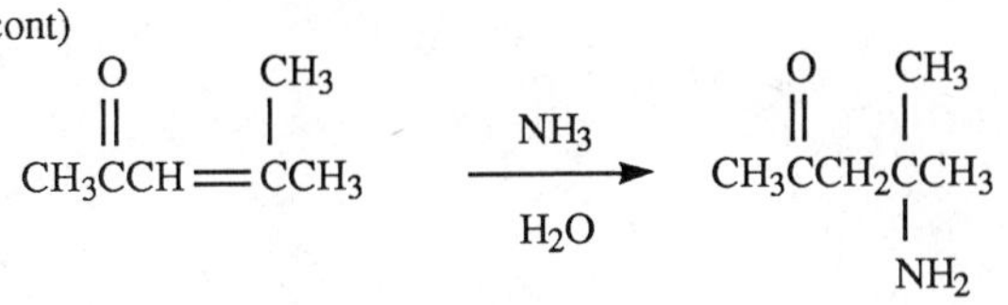

$$CH_3\overset{O}{\overset{||}{C}}CH=\overset{CH_3}{\overset{|}{C}}CH_3 \xrightarrow[H_2O]{NH_3} CH_3\overset{O}{\overset{||}{C}}CH_2\overset{CH_3}{\underset{NH_2}{\overset{|}{C}}}CH_3$$

C

(d)

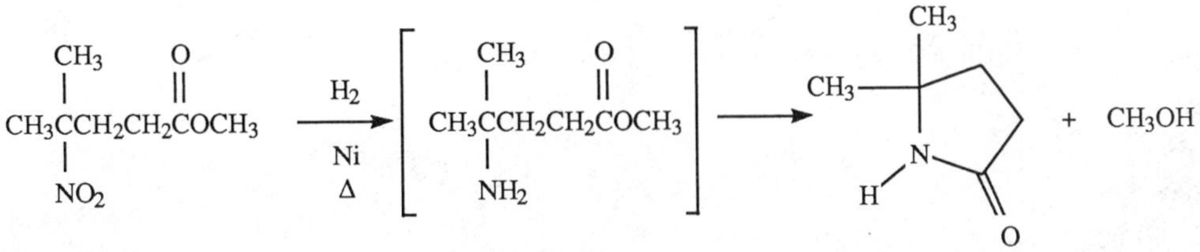

$$CH_3\overset{CH_3}{\underset{NO_2}{\overset{|}{C}}}CH_2CH_2\overset{O}{\overset{||}{C}}OCH_3 \xrightarrow[\Delta]{\underset{Ni}{H_2}} \left[CH_3\overset{CH_3}{\underset{NH_2}{\overset{|}{C}}}CH_2CH_2\overset{O}{\overset{||}{C}}OCH_3 \right] \longrightarrow$$ D + CH₃OH

(e)

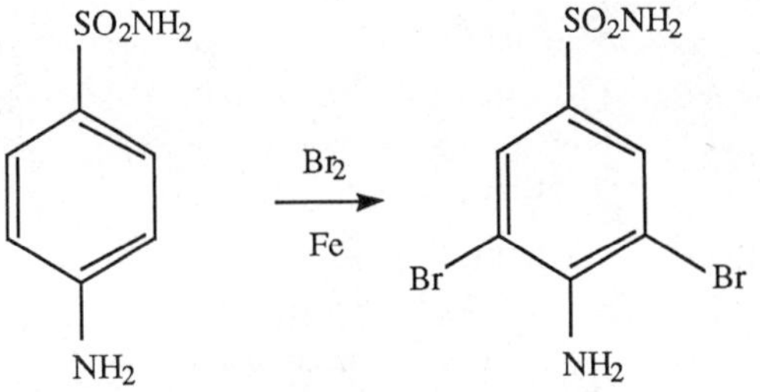

Br₂ / Fe

E

(f)

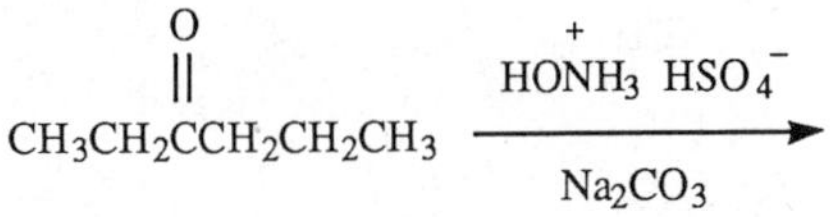

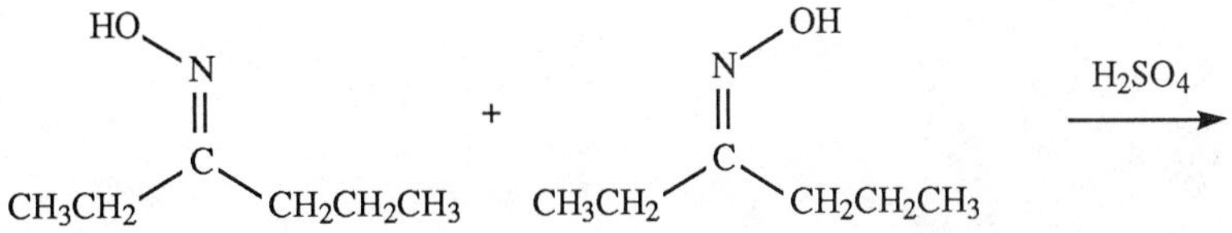

$$CH_3CH_2\overset{O}{\overset{||}{C}}CH_2CH_2CH_3 \xrightarrow[Na_2CO_3]{HO\overset{+}{N}H_3\ HSO_4^-}$$

F   +   G $\xrightarrow{H_2SO_4}$

$$CH_3CH_2\overset{O}{\overset{||}{C}}NHCH_2CH_2CH_3 \quad + \quad CH_3CH_2CH_2\overset{O}{\overset{||}{C}}NHCH_2CH_3$$

H   I

22.27 (cont)

(g)
$$CH_3\overset{\overset{\displaystyle O}{\|}}{C}(CH_2)_4CH_3 \xrightarrow[\text{Ni}]{\text{NH}_3,\ \text{H}_2} CH_3\underset{\underset{\displaystyle NH_2}{|}}{C}H(CH_2)_4CH_3$$

J

(h)
$$CH_3CH_2CH_2NO_2 \xrightarrow[\displaystyle]{\overset{\overset{\displaystyle O}{\|}}{\text{HCH, NaOH}}} CH_3CH_2\underset{\underset{\displaystyle NO_2}{|}}{C}HCH_2OH \xrightarrow{\text{H}_2\text{SO}_4,\ \text{Fe}}$$

K

$$CH_3CH_2\underset{\underset{\displaystyle {}^+NH_3\ HSO_4{}^-}{|}}{C}HCH_2OH \xrightarrow{\text{Ca(OH)}_2} CH_3CH_2\underset{\underset{\displaystyle NH_2}{|}}{C}HCH_2OH$$

L   M

(i)

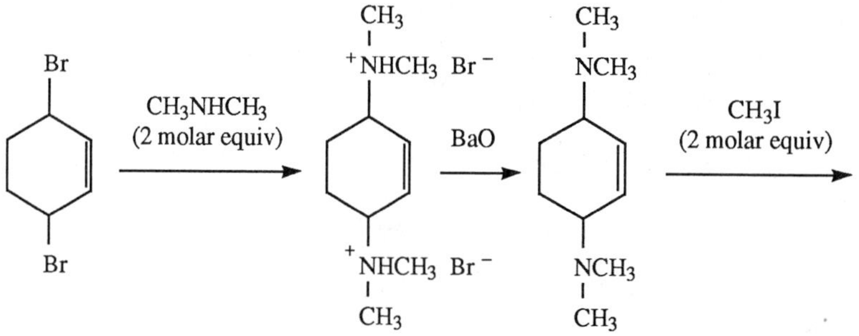

N   O

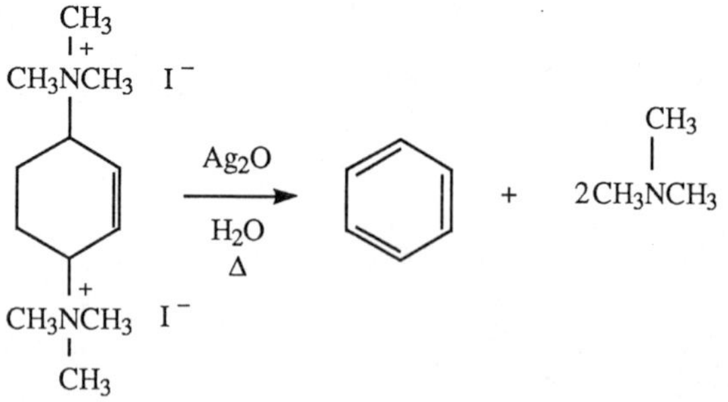

P   Q   R

22.27 (cont)

(j)

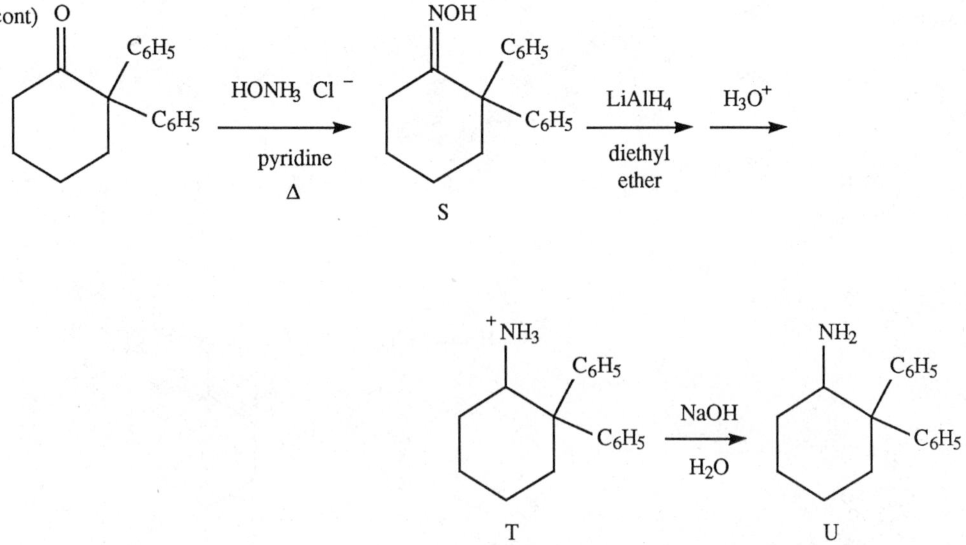

(k)

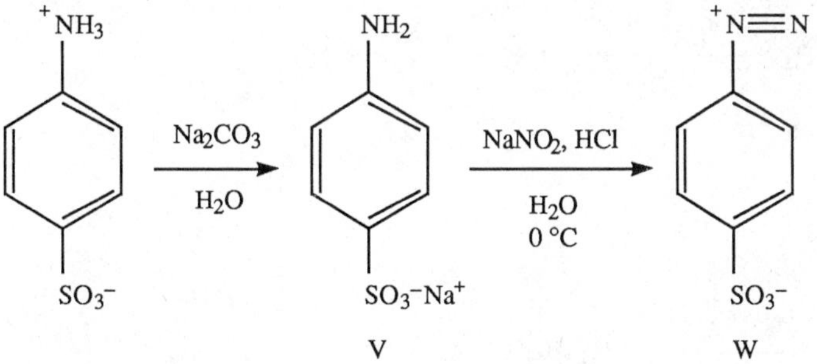

(l)

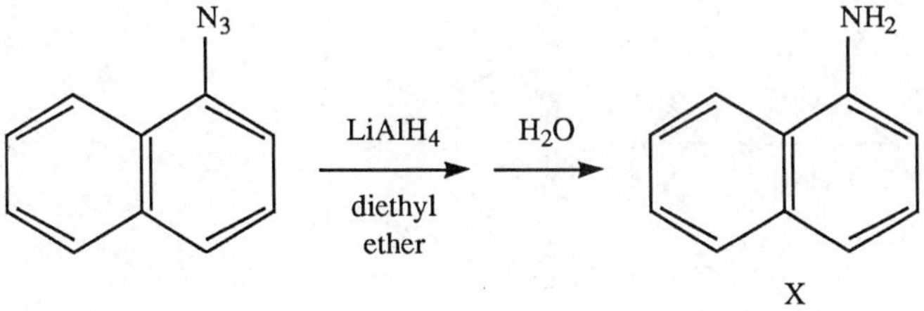

22.28

(a)

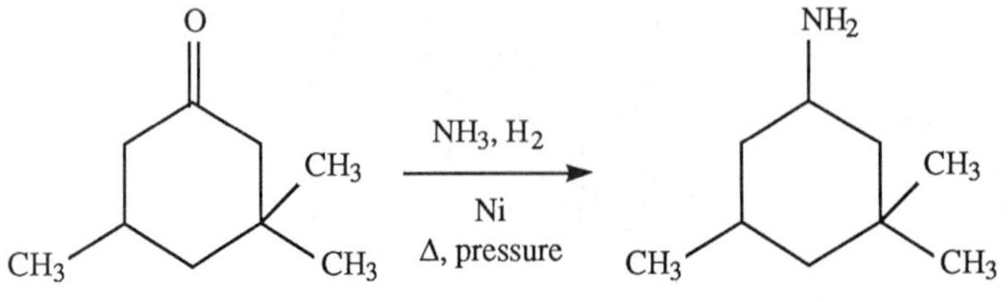

(b) CH_3CCH_3

(c)

(d) $CH_3CHCOCH_2CH_3$ $\xrightarrow[\text{dimethyl sulfoxide}]{NaNO_2}$ $CH_3CHCOCH_2CH_3$
      | |
      Cl NO_2

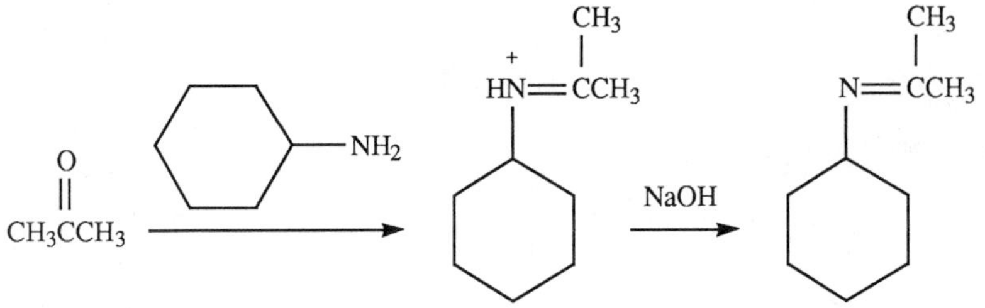

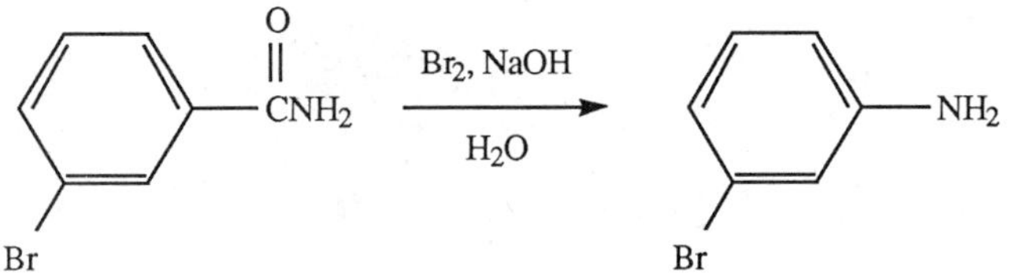

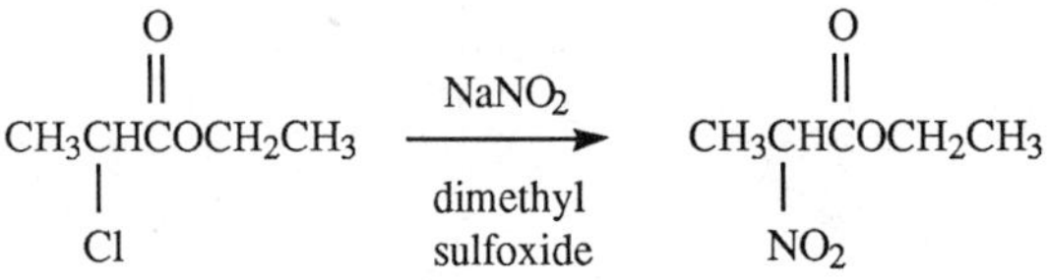

                                                              E

(e)

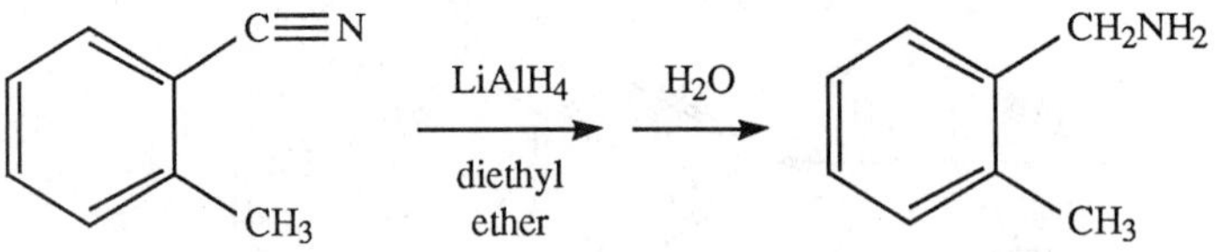

(f) CH_3CNH_2 $\xrightarrow[\text{H}_2\text{O}]{Br_2, NaOH}$ CH_3NH_2 + HCO_3^-

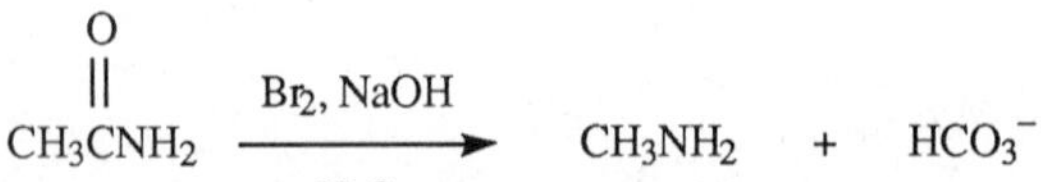

                                        G

22.28 (cont)

(g)

(h) $CH_3CH_2CHCH_2CH_2CH_3$ $\xrightarrow[\text{pyridine}]{\text{TsCl}}$ $CH_3CH_2CHCH_2CH_2CH_3$ $\xrightarrow{CH_3NHCH_3}$ $\xrightarrow{NaOH}$
                            | |
                           OH   OTs

                                                                                          J

$CH_3CH_2CHCH_2CH_2CH_3$ $\xrightarrow[\text{acetonitrile}]{CH_3I}$ $CH_3CH_2CHCH_2CH_2CH_3$
            | |+
       CH_3NCH_3 $CH_3NCH_3 \ \ I^-$
            K |
                                                                              CH_3

                                                                                L

(i) $Br(CH_2)_{10}Br$ + CH_3NCH_3 $\xrightarrow[\substack{\text{methanol}\\25\,°C\\2\text{ weeks}}]{}$ $Br^- \ \ CH_3N(CH_2)_{10}NCH_3 \ \ Br^-$ $\xrightarrow[\text{H}_2\text{O}]{Ag_2O}$
                              | +| |+
                            CH_3 CH_3 CH_3

      (1 equiv)   (2 equiv)   M

      $HO^- \ \ CH_3N(CH_2)_{10}NCH_3 \ \ OH^-$ $\xrightarrow{\Delta}$ $CH_2{=}CH(CH_2)_6CH{=}CH_2$ + $2\,CH_3NCH_3$
              +| |+ |
            CH_3 CH_3 CH_3

                    N   O

(j)

22.28 (cont)

(k)

$$\xrightarrow[\substack{\text{dimethyl} \\ \text{sulfoxide}}]{\text{NaNO}_2}$$

Q

(l)

$$\xrightarrow[\substack{\text{dichloromethane} \\ \text{18 h}}]{}$$

R

22.29

(a)

$$\xleftarrow{}$$

$$\text{Na}^+$$
$$\underset{\underset{O}{\|}}{\underset{-}{}}\,\underset{\underset{O}{\|}}{}$$
$$CH_3CCHCOCH_2CH_3$$

$$-CH=CHNO_2 \xleftarrow{\text{HCl}} \xleftarrow[\text{H}_2\text{O}]{\text{CH}_3\text{NO}_2, \text{NaOH}}$$

(b) $\underset{\substack{| \\ CH_3}}{H_2NCH_2CHCH_2CH_2NH_2} \xleftarrow{\text{Br}_2, \text{NaOH}} \underset{\substack{\| \quad | \quad \| \\ O \quad CH_3 \quad O}}{H_2NCCH_2CHCH_2CH_2CNH_2} \xleftarrow{\text{NH}_3 \text{ (excess)}}$

$\underset{\substack{\| \quad | \quad \| \\ O \quad CH_3 \quad O}}{ClCCH_2CHCH_2CH_2CCl} \xleftarrow{\text{SOCl}_2 \text{ (2 molar equiv)}} \underset{\substack{\| \quad | \quad \| \\ O \quad CH_3 \quad O}}{HOCCH_2CHCH_2CH_2COH}$

22.29 (cont)

(c) $CH_3CH_2\overset{\overset{\displaystyle O}{\|}}{C}NHCH_2CH_3$ $\xleftarrow[\Delta]{H_2SO_4}$ $CH_3CH_2\overset{\overset{\displaystyle NOH}{\|}}{C}CH_2CH_3$ $\xleftarrow[Na_2CO_3]{\overset{+}{HONH_3}\ Cl^-}$ $CH_3CH_2\overset{\overset{\displaystyle O}{\|}}{C}CH_2CH_3$

(d) 1. Ph—$CH_2CH_2\overset{\underset{\displaystyle NH_2}{|}}{C}H\,CH_2\overset{\underset{\displaystyle NH_2}{|}}{C}HCH_3$ $\xleftarrow[Ni]{\substack{NH_3,\ H_2\\(2\ molar\ equiv)}}$ Ph—$CH_2CH_2\overset{\overset{\displaystyle O}{\|}}{C}CH_2\overset{\overset{\displaystyle O}{\|}}{C}CH_3$

2. Ph—$CH_2CH_2\overset{\underset{\displaystyle NH_2}{|}}{C}H\,CH_2\overset{\underset{\displaystyle NH_2}{|}}{C}HCH_3$ $\xleftarrow[\substack{diethyl\\ether}]{H_2O}$ $\xleftarrow{LiAlH_4}$

Ph—$CH_2CH_2\overset{\underset{\displaystyle N_3}{|}}{C}HCH_2\overset{\underset{\displaystyle N_3}{|}}{C}HCH_3$ $\xleftarrow{NaN_3}$ Ph—$CH_2CH_2\overset{\underset{\displaystyle OTs}{|}}{C}HCH_2\overset{\underset{\displaystyle OTs}{|}}{C}HCH_3$ $\xleftarrow[pyridine]{TsCl}$

Ph—$CH_2CH_2\overset{\underset{\displaystyle OH}{|}}{C}HCH_2\overset{\underset{\displaystyle OH}{|}}{C}HCH_3$ $\xleftarrow[methanol]{NaBH_4}$ Ph—$CH_2CH_2\overset{\overset{\displaystyle O}{\|}}{C}CH_2\overset{\overset{\displaystyle O}{\|}}{C}CH_3$

(e) 1. Ph—$CH_2OCH_2CH_2CH_2NH_2$ $\xleftarrow[Ni]{NH_3,\ H_2}$ Ph—$CH_2OCH_2CH_2\overset{\overset{\displaystyle O}{\|}}{C}H$ $\xleftarrow[dichloromethane]{\text{pyridinium } CrO_3Cl^-}$

Ph—$CH_2OCH_2CH_2CH_2OH$ $\xleftarrow{Ph\text{-}CH_2Cl}$ $Na^{+-}OCH_2CH_2CH_2OH$ $\xleftarrow[\substack{(1\ equiv)}]{NaH}$ $HOCH_2CH_2CH_2OH$

2. Ph—$CH_2OCH_2CH_2CH_2NH_2$ $\xleftarrow[Pt]{H_2}$ Ph—$CH_2OCH_2CH_2CH_2N_3$ $\xleftarrow[\substack{dimethyl\\sulfoxide}]{NaN_3}$

22.29 (cont)

(e) 2. (cont)

$$\underset{\text{pyridine}}{\overset{\text{PBr}_3}{\longleftarrow}}$$

[from (e) 1]

22.30

1.

$$\underset{\text{carbon tetrachloride}}{\overset{\text{BrCH(COCH}_2\text{CH}_3)_2}{\longrightarrow}}$$

$$\underset{\text{ethanol}}{\overset{\text{CH}_3\text{CH}_2\text{ONa}}{\longrightarrow}}$$

$$\overset{\text{BrCH}_2\text{CH}_2\text{SCH}_3}{\longrightarrow}$$

$$\underset{\Delta}{\overset{\text{H}_2\text{O, H}_2\text{SO}_4}{\longrightarrow}}$$

$$+ \quad \text{CH}_3\text{SCH}_2\text{CH}_2\overset{\text{O}}{\overset{\|}{\text{CHCOH}}} \quad \overset{\text{N}}{\longrightarrow} \quad \text{CH}_3\text{SCH}_2\text{CH}_2\overset{\text{O}}{\overset{\|}{\text{CHCO}^-}}$$

$$\underset{\text{}^+\text{NH}_3 \ \text{HSO}_4^-}{} \qquad \underset{\substack{\text{}^+\text{NH}_3 \\ \text{methionine}}}{}$$

2.

$$\overset{\text{CH}_2\text{Br}}{\longrightarrow}$$

$$\underset{\Delta}{\overset{\text{H}_2\text{O, H}_2\text{SO}_4}{\longrightarrow}}$$

22.30 (cont)

2. (cont)

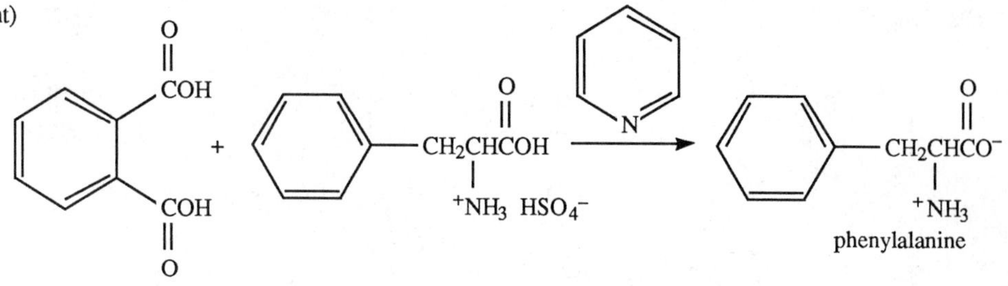

phenylalanine

3.

aspartic acid

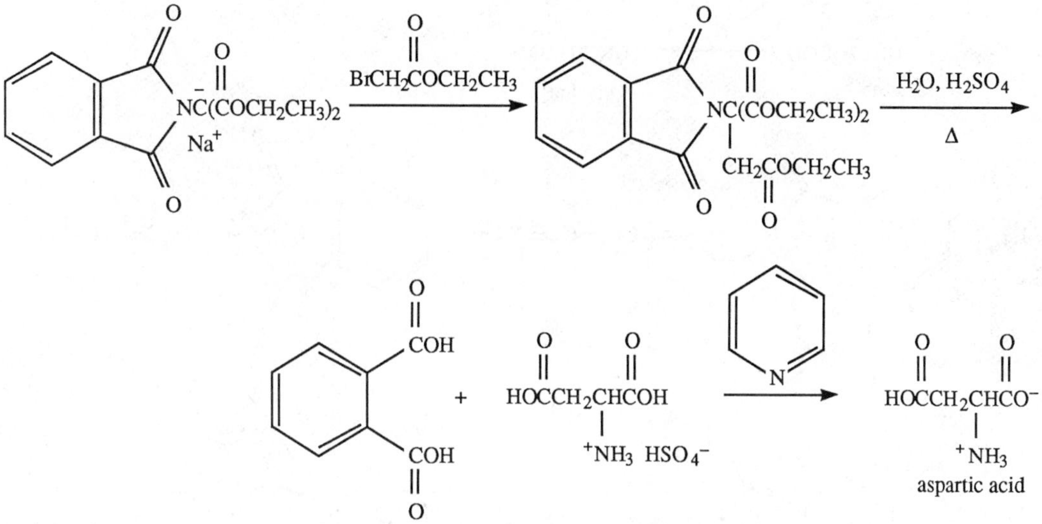

22.31

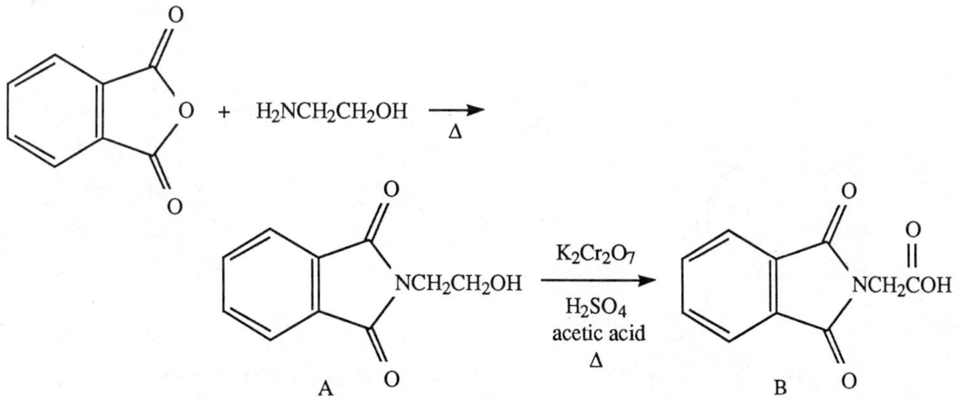

22.31 (cont)

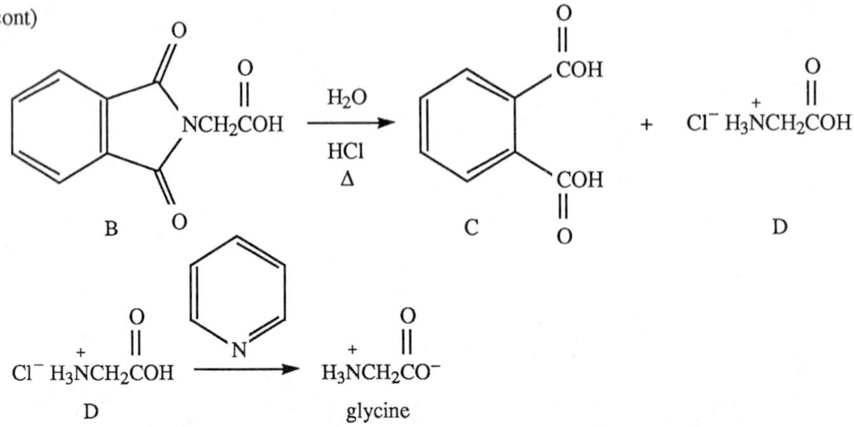
NCH2COH
H2O
HCl
Δ
COH
COH
+
Cl⁻ H3NCH2COH
B
C
D
Cl⁻ H3NCH2COH
D
H3NCH2CO⁻
glycine

22.32

1.

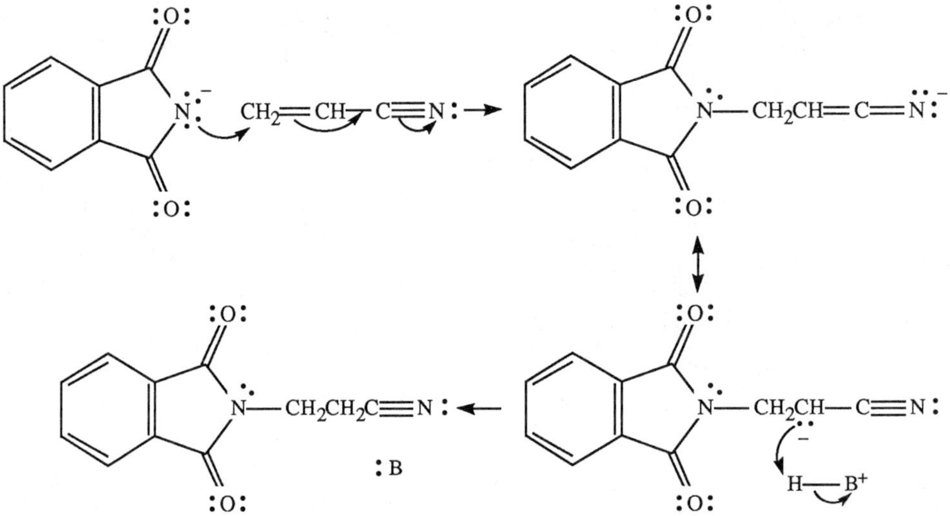
CH2=CH–C≡N
N–CH2CH=C=N
N–CH2CH–C≡N
H–B+
N–CH2CH2C≡N
:B

2.

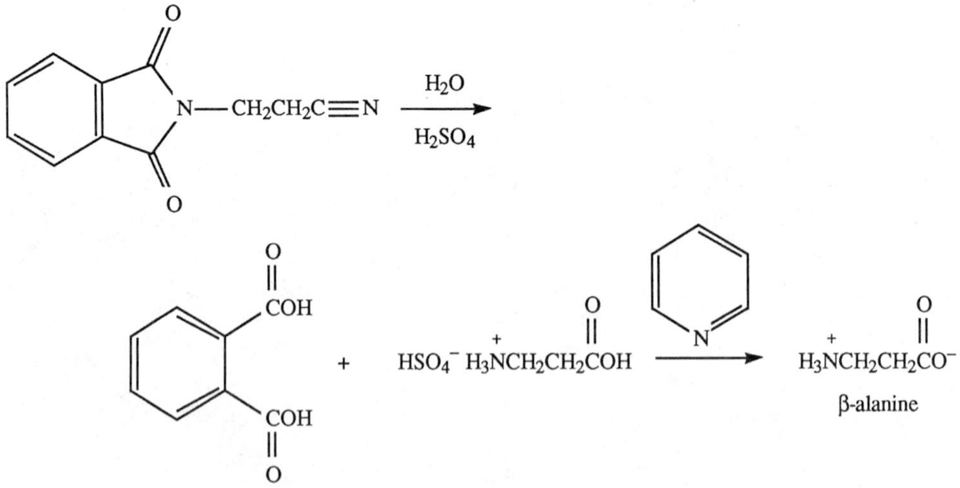
N–CH2CH2C≡N
H2O
H2SO4
COH
COH
+
HSO4⁻ H3NCH2CH2COH
H3NCH2CH2CO⁻
β-alanine

22.33

22.33 (cont)

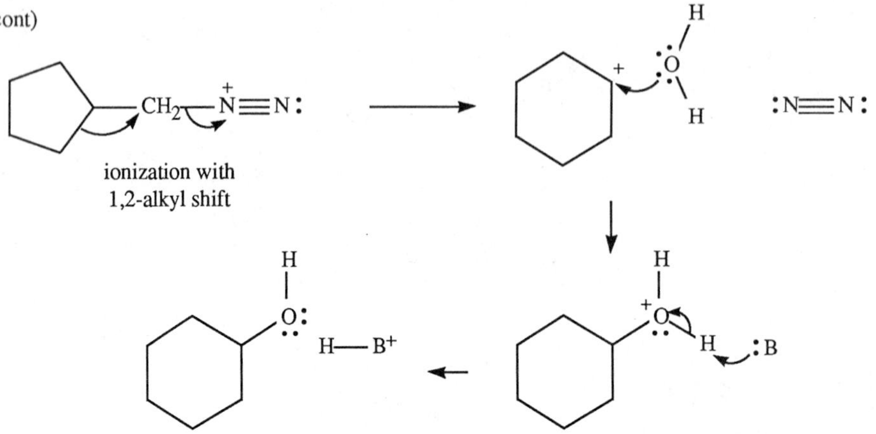

ionization with
1,2-alkyl shift

22.34

(CH₃C)₂O

pyridine

A

HC≡N

C

H₂

catalyst
acetic acid

B

NaNO₂
HCl
H₂O

D

NaOH   H₃O⁺

H₂O

E

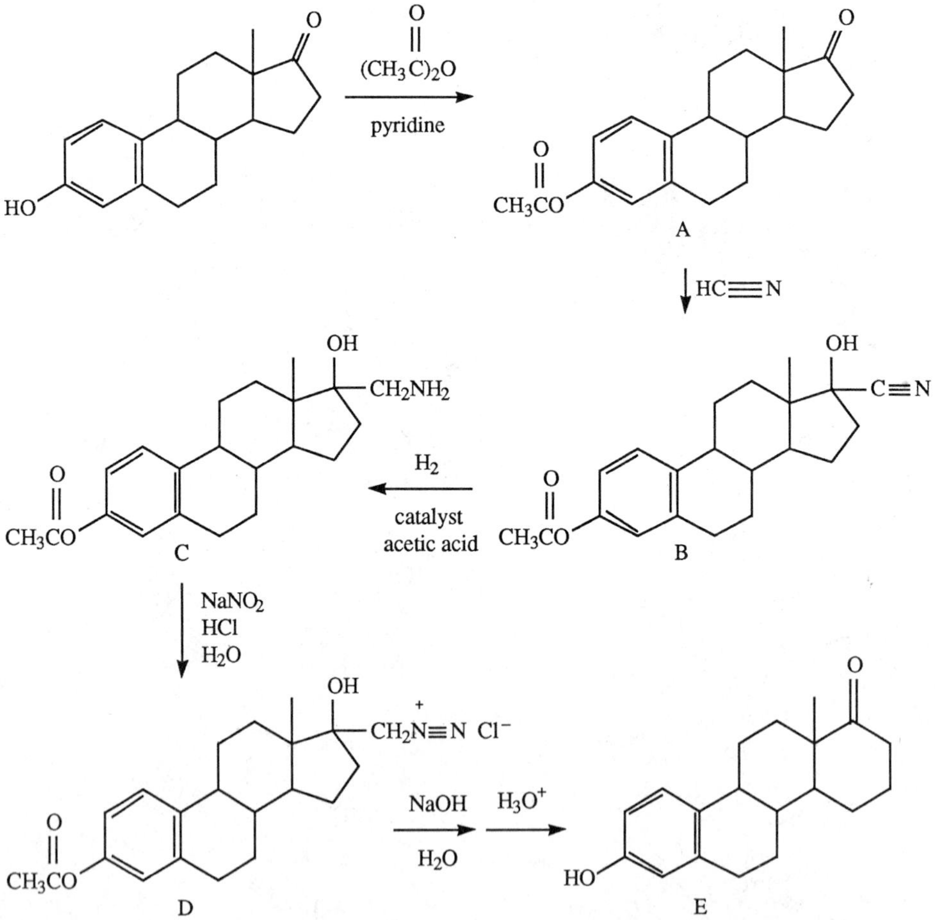

(Note that the acetyl group is
also removed in the last step.)

22.34 (cont)

<u>Mechanism of ring expansion</u>

(See the previous problem for the mechanism of the formation of the diazonium ion.)

22.35 $CH_3(CH_2)_5NH_2$ +

water soluble   water-insoluble precipitate

22.35 (cont)

$(CH_3CH_2CH_2)_2NH$ + [benzenesulfonyl chloride] $\xrightarrow[\text{H}_2\text{O}]{\text{NaOH}}$

[benzenesulfonamide] $SN(CH_2CH_2CH_3)_2$ $\xrightarrow[\text{to pH 3}]{\text{HCl}}$ no change

water-insoluble precipitate

$(CH_3CH_2)_3N$ + [benzenesulfonyl chloride] $\xrightarrow[\text{H}_2\text{O}]{\text{NaOH}}$

[benzene]SO^-Na^+ + $(CH_3CH_2)_3N$ $\xrightarrow[\text{to pH 3}]{\text{HCl}}$ $(CH_3CH_2)_3\overset{+}{N}H \; Cl^-$

water soluble

pK_a of conjugate acid is –0.6

A primary amine is converted into a sulfonamide that has an acidic hydrogen atom on the nitrogen and thus is deprotonated by the base. The salt of the sulfonamide is soluble in water, but the solid sulfonamide precipitates when the solution is acidified.

The sulfonamide of a secondary amine has no acidic hydrogen atom on the nitrogen and is not affected by the basic solution. We observe a precipitate in the test tube, and no change is seen when the solution is acidified.

A tertiary amine does not react with benzenesulfonyl chloride, which it converted slowly by base into sodium benzenesulfonate, a water soluble salt. What we observe depends upon the physical properties of the amine. If the amine has a low molecular weight and is soluble in water, we will have a clear basic solution that remains clear upon acidification. If the amine is not soluble in water, we will see the amine as a second phase in base but it will dissolve when we acidify the solution. The pK_a of the conjugate acid of sodium benzenesulfonate is –0.6. The salt remains in solution at a pH that will protonate an amine. Benzenesulfonic acid is also quite soluble in water, so no precipitate is seen under acidic conditions.

22.36

A
(mixture of stereoisomers)

22.36 (cont)

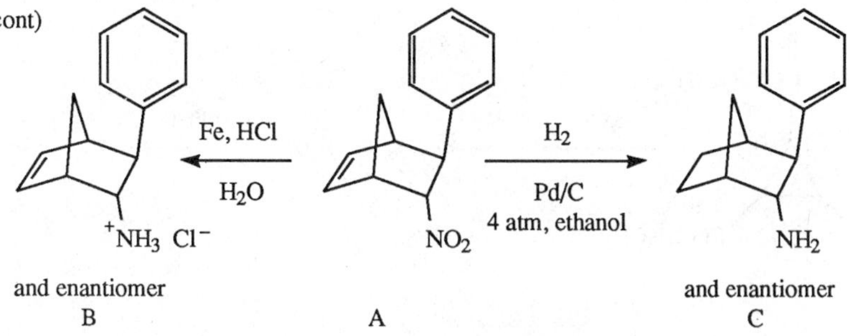
Fe, HCl
H_2O
$^+NH_3$ Cl^-
and enantiomer
B
NO_2
A
H_2
Pd/C
4 atm, ethanol
NH_2
and enantiomer
C

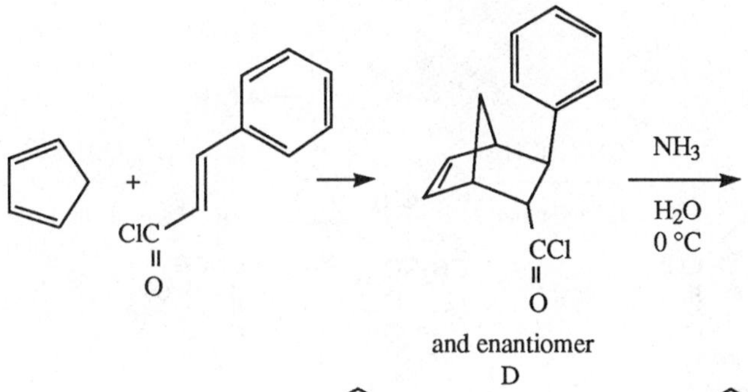
+
ClC
O
CCl
O
and enantiomer
D
NH_3
H_2O
0 °C

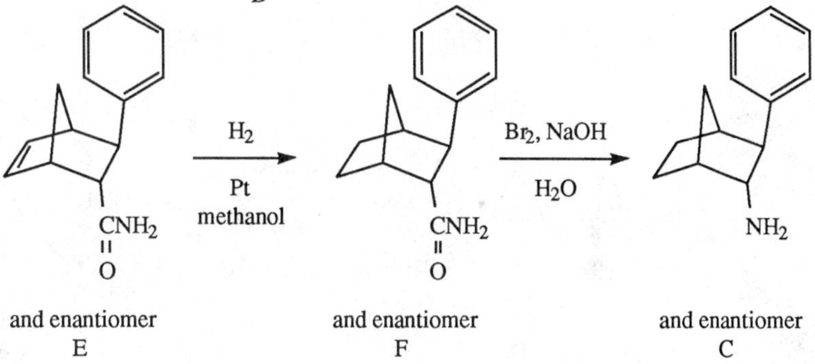
CNH_2
O
and enantiomer
E
H_2
Pt
methanol
CNH_2
O
and enantiomer
F
Br_2, NaOH
H_2O
NH_2
and enantiomer
C

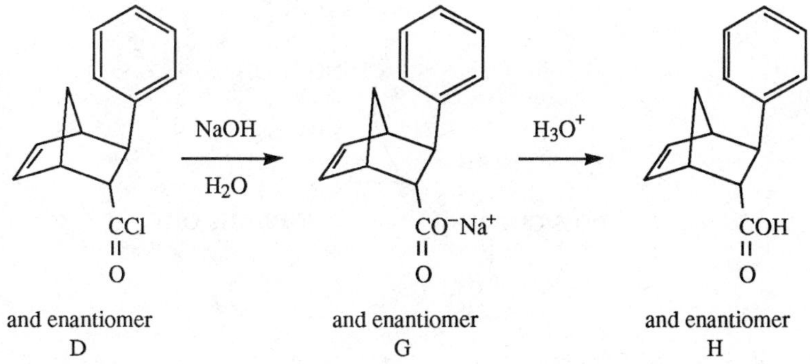
CCl
O
and enantiomer
D
NaOH
H_2O
CO^-Na^+
O
and enantiomer
G
H_3O^+
COH
O
and enantiomer
H

22.37

(a)

22.37 (cont)

(b)

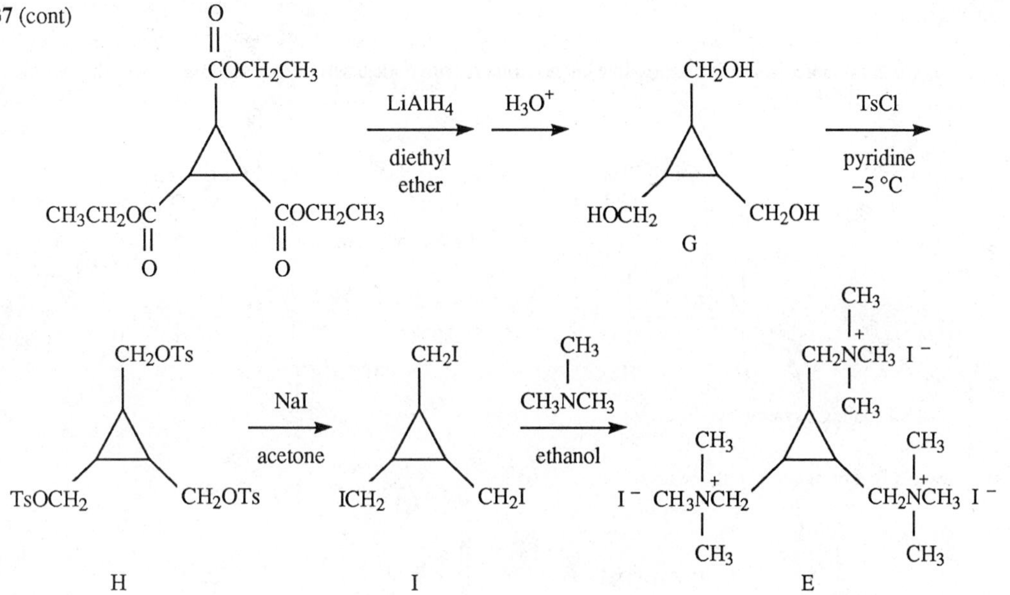

(c)

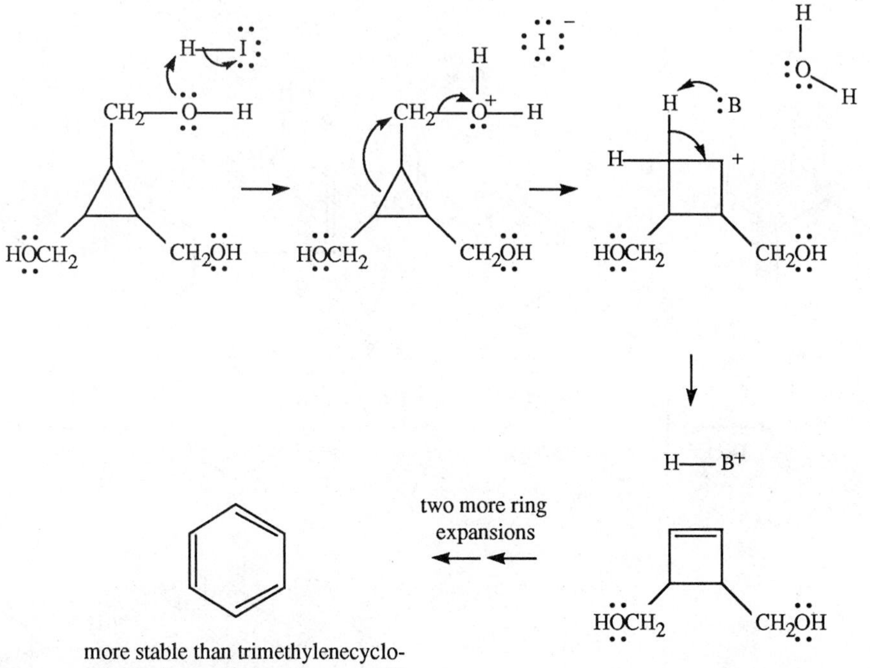

more stable than trimethylenecyclo-
propane because of the decrease of
ring strain and the gain of aromaticity;
care must be taken to prevent the
formation of carbocations.

22.37 (cont)

(d)   Cis-trans stereoisomerism is possible but no isomers with optical activity are seen because of the symmetry of the molecule.

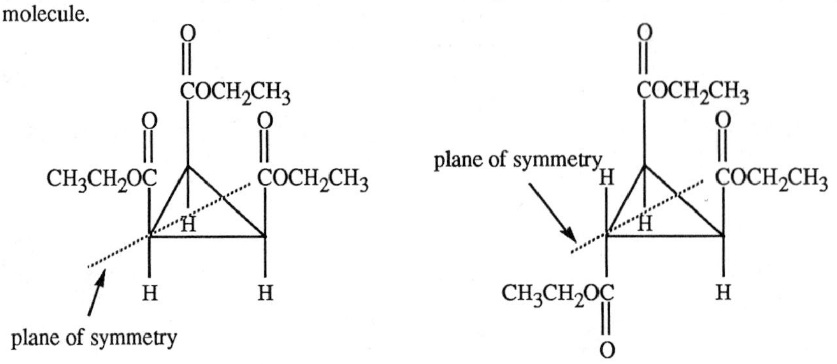

plane of symmetry

(e)   An E_2 elimination could be used.

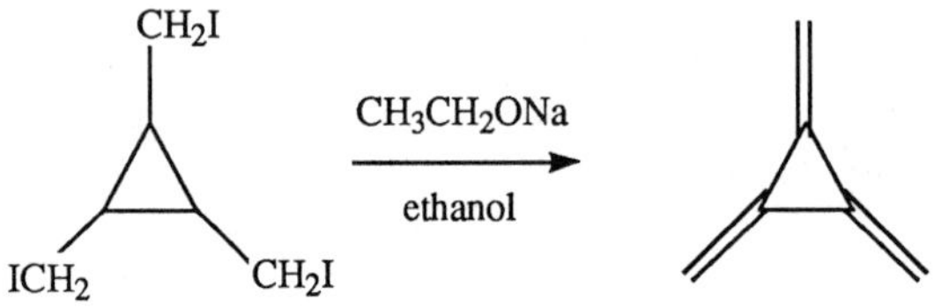

$$CH_3CH_2ONa$$

ethanol

22.38

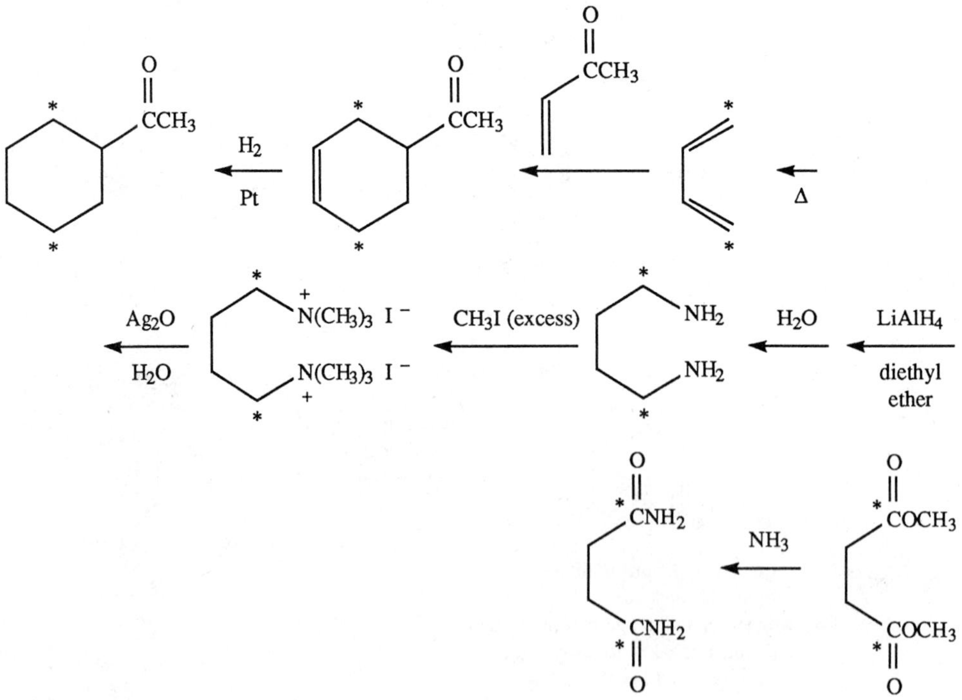

H_2

Pt

Ag_2O

H_2O

CH_3I (excess)

H_2O

$LiAlH_4$

diethyl
ether

NH_3

22.39

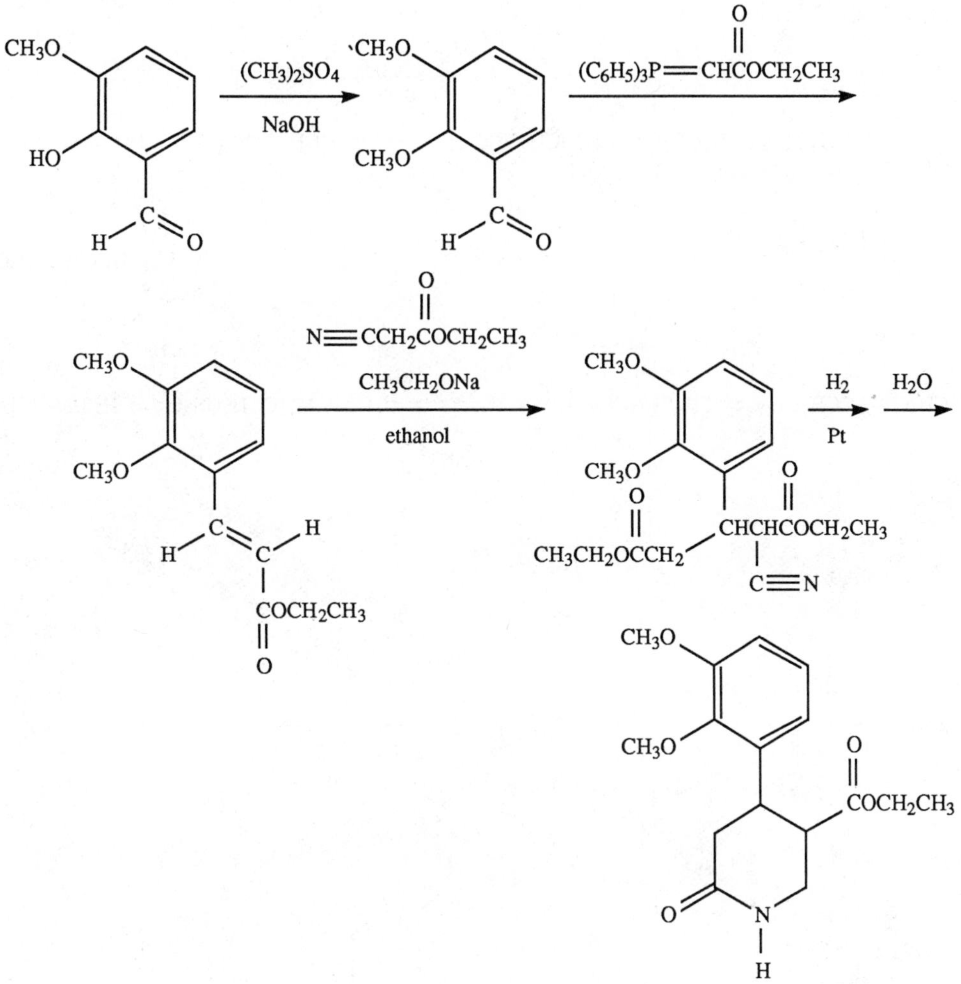

22.40

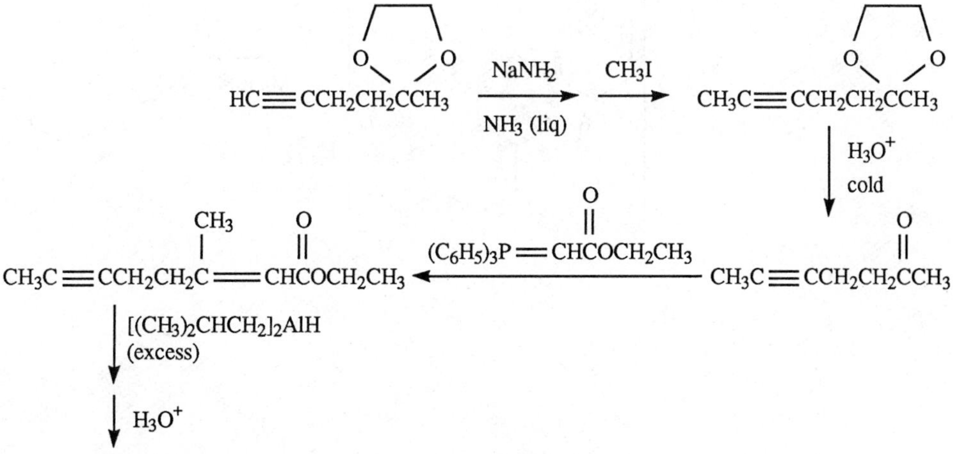

22.40 (cont)

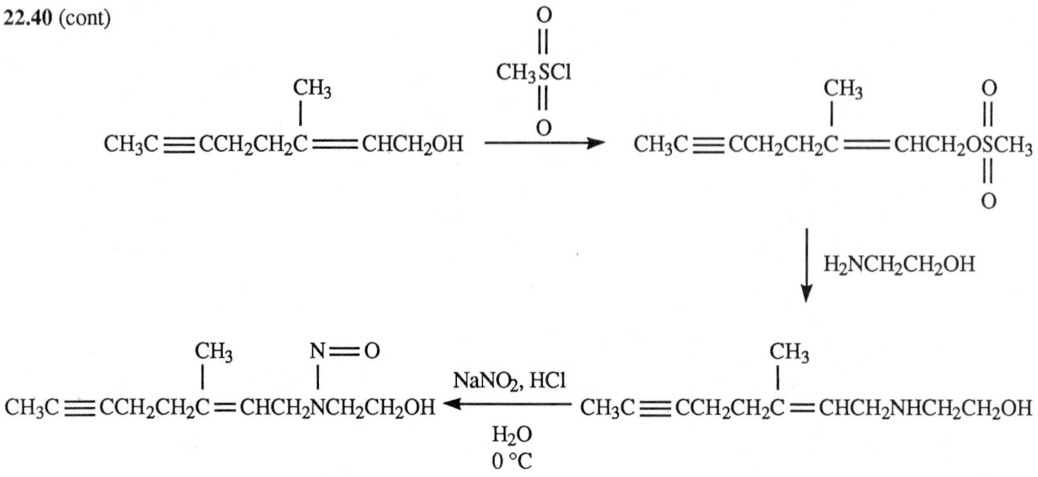

$$CH_3C\equiv CCH_2CH_2\underset{\underset{CH_3}{|}}{C}=CHCH_2OH \quad \xrightarrow{\overset{\overset{O}{\|}}{\underset{\underset{O}{\|}}{CH_3SCl}}} \quad CH_3C\equiv CCH_2CH_2\underset{\underset{CH_3}{|}}{C}=CHCH_2O\overset{\overset{O}{\|}}{\underset{\underset{O}{\|}}{S}}CH_3$$

$$\Big\downarrow \text{H}_2\text{NCH}_2\text{CH}_2\text{OH}$$

$$CH_3C\equiv CCH_2CH_2\underset{\underset{CH_3}{|}}{C}=CH\underset{\underset{N=O}{|}}{C}H_2NCH_2CH_2OH \quad \xleftarrow[\substack{H_2O \\ 0\,^{\circ}C}]{NaNO_2,\ HCl} \quad CH_3C\equiv CCH_2CH_2\underset{\underset{CH_3}{|}}{C}=CHCH_2NHCH_2CH_2OH$$

22.41

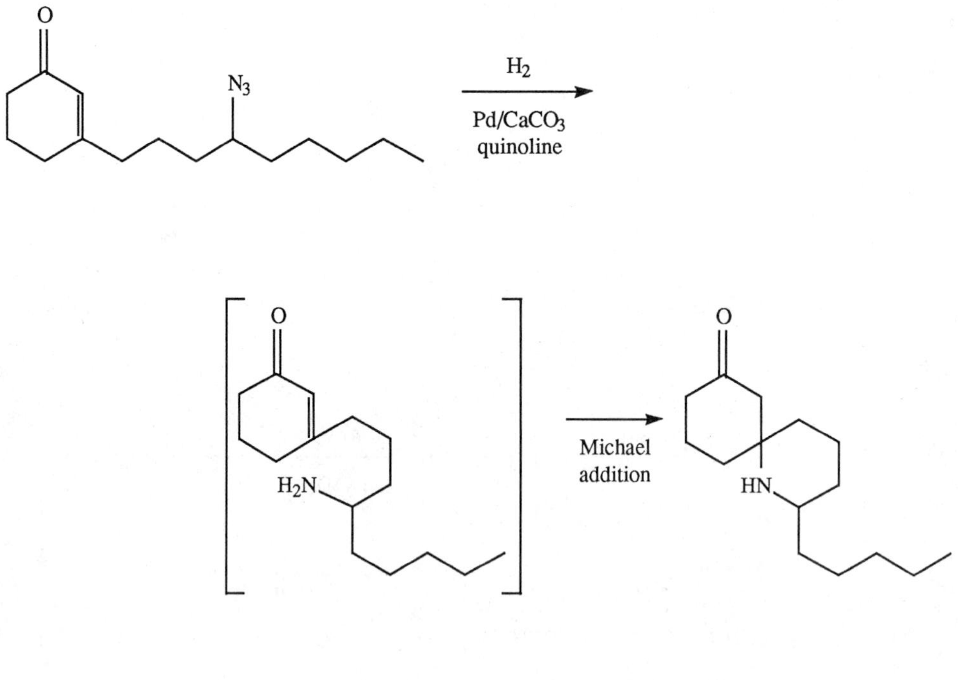

22.41 (cont)

22.42   The analysis of the proton magnetic resonance spectrum for Compound B is given below:

δ 3.2	(2H, singlet, —N$\underline{H}_2$)
δ 3.4	(3H, singlet, —OC$\underline{H}_3$)
δ 6.0 – 6.4	(4H, distorted para substitution pattern, Ar$\underline{H}$)

The distorted aromatic resonances and the relatively high chemical shift values tells us that the two groups para to each other on the aromatic ring are both electron donating. Compound B is *p*-methoxyaniline. Since Compound B is made by reduction of Compound A using reducing agents such as tin in hydrochloric acid, Compound A must have a nitro group in place of the amino group. Compound A is *p*-nitroanisole (*p*-methoxynitrobenzene).

Compound A   Compound B

22.43 The proton magnetic resonance spectrum of Compound C, $C_{10}H_{15}N$, shows a typical ethyl splitting pattern (a triplet at δ 1.1 and a quartet at δ 3.1). The chemical shift of the quartet tells us that the ethyl group is bonded to the nitrogen atom. There are also protons absorbing in the aromatic region (δ 6.0 – δ 6.9). The integration is 5:4:6 and tells us that there must be two ethyl groups bonded to the nitrogen. Compound C is *N,N*-diethylaniline.

$$CH_2CH_3$$
$$|$$
$$\text{—NCH}_2CH_3$$

The proton magnetic resonance spectrum is summarized below:

δ 1.1	(6H, triplet, *J* 7 Hz, —CH₂C<u>H</u>₃)
δ 3.1	(4H, quartet, *J* 7 Hz, —NC<u>H</u>₂CH₃)
δ 6.0 – 6.9	(5H, Ar<u>H</u>)

22.44

22.8 ppm   34.4 ppm

$$CH_3 \longrightarrow \overset{\overset{O}{\|}}{C} \longrightarrow NHCH_2CH_3$$

171.0 ppm   14.6 ppm

22.45 Compound D, C_3H_9N, has no units of unsaturation. The appearance of only two bands in the C-13 nuclear magnetic resonance spectrum tells us that Compound D is symmetrical. Compound D is isopropylamine (2-aminopropane).

26.2 ppm

$$CH_3CHCH_3$$
$$|$$
$$NH_2 \qquad 42.8 \text{ ppm}$$

22.46 Compound E, $C_8H_{11}N$, has four units of unsaturation, which suggests the presence of an aromatic ring. This is confirmed by the presence of bands in the C-13 spectrum in the aromatic region. The solubility properties of Compound E tell us that it is an amine and the lack of reactivity with benzenesulfonyl chloride tells us that it is a tertiary amine. Six of the eight carbon atoms in Compound E are the benzene ring, therefore, the remaining carbon atoms must be two methyl groups. Compound E is *N,N*-dimethylaniline.

23

The Chemistry of Aromatic Compounds II. Synthetic Transformations

23.1

(a)

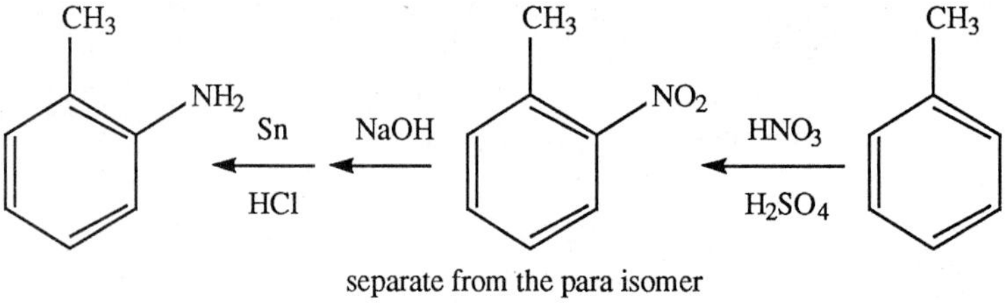

separate from the para isomer

(b)

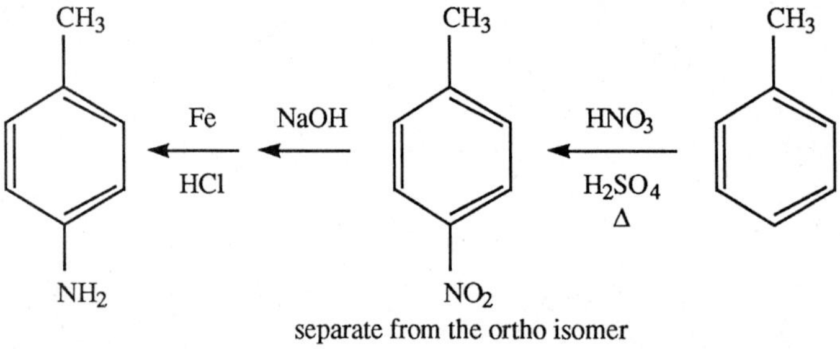

separate from the ortho isomer

(c)

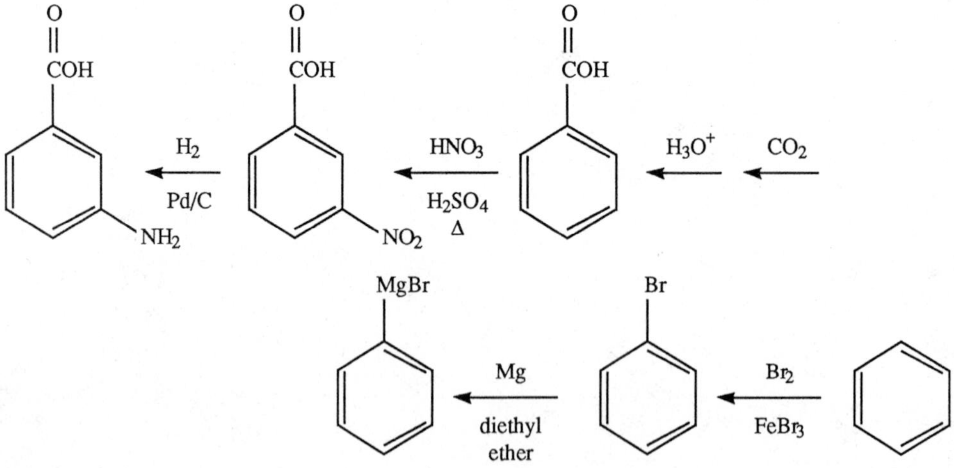

Concept Map 23.1   Diazonium ions in synthesis.

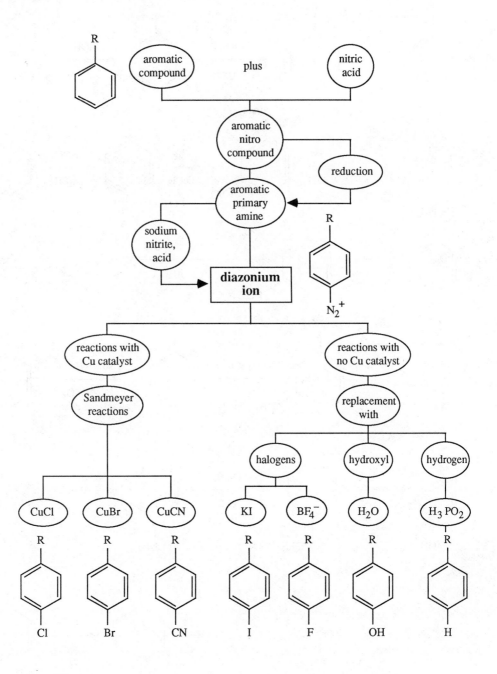

23.2

(a)

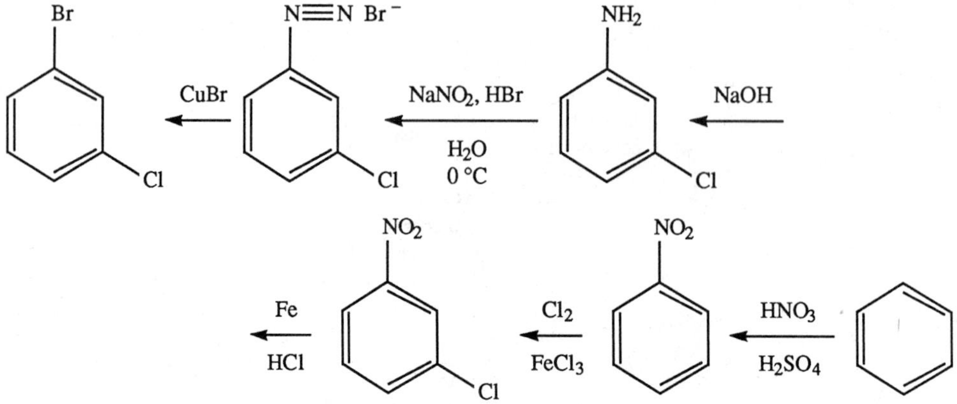
Br
Cl
CuBr
N≡N Br⁻
Cl
NaNO₂, HBr
H₂O
0 °C
NH₂
Cl
NaOH
NO₂
Cl
Fe
HCl
Cl₂
FeCl₃
NO₂
HNO₃
H₂SO₄

(b)

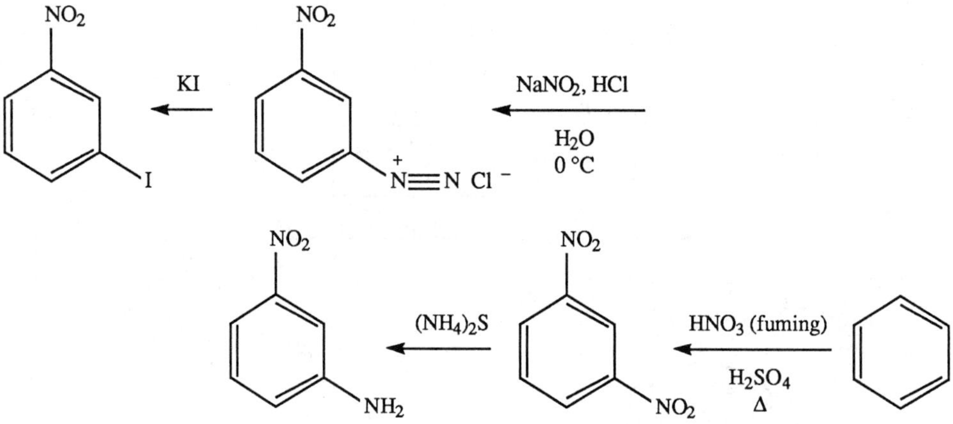
NO₂
I
KI
NO₂
N≡N Cl⁻
NaNO₂, HCl
H₂O
0 °C
NO₂
NH₂
(NH₄)₂S
NO₂
NO₂
HNO₃ (fuming)
H₂SO₄
Δ

(c)

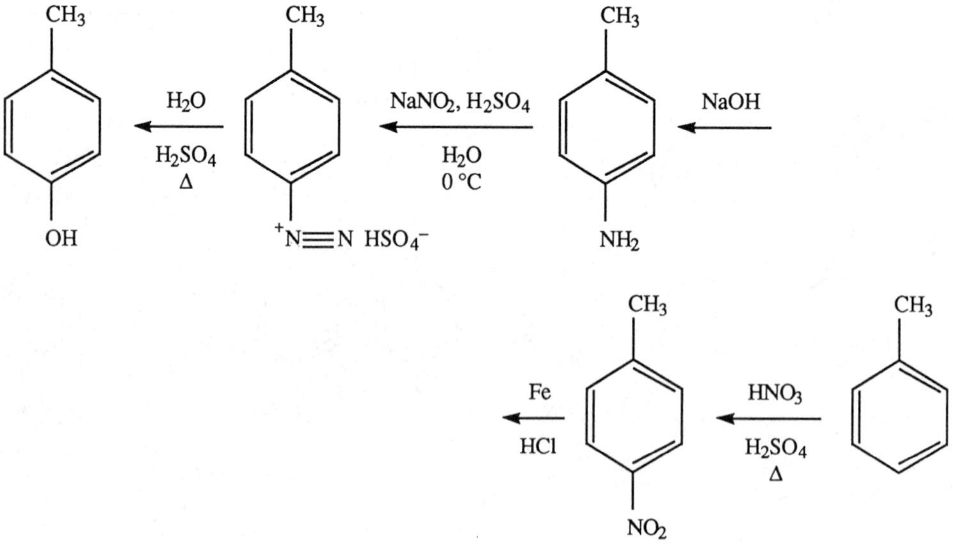
CH₃
OH
H₂O
H₂SO₄
Δ
CH₃
N≡N HSO₄⁻
NaNO₂, H₂SO₄
H₂O
0 °C
CH₃
NH₂
NaOH
CH₃
NO₂
Fe
HCl
HNO₃
H₂SO₄
Δ
CH₃

23.2 (cont)

(d)

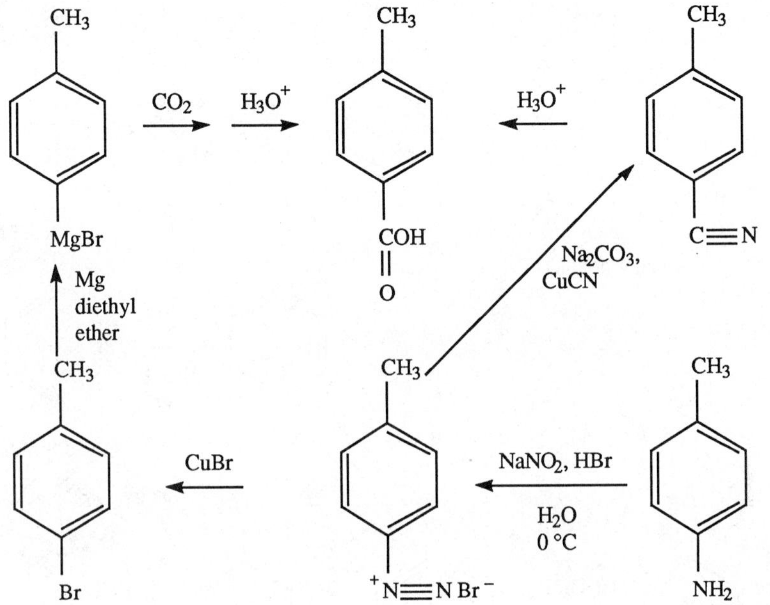

[from (c)]

p-Bromotoluene can also be synthesized by treating toluene with bromine the the presence of ferric bromide, but it will be contaminated with *o*-bromotoluene. Separation of *o*-bromotoluene from *p*-bromotoluene is difficult because they differ by only 3° in boiling point. *p*-Nitrotoluene will also have the ortho isomer as a contaminant, but the two nitrotoluenes are easy to separate because the ortho isomer is a liquid at room temperature and the para isomer is a solid.

(e)

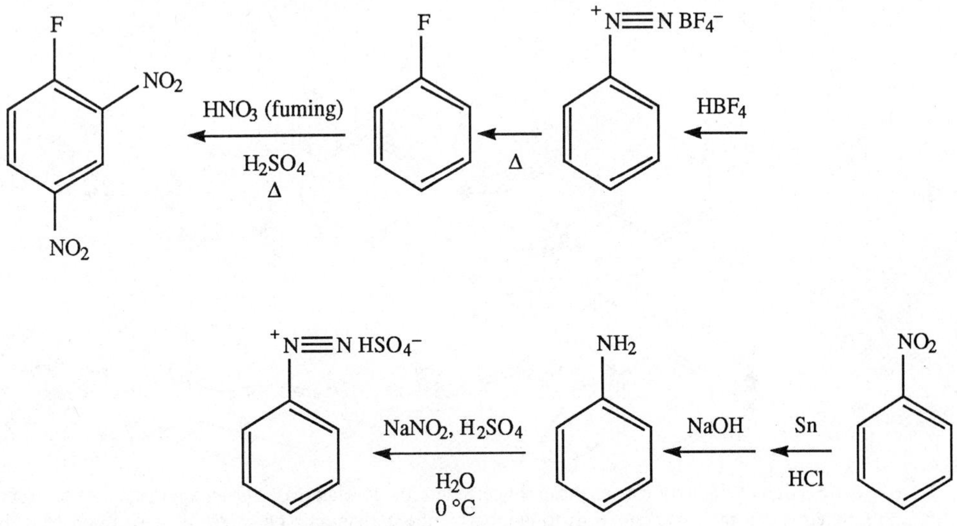

[from (a)]

23.2 (cont)

(f)

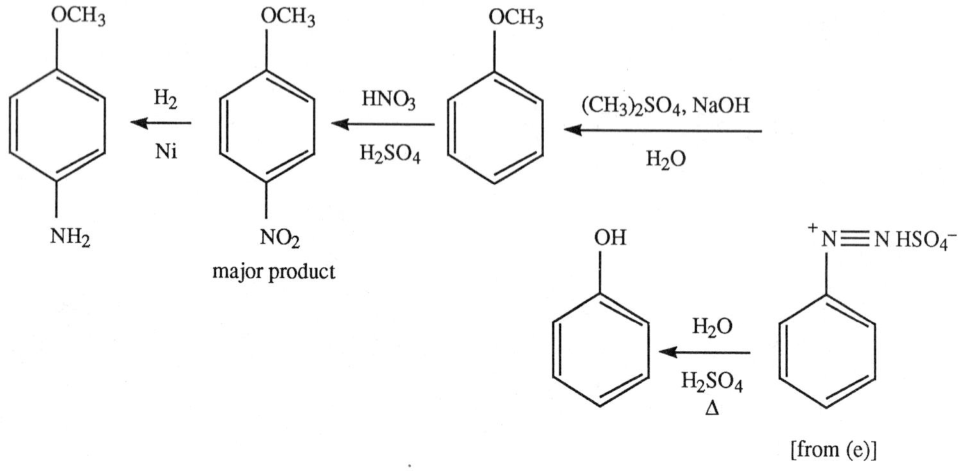

major product

[from (e)]

(g)

major product

[from (f)]

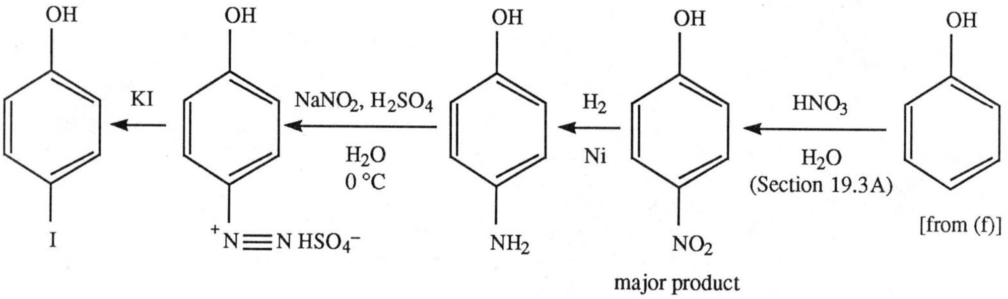

23.3

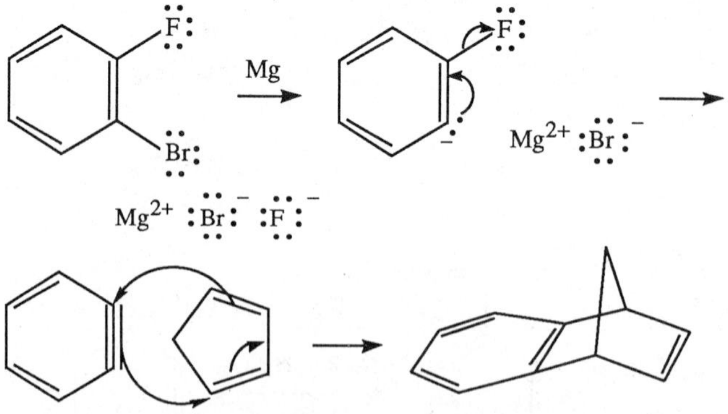

(See Section 13.6A for the mechanism suggested for the formation of a Grignard reagent. The structure of the Grignard reagent is written so as to emphasize the carbanionic character of the carbon atom bonded to magnesium.)

Concept Map 23.2   Reactions of diazonium ions.

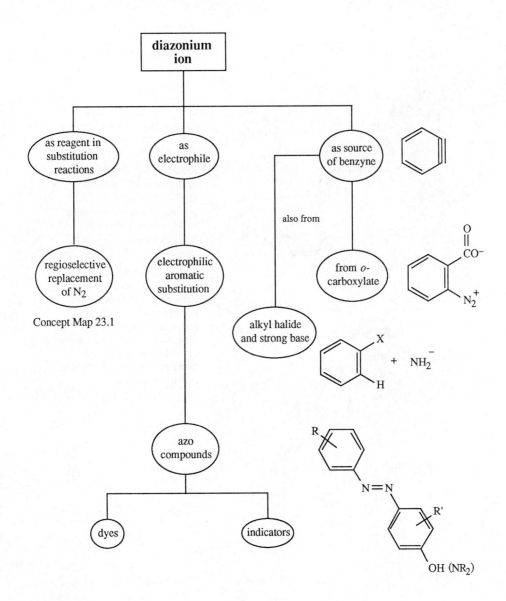

Concept Map 23.1

23.4

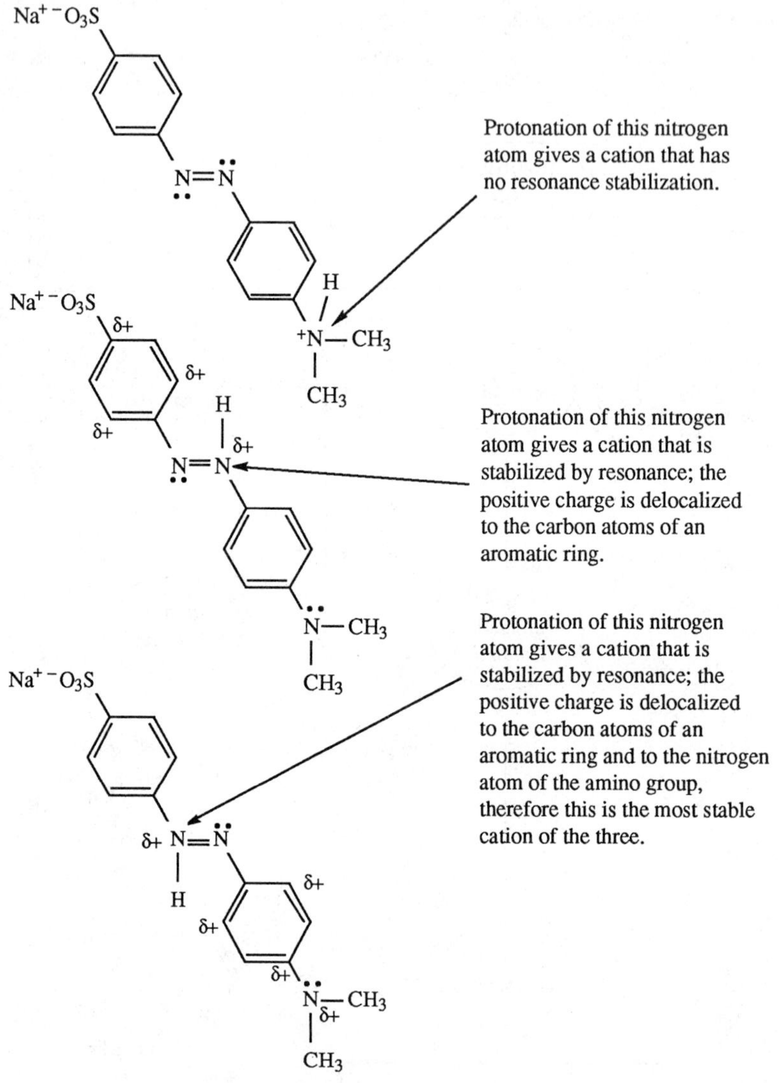
Na⁺ ⁻O₃S
N=N
H
⁺N—CH₃
CH₃
Protonation of this nitrogen atom gives a cation that has no resonance stabilization.
Na⁺ ⁻O₃S
δ+
δ+
δ+
H
δ+
N=N
N—CH₃
CH₃
Protonation of this nitrogen atom gives a cation that is stabilized by resonance; the positive charge is delocalized to the carbon atoms of an aromatic ring.
Na⁺ ⁻O₃S
δ+ N=N
H
δ+
δ+
δ+
N—CH₃
δ+
CH₃
Protonation of this nitrogen atom gives a cation that is stabilized by resonance; the positive charge is delocalized to the carbon atoms of an aromatic ring and to the nitrogen atom of the amino group, therefore this is the most stable cation of the three.

23.5

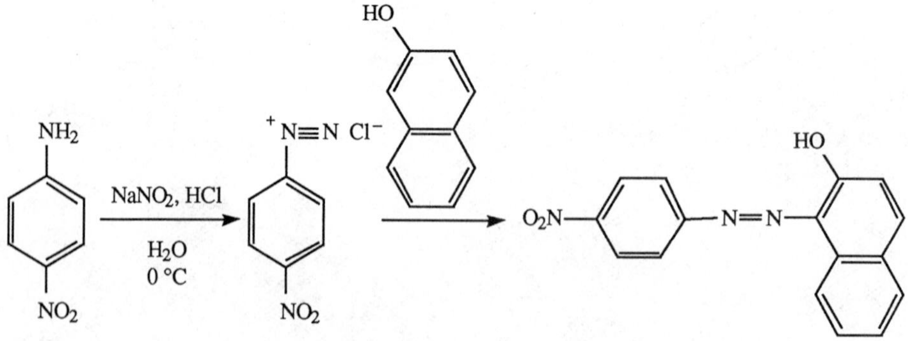
NH₂
NO₂
NaNO₂, HCl
H₂O
0 °C
⁺N≡N Cl⁻
NO₂
HO
O₂N—
—N=N—
HO

23.6

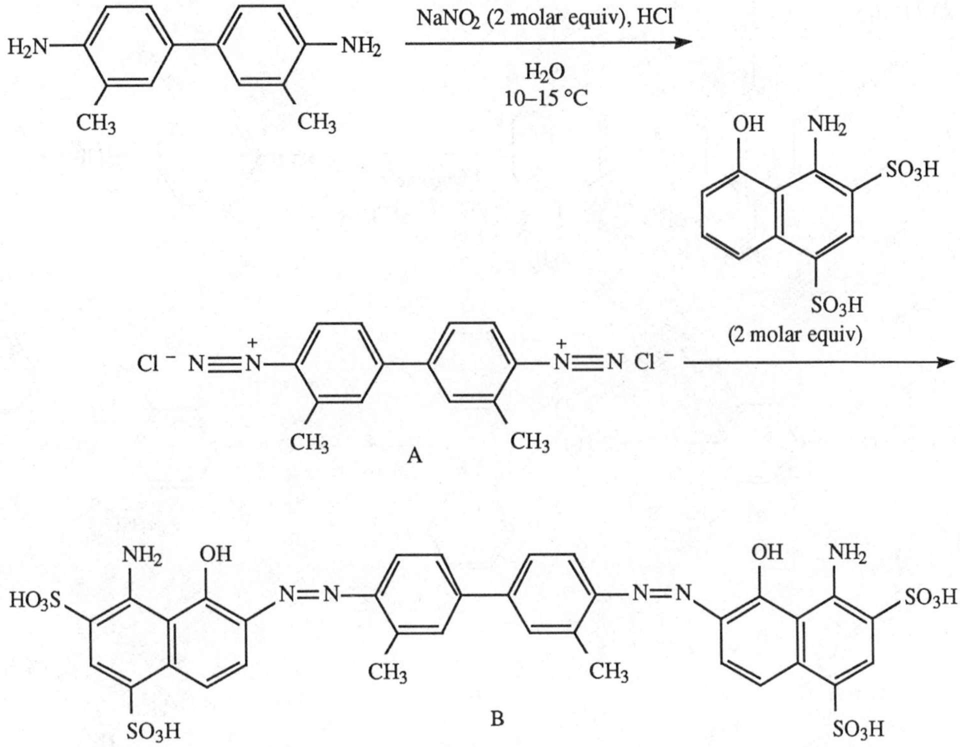

H2N
NH2
CH3
CH3
NaNO2 (2 molar equiv), HCl
H2O
10–15 °C
OH
NH2
SO3H
SO3H
(2 molar equiv)
Cl⁻ N≡N⁺
N⁺≡N Cl⁻
A
NH2
OH
HO3S
N=N
SO3H
B

23.7

(a)

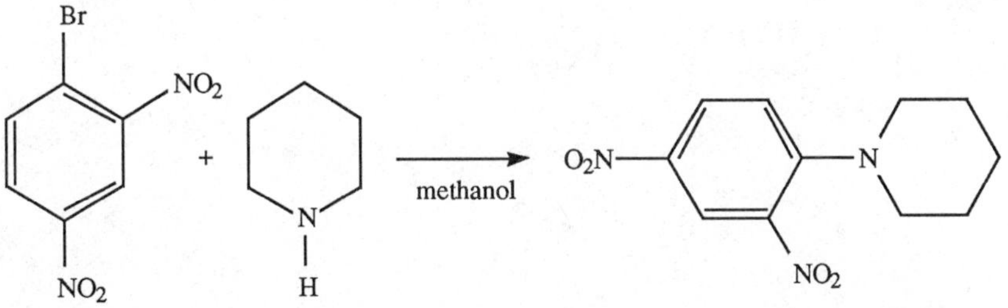

Br
NO2
NO2
+
N
H
methanol
O2N
N
NO2

(b)

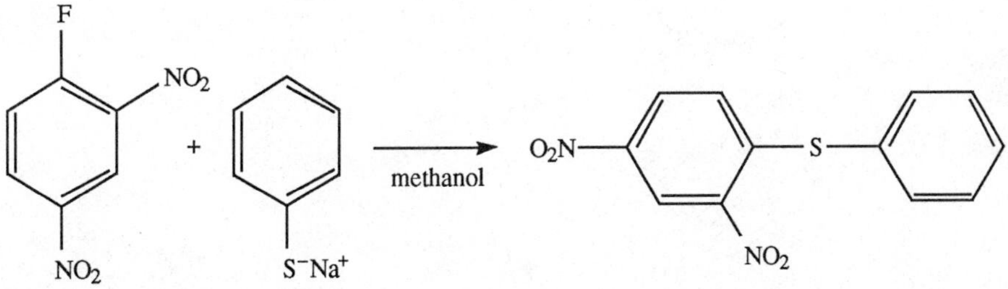

F
NO2
NO2
+
S⁻Na⁺
methanol
O2N
S
NO2

23.7 (cont)

(c)

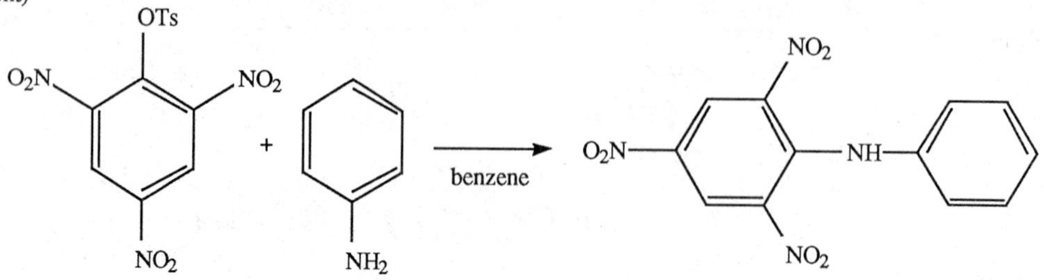
OTs
O_2N
NO_2
+
NO_2
NH_2
benzene
NO_2
O_2N
NH
NO_2

(d)

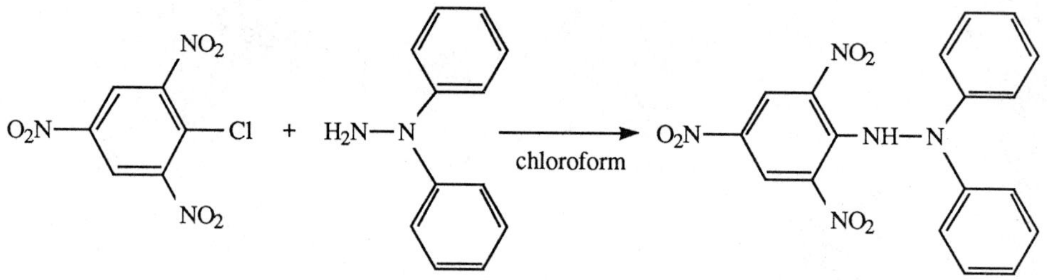
NO_2
O_2N
Cl
+
$H_2N—N$
NO_2
chloroform
NO_2
O_2N
NH—N
NO_2

(e)

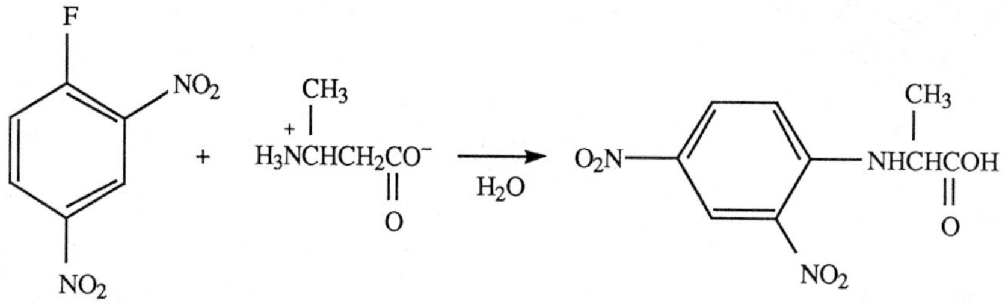
F
NO_2
+
CH_3
$H_3\overset{+}{N}CHCH_2CO^-$
O
NO_2
H_2O
O_2N
CH_3
NHCHCOH
O
NO_2

Concept Map 23.3   Nucleophilic aromatic substitution.

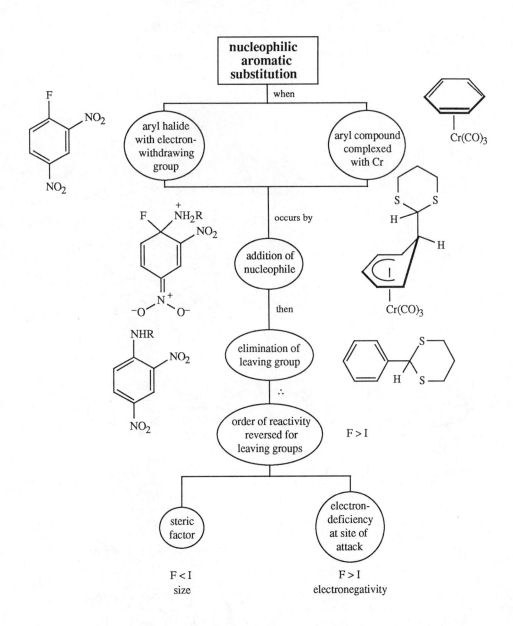

23.8

(a)

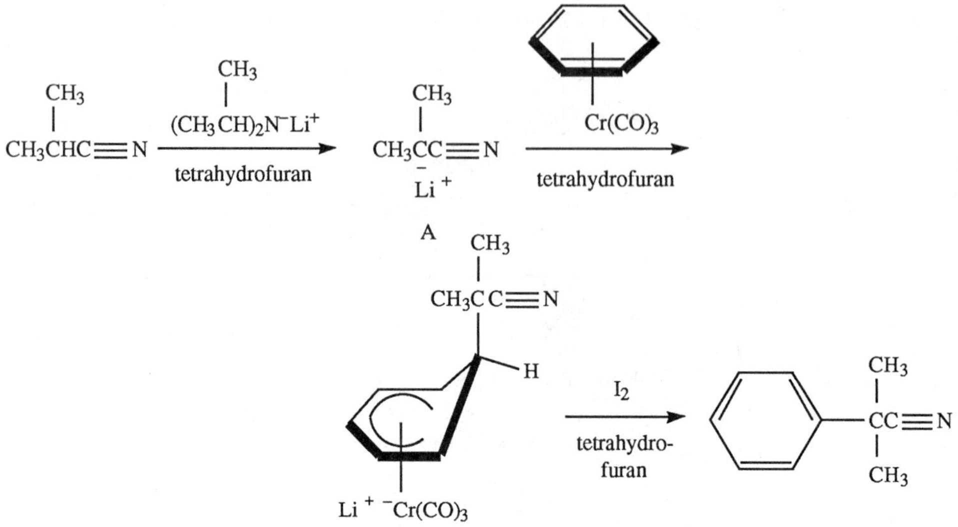

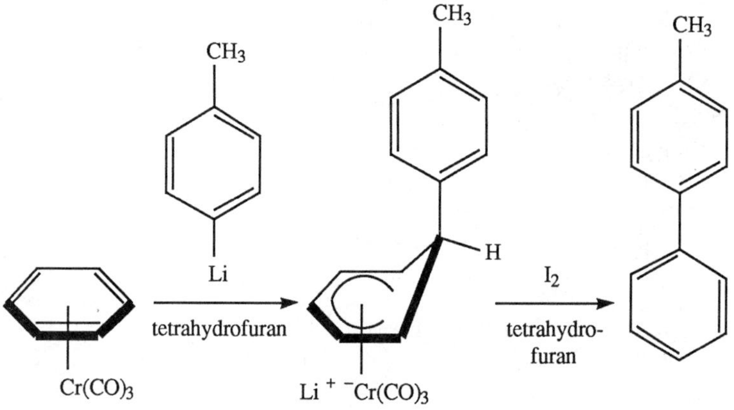

(b)

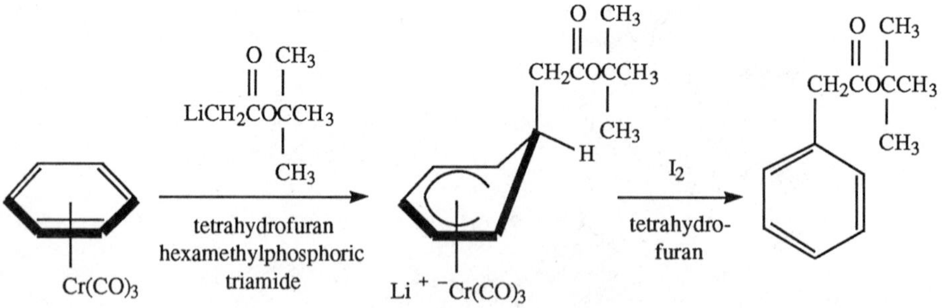

(c)

23.8 (cont)

(d)

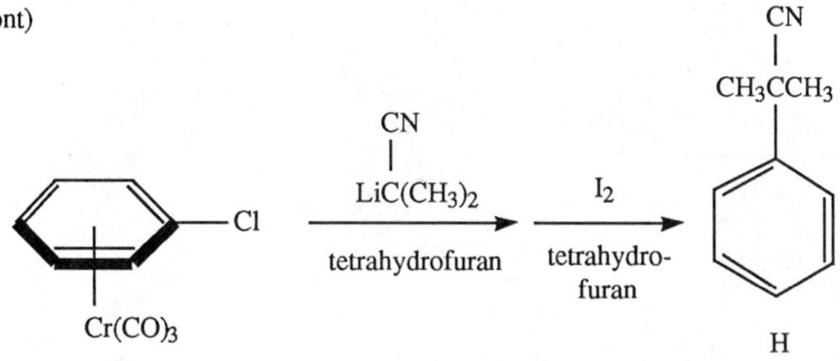

23.9

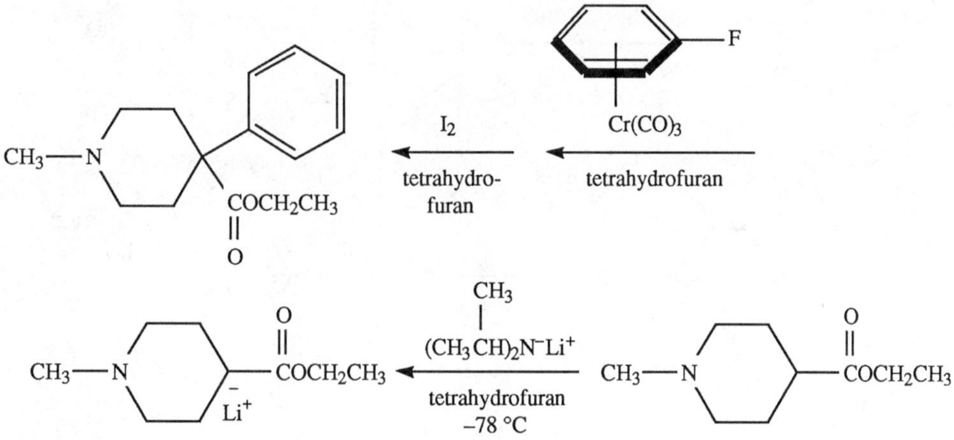

23.10

(a)

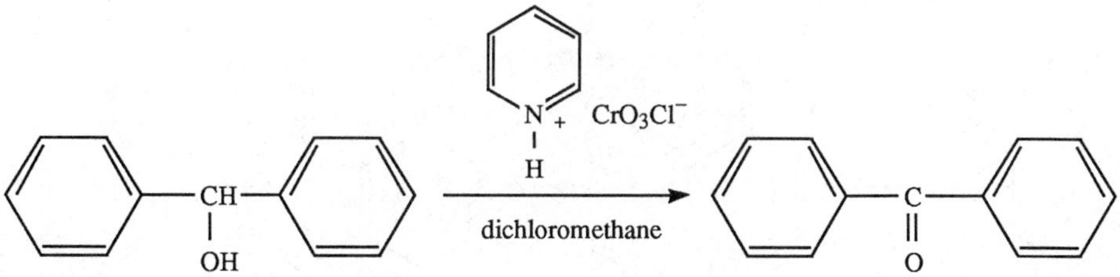

(b)

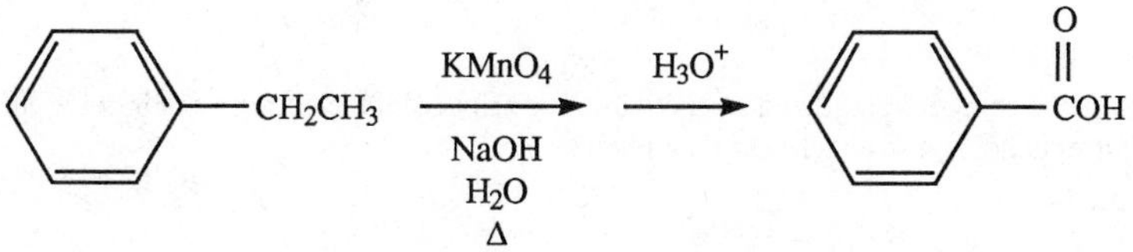

23.10 (cont)

(c)

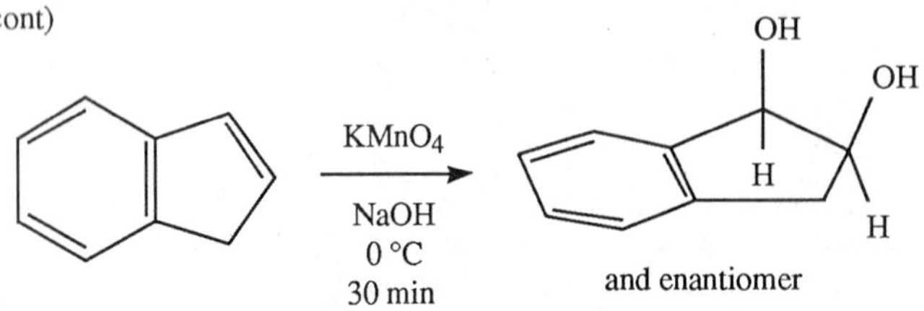

and enantiomer

(d)

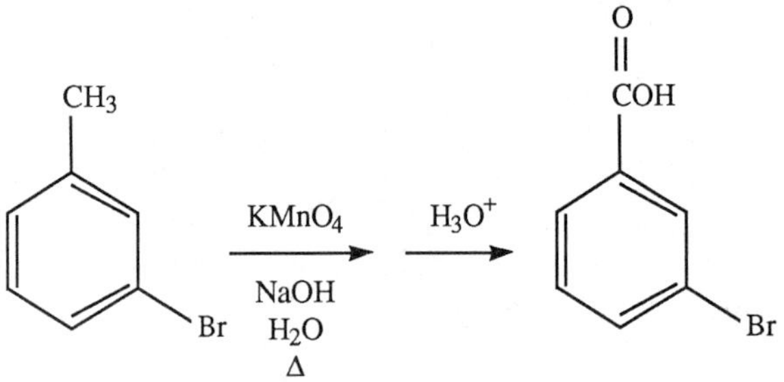

(e)

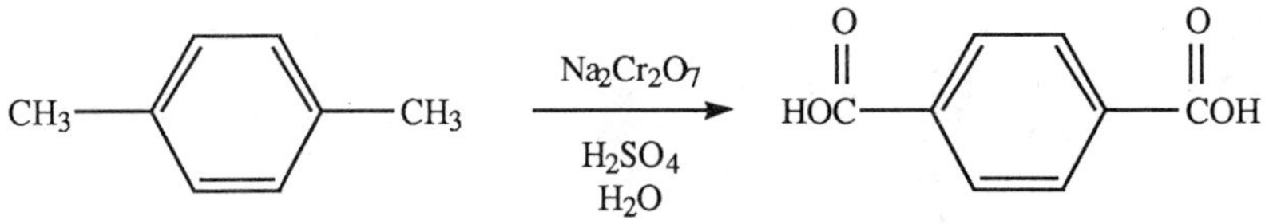

23.11   (a)   Resonance contributors
            1.   *m*-nitrophenolate anion

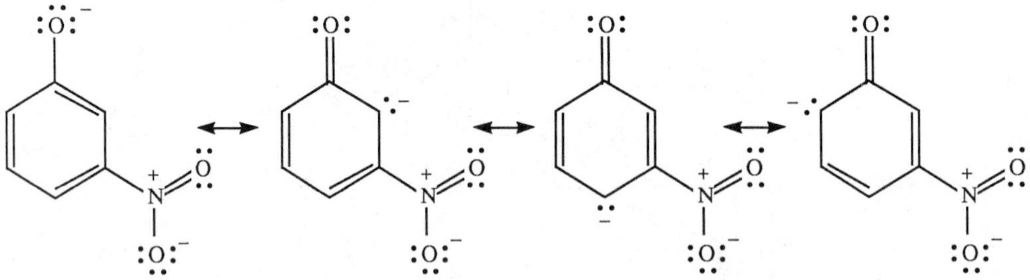

The negative charge is stabilized by the inductive effect of the electron-withdrawing nitro group, so *m*-nitrophenol is a stronger acid than phenol.

23.11 (cont)

2.   *p*-nitrophenolate anion

resonance contributor
in which the negative
charge is delocalized
to the oxygen atom of
the nitro group

The negative charge of the *p*-nitrophenolate anion is stabilized not only by the inductive effect but by a resonance effect. Additional delocalization of charge, beyond what is available in the phenolate anion and in the *m*-nitrophenolate anion, is possible for the *p*-nitrophenolate anion. This extra stabilization of the conjugate base is reflected in the greater acidity of *p*-nitrophenol.

(b)

$pK_a\,0.25$ $pK_a\,6.4$

Picric acid is a sufficiently strong acid to protonate the weak base bicarbonate anion. Although it is a phenol, the effect of the three nitro groups stabilizes the conjugate base of picric acid and thus increases the acidity of the compound so that it is more acidic than a simple carboxylic acid.

23.12 A carboxylic acid can usually be distinguished from a phenol by reaction with sodium bicarbonate, as long as the pK_a of the phenol is greater than that of sodium bicarbonate. For example:

insoluble in water   soluble in water   insoluble in water

Aqueous sodium bicarbonate dissolves *p*-chlorobenzoic acid with the evolution of carbon dioxide gas. The carboxylic acid will be reprecipitated by strong acid. The phenol should not be soluble in sodium bicarbonate solution but should be soluble in a stronger base such as sodium hydroxide.

insoluble in water

insoluble in water   soluble in water   insoluble in water

23.13

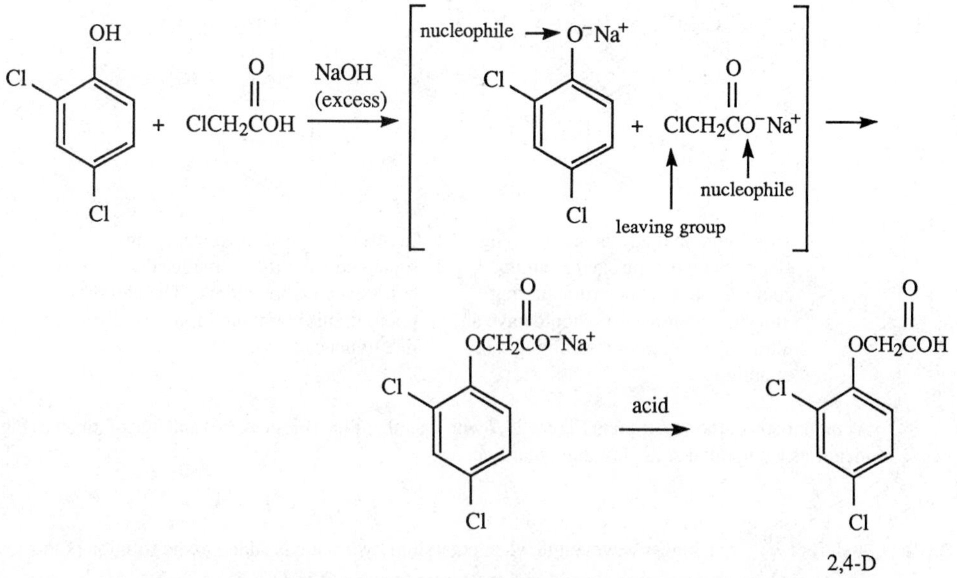

2,4-D

The phenolate anion is a stronger base and a better nucleophile than the chloroacetate anion. The oxygen atom of the phenolate anion will displace the chloride leaving group of the chloroacetate anion. Aromatic nucleophilic substitution reactions require strongly electron-withdrawing substituents on the benzene ring to stablize the negative charge in the transition state. The chlorine atoms on the aromatic ring of the dichlorophenolate ion are not displaced in nucleophilic substitution reactions.

23.14   (1)

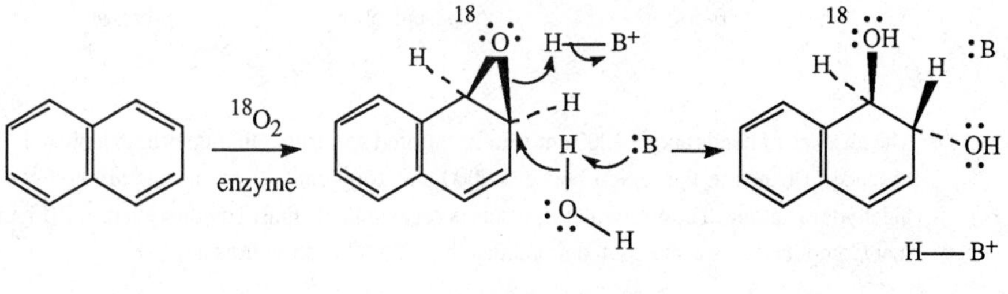

(2)

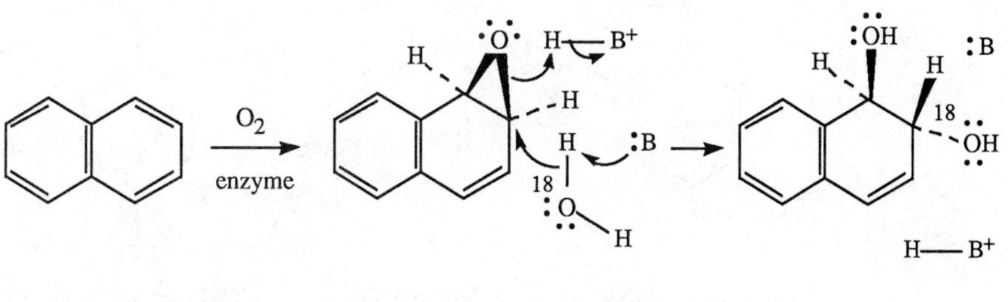

23.15

$$CH_3CH_2\overset{\bullet\bullet}{N}CH_2CH_3$$

$$CH_2CH_2\overset{\bullet\bullet}{N}CH_3$$
$$\underset{\displaystyle CH_3}{|}$$

This compound has the nonbonding electrons on the nitrogen atom in conjugation with the aromatic ring. It is an arylamine and should have an ultraviolet spectrum resembling that of aniline.

In this compound the nitrogen atom is separated from the aromatic ring by sp^3-hybridized carbon atoms. The chromophore in this compound is an alkylbenzene, like toluene.

A comparison of the spectrum in Figure 23.7 with that of aniline (Figure 23.6) and that of toluene (Figure 23.3) identifies Compound X as *N,N*-diethylaniline.

23.16   A shift of λ_{max} to a higher wavelength when potassium hydroxide is added to the solution points to a phenol. There are three possible phenols with the molecular formula C_7H_8O.

OH
   CH$_2$

OH

   CH$_2$

OH

CH$_2$

o-cresol   *m*-cresol   *p*-cresol

23.17   The absence of bands above 3000 cm^{-1} in the infrared spectrum eliminates both a phenol and an alcohol. The absence of bands in the region between 2000 and 1600 cm^{-1} eliminates a carbonyl compound such as an aldehyde or ketone. The ultraviolet spectrum is very similar to that of methoxybenzene (Figure 23.4) suggesting that Compound Z is a substituted aromatic ether. Possible structures are:

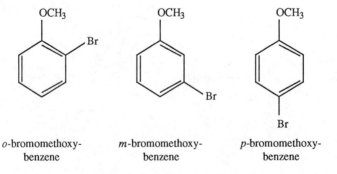

OCH$_3$
   Br

OCH$_3$

   Br

OCH$_3$

Br

o-bromomethoxy-   *m*-bromomethoxy-   *p*-bromomethoxy-
benzene   benzene   benzene

23.18

(a)

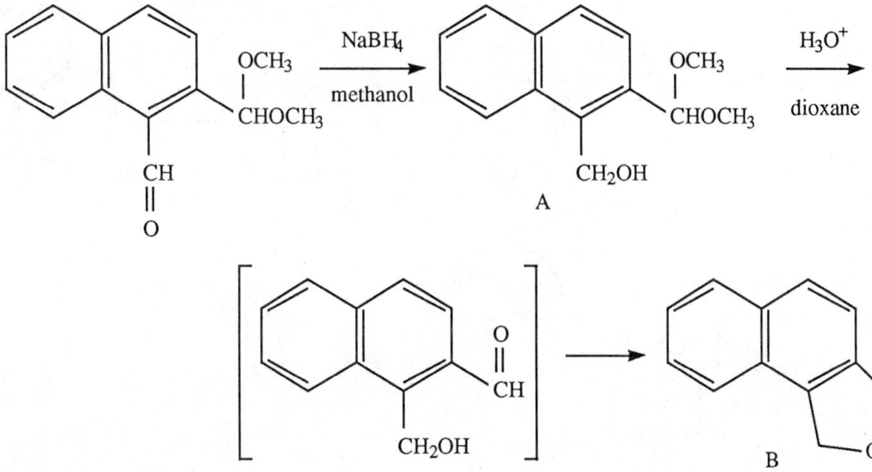
OCH3
CHOCH3
CH
O
NaBH4
methanol
OCH3
CHOCH3
CH2OH
A
H3O+
dioxane
O
CH
CH2OH
OH
H
O
B

(b)

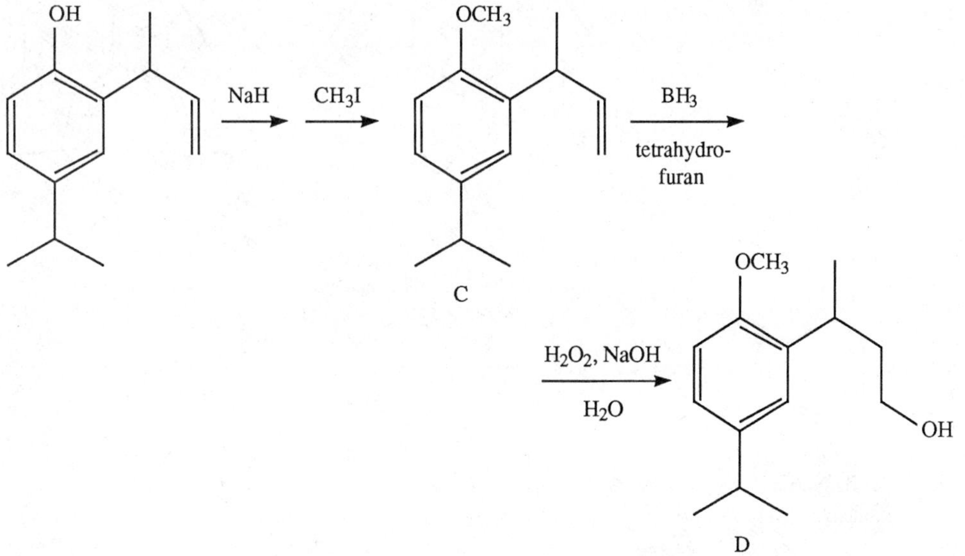
OH
NaH
CH3I
OCH3
C
BH3
tetrahydro-
furan
H2O2, NaOH
H2O
OCH3
OH
D

(c)

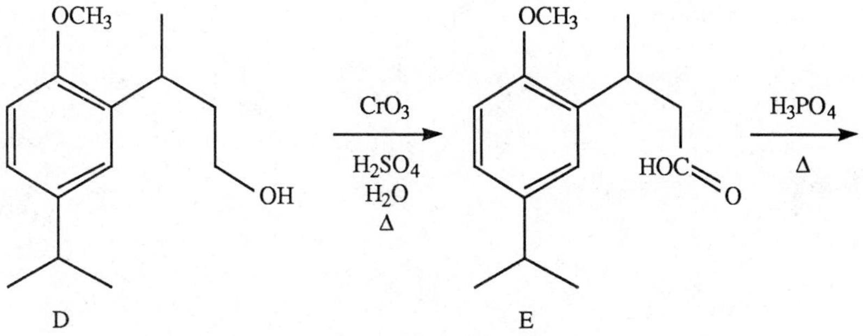
OCH3
OH
D
CrO3
H2SO4
H2O
Δ
OCH3
HOC
O
E
H3PO4
Δ

23.18 (cont)

(c) (cont)

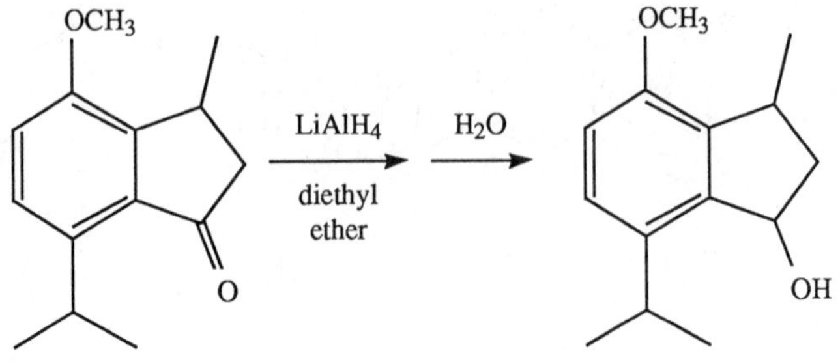

F

LiAlH₄ / diethyl ether   H₂O

G

(d)

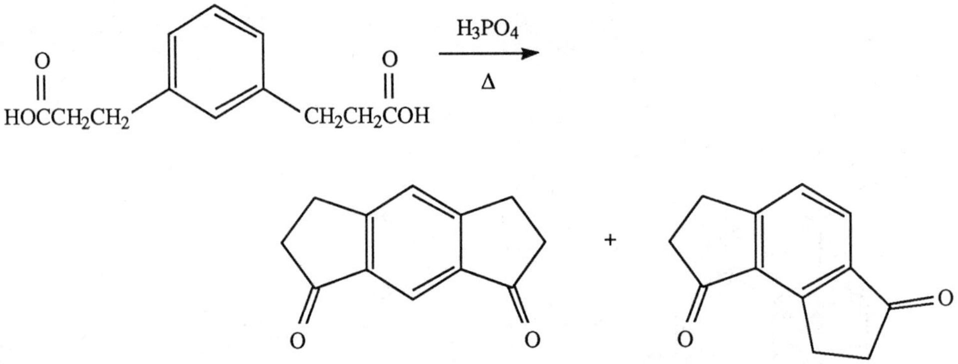

H₃PO₄ / Δ

H   +   I

(e)

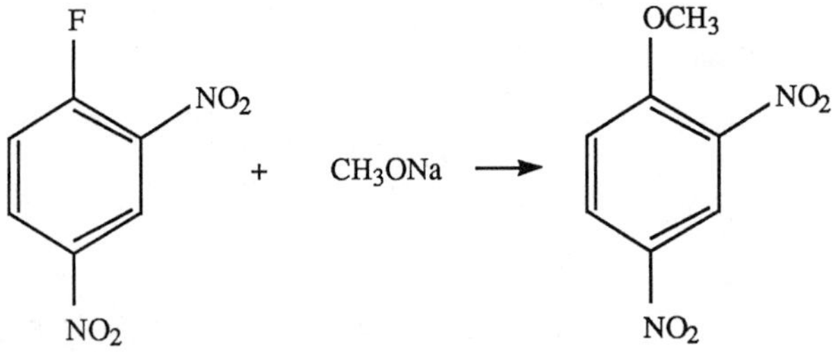

+   CH₃ONa

J

(f)

KMnO₄ / H₂O / Δ

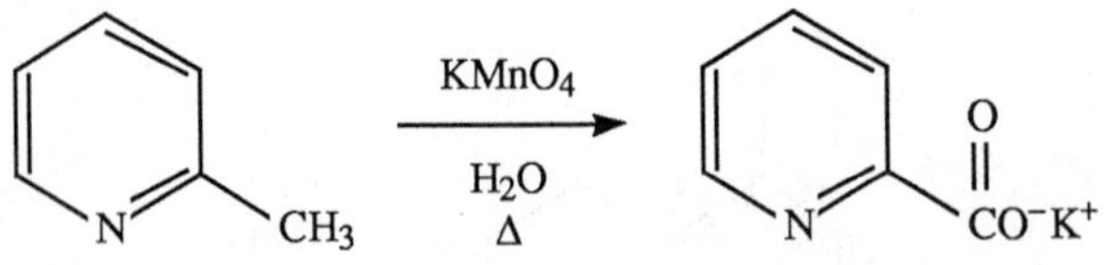

K

23.18 (cont)

(g)

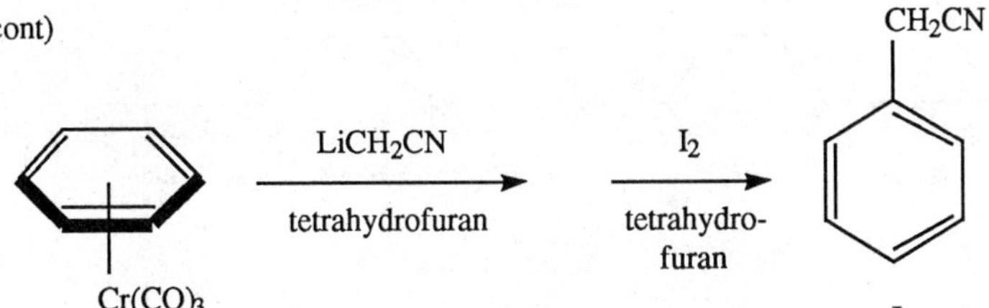

(h)

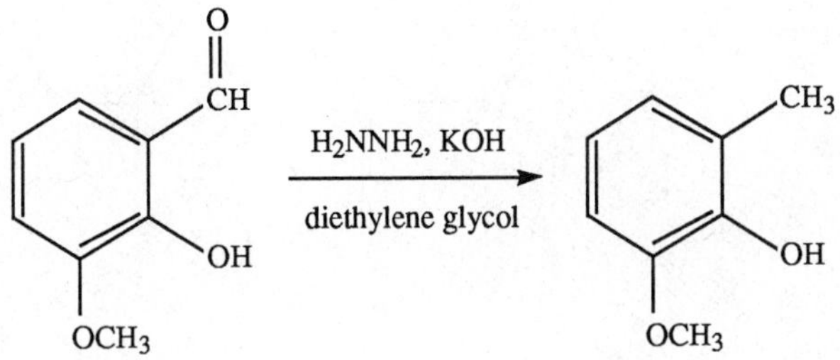

(i)

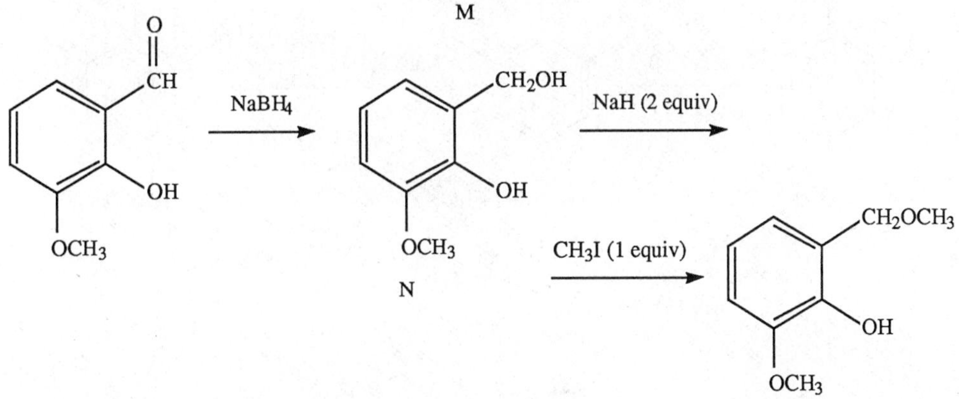

(an alkoxide ion is a better
nucleophile than a phenolate ion)

(j) $CH_2(COCH_2CH_3)_2$ $\xrightarrow[\text{diethyl ether}]{Na}$ $Na^+ {}^-CH(COCH_2CH_3)_2$

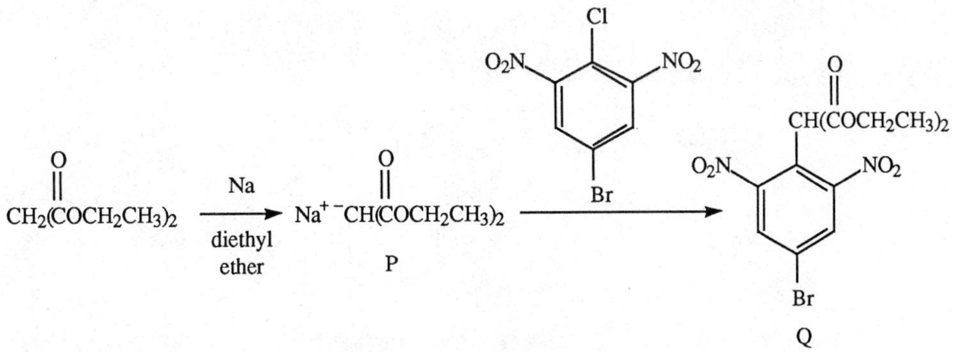

Chloride ion is a better leaving group than bromide ion for nucleophilic substitution reactions on aromatic rings (Section 23.3A). The location of the chlorine atom ortho to the nitro group is also essential to the reaction.

23.18 (cont)

(k)

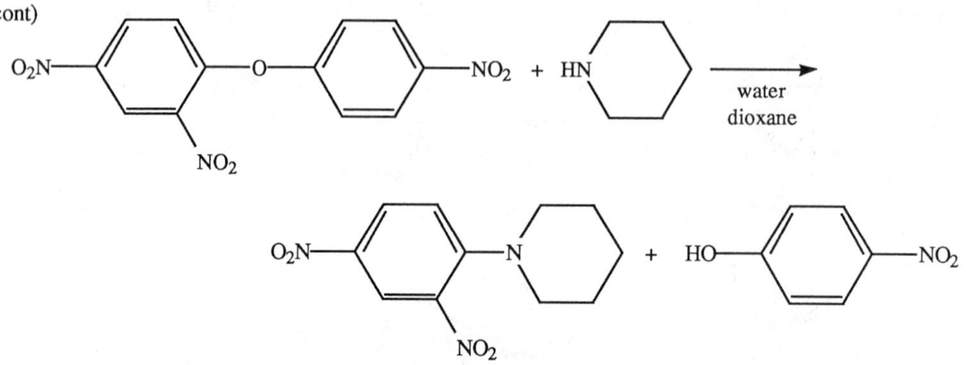

R   S

(l)

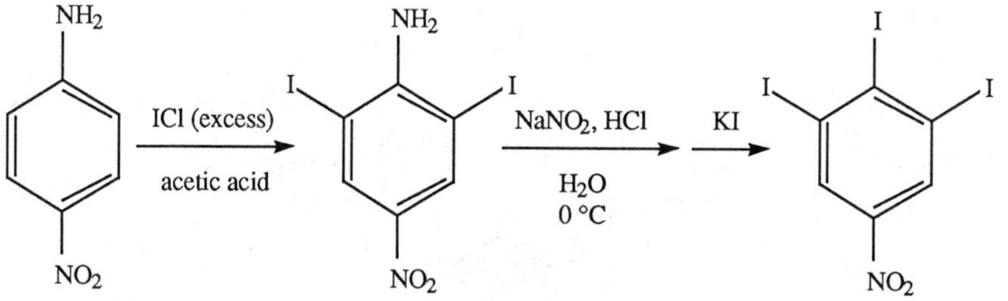

T   U

(m

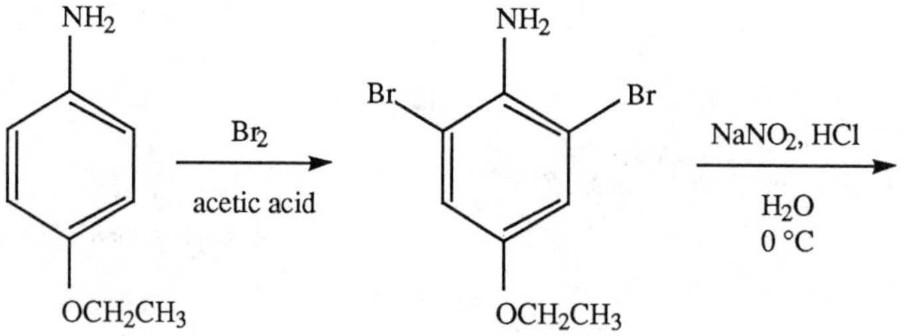

V

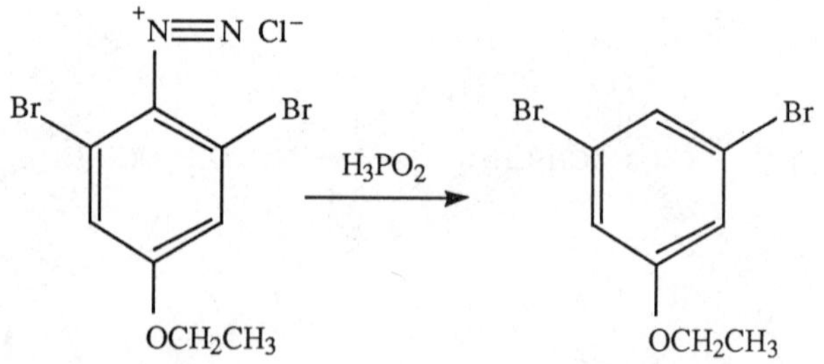

W   X

23.18 (cont)

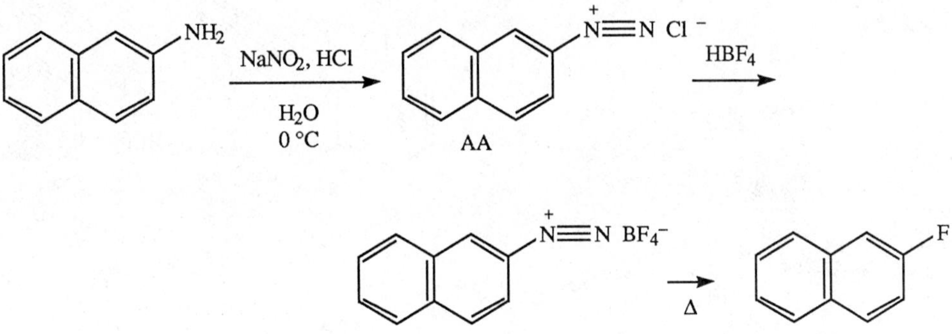

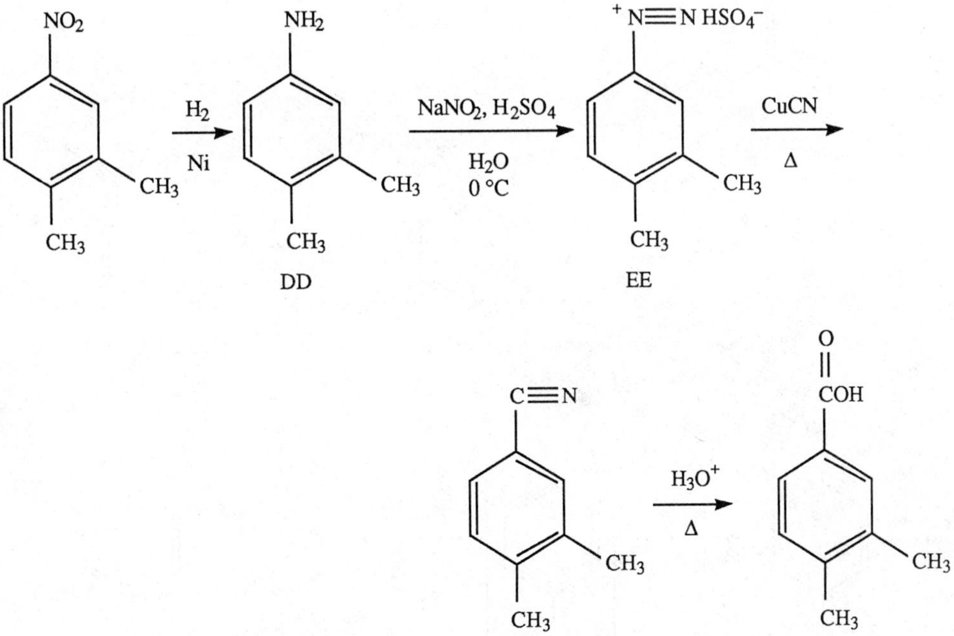

(n)

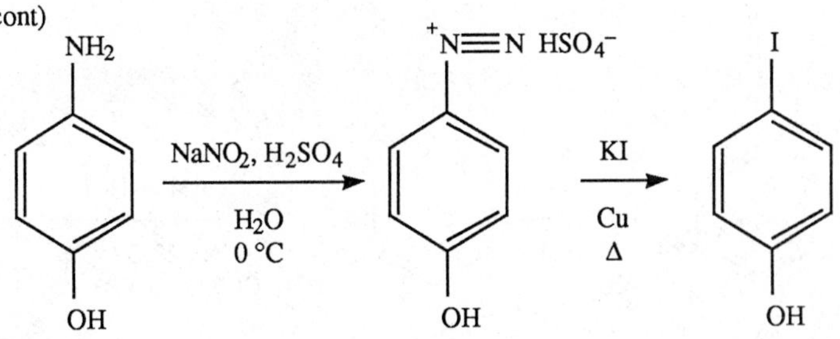

(o)

(p)

23.19

(a)

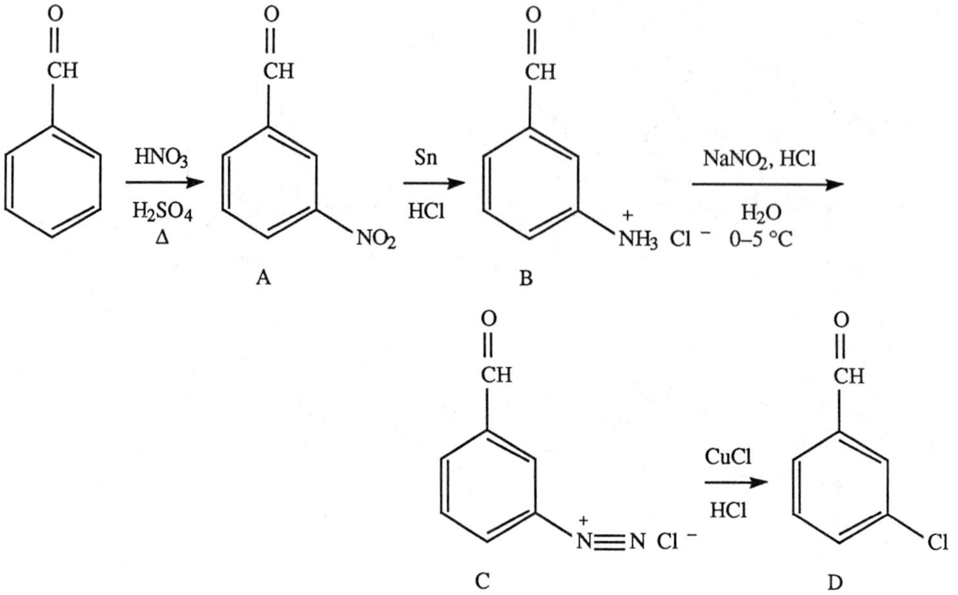
CH
HNO3
H2SO4
Δ
NO2
A
Sn
HCl
NH3 Cl⁻
B
NaNO2, HCl
H2O
0–5 °C
N≡N Cl⁻
C
CuCl
HCl
Cl
D

(b)

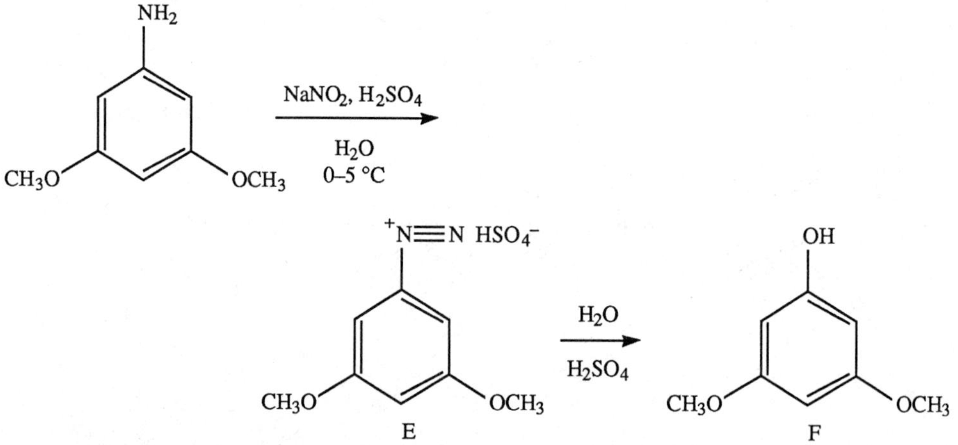
NH2
CH3O
OCH3
NaNO2, H2SO4
H2O
0–5 °C
N≡N HSO4⁻
E
H2O
H2SO4
OH
F

(c)

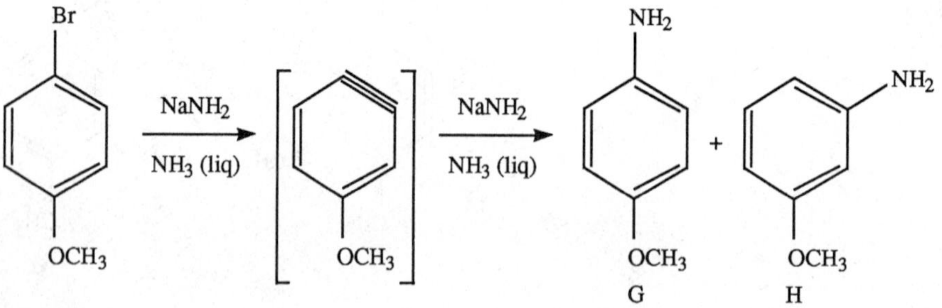
Br
OCH3
NaNH2
NH3 (liq)
NH2
G
H

23.19 (cont)

(d)

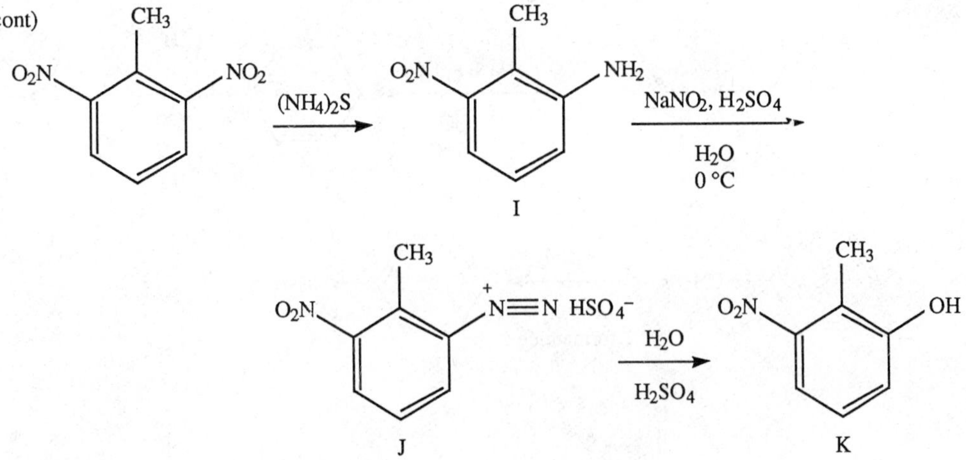

I

J

K

(e)

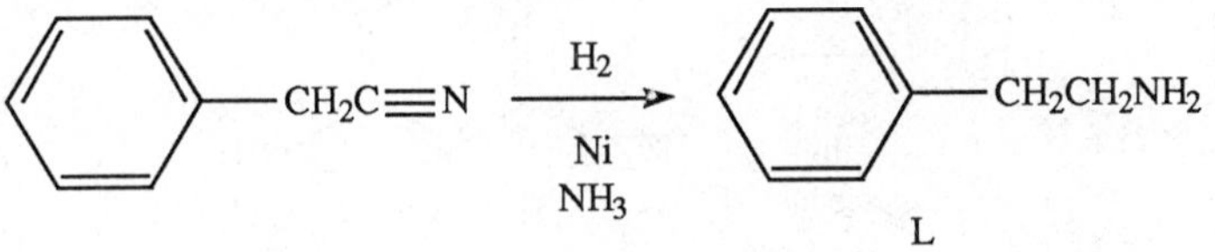

L

(f)

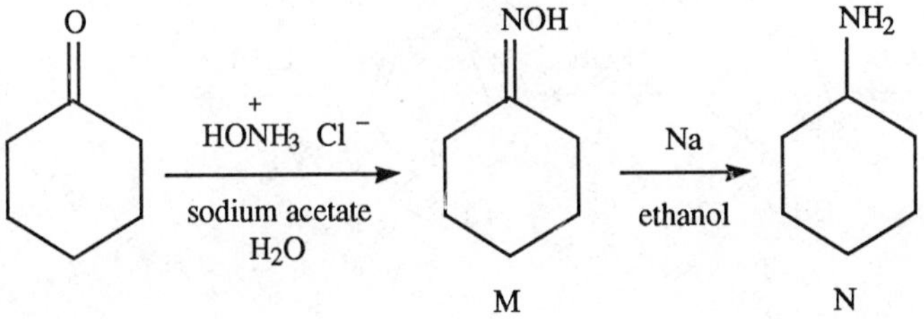

M

N

(g)

O

P

methyl red

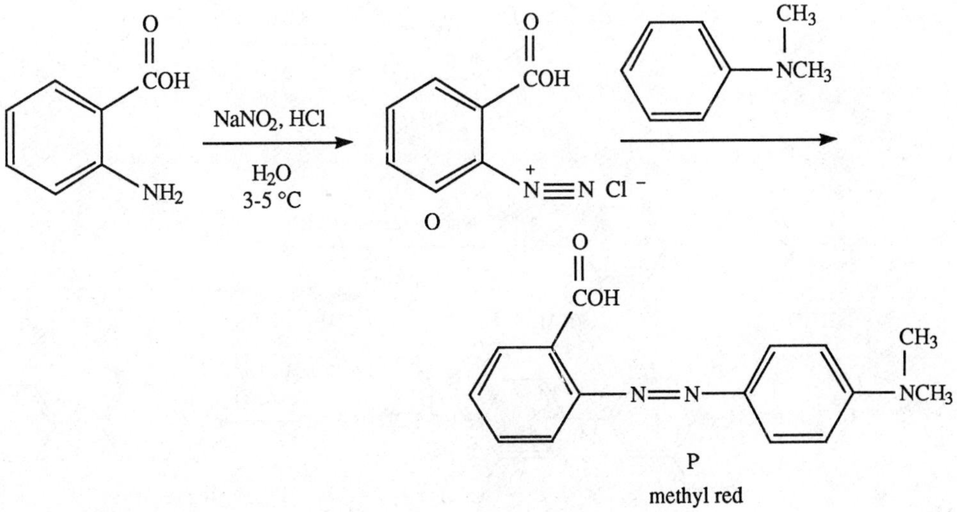

23.19 (cont)

(h)

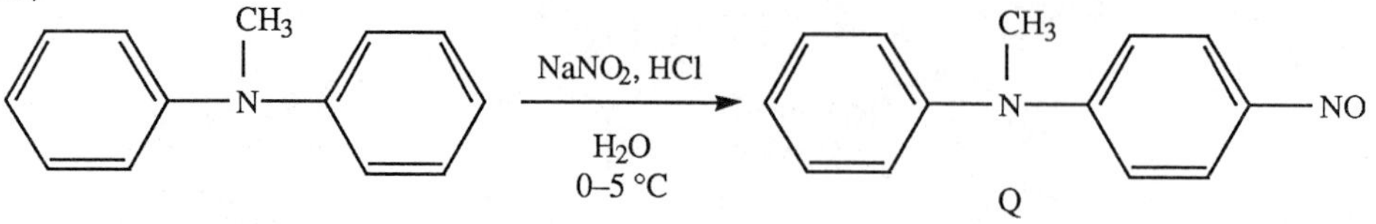
CH3
N
NaNO2, HCl
H2O
0–5 °C
CH3
N
NO
Q

(i)

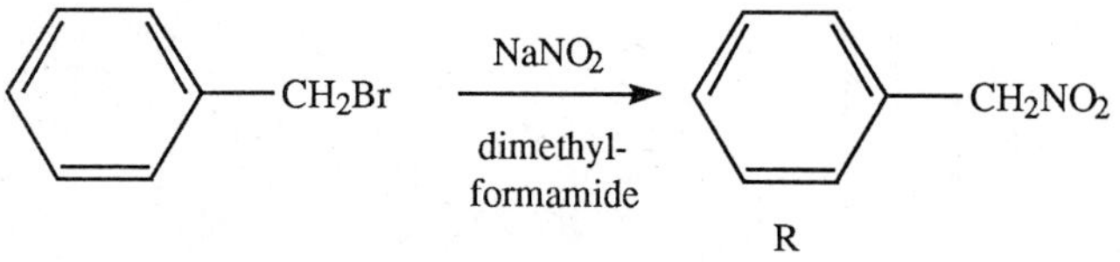
CH2Br
NaNO2
dimethyl-
formamide
CH2NO2
R

(j)

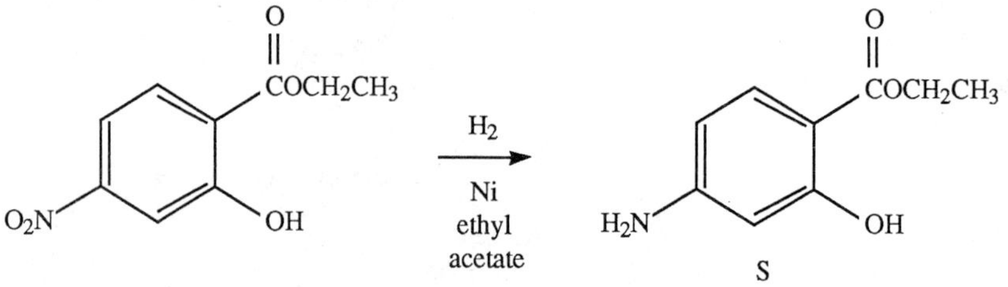
O
COCH2CH3
O2N
OH
H2
Ni
ethyl
acetate
O
COCH2CH3
H2N
OH
S

(k)

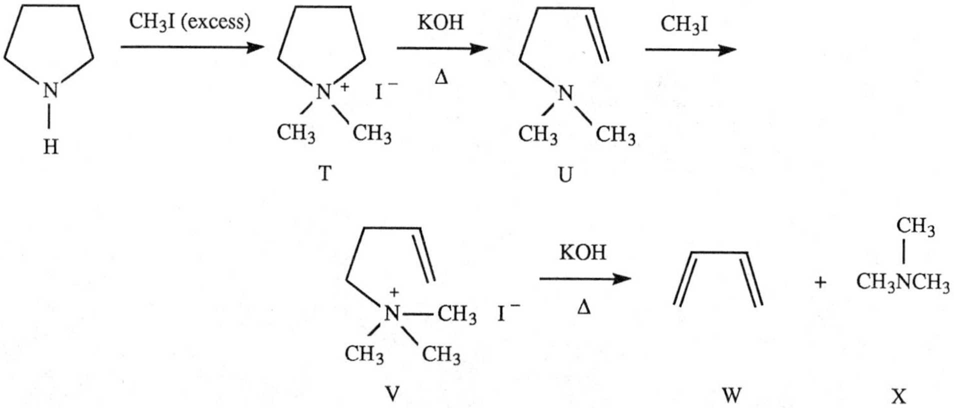
N
H
CH3I (excess)
N+ I−
CH3 CH3
T
KOH
Δ
N
CH3 CH3
U
CH3I
N—CH3 I−
CH3 CH3
V
KOH
Δ
W
+
CH3
CH3NCH3
X

(l)

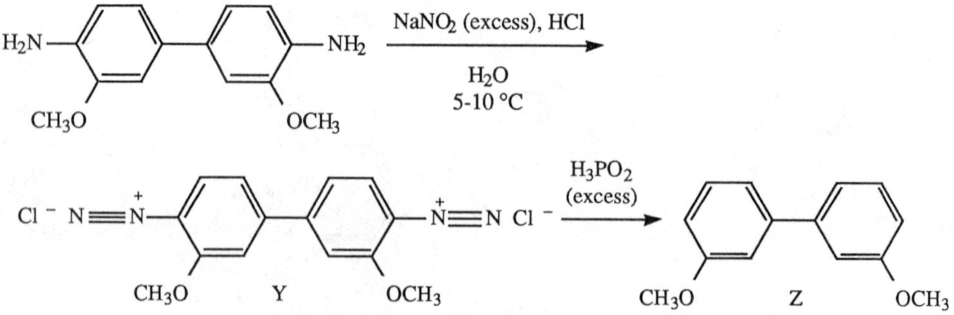
H2N
NH2
CH3O
OCH3
NaNO2 (excess), HCl
H2O
5-10 °C
Cl− N≡N
N≡N Cl−
CH3O
Y
OCH3
H3PO2
(excess)
CH3O
Z
OCH3

23.20

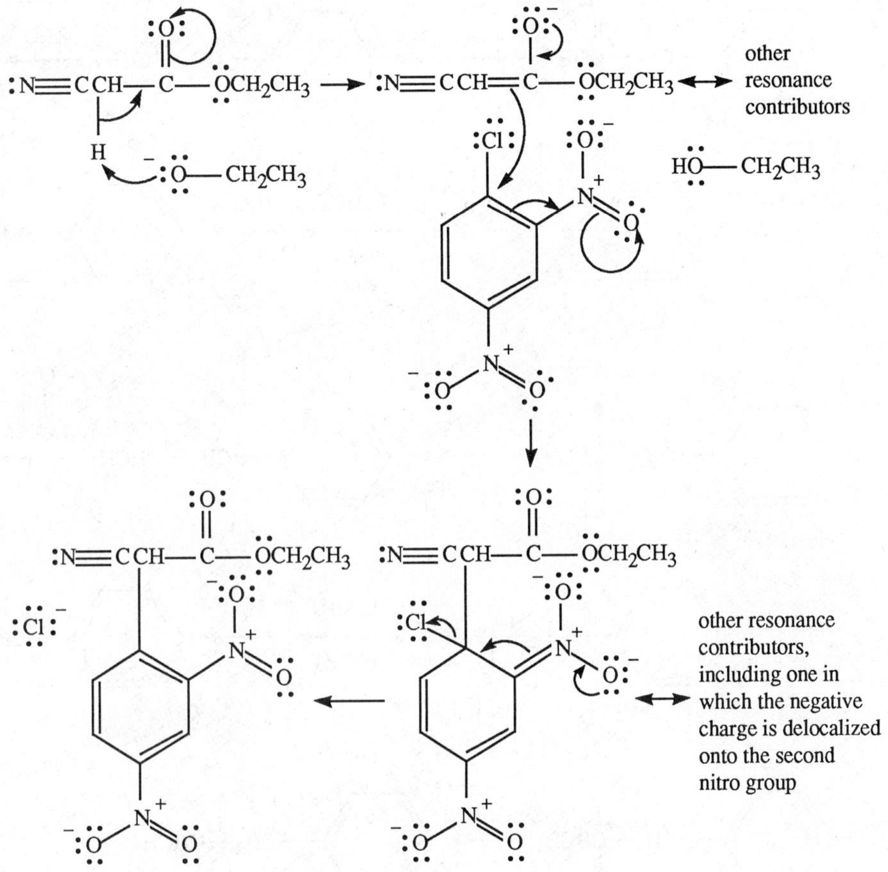
other
resonance
contributors
other resonance
contributors,
including one in
which the negative
charge is delocalized
onto the second
nitro group

23.21

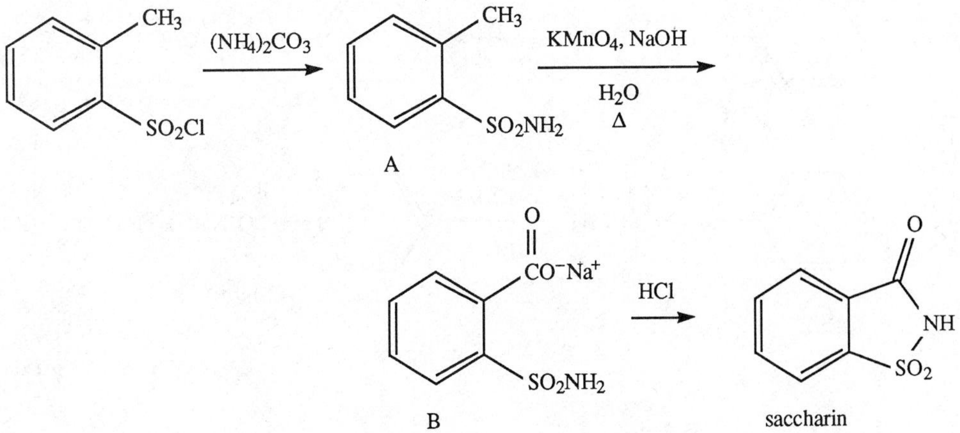
$(NH_4)_2CO_3$
$KMnO_4$, NaOH
H_2O
Δ
A
HCl
B
saccharin

23.22

(a)

23.22 (cont)

(b)

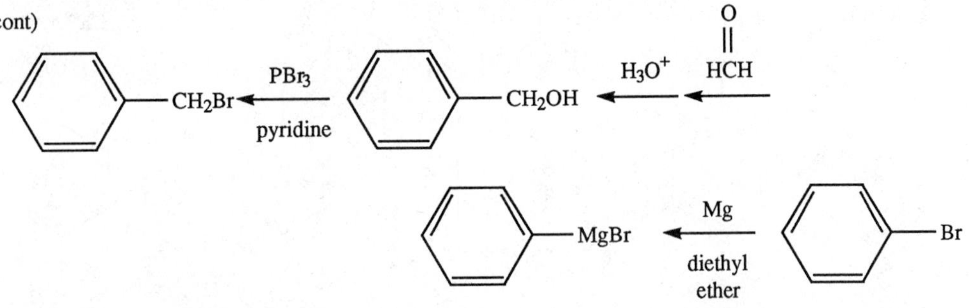

1.

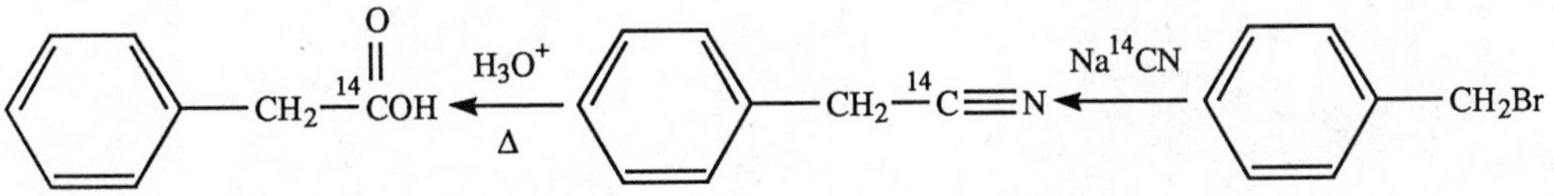

2.

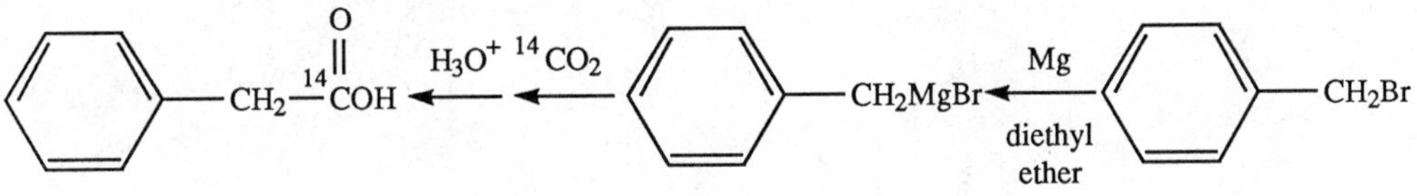

23.23

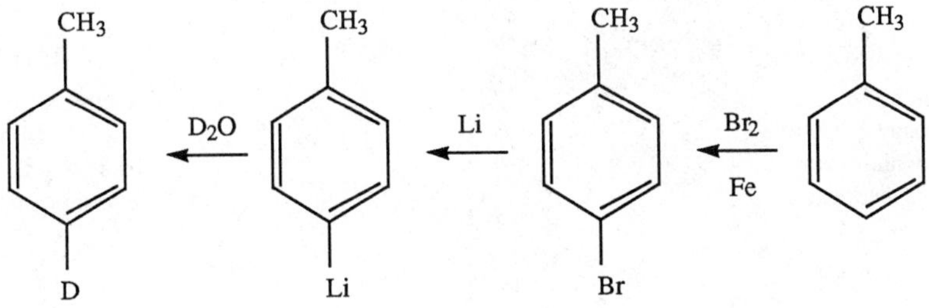

separate from
ortho isomer

most acidic   least acidic

23.24   (a)

carboxylic
acid

phenol

alcohol

amine

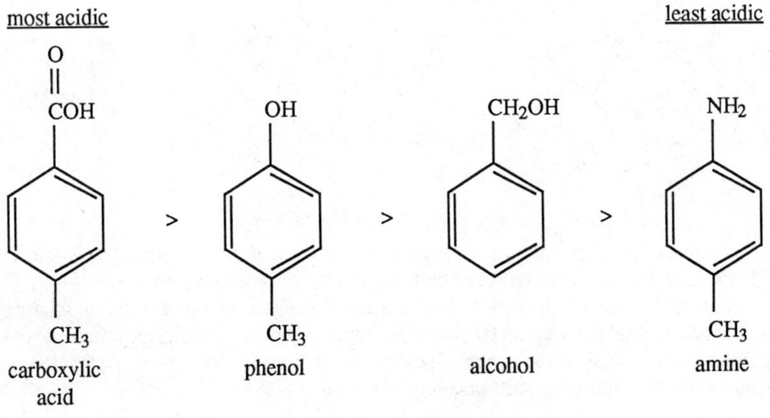

23.24 (cont)   most acidic   least acidic

(b)

OH
>
OH
>
OH
>
OH

NO₂   F   CH₃

nitro group is
strongest electron-
withdrawing
substituent

(c)

O
‖
COH
>
O
‖
COH
>
O
‖
COH
>
O
‖
COH

NO₂   Cl   OCH₃

nitro group is   methoxy group
strongest electron-   is electron-
withdrawing   donating in the
substituent   para position

23.25

The reaction is a nucleophilic aromatic substitution reaction. The relative rates of reaction for the different nucleophiles reflect the relative nucleophilicities of the different species.

Methoxide is the best nucleophile in the series because it is by far the strongest base. The thiophenoxide anion, though a weaker base than the phenolate anion, is the better nucleophile because of the polarizability of the sulfur atom. Aniline is the weakest nucleophile. It is a weak base, meaning that the nonbonding electrons are not easily available for bond formation.

23.26

For Y: ——OCH$_3$ >   ——CH$_3$ >   ——H   >   ——Cl   >   ——NO$_2$

      fastest   slowest

The order of reactivity depends on the nucleophilicity of the nitrogen atom in the substituted aniline, which in turn is dependent on the availability of the nonbonding electrons. Electron-donating groups at the para position would increase the rate, and electron-withdrawing groups would decrease the rate of the reaction.

The methoxyl group is strongly electron donating by resonance.

The methyl group is weakly electron donating by induction.

The chlorine is electron withdrawing by induction.

The nitro group is strongly electron withdrawing by resonance.

23.27

(a)

(b)

23.27 (cont)

(b) (cont)

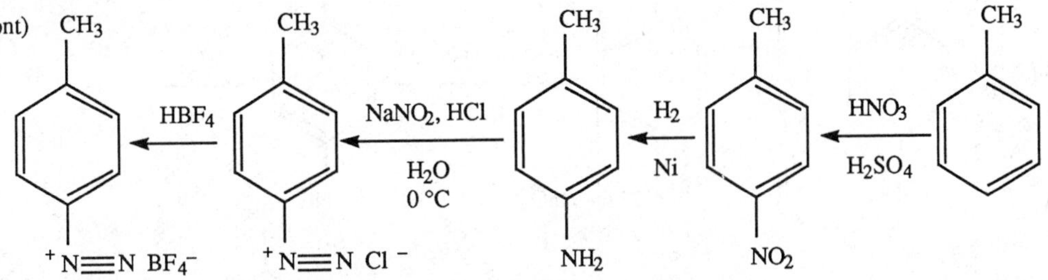

(c)

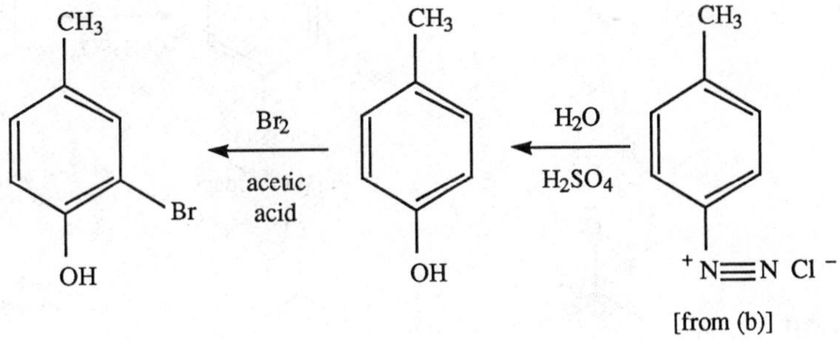

[from (b)]

(d)

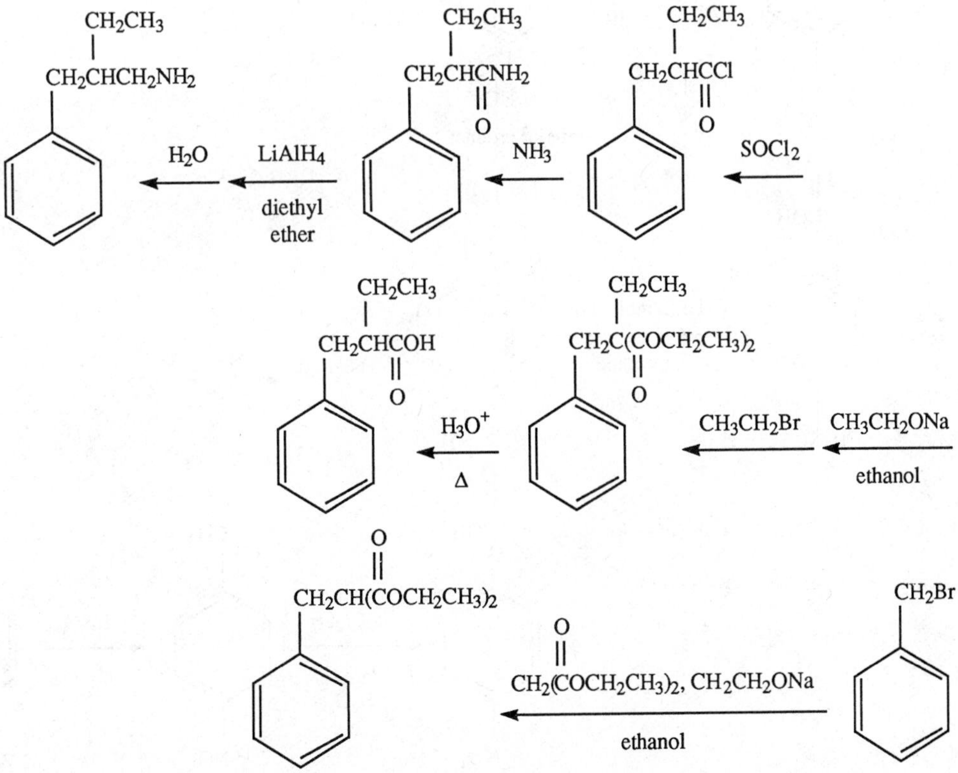

23.27 (cont)

(e)

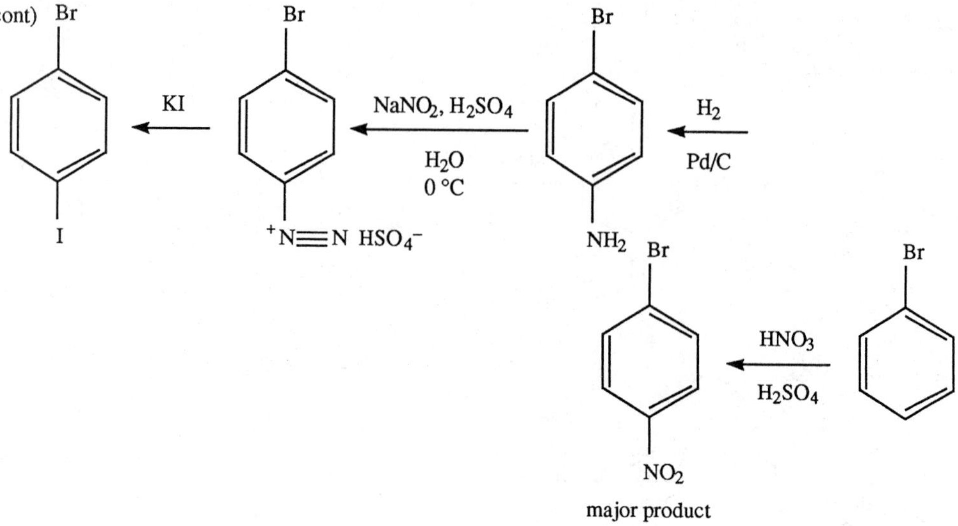

(f)

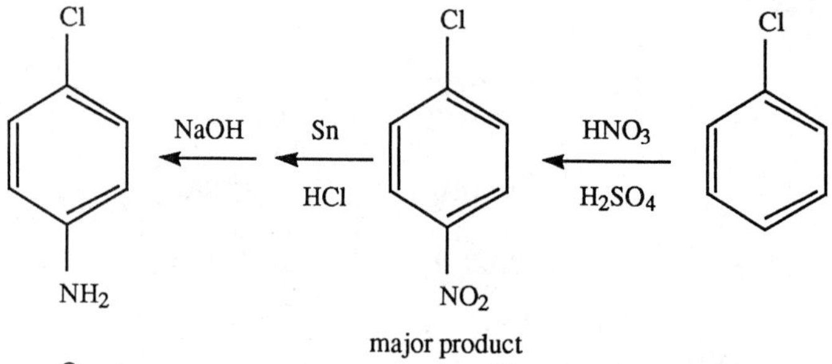

(g)

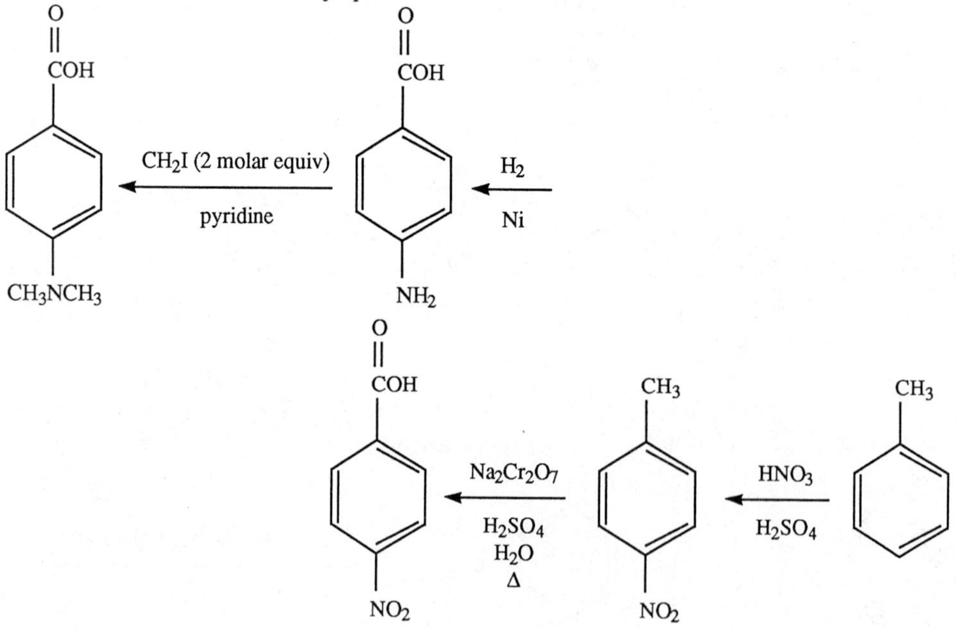

23.28   This reaction goes by way of a tetrahedral anionic intermediate.

Delocalization of the negative charge onto the para nitro group requires that the p orbitals on nitrogen, oxygen and carbon be parallel to each other.

As the R group gets larger [$H > CH_3 > (CH_3)_3C$], the nitro group can no longer remain in the plane of the ring, decreasing the p orbital overlap, and the stabilization of the intermediate anion. The large R groups ortho to the nitro group are said to cause **steric inhibition** of resonance, increasing the energy of activation, which in turn leads to a decrease in the rate of the reaction.

23.29   Iron has two electrons in the $4s$ orbital and six electrons in the $3d$ orbitals in its outermost shell. Iron(II) has lost the two electrons from the $4s$ orbital and has six electrons in the $3d$ orbitals. When it bonds to two cyclopentadienyl anions, each of which has six π electrons, iron has a total of eighteen electrons in its outermost shell, giving it the electronic configuration of the inert gas krypton, and leading to the stable compound ferrocene. Iron(III) has only five electrons in the $3d$ orbitals. Bonding with two cyclopentadienyl anions does not give iron(III) a filled shell, and, therefore, no stable compound forms.

23.30   The infrared spectrum tells us that Compound A is an alcohol (3329 cm^{-1} is the —O—H stretching frequency and 1046 cm^{-1} is the C—O stretching frequency) and Compound B is an ether (1245 and 1049 cm^{-1} are the C–O stretching frequencies). The proton magnetic resonance spectra are analyzed below:

Compound A:

δ 2.5	(2H, triplet, *J* 7 Hz, —CH$_2$CH$_2$—)
δ 3.4	(2H, triplet, *J* 7 Hz, —OCH$_2$CH$_2$—)
δ 3.7	(1H, singlet, —OH)
δ 6.7	(5H, singlet, ArH)

Compound B:

δ 1.3	(3H, triplet, *J* 7 Hz, —CH$_2$CH$_3$)
δ 3.7	(2H, quartet, *J* 7 Hz, —OCH$_2$CH$_3$)
δ 6.3 – 7.0	(5H, multiplet, ArH)

The fragments for both Compound A and Compound B add up to C$_8$H$_{10}$O, which corresponds to a molecular weight of 122, which is where the molecular ion of each compound appears in the mass spectrum. We thus have the molecular formulas for Compounds A and B. Compound A is 2-phenylethanol. The base peak for Compound A, *m/z* 91, is the tropylium ion, which results from the following fragmentation.

Compound A
2-phenylethanol

M$^{+}_{\bullet}$
m/z 122

tropylium ion
m/z 91

From the infrared spectrum we know that Compound B is an ether. The quartet at δ 3.7, and the triplet at δ 1.3 in the proton magnetic resonance spectrum tells us that there is an isolated ethyl group in which the methylene group is attached to an oxygen atom. Compound B is ethyl phenyl ether (ethoxybenzene). The base peak in Compound B, *m/z* 94, is 28 mass units less than the molecular ion and corresponds to a rearrangement in which ethylene, C$_2$H$_4$, is lost.

23.30 (cont)

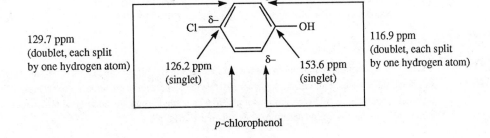

Compound B
ethyl phenyl ether
ethoxybenzene

M $\overset{+}{\underset{\bullet}{}}$
m/z 122

m/z 94

23.31   The increase in λ_{max} in the ultraviolet spectra in basic solution compared with the spectrum in acidic solution tells us that Compound C is a phenol. The presence of the hydroxyl group is confirmed by the broad band in the infrared spectrum at 3400 cm^{-1}.

The proton magnetic resonance data is summarized below:

δ 5.6   (1H, singlet, ArO$\underline{\text{H}}$)
δ 6.1 – 6.9   (4H, para substitution pattern, Ar$\underline{\text{H}}$)

The data so far tell us that Compound C is a para-substituted phenol. The C-13 nuclear magnetic resonance spectrum has only four bands instead of six in the aromatic region, suggesting that the benzene ring is symmetrically substituted, such as in para substitution.

The mass spectrum shows two molecular ions at *m/z* 128 and 130, with relative intensities of 3:1. The relative intensities suggest that Compound C contains chlorine. Putting together all the information gives us *p*-chlorophenol for the identity of Compound C.

The C-13 nuclear magnetic resonance spectrum is analyzed below.

129.7 ppm
(doublet, each split
by one hydrogen atom)

126.2 ppm
(singlet)

153.6 ppm
(singlet)

116.9 ppm
(doublet, each split
by one hydrogen atom)

p-chlorophenol

23.32   Integration of the bands in the proton magnetic resonance spectrum gives a ratio of 4:2:2:3. Disappearance of the broad band at δ 3.2 when Compound D is shaken with D_2O tells us that those hydrogens must be acidic protons attached to an oxygen or a nitrogen atom. The decrease in λ_{max} in the ultraviolet spectra in acid solution compared with basic solution tells us that Compound D is an aryl amine. The two bands in the infrared spectrum at 3480 and 3376 cm-1 point to the presence of a primary amine and the band at 1621 cm^{-1} confirms the presence of the aromatic ring.

The proton magnetic resonance data is summarized below:

δ 1.1	(3H, triplet, *J* 7 Hz, —CH$_2$C<u>H</u>$_3$)
δ 2.3	(2H, quartet, *J* 7 Hz, ArC<u>H</u>$_2$CH$_3$)
δ 3.2	(2H, broad singlet, —N<u>H</u>$_2$)
δ 6.0 – 6.7	(4H, Ar<u>H</u>)

The data so far tells us that Compound D is an ethylaniline. The aromatic region in the proton magnetic resonance spectrum does not show the typical splitting pattern for para substitution, therefore Compound D must be either ortho or meta substituted. The C-13 nuclear magnetic resonance spectrum has eight bands, which also tells us that Compound D cannot be *p*-ethylaniline, which should have only six bands in the C-13 spectrum because of the symmetry of the molecule. The data, however, does not allow us to distinguish between *o*-ethylaniline and *m*-ethylaniline.

23.33   The absorption bands reported for the first compound are comparable to those for an alkylbenzene (compare with the spectrum of toluene, Figure 23.3). This suggests that the double bond in this compound is not conjugated with the aromatic ring. The first compound is 3-phenylpropene. In the spectrum of the second compound, the absorption bands are shifted to longer wavelengths and are also more intense, suggesting that this compound has a more extensive system of conjugation than the first one. The second compound is, therefore, 1-phenylpropene.

3-phenylpropene

no conjugation of the
double bond with the ring

1-phenylpropene

double bond in con-
jugation with the ring

23.34   (a)   The highest wavelength of maximum absorption for Vitamin B$_6$ shifts from 292 nm to 315 nm when the solution is made basic. This change suggests that the vitamin contains a phenolic hydroxyl group. When the acidic proton of the hydroxyl group is replaced by a methyl group, the compound no longer ionizes at high pH and the absorption spectrum does not change with pH.

23.33 (cont)

(b) The structure of Vitamin B$_6$ can be derived from its molecular formula, $C_6H_{11}NO_3$, and from the structures of the oxidation products of its methyl ether.

dicarboxylic acid
$C_9H_9NO_5$

precursor of
the lactone
$C_9H_{11}NO_4$

lactone

Vitamin B$_6$
$C_8H_{11}NO_3$

methyl ether
of Vitamin B$_6$
$C_9H_{13}NO_3$

The side chains that are oxidized are the ones already bearing a hydroxyl group. Oxidation of an alcohol is easier than oxidation of an alkyl group. The reaction conditions, potassium permanganate at room temperature, are not vigorous enough to oxidize the alkyl group. The equations outlining the transformation of Vitamin B$_6$ are given on the previous page.

(c) Possible models for Vitamin B$_6$ should have a pyridine ring with a hydroxyl substituent. Alkyl substituents at different positions on the ring would also be helpful. Ultraviolet spectra of the model compounds should be taken under both acidic and basic conditions. Structures of some possible models are:

23.35

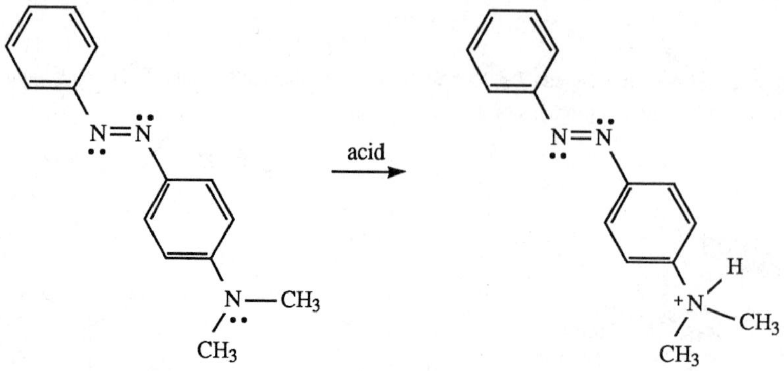

butter yellow
λ_{max} 408 nm

λ_{max} 320 nm

Protonation of the nonbonding electrons on
the amino group decreases the conjugation
in the chromophore and, therefore, the
wavelength of maxiumu absorption

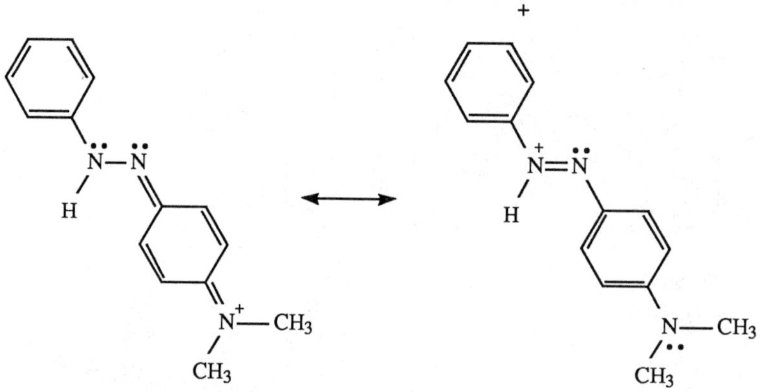

The chromophore in this
species has quinoid character.

λ_{max} 510 nm

24

The Chemistry of Heterocyclic Compounds

24.1 (a) 3-aminopyridine

(c) ethyl 1-pyrrolecarboxylate

(e) 3-methyl-2-isoxazoline
(Note that the number 2 defines the position of the double bond.)

(g) 4,4-dimethylazetidinone

(i) 3-methylfuran

(k) 1,4-dimethylpyrazole

(m) 2,3-diphenyloxazole

(b) 2,5-dimethylthiophene

(d) 4-ethylisoquinoline

(f) 1-ethylpyridine

(h) 3-indolecarboxylic acid

(j) 8-hydroxyquinoline

(l) 3-thiophenecarboxamide

24.2 (a)

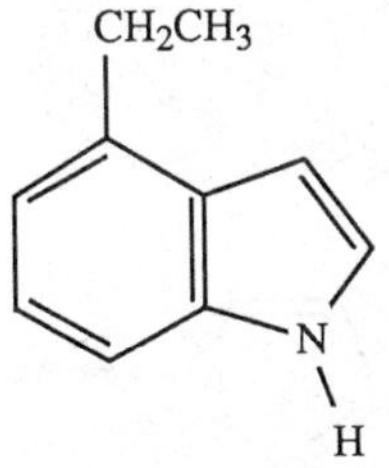

(b)

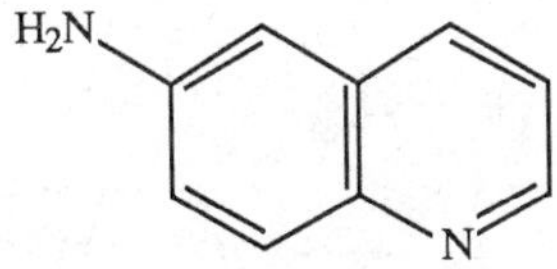

(c)

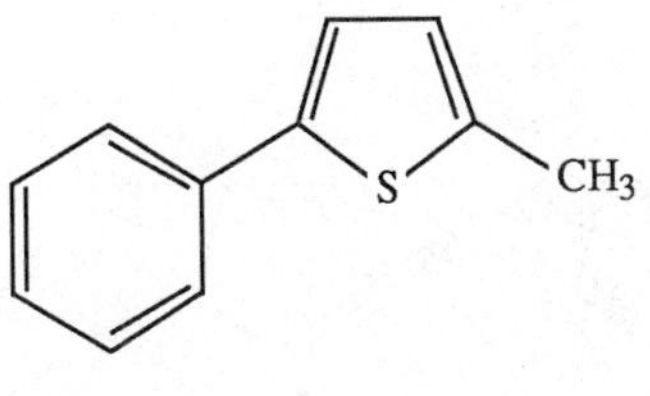

(d)

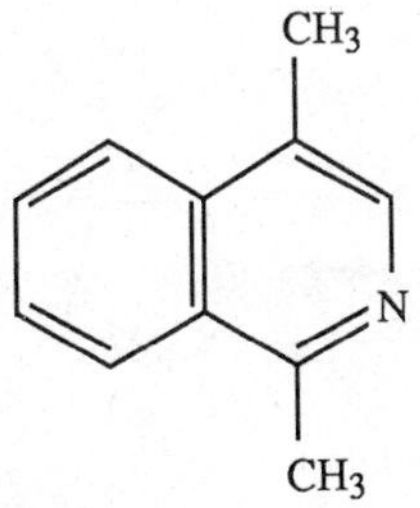

24.2 (cont)

(e)

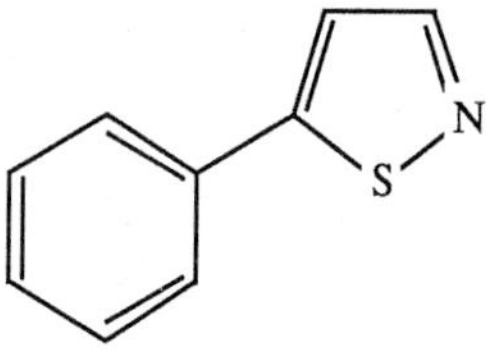

(f)

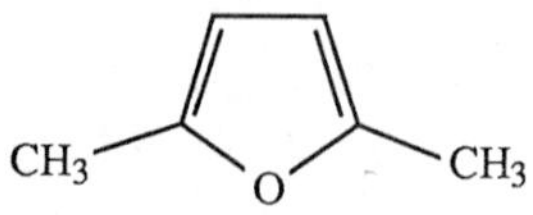

(g)

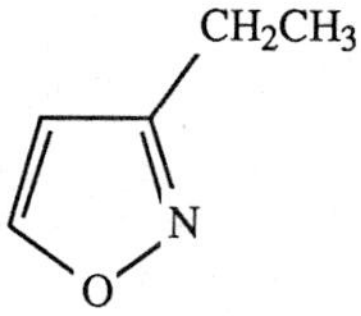

(h)

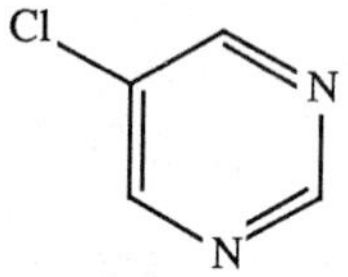

(i)

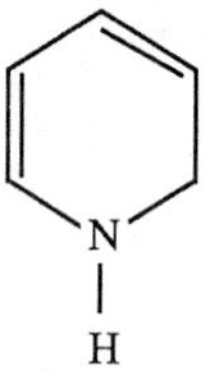

(j)

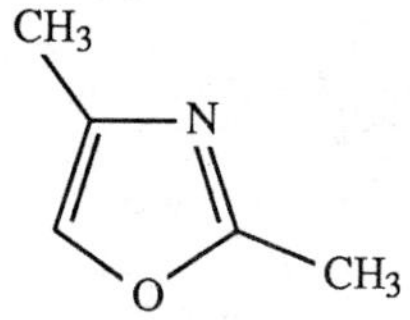

(k)

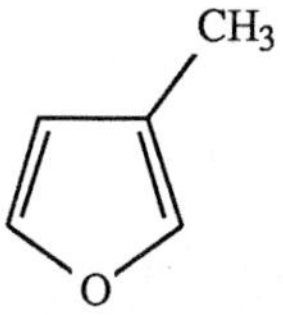

(l)

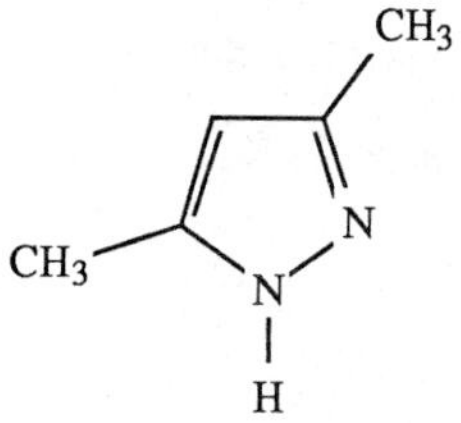

(m)

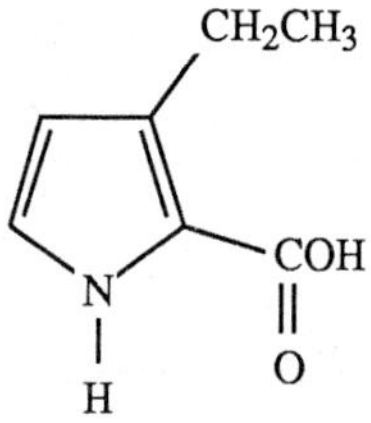

24.3

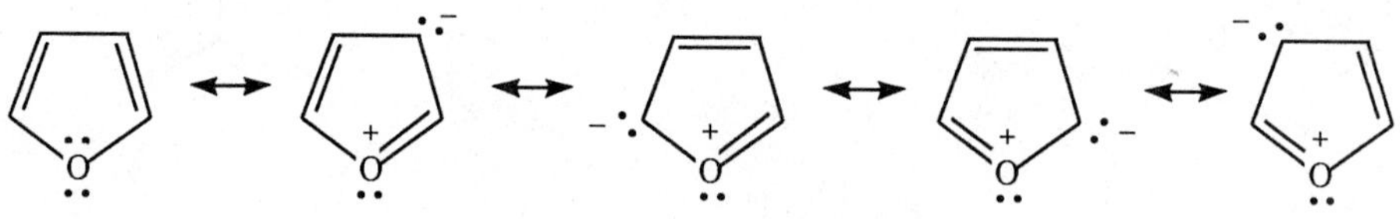

major

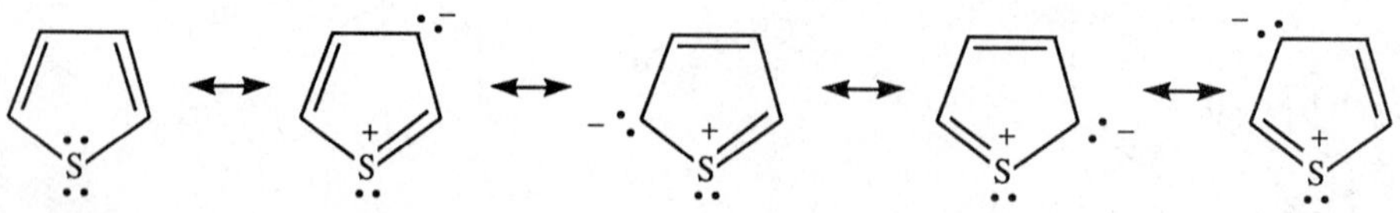

major

24.3 (cont)

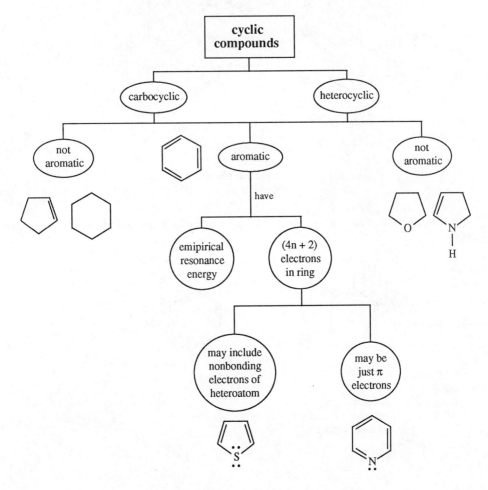

major

major

- -

Concept Map 24.1   Classification of cyclic compounds.

24.4   (a)   The basicity of an amine is determined by the availability of the nonbonding electrons on the nitrogen atom. The pK_a values for their conjugate acids indicate that 2-methylpyridine is a slightly stronger base than pyridine, and 3-nitropyridine is a much weaker base than pyridine. The methyl group is weakly electron donating; the nitro group is strongly electron withdrawing.

pK_a of con-
jugate acid   6.0   5.2

strongest base;
inductive effect of
the methyl group

pK_a of con-
jugate acid

8.0

weakest base;
inductive and
resonance effect of
the nitro group

(b)   The arguments used in part (a) of this problem can be applied to the compounds in this series also.

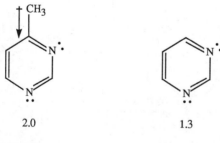

pK_a of con-
jugate acid   2.0   1.3

strongest base;
inductive effect of
the methyl group

24.4 (cont

(b) (cont)

pK$_a$ of con-
jugate acid

-0.68

weakest base;
inductive and
resonance effect of
the carbonyl group

(c)   The cyano group withdraws electrons from the ring and from the nitrogen atom of the substituted pyridines. Both inductive and resonance effects operate. The basicity of the pyridines appears to depend on the distance of the cyano group from the nitrogen atom in the ring.

pK$_a$ of con-   1.9   1.5 -0.3
jugate acid

strongest base;   weakest base;
cyano group farthest   cyano group closest
from the nitrogen   to the nitrogen
atom in the ring   atom in the ring

24.5

(a) $CH_3CCH_2CH_2CCH_3$ $\xrightarrow[\Delta]{P_4S_{10}}$

(b) $HCCH_2CH_2CH$ $\xrightarrow{HCl}$

(c) $CH_3CCH_2CH_2CCH_3$ $\xrightarrow[\Delta]{P_4Se_{10}}$

selenium is in the same
family as oxygen and sulfur

24.5 (cont)

(d)

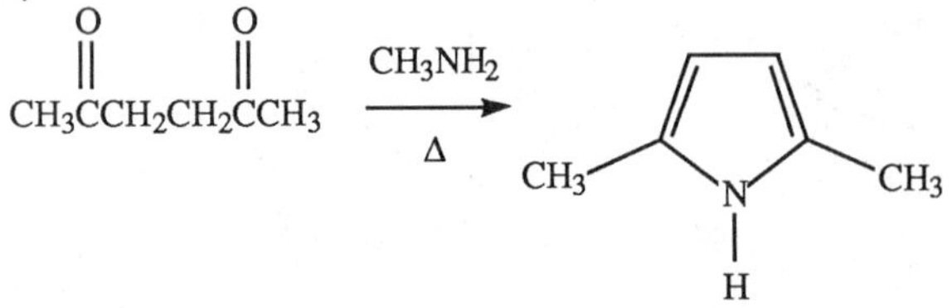

24.6

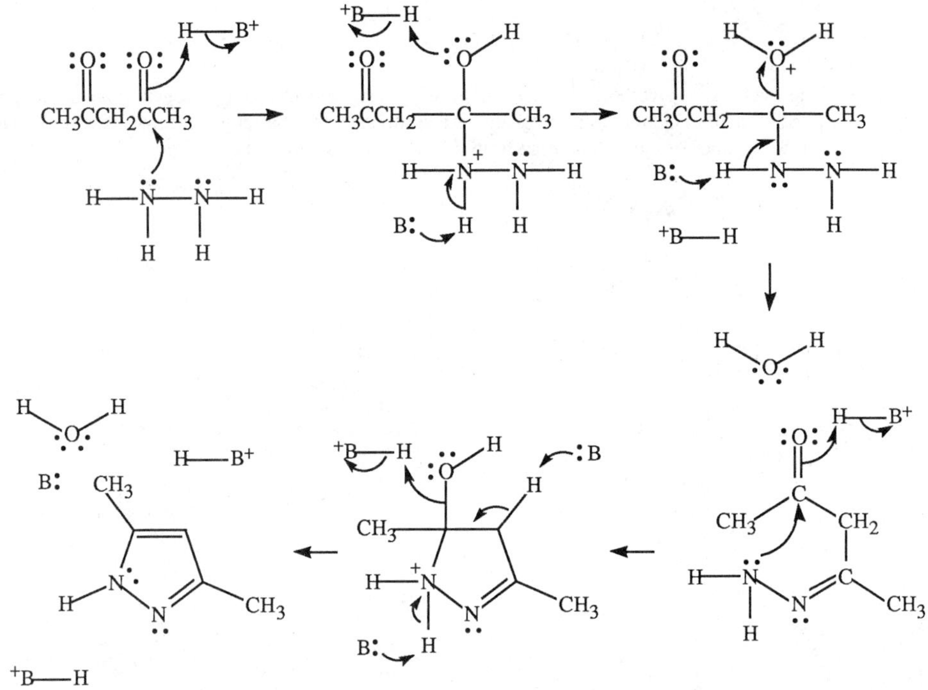

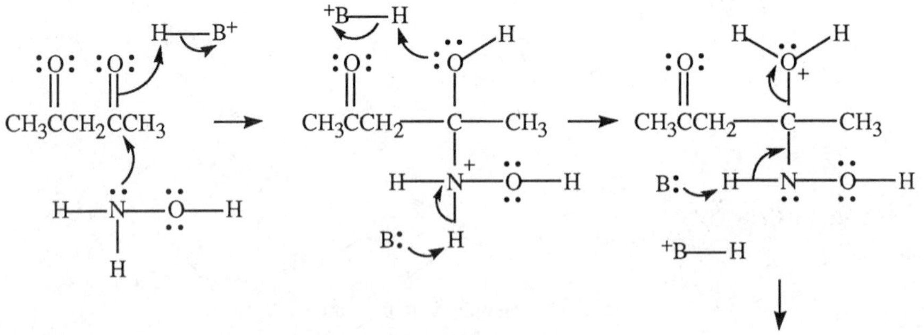

24.6 (cont)

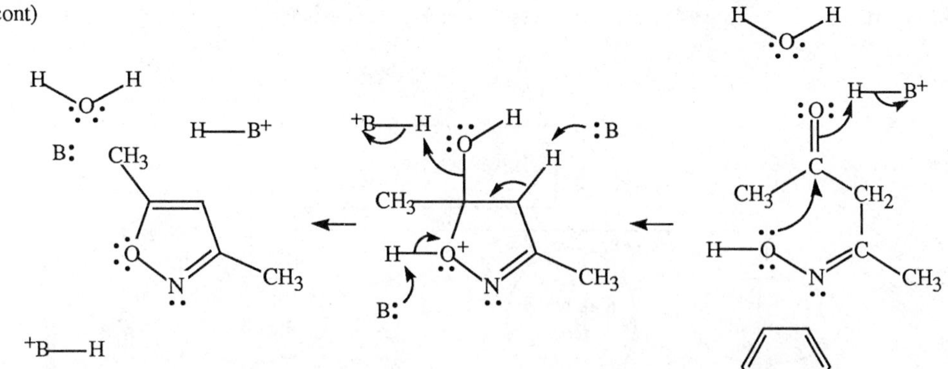

H—B⁺

B: CH₃

H—B⁺

⁺B—H   ⁺B—H   :B

CH₃

CH₃

CH₃

B:

⁺B—H

CH₃

CH₃   CH₂

CH₃

24.7

(a) CCH_2C $HONH_3 Cl^-$, NaOH Δ

(b) $C-C$ + CH_3CHCH CH_3 $CH_3CO^{-+}NH_4$ acetic acid Δ $CHCH_3$ CH_3 H

(c) $HCCH_2Br$ + CNH_2 NH   ethanol Δ

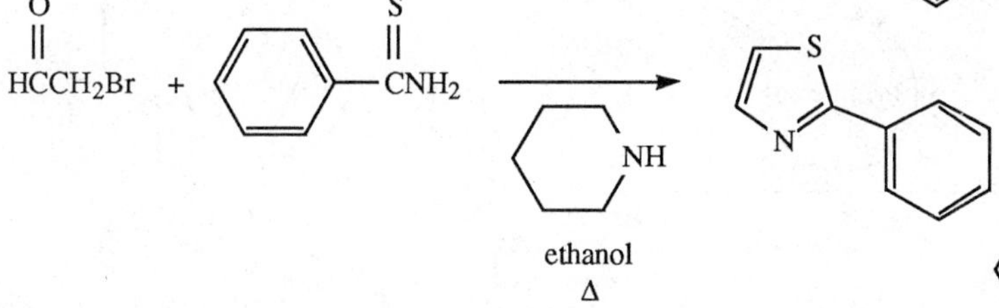

(d) CCH_2CCH_3 + $NHNH_2$ Δ CH_3 less hindered carbonyl group

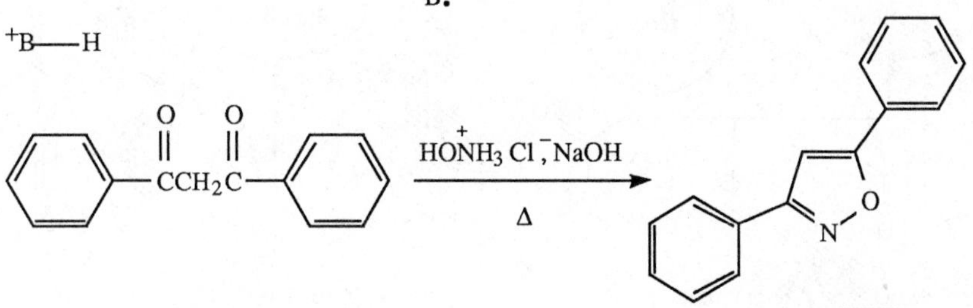

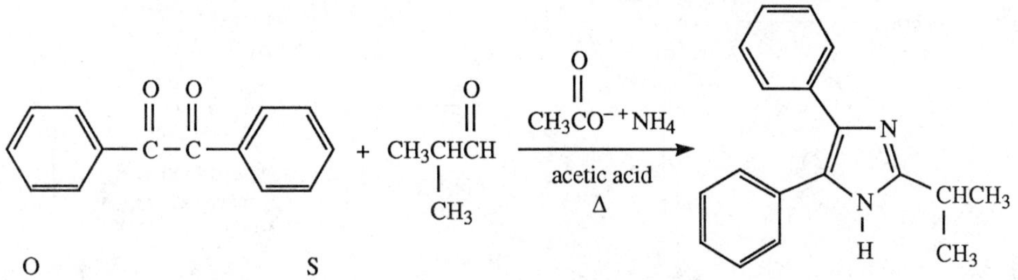

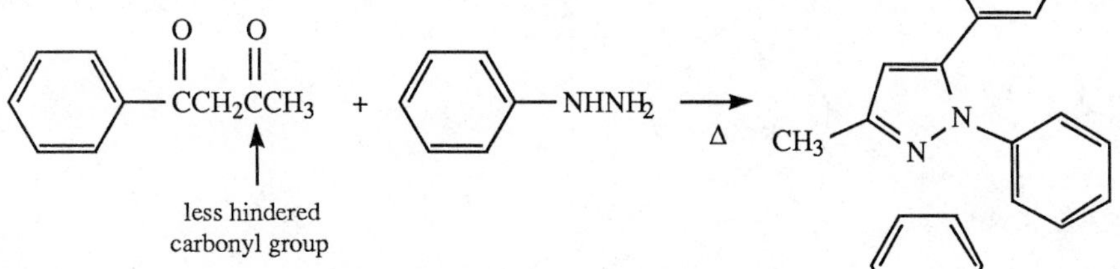

(e) CCH_2CCH_3 $N_2HNH_3 HSO_4^-$, NaOH Δ CH_3 H

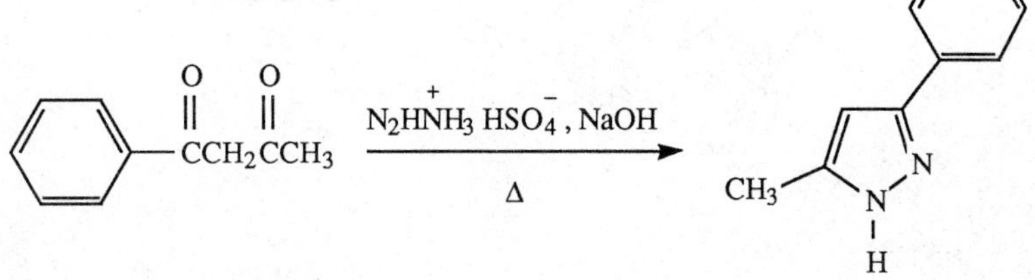

Concept Map 24.2   Synthesis of heterocycles from carbonyl compounds.

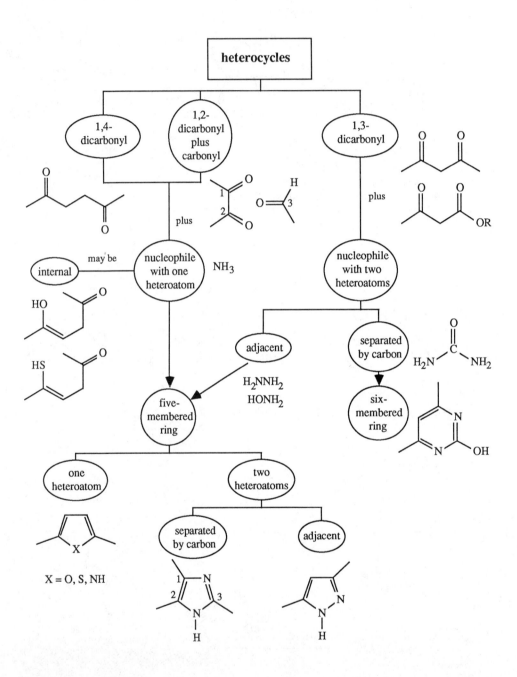

24.8

(a)

$$CH_3CCH_2COCH_2CH_3 + H_2NCNH_2 \xrightarrow[\text{ethanol}]{\text{HCl}}$$

(with the two left carbonyls shown as O double bonds, and urea H₂NC(=O)NH₂; product pyrimidine bearing OH, N, CH₃, N, OH)

(b)

$$CH_3CCH_2CCH_3 + CH_3ONHCNH_2 \xrightarrow[\text{ethanol}]{\text{HCl}}$$

(product pyrimidine bearing CH₃, OCH₃, N, CH₃, N, O)

(c)

$$\text{C}_6H_5\text{-CCH}_2\text{CCH}_3 + H_2NCNH_2 \xrightarrow[\text{ethanol}]{\text{HCl}}$$

(product pyrimidine bearing CH₃, N, phenyl, N, OH)

24.9

(a)

$$\text{thiophene} + CH_3CH_2CCl \xrightarrow{(CH_3CH_2)_2O\cdot BF_3}$$

(product A: 2-thienyl—C(=O)CH₂CH₃)

A

(b)

$$\text{(N-methylpyrrole)} + HNO_3 \xrightarrow{\text{acetic anhydride}}$$

(product B: N-methylpyrrole—NO₂)

B

(c)

$$\text{(2-bromothiophene)} \xrightarrow[\text{ether}]{\text{Mg}} \text{(2-thienyl—MgBr)}$$

C

$$\xrightarrow{CO_2}$$

(D: 2-thienyl—$CO^-Mg^{2+}Br^-$)

D

$$\xrightarrow{H_3O^+}$$

(E: 2-thienyl—COH)

E

24.9 (cont)

(d)

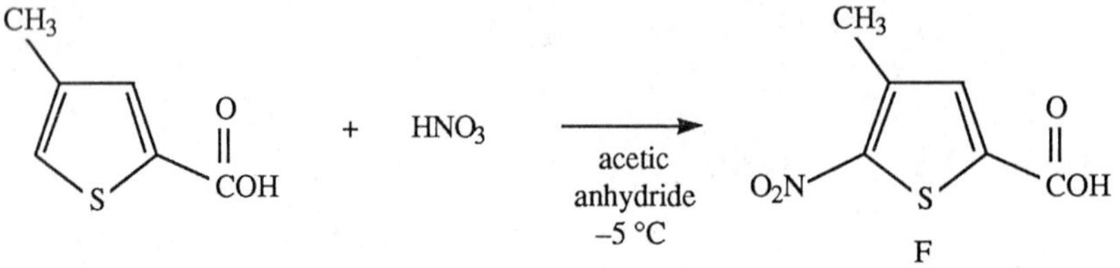

(e)

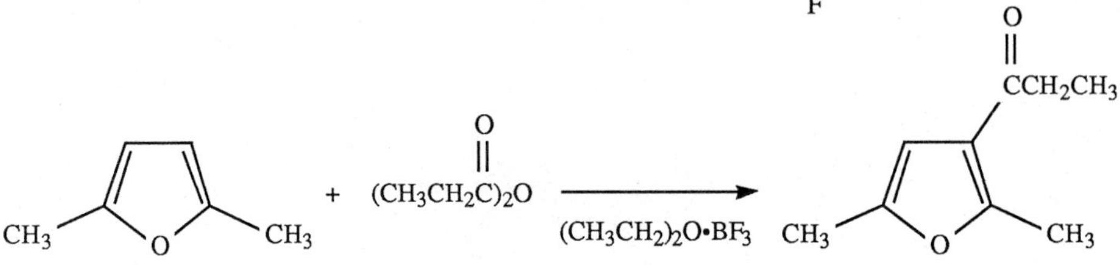

(f)

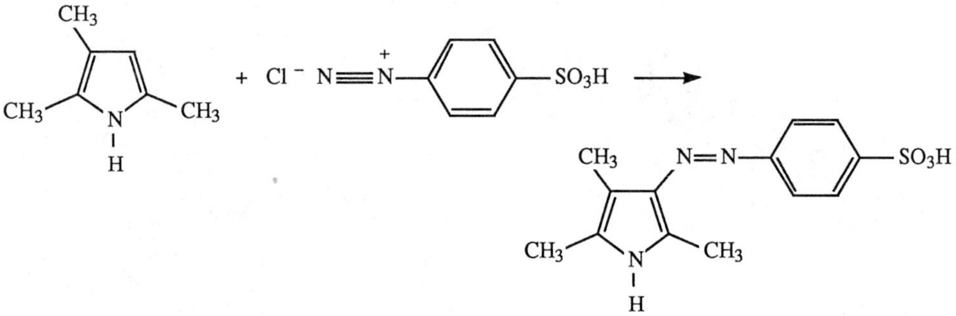

(g)

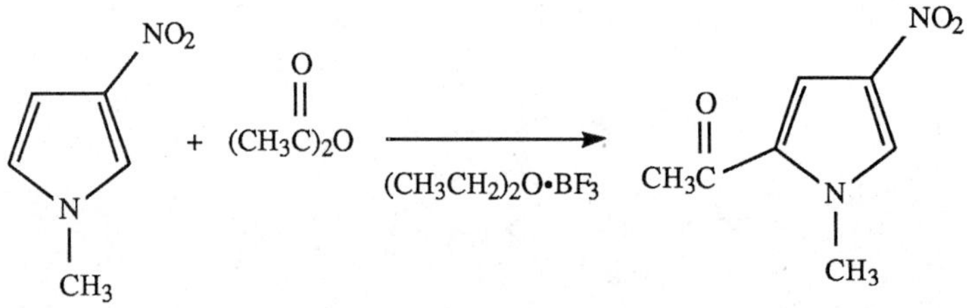

(h)

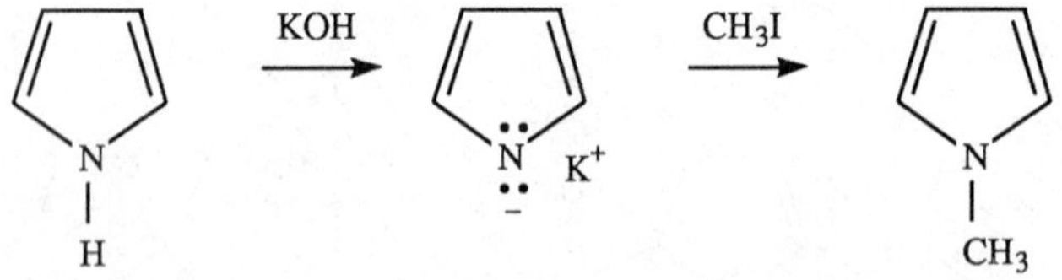

24.9 (cont)

(i)

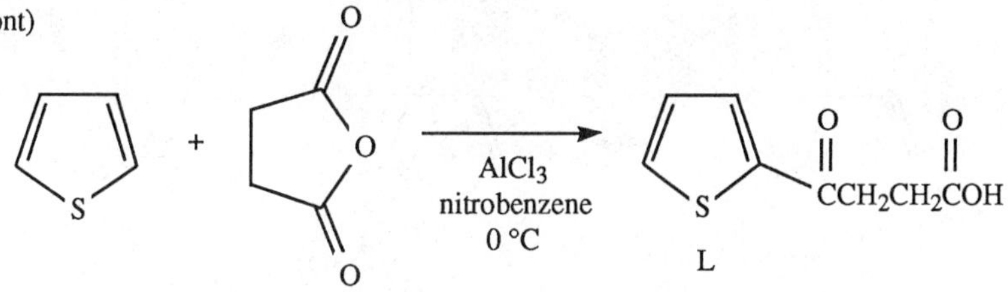

(j)

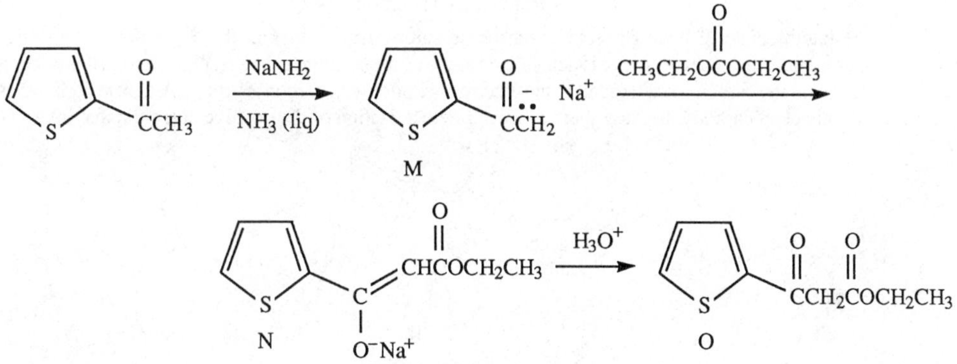

(k)

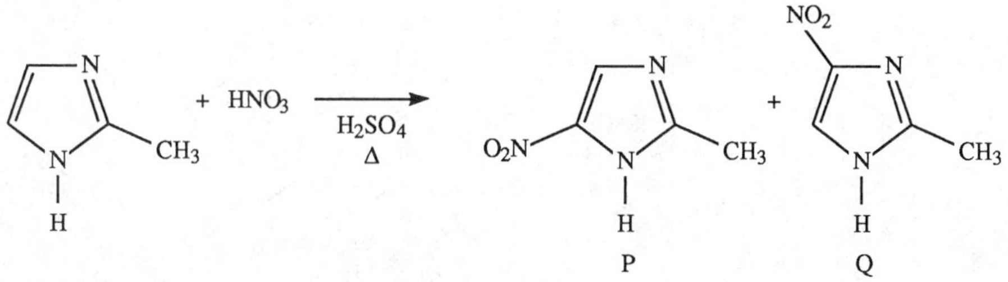

24.10   Three possible intermediates can form:

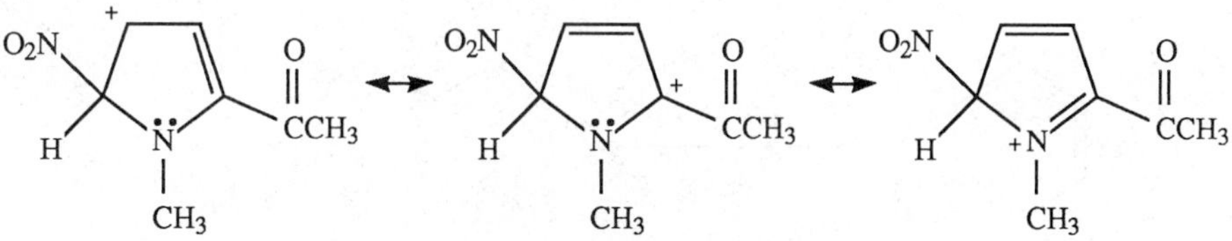

from attack at carbon 5

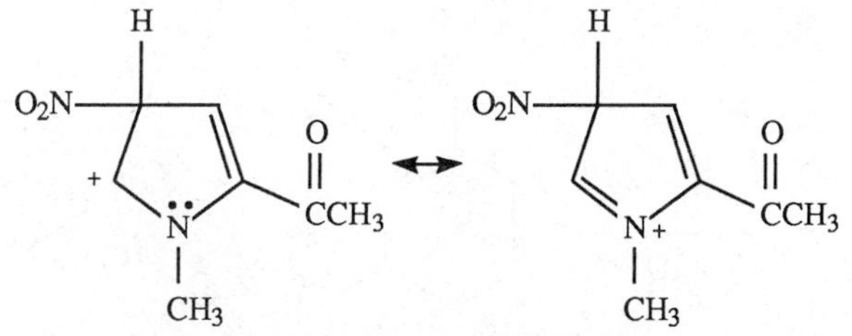

from attack at carbon 4

24.10 (cont)

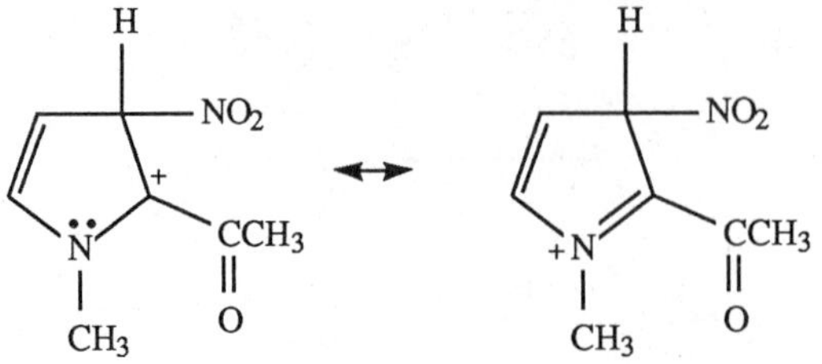

from attack at carbon 3

The intermediate of lowest energy is the one resulting from attack at the 4 position of the ring. In this intermediate the positive charge is delocalized to the nitrogen atom and no resonance contributor has a positive charge on the carbon atom bonded to the electron-withdrawing carbonyl group. Attack at carbon 5 is favored over attack at carbon 3 because there is greater delocalization of the positive charge in that intermediate.

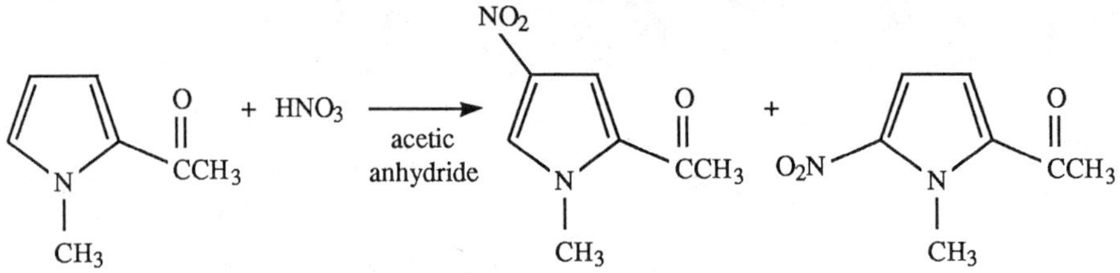

major product

24.11

(a)

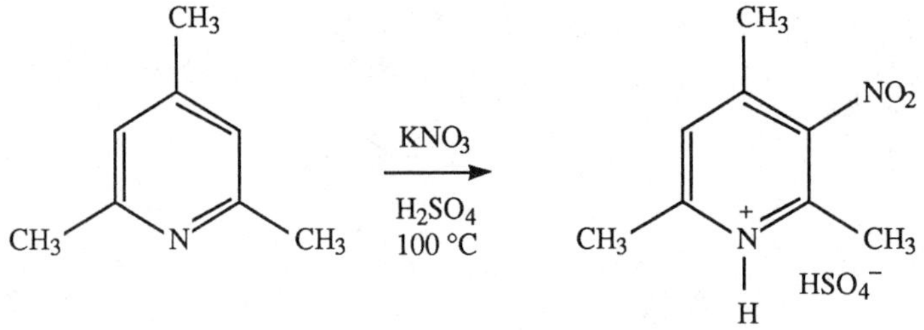

A

(b)

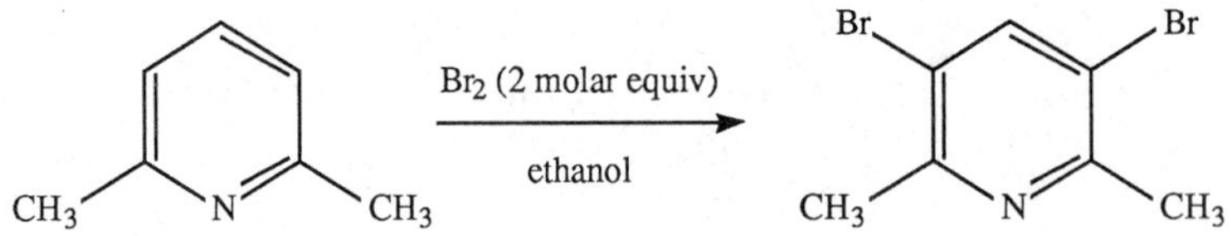

B

(c)

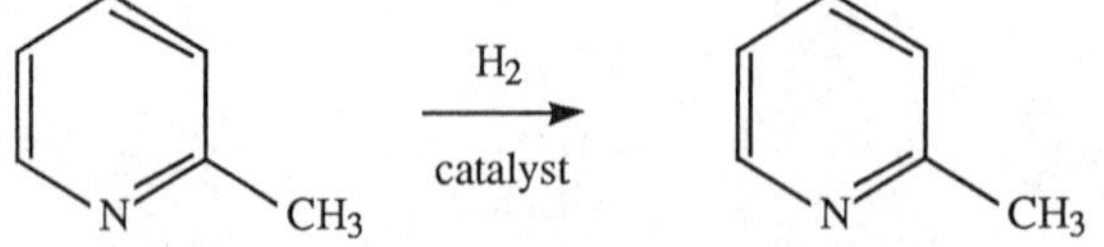

C

24.11 (cont)

(d)

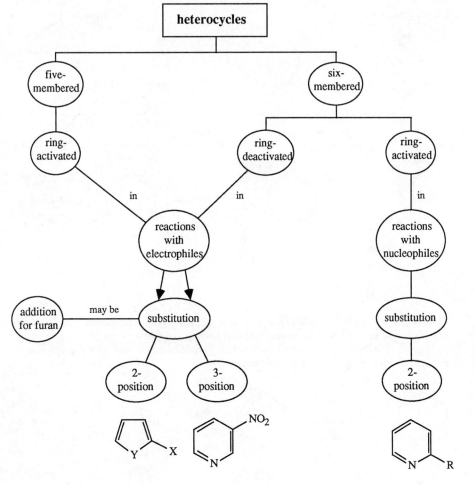

Concept Map 24.3   Substitution reactions of heterocycles.

24.12

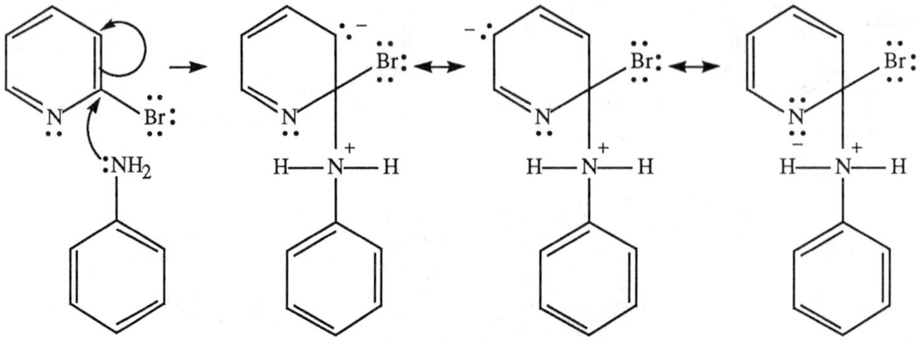

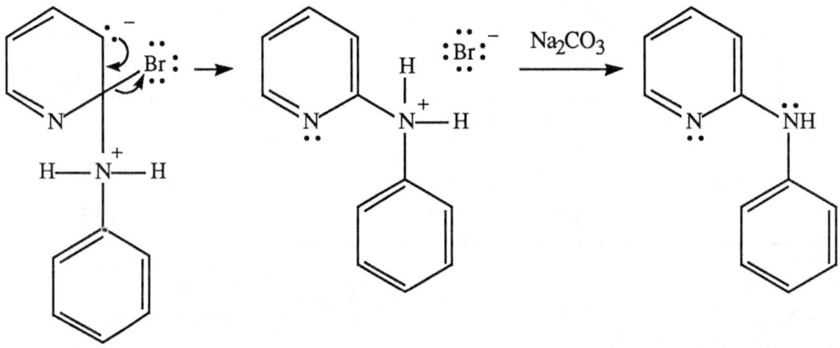

Concept Map 24.3   (see p. 793)

24.13

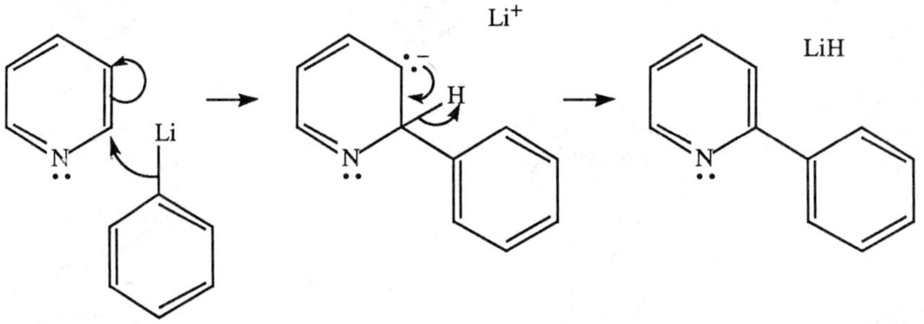

The reaction starts with attack of the organolithium reagent (a very strong nucleophile) on the carbon adjacent to the nitrogen atom to give a resonance-stabilized intermediate (see Problem 24.12). The leaving group is hydride ion. The driving force for the loss of hydride ion in this reaction and in the reaction of amide ion with pyridine is the recovery of aromaticity of the pyridine ring.

24.14

(a)

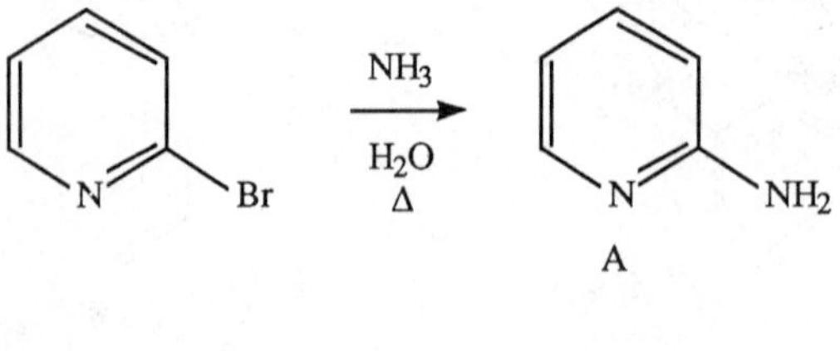

24.14 (cont)

(b)

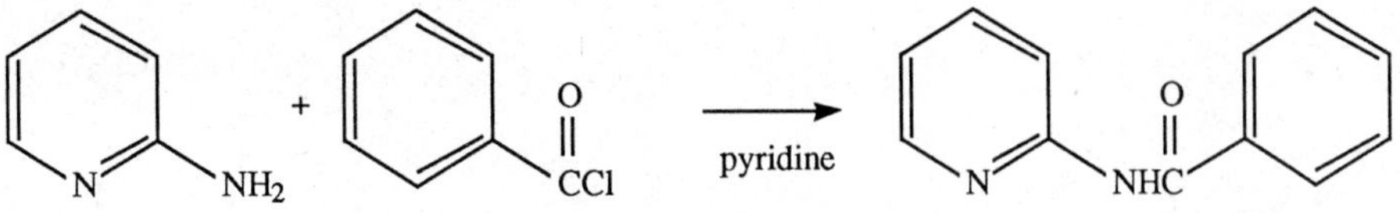

B

(c)

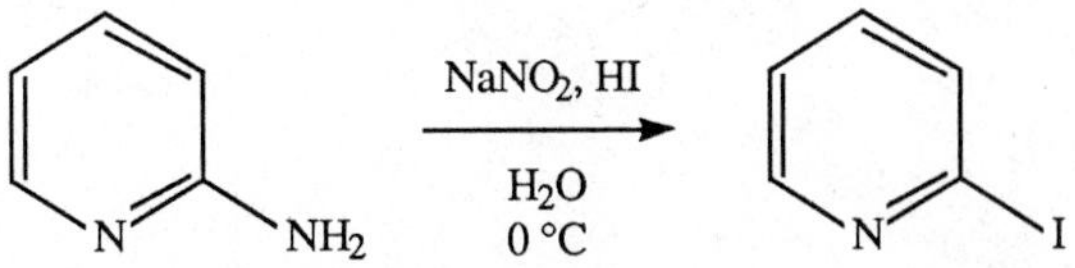

C

(d)

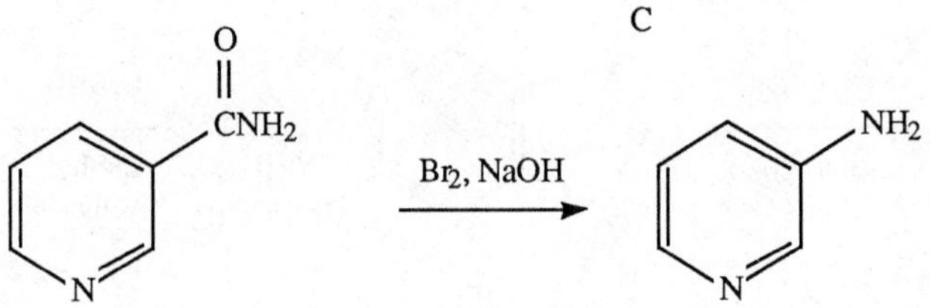

D

(e)

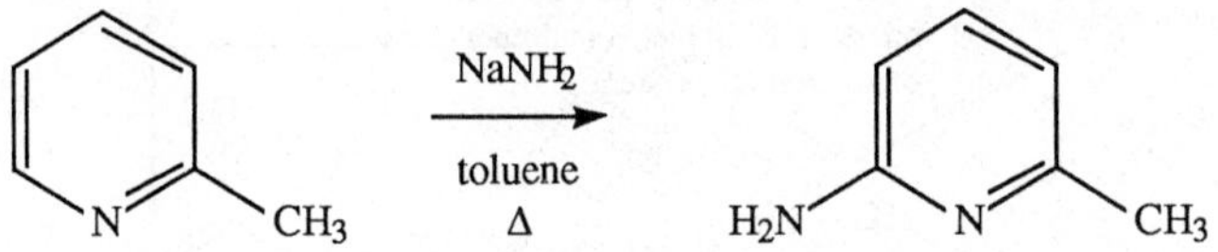

E

(f)

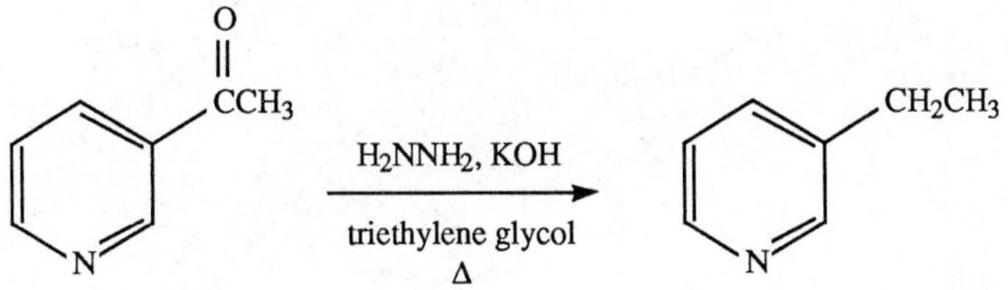

F

(g)

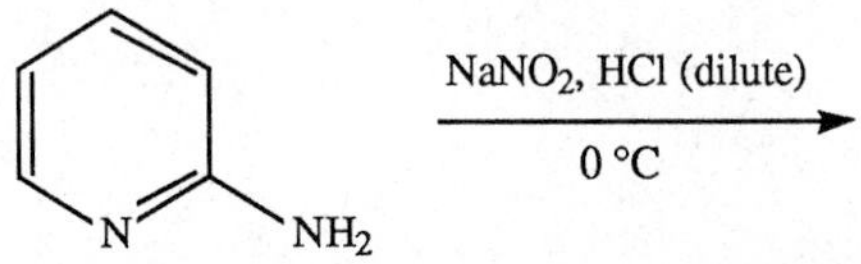

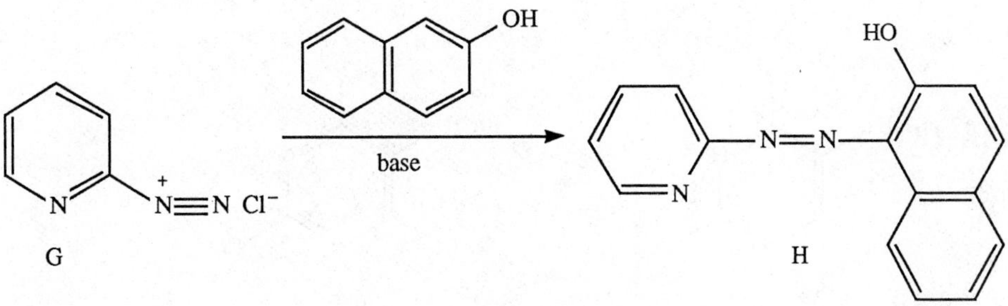

G   H

24.15

(a)

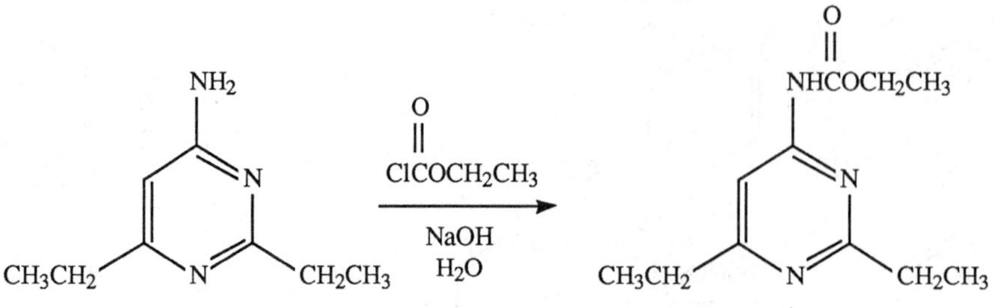

A

(b)

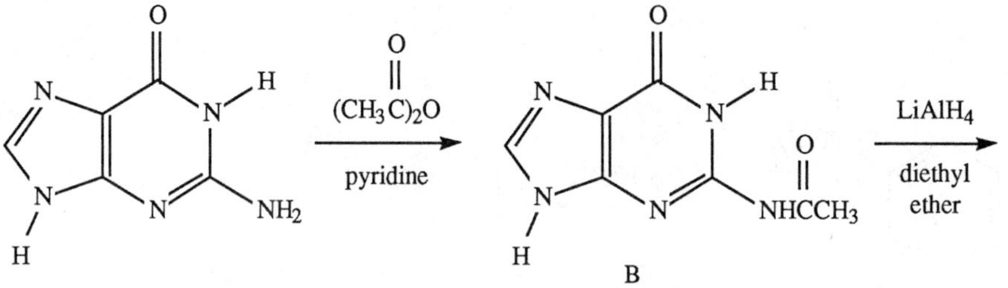

B

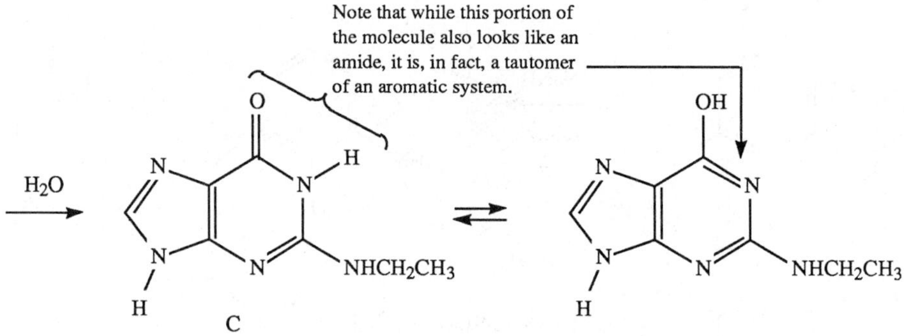
Note that while this portion of the molecule also looks like an amide, it is, in fact, a tautomer of an aromatic system.

C

(c)

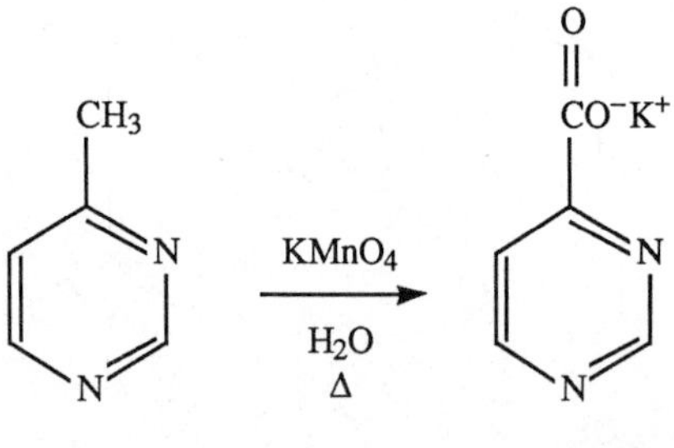

D

(d)

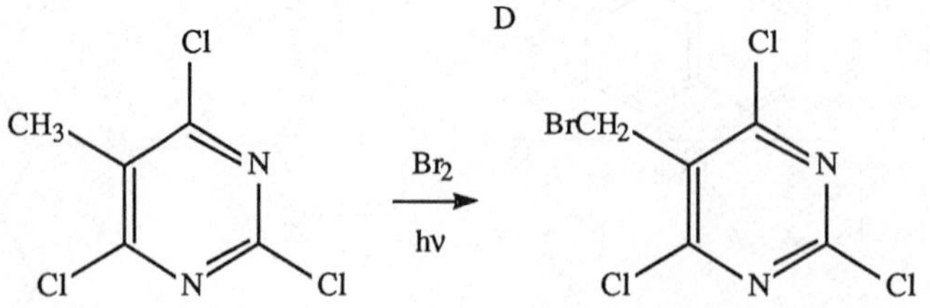

E

24.15 (cont)

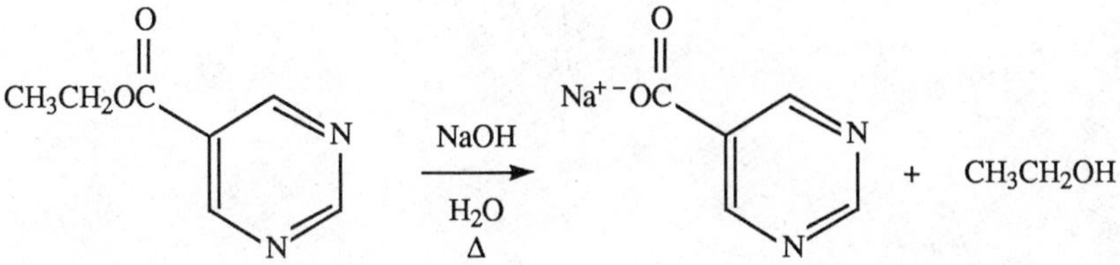

(e)

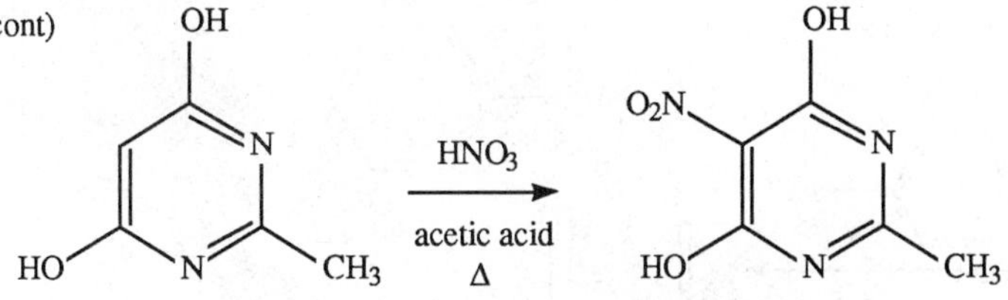

(f)

(g)

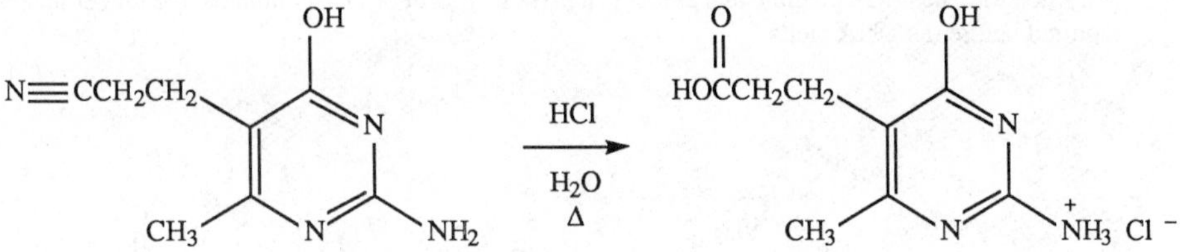

(h)

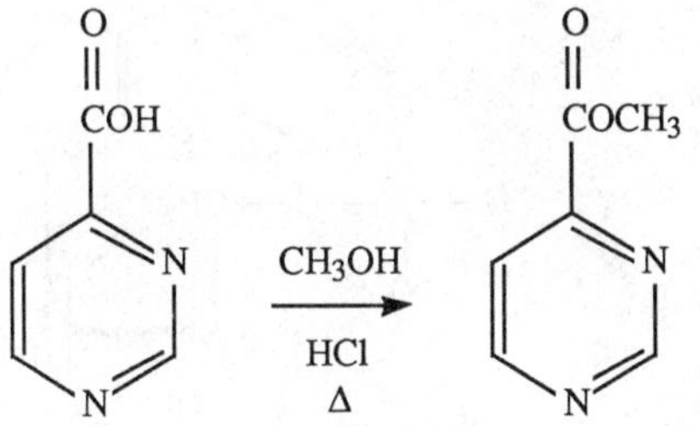

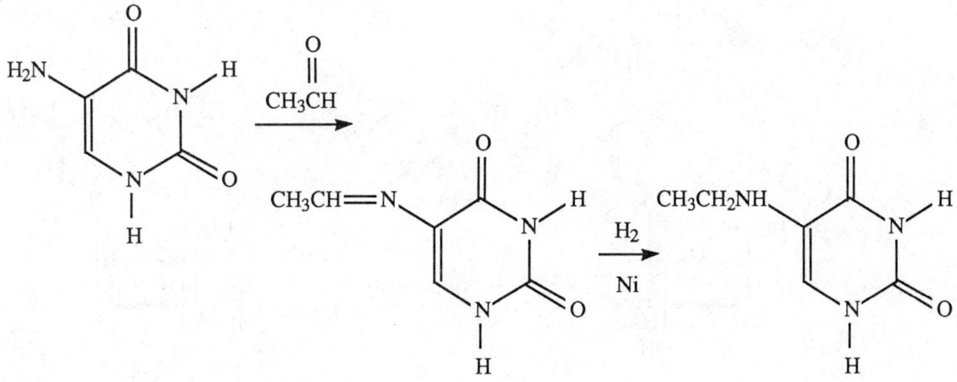

(i)

24.15 (cont)

(j)

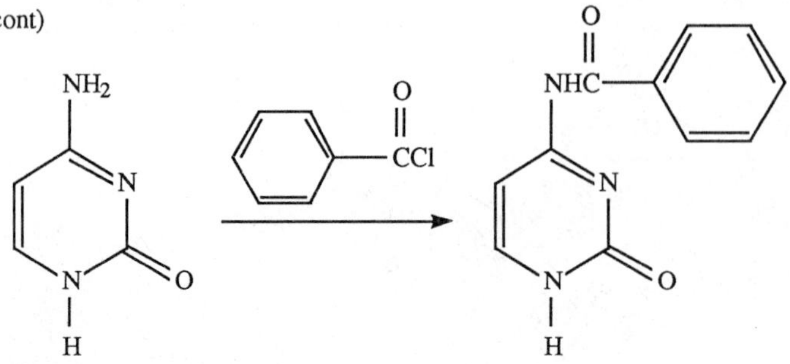

24.16   The adenine-thymine pair is held together by two hydrogen bonds, and the guanine-cytosine pair is held together by three hydrogen bonds. It takes more energy to pull apart a guanine-cytosine pair than an adenine-thymine pair, therefore, the more guanine and cytosine a particular form of DNA contains, the higher the temperature required before the DNA melts.

24.17   (a)

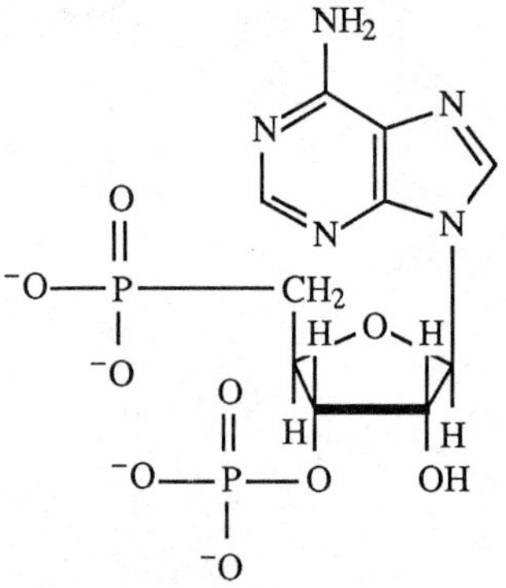

(b)

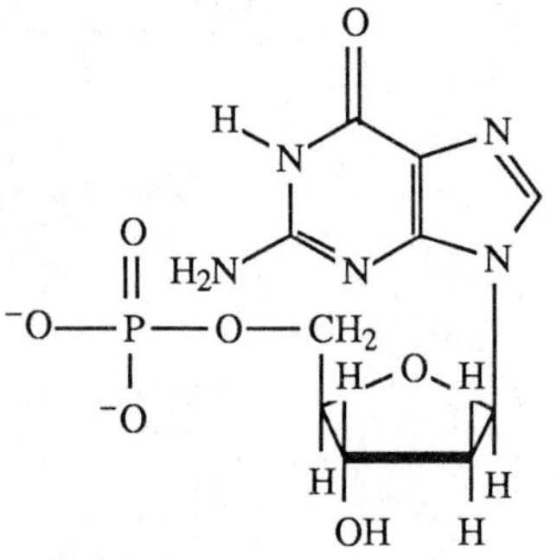

(c)

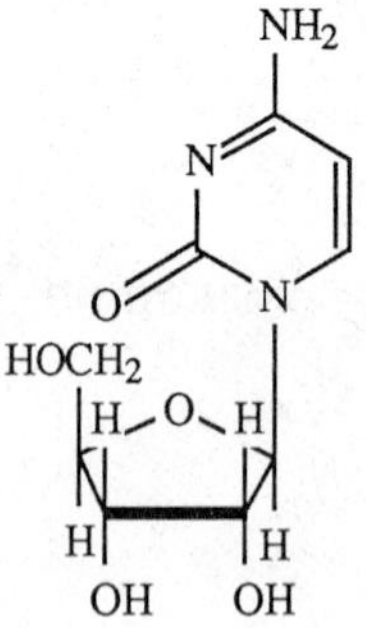

(d)

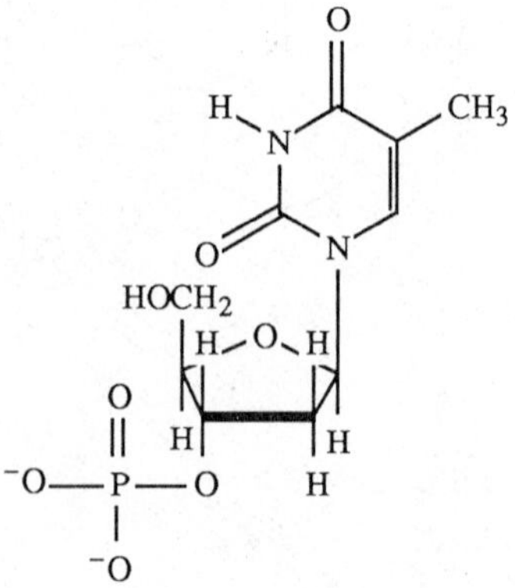

24.17 (cont)

(e)

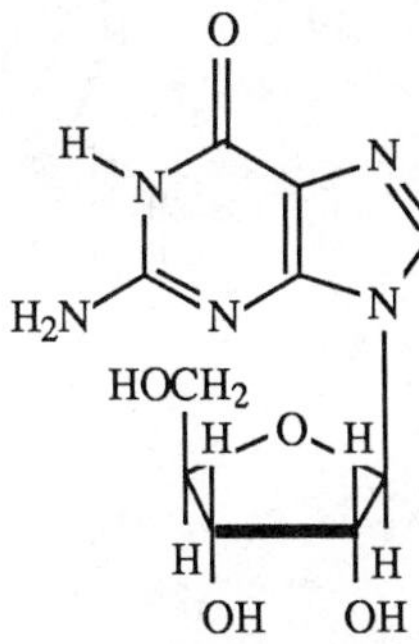

(f)

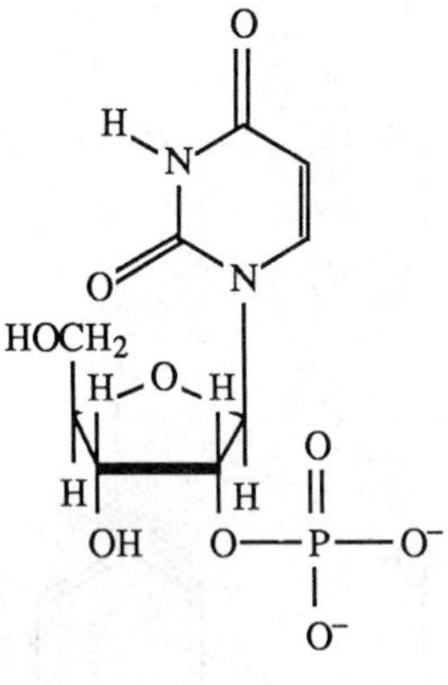

24.18

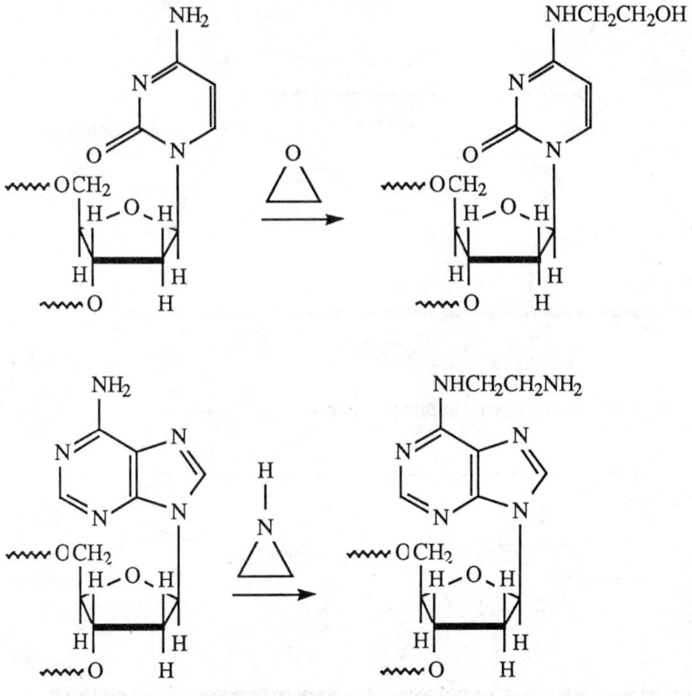

The heterocycles with three-membered rings are vulnerable to nucleophilic attack. The bases in DNA contain nucleophilic sites, such as amino groups. Reaction of an amino group with an oxirane or an aziridine creates a new functionality at the sites in DNA that were previously involved in the hydrogen bonding so important to base pairing. Oxiranes and aziridines, therefore, disrupt the processes necessary for the transmission of genetic information and cause mutations.

24.19

$$\text{(S)-ethanol-1-}d + \text{nicotinamide} + H_2O \xrightarrow[\text{dehydrogenase}]{\text{yeast alcohol}} CH_3CD + \text{product} + H_3O^+$$

(S)-ethanol-1-d   R

$$\text{(R)-ethanol-1-}d + \text{nicotinamide} + H_2O \xrightarrow[\text{dehydrogenase}]{\text{yeast alcohol}} CH_3CH + \text{product} + H_3O^+$$

(R)-ethanol-1-d   R

24.20

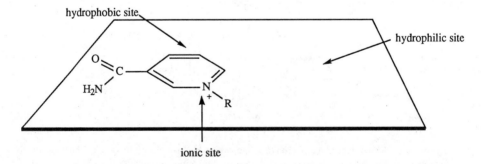

Nicotinamide fits the active site of the enzyme in this orientation.

Nicotinamide does not fit the active site of the enzyme in this orientation.

The two orientations of the nicotinamide molecules as drawn are enantiomers of each other and are distinguished from each other by the enzyme.

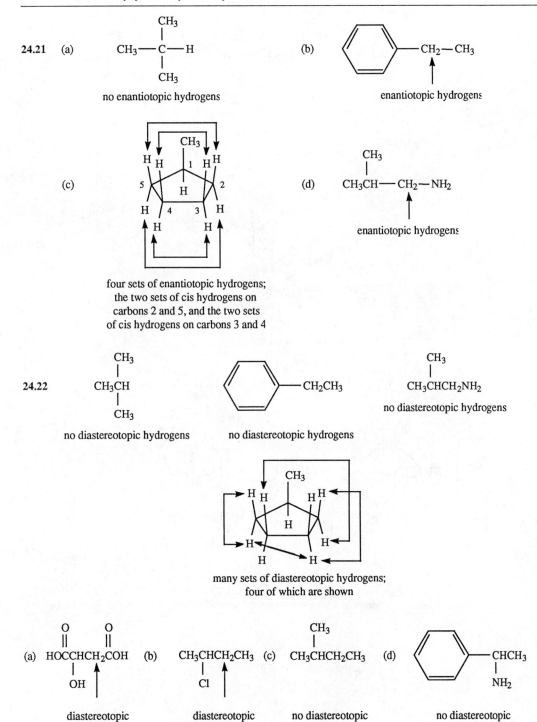

24.21   (a)   no enantiotopic hydrogens

(b)   enantiotopic hydrogens

(c)   four sets of enantiotopic hydrogens;
the two sets of cis hydrogens on
carbons 2 and 5, and the two sets
of cis hydrogens on carbons 3 and 4

(d)   enantiotopic hydrogens

24.22   no diasterotopic hydrogens

no diasterotopic hydrogens

no diasterotopic hydrogens

many sets of diastereotopic hydrogens;
four of which are shown

(a)   diastereotopic hydrogens

(b)   diastereotopic hydrogens

(c)   no diastereotopic hydrogens

(d)   no diastereotopic hydrogens

24.23

One hydrogen atom from testosterone goes to the solvent, the other to the back face of the nicotinamide group.

24.24

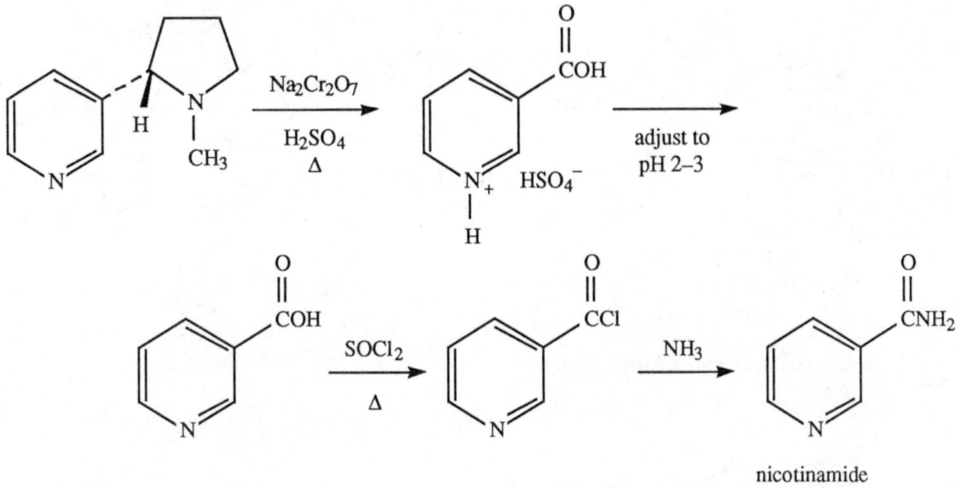

Na₂Cr₂O₇ / H₂SO₄ / Δ

adjust to pH 2–3

nicotinamide

SOCl₂ / Δ

NH₃

24.25

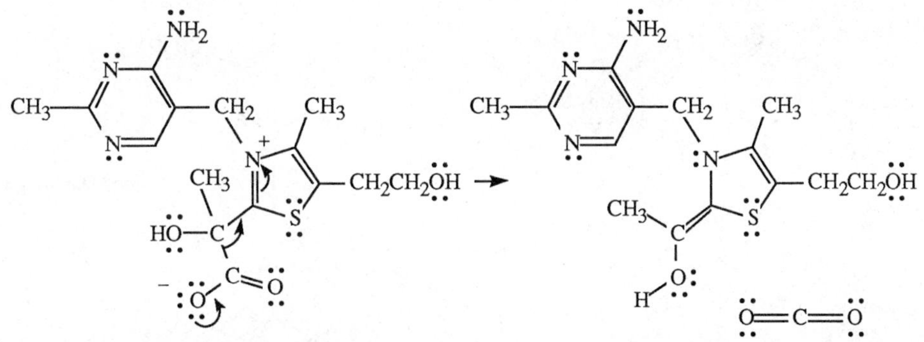

intermediate from the addition of
the ylid from thymine to pyruvic acid

Decarboxylation occurs easily because the carbanion that is formed is stabilized by delocalization of charge to the positively charged nitrogen atom of the ylid. (See Problem 24.26 for the resonance contributors of the decarboxylation product.)

24.26

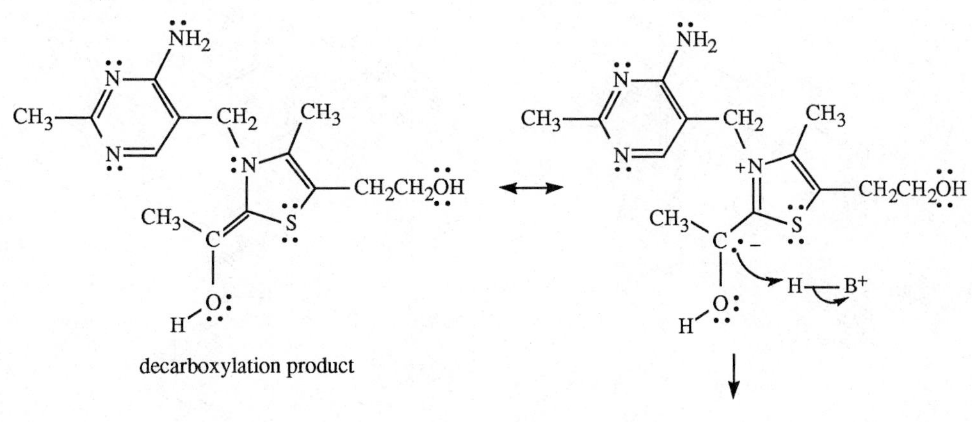

decarboxylation product

24.26 (cont)

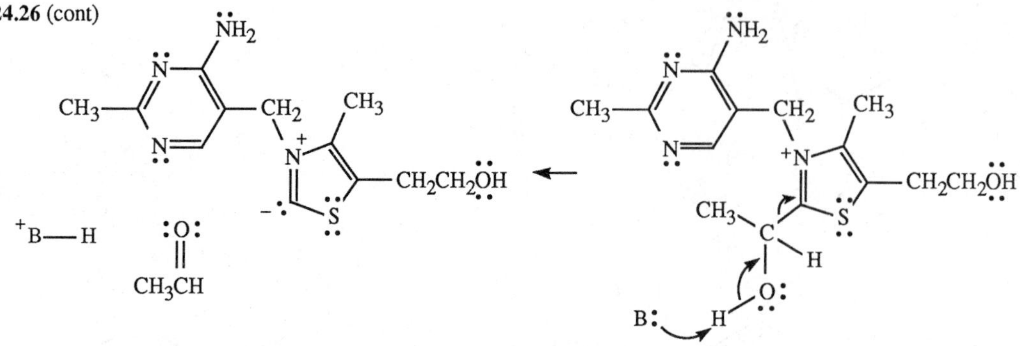

This resembles a retroaldol reaction.

24.27

(a)

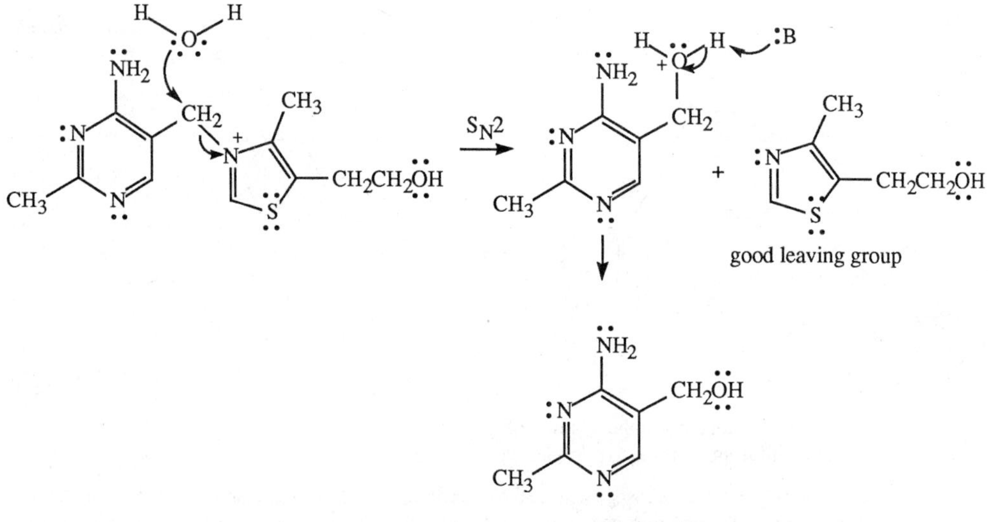

good leaving group

(b)

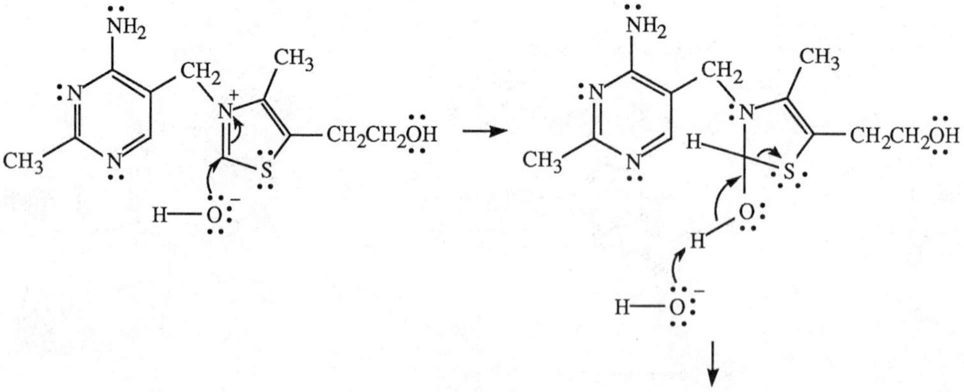

24.27 (cont)

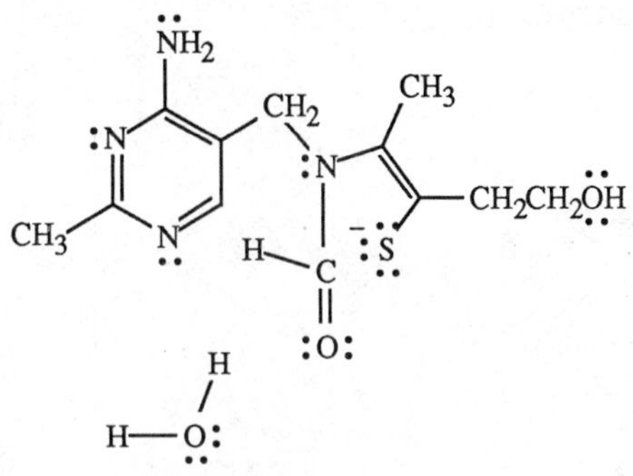

24.28   dehydratase   +   2H$_2$O

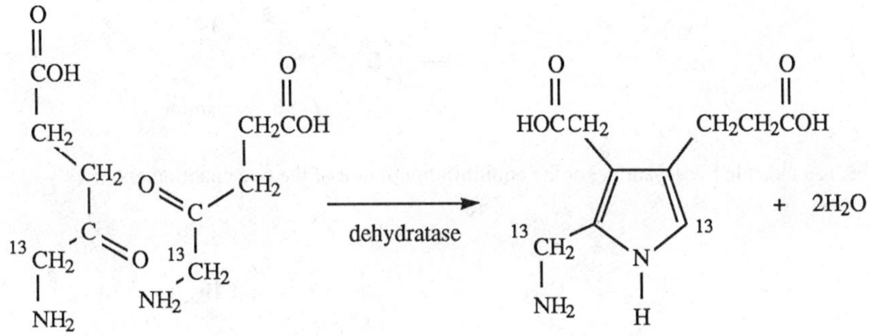

24.29

(–)-hyoscyamine

planar enolate ion; protonation
on top or bottom equally likely

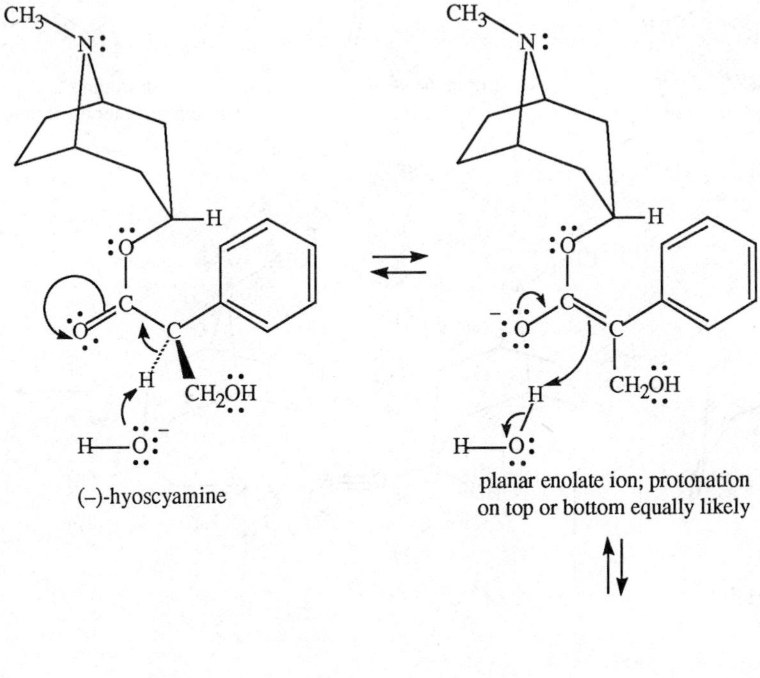

24.29 (cont)

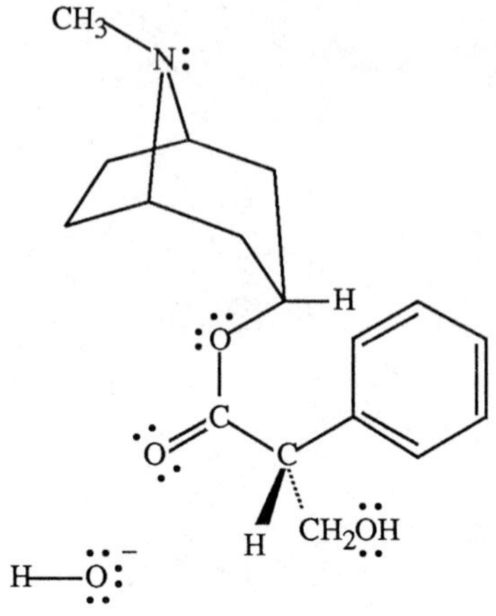

(+)-hyoscyamine

Atropine is a racemic hyoscyamine, or the equilibrium mixture of the two enantiomers.

24.30

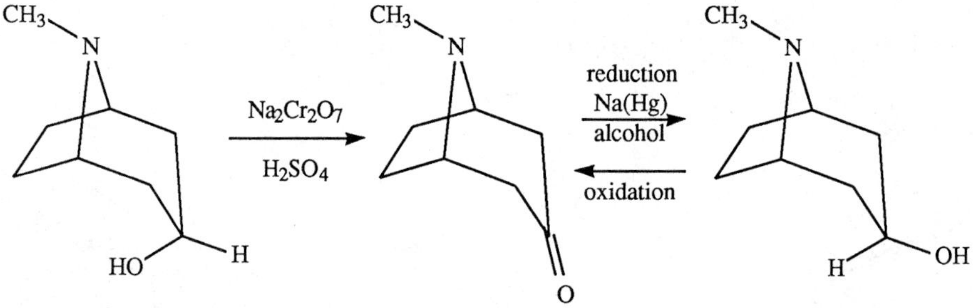

tropine   tropinone   ψ-tropine
                                                         a stereoisomer of tropine

24.31

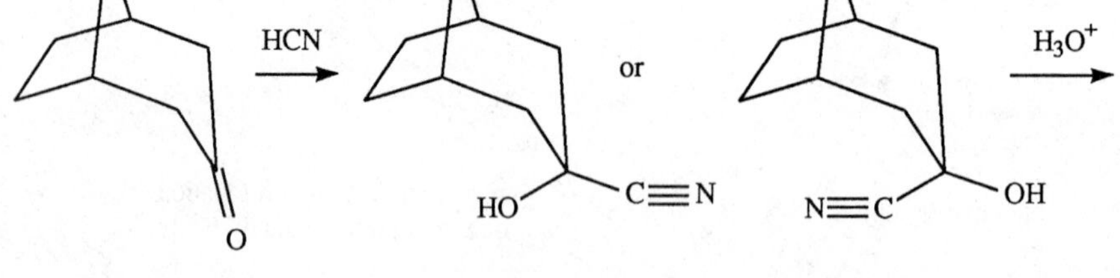

24.31 (cont)

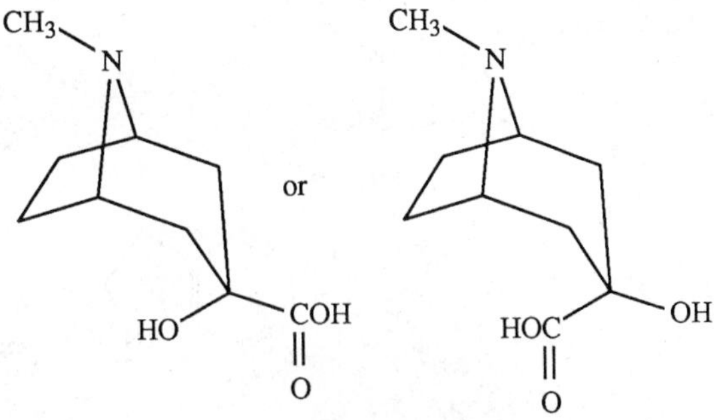

α-ecgonine

24.32

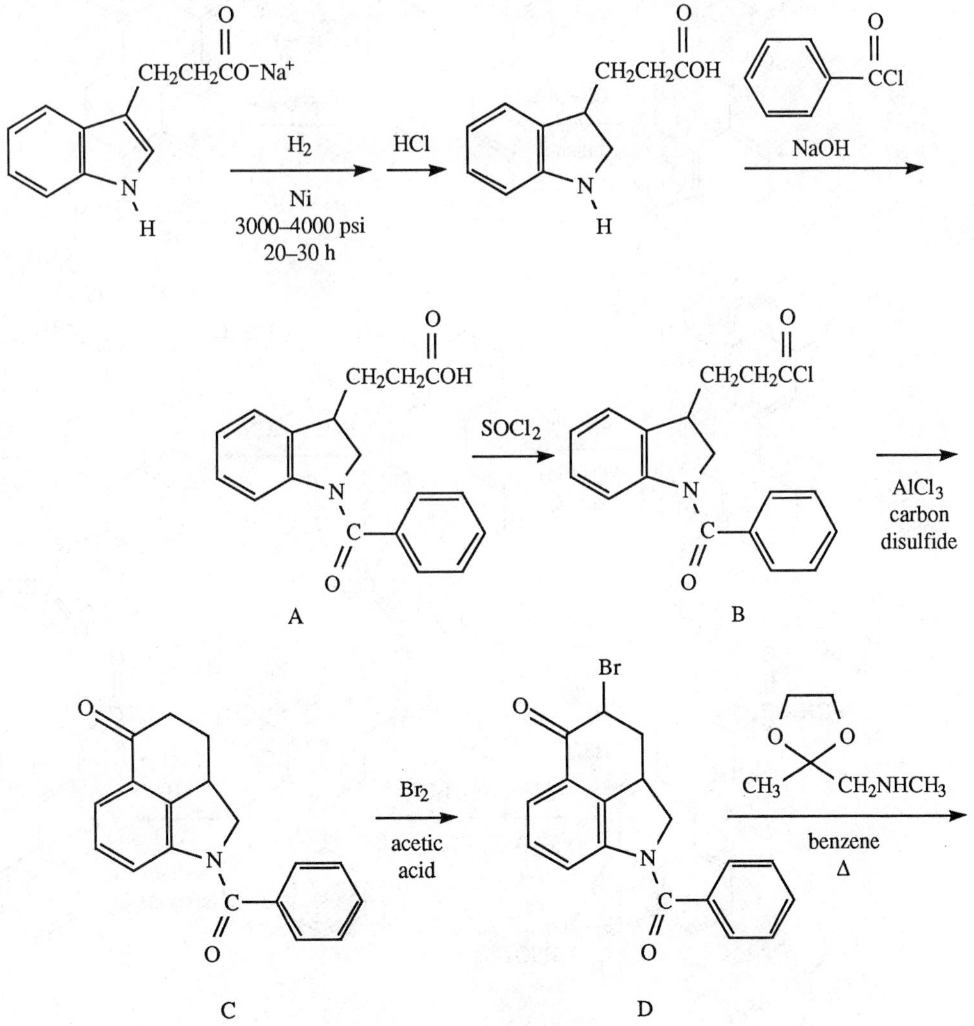

A

B

C

D

24.32 (cont)

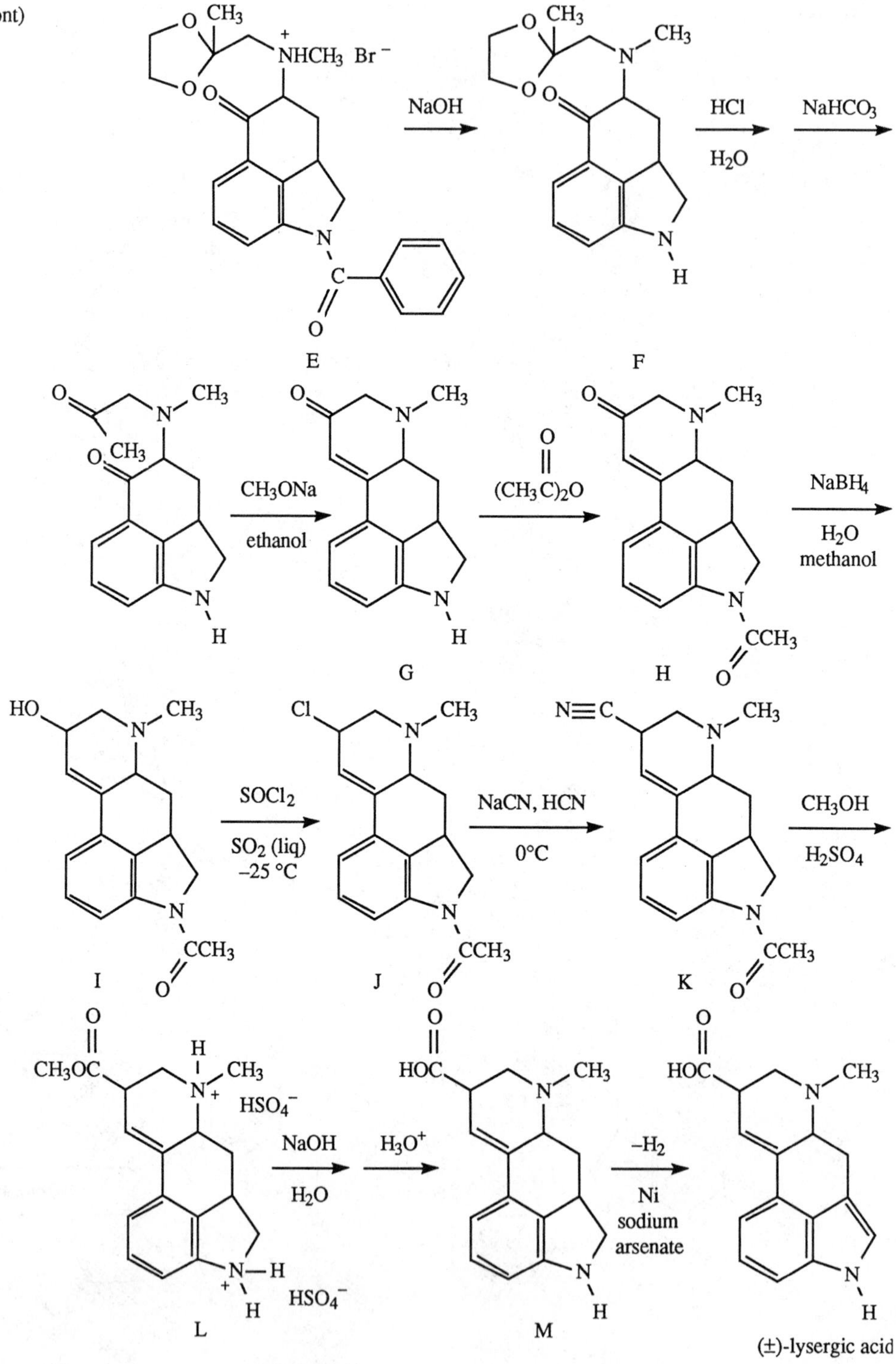

E

→ NaOH →

F

→ HCl / H₂O → NaHCO₃ →

G

CH₃ONa / ethanol →

H

(CH₃C)₂O →

NaBH₄ / H₂O / methanol →

I

SOCl₂ / SO₂ (liq) / −25 °C →

J

NaCN, HCN / 0°C →

K

CH₃OH / H₂SO₄ →

L

NaOH / H₂O →

H₃O⁺ →

M

−H₂ / Ni sodium arsenate →

(±)-lysergic acid

24.33

24.33 (cont)

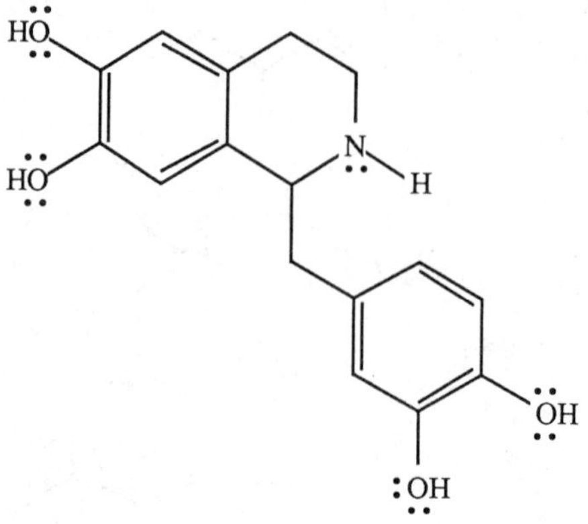

24.34

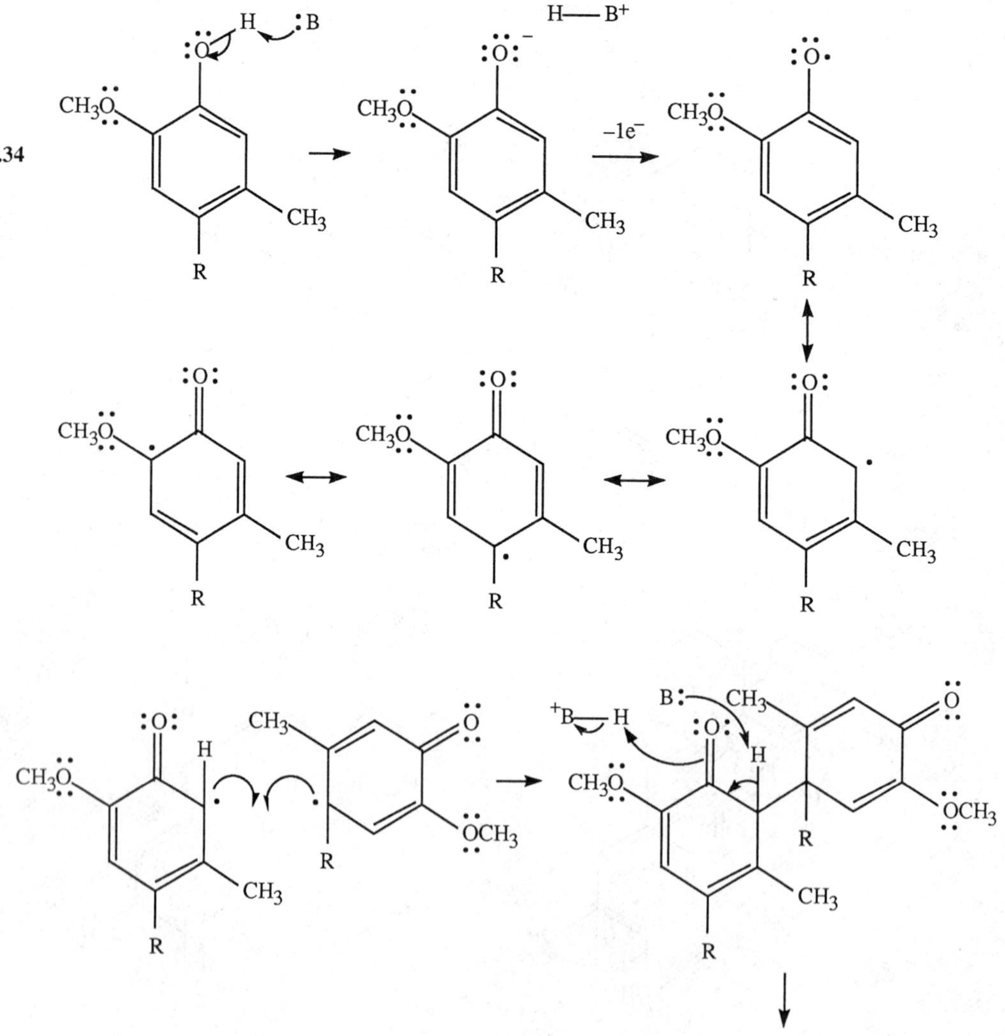

24.34 (cont)

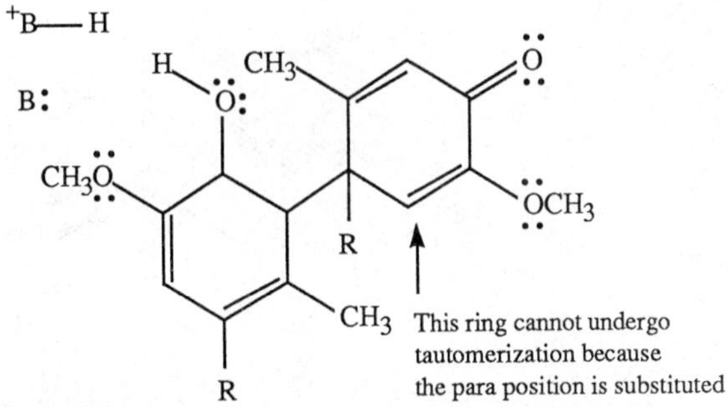

This ring cannot undergo tautomerization because the para position is substituted

24.35

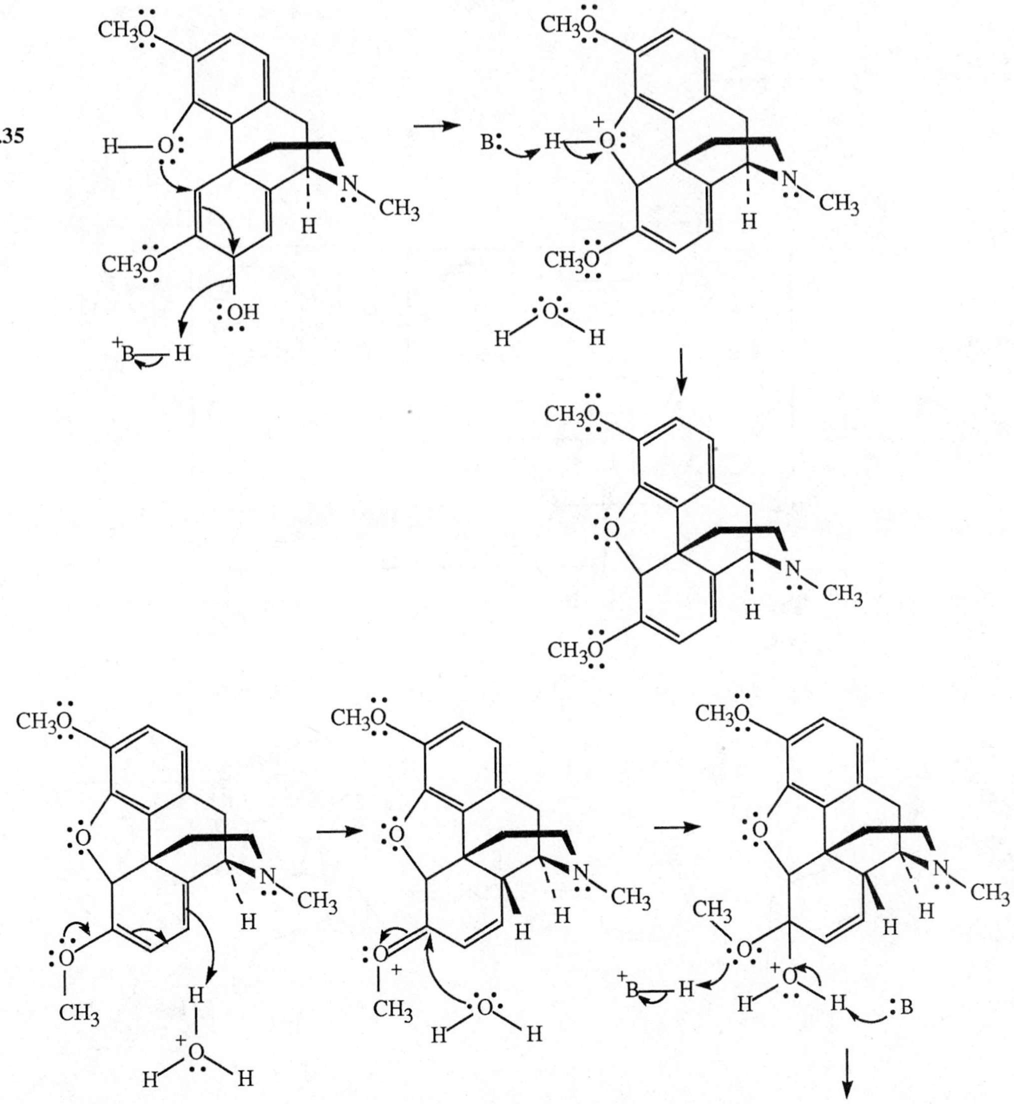

24.35 (cont)

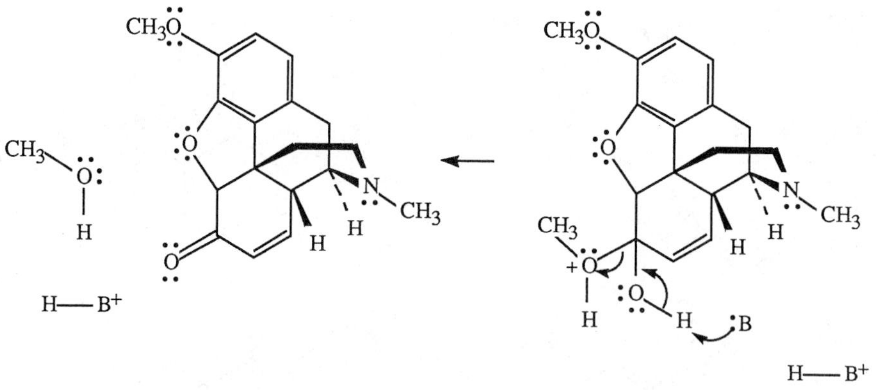

24.36

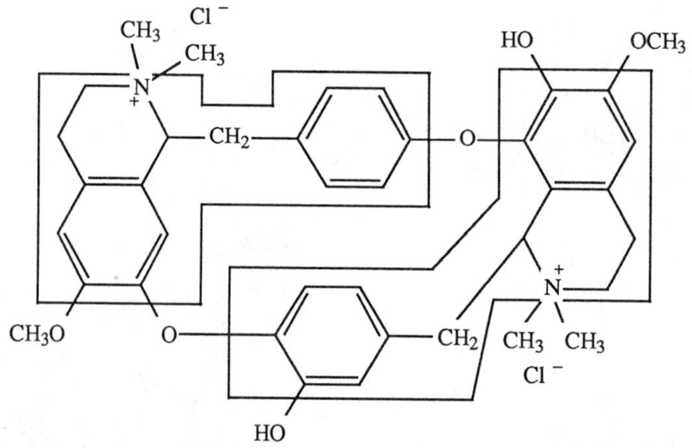

24.37

(a)

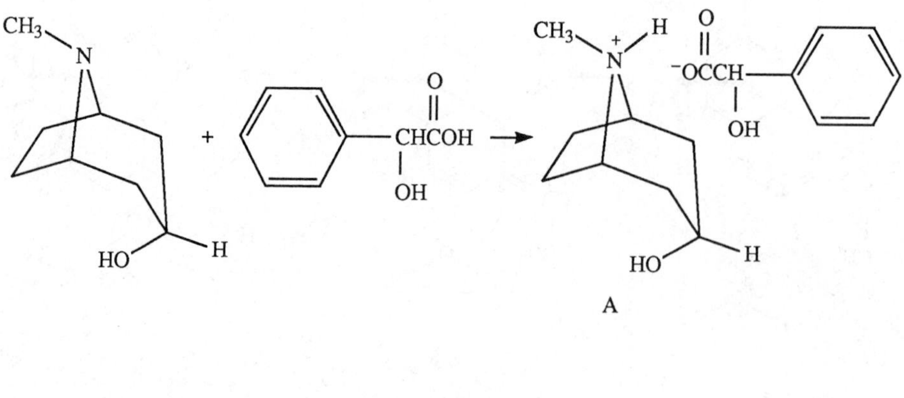

A

24.37 (cont)

(b)

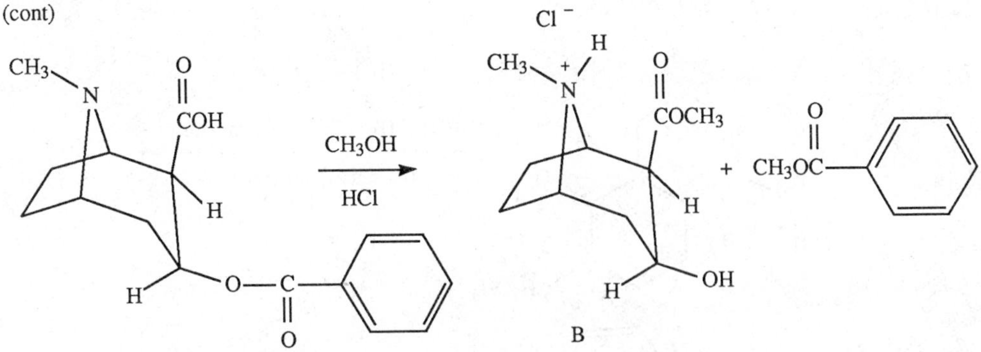

(c)

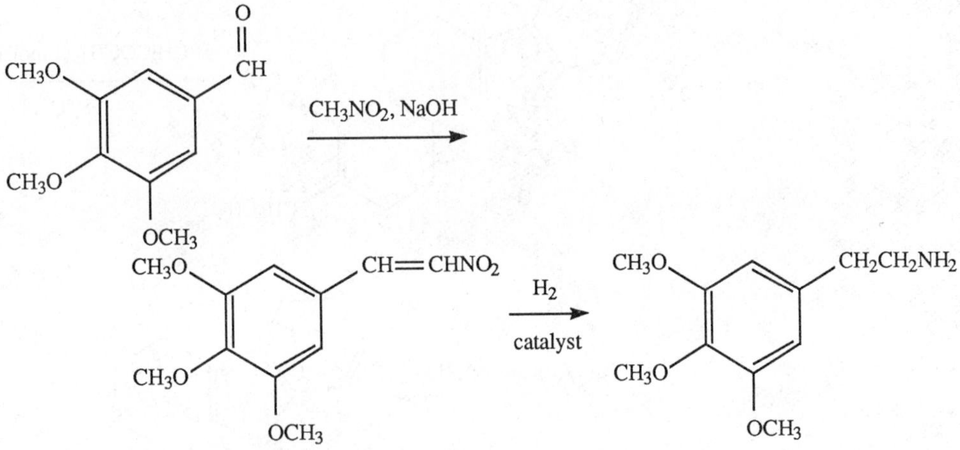

(d)

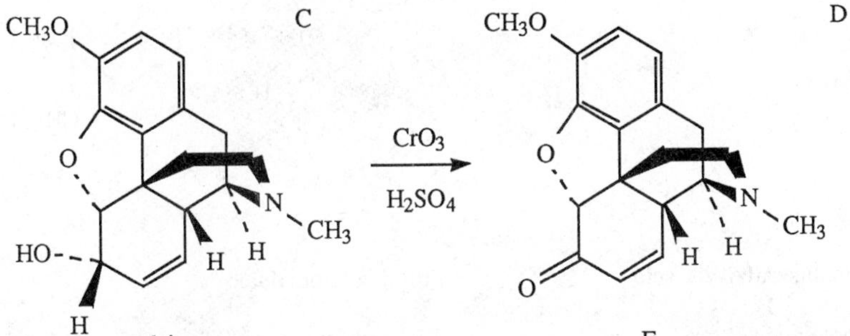

codeine

(e)

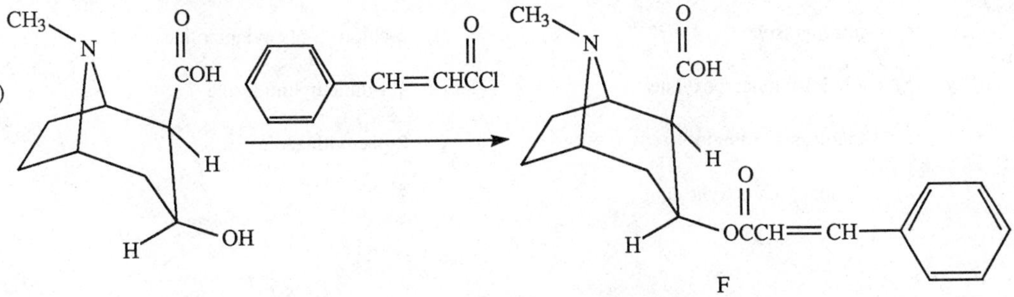

24.37 (cont)

(f)

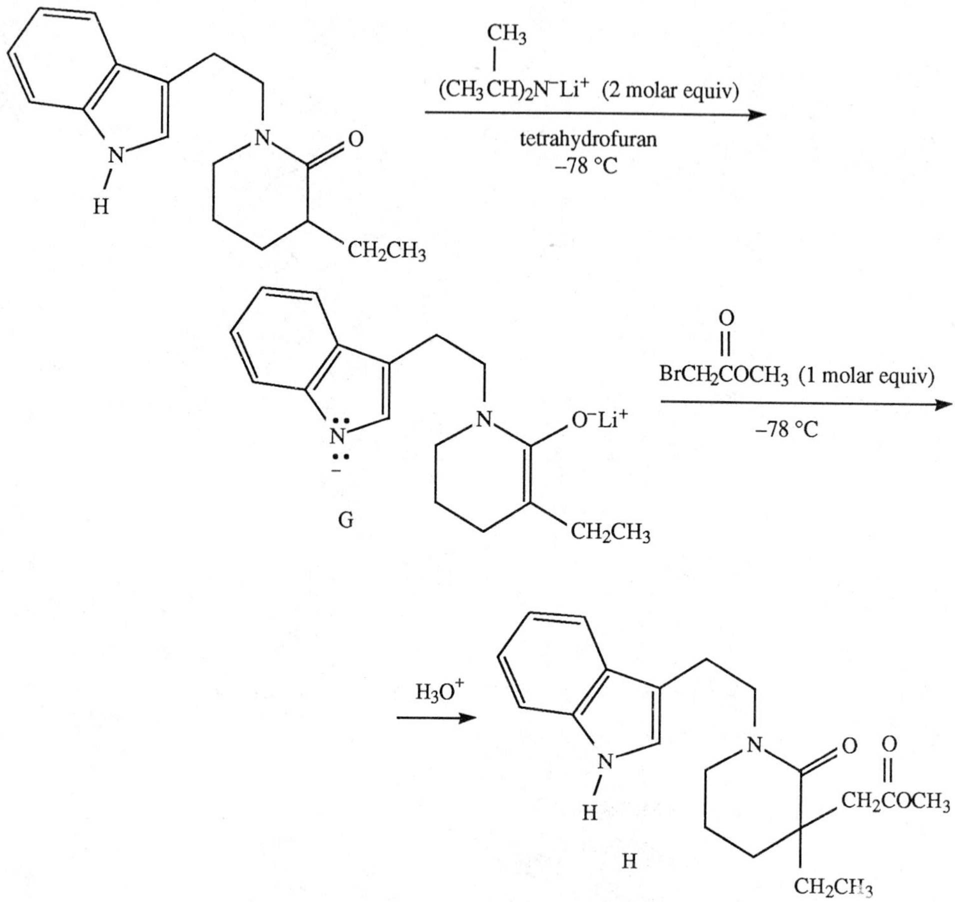

24.38 (a) 3-isoquinolinecarboxylic acid

(c) 2-aminopyrimidine

(e) 4-nitropyrazole

(g) methyl 3-furancarboxylate

(i) 3-indolecarboxamide

(k) 2-thiophenecarboxylic acid

(b) 3-chloropyridine

(d) 3-phenylisoxazolidine

(f) 6-chloro-3-phenylquinoline

(h) 1,4-dimethylimidazole

(j) 2-phenylthiazole

24.39

(a)

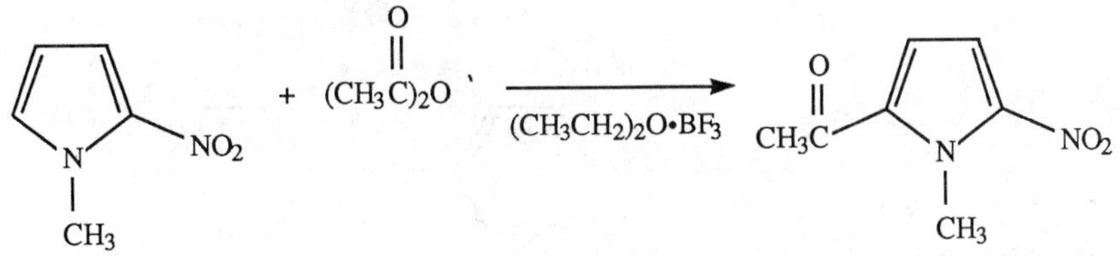

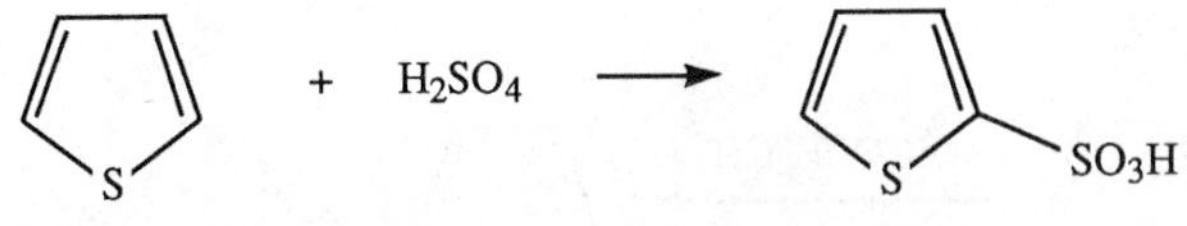

(b)

(c)

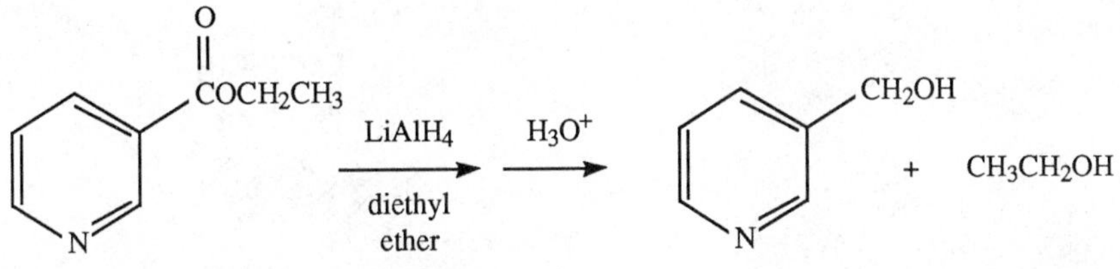

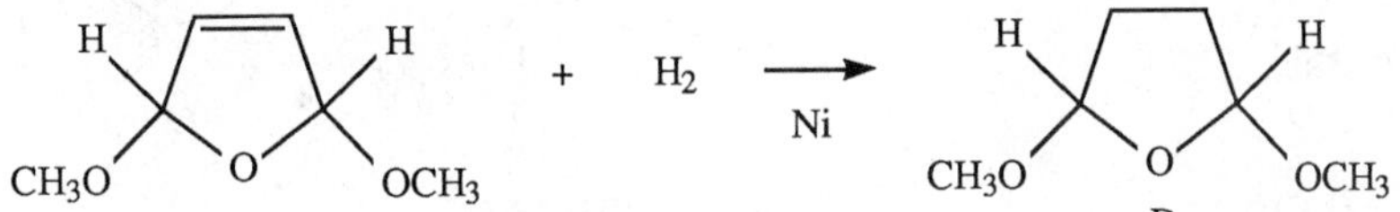

(d)

(e)

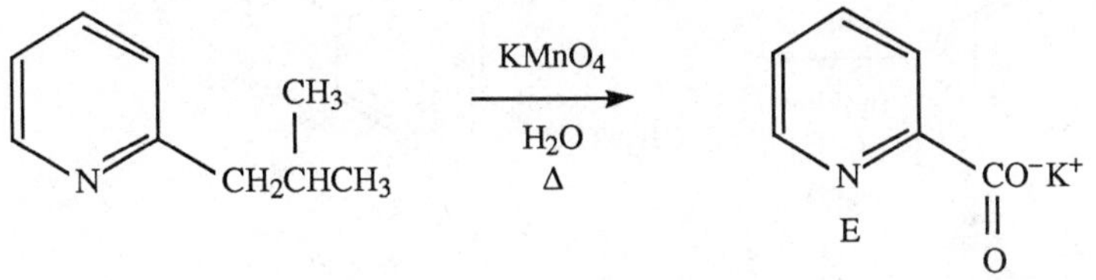

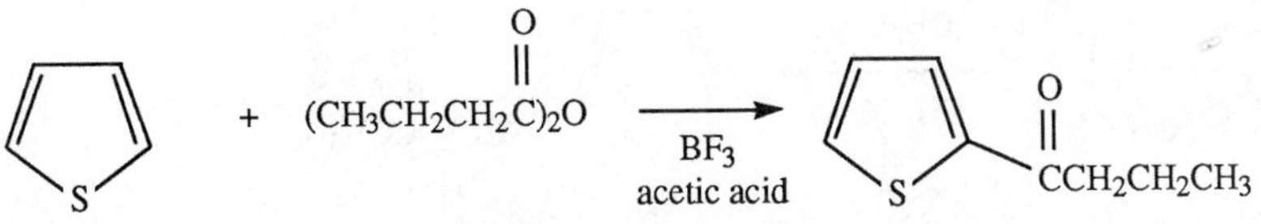

(f)

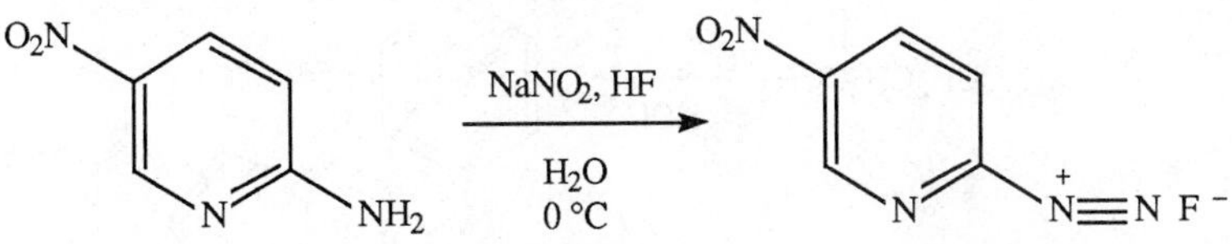

(g)

24.39 (cont)

(h) 4-chloropyridine + aniline → N-phenyl-4-aminopyridine

H

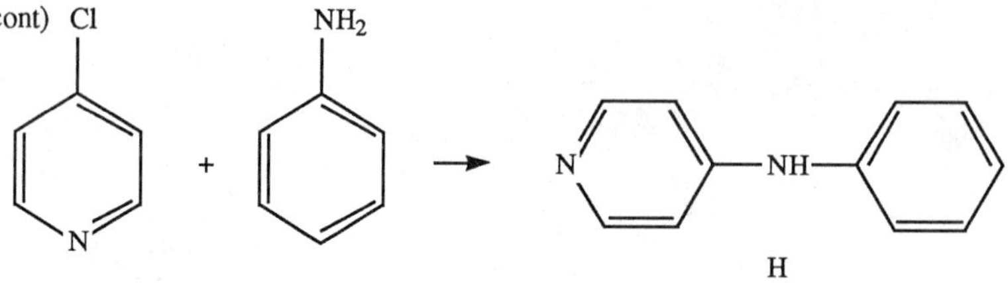

(i) pyrrole $\xrightarrow{\text{KOH}}$ pyrrolide K^+ (**I**) $\xrightarrow{\text{ClCOCH}_2\text{CH}_3}$ 1-propanoylpyrrole

J

(j) 2-acetylthiophene $\xrightarrow{\text{Cl}_2,\ \text{NaOH}}$ CHCl_3 + thiophene-2-CO^-Na^+ (**K**) $\xrightarrow{\text{H}_3\text{O}^+}$ thiophene-2-CO^-Na^+ (**L**)

(k) 2,4-diethoxypyridine $\xrightarrow[\substack{\text{pyridine}\\ 0\,°\text{C}}]{\text{Br}_2}$ 3,5-dibromo-2,4-diethoxypyridine

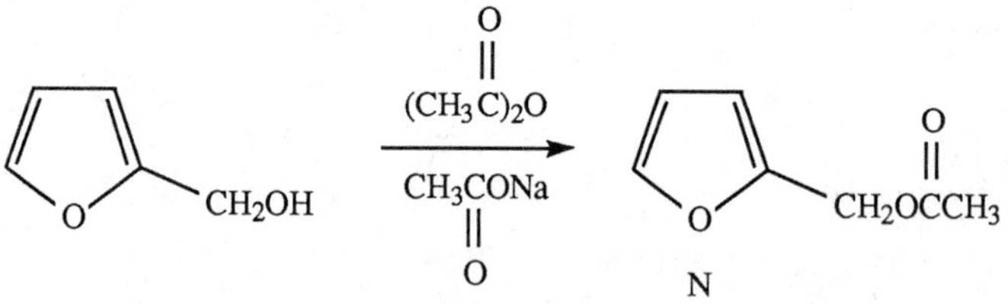

M

(l) furfuryl alcohol $\xrightarrow[\substack{\text{CH}_3\text{CONa}}]{(\text{CH}_3\text{C})_2\text{O}}$ furfuryl acetate

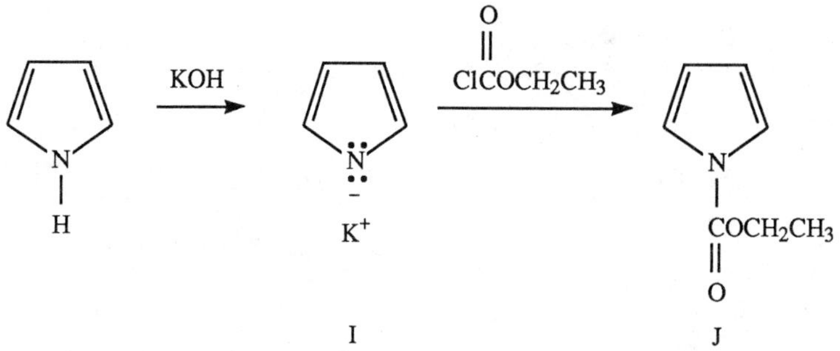

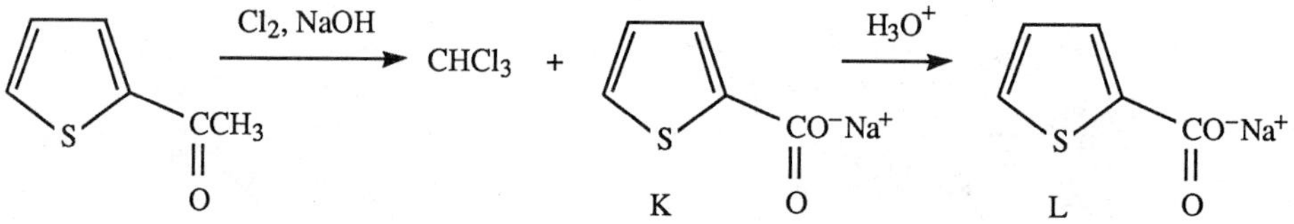

N

(m) 2-aminopyridine + $\text{CH}_3\text{CCH}_2\text{CH}_2\text{CCH}_3$ $\xrightarrow{\text{HCl}}$ 2-(pyrrol-1-yl)pyridine

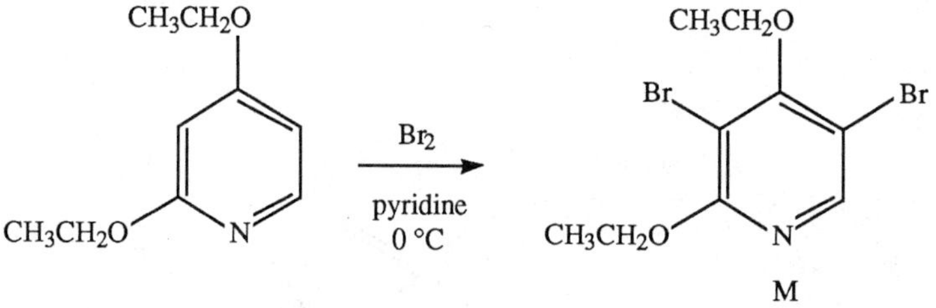

O

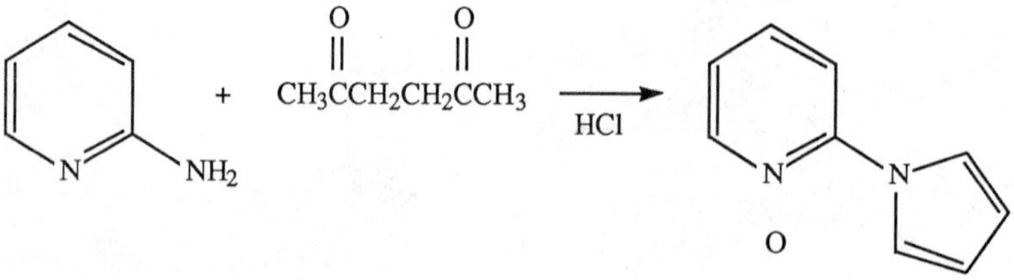

24.39 (cont)

(n)

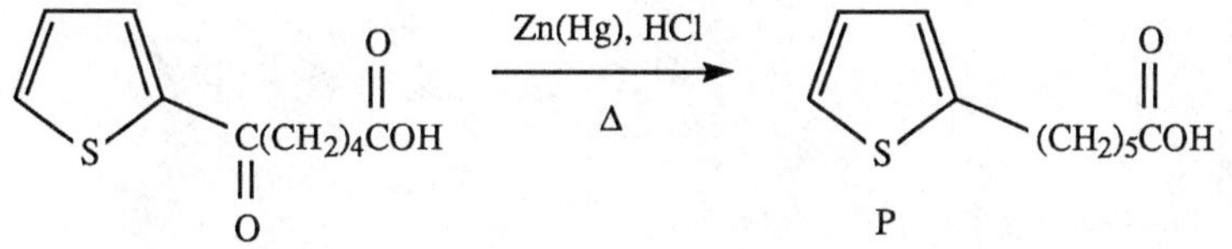

(o)

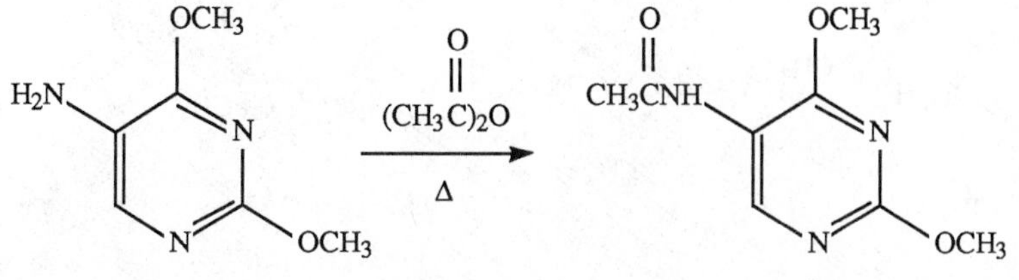

(p)

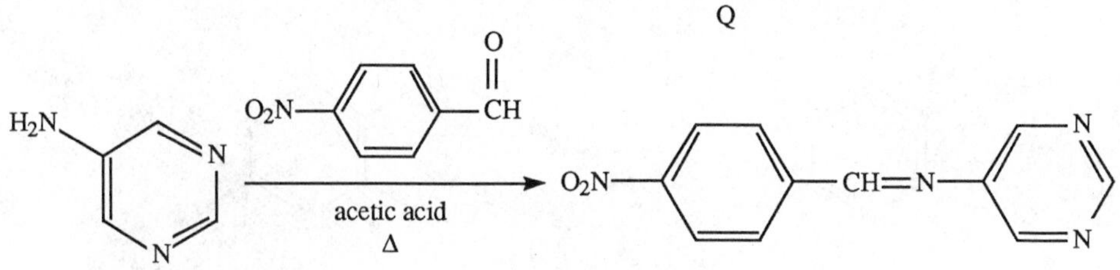

(q)

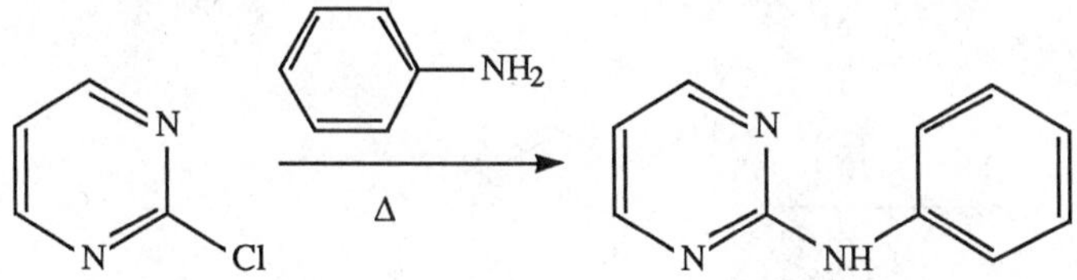

(r)

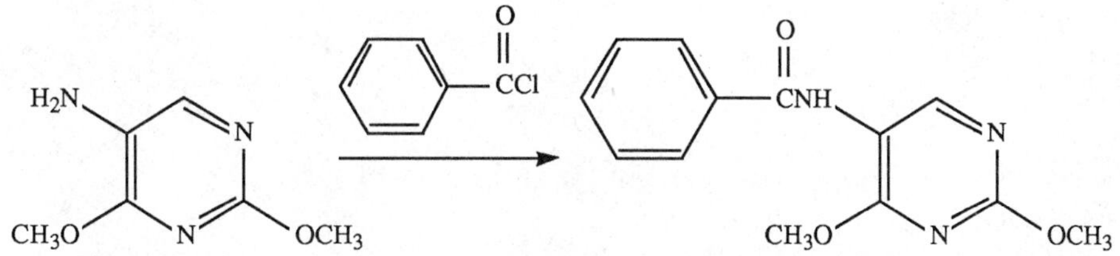

(s)

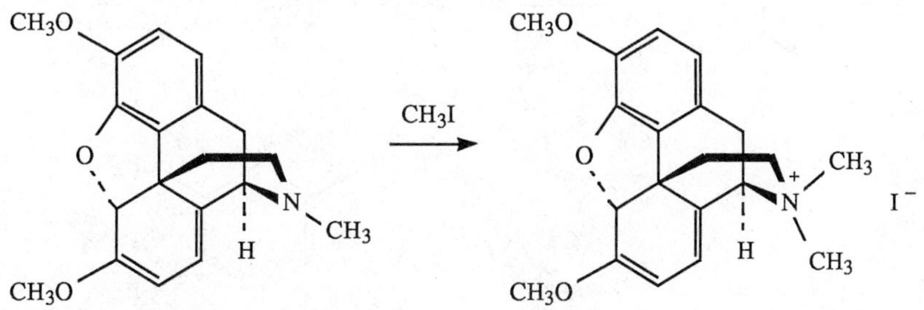

24.39 (cont)

(t)

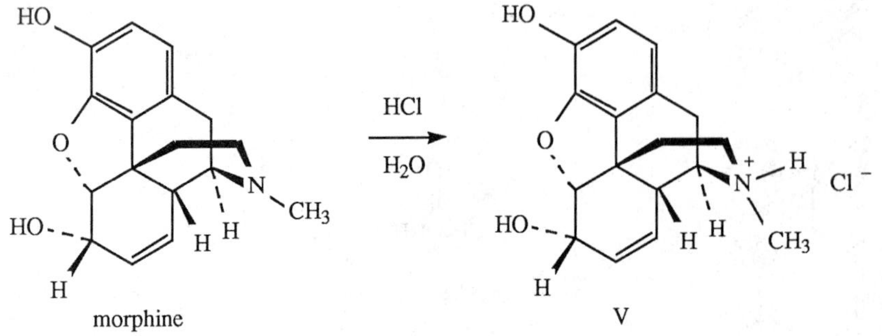

morphine   V

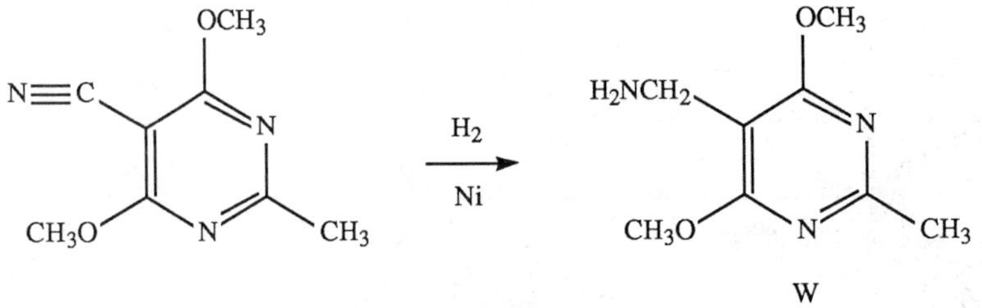

(u)

W

(v)

X

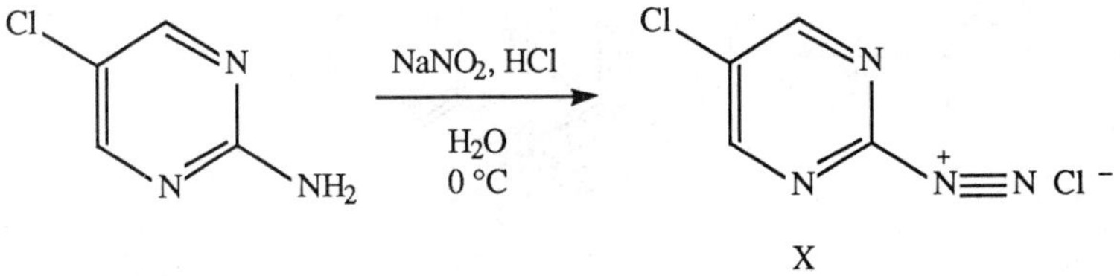

24.40

(a)

A

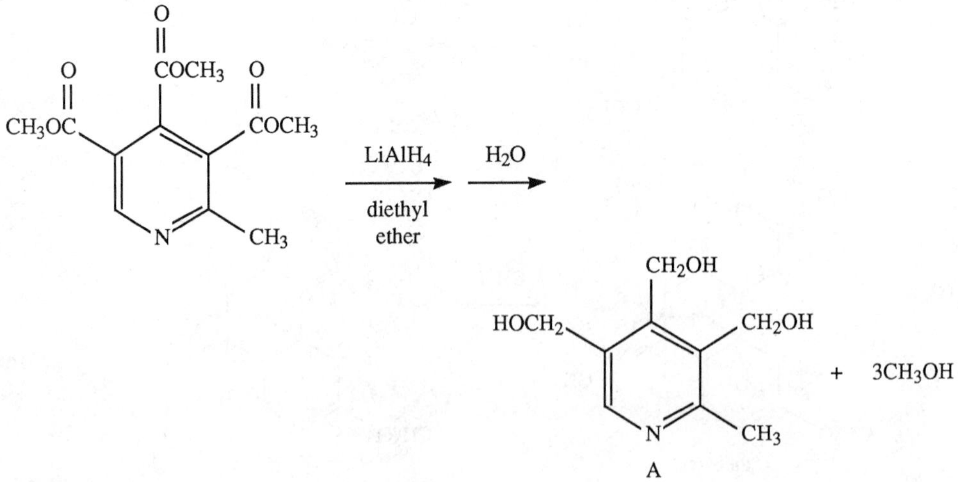

24.40 (cont)

(b)

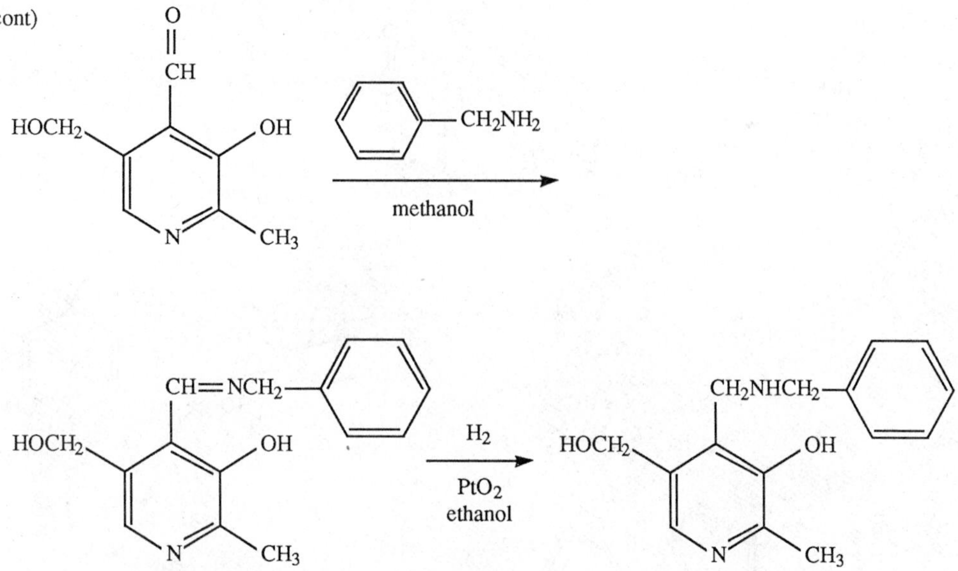

(c)

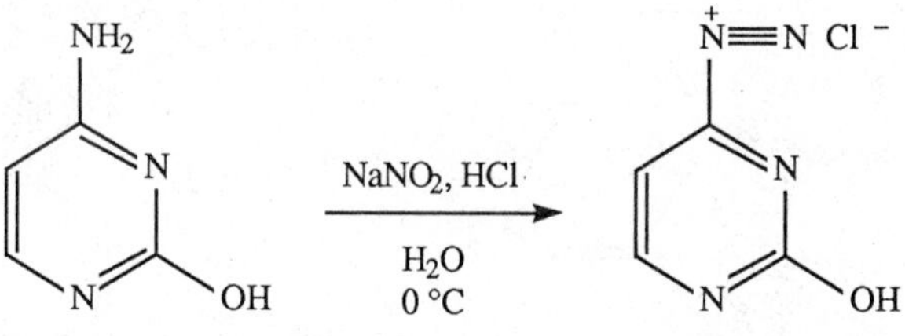

(d)

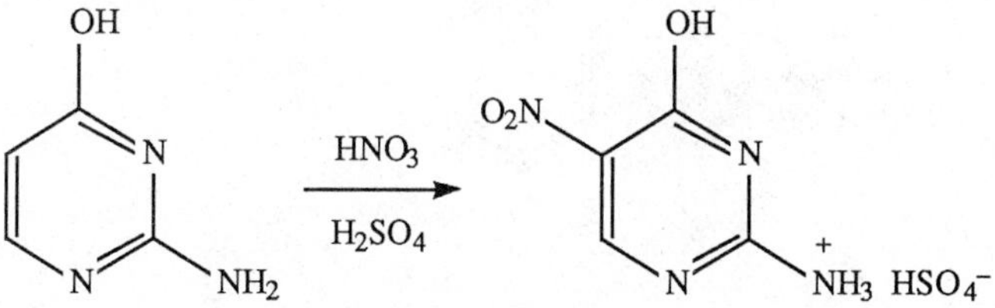

(e)

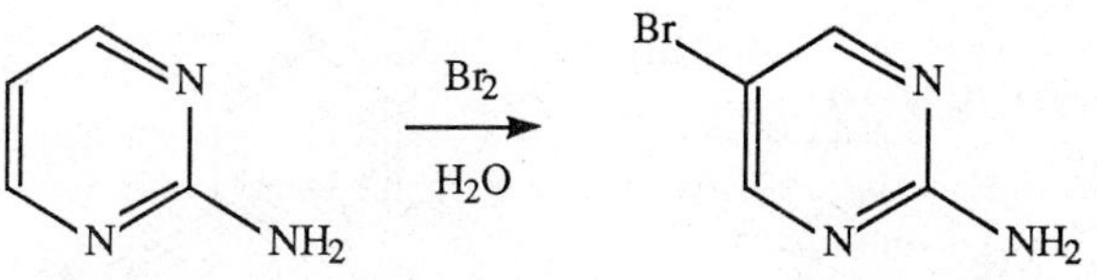

24.40 (cont)

(f)

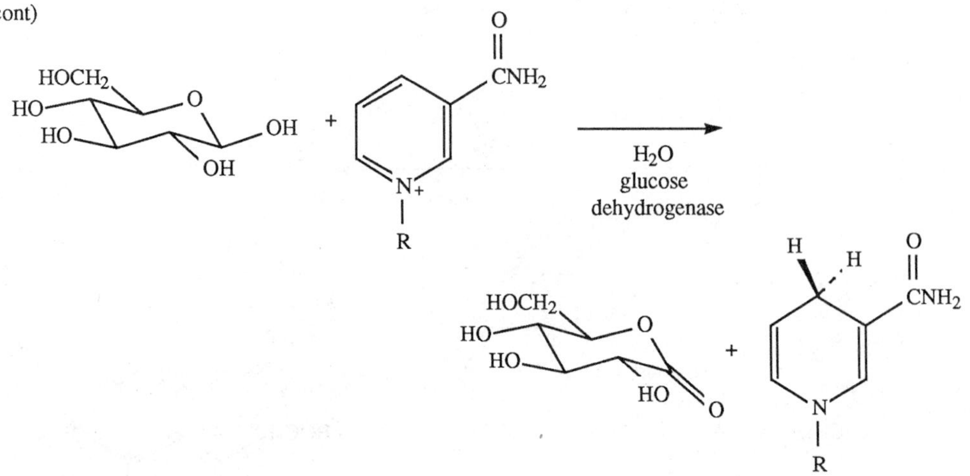

G

(g)

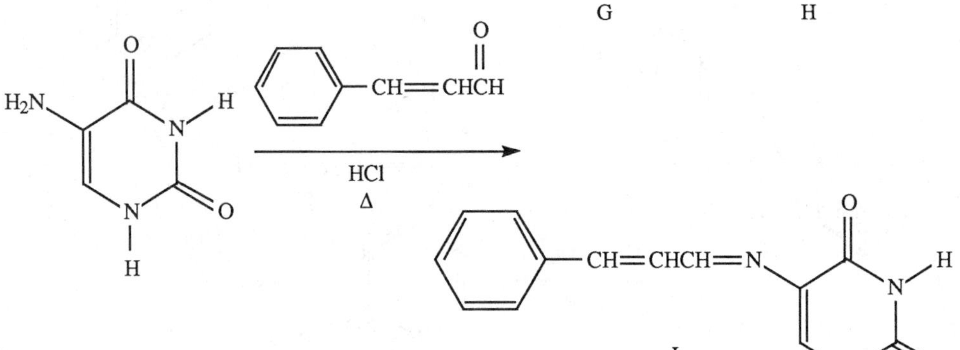

I

(h)

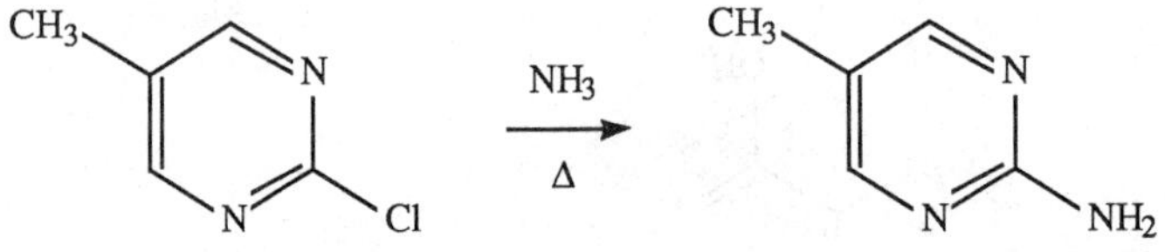

J

(i)

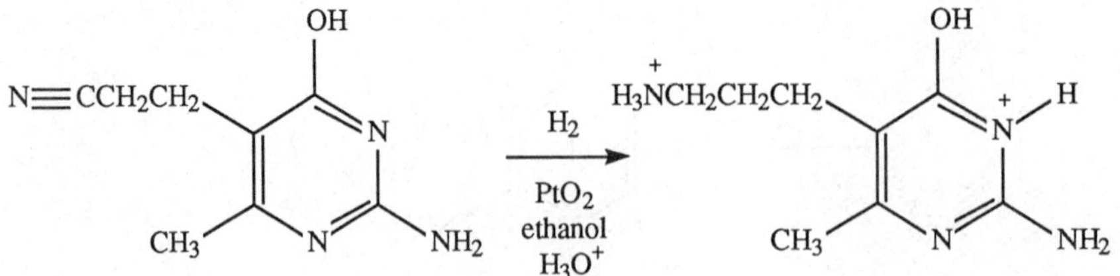

K

24.40 (cont)

(j)

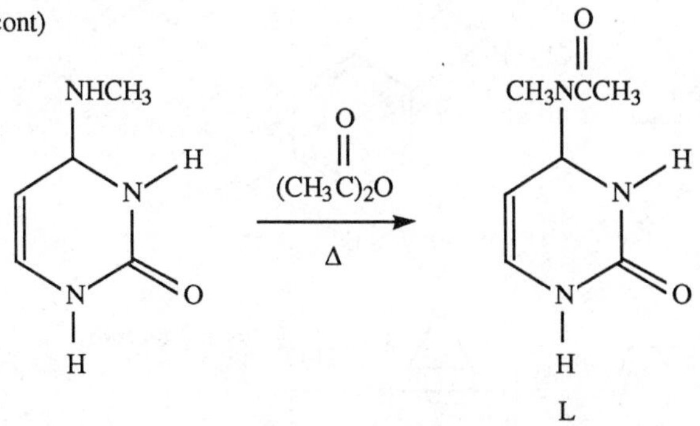

(k)

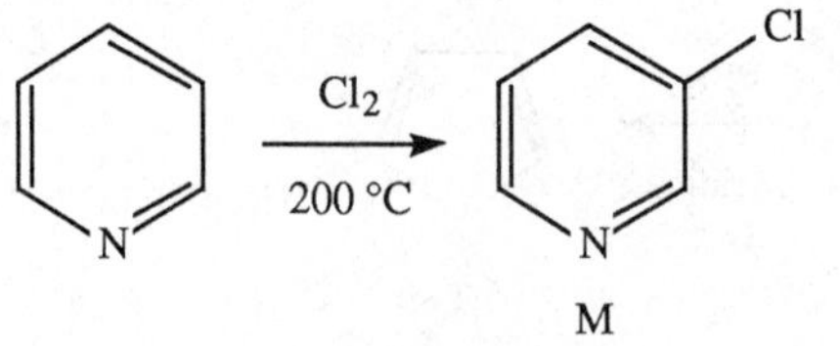

(l)

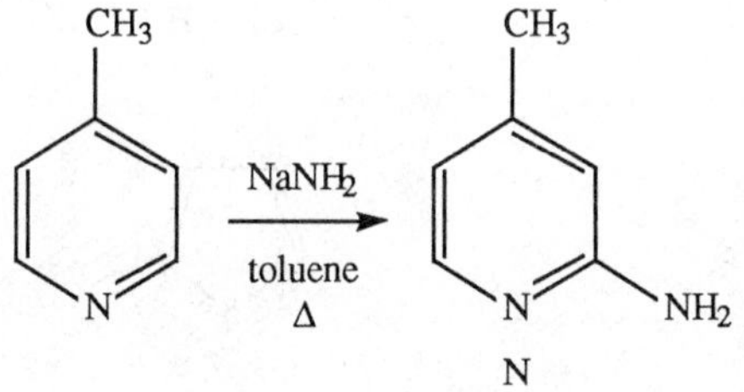

(m)

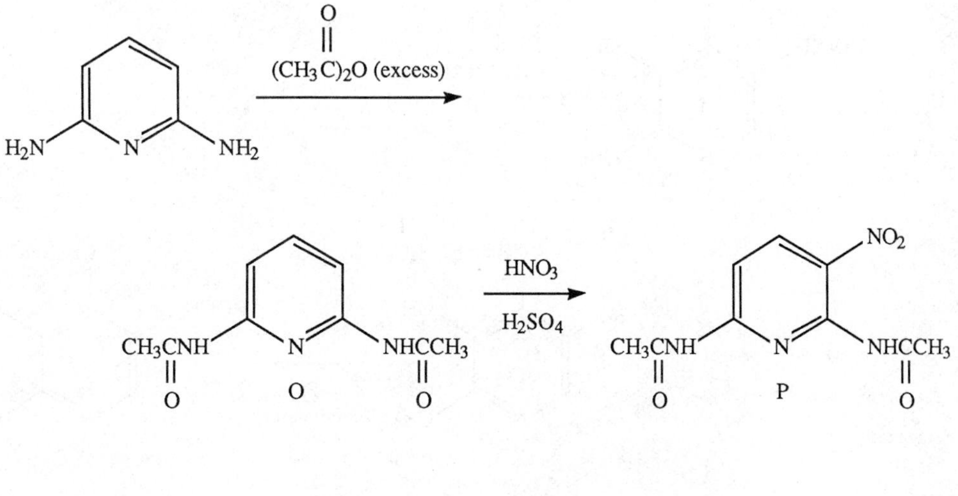

24.40 (cont)

(n)

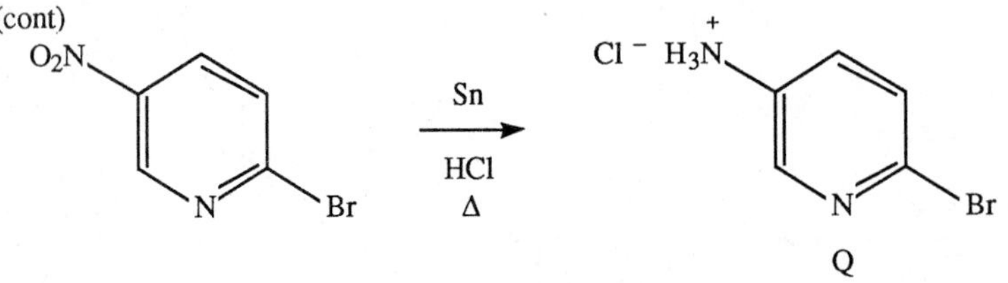

(o)

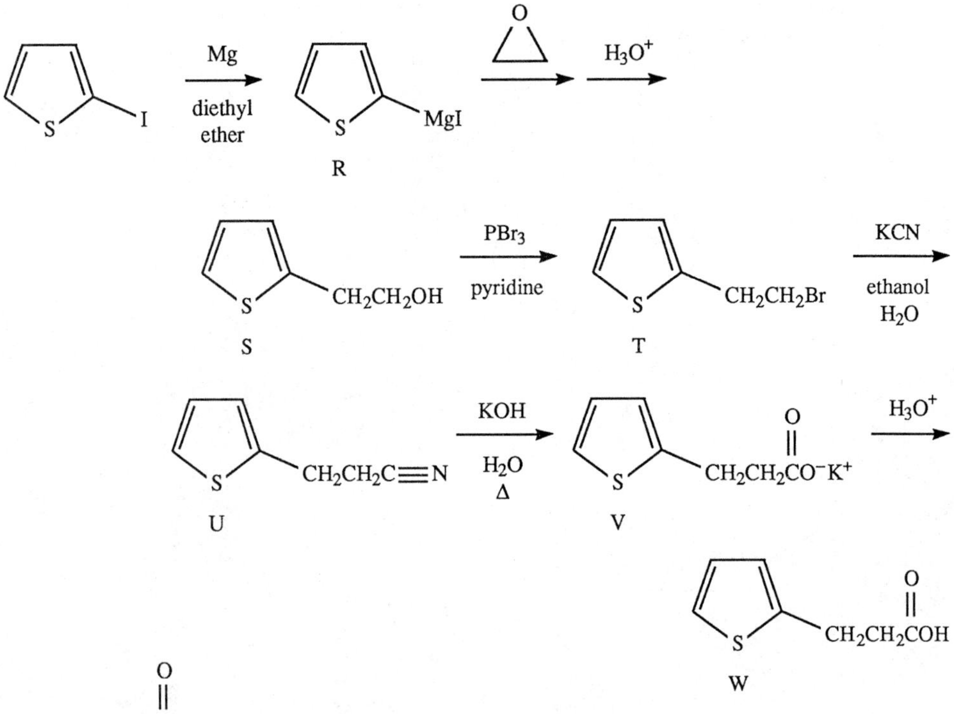

(p)

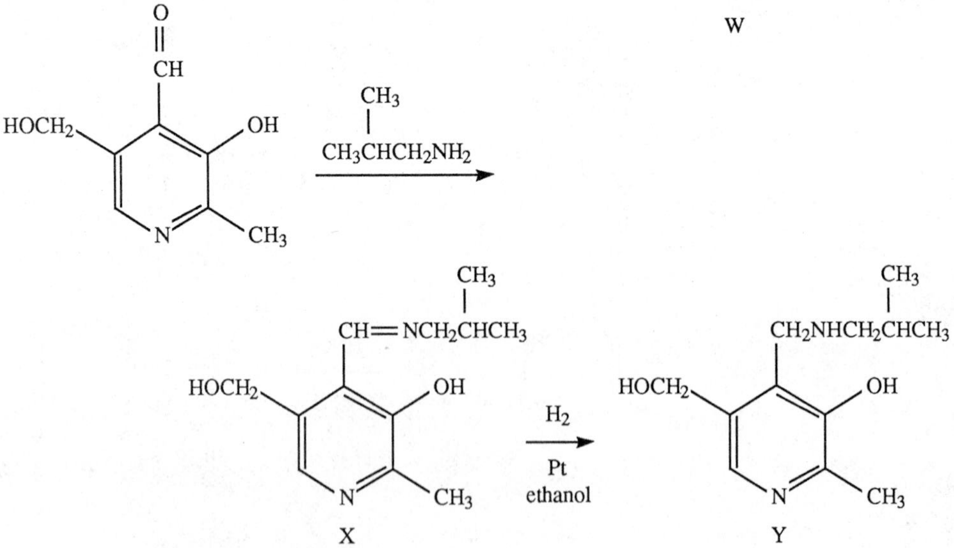

24.40 (cont)

(q)

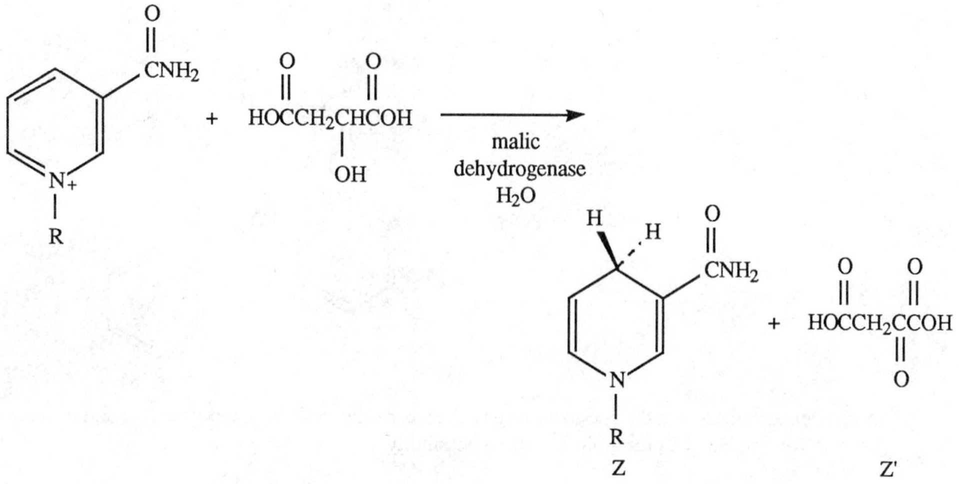

24.41

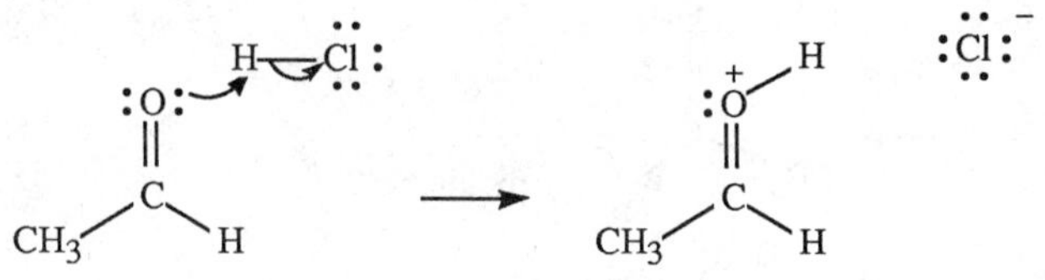

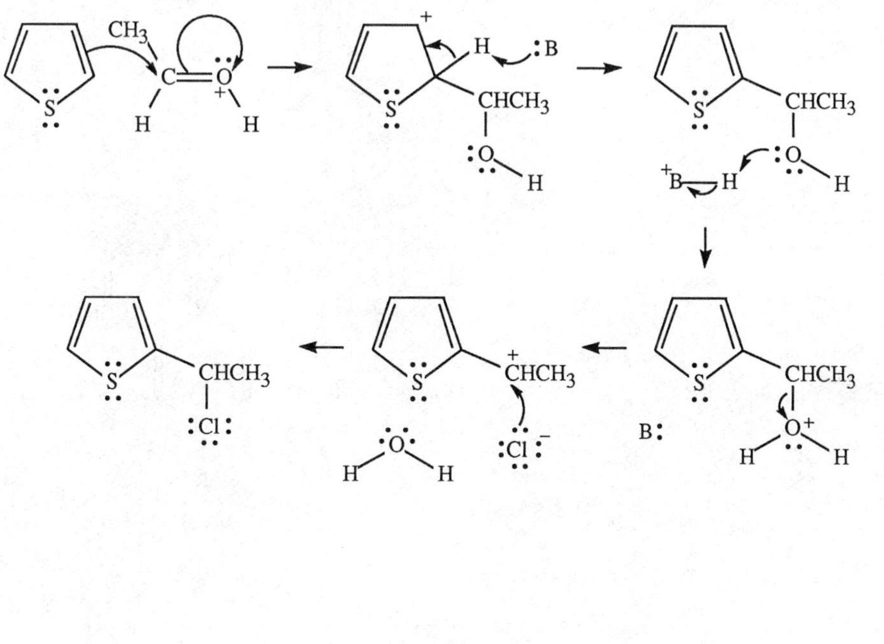

24.41 (cont)

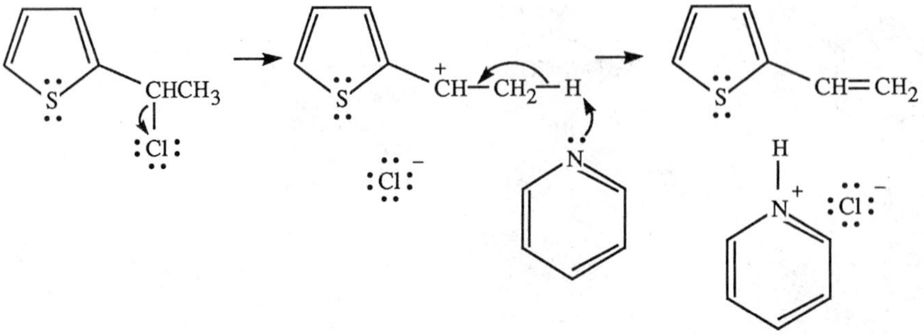

A carbocation adjacent to the thiophene ring is like a benzylic cation in stability so S_N1 and E_1 reactions have been shown. S_N2 and E_2 reactions are also a possibility.

24.42

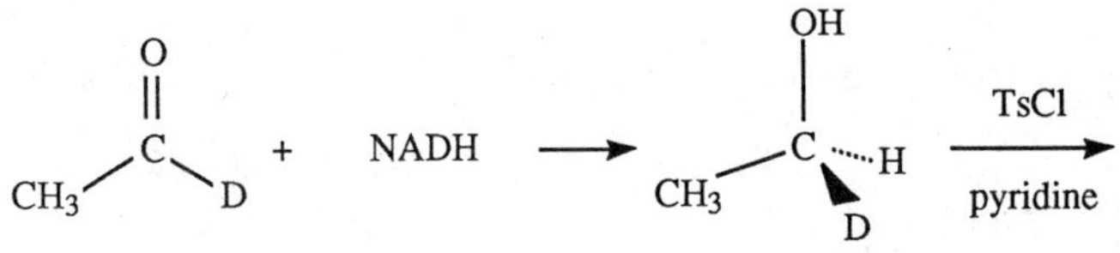

A
(*S*)-ethanol-1-*d*

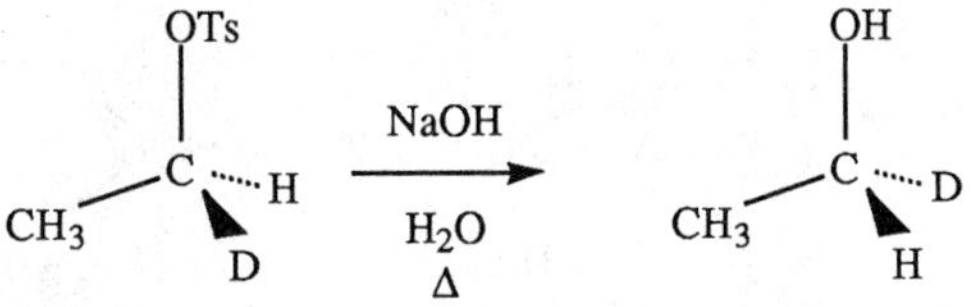

B

C
(*R*)-ethanol-1-*d*

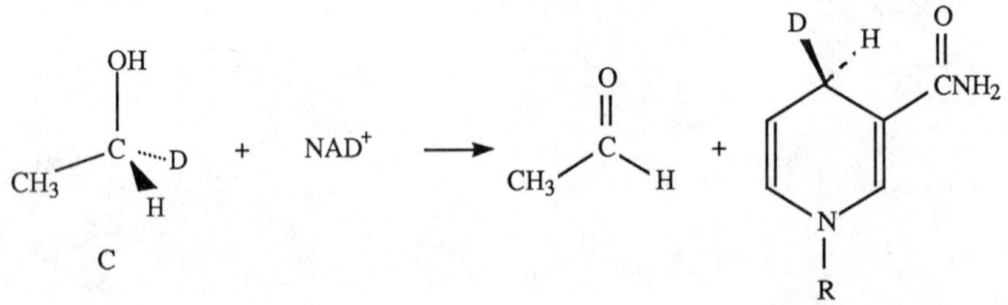

C

NADD

24.42 (cont)

Alcohol A is converted to tosylate B with retention of configuration. The reaction of the tosylate B with hydroxide ion occurs by an S_N2 reaction with inversion of configuration. Thus an optically active alcohol is converted into its enantiomer. When the two alcohols, A and C, are oxidized with NAD^+, deuterium is removed from C and hydrogen from A.

These reactions are the reverse of those used to prepare the alcohols by enzymatic reductions. Reduction of acetaldehyde-1-*d* by NADH gave Alcohol A, while that of acetaldehyde by NADD gave Alcohol C (see p.~ in text).

The enzymatic oxidation-reduction reactions are stereoselective, so the proof that the enantiomeric alcohols lose deuterium and hydrogen selectively in the experiments outlined above also proves that the original reduction reactions give rise to enantiomeric ethanol-1-*d* species.

24.43

4-methyl-imidazole	imidazole	benzimidazole	4-nitro-imidazole
least acidic			most acidic

| pK_a values | 7.5 | 7.0 | 5.5 | −0.1 |

The acidity of each of these species is related to the basicity of the conjugate base, which, in turn, is related to the availability of nonbonding electrons on a nitrogen atom. In all of the imidazoles, basicity is derived from both of the nitrogen atoms in the ring because electrons from both are involved in stabilization of the conjugate acid by delocalization of charge. For example:

protonation at N-3	stabilization of the cation by electrons from N-1

24.43 (cont)

We expect 4-methylimidazole to be more basic than imidazole because the methyl group is electron donating. In benzimidazole the nonbonding electrons are delocalized to the aromatic ring, making them less available at the nitrogen atoms.

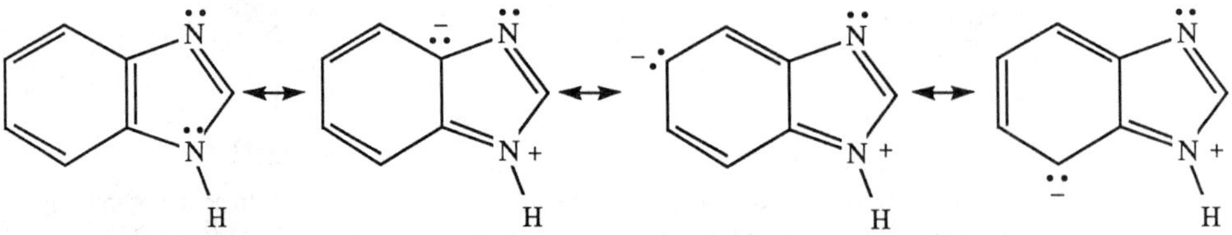

In 4-nitroimidazole, the strong electron-withdrawing inductive effect of the nitro group combines with the resonance effect to greatly reduce electron density in the ring and at the nitrogen atoms.

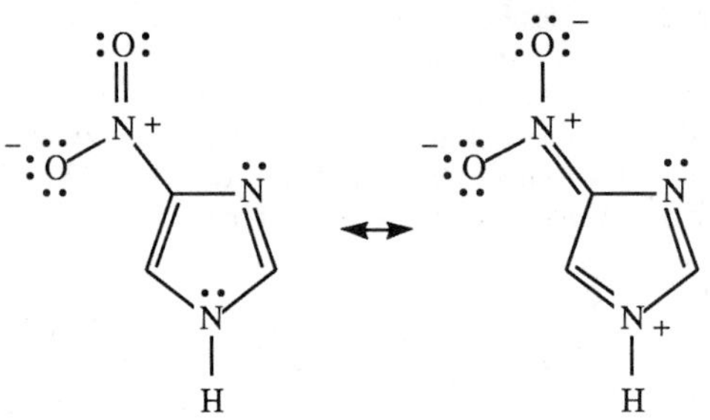

24.44

(a)

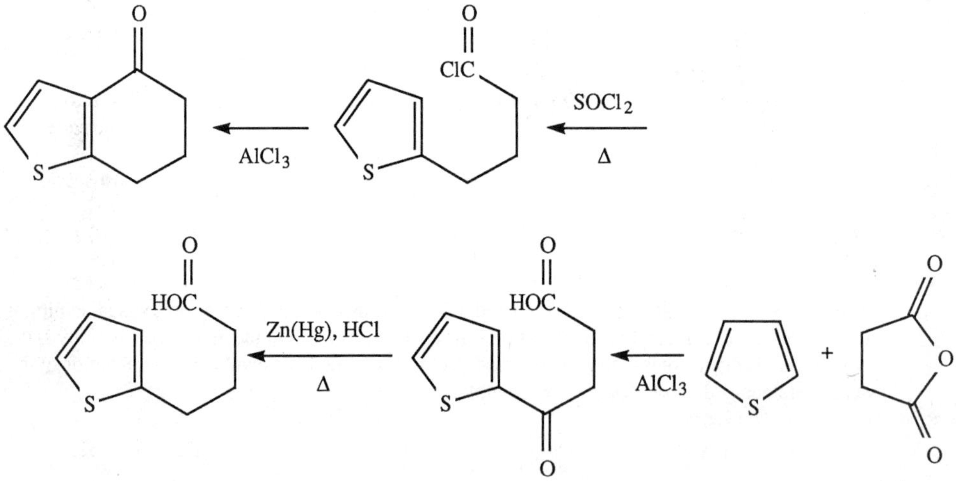

(b)

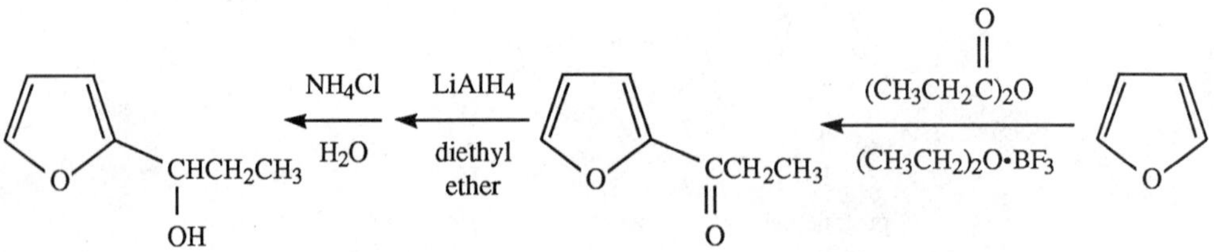

24.44 (cont)

(c)

$$\text{Cl}\quad\xleftarrow[\text{FeCl}_3]{\text{Cl}_2}\quad\text{(2-aminopyridine)}\xleftarrow[\substack{\text{toluene}\\\Delta}]{\text{NaNH}_2}\quad\text{(pyridine)}$$

(d)

$$\xleftarrow{\text{H}_3\text{O}^+}\quad\xleftarrow[\text{(excess)}]{\text{C}_6\text{H}_5\text{-Li}}$$

$$\text{HOC}\ (\overset{\|}{\text{O}})\ \cdots\ \text{COH}\ (\overset{\|}{\text{O}})\quad\xleftarrow[\substack{\text{H}_2\text{SO}_4\\\Delta}]{\text{K}_2\text{Cr}_2\text{O}_7}\quad\text{CH}_3\cdots\text{N}\cdots\text{CH}_3$$

(e)

$$\text{CNH}_2\ (\overset{\|}{\text{O}}),\ \text{CH}_3\quad\xleftarrow[\Delta]{\text{NH}_3}\quad\text{COCH}_2\text{CH}_3\ (\overset{\|}{\text{O}}),\ \text{CH}_3\quad\xleftarrow[(\text{CH}_3\text{CH}_2)_2\text{O}\cdot\text{BF}_3]{\overset{\text{O}}{\overset{\|}{\text{ClCOCH}_2\text{CH}_3}}}$$

$$\text{(N-CH}_3\text{ pyrrole)}\quad\xleftarrow[]{\text{CH}_3\text{I}\qquad\text{NaOH}}\quad\text{(pyrrole, N-H)}$$

(f)

$$\text{CCH}_2\text{Br}\ (\overset{\|}{\text{O}})\quad\xleftarrow[\substack{\text{acetic}\\\text{acid}}]{\text{Br}_2}\quad\text{CCH}_3\ (\overset{\|}{\text{O}})\quad\xleftarrow[\text{AlCl}_3]{\overset{\text{O}}{\overset{\|}{\text{CH}_3\text{CCl}}}}\quad\text{(thiophene)}$$

(g)

$$\text{CH}_3$$
$$\text{CCH}_2\text{C(COCH}_2\text{CH}_3)_2\ (\overset{\|}{\text{O}}\ \overset{\|}{\text{O}})\quad\xleftarrow[]{\text{[from (f)]}}\quad\text{CCH}_2\text{Br}\ (\overset{\|}{\text{O}})\quad\xleftarrow[\substack{\textit{tert}\text{-butyl}\\\text{alcohol}}]{\substack{\text{CH}_3\\|\\\text{CH}_3\text{COK}\\|\\\text{CH}_3}}$$

24.44 (cont)

(g) (cont)

$$\underset{CH_3CH(COCH_2CH_3)_2}{\overset{O}{\|}} \xleftarrow[\text{ethanol}]{CH_3I \quad CH_3CH_2ONa} \underset{CH_2(COCH_2CH_3)_2}{\overset{O}{\|}}$$

(h)

Thiophene-CCH₂CHCH₂-phenyl with COH and O groups $\xleftarrow[\substack{H_2O \\ \Delta}]{H_2SO_4}$

Thiophene-CCH₂C(COCH₂CH₃)₂ with CH₂-phenyl group $\xleftarrow{\text{[from (f)]}}$ Thiophene-CCH₂Br

$$\xleftarrow[\substack{\textit{tert}\text{-butyl} \\ \text{alcohol}}]{\substack{CH_3 \\ | \\ CH_3COK \\ | \\ CH_3}}$$

phenyl-CH₂C⁻(COCH₂CH₃)₂ K⁺

phenyl-CH₂CH(COCH₂CH₃)₂ $\xleftarrow{\overset{O}{\|} Na^{+-}CH(COCH_2CH_3)_2}$ phenyl-CH₂Br

24.45

aniline-NHNH₂ + $CH_3CH{=}CHCOH$ (with O) $\longrightarrow$
$$\left[\underset{\substack{H-N \quad \quad N-H \\ \\ \text{phenyl}}}{\overset{\substack{CH_3 \\ | \\ CH-CH_2COH \overset{O}{\|}}}{}} \right] \longrightarrow$$
5-methyl-2-phenyl-pyrazolidinone   C₁₀H₁₂N₂O

product from nucleophilic
addition to α,β-unsaturated
compound

24.45 (cont)

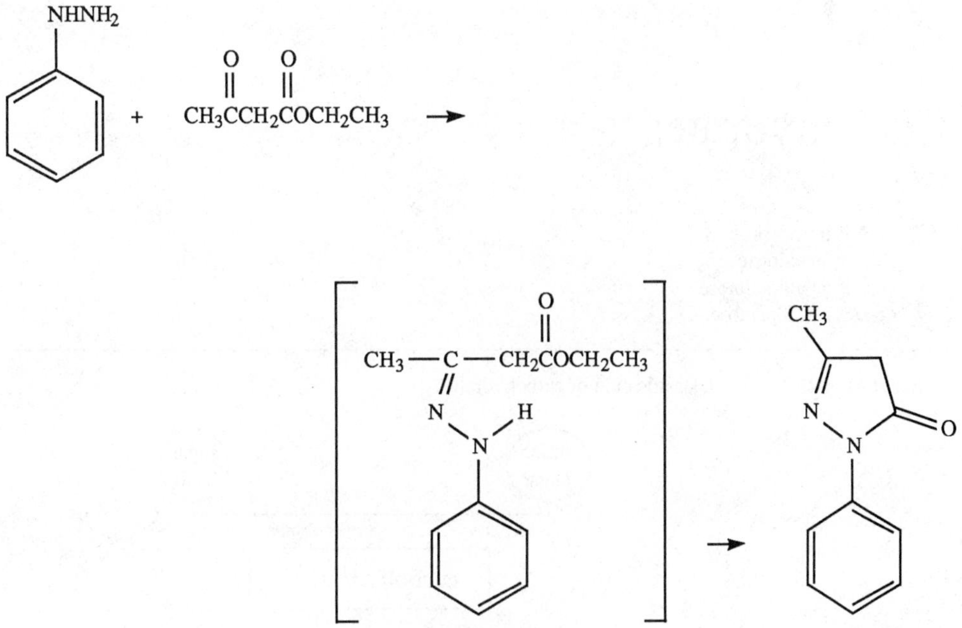

product from nucleophilic
addition to carbonyl of ketone
followed by loss of water to
give a phenylhydrazone

$C_{10}H_{10}N_2O$

25

Carbohydrates

25.1 A is a ketotriose
B is an aldotriose
C is an aldotetraose
D is a ketopentose

Concept Map 25.1   Classification of carbohydrates.

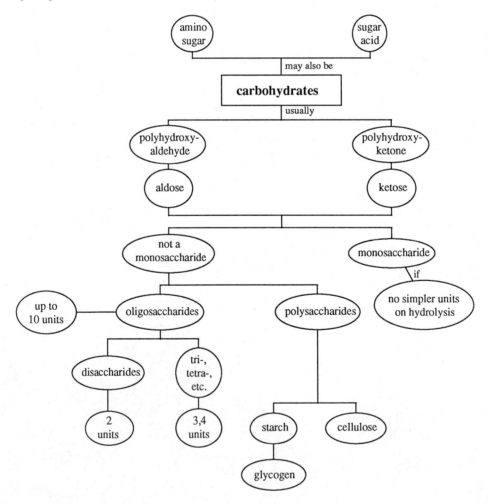

25.2   The configuration of (–)-lactic acid has been established as being *R* (Section 25.2A). Once we write the correct stereochemical structure for (*R*)-(–)-lactic acid, we can assign configuration to (–)-1-buten-3-ol and (–)-2-butanol also, because the reactions used for the interconversions do not break bonds to the stereocenter.

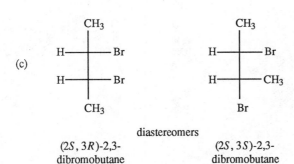

25.3   **(a)**

enantiomers

(*S*)-2-bromobutane   (*R*)-2-bromobutane

(b)

same
(*R*)-lactic acid

(c)

diastereomers

(2*S*, 3*R*)-2,3-   (2*S*, 3*S*)-2,3-
dibromobutane   dibromobutane

25.3 (cont)

(d)

O
‖
COH
H————OH
HO————H
COH
‖
O

H   O
     ‖
HO————COH
HO————COH
     ‖
H   O

diastereomers

(2*R*, 3*R*)-2,3-
dihydroxy
butanedioic acid

(2*R*, 3*S*)-2,3-
dihydroxy
butanedioic acid

25.4   D-(+)-glyceraldehyde ≡ (*R*)-(+)-glyceraldehyde
D-(−)-lactic acid ≡ (*R*)-(−)-lactic acid
D-(−)-tartaric acid ≡ (2*S*, 3*S*)-(−)-2,3-dihydroxybutanedioic acid
D-(−)-ribose ≡ (2*R*, 3*R*, 4*R*)-(−)-2,3,4,5-tetrahydroxypentanal
D-(+)-glucose ≡ (2*R*, 3*S*, 4*R*, 5*R*)-(+)-2,3,4,5-pentahydroxyhexanal
D-(−)-fructose ≡ (3*S*, 4*R*, 5*R*)-(−)-1,3,4,5,6-pentahydroxy-2-hexanone
L-(−)-glyceraldehyde ≡ (*S*)-(−)-glyceraldehyde
L-(+)-lactic acid ≡ (*S*)-(+)-lactic acid
L-(+)-tartaric acid ≡ (2*R*, 3*R*)-(+)-2,3-dihydroxybutanedioic acid
L-(+)-alanine ≡ (*S*)-(+)-alanine
L-(+)-arabinose ≡ (2*R*, 3*S*, 4*S*)-(+)-2,3,4,5-tetrahydroxypentanal

25.5

O
‖
COH
H————OH
H————OH
COH
‖
O

meso-tartaric acid

O   O
‖   ‖
HOC   COH
    \   /
  H⋯C————C⋯H
   HO   OH

The plane of symmetry bisects
the central carbon-carbon bond

25.6

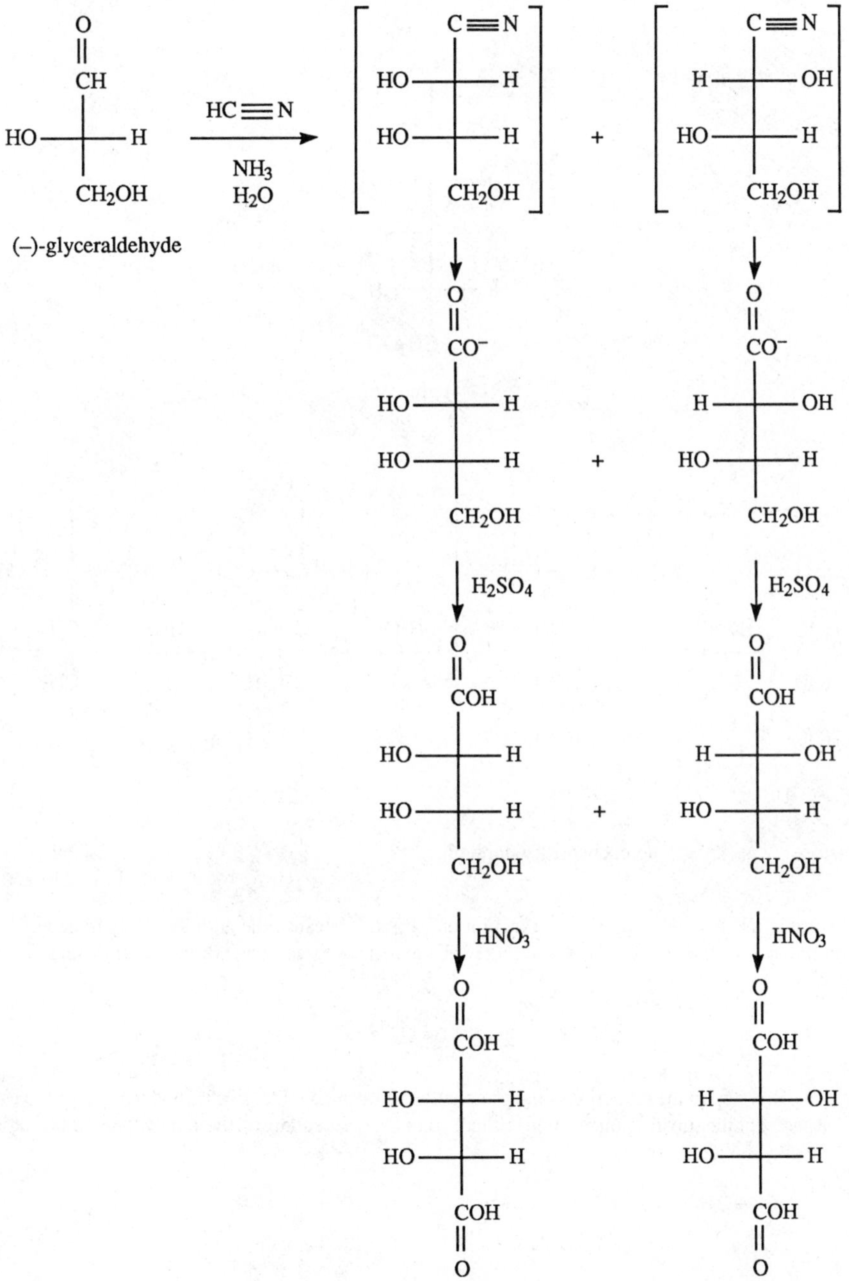

$(-)$-glyceraldehyde

$$\underset{\substack{\text{HC}\equiv\text{N}\\[2pt]\text{NH}_3\\\text{H}_2\text{O}}}{\longrightarrow}$$

$(2S, 3R)$-2,3- dihydroxy-
butanedioic acid

meso-tartaric acid
optically inactive

$(2R, 3R)$-2,3- dihydroxy-
butanedioic acid

L-$(+)$-tartaric acid
optically active

$$\underset{\text{L-(-)-glucose}}{\begin{array}{c} \overset{\displaystyle O}{\overset{\displaystyle \|}{CH}} \\ \text{HO}\text{---}\!\!\!\!\!\!\overset{}{}\!\!\!\!\!\!\text{---H} \\ \text{H}\text{---}\!\!\!\!\!\!\!\!\!\!\!\!\text{---OH} \\ \text{HO}\text{---}\!\!\!\!\!\!\!\!\!\!\!\!\text{---H} \\ \text{HO}\text{---}\!\!\!\!\!\!\!\!\!\!\!\!\text{---H} \\ CH_2OH \end{array}}$$

25.7

25.8

fructose   no stereochemistry implied   2-methyl-
hexanoic acid

Fructose is a ketose with the ketone function on carbon 2 because the addition of hydrogen cyanide to the carbonyl and the hydrolysis of the resulting nitrile give a seven-carbon carboxylic acid with branching at the second carbon atom.

25.9 (a) The first step in the reduction is the protonation of the alcohol to give a good leaving group followed by a nucleophilic substitution reaction (either S_N1 or S_N2, depending on the nature of the carbon atom) to form an alkyl halide.

$$R\text{---}OH \;+\; HI \longrightarrow R\text{---}I \;+\; H_2O$$

(b)

25.9   (b) (cont)

The rest of the reaction must be a free radical chain reaction.

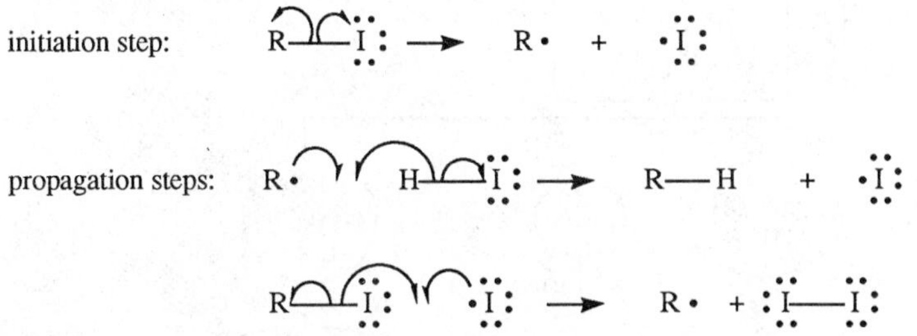

initiation step:

propagation steps:

- -

Concept Map 25.2   (see p. 836)

- -

25.10

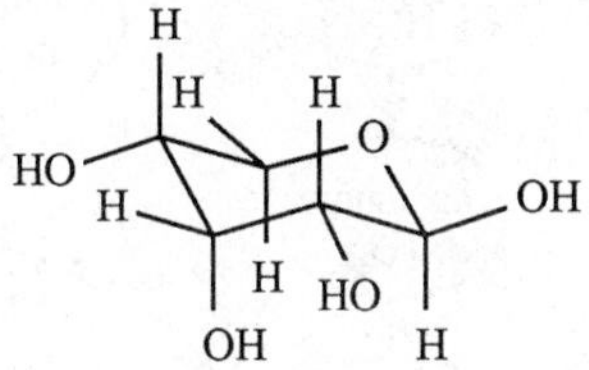

β-D-ribopyranose

25.11

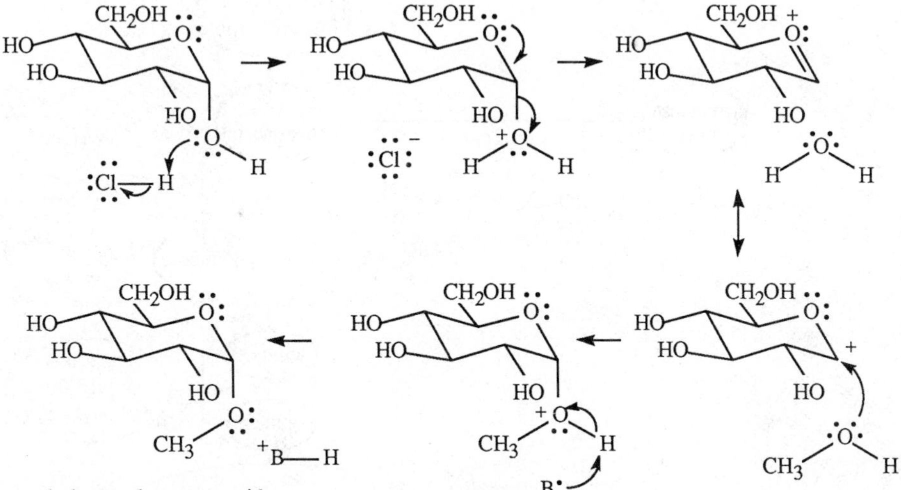

methyl α-D-glucopyranoside

Both α- and β-glucopyranose give the same carbocation intermediate. The intermediate reacts with methanol to give either the α-glucopyranoside or the β-glucopyranoside, depending on which side of the carbocation is attacked by the alcohol. The composition of the product mixture is determined by the relative stabilities of the two methyl glucopyranosides.

Concept Map 25.2   Structure of glucose.

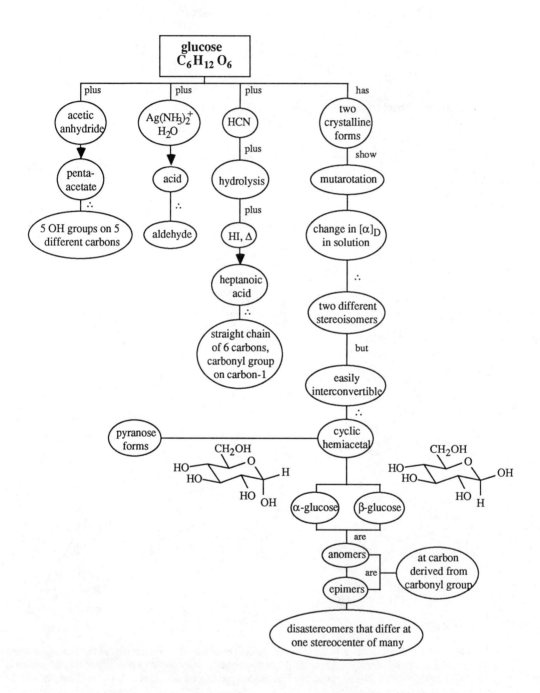

25.12

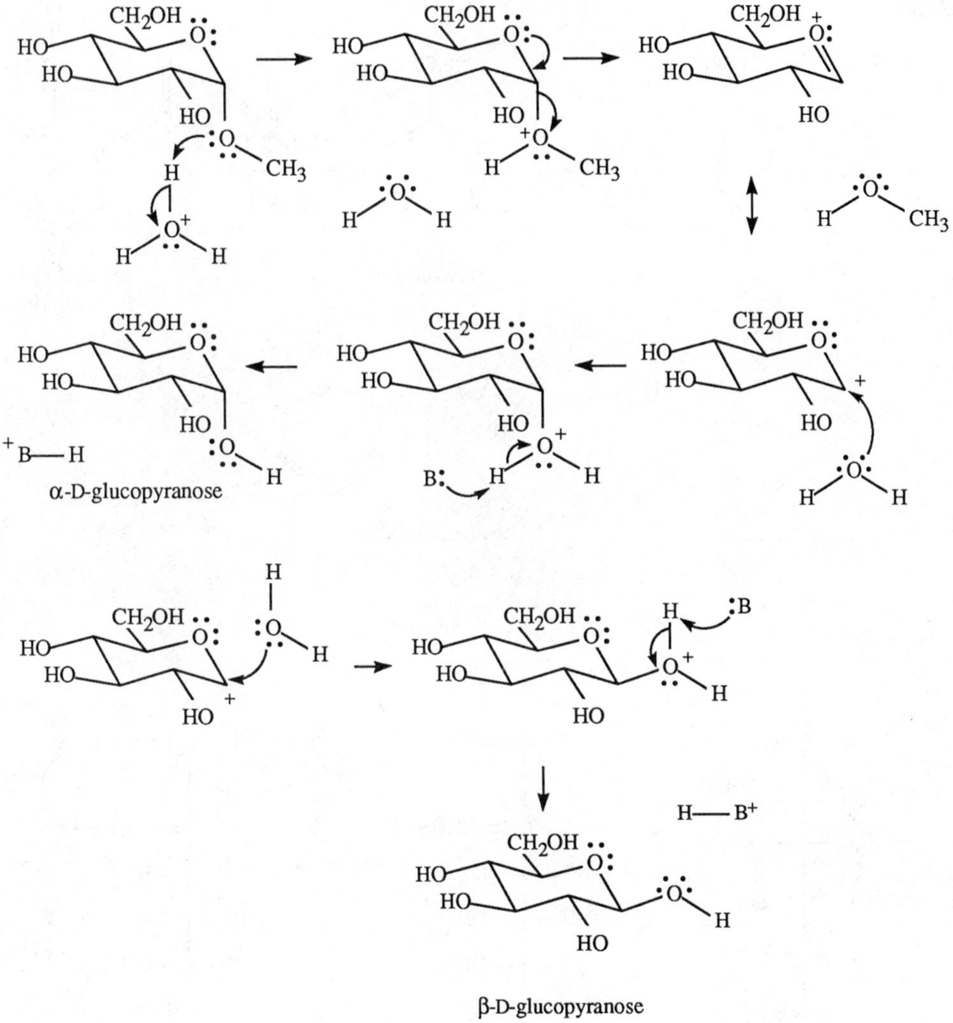

α-D-glucopyranose

β-D-glucopyranose

25.13

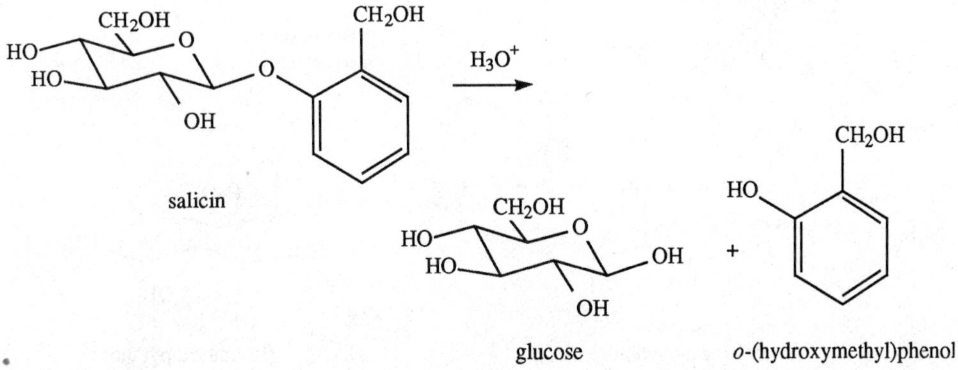

salicin

glucose

o-(hydroxymethyl)phenol

25.13 (cont)

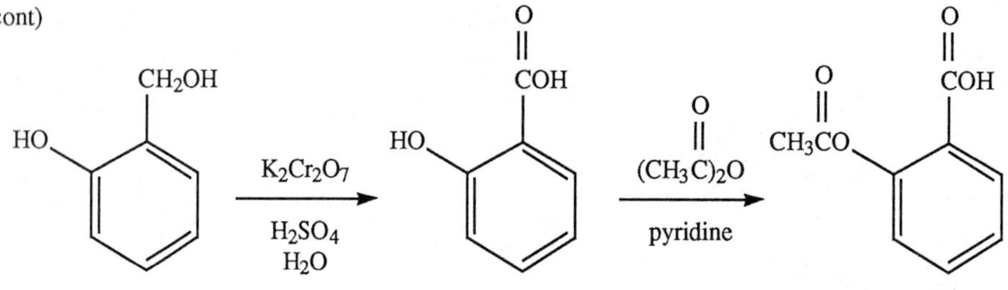

salicylic acid

acetylsalicylic acid
aspirin

25.14

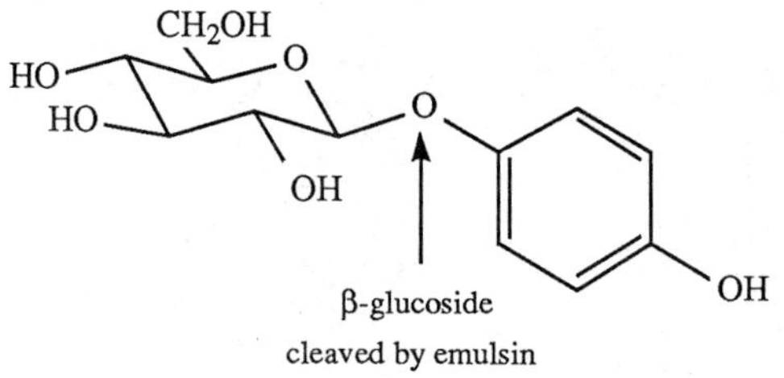

β-glucoside
cleaved by emulsin

arbutin

25.15

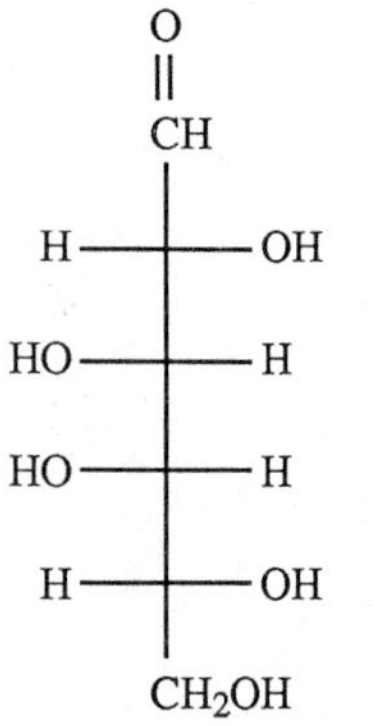

D-(+)-galactose

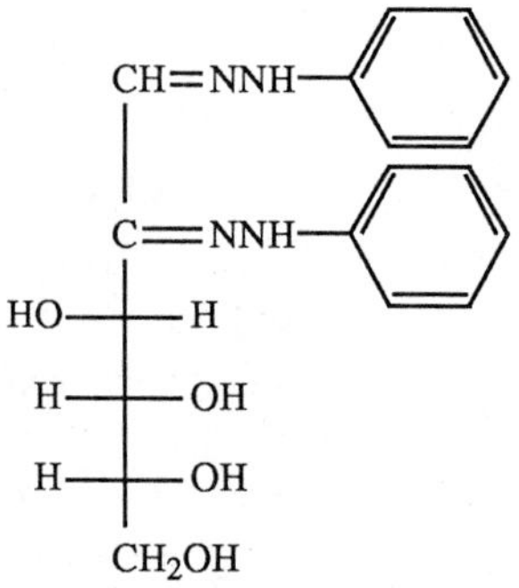

glucosazone

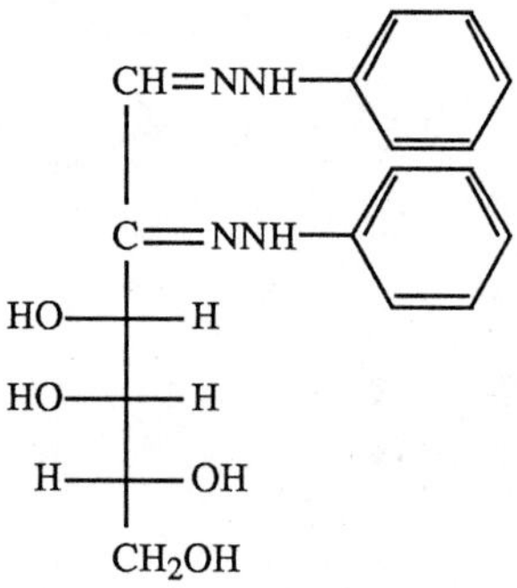

galactosazone

diastereomers
also epimers

25.16

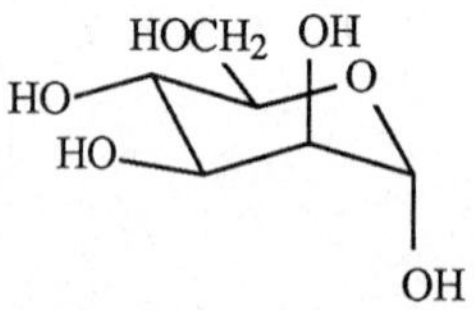

α-D-mannopyranose

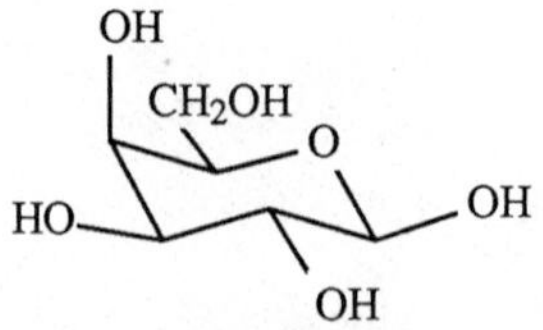

β-D-galactopyranose

25.17   (a)

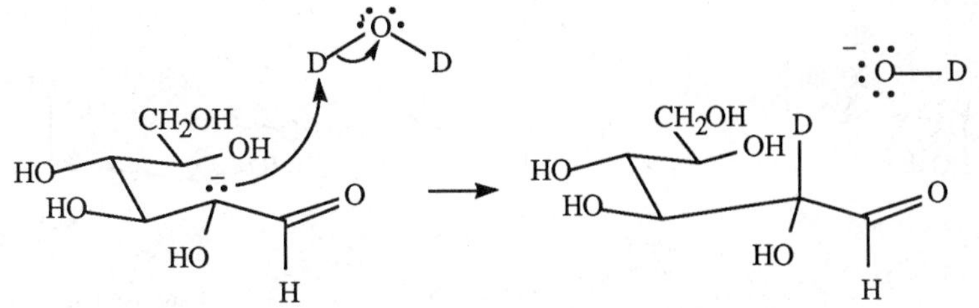

glucose with
deuterium at C-2

The carbanion at C-2 reacts with D_2O to give a deuterium atom bonded to carbon at that position in glucose or mannose.

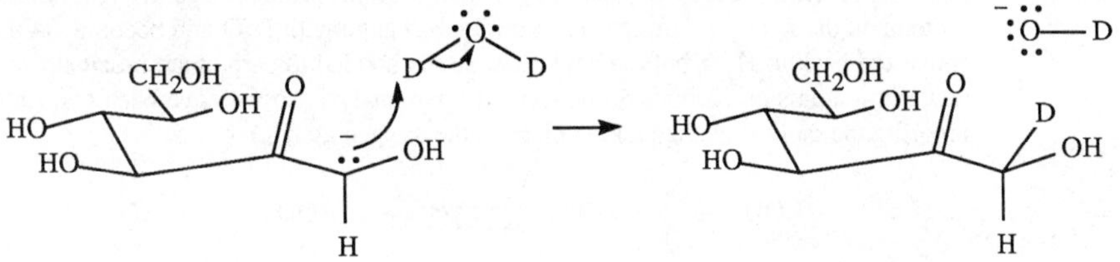

fructose with
deuterium at C-1

The carbanion at C-1 gives fructose with deuterium bonded to C-1.

(b) Fructose can pick up more than one deuterium atom because it can enolize again losing a hydrogen atom from C-1. This enolate anion can be deuterated again at C-1.

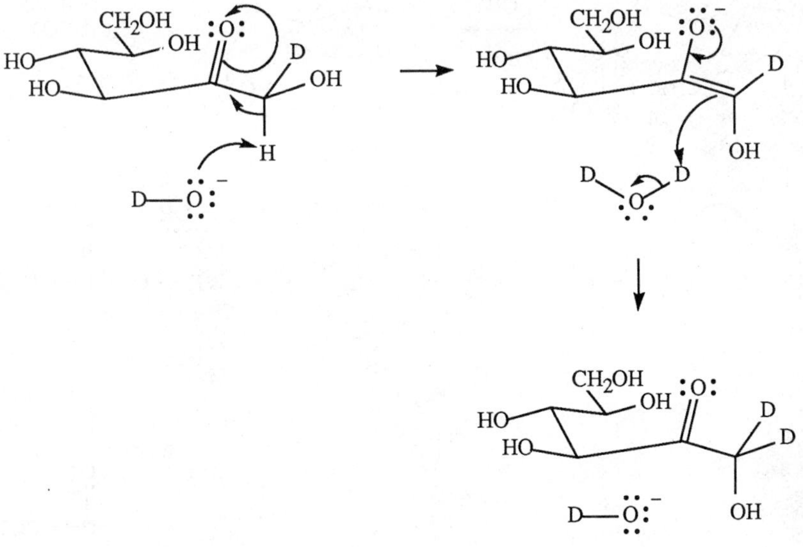

25.17   (b) (cont)

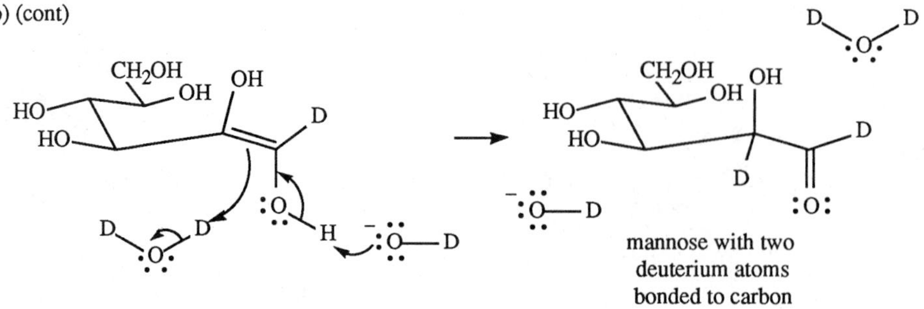

mannose with two
deuterium atoms
bonded to carbon

The presence of more than one deuterium atom, on the average, in mannose suggests that some of the mannose is coming from fructose. (In the answer to this problem we have ignored the fact that all of the protons on the hydroxyl groups very rapidly exchange with D_2O and become —OD groups. They are converted back to H_2O, but the deuterium atoms bonded to carbon are much slower to exchange. The deuterium atoms on carbon can be located upon analysis so we have been concentrating on them and ignoring the rapid exchange taking place at the oxygen atoms.)

$$ROH \;+\; D_2O \;\rightleftharpoons\; ROD \;+\; HOD$$

$$ROD \;+\; H_2O \;\rightleftharpoons\; ROH \;+\; HOD$$

25.18

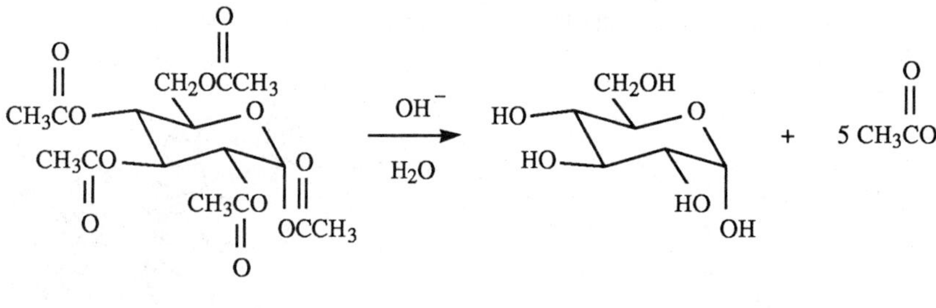

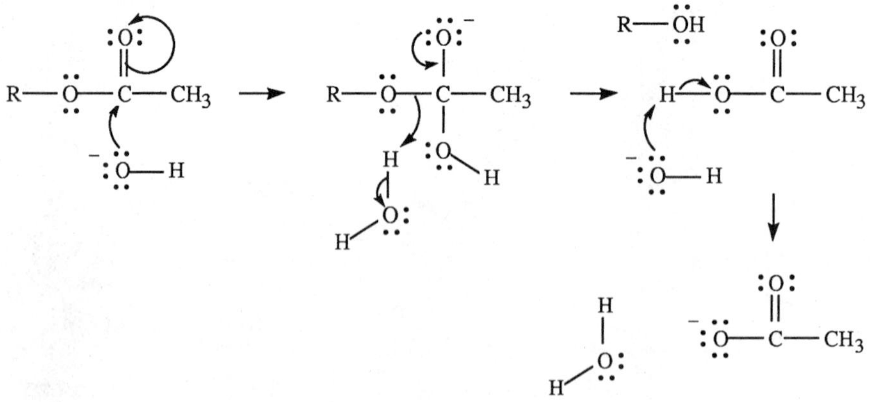

25.19

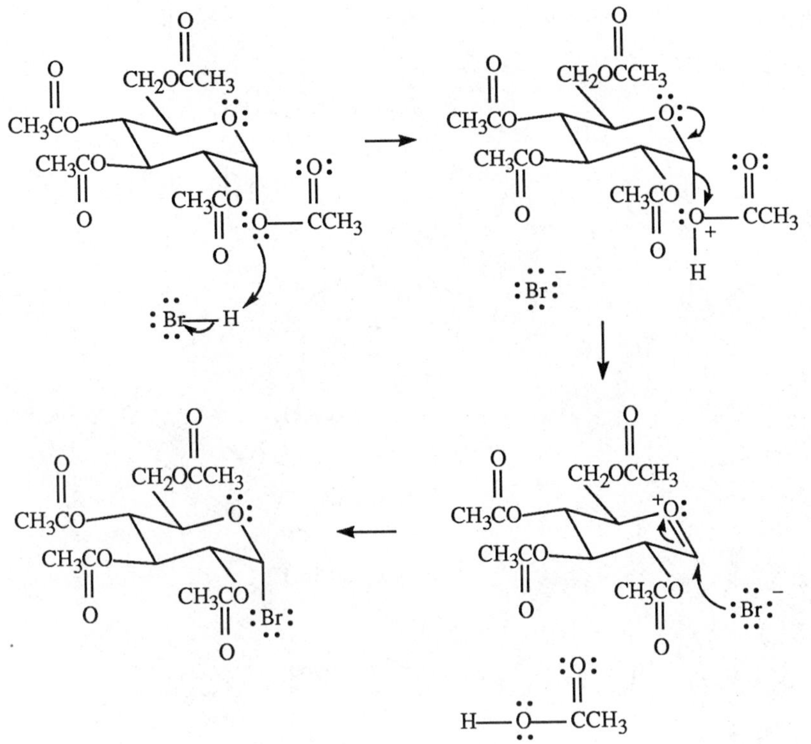

25.20   Acid hydrolysis would also cleave the glycosidic linkage between glucose and the phenol.

25.21

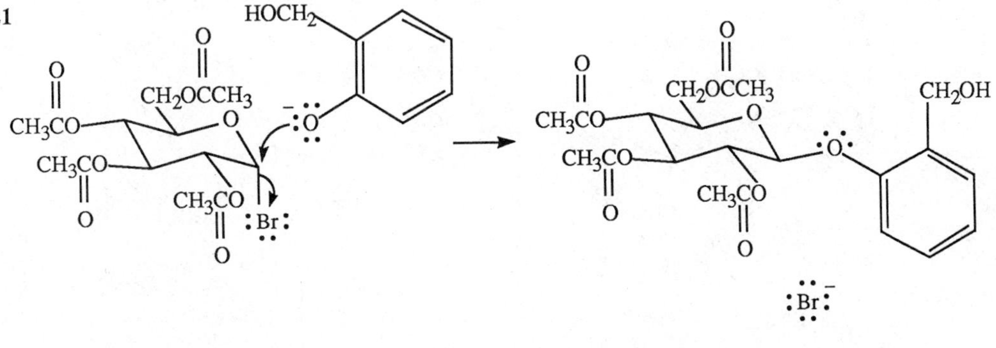

The reaction is an S_N2 reaction. It takes place with inversion of configuration.

25.22   Hydrolysis at C-1 gives a resonance-stabilized carbocation.

resonance-stabilized
cation

Cleavage of the linkage at C-2 gives a secondary carbocation with no possibility of delocalization of charge.

25.22 (cont)

secondary
carbocation at C-2
no delocalization
of charge possible

25.23 $R-\overset{..}{\underset{..}{O}}-CH_3$ $\longrightarrow$ $R-\overset{+}{\underset{|}{\underset{H}{O}}}-CH_3$ $\longrightarrow$ $R-\overset{..}{\underset{..}{O}}-H$ $+$ $CH_3-\overset{..}{\underset{..}{I}}:$

The concentrated acid protonates the ether, converting it to an oxonium ion with a good leaving group. Iodide ion displaces the leaving group. In dilute hydrochloric acid, water molecules compete with the ether for protons. The concentration of oxonium ions is much lower in dilute acid than in the concentrated acid. In addition, chloride ion is a weaker nucleophile than iodide ion, so the S_N2 reaction does not occur as readily. Therefore, it is possible to hydrolyze a glycoside linkage in dilute acid and leave untouched methyl ether functions at other hydroxyl groups in a sugar.

25.24

D-allose   allaric acid

25.25

25.25 (cont)

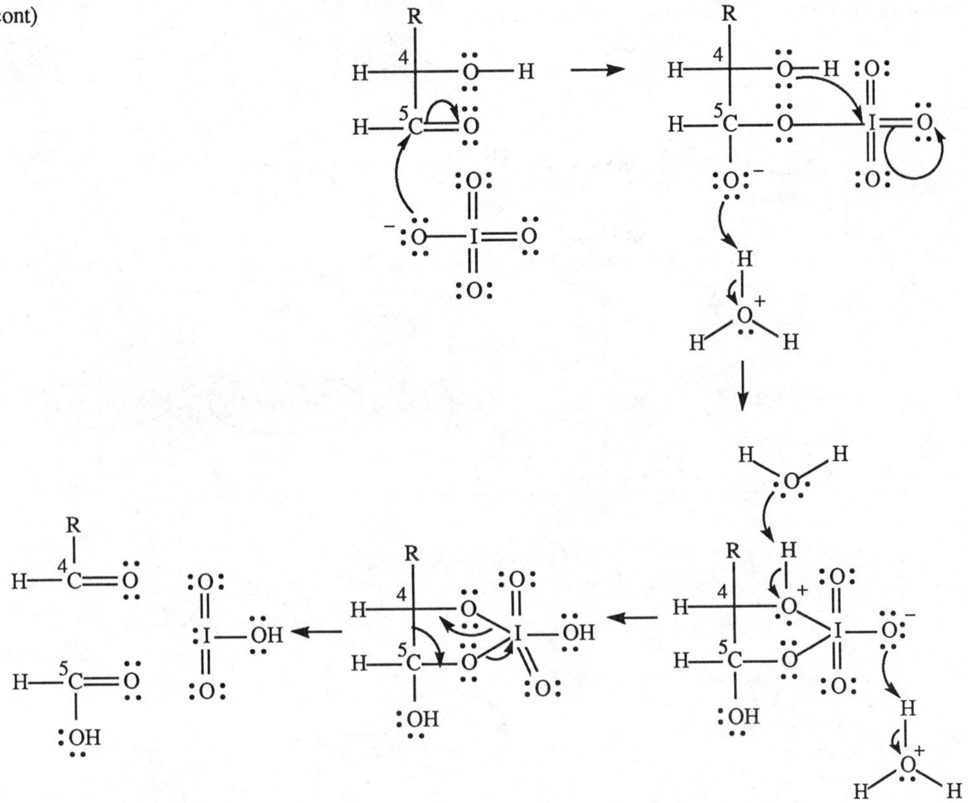

25.26

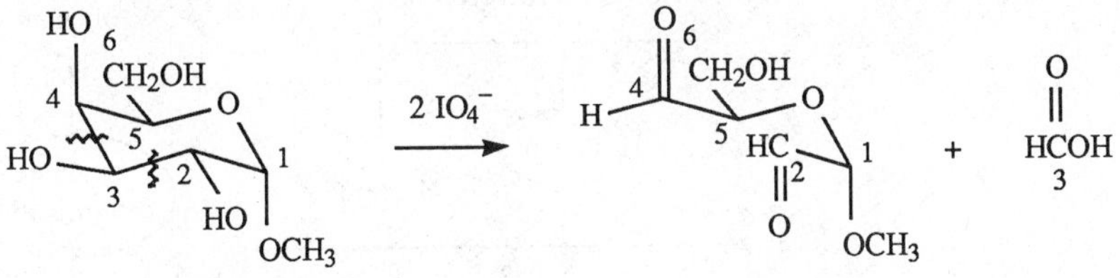

methyl α-D-galactopyranoside

25.27 $SrCO_3$ (s) $\rightleftharpoons$ [$SrCO_3$ (aq)] $\longrightarrow$ Sr^{2+} + CO_3^{2-} $\xrightarrow{H_3O^+}$

$$H_2O \ + \ HCO_3^- \ \xrightarrow{\ H_3O^+\ } \ H_2CO_3 \ \longrightarrow \ H_2O \ + \ CO_2$$

Strontium carbonate is a source of a low concentration of carbonate ions, which are basic and react with acid to give carbonic acid, and ultimately carbon dioxide. The oxidation of the carbohydrate with bromine in water has HBr as one of the products. If the acid were not neutralized, it would catalyze the hydrolysis of the acetal linkage at the anomeric carbon atom. The stereochemistry of carbon 1 in the product of the oxidation reaction would be prematurely lost.

Concept Map 25.3   Reactions of monosaccharides.

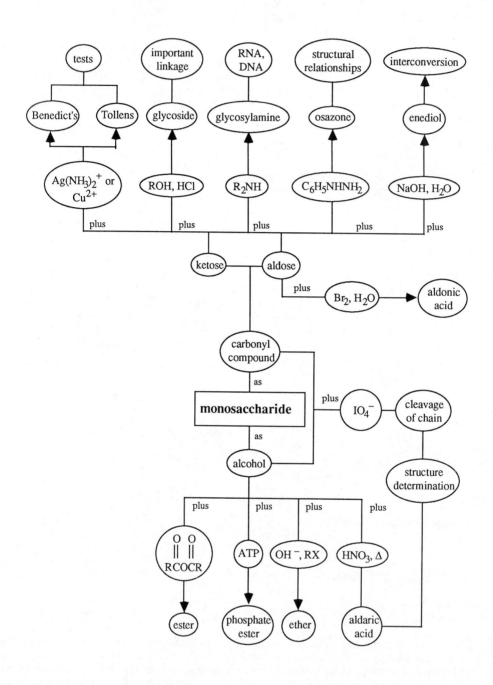

25.28

(a)

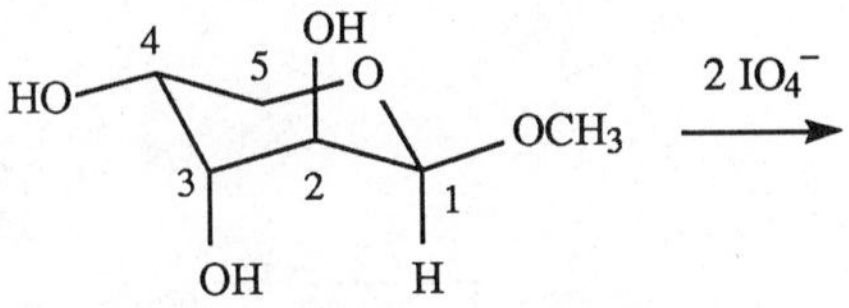

methyl β-D-arabinopyranoside

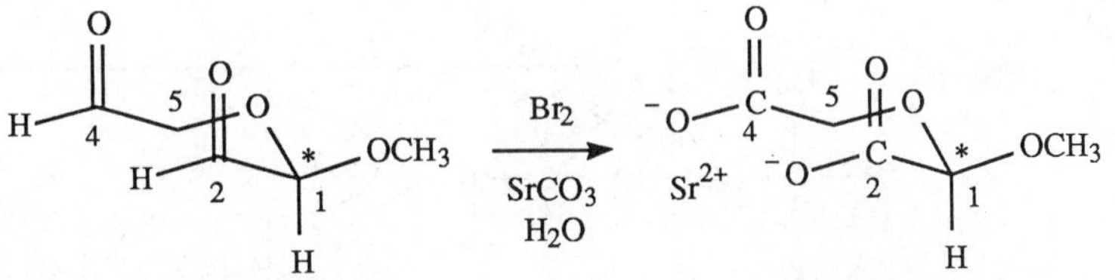

dialdehyde with
one stereocenter

strontium salt of the diacid;
C-1 is the only stereocenter

(b)   The strontium salt of the diacid obtained from the α–pyranoside would be the enantiomer of the strontium salt shown in part (a).

25.29   Excess sodium borohydride would reduce the aldose to the alcohol.

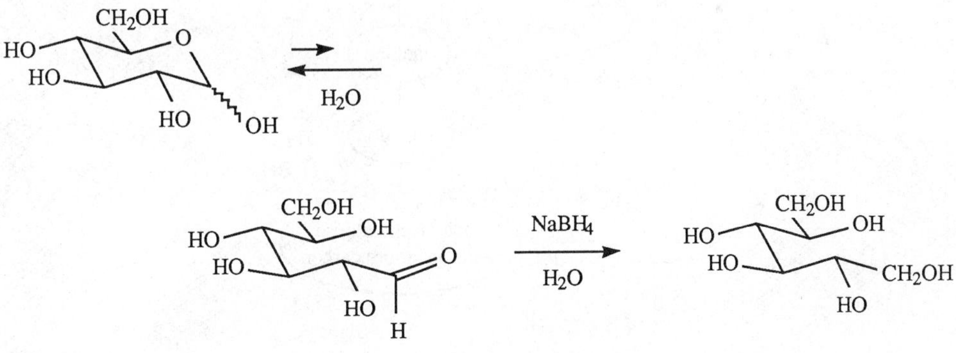

Concept Map 25.4   The Kiliani-Fischer synthesis.

25.30

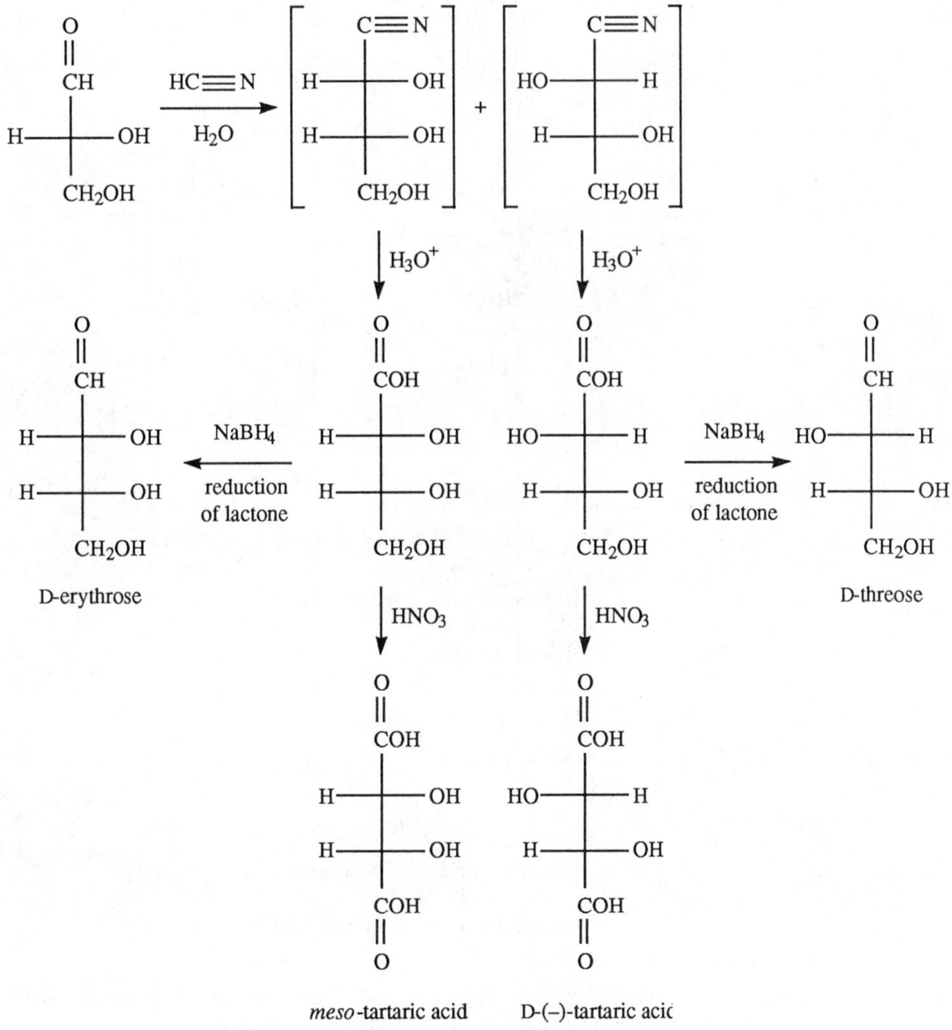

O
CH
H
OH
CH2OH
HC≡N
H2O
C≡N
H
OH
H
OH
CH2OH
+
C≡N
HO
H
H
OH
CH2OH
H3O+
O
COH
NaBH4
reduction of lactone
D-erythrose
D-threose
HNO3
meso-tartaric acid
D-(–)-tartaric acic

25.31

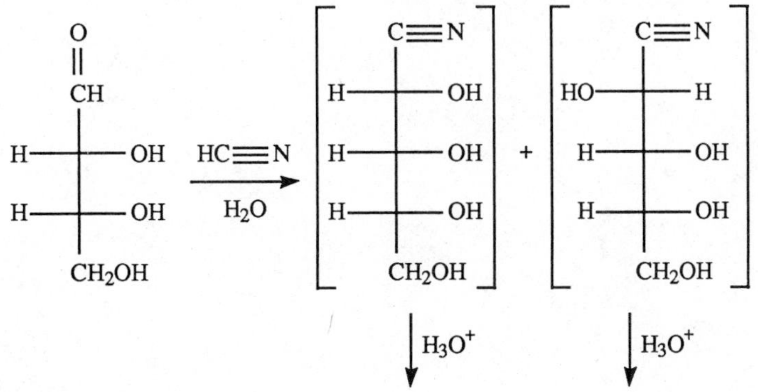

O
CH
H
OH
H
OH
CH2OH
HC≡N
H2O
C≡N
+
HO
H
CH2OH
H3O+

25.31 (cont)

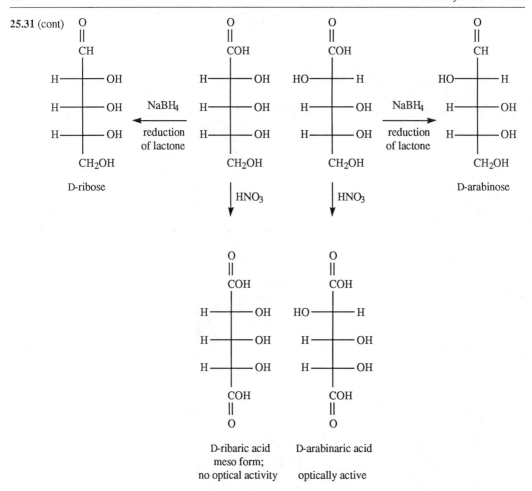

D-Ribose must have all of the OH groups on the right in the Fischer projection formula while D-arabinose must be the epimer of D-ribose at the new stereocenter created in this synthesis.

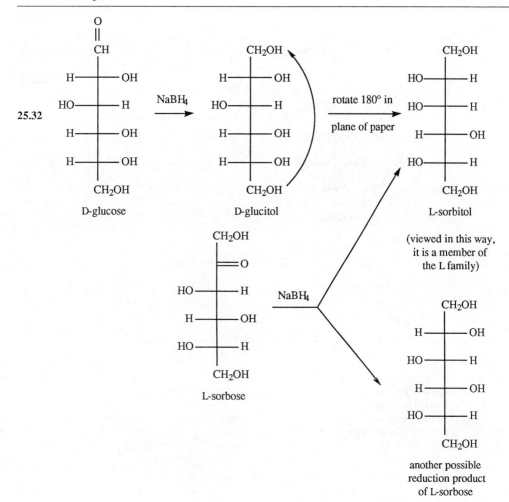

25.32

D-glucose →NaBH₄→ D-glucitol →rotate 180° in plane of paper→ L-sorbitol

(viewed in this way, it is a member of the L family)

L-sorbose →NaBH₄→ another possible reduction product of L-sorbose

25.33   sucrose →H₂O, CH₃COH→

25.33 (cont)

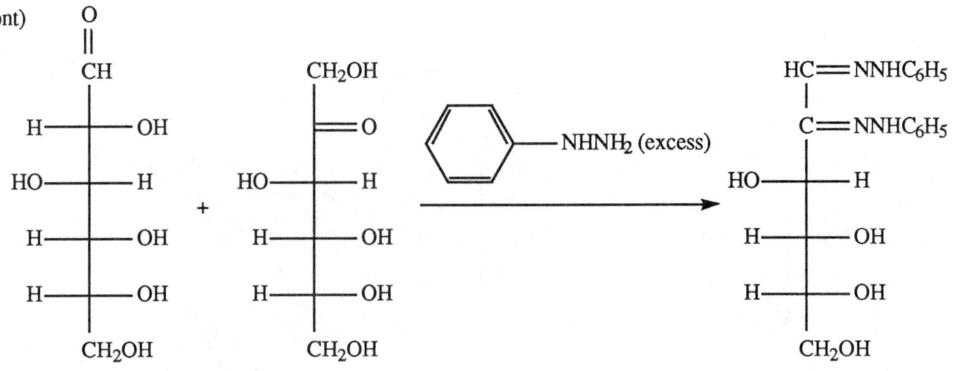

glucose   fructose   glucosazone

25.34

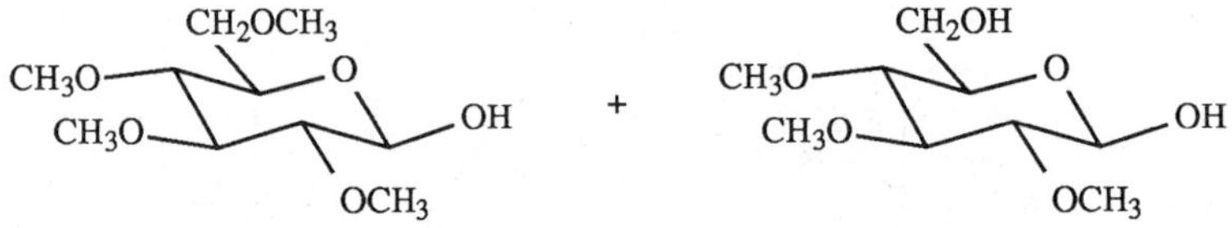

2,3,4,6-*O*-tetramethyl-   +   2,3,4,-*O*-trimethyl-
D-glucopyranose   D-glucopyranose
(therefore hydroxyl group   (therefore hydroxyl group
at C-1 in glycosidic linkage)   at C-6 in glycosidic linkage;
                                             anomeric carbon atom free
                                             because gentiobiose is
                                             a reducing sugar)

$\uparrow$ H$_3$O$^+$

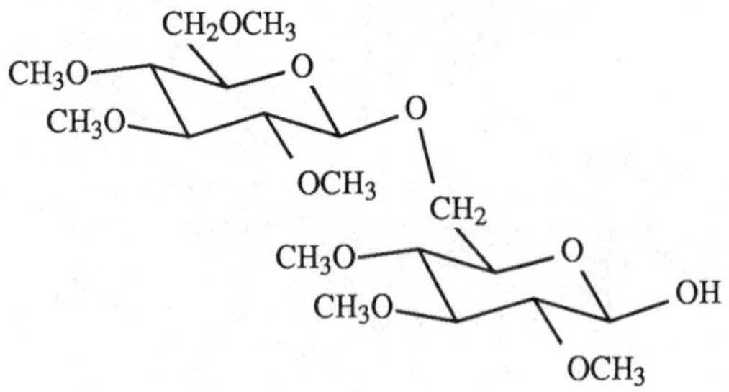

methylated gentiobiose

$\uparrow$ methylation

25.34 (cont)

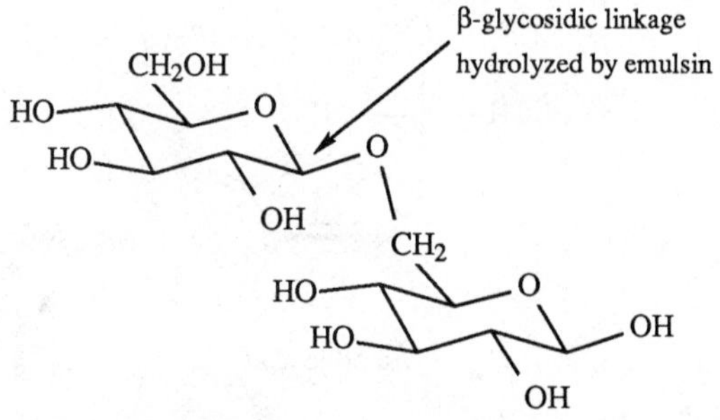

β-glycosidic linkage
hydrolyzed by emulsin

O-β-D-glucopyranosyl-(1 → 6)-D-glucopyranose

25.35

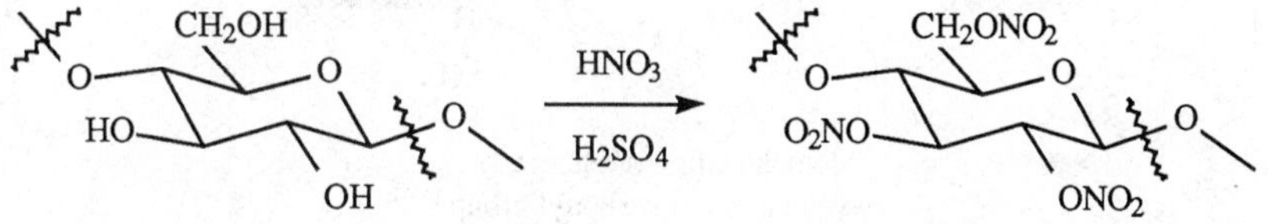

$$\xrightarrow[\text{H}_2\text{SO}_4]{\text{HNO}_3}$$

portion of cellulose   portion of cellulose trinitrate

25.36

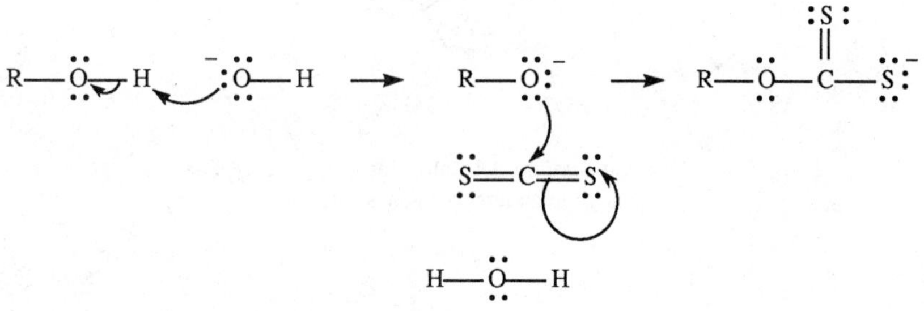

25.37

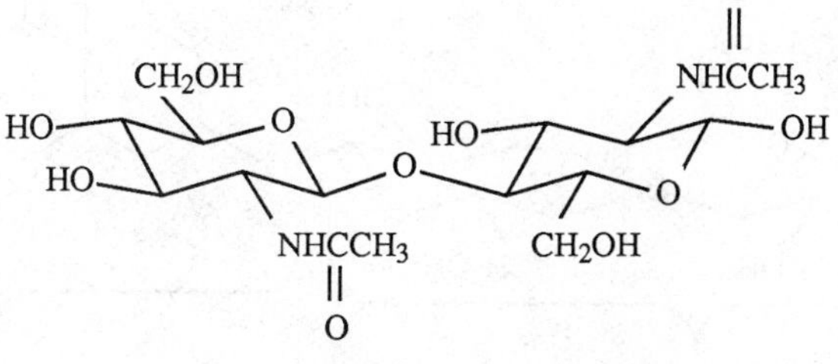

chitobiose

25.38

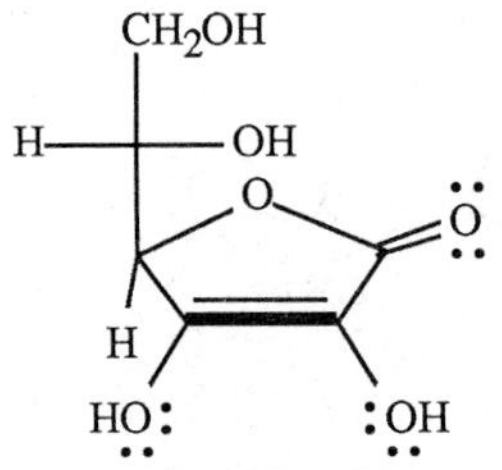

ascorbic acid

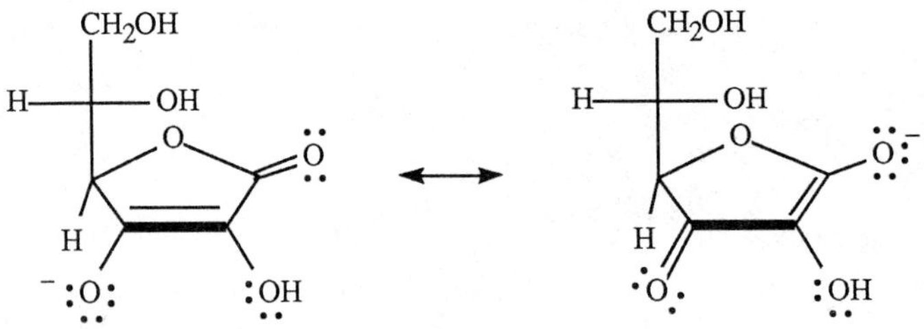

delocalization of charge to
oxygen atom of carbonyl group

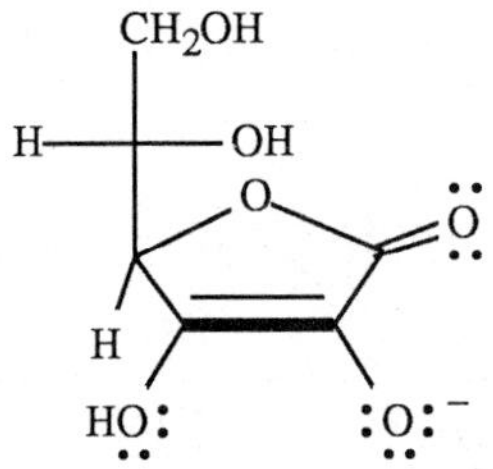

no possible delocalization of
charge to another oxygen atom

25.39

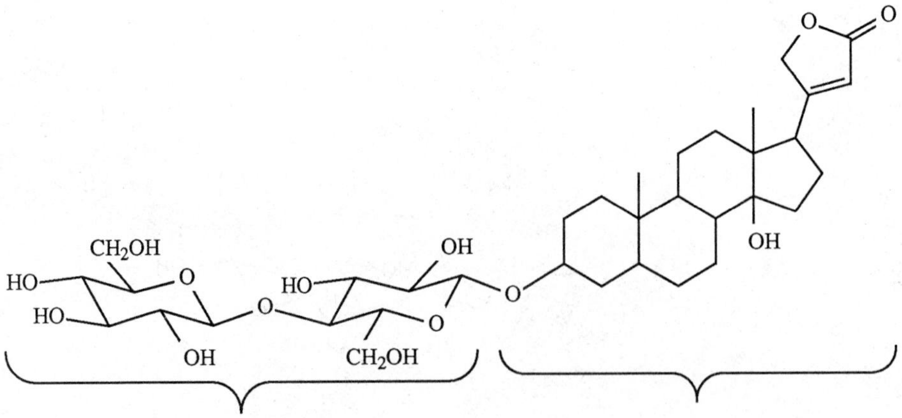

cellobiose   digitoxigenin

25.40   (a)

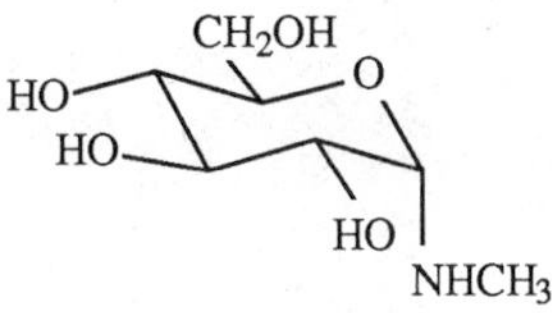

(b)

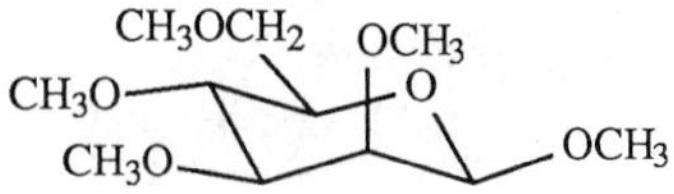

(c)

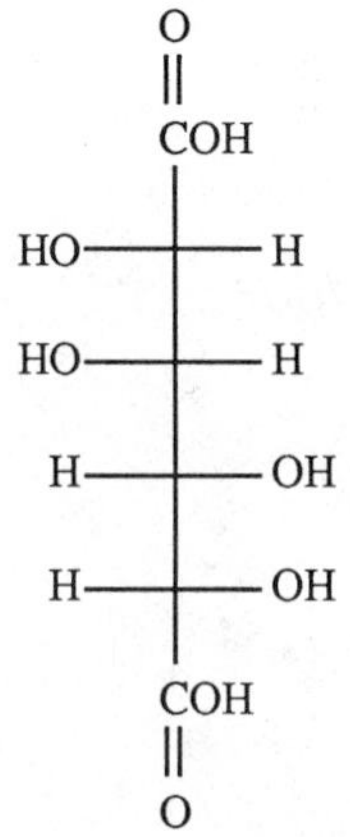

(d)

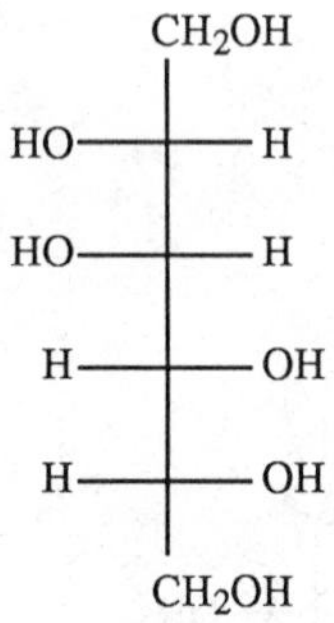

(e)

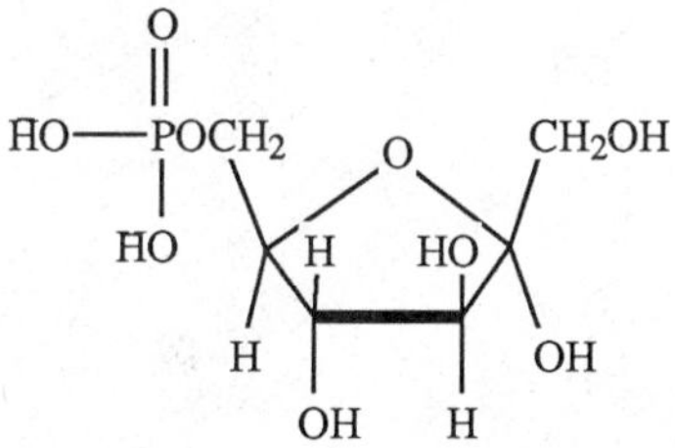

(f)

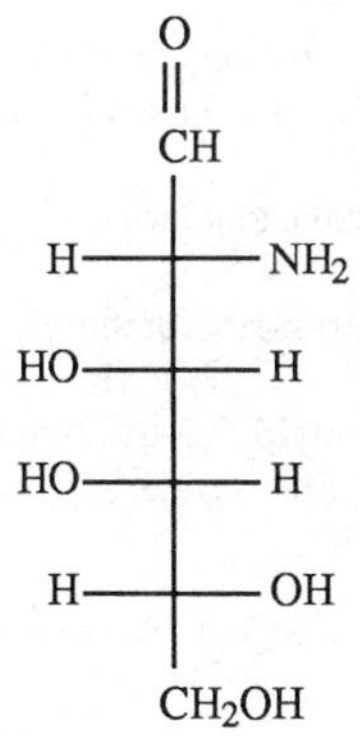

or

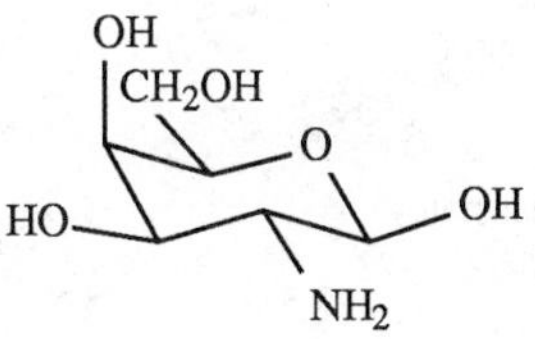

25.40 (cont)

(g)

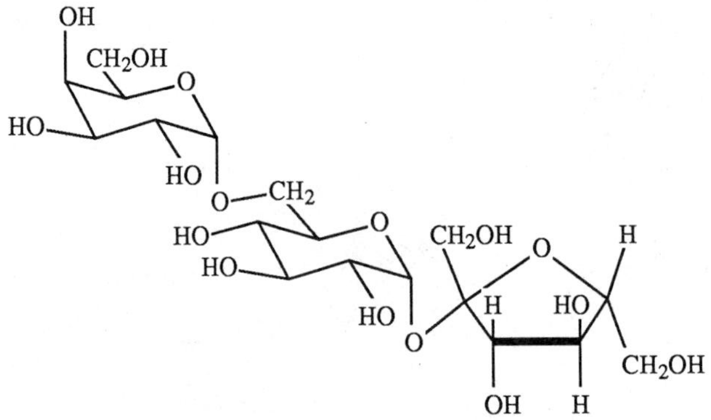

(h)

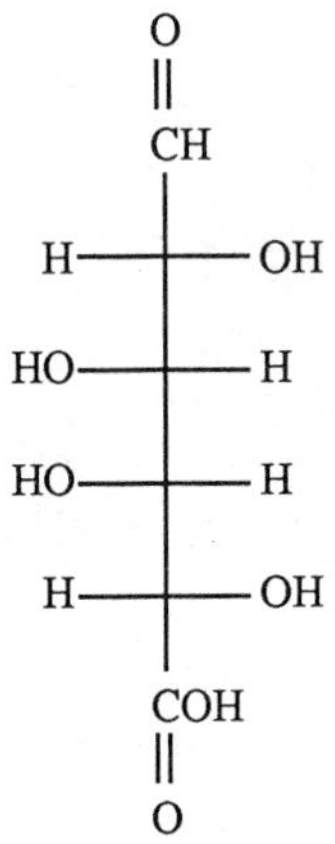

or

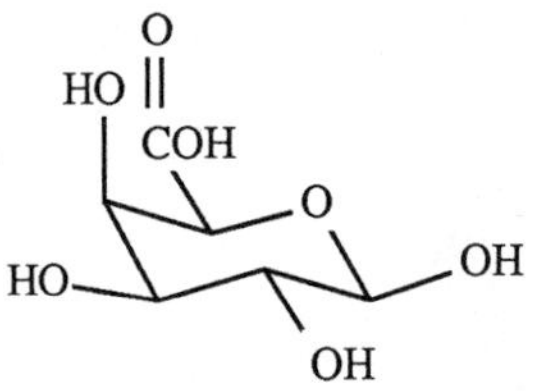

25.41 (a) *N*-acetyl-2-amino-2-deoxy-D-galactose
2-acetamido-2-deoxy-D-galactose

(b) α-D-glucopyranose-1-phosphate
(shown in ionized form)

(c) methyl β-D-2-deoxyribofuranoside

(d) δ-galactonolactone

(e) L-galactose

(f) *N*-(β-D-ribofuranosyl)dimethylamine

(g) α-D-galactopyranose pentaacetate

(h) methyl 2,3,5-tri-*O*-methyl-β-D-
ribofuranoside

25.42

(a)

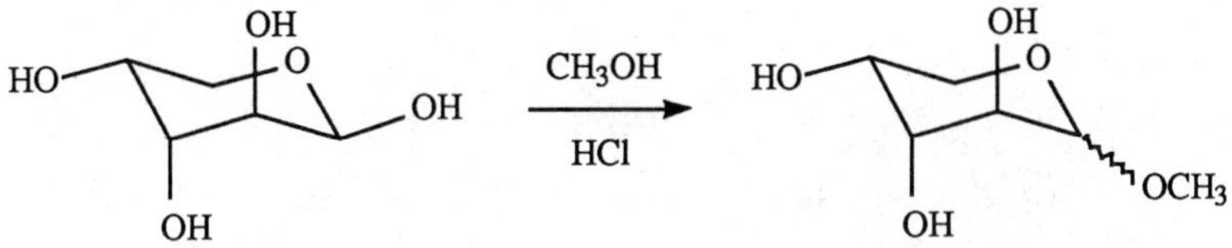

shown as β-D-arabinopyranose

mixture of methyl α- and
β-D-arabinopyranosides

(b)

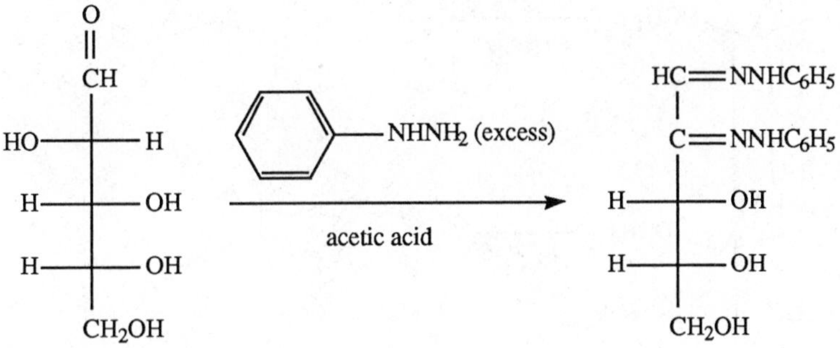

(c)

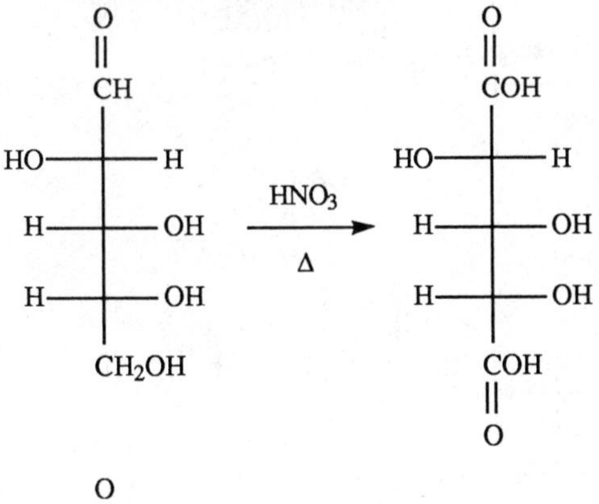

(d)

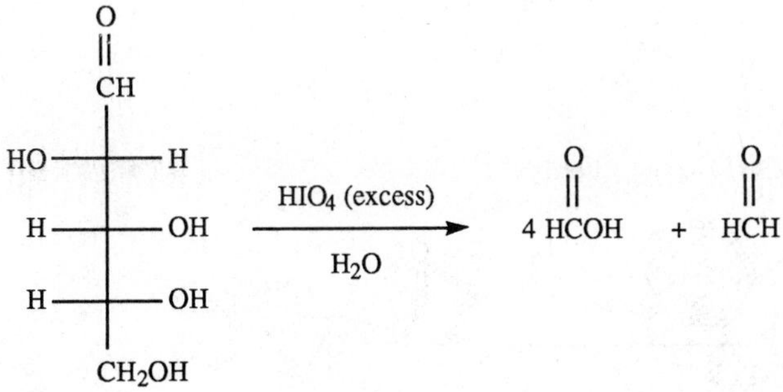

25.42 (cont)

(e)

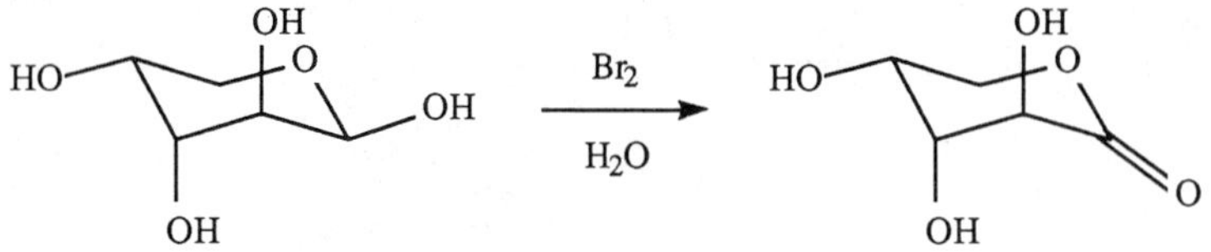

$$\xrightarrow[\text{H}_2\text{O}]{\text{Br}_2}$$

(f)

$$\xrightarrow[\text{NaOH}]{(\text{CH}_3)_2\text{SO}_4}$$

(g)

$$\xrightarrow[\text{HCl}]{\text{H}_2\text{O}}$$

(h)

$$\xrightarrow[\text{pyridine}]{(\text{CH}_3\text{C})_2\text{O (excess)}}$$

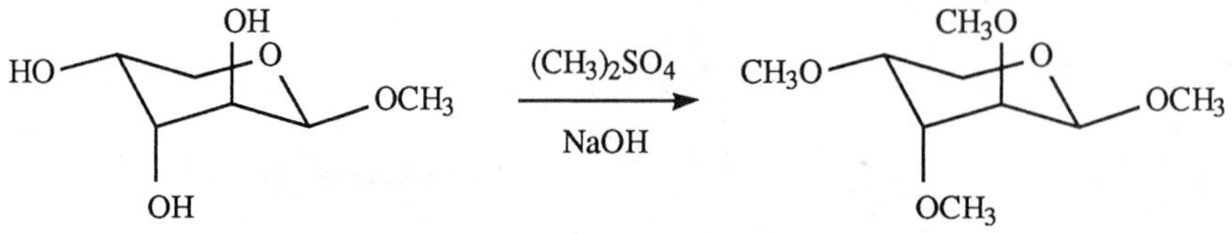

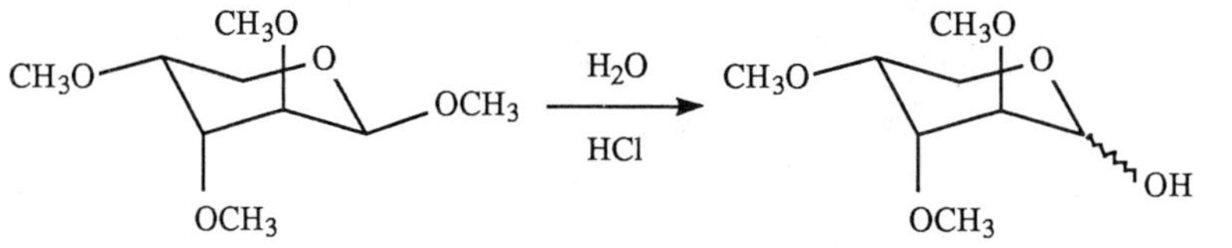

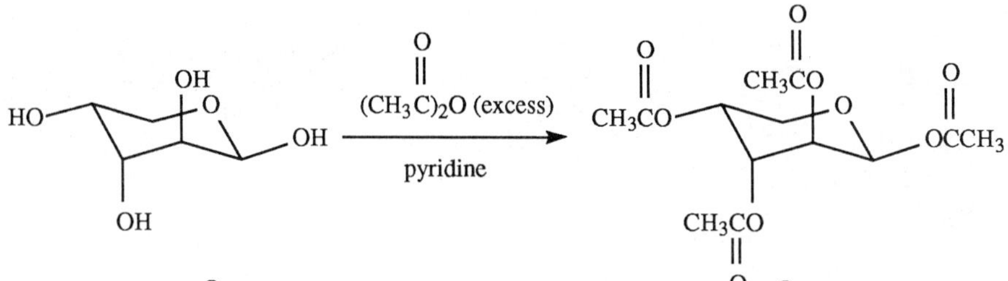

(i)

$$\xrightarrow[\text{cold}]{\text{HBr}}$$

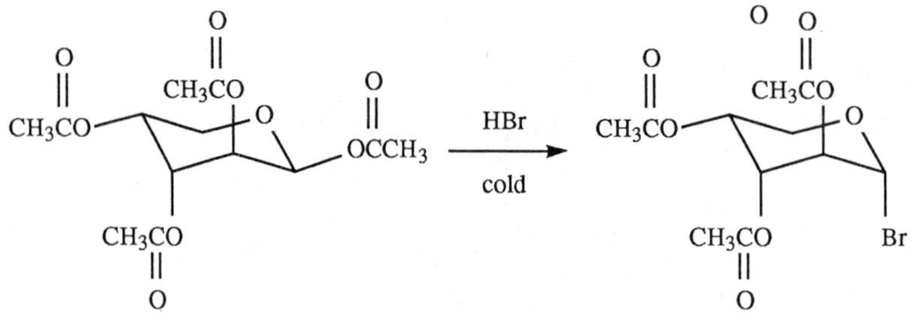

(j)

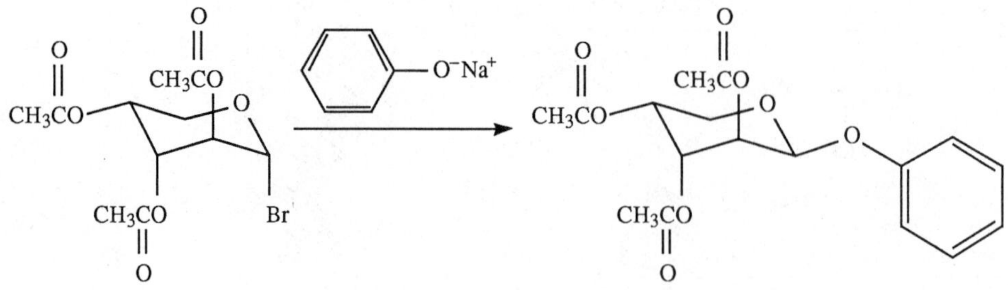

25.42 (cont)

(k)

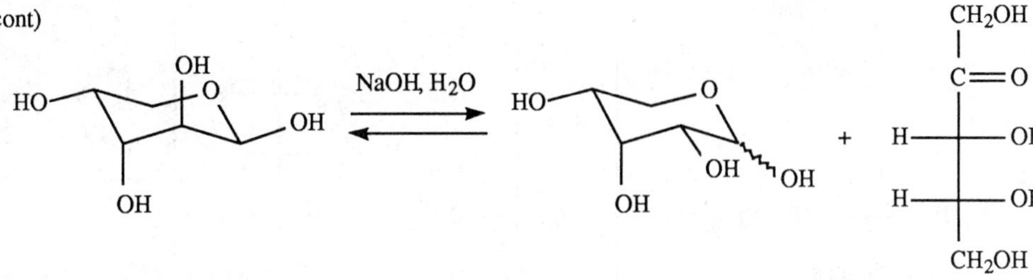

Note: D-arabinose will also be present in the equilibrium mixture.

(l)

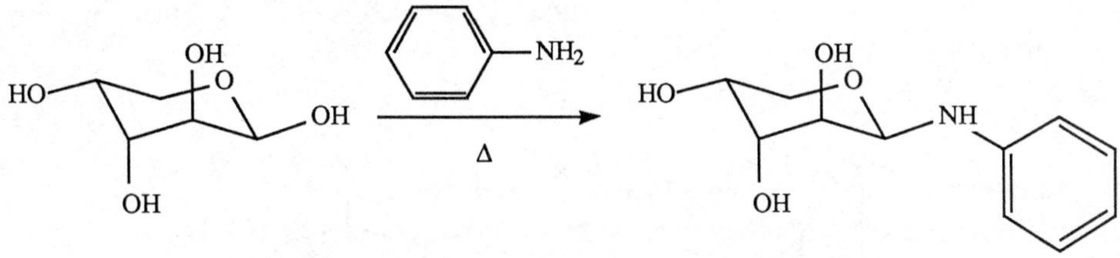

(m)

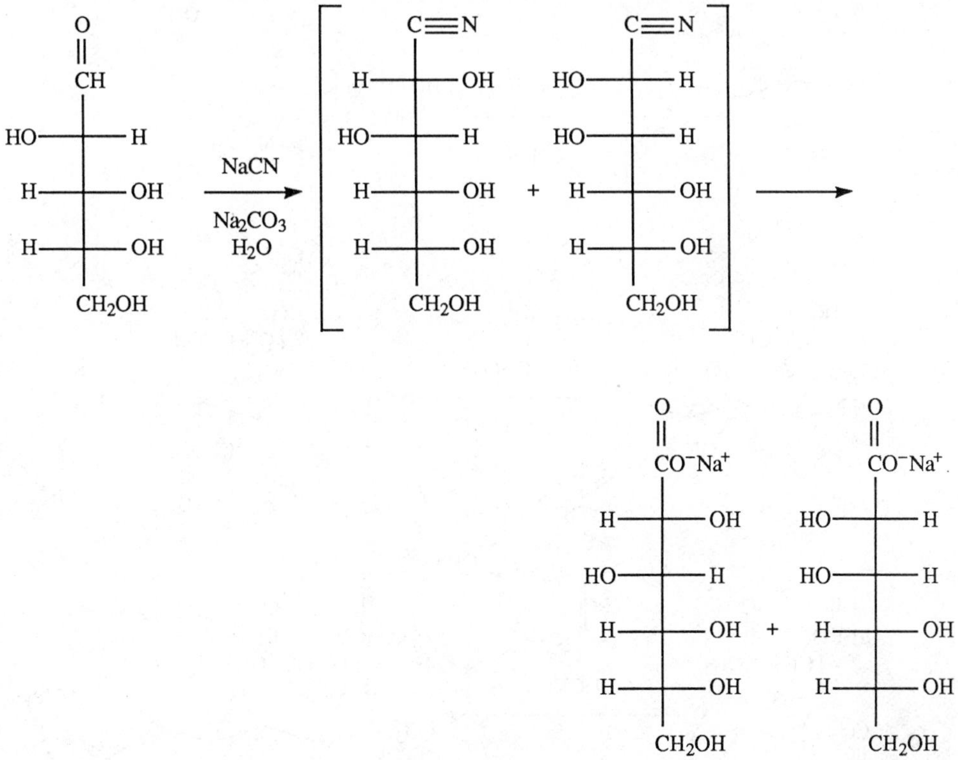

25.43

(a)

(b)

25.43 (cont)

(c)

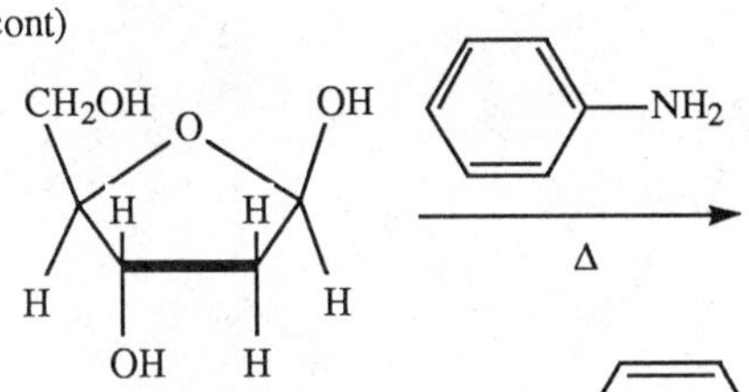

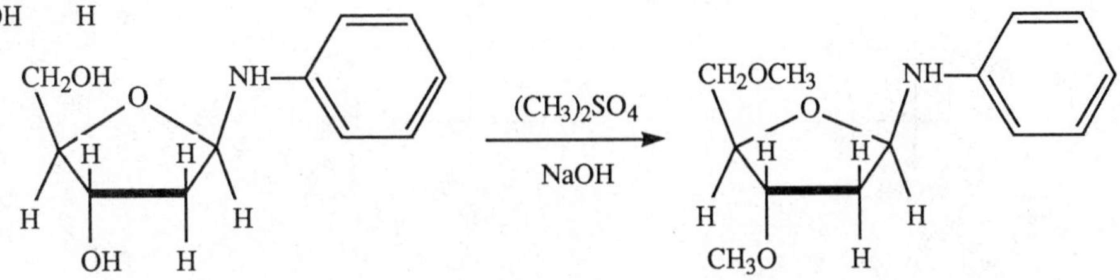

(d)

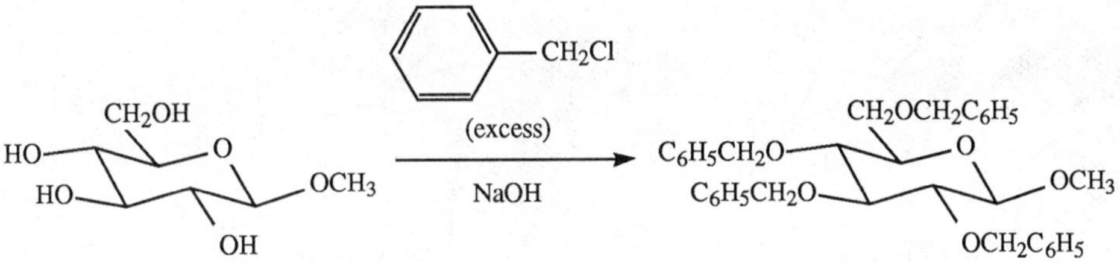

(e)

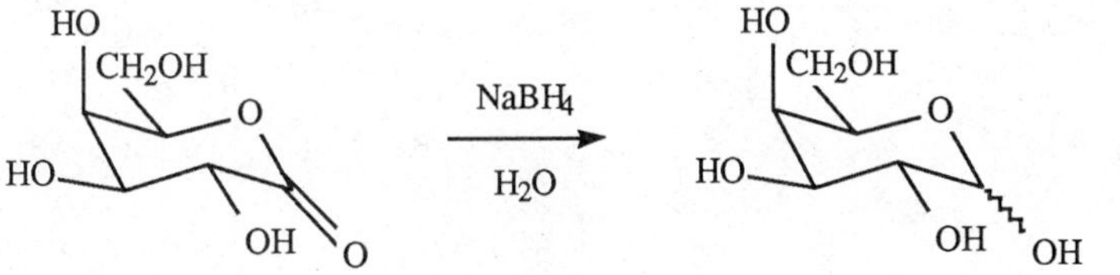

(f)

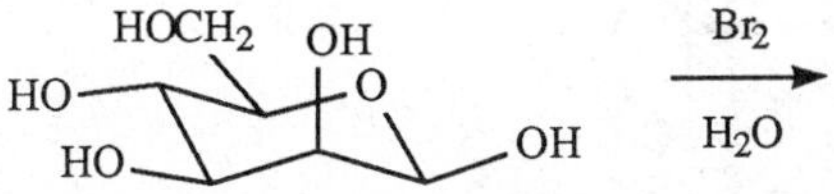

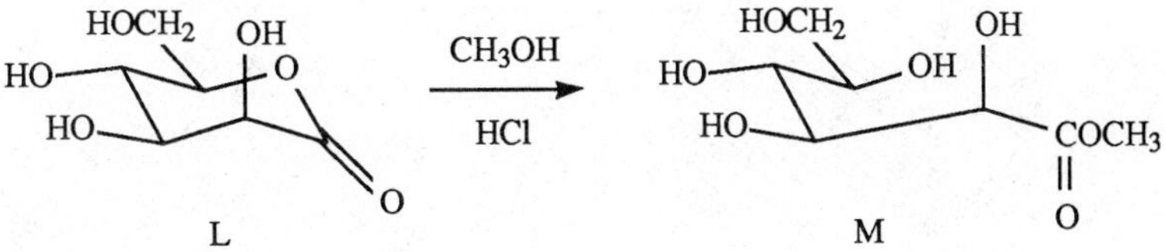

25.43 (cont)

(g)

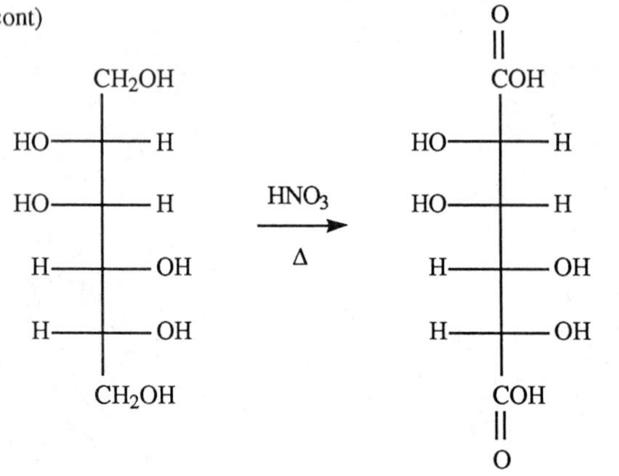

(h)

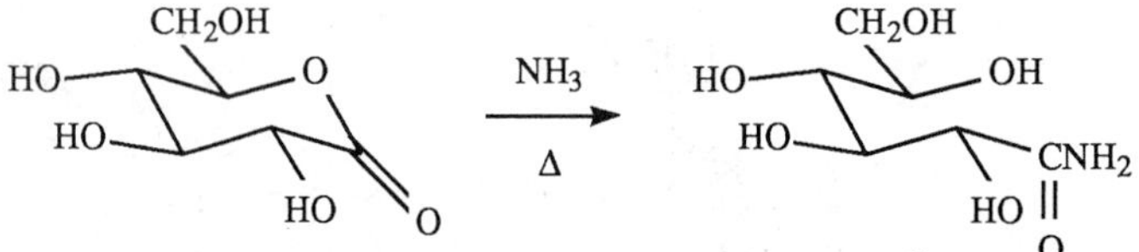

(i)

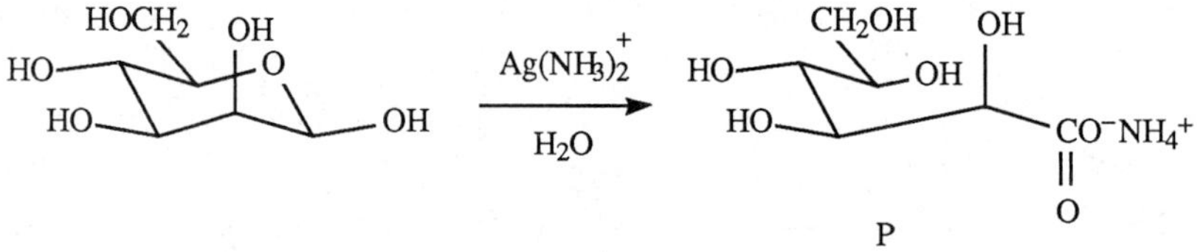

(j)

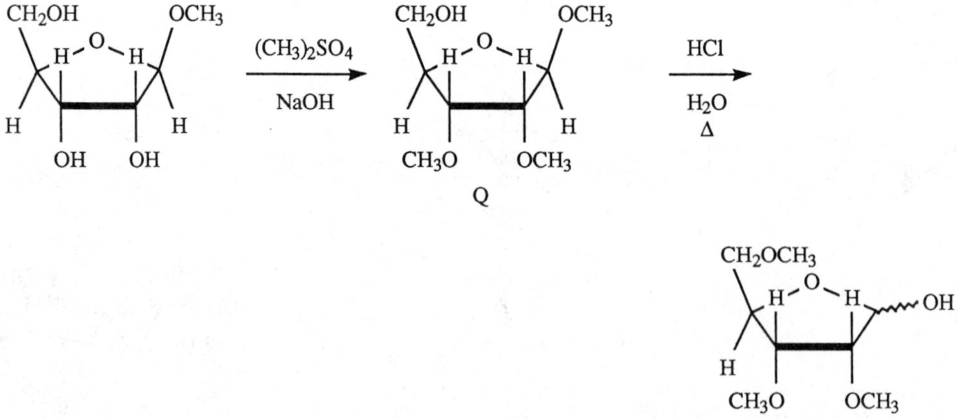

25.43 (cont)

(k)

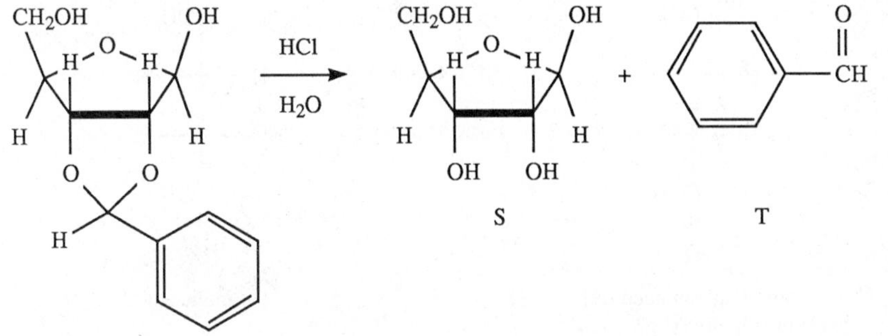

25.44

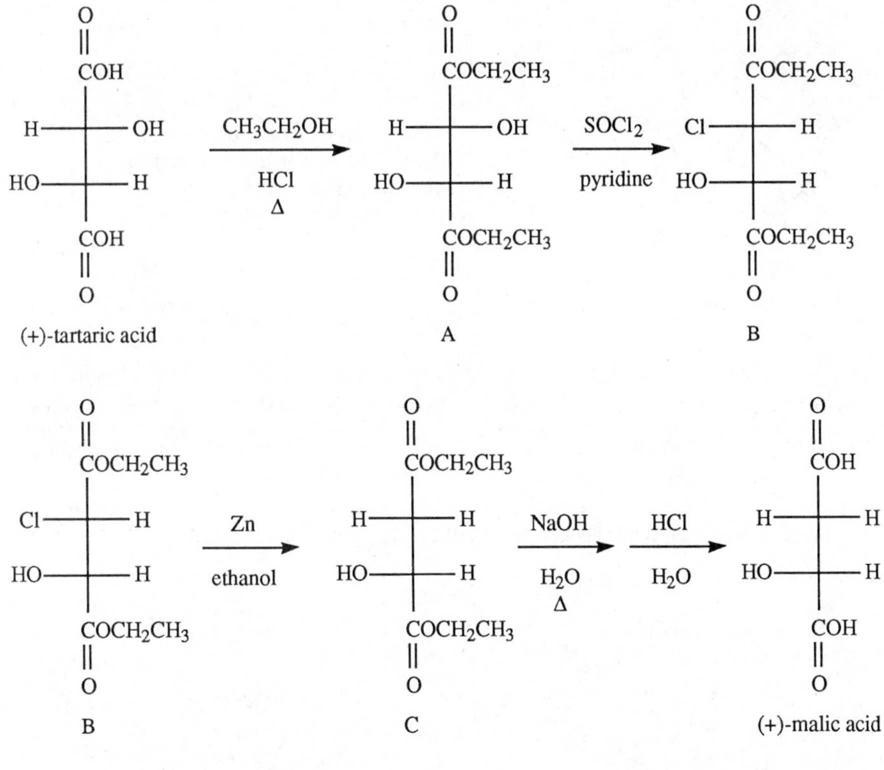

(+)-tartaric acid   A   B

B   C   (+)-malic acid

Note: Conversion of either hydroxyl group in (+)-tartaric acid to the chloro compound and reduction of the carbon-chlorine bond gives (+)-malic acid.

25.44 (cont)

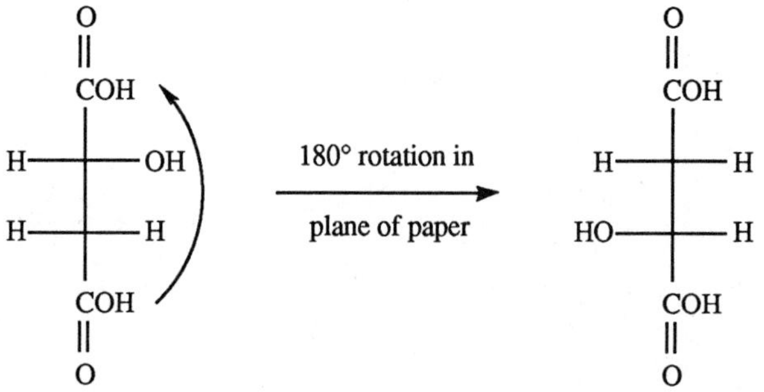

compound obtained if the
lower hydroxyl group in
(+)-tartaric acid is replaced
by hydrogen

180° rotation in

plane of paper

(+)-malic acid

25.45

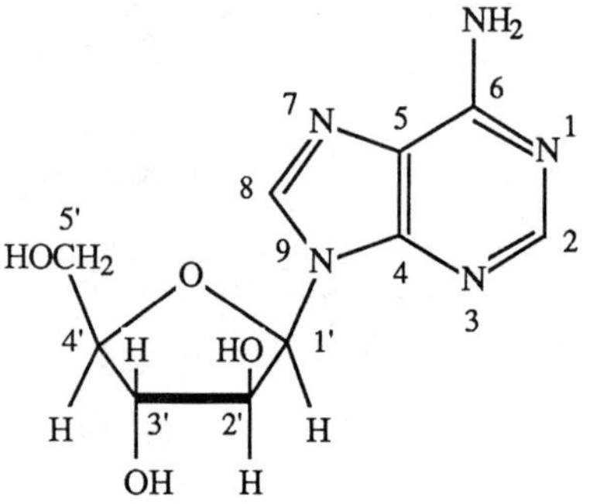

Vira-A

(Arabinose is epimeric with ribose at
C-2, the 2' position in the sugar residue,
so the structure of Vira-A is easily
derived from that of adenosine.)

25.46

(a)

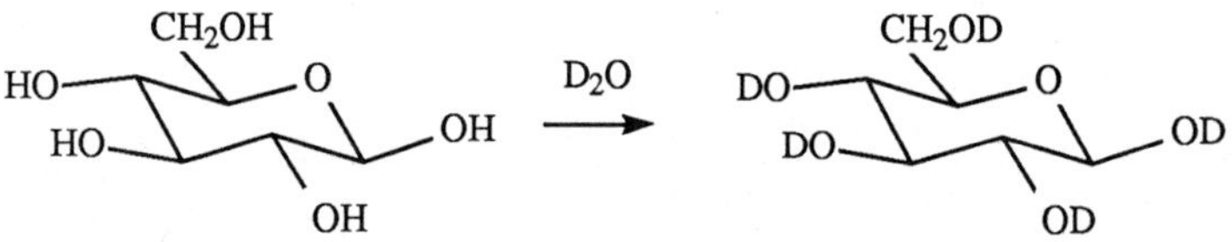

(Exchange of hydrogen atoms bonded to oxygen and nitrogen occurs rapidly.)

(b)

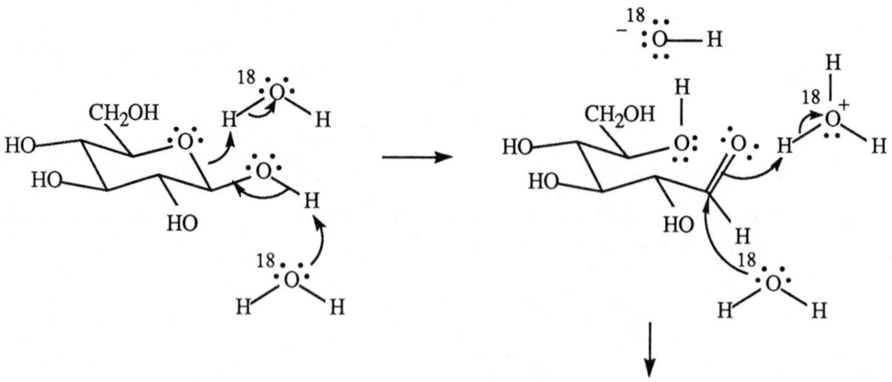

25.46

(b) (cont)

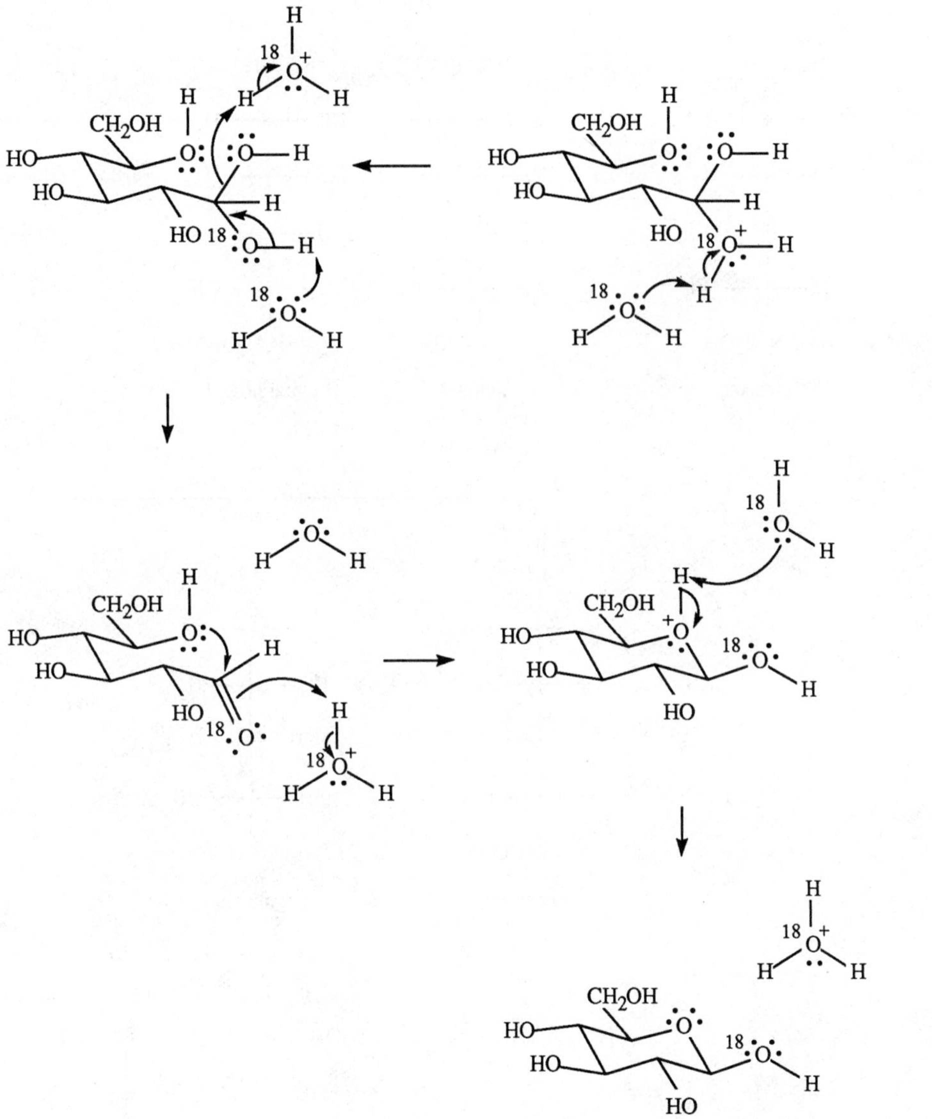

Exchange of oxygen atoms occurs at the anomeric carbon atom by way of the open-chain form of glucose.

25.47

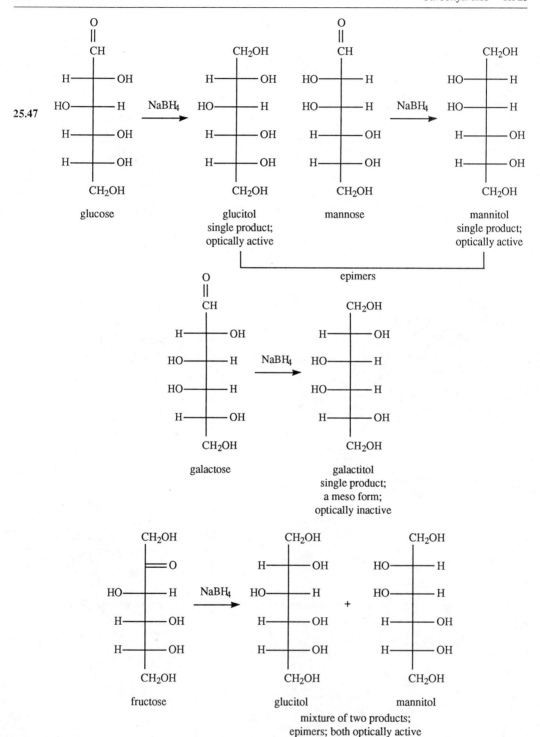

25.48

1,2,5,6-di-*O*-isopropylene-D-glucofuranose

a similar series
of steps at
→ → → →
C-5 and C-6

25.49   isomaltose $\xrightarrow[\text{or maltase}]{H_3O^+}$ 2

CH$_2$OH / glucose structure

glucose

isomaltose $\xrightarrow[\text{H}_2\text{O}]{\text{Br}_2}$ isomaltobionic acid lactone $\xrightarrow[\text{NaOH}]{\text{(CH}_3)_2\text{SO}_4}$

O-octamethylisomaltobionic acid methyl ester $\xrightarrow{H_3O^+}$

2,3,4,5-tetra-
O-methyl-D-gluconic acid

+

2,3,4,6-tetra-
O-methyl-D-glucopyranose

Only the hydroxyl group at C-6 of one glucopyranose unit and at C-1 of the other one are free in the hydrolysis products of the methyl ester of *O*-octamethylisomaltobionic acid, therefore those two carbon atoms must have been attached to each other in a glycosidic linkage. Hydrolysis of the disaccharide by maltase means that there is an α-glycosidic linkage between the two units. This information enables us to reconstruct the structure of isomaltose.

O-octamethylisomaltobionic acid methyl ester $\xleftarrow[\text{NaOH}]{\text{(CH}_3)_2\text{SO}_4}$ isomaltobionic acid lactone

$\uparrow$ $\xrightarrow[\text{H}_2\text{O}]{\text{Br}_2}$ potential aldehyde, therefore isomaltose is a reducing sugar, shown as the β-anomer

O-α-D-glucopyranosyl-(1 → 6)-D-glucopyranose

25.50   (a)

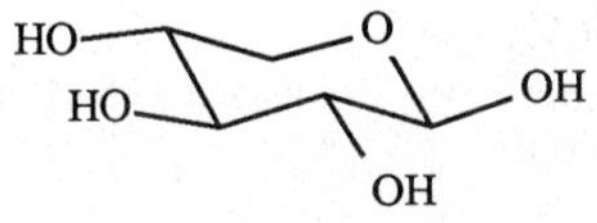

xylose
β-D-xylopyranose

(b)   The glucose unit has the free hemiacetal function.

(c)

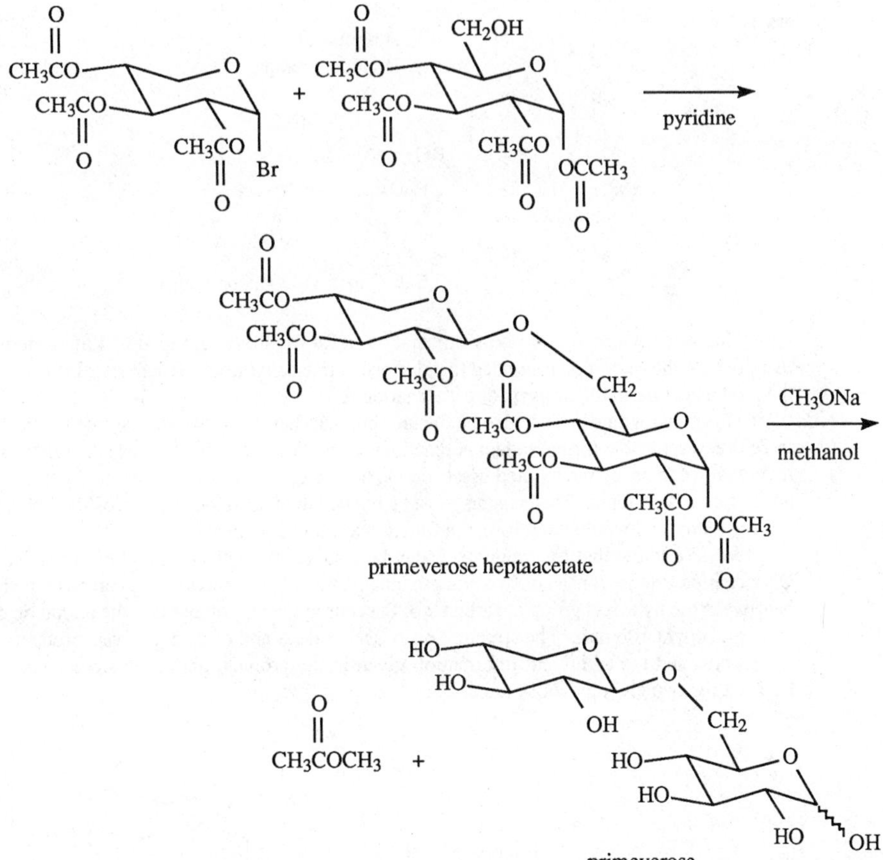

primeverose heptaacetate

CH_3COCH_3 +

primeverose

(d)

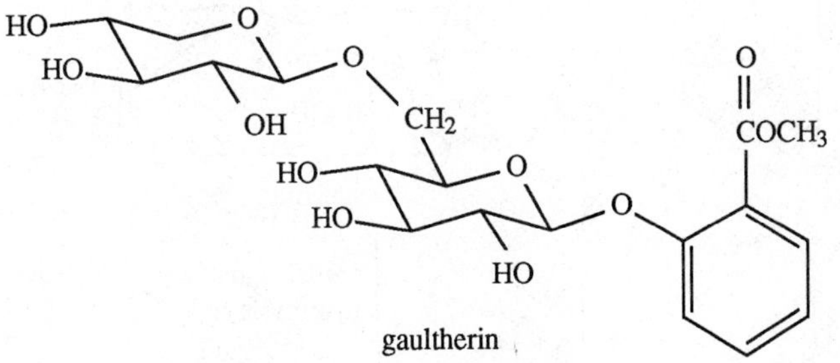

gaultherin

25.51   (a)   Trisaccharide A $\xrightarrow[\text{NaOH}]{(CH_3)_2SO_4}$ compound with 11 methyl groups $\xrightarrow{H_3O^+}$

2,3,4,6-tetra-*O*-
methyl-D-glucopyranose

+

2,3,4,6-tetra-*O*-
methyl-D-galactopyranose

3,6-di-*O*-methyl-D-glucopyranose

+

CH_3OH

(from hydrolysis of
the methyl glycoside
of trisaccharide A)

This glucopyranose is the only one that has a free hydroxyl group at C-4 after methylation, and must, therefore, be the sugar that is bonded to galactose in trisaccharide A. Otherwise, lactose would not be one of the products of partial hydrolysis of trisaccharide A.

This glucopyranose must also be the unit in trisaccharide A that has the free hemiacetal linkage and is, therefore, responsible for the reducing reactions of A. We know this because the other two sugars from the hydrolysis of the fully methylated trisaccharide have methyl groups at all of the hydroxyl groups except the anomeric carbon atoms. These sugars must be bonded to other sugars by glycosidic linkages at the anomeric carbon atoms, and would, therefore, not function as reducing agents.

We also know that trisaccharide A can be cleaved into galactose and disaccharide B by an enzyme. Disaccharide B must consist of two glucose units. The only connection that can be made between the sugars is between the hydroxyl group at carbon 2 of the central glucopyranose and the anomeric carbon atom of the other glucopyranose unit. The structure of trisaccharide A and of its hydrolysis products follow from these deductions, and the additional information given in the problem about the stereochemistry of the linkage between the two glucopyranose units.

partial acid
hydrolysis

25.51   (a) (cont)

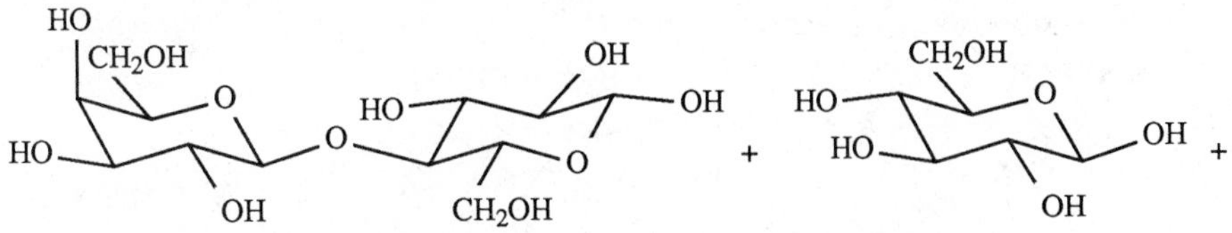

lactose   +   glucose   +

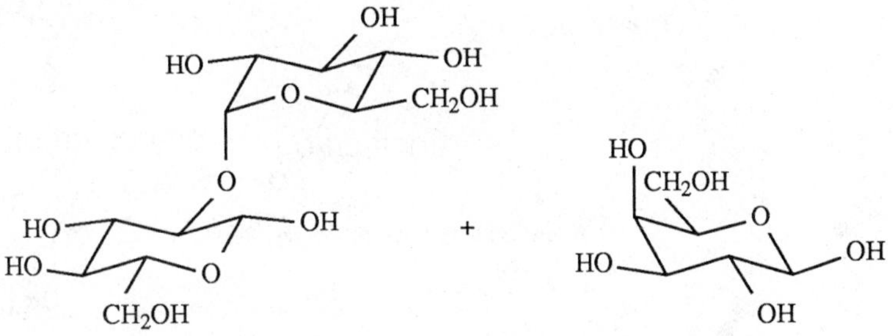

disaccharide B   +   galactose

formic acid

Reduction of the aldehyde function by $NaBH_4$ opens the middle ring to give this alcohol.

(b)

formic acid

formaldehyde

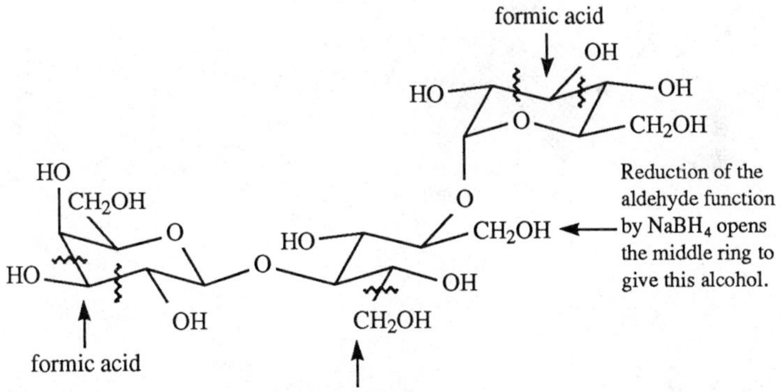

25.52

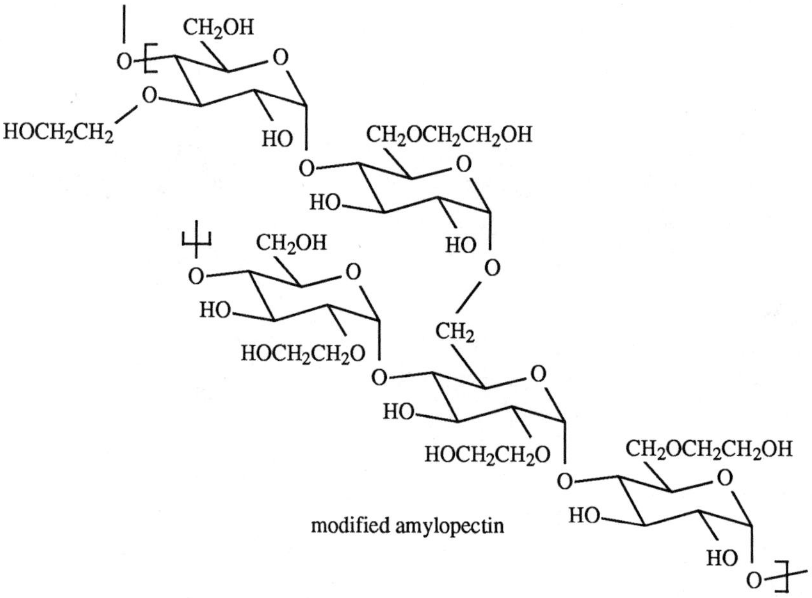

modified amylopectin

25.53

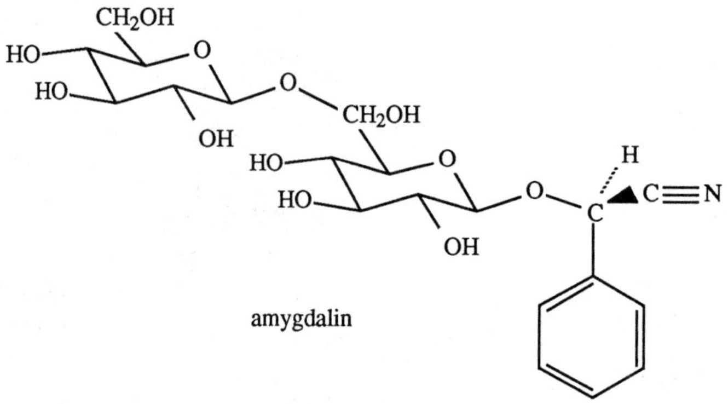

amygdalin

25.54

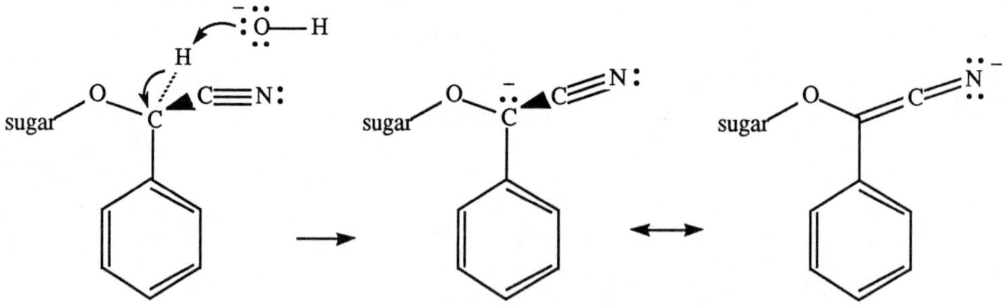

Loss of a proton gives a resonance-
stabilized carbanion. The carbanion
does not retain stereochemistry and
can be protonated on either side.

25.54 (cont)

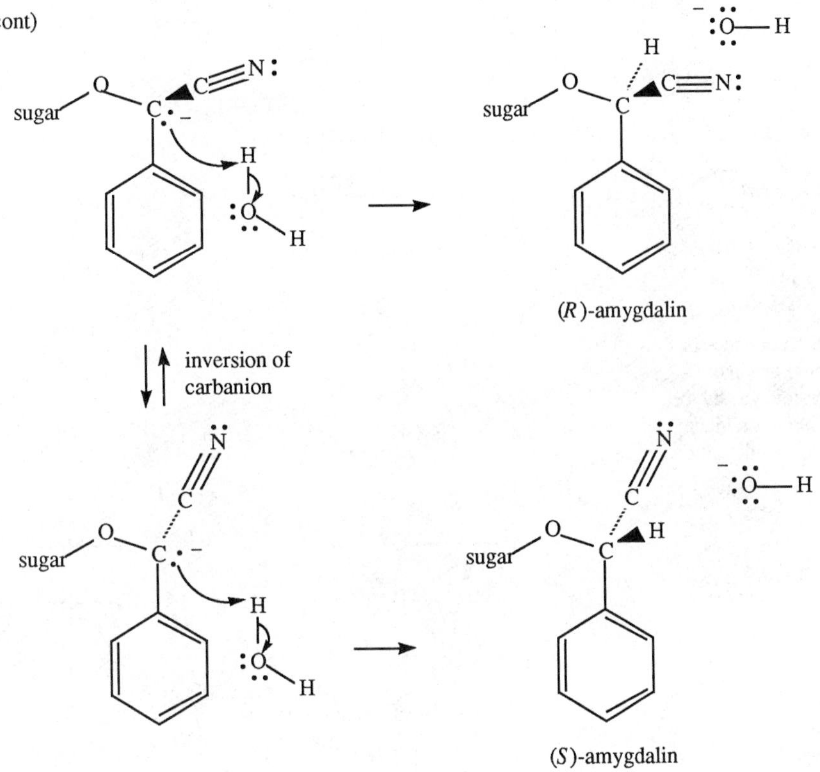

(*R*)-amygdalin

inversion of
carbanion

(*S*)-amygdalin

25.55   The reaction is an aldol condensation reaction between two glyceraldehyde molecules. The enediol intermediate that is necessary for the condensation reaction is formed faster from dihydroxyacetone (in which
deprotonation occurs at an O–H bond) than from glyceraldehyde (in which deprotonation must occur at a C–H
bond). The stereochemistry of the products is determined by the side of the carbonyl group that is attacked by
the enediol. The reaction is shown as taking place with the polar and bulky hydroxyl group of the enediol
pointing away from the oxygen atom of the carbonyl group undergoing nucleophilic attack.

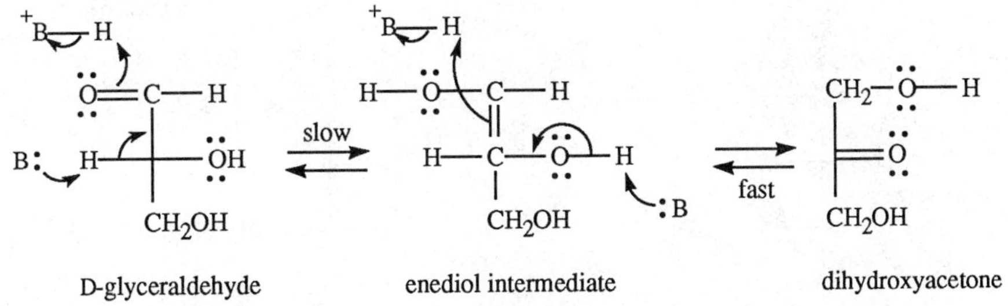

D-glyceraldehyde   enediol intermediate   dihydroxyacetone

25.55 (cont)

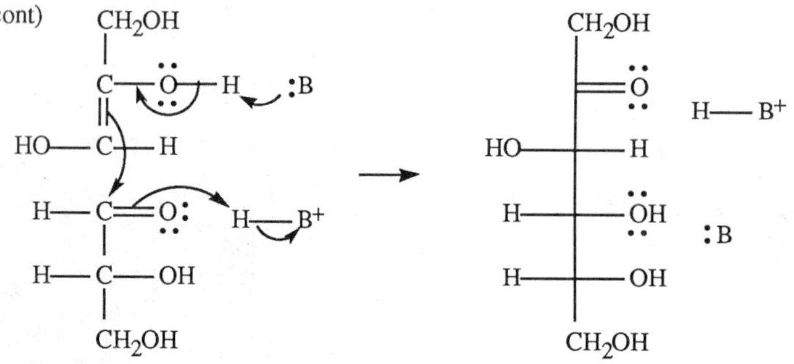

attack on one face of the
carbonyl group of D-
glyceraldehyde by the
enediol intermediate;
an aldol condensation

D-fructose

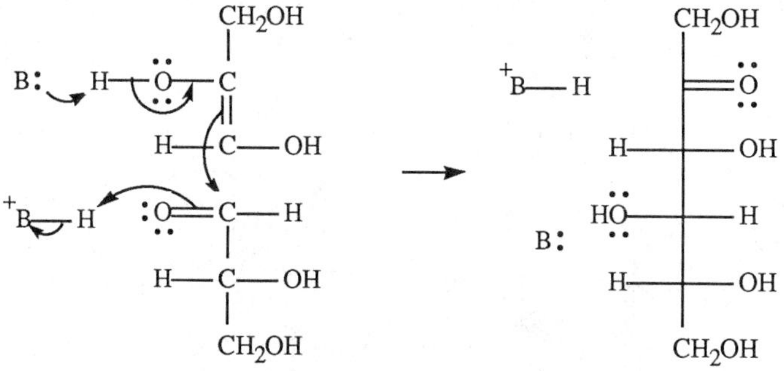

attack on the other face
of the carbonyl group of
D-glyceraldehyde by the
enediol intermediate

D-sorbose

25.56

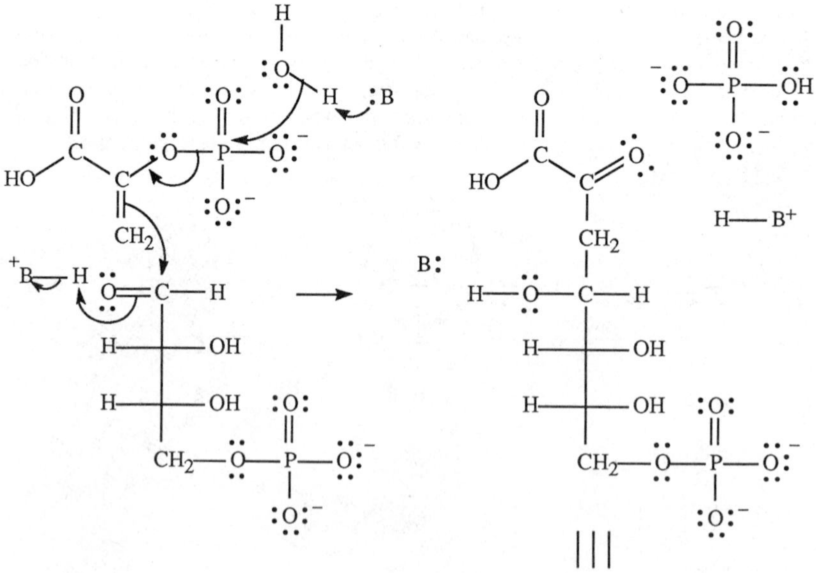

25.56 (cont)

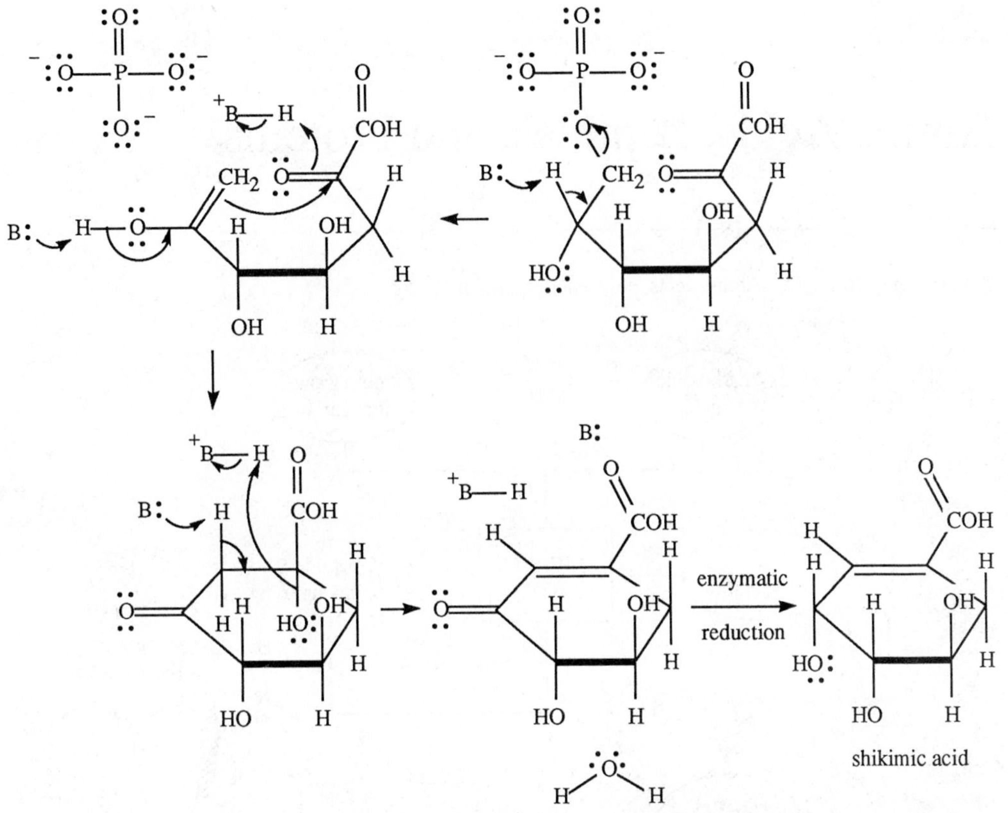

The transformations shown above occur in the presence of enzymes, and many bonds are shown being broken and made in one step. The stereoselectivity of an enzymatic reaction is demonstrated by the first transformation. In the laboratory, a mixture of diastereomers would be expected.

26

Amino Acids, Peptides, and Proteins

Concept Map 26.1   Amino acids, polypeptides, proteins.

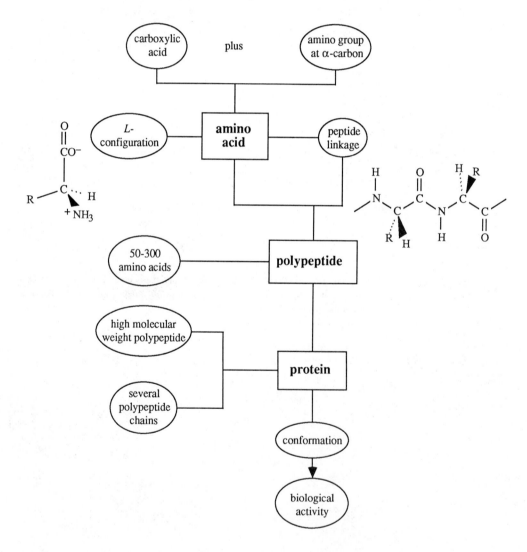

Concept Map 26.2   Acid-base properties of amino acids.

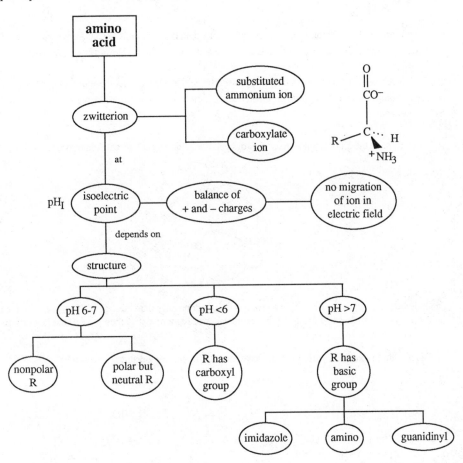

26.1   (a)

$$\underset{\underset{+NH_3}{|}}{HOCCH_2CHCO^-} \overset{OH^-}{\underset{H_3O^+}{\rightleftharpoons}} \underset{\underset{+NH_3}{|}}{^-OCCH_2CHCO^-}$$

aspartic acid

$$\underset{\underset{+NH_3}{|}}{HOCCH_2CH_2CHCO^-} \overset{OH^-}{\underset{H_3O^+}{\rightleftharpoons}} \underset{\underset{+NH_3}{|}}{^-OCCH_2CH_2CHCO^-}$$

glutamic acid

26.1 (cont)

The carboxylic acid groups on carbon 3 of aspartic acid is more acidic because it is closer to the electron-withdrawing effect of the ammonium ion on carbon 2 than is the acid group on carbon 4 of glutamic acid.

(b)   The carboxylic acid group on carbon 3 in aspartic acid is more acidic, therefore it has a lower pK_a. The pH required to keep that group protonated and the molecule in the balanced zwitterionic state would also be lower.

26.2   All of them have the *S* configuration except for cysteine, which has the *R* configuration.

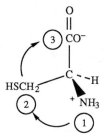

In all other amino acids the carboxylate group is of higher priority than the rest of the chain. The presence of sulfur in cysteine makes that group on the stereocenter of higher priority than the carboxylate group.

26.3   Isoleucine, threonine and hydroxyproline have more than one stereocenter.

Ile	Thr	Hypro
configuration not specified at carbon 3	(2*S*, 3*R*)-2-amino-3-hydroxybutanoic acid	(2*S*, 4*S*)-4-hydroxyproline

26.4   (a) $^-OCCHCH_2CH_2CH_2CH_2NH_2$ + H——B$^+$ $\rightleftarrows$ $^-OCCHCH_2CH_2CH_2CH_2NH_3^+$ + :B

with $^+NH_3$ groups below, labeled lysine

When lysine is protonated at the ε-amino group, the resulting positive charge is localized at that nitrogen atom.

26.4 (cont)

$$\overset{O}{\underset{||}{}}\overset{NH}{\underset{||}{}}$$

$$^-OCCHCH_2CH_2CH_2NHCNH_2 \; + \; H\!-\!B^+ \;\rightleftharpoons\; {}^-OCCHCH_2CH_2CH_2NHCNH_2 \; + \; B$$

with $^+NH_3$ below the first carbon (arginine) and $\overset{+}{N}H_2$ on the second structure.

arginine

$$^-OCCHCH_2CH_2CH_2\overset{+}{N}H\!=\!CNH_2 \;\longleftrightarrow\; {}^-OCCHCH_2CH_2CH_2NHC\!=\!\overset{+}{N}H_2$$

with O (C=O), NH_2, and $^+NH_3$ substituents as drawn.

When the guanidinyl group in arginine is protonated, the resulting cation is stabilized by delocalization of the charge to three nitrogen atoms. The guanidinyl group is, therefore, more basic, more easily protonated, than an amino group. The conjugate acid of guanidine, with a pK_a of 13.2, is a weaker acid than the conjugate acid of a primary amine, p$K_a \sim 10.5$.

(b)

tryptophan

indole ring
system

The nonbonding electrons on the nitrogen atom in the indole ring are part of an aromatic sextet, and are, therefore, not available for protonation. Protonation of that nitrogen atom would lead to loss of aromaticity, so the difference in energy between indole and its conjugate acid would be very large.

26.5

(a)

26.5 (cont)

(b)

$$\underset{\overset{+}{N}H_3}{\overset{\overset{O}{\parallel}}{\underset{H}{\overset{CO^-}{\underset{|}{C}}}}} H \xrightarrow[\substack{0\,°C}]{NaNO_2,\,HCl} \underset{OH}{\overset{\overset{O}{\parallel}}{\underset{H}{\overset{COH}{\underset{|}{C}}}}} H \; + \; \underset{Cl}{\overset{\overset{O}{\parallel}}{\underset{H}{\overset{COH}{\underset{|}{C}}}}} H \; + \; N_2\uparrow$$

(c)

A phenyl—CH$_2$ group attached to a carbon bearing CO$^-$ and $\overset{+}{N}H_3$ reacts with benzoyl chloride (C$_6$H$_5$COCl) and NaOH to give **A**, an N-benzoyl derivative with CO$^-$ and NHC(=O)C$_6$H$_5$ groups, which then reacts with SOCl$_2$.

A

$$\xrightarrow{SOCl_2}$$

B (the acid chloride, CCl with NHC=O benzoyl group) $\xrightarrow{NH_3}$ **C** (the amide, CNH$_2$ with NHC=O benzoyl group)

(d)

$$\underset{\overset{+}{N}H_3}{\overset{\overset{O}{\parallel}}{\underset{HSCH_2}{\overset{CO^-}{\underset{|}{C}}}}} H \xrightarrow[\substack{NaOH}]{CH_3I\,(excess)} \underset{\overset{+}{N}(CH_3)_3\;Cl^-}{\overset{\overset{O}{\parallel}}{\underset{CH_3SCH_2}{\overset{COCH_3}{\underset{|}{C}}}}} H$$

(e)

$$\underset{\overset{+}{N}H_3}{\overset{\overset{O}{\parallel}}{\underset{HOCH_2}{\overset{CO^-}{\underset{|}{C}}}}} H \xrightarrow[\substack{dichloromethane}]{\substack{\text{pyridinium}\ CrO_3Cl^-}} \underset{\overset{+}{N}H_3}{\overset{\overset{O}{\parallel}}{\underset{HC}{\overset{CO^-}{\underset{|}{C}}}}} H$$

26.5 (cont)

(f)

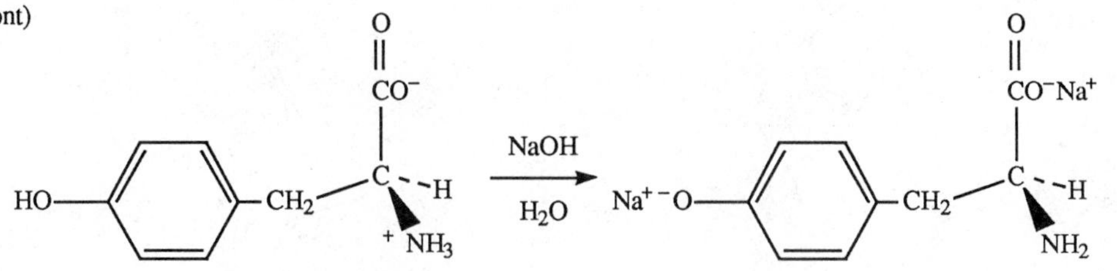

(g)

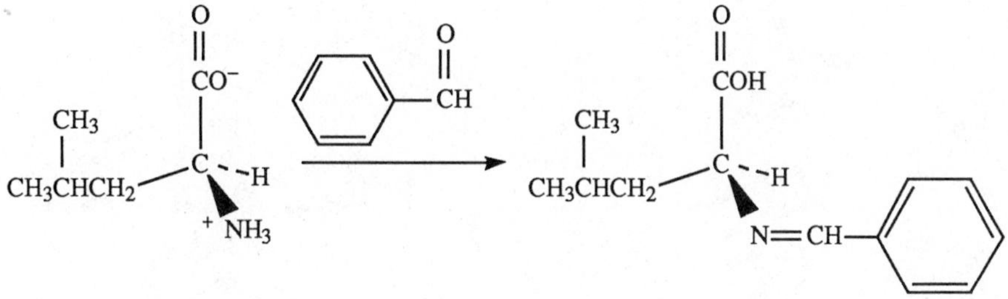

(h)

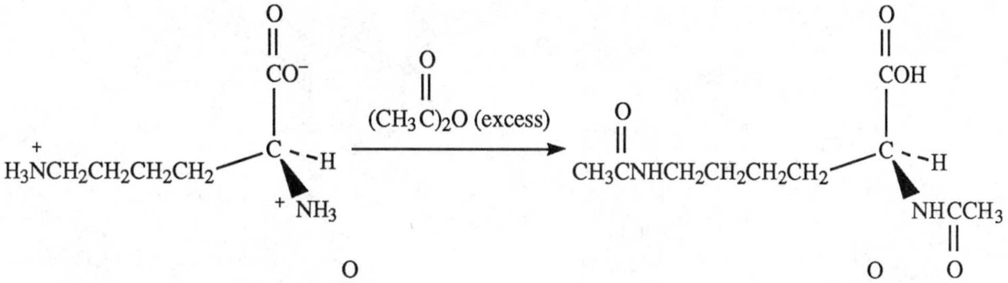

(i)

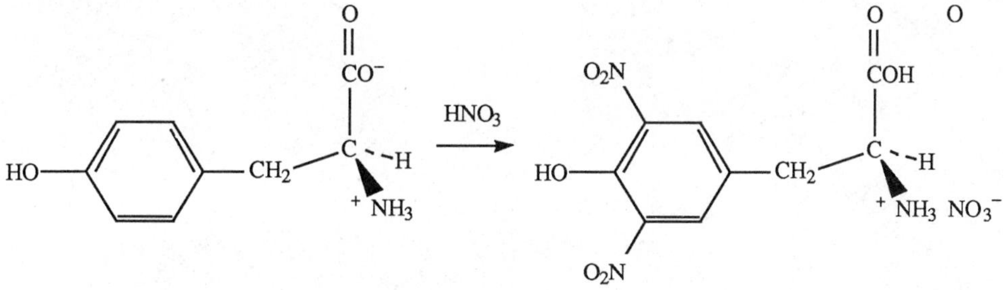

26.6

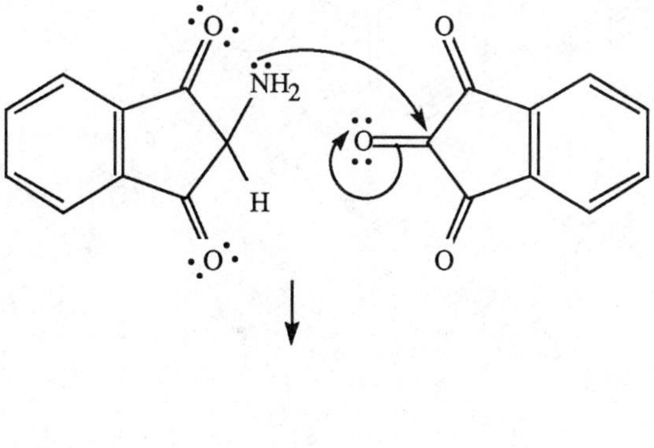

26.6 (cont)

26.7   One possible peptide is:

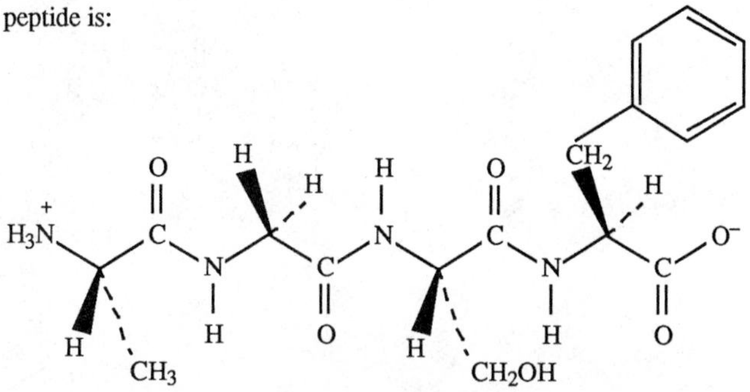

Ala-Gly-Ser-Phe

N-terminal   *C*-terminal
amino acid   amino acid

The other twenty-two, besides the one above and Ser-Ala-Phe-Gly, are:

Ala-Gly-Phe-Ser	Ser-Ala-Gly-Phe
Ala-Ser-Gly-Phe	Ser-Gly-Ala-Phe
Ala-Ser-Phe-Gly	Ser-Gly-Phe-Ala
Ala-Phe-Ser-Gly	Ser-Phe-Ala-Gly
Ala-Phe-Gly-Ser	Ser-Phe-Gly-Ala
Gly-Ala-Ser-Phe	Phe-Ala-Ser-Gly
Gly-Ala-Phe-Ser	Phe-Ala-Gly-Ser
Gly-Ser-Ala-Phe	Phe-Ser-Gly-Ala
Gly-Ser-Phe-Ala	Phe-Ser-Ala-Gly
Gly-Phe-Ala-Ser	Phe-Gly-Ser-Ala
Gly-Phe-Ser-Ala	Phe-Gly-Ala-Ser

N-terminal	*C*-terminal	*N*-terminal	*C*-terminal
amino acid	amino acid	amino acid	amino acid

26.8

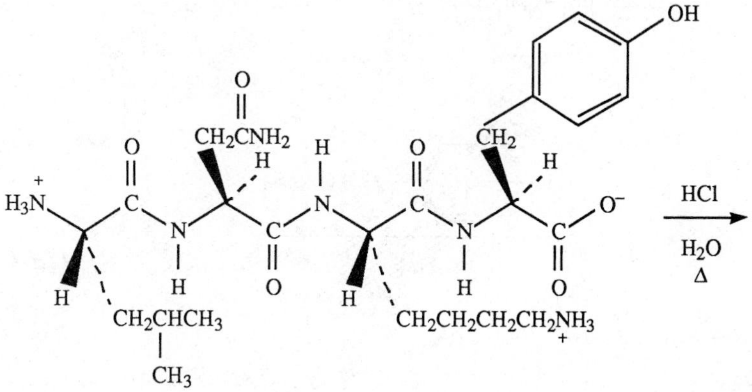

26.8 (cont)

$$CH_3CHCH_2 \overset{\displaystyle CH_3}{\underset{\overset{|}{\overset{+}{N}H_3\ Cl^-}}{\overset{\overset{\displaystyle O}{\overset{\|}{COH}}}{\overset{|}{C}}}} H \quad + \quad HOCCH_2 \overset{\overset{\displaystyle O}{\overset{\|}{COH}}}{\underset{\overset{+}{N}H_3\ Cl^-}{\overset{|}{C}}} H \quad + \quad \cdots$$

Leu   Asp

Tyr

The hydrolysis was done in strong acid, therefore all amino acids are shown in their fully protonated forms.

26.9   (a)   Gly-Phe-Thr-Lys will have $pH_I > 6$; it will move to the negative pole in an electric field at pH 6.

(b)   Tyr-Ala-Val-Asn will have $pH_I \sim 6$; it will not move in an electric field at pH 6.

(c)   Trp-Glu-Leu will have $pH_I < 6$; it will move to the positive pole in an electric field at pH 6.

(d)   Pro-Hypro-Gly will have $pH_I \sim 6$; it will not move in an electric field at pH 6.

26.10   *N*-terminal lysine

$$+ \quad Cl^-\ H_3NCH_2CH_2CH_2CH_2 \overset{\overset{\displaystyle O}{\overset{\|}{COH}}}{\underset{\overset{+}{N}H_3\ Cl^-}{\overset{|}{C}}} H \quad + \quad NH_4^+\ Cl^-$$

Lys   ammonium chloride
                                        (from the hydrolysis of
                                        the amide group in Asn)

$F-\!\!\!\!\bigcirc\!\!\!\!-NO_2$ (excess), base

26.10 (cont)

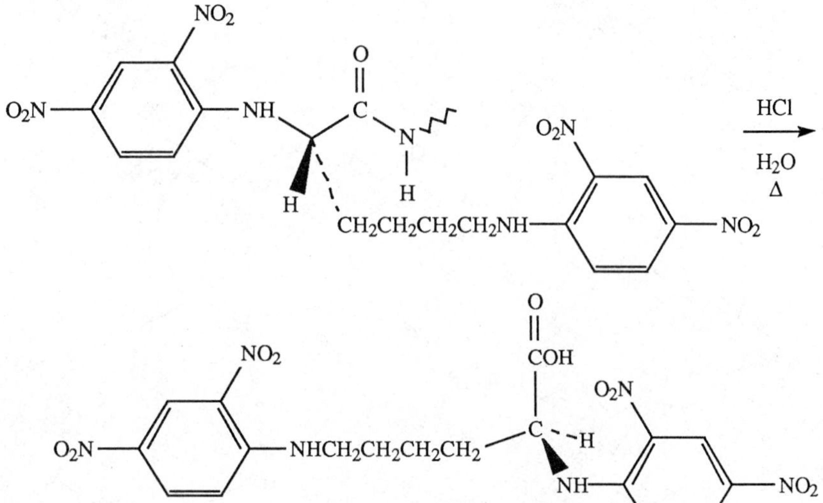

The *N*-terminal lysine
will be doubly labeled.

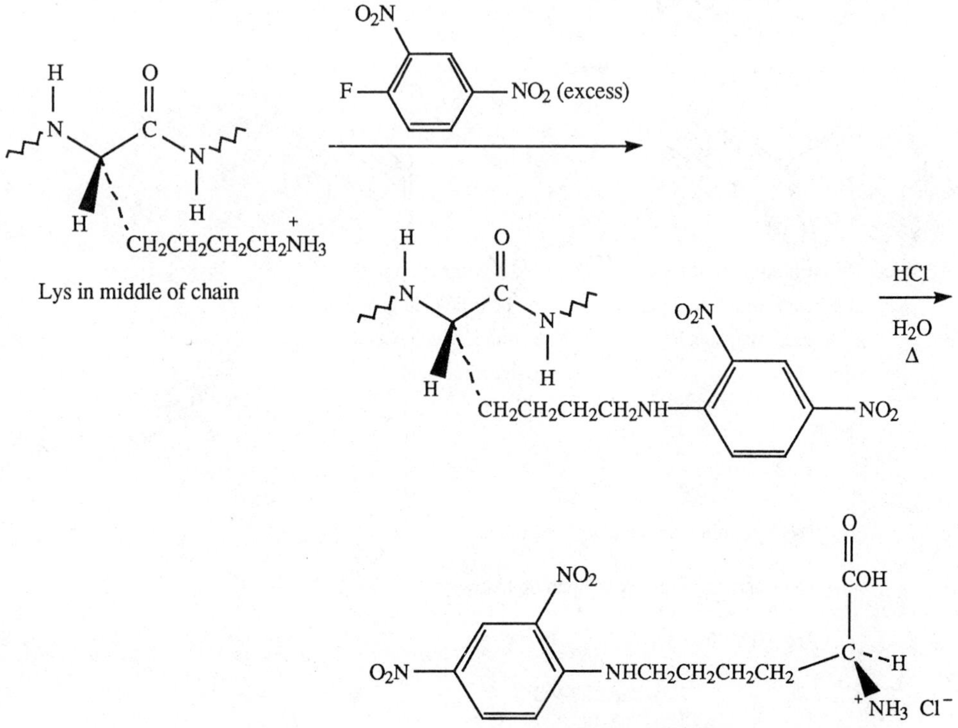

Lys in middle of chain

The lysine from the middle of the chain
will have a single dinitrophenyl label.

26.11

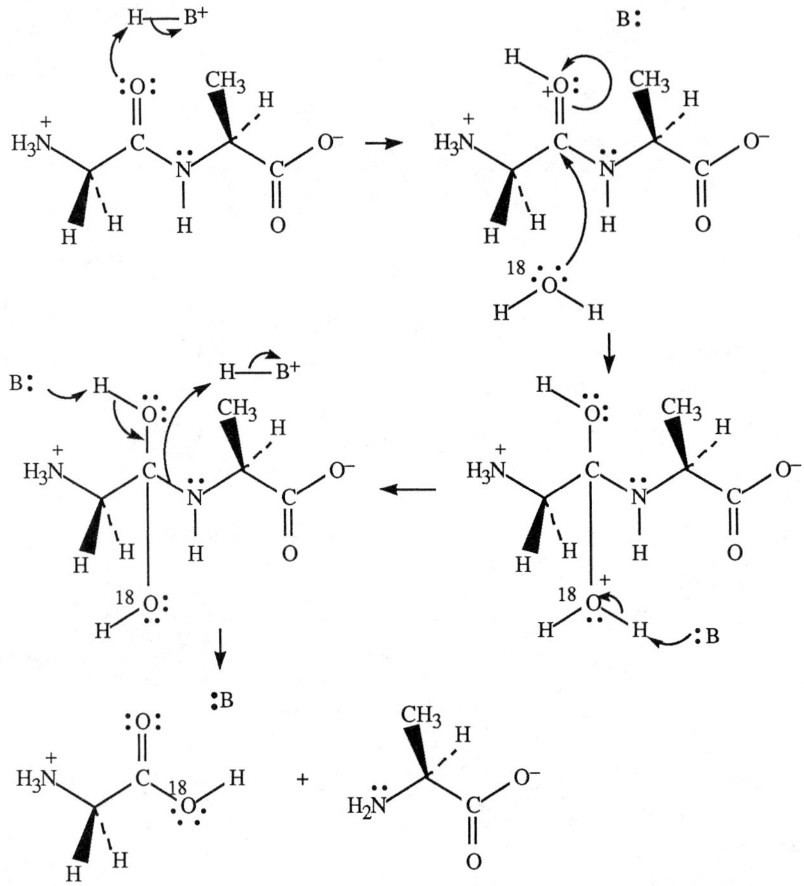

The carboxylic acid function
of glycine, which was part of
a peptide linkage, is labeled
with ^{18}O.

No ^{18}O turns up in the
carboxylic acid group of
alanine, which was the
C-terminal acid.

26.12   (a)   The *N*-terminal amino acid is arginine

(b)   The fragments are shown lined up below:

Arg–Gly–Pro
         Pro–Pro
              Pro–Phe–Ile–Val
              Pro–Phe–Ile

Therefore the structure of the peptide is: Arg–Gly–Pro–Pro–Phe–Ile–Val

Concept Map 26.3   Proof of structure of peptides and proteins.

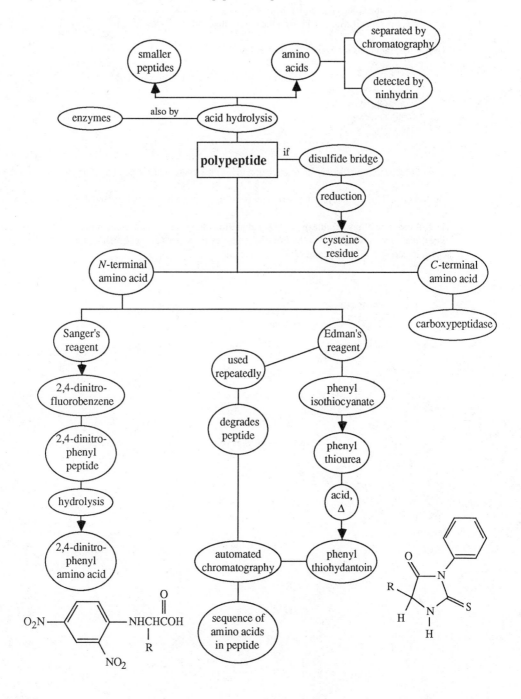

26.13   (a)   Cysteine is the *N*-terminal amino acid.
Glycinamide is the *C*-terminal amino acid.
Arginine is the second amino acid from the *C*-terminal end.

The peptide fragments are shown lined up below:

                        Cys–Pro–Arg–Gly (peptide of undetermined order*)
                Asp–Cys
            Glu–Asp–Cys
        Phe–Glu–Asp
        Phe–Glu
Cys–Try–Phe

*The information from the trypsin degradation gives the structure of the peptide of undetermined order.

Cys–Try–Phe–Glu–Asp–Cys—Pro–Arg–Gly is the order of amino acids from the degradation work.

(b)   The full structure of the peptide must show a disulfide bridge and three amide groups to account for the three moles of ammonia lost in the hydrolysis. There are three free carboxylic acid groups in the structure shown in part (a): Glu, Asp, and the *C*-terminal Gly.

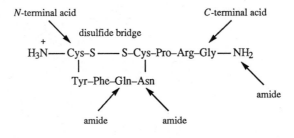

arginine vasopressin

Concept Map 26.4   Protection of functional groups in peptide synthesis.

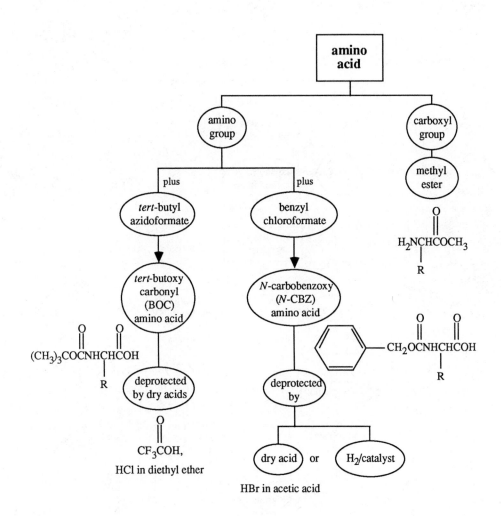

Concept Map 26.5   Activation of the carbonyl group in peptide synthesis.

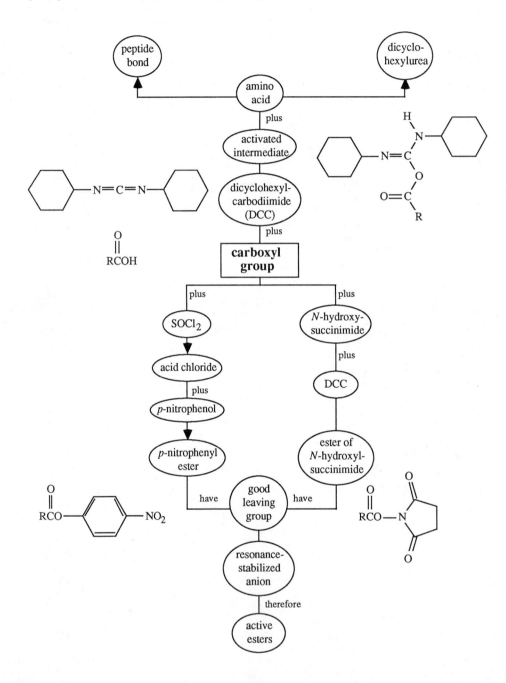

26.14

(a)

$$CH_3CO-\underset{\substack{|\\CH_3}}{\overset{\substack{CH_3\\|}}{C}}-\overset{O}{\overset{||}{C}}NHCH_2\overset{O}{\overset{||}{C}}NHCH_2\overset{O}{\overset{||}{C}}OH \quad \xrightarrow[\substack{dichloro-\\methane}]{\overset{O}{\overset{||}{CF_3COH}}}$$

$$\overset{O}{\overset{||}{CF_3CO^-}} \; \overset{+}{H_3}NCH_2\overset{O}{\overset{||}{C}}NHCH_2\overset{O}{\overset{||}{C}}OH \; + \; CH_3C\overset{CH_3}{\overset{|}{=}}CH_2 \; + \; CO_2$$

A

(b)

$$\underset{\overset{|}{\underset{NH_3}{+}}}{CH_3CHCH_2}\overset{CH_3}{\underset{|}{\overset{O}{\overset{||}{CO^-}}}}C{\cdots}H \quad + \quad \text{C}_6\text{H}_5\text{—CH}_2\overset{O}{\overset{||}{O}}\text{C}\text{Cl} \longrightarrow$$

B

$$\xrightarrow{\text{HO—N(succinimide)}, \text{ DCC}}$$

C

(c)

$$CH_3CH\overset{CH_3}{\underset{\overset{|}{\underset{NH_3}{+}}}{\overset{|}{C}}}{\cdots}H \; + \; CH_3CO\overset{CH_3}{\underset{\overset{|}{CH_3}}{\overset{|}{|}}}\overset{O}{\overset{||}{}}CN_3 \longrightarrow$$

D

26.14 (cont)

(d)

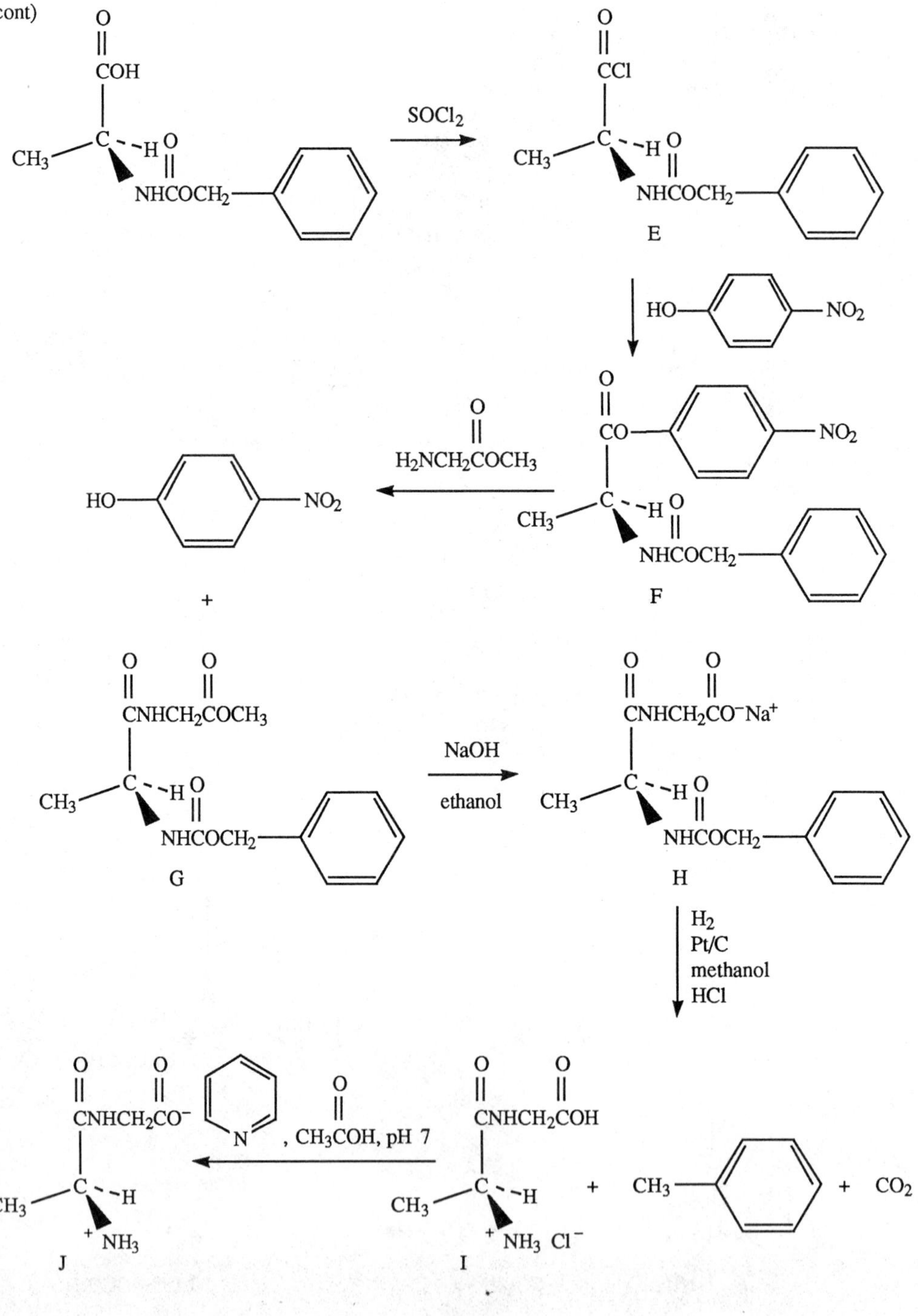

COH
SOCl2
CCl
E
HO
NO2
CO
H2NCH2COCH3
F
CNHCH2COCH3
G
NaOH
ethanol
CNHCH2CO−Na+
H
H2
Pt/C
methanol
HCl
CNHCH2CO−
, CH3COH, pH 7
CNHCH2COH
NH3 Cl−
J
I
CO2

26.15   The phenolic hydroxyl group can also behave as a nucleophile if unprotected, especially if the reaction is run in base. One possible complication in the synthesis of methionine enkephalin might be:

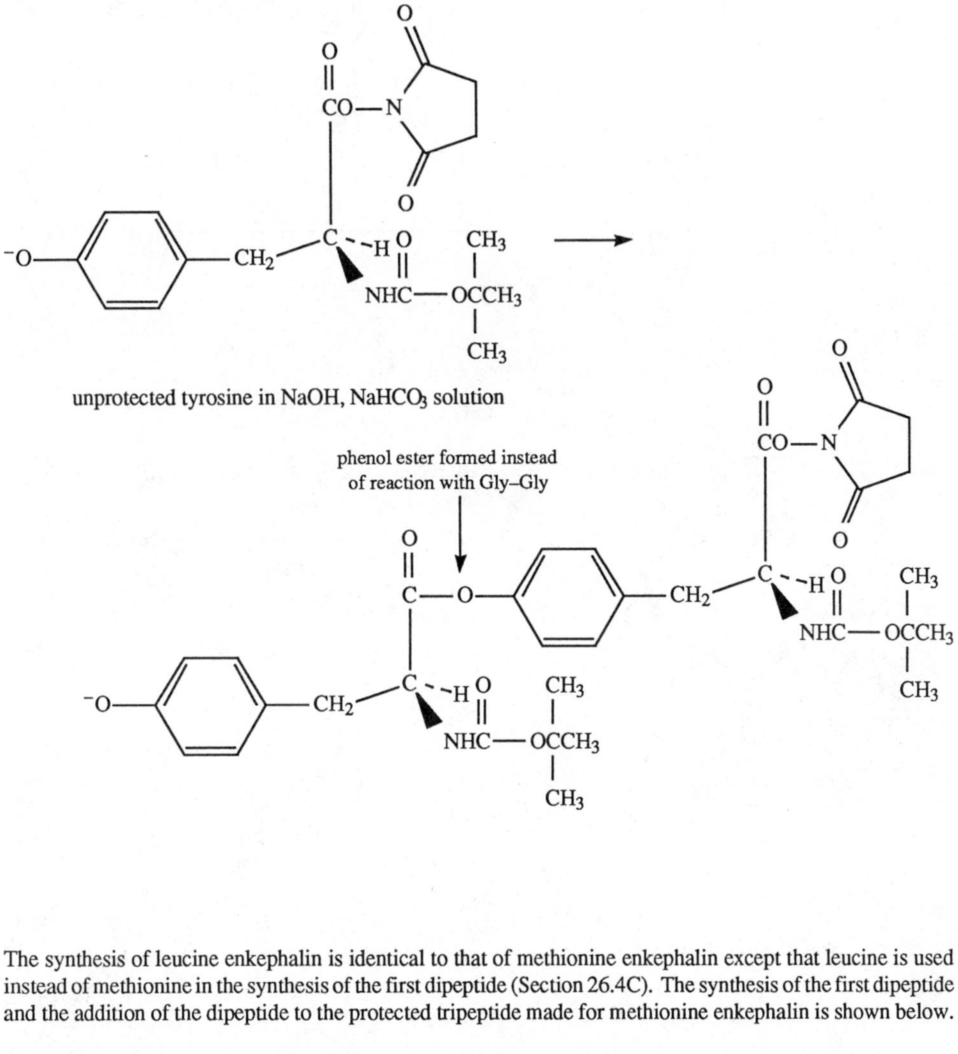

unprotected tyrosine in NaOH, NaHCO₃ solution

phenol ester formed instead
of reaction with Gly–Gly

26.16   The synthesis of leucine enkephalin is identical to that of methionine enkephalin except that leucine is used instead of methionine in the synthesis of the first dipeptide (Section 26.4C). The synthesis of the first dipeptide and the addition of the dipeptide to the protected tripeptide made for methionine enkephalin is shown below.

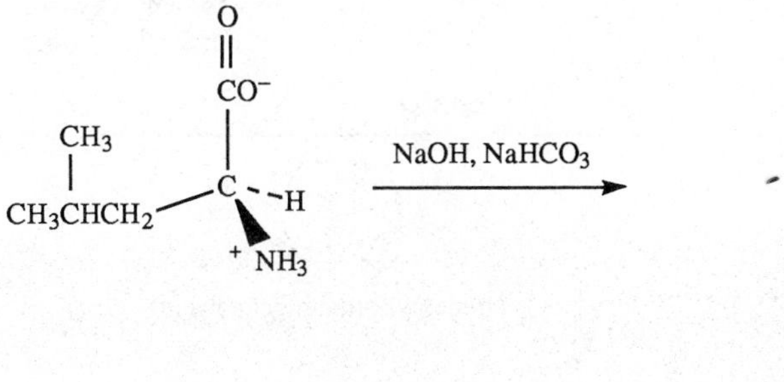

NaOH, NaHCO₃

26.16 (cont)

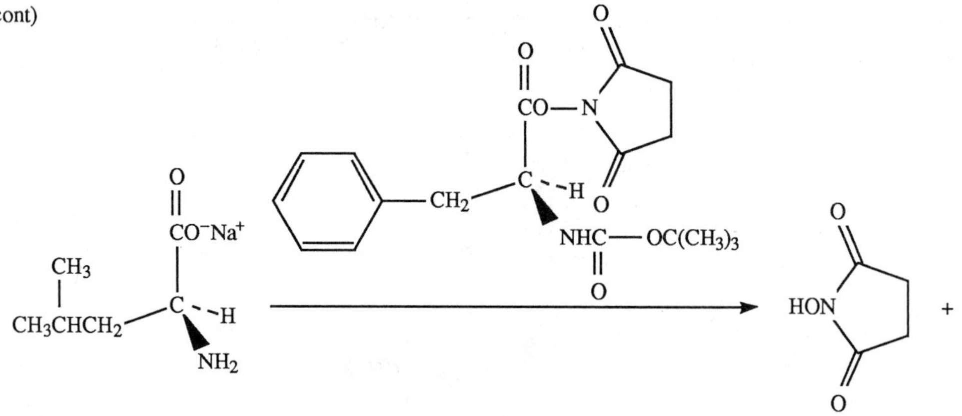

CO⁻Na⁺
CH3
CH3CHCH2
NH2
CH2
CO—N
NHC—OC(CH3)3
HON
+

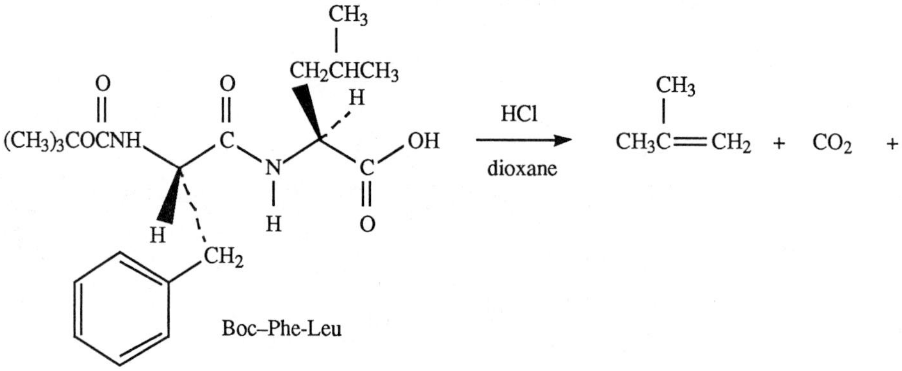

(CH3)3COCNH
CH2CHCH3
CH3
OH
CH2
HCl
dioxane
CH3C═CH2 + CO2 +
Boc–Phe-Leu

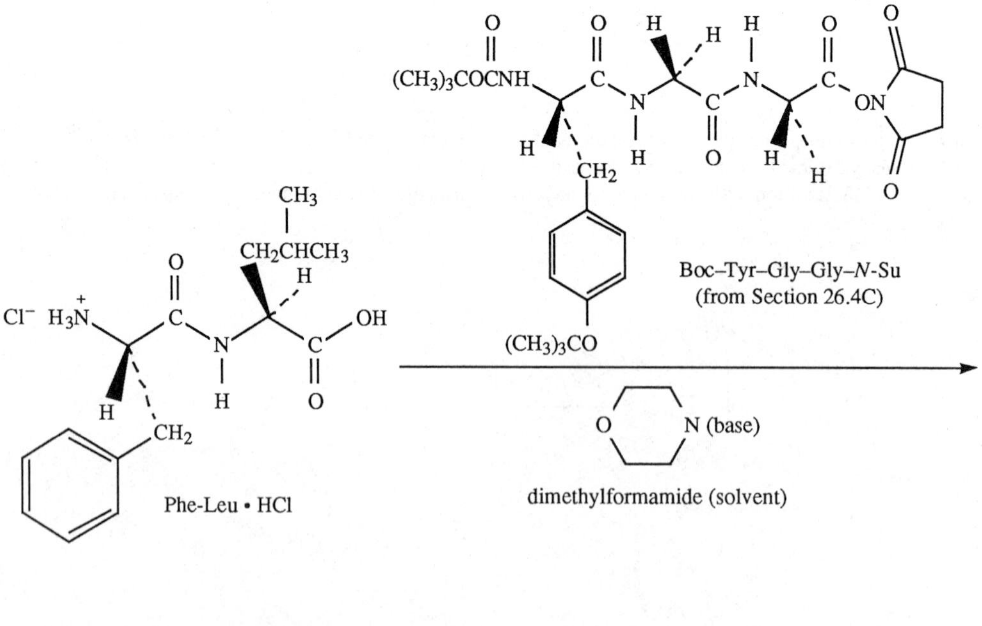

Cl⁻ H3N
CH2CHCH3
CH3
OH
CH2
Phe-Leu • HCl
(CH3)3COCNH
ON
(CH3)3CO
Boc–Tyr–Gly–Gly–N-Su
(from Section 26.4C)
N (base)
dimethylformamide (solvent)

26.16 (cont)

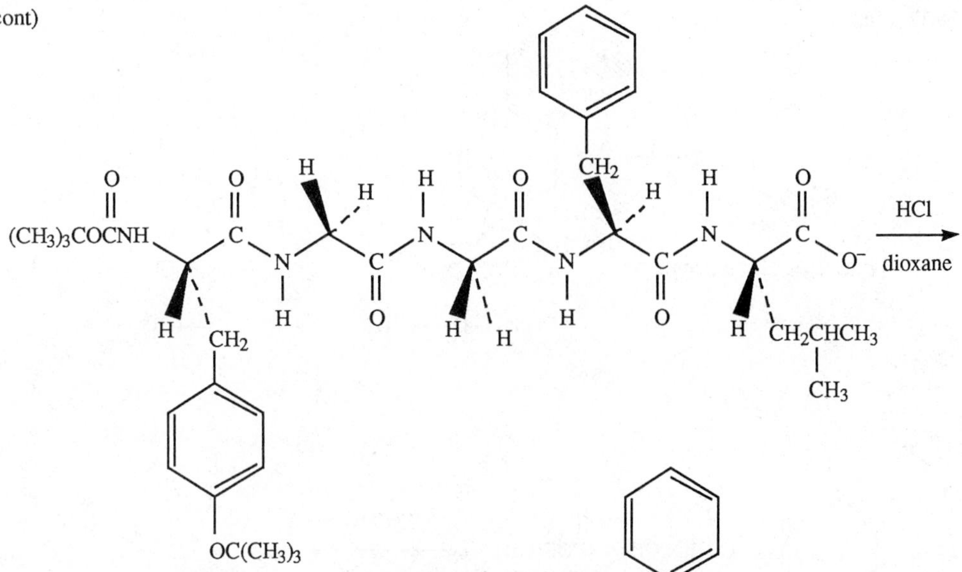

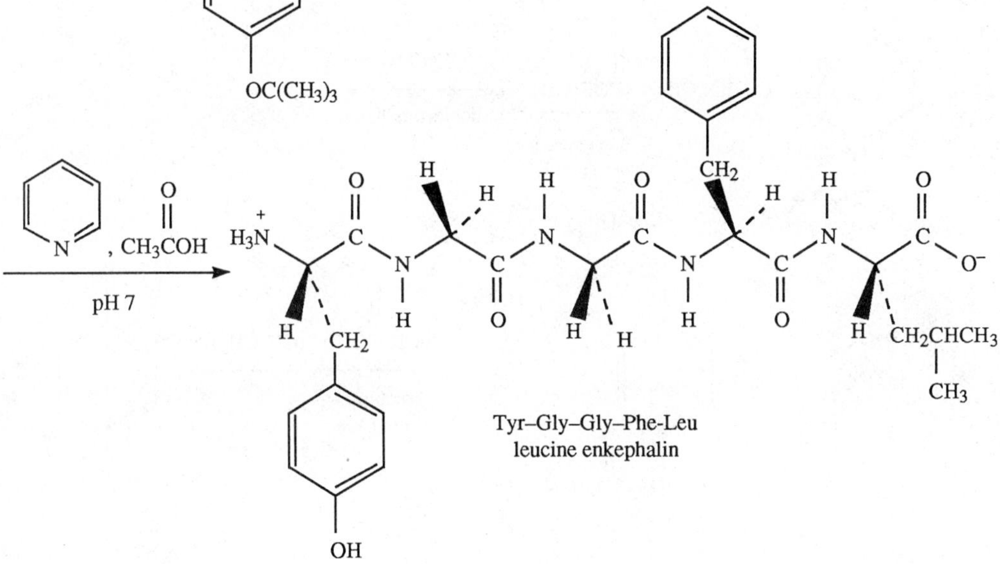

Tyr–Gly–Gly–Phe-Leu
leucine enkephalin

26.17

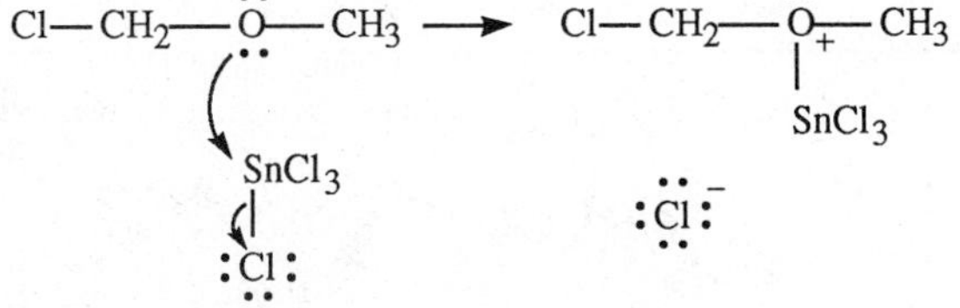

26.17 (cont)

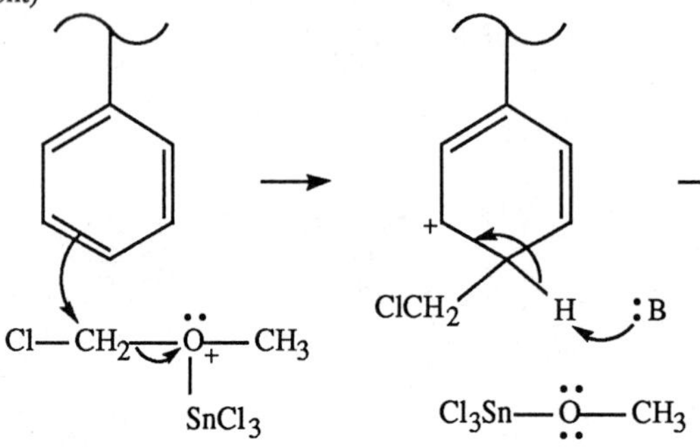

26.18

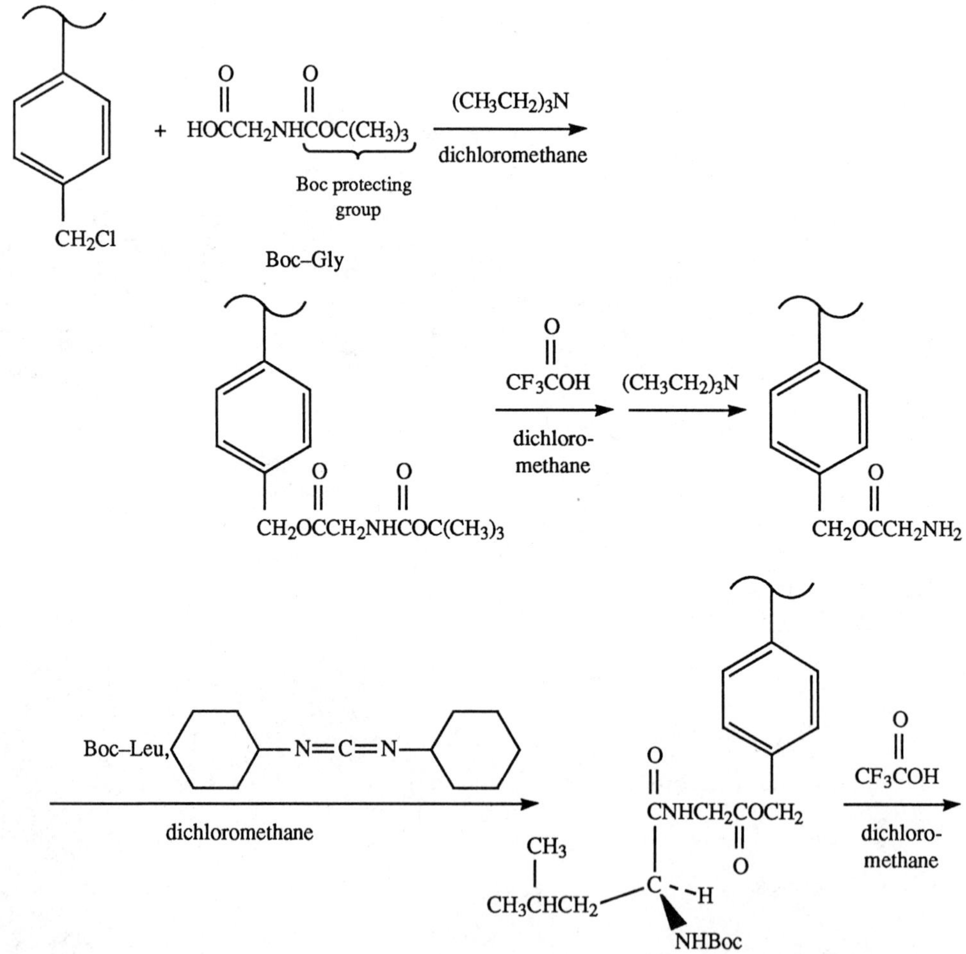

Boc protecting
group

Boc–Gly

26.18 (cont)

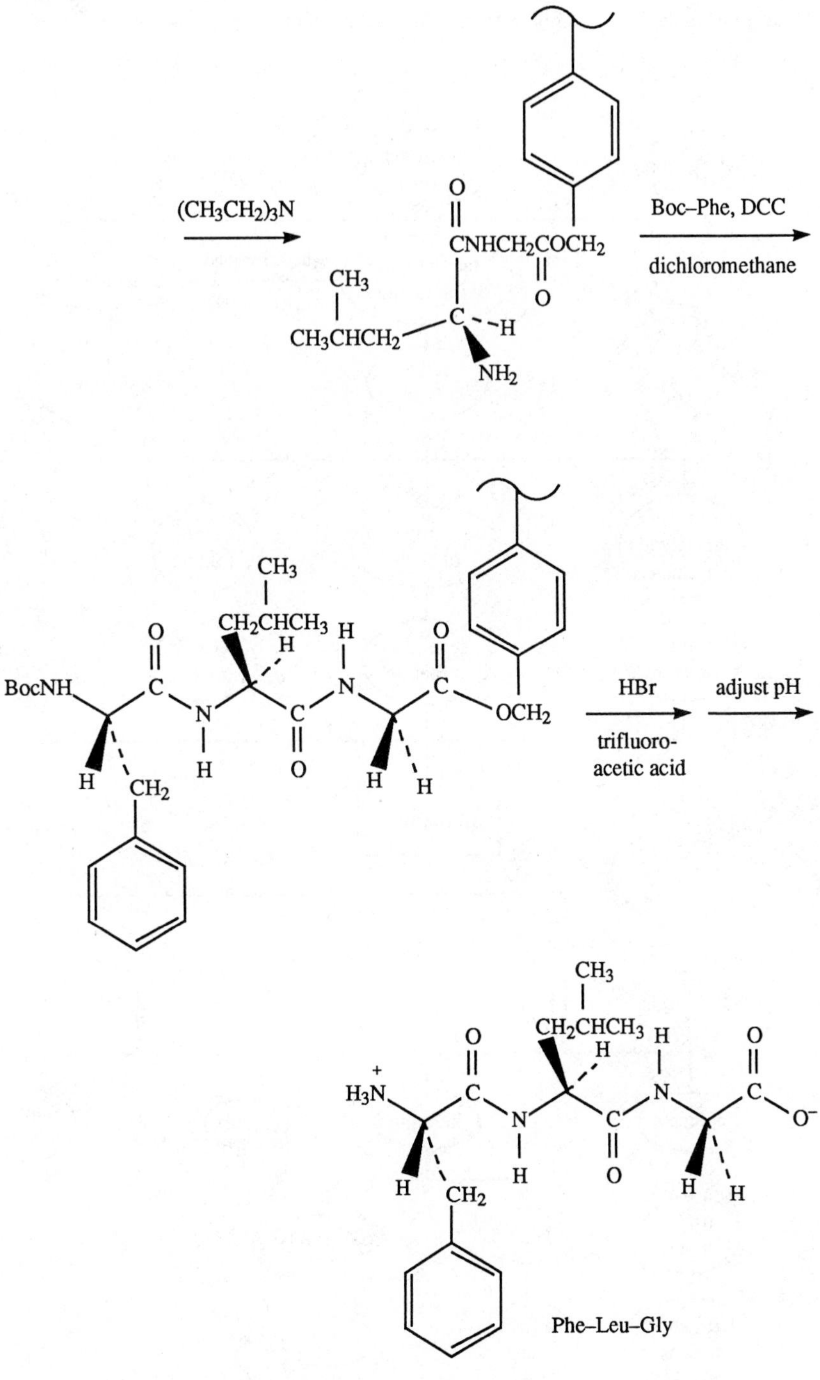

$(CH_3CH_2)_3N$

$\xrightarrow{\hspace{3cm}}$

CH₃CHCH₂—C—H structure with NH₂, CNHCH₂COCH₂ attached to aromatic ring

$\xrightarrow[\text{dichloromethane}]{\text{Boc–Phe, DCC}}$

BocNH peptide structure

$\xrightarrow[\substack{\text{trifluoro-}\\\text{acetic acid}}]{\text{HBr}}$ $\xrightarrow{\text{adjust pH}}$

Phe–Leu–Gly

Concept Map 26.6   Conformation and structure in proteins.

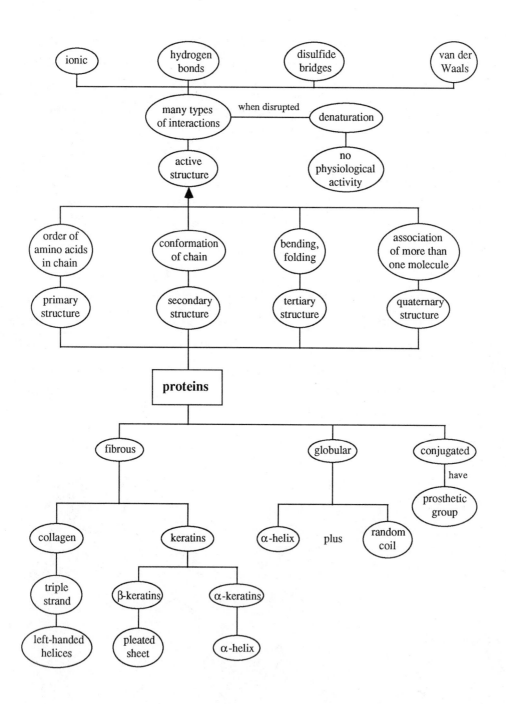

26.19 (a)

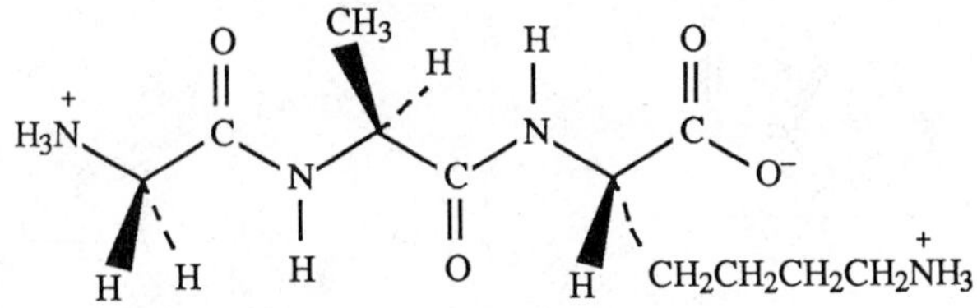

(shown at pH 6)

(b)

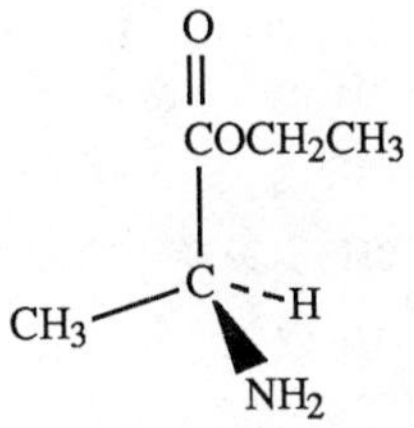

(c)

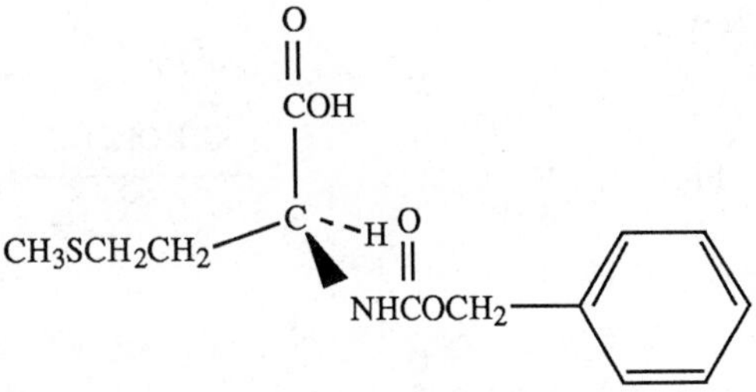

(d)

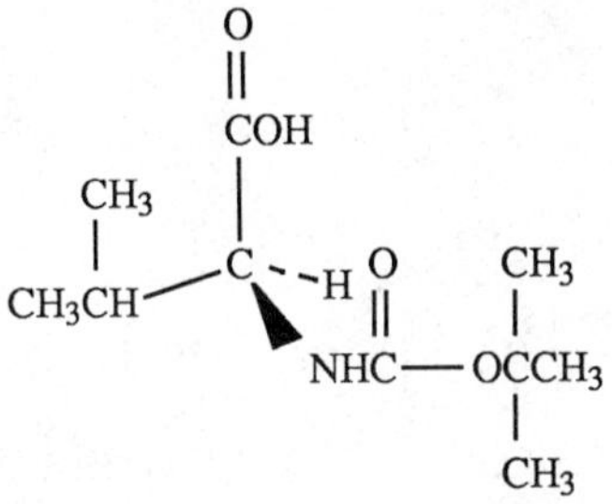

(e)

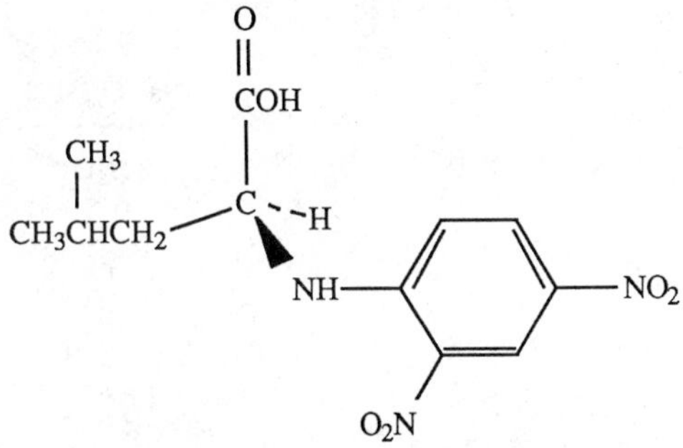

(f)

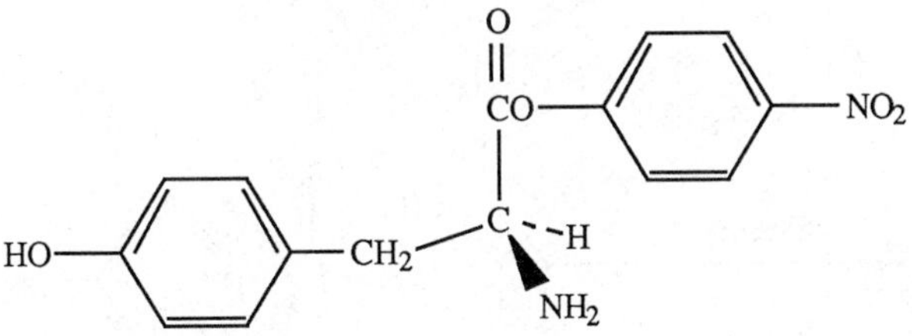

26.20 (a) *N*-2,4-dinitrophenylphenylalanine

(b) glycylcysteinylhistidine

(c) *N*-benzoylglycine; hippuric acid

(d) *p*-nitrophenyl alaninate

(e) *N-tert*-butoxycarbonylleucine

(f) methyl cysteinate

26.21

(a)

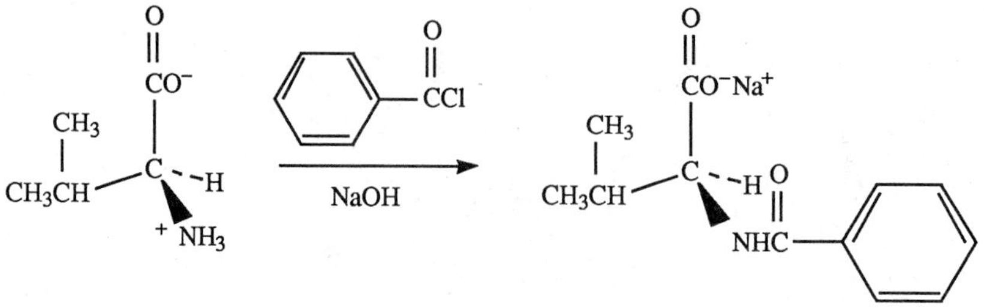

$$\xrightarrow{\text{NaOH}}$$

reagent: benzoyl chloride (PhCOCl)

(b)

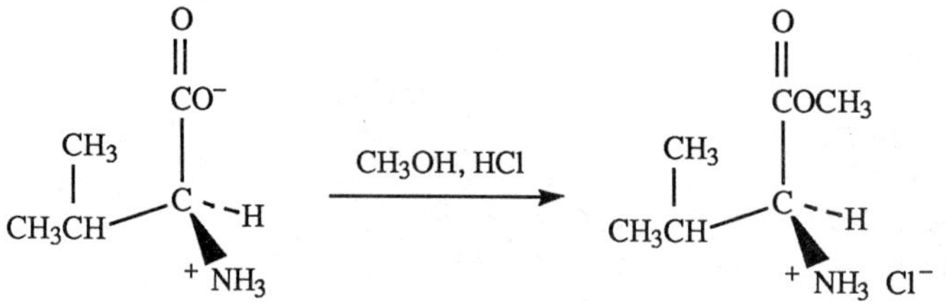

$$\xrightarrow{\text{CH}_3\text{OH, HCl}}$$

(c)

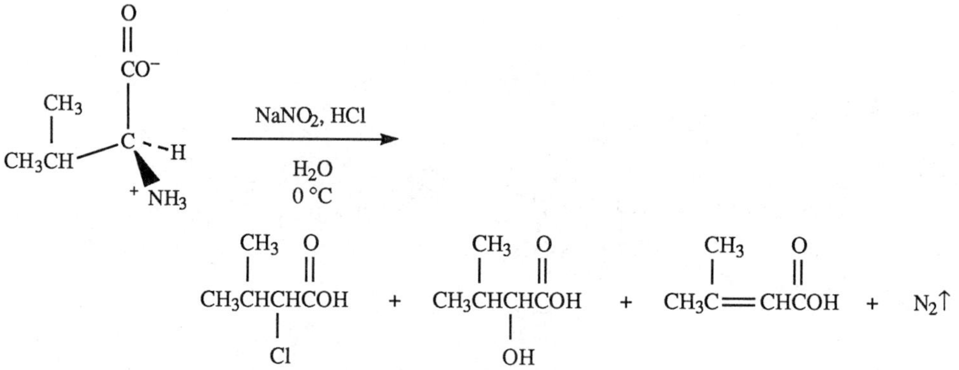

$$\xrightarrow[\substack{\text{H}_2\text{O} \\ 0\,^\circ\text{C}}]{\text{NaNO}_2,\ \text{HCl}}$$

$$\underset{\underset{\text{Cl}}{|}}{\overset{\overset{\text{CH}_3}{|}}{\text{CH}_3\text{CHCHCOH}}} \ \overset{\text{O}}{\overset{\|}{}} \quad + \quad \underset{\underset{\text{OH}}{|}}{\overset{\overset{\text{CH}_3}{|}}{\text{CH}_3\text{CHCHCOH}}} \ \overset{\text{O}}{\overset{\|}{}} \quad + \quad \overset{\overset{\text{CH}_3}{|}}{\text{CH}_3\text{C}}\!\!=\!\!\text{CHCOH} \ \overset{\text{O}}{\overset{\|}{}} \quad + \quad \text{N}_2\uparrow$$

(d)

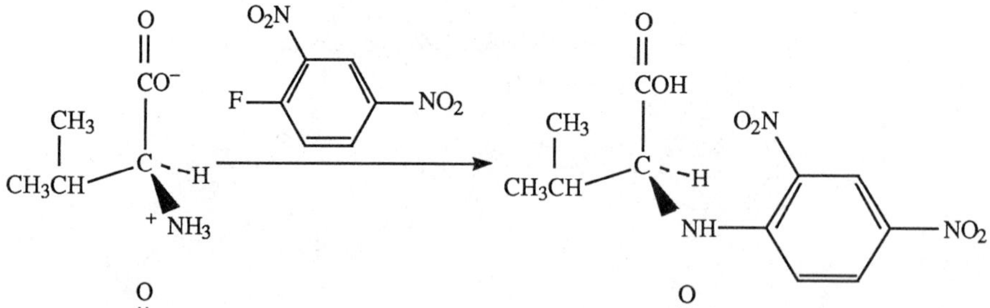

reagent: 2,4-dinitrofluorobenzene (O_2N, F, NO_2)

(e)

reagent: phenyl isothiocyanate ($N{=}C{=}S$)

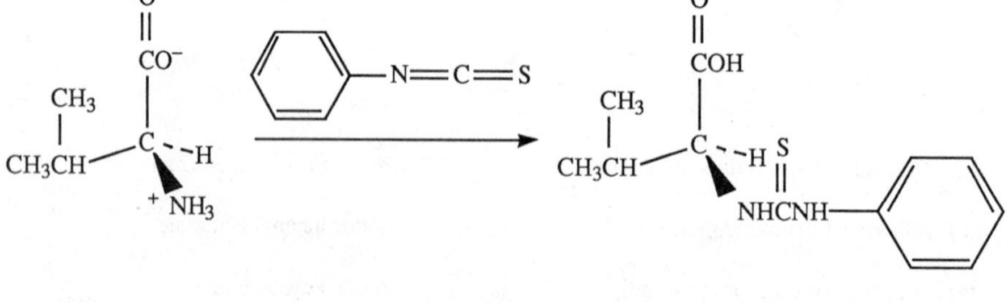

26.21 (cont)

(f)

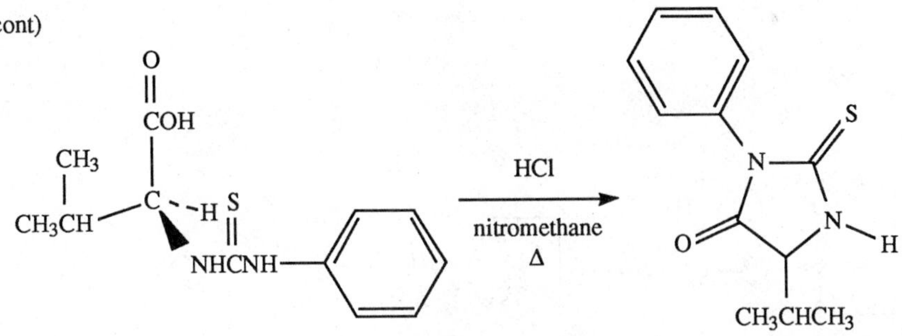

(g)

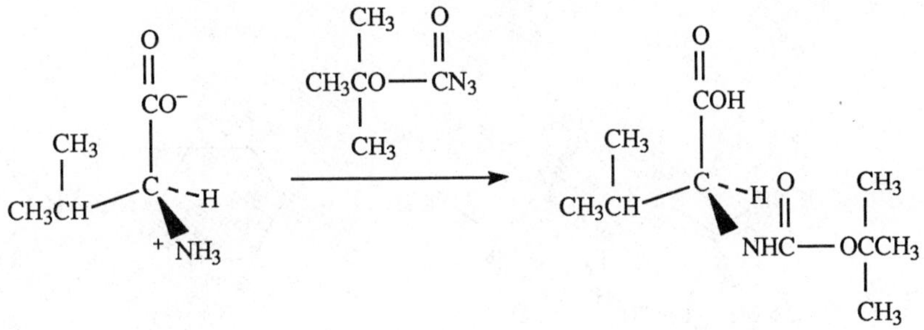

(h)

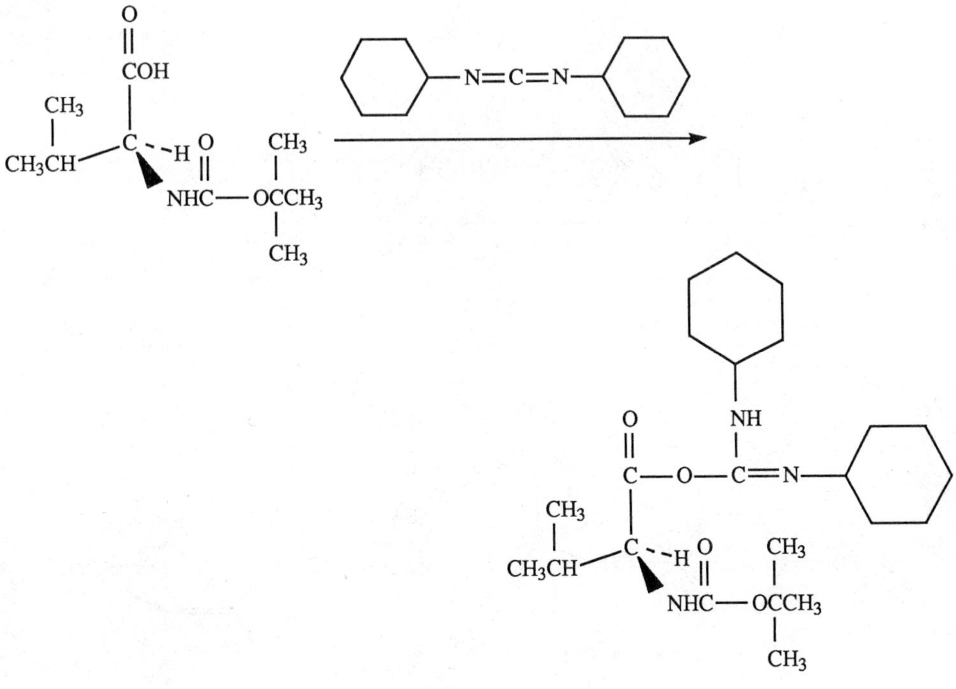

26.21 (cont)

(i)

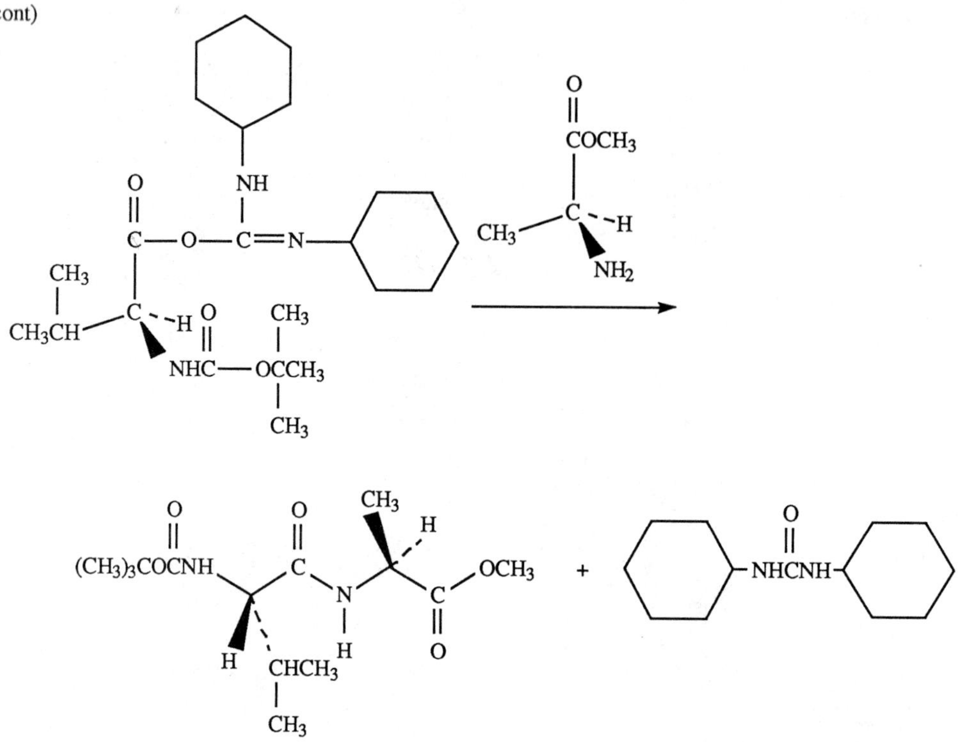

(j)

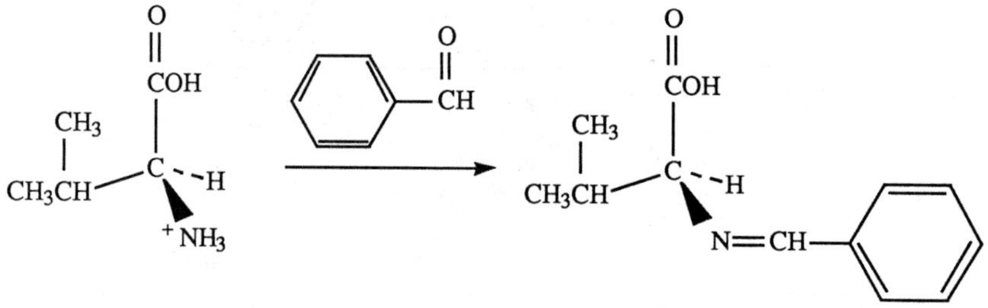

26.22

(a)

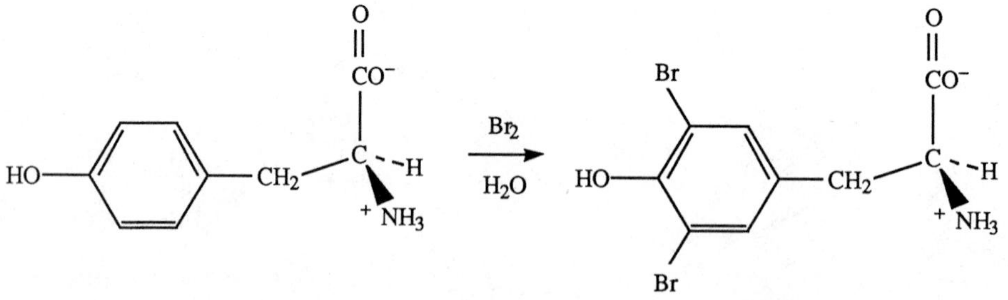

26.22 (cont)

(b)

$$
\text{H}_3\overset{+}{\text{N}}\text{CH}_2\text{CH}_2\text{CH}_2\text{CH}_2-\overset{\overset{\displaystyle\text{CO}^-}{\overset{\displaystyle\|}{\text{O}}}}{\underset{\overset{+}{\text{NH}_3}}{\text{C}}}\!\!-\text{H} \quad\xrightarrow[\substack{\text{H}_2\text{O}\\ 0\,°\text{C}}]{\text{NaNO}_2,\ \text{HCl}}\quad \left[\ \overset{+}{\text{N}_2}\text{CH}_2\text{CH}_2\text{CH}_2\text{CH}_2-\overset{\overset{\displaystyle\text{CO}^-}{\overset{\displaystyle\|}{\text{O}}}}{\underset{\overset{+}{\text{N}_2}}{\text{C}}}\!\!-\text{H}\ \right] \longrightarrow
$$

$$
2\text{N}_2\!\uparrow\ +\ \text{HOCH}_2\text{CH}_2\text{CH}_2\text{CH}_2\overset{\overset{\displaystyle\text{O}}{\|}}{\underset{\text{OH}}{\text{CHCOH}}} \ +\ \text{HOCH}_2\text{CH}_2\text{CH}_2\text{CH}_2\overset{\overset{\displaystyle\text{O}}{\|}}{\underset{\text{Cl}}{\text{CHCOH}}}\ +
$$

$$
\text{HOCH}_2\text{CH}_2\text{CH}_2\text{CH}\!\!=\!\!\text{CH}\overset{\overset{\displaystyle\text{O}}{\|}}{\text{COH}}\ +\quad
$$

other combinations of
displacements and
eliminations at both
diazonium ions

(c)

$$
\text{CH}_3-\overset{\overset{\displaystyle\text{CO}^-}{\overset{\displaystyle\|}{\text{O}}}}{\underset{\overset{+}{\text{NH}_3}}{\text{C}}}\!\!-\text{H} \quad\xrightarrow{(\text{CH}_3\text{C})_2\text{O}}\quad \text{CH}_3-\overset{\overset{\displaystyle\text{COH}}{\overset{\displaystyle\|}{\text{O}}}}{\underset{\underset{\overset{\displaystyle\|}{\text{O}}}{\text{NHCCH}_3}}{\text{C}}}\!\!-\text{H}
$$

(d)

$$
\text{CH}_3\text{CHCH}_2-\underset{\text{CH}_3}{\underset{|}{}}\overset{\overset{\displaystyle\text{CO}^-}{\overset{\displaystyle\|}{\text{O}}}}{\underset{\overset{+}{\text{NH}_3}}{\text{C}}}\!\!-\text{H} \quad\xrightarrow{\ \ \ \ }\quad \text{CH}_3\text{CHCH}_2-\underset{\text{CH}_3}{\underset{|}{}}\overset{\overset{\displaystyle\text{COH}}{\overset{\displaystyle\|}{\text{O}}}}{\underset{\underset{\overset{\displaystyle\|}{\text{O}}}{\text{NHC}}}{\text{C}}}\!\!-\text{H}
$$

(e)

$$
\text{HOCCH}_2-\underset{\overset{\displaystyle\|}{\text{O}}}{}\overset{\overset{\displaystyle\text{CO}^-}{\overset{\displaystyle\|}{\text{O}}}}{\underset{\overset{+}{\text{NH}_3}}{\text{C}}}\!\!-\text{H} \quad\xrightarrow{\text{CH}_3\text{CH}_2\text{OH (excess), HCl}}\quad \text{CH}_3\text{CH}_2\text{OCCH}_2-\underset{\overset{\displaystyle\|}{\text{O}}}{}\overset{\overset{\displaystyle\text{COCH}_2\text{CH}_3}{\overset{\displaystyle\|}{\text{O}}}}{\underset{\overset{+}{\text{NH}_3}\ \text{Cl}^-}{\text{C}}}\!\!-\text{H}
$$

26.22 (cont)

(f)

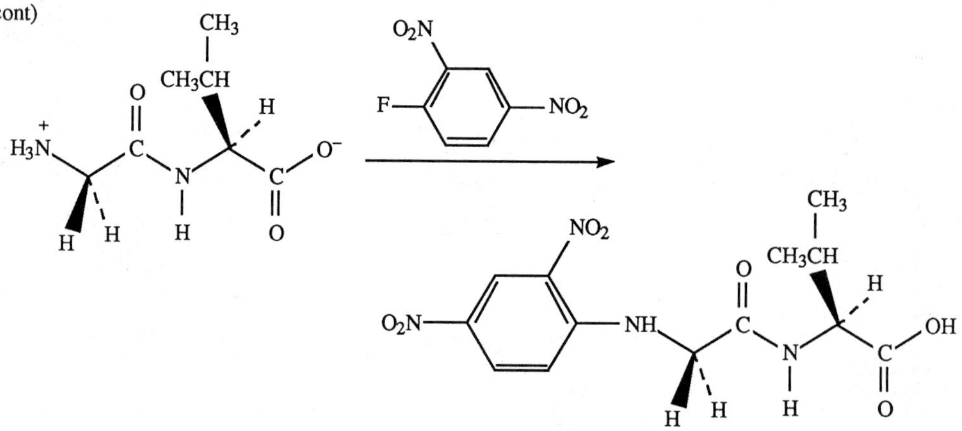

(g)   product of (f) 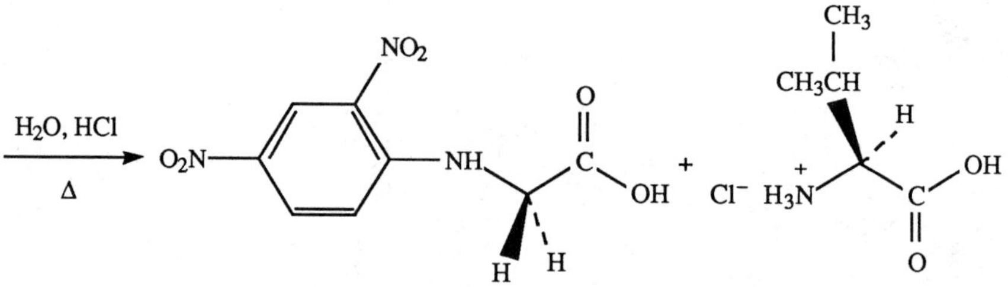 $\xrightarrow[\Delta]{\text{H}_2\text{O, HCl}}$

(h)

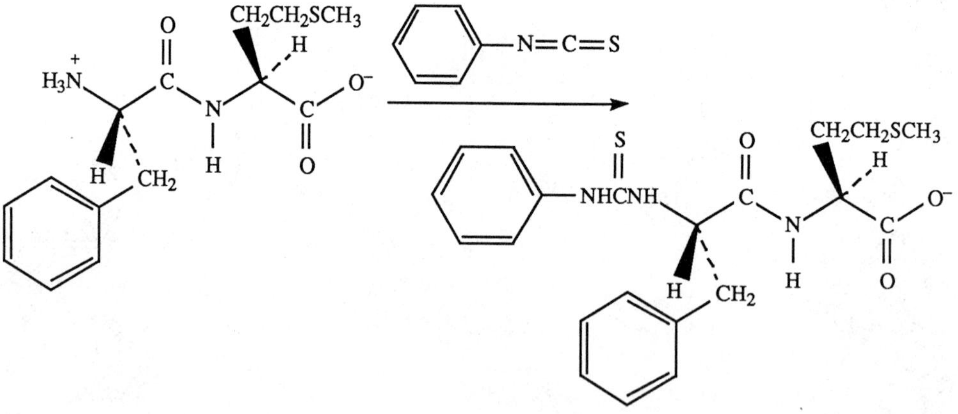

26.22 (cont)

(i)   product of (h) $\xrightarrow[\substack{\text{nitromethane} \\ \Delta}]{\text{HCl}}$

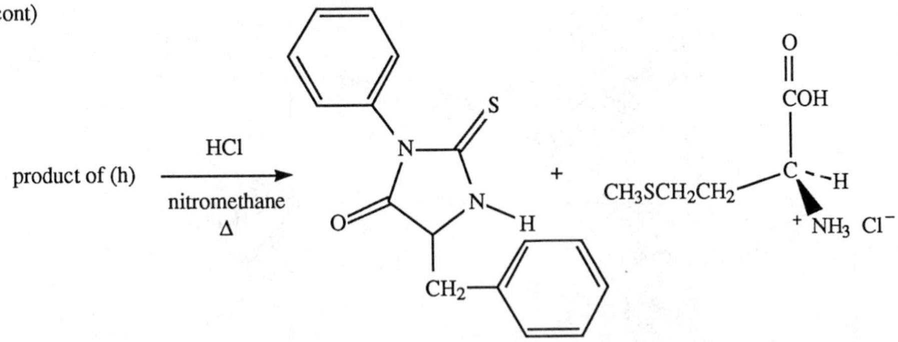

(j)

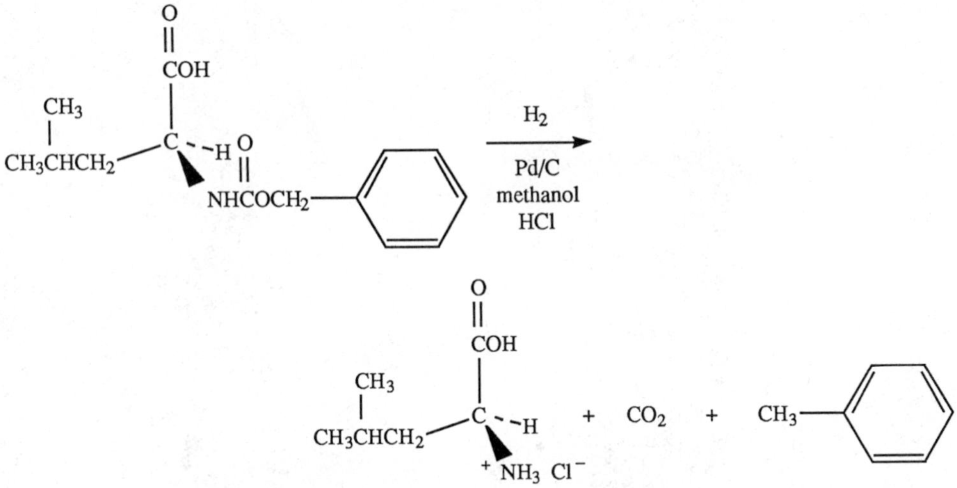

$$\xrightarrow[\substack{\text{Pd/C} \\ \text{methanol} \\ \text{HCl}}]{\text{H}_2}$$

$+ \quad CO_2 \quad + \quad CH_3-$

(k) $\xrightarrow[\text{dioxane}]{\text{HCl}}$

$+ \quad CO_2 \quad + \quad CH_3C\!=\!CH_2$

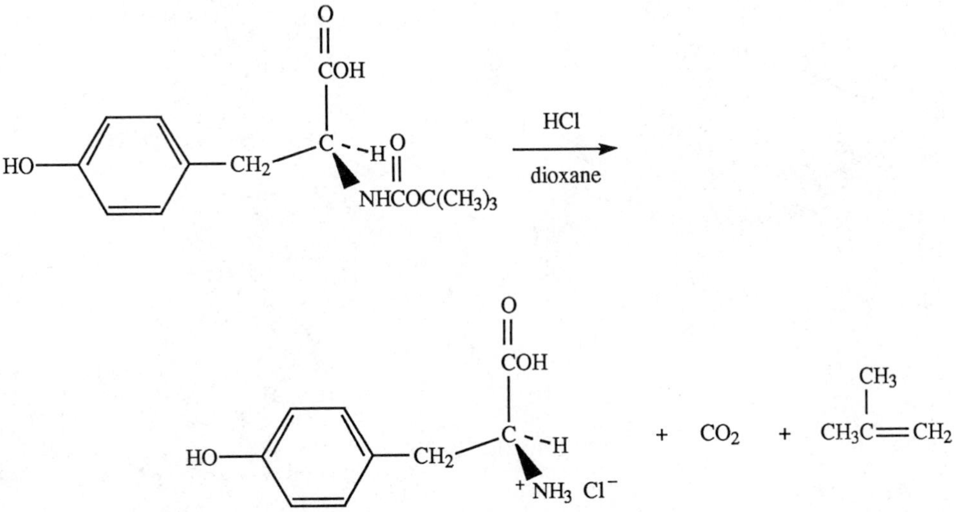

26.22 (cont)

(l)

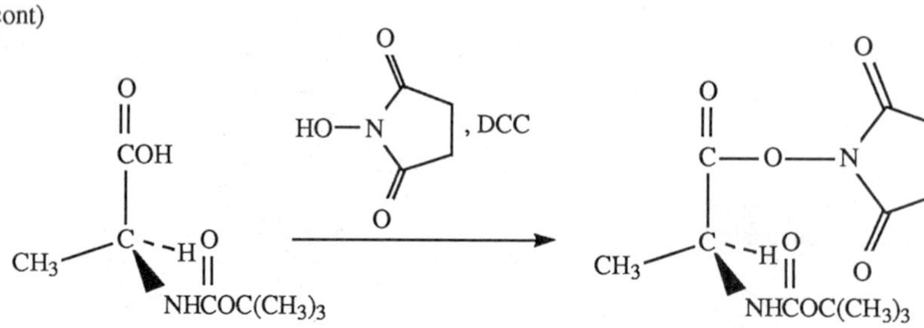

26.23

(a)

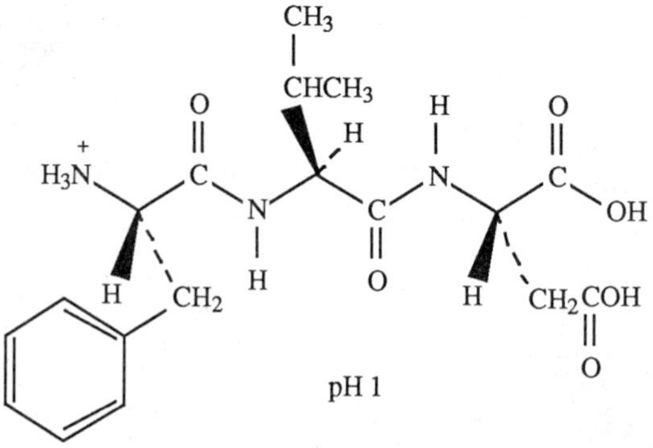

pH 1

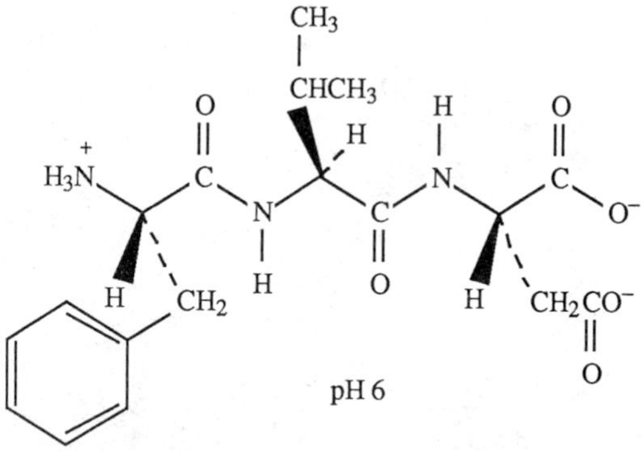

pH 6

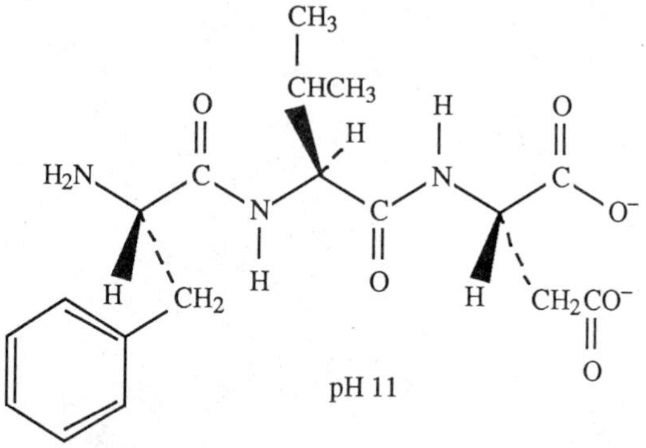

pH 11

(b)

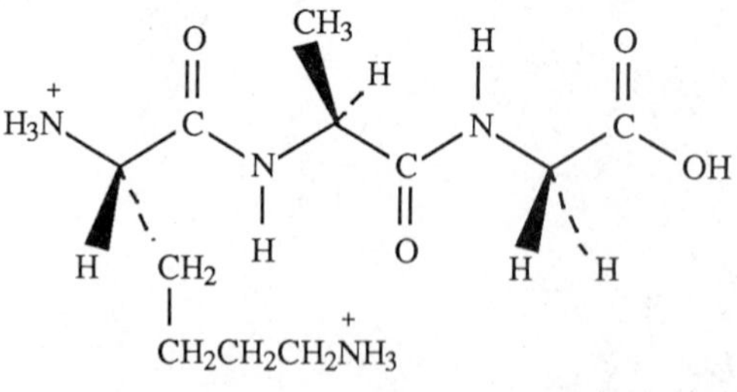

pH 1

26.23 (b) (cont)

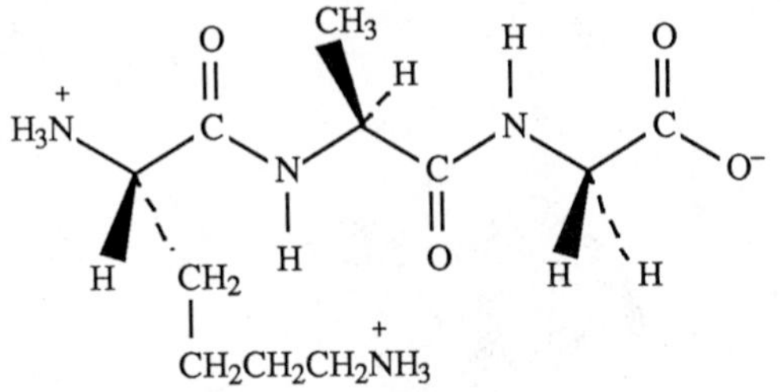

$H_3\overset{+}{N}$ ... C(=O) ... N(H) ... (CH_3)(H) ... C(=O) ... N(H) ... C(=O) ... O^-

H   CH_2
CH_2CH_2CH_2$\overset{+}{N}H_3$

pH 6

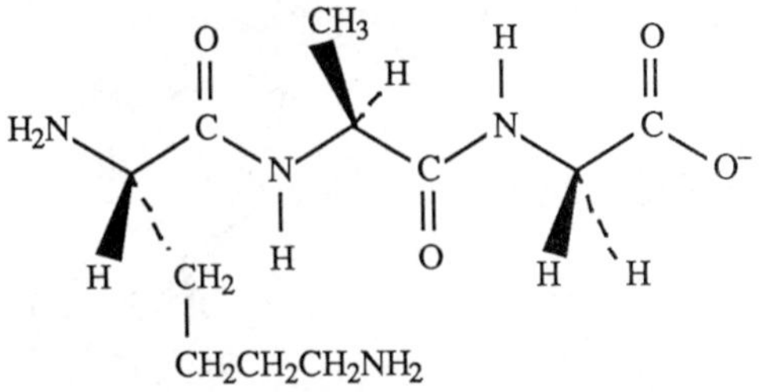

H_2N ... C(=O) ... N(H) ... (CH_3)(H) ... C(=O) ... N(H) ... C(=O) ... O^-

H   CH_2
CH_2CH_2CH_2NH_2

pH 11

(c)

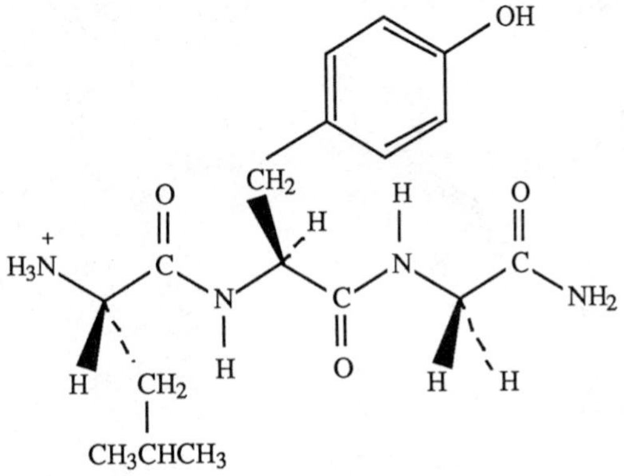

OH

CH_2

$H_3\overset{+}{N}$ ... C(=O) ... N(H) ... H ... C(=O) ... N(H) ... C(=O) ... NH_2

H   CH_2
CH_3CHCH_3

pH 1

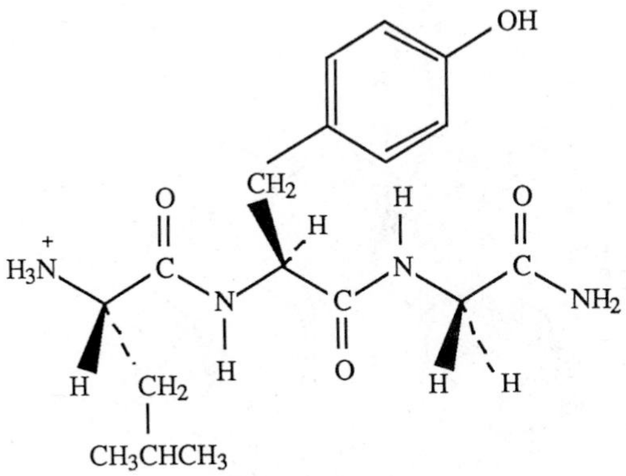

OH

CH_2

$H_3\overset{+}{N}$ ... C(=O) ... N(H) ... H ... C(=O) ... N(H) ... C(=O) ... NH_2

H   CH_2
CH_3CHCH_3

pH 6

O^-

CH_2

H_2N ... C(=O) ... N(H) ... H ... C(=O) ... N(H) ... C(=O) ... NH_2

H   CH_2
CH_3CHCH_3

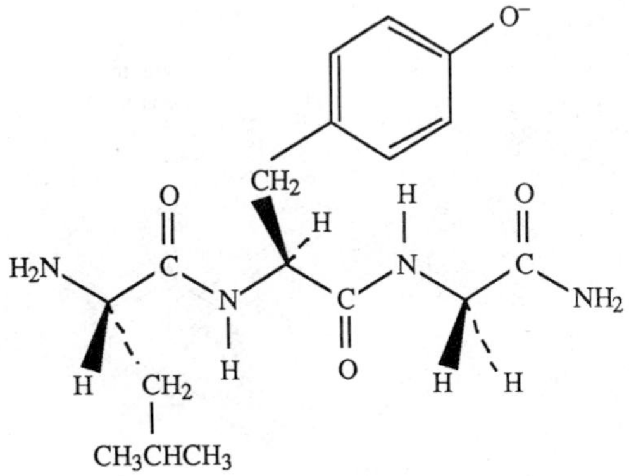

pH 11

26.23 (cont)

(d)

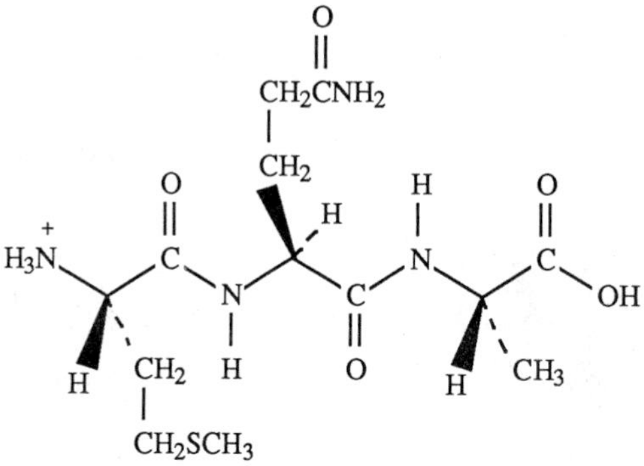

pH 1

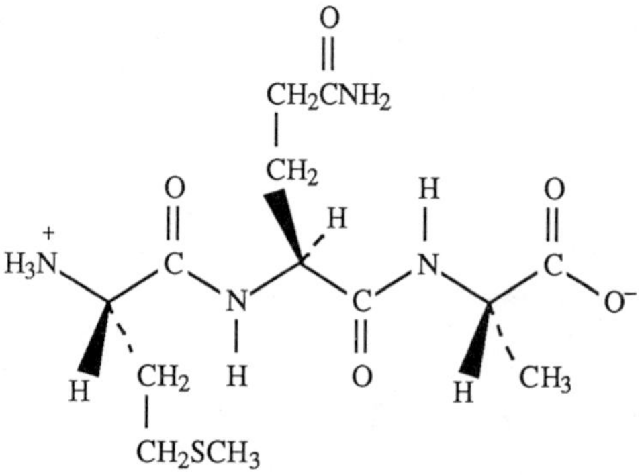

pH 6

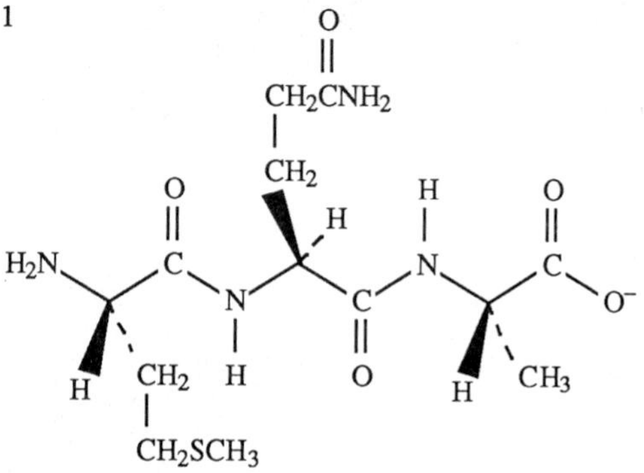

pH 11

26.24

Note that the configuration
at this stereocenter is reversed

(a)

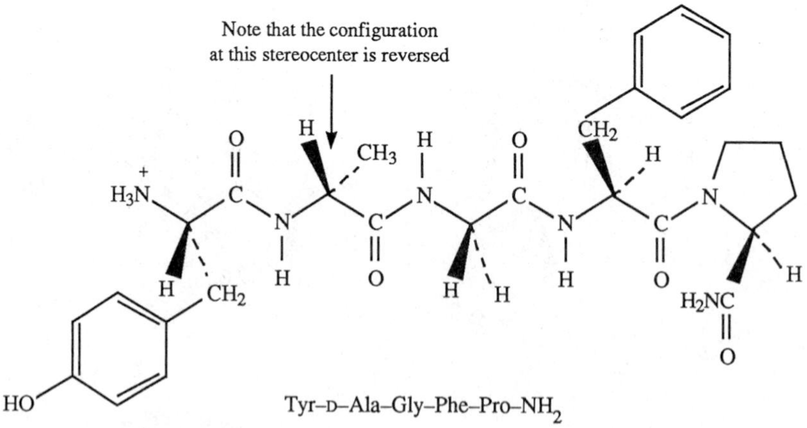

Tyr–D–Ala–Gly–Phe–Pro–NH$_2$

26.24 (cont)

(b)

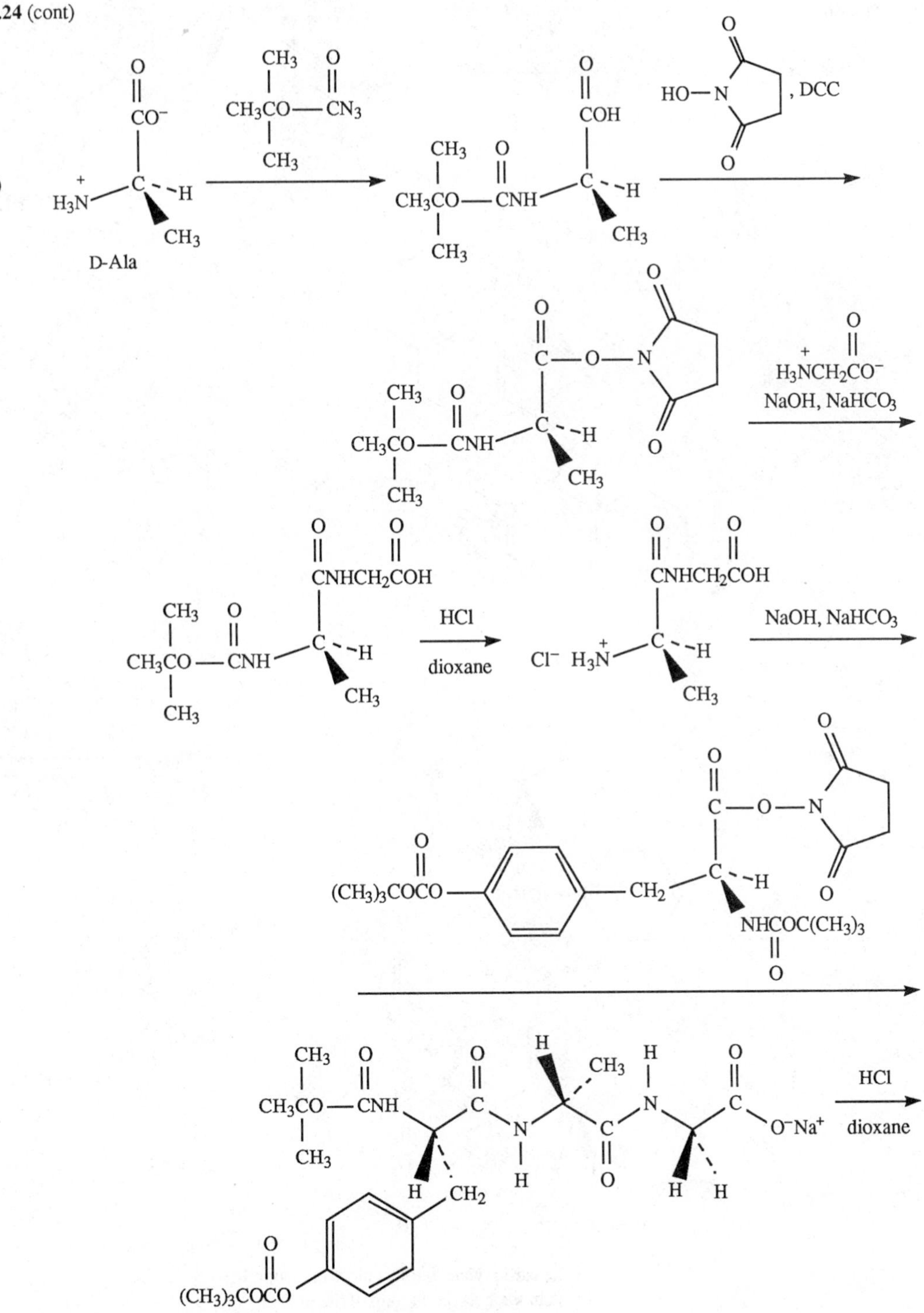

26.24 (cont)

(b) (cont)

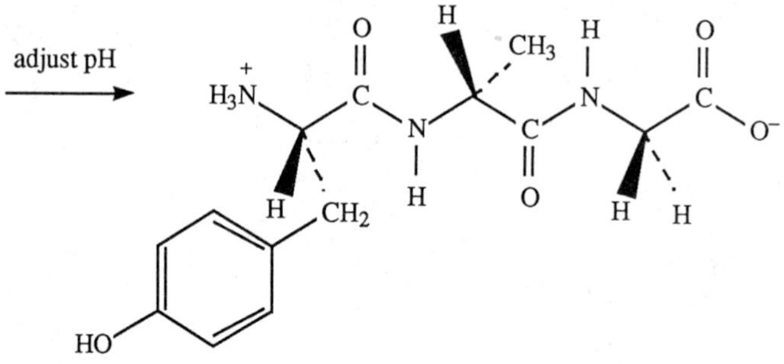
adjust pH
$H_3\overset{+}{N}$
CH$_3$
CH$_2$
HO
O$^-$

26.25

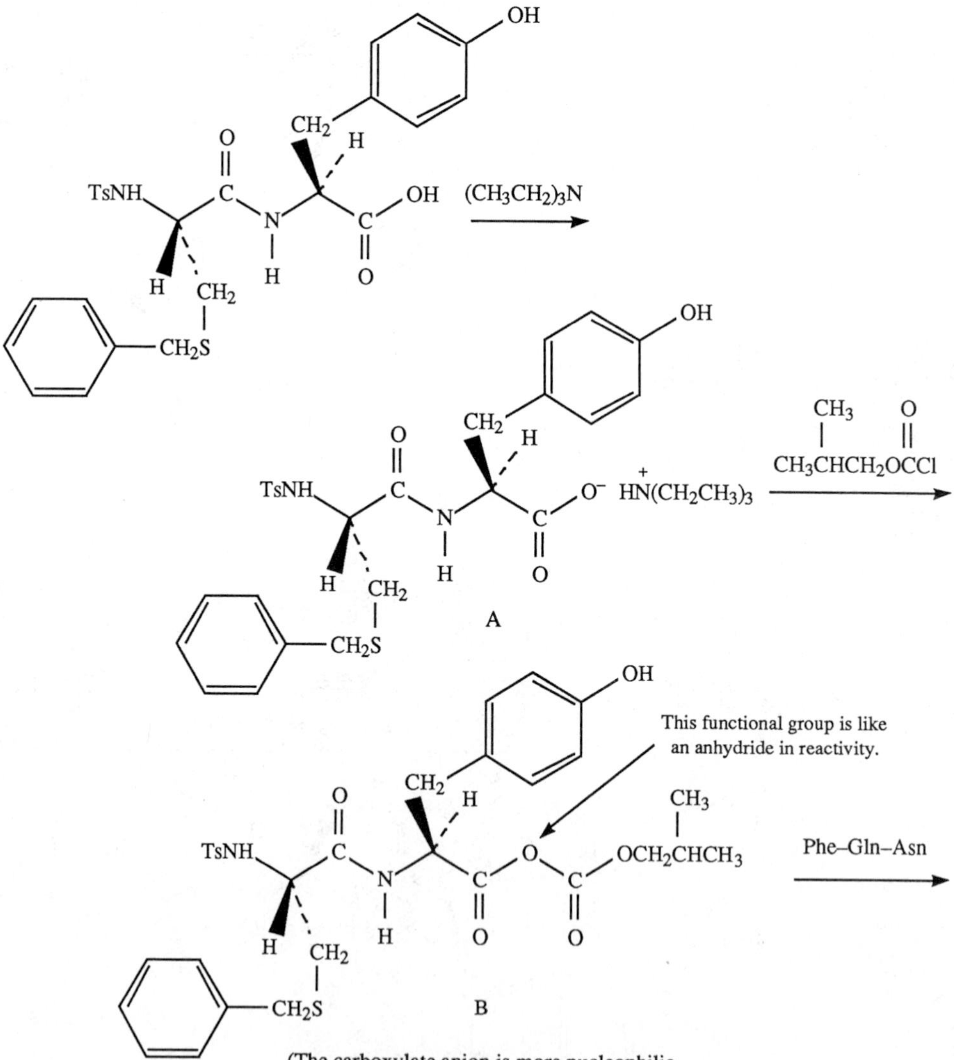
TsNH
OH
$(CH_3CH_2)_3N$
CH$_2$S
O$^-$ $H\overset{+}{N}(CH_2CH_3)_3$
A
CH$_3$
CH_3CHCH_2OCCl
This functional group is like an anhydride in reactivity.
OCH_2CHCH_3
Phe–Gln–Asn
B

(The carboxylate anion is more nucleophilic than the hydroxyl group of the phenol.)

26.25 (cont)

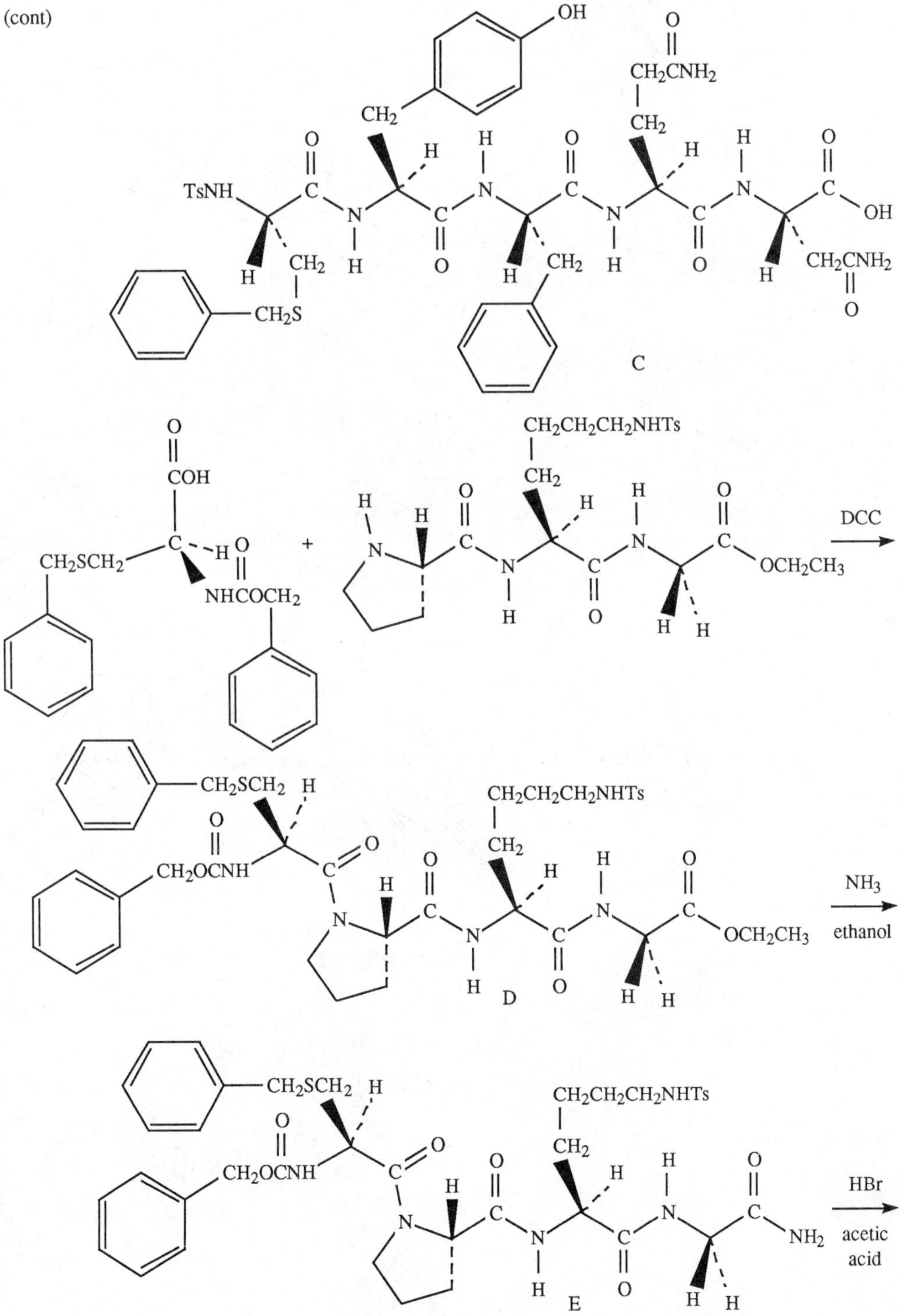

C

DCC

D

NH₃
ethanol

E

HBr
acetic
acid

26.25 (cont)

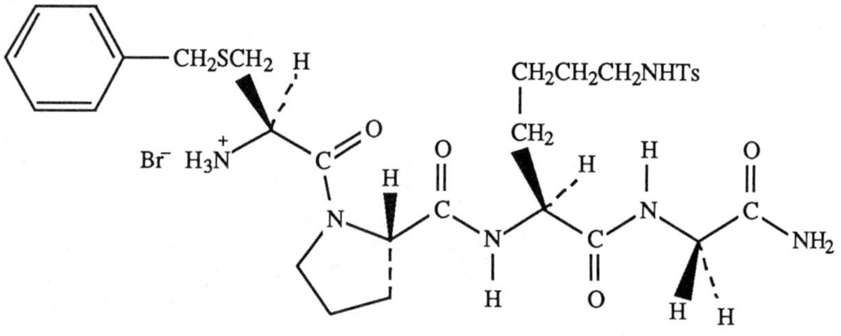

F

C + F

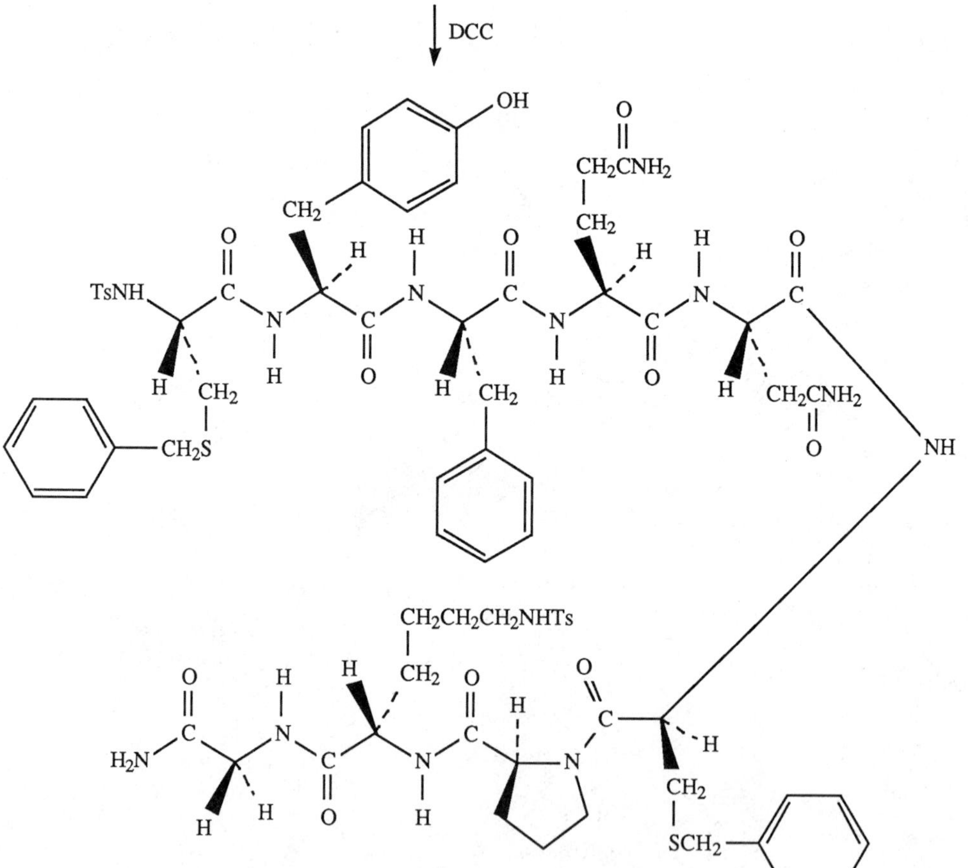

↓ DCC

G

↓ Na
NH₃ (liq)

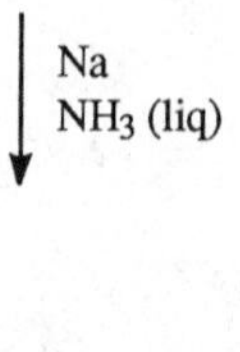

26.25 (cont)   Cys–Tyr–Phe–Gln–Asn–Cys–Pro–Lys–Gly—NH$_2$

N-terminal end   *C*-terminal end

H

air oxidation ↓

Phe——————Gln
  /   \
Tyr   Asn
  | |
Cys—— S —— S — Cys—Pro—Lys—Gly—NH$_2$

N-terminal end   *C*-terminal end

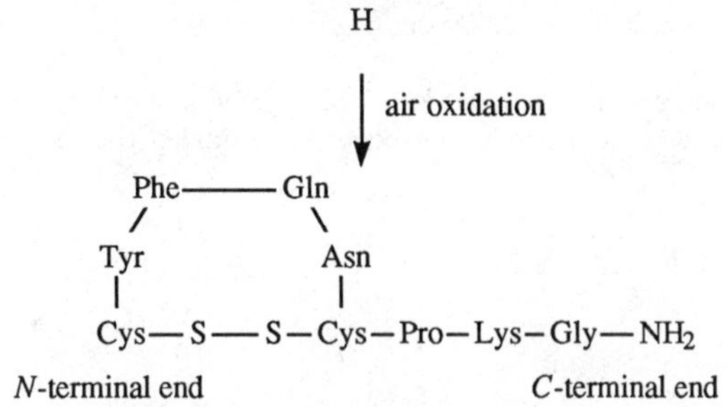

I

lysine vasopressin

26.26

(a)

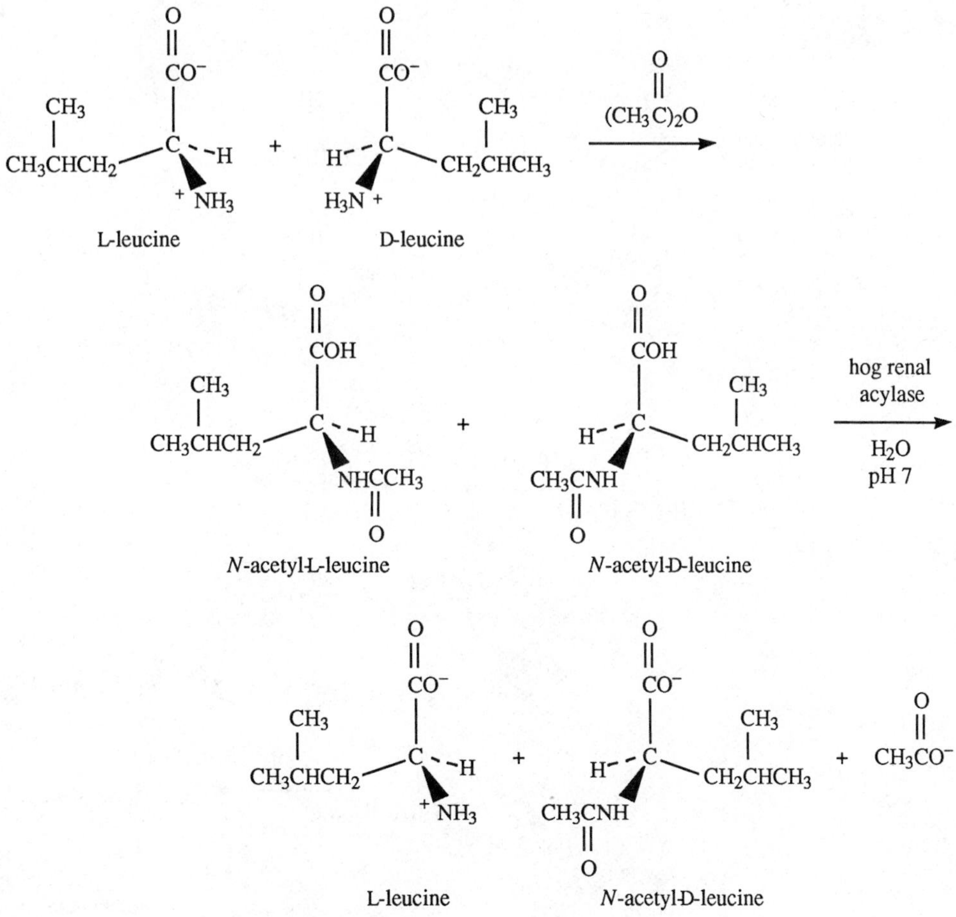

L-leucine   D-leucine $(CH_3C)_2O$

N-acetyl-L-leucine   *N*-acetyl-D-leucine   hog renal acylase H_2O pH 7

L-leucine   *N*-acetyl-D-leucine   + CH_3CO^-

26.26 (cont)

(b)   At pH 6, L-leucine will be its most insoluble as an internal salt or zwitterion. If the pH of the system is adjusted to 6, we can expect L-leucine to crystallize out.

The pH of the solution would have to much lower for *N*-acetyl-D-leucine to precipitate, because the basicity of the amino group has been eliminated by conversion to an amide. The compound has the acidity of an alkyl carboxylic acid.

mixture from part (a) $\xrightarrow[\text{to pH 6}]{H_3O^+}$

$$
\begin{array}{c}
\text{O} \\
\| \\
\text{CO}^- \\
| \\
\text{CH}_3\text{CHCH}_2\!-\!\!\overset{\displaystyle\text{CH}_3}{\underset{+\text{NH}_3}{\overset{|}{\text{C}}}}\!-\!-\text{H}
\end{array}
$$

precipitate;
filter out of solution

solution remaining $\xrightarrow[\text{to pH 1}]{H_3O^+}$

$$
\begin{array}{c}
\text{O} \\
\| \\
\text{COH} \\
| \\
\text{H}-\!-\overset{\displaystyle}{\underset{\text{CH}_3\text{CNH}}{\text{C}}}\!-\!\text{CH}_2\text{CHCH}_3 \\
\| \\
\text{O}
\end{array}
$$

precipitate;
filter out of solution

(c)

$$
\begin{array}{c}
\text{O} \\
\| \\
\text{COH} \\
| \\
\text{H}-\!-\overset{\displaystyle\text{CH}_3}{\underset{\text{CH}_3\text{CNH}}{\text{C}}}\!-\!\text{CH}_2\text{CHCH}_3 \\
\| \\
\text{O}
\end{array}
\qquad \xrightarrow[\Delta]{H_2O,\ HCl}
$$

$$
\begin{array}{c}
\text{O} \\
\| \\
\text{COH} \\
| \\
\text{H}-\!-\overset{\displaystyle\text{CH}_3}{\underset{\text{Cl}^-\ \text{H}_3\text{N}^+}{\text{C}}}\!-\!\text{CH}_2\text{CHCH}_3
\end{array}
\qquad \xrightarrow[\text{to pH 6}]{OH^-}
\qquad
\begin{array}{c}
\text{O} \\
\| \\
\text{CO}^- \\
| \\
\text{H}-\!-\overset{\displaystyle\text{CH}_3}{\underset{\text{H}_3\text{N}^+}{\text{C}}}\!-\!\text{CH}_2\text{CHCH}_3
\end{array}
$$

precipitate;
filter out of solution

26.27

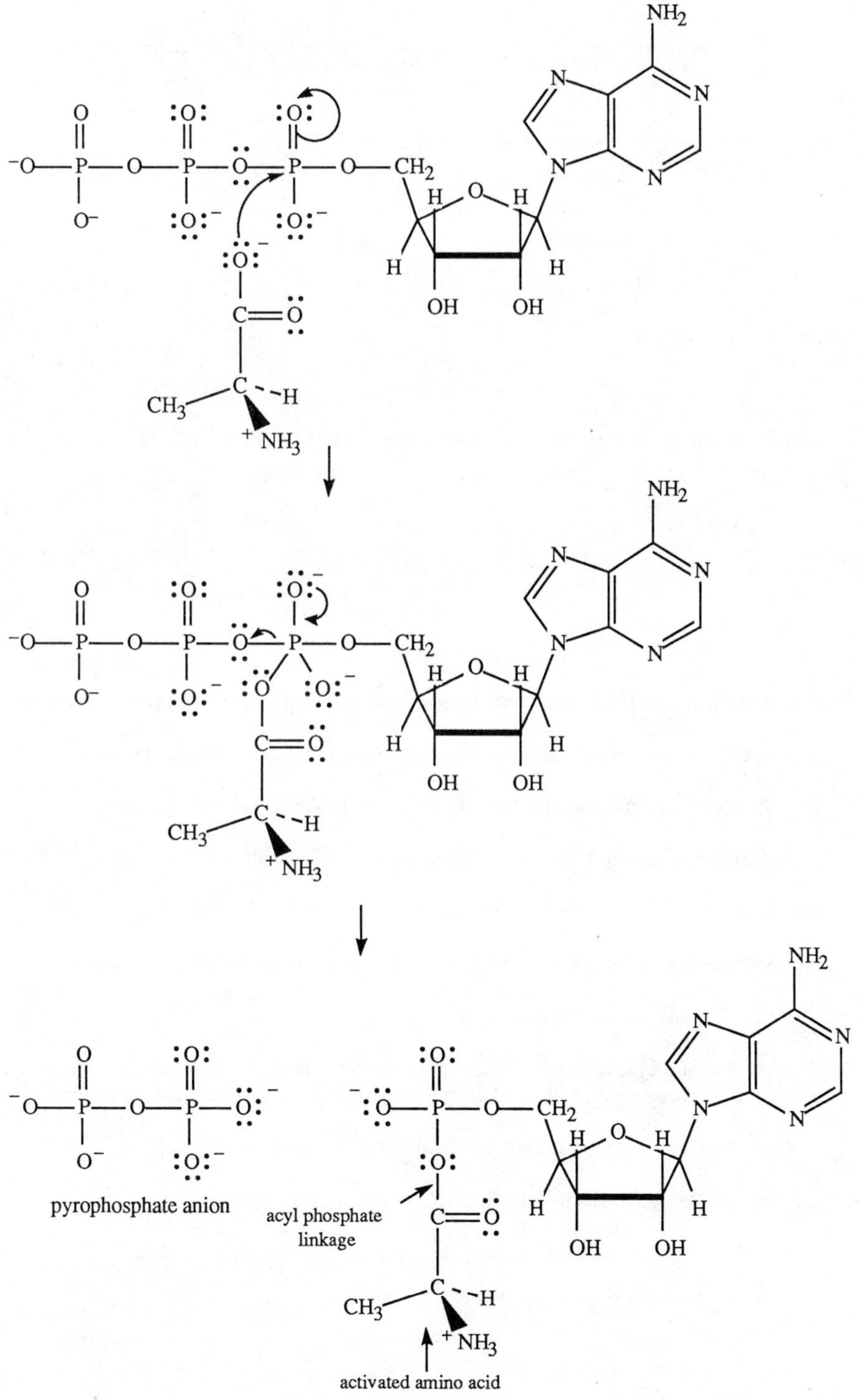

pyrophosphate anion

acyl phosphate
linkage

activated amino acid

26.28   xenopsin $\xrightarrow[\text{trypsin}]{}$ ⌐Glu—Gly—Lys—Arg   +   ⌐Glu—Gly—Lys—Arg—Pro—Trp

xenopsin $\xrightarrow[\text{chymotrypsin}]{}$ same peptides as above   +   Ile—Leu

N-terminal amino acid is ⌐Glu

C-terminal amino acid is Leu

Therefore, xenopsin is ⌐Glu—Gly—Lys—Arg—Pro—Trp—Ile—Leu

26.29   Combining the fragments from trypsin and chymotrypsin digestion gives the following three segments.

A peptide containing 17 amino acids from the *N*-terminal end of viscotoxin A$_2$:

Lys–Ser–Cys–Cys–Pro–Asn–Thr–Thr–Gly–Arg–Asn–Ile–Tyr–Asn–Thr–Cys–Arg

A peptide containing 17 amino acids from the *C*-terminal end:

Ser–Gly–Cys–Lys–Ile–Ile–Ser–Ala–Ser–Thr–Cys–Pro–Ser–Tyr–Pro–Asp–Lys

A 12 amino acid peptide; no overlap with either of the *N*- or *C*-terminal fragments:

Phe–Gly–Gly–Gly–Ser–Arg–Glu–Val–Cys–Ala–Ser–Leu

The three peptides add up to 46 amino acids, the total number contained in viscotoxin A$_2$. The complete order of amino acids in viscotoxin A$_2$ is thus the order that exists in the sequence of the three peptide fragments.

Lys–Ser–Cys–Cys–Pro–Asn–Thr–Thr–Gly–Arg–Asn–Ile–Tyr–Asn–Thr
                                                                         |
Ser–Leu–Ser–Ala–Cys–Val–Glu–Arg–Ser–Gly–Gly–Gly–Phe–Arg–Cys
|
Gly–Cys–Lys–Ile–Ile–Ser–Ala–Ser–Thr–Cys–Pro–Ser–Tyr–Pro–Asp–Lys

26.30

(a)

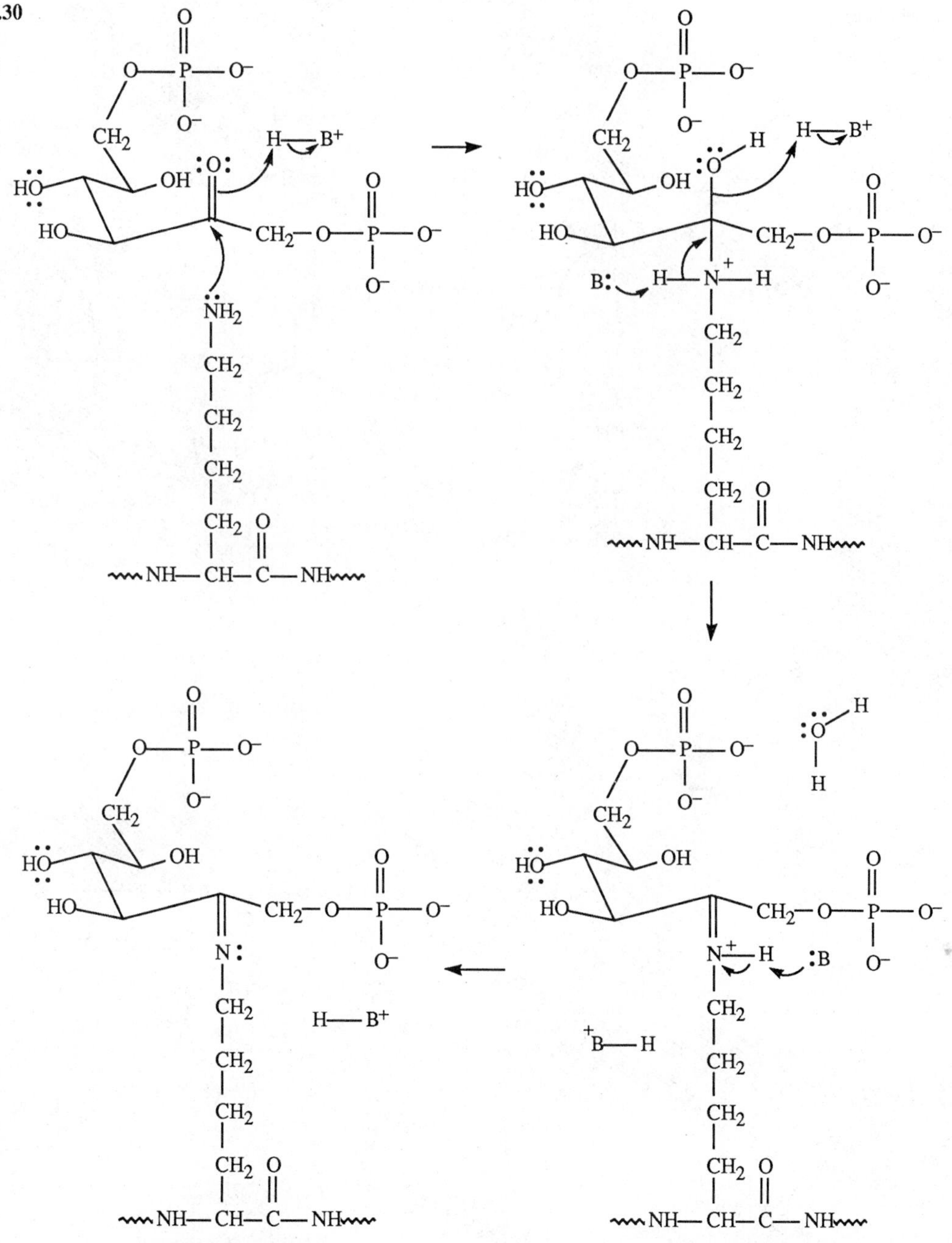

Schiff base at the
active site of the enzyme

26.30 (cont)

(b)

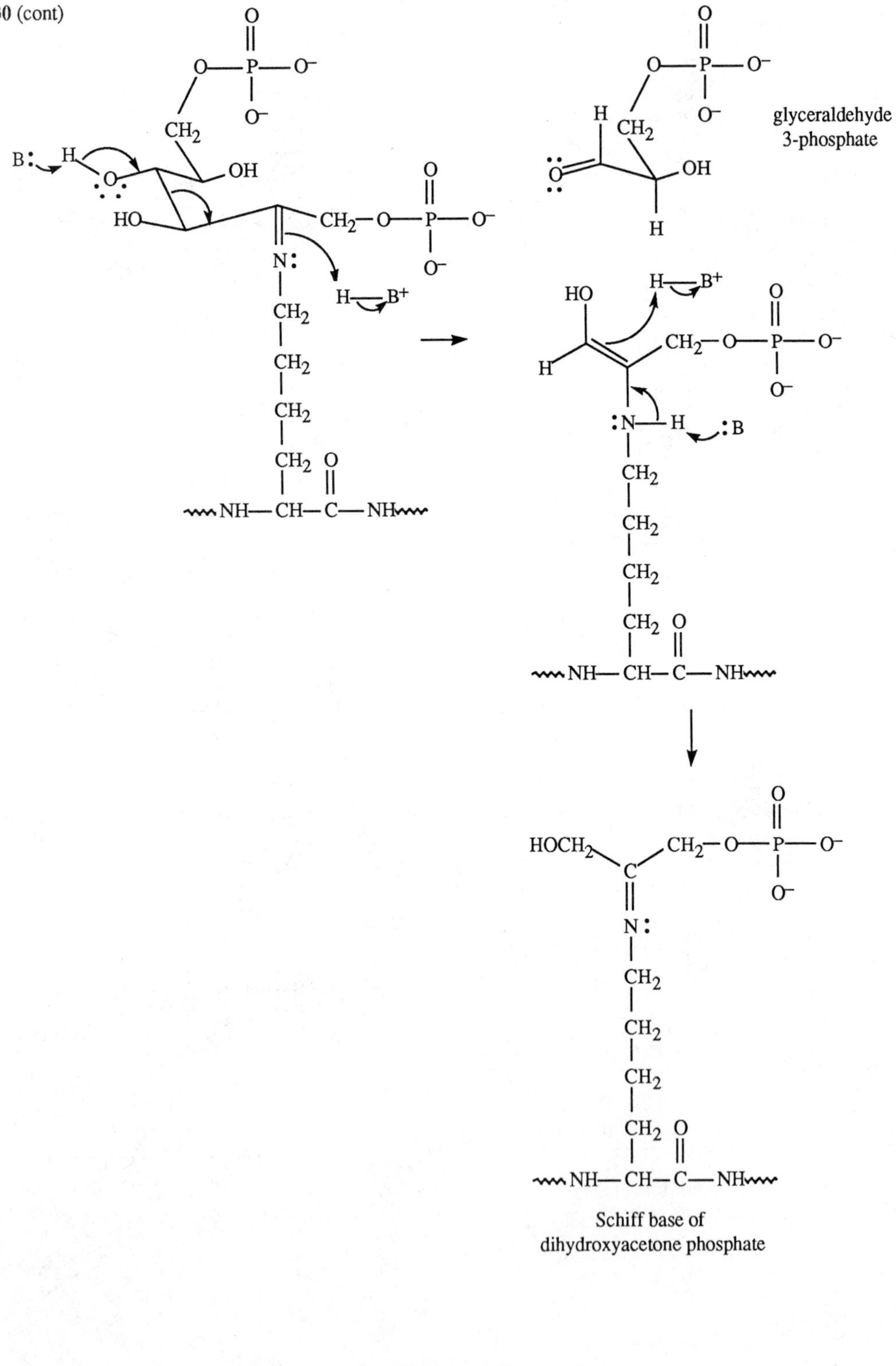

glyceraldehyde
3-phosphate

Schiff base of
dihydroxyacetone phosphate

26.30 (cont)

(c)

$$HOCH_2-\overset{\displaystyle N}{\underset{\displaystyle |}{\overset{\displaystyle ||}{C}}}-CH_2-O-\overset{\displaystyle O}{\underset{\displaystyle |}{\overset{\displaystyle ||}{P}}}-O^-$$

with O^- on phosphorus and the nitrogen chain:

N
|
CH_2
|
CH_2
|
CH_2
|
CH_2

attached to the ring structure with $\overset{H}{\underset{\ }{C}}$ bonded to N–H and C=O.

$$\xrightarrow[\text{H}_2\text{O}]{\text{NaBH}_4 \quad \text{hydrolysis}}$$

$$\underset{\displaystyle CH_2OH}{\overset{\displaystyle CH_2OH}{\overset{\displaystyle |}{CHNHCH_2CH_2CH_2CH_2}}}-\overset{\overset{\displaystyle O}{\overset{\displaystyle ||}{CO^-}}}{\underset{\displaystyle \overset{+}{N}H_3}{C}}\text{--}H$$

lysine derivative of
dihydroxyacetone

26.31   Peptide P contains 11 amino acids:

Arg, Gln (2), Gly, Leu, Lys, Met, Phe (2), Pro (2)

Peptide A contains Arg, Gln (2), Lys, Phe, Pro (2), therefore Peptide B must contain Gly, Leu, Met, Phe.

$$\text{Peptide P (and Peptide A)} \xrightarrow{\text{Edman's reagent}} \begin{array}{l}\text{phenylthiohydantoins of}\\ \text{Arg, Pro, Lys, Pro, in that order}\end{array}$$

therefore the *N*-terminal amino acid of Peptide P (and Peptide A) is Arg. Peptide A contains the amino acids from the *N*-terminal end of P, and Peptide B must contain the amino acids from the *C*-terminal end of P.

$$\text{Peptide A} \xrightarrow{\text{carboxypeptidase}} \text{1) Phe; 2) Gln}$$

therefore the *C*-terminal amino acid of A is Phe. Peptide A must have the structure Arg–Pro–Lys–Pro–Gln–Gln–Phe.

$$\text{Peptide B} \xrightarrow{\text{Edman's reagent}} \begin{array}{l}\text{phenylthiohydantoins of}\\ \text{Phe, Gly, Leu, in that order}\end{array}$$

therefore Peptide B must have the structure Phe–Gly–Leu–Met.

$$\text{Peptide P} \xrightarrow{\text{carboxypeptidase}} \text{no amino acid}$$

therefore the *C*-terminal amino acid of P does not have a free carboxylic acid group.

26.31 (cont)

Peptide P $\xrightarrow{\text{HCl (0.03 M)}}$ Peptide C

(hydrolysis of
amide bonds)

Peptide C $\xrightarrow{\text{carboxypeptidase}}$ Gly, Met, Leu, Phe (order not determined)

therefore Peptide C has a *C*-terminal amino acid with a free carboxylic acid group. The same amino acids as those present in Peptides B and P are released to the solution, so Peptide C has the same sequence of amino acids as Peptide P. Both have Met as the *C*-terminal amino acid. In Peptide P, the *C*-terminal acid exists as the amide, Met—NH$_2$.

These conclusions can be summarized as answers to the subsections of the problem.

(a)   The *N*-terminal amino acid of Peptide P is Arg.

(b)   The *C*-terminal amino acid of Peptide P is Met.

(c)   The sequence of amino acids in Peptide A is Arg–Pro–Lys–Pro–Gln–Gln–Phe.

(d)   The sequence of amino acids in Peptide B is Phe–Gly–Leu–Met.

(e)   The complete structure of Peptide B is Phe–Gly–Leu–Met—NH$_2$.

(f)   The sequence of amino acids in Peptide C is Arg–Pro–Lys–Pro–Gln–Gln–Phe–Phe–Gly–Leu–Met.

(g)   The complete structure of Peptide P is Arg–Pro–Lys–Pro–Gln–Gln–Phe–Phe–Gly–Leu–Met—NH$_2$.

26.32   The solid state synthesis of Peptide P:

(a)

26.32 (cont)

(b)

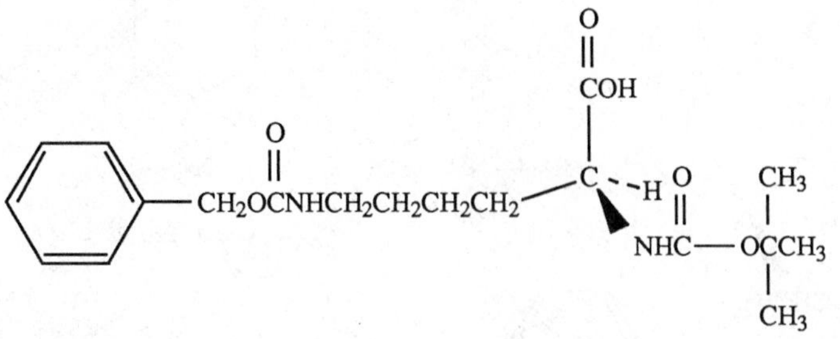

(c)

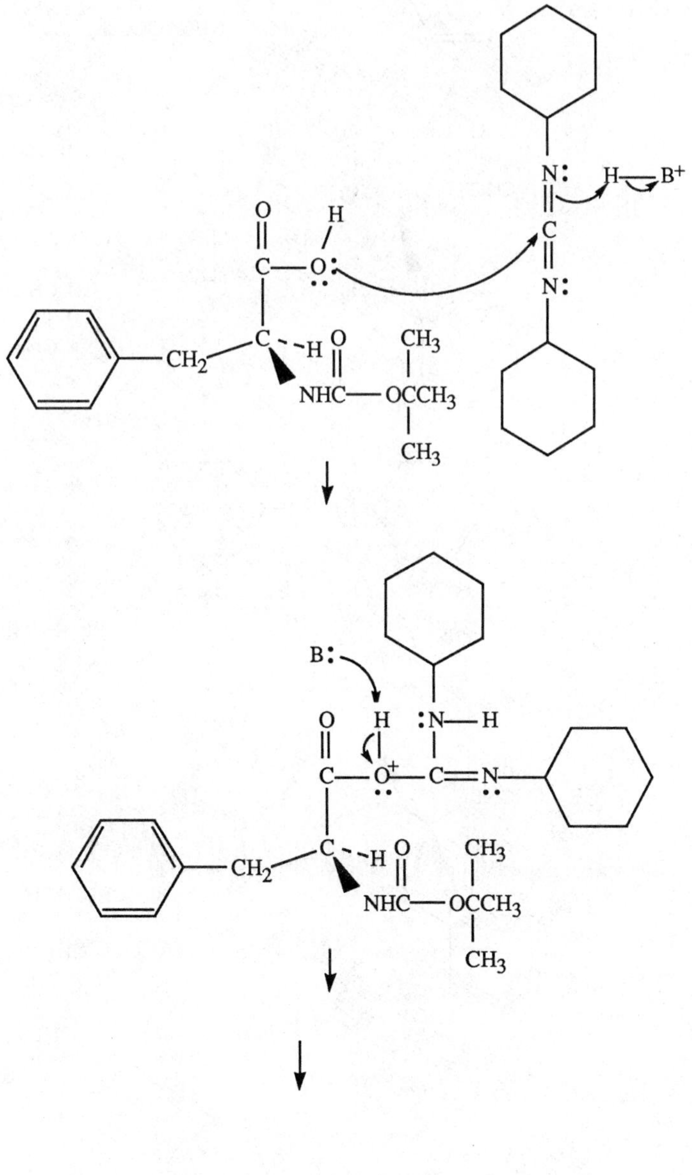

26.32 (c) (cont)

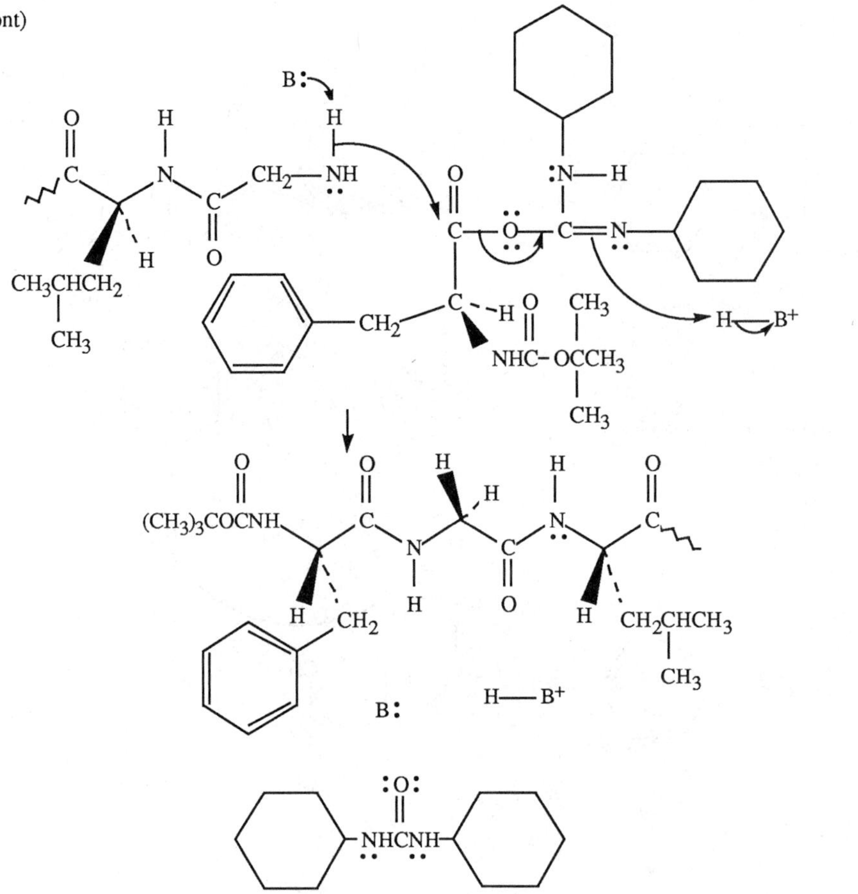

(d)

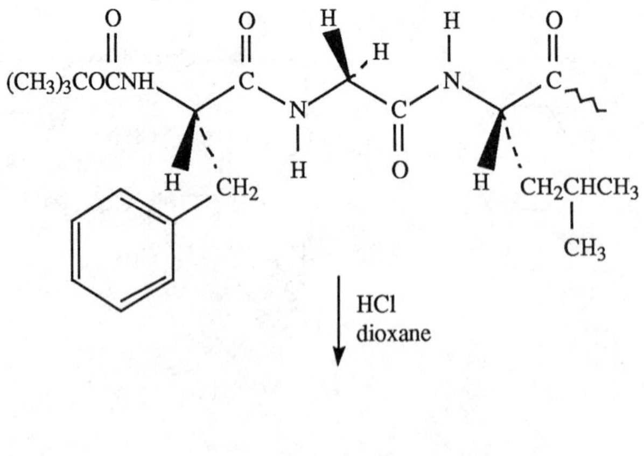

HCl
dioxane

26.32 (d) (cont)

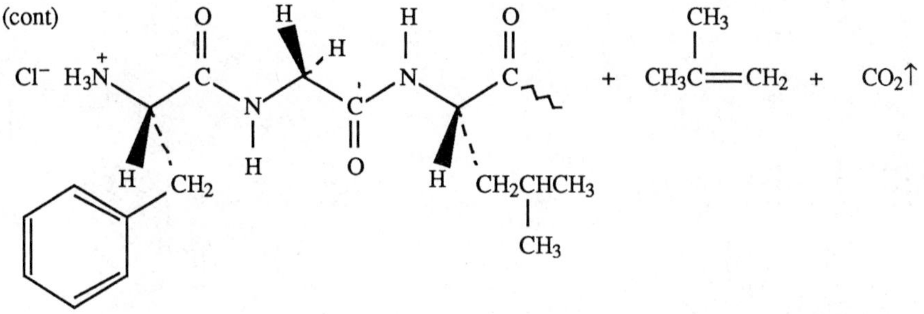

(e)

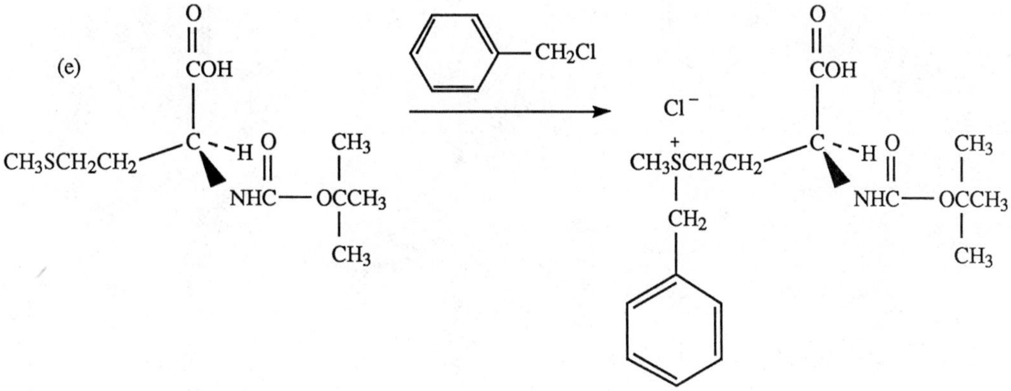

26.33

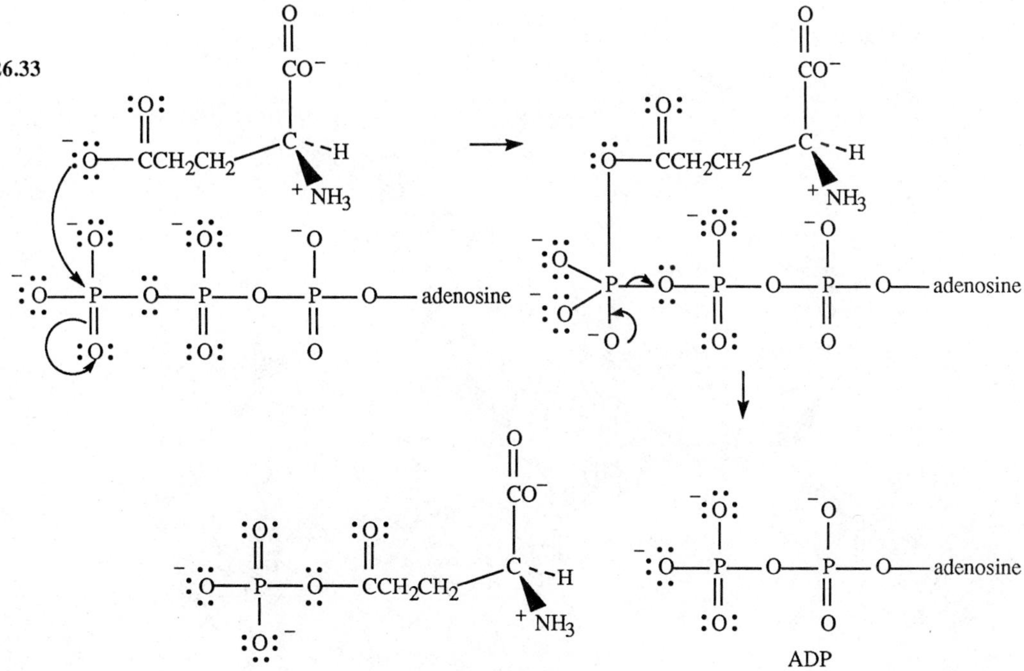

ADP

26.33 (cont)

26.33 (cont)

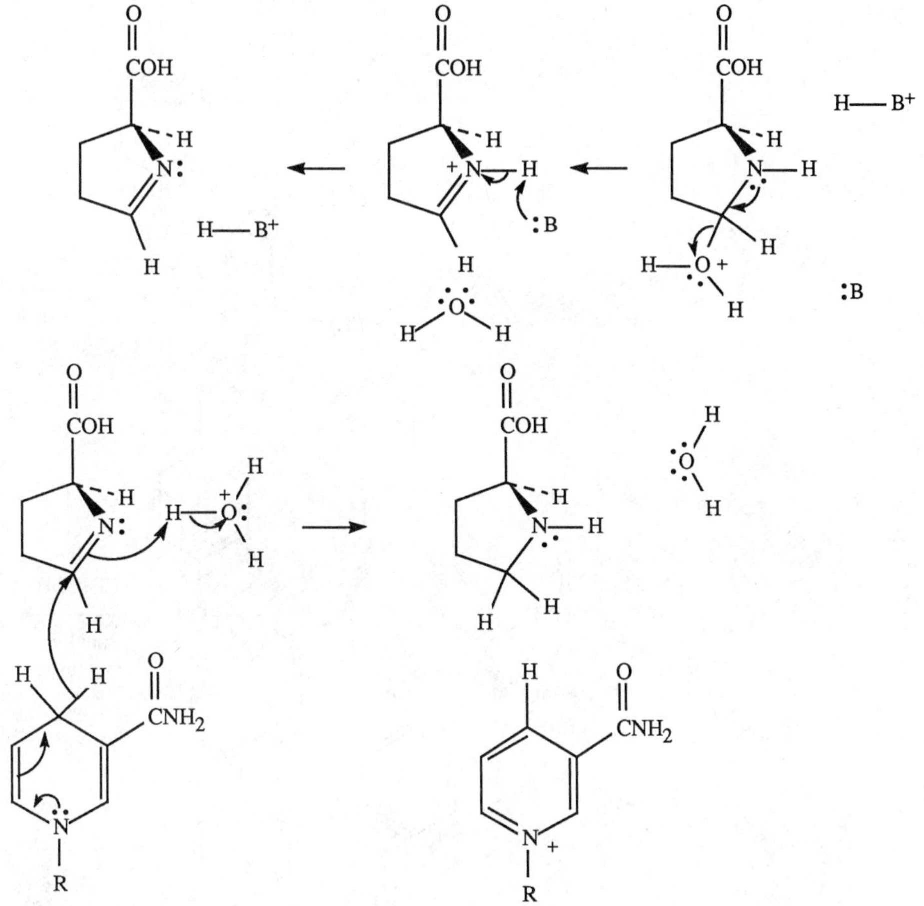

26.34

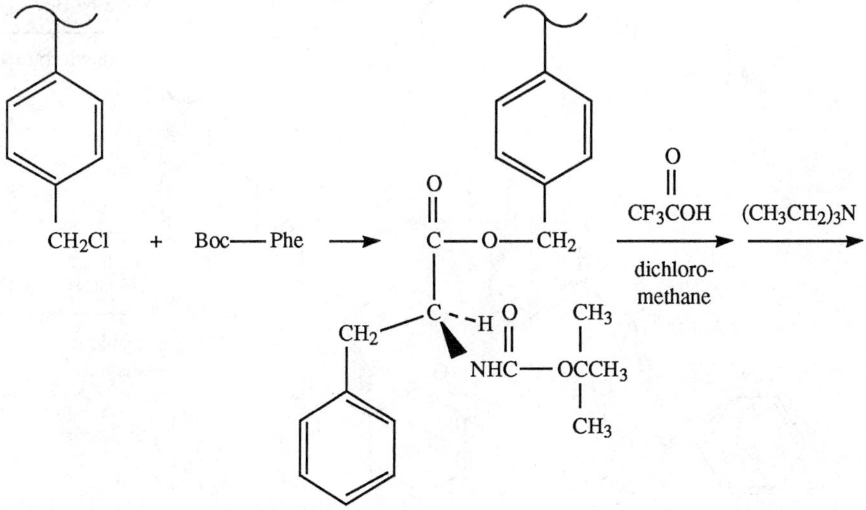

26.34 (cont)

$$\xrightarrow[\text{dichloromethane}]{\text{Boc}—\text{Gly, DCC}}$$

$$\xrightarrow[\substack{\text{dichloro-}\\\text{methane}}]{\overset{\displaystyle O}{\overset{\|}{\text{CF}_3\text{COH}}} \quad (\text{CH}_3\text{CH}_2)_3\text{N}}$$

$$\xrightarrow[\text{dichloromethane}]{S\text{–Bz–}N\text{–Boc–Cys, DCC}}$$

$$\xrightarrow[\substack{\text{dichloro-}\\\text{methane}}]{\overset{\displaystyle O}{\overset{\|}{\text{CF}_3\text{COH}}} \quad (\text{CH}_3\text{CH}_2)_3\text{N}}$$

26.34 (cont)

26.34 (cont)

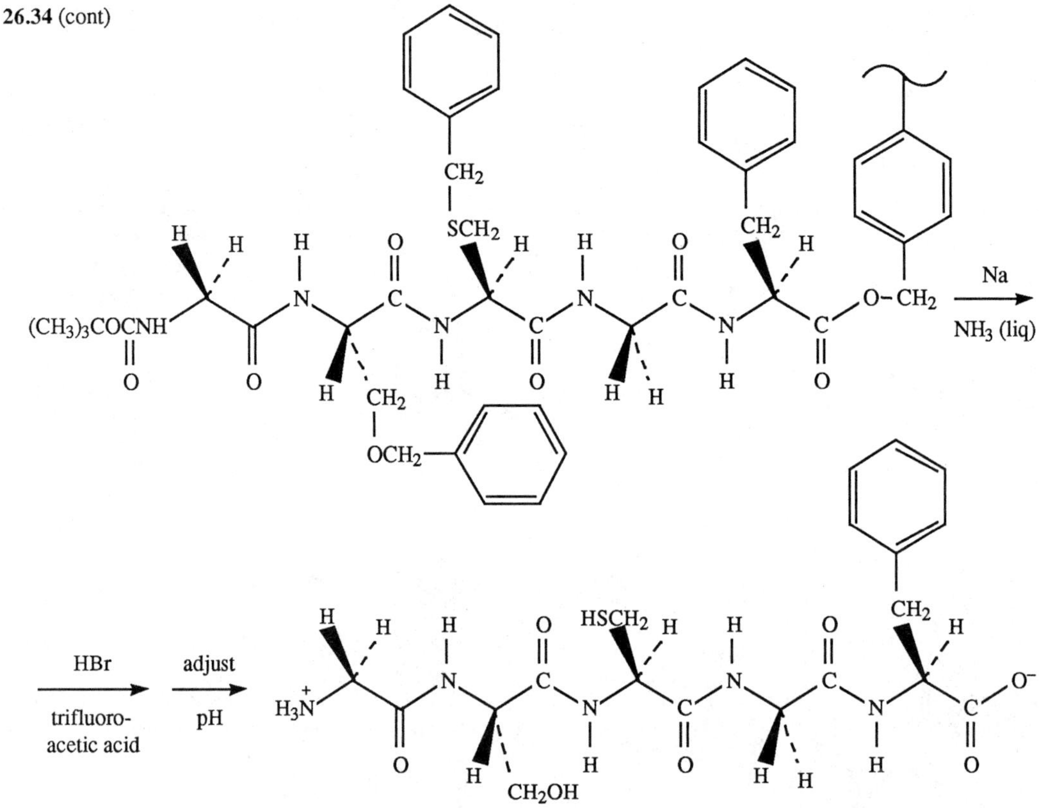

Gly–Ser–Cys–Gly–Phe

26.35 (a) β-Carboxyaspartic acid is a malonic acid derivative. Malonic acid is stable when it is in basic solution, but decarboxylates when it is heated with acid. β-Carboxyaspartic acid, therefore, survives basic hydrolysis of a peptide better than it survives acid-catalyzed hydrolysis. The decarboxylation product of β-carboxyaspartic acid is aspartic acid, which is then seen as the product of acid-catalyzed hydrolysis.

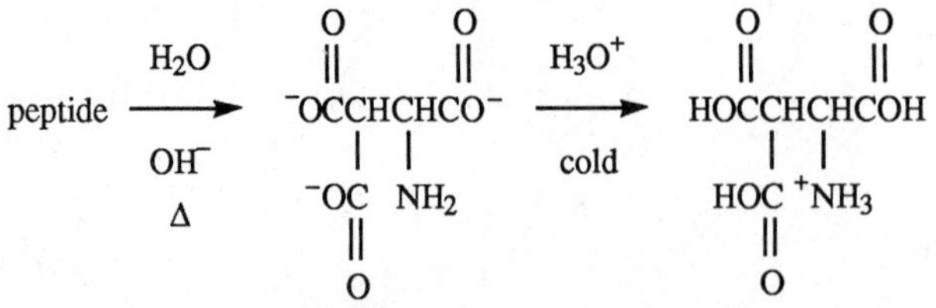

stable towards
decarboxylation

β-carboxyaspartic
acid (at low pH)

26.35 (a) (cont)

peptide $\xrightarrow[\Delta]{\underset{H_3O^+}{H_2O}}$ $\left[\begin{array}{c} \underset{O}{\overset{O}{\|}}\ \ \ \underset{O}{\overset{O}{\|}} \\ HOCCHCHCOH \\ | \quad\quad | \\ HOC \ \ ^+NH_3 \\ \overset{\|}{O} \end{array} \right]$ $\longrightarrow$ $CO_2\uparrow$ $+$ $\underset{\underset{^+NH_3}{|}}{\overset{\overset{O}{\|}\quad\quad\overset{O}{\|}}{HOCCH_2CHCOH}}$

aspartic acid
(at low pH)

(b) $CH_2(COCH_3)_2$ $\xrightarrow[\text{methanol}]{CH_3ONa}$ $Na^+\ ^-CH(COCH_3)_2$ $\xrightarrow{ClCH_2COCH_3}$ $CH_3OCCH_2CH(COCH_3)_2$

                                                          A   B

$CH_3OCCH_2CH(COCH_3)_2$ $\xrightarrow[\underset{\Delta}{KOH}]{\text{—CH}_2OH}$

     B

$-CH_2OCCH_2CH(COCH_2-\ )_2$ $\xrightarrow[\substack{\text{carbon}\\ \text{tetrachloride}\\ \text{dark}}]{Br_2}$

     C

$-CH_2OCCH_2C(COCH_2-\ )_2$ $\xrightarrow[\text{diethyl ether}]{(CH_3CH_2)_3N}$

         Br

     D

$-CH_2OCCH=C(COCH_2-\ )_2$

     E

$-CH_2OCCH=C(COCH_2-\ )_2$ $\xrightarrow[\text{tetrahydrofuran}]{HN_3}$

     E

26.35   (b) (cont)

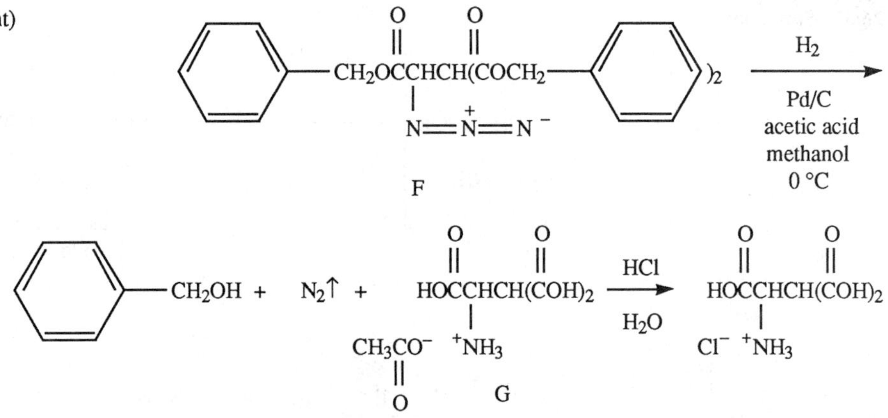

(c)   Free radical substitution of the benzylic hydrogen atoms of the benzyl ethers can occur with Br_2 in the presence of light.

(d)   The ester groups on the carbon-carbon double bond are electron withdrawing. The acid adds to give the more stable carbocation, the one with the positive charge on the carbon atom with only one ester group.

26.36

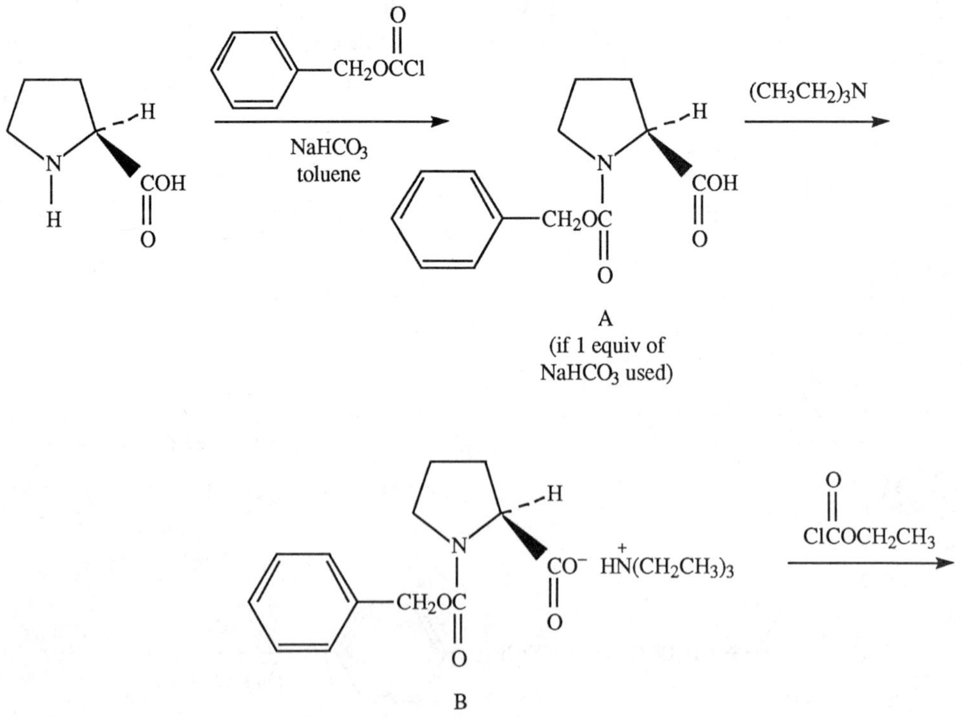

26.36 (cont)

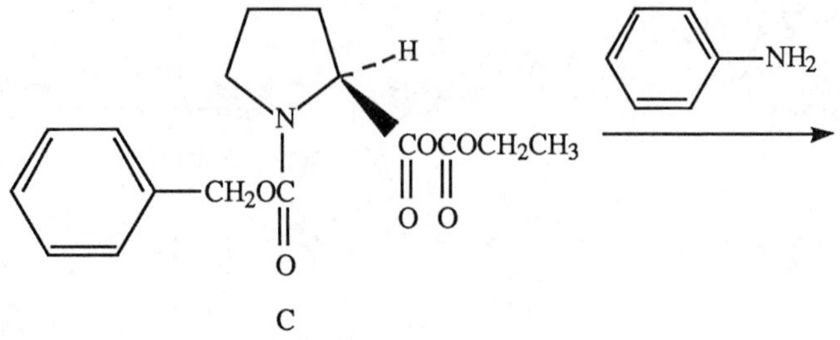

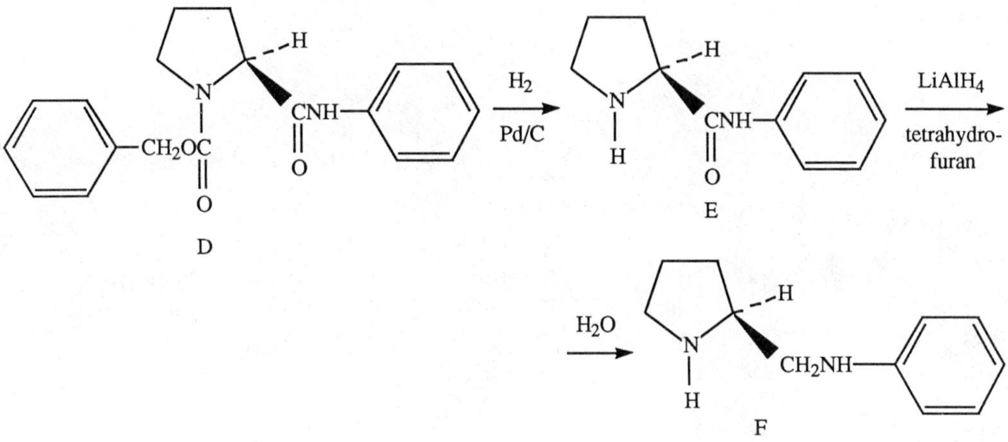

C

D

E

F

26.37

A

B

C

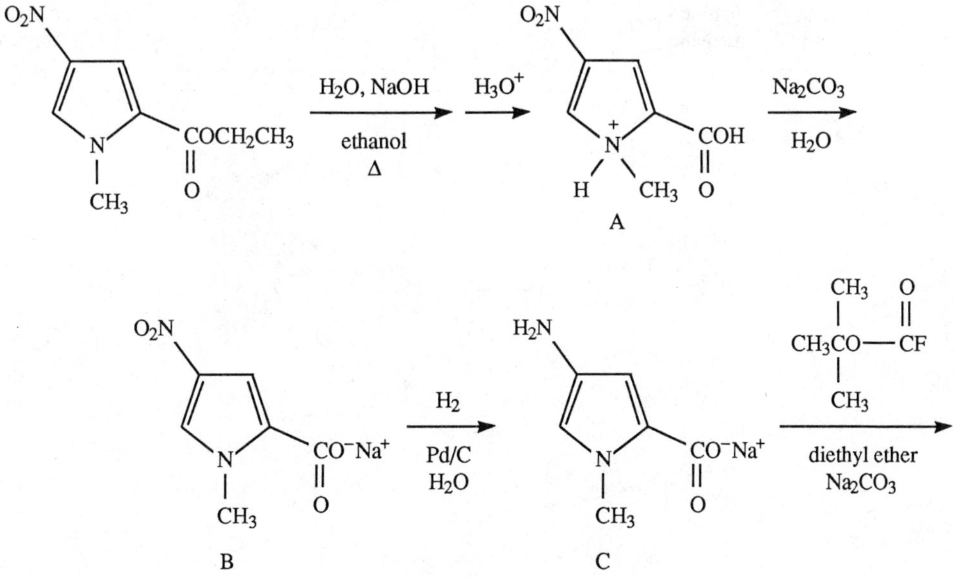

26.37 (cont)

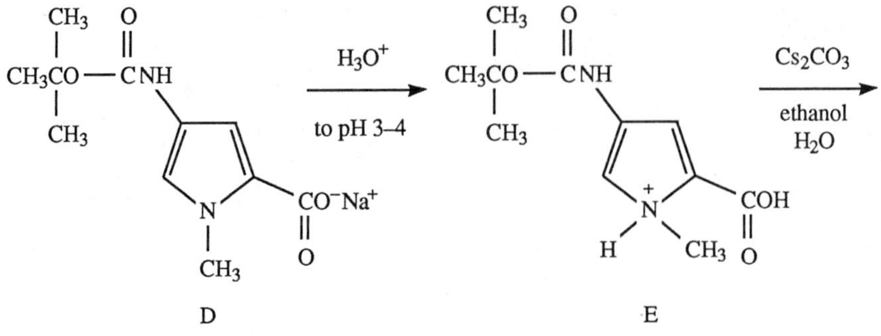

H_3O^+

to pH 3–4

D

Cs_2CO_3

ethanol

H_2O

E

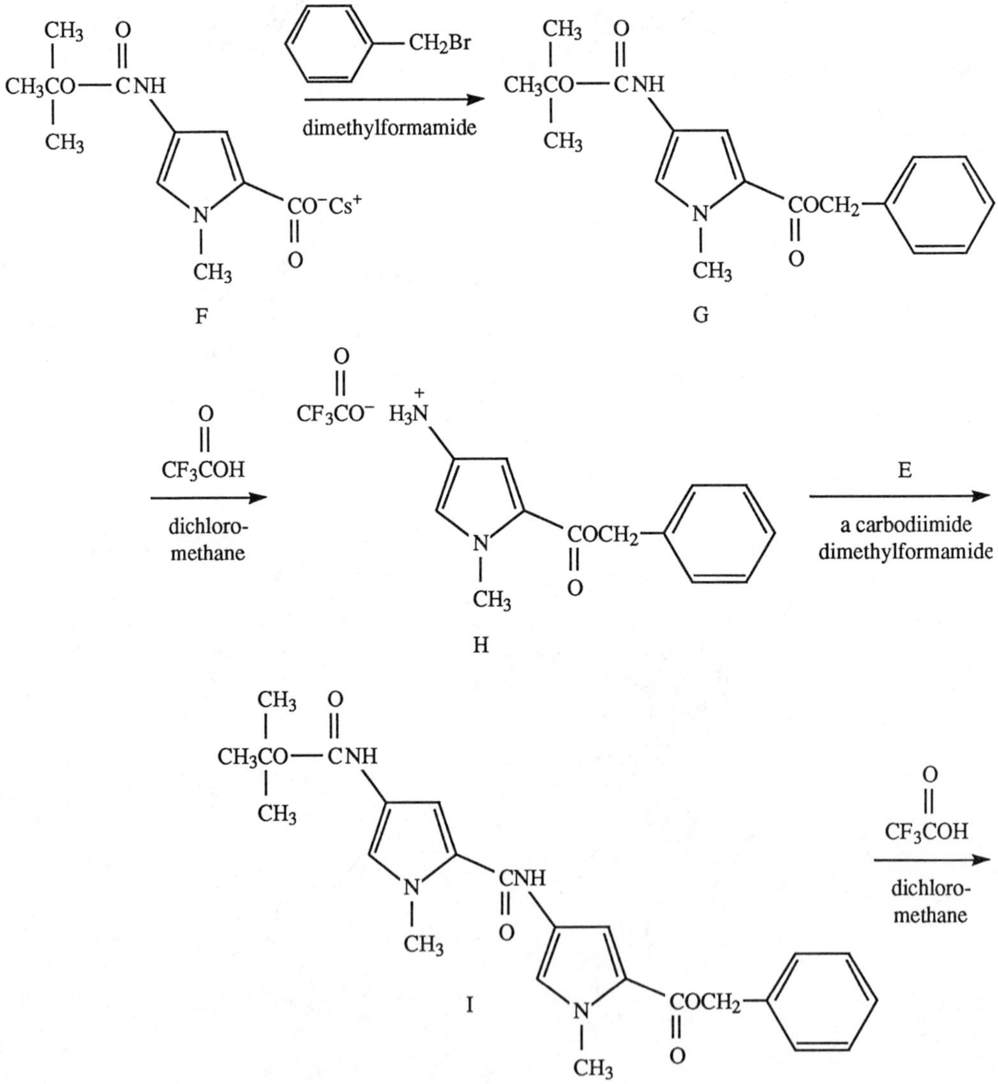

CH_2Br

dimethylformamide

F

G

CF_3COH

dichloro-
methane

CF_3CO^- H_3N^+

H

E

a carbodiimide
dimethylformamide

I

CF_3COH

dichloro-
methane

26.37 (cont)

26.37 (cont)

$$\xrightarrow[\substack{\text{dioxane}\\ \text{pH 4}\\ \text{H}_2\text{O}}]{\text{HCl}} \text{Distamycin A}$$

27

Macromolecular Chemistry

Concept Map 27.1   Polymers.

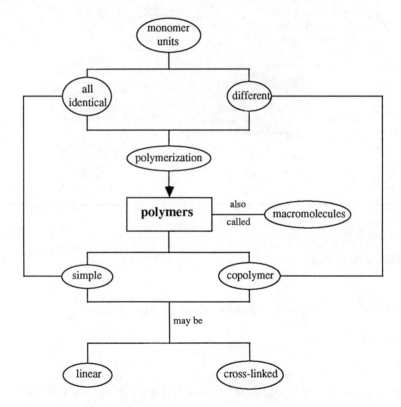

Concept Map 27.2   Properties of polymers.

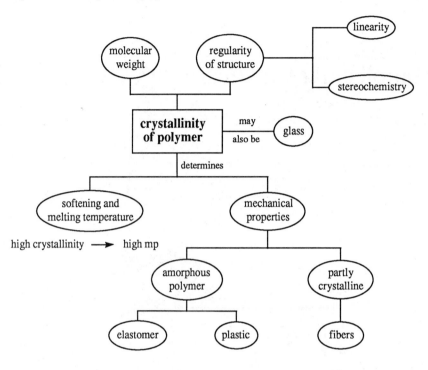

- -

27.1   Formation of the initiator:

Initiation step:

Chain-propagation step:

27.2   (a) 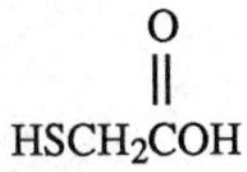 $\overset{\overset{\displaystyle O}{\|}}{HSCH_2COH}$

(b) 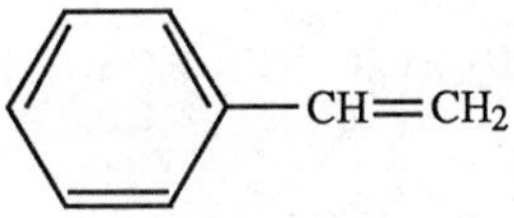 $-CH=CH_2$

(c) $CH_2=CHCH_3$

(d)  $O=C=N-$ $-N=C=O$

an isocyanate

(e) $CH_2=\overset{\overset{\displaystyle CH_3}{|}}{\underset{\underset{\displaystyle O}{\|}}{C}COCH_3}$

+

$HOCH_2CH_2OH$

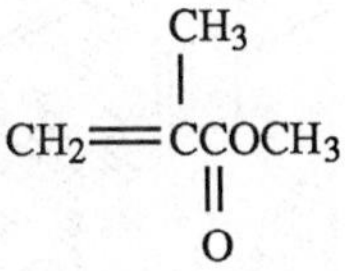

27.3   (a) 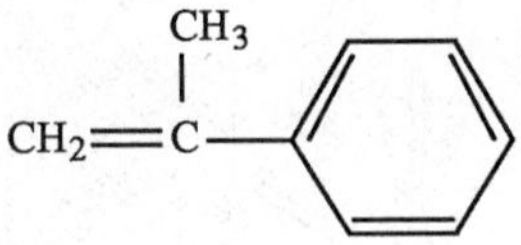 $CH_2=\overset{\overset{\displaystyle CH_3}{|}}{C}-$ and $N\equiv CCH=CHC\equiv N$; a copolymer

(b) $CH_2=CCl_2$; a simple polymer

(c) 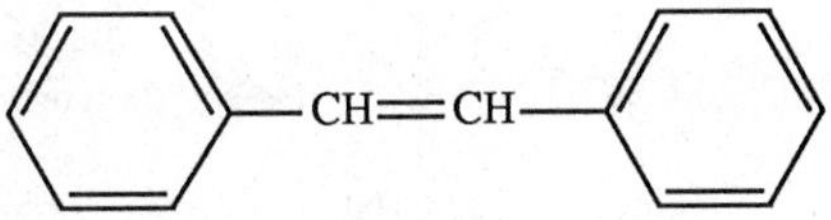 $-CH=CH-$ and 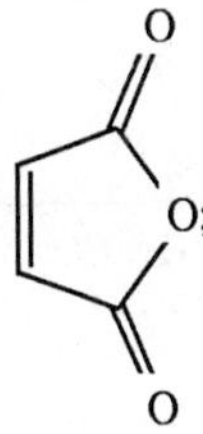 ; a copolymer

(d) 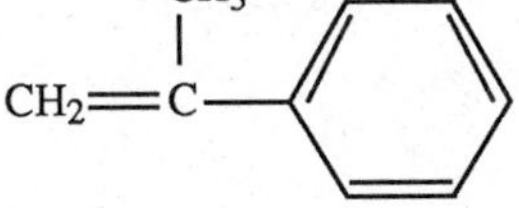 $CH_2=\overset{\overset{\displaystyle CH_3}{|}}{C}-$ and $CH_2=\overset{\overset{\displaystyle CH_3}{|}}{C}C\equiv N$; a copolymer

27.4

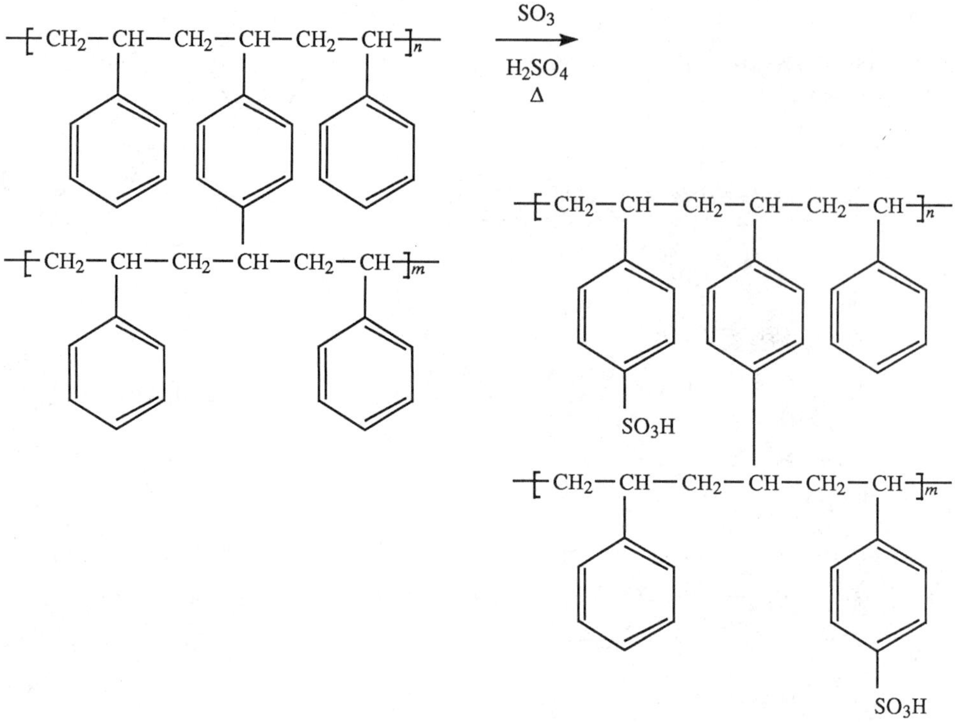
SO3
H2SO4
Δ
SO3H
SO3H

27.5

(a)

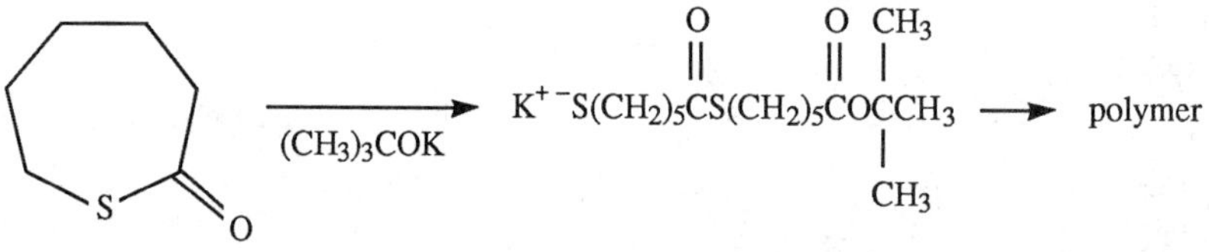
(CH3)3COK
K+ −S(CH2)5CS(CH2)5COCCH3
polymer

(b)

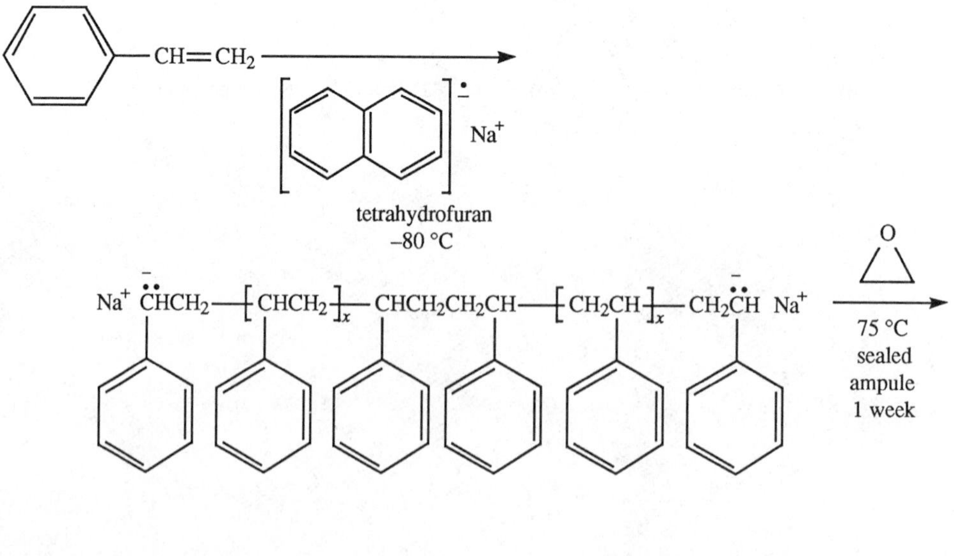
CH=CH2
Na+
tetrahydrofuran
−80 °C
75 °C
sealed
ampule
1 week

27.5 (cont)

(b) (cont)

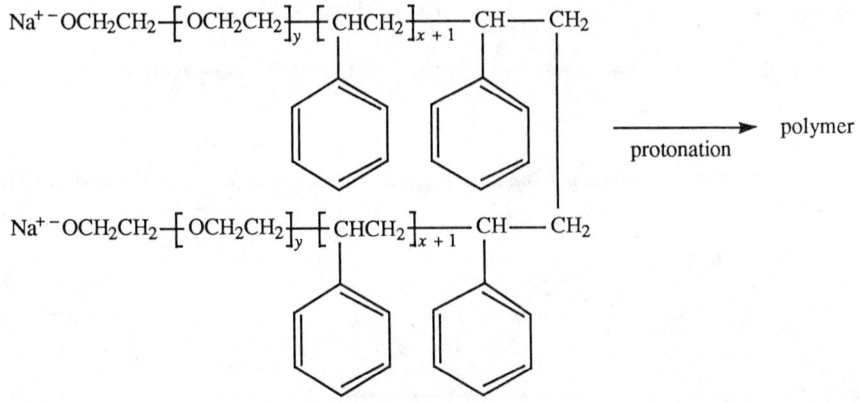

$$Na^{+\,-}OCH_2CH_2 \left[OCH_2CH_2 \right]_y \left[CHCH_2 \right]_{x+1} CH \!\!-\!\! CH_2$$

$$\xrightarrow[\text{protonation}]{} \text{polymer}$$

$$Na^{+\,-}OCH_2CH_2 \left[OCH_2CH_2 \right]_y \left[CHCH_2 \right]_{x+1} CH \!\!-\!\! CH_2$$

(c) $CH_2\!=\!CHCl$ + $CH_2\!=\!CHCOCH_3$

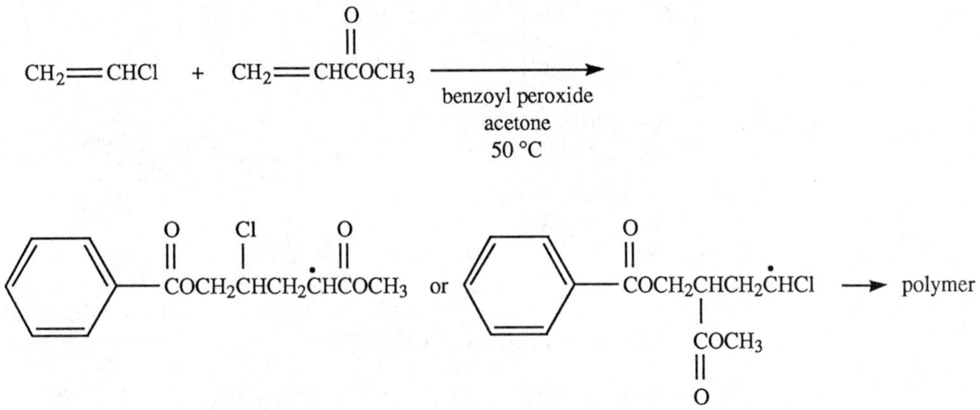

$$\xrightarrow[\substack{\text{benzoyl peroxide} \\ \text{acetone} \\ 50\,°C}]{}$$

C$_6$H$_5$—$COCH_2CHCH_2\overset{\bullet}{C}HCOCH_3$ (Cl on middle carbon)   or   C$_6$H$_5$—$COCH_2CHCH_2\overset{\bullet}{C}HCl$ ($COCH_3$ substituent) → polymer

(d)   C$_6$H$_5$—$CH\!=\!CH_2$ + maleic anhydride

$$\xrightarrow[\substack{\text{benzoyl peroxide} \\ \text{benzene} \\ 80\,°C}]{}$$

C$_6$H$_5$—$COCH_2CH$— (succinic anhydride ring with $\bullet$ and phenyl)   or   C$_6$H$_5$—CO—O— (succinic anhydride ring with $CH_2\overset{\bullet}{C}H$ phenyl) → polymer

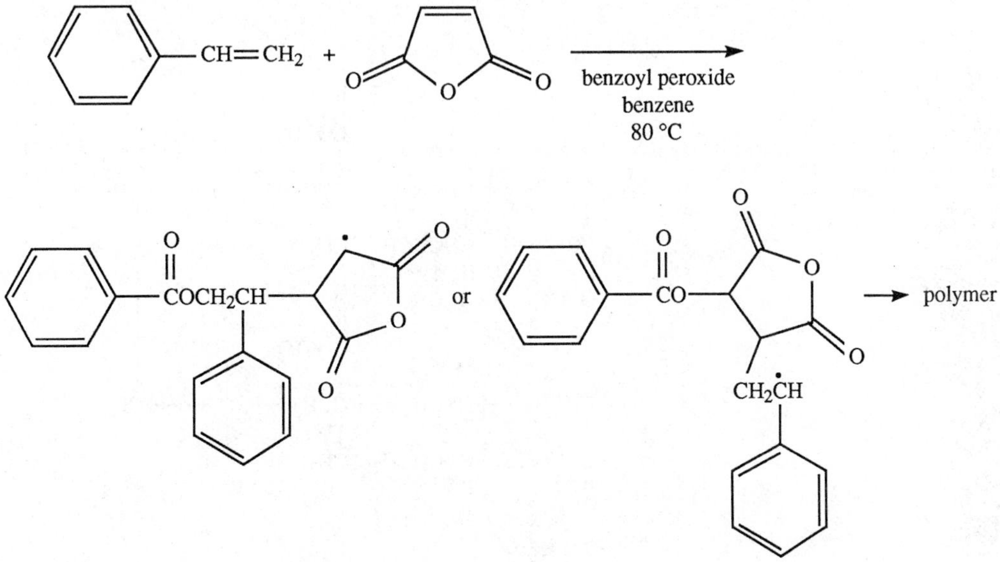

27.5 (cont)

(e)

$$CH_3CH\overset{O}{\overset{/\ \backslash}{-}}CH_2 \xrightarrow[\Delta]{KOH} HOCH_2\overset{CH_3}{\underset{|}{C}}HOCH_2\overset{CH_3}{\underset{|}{C}}HO^-K^+ \longrightarrow \text{polymer}$$

This is an S_N2 substitution reaction. The nucleophile reacts at the less hindered carbon atom of the oxirane.

27.6

$$CH_2{=}\overset{CH_3}{\underset{\underset{O}{||}}{\underset{|}{C}}}COCH_3 \xrightarrow{\hspace{3cm}}$$

tetrahydrofuran
−80 °C

$$Na^{+-}\underset{\underset{CH_3}{|}}{\overset{\overset{O}{||}}{\underset{|}{C}}CH_2}\left[\underset{\underset{CH_3}{|}}{\overset{\overset{O}{||}}{\underset{|}{C}}CH_2}\right]_x\underset{\underset{CH_3}{|}}{\overset{\overset{O}{||}}{\underset{|}{C}}CH_2CH_2}C{-}\left[\underset{\underset{CH_3}{|}}{\overset{\overset{O}{||}}{\underset{|}{C}}H_2C}\right]_x\underset{\underset{CH_3}{|}}{\overset{\overset{O}{||}}{\underset{|}{C}}H_2C}{}^-Na^+ \xrightarrow{CH_3\overset{CH_3}{\underset{|}{C}}{=}CH\overset{O}{\underset{||}{C}}O\overset{CH_3}{\underset{|}{C}}HCH_3}$$

$$\xrightarrow{CH_3OH}$$

27.6 (cont)

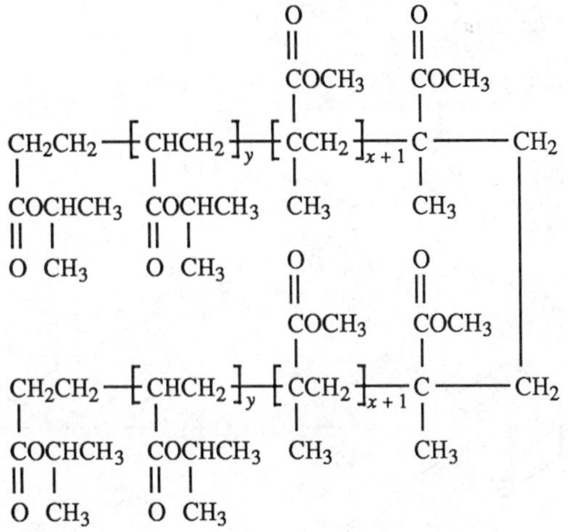

$$
\begin{array}{c}
\hspace{3cm} \overset{O}{\overset{||}{COCH_3}} \quad \overset{O}{\overset{||}{COCH_3}} \\
CH_2CH_2 \!-\!\!\left[CHCH_2 \right]_{\!y}\!\!\left[CCH_2 \right]_{\!x+1}\!\!-\!C\!-\!\!-\!CH_2 \\
\hspace{2cm} |\hspace{2.2cm}|\hspace{2.2cm}|\hspace{1.2cm}| \\
\hspace{1.5cm} COCHCH_3 \quad COCHCH_3 \quad CH_3 \quad CH_3 \\
\hspace{1.5cm} ||\ | \hspace{2cm} ||\ | \\
\hspace{1.5cm} O\ CH_3 \hspace{1.5cm} O\ CH_3
\end{array}
$$

A block copolymer can also be made by starting with isopropyl acrylate and then adding methyl methacrylate.

27.7 $AlCl_3 + CH_3C\!-\!\ddot{Cl}\!: \longrightarrow CH_3C\!\!\equiv\!\!\overset{+}{O}\!: + \ ^-AlCl_4$

initiation → propagation

polymer ←

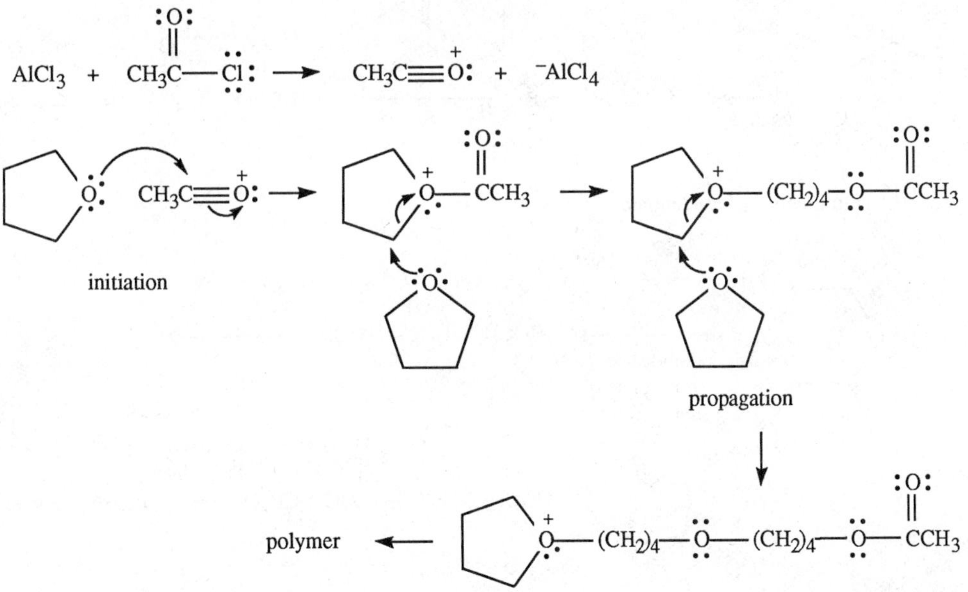

27.8

(a)

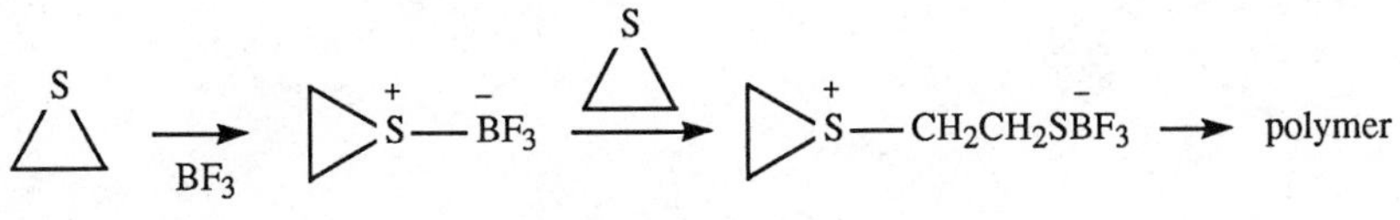

$$\triangle \ \xrightarrow{BF_3}\ \overset{+}{S}\!-\!\overset{-}{BF_3}\ \xrightarrow{\ \triangle\ } \ \overset{+}{S}\!-\!CH_2CH_2SBF_3 \ \longrightarrow \ \text{polymer}$$

27.8 (cont)

(b)

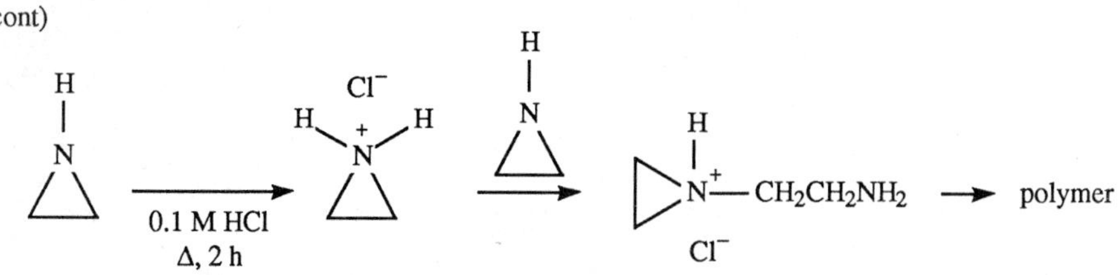

(c)

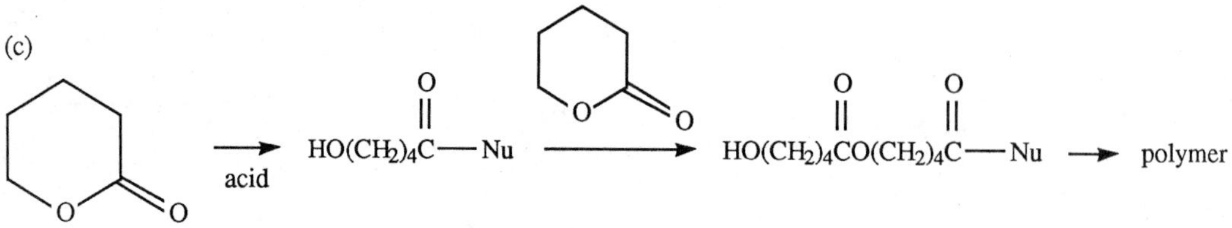

(Nu represents the conjugate base of the acid initiator.)

(d)

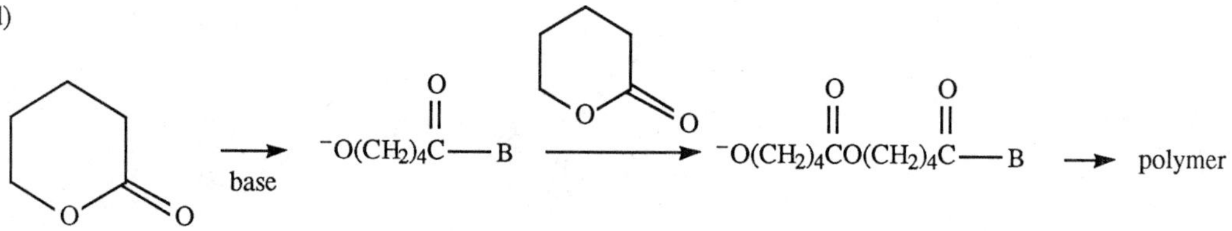

(B represents the basic initiator.)

(e)

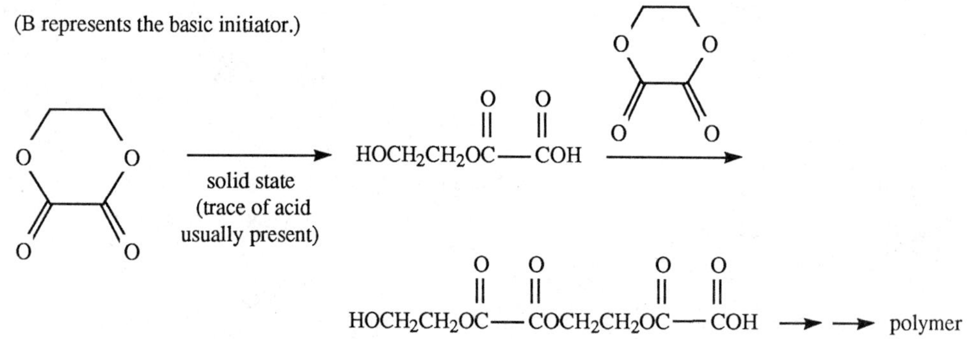

(f)

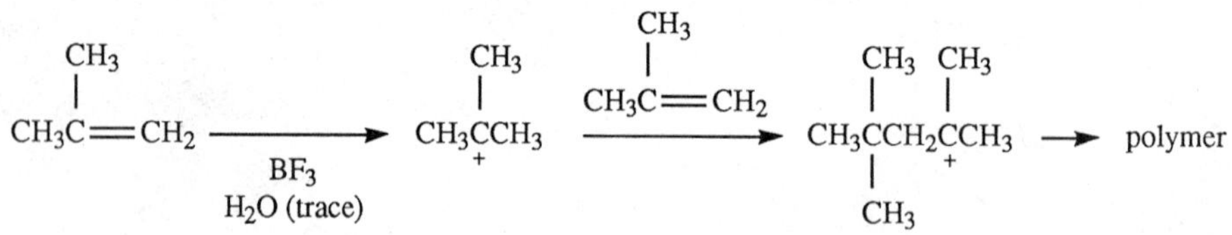

27.9

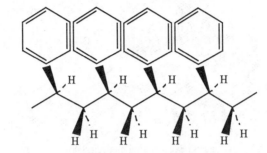

isotactic polystyrene

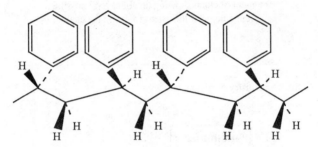

syndiotactic polystyrene

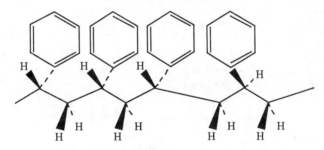

atactic polystyrene

fragment of the chain, showing only one of
several possible orientations of the phenyl
groups with respect to each other

27.10   In order for a polymer to be highly crystalline, it has to be highly ordered so it can pack easily into a crystal structure. Isotactic and syndiotactic polymers, with their very regular primary structure, have strong interactions between chains and crystallize, but atactic polymers with their much less regular structure do not have as good interactions between chains and will be more amorphous.

27.11

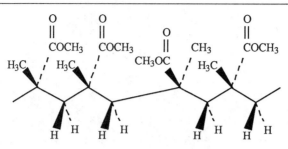

atactic poly(methyl methacrylate)

fragment of chain showing one of several possible
arrangements of the ester groups on the chain

Concept Map 27.3   Stereochemistry of polymers.

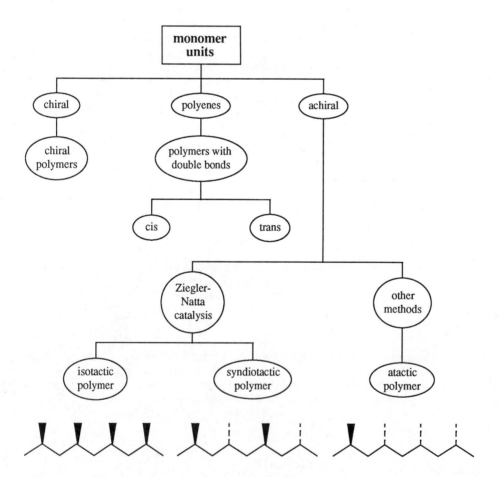

27.12

(a)

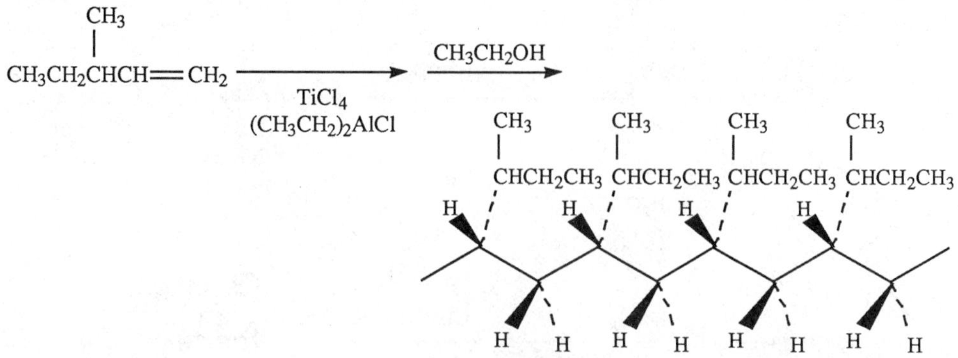

$$CH_2{=}CHCH_2CH_3CH_2CHCH{=}CH_2$$

with CH₃ substituent, reaction:

$CH_3CH_2CHCH{=}CH_2$ (with CH₃ group) $\xrightarrow[\text{(CH}_3\text{CH}_2)_2\text{AlCl}]{\text{TiCl}_4}$ $\xrightarrow{CH_3CH_2OH}$

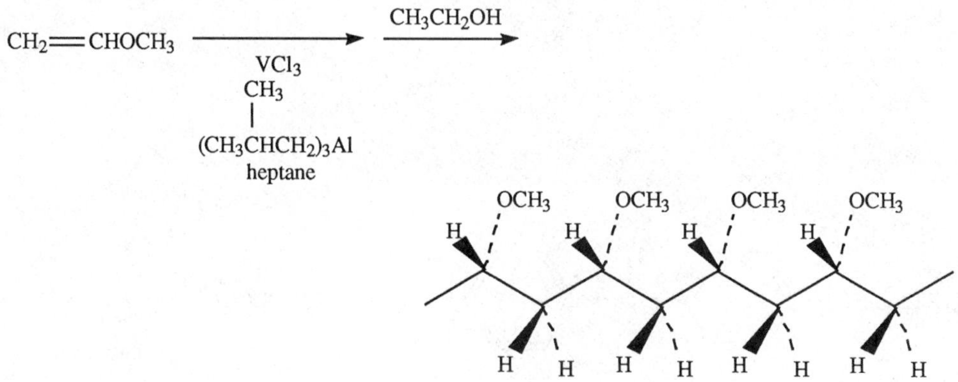

(b)

$CH_2{=}CHOCH_3$ $\xrightarrow[\begin{array}{c}\text{VCl}_3\\ \text{(CH}_3\text{CHCH}_2)_3\text{Al}\\ \text{heptane}\end{array}]{}$ $\xrightarrow{CH_3CH_2OH}$

(c)

$CH_2{=}CH_2$ + (2-methylpropene type: $C{=}C$ with two CH_3 and two H) $\xrightarrow[\begin{array}{c}\text{VCl}_3\\ \text{(CH}_3\text{CHCH}_2)_2\text{AlCl}\\ \text{heptane}\\ -30\ ^\circ\text{C}\end{array}]{}$ $\xrightarrow{CH_3CH_2OH}$

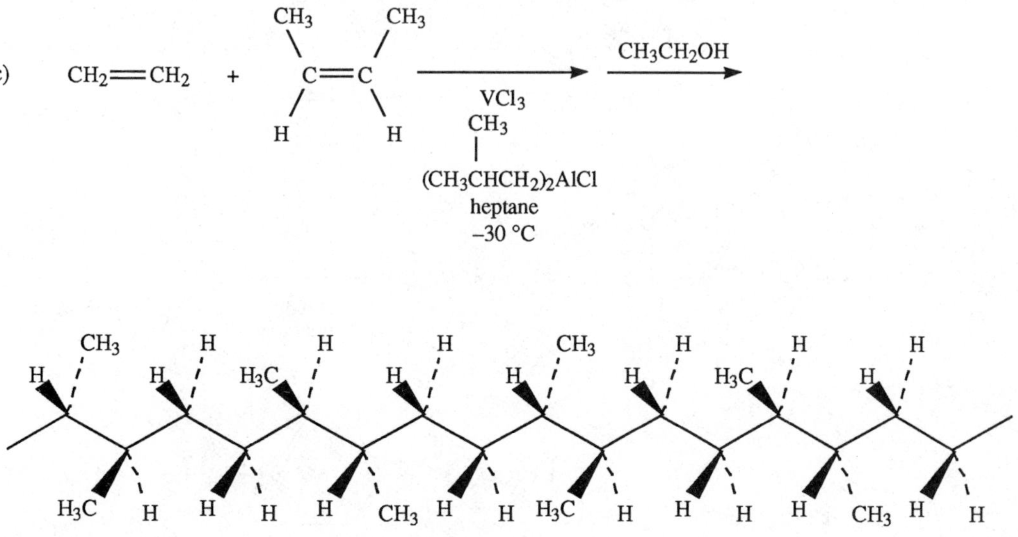

27.12

(d)

$$CH_2{=}\underset{\underset{O}{\|}}{\overset{\overset{CH_3}{|}}{C}}COCH_3 \quad \xrightarrow[\substack{VCl_3 \\ CH_3 \\ | \\ (CH_3CHCH_2)_3Al \\ heptane}]{CH_3CH_2OH} \quad \underset{\underset{\underset{O}{\|}}{COCH_3}}{\overset{\overset{CH_3}{|}}{-\!\!\left[CH_2C\right]_n\!\!-}}$$

(e)

$$CH_3CH_2CH{=}CH_2 \quad \xrightarrow[\substack{FeCl_3 \\ (CH_3CH_2)_2AlCl}]{CH_3CH_2OH} \quad \underset{CH_2CH_3}{-\!\!\left[CH_2CH\right]_n\!\!-}$$

27.13

$$HO{-}\phi{-}\underset{\underset{CH_3}{|}}{\overset{\overset{CH_3}{|}}{C}}{-}\phi{-}OH \;+\; CH_3O\overset{\overset{O}{\|}}{C}OCH_3 \xrightarrow[\text{temperature}]{\text{high}}$$

$$-\!\!\left[O{-}\phi{-}\underset{CH_3}{\overset{CH_3}{C}}{-}\phi{-}O\overset{\overset{O}{\|}}{C}\right]_n\!\!- \;+\; 2n\ CH_3OH$$

$$CH_3O\overset{\overset{O}{\|}}{C}{-}\phi{-}\overset{\overset{O}{\|}}{C}OCH_3 \;+\; HO(CH_2)_4OH \xrightarrow[\text{temperature}]{\text{high}}$$

$$-\!\!\left[O(CH_2)_4O\overset{\overset{O}{\|}}{C}{-}\phi{-}\overset{\overset{O}{\|}}{C}\right]_n\!\!- \;+\; 2n\ CH_3OH$$

27.14 $H_2N(CH_2)_{10}COH \xrightarrow{\Delta} \left[NH(CH_2)_{10}C \right]_n + 2n\ H_2O$

(with C=O groups, i.e. $\overset{O}{\overset{\|}{C}}$)

The commonly available lactams contain from four to seven atoms in the ring. The lactam of this monomer would contain twelve atoms in the ring and would not be readily available.

27.15

(a) $ClCO(CH_2)_4OCCl + H_2N(CH_2)_6NH_2 \xrightarrow[\substack{NaOH \\ H_2O}]{} \left[NH(CH_2)_6NHCO(CH_2)_4OC \right]_n$

(b)

$+\ HOCH_2CH_2OH \xrightarrow[\substack{\text{dimethyl sulfoxide} \\ \text{4-methyl-2-pentantone} \\ 115\ °C}]{}$

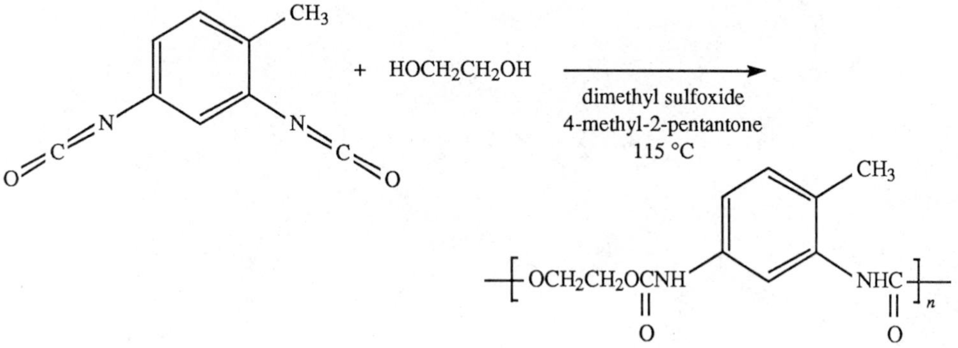

(c) $O=C=N(CH_2)_6N=C=O + HO(CH_2)_6OH \xrightarrow{185-195\ °C}$

$\left[O(CH_2)_4OC\,NH(CH_2)_6NHC \right]_n$

27.16

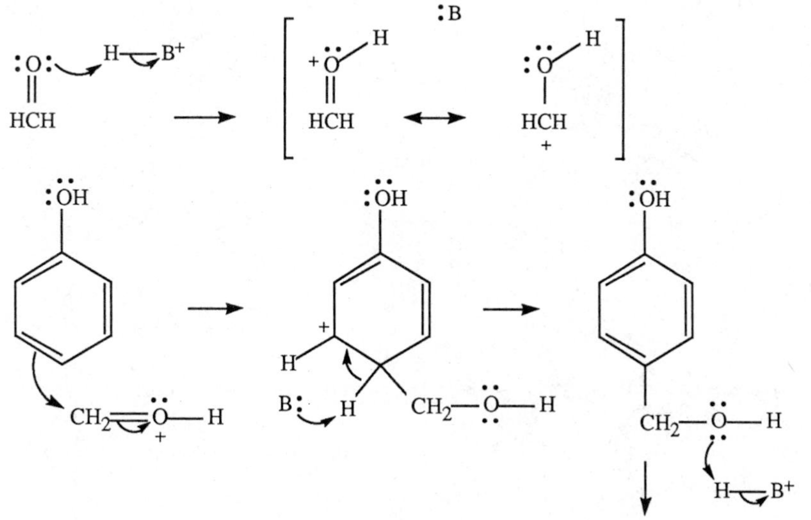

27.16 (cont)

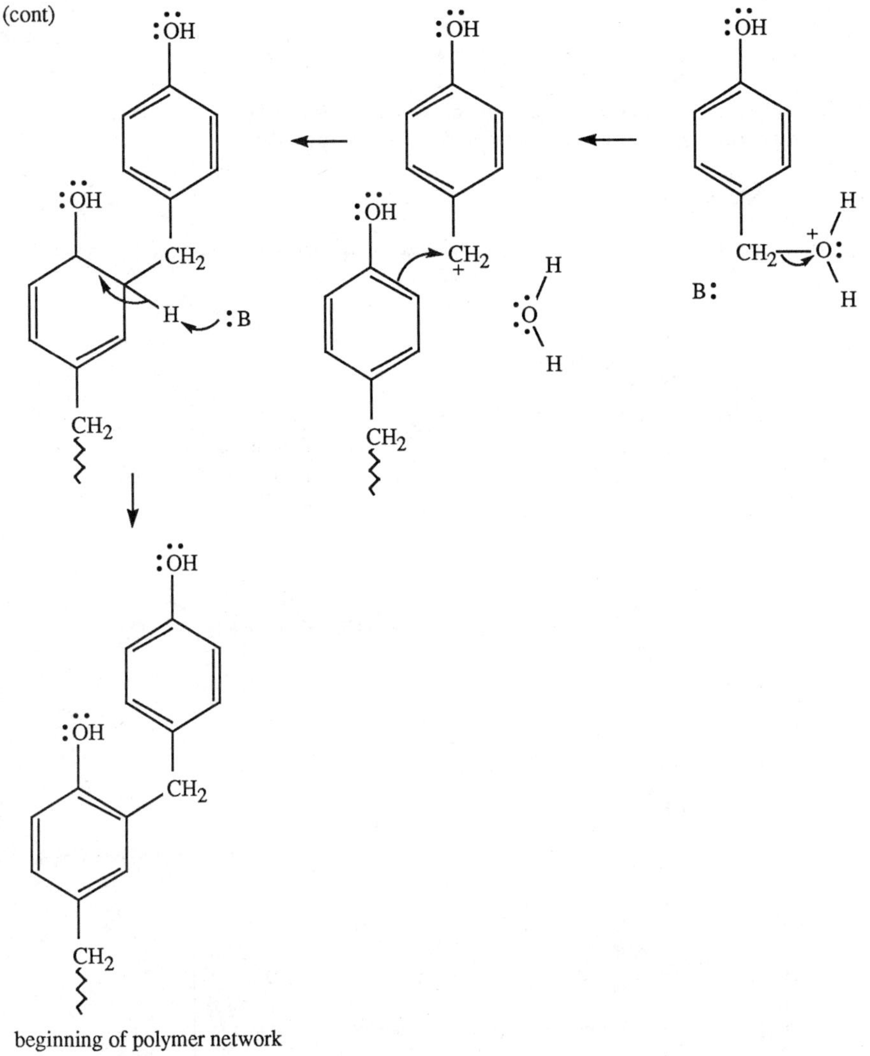

beginning of polymer network

27.17

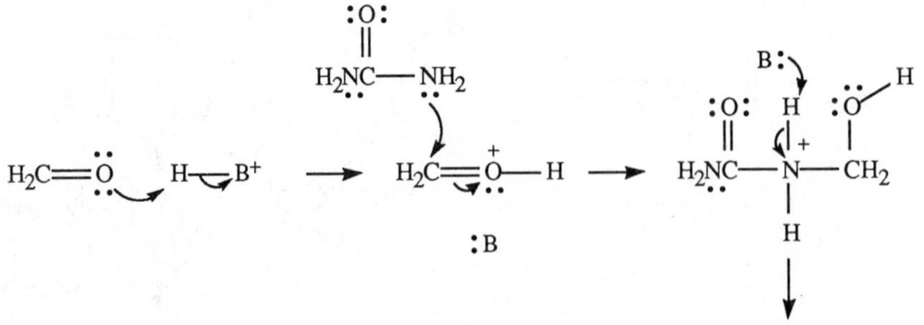

27.17 (cont)

Concept Map 27.4   Types of polymerization reactions.

27.18

27.18 (cont)

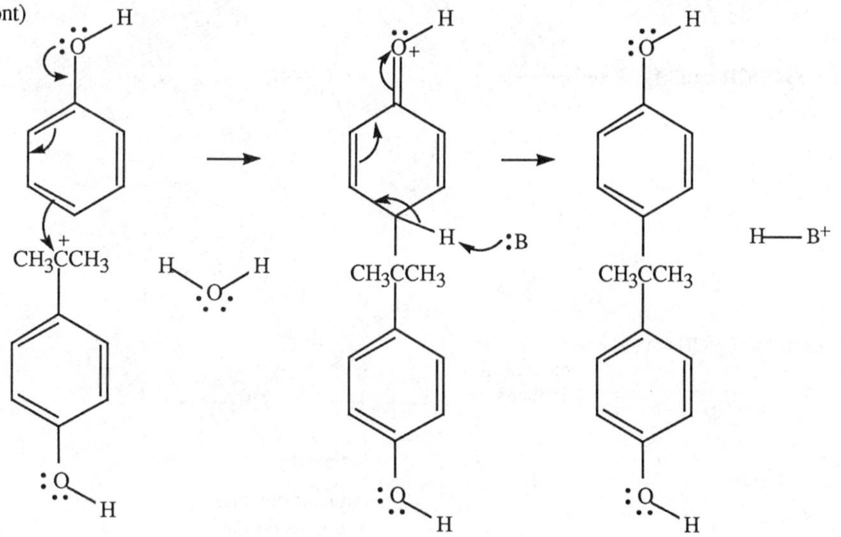

27.19 $CH_3CH{=}CH_2 \xrightarrow[\Delta]{Cl_2\,(g)} ClCH_2CH{=}CH_2 \xrightarrow[H_2O]{Cl_2} ClCH_2CH(Cl)CH_2OH$

free radical chlorination at the allylic position

electrophilic addition of chlorine to the double bond with water as the electrophile

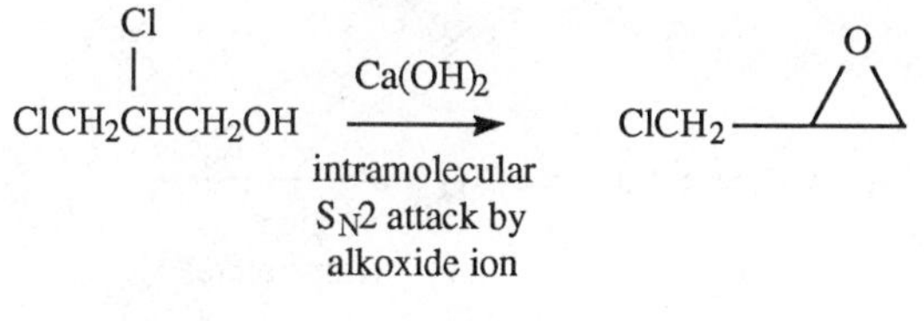
Ca(OH)2
intramolecular
SN2 attack by
alkoxide ion

27.20

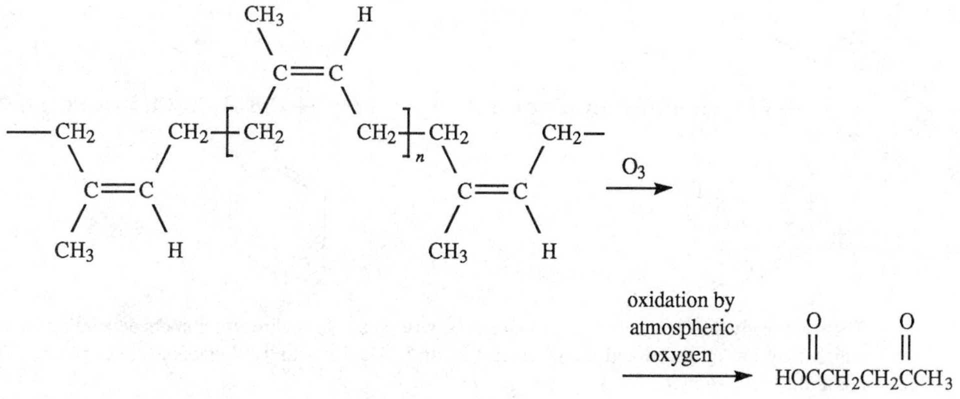
O3
oxidation by
atmospheric
oxygen

27.21 $CH_2\!\!=\!\!CHCH\!\!=\!\!CH_2$ $\xrightarrow[\text{(CH}_3\text{CH}_2)_3\text{Al}]{\text{TiBr}_4}$

27.22 $CH_3OCH\!\!=\!\!CHCH\!\!=\!\!CH_2$ $\xrightarrow[\text{initiator}]{\text{free radical}}$

amorphous polymer
cis double bonds

$CH_3OCH\!\!=\!\!CHCH\!\!=\!\!CH_2$ $\xrightarrow[\text{(CH}_3\text{CHCH}_2)_3\text{Al}]{\text{VCl}_3}$

crystalline polymer
trans double bonds

27.23 $CH\!\!=\!\!CH_2$ + $CH\!\!=\!\!CH_2$ + $CH_2\!\!=\!\!CHCH\!\!=\!\!CH_2$ $\xrightarrow[\text{temperature}]{\text{high}}$

There are many different copolymers that can be written for these three monomers depending on the order in which they are arranged and on whether 1,2- or 1,4-addition to 1,3-butadiene takes place. Two of the possibilities are shown.

27.24

(a)

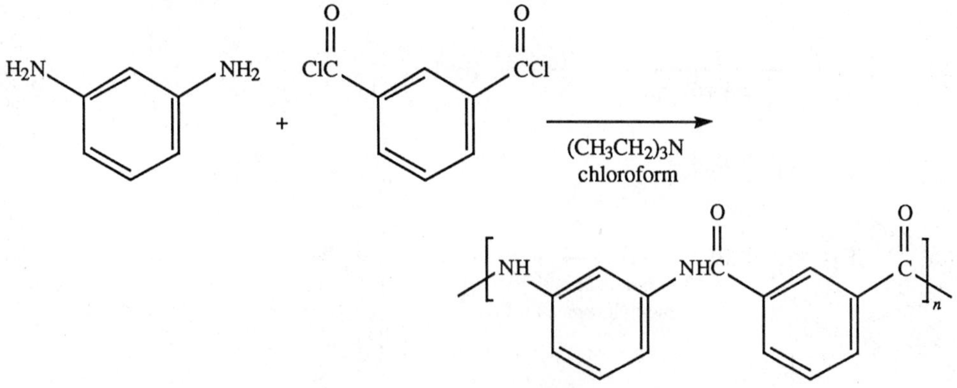

(b)

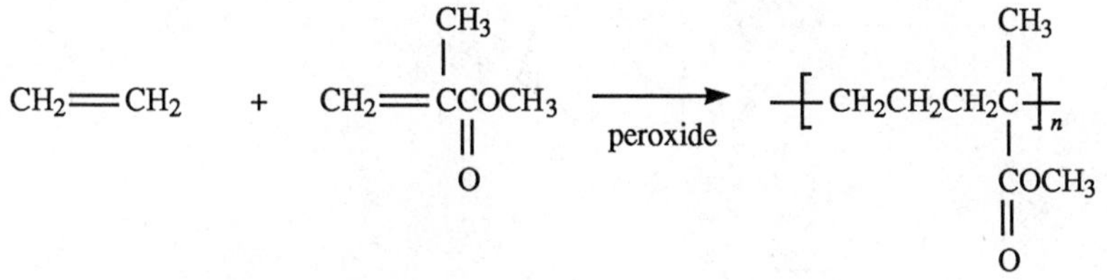

(c)

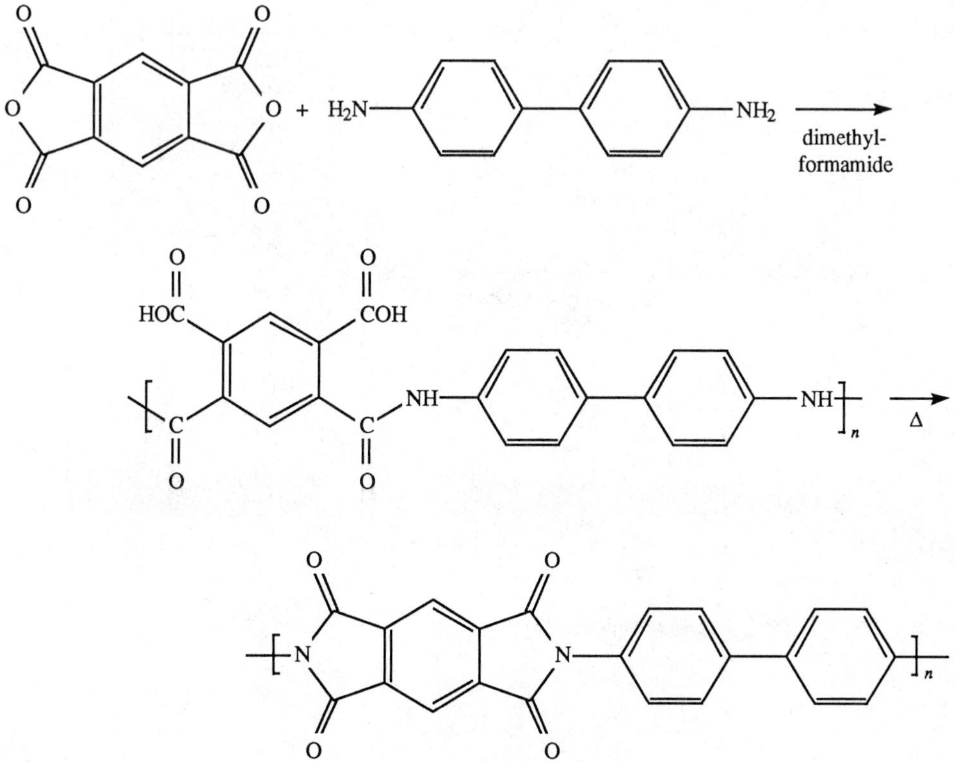

27.24 (cont)

(d)

$$\square\!-\!S \xrightarrow{(CH_3CH_2)_2O\cdot BF_3} \{SCH_2CH_2CH_2\}_n$$

(e)

$$CH_2\!=\!CCOCH_3 \xrightarrow[\substack{\text{free radical initiator} \\ \text{2-methylpropanenitrile}}]{} \{CH_2C\}_n$$

with substituents: N≡C (nitrile) on the monomer and product, and COCH$_3$ with O as the other group.

(f)

$$\xrightarrow[\substack{(CH_3CH_2)_2O\cdot BF_3 \\ \text{chloroform} \\ 20\,^{\circ}C}]{}$$

(g)

$$CH_3C\!=\!CH_2 \;+\; Cl\!-\!\!\bigcirc\!\!-\!CH\!=\!CH_2 \xrightarrow[\substack{SnCl_4 \\ \text{nitrobenzene} \\ 0\,^{\circ}C}]{} \{CH_2CHCH_2C\}_n$$

with CH$_3$ substituents and a p-chlorophenyl (Cl) ring on the product.

(h)

$$CH_2\!=\!CHOCH_2CHCH_3 \xrightarrow[(CH_3CH_2)_2O\cdot BF_3]{} \{CH_2CH\}_n$$

with CH$_3$ substituent on monomer; product side chain OCH$_2$CHCH$_3$ with CH$_3$.

(i)

$$CH_2\!=\!CHC\!\equiv\!N \xrightarrow[\substack{CaO \\ \text{dimethyl-} \\ \text{formamide} \\ 20\,^{\circ}C}]{} \;^-O\{CH_2CH\}_n CH_2\ddot{C}H \xrightarrow{HCl} HO\{CH_2CH\}_n CH_2CH_2$$

with C≡N groups on the side chains.

27.24 (cont)

(j)

$$CH_2=CHOCHCH_3 \xrightarrow[\substack{VCl_3 \\ (CH_3CHCH_2)_3Al \\ heptane}]{} \xrightarrow{CH_3CH_2OH}$$

with CH$_3$ on the CHOCHCH$_3$ group

(k)

$$CH_3O-\text{(benzene)}-CH=CH_2 \;+\; \text{(benzene)}-CH=CH_2 \xrightarrow[\substack{carbon\\tetrachloride\\nitrobenzene\\0\ ^\circ C}]{SnCl_4} \left[CHCH_2-CHCH_2\right]_n$$

with OCH$_3$ substituent

(l)

$$CH_2=CHCl \xrightarrow{benzoyl\ peroxide} \left[CH_2CH\right]_n$$
$$\hspace{3cm} | $$
$$\hspace{3cm} Cl$$

(m)

$$CH_2=CCOCH_3 \xrightarrow[\substack{methanol\\H_2O\\20\ ^\circ C}]{} \left[CH_2C\right]_n$$

with C≡N and COCH$_3$ (=O) groups

(n)

$$HO-\text{(benzene)}-\underset{\underset{CH_3}{|}}{\overset{\overset{CH_3}{|}}{C}}-\text{(benzene)}-OH \;+\; ClCCl\ (\text{C}=\text{O}) \xrightarrow[H_2O]{NaOH}$$

$$\left[O-\text{(benzene)}-\underset{\underset{CH_3}{|}}{\overset{\overset{CH_3}{|}}{C}}-\text{(benzene)}-OC\ (\text{C}=\text{O})\right]_n$$

27.24 (cont)

(o) $CH_2\!\!=\!\!CH_2$ + [cyclopentene] $\xrightarrow[\substack{VCl_4 \\ (hexyl)_3Al \\ heptane}]{}$ $\xrightarrow{CH_3CH_2OH}$ [polymer structure with $-CH_2CH_2-$ and cyclopentane ring]$_n$

(p) $CH_3CH_2OCH_2\!\!-\!\![oxetane\ with\ CH_2OCH_2CH_3]$ $\xrightarrow{(CH_3CH_2)_2O\cdot BF_3}$ $\left[OCH_2\underset{\substack{| \\ CH_2OCH_2CH_3}}{\overset{\substack{CH_2OCH_2CH_3 \\ |}}{C}}CH_2\right]_n$

(q)   [phenyl-substituted β-lactam (azetidinone) with N—H] $\xrightarrow[\substack{dimethyl \\ sulfoxide}]{pyrrolidide\ N^-K^+}$ $\left[NHCHCH_2\overset{O}{\overset{||}{C}}\right]_n NHCHCH_2\overset{O}{\overset{||}{C}}\!\!-\!\!N[pyrrolidine]$ (with phenyl substituents)

27.25   A cyano group or an ester group are electron withdrawing and stabilize a carbanionic intermediate by their inductive effects and also by resonance.

$$R'\!-\!CH_2\overset{-}{\ddot{C}}HC\!\!\equiv\!\!\ddot{N}: \quad\longleftrightarrow\quad R'\!-\!CH_2CH\!\!=\!\!C\!\!=\!\!\ddot{N}\!:^{-}$$

$$R'\!-\!CH_2\overset{-}{\ddot{C}}H\overset{\substack{:O: \\ ||}}{C}\!-\!\ddot{O}R \quad\longleftrightarrow\quad R'\!-\!CH_2CH\!\!=\!\!\overset{\substack{:\ddot{O}:^{-} \\ |}}{C}\!-\!\ddot{O}R$$

where R' is the initiator.

Radicals are also stabilized by groups such as the cyano group or a carbonyl group.

$$R'\!-\!CH_2\overset{\bullet}{C}HC\!\!\equiv\!\!\ddot{N}: \quad\longleftrightarrow\quad R'\!-\!CH_2CH\!\!=\!\!C\!\!=\!\!\overset{\bullet}{N}:$$

$$R'\!-\!CH_2\overset{\bullet}{C}H\overset{\substack{:\overset{\bullet}{O}: \\ ||}}{C}\!-\!\ddot{O}R \quad\longleftrightarrow\quad R'\!-\!CH_2CH\!\!=\!\!\overset{\substack{:\overset{\bullet}{O}: \\ |}}{C}\!-\!\ddot{O}R$$

Reactions of these monomers that go through anionic or radical intermediates have low energies of activation and proceed easily.

27.25 (cont)

Cationic intermediates are destabilized by the electron-withdrawing cyano and ester groups. Any resonance contributors that are written for such species delocalize the positive charge from carbon to the more electronegative nitrogen or oxygen atom.

$$\text{R'}\!-\!\overset{+}{\text{CH}_2\text{CHC}}\!\!\equiv\!\!\text{N}\!: \quad\longleftrightarrow\quad \text{R'}\!-\!\text{CH}_2\text{CH}\!\!=\!\!\text{C}\!\!=\!\!\overset{+}{\text{N}}\!:$$

$$\text{R'}\!-\!\text{CH}_2\overset{+}{\text{CHC}}\!-\!\overset{..}{\underset{..}{\text{OR}}} \quad\longleftrightarrow\quad \text{R'}\!-\!\text{CH}_2\text{CH}\!\!=\!\!\text{C}\!-\!\overset{..}{\underset{..}{\text{OR}}}$$

Such intermediates correspond to high-energy transition states and to high energies of activation for the reactions.

27.26 The rate of a reaction depends on the activation energy of the rate-determining step. The activation energy, in turn, is the difference in energy between the reagents and the transition state. In rationalizing most relative rates, it is necessary only to look at the relative stabilization of the transition states. The more stabilized the transition state is, in general, the lower the activation energy and the faster the reaction. However, if the energy of a reagent is lowered even more by stabilization, the activation energy will be larger, and the rate, correspondingly smaller.

In free radical polymerization the rate-determining step is the addition of an unstable radical to the alkene monomer. Substituents play an important role in stabilizing this radical, in which one species bears full radical character, and a less important role in stabilizing the transition state, in which radical character is distributed between two species. The consequences of this are illustrated graphically below.

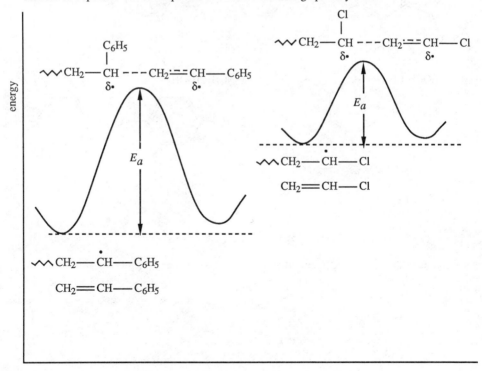

reaction coordinate

27.26 (cont)

      The phenyl group strongly stabilizes the reagent radical, the chlorine atom not as strongly. Even though the transition state for the reaction of the vinyl chloride may be higher in energy in absolute terms than the transition state for the reaction of styrene, the energy of activation for the reaction of vinyl chloride is smaller. The relative rates of the polymerization reactions thus reflect the relative stabilization of radicals by the substituents on the double bond.

27.27

(a)

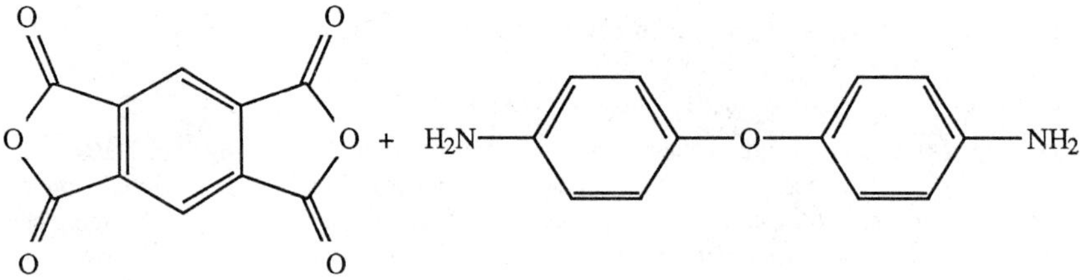

(b)

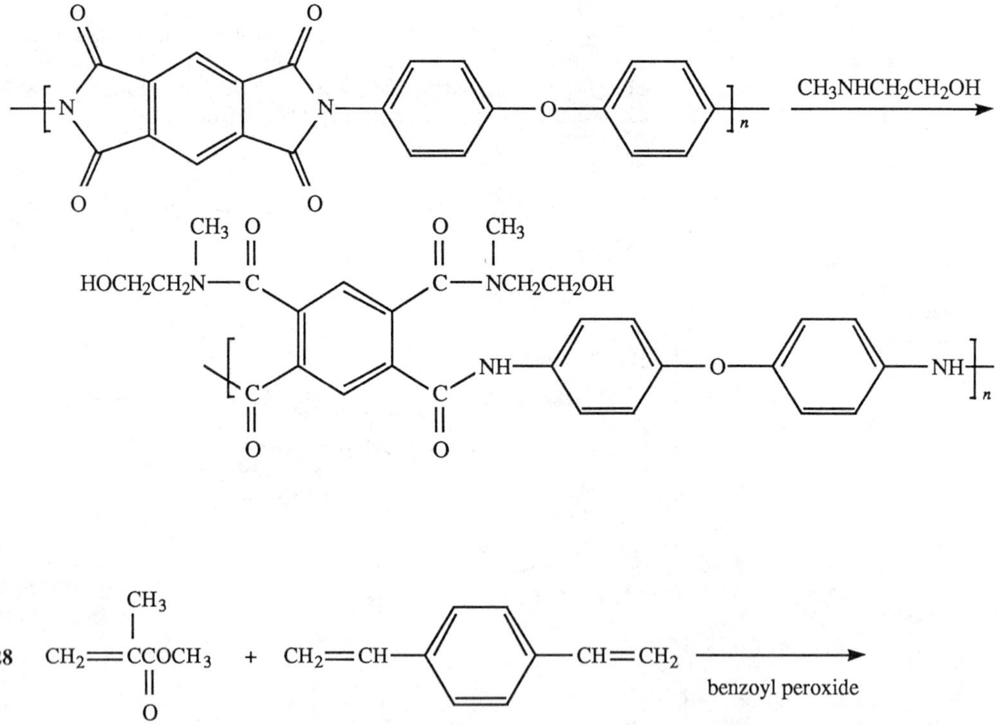

27.28

27.28 (cont)

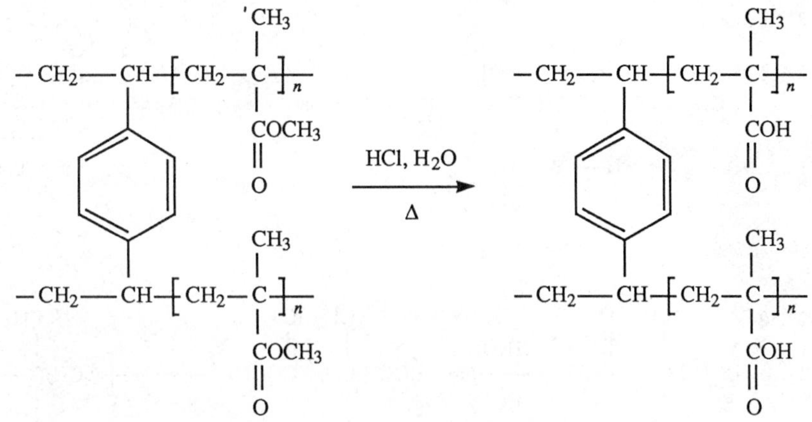

27.29

1.

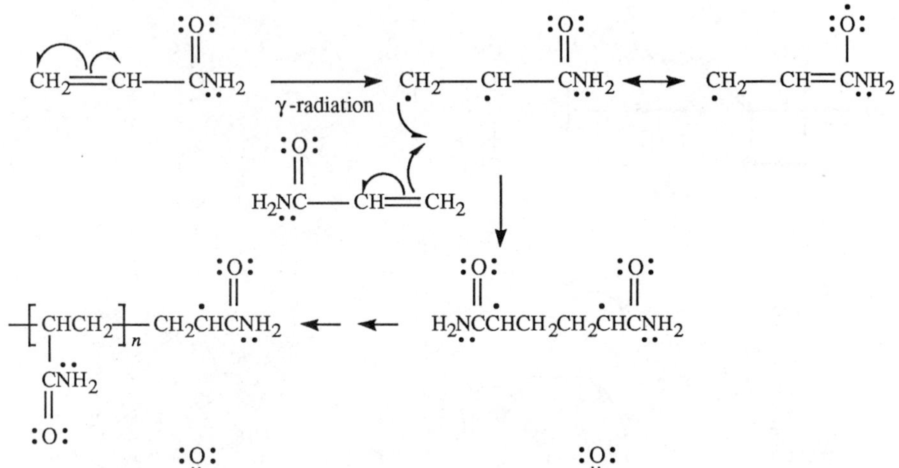

2.

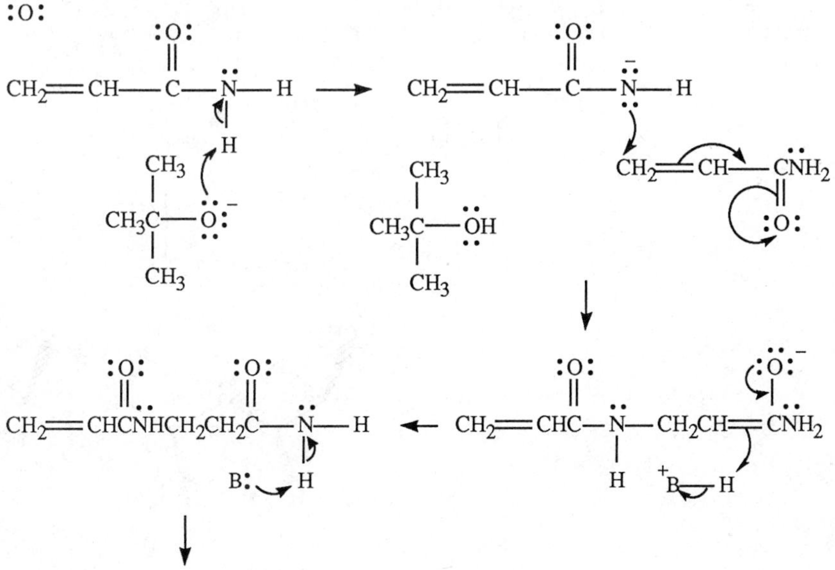

27.29 (cont)

2. (cont)

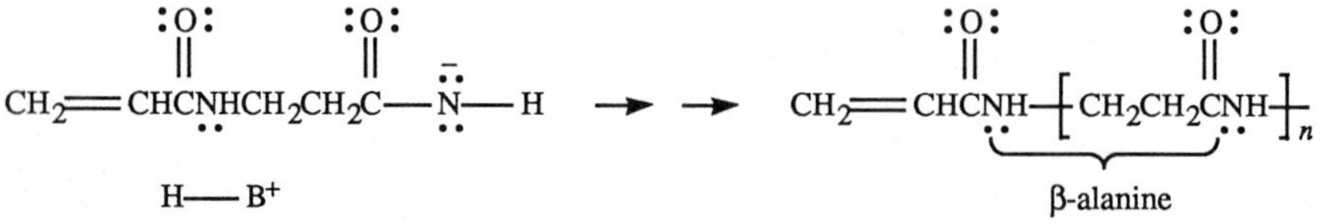

β-alanine

27.30

1.

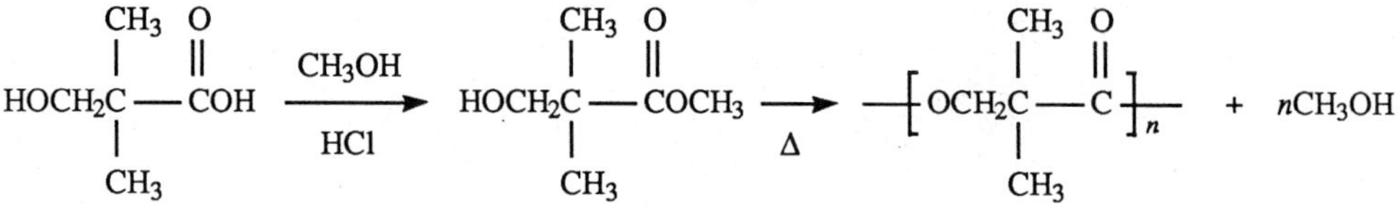

2.

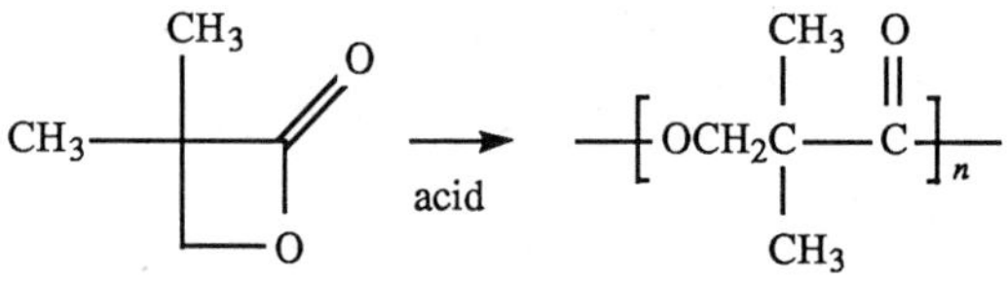

27.31

1.

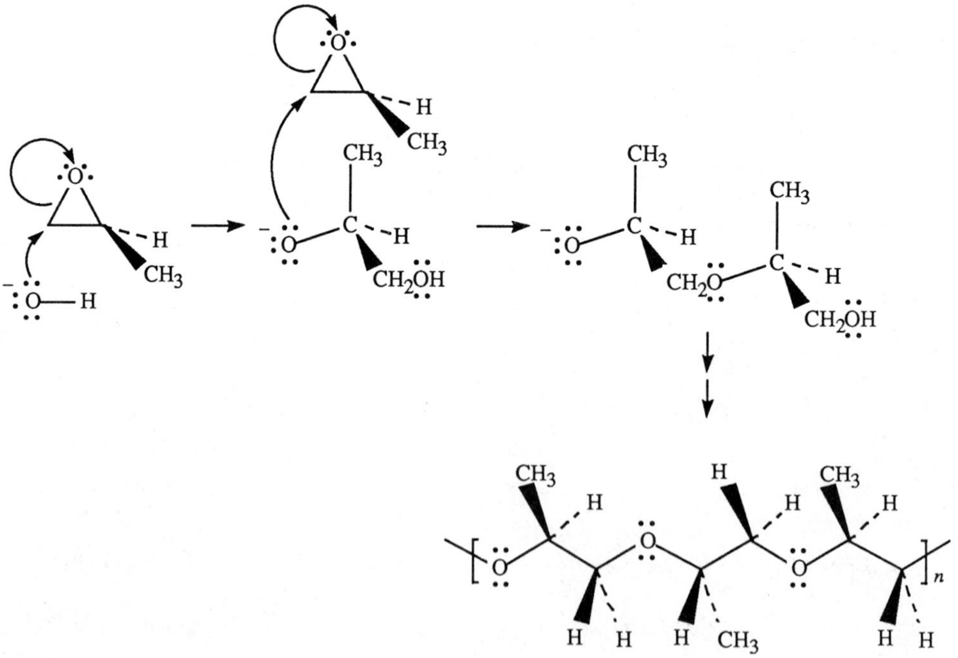

27.31 (cont)

2.

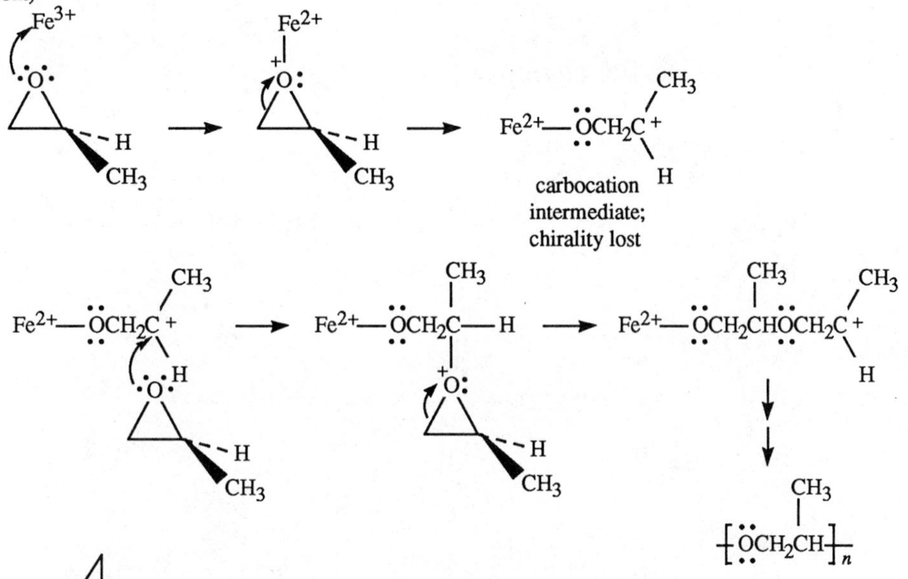

Fe^{2+}—ÖCH$_2$C$^+$

carbocation
intermediate;
chirality lost

$$\left[\!\!-\overset{\displaystyle CH_3}{\underset{}{\ddot{O}CH_2CH}}\!\!-\right]_n$$

27.32

(a)

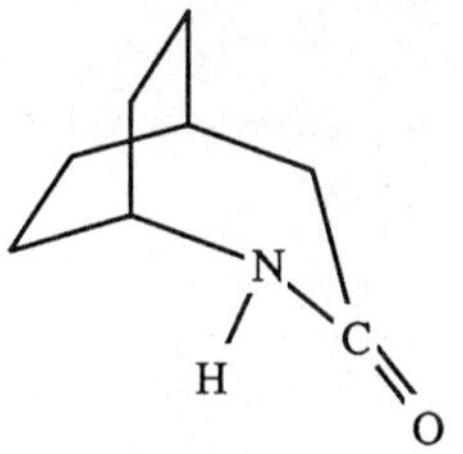

(b)

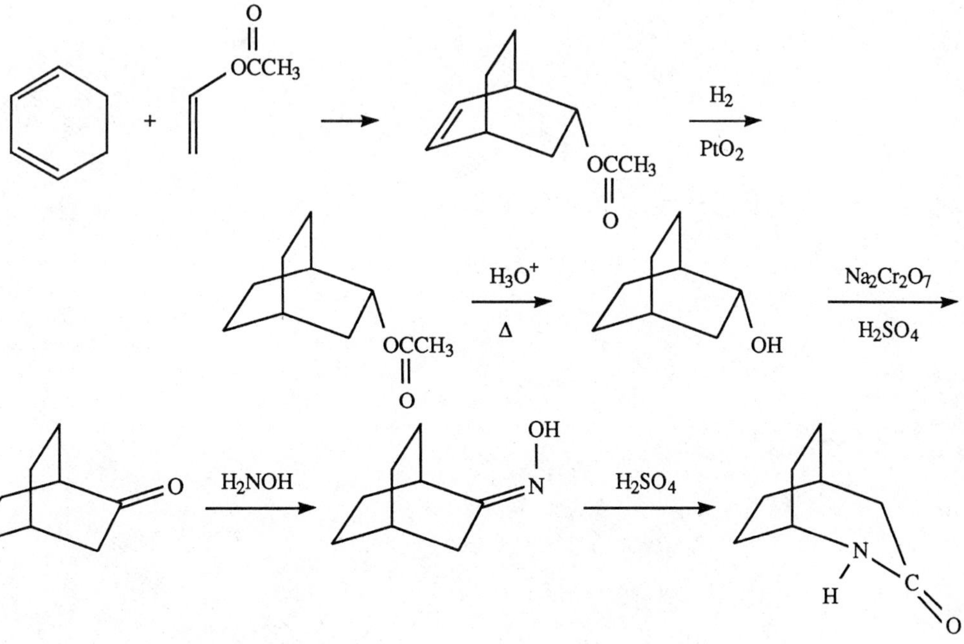

27.33

1.

$$HCH \longrightarrow (CH_3CH_2CH_2CH_2)_3\overset{+}{N}CH_2 - \overset{..}{\underset{..}{O}}: ^- \quad \overset{H}{\underset{H}{C}}=\overset{..}{\underset{..}{O}} \longrightarrow$$

:N(CH_2CH_2CH_2CH_3)_3

$$(CH_3CH_2CH_2CH_2)_3\overset{+}{N}CH_2 - \overset{..}{\underset{..}{O}}CH_2\overset{..}{\underset{..}{O}}: ^- \longrightarrow \longrightarrow (CH_3CH_2CH_2CH_2)_3\overset{+}{N}CH_2\overset{..}{\underset{..}{O}}\left[CH_2\overset{..}{\underset{..}{O}}\right]_n$$

2.

$$CH_2=CH-CH \longrightarrow :N\equiv C - \underset{CH=CH_2}{\overset{H}{C}} - \overset{..}{\underset{..}{O}}: ^- \quad \underset{\underset{CH_2}{\overset{\shortparallel}{CH}}}{\overset{H}{C}}=\overset{..}{\underset{..}{O}} \longrightarrow$$

$^-$:C≡N:

$$:N\equiv C - \underset{CH=CH_2}{\overset{H}{C}} - \overset{..}{\underset{..}{O}} - \underset{CH=CH_2}{\overset{H}{C}} - \overset{..}{\underset{..}{O}}: ^- \longrightarrow \longrightarrow :N\equiv C - \left[\underset{CH=CH_2}{C\overset{..}{\underset{..}{HO}}}\right]_n$$

28

Concerted Reactions

Concept Map 28.1   Concerted reactions.

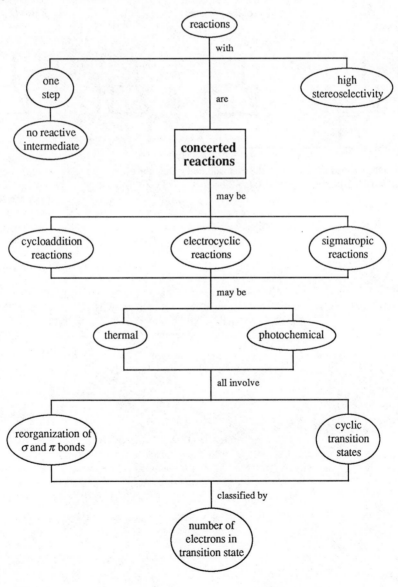

28.1

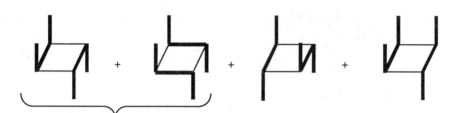

$\xrightarrow{\text{214 nm}}$

major component
of the mixture

from the cycloaddition
of two (Z)-2-butenes

from the cycloaddition
of two (Z)-2-butenes

from the cycloaddition
of two (E)-2-butenes

from the cycloaddition
of two (E)-2-butenes

from the cycloaddition
of one (E)-2-butene
with one (Z)-2-butene;
not seen as a product in
the early stages of the
photochemical reaction
of either (E)-2-butene
or (Z)-2-butene

28.2

CH₃   CH₂COCH₃ (with O above, ‖)
   C=C
H   H

$\xrightarrow[\text{diethyl ether}]{\text{LiAlD}_4}$ $\xrightarrow{\text{H}_3\text{O}^+}$

CH₃   CH₂CD₂OH
   C=C
H   H

A

$\xrightarrow[\text{pyridine}]{\text{TsCl}}$

CH₃   CH₂CD₂OTs
   C=C
H   H

B

$\xrightarrow[\text{acetone}]{\text{NaI}}$

CH₃   CH₂CD₂I
   C=C
H   H

C

$\xrightarrow[\text{methanol}]{\text{CH}_3\text{NCH}_3 \,(\text{with CH}_3 \text{ above})}$

28.2 (cont)

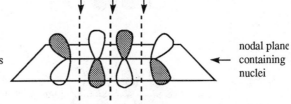

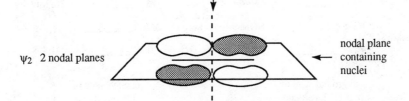

28.3 ψ_4 4 nodal planes

three perpendicular nodal planes

nodal plane containing nuclei

ψ_3 3 nodal planes

two perpendicular nodal planes

nodal plane containing nuclei

ψ_2 2 nodal planes

one perpendicular nodal plane

nodal plane containing nuclei

28.3 (cont)

ψ_1 1 nodal planes   nodal plane ← containing nuclei

28.4   The symmetry of a molecular orbital can be determined by looking at the symmetry of the atomic *p*-orbitals from which it is constructed since the molecular orbital retains that symmetry.

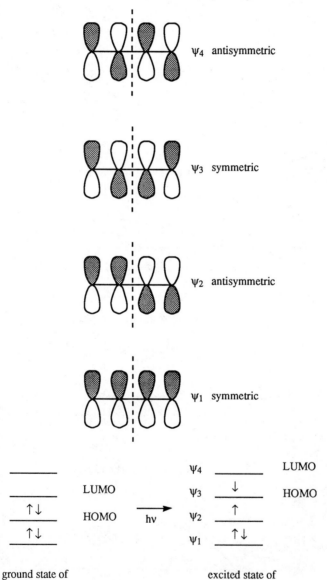

ψ_4 antisymmetric

ψ_3 symmetric

ψ_2 antisymmetric

ψ_1 symmetric

28.5

	ground state of 1,3-butadiene			excited state of 1,3-butadiene	
ψ_4	_____		ψ_4	_____	LUMO
ψ_3	_____	LUMO	ψ_3	↓	HOMO
ψ_2	↑↓	HOMO	ψ_2	↑	
ψ_1	↑↓		ψ_1	↑↓	

$\xrightarrow{h\nu}$

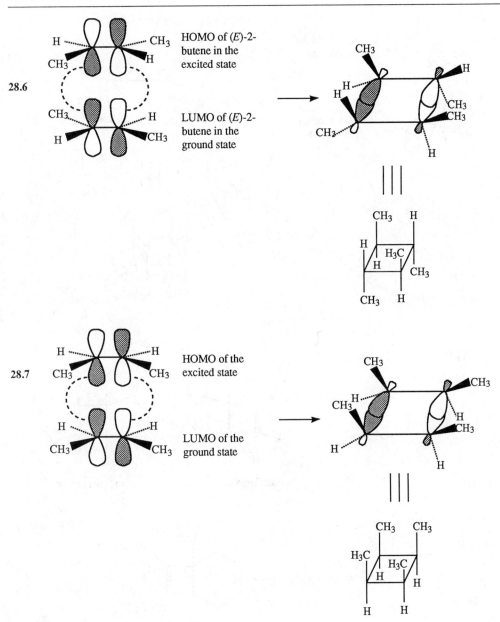

28.6

HOMO of (*E*)-2-butene in the excited state

LUMO of (*E*)-2-butene in the ground state

28.7

HOMO of the excited state

LUMO of the ground state

28.7 (cont)

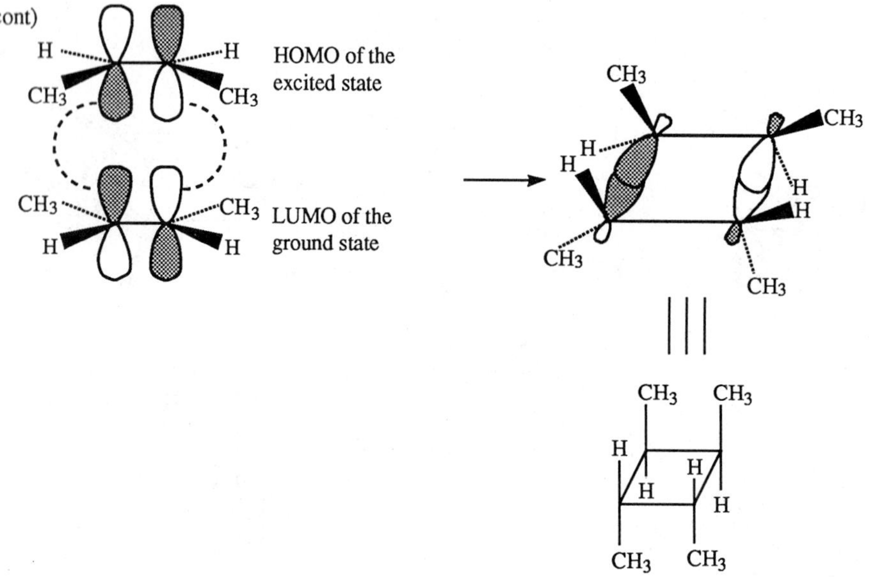
H
H
CH3
CH3
HOMO of the
excited state
CH3
CH3
H
H
LUMO of the
ground state
CH3
CH3
H
H
H
H
CH3
CH3
CH3
CH3
H
H
H
H
CH3
CH3

28.8

(a)

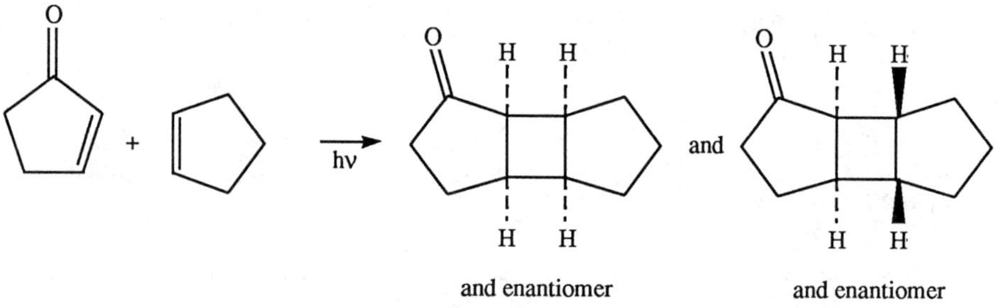
O
+
hν
O
H
H
H
H
and
O
H
H
H
H
and enantiomer
and enantiomer

(b)

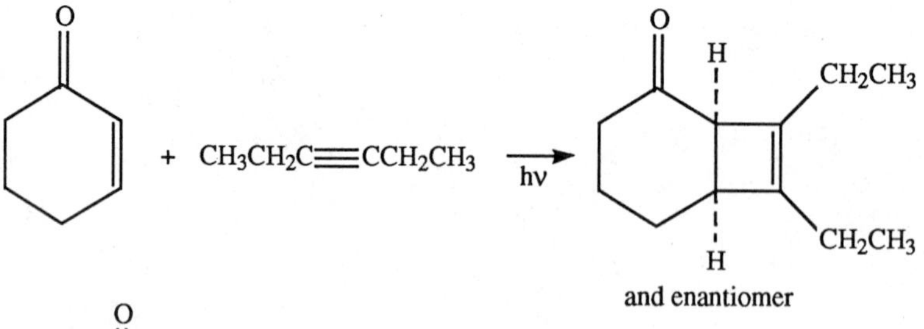
O
+ CH3CH2C≡CCH2CH3
hν
O
H
CH2CH3
CH2CH3
H
and enantiomer

(c)

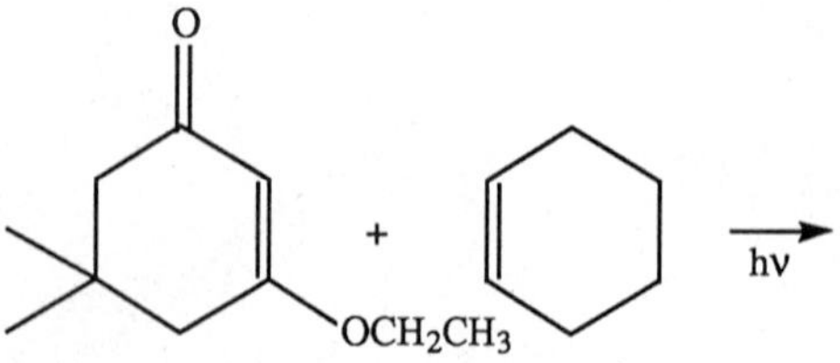
O
OCH2CH3
+
hν

28.8 (cont)

(c) (cont)

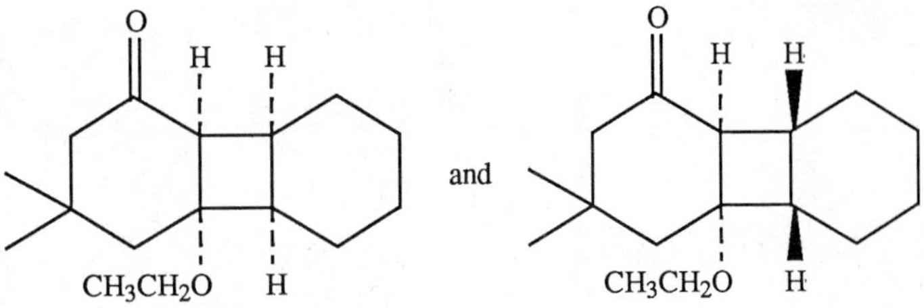

and enantiomer   and enantiomer

28.9

(a)

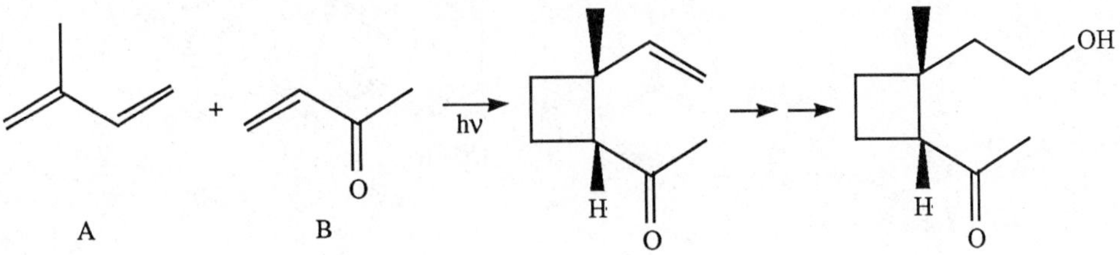

A   B   grandisol

(b)

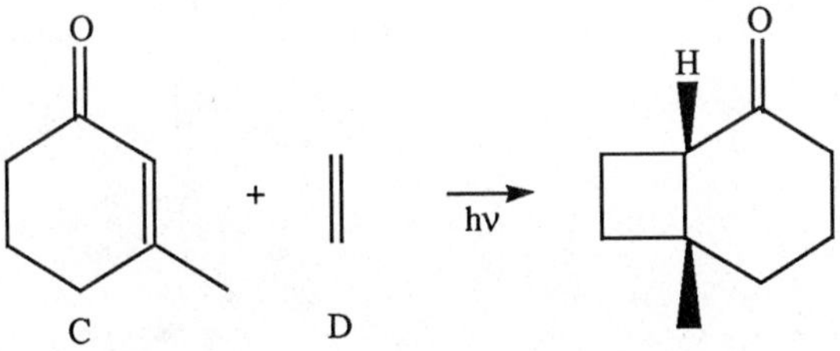

C   D

(c)

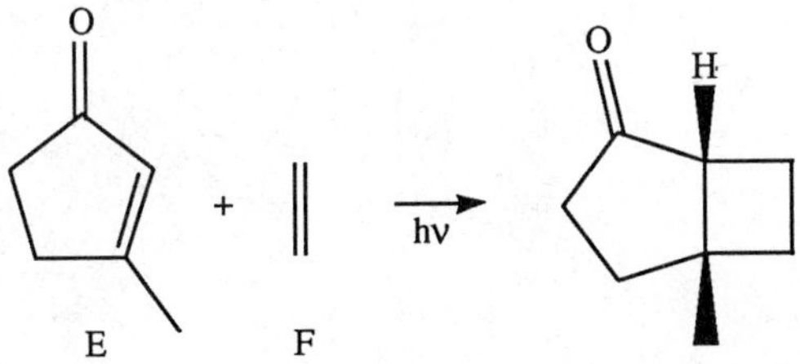

E   F

(d)

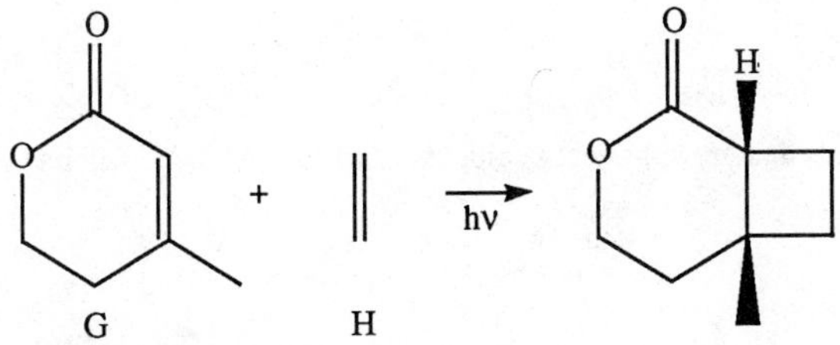

G   H

28.10

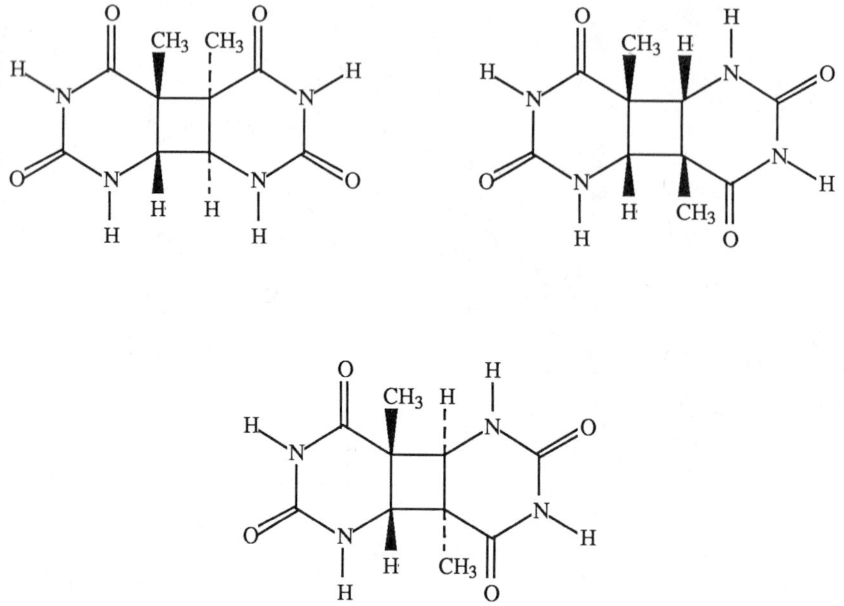

28.11

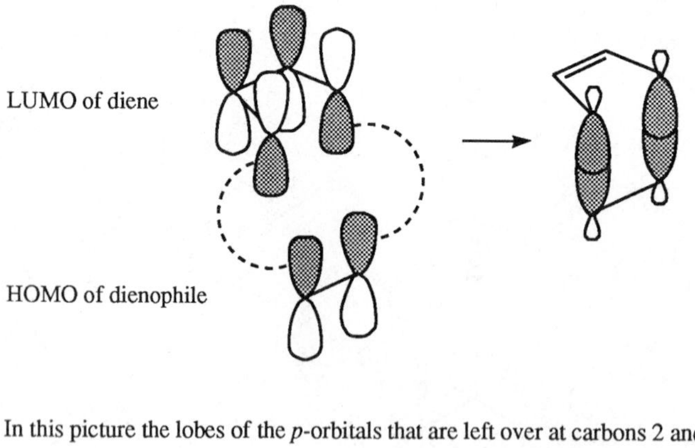

LUMO of diene

HOMO of dienophile

In this picture the lobes of the *p*-orbitals that are left over at carbons 2 and 3 of the diene will be bonding.

28.12

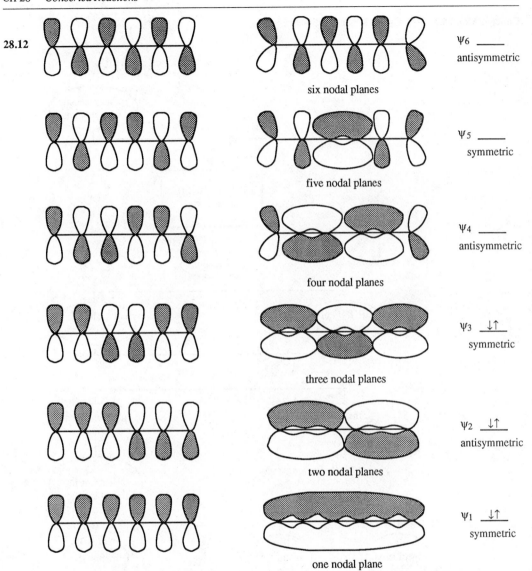

six nodal planes

five nodal planes

four nodal planes

three nodal planes

two nodal planes

one nodal plane

ψ_6 _____ antisymmetric

ψ_5 _____ symmetric

ψ_4 _____ antisymmetric

ψ_3 ↓↑ symmetric

ψ_2 ↓↑ antisymmetric

ψ_1 ↓↑ symmetric

The lower three molecular orbitals are bonding; the upper three are antibonding. The HOMO is ψ_3.

Concept Map 28.2   Cycloaddition reactions.

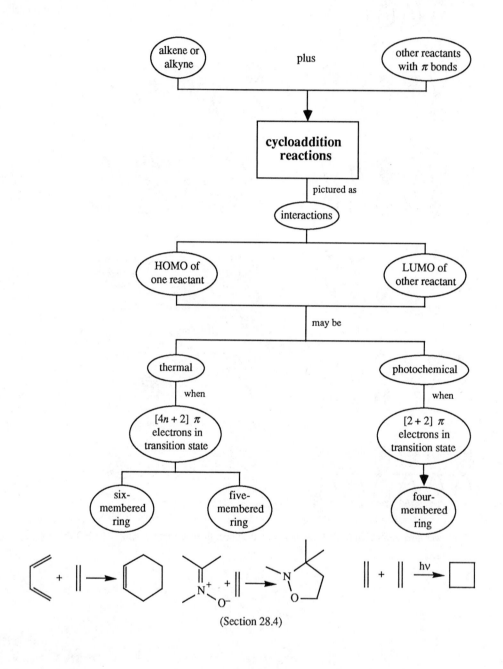

(Section 28.4)

28.13

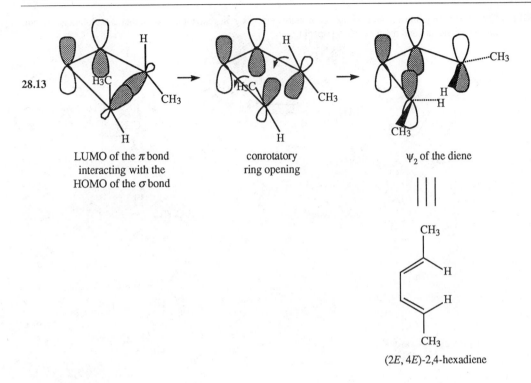

LUMO of the π bond
interacting with the
HOMO of the σ bond

conrotatory
ring opening

ψ₂ of the diene

(2E, 4E)-2,4-hexadiene

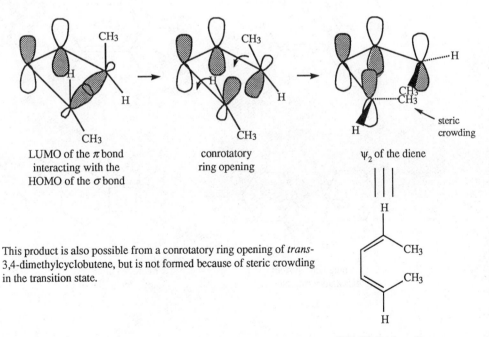

LUMO of the π bond
interacting with the
HOMO of the σ bond

conrotatory
ring opening

ψ₂ of the diene

steric
crowding

This product is also possible from a conrotatory ring opening of *trans*-3,4-dimethylcyclobutene, but is not formed because of steric crowding in the transition state.

(2Z, 4Z)-2,4-hexadiene

28.14   The observed equilibrium involves two conrotatory electrocyclic reactions of a butadiene-cyclobutene system, one a ring closure and the other a ring opening.

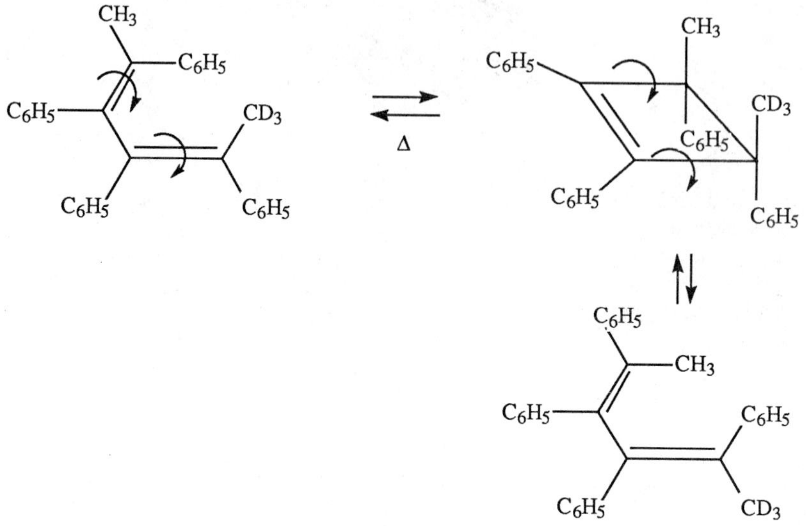

Other possible stereoisomers:

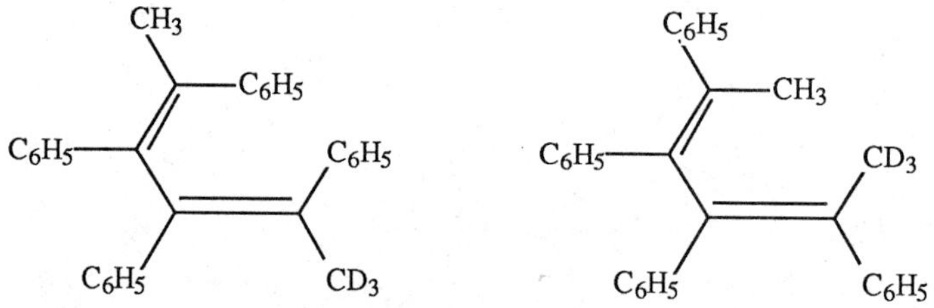

28.15

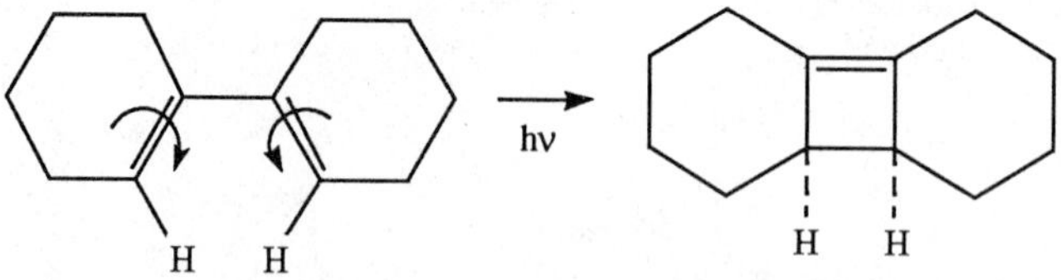

butadiene system
disrotatory ring closure

hydrogen atoms on
same side of the ring

28.16

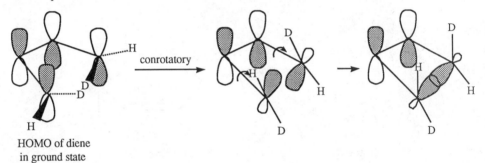

s-trans   *s*-cis   racemic mixture

Thermal process:

HOMO of diene
in ground state

conrotatory

Photochemical process:

HOMO of diene
in excited state

disrotatory

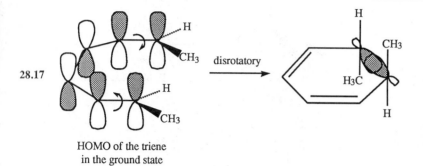

28.17

HOMO of the triene
in the ground state

(See Problem 28.12 for the molecular orbitals of the triene.)

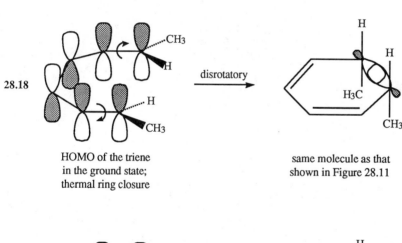

28.18

HOMO of the triene
in the ground state;
thermal ring closure

same molecule as that
shown in Figure 28.11

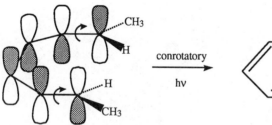

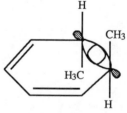

HOMO of the triene
in the excited state;
photochemical ring closure

enantiomer of the molecule
shown in Figure 28.12

Concept Map 28.3   Electrocyclic reactions.

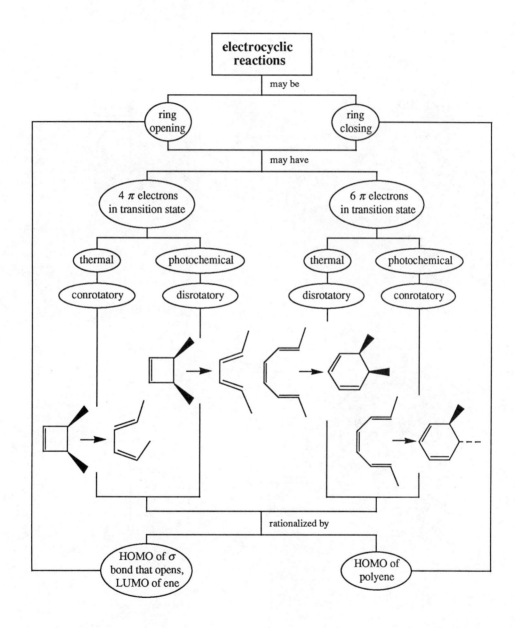

28.19   Homo of triene in the ground state

precalciferol   isopyrocalciferol

precalciferol   pyrocalciferol

28.20

allo-ocimene

hν
conrotatory

+

racemic mixture

Concept Map 28.4   Woodward-Hoffmann rules.

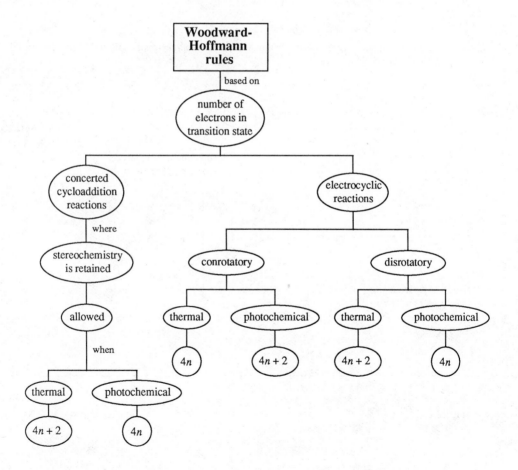

28.21

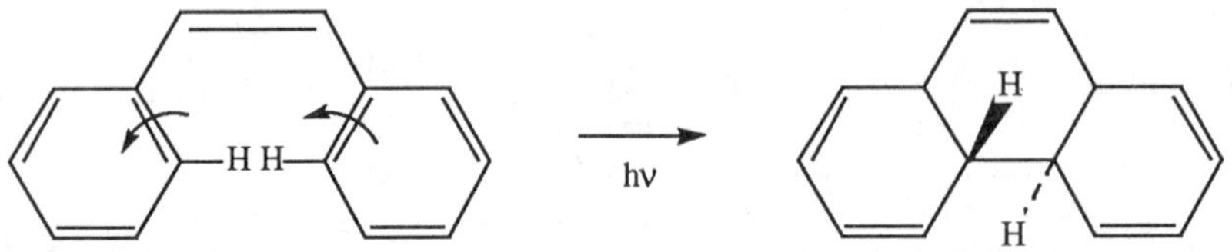

$4n + 2$ electrocyclic reaction; $n = 1$; conrotatory in the excited state; the hydrogen atoms at the ring junctions in dihydrophenanthrene will be trans to each other.

28.22

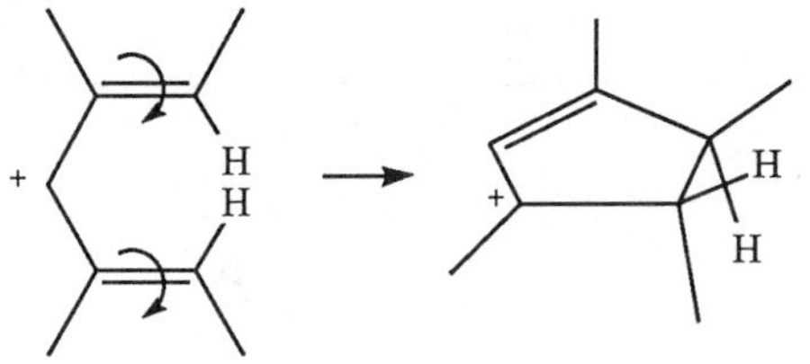

and enantiomer

$4n$ electrocyclic reaction, $n = 1$; conrotatory in ground state; the methyl groups on the new five-membered ring will be trans to each other.

28.23

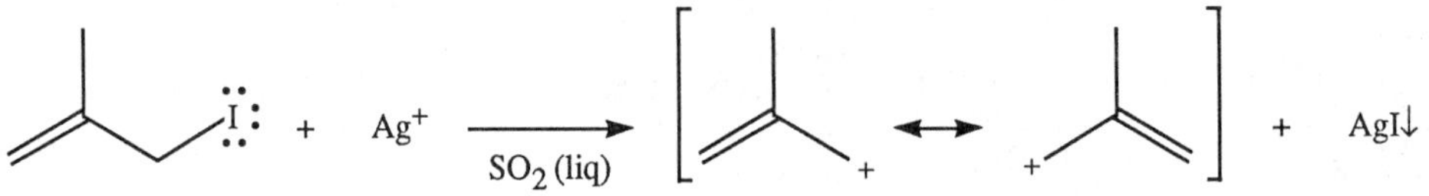

the dienophile is an allylic
cation with two π electrons

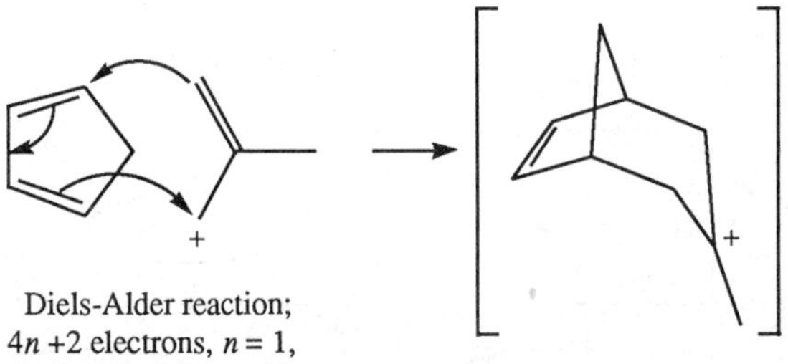

Diels-Alder reaction;
$4n + 2$ electrons, $n = 1$,
in the transition state

Concept Map 28.2   (see p. 960)

28.24

(a)

$$\underset{\substack{O \\ \parallel}}{CH_3CH} \quad + \quad \text{⟨phenyl⟩}-NHOH \quad + \quad \text{⟨phenyl⟩}-CH=CH_2 \quad \xrightarrow[60\ °C]{}$$

(structure: isoxazolidine ring with CH₃, two phenyl groups, O, N)

(b)

$$CH_2\overset{+}{=}N=\overset{-}{N} \text{(1 molar equiv)} \quad + \quad CH_2=CHCH=CH_2 \quad \longrightarrow$$
$$\text{diethyl ether}$$

(structure with CH=CH₂ and N=N ring) $\xrightarrow{\text{tautomerization}}$ (structure with CH=CH₂ and N–N, H)

(c)

$$\underset{\substack{| \\ \text{⟨phenyl⟩}}}{\overset{\substack{H \\ |}}{C}}=\overset{\substack{+ \\ \text{⟨phenyl⟩}}}{N} \quad + \quad CH_2=CHCH_2OH \quad \xrightarrow{70\ °C} \quad HOCH_2-\text{(isoxazolidine ring with phenyl, N, O)}$$
$$\underset{O^-}{}$$

(d)

$$CH_2\overset{+}{=}N=\overset{-}{N} \quad + \quad \underset{\substack{| \\ CH_3OC \\ \parallel \\ O}}{\overset{\substack{H \\ |}}{C}}=\underset{\substack{| \\ COCH_3 \\ \parallel \\ O}}{\overset{\substack{H \\ |}}{C}} \quad \longrightarrow$$

(pyrazoline product structure with COCH₃ groups)

and enantiomer

28.24 (cont)

(e) 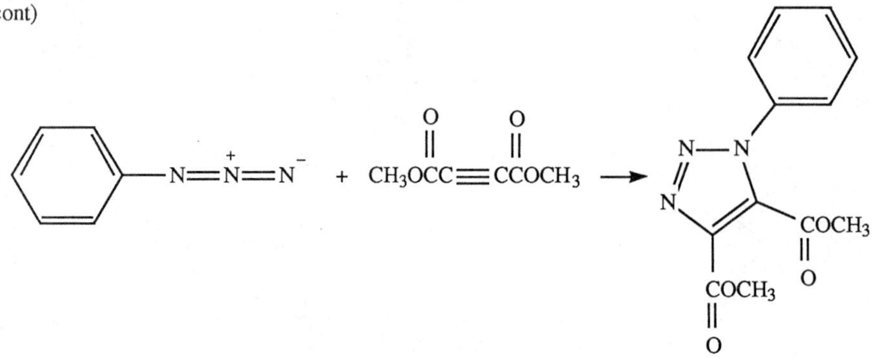

(f) 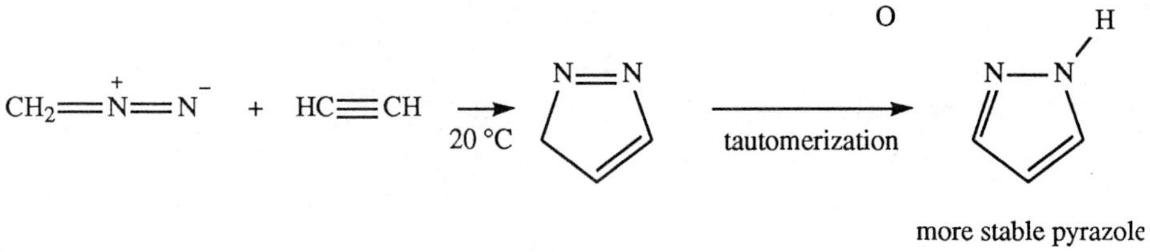
20 °C
tautomerization
more stable pyrazole

(g) 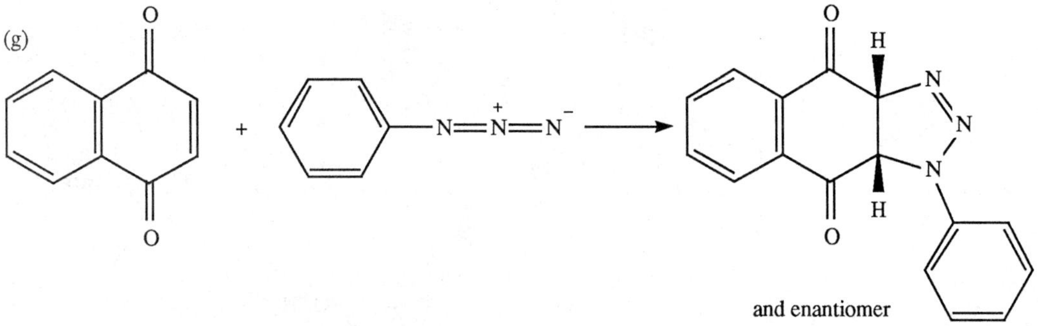
and enantiomer

(h) 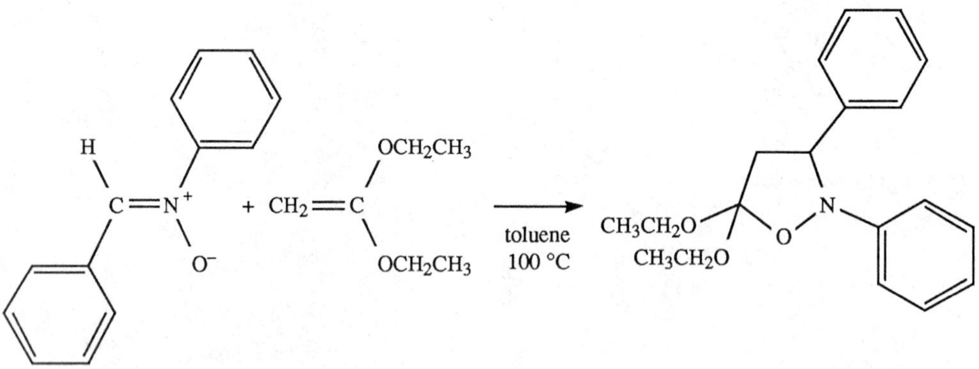
toluene
100 °C

28.24 (cont)

(i)

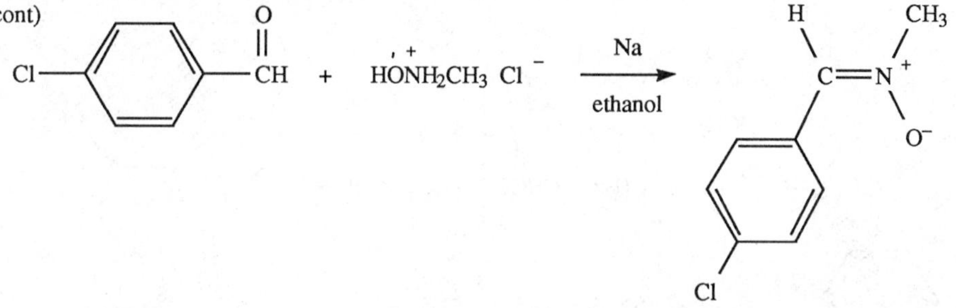

28.25

(a)

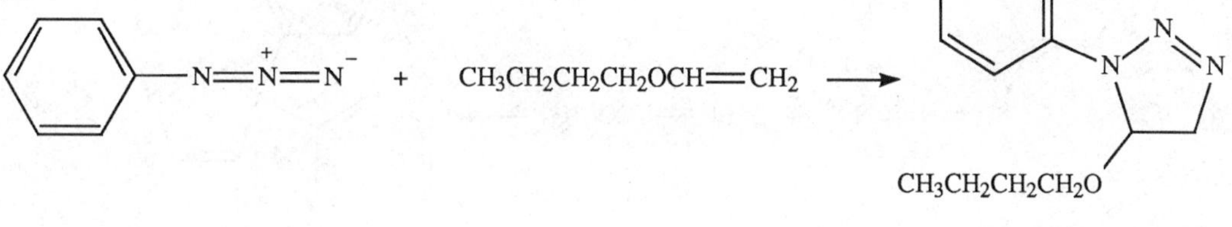

(b)

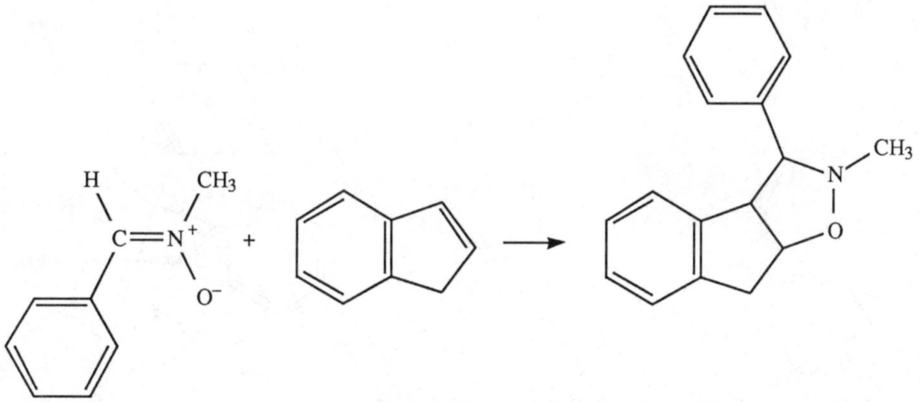

(c)

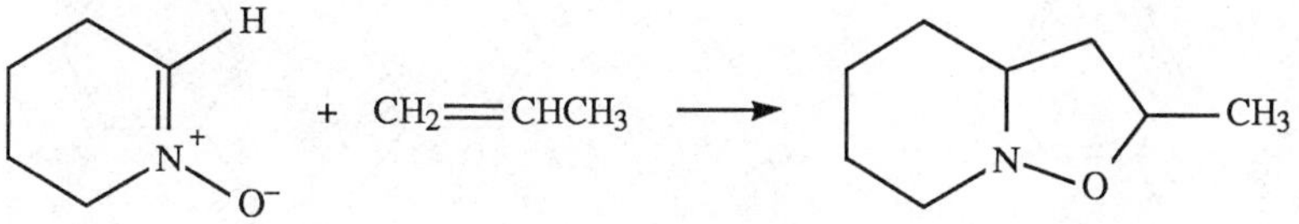

28.25 (cont)

(d)

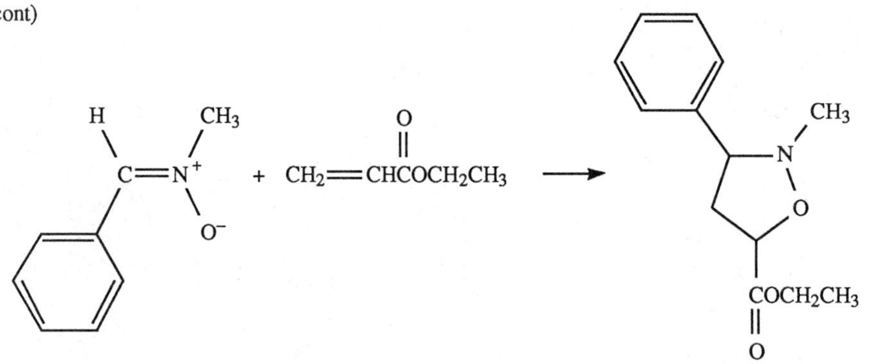

(e)

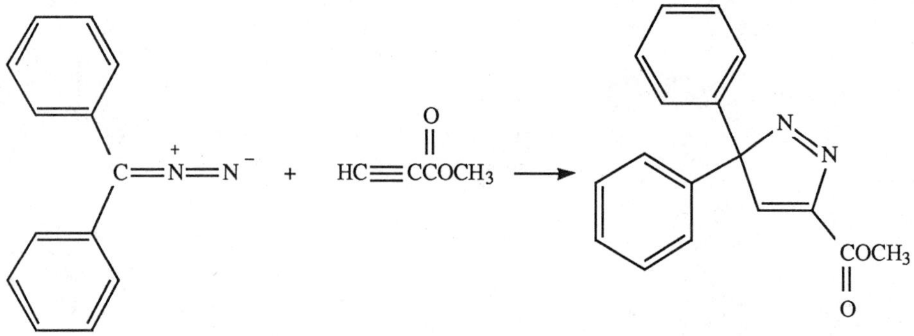

(f)

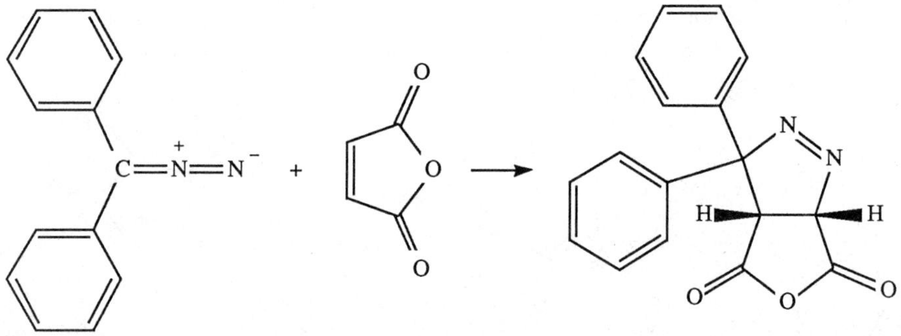

(g)

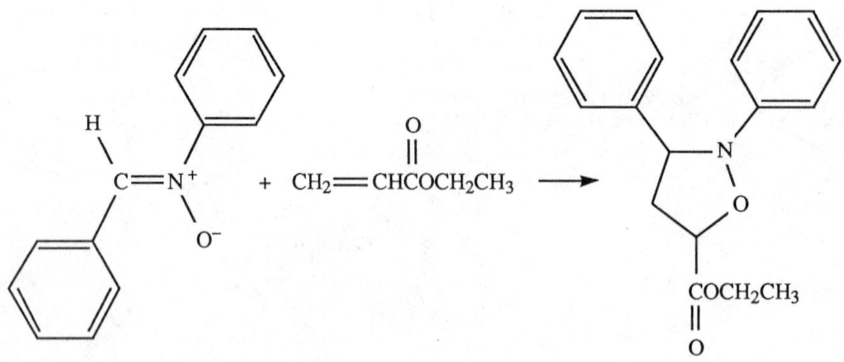

28.25 (cont)

(h) $CH_2{=}\overset{+}{N}{=}\overset{-}{N}$ + $CH_3CH_2CH_2CH_2OCH{=}CH_2$ $\longrightarrow$

$CH_3CH_2CH_2CH_2O$

28.26

	triene in the ground state			triene in the excited state
ψ_6 ____			ψ_6 ____	
ψ_5 ____			ψ_5 ____	LUMO
ψ_4 ____	LUMO		ψ_4 ↓	HOMO
ψ_3 ↑↓	HOMO	$\xrightarrow{h\nu}$	ψ_3 ↑	
ψ_2 ↑↓			ψ_2 ↑↓	
ψ_1 ↑↓			ψ_1 ↑↓	

(See Problem 28.12 for pictures of the orbitals.)

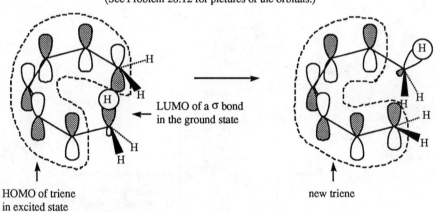

LUMO of a σ bond in the ground state

HOMO of triene in excited state

new triene

28.27

(a)

200 °C

28.27 (cont)

(b)

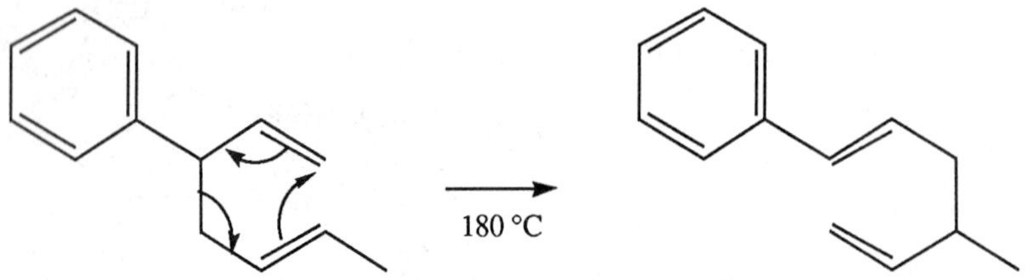
180 °C

(c)

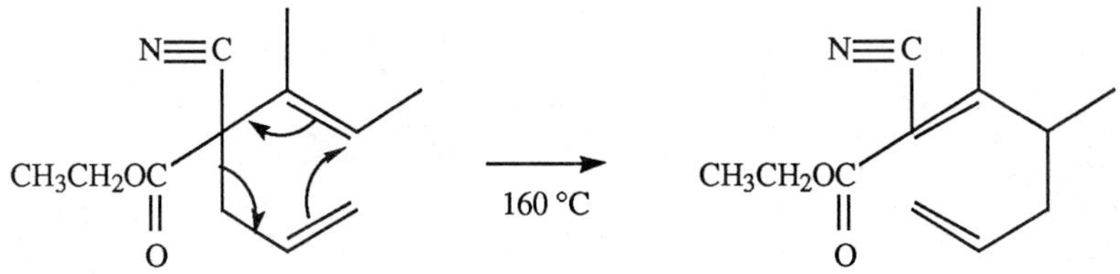
N≡C
CH3CH2OC
O
160 °C
N≡C
CH3CH2OC
O

(d)

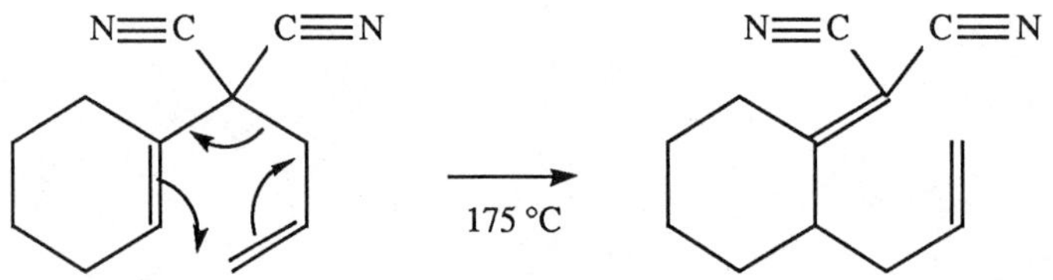
N≡C
C≡N
175 °C
N≡C
C≡N

28.28

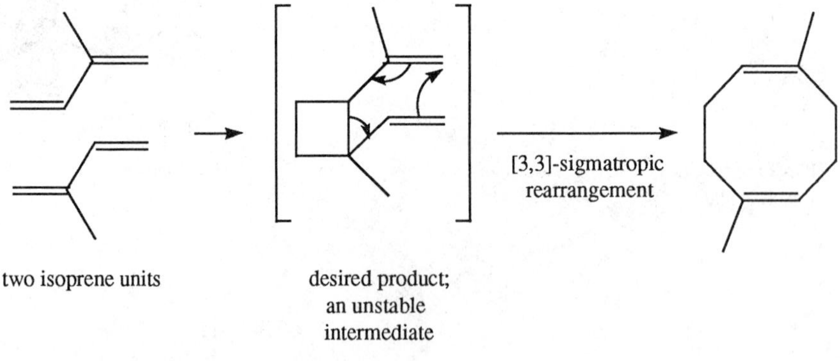
[3,3]-sigmatropic
rearrangement
two isoprene units
desired product;
an unstable
intermediate

The Cope rearrangement occurs easily because of relief of ring strain in going from a cyclobutane to a cyclooctadiene system.

Concept Map 28.5   Sigmatropic rearrangements.

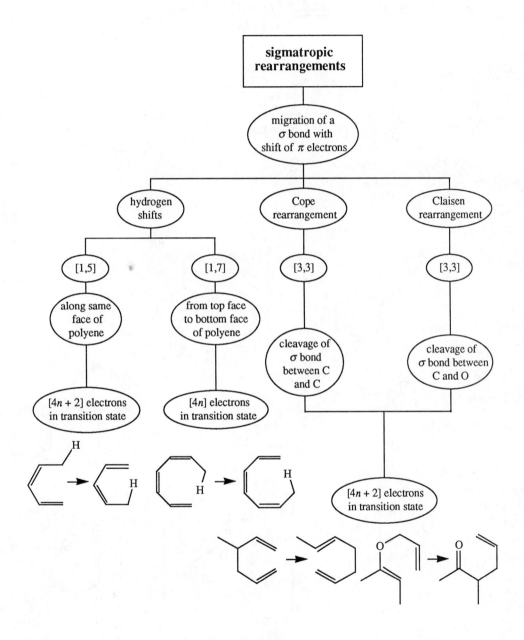

28.29

(a) $CH_2 = CHOCH_2CH = CH_2$ ≡ $\xrightarrow{255\ °C}$

(b) $\xrightarrow{175\ °C}$

(c) $\xrightarrow{\Delta}$ $\longrightarrow$

(d) $\xrightarrow{255\ °C}$ $\longrightarrow$

28.29 (cont)

(e)

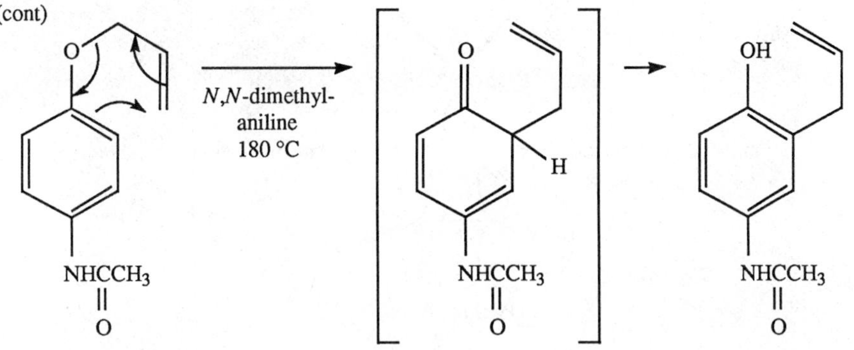

N,N-dimethyl-
aniline
180 °C

(f)

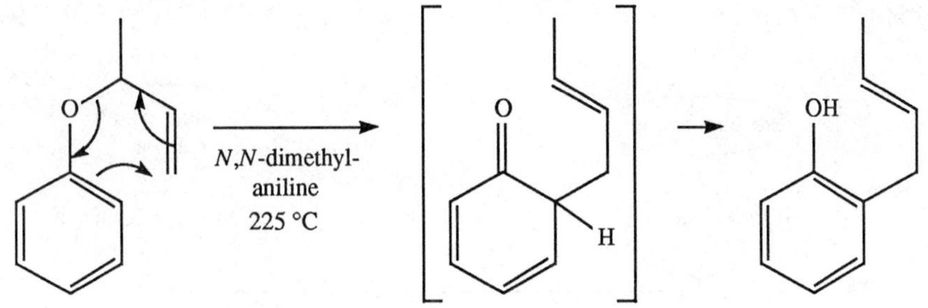

N,N-dimethyl-
aniline
225 °C

(g)

240 °C

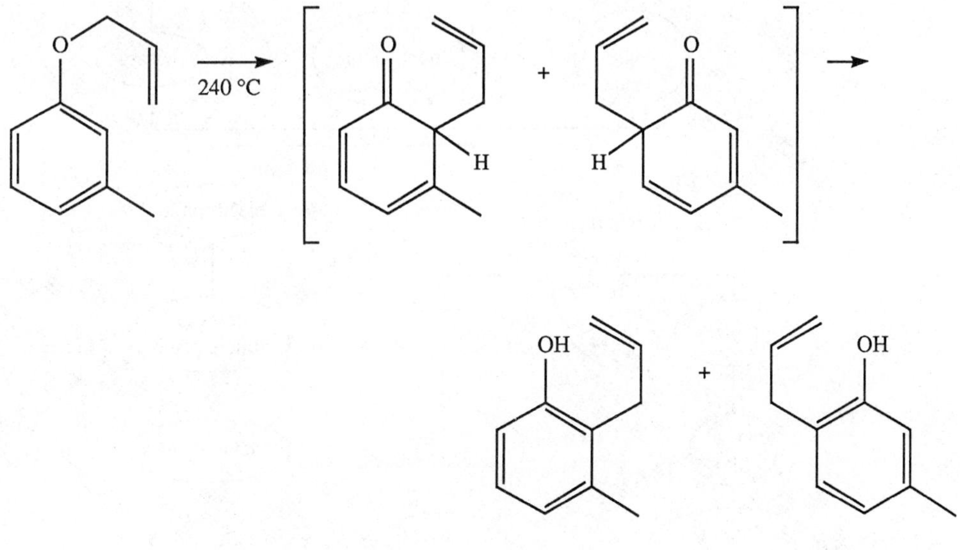

28.30

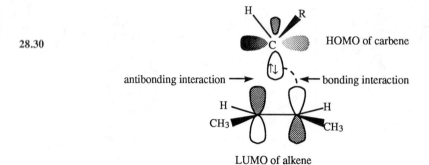

HOMO of carbene

antibonding interaction →   ← bonding interaction

LUMO of alkene

Concept Map 28.6   Carbenes.

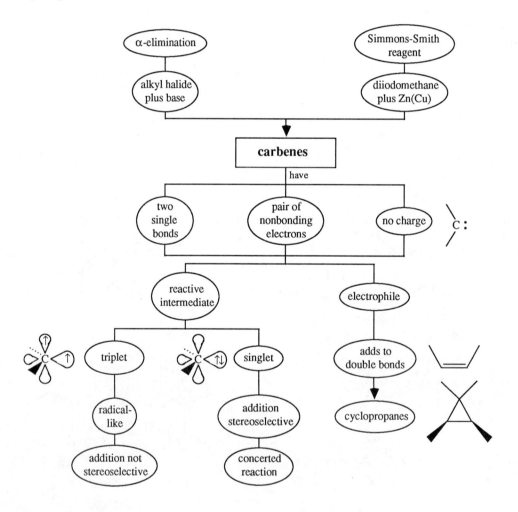

28.31

(a) CH_2I_2 $\xrightarrow[\text{diethyl ether}]{\text{Zn(Cu)}}$

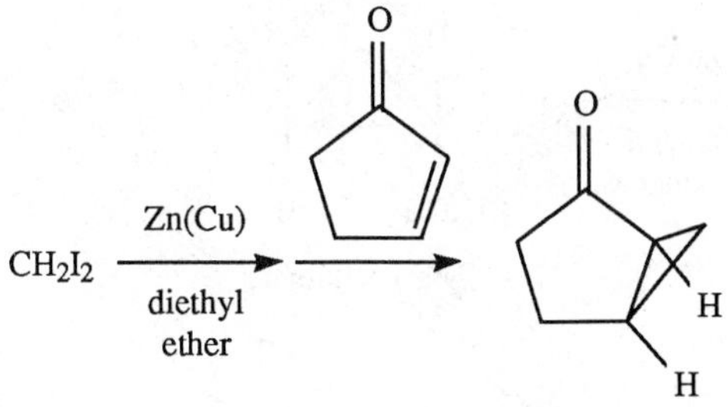

and enantiomer

(b) $(CH_3)_2C=C(CH_3)_2$ + $CHCl_3$ + $CH_3CO^-K^+$ (with CH_3 groups) $\xrightarrow{-10\,°C}$

(c)   cyclohexene + CH_2I_2 $\xrightarrow[\substack{\text{diethyl ether} \\ \Delta}]{\text{Zn(Cu)}}$

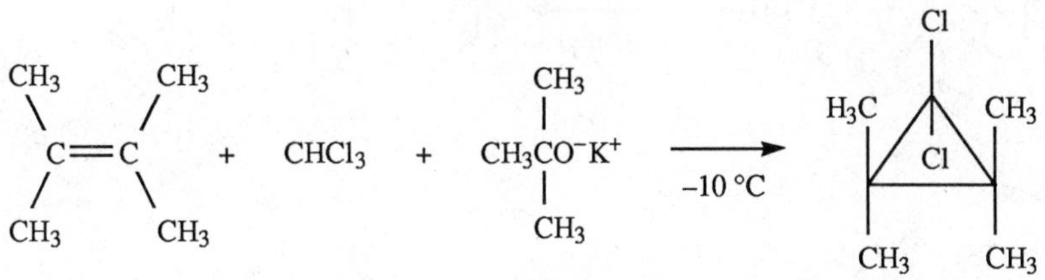

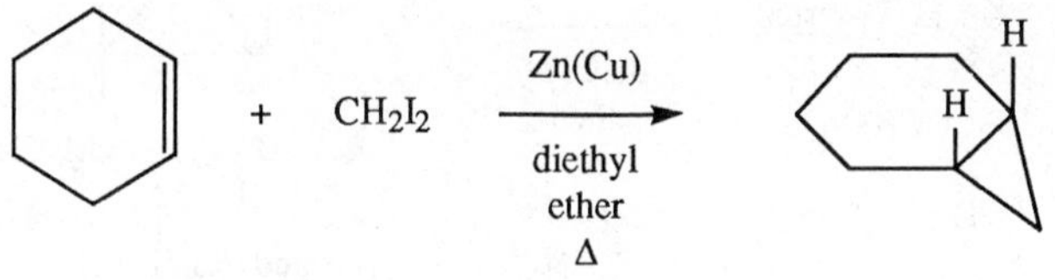

(d) $CH_2{=}CHOCH_2CH_3$ + $CHCl_3$ + $CH_3CO^-K^+$ $\xrightarrow{-10\,°C}$

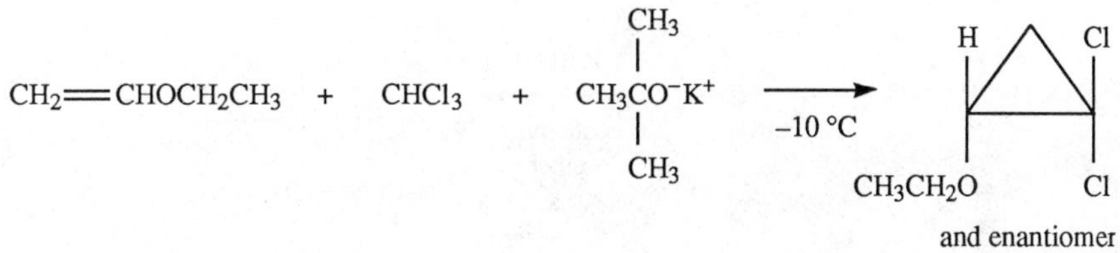

and enantiomer

(e) $CH_3(CH_2)_5CH{=}CH_2$ + CH_2I_2 $\xrightarrow[\substack{\text{diethyl ether} \\ \Delta}]{\text{Zn(Cu)}}$

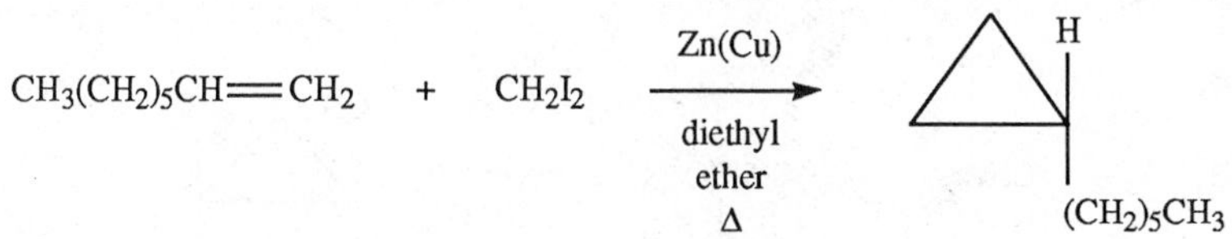

(f)   + CH_2I_2 $\xrightarrow[\substack{\text{diethyl ether} \\ \Delta}]{\text{Zn(Cu)}}$

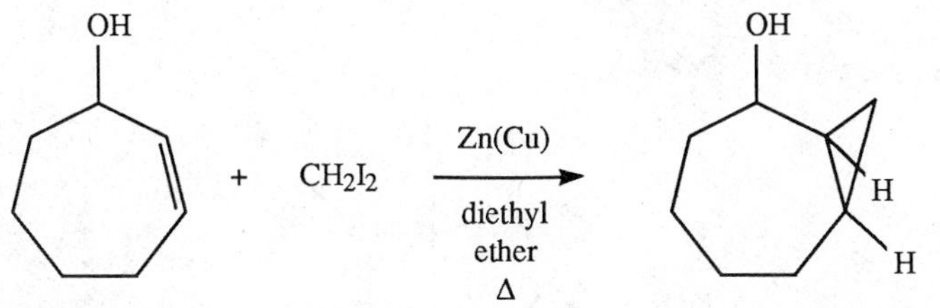

and enantiomer

28.31 (cont)

(g)

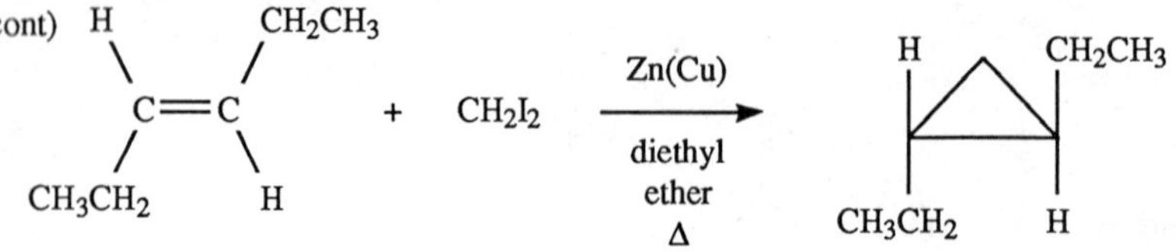

and enantiomer

(h)

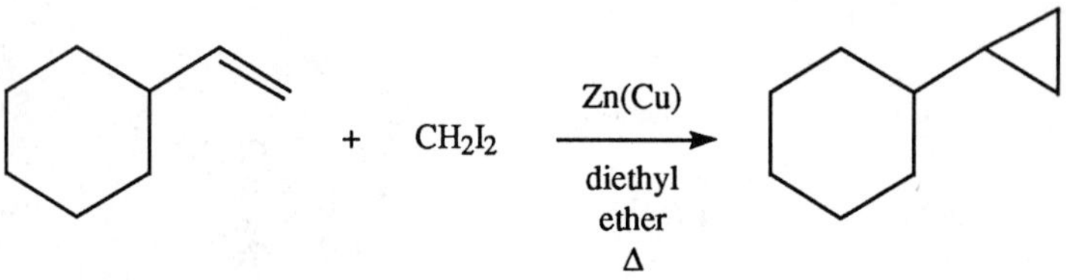

(i)

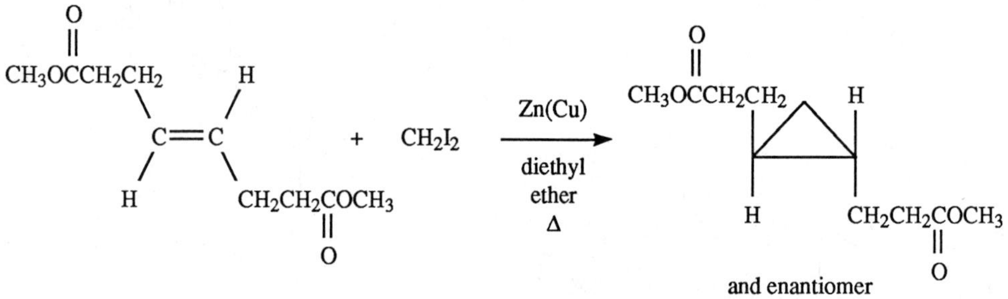

and enantiomer

(j)

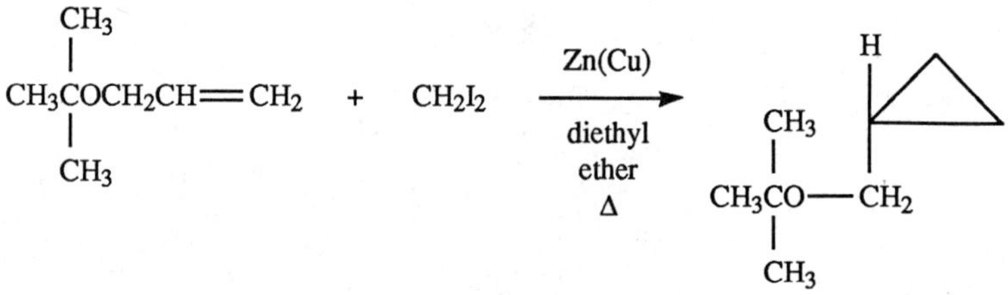

(k)

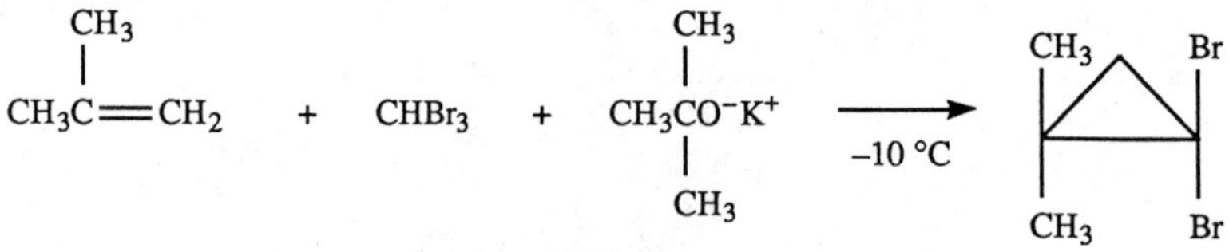

28.32

(a)

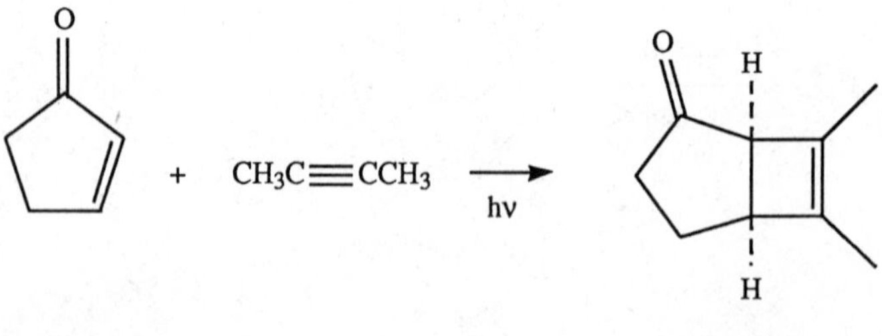

and enantiomer

28.32 (cont)

(b)

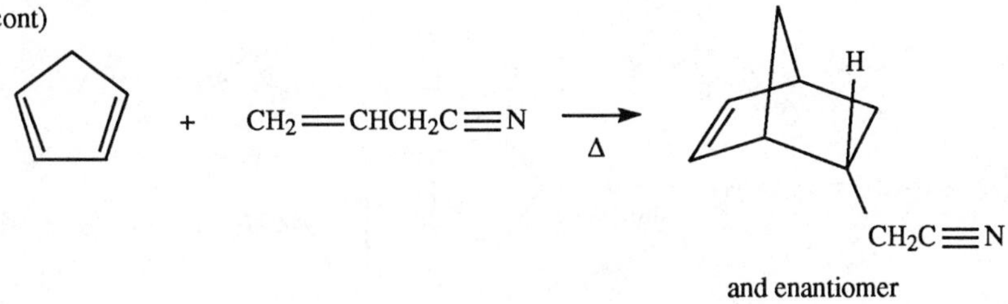

and enantiomer

(c)

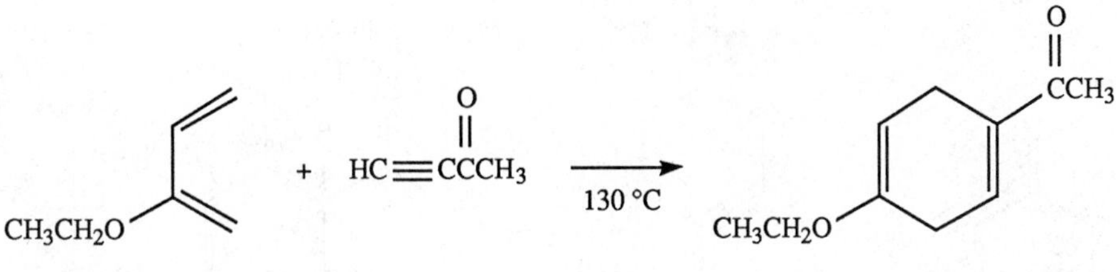

(d)

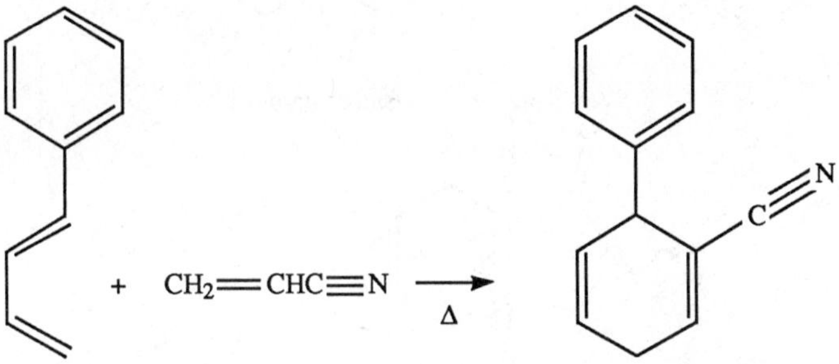

(e)

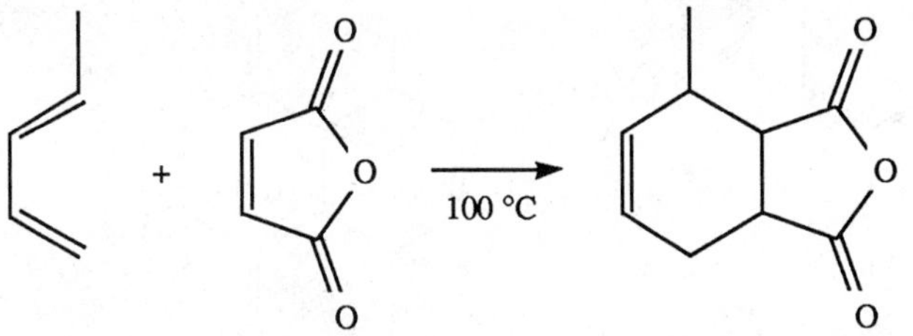

(f)

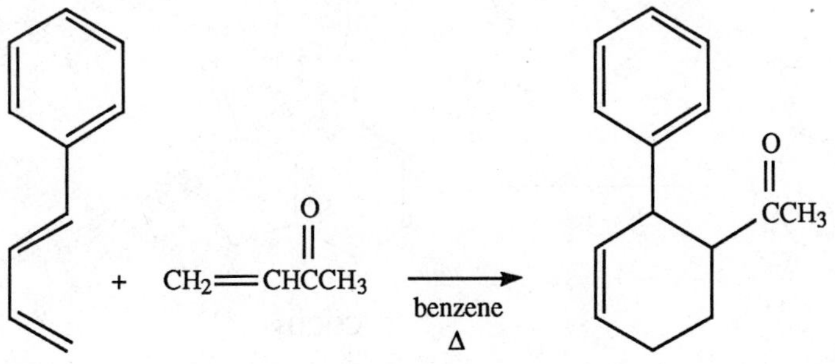

28.32 (cont)

(g)

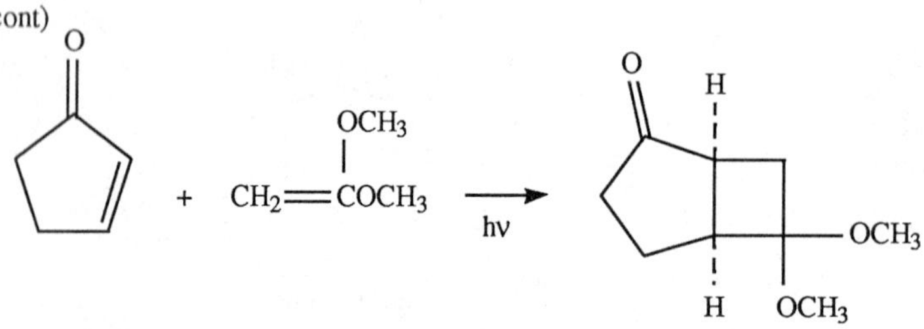

and enantiomer

(h)

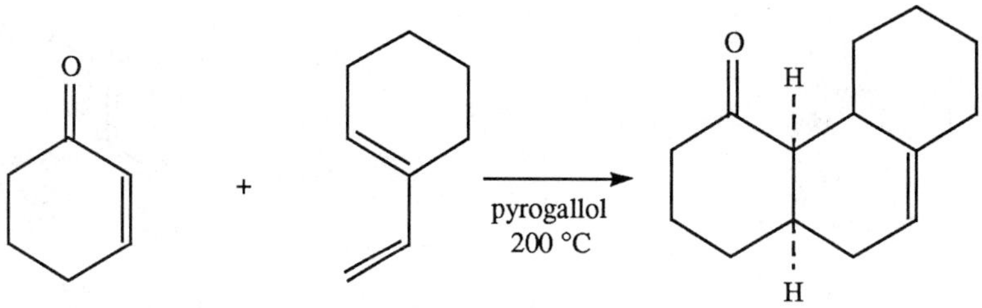

pyrogallol
200 °C

and enantiomer

(i)

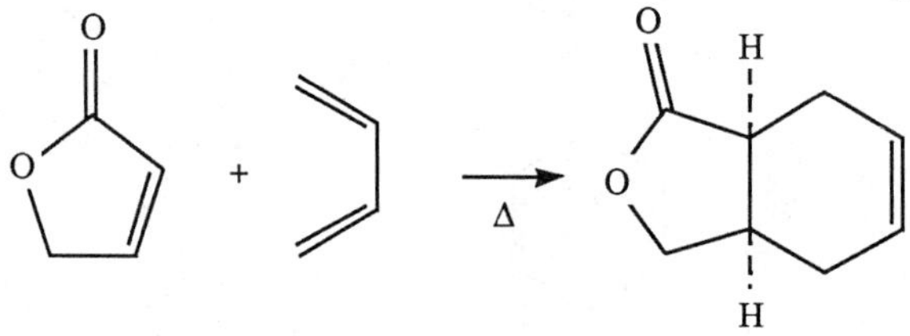

Δ

and enantiomer

(j)

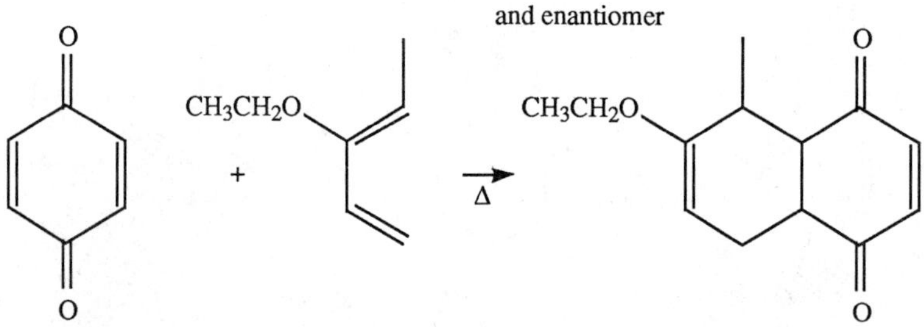

Δ

(k)

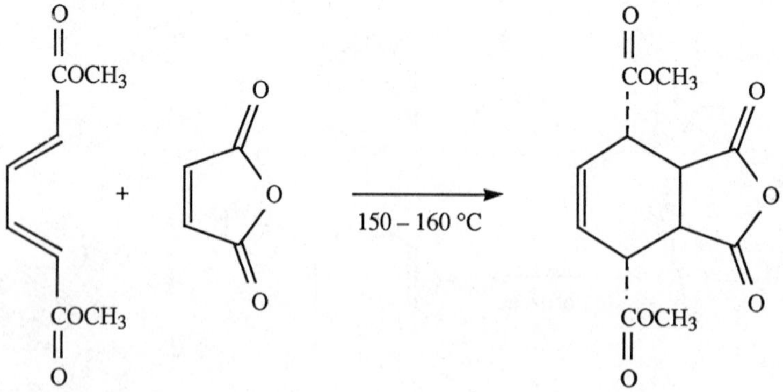

150 – 160 °C

28.32 (cont)

(l)

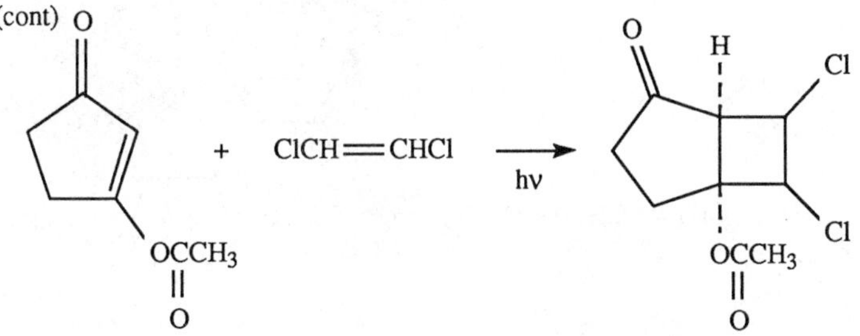

28.33

(a)

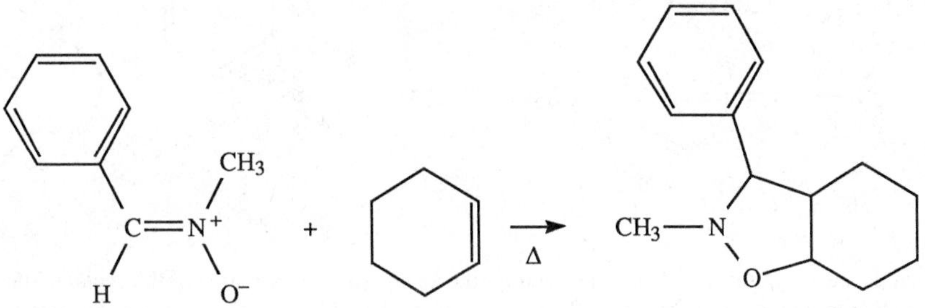

(b)

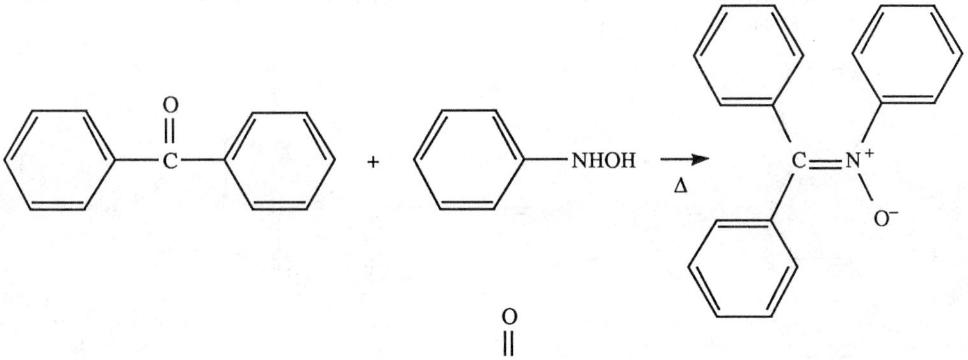

(c)

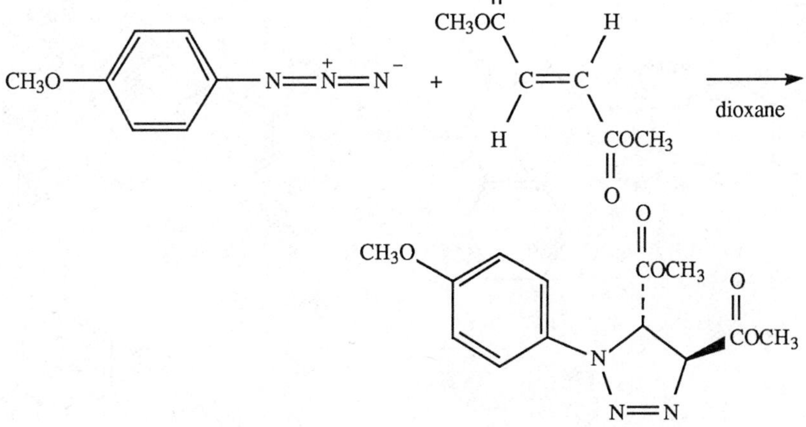

and enantiomer

28.33 (cont)

(d)

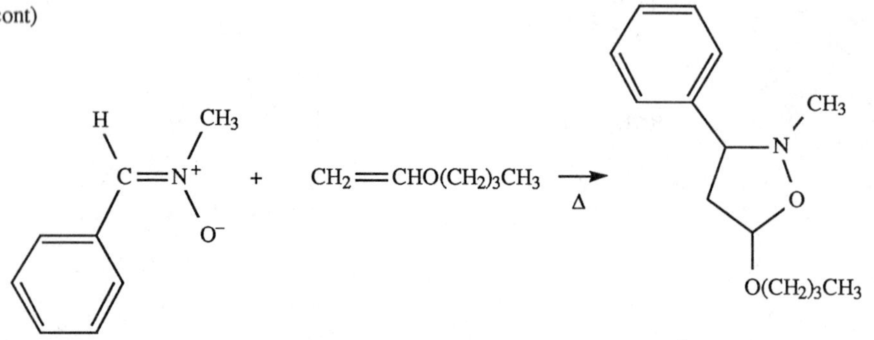

(e)

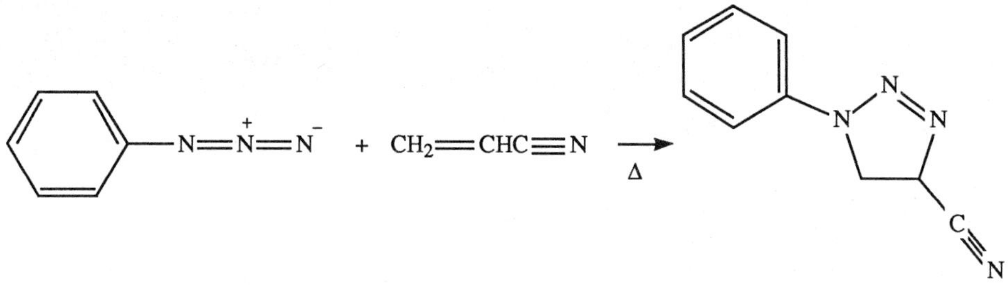

Note that the alkene in this problem is an electrophilic alkene so the orientation of the cycloaddition reaction is different from that seen in Problem 28.25 (a) in which phenyl azide adds to an alkene with an electron-donating group.

(f)

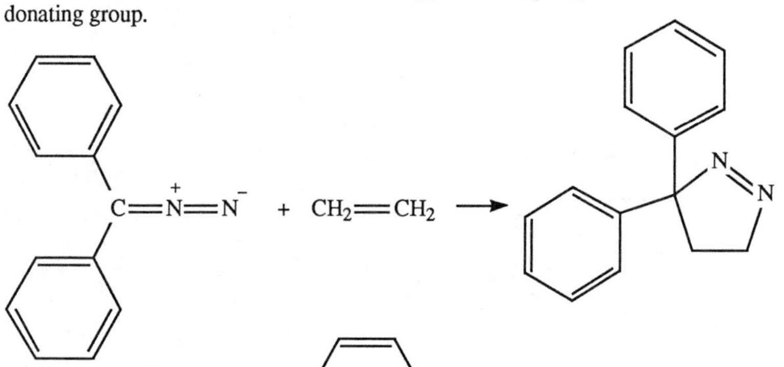

(g)

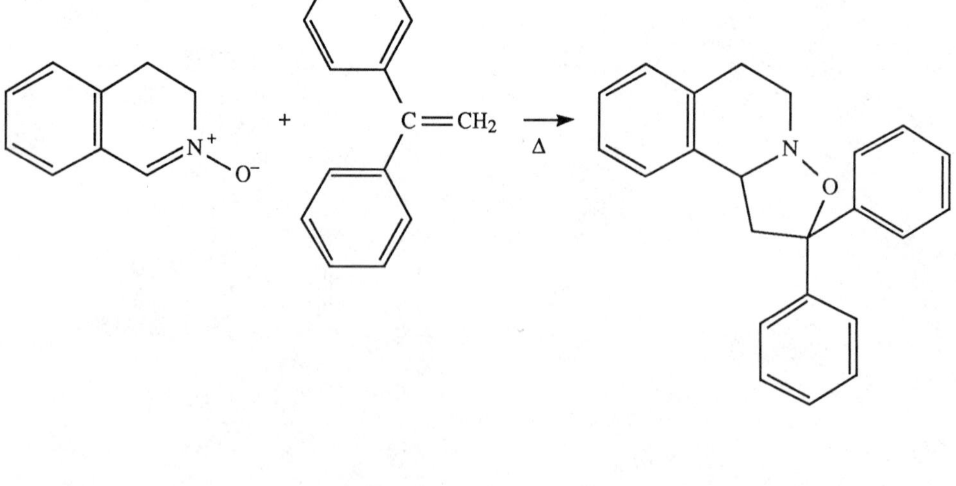

28.33 (cont)

(h)

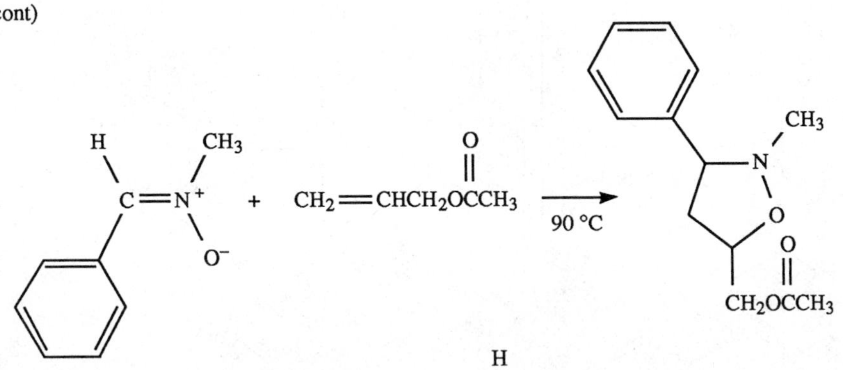

(i)

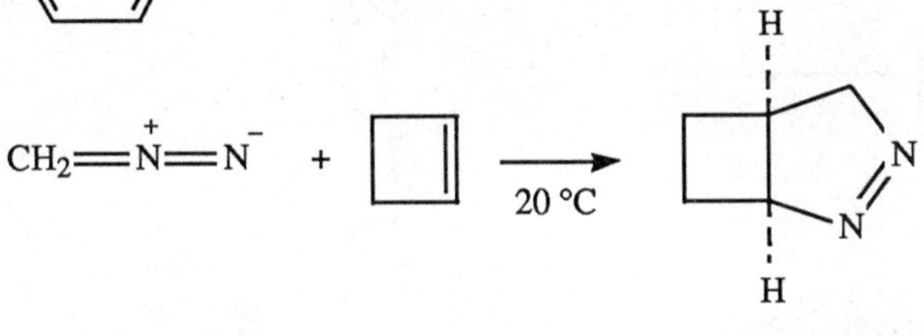

and enantiomer

28.34

(a)

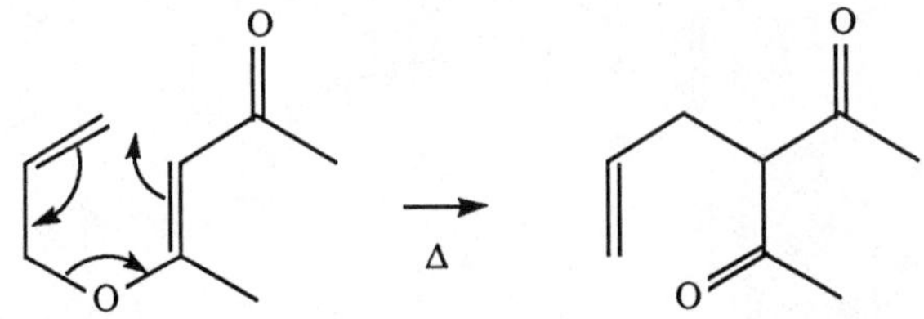

(b)

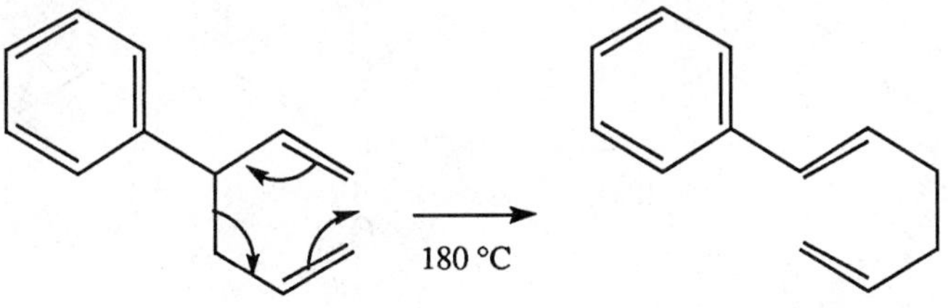

(c)

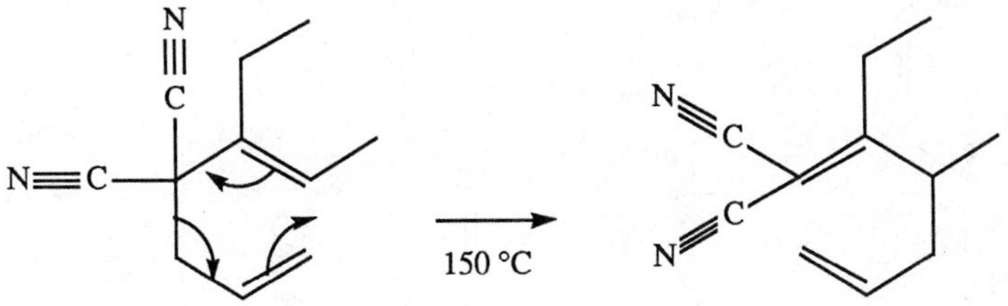

28.34 (cont)

(d)

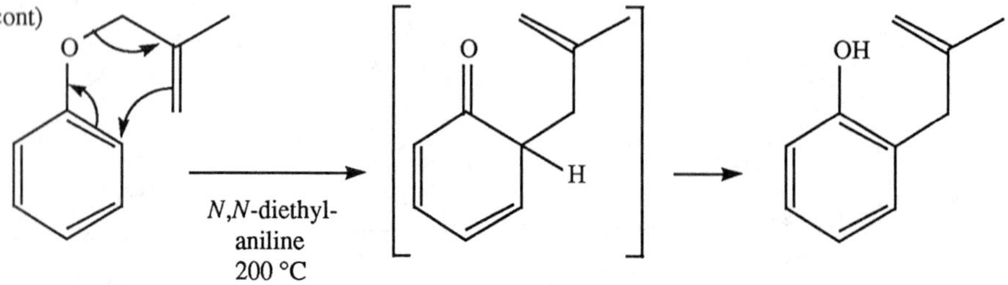
O
O
H
OH
N,N-diethyl-
aniline
200 °C

(e)

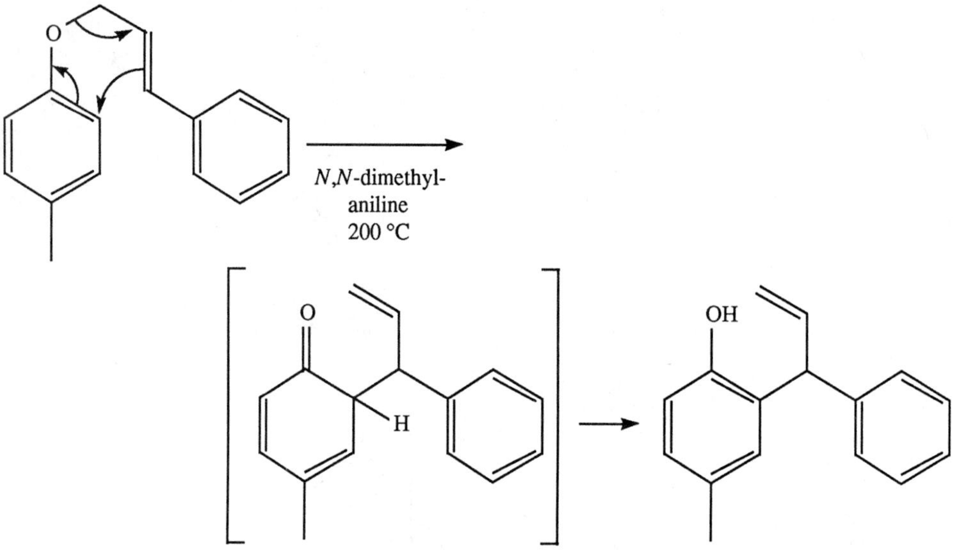
O
N,N-dimethyl-
aniline
200 °C
O
H
OH

(f)

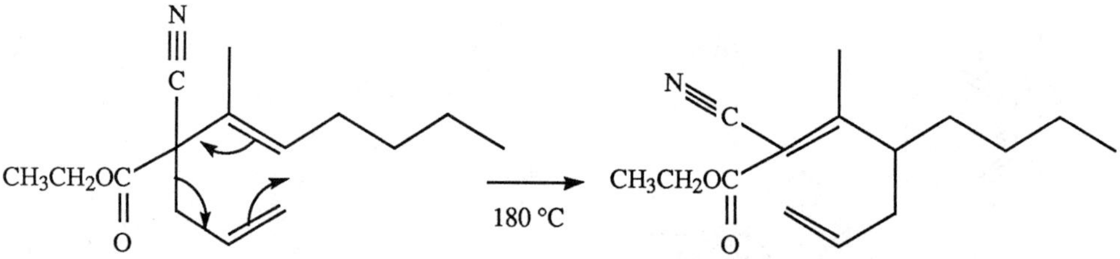
N
C
CH3CH2OC
O
180 °C
N
C
CH3CH2OC
O

(g)

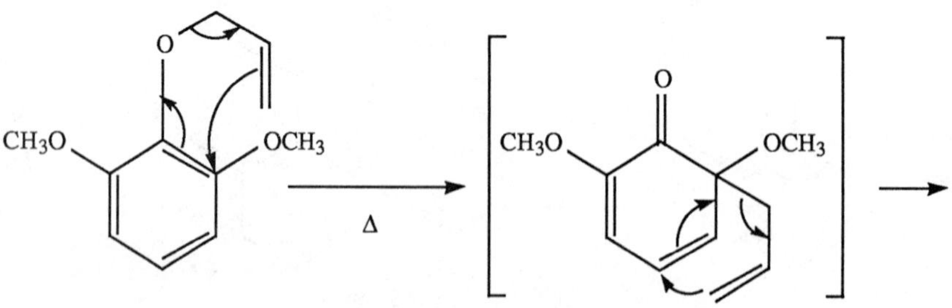
O
CH3O
OCH3
Δ
O
CH3O
OCH3

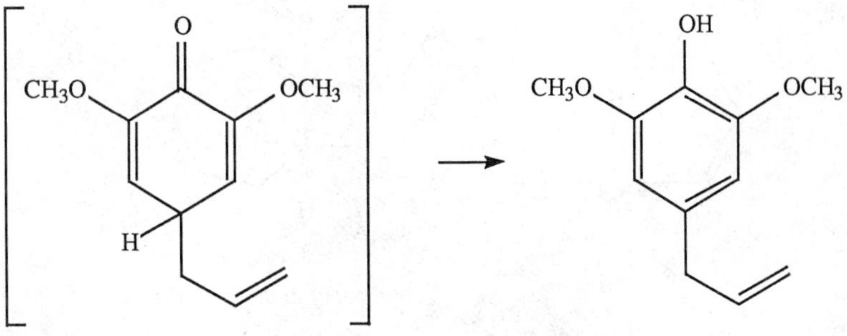

28.35

(a)

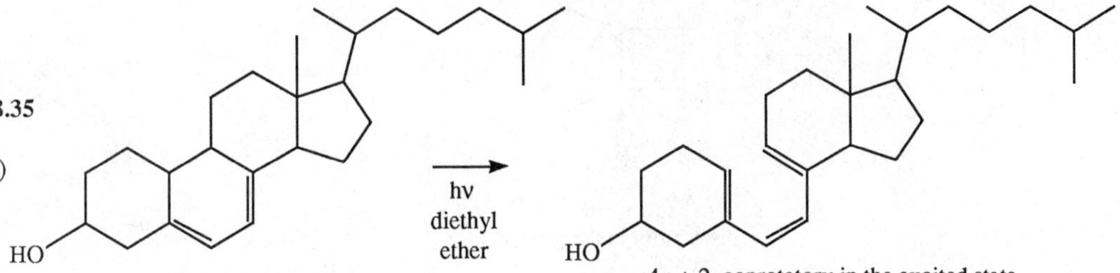

hν
diethyl
ether

$4n + 2$, conrotatory in the excited state

(b)

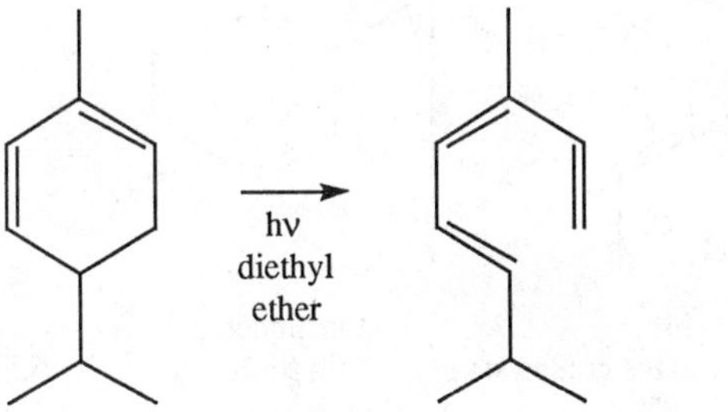

hν
diethyl
ether

$4n + 2$,
conrotatory in the excited state

(c)

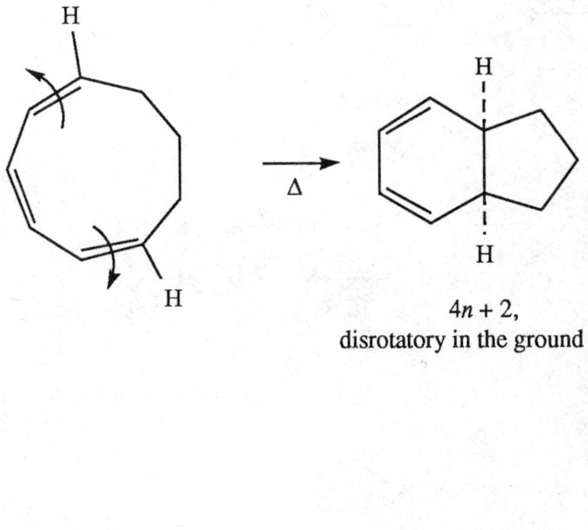

Δ

$4n + 2$,
disrotatory in the ground state

28.35 (cont)

(d)

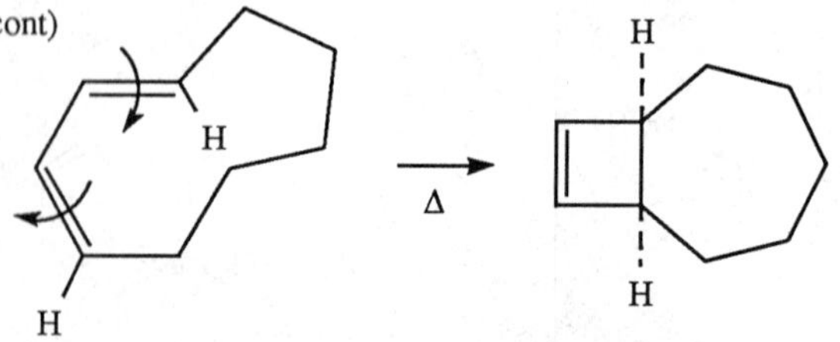

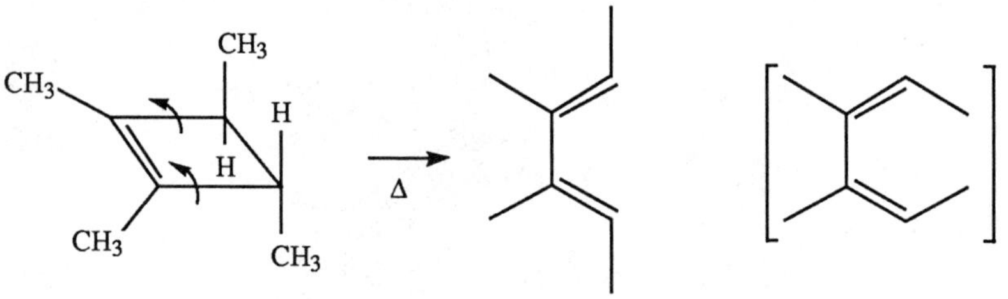

4*n*,
conrotatory in the ground state

(e)

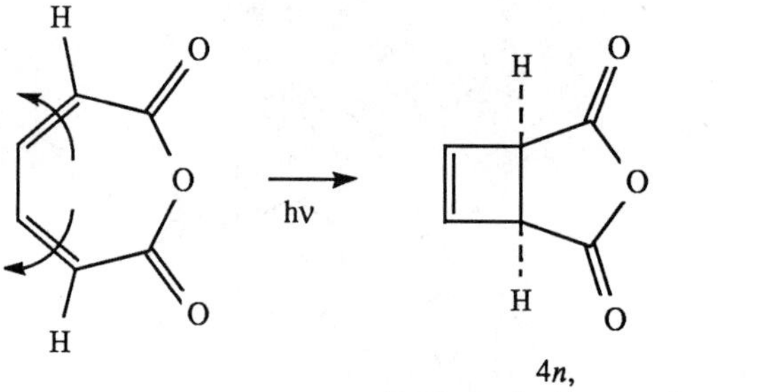

4*n*,
disrotatory in the ground state

(f)

4*n*,   too hindered
conrotatory in the ground state   to form

28.36

(a)

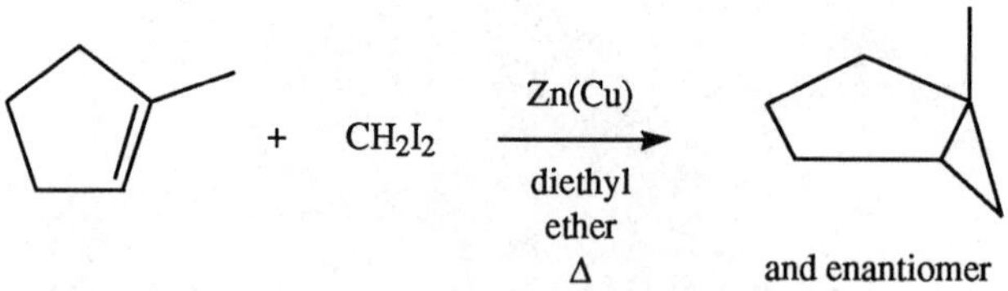

and enantiomer

(b)

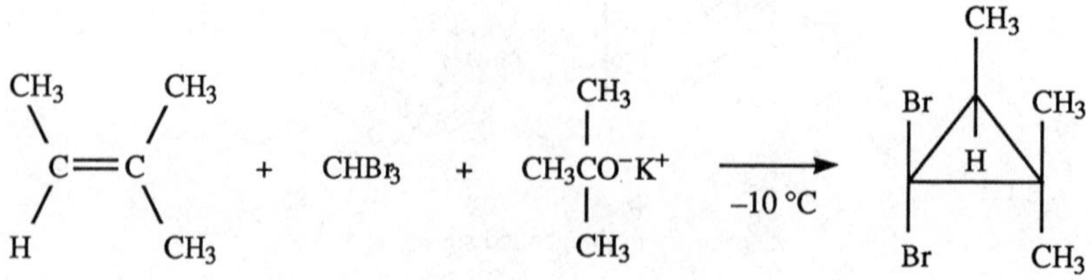

and enantiomer

28.36 (cont)

(c) $CH_2{=}CHCCH_3$ + CH_2I_2 $\xrightarrow[\substack{\text{diethyl}\\ \text{ether}\\ \Delta}]{\text{Zn(Cu)}}$

with the O double bond above CCH₃ on a cyclopropane ring bearing an H.

(d)

CH_3CH_2, CH_2CH_3 on a C=C with H, H   + CH_2I_2 $\xrightarrow[\substack{\text{diethyl}\\ \text{ether}\\ \Delta}]{\text{Zn(Cu)}}$ cyclopropane with CH_3CH_2, CH_2CH_3 and H, H

(e)

cyclopentene   + $CHBr_3$ + $CH_3CO^-K^+$ (with two CH_3 groups) $\longrightarrow$ bicyclic product with H, H and Br, Br

(f)

$CH_2{=}CCH_2CH_3$ (with CH_3)   + $CHCl_3$ + $CH_3CO^-K^+$ (with two CH_3 groups) $\xrightarrow[-10\,°C]{}$ cyclopropane with Cl, CH_2CH_3, Cl, CH_3

and enantiomer

28.37

(a) $Na^{14}C{\equiv}N$ +   (epoxide) $\longrightarrow$ $N{\equiv}^{14}CCH_2CH_2O^-Na^+$ $\xrightarrow[\Delta]{\text{HCl}}$

A

$HO^{14}CCH_2CH_2Cl$ $\xrightarrow[\substack{\text{diethyl}\\ \text{ether}}]{\text{LiAlH}_4}$ $Li^{+\,-}O^{14}CH_2CH_2CH_2Cl$ $\xrightarrow{}$ (phenol with OH)

B   **C**

(phenyl)$-OCH_2CH_2{}^{14}CH_2OH$ $\xrightarrow{\text{PBr}_3}$ (phenyl)$-OCH_2CH_2{}^{14}CH_2Br$ $\xrightarrow{\substack{CH_3\\ CH_3NCH_3}}$

D   **E**

28.37 (cont)

(a) (cont)

$$\text{C}_6\text{H}_5-\text{OCH}_2\text{CH}_2{}^{14}\text{CH}_2\overset{+}{\underset{\underset{\text{CH}_3}{|}}{\text{N}}}\text{CH}_3 \ \text{Br}^- \xrightarrow[\text{H}_2\text{O}]{\text{Ag}_2\text{O}}$$
(with CH₃ on top of N)

F

$$\text{C}_6\text{H}_5-\text{OCH}_2\text{CH}_2{}^{14}\text{CH}_2\overset{+}{\underset{\underset{\text{CH}_3}{|}}{\text{N}}}\text{CH}_3 \ \text{OH}^- \xrightarrow{\Delta} \text{C}_6\text{H}_5-\text{OCH}_2\text{CH}={}^{14}\text{CH}_2$$
(with CH₃ on top of N)

G

(b)   Dehydration of an alcohol would lead to rearrangements, giving the more highly substituted alkene.

(c)

OH

o-substituted: ${}^{14}\text{CH}_2\text{CH}=\text{CH}_2$ $\xrightarrow[\text{base}]{(\text{CH}_3)_2\text{SO}_4}$

OCH₃

o-substituted: ${}^{14}\text{CH}_2\text{CH}=\text{CH}_2$ H $\xrightarrow[\text{H}_2\text{O}]{\text{H}_2\text{O}_2,\ \text{HCOH}}$

OCH₃

o-substituted: ${}^{14}\text{CH}_2\underset{\underset{\text{OH}}{|}}{\text{CH}}-\underset{\underset{\text{OH}}{|}}{\text{CH}_2}$ I $\xrightarrow{\text{HIO}_4}$ $\overset{\text{O}}{\overset{||}{\text{HCH}}}$ J   +

OCH₃

o-substituted: ${}^{14}\text{CH}_2\overset{\text{O}}{\overset{||}{\text{CH}}}$ K $\xrightarrow[\text{H}_2\text{O}]{\text{Ag}_2\text{O}}$

OCH₃

o-substituted: ${}^{14}\text{CH}_2\overset{\text{O}}{\overset{||}{\text{C}}}\text{O}^-\text{Ag}^+$ + $\text{H}\overset{\text{O}}{\overset{||}{\text{C}}}\text{O}^-\text{Ag}^+$ $\xrightarrow{\text{HgO}}$ CO_2

L

(d)   No, the ¹⁴C label will be in the other product.

OCH₃

o-substituted: ${}^{14}\text{CH}_2\overset{\text{O}}{\overset{||}{\text{C}}}\text{OH}$

28.37 (cont)

(e)   Yes.

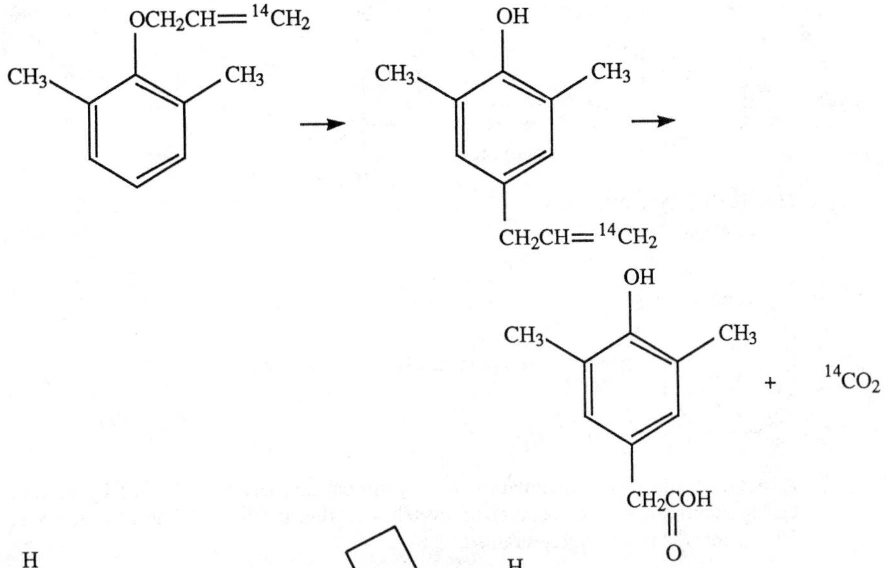

28.38

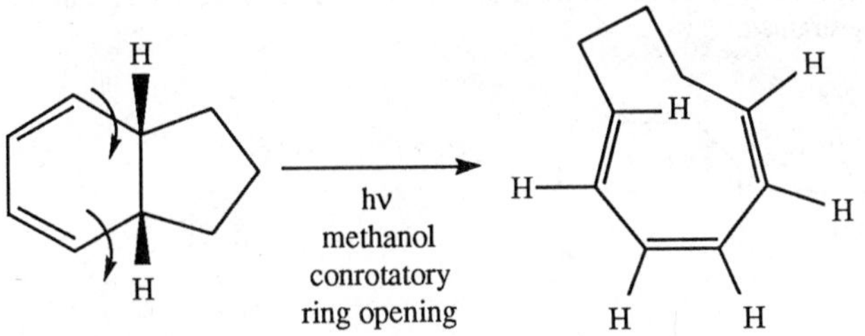

4n + 2 system
n = 1
A

B is (1Z, 3Z, 5E)-1,3,5 cyclononatriene.

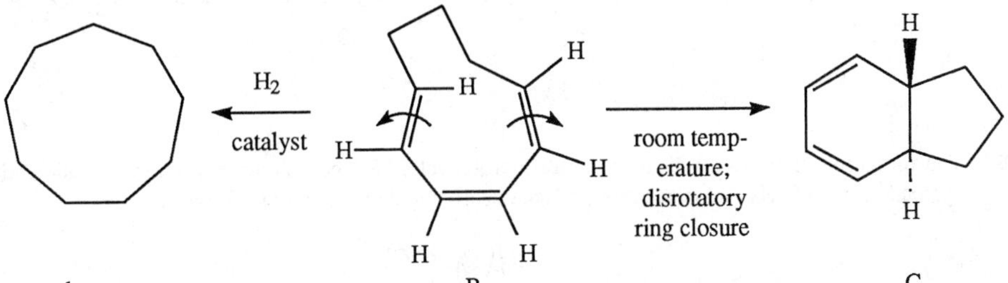

A λ_{max} of 290 nm (ε 2050) in the ultraviolet spectrum points to the presence of a conjugated π system. The presence of bands in the infrared spectrum of B at 1645, 975, 960, and 670 cm^{-1} tells us that B has carbon-carbon double bonds. The band at 1645 cm^{-1} is the C=C stretching frequency. The bands at 975, 960 and 670 cm^{-1} are the bending frequencies for the C—H bonds on the double bonds.

The bands in the proton magnetic resonance spectrum of B correspond to the chemical shifts of vinyl hydrogen atoms (δ 4.7 – 6.2, 6H), and allylic and alkyl hydrogen atoms (δ 0.6 – 2.6, 6H).

28.39

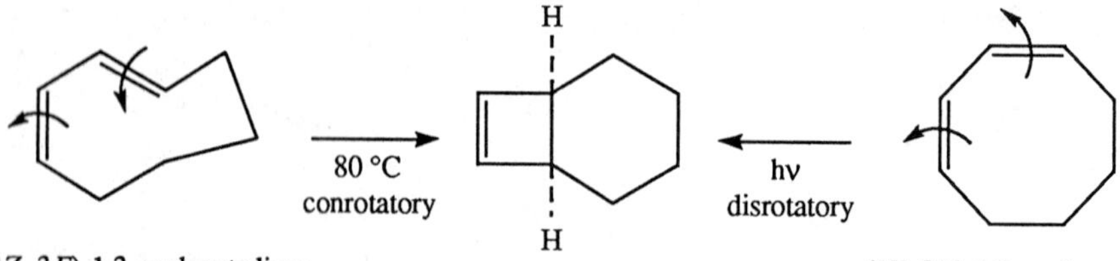

(1Z, 3E)-1,3-cyclooctadiene
4n π system, n = 1

(1Z, 3Z)-1,3-cyclooctadiene
4n π system, n = 1

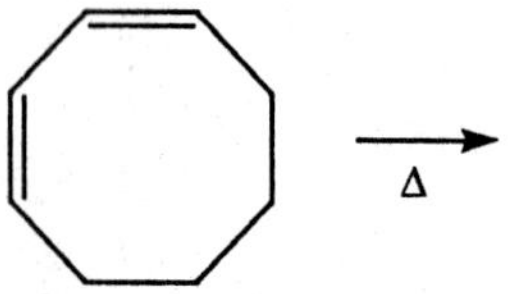

no product with a cyclobutane ring

A 4n π system undergoes a thermal conrotatory ring closure, but for the (1Z, 3Z)-1,3-cyclooctadiene, this would lead to a trans ring junction between the cyclohexane ring and the cyclobutane ring. Such a trans fusion of the four-membered ring is highly strained.

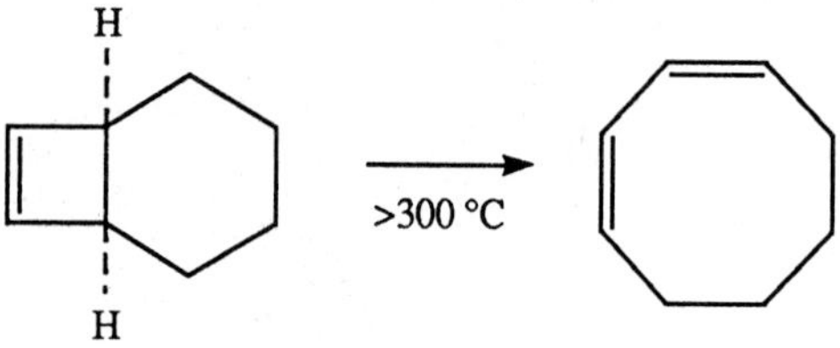

This cannot be a concerted reaction because the concerted thermal ring opening of a 4n π system would be conrotatory and would give the (1Z, 3E)-1,3-cyclooctadiene (the reverse of the first reaction shown above).

28.40   The formation of *trans*-5,6-dichloro-5,6-difluorobicyclo[2.2.1]hept-2-ene is a thermally allowed [4n + 2] cycloaddition (a Diels-Alder) reaction, which occurs with high stereoselectivity.

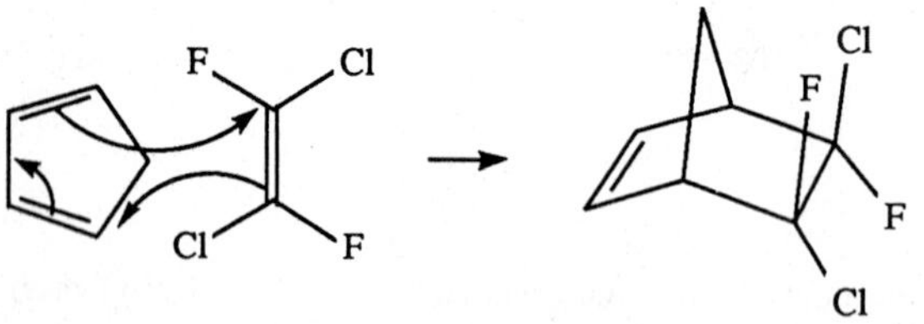

and enantiomer

The concerted thermal cycloaddition reaction for a system with 4n electrons is forbidden. The formation of 6,7-dichloro-6,7-difluorobicyclo[3.2.0]hept-2-ene goes by a radical mechanism, and gives mixtures of stereoisomers.

28.40 (cont)

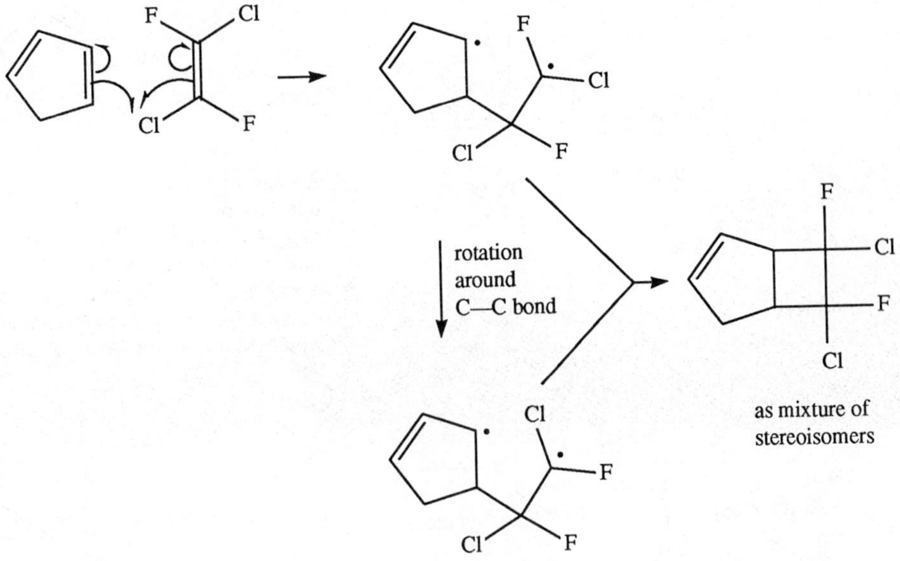

rotation
around
C—C bond

as mixture of
stereoisomers

28.41

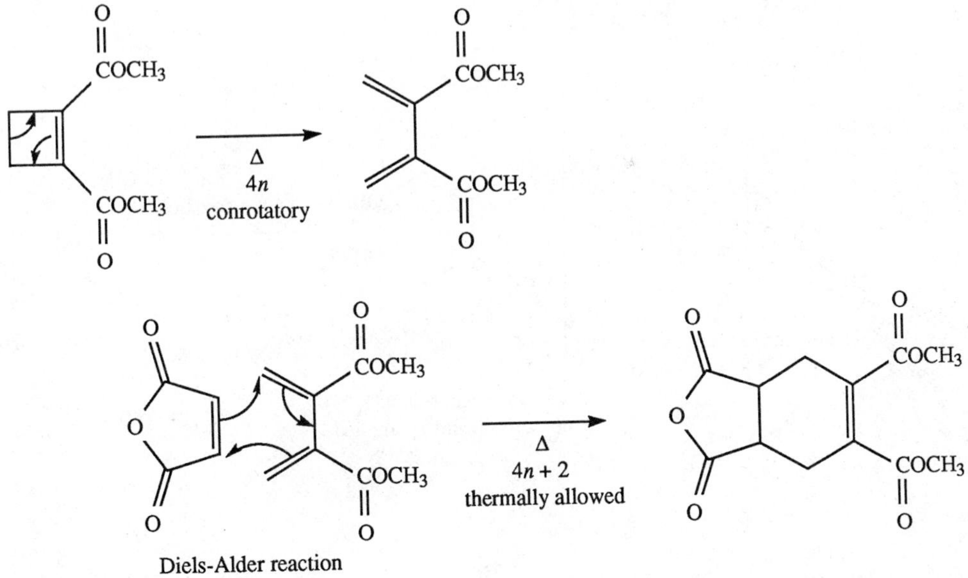

Δ
4n
conrotatory

Δ
4n + 2
thermally allowed

Diels-Alder reaction

28.42

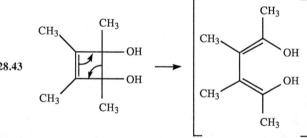

4n π system

A concerted reaction would produce a trans double bond in a six-membered ring. A six-membered ring containing a trans double bond is highly strained and unstable. Therefore, this cyclobutene does not open by a concerted reaction and would require high temperatures for reaction.

Δ
conrotatory

4n π system

A concerted reaction would produce cis double bonds in both rings. This cyclobutene will open easily by a concerted mechanism.

Δ
conrotatory

28.43

an unstable enol;
no stereochemistry implied

tautomerization

28.44

CH_3OCCH_2 CH_2COCH_3

$LiAlH_4$

diethyl ether

$Li^{+-}OCH_2CH_2$ $CH_2CH_2O^-Li^+$

TsCl

pyridine

H   H

A

28.44 (cont)

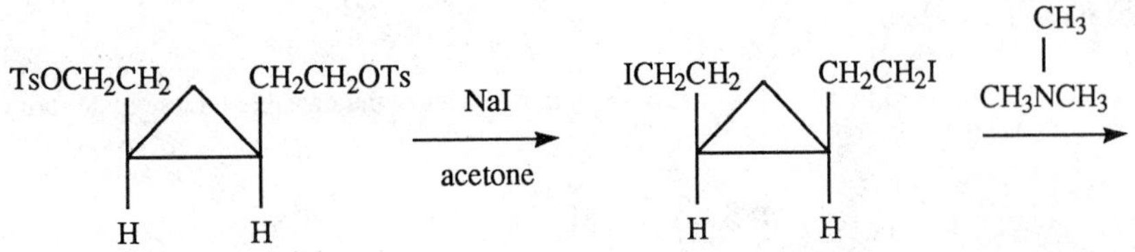

The ultraviolet spectrum
shows the absence of a
conjugated system.

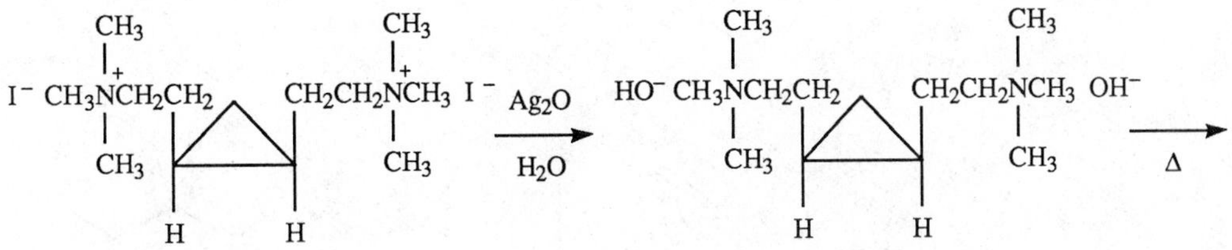

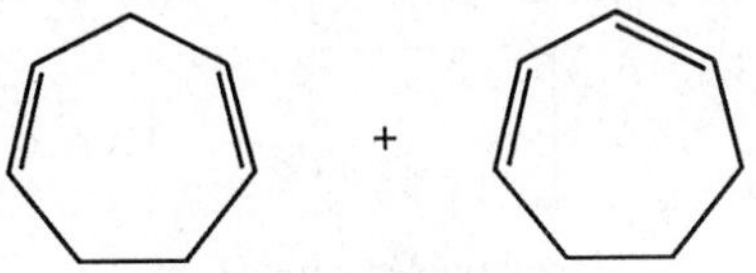

F   +   G

The ultraviolet spectrum
shows the presence of a
conjugated system.

F

G

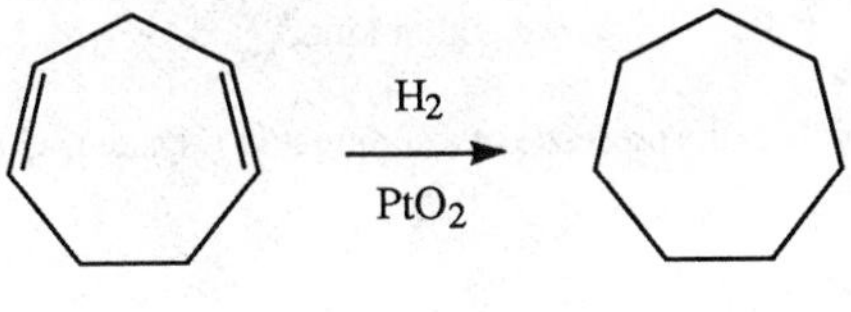

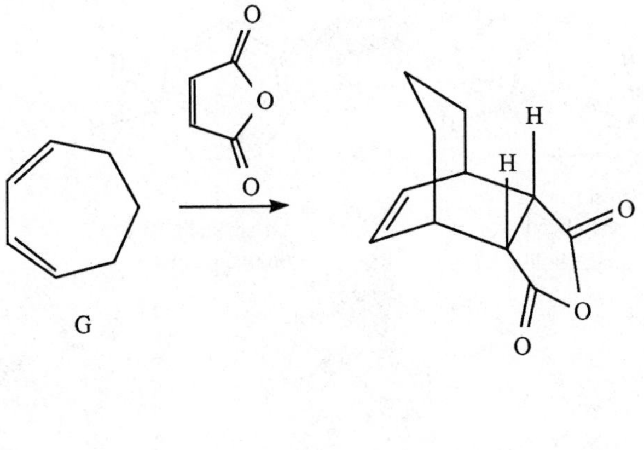

28.44 (cont)

F is formed by a [3,3] sigmatropic rearrangement of the unstable intermediate formed when compound E is heated.

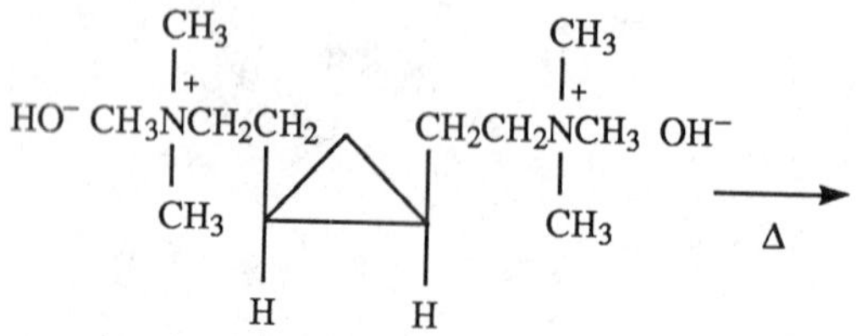

E

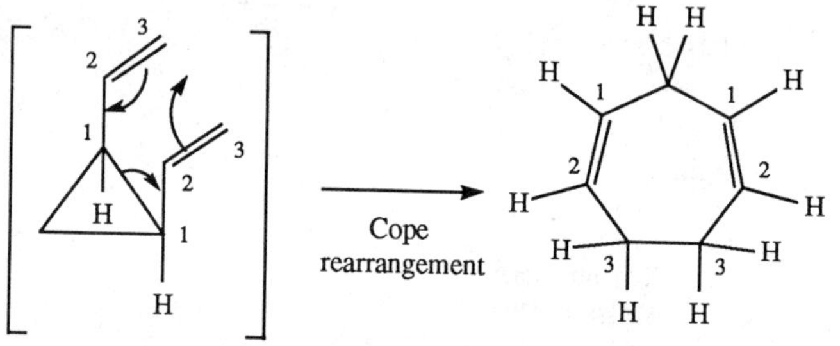

unstable intermediate

Cope
rearrangement

F

No bands in the infrared spectrum for terminal alkenes; bands in the proton magnetic resonance spectrum for 4 vinyl hydrogen atoms, 4 methylene hydrogen atoms of one kind, and 2 methylene hydrogen atoms of another kind.

A λ_{max} of 248 nm in the ultraviolet spectrum of C points to the presence of a conjugated π system. G is formed from F in the presence of strong base.

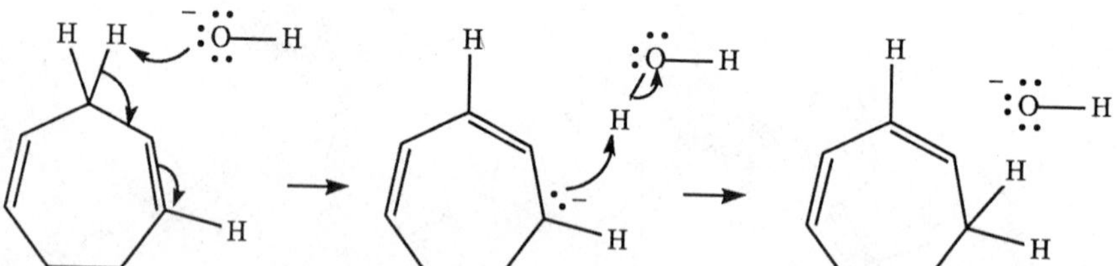

F

resonance-
stabilized
carbanion

G
more stable
conjugated diene

28.45

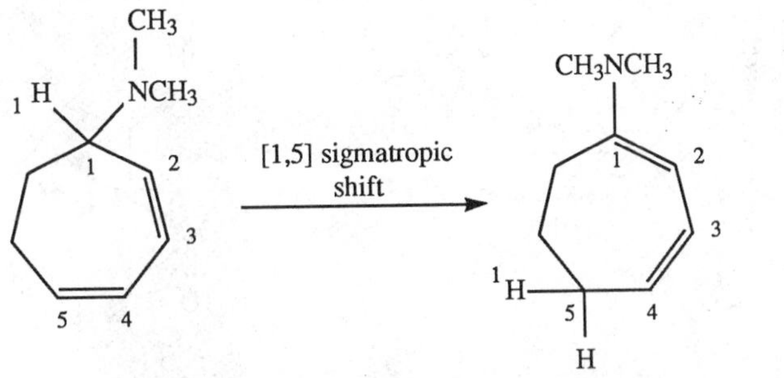

5-(*N*,*N*)-dimethylamino)-
1,3-cycloheptadiene

1-(*N*,*N*)-dimethylamino)-
1,3-cycloheptadiene

(The numbers on the structure
are to help identify the sigmatropic rearrangement and not
for nomenclature.)

28.46

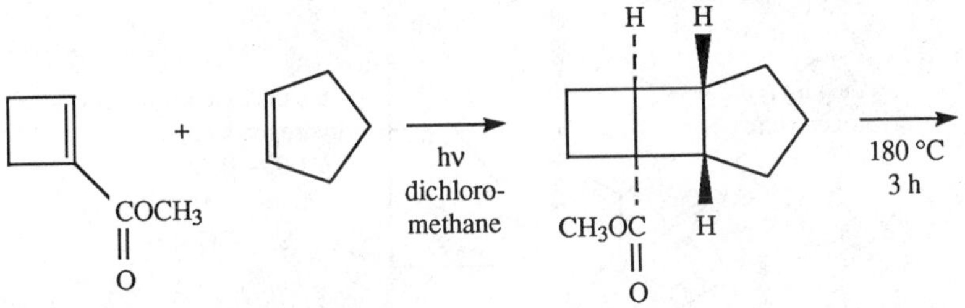

A
product of a [2 + 2]
cycloaddition reaction

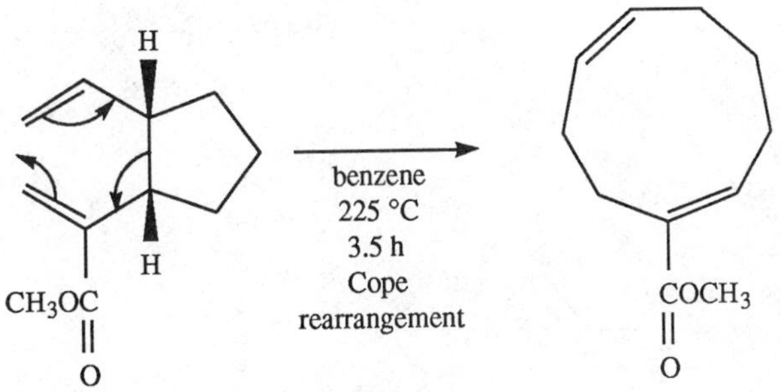

B
B is not the product of
a concerted reaction

C
product of the Cope
rearrangement of B

28.46 (cont)

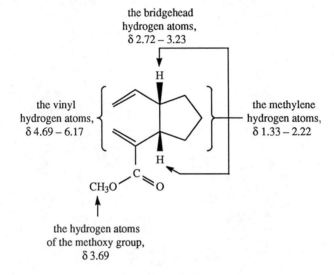

B

The assignment of the structure of B is supported by the results of the ozonolysis reaction (above) and by the proton magnetic resonance spectrum as shown below.

the bridgehead
hydrogen atoms,
δ 2.72 – 3.23

the vinyl
hydrogen atoms,
δ 4.69 – 6.17

the methylene
hydrogen atoms,
δ 1.33 – 2.22

the hydrogen atoms
of the methoxy group,
δ 3.69